Computing and Systems Technology Division 2014

Core Programming Area at the 2014 AIChE Annual Meeting

Atlanta, Georgia, USA
16-21 November 2014

Volume 1 of 2

ISBN: 978-1-5108-1255-0

Printed from e-media with permission by:

Curran Associates, Inc.
57 Morehouse Lane
Red Hook, NY 12571

Printed by Curran Associates, Inc. (2015)

For permission requests, please contact AIChE
at the address below.

AIChE
120 Wall Street, FL 23
New York, NY 10005-4020

Phone: (800) 242-4363
Fax: (203) 775-5177

www.aiche.org

Additional copies of this publication are available from:

Curran Associates, Inc.
57 Morehouse Lane
Red Hook, NY 12571 USA
Phone: 845-758-0400
Fax: 845-758-2634
Email: curran@proceedings.com
Web: www.proceedings.com

TABLE OF CONTENTS

VOLUME 1

VOLUME 2

392414 Overview of CAST Activities and Programming

Abstract:

An overview of activities of the Computing and Systems Technology (CAST) division, including conference technical programming, will be presented. The CAST Executive Committee consists of Mayuresh Kothare (Past Chair), Marianthhi Ierapetritou (Chair), Ray Adomaitis (1st Vice-Chair), Karl Schnelle (2nd Vice-Chair), Martha A. Grover (Secretary-Treasurer), and the following Directors: Christos Maravelias (2012-2014), Mario Eden (2012-2014), John Richards (2013-2015), Carl Laird (2013-2015), Andreas Linninger (2014-2016), Leo Chiang (2014-2016). The CAST Division Programming Chair is Nick Sahinidis (2014-2017).

377870 Synergistic Biomass and Natural Gas Process Design for Liquid Fuel Production with Reduced CO2 Emissions

Abstract:

For the past century, petroleum derived liquid hydrocarbons have been the predominant fuel for the transportation sector. However, concerns regarding scarce petroleum reserves and increasing greenhouse gas (GHG) emissions from fossil fuels urge advancements in alternative transportation fuels as well as production of liquid fuels from alternative carbon sources, such as natural gas (NG), coal, and biomass.

Liquid fuels derived from biomass can result in lower lifecycle GHG emissions relative to petroleum-derived fuels [1]. However, the supply of biomass-derived liquid fuel is limited by the sustainably available (SA) biomass feedstock and the relatively low (30-40 %) carbon conversion during standalone liquid fuel production. Biomass carbon conversion to liquid fuel can be increased by using supplemental non-carbon energy. However, such processes have to overcome the intermittent nature of most non-carbon energy sources as well as their high economic cost in the short-term [2]. The recent surge in shale gas reserves and its production, most notably in the U.S., has led to interest in utilizing NG as a bridge for the transition to a sustainable economy.

Towards reducing the CO_2 emissions associated with the transportation sector, we investigate the design of carbon and energy efficient processes for integrated biomass and natural gas (NG) conversion to liquid fuel. A process superstructure considering biomass conversion via gasification/Fischer-Tropsch (FT) synthesis and fast-hydropyrolsis/hydrodeoxygenation described in [3], and NG conversion via reforming followed by FT synthesis is established. Subsequently, a mixed integer nonlinear programming model (MINLP) is formulated to identify the process configurations that maximize the energy output as liquid fuel for different ratios of NG to biomass carbon feeds (δ_{ng}). For $1\ \% \leq \delta_{ng} \leq 150\ \%$, the optimal process configurations are capable of producing ~5-14 % more liquid fuel output than the combined fuel output of individual standalone processes converting the same amount of biomass and NG. This synergy originates from synthesizing additional liquid fuel by combining the residual biomass carbon with the excess hydrogen per carbon available from the NG feed. These integrated processes are also estimated to achieve up to 80 % reductions in greenhouse gas (GHG) emissions relative to petroleum-based fuels.

[1] Agrawal R, Singh NR. Synergistic Routes to Liquid Fuel for a Petroleum deprived future AIChE Journal 2009;55:1898.

[2] Singh N, Mallapragada D, Agrawal R, Tyner W. Economic analysis of novel synergistic biofuel (H2Bioil) processes. Biomass Conversion and Biorefinery 2012;2:141.

[3] Mallapragada DS, Agrawal R, Tawarmalani M. Synthesis of augmented biofuel processes using solar energy. AIChE Journal 2014.

383713 Higher-Order Inclusion Techniques for Guaranteed Parameter Estimation of Nonlinear Dynamic Systems

Abstract:

Among the available techniques to account for uncertainty in parameter estimation, guaranteed parameter estimation aims to determine all the parameter values of a model that are consistent with the measurements under given uncertainty scenarios [1]. Our focus in this presentation is on nonlinear dynamic systems and we consider the case that the uncertainty enters the estimation problem in the form of bounded measurement errors. Set-inversion algorithms, which approximate such parameter sets by a box partition at an arbitrary precision, were first developed for algebraic models in the early 1990s by Moore [2] and Jaulin and Walter [3] using interval analysis; and not long after were these algorithms extended to dynamic systems using ODE bounding techniques [4,5]. Nonetheless, seldom can any estimation problem with more than a few uncertain parameters be tackled with such algorithms. The main computational bottleneck for guaranteed parameter estimation in higher-dimensional dynamic systems appears to be the ability to compute tight bounds on parametric solutions of the dynamic system.

In the first part of the presentation, we review recent developments in ODE bounding techniques [6] based on Taylor models that enjoy higher-order convergence properties and we illustrate the benefits of these techniques for our problem. Next, we introduce optimization-based domain reduction techniques in order to enhance the convergence speed of the set-inversion algorithm as well as simple strategies that avoid recomputing the ODE bounds wherever possible. A challenging case study in anaerobic digestion is presented for a model describing complex liquid-gas transfer and pH self-regulation mechanisms and featuring multiple time scales. The results demonstrate that the proposed improvements allow tackling guaranteed parameter estimation in up to seven parameters within reasonable computational times.

References:

[1] Walter, E. (ed.) (1990). Parameter Identifications with Error Bound, *Mathematics & Computers in Simulation*, vol. 32. Elsevier.
[2] Moore, R.E. (1992). Parameter sets for bounded-error data. *Mathematics & Computers in Simulation*, vol. 34 (2), p. 113-119.
[3] Jaulin, L. and Walter, E. (1993). Set inversion via interval analysis for nonlinear bounded-error estimation. *Automatica*, vol. 29 (4), pp. 1053-1064.
[4] Jaulin, L. (2002). Nonlinear bounded-error state estimation of continuous-time systems. *Automatica*, vol. 38, pp. 1079-1082.
[5] Raissi, T., Ramdani, N. and Candau, Y. (2004). Set membership state and parameter estimation for systems described by nonlinear differential equations. *Automatica*, vol. 40, pp. 1771-1777.
[6] Villanueva, M.E., Houska, B. and Chachuat, B. (2013). Unified framework for the propagation of continuous-time enclosures for parametric nonlinear ODEs. *Journal of Global Optimization*, in revision.

Crystallization is one of the most important processes for the pharmaceutical industry, estimated to be over a one trillion dollar per year industry. One of the main issues associated with developing new products is the costs associated with finding out the best process parameters for operating the crystallization processes. Furthermore, it is noted that process modeling is an important area for crystallization systems research with such models allowing for novel designs and operating protocols to be developed systematically before they are tested experimentally [1].

The present work focuses on the modeling, simulation, and control of Ibuprofen crystal shape. Ibuprofen [2-(4-isobutyl-phenyl)-propionic acid) is commonly used as a nonsteroidal anti-inflammatory drug (NSAID) and belongs to the space group $P2_1/c$ with a unit cell of four molecules (two dimers) [2].

In order to model this process, several considerations must be made. Overall, this process must account for the growth rate, solubility of the continuous phase, nucleation rate (primary and secondary), and nucleation metastable zones [3]. In particular, experimental growth rate data is limited which is why we use kinetic Monte Carlo (KMC) simulations for this non-equilibrium process. KMC simulations are used to model the growth rate data in an area where growth rate data exists, and then can be used to estimate the real growth rates in nearby regions where data does not currently exist. For the KMC simulations, model parameters are modified while maintaining them in a physically meaningful range until satisfactory agreement between the calculated and the experimental growth rates are found. Work by Cano *et al.* measures the growth rates of the three independent faces, {001}, {011}, and {100} [4], however the amount of experimental data is limited and generating a model using this data alone is not feasible. Recently, Nguyen *et al.* experimentally measured the growth rates on the {001} and {011} faces as a function of the supersaturation [5]. They focused on crystallization under diffusion-limited conditions with supersaturation values ranging from 0.54 to 1.31, and these data are used in our modeling work. In order to generate a more complete model for Ibuprofen, modeling should be performed on all three faces when experimental data becomes available. Furthermore, crystallization from ethanol will be considered due to the fact that ample data is available to model the crystallization process in this solvent [3-8].

Additionally, the control of crystal shape is very important for crystallization processes since shape influences many of the crystal properties, such as flow properties, dissolution rates, and the bio-availability of the drug. The KMC simulations developed to model crystal growth are used alongside other data in order to model the crystal growth process for Ibuprofen in different types of crystallizers in order to estimate the shape and size distributions. The KMC simulations and modeling of the batch crystallization process will be developed in a similar fashion to our previous work [9-10]. Extensive simulations demonstrate the usefulness of the proposed model.

References:

[1] Winn D, Doherty MF, Modeling Crystal Shapes of Organic Materials Grown from Solution. AIChE Journal. 2000;46:1348-1367.
[2] McConnell JF, 2-(4-Isobuthlyphenyl) Propionic Acid. Crystal Structure Communication. 1974;78:73.
[3] Rashid A, White ET, Howes T, Litster JD, Marziano I, Crystallization Kinetics of Ibuprofen from Ethanol and Aqueous Ethanol. Chemical Engineering Transactions. 2011;24:631-636.

[4] Cano H, Gabas N, Canselier JP, Experimental study on the ibuprofen crystal growth morphology in solution. Journal of Crystal Growth. 2001;224:335-341.
[5] Nguyen TTH, Hammond RB, Roberts KJ, Precision measurement of the growth rate and mechanism of ibuprofen {001} and {011} as a function of crystallization environment. CrystEngComm. 2014; DOI: 10.1039/c4ce00097h
[6] Rashid A, White ET, Howes T, Litster JD, Marziano I, Growth rates of ibuprofen crystals grown from ethanol and aqueous ethanol. Chemical Engineering Research and Design. 2012;90:158-161.
[7] Rashid A, White ET, Howes ET, Liu L, Litster JD, Marziano I, Nucleation Kinetics for Ibuprofen Crystals Grown from Aqueous Ethanol. Proceedings of Chemeca 2011, New South Wales, Australia. September 2011.
[8] Rashid A, White ET, Howes T, Litster JD, Marziano I, Crystallizer Design Methods for Direct Crystallization of Ibuprofen as a Pharmaceutical Formulation. Proceedings of Chemeca 2013, Brisbane, Queensland, Australia. September 2013.
[9] Nayhouse M, Kwon JS, Christofides PD, Orkoulas G, Crystal shape modeling and control in protein crystal growth. Chemical Engineering Science. 2013;87:216-223.
[10] Kwon JS, Nayhouse M, Christofides PD, Orkoulas G, Modeling and control of protein crystal shape and size in batch crystallization. AIChE Journal. 2013;59:2317-2327.

361489 Parallel Distillation Calculations Using Openmp

Abstract:

With current CPU speeds reaching a practical maximum value, CPU manufacturers are now creating CPU's with multiple cores to allow for parallel processing. However, most process design simulators do not take advantage of parallelism. This paper presents speedup results for solving rigorous distillation columns by Newton's Method using OPENMP. A new thermodynamic engine was developed and used here which allows for parallel computation of thermodynamic properties such as the fugacity coefficients, enthalpies and also the associated temperature and composition derivatives. Strategies to avoid cache misses and keep relevant data in the L1 and L2 caches are discussed. Results are presented for the speedups obtained by calculating the thermodynamic properties, equation residuals and jacobian elements in parallel. The systems studied are rigorous superfractionator and demethanizer distillation columns.

388993 Data-Driven Models and Model Predictive Control Formulations for Load Leveling of Residential Buildings

Abstract:

Buildings account for about 60% of the electricity consumption in the United States [1]. While residential and commercial buildings have an approximately equal share, the impact of residential buildings on the grid is more pronounced, owing to the variability of their energy demand, which changes significantly during a day and between subsequent days. For example, the load placed on the grid by residential consumers is strongly influenced by weather and human activity patterns. Power producers and distributors deal with such demand fluctuations by bringing online additional, peaking, generation capacity, which typically consists of less efficient and more polluting power plants. Eliminating the use of peaking plants provides a strong incentive for mitigating variability in residential energy consumption.

The most significant residential energy consumer is the heating, ventilation and air conditioning (HVAC) system. In this paper, we investigate the potential for coordinated control of the HVAC systems in a residential community, with the purpose of achieving reductions in peak electricity demand, i.e. "leveling the residential load". Our work is based on experimental data obtained from a unique complement of 41 homes situated in Austin, Texas, for which power demand is sampled at high frequency using smart meters. Additionally, data concerning residential comfort preferences are collected using smart thermostats. We present a scheme for developing control-relevant, reduced-order models (ROMs) for each home, that tie energy consumption to outdoor dry bulb temperature and thermostat set points. A non-intrusive load monitoring technique is used to disaggregate A/C electricity consumption from whole-house electricity data reported by smart meters. We show that the proposed time-series models can use thermostat set points and outdoor dry bulb temperatures to accurately predict A/C loads.

We use the ROMs to construct a centralized model predictive control (MPC) scheme that manages energy use at the neighborhood level by altering the thermostat setpoints in individual homes. Simulation results confirm that the proposed controller can reduce the peak electricity load by pre-cooling homes and staggering the time A/C units turn on. Our approach is predicated on leveraging the thermal mass of each house and shifting the electricity peak demand of each individual house to a different time in order to level the overall load. On average the centralized MPC is able to reduce the peak load by 27% (15 kW) for the group of 41 houses without substantial increase in overall electricity consumption. We also show through simulations that decentralized control (whereby each house attempts to level its own peak) is less successful at leveling the overall load.

References

[1] U.S. Energy Information Administration, Annual energy review 2011 (2012).

373454 Using Measurement Uncertainty Information for Effective Model Development

Abstract:

Obtaining and handling uncertainty information is still a challenge in the Chemical Processing Industries (CPI). Typically, among the variables most affected by uncertainty, one typically finds the process outputs, comprising concentrations (main product and sub-products, reactants, etc.), measurements of quality properties (mechanical, chemical, etc.) or other relevant information about the end use of the product. With the increasing flexibility of processing units, these quantities can easily span different orders of magnitude and present rather different uncertainties associated with their measurements. This means that heteroscedasticity in the process outputs (Y's) is a rather prevalent feature in CPI, which must be properly managed and integrated in all the tasks that make use of process data. In this presentation we address this critical issue, by considering two sub-problems in which it can be decomposed: i) estimation of measurement uncertainty in heteroscedastic contexts; ii) integration of uncertainty in the outputs in model development. We present modelling approaches that describe the heteroscedastic behaviour of the Y-uncertainties in the operational domains of the measurements of interest and explore their interpolation and extrapolation capabilities and limitations. We also present the results of a Monte Carlo study where the added value of integrating different levels of information regarding the uncertainty affecting the output measurements is comparatively assessed. As a result of this study, valuable information is gathered on the benefits of collecting and using uncertainty information, which can be considered in decision making as a trade-off for the additional work the use of uncertainty information may encompass.

Dynamic Simulation, Sensitivity and Uncertainty Analysis of a Demonstration Scale Lignocellulosic Enzymatic Hydrolysis Process

Remus M. Prunescu, Technical University of Denmark, Kgs. Lyngby, Denmark
Gürkan Sin, Technical University of Denmark, Kgs. Lyngby, Denmark

Abstract

This study presents the uncertainty and sensitivity analysis of a lignocellulosic enzymatic hydrolysis model considering both model and feed parameters as sources of uncertainty. The dynamic model is parametrized for accommodating various types of biomass, and different enzymatic complexes, accounting a large number of parameters. The sensitivity analysis of model predictions with respect to model parameters is quantified by the delta mean square measure. By ranking the delta mean square, a reduced subset of parameters is found helping to identify the bottleneck of the model. The uncertainty analysis is carried for both model parameters and feed composition in order to assess the accuracy of the predictions. First, the model and feed parameters are sampled by Latin Hypercube Sampling (LHS) and then Monte Carlo simulations are run with the sampled values. Feed parameters are considered to be affected by non-zero mean noise because they are determined by a Near Infrared (NIR) instrument. LHS is performed on 2 parameters: the probability of the mean value and the probability of the standard deviation for each measurement. The Monte Carlo outputs are then analyzed by linear regression and the standardized regression coefficients (SRC) are computed for identifying the responsible parameters for model outputs precision. It is found that sugar yields are mostly sensitive to the composition of the enzymatic complex, and xylooligomers and glucose inhibition. pH is affected mostly by the amount of acetyl groups in the hemicellulose, while viscosity is sensitive to a few coefficients from its empirical equation.

Introduction

Biorefineries transform lignocellulosic agricultural wastes into products with higher added values following four major biochemical conversion steps: biomass pretreatment, enzymatic hydrolysis or liquefaction, fermentation, and purification [1]. Second generation bioethanol production technology had already reached commercial reality in 2012 [2] leading to the first plant commissioning in October 2013 by Beta Renewables in Crescentino, Italy [3], and followed by the first US plant named Project Liberty by POET-DSM, which started production in September 2014 in Iowa. Several other commercial size second generation bioethanol plants are expected to start operation in the near future: Abengoa Bioenergy (USA), DuPont (USA), Maabjerg Energy Concept (Denmark) etc.

During the pretreatment process, lignin is relocated and hemicellulose is partially hydrolyzed, allowing cellulose and the remaining hemicellulose to be exposed to enzymes for liquefaction. There are various types of pretreatment, out of which hydrothermal pretreatment with steam is seen as the best and cost-effective process [4], especially when the biorefinery is integrated with a power plant following the Integration Biomass Utilization System (IBUS) [1].

In the liquefaction phase, enzymes dissolve the pretreated lignocellulosic fibers in a high dry matter medium [5]. The enzymatic solution hydrolyses both hemicellulose and cellulose through a complex biochemical competitive conversion mechanism, which was thoroughly described and dynamically modeled in [6] with the following features: (1) Cellulose hydrolysis following the route: Cellulose $\rightarrow$ Cellobiose $\rightarrow$ Glucose and Cellulose $\rightarrow$ Glucose; (2) Hemicellulose hydrolysis: Xylan $\rightarrow$ Xylose; (3) Inhibition from sugar production; (4) Acetic acid production during hemicellulose hydrolysis; (5) Enzymatic complex parametrization; (6) Plug flow transport phenomena for the first hours of liquefaction; (7) Suitable for continuous and batch reactors; (8) pH calculator; (9) Viscosity calculator; (10) pH and temperature dependency of reaction rates; (11) Langmuir type adsorption of enzymes onto solids.

The model from [6] was analyzed in a continuous demonstration scale reactor for the first 7 – 10 hours of liquefaction. In this new study the hemicellulose hydrolysis route is extended with xylooligomers production because of their strong inhibition capabilities [7]. The process is then analyzed in batch mode for 200 h of enzymatic hydrolysis. It is found that the sensitivity of model parameters change in time: in the first hours of liquefaction some reactions are more active than others, leading to a high sensitivity on endo-exo type cellulase, β-xylosidase, and xylooligomers inhibition; as liquefaction progresses, the xylooligomers pool is reduced and their inhibition becomes less important being replaced by glucose inhibition and xylooligomers hydrolysis.

The uncertainty of model predictions with respect to model parameters also change in time, i.e. a high uncertainty in the first hours of liquefaction, which is gradually reduced as substrates are consumed. The uncertainty with respect to feed parameters has an opposite behavior as it gradually increases as the hydrolysis process progresses.

Dynamic Simulation

Model parameters are set as in [8] and [6] with similar values for xylan to xylose conversion routes. A simulation is prepared for 200 h, which is higher than a normal enzymatic hydrolysis time. The initial conditions are listed in Table 1. In order to increase the cost efficiency of the process, liquefaction occurs at high dry matter values [5], i.e. 20 % to 40 %. In this simulation scenario, water content is set to 600 g/kg corresponding to a dry matter of 40 %. Fibers are soaked with acetic acid from the pretreatment process and base is added in order to bring the pH to the enzymatic optimal value of 5 units. The enzyme dosage or concentration corresponds to the optimal value indicated in [9], i.e. 4 g per 100 g of cellulose. It is assumed that the pretreated fibers have no sugar content, i.e. cellobiose, glucose, xylooligomers and xylose concentrations are set to 0. The hold-up of the batch reactor does not matter in this scenario because there are no inflows, nor outflows from the tank.

Table 1: Initial conditions for dynamic simulation. The enzyme dosage is set according to [9], i.e 4 g per 100 g of cellulose. The composition of water is set such that a dry matter of 40 % is obtained.

(a) Initial states.

State	Value	Unit
Cellulose	170	g/kg
Xylan	50	g/kg
Lignin	140	g/kg
Ash	10	g/kg
Acid	7	g/kg
Cellobiose	0	g/kg
Glucose	0	g/kg
Xylooligomers	0	g/kg
Xylose	0	g/kg
Furfural	4	g/kg
Base	3	g/kg
Enzymes	7	g/kg
Water	600	g/kg
Other	10	g/kg

(b) Dry matter, pH, and viscosity.

Calculated	Value	Unit
Dry matter	40	%
pH	5	-
Viscosity	2.44	g/(m s)

Figure 1 displays the simulation results: cellulose and xylan decomposition in the top left plot, glucose and xylose yields with intermediate cellobiose and xylooligomers in the right top plot, and pH,

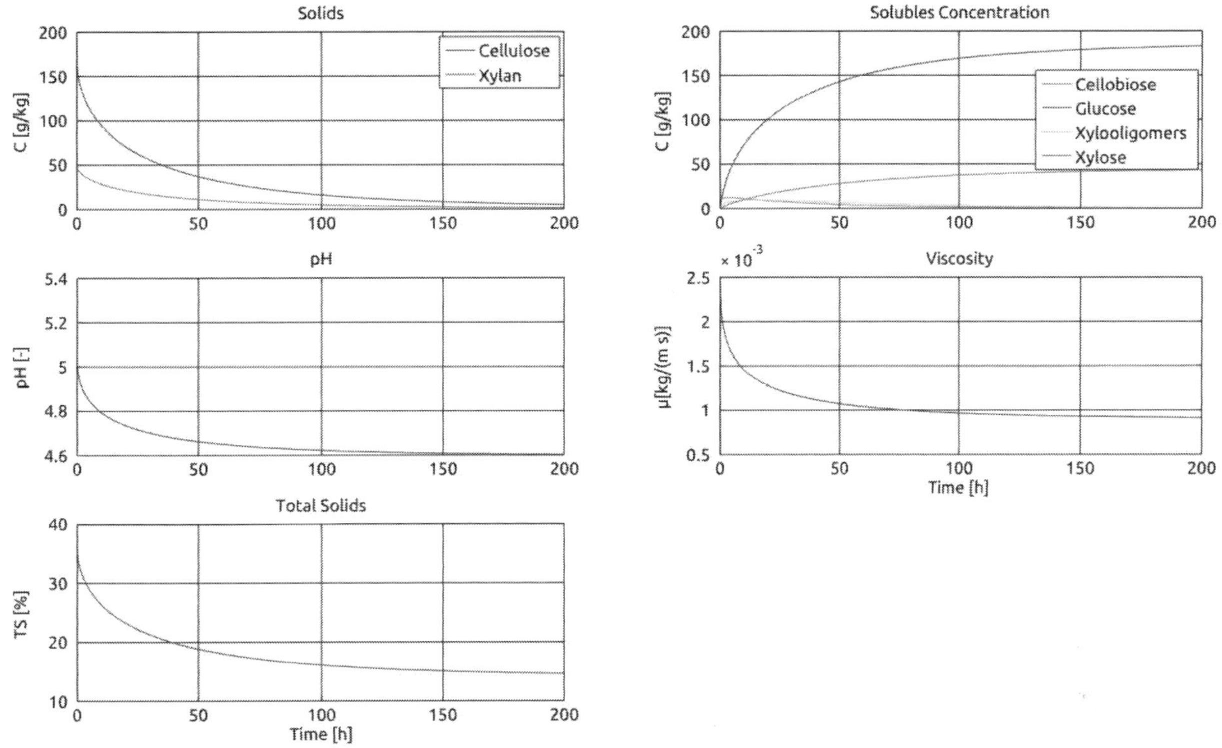

Figure 1: Dynamic simulation with model parameters from the literature.

viscosity and total solids in the remaining plots. Cellulose and xylan (solids) are hydrolyzed more rapidly in the first hours of liquefaction, and then all reaction rates gradually decrease due to product inhibition. Viscosity drops considerably in the first hours of liquefaction, from 2.5 g/(m s) to 1.5 g/(m s) after 10 h. pH also drops significantly in the first 10 h of liquefaction due to rapid xylan hydrolysis, which leads to acetic acid formation. In total, a drop of 0.4 pH units is recorded throughout the entire hydrolysis process. pH control would be necessary to keep the enzymatic activity close to optimality [10]. The solid content of the tank is reduced as cellulose and xylan get hydrolyzed. Lignin is not dissolved but rather only transported in solid state, so total solids cannot drop below 14 %. Lignin is recovered in the distillation and separation phase of the refining process, and sent to an evaporation unit where it forms bio-pellets. These bio-pellets are then co-burnt with coal in a nearby power plant for steam generation.

Sensitivity Analysis

In [11] it is shown that the model formulated by Kadam in [8], which has fewer parameters and is the basis of the model in this study, is over-parametrized. Many of the parameters are not significant for predicting the model outputs because not all involved phenomena have the same importance. E.g., when using an enzymatic complex with a high concentration of β-glucosidase, cellobiose to glucose becomes dominant and cellobiose concentration remains low leading to a minimal inhibitory effect on the other reactions. In this case, all parameters referring to cellobiose inhibition are expected to have a low sensitivity on model outputs and could be discarded from the model or set to a fixed value. This was the case from [6]. More than that, because the concentration of cellobiose stays low, the entire state could be dropped, thus achieving a reduction in model order and complexity.

The sensitivity analysis helps process understanding by quantifying the relation between model outputs and parameters, and ranks all involved phenomena with respect to their importance. Identifying a reduced subset of parameters also helps the model calibration procedure, which is computationally simplified and could run more often in a real industrial application.

The Delta Mean Square

A measure of sensitivity with respect to model parameters, and suitable for time varying signals, is the delta mean square δ_{ik}^{msqr} defined in [12]:

$$\delta_{ik}^{msqr} = \sqrt{\frac{1}{N} s_{nd,ik}^{\top} s_{nd,ik}} \tag{1}$$

where k is the parameter index, i is the model output index, N is the number of samples, and $s_{nd,ik}$ is a vector with the non dimensional sensitivity calculated in each sample:

$$s_{nd,ik} = \frac{\partial y_i}{\partial \theta_k} \frac{\theta_k}{sc_i} \tag{2}$$

$\partial y_i / \partial \theta_k$ represents the output variation with respect to a variation in parameter θ_k, and sc_i is a scaling factor with the same physical dimension as the corresponding observation in order to make this measure non dimensional. The scaling factor is chosen as the maximum value for that observation throughout the whole simulation time because some of the outputs, i.e. concentration of cellobiose and xylooligomers, are non-zero only throughout the first hours of liquefaction and a mean value would decrease their importance in a long simulation with many observation points:

$$sc_i = \max y_i(t) \tag{3}$$

All parameters are ranked according to δ_{ik}^{msqr} for each output i. As the sensitivity measure is non-dimensional, a cumulative variable is also defined as the sum of normalized sensitivities for a given parameter in all outputs. Because the model has to predict all defined outputs, the subset of significant parameters contains all parameters with a cumulative sensitivity above a threshold, which is set to 10 % of the maximum sensitivity.

Results

Figure 2 shows the sensitivity measure δ^{msqr} for cellobiose, glucose, xylooligomers and xylose. *Cellobiose sensitivity:* the most significant model parameters for cellobiose production are: β-glucosidase fraction from the enzymatic complex α_C^G; cellobiose to glucose reaction rate parameters from r_3, i.e. the reaction rate constant K_3, the overall inhibition term I_3, and glucose inhibition I_{G_3}; cellulose to cellobiose reaction rate constant K_1; endo-exo type cellulase fraction from the enzymatic mix α_C^E; and xylooligomers inhibition of r_1, i.e. I_{Xo_1}. These parameters are expected to affect cellobiose concentration as they appear either in cellobiose production r_1 or consumption r_3. β-glucosidase fraction is the most important factor as it directly influences the pool of cellobiose. Only inhibition terms I_{Xo_3}, I_{X_3}, I_{C_1}, I_{X_1} and I_{G_1} are not important in this scenario mostly because xylooligomers inhibition dominates in r_1, and the overall inhibition term I_3 and glucose inhibition I_{G_3} are more significant than the other individual inhibition terms from r_3.

Glucose sensitivity: glucose yield is sensitive to the following model parameters: cellulose to cellobiose reaction rate K_1 for creating a substrate for cellobiose to glucose reaction; fraction of exo-endo type cellulase α_C^E, which is directly involved in glucose production rates; xylooligomers inhibition I_{Xo_1}, which strongly inhibits glucose production; β-xylosidase fraction α_X^X for contributing to xylooligomers production and consumption; xylooligomers to xylose reaction rate constant K_6; xylooligomers to xylose overall inhibition term I_6; glucose inhibition of xylooligomers reduction I_{G_6}. The xylooligomers to xylose

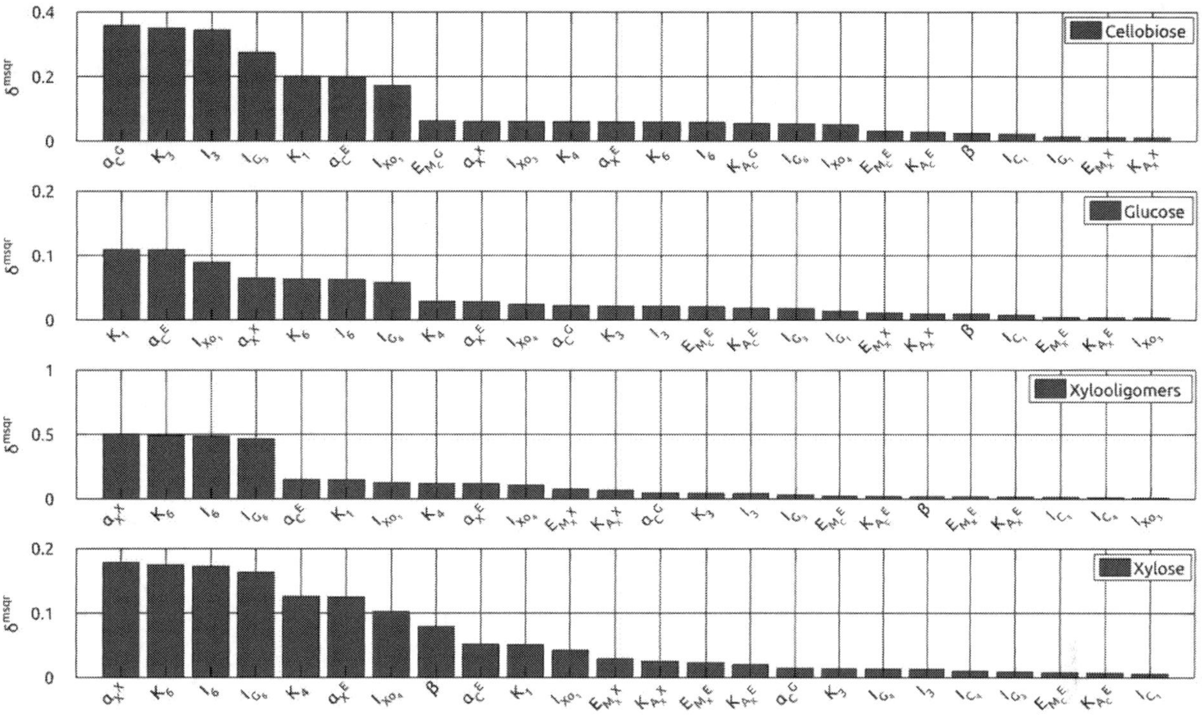

Figure 2: Sensitivity analysis: sorted δ_k^{msqr} for cellobiose, glucose, xylooligomers and xylose (top to bottom).

reaction parameters are important because they control the xylooligomers pool, which strongly inhibits glucose production, which is an important phenomena well captured by the model.

Xylooligomers sensitivity: the following model parameters are found to significantly influence xylooligomers production: xylooligomers to xylose parameters such as β-xylosidase fraction from the enzymatic complex α_X^X, the reaction rate constant K_6, the overall inhibition term I_6, and glucose inhibition I_{G_6}. β-xylosidase are directly involved in production and consumption of xylooligomers while the other parameters appear in the expression of r_6, which determines consumption of xylooligomers.

Xylose sensitivity: xylose yield is sensitive to model parameters affecting xylooligomers to xylose, and xylan to xylooligomers reaction rates: β-xylosidase fraction α_X^X, reaction rate K_6, the overall inhibition term I_6, glucose inhibition I_{G_6}, reaction rate constant K_4, endo-exo type xylanase α_X^E, xylooligomers inhibition I_{Xo_4}, and β_{Ac} indirectly through pH dependency.

pH and viscosity sensitivities: are summarized in the first two plots from Figure 3. pH is mostly sensitive to the β parameter, as expected because it is an indication of the amount of acetyl groups in hemicellulose and their hydrolysis lead to acetic acid production. Viscosity μ is sensitive to coefficients a_4, a_0, a_1 and a_2 from the relative viscosity equation from [13].

The cumulative δ^{msqr} sensitivity measure is plotted in the bottom part of Figure 3. In order to predict glucose and xylose yields, the pH and the viscosity of the medium, only a subset of approximately 23 parameters could be used. Most of these parameters relate to enzymatic complex composition, i.e. the α parameters, xylooligomers to xylose, and cellobiose to glucose reaction rates, which makes sense since these intermediate products with a higher degree of polymerization represent a step towards xylose and glucose production, and also strongly inhibit sugar production. For predicting the pH and viscosity of the mixture, β_{Ac}, and a_0, a_1, a_2, and a_4 would be enough.

This analysis identifies the bottleneck of the model predictions, i.e. the composition of the enzymatic complex and product inhibition, i.e. xylooligomers, cellobiose, glucose and xylose. The model can be reduced to 23 parameters and predict the sugar yields from the process, including pH and viscosity

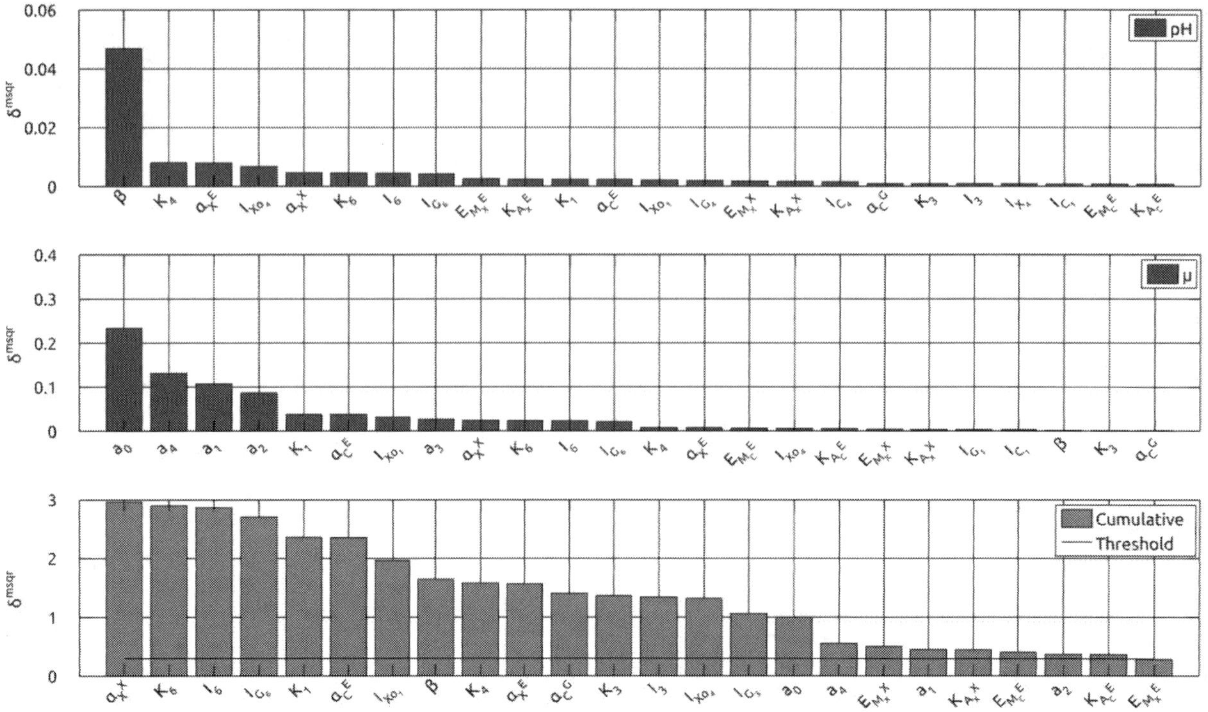

Figure 3: Sensitivity analysis: sorted δ_k^{msqr} for pH, viscosity and cumulative (top to bottom). The threshold indicates the most important parameter subset.

calculations.

Uncertainty Analysis of Model Predictions

The uncertainty analysis follows the standard Monte Carlo technique, which includes the following four steps [14]: (1) define parameters uncertainties with their range; (2) sampling of model and feed parameters using the Latin Hypercube Sampling technique (LHS); (3) run Monte Carlo simulations with sampled values; (4) evaluate results.

Sampling of Model and Feed Parameters

The uncertainty range for model parameters is based on expert information and is set to approximately 25 % like in [14]. Model parameters are considered uniformly distributed in the defined range. The feed parameters or the initial conditions for solids and acetic acid are typically measured with Near Infrared (NIR) equipment, which has a certain degree of uncertainty depending on how well the instrument is calibrated. This uncertainty is assumed to be normally distributed with mean value μ and standard deviation σ, such that it covers a range of 10 % uncertainty of nominal values:

$$\varepsilon \in N(\mu, \sigma) \tag{4}$$

where ε is the NIR measurement error, μ is the mean value and σ is the standard deviation. The mean value is assumed to be normally distributed while the standard deviation σ describes the measurement noise and is assumed to have a Gamma distribution parametrized in k and θ, i.e. a shape, and a scale parameter:

$$\mu \in N(0, \sigma_\mu) \qquad \sigma \in \Gamma(1, 2) \tag{5}$$

where σ_μ is set to cover 10 % uncertainty of the nominal values listed in Table 1.

Model and feed parameters are treated separately in the uncertainty analysis because feed parameters can shadow the effects of model parameters uncertainty. When treating model parameters uncertainty, the feed parameters are set constant as in Table 1.

Uncertainty Analysis with Respect to Model Parameters

Figure 4(a) displays the results after 800 Monte-Carlo simulations with sampled model parameters and constant feed parameters. The patch area identifies the 5th-95th percentile interval. The model has a higher uncertainty between 10 h and 80 h for glucose, cellobiose, xylooligomers and xylose, and gradually decreases as cellulose and xylan are depleted. It is expected to have a reduced uncertainty on sugar production as the substrate is consumed, near the end of the hydrolysis time, and a higher uncertainty in the first hours of liquefaction when the competitive conversion mechanism with product inhibition is more active due to a higher concentration of xylooligomers and cellobiose. The uncertainty on pH slowly increases as hydrolysis progresses due to the uncertainty on the amount of acetyl groups in hemicellulose, expressed as the β_{Ac} parameter. However, at the end of the liquefaction process, the overall uncertainty on pH reaches 0.1 units, which is considered acceptable. The viscosity uncertainty is slightly higher in the first hours of liquefaction due to the uncertainty on the model parameters that affect cellulose and hemicellulose decomposition. As the amount of solids is reduced, the uncertainty interval also slowly decreases.

Uncertainty Analysis with Respect to Feed Parameters

The case with feed parameters uncertainty is shown in Figure 4(b). As expected, the uncertainty interval slowly increases as the hydrolysis process progresses. Glucose has the highest uncertainty due to a higher variation in the initial concentration of cellulose. A 10 % uncertainty on xylan content translates to little uncertainty on pH time profile, and to a smaller uncertainty on xylose final yield. β_{Ac} has a higher degree of uncertainty on pH than xylan variation. The viscosity is not affected much, being influenced mainly by the initial total solids composition, which appears as a slightly higher uncertainty interval in the first hours of liquefaction.

Sensitivity Analysis - The Standardized Regression Coefficients (SRC)

In the last step of the methodology, in addition to inference statistics, a sensitivity analysis is also performed using linear regression of Monte Carlo simulations outputs, also known as the standardized regression coefficients (SRC) [15]. The linear model from Equation (6) is fitted using the least squares method on the model outputs y, i.e. concentrations of glucose, cellobiose, xylose, xylooligomers, pH and viscosity, for each Monte Carlo simulation:

$$y_{jk} = a + \sum_i b_i \theta_{ji} \tag{6}$$

where y_{jk} is the jth model output of simulation k, a and b_i are the linear model coefficients, and θ_{ji} is the ith model parameter from the jth sampled values.

The methodology aims at finding how much each model parameter contributes to the variance or uncertainty recorded in the model predictions. The relation between parameter and output variations is quantified by the β coefficient defined as:

$$\beta = \frac{\sigma_{\theta_i}}{\sigma_y} b_i \tag{7}$$

The standardized regression coefficients with respect to model parameters are ranked in Table 2 at different time samples, i.e. 1 h, and 200 h. These tables suggest that the coefficients change in time, which is expected due to the xylooligomers and cellobiose intermediate products.

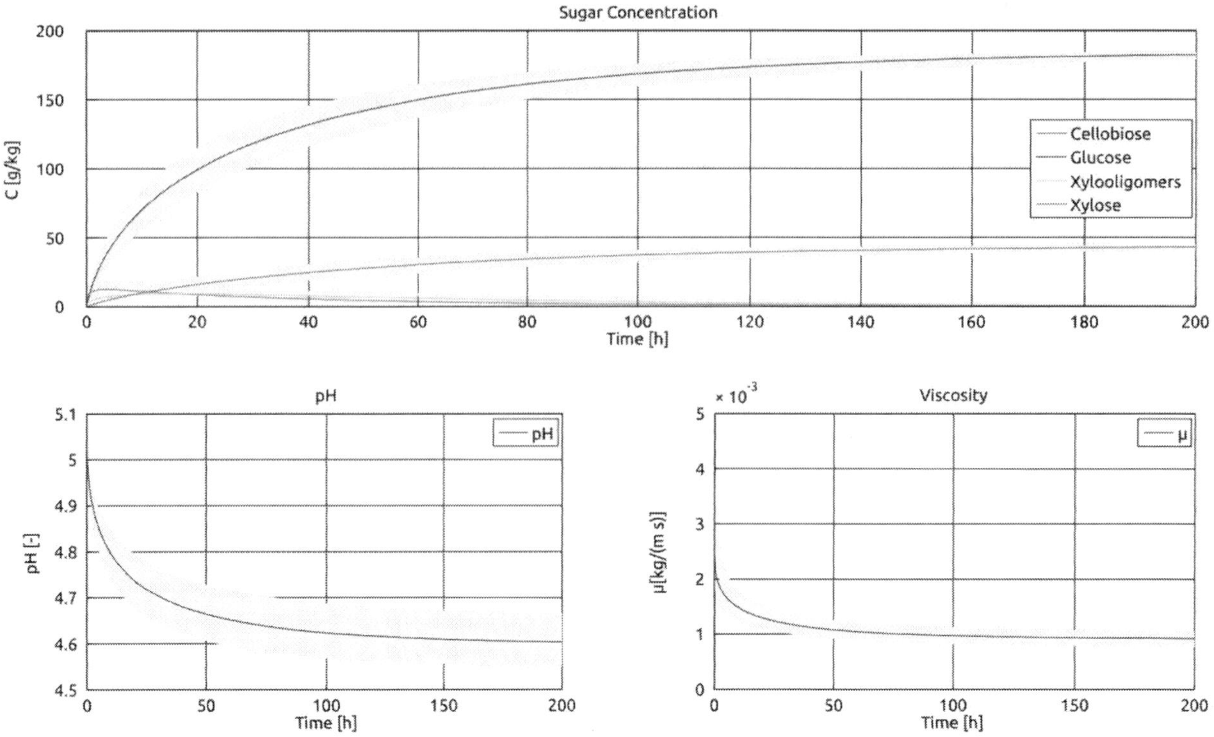

(a) Dynamic simulation with model parameters uncertainty. The patch area shows the 5th-95th percentile calculated after 800 Monte Carlo simulations.

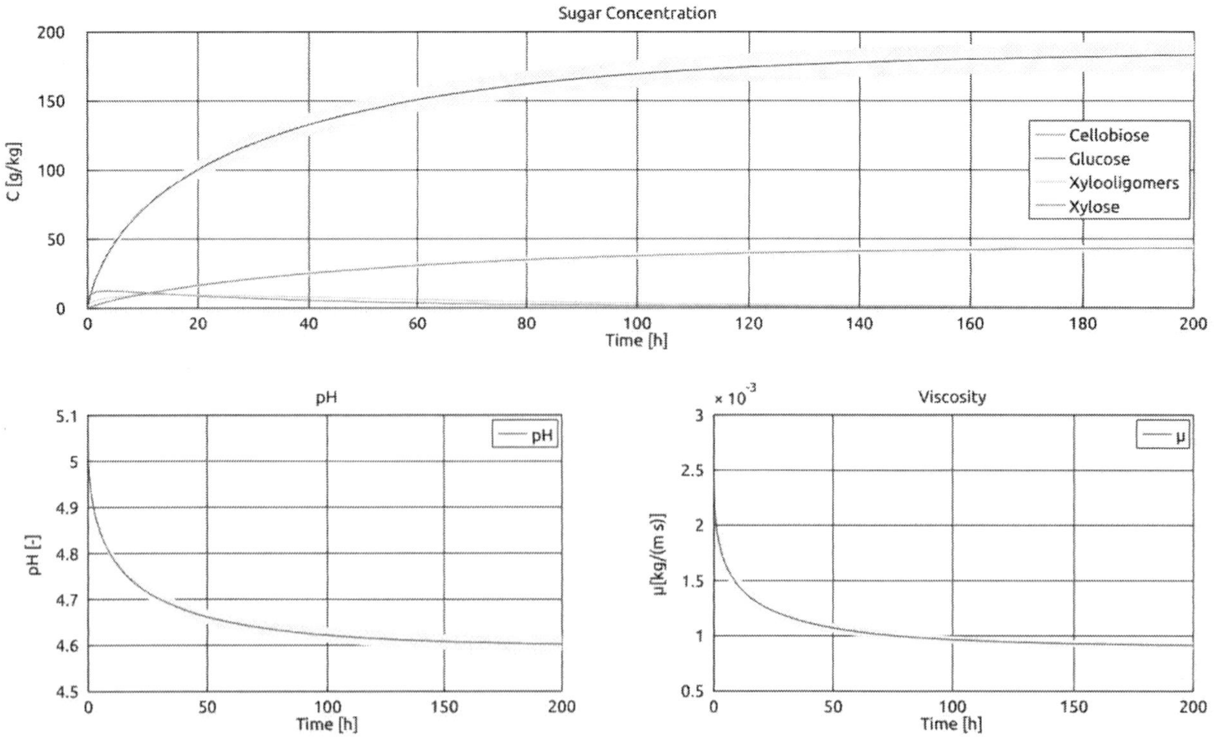

(b) Dynamic simulation with feed parameters uncertainty. The patch area shows the 5th-95th percentile calculated after 800 Monte Carlo simulations.

Figure 4: Dynamic simulation with model and feed parameters uncertainty.

(a) SRC coefficients for model parameters uncertainty at $t = 1\,\text{h}$.

θ	C_C	θ	C_G	θ	C_{Xo}	θ	C_X	θ	pH	θ	μ
α_C^E	0.48	α_C^E	0.45	K_4	0.58	α_X^X	0.41	β	−0.74	a_4	0.83
K_1	0.47	K_1	0.44	α_X^E	0.57	K_6	0.40	K_4	−0.40	a_2	0.20
I_{Xo_1}	0.38	I_{Xo_1}	0.34	I_{Xo_4}	0.47	I_6	−0.37	α_X^E	−0.39	a_0	0.20
α_C^G	−0.28	α_C^G	0.33	K_6	−0.22	α_X^E	0.33	I_{Xo_4}	−0.32	a_3	0.16
K_3	−0.27	K_3	0.32	α_X^X	−0.22	K_4	0.32	α_C^E	0.072	a_1	0.16

(b) SRC coefficients for model parameters uncertainty at $t = 200\,\text{h}$.

θ	C_C	θ	C_G	θ	C_{Xo}	θ	C_X	θ	pH	θ	μ
K_3	−0.37	α_C^E	0.43	α_X^X	−0.48	β	−0.45	β	−0.99	a_0	0.91
α_C^G	−0.37	K_1	0.40	I_{G_6}	−0.46	α_X^X	0.38	K_4	−0.044	a_1	0.34
I_{G_3}	−0.36	α_X^X	0.39	K_6	−0.46	I_{G_6}	0.37	α_X^E	−0.040	a_2	0.20
I_3	0.34	K_6	0.37	I_6	0.43	K_6	0.36	α_X^X	−0.032	a_4	0.073
α_X^X	−0.30	I_6	−0.36	K_4	−0.14	I_6	−0.35	I_{G_6}	−0.030	α_X^X	−0.032

Table 2: Standardized regression coefficients (SRC) or β coefficients.

Table 2(a) shows the coefficients after 1 h of liquefaction. Cellulose and glucose concentrations are mostly sensitive to the fraction of endo-exo type of cellulase, which is expected since these enzymes directly contribute to cellulose depolymerization and glucose formation. The reaction rate constant K_1 for cellulose to cellobiose, and xylooligomers inhibition on the same reaction rate are also important. β-glucosidase, and reaction rate constant K_3 for cellulose to glucose have a slightly lower importance. Concentration of xylooligomers is mostly sensitive to endo-exo type of xylanase, reaction rate constant K_4 for xylan to xylooligomers, and xylooligomers inhibition term I_{Xo_4}. Concentration of xylose is influenced by β-xylosidase, reaction rate constant K_6, and inhibition term I_6. pH is mostly affected by the acetyl groups parameter β_{Ac}, while viscosity remains high and only a_4 is significant. These results show that the endo-exo type of enzymes from the enzymatic complex, as well as xylooligomers inhibition are important factors in the first hours of liquefaction.

In Table 2(b) it is seen that, as glucose inhibition becomes dominant mostly due to to the fact that xylooligomers and cellobiose slowly drop in concentration, I_{G_6} is moving up the rankings of C_{Xo} and C_X. Enzymatic complex composition still remains in the first positions of the sensitivity rankings. pH remains sensitive to the acetyl groups concentration expressed in the β_{Ac} coefficient. As viscosity drops due to solids reduction, a_0 becomes the most significant parameter.

The standardized regression coefficients with respect to feed parameters are ranked in Table 3. Initial concentration of cellulose C_{C_S} is the most important parameter for cellobiose and glucose yields, which is expected. Initial concentration of xylan C_{X_S} is an important parameter for xylooligomers and xylose yields, as well as for the medium pH. Viscosity is sensitive to solid lignin C_{L_S} because lignin is neither produced, nor consumed during the hydrolysis process. A change in its concentration shifts the viscosity curve, directly affecting its value.

θ	C_C	θ	C_G	θ	C_{Xo}	θ	C_X	θ	pH	θ	μ
C_{C_S}	0.70	C_{C_S}	0.70	C_{X_S}	0.71	C_{X_S}	0.92	C_{X_S}	−0.99	C_{L_S}	0.57
C_W	−0.44	C_{X_S}	−0.49	C_{C_S}	0.33	C_{C_S}	−0.52	C_{C_S}	0.34	C_W	−0.39
C_{L_S}	−0.31	C_W	−0.41	C_{L_S}	−0.30	C_W	0.15	C_{L_S}	0.16	C_{X_S}	0.24
C_{X_S}	−0.18	C_{L_S}	−0.35	C_W	−0.28	C_{L_S}	−0.0081	C_W	0.064	C_{C_S}	0.10

Table 3: Standardized regression coefficients (SRC) or β coefficients for model output sensitivity with respect to feed parameters.

Conclusions

This report presents the sensitivity and uncertainty analysis of a liquefaction dynamic model. Two types of sensitivity analysis were conducted: one by calculating the delta mean square measure δ_{ik}^{msqr} for each output with respect to each model parameter; and another one by computing the standardized regression coefficients (SRC). Both analysis showed that the cellobiose, glucose, xylooligomers, and xylose, as well as the pH and viscosity of the medium are mostly sensitive to the enzymatic complex composition. Product inhibition by xylooligomers is mostly influential in the first hours of liquefaction, and as these high degree of polymerization sugars get depleted in time, glucose inhibition takes over. The model could be simplified to 23 parameters by choosing the most significant parameters above a threshold from the cumulative delta mean square measure.

The uncertainty analysis helped identifying the accuracy of the model considering 25 % uncertainty in model parameters and 10 % uncertainty in feed parameters. Model parameters uncertainty affects the process in its first hours and gradually decreases as the liquefaction progresses in time. Feed parameters had an opposite behavior, the uncertainty slowly increasing as the hydrolysis advanced in time.

This analysis is valuable for identifying and understanding the bottlenecks of a liquefaction process. The enzymatic complex should be identified prior to using the model in an industrial setup. The model could also be used to optimize the composition of the enzymatic solution, as well as the enzyme dosage. These issues are subject to future studies.

References

[1] J. Larsen et al. "The IBUS Process – Lignocellulosic Bioethanol Close to a Commercial Reality". In: *Chemical Engineering & Technology* 31.5 (May 2008), pp. 765–772. DOI: 10.1002/ceat.200800048.

[2] J. Larsen, M. Ø. Haven, and L. Thirup. "Inbicon makes lignocellulosic ethanol a commercial reality". In: *Biomass and Bioenergy* 46 (Nov. 2012), pp. 36–45. DOI: 10.1016/j.biombioe.2012.03.033.

[3] Novozymes A/S. "Beta Renewables opens biofuels plant in Italy". In: *Focus on Catalysts* 2013.12 (Dec. 2013), p. 6. DOI: 10.1016/S1351-4180(13)70461-6.

[4] D. Chiaramonti et al. "Review of pretreatment processes for lignocellulosic ethanol production, and development of an innovative method". In: *Biomass and Bioenergy* 46 (Nov. 2012), pp. 25–35. DOI: 10.1016/j.biombioe.2012.04.020.

[5] H. Jørgensen et al. "Liquefaction of lignocellulose at high solids concentrations". In: *Biotechnology and Bioengineering* 96.5 (2007), pp. 862–870. DOI: 10.1002/bit.

[6] R. M. Prunescu and G. Sin. “Dynamic modeling and validation of a lignocellulosic enzymatic hydrolysis process–a demonstration scale study.” In: *Bioresource technology* 150 (Dec. 2013), pp. 393–403. DOI: 10.1016/j.biortech.2013.10.029.

[7] Q. Qing, B. Yang, and C. E. Wyman. “Xylooligomers are strong inhibitors of cellulose hydrolysis by enzymes.” In: *Bioresource technology* 101.24 (Dec. 2010), pp. 9624–9630. DOI: 10.1016/j.biortech.2010.06.137.

[8] K. L. Kadam, E. C. Rydholm, and J. D. McMillan. “Development and validation of a kinetic model for enzymatic saccharification of lignocellulosic biomass.” In: *Biotechnology progress* 20.3 (2004), pp. 698–705. DOI: 10.1021/bp034316x.

[9] Novozymes A/S. *Cellulosic ethanol Novozymes Cellic CTec3 - Secure your plant’s lowest total cost.* Tech. rep. 2012, pp. 1–6.

[10] R. M. Prunescu, M. Blanke, and G. Sin. “Modelling and L1 adaptive control of pH in bioethanol enzymatic process”. In: *Proceedings of the 2013 American Control Conference.* 2013, pp. 2–9.

[11] G. Sin, A. S. Meyer, and K. V. Gernaey. “Assessing reliability of cellulose hydrolysis models to support biofuel process design—Identifiability and uncertainty analysis”. In: *Computers & Chemical Engineering* 34.9 (Sept. 2010), pp. 1385–1392. DOI: 10.1016/j.compchemeng.2010.02.012.

[12] R. Brun, P. Reichert, and H. Künsch. “Practical identifiability analysis of large environmental simulation models”. In: *Water Resources Research* 37.4 (2001), pp. 1015–1030.

[13] D. G. Thomas. “Transport characteristics of suspension: VIII. A note on the viscosity of Newtonian suspensions of uniform spherical particles”. In: *Journal of Colloid Science* 20.3 (1965), pp. 267–277.

[14] R. Morales-Rodriguez et al. “A framework for model-based optimization of bioprocesses under uncertainty: Lignocellulosic ethanol production case”. In: *Computers & Chemical Engineering* 42 (July 2012), pp. 115–129. DOI: 10.1016/j.compchemeng.2011.12.004.

[15] J. Helton and F. Davis. “Latin hypercube sampling and the propagation of uncertainty in analyses of complex systems”. In: *Reliability Engineering & System Safety* 81.1 (July 2003), pp. 23–69. DOI: 10.1016/S0951-8320(03)00058-9.

The Wind Integration National Dataset (WIND) Toolkit: wind power forecasts and production time series

Authors: Caroline Draxl, Dan Getman, Wesley Jones, Jim McCaa, Bri-Mathias Hodge

Regional wind integration studies in the United States require detailed wind power output data at many locations to perform simulations of how the power system will operate under high penetration scenarios. The wind datasets that serve as inputs into the study must realistically reflect the ramping characteristics, spatial and temporal correlations, and capacity factors of the simulated wind plants, as well as being time synchronized with available load profiles. The WIND Toolkit described here fulfills these requirements as a state-of-the-art national (US) wind resource, power production and forecast dataset.

The Weather Research and Forecasting (WRF) model was run at 2km by 2km spatial resolution and 5 minute temporal resolution over the entire continental United States for the years 2007 - 2013 to produce this unprecedented dataset, consisting of over 1 petabyte of model output. As would be expected, this was an extremely computationally expensive and required the use of the NREL Peregrine High Performance Computing (HPC) environment. Three different datasets were created: a meteorological dataset, a power production dataset, and the forecasts dataset. The power production dataset was created by utilizing the information provided in the meteorological dataset for a subset of data that includes over 126,000 onshore and offshore wind power production sites. Wind speed was converted to power through a multistep process that included: bias removal from wind speeds, wind speed adjustment for wakes, application of power curves, and statistical adjustment to power. Appropriate power curves were used for each of the categories considered, including offshore and class 1 – 4 wind sites.

State-of-the-art forecast accuracy was mimicked by reforecasting the 2007-2013 period using standard industry techniques. WRF was run on a nationwide 6-km grid for these 7 years to produce hourly output of barometric pressure, wind speed and direction (at 100 m), relative humidity, temperature, and air density. The simulated operational power output and forecasts are freely available for 1-hour, 4-hour, 6-hour, and day-ahead forecast horizons via an online data interface for all of the selected production sites.

The Economic Potential of Process Stoichiometries in Incomplete Markets

Fernando Garcia, Fanghui Fan, and Jeffrey Kantor*

May 13, 2014

Abstract

Modern analysis and design of chemical processes involves the integration of financial and plant operations. A key aspect in optimizing process goals is the attainment of economic and thermodynamic feasibility. In previous work, we proposed a process model for the underlying reactions, stoichiometry, mass and energy balances of a complex chemical process by imposing a minimal set of assumptions regarding the uncertain nature of price dynamics in the commodity markets [5]. Employing a framework developed by Shinnar [9, 8], Hildebrandt and Glasser [3, 2, 1] for chemical stoichiometry and thermodynamics we establish a modern stochastic interpretation of the concept of "economic potential" first proposed by Douglas [4].

A process model is formulated using the idea of a "simple process" [3, 2, 1], which combines basic chemical stoichiometry and thermodynamics to determine its feasibility. We then construct an investment portfolio comprised of commodities and risk-free assets to pose an investment problem, which calculates the "fair" price for leasing a unit of capacity of this chemical process. A focal point of this work is to illustrate how to obtain arbitrage-free scenario trees for stochastic commodity prices [6, 7]. Because the underlying assets cannot span the dynamics of this portfolio the market is "incomplete", which results in the existence of a range of prices. A linear programming problem is then solved to obtain the upper and lower bounds to the prices of this process lease or what in quantitative finance is known as the bid-ask price [10]. Tighter bounds are established by introducing second order stochastic dominance – the natural generalization of conditional value at risk – to measure preferences of risk-averse operators.

These markets are characterized for subjecting their participants to some degree of financial risk, which in turn may represent monetary losses. A key point of this study will be to determine efficient ways to hedge this risk and minimize the likelihood of incurring a deficit.

Keywords: Process stoichiometry, valuation, incomplete markets.

References

[1] David Glasser Bilal Patel, Diane Hildebrandt and Brendon Hausberger. Thermodynamics analysis of processes. 1. implications of work integration. *Industrial & Engineering Chemistry Research*, 44:3529–3537, 2005.

[2] David Glasser Diane Hildebrandt, Brendon Hausberger and Bilal Patel. Synthesis and integration of chemical processes from a mass, energy, and entropy perspective. *Industrial & Engineering Chemistry Research*, 46(25):8756–8766, 11 2007.

[3] David Glasser Diane Hildebrandt, Brendon Hausberger and Bilal Patel. Producing transportation fuels with less work. *Science*, 323(5922):1680–1681, Mar 2009.

[4] James M. Douglas. *Conceptual Design of Chemical Processes*. McGraw-Hill, 1988.

[5] Fernando Garcia and Jeffrey C. Kantor. The economic potential of process stoichiometries in incomplete markets. 2012.

[6] Alois Geyer, Michael Hanke, and Alex Weissensteiner. No-arbitrage conditions, scenario trees and multi-asset financial optimization. *European Journal of Operational Research*, 206:609–613, 2010.

[7] Pieter Klaasen. Generating scenario trees for multistage decision problems. *Management Science*, 48(11):1512–1516, November 2002.

[8] Reuel Shinnar. Thermodynamic analysis in chemical process and reactor design. *Chemical Engineering Science*, 43(8):2303–2318, 1988.

[9] Reuel Shinnar, Dan Shapira, and Shlomlth Zakal. Thermochemical and hybrid cycles for hydrogen production. a differential economic comparison with electrolysis. *Industrial and Engineering Chemical Process Design & Development*, 20:581–593, 1981.

[10] Aleš Černý. *Mathematical Techniques in Finance: Tools for Incomplete Markets*. Princeton University Press, 2nd edition, 2010.

*Author to whom correspondence should be addressed: `Kantor.1@nd.edu`

New criteria for evaluating variable selection performance

Zi Xiu Wang[1], Qinghua He[2], and Jin Wang[1]
[1]Department of Chemical Engineering, Auburn University, Auburn, AL 36849
[2]Department of Chemical Engineering, Tuskegee University, Tuskegee, AL 36088

Data-driven soft sensors have been widely used in both academic research and industrial applications for predicting hard-to-measure variables or replacing physical sensors to reduce cost. It has been shown that the performance of these data-driven soft sensors could be greatly improved by selecting only the vital variables that strongly affect the primary variables, rather than using all the available process variables. Extensive researches have been conducted to develop algorithms to select the relevant variables. However, the existing criteria, such as sensitivity, specificity, accuracy, and geometric mean of sensitivity and specificity, for evaluating variable selection methods have a major drawback: not applicable unless the ground truth of the data is known, which is most likely the case for the industrial applications. For such cases, soft sensor prediction performance becomes the most commonly used criterion to indirectly evaluate variable selection performance.

When the ground truth is not available, how consistent the variable selection results from different sampling runs can provide important information on the robustness of the variable selection methods. In this work, we proposed several indices to evaluate the variable selection performance. Some are directly based on the probability of variables being selected, and others are based on information entropy associated with the selection frequency. The proposed indices are evaluated with several case studies and compared with the ones mentioned above, such as sensitivity and specificity.

Session: Advances in Data Analysis: Theory and Applications

14AIChE Annual Meeting, Atlanta, GA

Detection of Causality between Industrial Alarm Data based on Transfer Entropy

(No. 103f, Paper382963)

Fan Yang, Weijun Yu

National Tsinghua Laboratory for Information Science and Technology

Department of Automation, Tsinghua University

- ***Importance of Detection of Causality***
- ***Basic Definition of Transfer Entropy***
- ***TE Method based on Alarm Data***
- ***Case Study***

Importance of Detection of Causality

The number of alarm variables becomes larger:

- Modern industrial processes become more and more complex.
- More physical quantities become measureable and monitored by DCS. We can easily get corresponding alarm variables by configurations in monitoring systems.
- There are complex interactions between different parts of the process. Once an abnormal situation occurs at some place in the process, the fault may spread quickly to other parts.

Under such situation, operators usually cannot identify the type of the fault, find the root cause and take effective actions in time, leading to serious consequence.

Importance of Detection of Causality

What can be done if we obtain the causality between alarm variables:

- We may find the propagation path of the fault and help operators find the root of the fault.
- Operators can take effect actions immediately, which will reduce the amount of serious accidents.

Thus the detection of the causality between alarm data becomes very important.

- ***Importance of Detection of Causality***
- ***<u>Basic Definition of Transfer Entropy</u>***
- ***TE Method based on Alarm Data***
- ***Case Study***

Basic Definition of Transfer Entropy

- In 2000, Schreiber proposed the definition of Transfer Entropy (TE) to describe the causality between variables.
- The basic definition of TE:

$$TE_{J\to I} = \sum p(i_{t+1}, i_t^{(k)}, j_t^{(l)}) \log_2 \frac{p(i_{t+1}|i_t^{(k)}, j_t^{(l)})}{p(i_{t+1}|i_t^{(k)})}$$

- If there exist causality from *J* to *I*, it will be helpful to predict the value of *I* using the historical data of *J*, then the value of the numerator should be larger than the value of the denominator and the value of the whole calculation equation;
- If there is no causality from *J* to *I*, the value of the whole equation should be close to 0.

References

[1] Schreiber T. Measuring information transfer. *Physical Review Letters*, 2000, 85(2): 461.
[2] Duan P, Yang F, Chen T, Shah SL. Direct causality detection via the transfer entropy approach. *IEEE Transactions on Control Systems Technology*, 2013, 21(6): 2052-2066
[3] Duan P, Yang F, Chen T, Shah SL. Transfer zero-entropy and its application for capturing cause and effect relationship between variables. *IEEE Transactions on Control Systems Technology*, DOI: 10.1109/TCST.2014.2345095

Basic Definition of Transfer Entropy

We suggest using alarm data directly other than using continuous time series:

- Continuous series can describe the whole characteristic of the process but it is very computational costly
- In most cases, we put more concern in the regular patterns under abnormal situations other than the dynamic relationships under normal situations, so it is unnecessary to take all cases into account
- Some alarm variables themselves are not generated by continuous variables, such as switch variables and state variables

Basic Definition of Transfer Entropy

Generally speaking, the value of TE will not be exact 0 because of noise, so we need a threshold:

- We use Monte Carlo method with surrogate data:

a. We keep the number of alarms in series J but change the locations of these alarms randomly, which can get surrogate series J'.

b. We then calculate TE using I and J' and get a result.

c. We can repeat the procedure some times and use the average μ_λ as the threshold.

d. When the calculation result is greater than the threshold by some σ_λ, we can conclude that there exists significant causality from J to I.

- ***Importance of Detection of Causality***
- ***Basic Definition of Transfer Entropy***
- ***<u>TE Method based on Alarm Data</u>***
- ***Case Study***

TE method based on Alarm Data

<u>Alarm series or event:</u>

•Each digit in the series is 0 or 1, where 0 represents that there is no alarm at present and 1 represents that there is an alarm at present.

•There are different characteristic of two-value series in different areas: First, electrical signals in the neuroscience area always exist in very short time, so the series is very sparse and people put more emphasis in the happening time of the signal; while alarms in the industrial process often last for some time and once an abnormal situation occurs, maybe the alarms here will be intensive.
Second, the electrical signals in the neuroscience are usually clean, while there are often some false alarms and missed alarms in the industrial process because of noise, as shown in Fig. 1.

<u>Thus we need some data preprocessing before using industrial data for analysis.</u>

TE method based on Alarm Data

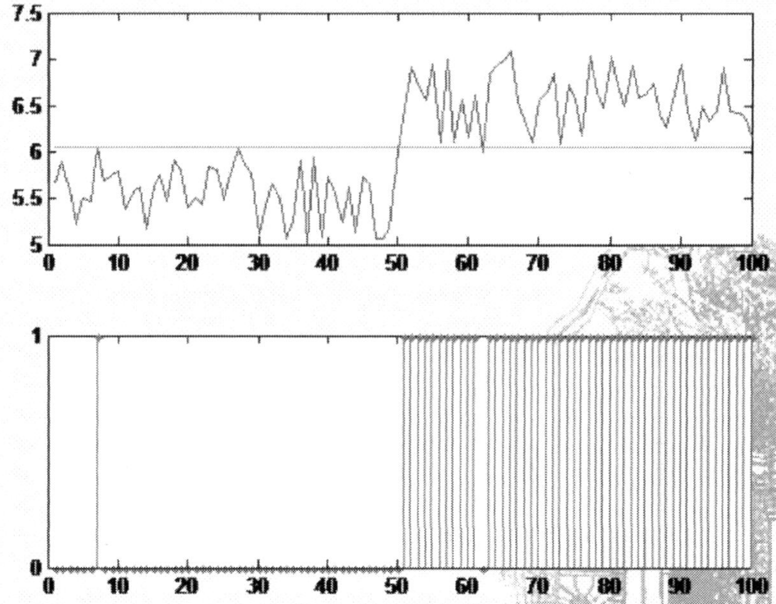

Fig. 1 False alarms and missing alarms generated by noise
(false alarm at *t*=7, missing alarm at *t*=62)

TE method based on Alarm Data

Data preprocessing:

- Get rid of those alarms with no other alarms before or after them in a quite long time

- Put some more signals around those alarms which last for a long time before recovering.

Thus we can reduce the number of false alarms and missed alarms.

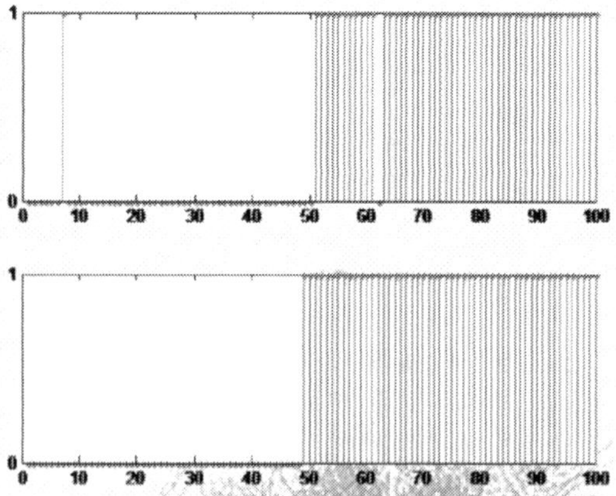

Fig. 2 False and missed alarms removed by preprocessing

TE method based on Alarm Data

The procedure of the method to detect causality between industrial alarm data based on TE is shown as follows:

- a. Get the original series;
- b. Do the preprocessing and get series *I*, *J*;
- c. Calculate the TE result from *J* to *I*;
- d. Calculate the threshold:
  - d1. Change *J* and get surrogate data *J'*;
  - d2. Calculate the TE result from *J'* to *I*;
  - d3. Repeat d1, d2 and use the average result as the threshold;
- e. Compare the result in c with the threshold in d.

If the result is larger than the threshold, there exists causality from *J* to *I*;
If the result is smaller than the threshold, there doesn't exist causality from *J* to *I*.

- ***Importance of Detection of Causality***
- ***Basic Definition of Transfer Entropy***
- ***TE Method based on Alarm Data***
- ***Case Study***

Case Study

The first case is described by a numerical equation as follows:

$$\begin{cases} Y_{k+1} = 0.8X_k + 0.5Y_k + v_{1k} \\ Z_k + 1 = 0.6Y_k + v_{2k} \end{cases} \quad (1)$$

Convert these variables into two-value alarm series with appropriate thresholds:

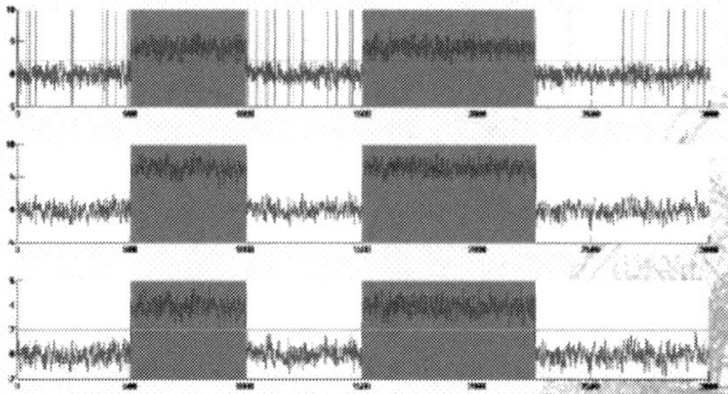

Fig. 3 Measured values of *X*, *Y*, *Z* and corresponding alarm series

We can know from equation (1) that there exists direct causality from *X* to *Y* and *Y* to *Z*, and indirect causality from *X* to *Z*. But there is no causality from *Y* to *X*, *Z* to *Y* or *Z* to *X*.

15

Case Study

If we calculate the TE result using original alarm series, we can get the result in table 1.

Table 1 Directing calculated TE results and corresponding thresholds

$TE_{col \to row}$	X	Y	Z
X	NA	0.0087(0.0005)	0.0122(0.0009)
Y	0.0978(0.0009)	NA	0.0227(0.0004)
Z	0.0934(0.0005)	0.0019(0.0011)	NA

We can see that the results in table 1 don't fit equation (1) very well.

If we do the preprocessing first, the result is shown in table 2.

Table 2 Calculated TE results after pre-processing and corresponding thresholds

$TE_{col \to row}$	X	Y	Z
X	NA	0.0120(0.0007)	0.0120(0.0011)
Y	0.0009(0.0010)	NA	0.0146(0.0010)
Z	0	0	NA

We can see that the result indicates the causality from *X* to *Y*, *Y* to *Z* and *X* to *Z*, which fits equation (1) well.

16

Case Study

We choose some typical input-output variables of Tennessee Eastman Process (TEP). The chosen input-output variables are:

- Feed A valve input and feed A output;
- Feed A valve input and feed D output;
- Reactor cooling water flow valve input and reactor temperature output.

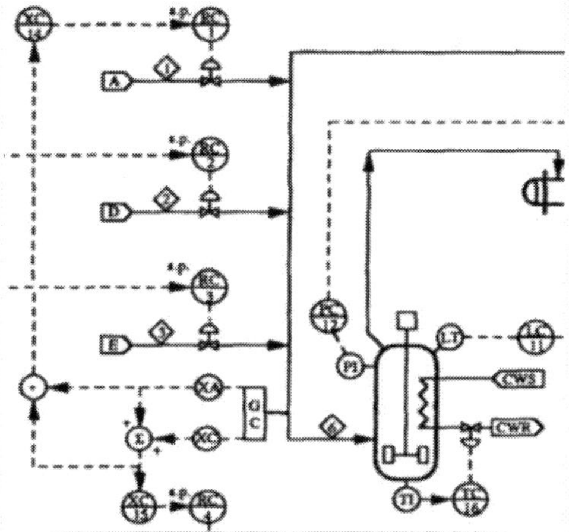

Fig. 4 Portion of TEP flowsheet and control loops

Case Study

Table 3 Calculated results of TEP

Groups	$TE_{I\to O}$	$TE_{O\to I}$
1	0.0027(0.0002)	0.0026(0.0003)
2	0.0001(0.0000)	0
3	0.0007(0.0001)	0.0001(0.0003)

We can see that the results are consistent with our expectation, which means the method is effective.

Recent Work

- Symbolic transfer entropy

19

Acknowledgments

- National Natural Science Foundation of China
- National High-Tech R&D Program of China
- Tsinghua University Initiative Scientific Research Program

Thanks for your attention!

Email: yangfan@tsinghua.edu.cn

20

386916 Subset Selection in Multiple Linear Regression

Abstract:

The ALAMO methodology has been developed recently to address the problem of discovering simple algebraic models from data obtained from simulations or experiments [1]. An important part of building these surrogate models is selecting the best subset from a large number of linear and nonlinear functions of explanatory variables. The selected subset will balance model complexity with the goodness-of-fit of the model in order to uncover underlying physical relationships instead of overfitting to the noise in the data.

The purpose of this paper is to present a systematic analysis of new and existing approaches to the subset selection problem encountered in ALAMO. The same problem of subset selection arises naturally in a variety of applications and has been the subject of study in the machine learning and statistics literatures [2, 3]. Yet, an effective solution approach to this problem is still elusive due to its highly combinatorial and nonlinear nature. It is often the practice to use greedy step-wise heuristics to produce a good fitting subset of regression variables [2]. These heuristics typically use different model fitness metrics, including Akaike's Information Criterion and Mallows' Cp, in order to define a stopping point. We compare these heuristic stepwise algorithms, exhaustive search algorithms [3], and newly proposed direct optimization of integer programming formulations for several different model selection criteria. For this purpose, we use a large test set with problems from a variety of applications.

References

[1] Cozad, A., N. V. Sahinidis, and D. C. Miller, Automatic learning of algebraic models for optimization, AIChE Journal, **60**, 2211-2227, 2014.

[2] Miller, A. J. (1990). *Subset selection in regression*. London [England]: Chapman and Hall.

[3] Furnival, G.M. and R. W. Wilson (1974). *Regression by leaps and bounds*, Technometrics, **16**, 499-511.

There is an increasing demand for systematic methods to determine the causal relationships among variables in complex systems. These causal relationships allow for developing complex models for use in the design, optimization and control of the systems, and detecting the hazard propagation pathways needed in risk management [1-4]. Causal relation determination can be made using expert knowledge, tentative mathematical models, and/or historical data [5]. In the past decade, the research community has focused particularly on the last approach. Besides various heuristic and hybrid methods [6], the existing data-based causal relationship determination methods are mainly of two types: score and search methods, and conditional independence tests [7]. Despite many advantages of these existing methods, finding directed causal structures with higher accuracy and lower computational cost is still an open problem.

In this work, a novel data-based method of determining direct causal relationships among variables of a process is presented. Of particular interest is to overcome the challenges imposed by different types of cyclic causalities present in modern industrial process systems. The cyclic causalities are mainly due to control and process bidirectional relationships. A new measure of conditional independence that finds the skeleton of undirected structure and a new technique that directs causal arrows and has the capability to capture highly non-linear interactions are introduced. Such a directed causal structure is also useful to identify hidden variables within a system. The application and performance of the proposed method in detecting and diagnosing process system abnormal operation and propagation of hazard will be shown by simulating several process examples.

References

[1]. Ahooyi, T. M.; Soroush, M.; Arbogast, J. E.; Seider, W. D.; Oktem, U. G. Maximum-likelihood maximum-entropy constrained probability density function estimation for prediction of rare events. *AIChE Journal* **2014**, *60* (3), 1013-1026.

[2]. Mohseni Ahooyi, T.; Abrogast, J. E.; Oktem, U.; Seider, W. D.; Soroush, M. Estimation of Complete Discrete Multivariate Probability Distributions from Scarce Data with Application to Risk Assessment and Fault Detection. *Industrial & Engineering Chemistry Research* **2014**, *53* (18), 7538-7547.

[3]. Mohseni Ahooyi, T.; Abrogast, J. E.; Soroush, M. Rolling Pin Method: a Novel Efficient General Method of Joint Probability Modeling. *Industrial & Engineering Chemistry Research, accepted*, DOI: 10.1021/ie503584q.

[4]. Mohseni Ahooyi, T.; Abrogast, J. E.; Soroush, M. Applications of the Rolling Pin Method: I) an Efficient Alternative to Bayesian Network Modeling and Inference. *Industrial & Engineering Chemistry Research*, accepted, DOI: 10.1021/ie503585m.

[5]. Brown, L. E.; Tsamardinos, I.; Aliferis, C. F. A comparison of novel and state-of-the-art polynomial Bayesian network learning algorithms. *In Proceedings of the Twentieth National Conference On Artificial Intelligence*, Veloso, M. M.; Kambhampati, S. (eds). 2, AAAI Press, 739–745 (**2005**).

[6]. Shaughnessy, P.; and Livingston, G. Evaluating the causal explanatory value of Bayesian network structure learning algorithms. Research paper, Department of Computer Science, University of Massachusetts Lowell (**2005**).

Multi-Objective Optimization for Solid Waste Management Systems

José Ezequiel Santibañez-Aguilar,[1] Juan Martínez-Gómez,[1] José María Ponce-Ortega,[1] Fabricio Nápoles-Rivera,[1] Medardo Serna-González[1], and Mahmoud M. El-Hawagi[2, 3]

[1]*Chemical Engineering Department, Universidad Michocana de San Nicolás de Hidalgo, Morelia, Michoacán, México, 58060*

[2]*Chemical Engineering Department, Texas A&M University, College Station, TX, USA, 77843*

[3]*Adjunct Faculty at the Chemical and Materials Engineering Department, King Abdulaziz University, Jeddah, Saudi Arabia.*

Several options for reusing the municipal solid waste have been reported; however, integral approaches for optimizing the entire supply chains associated to the municipal solid waste have not been considered. Supply chains associated to waste management might help to reduce the waste generation in an environmentally and economically sustainable way. Particularly, multi-objective optimization approaches accounting for economic, environmental, social and safety issues in the entire life cycle have not been reported. Therefore, this work presents a mathematical programming model to the optimal planning for reusing municipal solid waste taking into account economic, environmental, social and safety issues simultaneously in a distributed framework. The optimization model is able to select the processing technologies, consumers, cities producing wastes, amount of recycled waste, products and location of processing facilities.

The optimization problem is formulated as a multi-objective problem that considers three different objectives: the net annual profit, the amount of reused wastes and the total fatalities generated with the considered risks. The addressed problem considers several cities with a given municipal solid waste production, these wastes can be utilized to obtain different useful products like energy, recycled material, olefins, fuels, etc. The waste is stored in several disposition centers or dumps; these dumps can affect the population's safety in different ways. The first one is the fatalities associated for consuming polluted water, since the pollutants in dumps can be transported through leaching from the surface to shallow well that can serve as water source to the population. In addition, the risk is associated the intoxication by the gas emissions caused by the burning of a given dump; this risk is taken into account by the implementation of a gas dispersion model. One of the main considerations is that each city is divided in several grids named sites, each one of these sites have different waste production, population and may have processing plants and dumps, this division is accomplished because each dump can affect in different magnitude each one of the sites, since each site is located at different distance from the dump considered.

The economic objective function is the net annual profit and considers all the costs of the entire supply chain; in this context, the net annual profit is the revenue from sales minus the costs for operation, capital, transportation, disposing and separating.

$$\begin{aligned} NETPROFIT = {} & REVENUE - OPCOST - CAPCOST^{total} \\ & - TRANSPCOST - DISPCOST - SEPCOST \end{aligned} \quad (1)$$

The safety objective function is based on the total fatalities generated by intoxication and leaching. For the fatalities for intoxication, a frequency factor ($f^{frequency}_{factor}$) to obtain the number of times that the landfill is burned yearly is considered. The total fatalities are equal to the sum of the total fatalities ($F_{k,s}^{leaching\ fatalities}$) for leaching plus the total fatalities for intoxication ($F_{g,d,a}^{intoxication\ fatalities}$).

$$F^{total\ fatalities} = f^{frequency\ factor} \sum_g \sum_d \sum_a F_{g,d,a}^{intoxication\ fatalities} + \sum_k \sum_s F_{k,s}^{leaching\ fatalities} \tag{2}$$

The total consumed waste is a way to take into account the environmental impact although there are many forms to assess the environmental impact; this is an adequate way for the case of the municipal solid waste because one of the main impacts is for the huge amount of waste that is disposed in landfills. In this way, the total consumed waste is equal to the total waste that is sent to the processing facilities from dumps divided by the total waste that is produced in the each site of all cities.

$$WASTECONS = \sum_{wn \in WASTES} \sum_{d \in DUMPS} W_{wn,d}^{out\ dump} \Big/ \sum_{wn \in WASTES} \sum_{s \in SITES} W_{wn,s}^{produced} \tag{3}$$

A case study for planning the waste management in the city of Morelia in Mexico was considered. Figure 1 shows the tradeoffs between the number of fatalities and the net annual profit; notice the number of fatalities can decrease until 21 % from the maximum number of fatalities while the change in the profit is about 19 %. It is important to note that the fatalities for leaching are the most important in the proposed case study. Furthermore, Figure 1 shows that the consumed waste is almost constant when the Pareto curve is carried out with the net annual profit and the number of fatalities, and this way the number of fatalities is not function of the consumed waste. Results show that it is possible to implement a distributed processing system to reuse municipal solid waste in an economically attractive way. In addition, results can be used for governments to make decisions about the waste disposal and define the amount of waste that must be reused to obtain several products. It should be noted that results include the supply chain configuration.

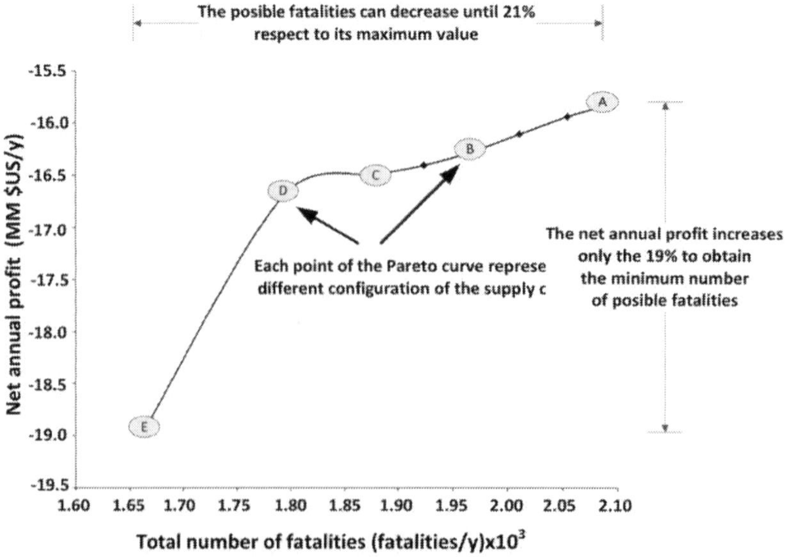

Figure 1. Pareto curve between the net annual profit and the number of fatalities.

384303 Global Optimization for Sustainable Design and Synthesis of Algae Processing Network for CO2 Mitigation and Biofuel Production Using Life Cycle Optimization

Abstract:

Renewable biofuel is considered as a promising fossil fuel substitute due to its potential for large-scale production and carbon mitigation. Among various biomass feedstocks with increasing attention in the recent decades, algal biomass offers significant advantages over conventional energy crops [1]. Algae have high area productivity and accumulate lipid materials up to 80% of its dry cell weight. They are able to utilize non-arable land and waste water for cultivation. Additionally, the hydrocarbon algal biofuels converted via hydroprocessing accommodate the specifications of drop-in fuels, which can be directly incorporated into the existing infrastructure [2]. In order to reduce environmental impacts and improve the overall economic performance, systematic modeling and optimization frameworks can be utilized to identify and assess the sustainable design and synthesis of a processing network [3-5]. Therefore, applying such a framework to the algae processing network provides valuable insights into the technology pathways in the context of sustainability. The goal of this research is to identify and assess the optimal sustainable design and synthesis of an algae processing network with both economic and environmental concerns.

In this work, we propose a superstructure of an algae processing network including 7,800 processing routes and 11 processing sections, namely cultivation, harvesting, primary dewatering, secondary dewatering, storage, cell disruption & hydrothermal liquefaction, lipid extraction, upgrading, remnant treatment, electricity generation, and steam generation. Feed gas, off-gas, makeup nutrients, and recycled materials are fed into the cultivation system and mature algae product from cultivation is concentrated by several technologies. First, 10% of the dilute slurry is harvested by auto flocculation and dissolved air flotation. The water content is further reduced by primary and secondary dewatering. Most of the reduced materials are recycled to the cultivation facility. In order to keep downstream facilities operating continuously, we employ a storage tank to temporarily store part of the enriched algae biomass for the production at night [6]. Next, the enriched microalgae paste undergoes cell disruption and lipid extraction or HTL so that most of the lipid materials are separated from algal cells. The lipid materials are then upgraded to biodiesel or renewable diesel depending on the specific conversion technologies. The rest of the algal body or remnant is fed to an anaerobic digester and decomposed into biogas, aqueous nutrients, and unreacted solids. Additionally, we couple electricity and steam generation into the network to satisfy the onsite energy consumption and reuse the resulting off-gas.

Based on this superstructure, a multi-objective mixed-integer nonlinear programming (MINLP) model for the sustainable design and synthesis is developed following the life cycle optimization frame work that integrates the multi-objective superstructure optimization scheme with life cycle assessment and techno-economic analysis [3, 4, 7-10]. The model consists of two objective functions and simultaneously minimize the unit annualized cost and the unit Global Warming Potential (GWP). As the final biofuel product can be biodiesel or renewable diesel depending on the specific upgrading technology, the unit objective functions are associated with the unit gasoline gallon equivalent (GGE) of the specific biofuel produced. To efficiently solve the problem, we propose a global optimization strategy that integrates a branch-and-refine algorithm based on successive piecewise linear approximations and an exact parametric algorithm based on Newton's method [11-15]. The multi-objective optimization problem results in two Pareto-optimal curves for biofuel production and biological carbon sequestration, respectively. Three optimal processing routes are identified. From the biofuel production perspective, the most environmentally sustainable processing route achieved a unit GWP of 16.520 kg CO2-eq/GGE and a unit annualized cost of $9.712/GGE, while the most economical processing route resulted in a unit annualized cost of $7.017/GGE and a unit GWP of 26.791 kg CO2-eq/GGE. If the algae processing network is utilized for biological carbon sequestration, the unit carbon sequestration and utilization cost ranges from $10.43 /t of CO2 to $1.64 /t of CO2, corresponding to unit GWP's from 352.41 kg CO2-eq/t of CO2 to 412.90 kg CO2-eq/t of CO2, respectively.

References

[1] J. Ferrell and V. Sarisky-Reed, "National algal biofuels technology roadmap," US Department of Energy College Park, MD2010.

[2] D. Yue, F. You, and S. W. Snyder, "Biomass-to-bioenergy and biofuel supply chain optimization: Overview, key issues and challenges," *Computers & Chemical Engineering,* 2013.

[3] D. J. Yue, M. A. Kim, and F. Q. You, "Design of Sustainable Product Systems and Supply Chains with Life Cycle Optimization Based on Functional Unit: General Modeling Framework, Mixed-Integer Nonlinear Programming Algorithms and Case Study on Hydrocarbon Biofuels," *Acs Sustainable Chemistry & Engineering,* vol. 1, pp. 1003-1014, Aug 2013.

[4] B. Wang, B. H. Gebreslassie, and F. Q. You, "Sustainable design and synthesis of hydrocarbon biorefinery via gasification pathway: Integrated life cycle assessment and technoeconomic analysis with multiobjective superstructure optimization," *Computers & Chemical Engineering,* vol. 52, pp. 55-76, May 10 2013.

[5] B. H. Gebreslassie, R. Waymire, and F. You, "Sustainable design and synthesis of algae-based biorefinery for simultaneous hydrocarbon biofuel production and carbon sequestration," *AIChE Journal,* vol. 59, pp. 1599-1621, 2013.

[6] J. Gong and F. You, "Optimal Design and Synthesis of Algal Biorefinery Processes for Biological Carbon Sequestration and Utilization with Zero Direct Greenhouse Gas Emissions: MINLP Model and Global Optimization Algorithm," *Industrial & Engineering Chemistry Research,* vol. 53, pp. 1563-1579, 2014/01/29 2014.

[7] F. Q. You, L. Tao, D. J. Graziano, and S. W. Snyder, "Optimal design of sustainable cellulosic biofuel supply chains: Multiobjective optimization coupled with life cycle assessment and input-output analysis," *Aiche Journal,* vol. 58, pp. 1157-1180, Apr 2012.

[8] F. Q. You and B. Wang, "Life Cycle Optimization of Biomass-to-Liquid Supply Chains with Distributed-Centralized Processing Networks," *Industrial & Engineering Chemistry Research,* vol. 50, pp. 10102-10127, Sep 7

2011.

[9] B. H. Gebreslassie, M. Slivinsky, B. L. Wang, and F. Q. You, "Life cycle optimization for sustainable design and operations of hydrocarbon biorefinery via fast pyrolysis, hydrotreating and hydrocracking," *Computers & Chemical Engineering,* vol. 50, pp. 71-91, Mar 5 2013.

[10] Q. Zhang, J. Gong, M. Skwarczek, D. Yue, and F. You, "Sustainable process design and synthesis of hydrocarbon biorefinery through fast pyrolysis and hydroprocessing," *AIChE Journal,* vol. 60, pp. 980-994, 2014.

[11] F. Q. You and I. E. Grossmann, "Stochastic Inventory Management for Tactical Process Planning Under Uncertainties: MINLP Models and Algorithms," *Aiche Journal,* vol. 57, pp. 1250-1277, May 2011.

[12] Z. X. Zhong and F. Q. You, "Globally convergent exact and inexact parametric algorithms for solving large-scale mixed-integer fractional programs and applications in process systems engineering," *Computers & Chemical Engineering,* vol. 61, pp. 90-101, Feb 11 2014.

[13] F. Q. You, J. M. Pinto, I. E. Grossmann, and L. Megan, "Optimal Distribution-Inventory Planning of Industrial Gases. II. MINLP Models and Algorithms for Stochastic Cases," *Industrial & Engineering Chemistry Research,* vol. 50, pp. 2928-2945, Mar 2 2011.

[14] D. J. Yue and F. Q. You, "Planning and Scheduling of Flexible Process Networks Under Uncertainty with Stochastic Inventory: MINLP Models and Algorithm," *Aiche Journal,* vol. 59, pp. 1511-1532, May 2013.

[15] F. Q. You, P. M. Castro, and I. E. Grossmann, "Dinkelbach's algorithm as an efficient method to solve a class of MINLP models for large-scale cyclic scheduling problems," *Computers & Chemical Engineering,* vol. 33, pp. 1879-1889, Nov 12 2009.

391695 Towards Integration of Life Cycle Assessment within a Multi Objective Supply Chain Optimization Modeling Framework

Abstract:

This paper presents advancements in integration of multiple environmental indicators (CO2e, Human health, Ecosystem health) based on life cycle assessment methods within the framework of supply chain modeling and optimization. We present results related to the optimal design of supply chain networks with consideration of environmental and economic objectives. The environmental objective is measured through life cycle assessment indicators. IMPACT 2002+ methodology is used to evaluate the environmental scores. Global warming, human health damages and ecotoxicity impacts are considered for evaulating the environmental impacts. On the other hand, minimization of total production cost and maximization of net present value of the supply chain is another goal. A multi objective mixed integer linear programing (MILP) model is developed for optimal design of green supply chain. Multi objective optimization is accomplished through non dominated sorted genetic algorithm. The resulting Pareto-optimal curves demonstrate the tradeoff between economic and environmental pathways. This method considers life cycle emissions from cradle to gate for each process and suggests alternatives technologies for the reduction of these pollutants while considering financial implications. The method can be used to derive optimal solutions for scheduling and planning or could be used to design future green supply chains. The methodology is demonstrated via several case studies.

383858 Optimal Design of Microgrids with Innovative CHP Systems: Integration of Process Optimisation and Life Cycle Assessment

Abstract:

Optimal design of microgrids with innovative CHP systems: integration of process optimisation and life cycle assessment

Di Zhang, Sara Evangelisti, Paola Lettieri, Lazaros G. Papageorgiou

Department of Chemical Engineering, University College London, Torrington Place, London WC1E 7JE, UK

As an alternative to current centralised energy generation, microgrids are adopted to provide local energy with lower energy expenses and gas emissions by utilising distributed energy resources (DER). DER is defined as local resources related to the energy system, such as generators, boilers and energy storage devices. Micro Combined Heat and Power (micro CHP) units simultaneously produce heat and power for a single building. In order to meet renewable energy target in the UK, the installation of CHP units is considered because of their flexibility, reliability and safer operating conditions with high overall efficiencies. Several micro CHP technologies have been developed recently for applications at domestic scale, including fuel cell, micro turbine, internal combustion engine and sterling engine. The optimal design of DERs within microgrid plays an important role in promoting the penetration of microgrid. It attracts money research attentions from the economic view.

Different tools exist to evaluate the environmental impacts of a process or a service. Life Cycle Assessment (LCA) is recognised as one of the best tools available. Moreover, LCA is one of the most developed and widely used environmental assessment tools for comparing alternative technologies when the location of the activity is defined. Studies have been published on carbon footprint of micro – CHP units rather than a holistic environmental appraisal. A few studies have focused on the environmental impacts of different micro-CHP technologies analysing the functioning of single units rather than microgrids.

Limited studies in the literature have been focused on coupling environmental and economic sustainability in a multi-objective optimisation model: the majority of them focused on the operation of a single unit. To our knowledge, no studies have been published on the optimal design of microgrids considering environmental and economic metrics. In this work, the optimal design of microgrids with innovative CHP units is addressed while involving both concerns. A multi-objective modelling framework for the optimal design is presented. The trade-off between the two conflicting objectives is obtained by solving the proposed model using the ε-constraint method.

This work considers a general microgrid, which involves several participant sites of different building types. Macrogrid is available to provide electricity to the participant in the microgrid and extra electricity can also be sold back to the macrogrid when it benefits. The candidate technologies include CHP generators (with different capacities and heat-to-power ratios), boilers, thermal storage and a macrogrid power connection. Micro CHP technologies considered in this work are Proton exchange membrane (PEM) Fuel Cell, Solid Oxide Fuel Cell (SOFC), Internal Combustion Engine (ICE) and Stirling Engine (SE). Turn-key costs of CHP generators are based on the CHP types as well as the capacity range. The microgrid and the macrogrid are interacted and constrained through exporting or importing electricity.

The goal of the LCA in this work is to estimate the environmental impacts of four different CHP systems with additional thermal energy storage (TES) working in a distributed generation network. The life span considered is 15 years. The impact categories considered in this study are global warming potential (GWP) and acidification potential (AP). The output of LCA is used as the input for the optimisation model. A multi-objective mixed-integer linear programming (MILP) approach is developed for this problem. The key decision variables include unit capacities and resources utilised. They are determined by minimise the total equivalent annualised cost (EAC), GWP and AP of all participants, subject to equipment capacity constraints, CHP ramp limit constraints, energy demand constraints, CHP selection constraints, thermal storage constraints and transfer price level constraints.

The proposed model is implemented for different microgrid modes. For each mode, optimal design of the microgrid is obtained with (a) the candidate technologies selected and their capacities, (b) energy resources consumed, (c) energy production plan, (d) thermal energy storage plan. The trade-off between the economic and environmental objectives is analysed with a set of Pareto-optimal solutions.

Keywords: microgrids; life cycle assessment; multi-objective optimisation; ε-constraint method

Application of sustainability footprint and multicriteria optimization for the choice of sustainable chemical processes

AIChE Annual Meeting, Atlanta, GA, November 16-21, 2014
For submission to "Advances in Life Cycle Optimization for Process Development"
Rajib Mukherjee[1], Debalina Sengupta[2], Subhas K. Sikdar[3]

1. *Vishwamitra Research Institute, Chicago, IL*
2. *Texas A &M University, College Station, TX*
3. *NRMRL, US EPA, Cincinnati, OH,*

Abstract

A chemical product can be obtained by different process pathways involving varying amounts and types of resources, utilities, product, and byproduct formation. Evidently, the sustainability in terms of economic, environmental and social impacts of these processes will vary relatively. When such competing process options are presented, and compared to an existing process pathway, it is necessary to evaluate and identify the most sustainable process among these competing options. Sustainability of a chemical process is generally evaluated with indicators that need process and chemical properties data. These indicators individually reflect the impacts of the process on the respective areas of sustainability. In order to choose among several alternative processes an overall comparative analysis is essential.

To make a decision on the most sustainable process, the indicators need to be rationally aggregated into a single index. In this research, the suite of sustainability indicators includes traditional economic indicators to represent the economic dimension, environmental indicators, and human health indicators to represent the social dimension. The environmental and human health indicators were computed using the TRACI impact characterization method. Process flow and economic data were used to compute the indicators in these categories. These indicators were then aggregated to compute the sustainability footprint (D_e) of each process and evaluate the relative sustainability of the competing processes.

This method of process selection is applied to five competing processes for methanol production from carbon dioxide. These are novel chemical processes which involve carbon dioxide as raw material, and a uniform industrial scale capacity for production of methanol. The processes differ in the amount of raw material required, byproducts formed, and utilities used. Based on process data, these processes are compared using the sustainability footprint method. This alone should give us information to evaluate the best process as the one which has the minimum sustainability footprint, D_e.

However, we wanted to see if the minimization of the sustainability footprint, D_e, can be used as an objective function for a mixed integer optimization problem. The aim in solving the multi-criteria problem will be to select the most sustainable option among the five competing processes, and also determine the optimum values of process parameters which will enable the same.

Based on our results, this research will demonstrate how an aggregate sustainability footprint index can be utilized to compare between competing processes, as well as how it can be integrated into a mixed integer problem formulation to determine the optimum process conditions which help in determining the most sustainable process among competing options.

Sustainable Biorefinery Supply Network using a Concept of Eco- and Total Profit – Case Study of the European Union

Lidija Čuček,[a] Mariano Martín,[b] Ignacio E. Grossmann,[c] Zdravko Kravanja[a]

[a]*Faculty of Chemistry and Chemical Engineering, University of Maribor, Smetanova ul. 17, Maribor, Slovenia, lidija.cucek@um.si, zdravko.kravanja@um.si*

[b]*Department of Chemical Engineering, University of Salamanca, Salamanca, Spain, mariano.m3@usal.es*

[c]*Department of Chemical Engineering, Carnegie Mellon University, Pittsburgh, USA, grossmann@cmu.edu*

Abstract

Nowadays the shortages of energy and water resources, the increasing prices, the demand for and dependence on fossil fuels and a growing environmental consciousness, have become societal priorities. These concerns have supported the development of alternative transportation fuels and the use of renewable sources of energy. Thus, governments across the globe have set targets on fossil fuels' substitutions over the next decade. In order to be able to meet these targets, not only process designs for the productions of alternative fuels are needed, but also the development of the networks which account for all the stages from the energy source to the final product. Biomass, in particular, is a challenging resource because of its seasonality, its degradation over time, its low density, regional availability, growth yield, and the large area required to grow it. However, biomass can provide a reasonable alternative to crude-based gasoline and diesel in the short and medium terms, although the competitiveness of biomass as an energy source is highly dependent on its supply chain. Lately, several studies have extended the supply-chain studies to actual regions across the US, but only for a limited number of raw materials and fuels, focusing mainly on lignocellulosic resources and in most of the cases only for bioethanol production, although FT-products have also been accounted for [1,2].

This contribution deals with a sustainable large-scale biorefinery supply network by considering both economic and environmental objectives. A biorefinery's supply network accounts for different biomass sources (algae, lignocellulosic materials, corn, wheat, cooking oil), conversion technologies (biochemical, thermochemical, thermo-biochemical, oil extraction, transesterification) and includes first, second, and third generations of biofuels, such as bioethanol, biodiesel, hydrogen, Fischer-Tropsch diesel, and green gasoline [3] accounting for water consumption and availability and energy integration within the network. Multi-period synthesis of an optimally-integrated biorefinery supply network [4] was performed through maximization of the economic objective, and applied on the case study of the European Union [5].

In this contribution the synthesis has now been extended to simultaneously accounting for environmental impact. The concept of Eco- and Total profit [6] was used for this purpose, which offers two important advantages over conventional multi-objective approaches: i) the concept accounts for both, burdening and unburdening effects [6] on the environment, as Eco-Profit is defined as the difference between Eco-Benefit (unburdening) and Eco-Cost (burdening), expressed in monetary terms. ii) by applying the concept where multi-objective solutions are reduced into single-objective ones as Total profit is composed of both, economic and Eco-profits. This enables the obtaining of sustainable solutions directly, which is especially important when solving supply network problems at the continental level when the models are huge. The solutions obtained for the case study of the European Union are those alternatives that would unburden the environment the most while having significant profit.

Acknowledgments

The authors acknowledge the financial support from the Slovenian Research Agency (Programs P2-0032 and P2-0377) and Centre for Advanced Process Decision-making.

References

[1] Elia, J.A., Baliban, R.C., Floudas, C.A. Nationwide energy supply chain analysis for hybrid feedstock processes with significant CO_2 emissions reduction. AIChE Journal. 2012; 58(7), 2142-2154.

[2] Marvin, A. W., Schmidt, L. D., Benjaafar, S., Tiffany, D. G., & Daoutidis, P. Economic Optimization of a Lignocellulosic Biomass-to-Ethanol Supply Chain. Chemical Engineering Science. 2012;67, 68-79.

[3] Martín M, Grossmann IE. On the systematic synthesis of sustainable biorefineries. Industrial & Engineering Chemistry Research. 2013;52(9):3044-64.

[4] Čuček L, Martín M, Grossmann IE, Kravanja Z. Multi-period Synthesis of Optimally-Integrated Biomass and Bioenergy Supply Network. Computers and Chemical Engineering. 2014:doi: 10.1016/j.compchemeng.2014.02.020.

[5] Čuček L, Martín M, Grossmann IE, Kravanja Z. Large-Scale Biorefinery Supply Network – Case Study of the European Union. In: J. J. Klemeš, P. S. Varbanov, P. Y. Liew, editor. Computer Aided Chemical Engineering: Elsevier; 2014, accepted for publication.

[6] Kravanja Z, Čuček L. Multi-objective optimisation for generating sustainable solutions considering total effects on the environment. Applied Energy. 2013;101:67-80.

Planning and Scheduling Industrial Waste Management Using Knowledge Based Lagrangean Decomposition

Elisabet Capón-García[a], Edrisi Muñoz[b], José Miguel Laínez[c], Antonio Espuña[d], Konrad Hungerbühler[a], and Luis Puigjaner[d]

[a]*Chemistry and Applied Biosciences, Institute for Chemical- and Bioengineering, Swiss Federal Institute of Technology, Zurich (ETHZ)*
[b]*Information Technologies, Centro de Investigacion en Matematicas A.C., Guanaguato, Mexico,*
[c]*Department of Industrial & Systems Engineering, University at Buffalo, Buffalo, NY,*
[d]*Chemical Engineering Department, Universitat Politècnica de Catalunya - ETSEIB, Barcelona, Spain*

Waste liquid effluents originated from industrial activities contain contaminants and certain species, such as organic solvents, with material and energy recovery potential. Therefore, not only does industrial waste management stand as an end-of-pipe problem, but also as a potential origin of resources, which can result insignificant benefits if material and energy integration strategies are used. The simultaneous planning and scheduling of industrial waste management allows improved resource utilization and reduced environmental impact. However, rigorous mathematical models usually result in large scale problems. In order to tackle the problem complexity, decomposition techniques based on information flows between a master and a set of sub-problems are widely applied (Maravelias and Sung, 2009, Calfa et al., 2013).

In this work, we exploit this capability of ontologies to address the optimal integration of planning and scheduling using a Lagrangean decomposition approach. Thus, ontologies improve information sharing and communication in the enterprise and can even represent holistic mathematical models facilitating the use of analytic tools and providing higher flexibility for model building. Therefore, the scheduling/planning sub-problems are created for the waste management facility and their dual solution information is shared by means of the ontological framework. Furthermore, this work presents a multi-period waste management multi-objective optimization, considering economic and environmental issues. The behavior of waste treatment units is included in the optimization problem as black-box models based on industrial practice.

A case study of the planning and scheduling of industrial waste facilities is presented to show the advantages and limitations of the proposed approach in a multi-objective environment considering economic and environmental issues using the decomposition strategy. The proposed framework leads to reduced computation effort and better solutions in terms of solution quality, since waste stream scheduling is integrated in decision-making.

Acknowledgements
Financial support received from the Spanish "Ministerio de Economía y Competitividad" and the European Regional Development Fund, both funding the research Project

SIGERA (DPI2012-37154-C02-01), and the "Agència de Gestió d'Ajuts Universitaris i de Recerca (2014SGR1092)" is thankfully acknowledged.

References

Calfa, B. A., Agarwal, A., Grossmann, I. E., Wassick, J. M., 2013. Hybrid bilevel-lagrangean decomposition scheme for the integration of planning and scheduling of a network of batch plants. Industrial & Engineering Chemistry Research 52 (5), 2152-2167.

Maravelias, C. T., Sung, C., 2009. Integration of production planning and scheduling: Overview, challenges and opportunities. Computers & Chemical Engineering 33 (12), 1919-1930.

391026 Identifying the Preferred Subset of Alternatives for Environmental Improvements Via an MILP Approach Based on the Analytic Hierarchy Process

Abstract:

The combined use of multi-objective optimization (MOO) and LCA has recently gained wider interest in process systems engineering. This approach provides as output a set of Pareto solutions that represent the optimal trade-off between the economic and environmental concerns considered in the analysis. From this set of optimal alternatives, decision-makers should identify the ones that better fulfill their preferences. Generating a large and representative enough subset of Pareto points to aid decision-making is challenging. This task is particularly difficult in problems with a large number of (environmental) objectives, as is the case when incorporating LCA principles in MOO.

In this work we present a mixed-integer linear programming method that simplifies MOO problems by concentrating on determining only a reduced number of Pareto points that are particularly appealing. Our approach, which relies on the analytic hierarchy process (AHP), identifies a set of weights to be assigned to the environmental objectives so as to translate them into a single aggregated indicator that reflects to the maximum extent possible the decision-makers' preferences. For every such combination of weights, we solve a single objective problem that optimizes the corresponding weighted sum of objectives, thereby generating only a subset of alternatives (Pareto solutions) reflecting preferences with a large degree of consistency. We illustrate the capabilities of our approach through its application to the design of supply chains for bioethanol production.

Verifying Differentiability of Solutions of Nonsmooth Dynamic Systems

Kamil A. Khan, Ali M. Sahlodin, and Paul I. Barton
Process Systems Engineering Laboratory, Massachusetts Institute of Technology, Cambridge, MA

Nonsmoothness can enter dynamic systems through phenomena such as thermodynamic phase changes and flow regime transitions. We consider the broad class of nonsmooth dynamic systems that can be represented as solutions of parametric ordinary differential equations (ODEs) with right-hand side (RHS) functions that are finite compositions of analytic functions and absolute-value functions. As shown in [2], state variables of these systems may be nondifferentiable with respect to system parameters. In this case, when using optimization methods which require first-order sensitivity information, generalized derivatives [1,2] must be computed. These generalized derivatives are significantly more difficult to furnish than the classical derivative.

Nevertheless, Clarke [1] shows that it is possible for state variables to be differentiable with respect to system parameters, even when nonsmooth domain points of the RHS function are visited along an ODE solution trajectory. Moreover, Clarke provides a sufficient condition for this differentiability: namely, that nonsmooth domain points of the RHS function be reached only at a set of times with zero Lebesgue measure.

By exploiting the structure of the considered ODE RHS functions, we show that Clarke's sufficient condition for differentiability can be recast in a numerically tractable form [3]. Firstly, we show that Clarke's sufficient condition is satisfied if there is no "valley-tracing mode", which is a nonzero duration on which the argument of some absolute-value function in the RHS is identically zero. Secondly, we show that any valley-tracing mode must begin either at the initial time, or when some other absolute-value function in the RHS switches between its two linear pieces. This condition can be detected numerically using standard event-detection techniques [5], and can sometimes be shown *a priori* never to hold. If no valley-tracing mode is detected for a particular parameter value, then a classical derivative exists, and may be computed using classical ODE sensitivity theory. Numerical examples are presented for illustration, in which nonsmooth flow systems are optimized [4].

[1] F. H. Clarke, *Optimization and Nonsmooth Analysis*, John Wiley & Sons, Inc., New York, 1983.

[2] K. A. Khan and P. I. Barton, Generalized derivatives for solutions of parametric ordinary differential equations with non-differentiable right-hand sides, J. Optimiz. Theory App., In press (2014), DOI: 10.1007/s10957-014-0539-1.

[3] K. A. Khan and P. I. Barton, Switching behavior of solutions of ordinary differential equations with nonsmooth right-hand sides, In preparation (2014).

[4] K. A. Khan, A. M. Sahlodin, M. Yunt, and P. I. Barton, Parametric sensitivity analysis of dynamic systems using nonsmooth analysis: II. Computation. In preparation (2014).

[5] T. Park and P. I. Barton, State event location in differential-algebraic models, ACM T. Model. Comput. S., 6 (1996) 137-165.

385111 Robust Online Optimization for Grade Transitions in Polyethylene Solution Polymerization

Abstract:

There are various grades of linear low-density polyethylene (LLDPE) tailored to different applications, with each grade defined by specifications of product properties such as melt index and density. Typically, several different grades are produced in the same production line. In certain instances, complex transitions rely heavily on operator/expert experience. Given the large market of LLDPE and the current experience-based transitions, there is a need, and also room for improvement, to perform transitions and to change operation conditions in a smarter way such that the transition time as well as the off-grade product could be minimized.

In this study, we consider a mathematical model capturing the dynamics of the solution polymerization process carried out in a CSTR. Besides the mass and heat balance equations, the whole model adopts the molecular weight moment model for the prediction of product properties and incorporates a simple, yet accurate, vapor-liquid equilibrium (VLE) model derived from rigorous calculation. As described in Shi et al. (2013), the model is validated via input step responses and then simultaneous dynamic optimization approach is applied to solve the problem. Two optimization formulations, single stage and multistage optimization, have been developed to deal with single-value specification and specification bands of product properties, respectively. The results show significant improvement in transition times and off-grade production compared with step change simulation. Also, the multistage formulation designed for problems with specification bands outperforms the single stage formulation as the former one leads to more aggressive control profiles and allows oscillations within the bands. Another advantage of multistage formulation is that it's capable to minimize the transition time and off-grade production directly.

The offline dynamic optimization demonstrates the potential in dealing with grade transition problems, but its performance can deteriorate in the presence of uncertainties, disturbances and model mismatch. In order to assess the impact, different transitions under various uncertainty levels are performed and robust optimization strategies for grade transitions are taken into account to obtain optimal policies which can be applied to systems with different uncertainty levels. This talk describes two complementary approaches to deal with these uncertainties. The first deals with the development of a robust, multi-scenario optimization approach. As detailed in Huang et al. (2009) and Lucia, Finkler and Engell (2013), a multi-scenario dynamic optimization problem is posed and solved on-line. Next, we consider a model-based state estimator that provides an estimate of the uncertainties to feed into the model-based on-line optimizer. Finally, we develop an optimization formulation that combines both strategies and we demonstrate improved performance of this synergistic strategy, over implementations of either strategy by itself.

References

Shi, J., Hamdan, I, Munjal, S., & Biegler, L. T. (2013) Model-Based Control for Optimal Transitions in Polyethylene Solution Polymerization. AIChE Annual Meeting. Presentation.

Lucia, S., Finkler, T., & Engell, S. (2013). Multi-stage nonlinear model predictive control applied to a semi-batch polymerization reactor under uncertainty.Journal of Process Control, 23(9), 1306-1319.

Huang, R., Patwardhan, S. C., & Biegler, L. T. (2009). Multi-scenario-based robust nonlinear model predictive control with first principle models. Computer Aided Chemical Engineering, 27, 1293-1298.

Dynamic Model of a Smart Refractory Brick for Gasifier Applications

Qiao Huang, Debangsu Bhattacharyya

Department of Chemical Engineering, West Virginia University, Morgantown, WV 26506, USA

Edward M. Sabolsky

Department of Mechanical & Aerospace Engineering, West Virginia University, Morgantown, WV 26506, USA

The operating temperature is the most important variable in an entrained-flow gasifier. A lower temperature results in lower carbon conversion and can lead to disruption in the flow of molten slag. A higher temperature can result in reduced life of the refractory leading to significant capital cost and undesired downtime. However, the current state of the art in measurement technology is not adequate for measuring the temperature profile inside the entrained flow gasifiers reliably mainly because of very high temperature and extreme harsh conditions. As a result, the thermocouples that are inserted into the gasifier through open access ports within the refractory survive for only a few weeks. To circumvent this problem, a novel smart refractory is being developed at West Virginia University. This refractory contains embedded sensors throughout the volume of high-chromia refractory brick. Since these sensors are embedded, they do not negatively impact the intrinsic properties of the refractory. However, in course of operation, the slag will penetrate into the smart refractory. When the slag penetrates into a location, the material properties will be that of a mixture of slag and refractory and are expected to change considerably as the material properties of the refractory and the slag are very different in terms of thermal and electrical characteristics. Therefore, characterization of these sensors by considering the data from the pristine bricks would not be sufficient for satisfactory estimation of the variables of interest. A model-based approach can be very useful for satisfactory estimation. Other than estimating the temperature profile, the model-based approach can be used for estimating the change in the material properties and can therefore be used for estimating the extent of slag penetration.

With this motivation, a rigorous, first-principles, dynamic model of the smart refractory brick has been developed. The thermal model has been developed by considering properties of the pristine smart refractory bricks as well as that of the slag-infiltrated brick. Using in-house experimental data and data from the open literature, correlations have been developed for thermal conductivity, heat capacity, and emissivity for the pristine and the slag-infiltrated brick. Models of the interdigitated capacitors as well as the thermistors have been developed by considering the location and the geometries of the embedded sensors. Finally, the raw, temporal measurements from the embedded sensors are used in a model-based estimation framework to obtain the satisfactory estimate of the variables of interest.

The dynamic, distributed model including the electrical systems model is developed in Aspen Custom Modeler for easy integration with other Aspen Engineering Suite models while the estimator algorithms are developed in MATLAB. The co-simulation using Aspen Plus Dynamics and MATLAB is accomplished by using the available commercial link.

Optimization of Dynamic Systems with Linear Programs Embedded

Jose A. Gomez[a], Kai Höffner[a], Paul I. Barton[a]

[a] *Process Systems Engineering Laboratory, Massachusetts Institute of Technology, Cambridge MA 02139, USA*

The acceleration in the process of genome sequencing in recent years has increased the availability of metabolic network reconstructions for a variety of species. These genome-based networks can be used within the framework of flux balance analysis (FBA) to predict growth and uptake rates accurately [1]. These rates can be used for the simulation of dynamic biological systems (DFBA). DFBA results in a system of ordinary differential equations (ODEs) with linear programs (LPs) embedded [2]. The LPs are said to be embedded because their solution depend on the differential states whereas the right-hand sides of the differential equations depend on the solutions of the LPs. For community simulations, each species is modeled by a different LP.

A broad range of dynamic optimization problems for this class of systems can be formulated [3]. Some examples include maximizing the concentration of lipids in microalgae for biofuel production or minimizing the concentration of contaminants in wastewater being treated biologically. The solution of these optimization problems is hindered by several difficulties.

First, the solution of each of the embedded LPs is often not a singleton. For the ODE system to be well-defined, the fluxes that influence the right-hand side of the differential equations must be unique. In the context of this work, this is achieved by performing lexicographic optimization where subsequent LPs are solved hierarchically to yield unique fluxes [2].

Another difficulty is introduced because not all elements of the set of optimization parameter values lead to feasible points. Only parameter values for which all LPs in the system are feasible for all simulation time are part of the feasible set for optimization. Therefore, no explicit representation of the set of feasible points in the parameter space exists. We use penalty methods to deal with this difficulty by transforming the constrained optimization problem into an unconstrained optimization problem. As proposed in [2, 3], an extended ODE system is defined such that the solution of this extended ODE exists for all parameter values. In addition, different penalty function formulations are presented in this work.

Next, an objective function dependent on the final values of the states and their trajectories is continuous given some mild assumptions but usually nondifferentiable. In addition, objective functions are usually nonconvex. This work is only concerned with local optimization. Bundle algorithms are the most common choice when performing local nonsmooth optimization [4]. A bundle solver from the several options presented by Karmitsa *et al.* is implemented [5]. These solvers require for each iterate the function value and an element of the generalized gradient. An element of the generalized gradient is calculated using Nesterov's lexicographic derivatives by incorporating into the ODE system a small number of LPs [3, 6].

Email addresses: **jagomezr@mit.edu** (Jose A. Gomez), **hoeffner@mit.edu** (Kai Höffner), **pib@mit.edu** (Paul I. Barton)

An efficient numerical implementation is illustrated through a large-scale dynamic flux balance analysis example.

Keywords: Dynamic flux balance analysis, linear programming, nonsmooth optimization, nonsmooth dynamic systems, lexicographic derivative, lexicographic optimization, penalty methods, extended ordinary differential equations systems.

References

[1] Jeffrey D. Orth, Ines Thiele, and Bernhard O. Palsson. What is flux balance analysis? *Nature Biotechnology*, 28(3):245, 2010.

[2] SM Harwood, K Höffner, and PI Barton. Solution of ordinary differential equations with a linear program embedded: the right-hand side case. Submitted.

[3] K Höffner, KA Khan, and PI Barton. Generalized derivatives of dynamic systems with a linear program embedded: Strictly feasible case. 2014. Preprint submitted to Elsevier.

[4] MM Mäkelä. Survey of bundle methods for nonsmooth optimization. *Optimization Methods and Software*, 17(1):1, 2001.

[5] N Karmitsa, ABagirov, MM Mäkelä. Empirical and Theoretical Comparisons of Several Nonsmooth Minimization Methods and Software. *Turku Centre for Computer Science Technical Report*, 959, 2009.

[6] Y Nesterov. Lexicographic differentiation of nonsmooth functions. *Mathematical Programming*, 104(2-3):669, 2005.

Centre for Process Systems Engineering

Imperial College London

A framework for Model Reduction, State Estimation and Multi-Parametric Model Predictive Control in Anaesthesia

Ioana Nascu, Romain S. C. Lambert, Efstratios N. Pistikopoulos

erc MOBILE

What is anaesthesia?

- Muscle relaxation
  induced to prevent unwanted movement or muscle tone
- Analgesia
  blocking the conscious sensation of pain
- Hypnosis
  a general term which indicates unconsciousness and absence of post operative recall of events

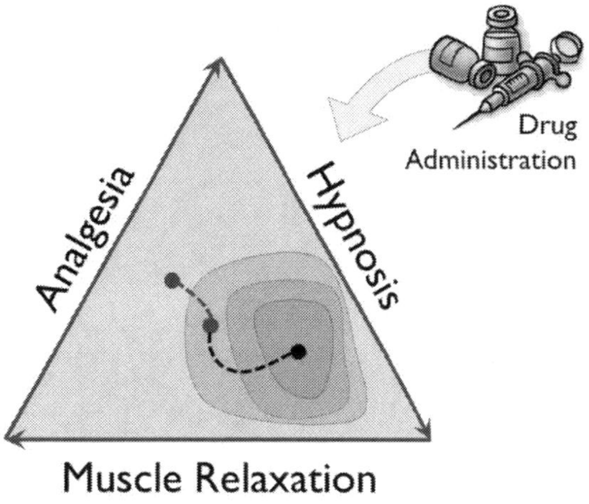

erc MOBILE

Anaesthesia – Current Practice

Anaesthetist provides hypnosis, analgesia, amnesia and muscle relaxation, while maintaining the vital functions of the patient

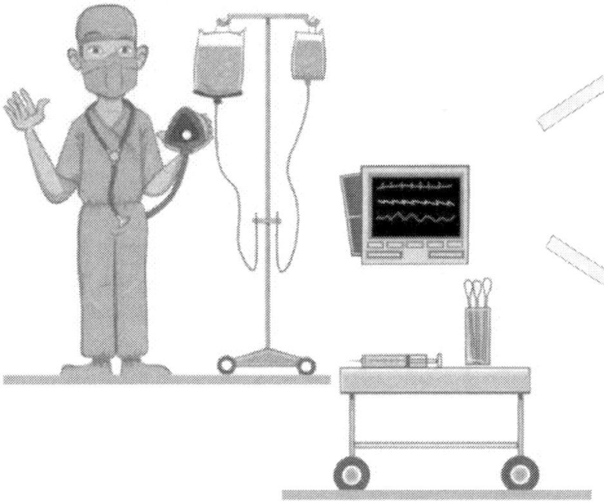

Intravenous Anaesthesia

Volatile Anaesthesia

erc MOBILE

Centre for Process Systems Engineering
Imperial College London

Anaesthesia – Current Practice

Anaesthetist provides hypnosis, analgesia, amnesia and muscle relaxation, while maintaining the vital functions of the patient

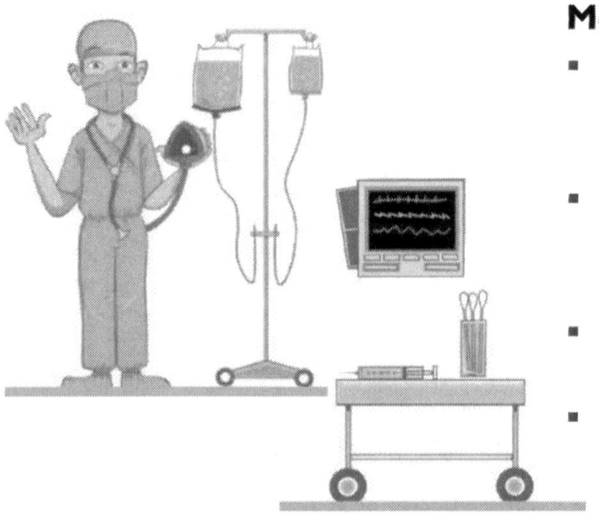

Motivation for control

- Support anaesthetist during critical situations
  - Automate drug delivery system
- Safety of the patient
  - Reduced side-effects by optimising the drug infusion rates
- Teaching tool for insights and understanding
- Testing platform

erc MOBILE

Framework for multi-parametric/explicit model predictive control

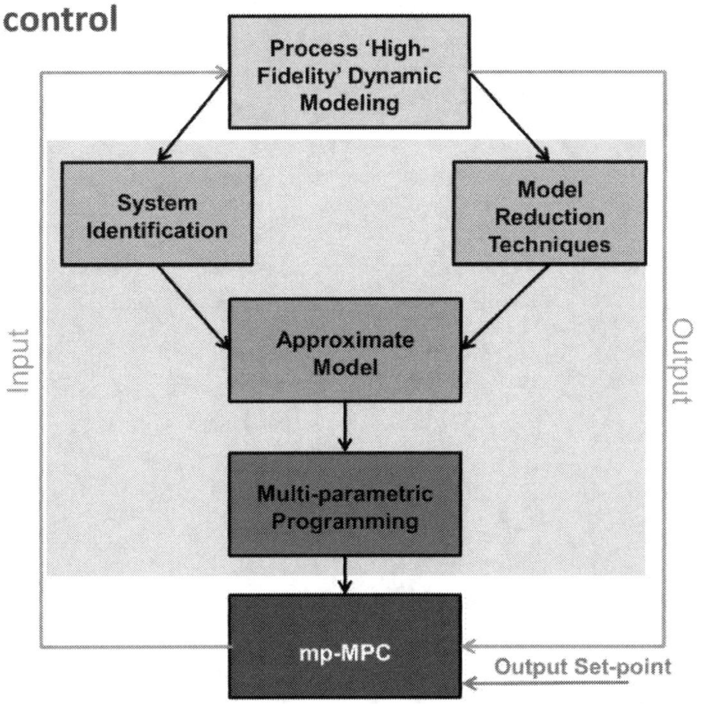

* Pistikopoulos, AIChE Journal, Perspectives in multiparametric programming and explicit model predi control, AIChE Journal. 55(8), 1918-1925 (2007)

erc MOBILE

Framework for multi-parametric/explicit model predictive control

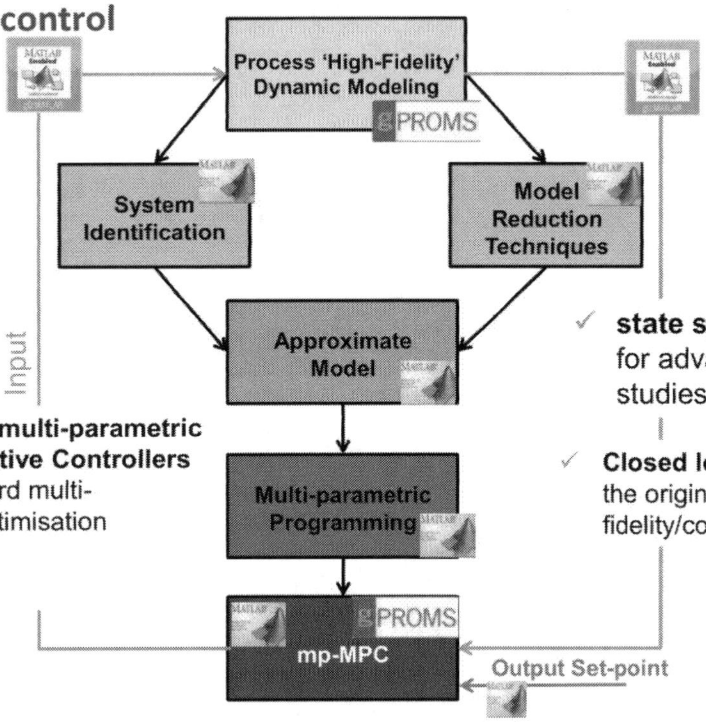

- Design of the **multi-parametric Model Predictive Controllers**
- Use of standard multi-parametric optimisation techniques

- **state space model** for advanced control studies
- **Closed loop validation** on the original high fidelity/complexity model

erc MOBILE

Framework for multi-parametric/explicit model predictive control

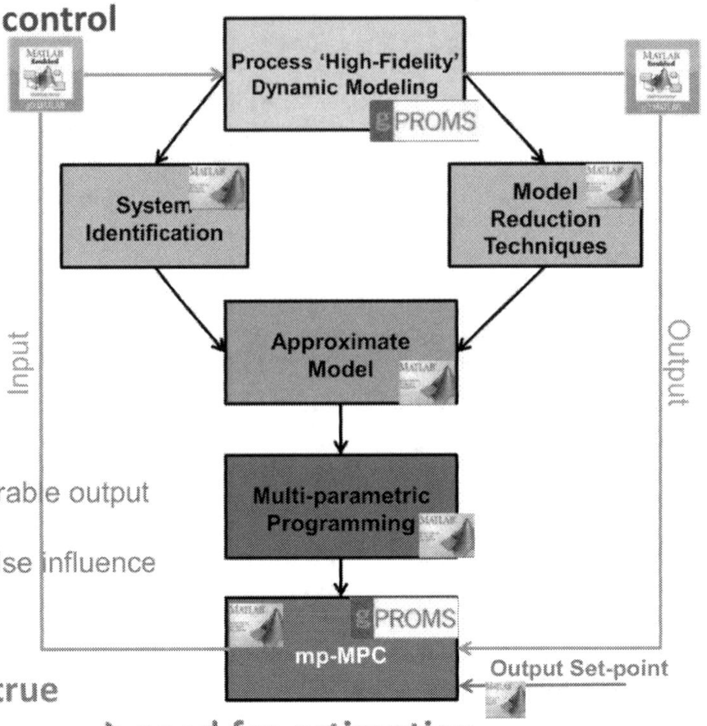

Assumptions:

- Clear measurable output
- Not much noise influence

may not be true

→ need for estimation

Centre for Process Systems Engineering
Imperial College London

Moving Horizon Estimation (MHE) – at a glance

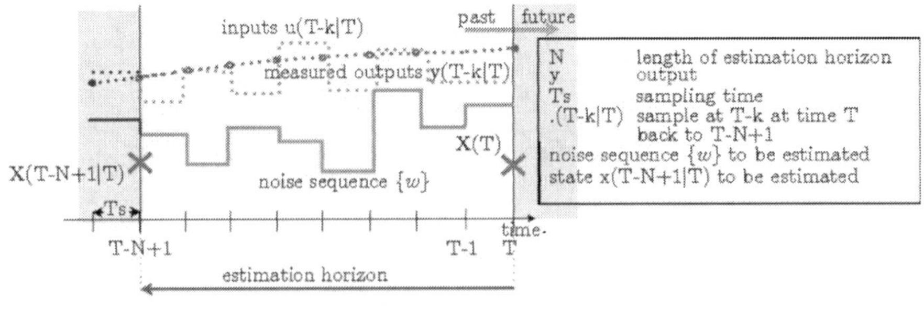

- **Model-based state estimator**
- **Obtains current state estimate x_T**
- **Main advantage: incorporates system constraints**
- **MHE is dual to MPC: *backwards MPC***

•Darby, M.L. and M. Nikolaou, *A parametric programming approach to moving-horizon state estimation.* Automatica, 2007
•Rawlings, Mayne, Model Predictive Control: Theory and Design, Nob Hill Publishing, 2009
•Rao, Moving Horizon Strategies for the Constrained Monitoring and Control of Nonlinear DiscreteTime Systems, Ph.D. thesis, 200

Centre for Process Systems Engineering

Imperial College London

Framework for multi-parametric/explicit model predictive control

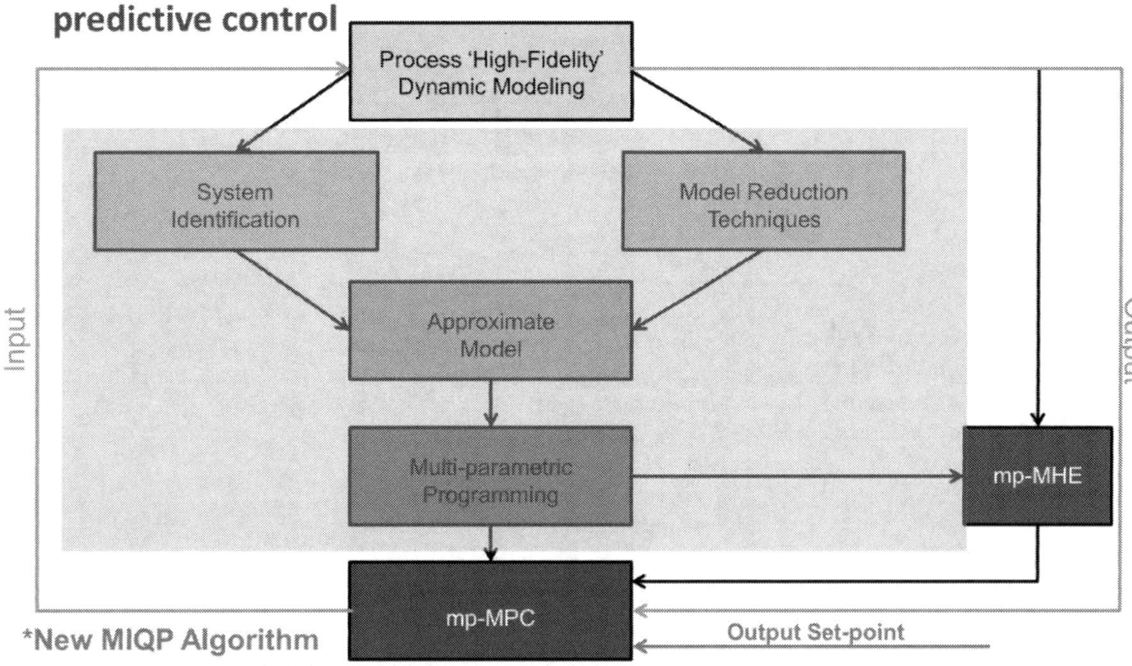

*New MIQP Algorithm

•Voelker, A., K. Kouramas, and E.N. Pistikopoulos. *Simultaneous constrained moving horizon state estimation and model predictive control by multi-parametric programming*. 2010
•Lambert, R.S.C., I. Nascu, and E.N. Pistikopoulos. *Simultaneous reduced order multi-parametric moving horizon estimation and model based control*. 2013
•Oberdieck, R. & E.N. Pistikopoulos, Automatica (submitted) 2014; Pistikopoulos (Session nr, 136)

erc MOBILE

Centre for Process Systems Engineering

Imperial College London

Anaesthesia - Input/Output Control System

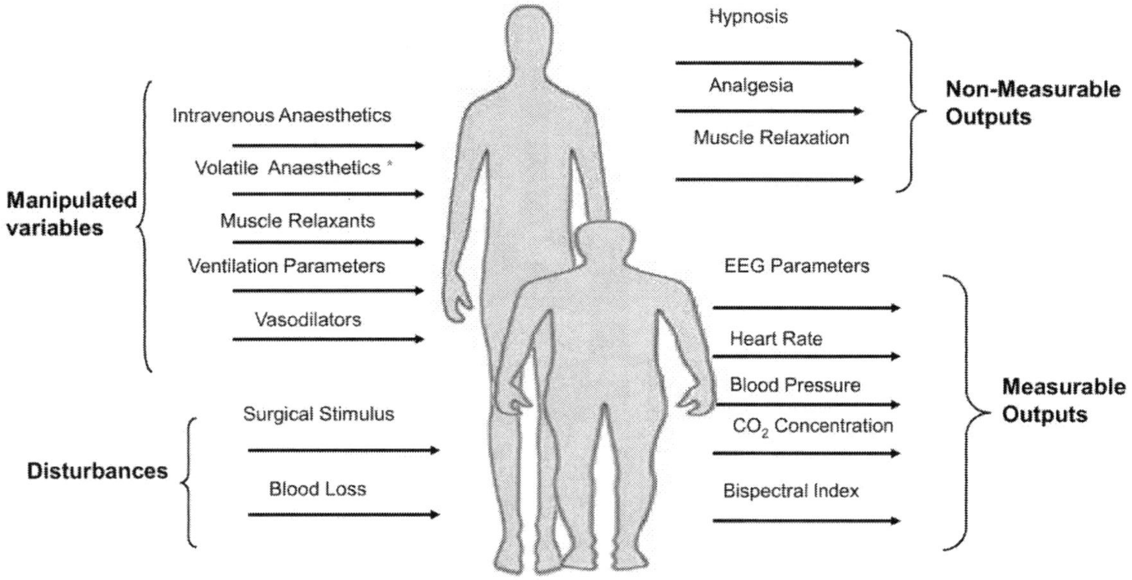

* Krieger, Pistikopoulos et al., A novel physiological based compartmental model for volatile anaesthesia, IEEE Transactions on Biomedical Engineering, 2014

erc MOBILE

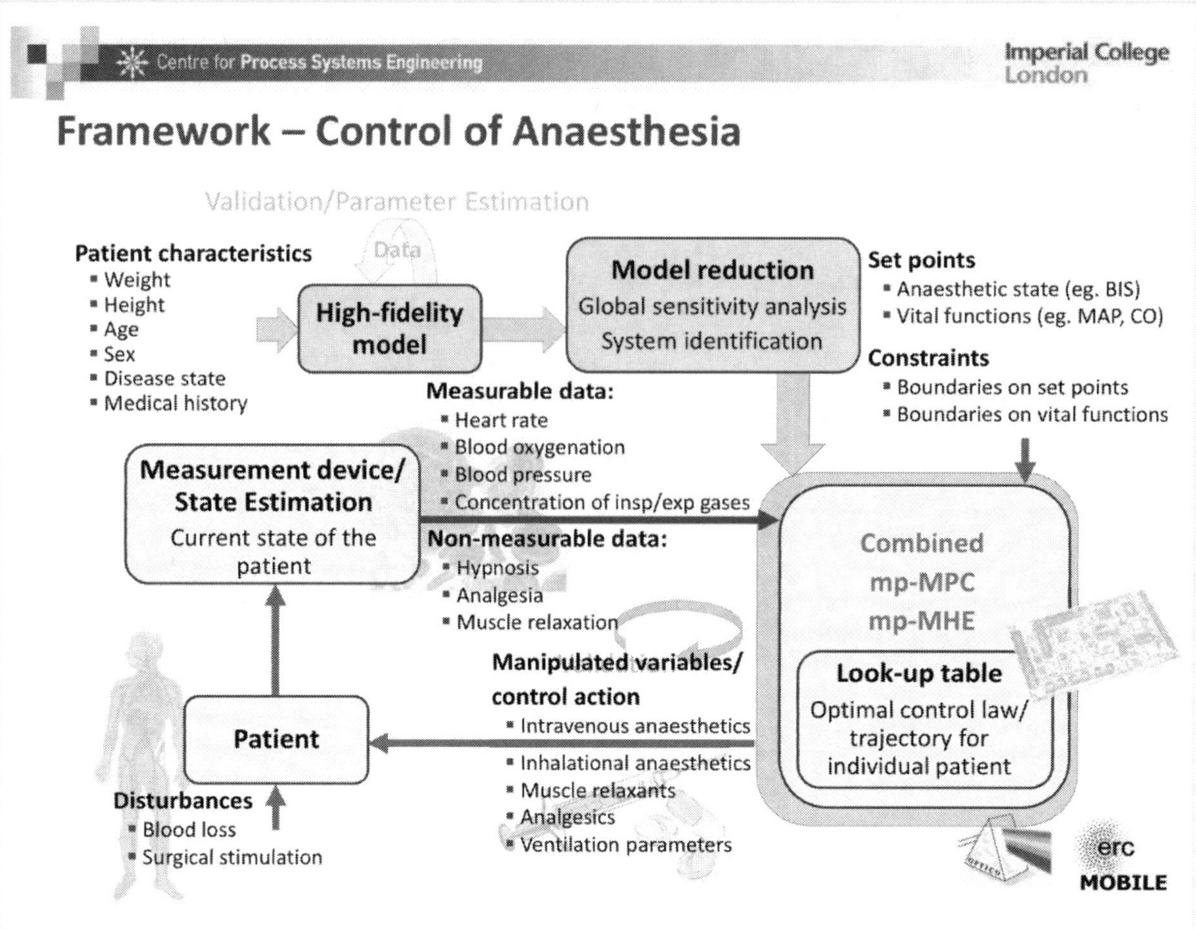
Centre for Process Systems Engineering
Imperial College London
Framework – Control of Anaesthesia
Validation/Parameter Estimation
Data
Patient characteristics
Weight
Height
Age
Sex
Disease state
Medical history
High-fidelity model
Model reduction
Global sensitivity analysis
System identification
Set points
Anaesthetic state (eg. BIS)
Vital functions (eg. MAP, CO)
Constraints
Boundaries on set points
Boundaries on vital functions
Measurable data:
Heart rate
Blood oxygenation
Blood pressure
Concentration of insp/exp gases
Measurement device/ State Estimation
Current state of the patient
Non-measurable data:
Hypnosis
Analgesia
Muscle relaxation
Combined
mp-MPC
mp-MHE
Look-up table
Optimal control law/ trajectory for individual patient
Manipulated variables/ control action
Intravenous anaesthetics
Inhalational anaesthetics
Muscle relaxants
Analgesics
Ventilation parameters
Patient
Disturbances
Blood loss
Surgical stimulation
erc
MOBILE

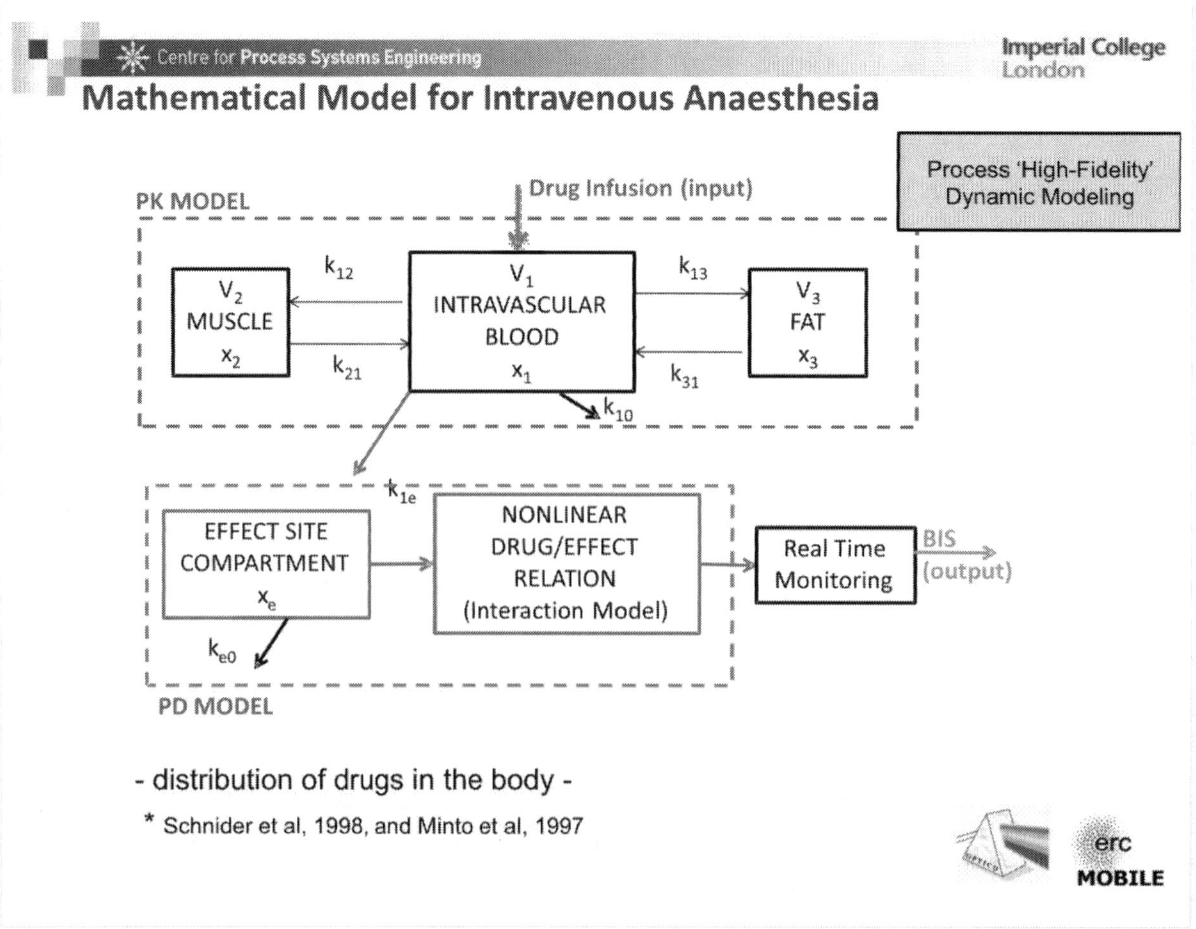
Centre for Process Systems Engineering
Imperial College London
Mathematical Model for Intravenous Anaesthesia
Process 'High-Fidelity' Dynamic Modeling
PK MODEL
Drug Infusion (input)
V2
MUSCLE
x2
k12
k21
V1
INTRAVASCULAR BLOOD
x1
k13
k31
V3
FAT
x3
k10
k1e
EFFECT SITE COMPARTMENT
xe
ke0
NONLINEAR DRUG/EFFECT RELATION (Interaction Model)
Real Time Monitoring
BIS
(output)
PD MODEL
- distribution of drugs in the body -
* Schnider et al, 1998, and Minto et al, 1997
erc
MOBILE

Centre for Process Systems Engineering

Imperial College London

Mathematical Model for Intravenous Anaesthesia
- Pharmacokinetic Model-

Process 'High-Fidelity' Dynamic Modeling

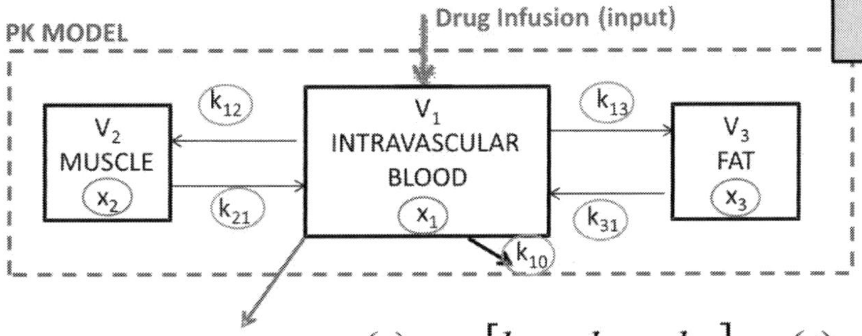

x – drug concentration

k – drug transfer frequency

$$\dot{x}_1(t) = -[k_{10} + k_{12} + k_{13}] \cdot x_1(t) + k_{21} \cdot x_2(t) + k_{31} \cdot x_3(t) + u(t)$$

$$\dot{x}_2(t) = k_{12} \cdot x_1(t) - k_{21} \cdot x_2(t)$$

$$\dot{x}_3(t) = k_{13} \cdot x_1(t) - k_{31} \cdot x_3(t)$$

LINEAR MODEL

erc MOBILE

Centre for Process Systems Engineering

Imperial College London

Mathematical Model for Intravenous Anaesthesia
- Pharmacodynamic Model-

Process 'High-Fidelity' Dynamic Modeling

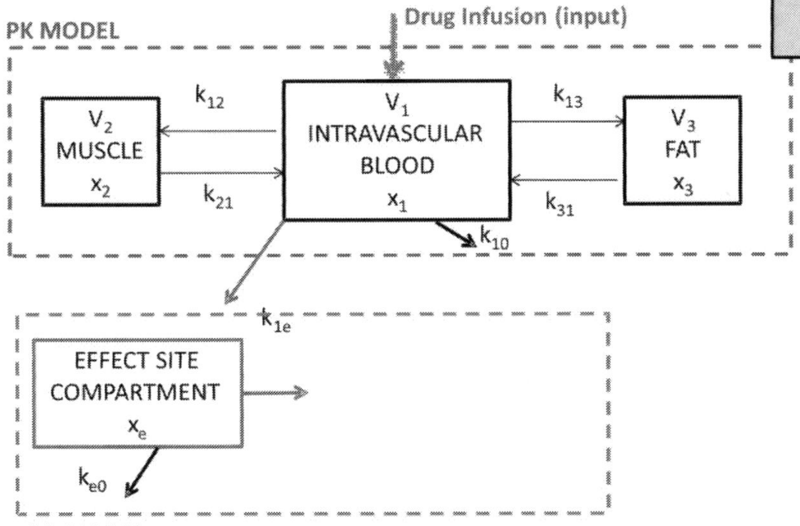

$$\dot{C}_e(t) = k_{e0} \cdot (C_e(t) - C_p(t))$$

erc MOBILE

Centre for Process Systems Engineering

Imperial College London

Mathematical Model for Intravenous Anaesthesia - Pharmacodynamic Model-

Process 'High-Fidelity' Dynamic Modeling

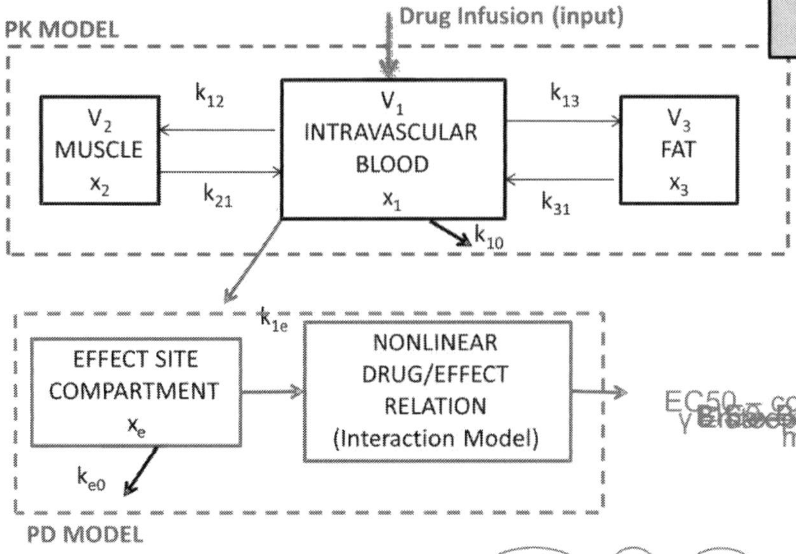

EC50 = concentration at 50% maximum effect

$$BIS(t) = E_0 - E_{max} \cdot \frac{C_e(t)^{\gamma}}{C_e(t)^{\gamma} + EC_{50}^{\gamma}}$$

NONLINEAR MODEL

erc MOBILE

Centre for Process Systems Engineering

Imperial College London

Mathematical Model for Intravenous Anaesthesia - Pharmacodynamic Model-

Process 'High-Fidelity' Dynamic Modeling

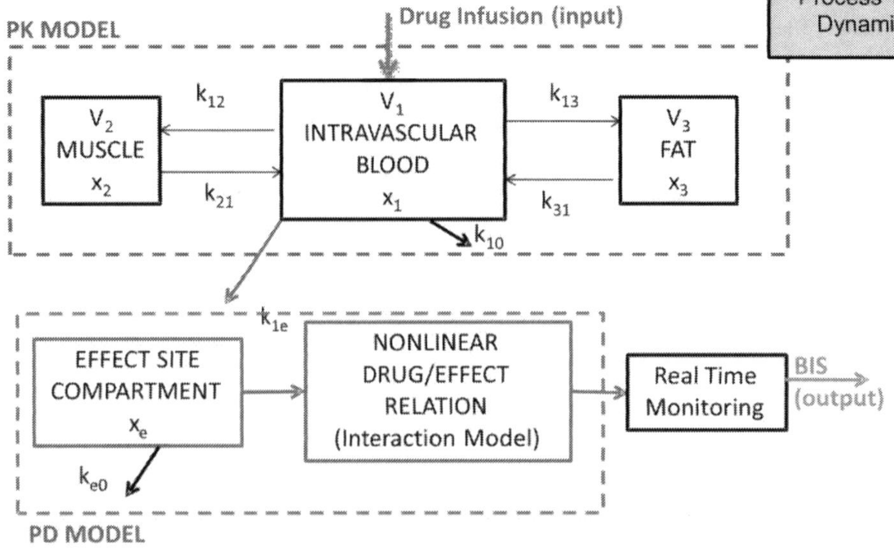

erc MOBILE

Centre for Process Systems Engineering

Imperial College London

Mathematical Model for Volatile Anaesthesia - Pharmacokinetic Model-

Process 'High-Fidelity' Dynamic Modeling

Body compartment

$$V_{b,i}\frac{dC_{b,i}}{dt} = Q_i(C_a - C_{b,i}) - Q_i(\lambda_i C_{b,i} - C_{t,i})$$

$$V_{t,i}\frac{dC_{t,i}}{dt} = Q_i(\lambda_i C_{b,i} - C_{t,i}) \Leftarrow u_{t,i}$$

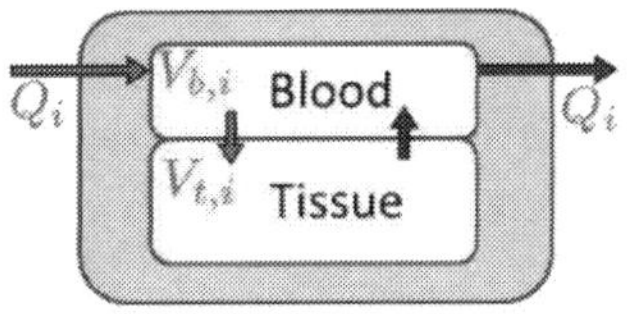

$$C_{t,i}|_{eq} = \lambda_i C_{b,i}$$

Lungs

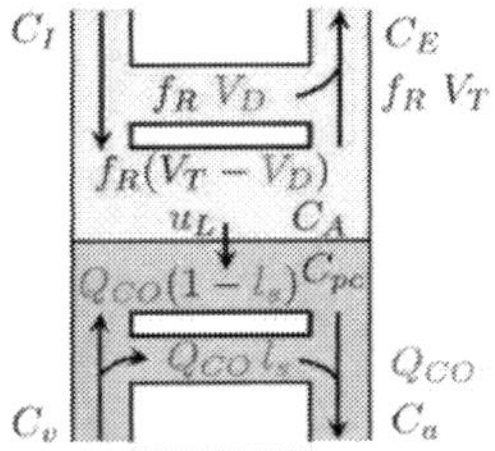

$$V_L \frac{dC_A}{dt} = f_R(V_T - V_D)(C_I - C_A) - Q_{CO}(1 - l_s)(\lambda C_A - C_v)$$

$\Uparrow u_L$

$$V_L = 11.97 \cdot \exp(-0.097 \cdot \text{BMI}) + 0.46$$

Blood pools

$$C_a Q_{CO} = C_{pc} Q_{CO}(1 - l_s) + C_v Q_{CO} l_s$$

$$C_v = \frac{\sum_i Q_i C_{b,i}}{Q_{CO}}$$

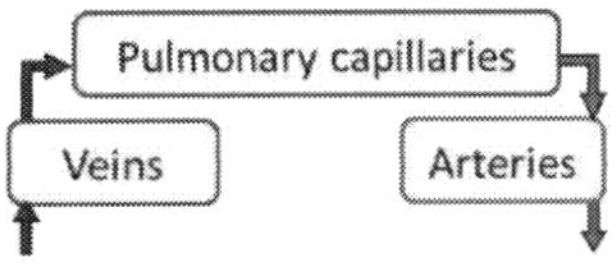

[1] Krieger, A., et al. (2011). 21st European Symposium on Computer Aided Process Engineering, volume 29, 1495-1499. Elsevier.

18

erc MOBILE

Centre for Process Systems Engineering

Imperial College London

Mathematical Model for Volatile Anaesthesia - Pharmacodynamic Model-

Process 'High-Fidelity' Dynamic Modeling

Hypnotic effect of the anesthetic agent

measured by the Bispectral Index (BIS)

$$\frac{dC_e}{dt} = k_{e0}(C - C_e)$$

$$BIS = BIS_0 + (BIS_{max} - BIS_0)\frac{C_e^{\gamma}}{C_e^{\gamma} + C_{50}{}^{\gamma}}$$

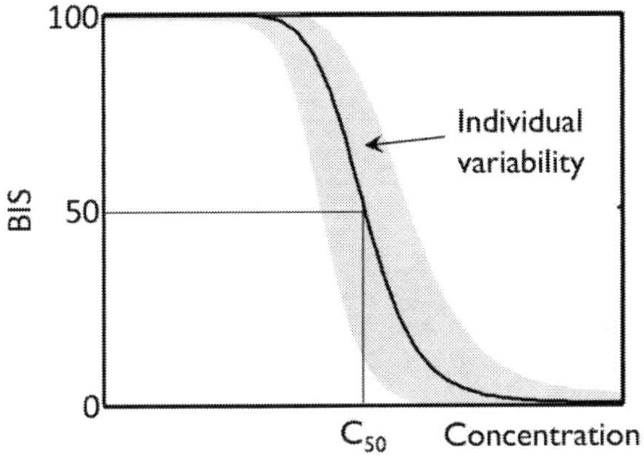

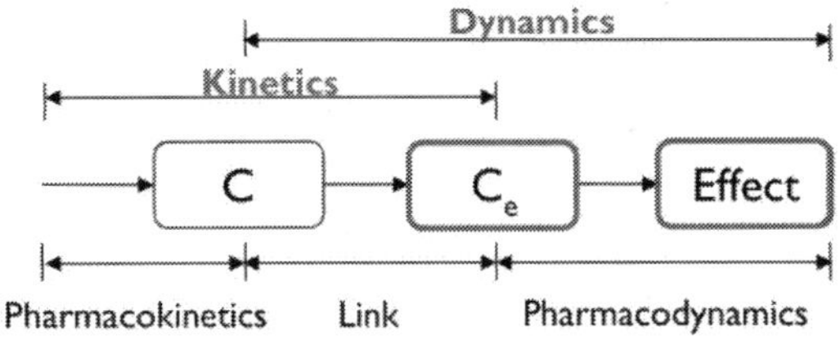

18

erc MOBILE

Mathematical Model for Anaesthesia

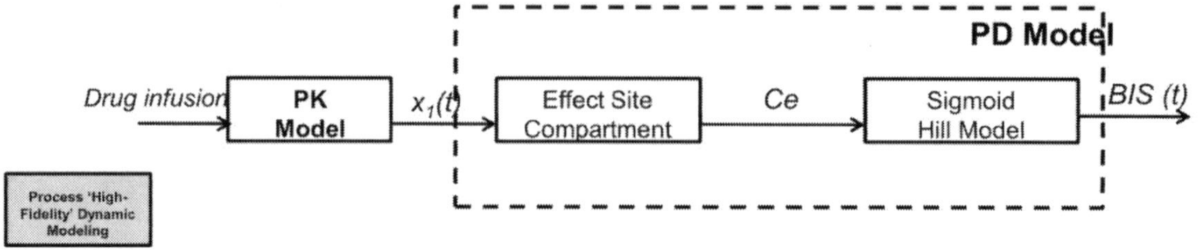

Intravenous Anaesthesia

- **Pharmacokinetic model**

$$\dot{x}_1(t) = -[k_{10} + k_{12} + k_{13}] \cdot x_1(t) + k_{21} \cdot x_2(t) + k_{31} \cdot x_3(t) + u(t)$$
$$\dot{x}_2(t) = k_{12} \cdot x_1(t) - k_{21} \cdot x_2(t)$$
$$\dot{x}_3(t) = k_{13} \cdot x_1(t) - k_{31} \cdot x_3(t)$$

- **Pharmacodynamic model**

$$\dot{C}_e(t) = k_{e0} \cdot (C_e(t) - C_p(t))$$

$$BIS(t) = E_0 - E_{\max} \cdot \frac{C_e(t)^\gamma}{C_e(t)^\gamma + EC_{50}^\gamma}$$

Volatile Anaesthesia

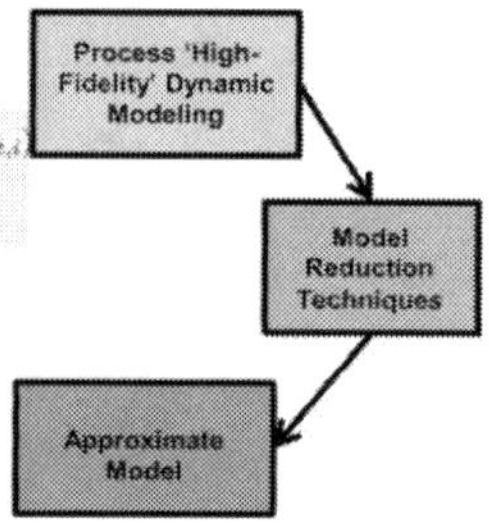

$$V_{b,i}\frac{dC_{b,i}}{dt} = Q_i(C_a - C_{b,i}) - Q_i(\lambda_i C_{b,i} - C_{t,i})$$
$$V_{t,i}\frac{dC_{t,i}}{dt} = Q_i(\lambda_i C_{b,i} - C_{t,i})$$
$$V_L\frac{dC_A}{dt} = f_R(V_T - V_D)(C_I - C_A) - Q_{CO}(1 - l_s)(\lambda C_A - C_v)$$
$$V_L = 11.97 \cdot \exp(-0.097 \cdot BMI) + 0.46$$
$$C_a Q_{CO} = C_{pc} Q_{CO}(1 - l_s) + C_v Q_{CO} l_s$$
$$C_v = \frac{\sum_i Q_i C_{b,i}}{Q_{CO}}$$

$$\frac{dC_e}{dt} = k_{e0}(C - C_e)$$

$$BIS = BIS_0 + (BIS_{max} - BIS_0)\frac{C_e^\gamma}{C_e^\gamma + C_{50}{}^\gamma}$$

erc MOBILE

Centre for Process Systems Engineering
Imperial College London

Framework – Control of Anaesthesia

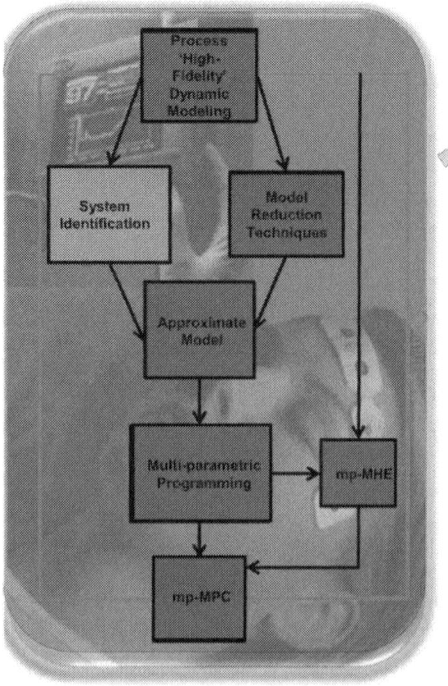

Volatile Anaesthesia -reduced model- → **Nonlinearity compensation -invHill-**

OPTICO erc MOBILE

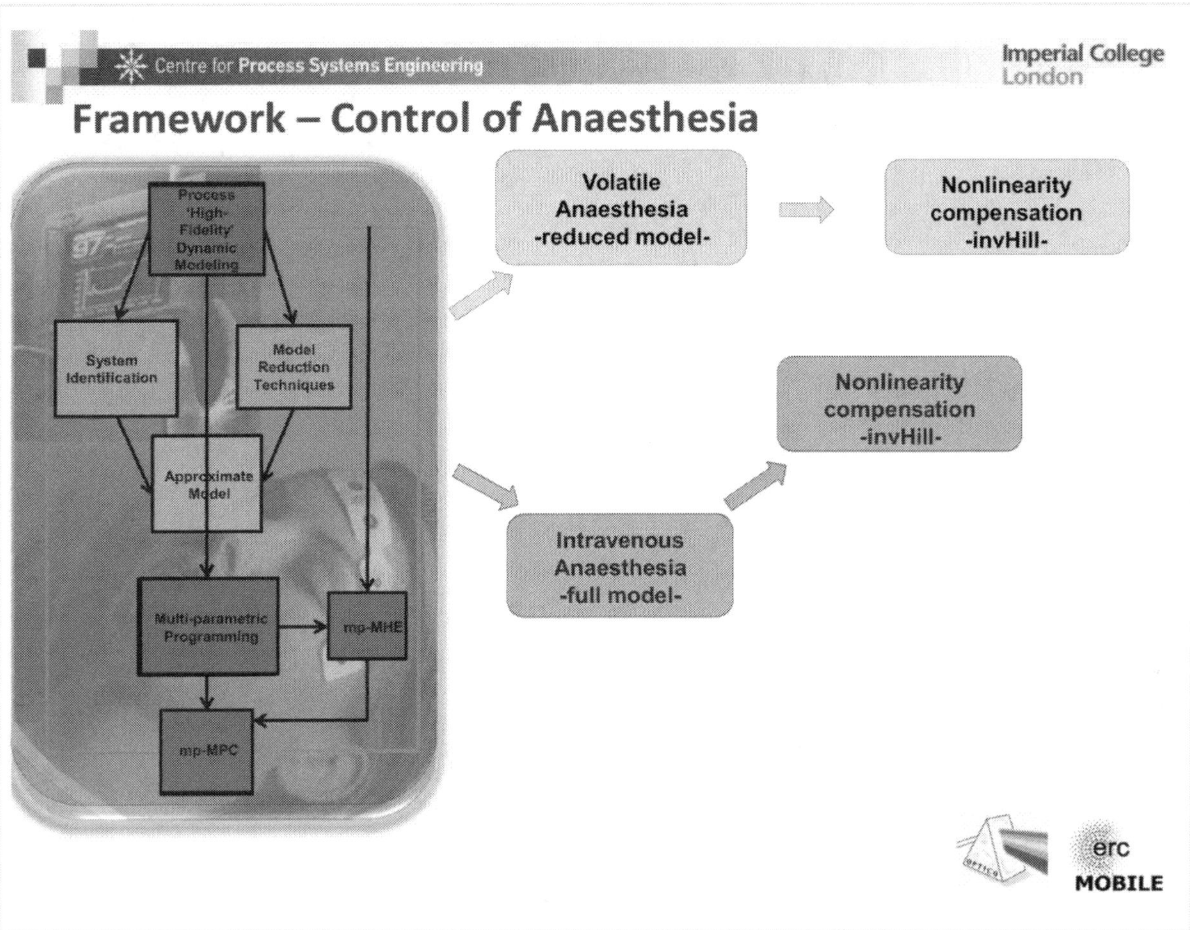
Centre for Process Systems Engineering
Imperial College London
Framework – Control of Anaesthesia
Process 'High-Fidelity' Dynamic Modeling
System Identification
Model Reduction Techniques
Approximate Model
Multi-parametric Programming
mp-MHE
mp-MPC
Volatile Anaesthesia -reduced model-
Nonlinearity compensation -invHill-
Nonlinearity compensation -invHill-
Intravenous Anaesthesia -full model-
erc
MOBILE

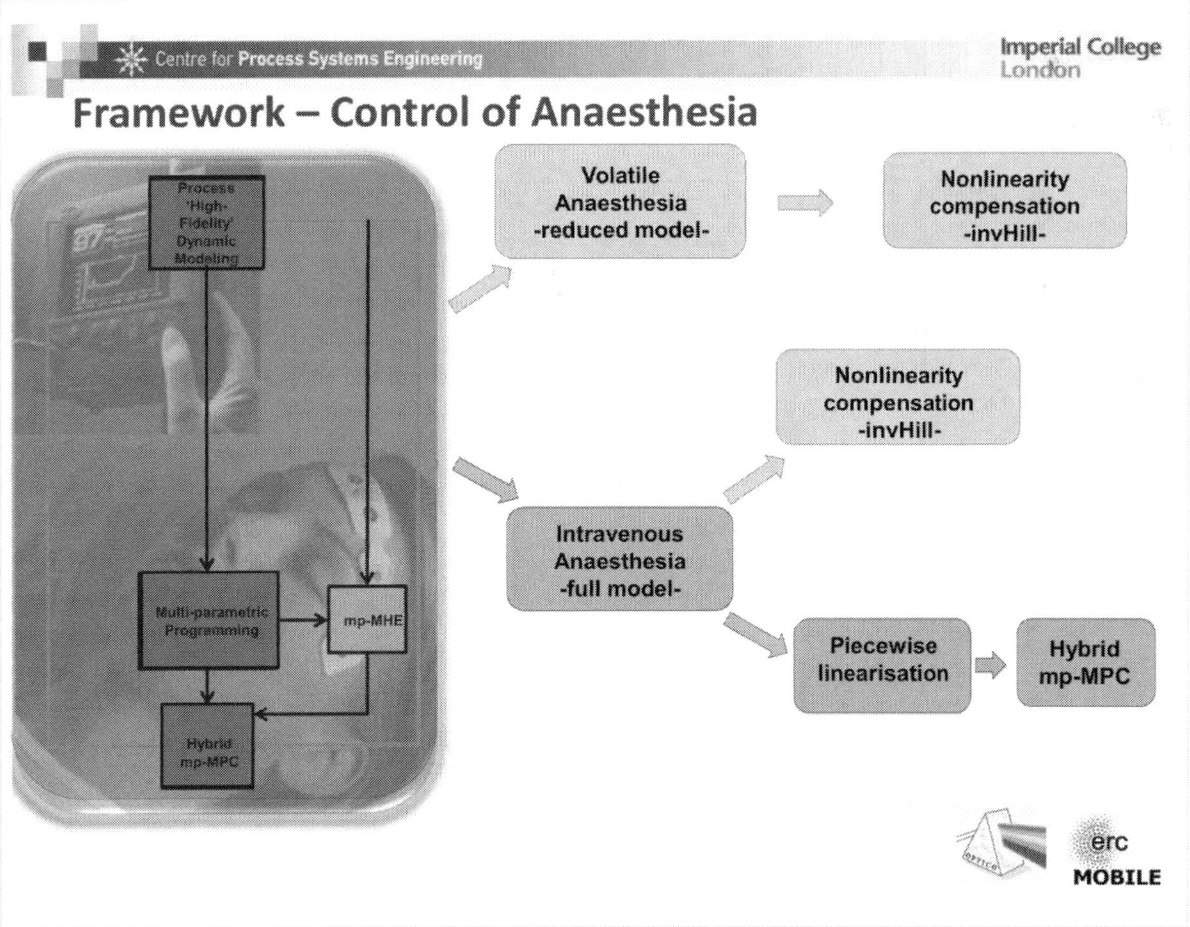
Centre for Process Systems Engineering
Imperial College London
Framework – Control of Anaesthesia
Process 'High-Fidelity' Dynamic Modeling
Multi-parametric Programming
mp-MHE
Hybrid mp-MPC
Volatile Anaesthesia -reduced model-
Nonlinearity compensation -invHill-
Nonlinearity compensation -invHill-
Intravenous Anaesthesia -full model-
Piecewise linearisation
Hybrid mp-MPC
erc
MOBILE

Centre for Process Systems Engineering — Imperial College London

Intravenous Anaesthesia – Control Problem

- Aim: automatic regulation of Depth of Anaesthesia (DOA);
- **Inputs**
  - **Intravenous: Propofol**
  - **Volatile: Isoflurane**

✓ acts on hypnotic state of brain;
✓ Propofol, Isoflurane = f(weight, height, age, gender, drug resistance)

- **Output #1: BIS;**
- measure of brain electrical activity, by means of frequency domain analysis of EEG;
- BIS = f(consciousness)
- BIS: 0 - 100 %

BIS = 0 – flat EEG
BIS = 100 – fully conscious patient

BIS = 40 – 60
general anaesthesia

erc MOBILE

Centre for Process Systems Engineering — Imperial College London

Anaesthesia – Patient Data

Patient	Age	Height (cm)	Weight (Kg)	Gender	EC_{50}	E_0	γ
1	40	163	54	M	6.33	98.8	2.24
2	36	163	50	M	6.76	98.6	4.29
3	28	164	52	M	8.44	91.2	4.1
4	50	163	83	M	6.44	95.9	2.18
5	28	164	60	F	4.93	94.7	2.46
6	43	163	59	M	12.0	90.2	2.42
7	37	187	75	F	8.02	92.0	2.1
8	38	174	80	M	6.56	95.5	4.12
9	41	170	70	M	6.15	89.2	6.89
10	37	167	58	M	13.7	83.1	1.65
11	42	179	78	F	4.82	91.8	1.85
12	34	172	58	M	4.95	96.2	1.84
Mean	38	169	65	M	7.42	93.1	3

Ionescu, C.M., I. Nascu, and R. De Keyser. *Towards a multivariable model for controlling the depth of anaesthesia using propofol and Remifentanil.* 2012

erc MOBILE

Framework – Control of Anaesthesia

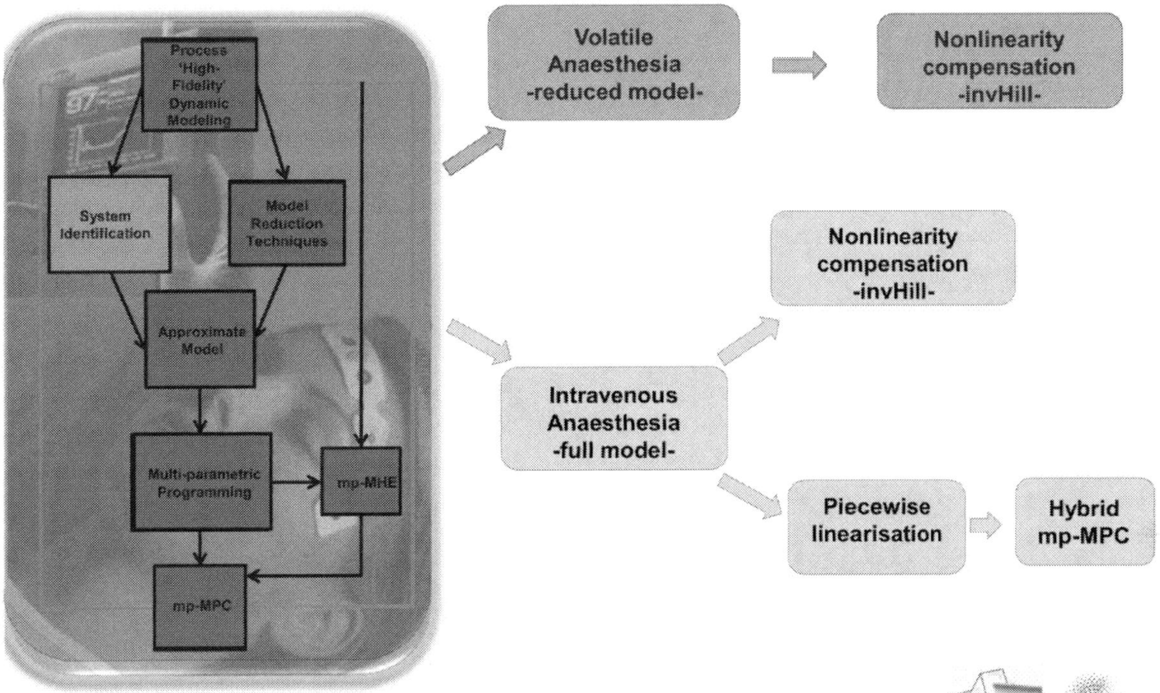

Simultaneous mp-MPC and mp-MHE

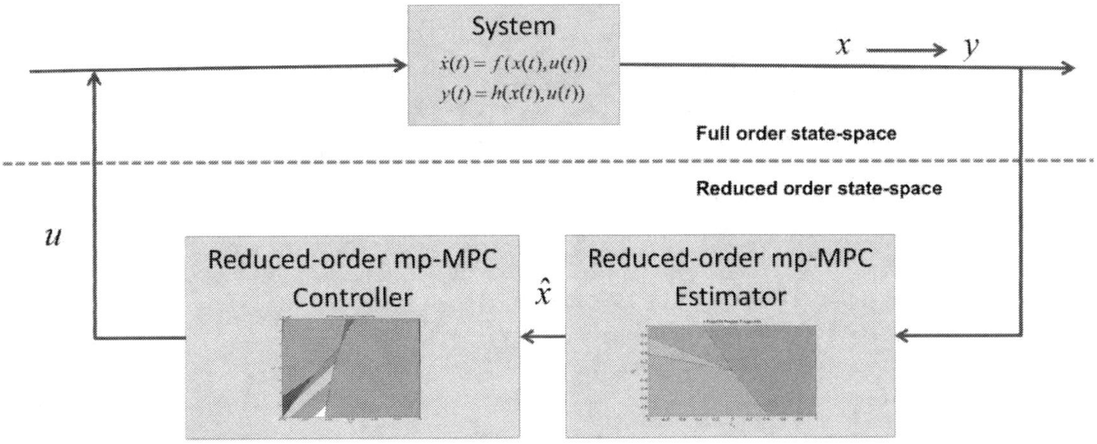

- Advantages:
  - reduction in computational complexity, necessary for tractability of mp-MPC and mp-MHE

Example – *Volatile anaesthesia*

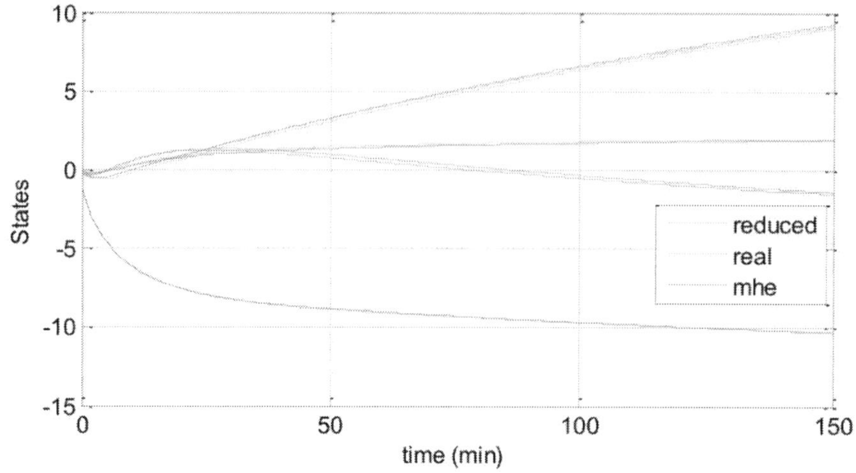

- Comparison of actual and estimated reduced order state and mhe
- 4 states

Example – *Volatile anaesthesia*

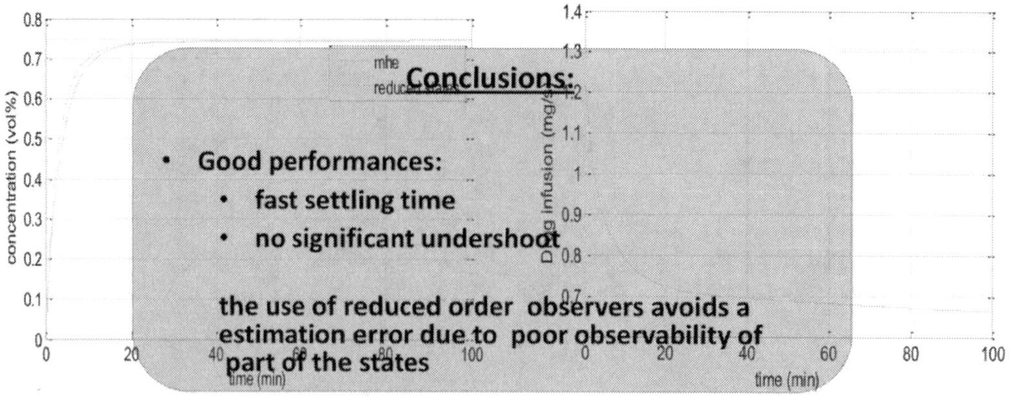

Conclusions:

- Good performances:
  - fast settling time
  - no significant undershoot

the use of reduced order observers avoids a estimation error due to poor observability of part of the states

- Close loop simulation of a set-point change operated through simultaneous mp-MHE and mp-MPC
- Evolution of the control input variable

Framework – Control of Anaesthesia

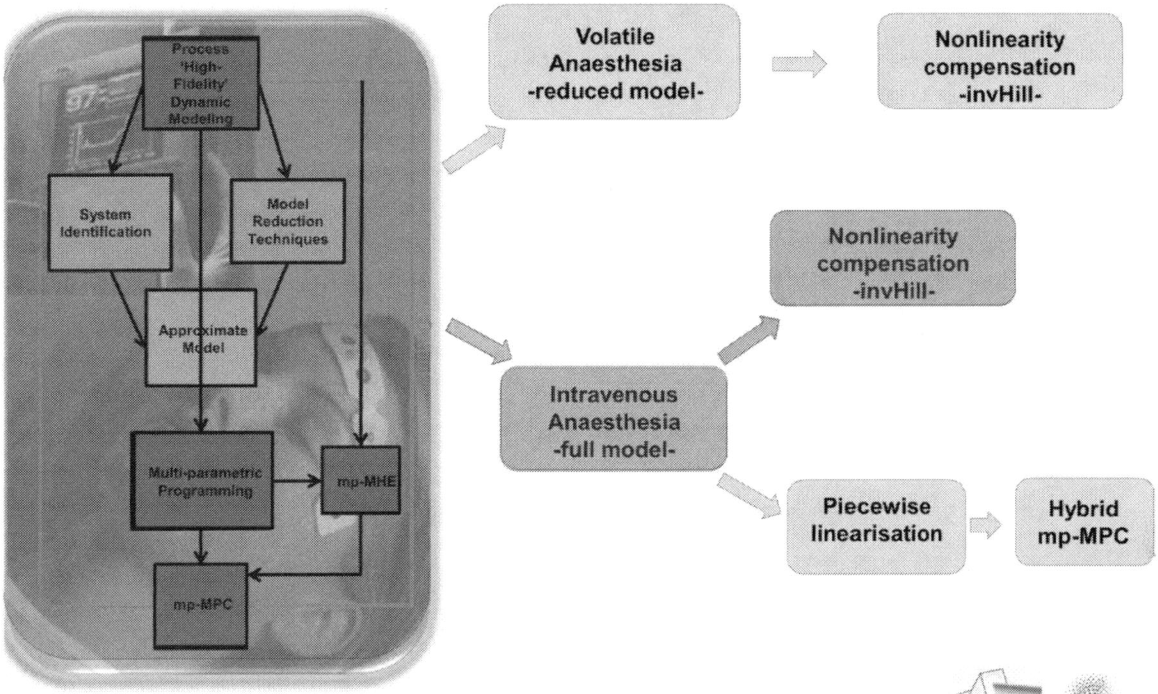

Intravenous Anaesthesia- Simultaneous mp-MHE and mp-MPC

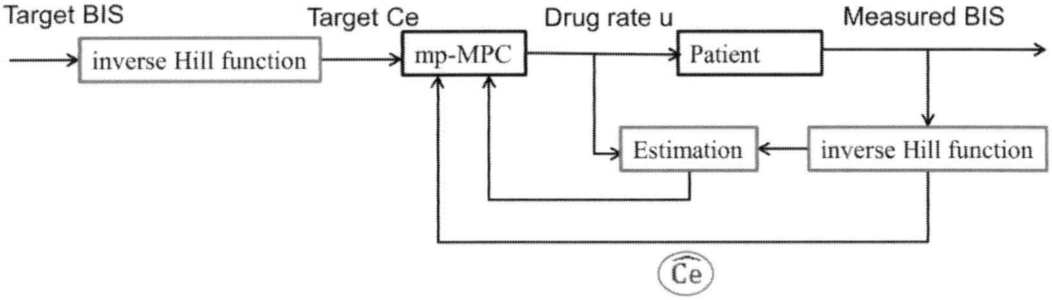

Estimation - Kalman filter
- MHE

Centre for Process Systems Engineering

Imperial College London

Intravenous Anaesthesia- Simultaneous mp-MHE and mp-MPC

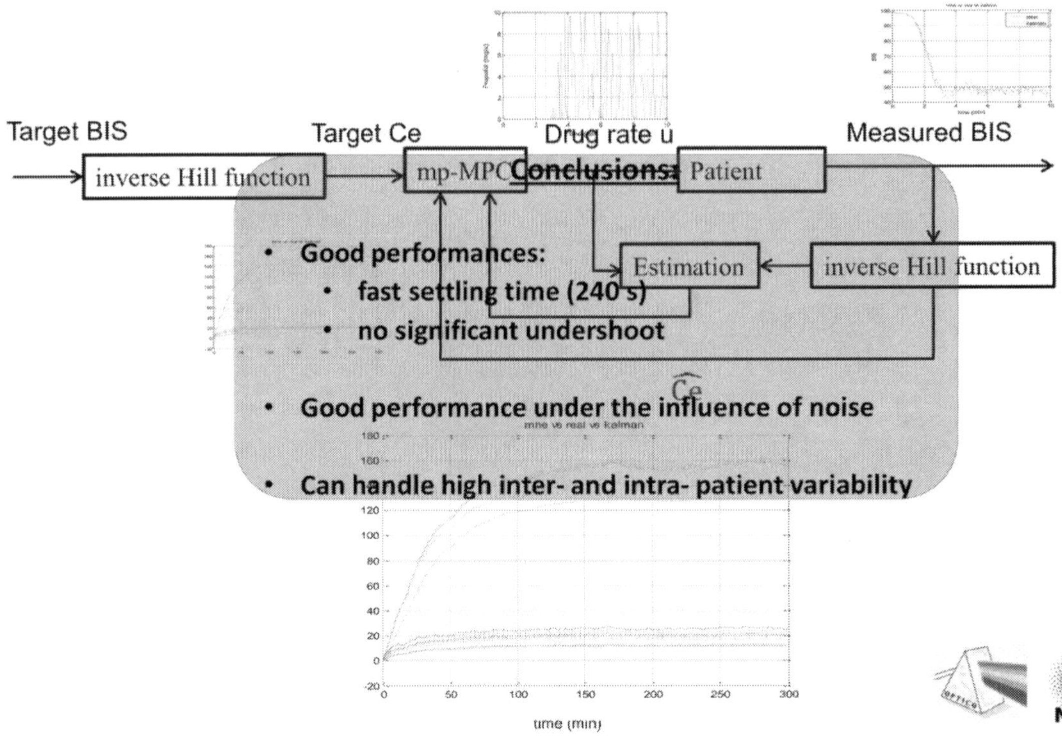

Conclusions:

- **Good performances:**
  - **fast settling time (240 s)**
  - **no significant undershoot**
- **Good performance under the influence of noise**
- **Can handle high inter- and intra- patient variability**

erc MOBILE

Centre for Process Systems Engineering

Imperial College London

Framework – Control of Anaesthesia

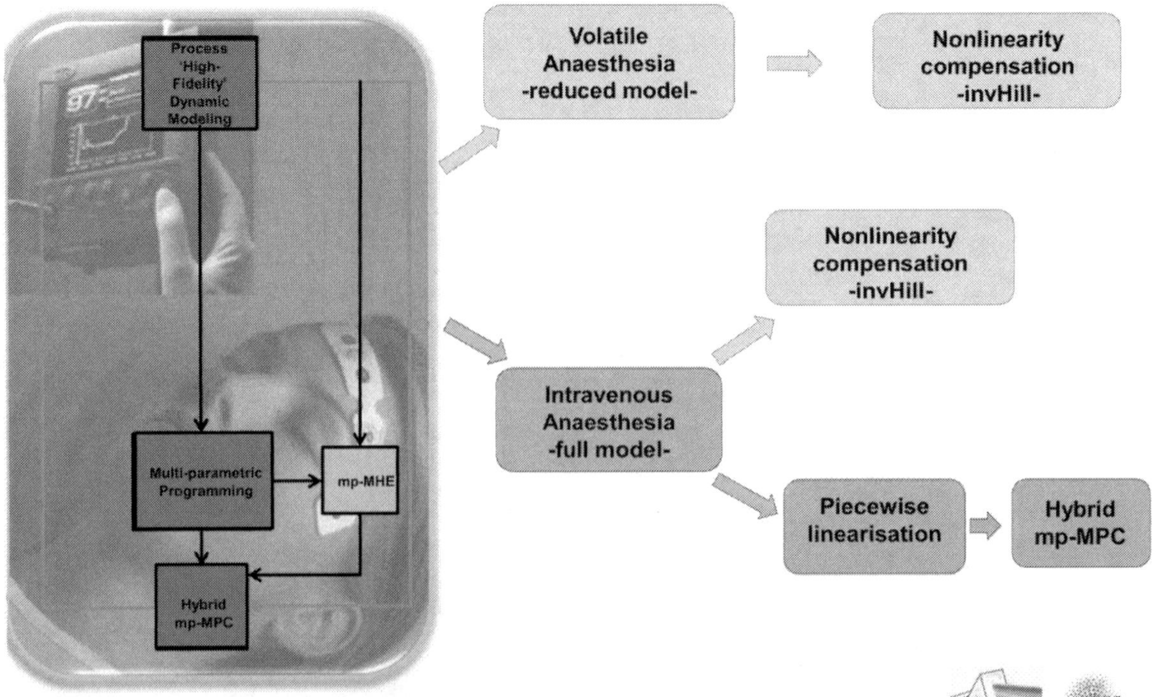

erc MOBILE

Intravenous Anaesthesia- Hybrid Control*

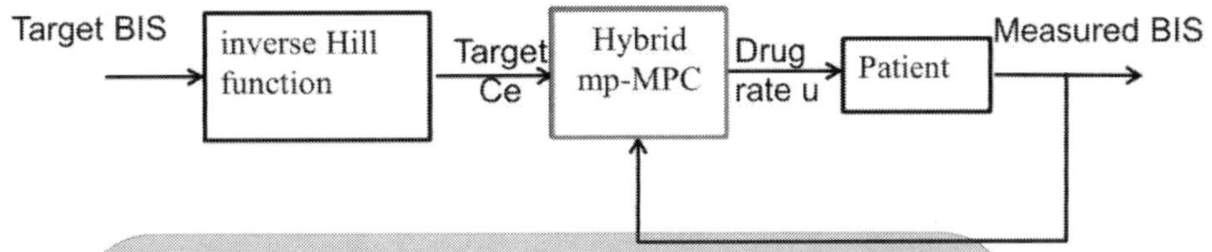

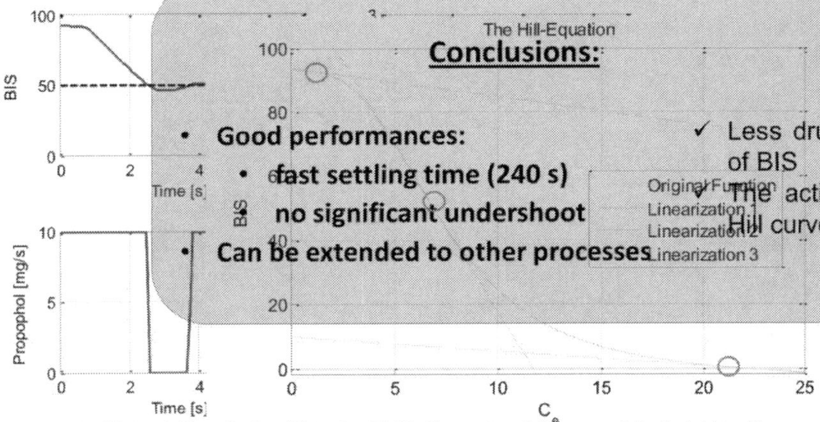

Conclusions:

- **Good performances:**
  - **fast settling time (240 s)**
  - **no significant undershoot**
- **Can be extended to other processes**

✓ Less drug infusion with decrease of BIS

✓ The active part of the piecewise Hill curve linearisation

*Nascu, I., Richard Oberdiecka, Efstratios N. Pistikopoulos A framework for hybrid multi-parametric model-predictive control with application to volatile anaesthesia, submitted, ESCAPE 2015

erc MOBILE

Concluding Remarks & Ongoing Research

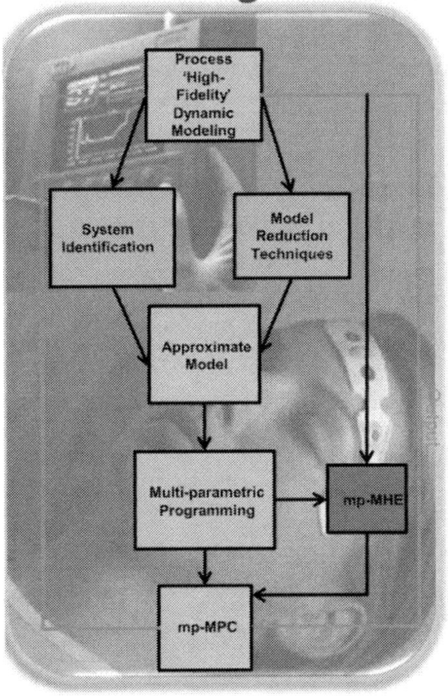

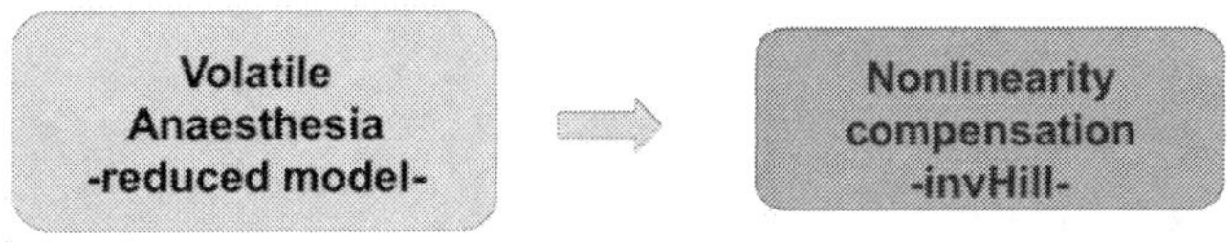

✓ The estimators provide sufficiently accurate information to reach the set points based only on the measured information

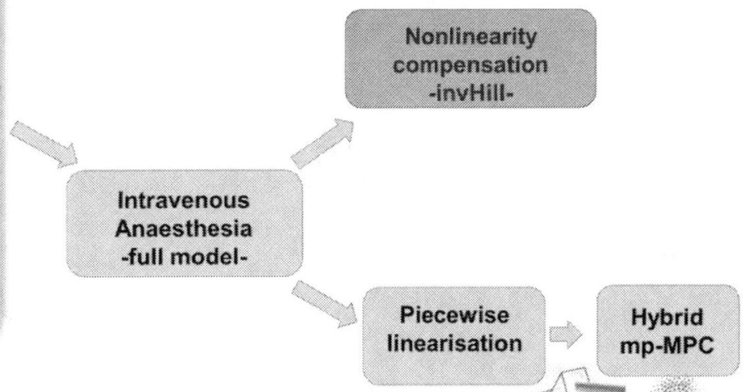

erc MOBILE

Concluding Remarks & Ongoing Research

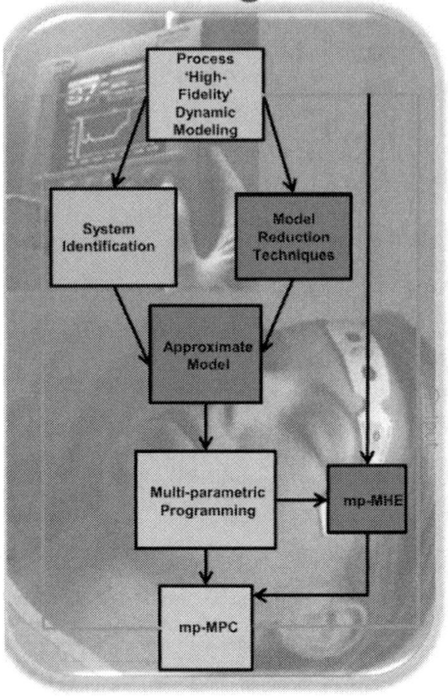

Volatile Anaesthesia -reduced model- → **Nonlinearity compensation -invHill-**

✓ Reduced order models is important for simultaneous mp-MHE and mp-MPC

Intravenous Anaesthesia -full model- → **Nonlinearity compensation -invHill-**

Intravenous Anaesthesia -full model- → **Piecewise linearisation** → **Hybrid mp-MPC**

erc MOBILE

Centre for Process Systems Engineering
Imperial College London

Concluding Remarks & **Ongoing Research**

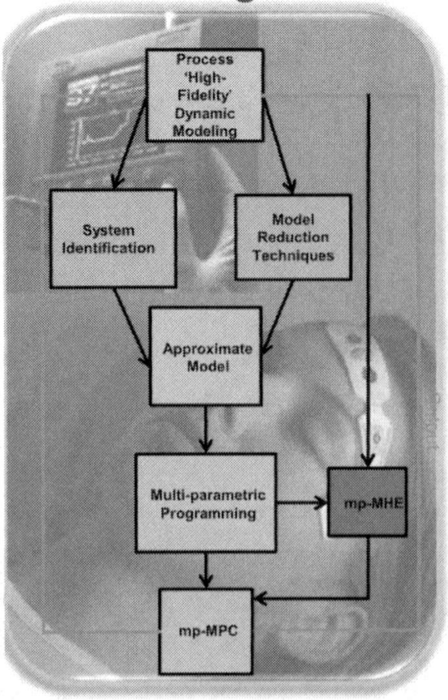

Volatile Anaesthesia -reduced model- → **Nonlinearity compensation -invHill-**

robust controllers to hedge against the estimation error by incorporating model reduction error metrics ir the calculation of mp-MHE error

Intravenous Anaesthesia -full model- → **Nonlinearity compensation -invHill-**

Intravenous Anaesthesia -full model- → **Piecewise linearisation** → **Hybrid mp-MPC**

erc MOBILE

Concluding Remarks & **Ongoing Research**

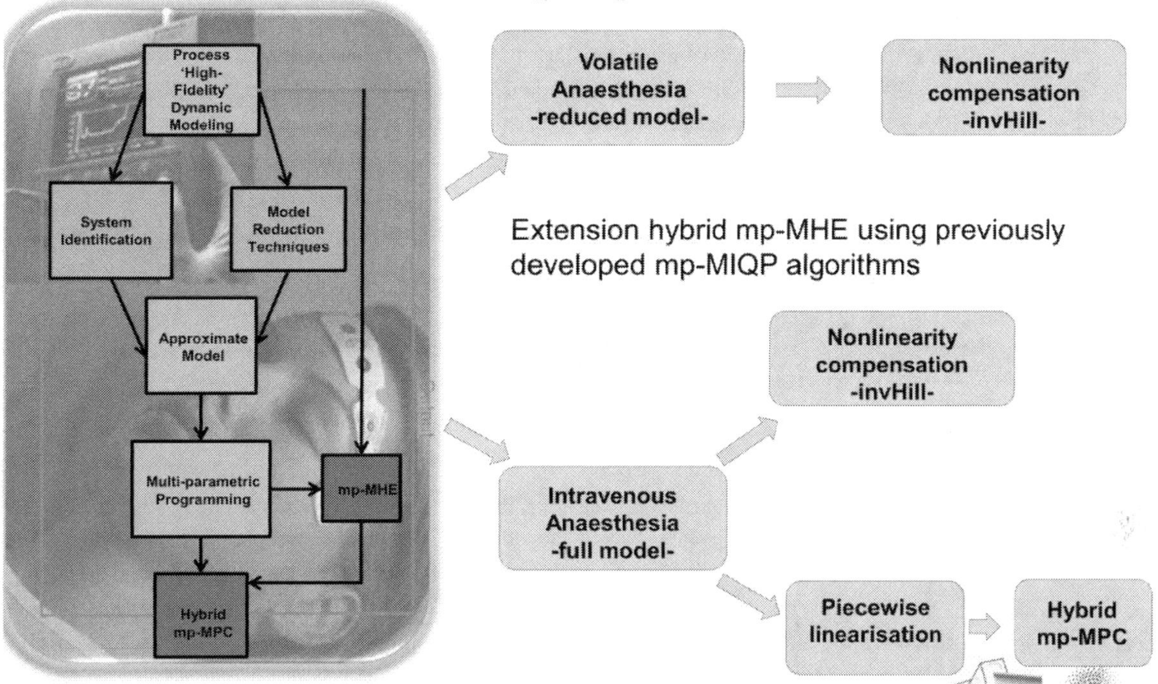

erc MOBILE

Centre for Process Systems Engineering

Imperial College London

A framework for model reduction, State Estimation and Multi-Parametric Model Predictive Control in Anaesthesia

Ioana Nascu, Romain S. C. Lambert, Efstratios N. Pistikopoulos

erc MOBILE

Acknowledgements

- OPTICO project (G.A. No 280813) for "Model-based Optimization and Control for Process-Intensification in Chemical and Biopharmaceutical Processes"
- EPSRC - GR/T02560/01, EP/E047017/1
- MOBILE, ERC Advanced Grant, No: 226462
- Mr. Richard Oberdieck

377460 Dynamic Feasibility Analysis of Black Box Processes

Abstract:

The concept of process feasibility describes the ability of a process to remain operable and satisfy all product specifications in the presence of uncertainty. This uncertainty can be introduced by variation in process inputs, error in estimated model parameters, or variability in process design and operating conditions. A rigorous mathematical formulation describing process feasibility was developed in the 1980s, with the goal of solving optimal design under uncertainty problems.[1,2] This feasibility test problem is an optimization problem, which seeks to minimize the violation of process constraints over a set of uncertain parameters, state variables and control variables for a given process design. To date, approaches to solving the feasibility test problem have generally involved deterministic optimization solvers. These require explicitly known constraints and also make certain assumptions regarding the convexity of the feasible region.[3] However, recent developments in derivative-free optimization have enabled the determination of feasible regions for processes with black box constraints, without making inferences regarding the convexity of the feasible region.[4,5] This is useful in addressing the feasibility analysis of integrated process models, which often contain a large number of potentially complex equations. In these applications, closed-form expressions for the constraints may not be available. In addition, the feasible region for these processes can have complex shapes, being nonconvex or even disjoint in nature.

This work is concerned with surrogate-based optimization methods, which involve the identification of a surrogate function to represent a complex process model. The surrogate function is less complex and faster to evaluate that the original model.[6] The process is then optimized using the surrogate representation rather than the complex process model. In previous work, the ability of surrogate-based methods to identify complex feasible regions has been demonstrated.[4] In addition, surrogate-based feasibility analysis has been applied to pharmaceutical processes to identify their design space during steady state operation.[7] In practice, many pharmaceutical manufacturing processes are inherently time-variant. As such, the feasible operating region may not be constant with respect to time. In the current work, a method is proposed for the surrogate-based feasibility analysis of dynamic processes, where the shape and size of the feasible region may vary over time. This methodology involves the development of a response surface representation of the feasibility function followed by the refinement of the response surface using an adaptive sampling strategy. The feasible region is identified using an iterative optimization process in which new points are added to the response surface based on their anticipated ability to improve upon the current model. The response surface is developed using Kriging, an interpolating black box modeling methodology, which allows determination of prediction variance at each sampled point. The prediction variance informs an expected improvement function, which is used to direct sampling towards points on the boundary of the feasible region. The adaptive sampling process continues until the expected improvement is below a threshold value, at which time the feasibility surrogate function is identified and used to determine the feasible region for the process. The proposed method is demonstrated on a pharmaceutical manufacturing process described by a complex process model with black box constraints.

1. Swaney RE, Grossmann, I.E. An Index for Operational Flexibility in Chemical Process Design Part I: Formulation and Theory. *Aiche J.* 1985;36.

2. Swaney RE, Grossmann, I.E. An Index for Operational Flexibility in Chemical Process Design}Part II: Computational Algorithms. *Aiche J.* 1985;31.

3. Biegler LT, Grossmann IE, Westerberg AW. Systematic Methods of Chemical Process Design. Upper Saddle River, NJ: Prentice Hall; 1997.

4. Boukouvala F, Ierapetritou MG. Feasibility analysis of black-box processes using an adaptive sampling Kriging-based method. *Comput Chem Eng.* 2012;36:358-368.

5. Boukouvala F, Ierapetritou M. Simulation-Based Derivative-Free Optimization for Computationally Expensive Function. Paper presented at: AIChE Annual Meeting2012; Pittsburgh, PA.

6. Jones DR. A taxonomy of global optimization methods based on response surfaces. *J Global Optim.* 2001;21(4):345-383.

7. Boukouvala F, Muzzio FJ, Ierapetritou MG. Design Space of Pharmaceutical Processes Using Data-Driven-Based Methods. *J Pharm Innov.* 2010;5(3):119-137.

378224 Dynamic Modeling and Optimization of an Industrial Super-Staged Argon Plant

Abstract:

Air separation units (ASUs) are power-intensive processes with major operation costs due to electricity consumption. Industrial gas producers in US consume over $ 1 billion dollars' worth of electricity annually (Chen et al., 2010). Traditionally, ASUs had infrequent changes in production rate and operation conditions, due to a relatively steady electricity price and customer demand. However, the situation has changed after electricity price deregulation, and operation of ASUs is subject to time-of-day or real-time electricity price fluctuations and varying customer demand. Frequent changes in operating conditions and/or startup/shutdown would be required to gain economic benefits from utility price fluctuations (Zhu et al., 2001). In other words, the profitability of air separation plants is highly sensitive to operation policies that utilize smart energy management and plant designs that allow flexible operation (Mitra et al., 2012). Furthermore, the tight energy and material integration and strict operational constraints also pose operation/design challenges and research opportunities. These process characteristics of ASUs call for an optimization-based framework to systematically explore both operation and design to obtain improvements in process performance.

Systematic approaches to the design of dynamically operable plants have been the subject of many research studies over the past three decades. Studies on operation and design of ASUs in the literature are relatively limited, with one or more of the following considerations: (1) dynamic behavior (i.e. through implementing rigorous dynamic models), (2) design modification, (3) uncertainty and disturbances, (4) multi-period operation policy and (5) control structure (e.g. Zhu et al., 2010; Mitra et al., 2012). In our study, we aim to address integrated design and operation problems under uncertainty for a super-staged argon system with emphasis on the cold-box through a rigorous dynamic model in an optimization-based framework.

The proposed strategy relies on an accurate and yet robust model that can be solved efficiently in the optimization studies. Two modeling approaches were investigated. First principles models that capture key process phenomena were first developed with columns modeled in a stage-wise manner. The posed integrated model was then validated using available plant information through parameter estimation to ensure that the model accurately represents the plant. Desired prediction accuracy was achieved after the model validation with negligible discrepancies in key process variables. However, the stage-wise model results in a large scale differential-algebraic equation system, which poses challenges in dynamic optimization studies with either the simultaneous or sequential approach. Hence, model reduction was conducted through an orthogonal collocation approach (Stewart et al., 1985; Swartz and Stewart, 1986). Significant model reduction can be achieved without losing desired prediction accuracy, and promising improvements in the solution time have been observed in simulation and optimization studies performed. The orthogonal collocation approach can also be applied to model the primary heat exchanger in the air separation plant. In our study, gPROMS is the main modeling platform. While solving the dynamic optimization problems, both the sequential approach (through gOPT) and simultaneous approach (through AMPL) are applied, with an in-house developed program MLDO used to convert the gPROMS model into AMPL code.

In this paper, we present both full and reduced order modeling approaches, discuss and compare their computational performance, and illustrate their capability through both dynamic simulation and dynamic optimization.

References

Chen, Z., Henson, M., Belanger, P., and Megan, L. (2010). Nonlinear model predictive control of high purity distillation columns for cryogenic air separation. Control Systems Technology, IEEE Transactions on, 18(4), 811 –821.

Mitra, S., Grossmann, I. E., Pinto, J. M., and Arora, N. (2012). Robust Scheduling under Time-sensitive Electricity Prices for Continous Power-intensive Processes. In FOCAPO 2012: Foundations of Computer-Aided Process Operations, Savannah, Georgia January 8 - 11.

Stewart, W., Levien, K., and Morari, M. (1985). Simulation of fractionation by orthogonal collocation. Chemical Engineering Science, 40, 409 – 421.

Swartz, C.L.E., and Stewart, W. (1986). A collocation approach to distillation column design. AIChE Journal, 32, 1832-1838.

Zhu, G., Henson, M. A., and Megan, L. (2001). Low-order dynamic modeling of cryogenic distillation columns based on nonlinear wave phenomenon. Separation and Purification Technology, 24, 467 – 487.

Zhu, Y., Legg, S., and Laird, C. D. (2010). Optimal design of cryogenic air separation columns under uncertainty. Computers & Chemical Engineering, 34(9), 1377 – 1384.

386090 Extending the Scope of Mixed-Integer Nonlinear Programming Techniques to Black-Box Systems

Abstract:

First introduced in process systems engineering by Ignacio Grossmann in the late 70s, algebraic MINLP approaches are now a mainstream approach for many problems in process synthesis and process operations. In order to extend the scope and applicability of MINLP to simulation-based and experiment-based optimization, we have developed methodology to solve the following key problems:

1. Converting data and simulations to surrogate models that are accurate and as simple as possible
2. Exploitation of theoretical understanding in the building of empirical models
3. Discovery of highly nonlinear relationships that are hidden in data

This talk will discuss the above methodologies and their computational implementation in our ALAMO software.

Game Theory Based Distributed Temperature Control for Energy Saving of Smart Buildings

Zhaohui Zhang[1], Gang Li[1], Tim Salsbury[2] and S. Joe Qin[1,3],

[1]Mork Family Department of Chemical Engineering and Materials Science, University of Southern California, Los Angeles, CA,

[2]Johnson Controls, Milwaukee, WI

[3]Department of Electrical Engineering, University of Southern California, Los Angeles, CA

Abstract: Recently, model-free distributed temperature control for multiple smart buildings within a common interest community, such as a university or corporation, received increasing attention. From a system designer's perspective, it is desirable to have the group of buildings cooperate with each other in terms of their energy consumption schedule, therefore, the total benefit can be maximized, i.e. a most beneficial behavior can be achieved.

The total benefit consists of three parts: firstly, there is a peak demand charge for the community from utility companies when the real-time power load is beyond a given limit. Thus, the real-time total power load constraint for whole community is introduced. Once an anticipated threshold is exceeded, a penalty will be added for compensation and adjustment; secondly, we would like to minimize the total energy consumption cost, which is a function of both the real-time total power load and the time-of-use rate. Here, a time-of-use rate with higher cost during peak hours is considered, so that a flatter energy consumption curve over a day with lower peak and higher valley can be achieved; last but not least, there is a requirement for the real-time indoor temperature comfort level for each building, i.e. the desired temperature should not fall below a lower limit or above an upper limit. In this paper, a global utility function is designed to describe a weighted average of the above-mentioned three parts of benefit.

This global utility maximization problem can be solved by traditional convex optimization methods in a centralized way, where a central controller is utilized to handle all the buildings. However, it is of great importance to solve the problem in a distributed manner, which has several advantages over centralized way. Firstly, a distributed control strategy makes systems more reliable. In the case of controller breaks down for some buildings, other buildings' performance is not affected or only slightly affected in a distributed system. However, if a centralized controller breaks down, all buildings will be severely affected. Secondly, a distributed control system has a better expandability. The system can be constructed with a large scale and scattered in a large area. It provides a convenient infrastructure if a new building is built and becomes part of the existing control system. Furthermore, the optimization and computation load for each controller would be significantly reduced, since no one needs to do the computation for the whole system.

In this paper, the distributed global utility maximization problem is formulated as a series of sliding window games, with each building being treated as a player, the global utility being decomposed to each player's payoff, and each game being played over a control horizon from the current time step. In each game, we apply a distributed learning algorithm such that each building can learn to play a part of the optimal solution. In the game, each player maximizes its own payoff based on the action it played and the payoff it received, without knowing others. The algorithm can achieve a sub-optimal solution, i.e. the solution converges to the centralized optimal solution with a probability approaching to one.

The effectiveness of this method is demonstrated by simulation. Both fast convergence and sub-optimal property of the methodology can be illustrated. A flatter daily energy consumption curve and a lower peak to average ratio are also shown. Moreover, since the method utilizes sliding window games, pre-cooling effect during off-peak period and autonomous cooling discharging during on-peak period can be observed.

387165 Distributed Control of Steam Methane Reformer Temperatures in a Smart Manufacturing Framework

Abstract:

Steam methane reforming is a key process in the formation of syngas for hydrogen, methanol and ammonia production. The process takes place in a steam methane reformer (SMR), with the endothermic reforming reactions being carried out in catalyst-filled tubes placed in a gas-fired furnace. Typical furnaces can comprise hundreds of tubes and burners, and their dimensions are comparable to those of a four story building. The SMR is an energy-intensive process unit, and maximizing energy efficiency is of primary interest. SMR operation faces several significant disturbances, including weather (which affects heat losses to the environment) and potentially large changes in product demand.

Together with natural degradation of reforming catalyst activity, these disturbances have a strong effect on the spatial temperature distribution in the furnace. Such temperature imbalances have a negative impact on operating efficiency. Moreover, excessive temperatures can jeopardize the longevity of the tubes, and strict upper bounds must be enforced on the reformer temperature during operation. This is, however, a challenging task due to the distributed nature of the system and the difficulty of obtaining distributed temperature measurements (the latter associated with the extreme operating conditions and the complex geometry of the furnace).

In this paper, we present our results concerning the monitoring of temperature distribution in an industrial SMR furnace using a large array of infrared camera sensors, which produce a significant stream of data regarding the furnace temperature distribution. We reconstruct the furnace temperature distribution from infrared image data, and use this map as a basis for formulating a model predictive control problem for the distributed-parameter system, centered on minimizing the temperature imbalance in the furnace based on modulating the flow rate of fuel to the array of burners. We show an initial validation of this strategy using a detailed computational fluid dynamics model of the furnace.

To our knowledge, this work comprises of multiple firsts, including the only open-literature discussion of using massive scale image information for the monitoring of an industrial SMR, as well as the first application of distributed-parameter control of steam methane reforming. Furthermore, we report on the implementation of our monitoring and control algorithms in a readily-deployable smart-manufacturing computational infrastructure.

370146 Economic Model Predictive Control of Nonlinear Time-Delay Systems

Abstract:

Tightly integrating process economic optimization and process control has been a significant focus of research in the context of chemical process control for the last ten years. Recently, economic model predictive control (EMPC) has been proposed as a potential feedback control that merges economic optimization and control into a unified control methodology [1]-[4]. EMPC differs from standard tracking model predictive control in that it is formulated with a cost function or performance index that represents the process economics and not necessarily a quadratic cost function. Most of the theoretical work (e.g., rigorous closed-loop stability and performance analysis) on EMPC has been completed for systems described by finite-dimensional nonlinear ordinary differential equations (or finite-dimensional nonlinear discrete-time equations). While some work has been completed on formulating EMPC schemes for systems described by partial differential equations by reduced-order modeling (e.g., [4]), no work has been completed for applying EMPC to systems described by nonlinear difference differential equations (DDEs). To motivate the need for such work, it is important to point out that many chemical processes exhibit both nonlinearities due to complex reaction mechanisms and Arrhenius rate dependence and significant time delays due to transportation lags (e.g., flow through pipes), measurement delays, and control actuator delays. Moreover, to adequately capture both the delays and nonlinearities in the dynamic model used in the formulation of an EMPC a DDE model should be used.

In the present work, a two-mode EMPC is designed via Lyapunov-based techniques for systems described by nonlinear difference differential equations modeling state, control actuator, and measurement sensor delays (i.e., state, input, and output time-delays). First, the DDEs are written as an infinite-dimensional system in an appropriate Banach space and an explicit stabilizing controller is derived for system using a combination of Lyapunov-based and geometric techniques (e.g., [5]). Then, Lyapunov-based constraints are imposed in the EMPC problem which are based on the Lyapunov functional derived for the closed-loop DDE system under the explicit stabilizing controller. Under the first mode of operation, the EMPC may dictate a possibly time-varying operating policy, while under the second mode of operation, the computed input trajectory forces convergence of the closed-loop state to the desired set-point. Closed-loop stability under the proposed EMPC scheme is rigorously analyzed and conditions where closed-loop stability in the sense of boundedness of the closed-loop state trajectory (mode 1 operation of the EMPC) and convergence to the desired set-point (mode 2 operation of the EMPC) are given. Lastly, the EMPC scheme will be demonstrated using a chemical process example.

[1] Angeli D, Amrit R, Rawlings JB. On average performance and stability of economic model predictive control. IEEE Transactions on Automatic Control. 2012;57:1615-1626.
[2] Diehl M, Amrit R, Rawlings JB. A Lyapunov function for economic optimizing model predictive control. IEEE Transactions on Automatic Control. 2011;56:703-707.
[3] Huang R, Harinath E, Biegler LT. Lyapunov stability of economically oriented NMPC for cyclic processes. Journal of Process Control. 2011;21:501-509.
[4] Heidarinejad M, Liu J, Christofides PD. Economic model predictive control of nonlinear process systems using Lyapunov techniques. AIChE Journal. 2012;58:855-870.
[5] Lao L, Ellis M, Christofides PD. Economic model predictive control of transport-reaction processes. Industrial \& Engineering Chemistry Research, in press.
[6] Antoiades C, Christofides PD. Feedback control of nonlinear differential difference equation systems. Chemical Engineering Science. 1999;54:5677-5709.

378399 Supervisory Event-Triggered Control of Networked Process Systems

Abstract:

With the increased emphasis on smart plant operations in recent years, there has been an emerging shift towards the integration of wireless sensor networks in process monitoring and control systems. This shit brings significant economic and operational benefits but also introduces fundamental challenges due to the intrinsic limitations on the computational and communication resources of wireless devices. These limitations favor keeping the information exchange over the network to a minimum so as to conserve network resources, while maximizing process performance favors increased communication levels. An effort to address this conflict in the context of large-scale distributed process systems was initiated in [1] where a quasi-decentralized model-based networked control structure was developed with the aim of enforcing closed-loop stability with minimal communication requirements. A key idea was to embed, within each local control system, a set of predictive models that generate estimates of the states of the neighboring units when communication is suspended, and to update the models when communication is restored. Under the assumption of a fixed communication rate, an explicit characterization of the minimum allowable communication rate was obtained. In a subsequent study [2], a quasi-decentralized control strcture using a state-dependent communication logic was developed for nonlinear systems. The key idea there was to have each unit monitor the evolution of the local state and prompt the other units to send their measurements to update their models only when the unit is on the verge of instability. The local instability threshold for each unit was obtained based on a suitable local control Lyapunov function. Compared with the fixed communication rate approach, this approach allows the plant to adapt to changes in operating conditions by varying the communication rate; however, it can also lead to increased network utilization and the possibility of delays when all plant units attempt to access the network at the same time.

An alternative approach was proposed in [3] where a decentralized event-triggered communication strategy was presented. In this approach, each unit includes an additional model of its own dynamics and uses it to monitor the evolution of the model estimation error in the other plant units. When the error exceeds a certain threshold, the unit transmits its measurement updates to the other units in order to maintain closed-loop stability. The transmission thresholds in this case are determined based on a composite control Lyapunov function for the overall system. While this approach avoids the need for the different units to access the network at the same time, it focuses only on the overall plant stability without taking the performance of the individual subsystems into account. This can result in significant performance deterioration in situations where, for example, a transient local disturbance causes the local Lyapunov function to grow for some time but does not cause similar growth in the composite Lyapunov function (and thus does not trigger model updates). The lack of communication in this case will cause the local performance deterioration to go undetected and uncompensated for. The reduction in network utilization in this approach essentially comes at the expense of sacrificing local process performance.

To address the limitations of previous methods, we present in this work a supervisory event-triggered control framework for multi-unit dstributed plants subject to discrete communication between the local control systems. The objective is to balance the overall plant stability requirement with the local performance needs of the individual subsystems, while simultaneously optimizing the extent of information transfer between the distributed control systems. The proposed control structure is hierarchical in nature. It includes at the lower level a set of quasi-decentralized networked control systems that exchange state measurements at discrete times. Each control system relies on a model of the local subsystem as well as a set of models of the interconnected units, which are used to estimate the states of those units when communication is suspended. By monitoring both the local model estimation error and the evolution of the local control Lyapunov function, the local controller sends alarms to the higher-level supervisor in the event that either one breaches its specified threshold. The supervisor keeps track of the alarm signals reported by the various units and decides accordingly which models need to be updated (i.e., which subsystems should communicate with one another) at any given time. The supervisory logic takes into account the behavior of the local control Lyapunov functions as well as that of the composite control Lyapunov function for the overall plant in order to meet both the stability and performance requirements. The theoretical results are applied to a reactor-separator plant example, and the simulation results show the supervisory control strategy to be more effective than previous approaches using either an adaptive or a fully decentralized control strategy. The supervisory control approach achieves similar performance with less frequent model updates, and also takes the performance of the individual subsystems into explicitly into account.

References:

[1] Sun, Y. and N. H. El-Farra, "Quasi-decentralized model-based network control of process systems," *Comp. Chem. Eng.*, **32,** 2016-2029, 2008.

[2] Sun, Y. and N. H. El-Farra, "Quasi-decentralized networked process control using an adaptive communication policy," *Proceedings of American Control Conference*, pp. 2841-2846, Baltimore, MD, 2010.

[3] Garcia, E. and P. Antsaklis, "Decentralized model-based event-triggered control of networked systems," *Proceedings of American Control Conference*, pp. 6485--6490, Montreal, Canada, 2012.

380910 MPC-Based Supervisory Control of an Off-Gas Recovery Plant with Periodic Disturbances from Parallel Batch Reactors

Abstract:

Chemical Vapor Deposition (CVD) is one of the typical batch processes to produce pure silicon and thin film. While the CVD reactor can produce high-purity polysilicon, it requires a long processing time and thus defies mass production. In order to increase production rate, the number of CVD reactors working simultaneously needs to be increased. In case of polysilicon production, operators replace the deposited silicon to seed filament. This makes the reactors operating in a sequential and periodic manner. The off-gas from CVD reactors needs to be further separated and recovered to recycle back to the CVD feed. Owing to the inherent discontinuous and periodic nature of operating upstream reactors, the captured off-gas cannot reach a steady-state and is continuously transported to recovery units.

This study proposes an appropriate plant-wide control structure and model-predictive control scheme that allow for the effectively separation of the off-gas with non-steady periodic feed dynamics. One of the main control objectives is to minimize the fluctuation in the recovered flowrate to reduce the equipment size and energy consumption. The target process studied in this work produces 10,000 MT polysilicon per year and the number of CVD reactors is 29. In order to construct rigorous dynamic process models in Aspen Plus and Aspen Dynamics, CVD reaction kinetic models were constructed and their parameters were identified using genetic algorithm. Proportional-integral (PI) controllers were implemented for regulation lower-level inventory levels, and a multilayer control architecture where a centralized supervisory layer based on model predictive control (MPC) minimizes the recovered flow fluctuation, were designed using MATLAB/Simulink.

Using a rigorous plant-wide dynamic model, the performance of the developed control strategy is compared with fully decentralized classical PID schemes. Discrete mode of disturbances where some of the CVD reactors are shut down is also studied, and the benefits of using multivariable MPC approach as a supervisory RTO layer are shown and analyzed.

378619 Distributed Data-Based Fault Identification and Accommodation in Networked Process Systems

Abstract:

Control and supervision of large-scale process networks are important problems that encompass a host of fundamental and practical challenges. These problems have received significant attention in process control as evident by the large and growing body of research work over the past two decades (see, for example, [1]-[5] for some recent results and references). In addition to these advances, research efforts within process control have begun to address the new challenges arising from the integration of networked control systems in process operations. Efforts to address problems such as network resource constraints and real-time scheduling constraints were initiated in a number of studies and led to the development of resource-aware networked plant-wide control [6], state estimation [7] and communication scheduling methods [8] that enforce the desired stability and performance properties with minimal communication requirements. The resulting design methods, however, do not explicitly consider fault diagnosis or compensation in the control system design. These are important problems that require special attention in the context of large-scale process networks. Specifically, due to the interconnections between plant units through mass and energy flows and recycle, the adverse effects of local faults in a given subsystem may propagate to the rest of the plant and potentially cause failure at the plant level, if not explicitly accounted for. Timely identification and handling of faults are therefore essential capabilities that the networked plant-wide control system needs to have.

Motivated by these considerations, we present in this work a data-based framework for distributed fault identification and accommodation in networked process systems controlled over a resource-constrained communication medium. Initially, a quasi-decentralized networked control structure is designed to stabilize the plant in the absence of faults. The structure consists of a set of local model-based control systems that communicate with each other at discrete times. An explicit characterization of the networked closed-loop stability region is obtained in terms of the choice of the update period, the accuracy of the models, and the choice of controller design parameters. To address the fault identification problem, a set of local fault diagnosis agents are designed for the various subsystems. Each agent uses a moving-horizon parameter estimation scheme to estimate on-line the size of the local fault using the locally-sampled states and manipulated inputs. The resulting estimates are then transmitted to a higher-level supervisor that reconciles any potential discrepancies or ambiguities in the local fault diagnosis results which may be caused by the strong coupling between the dynamics of the individual subsystems as well as the presence of plant-model mismatch. Once the location and magnitude of the fault are identified, a number of possible fault accommodation strategies are devised to maintain closed-loop stability. These include updating the post-fault models and switching to a new set of stabilizing controller parameters. The selection of the appropriate fault accommodation strategy is made on the basis of the estimated fault magnitude and the characterization of the networked closed-loop stability region. Finally, the developed methodology is illustrated using a reactor-separator process example.

References:

[1] P. D. Christofides, J. Liu and D. M. de la Pena. *Networked and Distributed Predictive Control: Methods and Nonlinear Process Network Applications*. Springer-Verlag, London, 2011.

[2] B. T. Stewart, S. Wright and J. B. Rawlings, "Cooperative distributed model predictive control for nonlinear systems," *J. Process Control*, **21**, 698-704, 2011.

[3] K. R. Jillson and B. E. Ydstie, "Process networks with decentralized inventory and flow control," *J. Process Control*, **17**, 399-413, 2007.

[4] M. D. Tetiker, A. Artel, F. Teymour and A. Cinar, A., "Control of grade transitions in distributed chemical reactor networks: An agent-based approach," *Comp. Chem. Eng.*, **32**, 1984-1994, 2008.

[5] S. Jogwar, M. Baldea and P. Daoutidis, "Dynamics and control of process networks with large energy recycle, " *Ind. Eng. Chem. Res.*, **48**, 6087-6097, 2009.

[6] Y. Sun and N. H. El-Farra, "Quasi-decentralized model-based networked control of process systems," *Comp. Chem. Eng.*, **32**, 2016-2029, 2008.

[7] Y. Sun and N. H. El-Farra, "A quasi-decentralized approach for networked state estimation and control of process systems," *Ind. Eng. Chem. Res.*, **49**, 7957-7971, 2010.

[8] Y. Sun and N. H. El-Farra, "Resource-aware quasi-decentralized control of networked process systems over wireless sensor networks," *Chem. Eng. Sci.*, **69**, 93-106, 2012.

391594 Observer-Enhanced Distributed Moving Horizon State Estimation Subject to Communication Delays

Abstract:

The operation of large-scale complex processes for increased efficiency and profits is a challenging task which has attracted significant attention. In recent years, different types of networked control architectures have been developed for the control of large-scale processes. Among these control architectures, one promising approach is distributed model predictive control (DMPC) which becomes very popular due to its ability to deal with scale and interaction issues in large-scale complex processes. The existing DMPC algorithms can be broadly classified into non-cooperative and cooperative DMPC algorithms based on the cost function used in the local controller optimization problem. In a non-cooperative DMPC algorithm, each local controller optimizes a local cost function while in a cooperative DMPC algorithm, a local controller optimizes a global cost function. However, almost all of the above results are derived under the assumptions that the system states are available all the times or that a centralized state observer is available. These assumptions, however, either fail in many applications or are inconsistent with the distributed framework which is not favorable from a fault tolerance point of view. Therefore, it is desirable to develop state estimation schemes in the distributed framework.

In the literature, a majority of the existing results on state observer designs are derived in the centralized framework. For linear systems, Kalman filters and Luenberger observers are standard solutions. In the context of nonlinear systems, observer designs including high-gain observers for different specific classes of nonlinear systems are available, observers for systems with delayed measurements were also developed. It is worth noting that the capability of high-gain observers to be used in output feedback control designs has made high-gain observers very popular in output feedback control of nonlinear systems. In another line of work, moving horizon estimation (MHE) has become popular because of its ability to handle explicitly nonlinear systems and constraints on decision variables. In MHE, the state estimate is determined by solving online an optimization problem that minimizes the sum of squared errors. In order to have a finite dimensional optimization problem, the horizon (estimation window size into the past) of MHE is in general chosen to be finite. At a sampling time, when a new measurement is available, the oldest measurement in the estimation window is discarded, and the finite horizon optimization problem is solved again to get the new estimate of the state. In a recent work, a robust MHE scheme was developed which effectively integrates deterministic (high-gain) observers into the MHE framework. The resulting robust MHE scheme gives bounded estimation error and has a tunable convergence rate. This makes the robust MHE suitable for output feedback control system design and has been applied in the design of an output feedback Lyapunov-based MPC and an output feedback economic MPC.

Within the decentralized/distributed frameworks, there are some results developed in the context of linear systems and a few results for nonlinear systems in the framework of MHE. For linear systems, the results focus on decentralized deterministic observers and distributed Kalman filtering methods for sensor networks. Recently, in the framework of moving horizon estimation, distributed MHE (DMHE) schemes were also developed for constrained linear systems. For nonlinear systems, distributed moving horizon estimation schemes (DMHE) were developed with the local MHE designed based on a centralized model or the subsystem model. The above DMHE schemes inherit the advantages of classical MHE including the capability to handle nonlinearities, constraints and optimality. However, it is not easy to characterize the effects of bounded uncertainties.

In our recent work, an observer-enhanced DMHE design was developed for a class of nonlinear systems with bounded process uncertainties. In this DMHE, each subsystem MHE communicates with subsystems that it interacts with every sampling time. In the design of each subsystem MHE, an auxiliary deterministic nonlinear observer is taken advantage of to calculate a confidence region that contains the actual system state every sampling time. The subsystem MHE is only allowed to optimize its state estimate within the confidence region. This strategy was demonstrated to guarantee the convergence and ultimate boundedness properties of the estimation error. However, the above results were derived under the assumption that the communication between subsystems is flawless and there is no delay in the information transmission. In practice, this assumption may not hold especially when shared wireless communication network is used. Issues brought into the design by communication need to be carefully addressed.

Motivated by the above considerations, in this work we proposed a DMHE scheme that is able to handle time-varying communication delays. In the proposed design, a nonlinear observer-enhanced MHE is designed for each subsystem and the distributed MHEs are assumed to be able to communicate and exchange information with each other via a shared communication network which may introduce communication delays. To handle time-varying delays in the communication, an open-loop state predictor is designed for each subsystem to provide predictions of unavailable subsystem states. In the design of each predictor, the centralized system model is used. Based on the state predictions, an auxiliary nonlinear observer is used to generate a reference subsystem state estimate for each subsystem every sampling time. Based on the reference subsystem state estimate as well as the local output measurement, a confidence region is constructed for the actual state of a subsystem. A subsystem MHE is only allowed to optimize its state estimate within the corresponding confidence region at a sampling time. The proposed DMHE is proved to give decreasing and ultimately bounded estimation errors under the assumption that there is an upper bound on the time-varying delay. The theoretical results are illustrated via the application to a reactor-separator chemical process, and the proposed approach is shown to be superior to a DMHE approach without considering communication delays explicitly.

373599 Real-Time Economic Model Predictive Control of Nonlinear Systems

Abstract:

Economic model predictive control (EMPC) is a control methodology that unites feedback control and real-time process economic optimization (e.g., [1]-[5]). The optimization problem of EMPC consists of three main parts: an objective function that accounts for the process economics, process constraints including state and inputs constraints and other constraints like stability and performance constraints, and a dynamic model to predict the future evolution of the process (thus, be able to select the optimal input profile with respect to the economics over an operating horizon). Regardless of the implementation strategy of EMPC (i.e., centralized, distributed, or hierarchical), the computation time required to solve the optimization-based controller is non-zero in practice. The computation time may be significant or insignificant depending on the time constants of the process dynamics. When the computation time is significant, using an EMPC that does not account for the delay caused by the computation time may lead to unstable closed-loop operation and/or performance degradation. However, no theoretical work on the closed-loop stability properties of EMPC accounting for the computation delay as been completed.

To this end, EMPC for real-time implementation is considered in this work. Specifically, a Lyapunov-based EMPC (LEMPC) [4] explicitly accounting for computational delays is proposed. From a performance perspective, it may be advantageous to provide the EMPC with knowledge of the computation delay when they are significant. Thus, the EMPC is formulated with a model that treats the computational delay as an input time-delay and the average computation time is used to model the input time-delay. From a stability perspective, there is a (theoretical) maximum amount of time that the optimization problem solver may spend in computation and must return a control action by this maximum amount of time to ensure closed-loop stability. A rigorous bound on the maximum amount of computation time to ensure closed-loop is derived. The bound will be used to force the solver to return a control action by the maximum computational time required for stability. By the design of the LEMPC, the returned control action, which may be returned before the solver converges to a (local) solution, is guaranteed to be stabilizing. The proposed LEMPC is demonstrated on a chemical process example to show that closed-loop stability can be maintained in the presence of computation delay.

[1] Angeli D, Amrit R, Rawlings JB. On average performance and stability of economic model predictive control. IEEE Transactions on Automatic Control. 2012;57:1615-1626.
[2] Amrit R, Rawlings JB, Angeli D. Economic optimization using model predictive control with a terminal cost. Annual Reviews in Control. 2011;35:178-186.
[3] Huang R, Harinath E, Biegler LT. Lyapunov stability of economically oriented NMPC for cyclic processes. Journal of Process Control. 2011;21:501-509.
[4] Heidarinejad M, Liu J, Christofides PD. Economic model predictive control of nonlinear process systems using Lyapunov techniques. AIChE Journal. 2012;58:855-870.
[5] Ellis M, Durand H, Christofides PD. A tutorial review of economic model predictive control methods. Journal of Process Control, in press.

THE IMPACT OF ROOFING MATERIAL ON BUILDING ENERGY PERFORMANCE

Ali Badiee

CIVIL ENGINEERING

ABSTRACT

The way of using the water, electricity, natural gas, materials or in general "energy" more efficiently in all phases of residential and commercial building constructions have grown significantly in the last decade since the times when heating and cooling prices were a reasonably minor line item on a facility manager's budget are long gone. Oil costs, nevertheless lower than they were last year, remain high and tremendously unstable. Coal prices and natural gas are similarly on the growth. All of these instability and rises have gone to higher heating and cooling costs, and engineers are doing all of their efforts to keep them under control by using energy efficient building materials. An argument can be made that the emphasis on building energy efficiency has influenced the roofing manufacturing more than the other building industries. Several energy codes such as ASHRAE 90.1 (2010) and IECC (2012) have been developed, organizations, administrations created and guidelines established, all in the interest of demonstrating the matter of energy efficient roofing. The topic of this thesis will be to address the concern of roofing energy productivity and ranks different roofing materials in terms of energy efficiency. The yield of various roofing materials installed on five different UAB campus buildings have been survived, analyzed and evaluated in this study. Two parameters will be considered in this evaluation, primarily the energy consumption, and energy bills as a second factor. These data have been provided by the UAB Department of Energy and Facilities Management and monitored from 2007 to 2013, creating several spreadsheets. The energy utilities examined in this study are consisted of electricity, natural gas, domestic water and chilled water (where it was applicable) which measured separately in four different seasons during this particular period. It has been expected to observe a rooftop classification indicating green roof, white roof, river rock roof, concrete paver roof and black traditional roof respectively at the conclusion of this research.

Keywords: ASHRAE 90.1, IECC (2012), Green roof, White roof, Black roof, Concrete paver roof, River rock roof

392682 Biowall's Impact on Indoor Air Quality and Energy

Abstract:

The average American spends more than 90% of his or her time indoors where the air is 2-5 times more polluted than outdoors. A principal cause of indoor pollution and its associated health concerns are volatile organic compounds (VOCs), which are emitted by a wide array of household products. Taken as a whole, poor indoor air quality (IAQ) is estimated to cost the United States at least $40 billion annually due to related illness and lost productivity. The elderly and the young are especially at risk to illness due to lifestyle habits and weaker immune systems.

Energy efficient buildings typically seal the leakage paths for outside air infiltration but this can lead to poor IAQ as an unintended side effect. The Biowall is a new concept for maintaining high levels of indoor air quality and has the potential to reduce HVAC energy use by reducing outside air requirements for buildings. The device consists of a living plant filter that is integrated into the heating and cooling system of a residence or a small commercial building. This technology leverages the natural ability of plants to reduce CO_2 and remove VOCs.

To evaluate this concept, a living plant filter was designed and grown in a simulated building environment. The plant filter was integrated with a heating and cooling system and fully instrumented so precise measurements of air temperature, air humidity, air quality, and component-level energy consumption could be made. Various growth media, plant varieties, and watering strategies were evaluated to investigate options for improving a Biowall's performance. The experimental work involved introducing toluene, a common VOC in buildings, into a test chamber where precise measurements of the VOC decay rate could be made. Comparisons were made with the toluene removal rate, both with and without the Biowall.

Sustainable Buildings: Targets and Measurements

Godfried Augenbroe
Jason Brown
Georgia Institute of Technology School of Architecture

AIChE Annual Meeting
November 16-21, 2014
Atlanta, GA

Energy and emissions

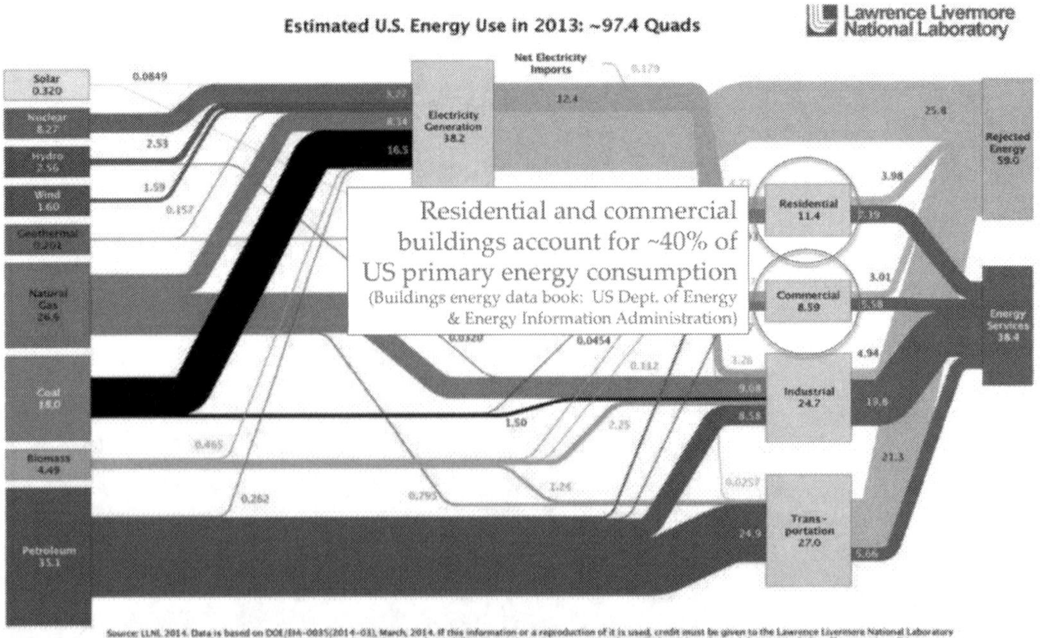

Energy and emissions

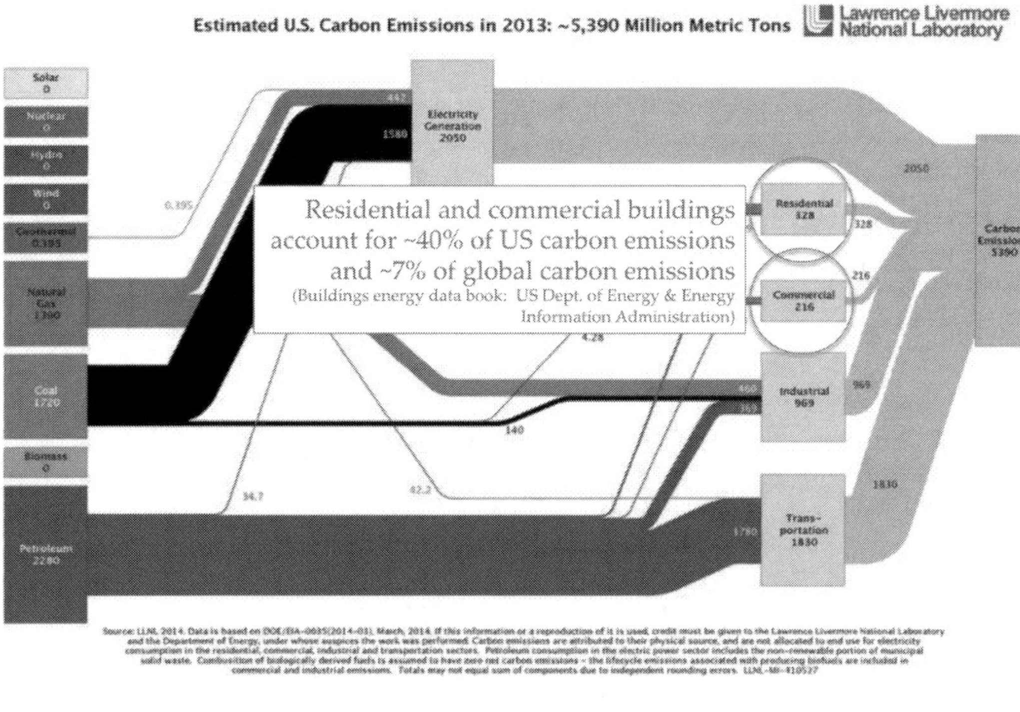

Energy and emissions

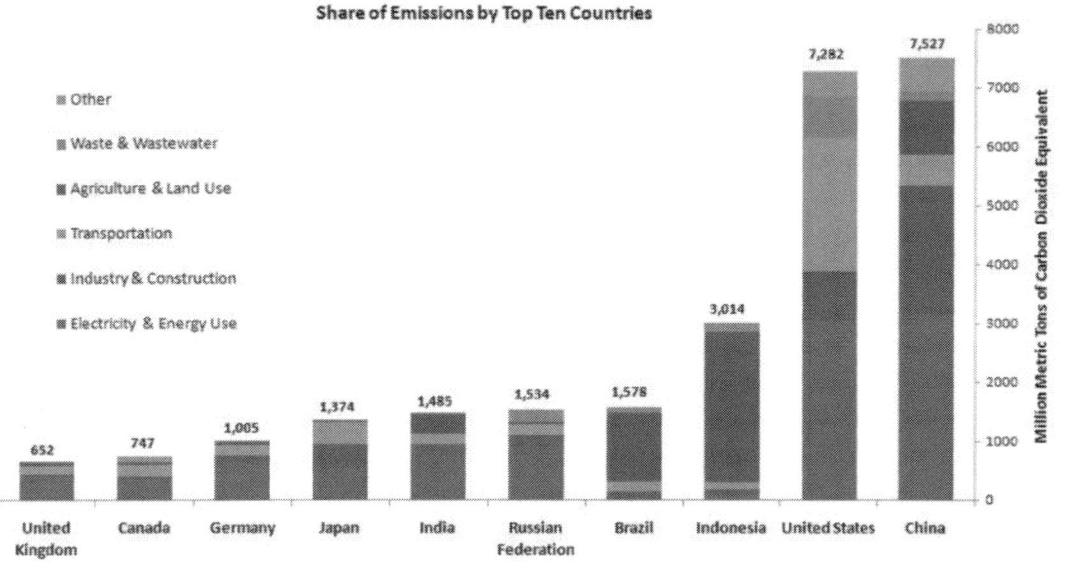

Source: Brown and Sovacool. 2011. *Climate Change and Global Energy Security*, Figure 2.1

Water

In 2005, US buildings used roughly 40,000 million gallons of water per day

This is roughly 10% of total US water consumption

Conservation measures

Commercial buildings end-uses, 2010

Residential buildings end-uses, 2010

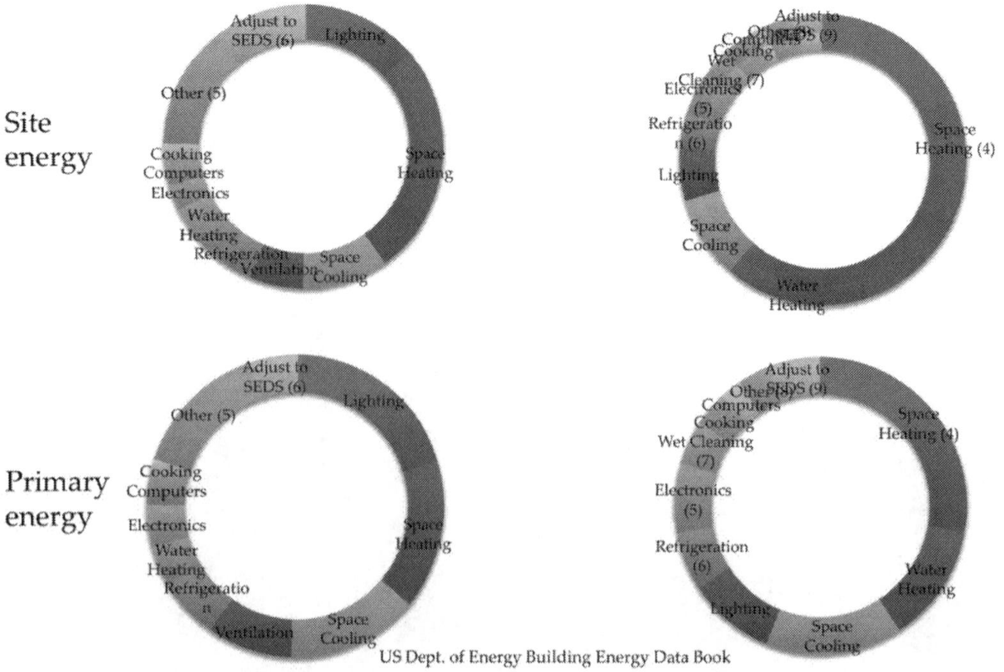

US Dept. of Energy Building Energy Data Book

Conservation measures

First things:

- Improvements to envelope: insulation, low U-value and low SHGC windows, etc.
- More efficient equipment: lighting, chillers, boilers

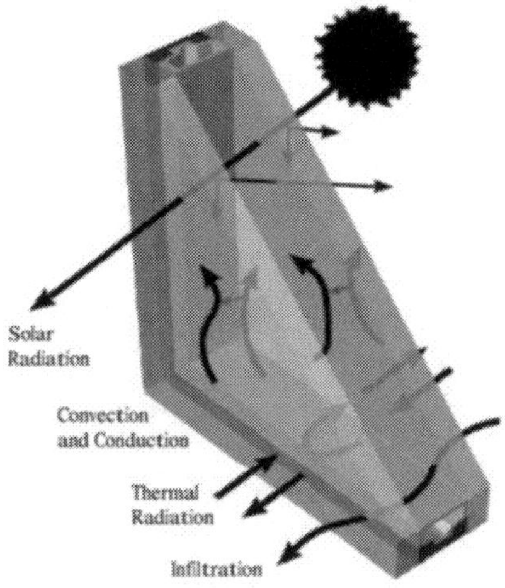

Conservation measures

Second: an integrated system

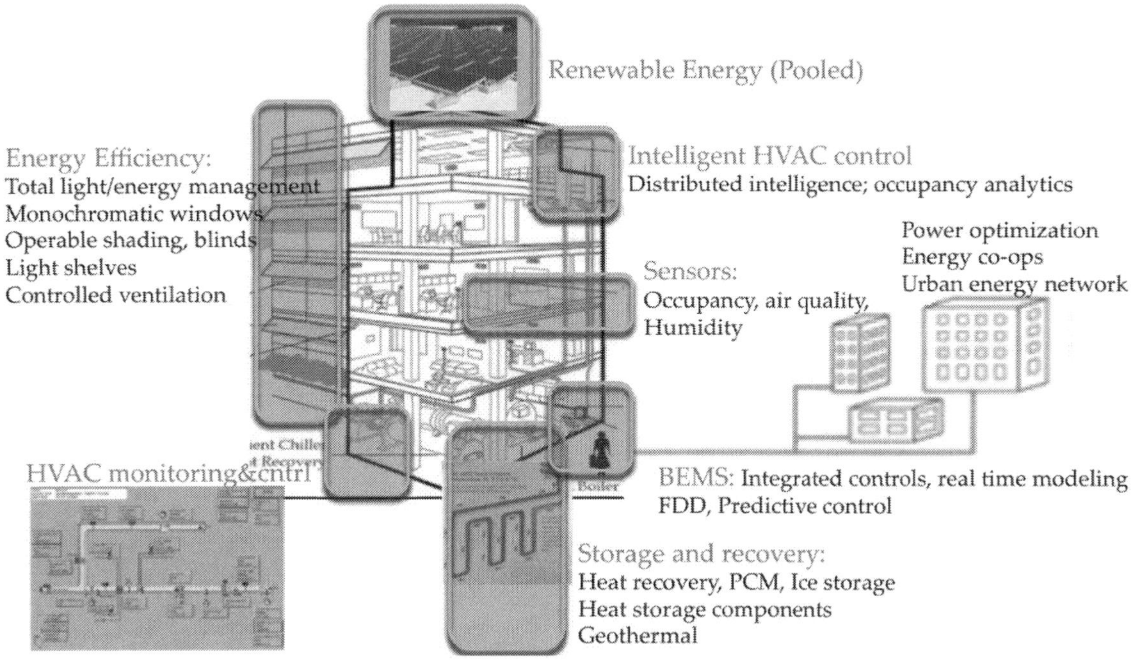

Targets

Goal: market-viable net-zero energy at building site by 2030

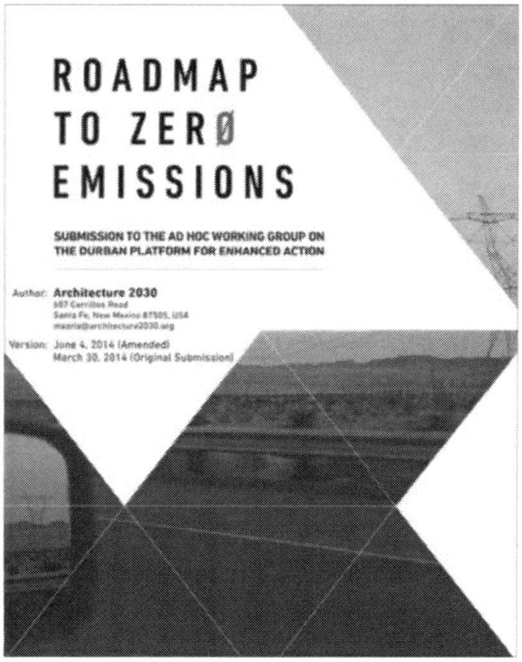

Goal: carbon-neutral buildings by 2030

Sustainability

Bruntland definition: sustainable development

> "…meets the needs of the present without compromising the ability of future generations to meet their own needs."

(World Commission On Environment and Development. (1987). Our common future. Oxford University Press, Oxford, United Kingdom.)

Daly's definition:

1. "harvesting rates [of renewable resources] do not exceed regeneration rates"
2. "nonrenewable resources are depleted at a rate equal to the rate of creation of renewable substitutes"
3. emission rates of substances harmful to life are less than the rate at which those substances are rendered harmless or are naturally recycled.

Daly, H.E. (1993) "Sustainable Growth: An Impossibility Theorem." In Daly, H.E. and Townsend, K.N., eds. (1993) *Valuing the Earth: Economics, Ecology, Ethics*. The MIT Press, Cambridge, Massachusetts.

Sustainability

Something is either sustainable or it is not

How do we assess the sustainability of buildings?

Building environmental assessment (BEA) systems

Building environmental assessment

In the spirit of, but more expansive than, life-cycle assessment

- Includes not just ecological impact, but social issues as well
- Ideally would be a full cradle-to-grave assessment of environmental and social effects

Actual practice deviates significantly from the ideal, however, due to:

- Difficulty in properly defining system boundaries
- Lack of data
- Time-consuming calculations even if the above are overcome
- AND: not all impacts can be verified

Building environmental assessment

As a result:

- A building's sustainability is not fully assessed
- Proxy or "feature-based" assessments stand in for difficult-to-measure outcomes

  e.g. giving points for locating a building next to public transit or providing showers and bike racks, etc.

- Such surrogates co-exist with more measurable, verifiable (at least partially) and a priori calculable estimates of outcomes

  e.g. estimates of energy consumption and ensuing emissions

- Scores are weighted and combined (sometimes in an ad-hoc way)
- A high "sustainability" score may not indicate good energy performance

  (e.g. Scofield, J.H. (2009) "Do LEED-certified buildings save energy? Not really..." Energy and Buildings 14, pp. 1386-1390)

- We only assess how less unsustainable a building is compared to a baseline

Building environmental assessment

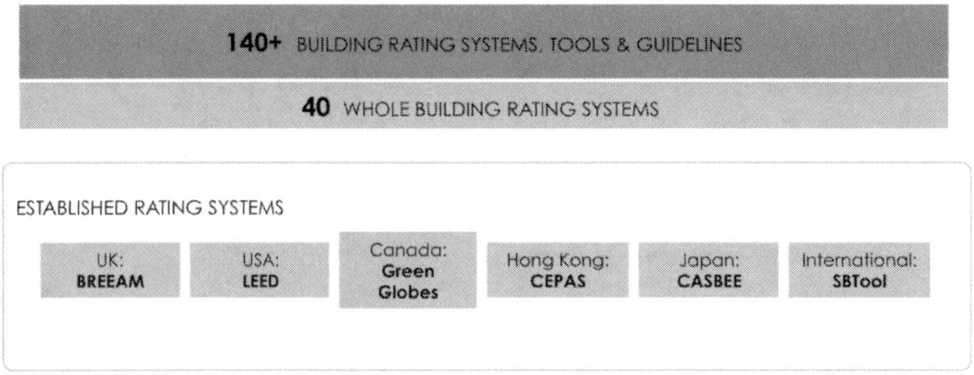

A great success of BEA systems is the creation of a sustainability or "green" marketplace

Building environmental assessment

LEED Green Building Rating System

LEED for New Construction 2009

EA Credit: Energy & Atmosphere

- **EA Prerequisite 2:** Minimum Energy Performance (Prerequisite)
  Need to show at least **10%** improvement (beyond that, earn extra points)
- **EA Credit 1:** Optimize Energy Performance (1-19 Points)
  More than **12%, up to 48%** (1-19 Points)

ANSI/**ASHRAE**/IESNA **Standard 90.1-2007**
"Energy Standard for Buildings Except Low-Rise Residential Buildings"

Use **Appendix G (Performance Rating Method)-Computer Simulation**

Envelope | **HVAC** | **Water Heating** | **Power** | **Other Equip.** | **Lighting**

Other options: Prescriptive Compliance Path
(*ASHRAE Advanced Energy Design Guide* or *Advanced Buildings™ Core Performance™ Guide*)

Building environmental assessment

LEED Green Building Rating System

ANSI/**ASHRAE**/IESNA **Standard 90.1-2007**
"Energy Standard for Buildings Except Low-Rise Residential Buildings"

1. Comply with the mandatory Provisions
   (a prescriptive path to compliance)

If a building cannot satisfy the prescriptive provisions (1),

2. Energy Cost Budget
   - An alternative to the prescriptive provisions
   - Evaluating the compliance of all proposed designs

If you need to rate the energy efficiency of building designs that exceed the requirements of this standard (i.e. LEED EA Scores),

3. **APPENDIX G: Performance Rating Method** (for LEED EA)
   - **Using a computer simulation model for the whole building project**
   - **Energy Cost Savings Percentage compared with the baseline building performance rating**

Envelope | **HVAC** | **Water Heating** | **Power** | **Other Equip.** | **Lighting**

Building environmental assessment

LEED Green Building Rating System

ANSI/**ASHRAE**/IESNA **Standard 90.1-2007**
"Energy Standard for Buildings Except Low-Rise Residential Buildings"

3. APPENDIX G: Performance Rating Method

- **Using a computer simulation model for the whole building project**
- **Energy Cost Savings Percentage compared with the baseline building performance rating**

$$\text{Percentage improvement (\%)} = 100 \times \frac{(\text{Baseline building performance} - \text{Proposed building performance})}{\text{Baseline building performance}}$$

Building environmental assessment

LEED Green Building Rating System

- **EA Credit 1: Optimize Energy Performance (1-19 Points)**

New Buildings	Existing Building Renovations	Points
12%	8%	1
14%	10%	2
16%	12%	3
18%	14%	4
20%	16%	5
22%	18%	6
24%	20%	7
26%	22%	8
28%	24%	9
30%	26%	10
32%	28%	11
34%	30%	12
36%	32%	13
38%	34%	14
40%	36%	15
42%	38%	16
44%	40%	17
46%	42%	18
48%	44%	19

*The minimum energy cost savings percentage for each point threshold

Building environmental assessment

Changes in LEEDv4 Green Building Rating System (coming in 2015)

- **Energy and atmosphere**: Energy consumption metering will be required; track and report performance for 5 years
- **Materials and resources**:
  - Includes verified LCA rather than proxies like % recycled content; early indications suggest LCA may become more business as usual, improving sustainability assessments
  - Incorporates hazard-based (chemical) ingredients reporting by a variety of pathways (e.g. Health Product Declaration)

Building environmental assessment

QSAS

QATAR SUSTAINABILITY ASSESSMENT SYSTEM

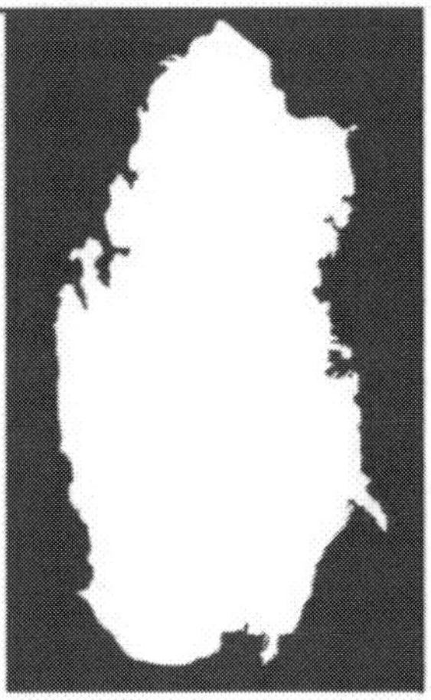

BARWA QSAS BQDRI

Building environmental assessment

QSAS Goals

[UC] The building's planning shall incorporate **urban** considerations.

[S] The building's development in relation to the existing **site** conditions shall be controlled.

[E] The building's depletion of fossil **energy** over its service life shall be controlled.

[W] The building's impact on the overall **water** resource shall be controlled.

[M] The impact of the building's use of **materials** on the environment shall be controlled.

[IE] The building's **indoor environment** shall be controlled.

[CE] The building's **cultural and economic value** shall be maintained or enhanced.

[MO] The building's **management and operations** plan shall be defined.

Building environmental assessment

QSAS Scoring

No	Category / Criteria	Weight
UC	Urban Connectivity	8.00%
UC.1	Load on Local Traffic Conditions	1.80%
UC.2	Pedestrian Pathways	1.08%
UC.3	Proximity to Amenities	0.69%
UC.4	Light Pollution	0.58%
UC.5	Noise Pollution	0.43%
UC.6	Public Transportation	1.30%
UC.7	Private Transportation	0.39%
UC.8	Sewer & Waterway Contamination	1.08%
UC.9	Shading of Adjacent Properties	0.65%
S	Site	9.00%
S.1	Ecological Value of Land	2.34%
S.2	Vegetation & Shading	1.05%
S.3	Desertification	1.75%
S.4	Rainwater Runoff	1.17%
S.5	Mixed Use	0.88%
S.6	Heat Island Effect	0.58%
S.7	Adverse Wind Conditions	0.88%
S.8	Acoustic Conditions	0.35%
E	Energy	24.00%
E.1	Energy Demand Performance	5.20%
E.2	Energy Delivery Performance	5.20%
E.3	Fossil Fuel Conservation	3.64%
E.4	CO2 Emissions	4.55%
E.5	NOx, SOx, & Particulate Matter	5.41%
W	Water	16.00%
W.1	Water Consumption	16.00%
M	Materials	8.00%
M.1	Regional Materials	1.85%
M.2	Responsible Sourcing of Materials	0.00%
M.3	Structure Reuse: On-site	0.91%
M.4	Materials Reuse: Off-site	1.54%
M.5	Recycled Materials	1.85%
M.6	Design for Disassembly	1.85%
M.7	Life Cycle Assessment (LCA)	0.00%
IE	Indoor Environment	14.00%
IE.1	Thermal Comfort	1.20%
IE.2	Low-Emitting Materials	1.60%
IE.3	Natural Ventilation	1.60%
IE.4	Mechanical Ventilation	1.60%
IE.5	Indoor Chemical & Pollutant Source Control	1.60%
IE.6	Views	1.20%
IE.7	Glare Control	1.20%
IE.8	Illumination Levels	1.20%
IE.9	Acoustic Quality	1.20%
IE.10	Daylight	1.60%
CE	Cultural & Economic Value	13.00%
CE.1	Heritage & Cultural Identity	8.67%
CE.2	Support of National Economy	4.33%
MO	Management & Operations	8.00%
MO.1	Commissioning Plan	2.67%
MO.2	Energy Use Sub-metering	0.89%
MO.3	Leak Detection	1.78%
MO.4	Organic Waste Management	0.00%
MO.5	Recycling Management	0.00%
MO.6	Intelligent Building Control System	2.67%
	Total	**100.00%**

Building environmental assessment

QSAS Scoring

- QSAS consists of six certification levels to denote a project's environmental impact
- Certification can only be achieved when the final score exceeds 0, earning 1, 2, 3, 4, 5, or 6 stars
- The highest certification level a building can reach is 6 stars

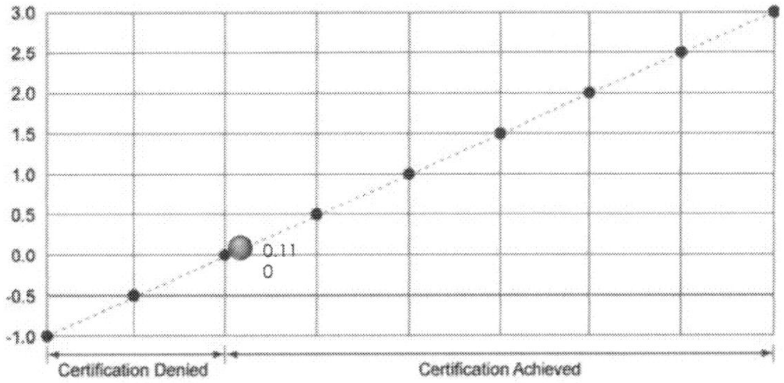

Score	Certification Level	QSAS Certification
X < 0	-	Certification denied
0.0 ≤ X ≤ 0.5	★	Certification achieved
0.5 < X ≤ 1.0	★★	
1.0 < X ≤ 1.5	★★★	
1.5 < X ≤ 2.0	★★★★	
2.0 < X ≤ 2.5	★★★★★	
2.5 < X ≤ 3.0	★★★★★★	

Building environmental assessment

QSAS Scoring: uses targets as baseline rather than % improvement over an ad-hoc baseline

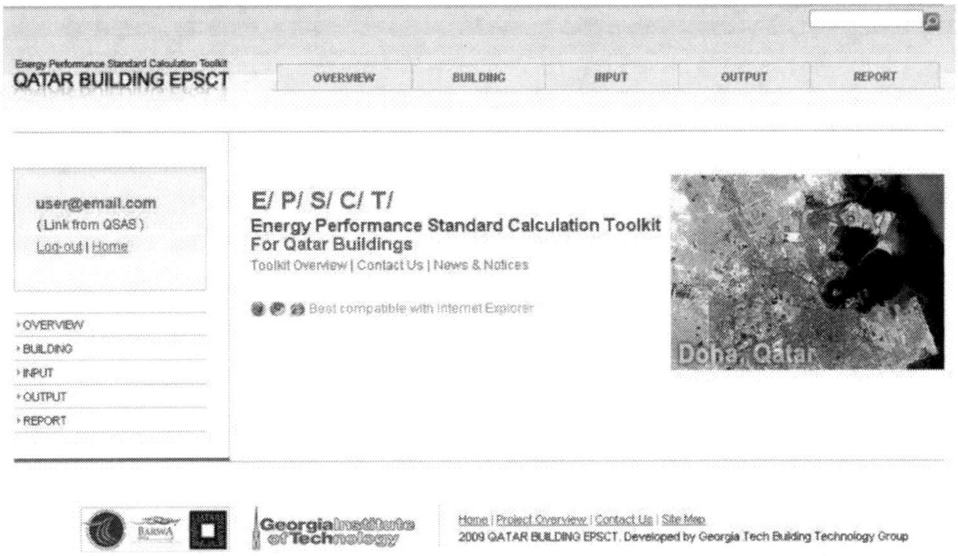

Building environmental assessment

QSAS Scoring: energy consumption assessment based on an international, *normative* standard (ISO 13790)

E_{design}: Calculated energy use

Calculated based on energy standard

E_{ref}: Reference Energy use

Energy consumption of a basket of buildings that represent local building practices.

$$EPC = \frac{E_{design}}{E_{ref}}$$

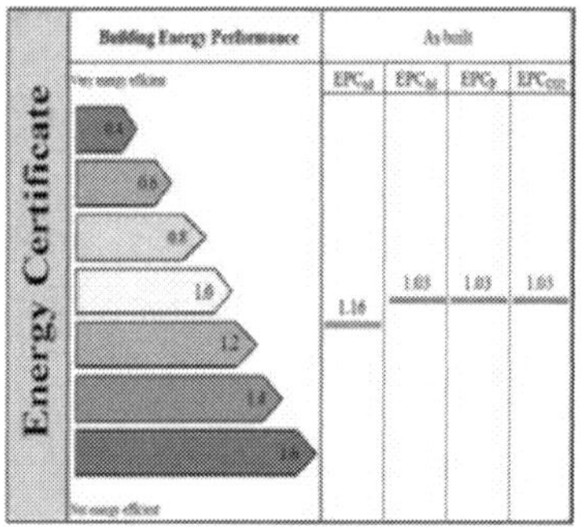

An example of how The EPC is used to "rank" a building

EPC ↓ **Energy Performance** ↑

Building environmental assessment

[E.1] Q_{nd}, EPC_{nd}

-Performance evaluation of thermal energy demand, based on calculation of heating & cooling demand
-Includes heat transmission, ventilation, solar heat gain, internal heat gain

Thermal Energy Need

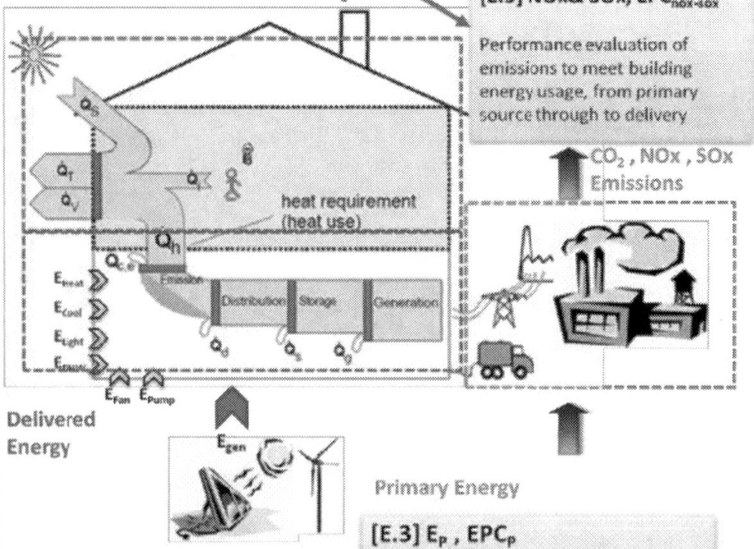

[E.4] CO_2, EPC_{CO2}
[E.5] NOx& SOx, $EPC_{nox\text{-}sox}$

Performance evaluation of emissions to meet building energy usage, from primary source through to delivery

CO_2, NOx, SOx Emissions

[E.2] E_{del}, EPC_{del}

-Performance evaluation of building energy consumption based on calculation of heating, cooling, lighting, DHW, receptacles, auxiliary energy consumption
-Consideration of all on site systems, including local energy generation and renewable energy systems

Delivered Energy

Primary Energy

[E.3] E_P, EPC_P

-Performance evaluation of primary energy use at the source, based on calculation of generation, distribution, district systems, delivery on site

Summary

Buildings are important components of the sustainability agenda

If sustainability is to be achieved, aggressive targets need to be met

Current assessments of sustainability vary in approach and are imperfect, but improving

One success of BEA systems is the creation of a sustainability or "green" marketplace

Backup: do LEED buildings save energy?

J.H. Scofield / Energy and Buildings 41 (2009) 1386–1390 1387

Table 1
Comparison of building-weighted mean site and source (case B only) energy intensities for LEED office sets with matched CBECS building sets (see text).

Case	LEED			CBECS			EUI reduction	LEED > CBECS	*T*-statistic	*p* value
	N	Mean	sdm	*N*	Mean	sdm				
Site energy intensity										
A	35	67.8	4.2	27	97.0	8.7	30%	12 (34%)	3.36	0.0019
B	32	67.6	4.4	26	98.9	9.0	32%	10 (31%)	3.53	0.0013
Source energy intensity										
B	32	178.3	13.2	26	227.5	18.9	22%	10 (31%)	2.53	0.0167

J.H. Scofield / Energy and Buildings 41 (2009) 1386–1390 1389

Table 2
Comparison of gsf-weighted mean site and source energy intensity (for case B) for LEED office sets with matched CBECS building sets (see text).

Case	LEED			CBECS			EUI reduce	LEED > CBECS	*T*-statistic	*p*
	gsf	Mean	sdm	gsf	Mean	sdm				
Site energy										
A	5,303,460	77.1	3.4	5,196,100	80.7	8.4	4%	68%	−0.40	0.69
B	3,835,620	77.6	3.5	3,776,100	86.6	9.3	10%	63%	−0.97	0.37
Source energy										
B	3,835,620	205.0	9.6	5,196,100	211.9	18.7	3%	64%	−0.35	0.73

Backup: abstract

The presentation starts with a discussion of the role of buildings in the global energy and water conservation agenda, well-established design measures in new buildings, and retrofits in existing buildings are reviewed, going from the low hanging fruit of energy and water efficiency measures, to improved management and control, and ultimately to the installation of renewables and recyclables. We compare measures especially on their cost per saving effectiveness, and discuss the feasibility of (net) zero-impact as the ultimate target of sustainable buildings.

The second part of the presentation focuses on building environmental assessment (BEA) systems. BEA systems measure the outcomes of buildings through a broad range of criteria, ranging from occupant health and productivity, energy and water conservation, material impacts, transport needs, etc. In the material impact criterion, they have recently started using LCA methods, or some reduced version of it when appropriate data is not available.

The primary aim of any BEA system is to encourage the design of green buildings and stimulate the market for sustainable construction components and materials. A properly managed set of guidelines, supported by an adequate BEA system, is the best instrument for local governments to regulate industries and to improve the sustainability of the built environment.

The current state of sustainability targets and assessment systems is critically analyzed.

373642 Sustainable Buildings

Abstract:

This unique session combines such a broad perspective of engineering practices in the face of contemporary society that we like to make time to promote audience interaction. If time allows or a speaker cannot make the session, a chair will prompt appropriate discussion in order to get feedback from the audience. We greatly appreciate your attendance and look forward to any comments you may have.

Avoided GHG Emissions Achieved By Using Architectural Window Film:
An LCA Compliant With New WBCSD Avoided Emissions Guidelines

Terrie Boguski, Harmony Environmental, LLC
Lauren Johnson, Eastman Chemical Company

Agenda

- Avoided Emissions Guidelines
- Architectural Window Film LCA
  - Product description
  - Goal and Scope
  - Results
  - Compliance with Avoided Emissions Guidelines
- Conclusions
- Questions

Avoided Emissions Guidelines

Avoided Emissions Guidelines

Image credit: ICCA and WBCSD, 2013

Avoided Emissions Guidelines

- Elements
  - Level in the Value Chain
  - Comparison Criteria
  - Calculation Methodology
  - Significance of Contribution

Architectural Window Film

Architectural Window Film

- Daylighting – +70% light transmittance
- Energy Savings – SHGC 0.41
- Transparent Appearance – 8% visible reflectance
- Fade Reduction – Blocks 99% of harmful UV radiation
- Uniform Appearance
- Noise Reduction

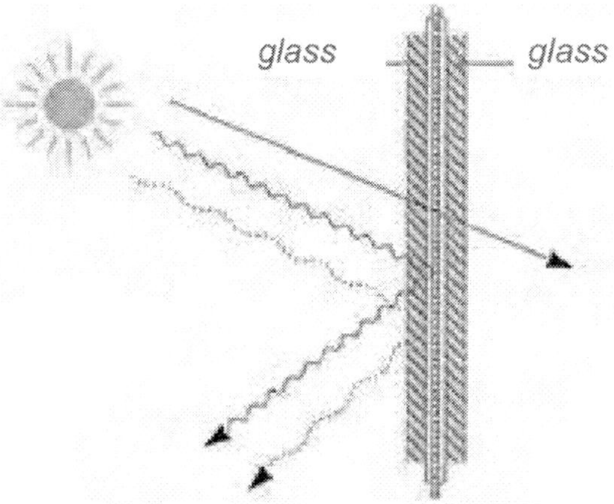

Architectural Window Film LCA

Architectural Window Film LCA

- Goal
  - Understand and document the environmental impact of using window film
  - Evaluate energy savings and avoided GHG emissions

Life Cycle Assessment
of Architectural Window Films

Prepared for
Eastman Chemical Company

by
Harmony Environmental, LLC
Olathe, KS

Harmony Environmental, LLC

January 22, 2014

Architectural Window Film LCA

- Scope
  - Cradle-to-Grave LCA of Window Film
    - 15 years of use
    - Follow ISO 14044
    - Follow the WBCSD Guidelines for Avoided Emissions

Architectural Window Film LCA

- LCA Boundaries

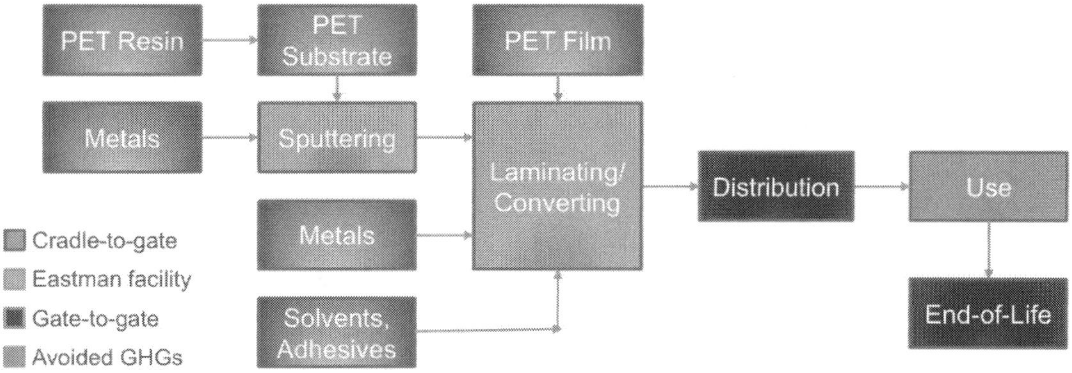

Architectural Window Film LCA

- Geographic coverage

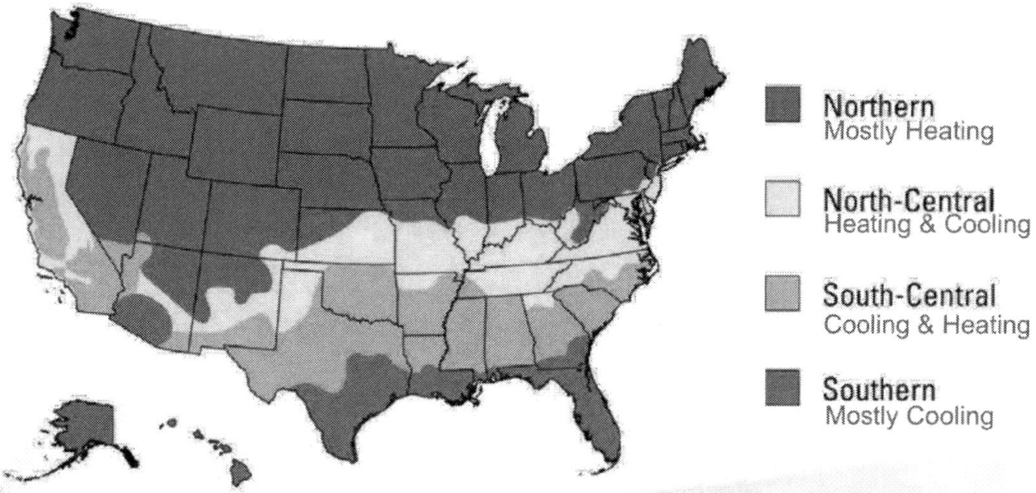

Adapted from US EPA Energy Star® Climate Zone Map and used per website permissions
http://www.energystar.gov/ia/products/windows_doors/Promotional_Map.pdf?a8d7-b26d

Architectural Window Film LCA

- Surrogates and Excluded Materials

Component	% of Mass of Product	Notes
Proprietary Adhesives and Coatings	13% to 15% depending on specific film	Estimated from MSDS component lists
Metals	<0.3% total metals	Surrogate data used for some metals
Silicone	<0.1%	Excluded

Architectural Window Film LCA

- Results

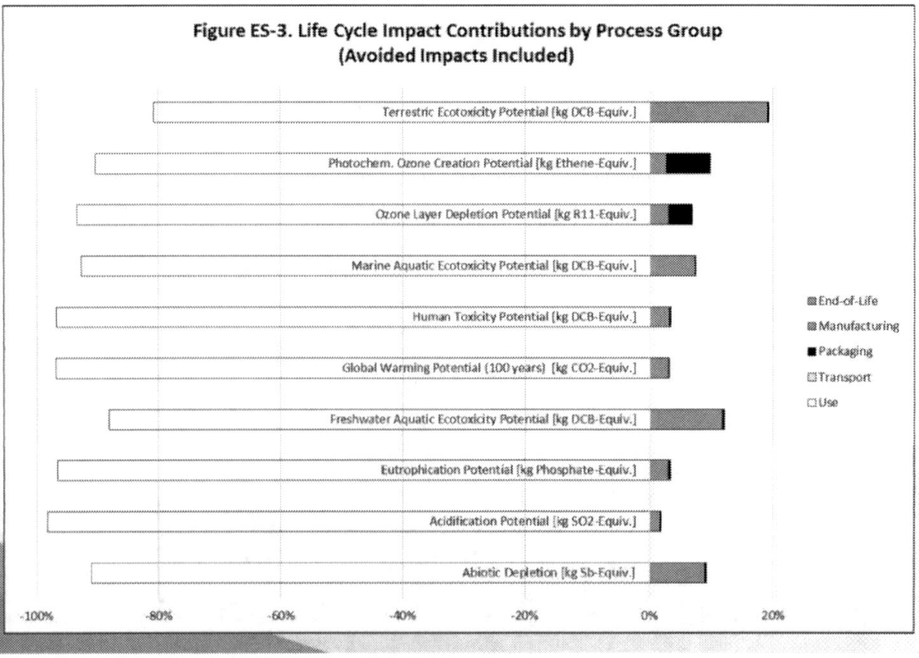

Figure ES-3. Life Cycle Impact Contributions by Process Group (Avoided Impacts Included)

Architectural Window Film LCA

- Results for GHG emissions

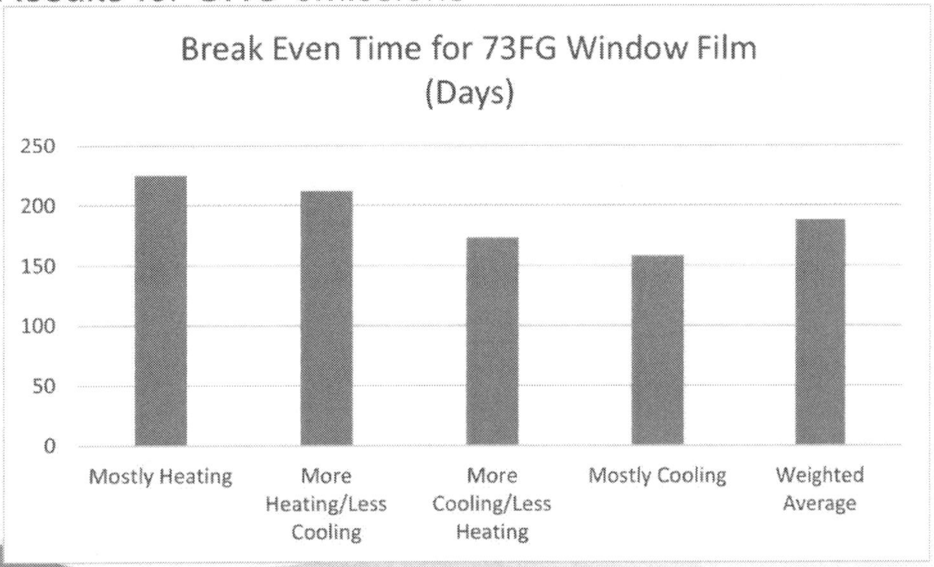

Architectural Window Film LCA

- Compliance with Avoided Emissions Guidelines
  - Level in the Value Chain
    - end-use level
  - Comparison Criteria
    - 15 years of window film use
    - compared fuel use for buildings with and without the window film applied to the windows.
    - all other building aspects are the same

Architectural Window Film LCA

- Compliance with Avoided Emissions Guidelines
  - Calculation Methodology
    - GHG reductions accumulated
      - over lifetime of 1 square meter of window film (0.87 square meters of window space)
  - Significance of Contribution
    - "fundamental"
      - provides the key technology
      - directly manufactures the end-use product.

Architectural Window Film LCA

- Conclusions
  - Benefits outweigh impacts
    - for all impact categories considered
  - Improvement Potential
    - film performance
    - sputtering process
    - greener electricity
    - metals/coatings with lower impacts
  - Provided meaningful results for consumers
    - Following the ICCA/WBCSD Avoided Emissions Guidelines

Thank you!

Lauren Johnson
Eastman Chemical Company
hjohnson@eastman.com
423-229-2289

EASTMAN

Terrie Boguski
President
Harmony Environmental, LLC
tboguski@harmonyenviro.com
913-780-3328

Harmony Environmental, LLC
Management • Analysis • Communication
Beyond Sustainability, Striving for Harmony

Avoided GHG Emissions Achieved By Using Architectural Window Film: An LCA Compliant With New ICCA/WBCSD Avoided Emissions Guidelines

This presentation will include an overview of the International Council of Chemical Associations (ICCA) and World Business Council for Sustainable Development (WBSCD) avoided emissions guidelines "Addressing the Avoided Emissions Challenge" and a compliant cradle-to-grave life cycle assessment study performed for architectural window film.

The goal of the life cycle assessment study was twofold: 1) to document the life cycle environmental impacts of architectural window films, and 2) to calculate the potential energy savings and avoided GHG emissions during use in compliance with new ICCA/WBSCD avoided emissions guidelines.

A full cradle-to-grave study was conducted covering the entire value chain of window film manufacture. Calculation of potential energy savings and avoided GHG emissions from use of the window film products is a key component of the project. Avoided emissions are the calculated differences in the life cycle emissions between two alternative solutions. For architectural films, this study compares the full life cycle of window films in buildings versus the alternative of the same windows in the same buildings without window films. Architectural window films are an optional upgrade to windows which may be factory installed or a retrofit. The impacts due to manufacturing, use and disposal of the building structure and window panes are identical between the two cases and are excluded from the scope. Potential energy savings and avoided GHG emissions calculations are made at the end-use level solely on whether or not window films are employed.

The results of the LCA show that the architectural window films studied have the potential to greatly improve building environmental performance by saving energy and avoiding emissions from the combustion of fuels used to provide heating and cooling. The environmental benefits from using architectural window films far outweigh the environmental impacts of making the film.

378735 Design of Robust Renewable Energy Supply Systems By Providing Visibility Based upon Improved Neural Network

Abstract:

Climate change and rising supply costs make us agree alternative energy sources should be introduced in replace of fossil fuels. Various renewable energy sources are thus investigated. It is clear that renewable energies become a practical contributor to energy supply system when we can manipulate their performances. However unpredictable nature of renewable energy outputs make it difficult to incorporate them into the main energy supply system directly. The wide implementation of renewable energies makes it even more necessary.

Alternatively the estimated output of renewable energy supply should be available before they are realized. When such estimation is available, the visibility of energy output would be greatly improved. In the end the portion of fossil fuels can be reduced.

This paper investigates existing forecasting methodologies such as neural network and the up-to-date linear / nonlinear time series methodologies are with the comparison of their performances. We propose a new optimized neural network based forecasting methodology for the case of photovoltaic power generation. The applicability of the proposed methodology is illustrated in 10KW real case in Korea.

Experimental Investigations on Brick Specimens made from Deinking Sludge

Shilpa Kulkarni [a 1], Vivek Kumar [b 1], Sanjeev K. Singh[2]., Mukesh C. Bansal [3]

[1]*Department of Paper Technology, Indian Institute of Technology- Roorkee, India,*
[2]*CSIR- Central Building Research Institute, Roorkee, India,*
[3]*Bharat Institute of Technology-Meerut, India*
[a]*shilpa.ips@gmail.com* [b]*vivekfpt@gmail.com* [2]sksingh_cbri@yahoo.co.in

Abstract:

The growing population, increasing pollution and decreasing land filling sites for the disposal of solid wastes possess potential problem. One such type of solid waste is the waste from recycled paper industries. It consists of organic and inorganic materials and also some heavy metals because of the writing or printing on the paper. However manufacturing process and raw material used plays a vital role in deciding the chemical composition of residue. Worldwide researchers are trying to use these wastes effectively and also economically into construction materials like bricks and titles. In this paper, deinking sludge, a recycled paper industry waste product, has been chemically and thermally characterized, in order to evaluate the possibility of its use as a construction materials. Characterization of materials is performed by using X- ray Fluorescence, X- ray Diffractometry, and Thermo gravimetric analysis for the determination of its composition, presence of crystalline material and to identify percentage of weight loss with change in temperature. Besides, the investigation also covers the brick making by utilizing the deinking sludge. The results indicate that the brick prepared from the combination of (paper sludge -cement) are light in weight. This novel construction material serves the objectives of reducing solid waste.

Keywords: Deinking Sludge *Compressive Strength *X- ray Fluorescence *Thermo gravimetric analysis *X- ray Diffractometry

1. Introduction

Nowadays there is an urgent need to manage the solid wastes as these are produced in huge quantities from various industries and agricultural sectors. The growing population, increasing pollution and decreasing land filling sites for the disposal of solid wastes possess potential problem. One such type of solid waste about which the present study is carried out is the waste from recycle paper industries. Deinking sludge is the by-product of paper industries which mainly consists of ink, fillers, fines, cellulose fibers *[1]*. The moisture content in deinking sludge normally varies from 60-75% (3, 4). Moreover manufacturing process, waste water treatment applied, and raw material used also play a vital role in deciding the chemical composition of residue *[2]*. It also possess problem of disposal because of its high volume production and decreasing land filling sites. Therefore utilizing these waste generated in huge quantities from recycle paper industry into some construction materials like bricks can address the problem of solid waste management. Moreover it can generate revenue also. Additionally, bricks manufactured from these wastes contain air pockets which make them light in weight *[3]*.

Worldwide researchers are trying to use these wastes effectively and also economically into some other developmental works like construction of homes, buildings, stadiums, dams and schools etc. Moreover it has been investigated that these wastes can be used for making bricks and can fulfill the ever growing demand of construction materials. It will not only save natural resources but also contribute to solid waste management. [4].

The main aim of present research work is to use recycle paper industry waste for making bricks with required properties. Solve the problem of disposal of solid waste and can revolutionize solid waste management.

2. Objectives of the study

This research aims is to evaluate the suitable mixing proportion in term of percentage of deinking sludge, Flyash, and Ordinary Portland Cement to achieve the required compressive strength. To achieve the above mentioned objectives, the investigation covers the following characterization: particle size analysis, specific gravity, moisture content and comparison of physical properties in terms of compressive strength, TGA analysis, FE-SEM image analysis.

3. Materials and methods

3.1 Materials used

Recycle paper industry waste called deinking sludge was obtained from one of the Recycled Paper Industry Ltd., Punjab, India. This sludge was received in grey color. Ordinary Portland cement (43-grade, Vikram Brand of Ultratech) conforming to IS: 12269 and Flyash (grade F) were also used.

3.2 Characterization of raw materials

Chemical analysis of recycled paper industry sludge and flyash were done by using Energy Dispersive X-ray Fluorescence Spectrometer (XRF, S4, Germany). X-ray diffraction pattern was recorded by model XRD- Philips X'Pert Pro with a scan rate of 2 degree/min. Particle size analysis was carried out by using Horiba's laser scattering diffraction Particle size distribution analyzer Partica LA-950. Thermogravimetry –differential analysis (ESTAR 6500 SII) was carried out to determine the thermal stability. Scanning electron microscope image analysis was recorded by using (Perking, Germany).

3.3 Experimental Investigation

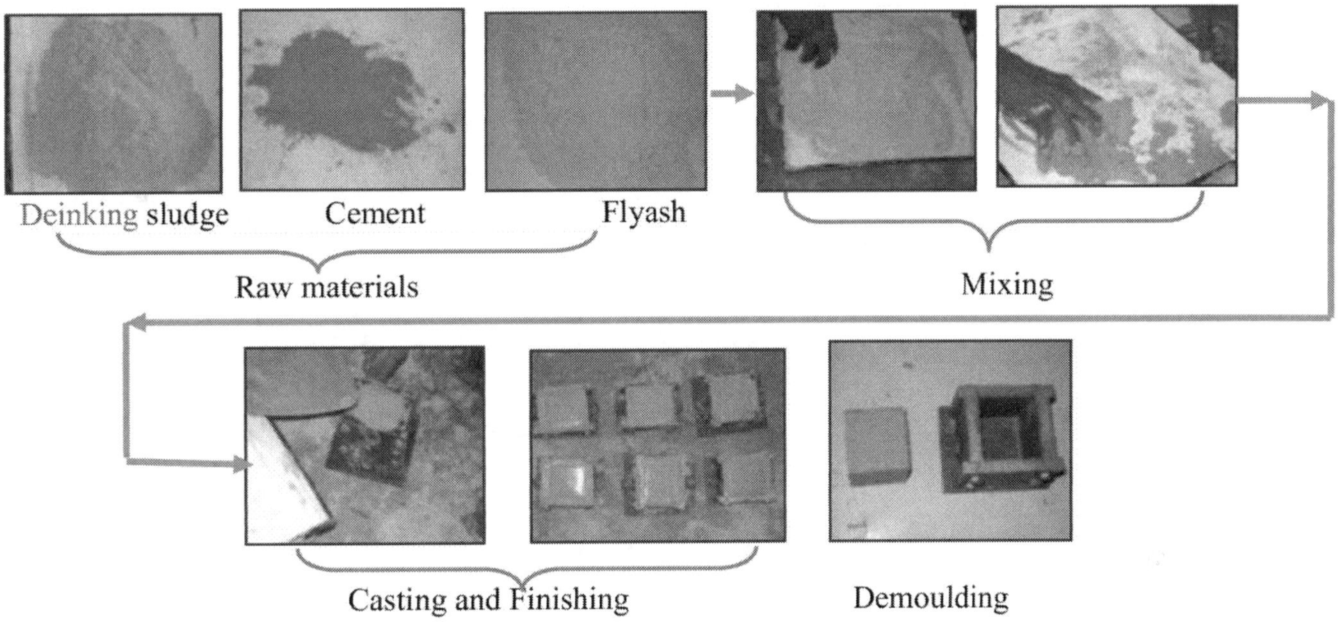

Figure 1: Experimental sequence for making specimens with deinking sludge

Deinking sludge was dried in oven at 60°C. After that, it was grinded and passed through the mesh sieve of 240 microns. The deinking sludge and flyash were mixed properly for 10 mins and agglomerates were left after that cement was sprayed uniformly over whole mixture and mixed it again for 10 mins. After that, water was added to the mixture and mixed well for 20 mins. When mixture was properly mixed then it was poured into the mould with three layers having better compaction. The mould size of 2 inches was used. Thereafter, finishing was carried out by using small trowel. Similarly all the different mix proportions were cast. After 48hrs, samples were demoulded carefully, because of its light weight and transfer into the tray. Moist curing was done for these specimens because deinking sludge is fibrous in nature. So it requires moist curing. Six numbers of cubes were caste for the determination of compressive strength testing. Different mix proportion represented in Table 2 and experimental sequence for fabricating the specimen is given in Figure 1.

Table 1: Chemical composition of Deinking Sludge, Flyash and Cement										
Oxides (%)	SiO_2	Al_2O_3	Fe_2O_3	CaO	K_2O	Na_2O	SO_2	MgO	TiO_2	Loss of ignition
Flyash	55.27	26.69	8.14	1.61	1.76	0.13	0.25	0.39	3.19	1.33
Deinking Sludge	15.50	8.90	1.17	62.43	0.17	0.38	1.62	3.44	1.64	-
Cement	19.32	5.84	5.02	64.29	-	-	2.37	0.83	-	4.0

4. Results and discussion

Chemical analysis of deinking was sludge shown in Table 1. According to the XRF analysis, the residue includes a large fraction of Ca in addition to Si, Al and Mg. This calcium most probably exists in the form of calcite ($CaCO_3$). The moisture content of deinking sludge is found to be 54 %.

Table 2: Mixing proportion of cubes specimen made with cement, flyash and deinking sludge with their compressive strength							
Mixing designation	Cement (g)	Flyash (g)	Deinking sludge (g)	Water (g)	Total	Compressive strength (KN)	
						28 days	56 days
M1	50	0	450	568	1068	1.2	1.8
M2	100	0	900	1330	2330	0.5	0.8
M3	100	900	0	308	1308	10.0	12.9
M4	150	1350	0	500	2000	13.4	15.3
M5	75	450	225	300	1050	1.1	1.8
M6	88	528	264	454	1334	1.4	2.1
M7	100	600	300	500	1500	7.4	10.8

From the above mentioned results, it was observed that these specimens required moist curing because deinking sludge was light weight material. M1 and M2 cubes that were made from deinking sludge only showed least compressive strengths. It was observed that it required some reactive material/adhesives/fillers/ binders for improving strength. While M3 and M4 made with flyash only showed better compressive strength as compared to other mixtures. However M5 and M6 showed lower compressive strength then M3, M4 and M7. The results showed that it required some increment in mixing proportion and additions of some reactive materials/binders/fillers/adhesives. Moreover M7 showed better strength then M1, M2, M5 and M6. From the overall results, it has been concluded that deinking sludge is a useful material for construction of light weighted bricks. Some reactive materials/binders/fillers/adhesives are required to improve the strength.

From the above mentioned different mixing proportion shown in Table 2, we made specimen cubes of 2 inches which is shown in Figure 2 (2a and 2b). While casting the cube specimens made with deinking sludge, flyash and cement and after demoulding, it became grey in colour. It also became hard and showed good compressive strength. While cubes specimens made with deinking sludge and cement only, the colour obtained after demoulding was yellow and cubes show poor compressive strength. During compressive strength testing of cubes specimen made with deinking sludge and cement only it was found that these specimens instead of breaking became compressed and also shows less compressive strength. While cubes specimens made with deinking sludge, flyash and cement broke easily and shows brittle nature. The result indicates that it requires some binder/reactive and filler materials for providing good strength.

Figure 2a: Deinking sludge + Flyash + Cement
(Cubes Specimens 2 inches)

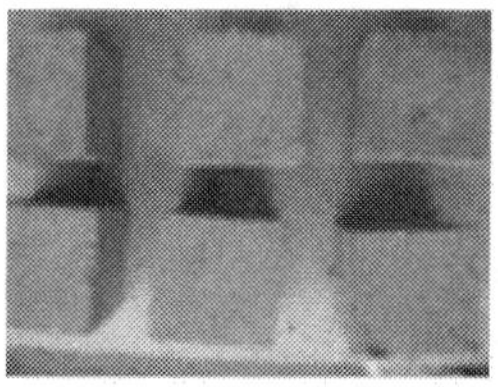

Figure 2b: Only Deinking sludge + Cement
(Cubes specimens, 2 inches)

Figure 2: Specimens made with addition of flyash shown in figure 2a and without addition of flyash shown in figure 2b

X-ray diffraction pattern was recorded with a scan rate of 2°/min and (0 to 100°) at 2θ angle with CuKa radiation (l = 1.542 A °) at 40 kV as given in Figure. 3. In this figure, it shows the intensities of the major and minor peaks for the minerals present. The major mineral phases in the specimens were calcium oxide , quartz (SiO_2), alumina (Al_2O_3) and hematite (Fe_2O_3).

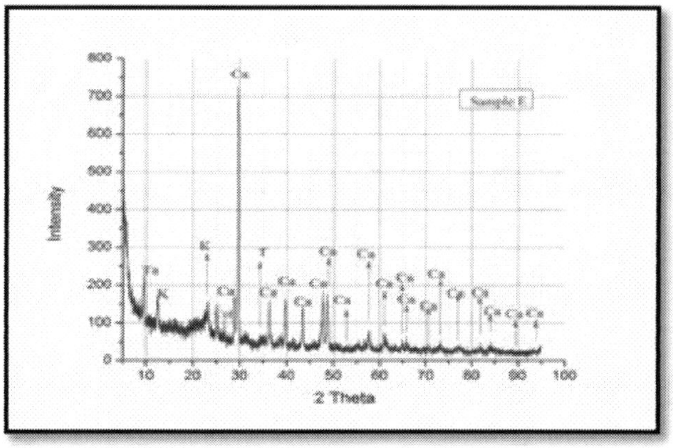

Figure 3: X-ray diffraction analysis of deinking sludge sample with continuous scanning from (0-95 °) at 2θ with intensity having (0-800 counts)

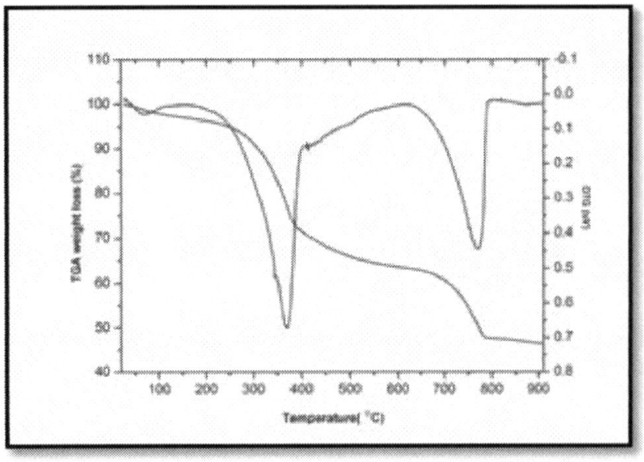

Figure 4: Thermal behavior of deinking sludge sample

Thermogravimetric Analysis (TGA) of deinking sludge was carried out to measure the amount and rate of change in the weight loss of a material as a function of temperature or time in a controlled atmosphere. Thermal behavior of deinking sludge for which the heating rate is 10 °C/ min with temperature from ambient to 900 °C in the presence of nitrogen supply of 200 ml/min **as** shown in Figure 4. The curve shows the loss in weight that occurred at different temperatures. This curve reveals the appearance of three distinct mass loss regions. The first loss (7.5%) occurred between 30°C and 280°C which is premature loss and could be attributed to the removal of superficial water molecules that may be present in the solid pores. The second loss occurred beyond 280 ° C where the material gets thermally degraded and the third mass loss beyond 600 ° C is due to combustion of solid organic matter present in deinking sludge.

Figure 5: FE-SEM analysis of deinking sludge sample

Scanning Electron Microscopy (SEM) images shown in Figure 5 for deinking sludge sample. In this figure, it indicates the presence of irregular pores and fibrous nature.

5 Conclusion

Deinking sludge brick's specimens, measuring cubes (2 inches) was molded in laboratory. The following conclusions can be drawn about specimens fabricates with deinking sludge:

- The specimen fabricates with paper sludge only **was** brittle in nature. Some binder is required.
- Recycling of such wastes by incorporating them into useful materials is a practical solution to reduce the solid waste.

- Innovative way for using waste material into useful products
- Further investigation is required in this area to make good quality/ useful product.
- Mix stuff, extra energy was needed for compaction
- Color of mix proportion obtained grey
- Proportion needed to be modified
- As the time of curing increases, specimen made with only paper sludge without flyash and cement it shrinks.

Acknowledgement

The financial assistance to this experimental study received from Ministry of Human Resource Department, Govt. of India, New Delhi.

References

1. Mohammad Ismail C, M.A. Ismail[1], S.K. Lau, Bala Muhammad, and Zaiton Majid' "*Fabrication of Bricks from Paper Sludge and Palm Oil Fuel Ash*", (1), 2010

2. Mucahit Sutcu, Sedat Akkurt "*Use of recycled paper processing residues in making porous brick with reduced thermal conductivity*", 35, 2625-2631, 2009

3. S. P. Raut, Rohant Sedmake, Sunil Dhunde, R.V. Ralegaonkar, Sachin A. Mandavganec," *Reuse of recycle paper mill waste in energy absorbing light weight bricks*", 27, 247-251, 2012.

4. S. Raut, R. Ralegaonkar, S.Mandavgane , "*Utilization of recycle paper mill residue and rice husk ash in production of lightweight bricks*", 4-7, 2012

388436 Multi-Objective Optimization of CHP Systems for Housing Complexes

Abstract:

Energy demand in the residential sector of Mexico is defined by the requirements of electricity and hot water for domestic use (HWS). These requirements may be satisfied by a cogeneration system (CHP). However, the implementation of such systems presents a particular problem.

On one hand, due to the nature of the domestic user, the energy demand is not uniform throughout the day and the season of the year, presenting differences between electricity demand and hot water demand. On the other hand, the costs of investment in such systems can become elevated to the domestic consumer. In addition, although the CHP systems have good efficiency compared to conventional power generation schemes, due to the use of fossil fuels continue to generate greenhouse gases emissions (GHG).

The above mentioned problems can be solved by using optimization models in the design stage. In this paper, the design of a CHP system, which consists of a diesel internal combustion and thermal storage tank, is presented. The thermal storage tank is used to smooth the gaps between electricity demand and hot water demand. The system interacts with the network of the electric company through a scheme in which the surplus power is sold or bought to cover missing energy demands.

The optimization model uses as objective functions to maximize the net present value (NPV) of the CHP system and minimize the generation of GHG emissions. As independent variables the schedule scheme operating internal combustion engine and the volume of thermal storage tank used. The partial load of the internal combustion engine and the temperature inside the thermal storage tank is restricted. The model has been solved by NLP models and genetic algorithms. Additionally, in order to make a comparison over conventional generation, the rate of primary energy savings (IPES) and avoided GHG emissions are calculated. Furthermore, the relationship between the temperature of hot water supply, the operation scheme of the primary drive and the volume of the thermal storage tank are shown.

A case study from the state of Michoacan in Mexico was considered to prove the proposed approach, this case study consists of a residential l complex of 1,110 houses; and the results show that there economically and environmentally attractive solutions with the implementation of the proposed approach.

Towards an automatic Hazop analysis

Heinz A Preisig NTNU - Trondheim, Norway heinz.preisig@chemeng.ntnu.no

Background

Hazop has a long history as an integral part of process design. Objective of an analysis is to catch all possible undesirable events in a plant design. So based on the simulated behaviour of the plant, one seeks if events occur that are undesirable given operating conditions and possible fault scenarios.

The conditions and possible faults are there many and thus as the hazop analysis aims at covering all alternatives, the combinatorial explosion threat is instantenuously aparent. Many suggestions have been made in the past not at least knowledge-based system were seen as the saviours in the 1980ties. Though the initial enthusiasm died out in the 90ties maybe based on the recognition that the complexity cannot be reduced with those instruments.

When working on fault detection an isolation, we developed a algorithm that let us compute the automaton representation of a plant driven by event-dynamic input signals and observed by a domain-observer, an event detection mechanism also called a quantiser. The algorithm is very simple and computes the automata very fast indeed also for large systems. Important is thereby to recognise that the computational complexity does not suffer of the combinatorial explosion in the context of the state. With regard to the state, the algorithm is linear.

Presentation

The hazop analysis defines first an operational domain for each process component / unit and all respective state variables. We then suggest systematically events that affect the unit both for all the connected units and additional exchange of extensive quantity of the unit with the environments. For the latter a standard example is a leak or an additional inflow of material or energy. We then compute for each of these inputs if there exists a piece in the boundary of the operation domain in which the gradient points outside. This indicates a possible undesirable transition out of the operational domain. Since these calculations are very simple indeed, we can cover a very large domain of input variations.

For the demonstration we introduce a standard toy system: the multiple coupled three-tank system, a standard casse for discrete-event dynamic research. Reason to use the example is that one has an easy insight into its mechanics. The increase in the model complexity that would occur when using a more complex case, does not really have a significant increase in the computational complexity of the algorithm.

Our aim is to connect our computer-aided modelling effort in which we use a very compact network description for the process model with this hazard analysis tool. We will use the graph representation to introduce systematically a large number of possible faults automatically. This will increase the possible domain of possible tested event scenarios greatly and should lead to a much more comprehensive coverage of the process analysis.

The drug product manufacturing sector is moving steadily and surely toward using continuous processes better controllability, higher quality and reduced manufacturing costs. With the advances in process analytical technologies and particle engineering, companies are already testing pilot scale production facilities. Although the regulatory agencies have not yet established clear guidelines for data monitoring, batch tracking and real-time release of drug products, it is expected that all data collected during production will play a role in meeting regulatory expectations. In addtion, the data collected during production will be vital in determining the process health, identifying process improvements and validating mechanistic models. It is imperative that a systematic framework be provided to manage information generated during manufacturing.

A workflow based framework has been developed for organizing the information associated with a production run and applied to a continuous tablet manufacturing pilot plant at Purdue University. The workflow for the production recipe is shown below.

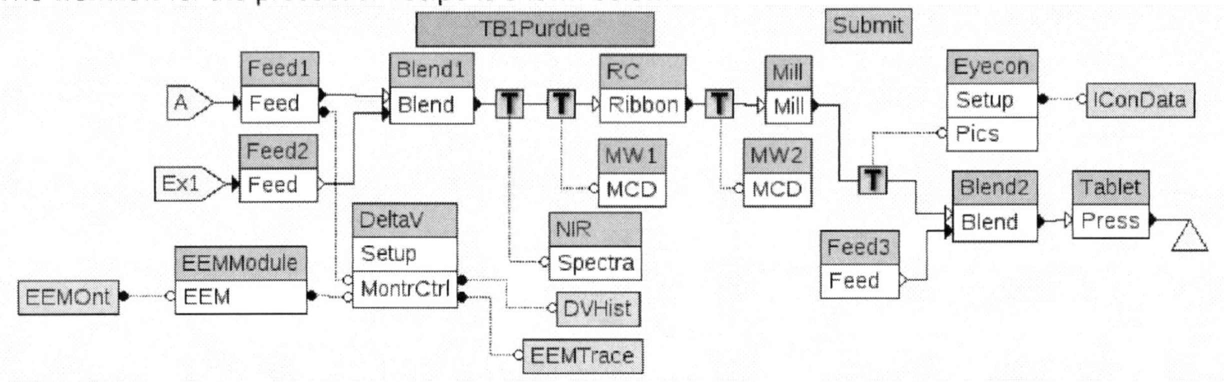

A workflow explicitly defines the relationship between the various operation via material flows and information flows, provides unique identities to all the data entered as parameters and through metadata defines the structure of all information generated during a run. While a data historian typically collects information about process variables, the data nodes shown above, such as, MCD, Spectra, IconData, provide access to raw information generated by instruments during a production run. Additionally, if an exceptional event management module is used during operation, a trace of all the actions taken can be stored as a node in the workflow, node EEMTrace shown above. If used, a model predictive control module can also be incorporated into a workflow with appropriate data trace collection nodes. In general, a workflow instance provides the full context about the a run and the unique references can be used for extracting desired information. The use of the workflow based framework in managing the data of the pilot plant and supporting data analysis functions will be illustrated.

374401 Model-Based Filtering of Large-Scale Datasets - a Biorefinery Application

Abstract:

Numerous sensors monitor large scale plants and generate an enormous amount of data that are most often overlooked. This paper presents a methodology for processing raw data in real-time in order to obtain more useful process parameters. The method relies on building an inferential sensor based on a dynamic model of the system, and is applied on a demonstration scale biorefinery. The study includes results from processing real-time industrial data.

In second generation biorefinery employing biochemical concepts, sugars are extracted from lignocellulosic agricultural wastes, which are then fermented in order to produce bioethanol. The biorefinery process consists of biomass pretreatment, fibers liquefaction or enzymatic hydrolysis, fermentation and separation steps [1]. In 2013 the first commercial size biorefinery started to produce 2nd generation bioethanol [2] and several more sites are under construction nowadays, soon to be commissioned.

Most biorefineries are monitored by trivial sensors, such as temperature, pressure, level, pH and flow measurements. The number of sensors in a typical commercial size plant can amount to several hundreds. It is impossible for process operators to monitor in real-time and make sense out of so many measurements. The operators normally extract historical values and post process them to calculate key parameters retrospectively, such as dry matter content in tanks, or mass and energy balances in different sections of the plant. In this way, sensors are fused together to produce fewer and better process indicators that help verify if the plant operates at desired optimal conditions or if the plant deviates from normal conditions, e.g. due to faults. Very often the raw historical data are filtered by trivial first or second order filters, and also downsampled in order to speed up calculations. Most filter parameters are found by cut and try design methods that eliminate noise.

A more advanced technique of filtering and sensor fusing can be accomplished using a dynamic model of the plant and design observer-like feedback loops in order to accommodate real measurements into the model. An overview of such filter designs can be seen in [3]. All these computations can be automatically performed fast enough in real-time nowadays, making real-time optimization and real-time fault detection possible. More than that, the dynamic model offers information that is not measured in reality, either because mounting specialized sensors in some points of the process can turn to be very expensive or because extracting samples from some reactors is nearly impossible, e.g. in steam pretreatment, the pressurized thermal reactor cannot be opened to extract samples and determine biomass composition.

The dynamic model comprises ordinary differential equations (ODEs) that describe mass and energy balances for each individual unit from the process. If transportation and mixing effects become important, computational fluid dynamics techniques are employed for describing them properly [4], e.g. the convection diffusion reaction and the heat convection diffusion equations on one axis are used in the thermal reactor, which is a very long cylindrical shaped tank [5]. Wherever steam exists, steam states are computed based on the IAPWS IF97 standard. Some units were also extended with fast pH and viscosity calculators, e.g. the liquefaction tank where enzymes are very sensitive to pH variations [6].

To transform the model into an observer or inferential sensor, innovations or estimation errors are constructed for each measured signal. The innovations are then multiplied by a gain matrix and fed back into the model. The gain matrix can be designed either by pole placement or through Kalman procedures, i.e. basic, extended or unscented Kalman filters depending on system nonlinearities. The purpose is to reject process and measurement noise, and to offer an accurate estimate of the plant future state. This study uses basic Kalman filters where matrix gains are computed offline because an industrial large scale setup is expected to have constant operating points.

The biorefinery inferential sensor is tested in real-time on a demonstration scale plant. Biomass composition in the pretreatment and liquefaction sections are estimated and compared to NIR and HPLC data, which were obtained offline in the laboratory. The estimates are in accordance with the laboratory analysis. This demonstrates that the observer is reliable for real-time optimization procedures or as soft measurements (as a soft sensor) for advanced control loops.

References:

[1] Larsen J, Haven M, Thirup L. Inbicon makes lignocellulosic ethanol a commercial reality. Biomass and Bioenergy, 2012; 46:36-45.

[2] Beta Renewables opens biofuels plant in Italy. Focus on Catalysts, 2013; 12:6.

[3] Khatibisepehr S, Huang B, Khare S. Design of inferential sensors in the process industry: A review of Bayesian methods. Journal of Process Control, 2013; 23:1575-1596.

[4] Egeland O, Gravdahl JT. Modeling and simulation for automatic control. 2002.

[5] Prunescu RM, Blanke M, Jensen JM, Sin G. Temperature Modelling of the Biomass Pretreatment Process. Proceedings of the 17th Nordic Process Control Workshop, 2012; 8-17.

[6] Prunescu RM, Sin G. Dynamic modeling and validation of a lignocellulosic enzymatic hydrolysis process - A demonstration scale study. Bioresource Technology, 2013; 150:393-403.

387773 Rational Approach for Designing Generic Droplet Based Systems Using Optimization Techniques

Abstract:

Droplet based microfluidic systems find applications in various areas, such as point of care diagnostics, drug discovery, development of organs on chip and so on. One of the primary challenges in designing droplet microfluidic systems is the lack of robust simulation and design algorithms. Microfluidic systems have complexities similar to pipeline networks. Droplet based microfluidic systems have additional difficulties due to droplet decisions at fluidic network junctions, making these systems challenging for computer aided rational design.

In the area of drug discovery, the current method for combinatorial drug screening of materials is to use robotic arm and well plates. The use of droplet microfluidic devices makes this process cheaper and faster. The formulation of an optimization problem to achieve a droplet microfluidic platform design for combinatorial sequencing of chemicals is proposed in this work.The optimization formulation results in a mixed integer nonlinear programming problem and is solved using mathematical programming and genetic algorithm approaches. The objective function is formulated in a manner that the final designs are robust to fabrication uncertainties. The results obtained by these methods demonstrate that it is possible to rationally design droplet based systems for generic applications

389960 Optimization of Air Usage in Activated Sludge Processes

Abstract:

Wastewater treatment plants are known to utilize excessive amounts of air in activated sludge tanks in order to maintain sufficient growth of biomass and proper mixing. This approach requires intensive usage of energy for the compression of air, and this comprises up to 60% of the energy burden of an entire wastewater treatment plant (Daw et al., 2012; Spellman, 2008). Usually, the rate of aeration is kept high in activated sludge tanks so as to minimize the effects of abrupt changes in influent conditions, such as composition, temperature or volume (Owen, 1982).

In the present study, an advanced activated sludge tank model is employed to assess the feasibility of using decreased aeration rates. The activated sludge tank model is a discretized dynamic PFR model, which was calibrated and validated with the historical data from the Stickney Water Reclamation Plant in Illinois. Consisting of a train 48 CSTR tanks, the discretized model allows individual manipulation of airflow rates at each CSTR, and is an expedient conduit for performing optimization and process control studies.

Initially, constrained nonlinear optimization was performed on the model with the objective of maintaining at least 6 mg/L of dissolved oxygen and less than 0.1 mg/L of ammonia in the effluent. This was mainly done by choosing a number of pivot CSTRs along the entire activated sludge tank. Aeration rates in the pivot CSTRs were designated as optimization variables, and the aeration rates of the remaining CSTRs were calculated by linear interpolation using the neighboring pivot CSTR tanks. This allows for the utilization of different aeration profiles at different portions of the activated sludge tank. Open-loop simulations with the optimized aeration profiles obtained from this strategy have shown that the total rate of aeration can be reduced by up to 40%.

As a further step, an agent-based control framework was implemented with the objective of real-time control of the aeration rates. Using a multi-agent configuration and a specific goal for effluent quality, agents manipulated the aeration rates in each CSTR, with the objective of utilizing the minimum amount of air. Contrary to the limitations imposed by the computational burden of constrained nonlinear programming, the agents were able to decrease the aeration relatively quickly by up to 50%, and maintain it at varying influent conditions. Several dynamic scenarios, including events such as storms, were also simulated to the test the aeration profiles obtained from both approaches.

References

Daw, J., Hallett, K., DeWolfe, J., Venner, I., 2012. Energy Efficiency Strategies for Municipal Wastewater Treatment Facilities. National Renewable Energy Laboratory, Technical Report NREL/TP-7A30-53341.

Owen, W.F., 1982. Energy in wastewater treatment. Prentice-Hall, Inc., Englewood Cliffs, NJ.

Spellman, F., 2008. Handbook of water and wastewater treatment plant operations. CRC Press.

ARTIFICIAL NEURAL NETWORKS FOR ENVIRONMENTAL AND BIOCHEMICAL MODELLING

D.A. Sarigiannis1, S.P. Karakitsios1
1Chemical Process and Energy Resources Institute, Centre for Research and Technology Hellas, Thessaloniki, Greece, 2Chemical Engineering, Aristotle University of Thessaloniki, Thessaloniki, Greece

Introduction

Artificial neural networks form a group of machine learning techniques, inspired by biological neurons. A neural network is a computer model whose architecture mimics the knowledge acquisition and organizational skills of the human brain. Specifically, ANNs consist of a number of interconnected processing elements, commonly referred to as neurons. The neurons are logically arranged into two or more layers and interact with each other via weighted connections. These scalar weights determine the nature and strength of the influence between the interconnected neurons. Each neuron is connected to all other neurons in the next layer. There is an input layer where data are presented to the neural network, and an output layer that holds the response of the network to the input. It is the intermediate layers, also known as hidden layers, which enable these networks to represent and compute complicated associations between patterns. Neural networks essentially learn through the adaptation of their connection weights.

The aim of the study is to present how ANNs may serve as efficient modelling schemes for tackling major environmental-health problems, providing key information (environmental or biochemical) at several stages of the source-to-effect continuum. This will be demonstrated in three different case studies, including the (a) assessment of exposure to benzene to gasoline station employees, (b) assessment of PM air pollution and (c) parameterization of physiology-based biokinetic (PBBK) models designed to estimate internal exposure to a large amount of industrial chemicals (xenobiotics) and pharmaceutical products.

Methodology

Main Methodological Concept

In their general form, ANNs refer to a parallel model architecture able to perform numerical calculations based on distributed processing. ANNs are composed from artificial neurons that operate according to a specified transfer function. The neurons interact with each other through the use of neuron connections called weight factors or simply weights. Positive or negative weights correspond to connections that propagate or suspend, respectively, signals from other neurons. ANNs are trained to perform a particular function by adjusting properly the values of these weights. For the needs of the present study, a generic multi-layer perceptron (MLP) network was utilized. This is a feed-forward artificial neural network model that maps sets of input data onto a set of appropriate output. It is a modification of the standard linear perceptron with three or more layers of neurons (nodes) and nonlinear activation functions. These architectures provide powerful models, since they have the ability to distinguish data that are either nonlinearly separable, or are separable by a hyperplane. Several training algorithms were applied and evaluated, but it was the Broyden-Fletcher-Goldfarb-Shanno (BFGS) quasi-Newton algorithm that proved to be the most efficient and was therefore used. The BFGS algorithm [1] is a method widely used to solve unconstrained nonlinear optimization problems. The selection of the BFGS quasi-Newton algorithm was consistent with the findings of Schlink et al. [2], who have recently reported the good performance of the quasi-Newton algorithm, in the case of Neural Networks that use the early stopping technique for the improvement of their generalization abilityA conceptual representation of a general MLP network is graphically illustrated in Figure 1. Such architectures provide more powerful models, since they have the ability to distinguish data that are either nonlinearly separable, or are separable by a hyperplane. It is noted that sigmoid units are employed with logistic as activation function at both the hidden and the output layers.

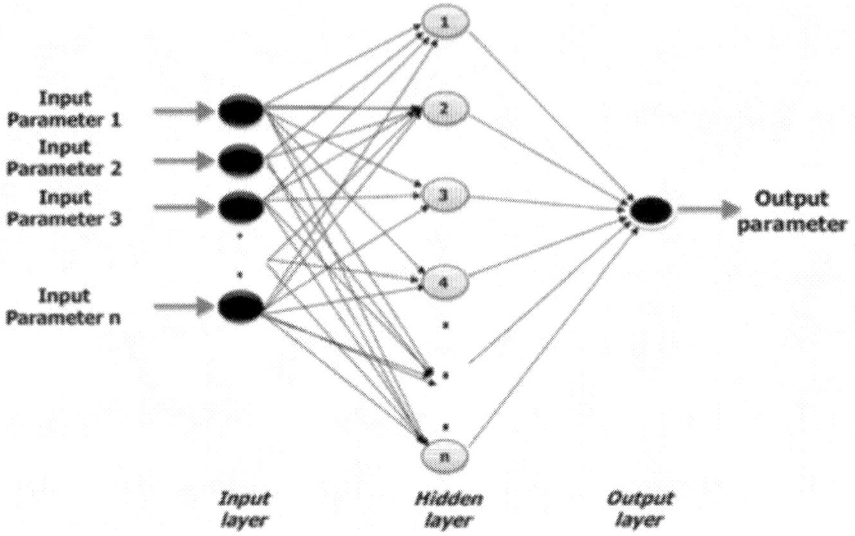

Figure 1. Graphical illustration of a generic ANN model

Beyond the basic architecture, the neural network design was optimized to meet the efficiency and accuracy needs of each case study. Several training algorithms were applied and evaluated, but it was the Broyden-Fletcher-Goldfarb-Shanno (BFGS) quasi-Newton algorithm that proved to be the most efficient and was therefore used. The BFGS algorithm is a method widely used to solve unconstrained nonlinear optimization problems. It is derived from Newton's method, a class of hill-climbing optimization technique that seeks the stationary point of a function, where the gradient is 0. Specifically, Newton's method assumes that the function can be locally approximated as a quadratic Taylor expansion in the region around the optimum and uses the first and second derivatives to find the stationary point.

Case studies tested

The applicability of ANN-expert systems to resolve complex problems in the environmental health risk assessment arena, is demonstrated in three different case studies, namely:

Prediction of exposure to benzene for gasoline station employees. The environmental health problem targeted in this study is exposure of gasoline filling station employees. Employees in fuel stations are exposed to high levels of gasoline vapours (including benzene) during refuelling, as well as high concentrations of emissions from tail pipe exhausts. Therefore, filling station employees are a vulnerable sector of the population because of their exposure during their 8 h shifts [3], especially considering the well known health effects focused mainly on increased risk of leukaemia after chronic exposure and neurotoxic effects after short-term exposure to high values of benzene vapour and other volatile organic chemicals present in this environmental setting. Detailed environmental and exposure data used in this study have already be described in detail by the authors [4, 5]. In the case of predicting exposure to benzene of gasoline station employees, an ANN with five input nodes was created for each gasoline station. The number of the input neurons corresponds to each input parameter, corresponding to i) the amount of gasoline hourly traded (affecting benzene emissions through evaporation and traffic into the gasoline station), ii) wind speed (affecting benzene concentrations through dispersion), iii) ambient temperature (affecting benzene emissions through evaporation and to a smaller extent dispersion), iv) traffic flow of the road in front of the filling station (benzene emissions close to the gasoline station) and v) benzene urban background concentration (affecting overall benzene levels). The hidden layer included ten nodes (found experimentally to provide the best results) and the output layer three (for each category of employees, meaning cashiers, miscellaneous activities and employees dealing with car refuelling). Finally, a third ANN was developed, using a mixed set of input and output data collected from both filling stations.

Prediction of atmospheric PMx concentrations. The second case study was the prediction of atmospheric PM concentrations in the city of Thessaloniki (Greece). Over the last couple of years, the use of biomass as heating source was allowed in Greece as a CO_2-neutral means of space heating in the large metropolitan areas of Athens and Thessaloniki affecting more than half of the country's population. At the same time the use of light heating diesel was heavily taxed. In the same period Greece faces a financial crisis with significant repercussions on the average household income. This resulted in a significant elevation of PM2.5 and PM10 concentrations during wintertime, systematically above 100 $\mu g/m^3$, while concentrations raised up to180 $\mu g/m^3$ (and even 220 $\mu g/m^3$ in extreme occasions) when favored by specific weather conditions. Last winter the Greek government revealed a plan for tackling acute pollution episodes in an attempt to tamper the adverse effects on public health. Thus, a forecasting model for PMx concentrations would be very useful for both increasing public awareness and ensuring cost-effective public health protection. In the specific study, data from 2 different classes of monitoring stations were used, one representative of urban background and one representative of typical traffic sites. On the whole, for each of the two groups of monitors the optimal ANN was designed and developed. The input layer of each ANN consisted of 8 neurons, corresponding to the following predictors: PM10 and PM2.5 concentrations of the previous day, wind speed and direction, temperature, humidity, precipitation and one categorical variable identifying the day of the week (weekday or weekend). The output layer consisted of 2 nodes, corresponding to the output parameters, namely the average daily PM10 and PM2.5 concentrations for each station. The second layer consisted of 14 neurons, experimentally found to yield the best results.

Prediction of biochemical parameters entering a PBBK model. The third case study is the application of ANNs on a completely different type of interactions between environmental stressors and human health. The fate of chemicals upon human exposure, is mathematically described by Physiology Based BioKinetic (PBBK) models. The latter are continuously gaining ground in regulatory toxicology, describing in quantitative terms the absorption, metabolism, distribution and elimination (ADME) processes in the human body, with a focus on the effective dose at the expected target site. A major problem for the inclusion of these models into the regulatory risk assessment process is the lack of generalization, which is mainly due the lack of proper parameterization of PBBK models for many compounds and especially for new chemicals. Chemical-specific input include tissue:blood partition coefficients, as well as metabolic parameters such as the maximal velocity (Vmax) and the Michaelis affinity constant (Km) or intrinsic clearance (Vmax/Km). These physicochemical parameters should have been obtained on the basis of independent measurements (*in vitro* and *in vivo*) or using algorithms in the valid domain of application, like quantitative structure–activity relationships (QSARs). QSAR models are regression or classification models used in the chemical and biological sciences and engineering for data poor chemicals, allowing us to predict biological properties related to a chemical compound on the basis of structural information. In this work, Abraham's [6] solvation equation was used for estimating biological properties. Abraham's model takes into account the excess molar refraction, the compound dipolarity/polarizability, the solute effective or summation hydrogen-bond acidity, the solute effective or summation hydrogen-bond basicity and the McGowan characteristic volume. These parameters comprised the input for the ANN platform we developed. Several network architectures were tested and optimal results were obtained using a twelve node hidden layer. For each chemical-specific biological property, a different ANN was used.

Results

Gasoline station employees exposure to benzene

The overall performance of the ANN applied in this case study performed excellent results (Figure 2). More in detail, the performance of the ANN applied to the urban filling station data is slightly better than the ANN applied to the rural one for the first two categories of employees (employees working in car refueling and employees doing miscellaneous activities).On the contrary, the performance referring to cashiers is worst in the urban station. This occurs because the indoor benzene concentration (where cashiers are exposed to) presents a time lag following the outdoor concentrations that determine the exposure of the other two categories of working people and is strongly affected by the overall benzene concentrations affecting the wider area.

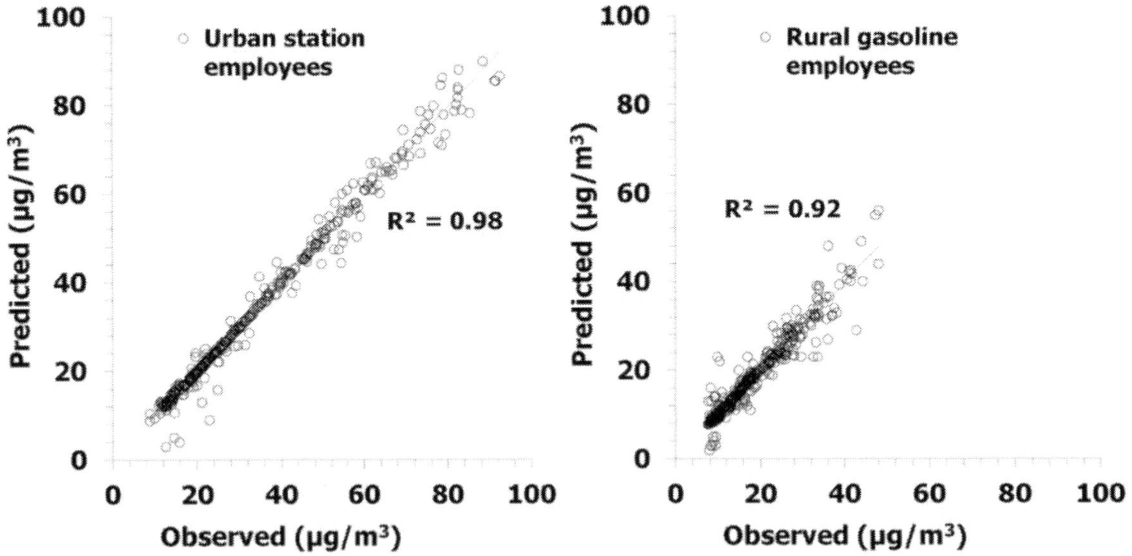

Figure 2. Dispersion among the observed and the predicted values by the ANN for the exposure of the employees of urban and rural filling station respectively

This is also the reason, that among the three categories of employees, the best performance of the model is for the miscellaneous activities employees. Exposure pattern of these employees is directly affected by the referred input parameters, located in the area of the station that does not appear to have acute changes in the observed benzene concentrations.

On the contrary, this occurs to the employees refueling cars during the refueling process, having as a consequence sometimes the quickly elevated levels of exposure to be hardly predicted with the same degree of accuracy. The results of the third ANN (employing data from both stations) were also very good (R^2=0.86). This reveals that the model in its general form and using these input parameters, is very well fitted in filling stations with different characteristics of the emission sources. Having in mind the simplicity of the model (one input parameter fewer) and the fact that for both the filling stations one model is required, for applications aiming to predict the exposure levels among employees from different filling stations, this is the proposed model.

Predicting ambient air PM concentrations

The model performed very well considering that the explanatory variables accounted for more than 90% of the observed values; R^2 scores of the training set for the traffic sites were equal to 0.90 and 0.97 for PM10 and PM2.5 respectively and for the urban background sites equal to 0.97 and 0.93. The overall performance of the developed ANN was also validated against an independent validation set, including measurements at different traffic and background sites within the city. The results of the validation of the ANN model are illustrated in figure 8. Considering the limited number of data used for training, the performance of the ANN (R^2 ranging between 0.68 and 0.78) is considered more than adequate.

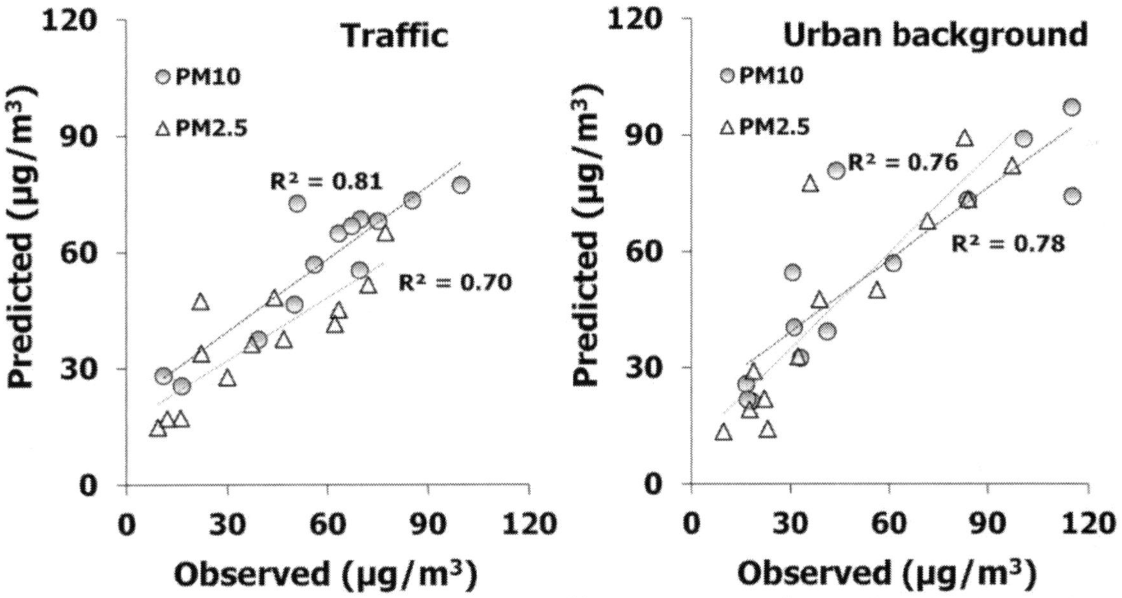

Figure 3. Observed vs. predicted values of PMx concentrations using the forecasting model

The good performance in the results of the exposure forecasting tool confirms the fact that ambient air concentrations are strongly influenced by meteorological conditions both directly and indirectly. A major advantage of the ANN model is its capability to capture the effect of parameters with complex influence on the PM concentration levels, such as temperature. Emission patterns tend to rise significantly when average daily temperature falls below 6-7 ^{0}C, especially emissions related with domestic heating. The fact that the model performs better in predicting PM2.5 levels confirms the correlation of temperature change with the change in residential heating emissions, especially those associated with biomass burning [7, 8]. In addition, temperature affects traffic emissions, by increasing the fraction of cold emissions [9] for typical vehicle routes. Wind speed and direction affect dispersion, diffusion and re-suspension of particles. Humidity affects the hydroscopic growth of particles [10, 11], while rainfall (wet deposition) is a dominant mechanism of atmospheric clearance [12]. The performance of the ANN was enhanced by distinguishing between weekends and weekdays in the overall model, incorporating in this way inter-day differences in emissions intensity and PMx composition. Using as input the ambient air concentrations of the previous day enhances the performance of the ANN, by providing an indication of the background concentrations, as well as the expected emission trends.

Prediction of biochemical parameters entering a PBBK model

A major breakthrough came from the use of Artificial Neural Networks coupled to Abraham's solvation equation for predicting biological/biochemical properties such as blood-tissue partition coefficients, Maximal Velocity (V_{max}) and Michaelis - Menten (K_m) constant.

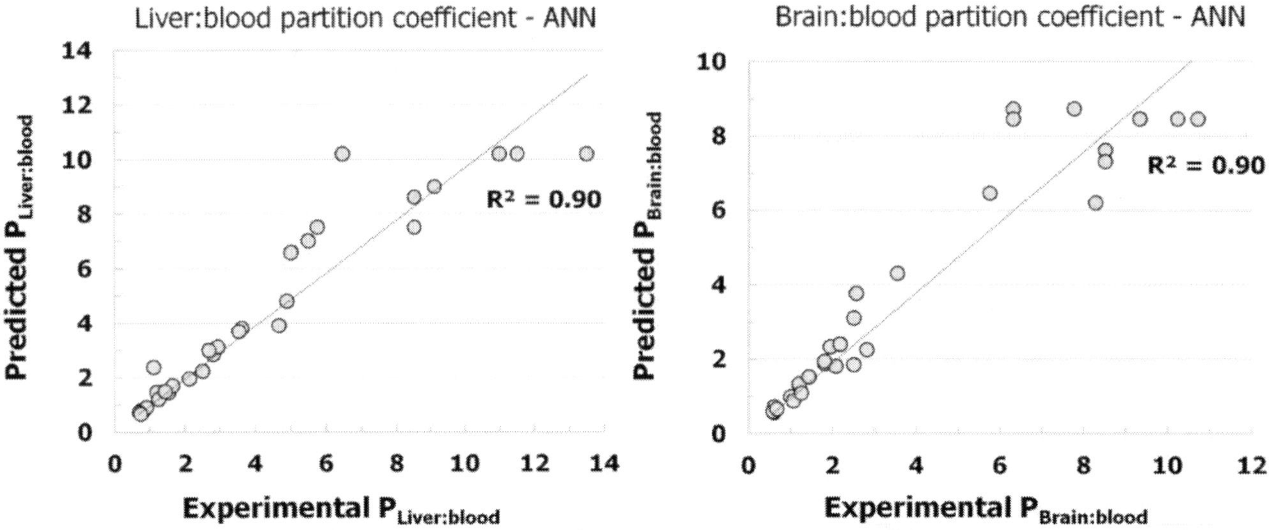

Figure 4. Experimental vs predicted liver:blood and kidney:blood partition coefficients

This was a remarkable advance, since till now, prediction capability of the Michaelis - Menten constant was rather poor; with our coupled ANN - Abraham's solvation equation method for the investigated group of chemicals R^2 went up to 0.7 (Figure 5), which is by far higher to any other existing methodology.

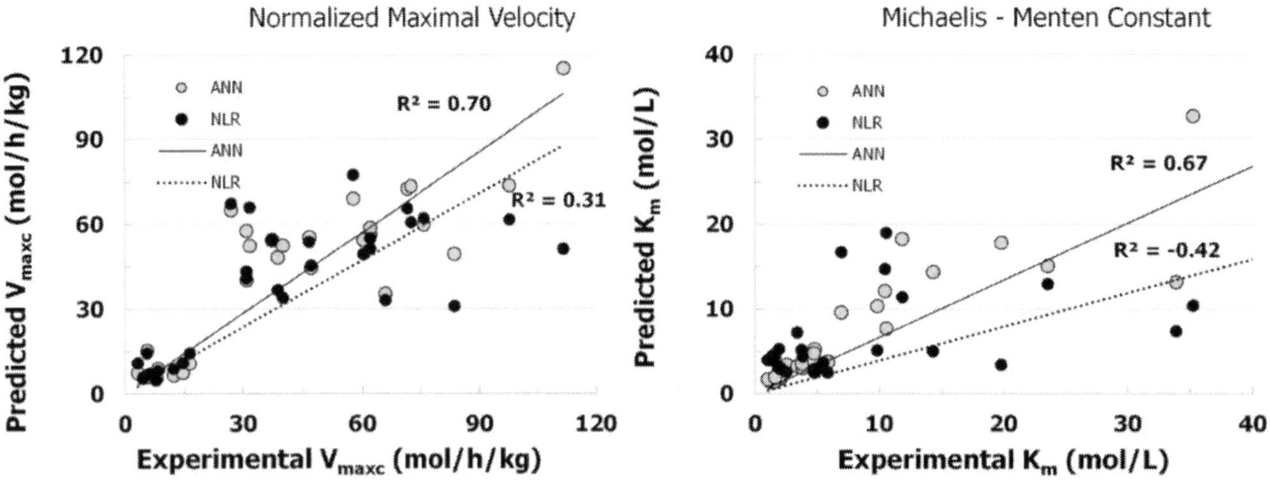

Figure 1. Experimental vs predicted Maximal Velocity (V_{max}) and Michaelis - Menten (K_m) constants for ANNs and Non-Linear Regression

The significantly improved performance of Abraham's equation and ANN combination can be ascribed to its capacity to represent mathematically the complex interactions of biochemical micro-processes, which are lumped into the metabolic and physicochemical parameters. Generally, the predictive capacity of the Abraham's model for new chemical entities is influenced by chemical nature of the training set molecules used for development of the model. In actual case, the test set molecules can be predicted well when these molecules are structurally very similar to the training set molecules. The reason is that the model has captured all features common to the training set molecules.

Concluding Remarks

The aim of this study was to demonstrate the applicability of ANNs to resolve different problems of the environmental health risk assessment continuum. These problems might refer to the prediction of macroscopic environmental parameters (such as environmental concentrations and personal exposure), as well as to essential physic-chemistry parameters for complex modelling systems such as biokinetic modelling. These applications serve different scopes. Assessment of exposure to occupational groups based on main environmental parameters, provides an assessment tool for rapid and robust evaluation of exposure to toxicants and eventually risk. Parameterization of complex biokinetic models, gives the opportunity of expanding the applicability of these models to a larger chemical space, creating a new era in the chemicals risk assessment.

In any case we need to keep in mind, that ANNs are not just "black boxes". The varying extent to which different models describe the underlying functional mechanisms of the processes of interest determines the varying level of the model empiricism. In general, more empirical, data-based models have a better practical value because of the non-linear complexity and interactions of the underlying process. On the other hand, mechanistically based models, are able to generate new knowledge, and to test hypothesis [13]. They rather facilitate us to capture the response of complex non-linear dynamic systems, where physical modelling is data-exhausting or computationally and time demanding.

However, the good performance of an ANN all the time relies on the previous knowledge of the physics and in turn the parameters that govern the process to be described. Considering a more integrative approach on resolving environmental health problems, ANN (or other type of expert systems) could serve as an intermediate step for applying a mechanistic approach; use of ANN for PBBK models parameterization is a typical example. Recent applications in integrated exposure biology include gene regulatory network inference, potentially capturing high-order and non linear dependencies between expression patterns [14]. In addition, the output of ANNs on environmental contamination [15] or exposure [4], could provide the input to more complex mechanistically based models, for further analysis of potential health implications. Overall, ANN models could provide the essential information on critical steps of the environmental health risk assessment process, either as stand-alone models, or as a part of a more integrative approach.

References

1. Dennis J.E.J., Schnabel R.B. (1996) Numerical Methods for Unconstrained Optimization and Nonlinear Equations. 1996: Soc for Industrial & Applied Math.
2. Schlink U., Dorling S., Pelikan E., Nunnari G., Cawley G., Junninen H., Greig A., Foxall R., Eben K., Chatterton T., Vondracek J., Richter M., Dostal M., Bertucco L., Kolehmainen M., Doyle M.

(2003). A rigorous inter-comparison of ground-level ozone predictions. Atmospheric Environment. 37(23), pp. 3237-3253.

3. Jo W.K., Song K.B. (2001). Exposure to volatile organic compounds for individuals with occupations associated with potential exposure to motor vehicle exhaust and/or gasoline vapor emissions. Sci Total Environ. 269(1-3), pp. 25-37.
4. Sarigiannis D.A., Karakitsios S.P., Gotti A., Papaloukas C.L., Kassomenos P.A., Pilidis G.A. (2009). Bayesian algorithm implementation in a real time exposure assessment model on benzene with calculation of associated cancer risks. Sensors. 9(2), pp. 731-755.
5. Karakitsios S.P., Papaloukas C.L., Kassomenos P.A., Pilidis G.A. (2007). Assessment and prediction of exposure to benzene of filling station employees. Atmospheric Environment. 41(40), pp. 9555-9569.
6. Abraham M.H. (1993). Application of solvation equations to chemical and biochemical processes. Pure and Applied Chemistry. 65(12), pp. 2503-2512.
7. Reisen F., Meyer C.P.M., Keywood M.D. (2013). Impact of biomass burning sources on seasonal aerosol air quality. Atmospheric Environment. 67, pp. 437-447.
8. Perrone M.G., Larsen B.R., Ferrero L., Sangiorgi G., De Gennaro G., Udisti R., Zangrando R., Gambaro A., Bolzacchini E. (2012). Sources of high PM2.5 concentrations in Milan, Northern Italy: Molecular marker data and CMB modelling. Science of the Total Environment. 414, pp. 343-355.
9. Ntziachristos L., Samaras Z. (2001). An empirical method for predicting exhaust emissions of regulated pollutants from future vehicle technologies. Atmospheric Environment. 35(11), pp. 1985-1999.
10. Sarigiannis D.A., Soulakellis N., Schäfer K., Tombrou M., Sifakis N.I., Assimakopoulos D., Lointier M., Dantou A., Saisana M. (2002). ICAROS: An integrated computational environment for the assimilation of environmental data and models for urban and regional air quality. Water, Air, & Soil Pollution: Focus. 2(5), pp. 641-654.
11. Sarigiannis D.A., Soulakellis N.A., Sifakis N.I. (2004). Information Fusion for Computational Assessment of Air Quality and Health Effects. Photogrammetric Engineering and Remote Sensing. 70(2), pp. 235-245.
12. Lohse K.A., Hope D., Sponseller R., Allen J.O., Grimm N.B. (2008). Atmospheric deposition of carbon and nutrients across an arid metropolitan area. Science of the Total Environment. 402(1), pp. 95-105.
13. Nestorov I.S., Hadjitodorov S.T., Petrov I., Rowland M. (1999). Empirical versus mechanistic modelling: comparison of an artificial neural network to a mechanistically based model for quantitative structure pharmacokinetic relationships of a homologous series of barbiturates. AAPS pharmSci [electronic resource]. 1(4), pp.
14. Grimaldi M., Visintainer R., Jurman G. (2011). RegnANN: Reverse Engineering Gene Networks using Artificial Neural Networks. PLoS One. 6(12), pp. e28646.
15. Sarigiannis D.A., Karakitsios S.P., Kermenidou M., Nikolaki S., Zikopoulos D., Semelidis S., Papagiannakis A., Tzimou R. (2014). Total exposure to airborne particulate matter in cities: The effect of biomass combustion. Science of The Total Environment. 493(0), pp. 795-805.

389057 Fault Detection in an Acid Gas Removal Unit of IGCC Plant Using Magnitude Ratio Algorithm

Abstract:

Tight carbon emission standards on power generation from fossil fuels have made researchers focus on plants with near-zero emissions. Integrated gasification combined cycle (IGCC) power plants with CO_2 capture are promising technologies for safe and clean power generation that enforce the environmental emission standards by treating the fuel gas in the acid gas removal (AGR) unit. Satisfying the environmental constraints depends crucially on the performance of AGR unit. Any faults that occur in the AGR unit can drive the process away from its nominal condition and may lead to violation of environmental constraints and hazardous consequences. An early detection and identification of the faults facilitate preventive actions for safe and optimal operations. Any abnormalities affect several process variables throughout the process which requires measurements of variables for symptoms to be observed. In order to detect and diagnose faults, crucial variables must be identified based on their economical and practical feasibility. An algorithmic approach for identifying the optimal number, type and location of the sensors for fault detection and diagnosis is useful, particularly for the large-scale, emerging fossil power plants with CO_2 capture.

The usefulness of magnitude ratio (MR) algorithm developed in our previous work[1], which enhances the diagnosis capability of signed directed graph (SDG) sensor placement algorithm, is demonstrated by implementing on three basic case studies. The algorithm uses all the variables from the process and assumes the magnitude ratio of each pair as a pseudo-sensor. Similar to SDG, pseudo-sensors have discrete states and are treated by symmetric difference operator in sensor network design framework. When used in SDG framework, the algorithm retains the properties of SDG and manages to add more information for improving diagnosis capability. The algorithm is implemented on the SELEXOL-based AGR unit to identify an optimal sensor network. The enhanced diagnosis capability of the MR algorithm observed in AGR unit is promising for further implementation of the algorithm in large and complex power plants.

Reference:

[1] P Mobed, J Maddala, D Bhattacharyya, R Rengaswamy, "On the Use of Magnitude Ratio in Sensor Placement Algorithms for Fault Detection and Diagnosis in Complex Energy Processes", *57th Annual ISA POWID Symposium*, Scottsdale, Arizona, USA, 2014.

Ontology-assisted Knowledge Acquisition and Management in Pharmaceutical Engineering

Miguel Francisco M. Remolona, Matthew F. Conway, Sriram Balasubramanian, Linxi Fan, Ziyan Feng, Tianhao Gu, Hyungtae Kim, Prasad M. Nirantar, Sarah Panda, Nithin R. Ranabothu, Neha Rastogi and Venkat Venkatasubramanian

Complex Resilient Intelligent Systems Laboratory
Department of Chemical Engineering
Columbia University
New York, NY 10025

Knowledge acquisition and management in pharmaceutical product development and manufacturing is a challenging problem, where one must deal with the generation, use, and management of massive amounts of diverse data, information, and knowledge. For example, a new drug application consists of 100,000-500,000 pages. In order to facilitate review and management of such massive information progress in two key tasks are required: information retrieval and knowledge management. This work focuses on advancing the state of the art in these two areas in pharmaceutical engineering using ontologies. Ontologies are a knowledge representation framework for storing and classifying information. We describe a prototype, called Holmes, which extracts the relevant information from scientific documents, classifies it using an ontology, and assists in semantic queries using a combination of algorithms and methods from computer vision, natural language processing, machine learning and database management. Holmes automatically extracts content such as text, equations, images, graphs, tables, etc., which are then processed separately to break down into consequent parts, classes and relations with each other. This information is then represented in a graphical database guided by the domain ontologies. We will present some results and discuss the next steps.

379302 A Methodology for the Simultaneous Design, Scheduling and Control of Multi-Product Processes Under Uncertainty

Abstract:

Chemical process industry has been reinforced time to time by development of methodologies that strengthen the various aspects of the industry. The competitive environment makes the industry vulnerable to non-optimal operations. The overall cost of plant has always been the classical measure of the optimality of the operations in the industry. This may include the capital cost, the operating cost, the cost associated with the control system, etc. An important section of the chemical industry produces various grades of products (multi-product) and scheduling of the production of these grades plays an important role in the economy of the operations. The presented work focuses on the development of a methodology that can address three aspects of the economy of multi-product plants; i.e. process design, control and scheduling. The approaches proposed in the literature to address this challenging problem can be classified in two groups: i) the sequential approach, which generates overall optimal solution by sequentially solving an optimization problem for each of the aspect separately; i.e. one of the three aspects at a time, and ii) the simultaneous approach, which adds more degrees of freedom by generating optimal solution of an optimization problem for all aspects simultaneously. Studies have been presented on integration of the aspects, where some of them present the integration of process design and control[1,4] ,while the other integrates scheduling and control[2] . There also have been studies which address integration of optimal transitions of grades with design and control with fixed sequence of production[3], while others have integrated design, scheduling and optimal control, where control decisions involve transition times[6] but not the controller tuning parameters. The work presented here involves the development of simultaneous design, scheduling and control (SDSC) approach. The integration of the three aspects is one of the novel features of the work, as sequence of production remains part of decision variables as opposed to fixed sequence considered in previous works[3]. Key contributions of the work also include determination of tuning parameters as decision variables and the integration of the three aspects under continuous influence of disturbances and uncertainty in product demands.

The optimization framework developed aims to minimize an overall cost function, which includes capital cost, operating cost, variability cost and transition cost, while taking into account various operational constraints. The decision variables consist of the design of process equipment's, the tuning parameters for the control system in place and the scheduling parameters. Scheduling parameters include optimal sequence of production and optimal transition trajectories. The proposed approach takes into account the influence of disturbances in the system by the identification of critical frequency from the disturbances, which is used to quantify the worst-case variability in the controlled variables. The uncertainty in the demands of products has also been addressed by creating critical demand scenarios with different probabilities of occurrence. These features foster the robustness of the approach to uncertainty. System stability analysis is performed at each optimal solution to ensure feasibility of the solution to achieve stable operation. To demonstrate the effectiveness of the SDSC approach, a multi-product continuous stirred-tank reactor (CSTR) system is considered[5], which produces 4 different grades of a product. A comparison study has been performed to prove an edge of simultaneous approach over the traditional sequential method and effect of demand uncertainty is demonstrated via a case study involving different demand scenarios.

References

(1) Flores-Tlacuahuac, A., Biegler, L. T. Simultaneous mixed-integer dynamic optimization for integrated design and control, Computers & Chemical Engineering, 31, 588–600, 2007

(2) Flores-Tlacuahuac, A., & Grossmann, I. E. Simultaneous cyclic scheduling and control of a multiproduct CSTR, Industrial & Engineering Chemistry Research, 45(20), 6175–6189, 2006.

(3) Flores-Tlacuahuac, A., Biegler, L. T. Integrated control and process design during optimal polymer grade transition operations, Computers and Chemical Engineering, 32 (11), 2823-2837, 2008.

(4) Ricardez-Sandoval, L. A., Douglas, P. L., & Budman, H. M. A methodology for the simultaneous design and control of large-scale systems under process parameter uncertainty, Computers and Chemical Engineering, 35(2), 307–318, 2011.

(5) Ricardez-Sandoval, L. A. Optimal design and control of dynamic systems under uncertainty: A probabilistic approach, Computers and Chemical Engineering, 43, 91–107, 2012.

(6) Terrazas-Moreno, S., Flores-Tlacuahuac, A., Grossmann, I.E. Simultaneous design, scheduling, and optimal control of a methyl-methacrylate continuous polymerization reactor, AIChE Journal. 54 (12), 3160–3170, 2008.

375346 Medium-Term Production Scheduling of a Large-Scale Steelmaking Continuous Casting Process Under Demand and Processing Time Uncertainty

Abstract:

Scheduling of steelmaking-continuous casting (SCC) processes is of major importance in iron and steel operations since the SCC process is often a bottleneck in iron and steel production.[1] Li et al.[2] developed a novel and effective unit-specific event based continuous time formulation for this process and extended the rolling-horizon approach[3-4] to decompose the entire MILP problem. The computational results show that the extended rolling horizon approach reduced the computational time and generated the same or better feasible solution than that without using the rolling horizon approach. This proposed model assumes that all the parameters are deterministic in nature.

However, several uncertainties such as demand fluctuation, and processing time uncertainty frequently happen during the realistic operations. In the presence of these uncertainties, the nominal schedule may often be suboptimal or even become infeasible. In general, two approaches can be used to address those uncertainties: reactive scheduling and preventative scheduling.[5] While reactive scheduling is a process to revise the generated schedule from nominal parameters when a disruption has occurred during the actual execution of the schedule, preventive scheduling seeks to accommodate future uncertainty at the scheduling stage. The uncertainty can be explicitly taken into account through preventive approaches such as two-stage stochastic programming, parametric programming, fuzzy programming, chance constraint programming, robust optimization techniques, and conditional value at risk.[5] Among these approaches, uncertain parameters are often represented by scenarios or non-scenarios. The detailed reviews on planning and scheduling under uncertainty can be referred to Li and Ierapetritou[6] and Verderame et al. [5].

In steelmaking and continuous casting process, reactive scheduling is often used for handling different types of disruptions such as machine breakdowns, rush orders, defect problems, and order cancellations.[7-9] Hou and Li[9] investigated comprehensively various disturbances in steel production dynamic environment and their effects. To address those disturbances, they proposed a general repair procedure and basic repair steps to provide an effective decision support for theory and practice of the integrated steel production rescheduling. Yu et al.[10] used fuzzy programming approach to address uncertain processing time in the steelmaking and continuous casting production process. The uncertain processing time was denoted by triangular fuzzy number.

In this paper, we first employ the robust optimization framework from Lin et al.,[11] Janak et al.,[12] Li et al.,[13] and Li et al.[14] to develop a deterministic robust counterpart optimization model for demand and processing time uncertainty during the steelmaking and continuous casting operations. The robust solution from the robust optimization framework is guaranteed to be feasible for the nominal parameters. The computational results show that the obtained schedule is more robust than the nominal schedule.

While the robust optimization framework aims at finding a single schedule that is immune to all the possible uncertainty realizations within a uncertainty set, two/multi-stage stochastic programming method provides flexibility of implementing different operational decisions after the realization of uncertainty. In this paper, a two-stage stochastic programming framework was also studied for the scheduling of steelmaking and continuous operations under demand and processing time uncertainty. Scenario based stochastic programming method was applied to the problem under investigation. To make the resulted stochastic programming problem computationally tractable, a novel scenario reduction method[15-16] has been applied to reduce the huge number of scenarios to a small set of representative realizations. The selected scenarios are incorporated into the stochastic programming model, which is further reformulated into its deterministic equivalent model and solved. Results demonstrate that the robust optimization-based solution is of comparable quality to the two-stage stochastic programming based solution.

References

[1] Tang, L. X., Lih, P. B., Liu, J. Y., Fang, L. Steel-making process scheduling using lagrangian relaxation, Int. J. Prod. Res., 2002, 40, 55-70.

[2] Li, J., Xiao, X., Tang, Q. H., Floudas, C. A. Production scheduling of a large-scale steelmaking continuous casting process via unit-specific event-based continuous-time models: Short-term and medium-term scheduling, Ind. Eng. Chem. Res., 2012, 51, 7300-7319.

[3] Lin, X., Floudas, C. A., Modi, S., Juhasz, N. M. Continuous-time optimization approach for medium-range production scheduling of a multiproduct batch plant, Ind. Eng. Chem. Res., 2002, 41, 3884-3906.

[4] Janak S. L., Floudas C. A., Kallrath J., Vormbrock N. Production scheduling of a large-scale industrial batch plant. II. Reactive scheduling, Ind. Eng. Chem. Res., 2006, 45, 8234-8252.

[5] Verderame, P. M., Elia, J. A., Li, J., & Floudas, C. A. Planning and scheduling under uncertainty: A review across multiple sections, Ind. Eng. Chem. Res., 2010, 49, 3993-4017.

[6] Li, Z., Ierapetritou, M. Processing scheduling under uncertainty: Review and challenges, Computers and Chemical Engineering, 2008, 32, 715-727.

[7] Tang, L. X., Wang, X. P. A predictive reactive scheduling method for color-coating production in steel industry, The International Journal of Advanced Manufacturing, 2008, 35, 633-645.

[8] Worapradya, K., Buranathiti, T. Production rescheduling based on stability under uncertainty for continuous slab casting, Proceedings of ASIMMOD, January 22-23, Bangkok, Thailand, 2009, 170-176.

[9] Hou, D. L., Li, T. K. Analysis of ramdom disturbances on shop floor in modern steel production dynamic environment. Procedia Engineering, 2012, 29, 663-667.

[10] Yu, S. P., Pang, X. F., Chai, T. Y., Zheng, B. L. Scheduling of steelmaking and continuous casting process under

processing time uncertainty. Control and Decision, 2009, 24, 10, 1-6.

[11] Lin, X., Janak, S. L., Floudas, C. A. A new robust optimization approach for scheduling under uncertainty: I. Bounded uncertainty. Computers and Chemical Engineering, 2004, 28, 1069-1085.

[12] Janak, S. L., Lin, X., Floudas, C. A. A new robust optimization approach for scheduling under uncertainty. II. Uncertainty with known probability distribution. Computers and Chemical Engineering, 2007, 31, 171-195.

[13] Li, Z., Ding, R., Floudas, C. A. A comparative theoretical and computational study on robust counterpart optimization: I. Robust linear optimization and robust mixed-integer linear optimization. Industrial and Engineering Chemistry Research, 2011, 50, 10567-10603.

[14] Li, Z., Tang, Q., Floudas, C. A. A comparative theoretical and computational study on robust counterpart optimization: II. Probabilistic guarantees on constraint satisfaction. Industrial and Engineering Chemistry Research, 2012, 51, 6769-6788.

[15] Li, Z., Floudas, C. A. Optimal scenario reduction framework based on distance of uncertainty distribution and output performance: I. Single reduction via mixed integer linear optimization, Computers and Chemical Engnineering, Under review, 2013a.

[16] Li, Z., Floudas, C. A. Optimal scenario reduction framework based on distance of uncertainty distribution and output performance: II. Sequential reduction, To be submitted, 2013b.

Robust Dynamic Real-Time Optimization of a Semi-Batch Emulsion-Copolymerization Process with Uncertainties in the Parameter Values

J.Puschke (Jennifer.Puschke@avt.rwth-aachen.de), A.Mitsos (Alexander.Mitsos@avt.rwth-aachen.de)

AVT – Process Systems Engineering (SVT), RWTH Aachen University
Turmstr. 46, 52064 Aachen, Germany

In industrial applications, emulsion polymerization is typically performed in semi-batch reactors according to a given recipe. A promising alternative to recipe-based operation is economic nonlinear model-predictive control (eNMPC) [8] to reduce the batch time and to increase productivity [2,4,10]. The controller determines the optimal trajectories of the manipulated variables in each sampling interval by a solution of a nonlinear dynamic optimization problem with an economical objective in real-time. However, the optimization problem is subject to multiple types of uncertainties, including model and input uncertainty, on top of the inevitable uncertainty resulting from imperfect knowledge of the initial state, which, under mild assumptions, can be represented as parametric uncertainties. Therefore, model-based control and optimization should be robust in the presence of parametric uncertainty, i.e., the solution has to satisfy the constraints (robust feasibility) and has to achieve at least a local optimum (robust optimality), despite imperfect knowledge on the parameters. With state-of-the-art problem formulations, algorithms and computers it is still an open research problem how to properly address robust feasibility and robust optimality in real-time in a unifying framework [6].[1]

To this end, the concept of a multi-model approach will be considered [5]. In the multi-model approach several process models are subject to optimization at once [3]. As a result, the optimization problem has many more constraints, thus making the eNMPC more robust. Despite the conservatism introduced by any robustification approach into the solution, it is expected, that the productivity (i.e., the space-time yield) is higher than the one resulting from commonly used recipe-based operation, which are more conservative by nature.

One model of the multi-models will have the estimated parameter values in the developed model structure and will be called nominal model. The other models are constructed using the same model structure but choosing different values for some important model parameters to reflect the solution variety over the uncertainty set in the parameter space as good as possible. Ideally, the models are chosen such that they correspond to the worst-cases with respect to robust feasibility and to a representative sample with respect to the objective to approximate robust optimality to the extent possible. Though neither robust feasibility nor robust optimality can be guaranteed by such a multi-model approach in general, we call the other models "critical" for convenience. The methodology presented is set up to approximate the yet computationally intractable correct solution of semi-infinite programming (SIP) problems which would guarantee robust feasibility if it is solved globally in the lower level [7, 9].

The difficulties of systematically deriving such critical models for a given nominal process model are sketched. Furthermore, a case study is presented, where the approach is applied to robustify an optimization of a semi-batch emulsion polymerization reactor as part of the European Union's

Seventh Framework Programme for research, technological development and demonstration under grant agreement no [280827] (COOPOL).

[1] COOPOL Deliverable 5.1: *Robust model-based control and dynamic real-time optimization methods,* Version dated January (2013).

[2] Eliçabe, G. E., Meira, G. R., Estimation and Control in Polymerization Reactors. A Review, Polymer Engineering and Science, 28(3), 121-135, (1988).

[3] Hartwich A. *Adaptive Numerical Methods for Dynamic Real-Time Optimization of Chemical Processes*. PhD thesis. RWTH Aachen, (2012).

[4] Helbig A., Abel O., Marquardt W., *Structural concepts for optimization based control of transient processes*. In F. Allgöwer and A. Zheng, editors, Nonlinear Model Predictive Control Workshop - Assessment and Future Directions, 26, 295–311. Birkhäuser Verlag, Basel, (2000).

[5] Huang R., Patwardhan S.C., Biegler L.T., *Multi-scenario-based robust nonlinear model predictive control with first principle models*, Computer Aided Chemical Engineering, Vol. 27, 1293-1298. (2009).

[6] Lucia S., Finkler T., Basak D., Engell S., *A new Robust NMPC Scheme and its Application to a semi-batch ReactorExample,* 8th IFAC Symposium on Advanced Control of Chemical Processes, Singapore, (2012)

[7] Mitsos A., Lemonidis P., Lee C.K., Barton P., *Relaxation-Based Bounds for Semi-Infinite Programs.* SIAM Journal on Optimization. Volume 19, p.77-113, (2008).

[8] Rawlings J.B., Amrit R., *Optimizing process economic performance using model predictive control.* In L. Magni, D. Raimondo and F. Allgöwer, editors, Nonlinear Model-Predictive Control: Towards New Challenging Applications, p.119–138. Springer-Verlag, (2009).

[9] Rossner N., King R., *Experimental Validation of Robust Process Design and Control Using Gaussian Mixture Densities,* Automatica accepted, (2013).

[10]Schei T. S., Ludvigsen H., Singstad P. and Foss B., *Operator Support System for Optimization of Suspension PVC Batch Polymerization Reactors,*In ECC 2001 (Industry Day), Porto, Portugal, (2001).

Variants to a Heuristic Knapsack Approach for Solving Pharmaceutical R&D Pipeline Management Problems

Brianna Christian, Selen Cremaschi
Department of Chemical Engineering, University of Tulsa, 800 South Tucker Drive, Tulsa, Ok, 74104, USA

With the rapid expiration of big selling pharmaceutical drug patents, pharmaceutical manufacturers are under pressure to improve the efficiency of the pharmaceutical research and development (R&D) pipeline. One of the ways to improve efficiency is to improve the management through scheduling and planning of clinical trials. The characteristics of the pharmaceutical R&D pipeline management problem include a set of new drug development projects, resources and expenditures associated with each project, and returns that will only be realized once the drugs reach market. To reach market the products are required to complete a stage-gate process where if a product fails to successfully complete a stage, the product does not continue through subsequent stages. Often the returns, the necessary resources, and whether or not each product will reach market are not known with certainty at the time of the pipeline planning. Decisions are made to maximize the returns by determining which projects to pursue and how to allocate resources to the chosen projects (Subramanian et al., 2001). The R&D pipeline management problem is challenging due to the endogenous nature of the uncertainty, i.e., decisions on whether or not to invest impact the realization of the uncertainty. The pharmaceutical R&D pipeline management problem can be formulated as a multistage stochastic program (Colvin and Maravelias, 2008). Literature on multistage stochastic programs (MSSPs) with endogenous uncertainty is limited, and often the available approaches can only successfully solve small problems.

Goel and Grossmann (2004a) presented an approximation method that searches the feasible solution space to find a "good" solution. The algorithm is applied to an offshore oil planning problem where the objective was to determine the optimal placement of production platforms given a limited investment budget. The decisions were whether or not to install well and production platforms. The uncertainty in the problem is the production amounts of each well and is realized after the installation of the platforms. Goel and Grossmann (2004b), considering the same offshore oil planning problem, proposed a Lagrangean duality based branch-and-bound approach. A valid lower bound is obtained by solving the Lagrange dual problem with relaxed non-anticipativity constraints. Colvin and Maravelias explored several formulation and solution procedures to handle larger instances of the pharmaceutical R&D pipeline management problem. One approach, a branch-and-cut algorithm (Colvin and Maravelias 2010), initially adds a percentage of non-anticipativity constraints and then iteratively adds constraints based on violations. More recently, Gupta and Grossmann (2014) proposed a Lagrangean decomposition scheme that utilizes a scenario grouping strategy. The scenario grouping strategy allows of partial decomposition of the full space model that may be useful in large problems. The scenario grouping strategy was applied to the oilfield planning problem from Goel and Grossmann (2004a) and a process network problems originally presented in Goel and Grossmann (2006). The process network problem consists of finding the optimal expansion and operation decisions for a fixed planning horizon. In this problem, the yields of the process equipment used for expansion are not known with certainty. Gupta and Grossman (2014) concluded that their scenario grouping strategy performed well on medium sized problems but that further work would be needed to show its effectiveness on larger problems.

In their previous work (Christian and Cremaschi, 2014), the authors presented a heuristic approach to solving the pharmaceutical R&D pipeline management problem using a series of knapsack problems. The knapsack solution approach yields implementable solutions, and was demonstrated to improve the solution times by several orders of magnitude for large instances of the R&D pipeline management problem. The solutions were within 3 % of the global optimal for small instances, while they degraded as the problem size increased. The degradation in the solutions with problem size is hypothesized to be a combination of the method's inability to see the entire planning horizon and methods inflexibility in knapsack sub-problem generation. Sub-problems for the original knapsack solution method are created at the realization of all uncertainties. The work presented in this talk explores several techniques to improve this knapsack heuristic approach.

The first approach relaxes the times at which the knapsack sub-problems are solved, allowing for knapsack sub-problems to be generated when there are sufficient resources available for decision making. Preliminary results show that the relaxed planning horizon solutions are identical or better to the solutions of the original heuristic approach for smaller instances of the pharmaceutical R&D pipeline management problem. In this talk, we will discuss how the flexible sub-problem generation affects the solution quality and time with different problem sizes.

The second approach implements a hedging approach where the knapsack problems are allowed to pack items in violation of the constraint used to prevent the greedy solution at the cost of penalizing the knapsack objective. The approach appears to be very sensitive to the parameters used to penalize the objective function. When penalty parameters are small the solution is prone to overscheduling yielding solutions that are worse than the original knapsack solution. When penalty parameters are too large the solution becomes identical to the knapsack solution. The talk will present how the solution quality changes with penalty parameters.

References

M. Colvin, C.T. Maravelias, 2008, A stochastic programming approach for clinical trial planning in new drug development, Computers & Chemical Engineering, 32, 11, 2626-2642.

M. Colvin, C. T. Maravelias, 2010, Modeling methods and a branch and cut algorithm for pharmaceutical clinical trial planning using stochastic programming, European Journal of Operational Research, 203, 1, 205-215.

Goel V, Grossmann IE. A stochastic programming approach to planning of offshore gas field developments under uncertainty in reserves. *Computers & Chemical Engineering.* 2004;28(8):1409-1429.

Goel V, Grossmann IE. A Lagrangean duality based branch and bound for solving linear stochastic programs with decision dependent uncertainty. In: Puigjaner L, Espuna A, eds. *European Symposium on Computer-Aided Process Engineering-15, 20A and 20B.* Vol 20a-20b. Amsterdam: Elsevier Science Bv; 2005:55-60.

V. Goel, I.E. Grossmann, 2006, A Class of stochastic programs with decision dependent uncertainty, Mathematical Programming, 108, 2/3, 355-394.

V. Gupta, I. E. Grossmann, 2011, Solution strategies for multistage stochastic programming with endogenous uncertainties.,Computers & Chemical Engineering, 35, 11, 2235-2247.

V. Gupta, I.E. Grossmann, 2014, A new decomposition algorithm for multistage stochastic programs with endogenous uncertainties. , Computers & Chemical Engineering, 62,0, 62-79.

Sandia Corporation. Pyomo <software.sandia.gov/trac/coopr> accessed on 11/10/2013.

D. Subramanian, J.F. Pekny, G.V. Reklaitis, 2001, A Simulation-Optimization Framework for Research and Development Pipeline Management, AIChE Journal, 47, 10, 2226-2242.

B. Christian, S. Cremaschi, A Quick Knapsack Heuristic Solution for Pharmaceutical R&D Pipeline Management Problems , ESCAPE 24.

A New Reactive Scheduling Methodology for Front-end Crude Oil Operations under Uncertainties of Tank Availability

Shujing Zhang and Qiang Xu

Dan F. Smith Department of Chemical Engineering

Lamar University, Beaumont, Texas 77710, USA

Abstract

Nowadays, the petroleum industry is confronted with volatile crude and product markets and strict environmental regulations, which result in low profit margins and intense competitions for refineries. Due to the ever deeper globalization, the supply chain of refineries is more vulnerable to many unexpected events, which will lead to interruptions of normal operations and incur adverse consequences such as failure to fulfill the customer orders, off-spec product generations, waste of raw materials/energy, or even unplanned facility shutdowns. Here, uncertainties refer to those abrupt incidents that deviate supply chain activities significantly apart from what are planned, which necessitate the rescheduling of future manufacturing activities. There are mainly two categories of rescheduling strategies from the literature review: proactive and reactive rescheduling. Reactive rescheduling puts more emphasis on solution quality instead of flexibility, and it makes amendments whenever the deviations actually become realized. However, proactive rescheduling usually seeks to generate an inherently flexible schedule which is robust in face of different uncertainties in the future and largely depends on probabilistic calculations. Proactive scheduling strategies in forms of robust programming, two-stage stochastic programming, fuzzy and parametric programming have been broadly reported, reactive scheduling methods, however, are still heavily dependent on heuristic tactics, which are criticized of incompleteness and lack of automation.

In this paper, a new reactive rescheduling methodology for short-term crude oil operations to handle various tank unavailability scenarios has been developed. A two-stage strategy will be implemented. On the first stage, a deterministic schedule is initially obtained based on a continuous-time global event model using nominal data provided/projected at the beginning of each scheduling time horizon. On the second stage, three different indicator binary variables are implemented respectively under any occurrence of tank unavailability to collect information from the current deterministic schedule. On basis of the obtained real-time information, rescheduling models are configured correspondingly by combining the first-stage scheduling model and amendment constraints. Depending on different circumstances, two alternative operation ways, i.e., evacuating incidental tank or not, will be studied. Note that when the incidental tank is evacuated, the left-over crudes from the tank are allowed to be sent to the downstream or other parallel tanks. Also, reactive scheduling is conducted under different evacuation times to obtain a Pareto frontier balancing between the evacuation time and their associated operation costs. Furthermore, two different strategies, full rescheduling *vs.* executed-task based partial rescheduling, are applied and compared in this study.

To demonstrate the efficacy of the proposed reactive scheduling methodology, several case studies are carried out. The results show that the proposed methodology can generate optimal rescheduling solutions seamlessly connected to the current completed scheduling operations and thus ensure the optimal operational continuity under impacts from uncertainties. Meanwhile, this methodology can systematically explore possible rescheduling actions for continuous crude process without any presumed heuristics.

379786 Identification-Based Optimization of Thermal Energy Storage Systems Under Uncertainty

Abstract:

The operation of a chemical process faces multiple uncertainties, including changes in feed quality, environmental conditions, and fluctuations in process demand. As a result, these variables should be specified at the design stage in terms of probability density functions, rather than as exact values, and the process design problem should be formulated and solved as an optimization problem under uncertainty. The typical approach for addressing this class of problems consists of sampling the probability distributions of the uncertain variables to generate a number of probability-weighted scenarios, which are then used to calculate an average value of the design objective function to be minimized. The number of scenarios increases substantially with the number of uncertain variables, which in turn increases the computational resources required to solve this problem.

Focusing in particular on the case of dynamical systems operating under uncertainty, we have recently [1] introduced the concept of identification-based optimization (IBO). The IBO principle relies on representing the uncertain variables as time varying pseudo-random multi-level signals (PRMS), which are imposed on the differential-algebraic equation (DAE) model of the system during the optimization iterations to efficiently sample the uncertainty space. The PRMSs are generated from the probability distributions of the uncertain variables using concepts from system identification theory. IBO thus converts the infinite dimensional optimization problem under uncertainty into a dynamic optimization problem.

Motivated by the need to provide rigorous design optimization solutions for thermal energy storage systems– whose operation is inherently transient and uncertain – in this paper provides a critical investigation of two numerical frameworks that are appropriate for solving the dynamic optimization problem associated with IBO. We first consider a full discretization approach [2], whereby the DAE model is converted using orthogonal collocation on finite elements. Then, we consider a sequential strategy, in which a DAE solver and a nonlinear programming algorithm (NLP) work in tandem. We propose initialization methods for both algorithms, focused specifically on IBO problems.

Finally, we demonstrate these concepts in the design of a thermal energy storage system based on phase-change materials, which is used for leveling the grid load of buildings [3]. We provide optimal design results, showing in the meantime that IBO is significantly more computationally efficient than scenario-based methods.

References

[1] S. Wang and M. Baldea. Identification-based optimization of dynamical systems design under uncertainty. Comput. Chem. Eng., 64(7):138-152, 2014.

[2] L.T. Biegler. Nonlinear Programming: Concepts, Algorithms, and Applications to Chemical Processes. SIAM, 2010.

[3] C. Sullivan, ConEd floats big boosts in incentives for energy storage. From http://www.governorswindenergycoalition.org/?p=7681, last retrieved May 11, 2014.

Bayesian Approach for Optimal Experimental Design of Thermo Gravimetric Analysis of CO_2 on Amine Sorbents

Jayashree Kalyanaraman, Matthew Realff, Yoshiaki Kawajiri

School of Chemical and Bio-molecular Engineering, Georgia Institute of Technology

AICHE Annual Meeting, Atlanta, Nov 17, 2014

Acknowledgement: US Department of Energy , Grant no: DE-FE007804

1

Application: CO_2 adsorption on amine sorbent loaded hollow fiber

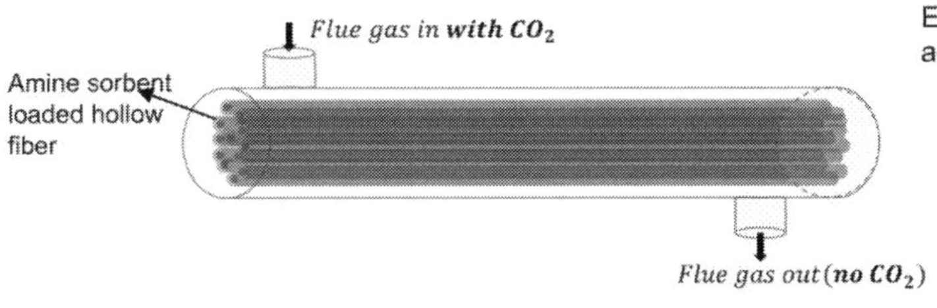

Exit gas CO_2 % varying with adsorption time - **Breakthrough curve**

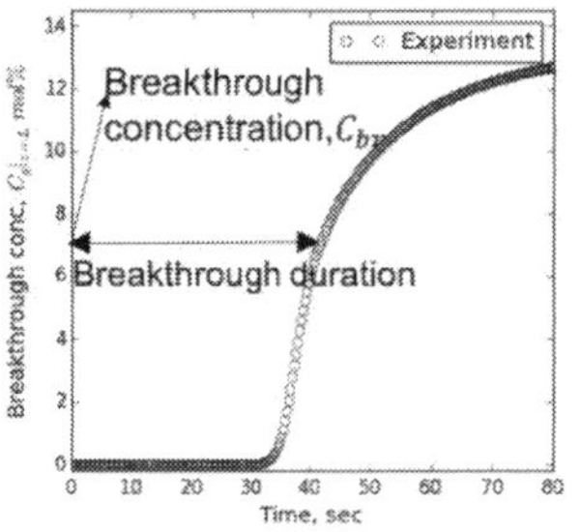

- Adsorption step run until exit CO_2 reaches maximum allowable concentration (breakthrough concentration)
- Breakthrough capacity q_{br}- Average sorbent loading through breakthrough time
  - Design variable related to process economics

q_{br} ↑ Capital and operating cost ↓

* Fan, Y. et al. Evaluation of CO2 adsorption dynamics of polymer/silica supported poly(ethylenimine) hollow fiber sorbents in rapid temperature swing adsorption, Int. J. Greenhouse Gas Control 2014, 21, 61-71.

2

Process Model Equations

Gas mass balance

$$\frac{\partial C_g}{\partial t} + u_g \frac{\partial C_g}{\partial z} + \frac{A_o \rho_f v_s}{V_{fs}} \frac{\partial q}{\partial t} = 0$$

Adsorption isotherm parameters

$$\theta = \{q_{m0}, \eta, A, B, \Delta H_0, b0\}$$

Sorbent mass balance

$$\frac{\partial q}{\partial t} = K_s(\psi q_{eq} - q_1)$$
$$+K_p((1-\psi)q_{eq} - q_2)$$

Adsorption isotherm equilibrium

$$q_{eq} = q_{max}\frac{bP}{(1+(bP)^n)^{1/n}}\ ;\ n = A + B\left(1 - \frac{T_0}{T}\right); q_{max} = q_{m0}e^{\eta\left(1-\frac{T}{T_0}\right)}$$
$$b = b_0 e^{\frac{\Delta H_0}{RT_0}\left(1-\frac{T_0}{T}\right)}$$

Gas heat balance

$$\rho_g C p_g \frac{\partial T_g}{\partial t} + u_g \frac{\partial T_g}{\partial z} + \frac{h_g A_o}{V_{fs}}(T_g - T_f) + q_{thm} = 0$$

Fiber heat balance

$$\rho_f C_{pf} \frac{\partial T_f}{\partial t} = \frac{\lambda_f}{r}\frac{\partial^2 T_f}{\partial z^2} + (-\Delta H_{ads})\frac{\partial q}{\partial t} v_s \rho_{sorb} + \frac{h_g A_o}{V_f}(T_g - T_f)$$

Thermogravimetric (TGA) expts

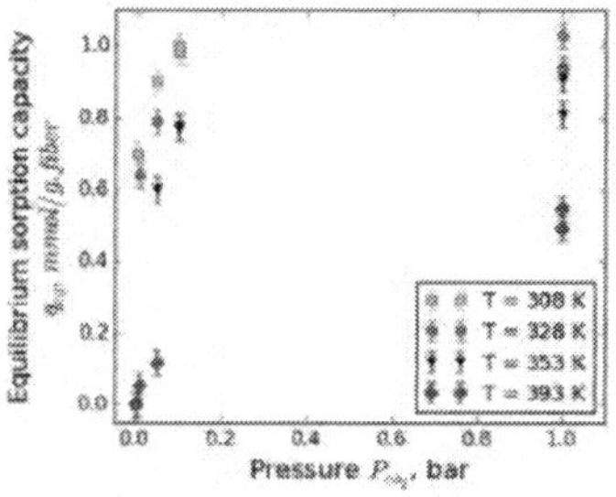

3

Source of uncertainties

Adsorption isotherm parameters

$$\boldsymbol{\theta} = \{\boldsymbol{q_{m0}}, \boldsymbol{\eta}, \boldsymbol{A}, \boldsymbol{B}, \boldsymbol{\Delta H_0}, \boldsymbol{b0}\}$$

Toth equilibrium model

Adsorption isotherm equilibrium

$$q_{eq} = q_{max}\frac{bP}{(1+(bP)^n)^{1/n}}\ ;\ n = A + B\left(1 - \frac{T_0}{T}\right); q_{max} = q_{m0}e^{\eta\left(1-\frac{T}{T_0}\right)}$$
$$b = b_0 e^{\frac{\Delta H_0}{RT_0}\left(1-\frac{T_0}{T}\right)}$$

Thermogravimetric (TGA) expts

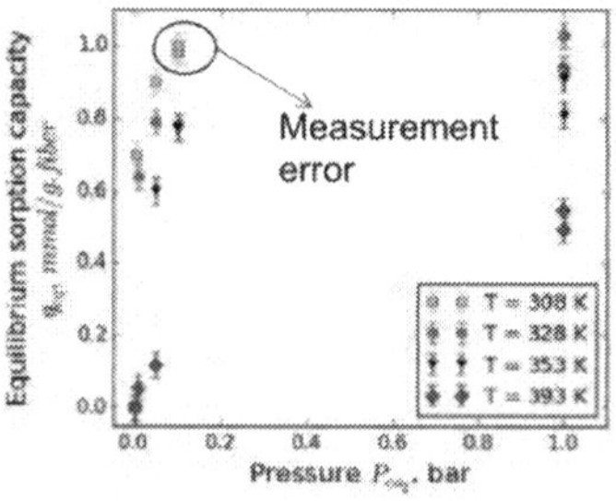

❑ **Parametric uncertainty**

Measurement error ⇨ Uncertainty/Error in θ estimate ⇨ Uncertainty/Error in prediction of breakthrough time and capacity q_{br}

4

Why is Uncertainty Quantification important?

❑ Adsorption step run until exit CO_2 reaches C_{br}

Process constraint: CO_2 concentration at exit < C_{br} during adsorption step

❖ **Robust design under uncertainty is needed**

Exit gas CO_2 % varying with adsorption time - **Breakthrough curve**

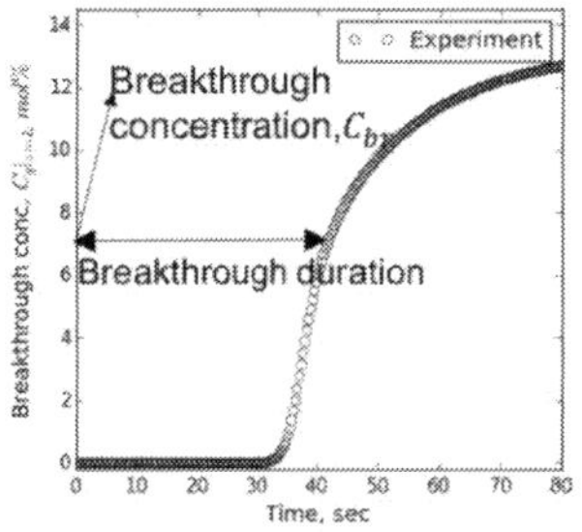

* Fan, Y. et al. Evaluation of CO2 adsorption dynamics of polymer/silica supported poly(ethylenimine) hollow fiber sorbents in rapid temperature swing adsorption, Int. J. Greenhouse Gas Control 2014, 21, 61-71.

5

Methods of Uncertainty Quantification (UQ)

Uncertainty Quantification: Evaluate the impact of parametric & model uncertainties on model prediction variables

Methods of UQ

Frequentist /Classical Approach

- Measurements as random variables
- Assumes Gaussian distribution
- Parameters are fixed

Bayesian Approach

- Parameters as random variables
- No assumption of distribution
- Includes prior knowledge of parameters

Non-probabilistic Approach

- Parameters are fixed
- Eg: Interval estimation, fuzzy logic
- Difficult to treat mathematically

6

Bayesian Approach of Uncertainty Quantification (UQ)

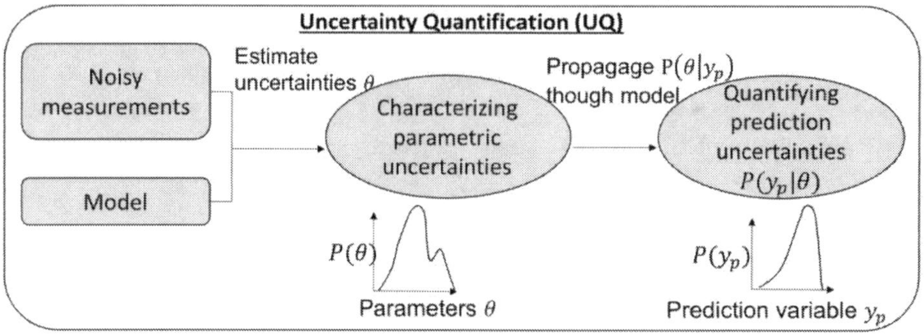

- Markov Chain Monte Carlo (MCMC) – 10 K- 1 M simulations for convergence
- ❑ **Adaptive Metropolis for faster convergence**

- Monte Carlo or Polynomial Chaos expansion – computationally expensive
- ❑ **Parallelized propagation using Monte Carlo**

7

Uncertainty Quantification (UQ) of CO_2 Adsorption

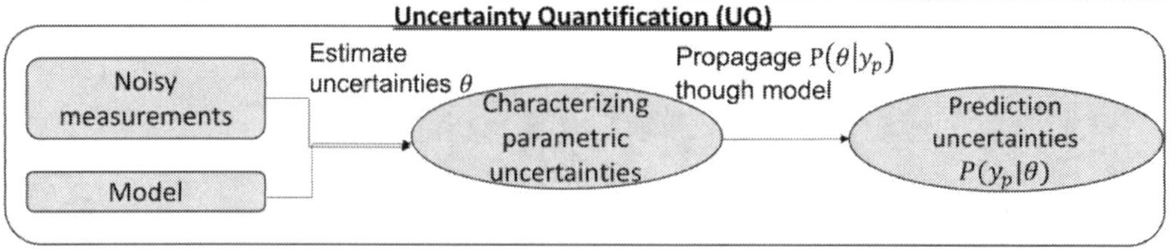

Thermogravimetric (TGA) expts of q_{eq}

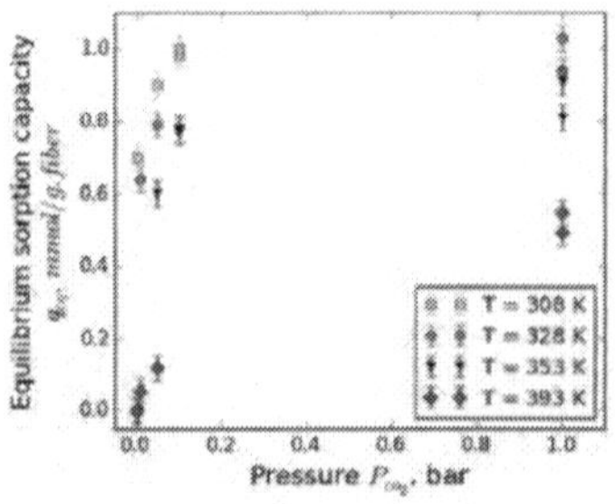

Isotherm parameters $P(\theta|q_{eq})$

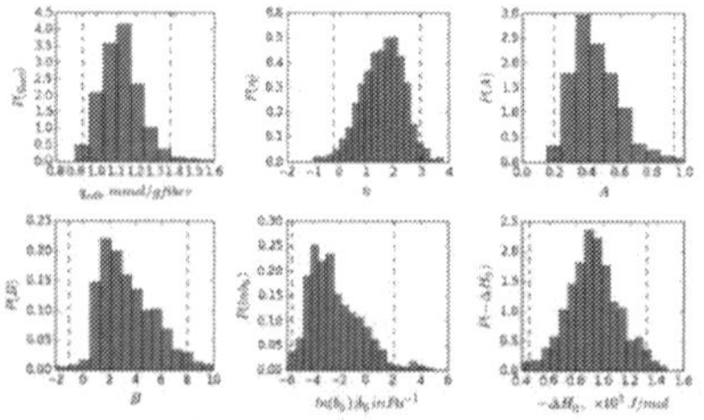

Prediction variable: Breakthrough capacity $P(q_{br}|q_{eq})$

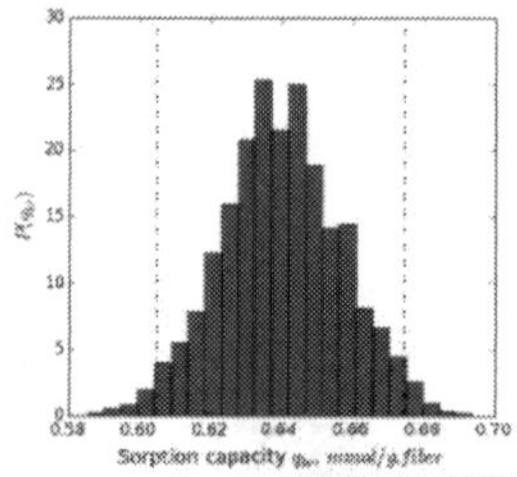

8

Uncertainty Quantification (UQ) of CO_2 Adsorption

Prediction variable: Breakthrough capacity $P(q_{br}|q_{eq})$

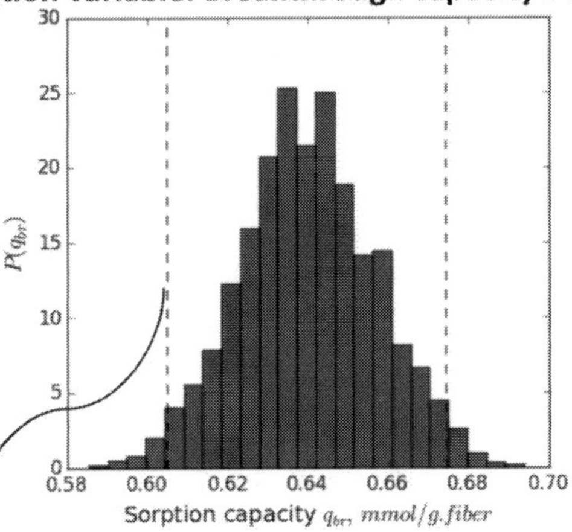

Dashed lines indicate 95% credible intervals

Prediction variable: Breakthrough capacity $P(q_{br}|q_{eq})$

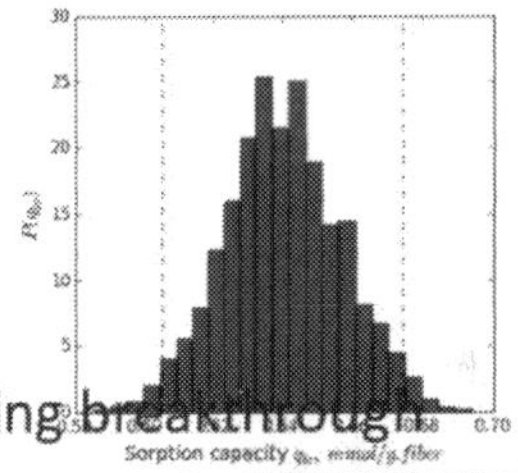

- ❑ Maximum sorption capacity obtainable without violating breakthrough concentration constraint = 0.605 mmol/g fiber

Uncertainty Reduction- Optimal Experimental Design (OED)

- ❑ Choose most efficient experiment that minimize time and resources
- ❑ Need to quantify uncertainty of process economic metric of y_p
  - ➢ Challenge : Computationally very expensive

Utility function : Value of information at d

$$U(d) = \int \log\left(\frac{P(y_p|\, y_{orig}, d)}{P(y_p|y_{orig})}\right) P(y_p|y_{orig}, d) P(y_{orig}|d) dy_p dy_{orig}$$

Log of pdf – Information content

Relative improvement in distribution of y_p after experiment at d

Utility function for Reduction of Prediction Uncertainty

$$U(d) = \int \log\left(\frac{P(y_p|\, y_{orig}, d)}{P(y_p|y_{orig})}\right) P(y_p|y_{orig}, d) P(y_{orig}|d)\, dy_p\, dy_{orig}$$

Evaluation

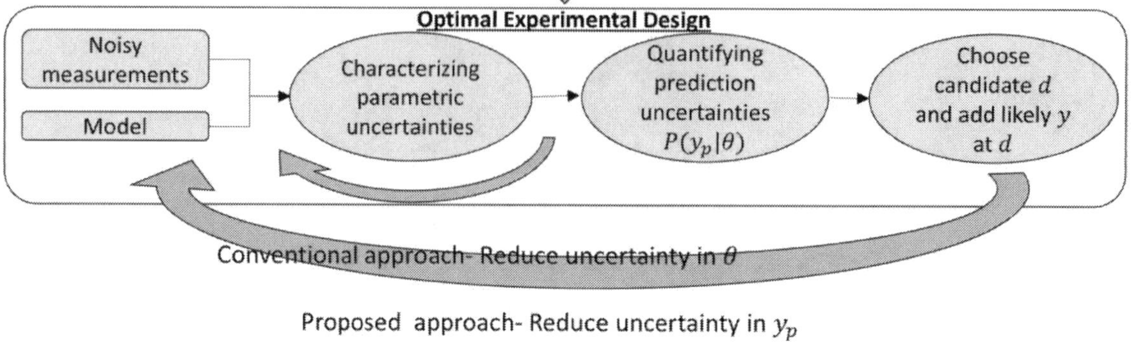

11 Georgia Institute of Technology

Utility function for Reduction of Prediction Uncertainty

$$U(d) = \int \log\left(\frac{P(y_p|\, y_{orig}, d)}{P(y_p|y_{orig})}\right) P(y_p|y_{orig}, d) P(y_{orig}|d)\, dy_p\, dy_{orig}$$

Evaluation costly

Modified Utility function- Less computationally expensive

$$U'(d) = \int \log\left(\frac{P(\theta_s|\, y_{orig}, d)}{P(\theta_s|y_{orig})}\right) P(\theta|y_{orig}, d) P(y_{orig}|d)\, d\theta\, dy_{orig}$$

Parameters with high sensitivity for y_p- found using Global Sensitivity Analysis

Relative improvement in distribution of θ after experiment at d

Global Sensitivity Analysis – Link between the parametric information to predictive information

12 Georgia Institute of Technology

Utility Function for Optimal Experiment Design

Modified Utility function surface $U'(d)$

$d = \{T, P_{co_2}\}$- Design conditions for experiment

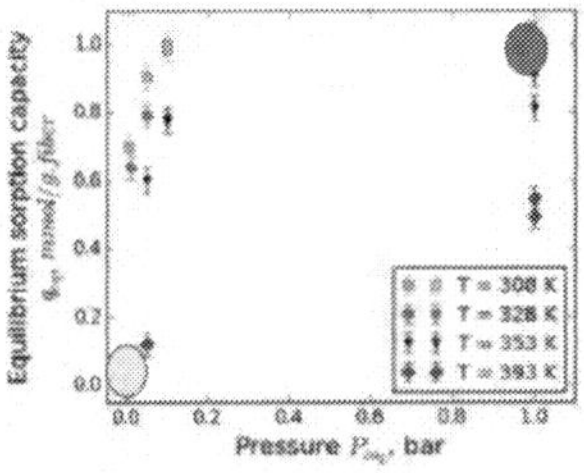

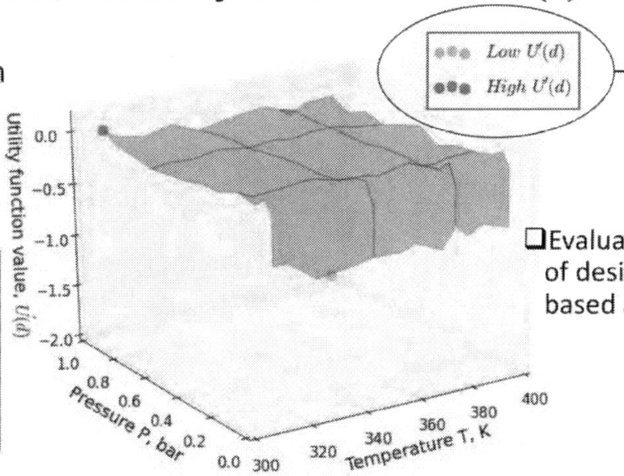

Two extreme cases chosen to verify $U'(d)$

❑Evaluated by enumeration of design space –Simulation based approach

* Rezeai, F. et al. Aminosilane-Grafted Polymer/Silica Hollow Fiber Adsorbents for CO2 Capture from Flue Gas, ACS 2013, 5, 3921-3931

13 Georgia Institute of Technology

Uncertainty Reduction by Optimal Experimental Design

Uncertainty comparison in Breakthrough capacity (design variable) prediction

Dashed lines indicate 95% credible intervals

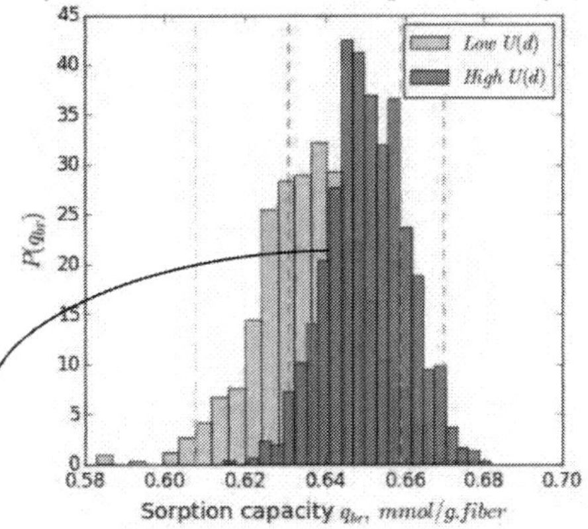

❑ Leads to ~ 5% reduction of net CO_2 capture cost *

* Kulkarni and Sholl, Analysis of equilibrium based TSA processes for direct CO_2 capture from air IECR (51), 2012

❑Breakthrough capacity considering uncertainty increased from 0.605 to 0.63 mmol/g fiber(5% increase) with one added data

14 Georgia Institute of Technology

Conclusions

- ➢Proposed computationally effective methodology for **optimal experimental design under uncertainty** for reducing prediction uncertainty
- ➢Utilized Global Sensitivity Analysis to determine information gain for prediction variable from parametric information gain

❑ Kalyanaraman, J. et al. Bayesian estimation of parametric uncertainties, quantification and reduction using optimal design of experiments for CO2 adsorption on amine sorbents, Computers and Chemical Engineering (submitted).

Thank YOU!

Prediction uncertainty comparison with data added at extreme utility

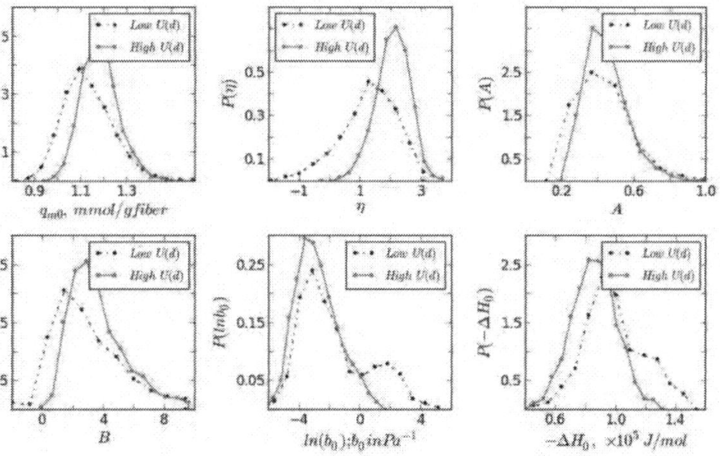

17 Georgia Institute of Technology

379124 Recent Advances in Chemical Production Scheduling

Abstract:

Despite the large number of publications in the area during the last two decades, there is no unified notation for chemical production scheduling. In this talk, we first develop a framework for the description of scheduling problems in the chemical industries. While building upon ideas used in discrete manufacturing, the proposed framework accounts for features such as material handling restrictions which make chemical production different. Second, we present a classification of the various modeling approaches that have been presented in the process systems engineering (PSE) literature and offer some critical insights. Third, we briefly review a series of mixed-integer programming (MIP) models we have recently developed to address limitations of existing approaches. We focus on a discrete-time formulation that allows us to address problems in all types of facilities, including facilities with different material handling constraints (mixing/splitting). Finally, we present an overview of two solution methods we recently developed, namely, a constraint propagation algorithm for the calculation of parameters that are used to generate four types of tightening constraints, and a series of reformulations. The proposed methods are applicable to a wide range of problems and lead to dramatic computational improvements.

363515 Stochastic Output Feedback Control of Nonlinear Systems with Probabilistic Uncertainties: Application to Control of Polymorphic Transformations in Batch Crystallization

Abstract:

Measurement noise, parametric and structural uncertainties, and exogenous disturbances are ubiquitous in complex dynamical systems. Uncertainties can lead to undesired variability of the system outputs and, as a result, a notable degradation of closed-loop performance. In this talk, a general framework will be presented for stochastic model predictive control (MPC) and state estimation of nonlinear systems with probabilistic, time-invariant uncertainties. Stochastic MPC with chance constraints (e.g., [1,2,3]) uses probabilistic descriptions of system uncertainties to realize acceptable levels of risk during system operation. Stochastic MPC enables shaping the probability distribution of system states, while ensuring the satisfaction of constraints with a desired probability level in a stochastic setting. This flexibility would have significant economic and safety implications for high performance operation of complex chemical and biological systems, which is often achieved in the vicinity of constraints. On the other hand, nonlinear estimation of the uncertain state variables enables output feedback application of stochastic MPC to (partly) suppress the effect of model mismatches and time-varying disturbances on control performance.

The key challenges in state estimation and control of nonlinear systems are the propagation of probabilistic uncertainties and reformulation of chance constraints in terms of computationally tractable expressions. In this work, generalized polynomial chaos (PC) theory [4,5] is used to present a probabilistic framework for stochastic nonlinear state estimation and control. The PC approach enables replacing the implicit mappings between the uncertain variables/parameters and system states with an expansion of orthogonal polynomials, whose statistical moments can be determined efficiently from the expansion coefficients (e.g., see [6,7,8]). Hence, PC expansions are computationally efficient surrogates for Monte Carlo-based approaches to perform uncertainty analysis for real-time estimation and control. A nonlinear Bayesian state estimation algorithm is presented that utilizes PC expansions to describe the evolution of state uncertainty, which is then used to maximize the posterior probability density function of the random state variables. The stochastic MPC approach is formulated to control the predicted probability densities of system states over a finite prediction horizon, while ensuring the satisfaction of constraints with a desired probability level. To obtain a computationally tractable stochastic optimal control problem, the individual chance constraints are transformed into explicitly convex second-order cone constraints for a general class of probability distributions with known mean and covariance [9].

The proposed stochastic output feedback control approach is applied for optimal control of the polymorphic transformation of L-glutamic acid in cooling batch crystallization [10]. Polymorphic transformation is of paramount importance in the specialty chemical and pharmaceutical industries, as such transformations can lead to undesired variations in the chemical and physical properties of the product crystals. The control problem in polymorphic transformations is to ensure consistent manufacturing of the desired polymorph in a stochastic setting due to time-invariant uncertainties in initial conditions and crystallization kinetic parameters. The simulation results indicate that the stochastic output feedback control approach enables effective shaping of the probability density of the performance index (i.e., the nucleation crystal mass to the seed crystal mass of crystals of the desired form). Stochastic MPC results in a significant reduction (approximately 30%) in the variance of probability density of the performance index, as compared to a nonlinear MPC approach that disregards the probabilistic system uncertainties. In addition, the chance constraints in stochastic MPC ensure that the system operation remains in the designated operating region defined in terms of state constraints, whereas in the nonlinear MPC approach the constraints are violated due to the system uncertainties.

[1] A. Schwarm and M. Nikolaou. Chance-constrained model predictive control. AIChE Journal, 45:1743-1752, 1999.

[2] D. Bernardini and A. Bemporad. Scenario-based model predictive control of stochastic constrained linear systems. In Proceedings of the 48thIEEE Conference on Decision and Control, 6333-6338, Shanghai, 2009.

[3] G. C. Calafiore and L. Fagiano. Robust model predictive control via scenario optimization. IEEE Transactions on Automatic Control, 58:219-224, 2013.

[4] N. Wiener. The homogeneous chaos. American Journal of Mathematics, 60:897-936, 1938.

[5] D. Xiu and G. E. Karniadakis. The Wiener-Askey polynomial chaos for stochastic differential equations. SIAM Journal of Scientific Computation, 24:619-644, 2002.

[6] Z. K. Nagy and R. D. Braatz. Distributional uncertainty analysis using power series and polynomial chaos expansions. Journal of Process Control, 17:229-240, 2007.

[7] J. Fisher and R. Bhattacharya. Linear quadratic regulation of systems with stochastic parameter uncertainties. Automatica, 45:2831-2841, 2011.

[8] A. Mesbah, S. Streif, R. Findeisen, and R. D. Braatz. Stochastic nonlinear model predictive control with probabilistic constraints. In Proceedings of the American Control Conference, Portland, 2014, in press.

[9] G. C. Calafiore and L. El Ghaoui. On distributionally robust chance-constrained linear programs. Journal of Optimization Theory and Application, 130:1-22, 2006.

[10] M. W. Hermanto, N. C. Kee, R. B. H. Tan , M. Chiu, and R. D. Braatz. Robust Bayesian estimation of kinetics for the polymorphic transformation of L-Glutamic acid crsytals. AIChE Journal, 54:3248-3259, 2008.

389951 Optimal Control of Uncertain Systems Using Dual Model Predictive Control (DMPC)

Abstract:

We present an adaptive model predictive controller (AMPC) for plants with uncertain parameters. Inspection of the objective function shows that the AMPC must include caution and probing in order to generate optimal controls and avoid singularities that may result when the model has close pole-zero cancellations. We show that DMPC avoids singularities, and provides performance than certainty equivalence AMPC by combing caution and probing in an optimal manner.

We consider linear ARX models in discrete time with Gaussian disturbances and quadratic performance cost. The least squares estimate provides the optimal estimate of the model at current time given past data. In certainty equivalence adaptive control this estimate is used directly to generate the controls by assuming that the estimate (being the optimal one) actually provides the best control. However, evaluation of the cost to go shows that optimal controls are functions of the current and future parameter-estimate error covariances. The current covariance matrix provides a rationale for caution, whereas the future covariance matrices can be reduced by probing. The optimal trade-off among control, caution, and probing is provided by the objective function.

In order to facilitate online solution of the DMPC problem we replace the recursive least-squares (Kalman) covariance update with an information matrix, which contains the same information, but provides the inverse. The nonlinear programming formulation remains nonlinear, but all nonlinear terms are bilinear. Although optimization problems constrained by bilinear equality constraints are not trivial to solve, there exist methods for solving this class of problem to global optimality that apply to the DMPC problem. We further simplify and speed up solution by exploiting the symmetry of the information matrix, reduce the number of variables through decomposition, and use of constraints for controls and outputs.

We demonstrate the application of DMPC to a finite impulse response (FIR) system. The advantage of the FIR is that the objective function reformulation is exact and the covariance predictions are explicit functions of the decision variables. Generalizations to Laguerre and ARX formulations follow but the computational expense is higher. The performance of the controller is compared with AMPC and fixed-parameter MPC through Monte Carlo simulations. We discuss the online solution of the optimization problem and the use of local instead of global solvers. The simulation examples show that the optimal excitation strategy provides better mean-square performance than approaches that rely on persistent excitation. Our controller excites the system just enough to obtain good estimates of the unknown parameters while minimizing the resulting adverse effect on output regulation. Simulation shows that the parameter estimates converge rapidly and that exact parameter estimates are obtained in the limit even though the excitation vanishes quickly.

373225 Assessing the Reliability of SOC Real-Time Optimization Methodologies

Abstract:

Increased industrial competition and the need for decreased energy consumption spent by production processes require the optimization of plant operations at optimum. The Real Time Optimization (RTO) technology is one of the best known and widely used techniques to improve profit and decrease the energy consumption of a process production. However, there is not a general consensus about the benefits of implementing this technology to increase the profit of process plants (Darby, Nikolaou, Jones, & Nicholson, 2011).

The classical way to design a RTO layer uses a first principles steady state model to describe the plant behavior and to optimize an economical objective function subject to this set of constraints. At each RTO iteration, some key model parameters are updated to reduce plant-model mismatch, using information provided by the plant measurements at actual operating points (Miletic and Marlin, 1998). The optimal economic set points, obtained by the RTO layer, are sent to Model Predictive Control (MPC) layer, which is responsible to keep the process under control at this target. A drawback related to this optimization approach is the low frequency of set point updates by the RTO layer. Since the RTO algorithm is performed under steady state conditions, the plant operates, in presence of disturbances, under suboptimum conditions until the detection of the next steady state. This has a clear disadvantage over other RTO methodologies, such as Economic MPC or Dynamic Real Time Optimization.

Considering this disadvantage, it is important that the control layer is more tightly coordinated with the RTO layer. In particular, the control layer must be robust regarding common disturbances affecting the plant profit. In other words the control layer should "obtain acceptable profit loss with constants set point values". That is the definition of self-optimizing control (SOC, Skogestad, 2000). The main idea behind the SOC is to choose a set of controlled variables that have set points values insensitive to disturbances, for instance, state variables that keep at active constraints despite presence of disturbances.

Several studies have developed methods to find optimal linear combinations of measured variables, which for artificial controlled variables that are insensitive to disturbances. These include the Exact Local method developed by Halvorsen and coworkers (2003), and the Null Space method by Alstad and Skogestad (2007), which uses the optimum output sensitivities to disturbances to find a matrix of measurements combinations. In this case, the artificial controlled variables present zero loss respect to the analyzed disturbances. Alstad and coworkers (2009) extended the Null space method, using extra measurements to reduce the loss assigned to measurement noise.

In this setting, the SOC methodology can be an alternative to reduce the low frequency of set points update in the classic RTO implementation (Jäschke and Skogestad, 2011). However, the practical implementation requires the solution of some challenges. The first one relates to the need to compute new solutions for large-scale nonlinear optimization problems for each disturbance analyzed, in order to compute the optimal sensibilities matrix necessary to the measurements combinations methods. An alternative for this problem could be to use the solver sIPOPT (Pirnay, López-Negrete and Biegler, 2012) to compute sensitivities with low computational cost, but this does not handle active set changes directly.

Another problem related to active set change is that it implies the need to modify the set of measured variables that must be controlled. For this reason, in the present work an algorithm is proposed to implement the SOC methodology in an MPC control, considering the choice of active constraints and linear combination of measured variables as controlled variables. The transformation matrix is updated in each RTO cycle, as are the manipulated set points and the MPC tuning. This is demonstrated on a detailed simulation of a multi-column distillation process. The results show better economic performance of the MPC with SOC compared with the classic MPC, and less change in the manipulated variables, thus indicating that it can be a good alternative to address the low frequency problem of the RTO algorithm.

REFERENCES

Alstad, V., & Skogestad, S. (2007). Null Space Method for Selecting Optimal Measurement Combinations as Controlled Variables. *Industrial & Engineering Chemistry Research*, *46*(3), 846–853. doi:10.1021/ie060285+

Alstad, V., Skogestad, S., & Hori, E. S. (2009). Optimal measurement combinations as controlled variables. *Journal of Process Control*, *19*(1), 138–148. doi:10.1016/j.jprocont.2008.01.002

Darby, M. L., Nikolaou, M., Jones, J., & Nicholson, D. (2011). RTO: An overview and assessment of current practice. *Journal of Process Control*, *21*(6), 874–884. doi:10.1016/j.jprocont.2011.03.009

Halvorsen, I. J., Skogestad, S., Morud, J. C., & Alstad, V. (2003). Optimal Selection of Controlled Variables †. *Industrial & Engineering Chemistry Research*, *42*(14), 3273–3284. doi:10.1021/ie020833t

Jäschke, J., & Skogestad, S. (2011). NCO tracking and self-optimizing control in the context of real-time optimization. *Journal of Process Control*, *21*(10), 1407–1416. doi:10.1016/j.jprocont.2011.07.001

Miletic, I. P., & Marlin, T. E. (1998). On-line Statistical Results Analysis in Real-Time Operations Optimization. *Industrial & Engineering Chemistry Research*, *37*(9), 3670–3684. doi:10.1021/ie9707376

Pirnay, H., López-Negrete, R., & Biegler, L. T. (2012). Optimal sensitivity based on IPOPT. *Mathematical Programming Computation*, *4*(4), 307–331. doi:10.1007/s12532-012-0043-2

Skogestad, S. (2000). Plantwide control: the search for the self-optimizing control structure. *Journal of Process Control*, *10*(5), 487–507. doi:10.1016/S0959-1524(00)00023-8

MPC-based scheduling of re-entrant line: some issues and extensions

Taeyeong Jung[1], Hong Jang[1] and Jay H. Lee[1]*

[1] Department of Biomolecular and Chemical Engineering, Korea Advanced Institute of Science and Technology, Deajeon, 305-701, Korea

Abstract

Model Predictive Control (MPC) – an on-line optimization-based control method - has been applied extensively in refineries and other chemical manufacturing processes. As its theory matured and computational capabilities improved, MPC's application domains expanded to include many new types of systems like mechanical devices and biomedical systems. In addition, it has also been proposed for solving certain scheduling problems in re-entrant manufacturing lines, like those found in semiconductor manufacturing fabs [1-3]. In these works, wafer flows in a fab have been modeled by the so called aggregated flow model (AFM), which is a continuous flow equation approximating what is inherently a discrete-event system. MPC is designed based on the AFM so as to follow a given 'fab-out' profile, which acts as a main setpoint trajectory, while maintaining desired inventory levels between processing steps. These studies demonstrated the effectiveness of MPC in controlling the output wafer quantity at a desired level on a simplified fab model called "Intel mini-fab" benchmark model [1].

As we were working with a major semiconductor manufacturer to implement this approach on their fabs, we found that there are a number of additional complicating factors not addressed in the previous studies. They include:

1. Production planning gives desired wafer quantities (of different device types) by a certain due date. This demand information from the planning layer must be translated into target fab-out quantity profiles. When multiple product types are involved, it is not clear how best to do this.
2. Significant uncertainties exist in a real fab environment, such as unexpected machine breakdowns/maintenances and last minute demand changes. While the feedback mechanism of MPC does help in counteracting them, they also call for continuous adjustment of the target production profiles as well as appropriate choices of inventory levels ('wafers-in-process' (WIPs)). Whereas too high levels of inventories will increase the cycle time and hurt the agility of the manufacturing, too low levels of inventories will hurt its robustness to deal with unexpected events.

3. Generally, semiconductor manufacturing lines involve hundreds of processing steps and total cycle time (from the start to the end) can exceed a month. As one of the major input for control is the raw wafer release into the line, it is important to use a large enough prediction and control horizon in the MPC in order to get a good performance. This, combined with the fact that AFM for a real fab is generally of very high order (as it needs to model the flow of multiple product types through hundreds of processing steps), puts significant burden on the computation.

In this work, we address the above-stated issues in applying MPC to fab-wide scheduling. In order to keep the complexity reasonable for thorough analysis, we also use the Intel mini-fab model for demonstration but we modify it to have a number of additional features like multiple product types and uncertain demands and process parameters. We propose a novel optimization formulation to relax the constant set-point assumption so that changing demands can be met more flexibly. In this optimization framework, objectives are to meet the demand requirement by the due date while keeping the production linearity in the system as much as possible. Various constraints including the buffer limit on each processing step and machine capacity utilization are also considered. The resulting scheduler not only suggests a future production plan on a (operator-)shift-by-shift basis but also solves the constraint-aware production optimization and the in-process inventory level control simultaneously at each scheduling instance. Simulations on the modified Intel mini-fab model will demonstrate that the proposed MPC-based scheduler automatically fixes the production plan under emergent machine breakdowns and so that the lost production is compensated for in the future operator shifts in order to meet the required demand by the due date. Additionally, as a way to keep the computational complexity manageable, we test several move blocking strategies to identify their advantages and disadvantages in this problem.

Keywords: economic model predictive control, dynamic real-time optimization, re-entrant line scheduling, move blocking strategy, Intel mini-fab benchmark model, aggregated flow model

Reference

[1] Vargas-Villamil, F. D., Rivera, D. E., & Kempf, K. G. (2003). A hierarchical approach to production control of reentrant semiconductor manufacturing lines. Control Systems Technology, IEEE Transactions on, 11(4), 578-587.

[2] Wang, W., & Rivera, D. E. (2008). Model predictive control for tactical decision-making in semiconductor manufacturing supply chain management. Control Systems Technology, IEEE Transactions on, 16(5), 841-855.

[3] Nandola, N. N., & Rivera, D. E. (2013). An improved formulation of hybrid model predictive control with application to production-inventory systems. Control Systems Technology, IEEE Transactions on, 21(1), 121-135.

STOCHASTIC MODEL PREDICTIVE CONTROL FOR OUTPUT TRACKING WITH BOUNDED CONTROL INPUTS

Bruno F. Santoro, University of São Paulo, São Paulo, Brazil
Darci Odloak, University of São Paulo, São Paulo, Brazil

Introduction

In the last 40 years, Model Predictive Control (MPC) has evolved from an industrial heuristic approach to a mature, well-based theory. From a practical standpoint, its use nowadays is widespread in the chemical industry, especially in the oil refining and petrochemical branches. On the other hand, from a theoretical point of view, many important stability questions have been addressed and solved [1]. Necessary and sufficient conditions for the stability of nominal controllers are completely understood.

An immediate consequence of this effort was the MPC extension to uncertain models, since plant-model mismatch can seriously deteriorate the performance of the closed-loop system. The earliest attempts to improve robustness of MPC, which date back from the 1990s, usually considered that the parameters describing the plant belong to a bounded set. This hypothesis allows worst-case analyses, which have been used to guarantee stability and feasibility but may be over conservative in terms of performance. More recently, there has been a trend to incorporate statistical knowledge about disturbances in MPC, reducing conservativeness without risking process safety.

The choice of a disturbance description is usually a trade-off between mathematical simplicity and realistic representation. Considering only linear models with independent and identically distributed noise, the main distinction is between multiplicative uncertainty, affecting matrices A, B or C, ([2], [3]) and additive noise, which directly affects states or outputs ([4], [5], [6]). The noise distribution is also an important hypothesis: if its support is bounded, standard robust MPC techniques, such as robust invariant sets, may be applied straightforwardly ([7],[8]).

Most of the literature so far only considers the regulator problem., The work in [9] is an exception as they present a set point tracking algorithm for output measures corrupted by noise but deterministic state evolution. However, their work does not handle the more general problem of tracking, in which the state is a random variable.

The main objective in this article is to track possibly time-varying set points for systems subject to additive process noise. It is known [10] that one way to achieve offset free tracking is to use a velocity description of the system, which means that Δu replaces u as the input. Also, in order to be implementable in practical situations, calculated control actions must respect its hard constraints. However, the situation is different concerning state constraints: relaxing some of these to a probabilistic counterpart is probably an opportunity window for some performance improvements, because robust model predictive control is usually based on worst-case considerations. One approach that takes advantage of this observation is [7], where the problem solved is to synthesize a strongly feasible MPC with probabilistic state constraints. This first predicted control move forces, at each sampling time, that the state on the next instant belongs to a "stochastic robust controlled invariant" (SRCI), which is a refinement of traditional robust invariant set: besides being invariant when faced with all possible disturbances, the probability of satisfying the state constraint in the future is also bounded. We incorporate this first-step approach to our optimization problem, defining probabilistic constraints relating states and set points.

Due to the presence of disturbances, the open-loop calculation of future control moves may cause render the control problem infeasible in the presence of state constraints [11]. Also, open-loop control could be excessively conservative when looking for inputs that stabilize the system for all possible disturbance realizations [1]. Therefore, it is advantageous to consider causal feedback policies, but the optimization is generally intractable if performed over the whole class of causal policies. A possible approach to find a suboptimal solution is to parameterize the input as an affine function of the state. As shown in [12], this strategy is equivalent to use a parameterization concerning the process disturbance. Moreover, the resulting optimization problem is convex and hence tractable. We propose a parameterization similar to [12] but regarding Δu and not u.

The structure of this paper is as follows: in the next Section we propose a formulation for a stochastic model predictive control for output tracking, with hard input constraints and probabilistic state constraints, based on noise affine feedback policies. In Section III, the controller performance is evaluated through simulation of a non-linear system. Finally, Section IV summarizes the results obtained so far and indicates next steps.

Controller Development

Consider a linear time-invariant system subject to process noise:

$$\begin{aligned} &x_p(k+1) = A_p x_p(k) + B_p u(k) + F_p w(k) \\ &y(k) = C_p x_p(k) \\ &x_p(k) \in \mathbb{R}^{nx}, u(k) \in \mathbb{R}^{nu}, y(k) \in \mathbb{R}^{ny}, w(k) \in \mathbb{R}^{nw} \end{aligned} \tag{1}$$

where {w(k)}} is a sequence of independent, uniformly distributed random variables with bounded support. Full-state measurement is assumed.

A possible velocity description to achieve offset-free tracking considers an augmented state defined as $x_v(k) = \left[\Delta x_p(k)^T \quad y(k)^T\right]^T$, where $\Delta x_p(k) = x_p(k) - x_p(k-1)$. Thus, it is clear that (1) is equivalent to

$$\begin{bmatrix} \Delta x_p(k+1) \\ y(k+1) \end{bmatrix} = \begin{bmatrix} A_p & 0 \\ C_p A_p & I_{ny} \end{bmatrix} \begin{bmatrix} \Delta x_p(k) \\ y(k) \end{bmatrix} + \begin{bmatrix} B_p \\ C_p B_p \end{bmatrix} \Delta u(k) + \begin{bmatrix} F_p \\ C_p F_p \end{bmatrix} \Delta w(k) \tag{2}$$

$$y(k) = \begin{bmatrix} 0 & I_{ny} \end{bmatrix} \begin{bmatrix} \Delta x_p(k) \\ y(k) \end{bmatrix}$$

$$\Leftrightarrow$$

$$\begin{aligned} &x_v(k+1) = A_v x_v(k) + B_v \Delta u(k) + F_v \Delta w(k) \\ &y(k) = C_v x_v(k) \end{aligned}$$

Note that the proposed incremental form causes $\Delta w(k)$ and $\Delta w(k-1)$ to be function of $w(k-1)$, therefore the process $\{\Delta w(k)\}$ is not composed of independent random variables.

Using this formulation, the problem that we would like to solve is

Problem 1
$$\min_{\Delta u(k),\cdots,\Delta u(k+m-1)} \mathbb{E}_{x(k)} \left[\sum_{i=1}^{p} \left\| y(k+i) - y_{sp} \right\|_Q^2 + \sum_{i=0}^{m-1} \left\| \Delta u(k+i) \right\|_R^2 \right] \tag{3}$$

subject to

$$\left| \Delta u(k+i) \right| \le \Delta u_{\max}, i = 0,\cdots,m-1 \tag{4}$$

$$u_{\min} \le u(k+i) \le u_{\max}, i = 0,\cdots,m-1 \tag{5}$$

$$\mathbb{P}\left[g_j^T x_v(k+1) \le h_j \right] \ge 1 - \alpha_j, \quad j = 1,\cdots,r; \tag{6}$$

Q and R are symmetric positive-definite tuning matrices, p is the prediction horizon, m is the control horizon, u_{min}, u_{max} and Δu_{max} are physical constraints on the inputs and control moves. $\mathbb{E}_{\mathfrak{X}(k)}[\bullet]$ is the conditional expectation given $\mathfrak{X}(k)$, which is the set of state observations up to time k, $\mathfrak{X}(k)=\{x_v(0),\cdots,x_v(k)\}$. $\mathbb{P}[\bullet]$ is the probability, g_j^T and h_j are parameters that define r linear constraints over the state, which have to be satisfied at a confidence level of $(1-\alpha_j)$.

Input Parameterization

Problem 1 is generally intractable if the optimization is performed over π, the whole class of causal policies. Inspired by the parameterization of inputs as a function of process disturbance proposed in [12], we propose an adaptation to the case of incremental input as follows:

$$\begin{bmatrix}\Delta u(k)\\ \Delta u(k+1)\\ \vdots \\ \Delta u(k+m-1)\end{bmatrix}=\begin{bmatrix}d(k)\\ d(k+1)\\ \vdots \\ d(k+m-1)\end{bmatrix}+\begin{bmatrix}0 & 0 & \cdots & 0 & 0\\ g(k+1,k) & 0 & \cdots & 0 & 0\\ \vdots & \vdots & \ddots & \vdots & 0\\ g(k+m-1,k) & g(k+m-1,k+1) & \dots & g(k+m-1,k+m-2) & 0\end{bmatrix}\begin{bmatrix}\Delta w(k)\\ \Delta w(k+1)\\ \vdots \\ \Delta w(k+m-1)\end{bmatrix}$$

$$\Delta\bar{U}(k)=\bar{D}(k)+\bar{G}(k)\Delta\bar{W}(k) \tag{7}$$

The decision variables of the optimization problem are the affine terms $d(k+i)\in\mathbb{R}^{nu}$ and the feedback gains $g(k+i,k+j)\in\mathbb{R}^{nu\times nx}$.

Objective Function

The objective function actually used in this work is

$$\min_{d,G}\mathbb{E}_{\mathfrak{X}(k)}\left[\sum_{i=1}^{p}\left\|y(k+i)-y_{tar}\right\|_Q^2+\sum_{i=0}^{m-1}\left\|\Delta u(k+i)\right\|_R^2\right]+V_O\left(y_{tar},y_{sp}\right) \tag{8}$$

The difference between (8) and (3) is the term $V_O(y_{tar},y_{sp})$, which is an offset function. We consider the same strategy of [13] to penalize the deviations between the real and artificial set points. y_{tar} is a new decision variable of the optimization problem, corresponding to an artificial reference that will be steered as close as possible to the true set point. It is advantageous to introduce an artificial set point since the original one may not be reachable when constraints are present and also it could not correspond to a steady state.

We consider that the prediction horizon is not necessarily equal to the control horizon, based on a dual controller framework, where the prediction of future inputs after the control horizon m is a simple feedback law:

$$\Delta u(k+j)=K\Delta w(k+j-1),\quad m<j<p \tag{9}$$

where $K\in\mathbb{R}^{nu\times nx}$ is calculated off-line such that (A_p+B_pK) is stable.

Due to the assumption of measured states, the expected value of the objective function may be expressed in terms of current state, future control moves and noise moments of first and second order.

Constraints

Bounds on the maximum allowed control moves follow naturally from the chosen parameterization, in a similar spirit to [6]. Let w_{max}, $w_{min} \in \mathbb{R}^{nw}$ be respectively the maximum and minimum values of the noise, which are finite due to the bounded support assumption. Note that the absolute values of w_{max} and w_{min} may be different, since the noise distribution is not necessarily symmetric and zero-mean. Therefore, a conservative version of (4) is given by

$$|d_i| + \|G_i\|_1 (w_{max} - w_{min}) \leq \Delta u_{max}, i = 1, \cdots, m \cdot nu \tag{10}$$

where d_i is the i-th element of vector $\bar{D}(k)$ and G_i is the i-th row of matrix $\bar{G}(k)$.

Regarding (5), it suffices to notice that future input values and control moves are related through

$$\begin{bmatrix} u(k) \\ u(k+1) \\ \vdots \\ u(k+m-1) \end{bmatrix} = \begin{bmatrix} u(k-1) \\ u(k-1) \\ \vdots \\ u(k-1) \end{bmatrix} + \underbrace{\begin{bmatrix} I & 0 & \cdots & 0 \\ I & I & \cdots & 0 \\ \vdots & \vdots & \ddots & \vdots \\ I & I & \cdots & I \end{bmatrix}}_{M} \begin{bmatrix} \Delta u(k) \\ \Delta u(k+1) \\ \vdots \\ \Delta u(k+m-1) \end{bmatrix} \tag{11}$$

Let

$$U_{max}(k) = \left[(u_{max} - u(k-1))^T \ \cdots \ (u_{max} - u(k-1))^T \right]; U_{min}(k) = \left[(u_{min} - u(k-1))^T \ \cdots \ (u_{min} - u(k-1))^T \right] \tag{12}$$

Then, (5) is equivalent to

$$U_{min}(k) \leq M \Delta \bar{U}(k) \leq U_{max}(k) \tag{13}$$

Considering the extreme values of the noise realizations, (13) is assured if

$$\begin{aligned} &M\bar{D}(k) + M\bar{G}^+(k)|\bar{W}_{max}| + M\bar{G}^-(k)|\bar{W}_{min}| \leq U_{max}(k) \\ &U_{min}(k) \leq M\bar{D}(k) - M\bar{G}^+(k)|\bar{W}_{max}| + M\bar{G}^-(k)|\bar{W}_{min}| \end{aligned} \tag{14}$$

where $\bar{G}^+(k)$ and $\bar{G}^-(k)$ are positive (element-wise) matrices such that $\bar{G}(k) = \bar{G}^+(k) - \bar{G}^-(k)$ and $\bar{W}_{max} = \left[w_{max}{}^T \ \cdots \ w_{max}{}^T \right]^T, \bar{W}_{min} = \left[w_{min}{}^T \ \cdots \ w_{min}{}^T \right]^T \in \mathbb{R}^{m \cdot nw}$

In order to accommodate constraints over the state and the artificial set point, let Ψ be an augmented state such that $\Psi(k) = \left[x_v(k)^T \ \ y_{tar}{}^T \right]^T$. In order to use standard results on invariant sets, it is also necessary to define a dynamic evolution for this augmented state. The dynamics is given by

$$\begin{bmatrix} x_v(k+1) \\ y_{tar}(k+1) \end{bmatrix} = \begin{bmatrix} A_v & 0 \\ 0 & I_{nx+ny} \end{bmatrix} \begin{bmatrix} x_v(k) \\ y_{tar}(k) \end{bmatrix} + \begin{bmatrix} B_v \\ 0 \end{bmatrix} \Delta u(k) + \begin{bmatrix} F_v \\ 0 \end{bmatrix} \Delta w(k) \tag{15}$$

Notice that $x_v(k)$ and $y_{tar}(k)$ may be related through constraints but not through their evolution in time, since $y_{tar}(k)$ assumes a constant, independent value at each sample time. Moreover, since we are assuming measured state, the value of $w(k-1)$ is known at time k. General probabilistic constraints over linear combinations of Ψ are given by

$$\mathbb{P}\left[g_j^T \Psi(k+1) \leq h_j \right] \geq 1 - \alpha_j, \quad j = 1, \cdots, r; \ k \in \mathbb{N}^* \tag{16}$$

Let Δ be an upper bound of the 1-norm of $w \in \mathfrak{W}$ and consider the conservative approach taken by [7], where constraints (16) are replaced with

$$\mathbb{P}\left[g_j^T \Psi(k+1) \leq h_j \mid \Psi(k)\right] \geq 1-\alpha_j, \quad j=1,\cdots,r; \;\; k \in \mathbb{N}^* \tag{17}$$

By partitioning g_j^T as $\left[g_{1,j}^T \;\; g_{2,j}^T\right]$, where each component corresponds to the distinct elements of Ψ and following the same reasoning as in [7], we conclude that (17) is equivalent to (18)

$$g_{1,j}^T\left(A_v x_v(k)+B_v \Delta u(k)-F_v w(k-1)\right)+g_{2,j}^T y_{tar}(k) \leq h_j-\mathbb{F}_{\left(g_{1,j}^T F_v\right) w}^{-1}\left(1-\alpha_j\right) \tag{18}$$

where $F_{g_j^T w}^{-1}(\cdot)$ is the left quantile function of the random variables $g_j^T F_v w$.

It is now possible to look for the MSRCI of system (15), subject to constraint (18). The result may be a polyhedron or, if it is not the case, a polyhedral approximation of the real set. It may be represented by $S_{SRCI}^* = \left\{\Psi \in \mathbb{R}^{2(nx+ny)} \mid \tilde{S}\Psi \leq \tilde{c}\right\}$ and matrix $\tilde{S}$ may be decomposed as $\tilde{S}=\left[\tilde{S}_1 \;\; \tilde{S}_2\right]$. Finally, since the first control move is not subject to feedback, the invariance constraint that should be added to the formulation sufficient to guarantee that $\Psi(k+1) \in S_{SRCI}^*, \forall w \in \mathfrak{W}$, is

$$\tilde{s}_{1,j}^T\left(A_v x_v(k)+B_v \Delta u(k)-F_v w(k-1)\right)+\tilde{s}_{2,j}^T y_{tar}(k) \leq \tilde{c}_j-\left\|\tilde{s}_{1,j}^T\right\|_\infty \left\|F_v\right\|_\infty \Delta, \quad j=1,\cdots,\tilde{r}; \tag{19}$$

where $\tilde{s}_{i,j}^T$ are the rows of $\tilde{S}_i$ (i = 1, 2). Notice that (19) is linear with respect to all the decision variables, there are no concerns about convexity issues and indeed the resulting problem is defined as follows:

Problem 2:

$$\min_{d,G} \mathbb{E}_{X(k)}\left[\sum_{i=1}^{p}\left\|y(k+i)-y_{tar}\right\|_Q^2+\sum_{i=0}^{m-1}\left\|\Delta u(k+i)\right\|_R^2\right]+V_O\left(y_{tar}, y_{sp}\right) \tag{20}$$

subject to (7), (10), (14), (18) and (19)

Feasibility and stability analysis

The assumption of bounded noise support allows one to use traditional robust control approaches to assure closed-loop stability. As previously defined, we have imposed that the state in the subsequent sampling time is included in a robust positive invariant set (19). Therefore, if the optimization problem is feasible at time k, it will remain feasible for all possible future noise realizations and the state is confined within the invariant set.

Simulation example

This section assesses the performance of the proposed controller in comparison to other possible control strategies, namely a finite horizon MPC and a truncated LQR, using a numerical example. The stochastic controller (Problem 2) was formulated using YALMIP [14] and solved with GUROBI® [15]. The finite horizon MPC controller is based on the same incremental description of the system as the stochastic controller, but ignores disturbances when calculating the predictions. The LQR controller steady state gain K_{LQR} was calculated using matrices Q and R. The inputs calculated with this gain were saturated before implementation considering the input and control move bounds.

The methodology of the simulation consists of 1000 Monte Carlo repetitions, with the same noise realization being provided to three controllers. Mean values for each sampling time are plotted to show that the outputs calculated by the stochastic controller converge in mean to the set point. Moreover, histograms of variables' distributions are presented to indicate that the proposed controller achieves lower output variances with less intense control moves.

The system considered here is an example adapted from [16], representing the control of a binary ethanol-water distillation column. The controlled variables are overhead ethanol mole fraction, side stream ethanol mole fraction and temperature of tray number 19 (ºC). The manipulated variables are the reflux flow rate (gpm), side stream product flow rate (gpm) and the reboiler stream pressure (psig), while the feed flow rate (gpm) and the feed temperature (ºC) are disturbances. Hereafter, the order of all variables is the same as in this paragraph.

The model presented in [16] is composed of first and second order transfer functions plus time delays, identified from experiments with a real column. Since our approach relies on the invariant set construction, it is not computationally feasible to consider systems with large state space descriptions. Therefore, we have simplified the original model by disregarding the time delays. The model was discretized using a sampling time of 1 minute.

Regarding system constraints, we imposed the following limits: $u_{min} = [0.068 \quad 0.00694 \quad 15.6]$, $u_{max} = [0.245 \quad 0.1 \quad 34]$ and $\Delta u_{max} = [0.02 \quad 0.02 \quad 5]$. Process noise is described by Normal distributions with zero-mean and standard deviation of 0.07 gpm for the feed flow rate and 8ºC for the temperature. To keep the realizations bounded, we have truncated the values at three standard deviations.

A soft constraint was added, forcing the overhead ethanol mole fraction (y_1) to be greater than its artificial set point with 90% of probability. In terms of set point tracking, the system was to be kept around the origin ($[0.7 \quad 0.52 \quad 92]$) for the first 100 minutes and in another steady state ($[0.64 \quad 0.4 \quad 95]$) for the second part. The same weights $Q = [10^2 \quad 10^2 \quad 10^{-2}]$ and $R = [10^{-1} \quad 10^{-1} \quad 10^{-3}]$ were used for all controllers, to standardize the relative importance of the variables. Prediction and control horizon were set at 30 and 3, respectively.

Figure 1: Mean of system outputs (Proposed controller in red, MPC in blue, LQR in black, set point in green)

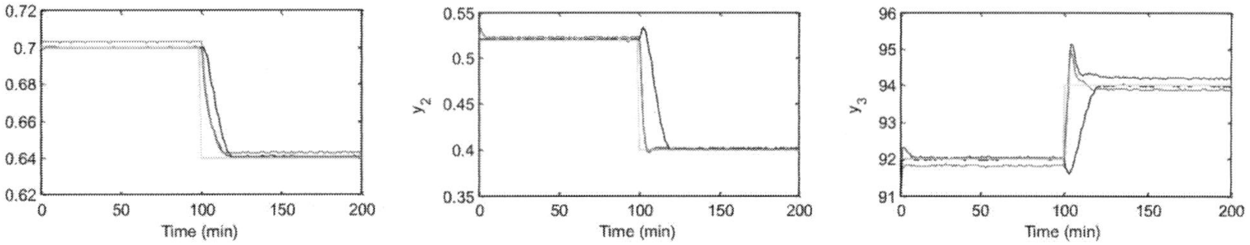

Figure 2: Mean of system inputs (Proposed controller in red, MPC in blue and LQR in black)

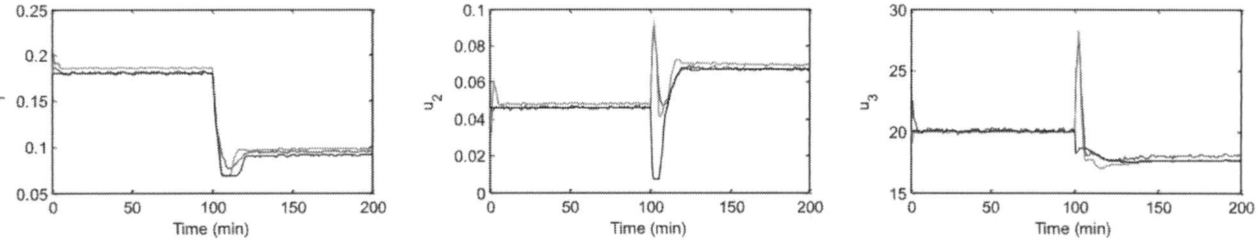

The noise realizations compensate each other since their distribution is symmetric. The main difference between the results of the proposed controller and literature benchmarks is the presence of a back-off in the overhead ethanol mole fraction (y_1) in relation to its set point, as a consequence of the soft constraint. In the first steady state of the simulation, from 10 to 100 min, there were, on average, 9.66% of violations of this constraint, and in the second steady state, from 120 to 200 min, the value was 9.91%. Since the violation percentage of the stochastic controller is close to the prescribed value of 10%, it shows that the approach is not excessively conservative.

It has been shown in Figure 1 that the inclusion of the soft constraint drives the system away from the set point. Even in the presence of such constraint, the proposed controller still outperforms the controllers that are oblivious to the constraint. As a performance metric, we propose the weighted sum of square errors using Q as weighting matrix. The average over all Monte Carlo repetitions of this index provides the following results: 6.9 for the proposed controller, 7.96 for the standard MPC and 20.81 for the truncated LQR. This is a consequence of the reduction of outputs variances, as shown in Figure 3 and Table 1.

Figure 3: Distribution of system outputs between 120 and 200 min for MPC (blue), the proposed controller (red), MPC (blue) and Truncated LQR (black). The green bar indicates the set point.

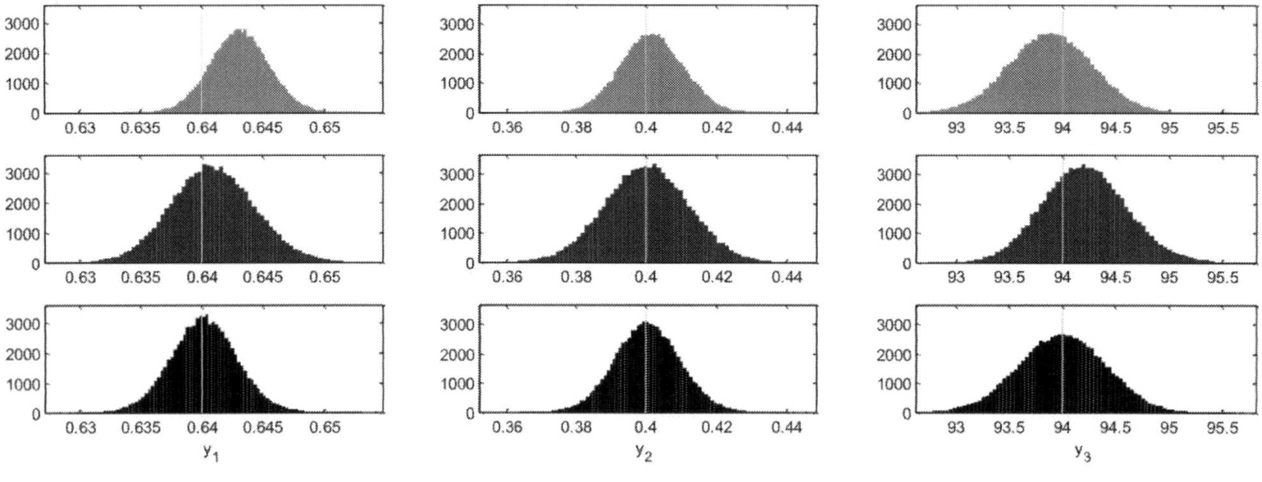

Table 1: Output standard deviations

	First steady state (10<k<100)			Second steady state (120<k<200)		
Outputs	Proposed controller	Standard MPC	Truncated LQR	Proposed controller	Standard MPC	Truncated LQR
y_1	0.0024	0.0038	0.0028	0.0024	0.0035	0.0027
y_2	0.0089	0.0123	0.0095	0.0089	0.012	0.0093
y_3	0.395	0.423	0.411	0.394	0.399	0.413

From Table 1, we verify that the proposed controller yields lower variances than the benchmarks in both steady states. The reduction is of about 30% when compared to the standard MPC and 10% to the truncated LQR. Since the steady states are far from the constraints, the truncated LQR actually corresponds to the optimal response in many time steps, however the bounds on Δu_{max} may turn it suboptimal, which justifies the slight gain in performance of the proposed approach.

Conclusions

This article presented the development of a stochastic predictive controller for output tracking. It was shown that it presents two main innovative contributions: a model representation following a velocity framework and probabilistic soft constraints regarding the states and the artificial set point. We have illustrated through a simulated example that the proposed controller is able to track set points and it outperforms conventional approaches by reducing outputs' variance.

References

1. Mayne, D.Q., Rawlings, J. B., Rao, C. V. and Scokaert, P. O. M. (2000), "Constrained model predictive control : Stability and optimality," *Automatica*, vol. 36, pp. 789–814.
2. Cannon, M., Kouvaritakis, B. and Wu, X. (2009), "Probabilistic Constrained MPC for Multiplicative and Additive Stochastic Uncertainty", *IEEE Trans. Automat. Contr.*, vol. 54, pp. 1626–1632.
3. Primbs, J. A. and Sung, C. H. (2009), "Stochastic Receding Horizon Control of Constrained Linear Systems With State and Control Multiplicative Noise", *IEEE Trans. Automat. Contr.*, vol. 54, pp. 221–230.
4. Cannon, M., Cheng, Q., Kouvaritakis, B. and Raković, S. V. (2012), "Stochastic tube MPC with state estimation", *Automatica*, vol. 48, pp. 536–541.
5. Ramponi, F., Chatterjee, D., Milias-Argeitis, A., Hokayem, P. and Lygeros, J. (2010), "Attaining Mean Square Boundedness of a Marginally Stable Stochastic Linear System With a Bounded Control Input", *IEEE Trans. Automat. Contr.*, vol. 55,pp. 2414–2418.
6. Hokayem, P., Cinquemani, E., Chatterjee, D., Ramponi, F. and Lygeros, J. (2012) "Stochastic receding horizon control with output feedback and bounded controls", *Automatica*, vol. 48 pp. 77–88.
7. Korda, M., Gondhalekar, R., Cigler, J. and Oldewurtel, F. (2011), "Strongly feasible stochastic model predictive control", *IEEE Conf. Decis. Control Eur. Control Conf.*, pp. 1245–1251.
8. Kouvaritakis, B., Cannon, M., Raković, S. V. and Cheng, Q. (2010), "Explicit use of probabilistic distributions in linear predictive control", *Automatica*, vol. 46, pp. 1719–1724.
9. Couchman, P. D., Cannon, M. and Kouvaritakis, B. (2006), "Stochastic MPC with inequality stability constraints", *Automatica*, vol. 42, pp. 2169–2174.
10. Maeder, U., Borrelli, F. and Morari, M. (2009), "Linear offset-free Model Predictive Control", *Automatica*, vol. 45, pp. 2214–2222.
11. Scokaert, P. O. M. and Mayne, D. Q. (1998), "Min-Max Feedback Model Predictive Control for Constrained Linear Systems", *IEEE Trans. Automat. Contr.*, vol. 43, pp. 1136–1142.
12. Goulart, P.J., Kerrigan, E.C. and Maciejowski, J.M. (2006), "Optimization over state feedback policies for robust control with constraints", *Automatica*, vol. 42, pp. 523–533.
13. Ferramosca, A., Limon, D., González, A.H., Alvarado, I. and Camacho, E.F. (2012), "Robust MPC for tracking zone regions based on nominal predictions", *J. Process Control*, vol. 22, pp. 1966–1974.
14. Löfberg, J. (2004), "YALMIP : A toolbox for modeling and optimization in MATLAB," in *IEEE International Symposium on Computer Aided Control Systems Design*, pp. 284–289.
15. Gurobi Optimization, Inc. (2013), "Gurobi Optimizer Reference Manual.".
16. Ogunnaike, B.A., Lemaire, J.P., Morari, M. and Ray, W.H. (1983), "Advanced Multivariable Control of a Pilot-Plant Distillation Column," *AIChE Journal*, vol. 29, pp. 632–640.

373029 Output Feedback Lyapunov Model Predictive Control of Stochastic Nonlinear Systems

Abstract:

Output Feedback Lyapunov Model Predictive Control of Stochastic Nonlinear Systems

Summary

The challenges of strong nonlinearities, system constraints, disturbances, and uncertainty are ubiquitous in the modern control landscape, from chemical reactors to fighter jet airframes.

In addressing these challenges, controllers based on Nonlinear Model Predictive Control (NMPC) [13] have enjoyed huge popularity due to their explicit constraint-handling ability [14], and ease of deployment. This controller anticipates the plant's response to control actions to devise the optimal future control plan, which is then repeatedly re-optimized as real-time information is incorporated. When NMPC is combined with Lyapunov analysis (LMPC) [15], the designer can guarantee feasibility and stability [12] prior to implementation, provided the system is within a known space of initial conditions.

However, without adequate modelling of disturbances, performance will be degraded. Since many process disturbances are inherently stochastic, control laws designed to attenuate stochastic disturbances can achieve better performance than conventional controllers based on 'worst case' bounded disturbance models.

In one approach, a stochastic LMPC framework is used to achieve optimal plant control. This controller accounts for system constraints as well as the distributions of probabilistic variables, but is also re-optimized as random variable sample paths are realized [8].

In [1], an LMPC formulation was developed for a class of stochastic nonlinear systems with affine input and additive vanishing noise. The paper first addressed the system's closed-loop stability and boundedness properties subject to discrete control implementation. The resulting system was stable in the sense that it had a known probability of being contained within given bounds using sufficiently small hold times. It was then shown that the Lyapunov function decay constraints were always feasible and that receding horizon implementation of the above scheme would provide stability in probability guarantees.

In many applications, the model's states are not measured (for technical or economic reasons), but can sometimes be inferred from the response of the system through the use of a state observer. For linear systems, the observer's design is exact, separable, and has asymptotically stable error dynamics [7]. Even for a stochastic linear system, filtering can typically yield accurate observers. In a nonlinear setting, techniques that rely on approximate local linearization have seen the most use [6], but the need for nonlinear observers with globally stable error has led to results on observer designs for completely feedback linearizable systems [5].

In [2] it was shown that systems in stochastic output feedback canonical form admit global state estimators. Using integrator backstepping and stochastic Lyapunov techniques, the connected system was stabilized in probability. The analysis also shows that the observer's error dynamics, which are in effect being afflicted by the random processes of the plant, can be designed to be asymptotically stable in probability as well.

In the present paper, we consider the problem of an output feedback Lyapuov-based MPC design that recognizes the stochastic nature of the system. To this end, we first define a state observer for systems diffiomorphic [4] to the specially structured stochastic output feedback form [3]. Then, we determine how the risk margins for stability and boundedness of the estimated states are affected by the observer dynamics, and how the dynamics of the controlled system under output feedback differ from a state feedback configuration. Lastly, we present a unified output feedback stochastic LMPC implementation. Simulation results are presented to illustrate the efficacy of the proposed results, using an isothermal reactor with sequential reaction kinetics.

References

[1] M. Mahmood, P. Mhaskar. Lyapunov-based model predictive control of stochastic nonlinear systems. *Automatica*, 48(9):2271-2276, 2012.

[2] H. Deng, M. Krstic. Output-feedback stochastic nonlinear stabilization. *IEEE Transactions on Automatic Control*, 44(2):328-333, 1999.

[3] S. Liu, Z. Jiang, J. Zhang. Global output-feedback stabilization for a class of stochastic non-minimum-phase nonlinear systems. *Automatica*, 44(8):1944-1957, 2008.

[4] Z. Pan. Differential geometric condition for feedback complete linearization of stochastic nonlinear system. *Automatica*, 37(1):145-149, 2001.

[5] A. Krener and W. Respondek. Nonlinear Observers with Linearizable Error Dynamics. *SIAM Journal of Control and Optimization*, 23(2):197-216, 1985.

[6] S. Julier and J. Uhlmann. Unscented Filtering and Nonlinear Estimation. *Proceedings of the IEEE*, 92(3):401-422, 2004.

[7] C. Chen, Linear Systems Theory and Design, 3rd Edition. *Oxford University Press*, New York. Chapter 8, Page 253, 1999.

[8] M. Cannon, B. Kouvaritakis, and X. Wu. Probabilistic Constrained MPC for Multiplicative and Additive Stochastic Uncertainty. *IEEE Transactions on Automatic Control*, 54(7):1626–1632, 2009.

[9] R. Chabour and M. Oumoun. On a universal formula for the stabilization of control stochastic nonlinear systems.

Stochastic Analysis and Applications, 17:359–368, 1999.

[10] H. Deng, M. Krstic, and R. J. Williams. Stabilization of stochastic nonlinear systems driven by noise of unknown covariance. *IEEE Transactions on Automatic Control*, 46(8):1237–1253, 2001.

[11] D. V. Hessem and O. Bosgra. Stochastic closed-loop model predictive control of continuous nonlinear chemical processes. *Journal of Process Control*, 16(3):225–241, 2006.

[12] M. Mahmood and P. Mhaskar. Enhanced stability regions for model predictive control of nonlinear process systems. *AIChE J.*, 54:1487–1498, 2008.

[13] D. Q. Mayne. Nonlinear model predictive control: An assessment. In *Proceedings of 5th International Conference on Chemical Process Control*, Pages 217–231, Tahoe City, CA, 1997.

[14] D. Q. Mayne, J. B. Rawlings, C. V. Rao, and P. O. M. Scokaert. Constrained model predictive control: Stability and optimality. *Automatica*, 36:789–814, 2000.

[15] P. Mhaskar, N. H. El-Farra, and P. D. Christofides. Techniques for uniting Lyapunov-based and model predictive control. In Rolf Findeisen, Frank Allguwer, and Lorenz Biegler, Editors, *Assessment and Future Directions of Nonlinear Model Predictive Control*, Volume 358 of *Lecture Notes in Control and Information Sciences*, Pages 77–91. Springer Berlin / Heidelberg, 2007.

Moving Horizon Estimation using Carleman Linearization and Sensitivity Analysis

Negar Hashemian, Antonios Armaou

In many processes, the existence of unmeasurable states necessitates the design of an observer. Moving Horizon Estimation (MHE) is a method to estimate the unknown states in systems exposed to noise signals and exogenous disturbances. In fact, the formulation of MHE is very similar to the dual of the optimization problem in model predictive control. The objective function appears in form of a summation of some stage cost functions and an initial penalty defined over a time horizon. By minimizing this cost function, we estimate the current state and update the last estimation of previous states. Additionally, moving horizon approximation method prevents the growth of the optimization problem in time through bounding the horizon. This approximation is based on the idea of using a cost function defined on a sliding window which considers only a finite number of time stages.

The most important advantage of MHE is that the formulation is applicable for constrained nonlinear/linear systems. This feature makes MHE superior to its competitors. For example, Kalman filter is an ideal state estimator for linear unconstrained systems. However, it fails in dealing with nonlinear or constrained problems which are unavoidable in chemical engineering systems. The alternative approach, extended Kalman filter, which uses linearization to handle the nonlinearities is also unsuccessful in comparison with MHE [1].

Since MHE requires solving a nonlinear optimization problem at each sampling time, it is computationally demanding. In this work, we present a method to accelerate the online computations using the sensitivity analysis. This method considers a continuous system with discrete measurements which is very common in chemical engineering systems. Then, it uses Carleman Linearization approach to represent the nonlinear system in a linear form. We assume the disturbances are constant in each time interval. This assumption enables us to solve the system analytically. Consequently, using the analytic solution, the gradient vector and Hessian matrix of the objective function are available. This information about the objective function considerably accelerates the optimization search and eliminates the errors associated with numerical approximations of derivatives.

To test the proposed method, we consider a Continuous Stirred Tank Reactor (CSTR) with a nonlinear model with an unknown disturbance in the flow rate. The simulation shows the third order Carleman linearized model can track the real state trajectories. Also, the results suggest that by providing the objective function gradient and Hessian, the simulation time decreases significantly, while the estimation remains precise.

Reference:

[1] E.L. Haseltine & J.B. Rawlings. "Critical evaluation of extended Kalman filtering and moving horizon estimation," *Ind. Eng. Chem. Res.*, vol. 44, pp. 2451-2460, 2005.

386128 Incorporating Measurement Lab Design Projects in an Undergraduate Chemical Engineering Laboratory Course

Abstract:

An initiative toward integrating process and product design concepts throughout the undergraduate chemical engineering curriculum has recently been undertaken at Vanderbilt University. In response to this challenge, an introduction to chemical process and product design concepts as part of the undergraduate laboratory experience has recently been implemented in the chemical engineering curriculum at Vanderbilt University. Specifically, half-semester "measurement lab" design projects have been included over the past two years in the junior chemical engineering laboratory course. The students work in groups on these projects, select their own project topic, design their experiments, collect data, perform analysis and present their results in a poster session at the end of the semester. Measurement lab sensors, analytical equipment, and software used for this project will be discussed. In addition, past example projects and their results will be highlighted. Finally, the impact of this effort on the senior chemical engineering capstone design projects will be discussed.

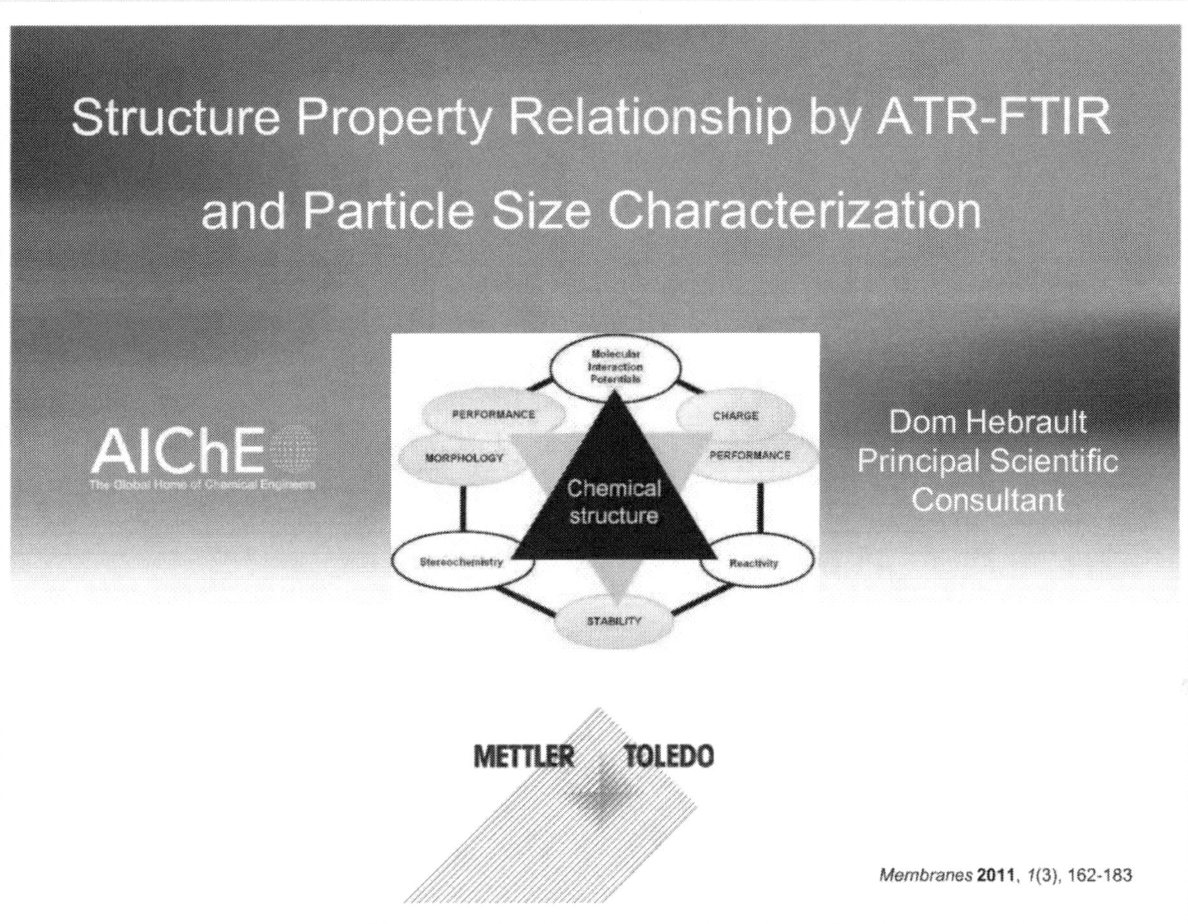
Structure Property Relationship by ATR-FTIR
and Particle Size Characterization
AIChE
The Global Home of Chemical Engineers
Molecular Interaction Potentials
PERFORMANCE
CHARGE
MORPHOLOGY
PERFORMANCE
Chemical structure
Stereochemistry
Reactivity
STABILITY
Dom Hebrault
Principal Scientific Consultant
METTLER TOLEDO
Membranes 2011, 1(3), 162-183

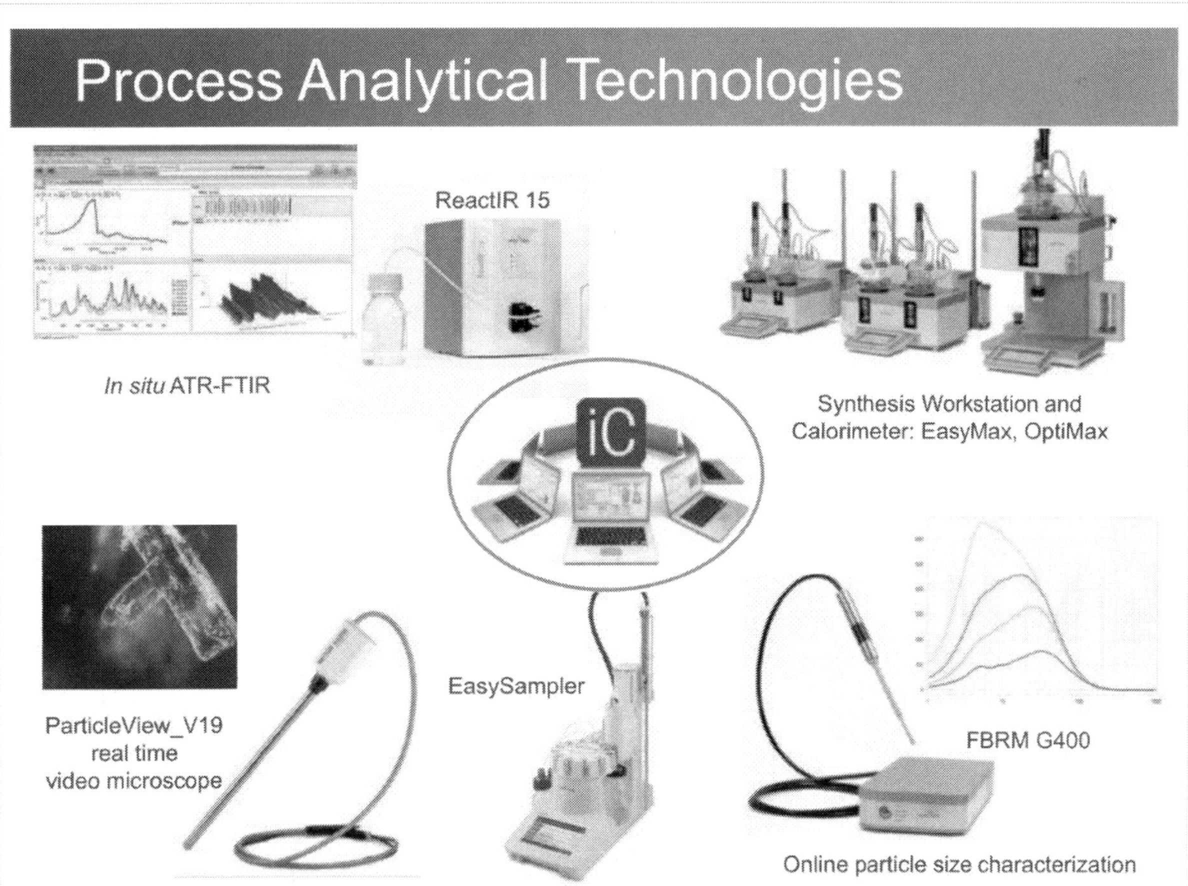
Process Analytical Technologies
ReactIR 15
In situ ATR-FTIR
Synthesis Workstation and Calorimeter: EasyMax, OptiMax
iC
ParticleView_V19 real time video microscope
EasySampler
FBRM G400
Online particle size characterization

Inline Mid-IR Monitoring

- Monitor chemistry *in situ*, under all reaction conditions
- "Movie" of the reaction
- Determine reaction evolution, outcome and pathway

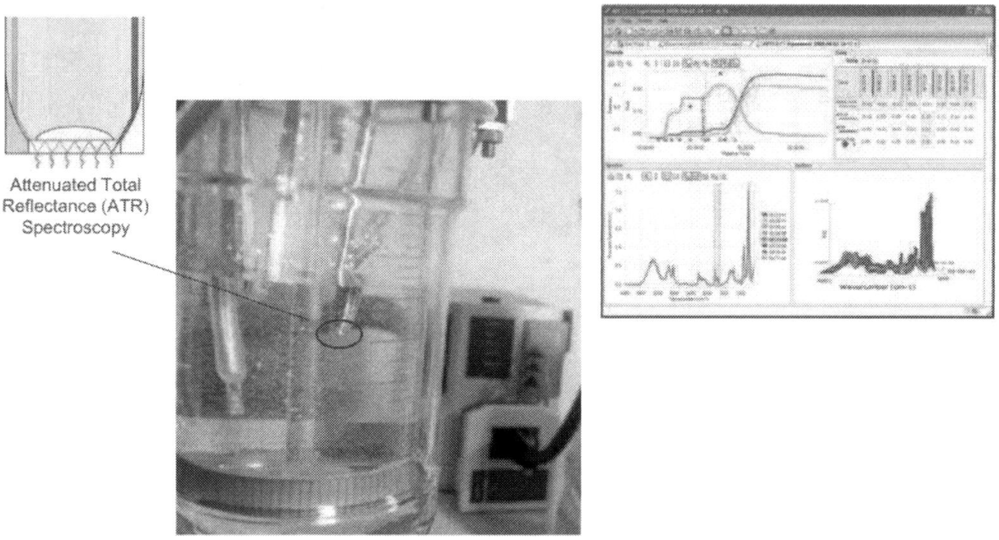

ReactIR Solutions from METTLER TOLEDO

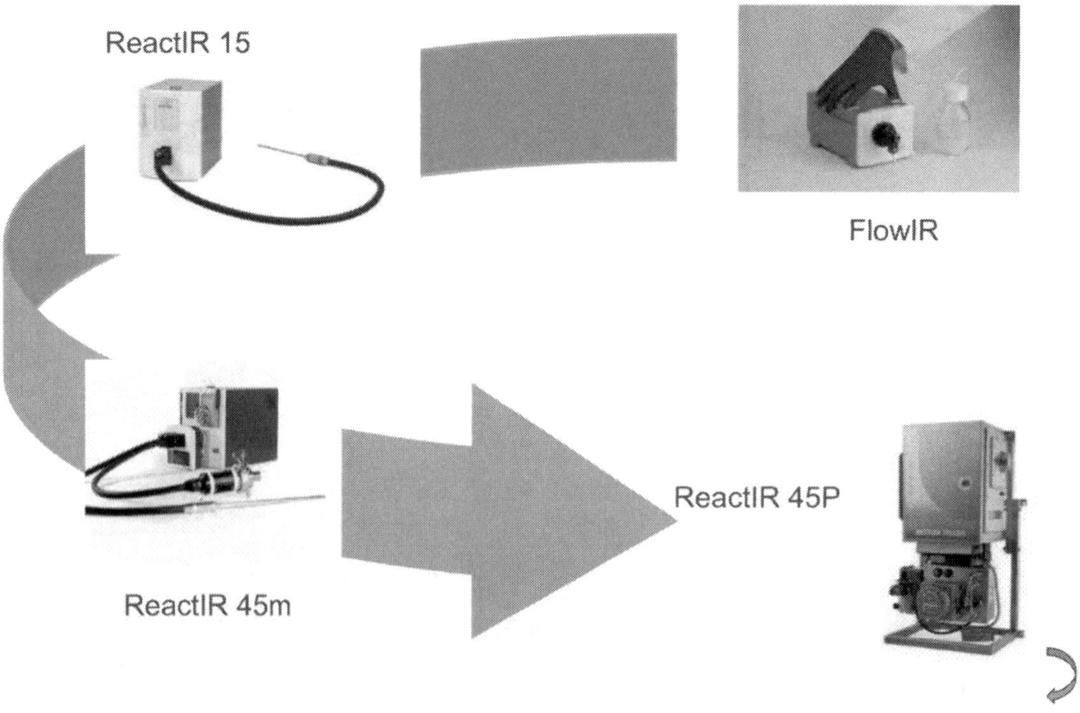

In Situ Particle Characterization Tools

FBRM® Technology
Focused **B**eam **R**eflectance **M**easurement

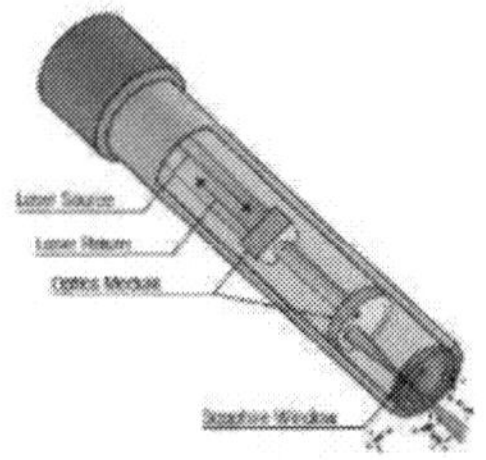
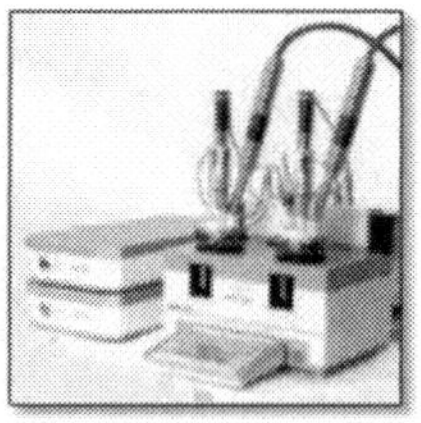
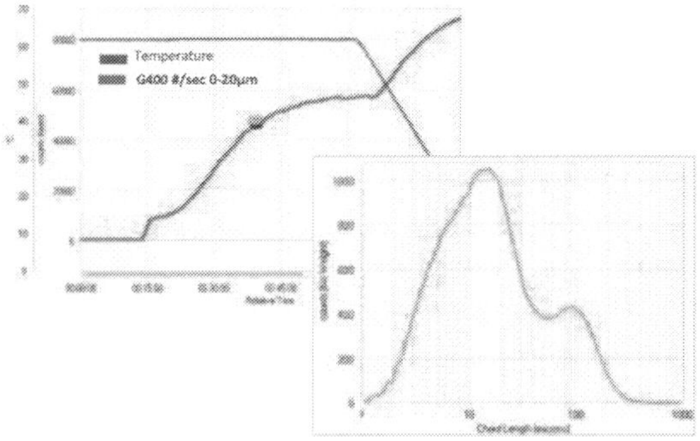

ParticleView™
Particle **V**ision and **M**easurement

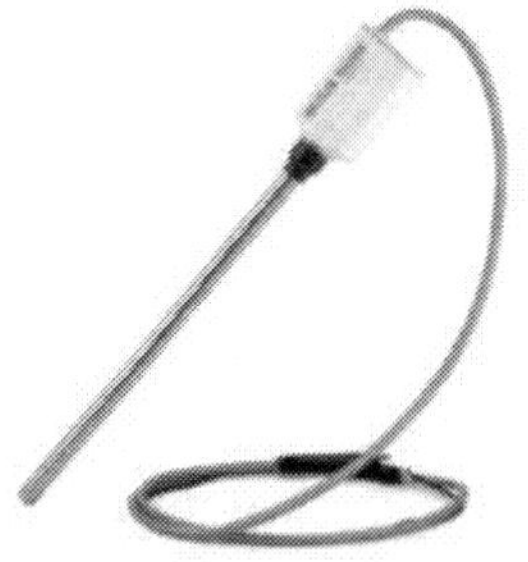
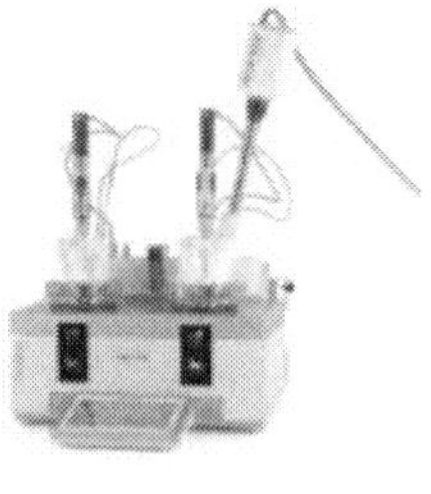
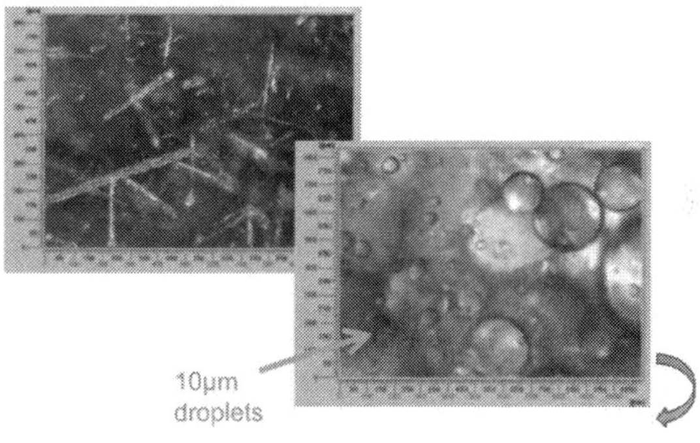

7

D2i or Data to information

iC Kinetics

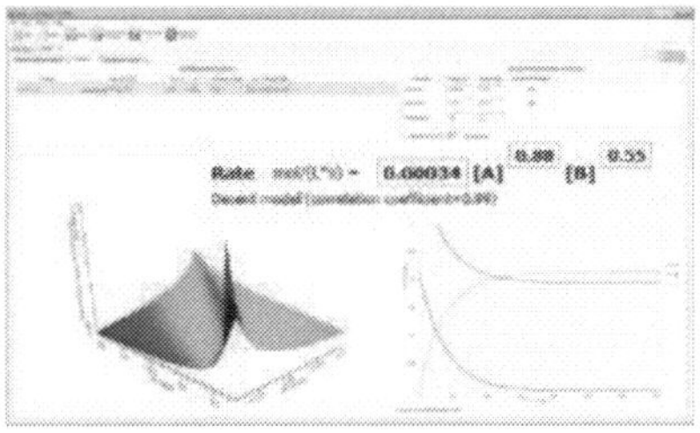

- A fast way to optimize chemistry
- Greater insight with fewer experiments
- Understand process robustness

ConcIRT Pro

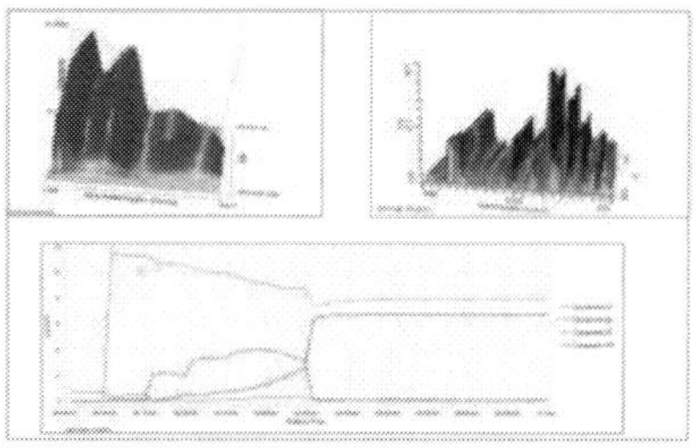

- Accurate trends for kinetics analysis by combining spectroscopy data from two techniques
- Improve process understanding by analyzing different batches simultaneously

iC PVM

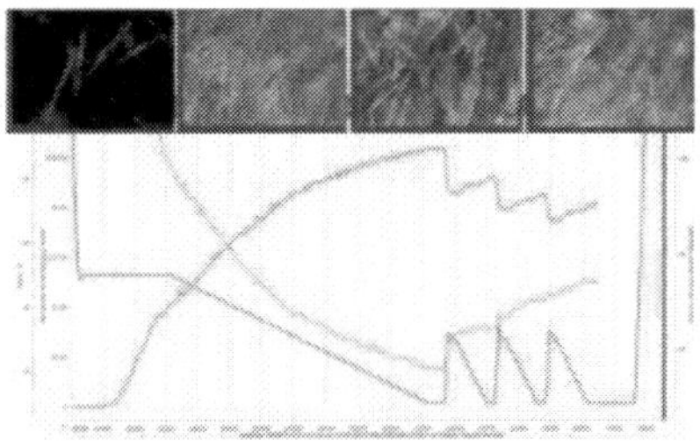

- Image-Based Process Analytics
- Enable simultaneous visual understanding and trending
- Trending based on particle size and concentration

iC Quant

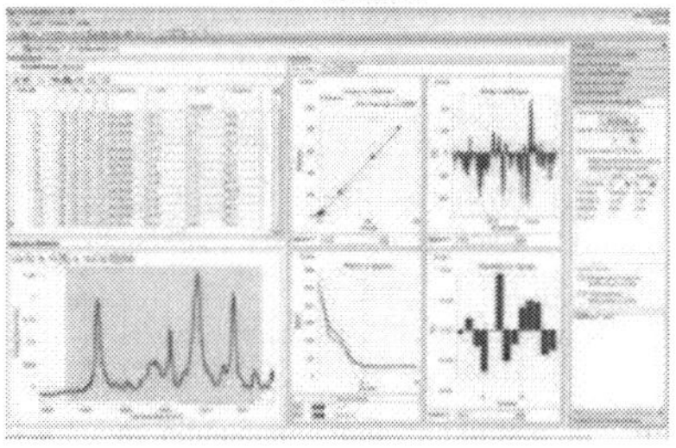

- Develop calibration models
- Predict concentration of key reaction species
- Understand reaction performance

Polymer Characterization Using SEC-FTIR

- Improve the quality of polyethylene (PE)
- Flow-FTIR coupled to GPC to deliver concentration for MW calculations, composition

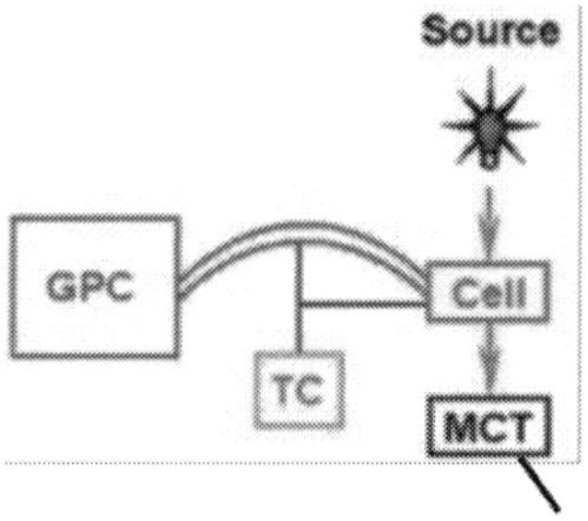

Online ATR-FTIR

GPC/FTIR interface with heated FTIR flow cell

FlowIR Technology

- Measurement of–CH_3 *versus* >CH_2 groups
- Chemometrics-based approach to short-chain branching, SBC (distribution, assay)

Chevron Phillips

Paul J. DesLauriers, D. C. Rohlfing, Chevron Phillips Chemical Company, *Macromolecular Symposia*, 282, 1, 136–149, 2009

Polymer Characterization Using SEC-FTIR

- Level of SCB → PE crystallinity, density, stress crack resistance
- Estimated using chemometrics for level of SBC as a function of MW

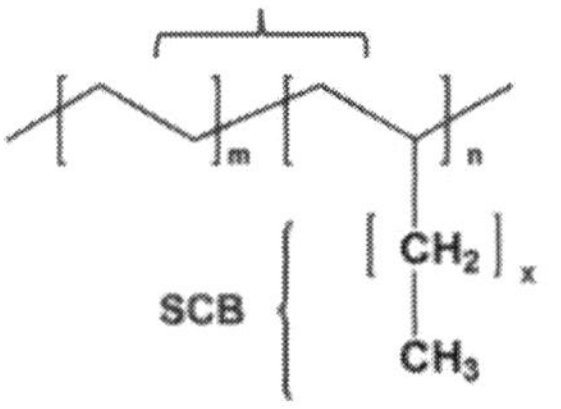

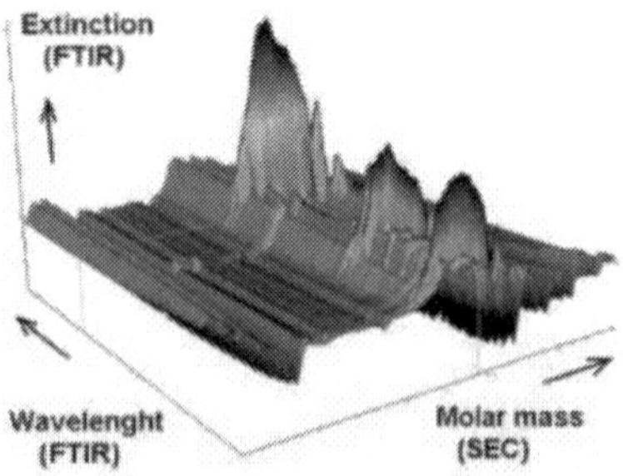

Plot of time-resolved FTIR spectra for PE

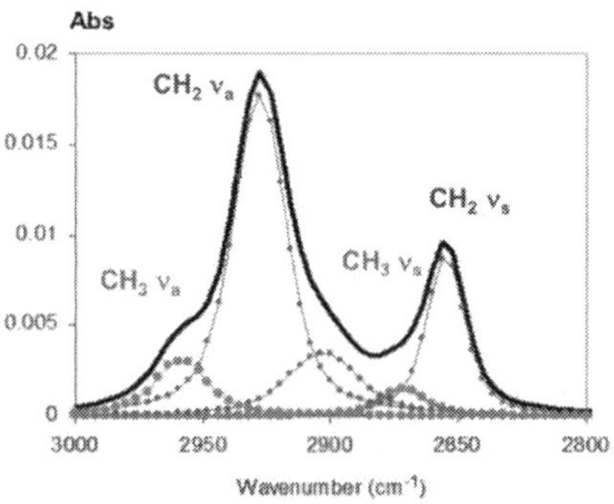

FTIR-based short chain branching (SBC) determination in PE

Chevron Phillips

Paul J. DesLauriers, D. C. Rohlfing, Chevron Phillips Chemical Company, *Macromolecular Symposia*, 282, 1, 136–149, 2009

Catalytic Ethylene Polymerization

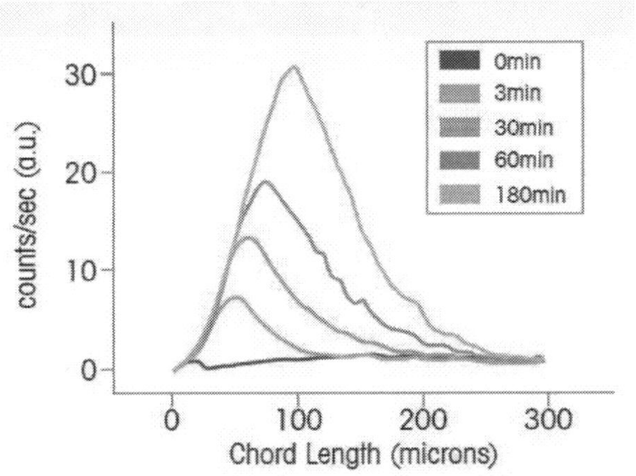

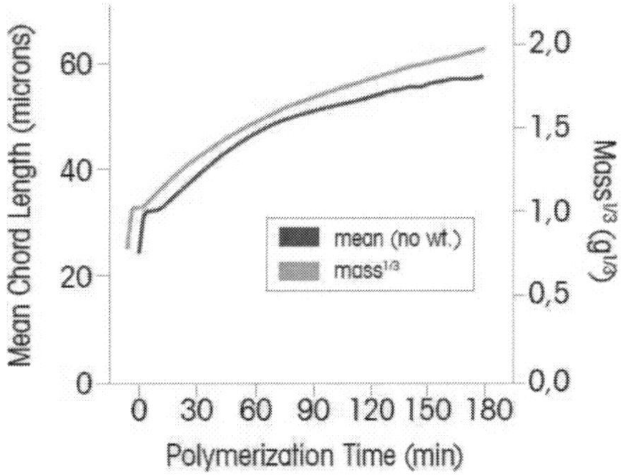

- FBRM, PVM to monitor polymer particle growth in catalytic ethylene slurry polymerizations
- FBRM mean size and mass uptake track polymerization in real-time

FMF Rainer Xalter, Rolf Mülhaupt, FMF Institute, University Freiburg, Germany, *Macromolecular Reaction Engineering, 4*, 1, 25–39, **2010**

Catalytic Ethylene Polymerization

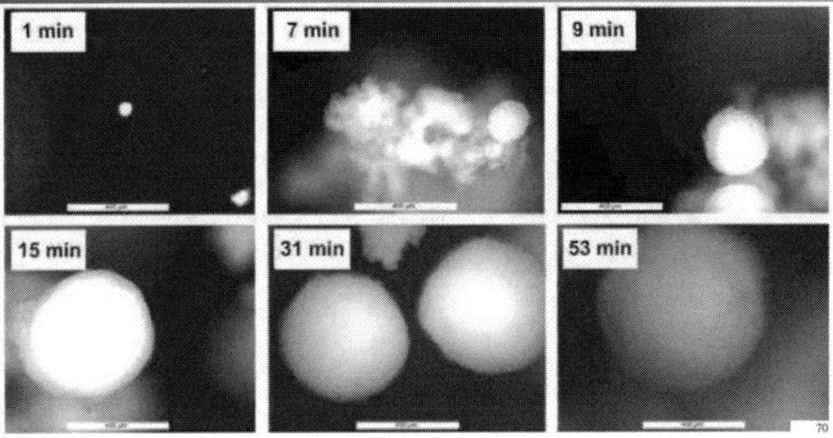

ParticleView *in situ* microscope provides visual confirmation of particle formation, growth, and little aggregation

Particle size influences PE physical properties

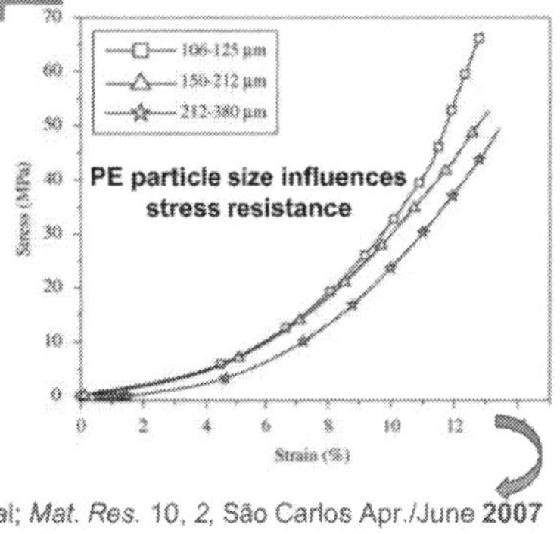

Gean Vitor Salmorial et.al; *Mat. Res.* 10, 2, São Carlos Apr./June **2007**

Polyurethane Production

Polyurethane production is complex

1. Exothermic reaction of polyol and isocyanate → polymer matrix
2. Blowing agent, in the matrix, converts liquid droplets into gas bubbles

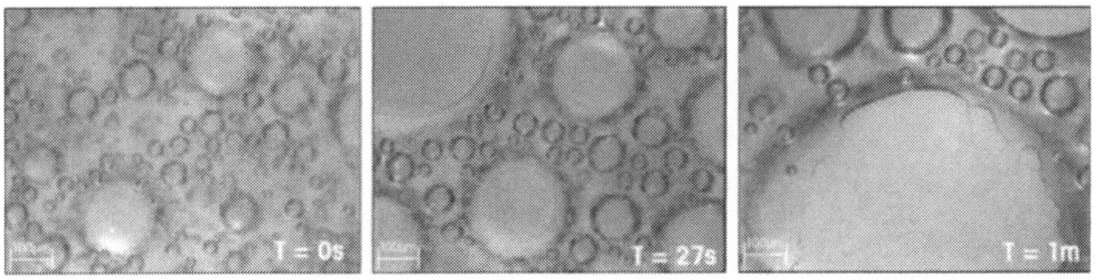

ParticleView *in situ* images

Coalescence < 1'; Unstable emulsion → Variation in foam size → Final product quality

<u>ParticleView</u>

- Monitor droplet distribution
- Screen surfactants
- Ensure more consistent and stable emulsion

Polyurethane Production

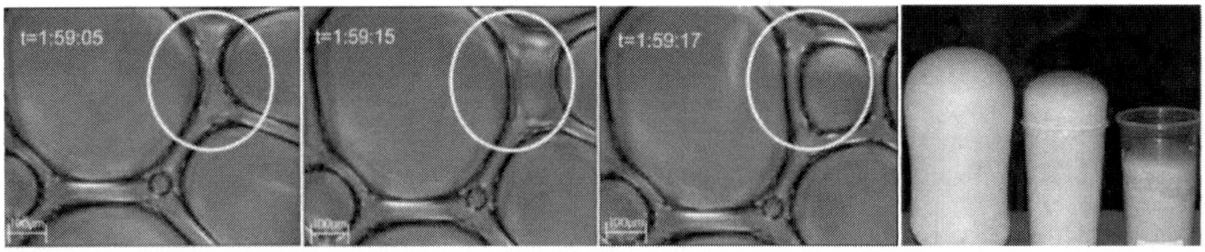

Spherical droplets → Non-spherical polyhedra (polymer struts), and trapped gas bubbles

ParticleView: Real-time monitoring of fast (seconds) bubble formation within polymer strut

Off-line analysis cannot capture fast dynamics and causes partial structure destruction

ParticleView provides a better understanding through *in situ* visual observation of the process with no sampling involved.

Acknowledgements

- James Rawlins and team, Mississippi Polymer Institute, University of Southern Mississippi, Hattiesburg, MS
- Paul J. DesLauriers, Chevron Phillips Chemical Company, Bartlesville, OK
- Rainer Xalter, Rolf Mülhaupt, FMF Institute, University Freiburg, Germany
- My colleagues in the reaction engineering team at Metter Toledo, Columbia, MD
- And YOU!

387597 Designing Products and Processes for Sustainability Using Greenscope: Application and Education

Abstract:

The U.S. Environmental Protection Agency has developed a beta software tool called GREENSCOPE (Gauging Reaction Effectiveness for the Environmental Sustainability of Chemistries with a multi-Objective Process Evaluator) to support sustainable processes and products. Evaluations of existing and new processes can be accomplished using GREENSCOPE with 140 indicators for objectives in four basis areas: environment, economics, energy, and efficiency. GREENSCOPE provides information on sustainability for processes, and through product/process decisions, about product sustainability.

Increasing sustainability can be approached through designing more sustainable products and educating those who design products and related facilities. As chemical engineers learn elements of product design and process systems engineering the opportunity exists to introduce sustainability concepts. The EPA's GREENSCOPE methodology and tool can be used to teach these concepts. This work describes example calculations of GREENSCOPE indicators for the oxidation of toluene and puts the indicators into context with best- and worst-case limits. The indicator calculations use data available from the process for the generation of oxidation products, thus transforming data into understandable information which describes sustainability. Targets for more sustainable process designs are understood, and enhancements can be considered to improve designs, either for real-world processes or for educational purposes.

390047 Development of QSPR Model Using 3D Molecular Descriptors Relating Solvent Structure to Crystal Morphology: An Efficiency Study of PCA, GA and Ann Algorithms

Abstract:

The paper outlines the development of a quantitative structure-property relationship (QSPR) that relates solvent structure to the morphology of ibuprofen crystals grown within that solvent. Morphology can be quantified by aspect ratio, and ibuprofen aspect ratio data was obtained for crystals grown in 16 different organic solvents. To provide a quantitative representation of the geometry optimized solvent molecules, 3D molecular descriptors were generated. Three different empirical force fields were used to estimate the three-dimensional structure of the solvent molecules while their effect on the developed models is analyzed. The descriptor data matrix, containing a multitude of descriptor types, is reduced in size, using Principal Component Analysis (PCA), Genetic algorithm (GA), Artificial Neural Network (ANN), and also combination of such algorithms for regression into linear models. The Singular Value Decomposition (SVD) procedure was used for PCA due to its increased stability in primary component prediction. The predictive capabilities of these models were also analyzed through means of internal and external validation methods.

391439 Advancements in Computer Aided Design of Task Specific Ionic Liquids

Abstract:

This presentation provides details on recent advancements related to computer-aided ionic liquid design method. In particular, consideration of environmental, ecological and health impacts along with thermophysical properties within the design framework is discussed. The proposed methodology is able to screen a variety of cations, anions and alkyl groups to arrive at optimal ionic liquid structures for a given application. Group contribution methods are used for ionic liquid property prediction while optimization methods are used to screen all possible structure. The designed ionic liquids are not only optimal for the given application but also have reduced environmental burden. The utility of the method is demonstrated through interesting case studies.

390590 Simulation and Optimization of Ethyl Benzene Production Using Aspen Plus Platform

Abstract:

Ethyl benzene is an aromatic hydrocarbon and is also important in the petrochemical industry and as an intermediate in the production of styrene, which is utilized for making polystyrene, it is a common plastic material. The study deals with optimization of the process for the production of ethyl benzene by liquid-phase benzene alkylation. This process involves the reaction of benzene with ethylene to form ethyl benzene. Ethylene reacts with ethyl benzene to form undesired product di-ethyl benzene, if the temperatures of a reactor or concentrations of ethylene are high. Di-ethyl benzene reacts with benzene to form ethyl benzene. Di-ethyl benzene is the highest-boiling component in the system; it comes out the bottom of two distillation columns. The recycling benzene is more expensive. The economic optimum steady-state design is developed that minimizes total annual cost. Therefore, it offers a classic case of an engineering invention and optimization of a procedure. The purpose of this study is to develop an optimum design for the ethyl benzene process considering the reactor size and the amount of benzene recycled. Optimization of the process led to an increase in the yield of ethyl benzene by 0.8832%.

Title: Mixed Integer Polynomial Optimisation

Author: Vivek Dua

Affiliation: Department of Chemical Engineering, Centre for Process Systems Engineering, University College London, Torrington Place, London WC1E 7JE, United Kingdom. E-mail: v.dua@ucl.ac.uk

Abstract

Mathematical modelling and model-based optimisation of chemical process systems have great potential for providing answers on how to optimally design and operate these systems. One key difficulty in not being able to fully exploit this potential is the presence of nonlinear terms in the mathematical models. This issue is further exacerbated when the model also includes integer variables, for incorporating structural choices. A generic formulation of such mathematical programming problems is given by Mixed Integer Nonlinear Programs (MINLP) (Grossmann, 2002).

Solving MINLPs is NP-hard and has created a huge interest for developing computationally efficient algorithms for obtaining its solution. New theoretical developments for solving MINLPs have pushed the boundaries of application of MINLPs to many areas in engineering and science. Several software tools are available to solve these problems, DICOPT (Viswanathan and Grossmann, 1990), MINOPT (MINOPT, 1998), BARON (Sahinidis, 1996), GloMIQO (Misener and Floudas, 2013), Alpha-ECP (Westerlund and Lundqvist, 2005) – to name a few. The reader is referred to recent survey papers by Belotti et al. (2013) and D'Ambrosio & Lodi (2011) presenting an overview of advances for solving MINLPs and the books (Biegler et al., 1997; Floudas, 1995; Nemhauser and Wolsey, 1988) for an introduction to the topic.

Gueddar and Dua (2012) reformulated MINLPs as a multi-parametric Nonlinear Program (mp-NLP) by first relaxing the binary variables as continuous variables and then treating them as parameters. An approximate solution of the mp-NLP is obtained and that solution was then used to estimate the solution at the terminal nodes of the Branch and Bound (B&B) tree and guide the search in the tree; integer variables are the branching variables in the tree. For details of multi-parametric programming the reader is referred to Dua and Pistikopoulos (1999), Pistikopoulos et al. (2007a,b), Pistikopoulos (2009) and Wittmann-Hohlbein and Pistikopoulos (2014).

In this work, the case when nonlinearities can be captured by polynomial functions alone is considered, an MINLP therefore becomes a Mixed Integer Polynomial Optimization (MIPOPT) problem. Patil et al. (2012) presented a Bernstein polynomial approach for solving such problems. Teles et al. (2013) proposed a discretization approach using bilinear terms as the building block. In this work, the integer variables are relaxed as continuous variables and then treated as parameters resulting in a multi-parametric polynomial program (mp3). An exact solution of the resulting mp3 can be obtained by exact multi-parametric nonlinear inversion of the optimality conditions, see for example Fotiou et al. (2007). The proposed approach hence does not require approximate solution of the mp-NLP followed by a tree search as in Gueddar and Dua (2012).

The optimality conditions for the mp3 are given by a set of simultaneous parametric polynomial equations which are solved analytically to give the parametric optimal solution as a function of the relaxed integer variables. Evaluation of the parametric optimal solution for integer variables fixed at their integer values followed by screening of the evaluated solutions gives the optimal solutions.

The proposed MIPOPT to mp3 reformulation followed by exact solution of the mp3 provides all the possible candidate solutions and does not use an iterative numerical method in the traditional sense to identify the global optimal solutions. Although the focus here is on the nonlinearities arising only from the polynomial terms, theoretical research work on obtaining analytical solution of equations incorporating other types of nonlinearities will further expand the applicability of the proposed approach where mixed integer programs were to be reformulated as multi-parametric programs. Additionally, many nonlinear functions can be approximated as polynomials and in those cases the proposed approach can be explored for obtaining approximate global optimal solutions.

References:

1. Belotti, P., Kirches, C., Leyffer, S., Linderoth, J., Luedtke J., Mahajan, A. (2013). Mixed-integer nonlinear optimization. Acta Numerica, 1-131.
2. Biegler, L.T., Grossmann, I.E. and Westerberg, A.W. (1997). Systematic Methods of Chemical Process Design. Prentice Hall.
3. D'Ambrosio, C. and Lodi, A. (2011). Mixed integer nonlinear programming tools: a practical overview. 4OR. 9(4), 329-349.
4. Dua, V., Pistikopoulos, E. N. (1999) Algorithms for the solution of multiparametric mixed-integer nonlinear optimization problems, Industrial & Engineering Chemistry Research, 38, 3976-3987.
5. Floudas, C.A. (1995). Nonlinear and Mixed-Integer Optimization: Fundamentals and Applications. Oxford University Press.
6. Fotiou, I.A., Rostalski, P., Parrilo, P.A. and Morari, M. (2007). Parametric optimization and optimal control using algebraic geometry methods. International Journal of Control, 79(11), 1340-1358.
7. Grossmann, I.E. (2002). Review of nonlinear mixed-integer and disjunctive programming Techniques. Optimization and Engineering, 3, 227–252.
8. Gueddar, T. and Dua, V. (2012). Approximate multi-parametric programming based B&B algorithm for MINLPs. Computers and Chemical Engineering, 42, 288-297.
9. MINOPT (1998). http://titan.princeton.edu/MINOPT
10. Misener, R. and Floudas, C.A. (2013). GloMIQO: Global mixed-integer quadratic optimizer, Journal of Global Optimization, 57, 3-50.
11. Nemhauser, G.L. and Wolsey, L.A. (1988). Integer and Combinatorial Optimization. Wiley.
12. Patil, B.V., Natraj, P.S.V. and Bhartiya, S. (2012). Global optimization of mixed-integer nonlinear (polynomial) programming problems: the Bernstein polynomial approach. Computing, 94, 325–343.
13. Pistikopoulos, E.N. (2009). Perspectives in multi-parametric programming and explicit model predicitive control. AIChE Journal, 55, 1918-1925.
14. Pistikopoulos, E.N., Georgiadis, M.C., & Dua, V. (2007a). Multi-Parametric Programming. Vol 1, Wiley-VCH.
15. Pistikopoulos, E.N,. Georgiadis, M.C., & Dua, V. (2007b). Multi-Parametric Model Based Control. Vol 2, Wiley-VCH.
16. Sahinidis, N.V. (1996). BARON: A general purpose global optimization software package, Journal of Global Optimization, 8, 201–205.
17. Teles, J.P., Castro, P.M., Matos, H.A. (2013). Univariate parameterization for global optimization of mixed-integer polynomial problems. European Journal of Operational Research, 229, 613–625.
18. Viswanathan, J. and Grossmann, I.E. (1990). A combined penalty function and outer-approximation method for MINLP optimization. Computers and Chemical Engineering, 14, 769–782.
19. Westerlund, T. and Lundqvist, K. (2005). Alpha-ECP, version 5.101: An interactive MINLP-solver based on the extended cutting plane method. Technical report 01-178-A, Process Design Laboratory at Abo Akademi University.
20. Wittmann-Hohlbein, M. and Pistikopoulos, E.N. (2014). Approximate solution of mp-MILP problems using piecewise affine relaxation of bilinear terms. Computers & Chemical Engineering, 61(11), 136–155.

Embedding complex simulations in flowsheet optimization problems: a reduced model trust region framework.

As multi-scale modeling and advanced simulation technologies mature, it is desirable to utilize their predictive power in process design and optimization. However, these models present unique challenges for traditional optimization frameworks. These models, such as those for complex fluid flow and molecular dynamics, span a very wide range of length and time scales. Simulations can be very computationally expensive, and derivative information may be unavailable. Through systematic use of reduced models, the proposed framework bridges the gap between time scales.

Consider the optimization of a process flowsheet in which one unit is represented with a complex model (CM). By constructing reduced models (RMs) that approximate the behavior of the CM and substituting these simpler models for the complex unit operation, one may simultaneously optimize the entire flowsheet with traditional, derivative-based techniques. Using trust region concepts from nonlinear programming, we construct a sequence of local RMs. If these reduced models are constructed in order to satisfy the "fully linear property" as described by Conn, Scheinberg, and Vicente,[1] convergence can be guaranteed under mild assumptions. Popular sample-based reduced model techniques such as polynomial regression and kriging are suitable choices for reduced models.

Previous work by Biegler et al. 2014[2] presented the above framework for inequality constrained problems. Although equality constraints may be eliminated for the optimization strategy, this process may still require many evaluations of the CM when the constraints are coupled across the process (e.g., in the presence of a recycle loop). This work discusses alternative algorithms using the filter method to handle feasibility. The filter methods are able to handle equality constraints directly.

Previous RM optimization methods use penalty functions to handle constraints. Penalty functions judge progress towards the optimum through a combined measure of objective decrease and feasibility. By contrast filter methods borrow concepts from multi-objective optimization and treat feasibility and the objective function separately. A filter will be more robust in handling equality constraints.

Two trust region filter RM optimization algorithms are presented. The first method assumes that the algorithm can obtain derivatives of the CM, even though they may be expensive. The second method assumes derivatives are unavailable.

The performance of these algorithms is demonstrated on the optimization of the classic ammonia synthesis process. The reactor is described by a system of differential algebraic equations. Reduced reactor models are constructed using kriging to represent extent of reaction and outlet temperature. The performances of both algorithms are analyzed and compared against the solution of the original model obtained using simultaneous collocation.

1. Conn, A. R., Scheinberg, K., & Vicente, L. N. (2009). *Introduction to Derivative-Free Optimization,* SIAM Publications (Philidelphia).

2. Biegler, L. T., Lang, Y. D., & Lin, W. (2014). Multi-scale optimization for process systems engineering. *Computers & Chemical Engineering,* 60, 17-30.

375107 Using the Finite Element Method to Model an Evaporating Droplet

Abstract:

Modeling an evaporating droplet with colloidal particles has many computational challenges. This solid-liquid-gas system requires a highly refined mesh especially at the drop contact line, where the evaporative flux is highest. The elliptic mesh method is used to update the mesh at every time step, accommodating the free liquid-gas interface and focusing the mesh towards the rapidly changing areas. The numerical model presented was compared with analytical solutions for specific cases.

384676 Transient CFD Simulation of Turbulent Structures in a Small Modular Nuclear Reactor

Abstract:

Preliminary steady flow thermal-hydraulic analysis on natural convection in small modular nuclear reactors (SMRs) identified regions of significant fluid-structure interactions [1,2 - URL]. The present work aimed at capturing the transient growth and propagation of eddies and vortex shedding in the hot leg riser of an SMR. Large eddy simulations (LES) were performed on regions where internal structures significantly altered flow patterns and were relevant to the structural integrity of the reactor pressure vessel. The commercial CFD suite STAR-CCM+ was utilized for this work.

385510 Higher-Order Inclusions of Nonlinear Systems By Chebyshev Models

Abstract:

Estimators that consist of a polynomial approximant and an interval remainder on the approximation error have been successfully used to create bounds for nonlinear systems, with applications to validated ODE integration and global optimization. Taylor models in particular are generated by creating a Taylor expansion of a function at a point, with the approximation error bounded for the given approximation range [1,2]. There is an arithmetic linked to Taylor models which allows for univariate and bivariate operations to be carried out and therefore Taylor models to be built up operation-wise for (sufficiently smooth) factorable functions. Nonetheless, it may appear unnatural to create the polynomial approximation at a point and then bound the approximation error over a range. Instead, the main focus of this presentation is on Chebyshev polynomials, which provide a polynomial approximation of a function over a range.

A Chebyshev expansion consists of a multilinear combination of Chebyshev polynomials similar to how a Taylor polynomial is made up of a multilinear combination of monomials [3]. We have developed an equivalent to Taylor model arithmetic in Chebyshev model arithmetic, which includes rules on how to carry out a number of common univariate operations along with bivariate addition and multiplication and steps such as bounding the range of a Chebyshev expansion. An important challenge in creating such an arithmetic arises in the multiplication of Chebyshev polynomials which creates two terms, whereas for monomials only one term is created. This proves to be a particular challenge for n-variate Chebyshev models as 2^n terms are created via multiplication. Besides, creating a Chebyshev expansion for intrinsic univariate functions proves to be more challenging than a Taylor expansion as the coefficients are calculated via integration as opposed to differentiation. Here, we have chosen to create a 'cheap' approximation via interpolation as we know the approximation error is no more than double that of the approximation created via integration. Moreover, we have developed a method for finding sharp remainder bounds for a subset of the univariate operations, which significantly reduces the width of the remainder term.

An implementation for Chebyshev models has been created using C++, which can deal with multivariate Chebyshev models and the commonly used operations. This can been used to compare the performance of Chebyshev models to Taylor models for function bounding, verified ODE integration, and bounding of the solutions of implicit equations. Our results show that we obtain tighter bounds with Chebyshev models in all the cases.

References:
[1] A. Neumaier (2002). Taylor forms - Use and limits. *Reliable Computing*, **9**(1):43-79.
[2] K. Makino; M. Berz (2003). Taylor models and other validated functional inclusion functions. *International Journal of Pure and Applied Mathematics*, **4**(4):379-456.
[3] L.N. Trefethen (2013). *Approximation Theory and Approximation Practice*. SIAM, Philadelphia.

377210 Recent Results and Insights into Using Graphics Processors (GPUs) for Attainable Regions Computations

Abstract:

The task of computing the limits of achievability for a system of reactions and associated kinetics finds special importance in the design of efficient chemical reactors, used in industrially relevant processes. Knowledge of these limits helps us to understand what potential improvements could be made to an existing reactor design, and further assists in setting performance targets for new, potentially novel, reactor designs.

For the past three decades, computation of these limits for chemical reactor networks specifically has been made possible by the idea of the *Attainable Region* (AR). The AR for a system of reactions represents the collection of all states achievable by the system for a given set of kinetics and initial conditions. The AR is represented as the set of all compositions achievable in a reactive system. It is commonly plotted as a convex polytope in n-dimensional composition space. The dimension and complexity of this space is determined by the number of components participating in the reactions as well as the number of independent reactions associated with the system. From this it follows that problems in AR theory revolve around understanding the nature of convex polytopes in higher dimensional spaces, understanding the characteristic structure of these regions, and also understanding how to compute the AR for a given set of component reactions, and feed points.

Recently, a large interest in AR theory has focussed towards developing novel computer algorithms which help to construct the AR via numerical means. Due to the higher dimensional and geometric nature of Attainable Regions, computation of the AR for a system of moderate complexity is often not easy to perform by hand. Rather, one must rely on automated AR construction techniques to assist in finding these regions.

Automated AR construction is often difficult in itself. The problem of determining a higher dimensional convex polytope, associated with a system of non-linear reaction kinetics and reactor models, is a complex procedure. Additionally, the large amount of data generated by the AR imposes a large computational workload even for problems of moderate complexity. Computation of the AR has hence only been achieved for relatively small systems, residing in two or three dimensions usually.

In roughly the same timeframe as the development of AR theory, rapid advances in computer architecture and high performance computing (HPC) have also been made. The increase in computational power has allowed researchers to investigate increasingly complex problems in science and engineering. More specialised technologies and techniques used in computer science have not always been widely adopted in mainstream chemical engineering research - HPC applications are often more widely noticed in fields of computational fluid dynamics (CFD) and catalysis rather than in the fields of computational reactor and process synthesis for instance.

More recently, the use of graphics processing units (GPUs) in general purpose computations has allowed for computationally demanding scientific discovery to be carried out on the computer. GPUs are highly powerful devices which are able to throughput a large number of floating point operations per cycle. Current GPUs are often considered to be an order of magnitude more powerful and are considered to be far more energy efficient when compared to traditional CPUs. It is for this reason that many HPC applications and computer codes have recently directed focus towards implementation on GPUs. Nevertheless, adoption of GPUs to computationally strenuous problems in chemical engineering has been predominantly in CFD-related fields - little research into the use of GPUs to reactor network synthesis problems has been achieved thus far. Unlike CPUs however, which are fast highly generalised processors, GPUs rely on parallel computation.

In this work, we highlight a number of advancements and insights gained, specifically to AR construction techniques, using GPUs to carry out computations. The computationally demanding tasks required by current AR construction algorithms warrants the use of GPUs. Recent results have shown how large gains in construction time and computational efficiency can be obtained using GPUs when the algorithm can be interpreted in a parallel manner. Dependent on the complexity and type of AR problem investigated, speed ups in construction time of 2x to 60x have already been observed. However, additional improvements and advances are still possible with adequate understanding of how to better exploit GPU performance in AR applications. This allows for more complex and industrially relevant systems to be considered which would have been computationally prohibitive in the past.

Furthermore, existing AR construction algorithms, which were in the past implemented on a serial CPU, may also be adapted for use with GPUs. The success and challenges of these implementations is also a topic of discussion in this work. A number of GPU implementations, both to existing algorithms, as well as to entirely new algorithms are currently being investigated. We wish to highlight a number of these issues in this work, discuss implementation challenges currently being faced with recent methods, and reflect on possible future directions using GPUs specifically in AR research.

The information generated by understanding the AR for a system may be valuable in the design of more efficient chemical reactors, however its computation has been limited to relatively small systems currently. The use of GPUs to accelerate AR computations has already shown to be beneficial, although greater performance may be gained with further research in this field. This may lead to more robust and computationally efficient AR construction methods in the future.

Fractional Calculus Based Models for a Photo-Catalytic Reactive System: Deterministic and Stochastic Approaches

Vicente Rico-Ramirez[a], Sergio Damian-Vazquez[a], Rasiel Toledo-Hernandez[a] and Urmila M. Diwekar[b]

[a]*Instituto Tecnologico de Celaya, Departamento de Ingenieria Quimica, Av. Tecnologico y Garcia Cubas S/N, Celaya, Guanajuato, Mexico, 38010*

[b]*Vishwamitra Research Institute, 368 56th Street, Clarendon Hills, IL, 60514, USA*

Fractional Calculus (1695)

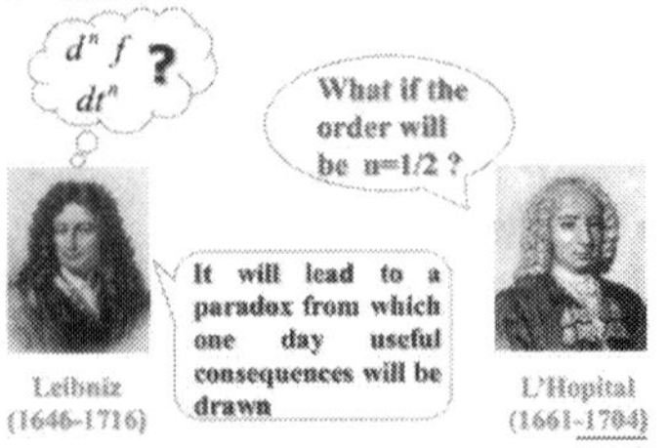

Fractional calculus is a generalization of ordinary differentiation and integration to an arbitrary NON INTEGER order. $\frac{d^n f}{dx^n}$

Basic Definitions

Fractional Integration Definition

$${}_0D_t^{-\alpha}F(t) = \frac{1}{\Gamma(\alpha)}\int_0^t \frac{F(\tau)}{(t-\tau)^{1-\alpha}}d\tau$$

$${}_0D_t^{-\alpha}F(t) = F(t) * \frac{t_+^{\alpha-1}}{\Gamma(\alpha)} \quad \text{Memory Effects}$$

Fractional Differentiation Definitions

Caputo Definition

$${}_0^CD_t^{\alpha}Y(t) = \frac{1}{\Gamma(1-\alpha)}\int_0^t \frac{Y'(\tau)}{(t-\tau)^{\alpha}}d\tau$$

Convolution integrals accounts for memory effects in dynamic systems

Numerical and Analytical Tools for Simulation and Optimization

A) Predictor-Corrector Technique for Integration of Fractional Differential Equations (FDE)

FDE $\quad {}_0^CD_t^{\alpha}y = f(t,y)$

This technique is a generalization of the Adams-Bashforth-Moulton Method for ordinary integration

B) Euler-Lagrange Optimality Conditions for Fractional Optimal Control Problems (FOCP)

Performance Index: System Dynamics

$$J(u) = H[x(a),x(b)] + \int_a^b F(x,u,t)\,dt \qquad {}_aD_t^{\alpha}x = G(x,u,t) \quad x(a) = x_a$$

Optimal Control **Adjoint Equations**

$$\frac{\partial F}{\partial u} + \lambda\frac{\partial G}{\partial u} = 0 \qquad {}_tD_T^{\alpha}\lambda = \frac{\partial F}{\partial x} + \lambda\frac{\partial G}{\partial x} \quad \lambda(T) = 0$$

State Equations

$${}_0D_t^{\alpha}x = G(x,u,t) \qquad x(0) = x_0$$

* The result is a two-point fractional boundary value problem

Previous Work: Modeling and Optimization

Modeling Capabilities: Fermentation and Other Biological Reactive Systems. Example: **Tequila Production Process**

Ordinary Model:	Fractional Model
$\frac{dx_1}{dt} = (\mu - D)x_1$	${}_0^CD_t^{\alpha_1}x_1 = k_c\,x_1\,x_2 - k_m x_1$
$\frac{dx_2}{dt} = -(Y_{xs} - Y_{ps})\mu x_1 - m_s x_1 - (-S_0 + x_2)D$	${}_0^CD_t^{\alpha_2}x_2 = -k_s\,x_1\,x_2$
$\frac{dx_3}{dt} = \alpha\mu x_1 - Dx_3$	${}_0^CD_t^{\alpha_3}x_3 = k_p\,x_1\,x_2$

$$\mu = \frac{\mu_{max}\,x_2}{k_s + x_2 + x_2^2 k_i}(1 - x_3 k_p)$$

Gradient Method for Solving a FOCP

* The two-point fractional boundary value problem is solved with a Matlab implementation of the gradient method for a FOCP, combined with the numerical integration technique for FDE's

A Photo-Catalytic Reactive System

Phenol Degradation

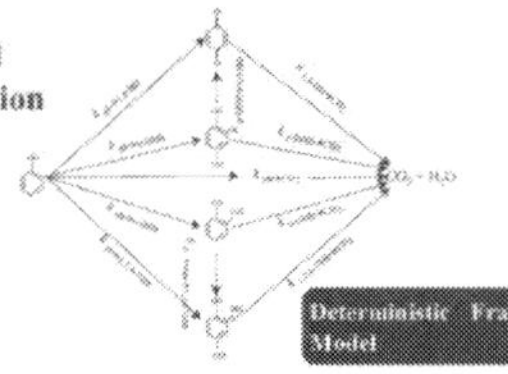

Ordinary Model

$$\frac{dC_{ph}}{dt} = -\frac{k_{ph}C_{ph}}{1 + k_{ph}C_{ph} + k_{p\text{-}DHB}C_{p\text{-}DHB} + k_{1,4\text{-}BQ}C_{1,4\text{-}BQ}}$$

$$\frac{dC_{p\text{-}DHB}}{dt} = \frac{k_{ph\to p\text{-}DHB}C_{ph} - k_{p\text{-}DHB\to p\text{-}CO_2}C_{p\text{-}DHB}}{1 + k_{ph}C_{ph} + k_{p\text{-}DHB}C_{p\text{-}DHB} + k_{1,4\text{-}BQ}C_{1,4\text{-}BQ}}$$

$$\frac{dC_{1,4\text{-}BQ}}{dt} = \frac{k_{ph\to 1,4\text{-}BQ}C_{ph} - k_{1,4\text{-}BQ\to p\text{-}CO_2}C_{1,4\text{-}BQ}}{1 + k_{ph}C_{ph} + k_{p\text{-}DHB}C_{p\text{-}DHB} + k_{1,4\text{-}BQ}C_{1,4\text{-}BQ}}$$

Deterministic Fractional Model

$${}_0^CD_t^{\alpha_1}C_{ph} = -k_1\,C_{ph}$$

$${}_0^CD_t^{\alpha_2}C_{p\text{-}DHB} = k_2\,C_{ph} - k_4 C_{p\text{-}DHB}$$

$${}_0^CD_t^{\alpha_3}C_{1,4\text{-}BQ} = k_3\,C_{ph} - k_5 C_{1,4\text{-}BQ}$$

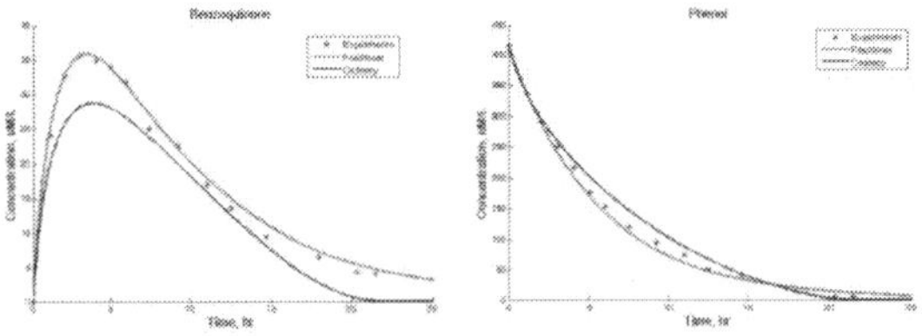

Ito Processes

Ordinary Calculus $\quad dx = a(x,t)dt + \sigma(x,t)dz \quad x(t_0) = x_0$

* It includes a deterministic component and a Wiener process (stochastic component, $dz = \in \sqrt{dt}$)
* A Wiener process satisfies the Markov property (distribution of future values depends only on its current value) and has independent increments

Fractional Ito Process

$${}_0X_t^{\alpha} = a(X,t)\,{}_0^CD_t^{\alpha}X + \sigma(X,t)\frac{dz}{dt}$$

* The deterministic component presents fractional dynamics and memory effects
* The stochastic component is still a Wiener process whose evolution depends only on the actual value of the stochastic variable

Stochastic Fractional Model

$${}_0C_{ph}^{\alpha_1} = {}_0^CD_t^{\alpha_1}C_{ph} + \sigma_1\frac{dz}{dt}$$

$${}_0C_{p\text{-}DHB}^{\alpha_2} = {}_0^CD_t^{\alpha_2}C_{p\text{-}DHB} + \sigma_2\frac{dz}{dt}$$

$${}_0C_{1,4\text{-}BQ}^{\alpha_3} = {}_0^CD_t^{\alpha_3}C_{1,4\text{-}BQ} + \sigma_3 \in_3 \frac{dz}{dt}$$

Stochastic Predictor-Corrector Integration Technique

SFDE $\quad {}_0y_t^{\alpha} = {}_0^CD_t^{\alpha}y + \sigma(y,t)\frac{dz}{dt} \qquad$ **where** $\quad {}_0^CD_t^{\alpha}y = f(t,y)$

Predictor Step

$$y_h^p(t_{n+1}) = \sum_{k=0}^{\alpha-1}\frac{t_{n+1}^k}{k!}y_0^k + \frac{1}{\Gamma(\alpha)}\sum_{j=0}^{n}d_{j,n+1}f(t_j,y_h(t_j)) + \frac{1}{\Gamma(\alpha)}\sum_{j=0}^{n}b_{j,n+1}\in_j\sigma(t_j,y_h(t_j))$$

Corrector Step

$$y_h(t_{n+1}) = \sum_{k=0}^{\alpha-1}\frac{t_{n+1}^k}{k!}y_0^k + \frac{h^{\alpha}}{\Gamma(\alpha+2)}\left[f(t_{n+1},y_h^p(t_{n+1})) + \sum_{j=0}^{n}a_{j,n+1}f(t_j,y_h(t_j))\right]$$
$$+ \frac{h^{\alpha}}{\Gamma(\alpha+2)}\sum_{j=0}^{n}a_{j,n+1}\in_j\sigma(t_j,y_h(t_j))$$

Weights

$$d_{j,n+1} = \frac{h^{\alpha}}{\alpha}\left[(n+1-j)^{\alpha} - (n-j)^{\alpha}\right] \quad b_{j,n+1} = \frac{h^{\alpha}}{\alpha}\left[(n+1-j)^{\alpha} - (n-j)^{\alpha}\right]$$

$$a_{j,n+1} = \begin{cases} n^{\alpha+1} - (n-\alpha)(n+1)^{\alpha} & \text{if } j = 0 \\ (n-j-2)^{\alpha+1} + (n-j)^{\alpha+1} - 2(n-j+1)^{\alpha+1} & \text{if } 1 < j < n \\ 1 & \text{if } j = n+1 \end{cases}$$

α_1	α_2	α_3
1.0000	0.8468	0.5600

k_1 $(hr^{-\alpha_1})$	k_2 $(hr^{-\alpha_2})$	k_3 $(hr^{-\alpha_3})$	k_4 $(hr^{-\alpha_2})$	k_5 $(hr^{-\alpha_3})$	σ_1	σ_2	σ_3
0.1486	0.0993	0.0601	0.000	0.3775	0.25	0.25	0.35

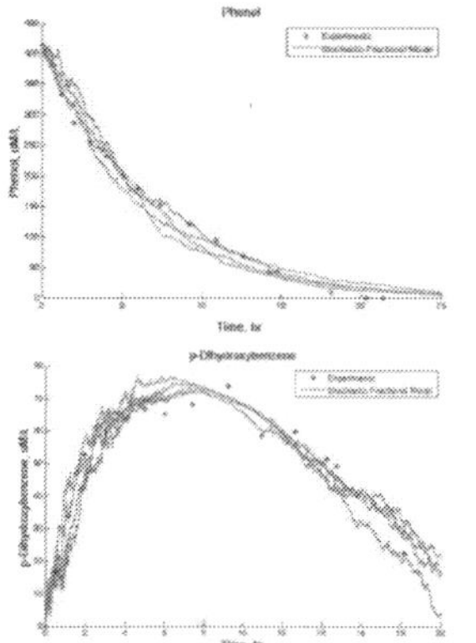

Summary

* Results have shown the potential of fractional calculus as a modeling tool for practical deterministic and stochastic applications in which memory effects are significant
* Some of our contributions to the area, apart from various fractional models, include the implementation of the gradient method for a FOCP, the predictor corrector technique for fractional stochastic models and the concept of a fractional Ito Process

Kinetic Theory for Granular Flow (KTGF) Revisited

Yujian Sun, Milorad P. Dudukovic*

Chemical Reaction Engineering Laboratory, Department of Energy, Environmental & Chemical Engineering, Washington University in St. Louis, St. Louis, MO 63124, USA

Abstract: Granular flows are ubiquitous in process engineering involving solid particles. Computational fluid dynamic models that treat the granular phase as a continuous fluid need constitutive models for closure of the stress tensor. The most commonly used models are derived from the kinetic theory for granular flow (KTGF), which was developed and popularized by many (e.g. Jenkins and Savage, 1983; Lun et al., 1984; Gidaspow, 1994). These traditional models assume that during an inelastic collision, only the normal component of the relative velocity is decreased, while the tangential component remains unaltered due to the assumption of "smooth" particles. In the present work, we assume that it is the velocity magnitude (speed) that is dampened due to an inelastic collision between two smooth or nearly smooth particles, rather than just the normal velocity component. Based on this assumption, new phenomenological models arise for solids pressure, shearing and bulk viscosity, thermal conductivity, and most importantly for the dissipation rate, which is found to be almost twice as large as the original one. Thus, by neglecting the effect of inelasticity on tangential velocities, the dissipation rate would be underestimated by almost one hundred percent. The proposed models reduce to those for dense gases (Chapman and Cowling, 1970) when the coefficient of restitution e is equal to one, as does Gidaspow (1994). The stress tensor now becomes asymmetric without the need to assume a different restitution coefficient in the tangential direction, and to add transport equations for angular velocity and rotational fluctuation energy, as indicated by Lun and Savage (1987) and Lun (1991). Therefore, the proposed models predict less underestimated dissipation rate without extra complexity. The significance of the current models and the comparison with more traditional ones are discussed.

Keywords: kinetic theory for granular flow (KTGF), phenomenological models, dissipation rate

References:

Chapman, S.; Cowling, T. G. *The Mathematical Theory of Non-Uniform Gases: An Account of the Kinetic Theory of Viscosity, Thermal Conduction, and Diffusion in Gases*; 3rd ed.; Cambridge University Press: Cambridge, 1970.

Gidaspow, D. *Multiphase Flow and Fluidization: Continuum and Kinetic Theory Descriptions*; Academic Press: New York, 1994.

*To whom correspondence should be addressed: dudu@seas.wustl.edu

Lun, C. K. K. Kinetic Theory for Granular Flow of Dense, Slightly Inelastic, Slightly Rough Spheres. *Journal of Fluid Mechanics* **1991**, *233*, 539.

Lun, C. K. K.; Savage, S. B. A Simple Kinetic Theory for Granular Flow of Rough, Inelastic, Spherical Particles. *Journal of Applied Mechanics* **1987**, *54*, 47.

Lun, C. K. K.; Savage, S. B.; Jeffrey, D. J.; Chepurniy, N. Kinetic Theories for Granular Flow: Inelastic Particles in Couette Flow and Slightly Inelastic Particles in a General Flowfield. *J. Fluid Mech.* **1984**, *140*, 223–256.

Jenkins, J. T.; Savage, S. B. A Theory for the Rapid Flow of Identical, Smooth, Nearly Elastic, Spherical Particles. *J. Fluid Mech.* **1983**, *130*, 187–202.

*To whom correspondence should be addressed: dudu@seas.wustl.edu

Influence of flue gas radiative properties, burner geometry and shadow effects in steam cracking furnace simulations

Yu Zhang[1,2], Carl M. Schietekat[1], Kevin M. Van Geem[1], Feng Qian[2], Guy B. Marin[1]

[1] Laboratory for Chemical Technology, Department of Chemical Engineering, Ghent University, Technologiepark 918, B-9052 Zwijnaarde, Belgium

[2] Key Laboratory of Advanced Control and Optimization for Chemical Processes of Ministry of Education, East China University of Science and Technology,130 Meilong Road, Shanghai200237, China

Introduction

Large fuel-fired furnaces have many applications in refining and (petro-)chemistry. In petrochemistry, one of the most important applications is in the hot section of steam cracking units for the production of olefins and aromatics. Multiple tubular reactors are suspended in these furnaces. The process is very energy-intensive as most reactions are highly endothermic, consuming about 8% of the total chemical industry energy use [1]. In order to improve the process' energy efficiency and the olefin selectivity, a lot of efforts have been made towards the rigorous simulation of steam cracking units. To this end detailed reaction networks describing the gas-phase radical chemistry [2] and coking models [3] have been implemented in one-dimensional, two-dimensional and three-dimensional reactor models [4,5]. To obtain accurate run length predictions with these reactor simulations a well-defined heat flux profile to the reactors is necessary, which can only be obtained by detailed furnace simulations with models describing combustion and radiative heat transfer.

In recent years, significant progress has been made in the modeling of steam cracking furnaces using computational fluid dynamics (CFD). However, furnace modeling can still be improved in three aspects. First, the flue gas radiative properties have been treated as gray in most of the works, which can lead to temperature underpredictions of more than 100 K [6]. Second, the detailed geometry of the burners is often omitted in the industrial furnace simulation, although it is accounted for during burner design. Hence, the effect of the burner geometry details on the heat flux to the reactors has never been assessed. Last, the so-called "shadow effect" [7] arising from the projected shadows between adjacent reactors always leads to significant heat flux non-uniformities. This phenomenon is important but is rarely considered in furnace simulations as often only a single reactor is simulated.

This work solves all three aforementioned issues in the simulation of an industrial naphtha cracking furnace equipped with long-flame floor burners. A "nine-band model" is developed from the exponential wide band model (EWBM) [8] and is validated against total emissvities calculated by Leckner's correlation [9] as well as heat flux profiles of a small test furnace. It is then used as a non-gray gas radiative property model and compared to the gray gas implementation of the Weighted Sum of Gray Gas Model (WSGGM) in the commercial software FLUENT 14.0. Three cases were studied to investigate the influence of gas properties and the burner geometry on the furnace simulation. Shadow effects are evaluated in all three cases by performing separate simulations for all 22 reactors in the furnace.

Mathematical models

A steam cracking reactor simulation software package COILSIM1D [5] is used in the present work. This software developed in in the Laboratory for Chemical Technology (LCT) contains a reaction network with more than 750 components and thousands of elementary reactions for steam cracking. By assuming an one-dimensional plug flow inside the cracking reactor, a set of continuity equations for the process gas species are solved simultaneously with the energy and the momentum equation. The detailed composition of complex feedstock like naphtha is obtained by molecular reconstruction method based on its commercial indices such as average molecular mass, PIONA analysis, specific density, etc. The software package SimCO embodied in COILSIM1D is used for this purpose. It provides two methods to do molecular reconstruction for naphtha, namely the Shannon entropy maximization or an Artificial Neutral Network (ANN) [10].

The furnace radiation section is divide into few millions of cells, on which the mass, momentum and energy conservation equations for a steady-state, compressible, reactive fluid are solved. Gravity is also taken into account since it is reported that it has some influences on the flow pattern of the flue gas [11]. The Renormalization Group (RNG) k-ε turbulence model is adopted to close the Reynolds Averaged Navier Stokes (RANS) equations. However, turbulence flow is greatly affected by the presence of a wall, while the k-ε model is valid primarily for turbulent core flows. This means that a near wall function is needed to describe the flow behavior close to a wall. In the present work, the standard wall function is chosen to reduce the computational cost because it does not require highly refined cells near the wall boundary.

For the modeling of combustion process, a two-step global reaction scheme derived by Westbrook and Dryer [12] is employed. In this reaction scheme, hydrocarbons are partially oxidized to carbon monoxide in the first step. The carbon monoxide then reacts with oxygen, yielding carbon dioxide. According to the fuel composition four reactions are included:

$$CH_4 + 1.5O_2 \rightarrow CO + 2H_2O \tag{1}$$

$$C_2H_4 + 2O_2 \rightarrow 2CO + 2H_2O \quad (2)$$

$$CO + 0.5O_2 \rightarrow CO_2 \quad (3)$$

$$H_2 + 0.5O_2 \rightarrow H_2O \quad (4)$$

An extended form of Eddy-Dissipation Model (EDM) known as finite-rate/eddy-dissipation model is adopted to account for the interaction between turbulent mixing and the combustion reactions. The minimum of the Arrhenius and eddy-dissipation reaction rates indicates whether the reaction or mixing is the rate determining step, and is used as the net reaction rate.

The Radiative Transfer Equation (RTE) is solve by the Discrete Ordinates (DO) model, resulting in equation (5).

$$\nabla \cdot (I(\vec{r},\vec{s})\vec{s}) = -(\kappa + \sigma_s)I(\vec{r},\vec{s}) + \kappa n^2 \frac{\sigma T^4}{\pi} + \frac{\sigma_s}{4\pi}\int_0^{4\pi} I(\vec{r},\vec{s}')\Phi(\vec{s},\vec{s}')d\Omega' \quad (5)$$

The change of radiative intensity I along a path in direction $\vec{s}$ (left-hand side) equals the decrease of intensity by the absorption and scattering away from direction $\vec{s}$ (first term on right-hand side) and increase of intensity due to the emission from the hot gases (second term) and the radiation scattered from other directions $\vec{s}'$ (third term). When performing a non-gray calculation, a number of spectral intensities are computed by solving RTEs for several wavelength bands separately. The total intensity over all bands is obtained from the sum of these spectral intensities.

Flue gas non-gray radiative properties

Solving the RTE for an absorbing-emitting gas mixture contained by opaque walls requires the knowledge of the gas radiative properties in terms of the absorption coefficient κ, which varies strongly and rapidly across the spectrum. Thus the gray gas model which uses averaged absorption coefficient over the entire spectrum is almost never a good one. To take the non-gray properties into account, the most successful wide band model, namely the Exponential Wide Band Model (EWBM) proposed by Edwards is adopted.

For the purpose of reducing computational cost, only 5 most important absorption bands out of 11 H_2O and CO_2 absorption bands in EWBM are taken into account in the present work. As shown in **Table 1**, two bands at 2.7 μm are overlapping to each other, hence the whole spectrum is split by four spectral ranges, yielding five so-called spectral windows in-between these absorption bands. So the number of RTEs need to be solved is nine, making a newly developed nine-band model.

The nine-band model is validated against EWBM and Leckner's correlation [9] by comparing the total emissivities of H_2O and CO_2 calculated from these models. The results are shown in **Figure 1** (left), the EWBM and the Leckner's correlation predicts almost same emissivity values for both H_2O and CO_2. Agreement between the EWBM and the nine-band model for CO_2 is excellent, while difference between the two models for H_2O is larger especially at lower temperature. This can be attributed to the omitting of some absorption bands in the nine-band model. Since the nine-band model is only inaccurate at temperatures below 1200 K, which is lower than the averaged furnace temperature, the effect of this inaccuracy on the furnace simulation results is quite limited. This statement is corroborated by **Figure 1** (right), where the average heat flux profile to all reactor coils in a small test furnace is depicted. The results using the nine-band model and the Leckner's correlation are almost identical, indicating that the nine-band model is suitable for the simulation of furnaces.

Table 1: Absorption bands of EWBM included in the nine-band model

Band name in EWBM	Nine-band model
H_2O – 71 µm	-
H_2O – 6.3 µm	Included
H_2O – 2.7 µm	Included
H_2O – 1.87 µm	-
H_2O – 1.38 µm	-
CO_2 – 15 µm	Included
CO_2 – 10.4 µm	-
CO_2 – 9.4 µm	-
CO_2 – 4.3 µm	Included
CO_2 – 2.7 µm	Included
CO_2 – 2.0 µm	-

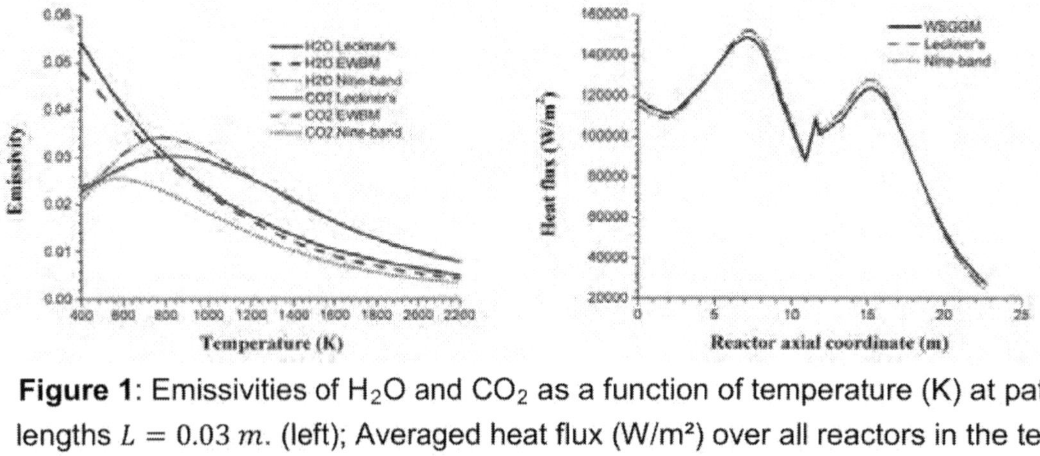

Figure 1: Emissivities of H_2O and CO_2 as a function of temperature (K) at path lengths $L = 0.03\,m$. (left); Averaged heat flux (W/m²) over all reactors in the test furnace as a function of the reactor axial coordinate (m). (right)

Results

Two furnace configurations were built to investigate the effect of modeling the burner details on the furnace and reactor simulations. In the first configuration, all the geometrical details such as stages and tips are modeled, while in the second configuration the burners are modeled as a rectangular inlet in the furnace floor. The gray implementation of WSGGM, which is widely used in steam cracking furnace simulation as gray gas model, is adopted to compare with the developed nine-band non-gray model. Three cases were studied: a 'gray gas radiation model and detailed burner geometry' case, a 'non-gray gas radiation model and detailed burner geometry' case and a 'non-gray gas radiation model and simplified burner geometry' case. They are referred to as "gray detailed", "non-gray detailed" and "non-gray simplified" respectively.

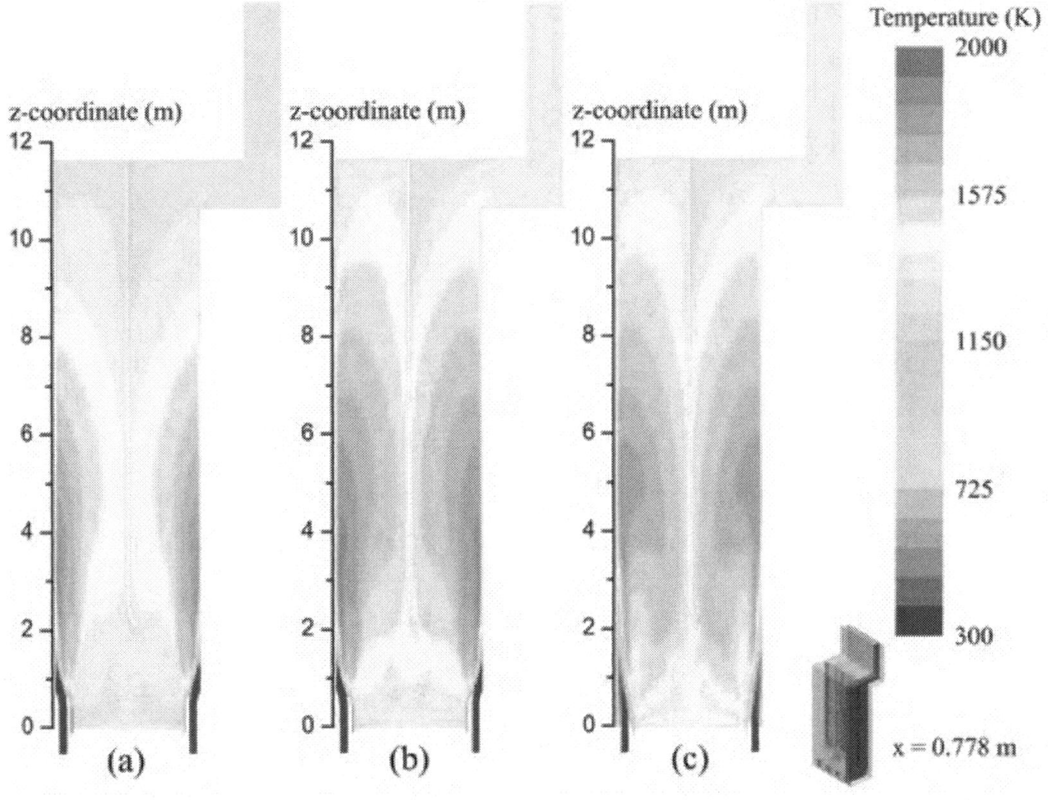

Figure 2: Temperature (K) field in a vertical cross section at x = 0.778 m: (a) gray detailed case; (b) non-gray detailed case; (c) non-gray simplified case.

The flue gas temperature contours of a vertical cross section at x = 0.778 m are depicted in **Figure 2**. Higher flame temperature is observer when non-gray gas model is applied, which is consistent with the work of Stefanidis et al. [13]. In spite of the big difference in flue gas temperature, flame shapes of the gray detailed and non-gray detailed cases very similar. On the other hand, simplification of the burner geometry gives a different temperature distribution with a much wider flame and a shift of the temperature maximum to higher positions. The results indicate that the radiative properties affect the mean flue gas temperature but have very limited influence on the combustion process.

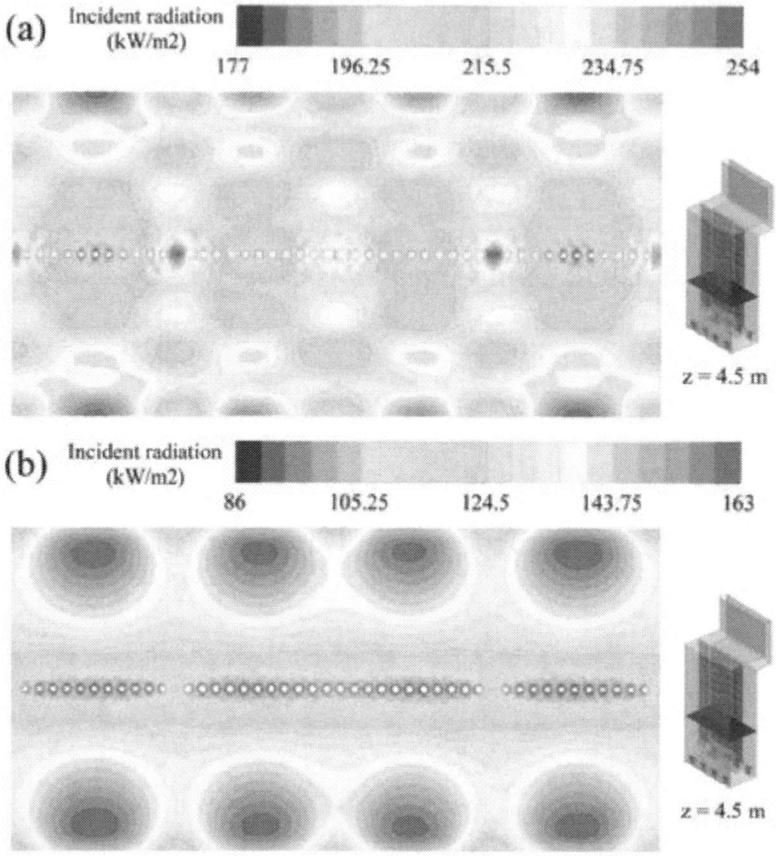

Figure 3: Incident radiation (kW/m²) field for the non-gray detailed case in a horizontal plane at z = 4.5 m in the: (a) first spectral window; (b) 2.7 μm band.

The bands in a non-gray model can be divided into two categories: the spectral windows and the absorbing/emitting bands. In the former category, the flue gas absorption coefficient is zero. Thus radiation emitted by the refractory walls with wavelengths in the spectral windows travels directly to the reactor coils. Radiation with wavelengths in the absorption bands, however, experiences an absorbing and re-emitting process by the flue gas before reaching the reactors. This leads to different radiative heat transfer behaviors in the spectral windows and the absorption bands. The band incident radiation in the first spectral window and the 2.7 μm band are depicted as an example in **Figure 3**. It is very clear that the location of maximum radiation intensity in the spectral windows, which is only determined by the refractory and reactor outer wall temperatures, is just right above the burners where the refectory wall temperature occurs. However, due to the absorption and redistribution of the radiation emitted from the refractory by flue gas, the maximum radiation intensity is found about 0.2 m away from the refractory. The total incident radiation distribution is a combination of these two categories.

Since a reactor simulation is performed for all 22 reactors individually, the variation of heat transfer to the different reactors and the resulting distribution of Coil Outlet Temperatures (COT) is studied and shown in **Figure 4**. The average value of the total transferred thermal power of the two non-gray cases is very similar, while those predicted by gray gas model are much higher. As a direct consequence of the heat transfer, COT difference between the reactors is up to 25.2 K, 29.1 K and 24.6 K for the gray detailed, non-gray detailed and non-gray simplified case respectively. A reduction of the non-uniformity of the heat transfer to the reactors could be obtained by e.g. altering the distribution of the fuel over the different burners or the mass flow rate over the different reactors.

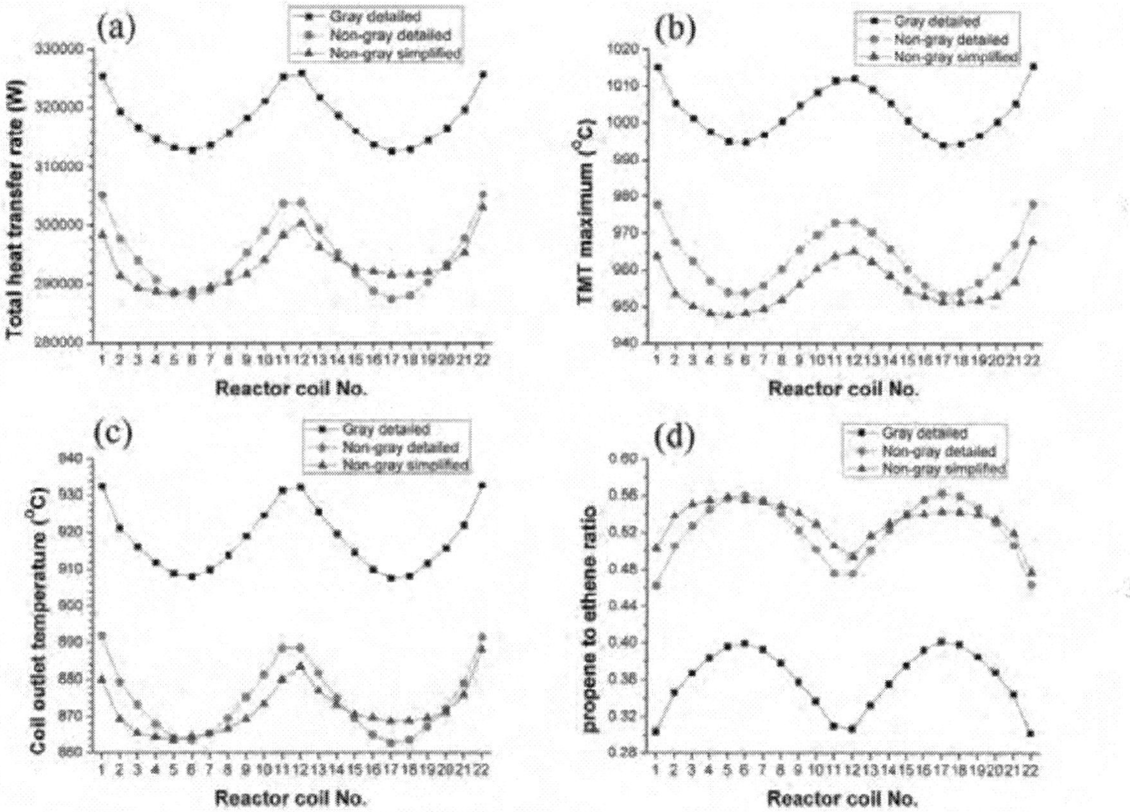

Figure 4: The distribution of (a) total transferred thermal power (kW), (b) maximum external tube metal temperature (K), (c) coil outlet temperature (K) and (d) propene-to-ethene ratio (wt%/wt%) over all 22 reactors

Conclusions

Simulation results show that the gray gas simulation overpredicts the flue gas temperature by about 70 K compared to the non-gray gas simulation, resulting in a 3.6% higher thermal efficiency and 44 K higher average Coil Outlet Temperature (COT). For the simulation with simplified burners, longer flames are simulated and more uniform heat flux profiles to the reactors are obtained. Hence, the tube metal temperature is underpredicted by about 20 K compared to the detailed burner simulation. Finally, shadow effects in the furnace cause strong differences in the total heat flux per reactor. The maximum COT difference between two reactors is about 29.1 K, leading to a difference in propene-over-ethene of 0.1 between two different U-coils in the furnace. This information can be utilized to unevenly distribute the flow rate of the naphtha feedstock individual reactors to obtain more uniform COTs and olefin yields.

Acknowledgment

The computational work was carried out using the STEVIN Supercomputer Infrastructure at Ghent University, funded by Ghent University, the Flemish Supercomputer Center (VSC), the Hercules Foundation and the Flemish Government – department EWI. The authors also acknowledge the financial support from Long Term Structural Methusalem Funding by the Flemish Government, the '111_ Project by the Chinese Government, and the China Scholarship Council (CSC).

References

1. Ren T, Patel M, Blok K. Olefins from conventional and heavy feedstocks: Energy use in steam cracking and alternative processes. *Energy.* 2006;31(4):425-451.
2. Van Geem KM, Reyniers M-F, Marin GB, Song J, Green WH, Matheu DM. Automatic reaction network generation using RMG for steam cracking of n-hexane. *AIChE Journal.* 2006;52(2):718-730.
3. Wauters S, Marin GB. Kinetic Modeling of Coke Formation during Steam Cracking. *Industrial & Engineering Chemistry Research.* 2002;41(10):2379-2391.
4. Schietekat CM, Van Cauwenberge DJ, Van Geem KM, Marin GB. Computational fluid dynamics-based design of finned steam cracking reactors. *AIChE Journal.* 2014;60(2):794-808.
5. Van Geem KM, Žajdlík R, Reyniers M-F, Marin GB. Dimensional analysis for scaling up and down steam cracking coils. *Chemical Engineering Journal.* 2007;134(1-3):3-10.
6. Modest MF. The Treatment of Nongray Properties in Radiative Heat Transfer: From Past to Present. *Journal of Heat Transfer.* 2013;135(6):061801-061801.
7. Heynderickx GJ, Cornelis GG, Froment GF. Circumferential tube skin temperature profiles in thermal cracking coils. *AIChE Journal.* 1992;38(12):1905-1912.
8. Edwards DK, Balakrishnan A. Thermal radiation by combustion gases. *International Journal of Heat and Mass Transfer.* 1973;16(1):25-40.
9. Leckner B. Spectral and total emissivity of water vapor and carbon dioxide. *Combustion and Flame.* 1972;19(1):33-48.
10. Pyl SP, Van Geem KM, Reyniers M-F, Marin GB. Molecular reconstruction of complex hydrocarbon mixtures: An application of principal component analysis. *AIChE Journal.* 2010;56(12):3174-3188.
11. Hassan G, Pourkashanian M, Ingham D, Ma L, Newman P, Odedra A. Predictions of CO and NOx emissions from steam cracking furnaces using GRI2.11 detailed reaction mechanism – A CFD investigation. *Computers & Chemical Engineering.* 2013;58:68-83.
12. Westbrook CK, Dryer FL. Simplified Reaction Mechanisms for the Oxidation of Hydrocarbon Fuels in Flames. *Combustion Science and Technology.* 1981;27(1-2):31-43.
13. Stefanidis GD, Merci B, Heynderickx GJ, Marin GB. Gray/nongray gas radiation modeling in steam cracker CFD calculations. *AIChE Journal.* 2007;53(7):1658-1669.

390483 Chematica: An Automatic Chemist for the XXI Century

Abstract:

The millions of reactions performed and compounds synthesized by organic chemists over the past two centuries connect to form a network larger than the metabolic networks of higher organisms and rivaling the complexity of the World Wide Web. Despite its apparent randomness, the network of chemistry has a well-defined, modular architecture and evolves in time according to trends that have not changed since the inception of the discipline, thus projecting into chemistry's future. Analysis of organic chemistry using the tools of graph theory, linguistics and machine learning enables the optimization of chemical syntheses including predictions of how unknown molecules can be efficiently made. All of these capabilities are embodied in a unified software environment called *Chematica*. *Chematica* plans synthesis of life-saving drugs, dyes and pigments, complex natural products and any other molecule one wishes to make in a matter of seconds vs. weeks to months by human experts. *Chematica* is the Deep Blue of chemistry, constantly self-learning and improving. It is already being deployed in and used by the American and E.U. chemical industries and also by several national security/intelligence agencies (in the context of chemical warfare prevention). Please come see the live demo of *Chematica* at the lecture!

Key References:

1. B.A. Grzybowski, K.J.M. Bishop, B. Kowalczyk, C.E. Wilmer The wired universe of organic chemistry *Nature Chemistry* 1, 31 (2009).

2. K.J.M. Bishop, R. Klajn & B.A. Grzybowski **The core and most useful molecules in organic chemistry** *Angew. Chem. Int. Ed.*, 45, 5348 (2006).

3. M. Fialkowski, K.J.M. Bishop, V. Chubukov, C.J. Campbell & B.A. Grzybowski, The architecture and evolution of organic chemistry, *Angew. Chem. Int. Ed.*, 44, 7263 (2005)

4. C.M. Gothard, S. Soh, N.A. Gothard, B. Kowalczyk, Y. Wei. B. Baytekin, B.A. Grzybowski Rewiring Chemistry: Algorithmic Discovery and Experimental Validation of One-Pot Reactions in the Network of Organic Chemistry. *Angew. Chem. Int. Ed.* 51, 7922 (2012).

5. M. Kowalik, C.M. Gothard, A.M. Drews, N.A. Gothard, A. Wieckiewicz, P.E. Fuller, B.A. Grzybowski, K.J.M. Bishop Parallel Optimization of Synthetic Pathways within the Network of Organic Chemistry. *Angew. Chem. Int. Ed.* 51, 7928 (2012).

6. P.E. Fuller, C.M. Gothard, N.A. Gothard, A. Wieckiewicz, B.A. Grzybowski Chemical Network Algorithms for the Risk Assessment and Management of Chemical Threats. *Angew. Chem. Int. Ed.* 51, 7933 (2012).

380229 Adjoint Model for the Efficient Calculation of Atmospheric Nucleation Sensitivities

Abstract:

Adjoints can be used for computationally efficient calculation of the model output gradient in both linear and nonlinear systems. While model gradients may be calculated in a forward run using finite differences, the computational time in this case increases with the dimension of the input space. During a reverse run, however, the model state is stored after certain steps in the instruction sequence and then passed back through a piecewise-differentiated sequence so that additional inputs have no additional computational cost. These reverse-mode adjoints have a wide range of applications including sensitivity study, stability analysis, and optimization. Here, we present one environmentally-relevant use with the adjoint model of an atmospheric ice nucleation code by Barahona and Nenes (ABN09). ABN09 calculates the change in nucleated ice crystal number with size and composition of seed material as well as atmospheric temperature and vertical velocity in a single run. Running this gradient model with inputs from a global climate model, we can see in which geographic regions ice crystals nucleate heterogeneously versus homogeneously and where these regimes are most sensitive to addition of ice nuclei. An attribution analysis is performed to see whether more variability in ice crystal number comes from inherent model bias or from input fluctuations. We test three heterogeneous nucleation spectra to describe the portion of ice-nucleating seed material within ABN09 to understand the effect on sensitivity of different model formulations.

389640 Multiscale Modeling with Dynamic Discrepancy

Abstract:

The dynamic discrepancy methodology is an approach to multi-scale modeling based on quantification of uncertainty in scale-bridging. It is a reduced order approach using stochastic functions to efficiently replace model variability removed when reducing problem complexity. It can be utilized in conjunction with rigorous reduced-order strategies such as the proper orthogonal decomposition. The stochastic character of the method leads to the opportunity for machine learning in reduced model training. A demonstration of the method on problems in modeling of carbon capture systems will be presented, using both hypothetical and real data sets.

366636 Multiscale Modelling of Polycaprolactone Self-Assembly in Water-Acetone Mixtures

Abstract:

Polycaprolactone (PCL) is an important polymer that finds relevant applications in the pharmaceutical and biomedical fields, often in the form of nanoparticles. These nanoparticles are generally obtained through solvent displacement, namely PCL is dissolved in a "good" solvent (e.g. acetone) and the solution is subsequently mixed with a "bad" solvent (e.g. water). While PCL molecules form in acetone (at moderate concentrations) a stable solution, upon mixing with water (with which PCL exhibits a lower compatibility), PCL molecules self-assemble, via a diffusion-limited aggregation process, into nanoparticles. In this work a multiscale model is developed and validated; the model accounts for all the key factors involved: (1) effect of mixing of water and acetone on the final nanoparticle characteristics, (2) polydispersity of the obtained nanoparticles and (3) complexity of the interactions between water, acetone and PCL molecules. For these three different key factors, three modeling approaches (acting on three different scales) are employed. The acetone-water mixing process is described through computational fluid dynamics, the PCL self-assembly into a polydisperse population of particles is described through a population balance model (directly solved within the computational fluid dynamics code), whereas the PCL molecule aggregation kernel is derived from the radius of gyrations and diffusion coefficients of PCL molecules, calculated in turn from molecular dynamics simulations.

Computational fluid dynamics simulations are carried out in Ansys Fluent 14, in which the population balance model is implemented via the quadrature method of moments with user-defined functions and scalars, while molecular dynamics simulations are carried out with GROMACS 4.6 (by employing the OPLS-AA force field together with the SPC model for PCL, acetone and water molecules).

Molecular dynamics simulations, carried out for PCL molecules of different molecular weights and in different water-acetone mixtures, highlight that PCL molecules behave very differently when they are in pure water or in water-acetone mixtures (notwithstanding the actual composition). In fact, as soon as some acetone is introduced into the system, it clusters around the PCL molecule, as evidenced by the enclosed figure (where acetone molecules are represented in light blue, whereas water molecules are represented in dark blue) drastically changing its behavior.

The values of radius of gyration obtained by molecular dynamics simulations are well described by the Flory law, whereas the diffusion coefficients obtained are well described by the Stokes-Einstein law. These two pieces of information are then combined together to quantify the PCL molecule aggregation kernel, that is in turn used to predict the final size distribution of PCL nanoparticles obtained by solvent displacement carried out in a confined impinging jets mixer, resulting in satisfactory agreement with experiments.

380242 Effect of Glycolic Acid on Hydration of Tricalcium Silicates – a Combined Experimental and Molecular-Scale Computational Study

Abstract:

Tricalcium silicate is the most abundant constituent of ordinary protland cement and its hydration is largely responsible for early strength development in concrete. Modern concrete contains not only cement but frequently also includes a variety of organic compounds that are used to modify one of any number of fluid or hardened properties. One class of such compounds is used to alter the rate of hydration. Glycolic acid, a well-known hydration inhibitor, is thought to interfere with calcium hydroxide precipitation, a by-product of tricalcium silicate hydration, and thus obstructs the rate of reaction. Little actual experimental evidence, however, in support of this or other hypothesis is available. A combined experimental and computational study has been done in an effort to elucidate this mechanism. Both a semi-empirical quantum mechanical (Parametric Method 3) and a force field method (CLAYFF) were employed for computational study. The recently developed semi-empirical force field CLAYFF considers flexibility of surface OH groups from calcium hydroxide and allows energy-momentum transfer between interfaces. The primary focus is to calculate the interaction energy (binding energy) of glycolic acid on the calcium hydroxide surface. X-ray, calorimetric and microstructural characterization were also used to elucidate the process and support the computational work.

390787 Biodetoxification By Lipid Emulsion Droplets: Coarse Grained Molecular Dynamics at the Oil/Water Interface

Abstract:

Drug intoxications cause 2 million emergency room visits each year and now lead to more accidental deaths than motor vehicle accidents. As specific antidotes do not always exist and often the particular intoxicant may not be readily identified, there has been increasing interest in broad-spectrum detoxification agents. Lipid droplets have exhibited considerable promise as resuscitation agents, and it is becoming increasingly clear that their multifunctionality imbues them with advantages beyond simply corporeal scavenging, or entrapment, of toxins. Their action is both direct stimulation of toxin-impaired function in critical organs such as the heart and indirect organ protection via local toxin sequestration and subsequent redistribution to 'storage' organs (e.g. adipose tissue). This indirect mechanism depends on the reversible uptake of drug molecules by lipid droplets. The requisite scavenging action, known as the 'lipid sink', is thought to be limited to acting on highly lipophilic toxins, with hydrophobic partitioning as the primary driving force. However, many of the compounds in question contain regions of considerable polarity or exist in a predominantly charged state at physiological pH. Hence, polar and electrostatic interactions are likely to play an important role in driving the sequestration of toxins. Furthermore, one-off clinical reports exist that credit lipid droplets with reversing toxicities due to drugs of low lipophilicity. Rational application of lipid biodetoxification and optimization of therapeutic formulations requires identification of the key physicochemical characteristics that govern association of toxin molecules with lipid droplets.

We have employed coarse-grained molecular dynamics simulations utilizing the MARTINI force field to develop molecular topologies for the multiphase system of interest (aqueous/oil system with phospholipid monolayer interface). The MARTINI scheme has also been used to represent the drug molecules of interest, ranging from local anesthetics to antidepressants and anticoagulants. In the case of the cardiotoxic and lipophilic local anesthetic bupivacaine, both enantiomers are sequestered within the phospholipid monolayer, where they are preferentially located near the phospholipid backbone. The presence of the anesthetic alters the monolayer structure and dynamics in a manner that depends on both the isomer and its charge. The highly lipophilic molecule brodifacoum causes considerable disruption of the monolayer, with the initial approach driven by hydrophobic partitioning and subsequent entrapment promoted by polar interactions between the toxin's hydroxycoumarin region and the phospholipid head group. The less lipophilic molecule acetaminophen also readily embeds within the lipid monolayer – even in its protonated form. If penetration into the oil core occurs, it does so over timescales greater than 200 ns and, in some cases, re-entry into the monolayer has been observed. In our initial studies, it appears that lipophilicity alone is not a good predictor of droplet association. A similar observation was made through an experimental study of drug uptake by commercial lipid emulsions.

388564 Numerical Analysis of the Water Depth of Open Raceway Ponds on Energy Consumption and Algal Productivity Using a Computational Fluid Dynamics Model

Abstract:

Some microalgal strains have an excellent potential for biofuel production and the treatment of wastewater by reducing nitrate, phosphate and organic matter in the wastewater. Open cultivation systems, especially open raceway ponds have received considerable attention for the commercial cultivation of microalgae for biofuel production because of their easier and cheaper setup than photobioreactors. Major limitations of open raceway ponds include poor light utilization by the cells, high evaporative losses of water, high diffusion of CO_2 to the atmosphere and requirement of large areas of land. The flow rate and the depth of water in a raceway pond affect the distribution of sunlight, CO_2 and nutrients for the algal production and also the energy consumption for the circulation of water in the pond. Therefore, the goal of this research is to use a 3D computational fluid dynamics model based on ANSYS FLUENT 15.0 to study the hydrodynamic effects of water depth and flow rate of the pond on energy consumption and algal productivity. The model will be used validated by the experimental data collected on a lab-scale raceway pond. The model can be used to optimize the design and analyze an open raceway pond at different scales.

386526 Computational Fluid Dynamics Modeling of Heat Transfer in Three Phases of Biomass Particles, Bed Materials and Gas during Fluidized Bed Gasification

Abstract:

Gasification has been identified as an energy-efficient, environmentally-friendly and economically-feasible technology to partially oxidize biomass into a gaseous mixture of syngas consisting of H_2, CO, CH_4 and CO_2. High-quality syngas can be further used to catalytically synthesize liquid fuels and produce hydrogen. Fluidized biomass gasification involves complicated heat transfer among the three phases of biomass particles, bed materials and fluidized gas. During fluidized bed gasification, a bed material such as sand is used as a heat carrier to directly supply heat to biomass particles. The biomass fluidized bed gasification process involves both particle-particle heat transfer and gas-particle heat transfer. There is limited research on heat transfer through particle-particle interaction during fluidized bed gasification. However, it is critical to understand how the biomass particle-bed particle interaction on the temperature profile of biomass particles in order to predict the yield and quality of syngas during fluidized bed gasification. In this research, a computational fluid dynamics (CFD) model is developed to analyze the behavior of three-phase, reactive gas-particle-particle flow with heat transfer via both gas-particle and particle-particle interaction during fluidized bed gasification of biomass. A commercial Fluent CFD software package is used as a basic platform for the computational model. A mathematical sub-model is developed to calculate the dynamic particle-particle heat transfer coefficient. A user defined function (UDF) is then developed to incorporate the sub-model into the Fluent CFD simulation platform. Experimental data are used to validate the CFD model with and without consideration of the particle-particle heat transfer. The model is used to predict the effects of biomass particle size, temperature, biomass/bed material ratio, biomass loading rate on the yield and composition of syngas using the CFD model.

The Impact of Radioactive Charging on the Microphysical Evolution and Transport of Radioactive Aerosols

Petros Vasilakos,[†] *Yong-ha Kim,*[†] *Sotira Yiacoumi,*[†] *Costas Tsouris,*[†,‡] *Athanasios Nenes*[†,*]
[†] Georgia Institute of Technology, Atlanta, Georgia 30332
[‡] Oak Ridge National Laboratory, Oak Ridge, Tennessee 37831-6181
*Email: athanasios.nenes@gatech.edu; Telephone: (404) 894-9225; (404) 894-9225

1. Introduction

Current atmospheric transport models used to study the transport and transformation of radioactive-particle debris do not take into consideration the increasing evidence that radioactive particles behave differently than non-radioactive ("cold") particles (Clement et al. 1995, Clement & Harrison 1992, Harrison & Carslaw 2003). Furthermore, predictions of radioactive debris fallout using information from extensive monitoring and conventional transport models alone are subject to considerable uncertainty and questions their use in the public response process if future nuclear reactor accidents or weapon testing accidents were to occur (Chesser et al. 2004, Lujanienė et al. 2007, Arimoto et al. 2005, Adeyini & Oladiran 2007, Yoshenko et al. 2006).

Aerosol dispersal models used in reactor accident scenarios do not consider all the microphysical aspects of aerosol atmospheric processing after emission (Yoshenko et al. 2006, Christoudias et al. 2013). State-of-the-art chemical transport models (CTMs) can address this limitation, as they quantitatively couple meteorological dynamics with chemical transformation of gas-phase and aerosol-phase pollutants to predict their behavior in the atmosphere. To accomplish this, CTMs, such as CMAQ and the Harvard GEOS-Chem model, have been coupled with key atmospheric processes (e.g., condensation/evaporation of semi-volatile species, coagulation of neutral particles, cloud processing, wet/dry deposition) which are then integrated over time. However, CTMs may omit important aspects of charged particle interactions that can play a major role during radioactive particle transport away from the radioactive sources. For example, a study of Chernobyl fallout over Tokyo reported that small radioactive particles, which had traveled half-way around the world, were surmised to have agglomerated with atmospheric components in the urban environment (Ooe et al. 1988).

The main reason why the atmospheric evolution of radioactive particles differs from other (neutral) atmospheric particles is that the former can develop charge spontaneously via direct ionization from radioactive decay and collision/adsorption of gas-phase ions (generated by the radioactive decay) from the gas-phase surrounding the particles (Harrison & Carslaw 2003, Clement et al. 1995, Clement & Harrison 1992). This charging may affect their transport, agglomeration/coagulation, and deposition processes. Currently, only charging mechanisms of radioactive aerosols, as well as a model to describe them, have been proposed. Few attempts have been made to incorporate the development of charge and its consequences in transport models for radioactive aerosols (Yoshenko et al. 2006, Clement et al. 1995, Clement & Harrison 1992). Neglecting these effects could have important implications, as prior work has shown that the charge developed during radioactive decay can be significant, and affect the predicted aerosol size distribution (Kim et al. 2013).

2. Objectives

The main objective of this paper is to present a comprehensive atmospheric modeling approach to provide a better understanding of the transport and deposition of radioactive debris in the environment. To investigate this, an augmented version of the TOMAS microphysics model (Adams & Seinfeld 2002) that accounts for radioactive charging is employed. The version used here has a resolution of 30 bins, covering particle sizes from 10 nm to 10 μm, meaning that the particle size distribution is split in 30 different sections, with each section having a lower and an upper limit on the particle diameter. All the particles are placed in each bin depending on their size, and considered to be internally mixed i.e. they have the same chemical composition. TOMAS accounts for all the relevant atmospheric processes of nucleation, condensation, coagulation, vertical mixing, cloud processing and deposition, in order to find the number concentration and size distribution of the modeled aerosol. To that end, only dry deposition, vertical mixing and coagulation were active (Figure 1). The model is referred to as a two-moment model (Adams & Seinfeld 2002) because it tracks two independent moments of the aerosol size distribution for each size bin – aerosol number and mass:

$$N_k = \int_{x_k}^{x_{k+1}} n_k(x)dx \quad (1)$$

$$M_k = \int_{x_k}^{x_{k+1}} x n_k(x)dx \quad (2)$$

where N_k, M_k are the total number and mass of aerosol in grid cell, $n_k(x)$ is the number of particles with masses included between $x + dx$, and x_k is the lowest boundary of the kth cell. The lowest and the largest boundaries of each cell are defined in terms of dry aerosol mass, in such a way that the largest boundary has double the mass of the lowest boundary. Using the above scheme, the ratio of M_k to N_k can vary with time, and greater computational speed is attained when gauged against single moment models, because of the coarser resolution required for the same accuracy (Adams & Seinfeld 2002).

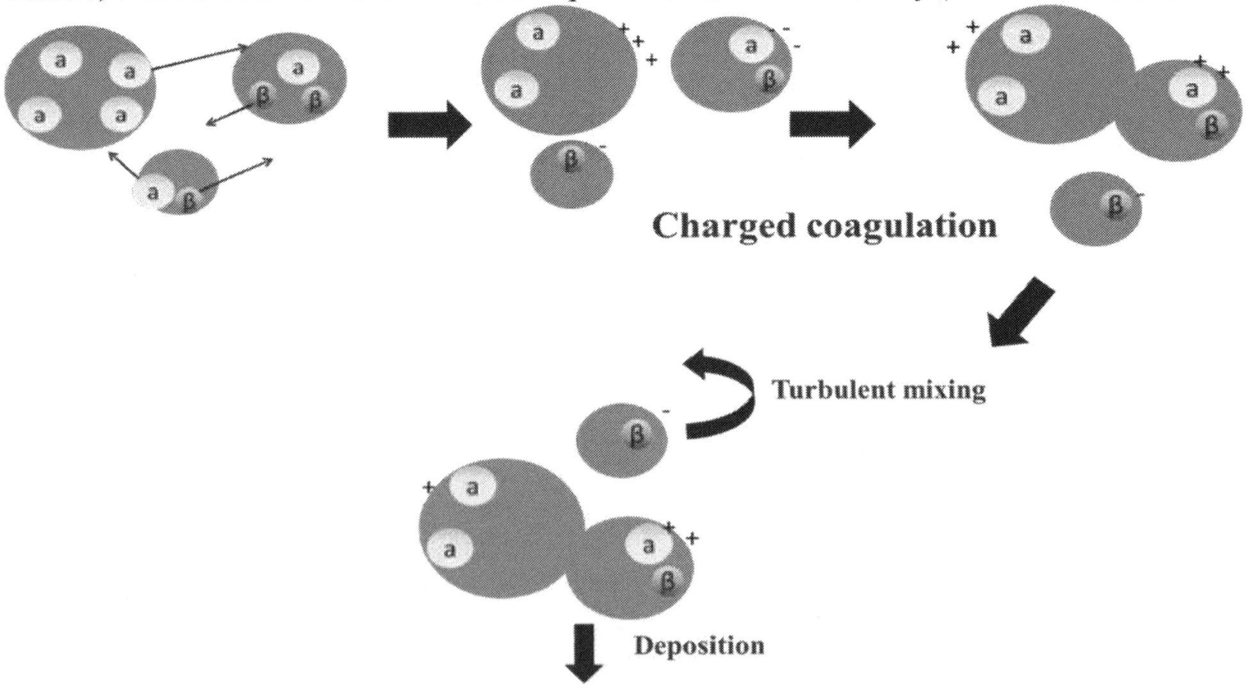

Figure 1. Schematic of the processes active in radioactive TOMAS

3. Model development and methodology

a. Average Correction Factor for Radioactivity-induced Charging

Radioactivity-induced charge can be accumulated on aerosol surfaces, but the number of the accumulated charges may be different for similar-size aerosols. The number concentration of aerosols with elementary charge j in cell k, $N_{k,j}$, can be predicted by assuming a Gaussian distribution (Clement et al, 1995).

$$N_{k,j} = \frac{\sum_k \sum_j N_{k,j}}{\sqrt{2\pi}\sigma} \exp\left(-\frac{(j-J)^2}{2\sigma^2}\right) \quad (3)$$

Coagulation rates of charged aerosols may differ from that of non-charged aerosols because of generation of electrostatic forces. The effects of electrostatic forces on Brownian coagulation can be included using the correction factor W, derived from the stability function (Spielman 1970), as

$$W_{k,i} = \frac{e^y - 1}{y} \quad (4)$$

with

$$y = \frac{j_k j_i e^2}{4\pi\varepsilon_0 \varepsilon (r_k + r_i) k_B T}$$

where e is the elementary charge, ε_0 is the vacuum permittivity, ε is the dielectric constant of the air, r is the aerosol radius, k_B is the Boltzmann constant, and T is the temperature. The average correction factor $\overline{W}$ proposed by Clement et al.(1995), which can consider the interaction of all charged aerosols, is:

$$\overline{W}_{k,i} = 1 + \frac{\sum_{j \neq 0} N_{k,j} N_{i,j} \left(W_{k,i}^{-1} - 1\right)}{\sum_j N_{k,j} \sum_j N_{i,j}} \quad (5)$$

The fractional term on the right hand side of eq. 3 is the average correction factor for charged aerosols in the grid cells. If repulsive electrostatic forces are predominant among aerosols, the sign of the fractional term is changed to minus and the average correction factor decreases below unity. The influences of radioactivity-induced charging on the collision can be corrected by multiplying the coagulation coefficient by eq. 3.

b. Coupling radioactivity-induced charging to TOMAS.

The TOMAS microphysics model is a moving sectional approach to track the 0th and 1st moments (number and mass) of aerosol size distributions (Adams & Seinfeld 2002, Lee & Adams 2012). The aerosol size distributions can be influenced by various microphysical processes such as deposition, vertical layer mixing, and coagulation. Particularly, in each cell, changes in number and mass concentrations, N and M by coagulation involving aerosol-aerosol interaction are given as

$$\begin{aligned}\frac{dN_k}{dt} &= 0.5K_{k-1,k-1}N_{k-1}{}^2 - K_{k,k}N_k{}^2 - N_k\sum_{i=k+1}^{I} K_{k,i}N_i + \psi_{k-1}\sum_{i=1}^{k-2} K_{k-1,i}M_i \\ &-\psi_{k-1}\sum_{i=1}^{k-1} K_{k,i}M_i - \frac{\psi_{k-1} - f_{k-1}}{2x_{k-1}}\xi\sum_{i=1}^{k-1} K_{k,i}M_i m_i - \frac{\psi_{k-1} - f_{k-1}}{2x_{k-1}}\xi\sum_{i=1}^{k-2} K_{k-1,i}M_i m_i\end{aligned} \quad (6)$$

$$\frac{dM_k}{dt} = K_{k-1,k-1}N_{k-1}M_{k-1} - K_{k,k}N_kM_k + N_k\sum_{i=1}^{k-1}K_{k,i}M_i - M_k\sum_{i=k+1}^{I}K_{k,i}N_i$$
$$+\psi_{k-1}x_k\sum_{i=1}^{k-2}K_{k-1,i}M_i - \psi_k x_{k+1}\sum_{i=1}^{k-1}K_{k,i}M_i + \frac{f_{k-1}}{2}\xi\sum_{i=1}^{k-2}K_{k-1,i}M_i m_i \tag{7}$$
$$-\frac{f_k}{2}\xi\sum_{i=1}^{k-1}K_{k,i}M_i m_i + \frac{\psi_k - f_k}{2x_k}\xi^3\sum_{i=1}^{k-1}K_{k,i}M_i m_i^2 - \frac{\psi_{k-1} - f_{k-1}}{2x_{k-1}}\xi^3\sum_{i=1}^{k-2}K_{k-1,i}M_i m_i^2$$

where i and k are integers to indicate the cells, K is the coagulation coefficient, I is the total number of the cells, ψ and f are the linear approximation parameters, ξ is the closure parameter determined by the cell spacing (here, $\xi = 1.0625$), x is the lowest dry mass boundary of the cell, m is the average particle mass of the cell, and t is the time.

In eqs. 5 and 6, the coagulation coefficient is the main parameter influencing coagulation rates of aerosols and it can be affected by several collision mechanisms involving interparticle forces and flow regimes. In TOMAS, it has been assumed that Brownian motion is the dominant collision mechanism (Adams & Seinfeld, 2002; Lee & Adams, 2012) and that the effects of electrostatic forces on the collision are negligible because, at equilibrium, the net charge of aerosols can become zero according to Boltzmann's law. Effects of non-continuum regimes on the collision are discussed by Dahneke (1983). A detailed description of TOMAS is available elsewhere (Adams & Seinfeld, 2002; Lee & Adams, 2012). The proposed scheme is presented in Figure 2.

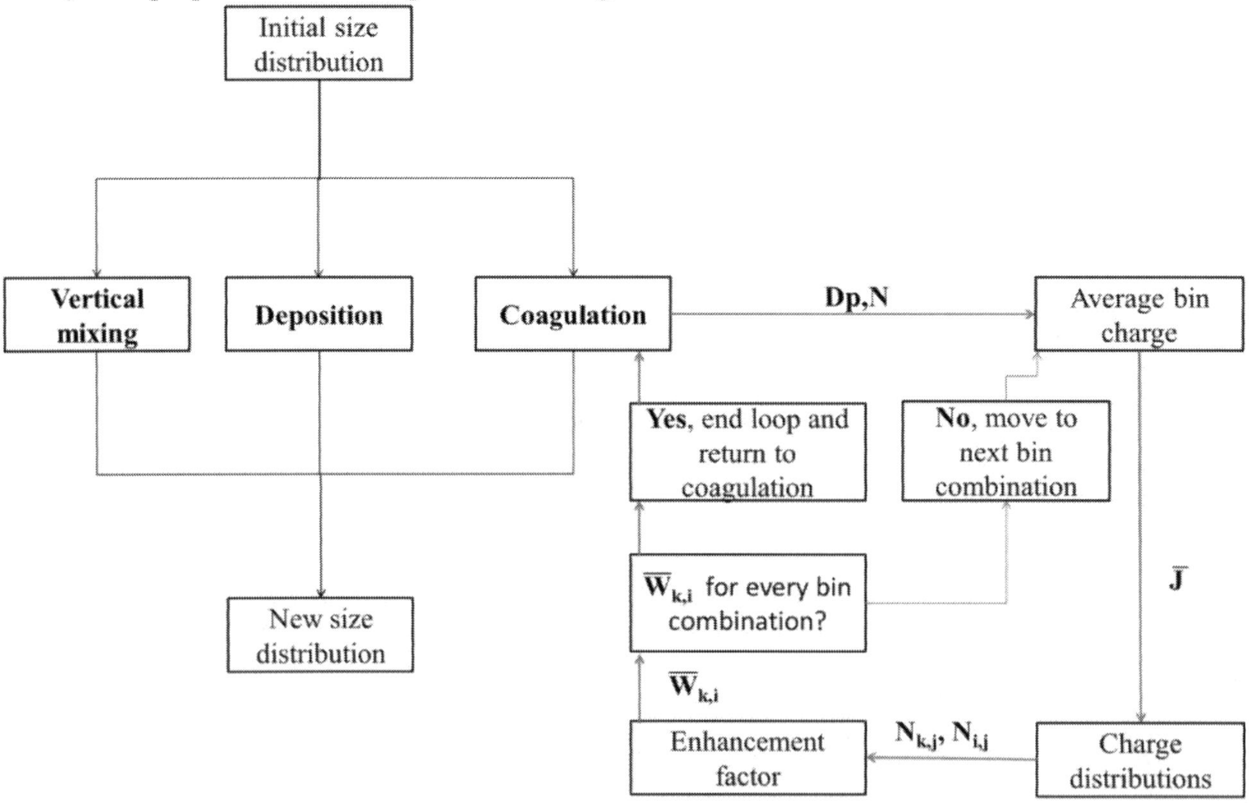

Figure 2. Flow diagram of the computational scheme. The number concentration and diameter of each bin are passed from the coagulation subroutine to the radioactive physics subroutine, where the average charge and charge distributions are calculated, and then used to estimate the enhancement factor, which is returned to the TOMAS coagulation module.

4. The effect of radioactive charging on the transport properties of aerosol

For all the cases simulated, three processes were accounted for; coagulation, vertical mixing (diffusion) and dry deposition. The simulation duration is set to 24 hours for iodine and 5 days for cesium. A higher simulation time was used for cesium, because of its higher half-life of 30.6 years compared to 8 days for iodine, leading to slower charging of the particles. For each case, simulations were carried out with two different configurations, one where the radioactivity effects are accounted for and another where they were ignored, and the results between those cases were compared. In both simulations, it was assumed that each of the five height levels had an initial particle population that followed a log-normal distribution with a single mode.

For the needs of the sensitivity analysis, the effect of radioactivity was gauged against various different initial radioactive particle distributions. The geometric mean diameter $\boldsymbol{D_g}$, the initial concentration of particles $\boldsymbol{N_0}$, and the standard geometric deviation $\boldsymbol{\sigma_g}$ are the parameters that define unique, single-mode, log-normal distribution.

Deposition characteristics of radioactive aerosols in the column consisting of five vertical layers were investigated using TOMAS with and without the average correction factor (hereafter radioactive and nonradioactive cases, respectively). Coagulation characteristics of radioactive aerosols in one layer were investigated first before analyzing their vertical transport. Then the aerosol size distributions and total mass residing in the layers were analyzed to elucidate effects of the radioactivity-induced charging on the particle lifetime.

Figure 3 shows the time evolution of the size distribution of ^{137}Cs aerosols at 400-m altitude, where the aerosol concentration was the highest in the 1000-m column. As time passed, the growth rates of ^{137}Cs aerosols became slower than those of nonradioactive aerosols. This could result from lower coagulation rates of ^{137}Cs aerosols. Because ^{137}Cs has lower decay rates, the diffusion charging could be enhanced, making most aerosols, including larger ones, negatively charged as shown in Figure 4. Gensdarmes et al. (2001) reported that ^{137}Cs aerosols can be negatively charged under conditions that the diffusion charging is prevailing. Thus, coagulation of ^{137}Cs aerosols could be slightly suppressed because of repulsive electrostatic forces, causing the accumulation of ^{137}Cs aerosols in this layer.

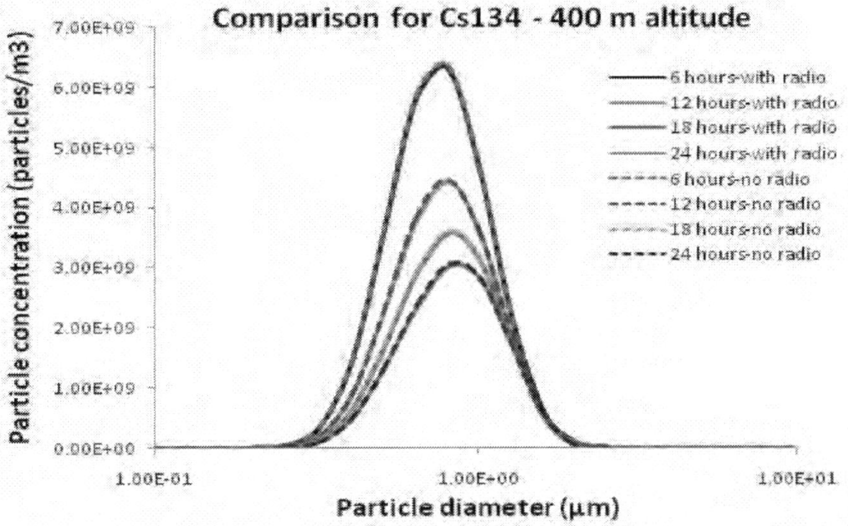

Figure 3. Evolution of the size distribution of ^{137}Cs aerosols at 400 m (t = 24 h, d_g = 0.6 μm, σ_g = 1.7, and $N_t = 10^{11}$ /m^3).

The time evolution of the size distribution of ^{131}I aerosols at 600-m altitude is presented in Figure 4. In contrast to ^{137}Cs aerosols, the growth rates of smaller ^{131}I aerosols were slightly faster, but those of larger ^{131}I aerosols were slower than those of nonradioactive aerosols. This could also be attributed to the results of different coagulation rates of ^{131}I aerosols, corrected by the radioactivity-induced charging, as shown in Figure 4. Thus, many ^{131}I particles reaching a certain size might not be deposited and were accumulated in this layer as shown in Figure 4.

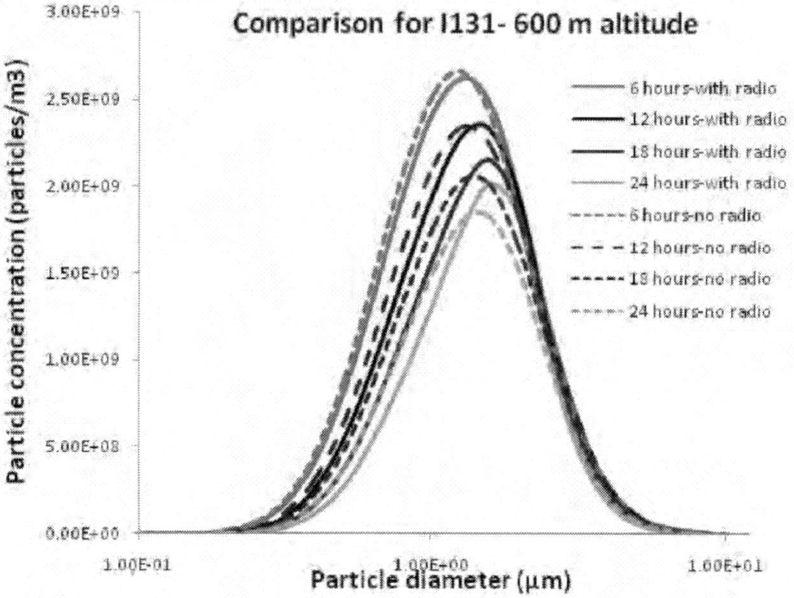

Figure 4. Evolution of the size distribution of ^{131}I aerosols at 600 m (t = 24 h, d_g = 1.0 μm, σ_g = 1.7, and $N_t = 10^{11}$ /m^3)

5. Investigating the importance of radioactive charging

The initial radioactive aerosol size distribution released from a radioactive source, can be important because it is highly related to the amount of the radiological debris deposited on the nearby ground. However, it is difficult to obtain the exact initial size distribution of radioactive aerosols. Thus, radioactivity-induced charging effects on the deposition of radioactive aerosols were investigated for various initial size distributions, using the column integrated mass ratio (**CIMR**) defined as:

$$CIMR = \frac{MR_t}{MNR_t} \qquad (8)$$

where $\boldsymbol{MR_t}$ and $\boldsymbol{MNR_t}$ are the total amounts of mass residing in the column after time t for the radioactive and nonradioactive cases, respectively.

Figure 5 shows CIMR for ^{137}Cs aerosols having various initial size distributions after 5 days, corresponding typical aerosol lifetime. The x-axis represents the initial geometric mean diameter while the y-axis represents the initial concentration as a function of the aerosol mean size. The contours represent the CIMR values. As the geometric diameter and total concentration of the initial aerosol size distribution increases, the CIMR also increases, indicating that less ^{137}Cs aerosols were deposited. The amount of the deposited ^{137}Cs aerosols could be reduced up to 30%. As noted in Figure 5, this might result from the accumulation of ^{137}Cs aerosols due to the suppression of coagulation by the radioactivity-induced charging of the aerosols. The geometric standard deviation of the initial aerosol size distribution was also found to affect CIMR.

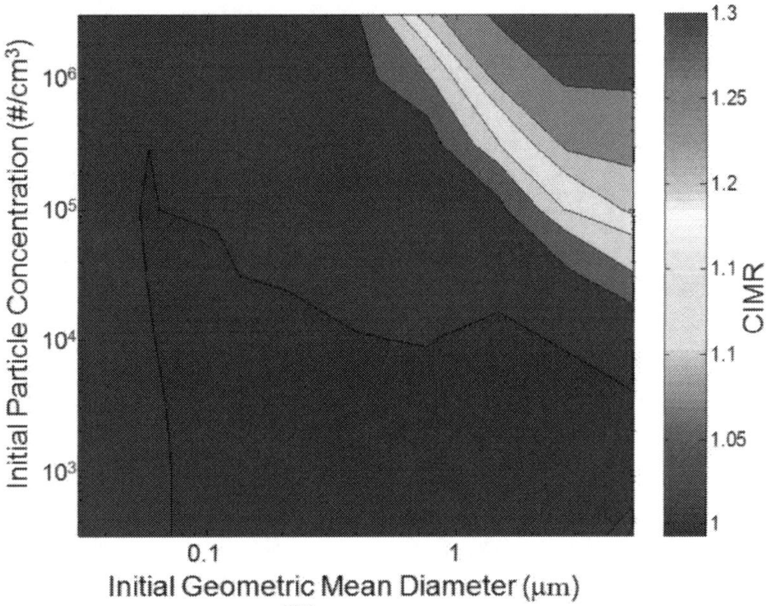

Figure 5. CIMR for ^{137}Cs aerosols after 5 days; $\sigma_g = 2.0$.

Figure 6 shows CIMR for ^{131}I aerosols after 1 day under various initial aerosol size distributions. Less ^{131}I aerosols were deposited, as the mean diameter or total concentration increased. In contrast to ^{137}Cs aerosols, the accumulation of many positive charges on ^{131}I aerosols resulted in the generation of strong electrostatic repulsive forces, severely hindering further growth of aerosols by coagulation in the column.

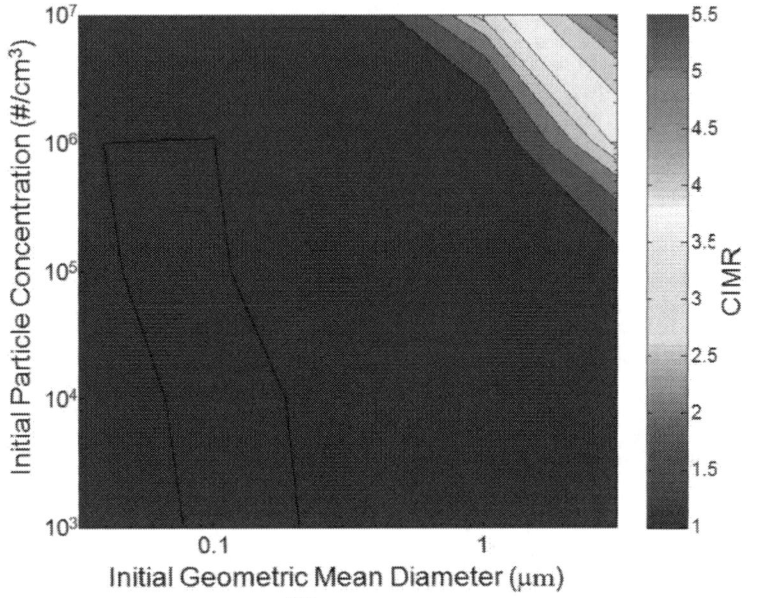

Figure 6. CIMR for ^{131}I particles after 1 day; $\sigma_g = 2.0$.

It was shown that for all cases, radioactivity may influence deposition characteristics of radioactive aerosols by electrostatic forces generated through radioactivity-induced charging. However, it should be noted that the average charge signs of ^{137}Cs and ^{131}I aerosols were opposite because of different dominating charging mechanisms. If both radioactive particles exist in the same column, the coagulation would be highly facilitated and thus the deposition rates of radioactive aerosols should be enhanced. This behavior suggests the importance of coupling the radioactivity effects to radioactive aerosol transport modeling and the necessity of the further development of models for multicomponent systems where various types of aerosols (e.g., background aerosols) exist.

REFERENCES

1) R.K. Chesser, M. Bondarkov, R.J. Baker, J.K Wickliffe, B.E. Rodgers, Reconstruction of radioactive plume characteristics along Chernobyl's Western trace, *J. Environ. Radioactivity*, **2004**, *71*, 147.
2) G. Lujanienė, V. Aninkevičius, V. Lujanas, Artificial radionucleotides in the atmosphere over Lithuania, *J. Environ. Radioactivity*, **2007**, doi:10.1016/j.jenvrad.2007.07.015.
3) Harrison, R. G., and K. S. Carslaw (2003), Ion-aerosol-cloud processes in the lower atmosphere,Rev. Geophys., 41, 1012, doi:10.1029/2002RG000114, 3.
4) R. Arimoto, J. L. Webb, M. Conley, Radioactive contamination of atmospheric dust over Southwestern New Mexico, *Atmospheric Environ.* **2005**, *39*, 4745.
5) M. O. Adeyini, E. O. Oladiran, Recent results on atmospheric radioactivity at Ibadan, Nigeria, *Radiation Measurements* **2006**, *41*, 330
6) V. I. Yoshenko, V. A. Kashparov, V. P. Protsak, S. M. Lundin, S. E. Levchuk, A. M. Kadygrib, S. I, Zvarich, X. V. Khomutinin, I. M. Maloshtan, V. P. Lanshin, M. V. Kovtun, J. Tschiervsch, Resuspension and redistribution of radionucleotides during grassland and forest fires in the Chernobyl exclusion zone. Part I: fire experiments, *J. Environ. Radioactivity* **2006**, *86*, 143.
7) V. I. Yoshenko, V. A. Kashparov, S. E. Levchuk, A. S. Glukhovskiy, Y. V. Khomutni, V. P. Protsak, S. M. Lundin, J. Tschiersch, Resuspension and redistribution of radionucleotides during grassland and forest fires in the Chernobyl exclusion zone. Part II: modeling the transport process, *J. Environ. Rad.* **2006**, *87*, 260.
8) H. Ooe, R. Seki, N. Ikeda, Particle-size distribution of fission products in airborne dust collected at Tsukuba from April to June 1986. J. Environ. Radioactivity 6(3), 219-223 (1988).
9) Christoudias, T., Proestos, Y., and Lelieveld, J.: Global risk from the atmospheric dispersion of radionuclides by nuclear power plant accidents in the coming decades, Atmos. Chem. Phys. Discuss., 13, 30287-30309, doi:10.5194/acpd-13-30287-2013, 2013.
10) Kim Y., Yiacoumi S. , Lee I., Mcfarlane J., Tsouris K., Influence of Radioactivity on Surface Charging and Aggregation Kinetics of Particles in the Atmosphere *Environmental Science & Technology,* **DOI:** 10.1021/es4047439, **2013**
11) Adams, P.J.; Seinfeld, J.H. Predicting global aerosol size distribution in general circulation models. *J. Geophys. Res.,* **2002**, *107(D19)*, 4370.
12) Clement, C.F.; Clement, R.A.; Harrison, R.G. Charge Distributions and Coagulation of Radioactive Aerosols. *J. Aerosol Sci.* **1995**, *26*, 1207-1225.
13) Clement, C.F.; Harrison, R.G. The Charging of Radioactive Aerosols. *J. Aerosol Sci.* **1992**, 23, 481-504.
14) Gensdarmes, F.; Boulaud, D.; Renoux, A. Electrical Charging of Radioactive Aerosols—Comparison of the Clement–Harrison Models with New Experiments. *J. Aerosol Sci.* **2001**, *32*, 1437-1458.

15) Yeh, H.C.; Newton, G.J.; Raabe, O.G.; Boor, D.R. Self-charging of 198 Au-labeled Monodisperse Gold Aerosols Studied with a Miniature Electrical Mobility Spectrometer. *J. Aerosol Sci.* **1976**, *7*, 245-253.
16) Fuchs, N.A. *The Mechanics of Aerosols*; Pergamon Press: New York, 1964.
17) Tzivion, S.; Feingold, G.; Levin, Z. An Efficient Numerical Solution to the Stochastic Collection Equation. *J. Atmos. Sci.* **1987**, 44, 3139-3149.
18) Masson, O.; Baeza, A.; Bieringer, J.; Brudecki, K.; Bucci, S.; Cappai, M.; Carvalho, F.P.; Connan, O.; Cosma, C.; Dalheimer, A.; Didier, D.; Depuydt, G.; De Geer, L.E.; De Vismes, A. ; Gini, L.; Groppi, F.; Gudnason, K.; Gurriaran, R.; Hainz, D.; Halldórsson, Ó.; Hammond, D.; Hanley, O.; Holeý, K.; Homoki, Zs.; Ioannidou, A.; Isajenko, K.; Jankovic, M.; Katzlberger, C.; Kettunen, M.; Kierepko, R.; Kontro, R.; Kwakman, P.J.M.; Lecomte, M.; Vintro, L.L.; Leppänen, A.-P.; Lind, B.; Lujaniene, G. ; Ginnity, P.M.; Mahon, C. M.; Malá, H.; Manenti, S. ; Manolopoulou, M.; Mattila, A.; Mauring, A.; Mietelski, J.W.; Møller, B.; Nielsen, S.P. ; Nikolic, J.; Overwater, R.M.W.; Pálsson, S. E.; Papastefanou, C.; Penev, I.; Pham, M.K.; Povinec, P.P.; Ramebäck, H.; Reis, M.C.; Ringer, W.; Rodriguez, A.; Rulík, P.; Saey, P.R.J.; Samsonov, V.; Schlosser, C.; Sgorbati, G.; Silobritiene, B.V.; Söderström, C.; Sogni, R.; Solier, L.; Sonck, M.; Steinhauser, G.; Steinkopff, T.; Steinmann, P.; Stoulos, S.; Sýkora, I.; Todorovic, D.; Tooloutalaie, N.; Tositti, L.; Tschiersch, J.; Ugron, A.; Vagena, E.; Vargas, A.; Wershofen, H.; Zhukova, O. Tracking of Airborne Radionuclides from the Damaged Fukushima Dai-Ichi Nuclear Reactors by European Networks. *Environ. Sci. Technol.* **2011**, *45*, 7670-7677.
19) Yoshida, N.; Kanda, J. Tracking the Fukushima Radionuclides. *Science* **2012**, *336*, 1115-1116.
20) Kauppinen, E.I.; Hillamo, R.E.; Aaltonen, S.H., Sinkko, K.T.S. Radioactivity Size Distributions of Ambient Aerosols in Helsinki, Finland, during May 1986 after the Chernobyl accident: Preliminary Report. *Environ. Sci. Technol.* **1986**, *20*, 1257-1259.
21) Simons, S. The Coagulation and Deposition of Radioactive Aerosols. *Ann. Nucl. Energy* **1981**, *8*, 287-294.
22) Mulpuru, S.R.; Pellow, M.D.; Cox, D.S.; Hunt, C.E.L.; Barrand, R.D. Characteristics of Radioactive Aerosols Generated from a Hot Nuclear Fuel Sample. *J. Aerosol Sci.* **1992**, *23*, S827-S830.
23) Lee, Y.; P. Adams. A fast and efficient version of the two-moment aerosol sectional (TOMAS) global aerosol microphysics model. *Aerosol. Sci. Technol.* **2012**, *46* (6), 678-689.
24) Kumar, S.; Ramkrishna, D. On the solution of population balance equations by discretization—I. A fixed pivot technique. *Chem. Eng. Sci.* **1996**, *51* (8): 1311-1332.

Electrochemical-Thermal Modeling of Large-Format Prismatic Li-ion Cells

Saeed Khaleghi Rahimian, Sun Ung Kim, Charles W. Monroe
Department of Chemical Engineering, University of Michigan, Ann Arbor 48109-2136, USA

Thermal management of large Large-format prismatic Li-ion cells is crucial due to their small surface-to-volume ratios, which can lead to local degradation of active materials within the battery, or catastrophic thermal runaway of the battery [1]. In order to predict the transient electrochemical-thermal response of a 15-Ah A123 prismatic pouch cell, a simple physics-based model has been developed. The 3D model is based on the Newman-Tobias porous-electrode theory [2] and coupling the charge and energy balances. Li ion transport inside the particles as well as liquid phase are not considered. A data-fitting approach based on dimensionless parameters predicts surface-average temperature, maximum and minimum temperature, as well as cell voltage. In spite of its simplicity with respect to the standard Dualfoil model [3], the proposed model prediction is within the experimental error as shown in Figure 1.

[1] T. M. Bandhauer, S. Garimella, and T. F. Fuller, *J.Electrochem. Soc.* 158, R1 (2011).

[2] J. S. Newman and C. W. Tobias, *J. Electrochem. Soc.* 109, 1183 (1962).

[3] M. Doyle, T. Fuller, and J. Newman, *J. Electrochem. Soc.*, 140, 1526 (1993).

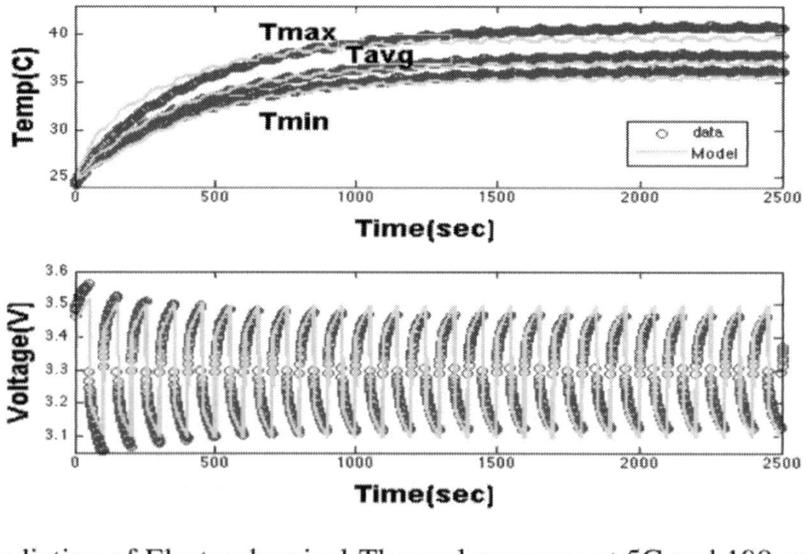

Figure 1, Prediction of Electrochemical-Thermal response at 5C and 100 sec period time

HEAT INTEGRATION OF CONTINUOUS STREAMS IN BATCH PLANTS

Jui-Yuan Lee [a,*], Esmael Seid [b], Thokozani Majozi [b]

[a] *Department of Chemical Engineering, National Tsing Hua University, 101, Section 2, Kuang-Fu Road, Taiwan, 30013, ROC*

[b] *School of Chemical and Metallurgical Engineering, University of the Witwatersrand, Johannesburg, 1 Jan Smuts Avenue, Braamfontein 2000, Johannesburg, South Africa*

E-mail address: juiyuan.lee@gmail.com (J.-Y. Lee).

Abstract

Batch processes are flexible allowing the production of different products within the same facility, and suitable for producing low volume, high value-added products such as pharmaceuticals and agrochemicals. The trend towards batch processing has necessitated the development of scheduling techniques. In addition to process scheduling, heat integration has been increasingly considered for batch plants to reduce external utility (e.g. steam and cooling water) requirements for tasks involving heating or cooling, such as endothermic and exothermic reactions. In this work, a mathematical technique for simultaneous process scheduling and heat integration of batch plants is presented. The formulation, based on a superstructure, aims to maximise the coincidence of availability of hot and cold process stream pairs with feasible temperature driving forces, whilst taking into account scheduling constraints. Heat integration during stream transfer can shorten the time required for heating and cooling in processing units, and is expected to enable higher production and lower utility consumption for batch plants. A case study is solved to demonstrate the application of the proposed mathematical model.

Keywords: Batch plants; Scheduling; Heat integration; Mathematical optimization.

Introduction

Batch processes have gained considerable interest because of their flexibility and adaptability to produce different products using the same facility. Such processes are suited for producing low volume and high value-added products, which are commonly found in industries such as agrochemicals, pharmaceuticals, polymers, foodstuffs and fine/specialty chemicals. A common feature of the various batch processing plants is the use of fossil fuels as the primary energy source. Reliance on fossil energy in turn causes serious problems to the environment. Stricter environmental regulations due to increased public awareness towards and the need for sustainable development have urged the process industries to seek alternative renewable energy sources and efficient end use of energy. Improved energy utilisation can be achieved through enhancements in plant machinery and process integration. The literature review focuses upon process integration methodologies developed for energy recovery in batch processes. Although process integration has been successfully applied for energy conservation for continuous processes, different techniques are required for batch processes, where time is also an important constraint in addition to temperature. The complexity of the design and operation of batch plants has led to the development of various methods, tools and mathematical models for energy minimisation.

While there have been many works addressing heat integration of batch plants, most of those only consider direct or indirect heat integration between processing units.[1] The opportunity for energy recovery through heat exchange between material streams during transfer was overlooked and remains to be exploited. This is the subject of the present study.

Problem Statement

The problem addressed in this paper can be briefly stated as follows. Given (i) production scheduling data, including equipment capacities, task durations, the time horizon of interest, the recipe for each product, raw material costs and product selling prices; (ii) operating temperatures for tasks, supply temperatures of raw materials and storage temperatures of final products; (iii) specific heat capacities of states and (iv) costs of external hot and cold utilities, it is desired to determine an optimal production schedule that maximizes the profit, which is defined as the difference between product revenue and utility cost. The following assumptions are made for the problem:

- Constant heat capacities
- Negligible heat losses for short-term storage
- Counter-current heat exchangers

Model Formulation

In this work, the mathematical modelling is based on the superstructure representations shown in Fig. 1. It can be seen that each processing unit can receive material from and send material to other units or storage. The input material may be heated or cooled in the processing unit before the task starts in order to meet the operating temperature; similarly, the output material may be heated or cooled in-situ before leaving for further processing or storage. To reduce the use of external hot and cold utilities, heat exchange between output (i.e. intermediate and product) streams (Fig. 1a) is considered, as well as heat exchange for raw material preheating (Fig. 1b). Note that output streams are to have heat exchange before entering next processing units or storage, while raw material streams are, after leaving storage. If there are two or more states produced in a processing unit, they are assumed to leave the unit at the same temperature and treated as separate streams for heat integration. Also, it is currently assumed that external utilities can only be used in processing units through jackets; no utility heaters or coolers are available for use. In addition to the necessary scheduling constraints, the mathematical model consists mainly of mass and energy balance equations and heat integration constraints. To simplify process operation for practical considerations, the one-to-one heat integration arrangement is assumed, in which a hot stream can only have heat exchange with one cold stream.

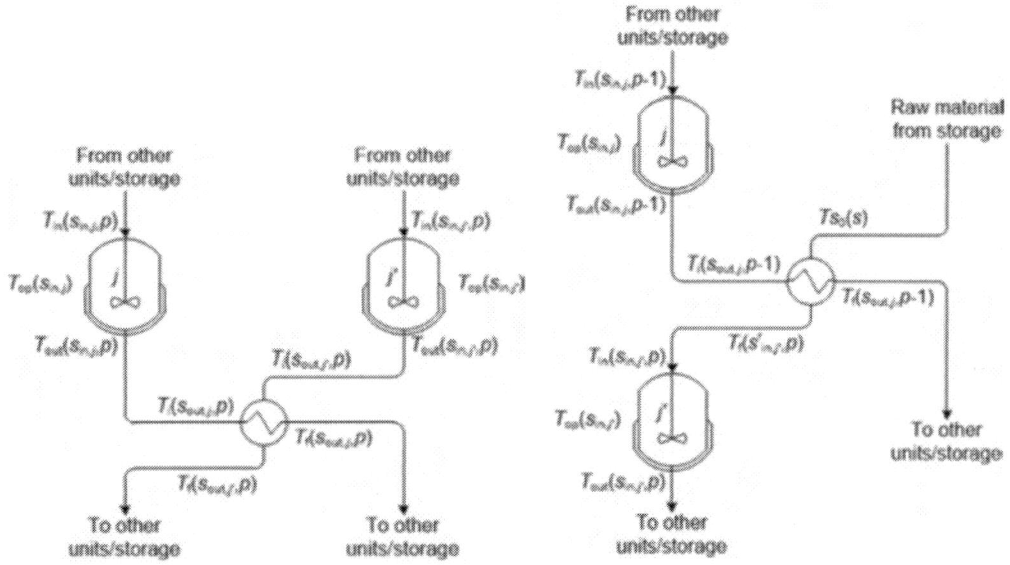

(a) (b)

Fig. 1. Superstructure for heat integration: (a) between outlet streams; (b) between outlet and raw material streams.

For profit maximisation, the objective function may be defined as the difference between the product revenue and the costs of external hot and cold utilities (i.e. steam and cooling water):

$$\max \sum_{s \in S^{\mathrm{P}}} Price(s) qs(s,p)\Big|_{p=|P|} - \sum_{p \in P} \sum_{s_{\mathrm{in},j} \in S_{\mathrm{in},j}^{\mathrm{eff}}} CST\left(q_{\mathrm{in}}^{\mathrm{h}}\left(s_{\mathrm{in},j},p\right)+q_{\mathrm{out}}^{\mathrm{h}}\left(s_{\mathrm{in},j},p\right)\right)$$

$$-\sum_{p \in P} \sum_{s_{\mathrm{in},j} \in S_{\mathrm{in},j}^{\mathrm{eff}}} CCW\left(q_{\mathrm{in}}^{\mathrm{c}}\left(s_{\mathrm{in},j},p\right)+q_{\mathrm{out}}^{\mathrm{c}}\left(s_{\mathrm{in},j},p\right)\right) \quad (1)$$

With the presence of bilinear terms (in energy balance equations) and the use of binary variables (mainly for process scheduling), the overall model is a mixed integer nonlinear programme (MINLP), for which global optimality cannot be guaranteed.

In the next section, the application of the proposed model is demonstrated through a modified literature case study. The model is implemented in the GAMS environment[2] with BARON[3] as the MINLP solver.

Illustrative Case Study

A multipurpose batch plant is to be scheduled to produce three products (S9-S11) from three raw materials (S1, S2 and S7) over a 16-h time horizon. The STN process representation is shown in Fig. 4, where states S3-S6 and S8 are all intermediates. This case study is adapted from Pinto et al.,[4] with heating and cooling requirements for materials included to explore heat integration opportunities. Tables 1 and 2 present the pertinent data for the case study. For simplicity, it is assumed that each state has a dedicated storage vessel and each processing task is carried out in a dedicated reactor. Note that the supply temperature in Table 2 refers to the initial temperature of a state in the storage vessel. Therefore, the supply temperatures of intermediates and products are taken as zero.

Table 1. Processing Data for the Case Study

Task	Unit	Duration (h)	Operating temperature (°C)	Capacity limits (kg)
1	R1	2	180	30-120
2	R2	2	40	20-80
3	R3	4	80	20-80
4	R4	2	120	50-200
5	R5	1	160	50-200
6	R6	1	100	35-140

It can be seen in Fig. 2 that unit R1 is to have heating for its input S1 and cooling for its output S4, as the supply temperature of S1 is lower than the operating temperature of R1, and the latter is higher than the operating temperature of R4. Similarly, the potential heating requirements for the inputs of R3 and R5 and the outputs of R2-R4 can be identified from Fig. 4, as well as the potential cooling requirements for the outputs of R5 and R6. However, uncertainties remain for R4 and R6. The input of R4 consists of S4 and S5, which are produced at different temperatures and can give a mixture temperature higher or lower than its operating temperature. In the case of R6, if the product storage temperature of S11 (80°C) is achieved by utility cooling in R5, then S8 will leave R5 at 80°C, which is lower than the operating temperature of R6 (100°C). However, if the product cooling is achieved by heat exchange, the final temperature of S8 may be higher than the operating temperature of R6. Therefore, for units R4 and R6, constraints (19)-(23) should be used.

Table 2. Material and Utility Data for the Case Study

State	Storage capacity (kg)	Initial inventory (kg)	Supply temperature (°C)	Product storage temperature (°C)	Heat capacity (kJ/kg °C)	Price ($/kg)
S1	UL	AA	80	-	2.5	-
S2	UL	AA	40	-	3.4	-
S3	120	0	0	-	3.3	-
S4	180	0	0	-	3.7	-
S5	120	0	0	-	4.0	-
S6	300	0	0	-	3.6	-
S7	UL	AA	60	-	3.5	-
S8	210	0	0	-	4.0	-
S9	UL	0	0	40	3.2	10
S10	UL	0	0	40	3.3	5
S11	UL	0	0	80	3.7	10

Material transfer time	10 min
Minimum temperature difference for heat transfer	10°C
Cooling water cost	$0.6/MJ
Cooling water inlet/outlet temperature	20/30°C
Heat capacity of cooling water	4.2 kJ/kg °C
Mass flow rate of cooling water	1800 kg/h
Steam cost	$3/MJ
Saturated steam temperature	200°C
Latent heat of steam	1940 kJ/kg
Mass flow rate of steam	40 kg/h

UL, unlimited; AA, available as/when required.

The hot and cold streams for heat exchange can also be identified from Fig. 2. There are three potential hot streams (S4outR1, S8outR5 and S11outR5) and five potential cold streams (S1inR1, S3outR2, S5outR3, S6outR4 and S7inR5). Note that the feed stream of R2 (S2inR2) and the product streams of R6 (S9outR6 and S10outR6) are not included; the former has no need for heating or cooling, while the storage temperatures of the latter can only be achieved with cooling water. Two scenarios based on fixed and variable batch sizes are considered in this case study. The objective is to maximize the profit given by the difference between product revenue and utility cost.

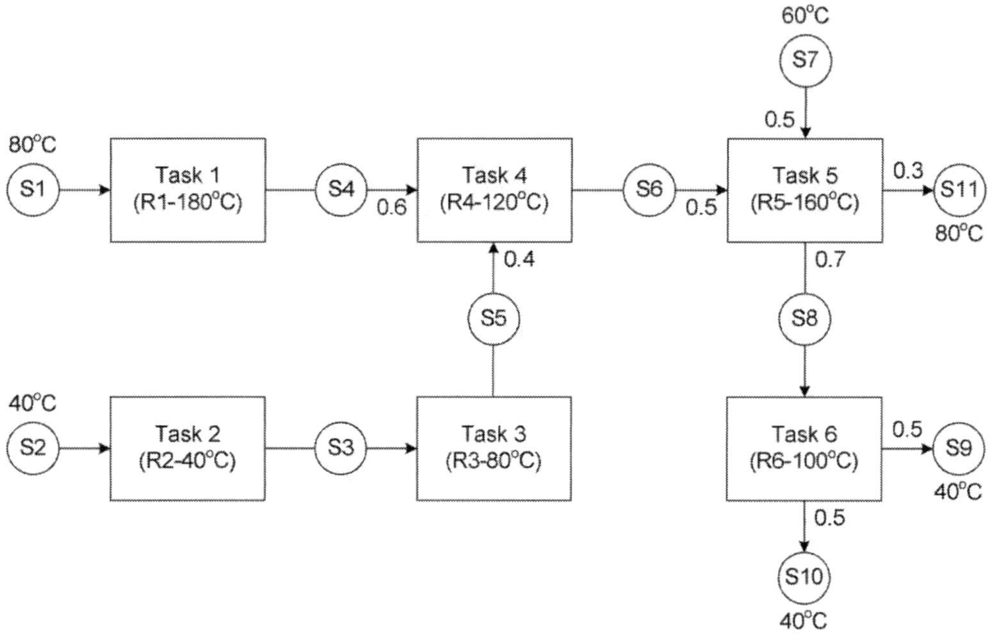

Figure 4. STN Representation for the Case Study

Batch operation is often carried out with fixed batch size in the process industries to minimize changes. In this scenario, the batch sizes for all processing units are fixed at the maximum capacities given in Table 1. Prior to the exploration of heat integration opportunities, the batch plant produces 140 kg of S9, 140 kg of S10 and 120 kg of S11 within the 16-h time horizon, while consuming 161.76 MJ of steam and 193.56 MJ of cooling water. This gives a profit of $2698.58, obtained from the proposed model by setting all the binary variables associated with heat exchange to zero. Fig. 3 shows the Gantt chart for the case without heat integration. Note that the heating and cooling required in R5 increase the material residence time to at least 2.47 h, which is much longer than the task duration of 1 h. Thus, there is not enough time to process another full-size batch in R5.

When heat exchange between process streams (three hot, S4outR1, S8outR5 and S11outR5; five cold, S1inR1, S3outR2, S5outR3, S6outR4 and S7inR5) is considered, the plant is able to produce 140 kg of S9, 140 kg of S10 and one batch more of S11 (i.e. 180 kg) within the same time horizon. The utility cost can be reduced to $430.68, giving a profit of $3469.32. This corresponds to a 28.6% improvement compared to the case without heat integration. Fig. 4 shows the optimal schedule with heat exchange between process streams. An overall comparison shows that heat integration of material transfer streams enables the batch plant to produce a larger quantity of products with much less consumption of external utilities.

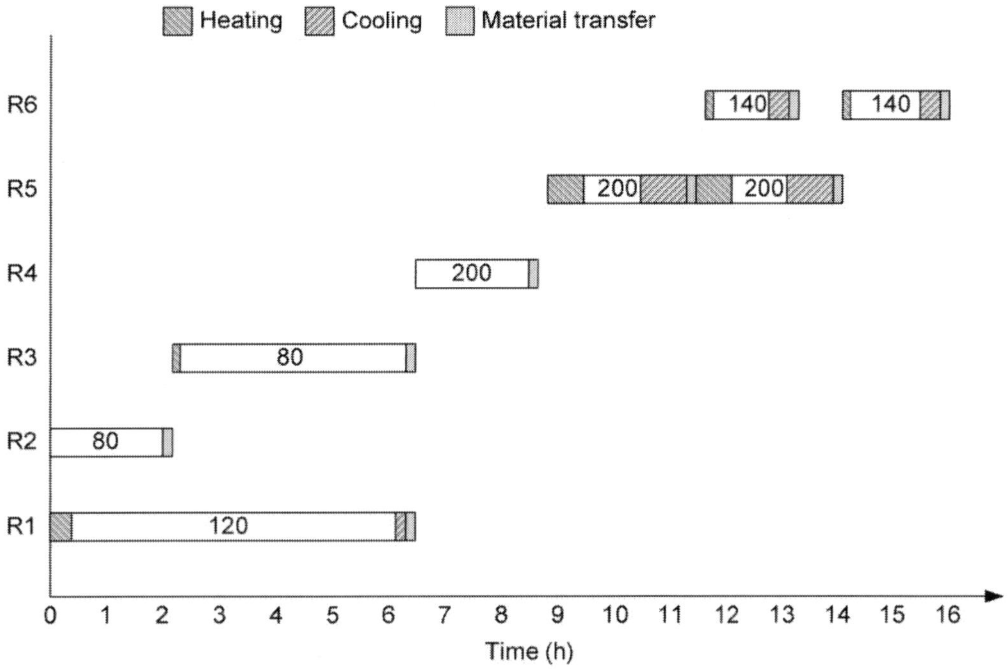

Figure 3. Gantt Chart for the Case without Heat Integration

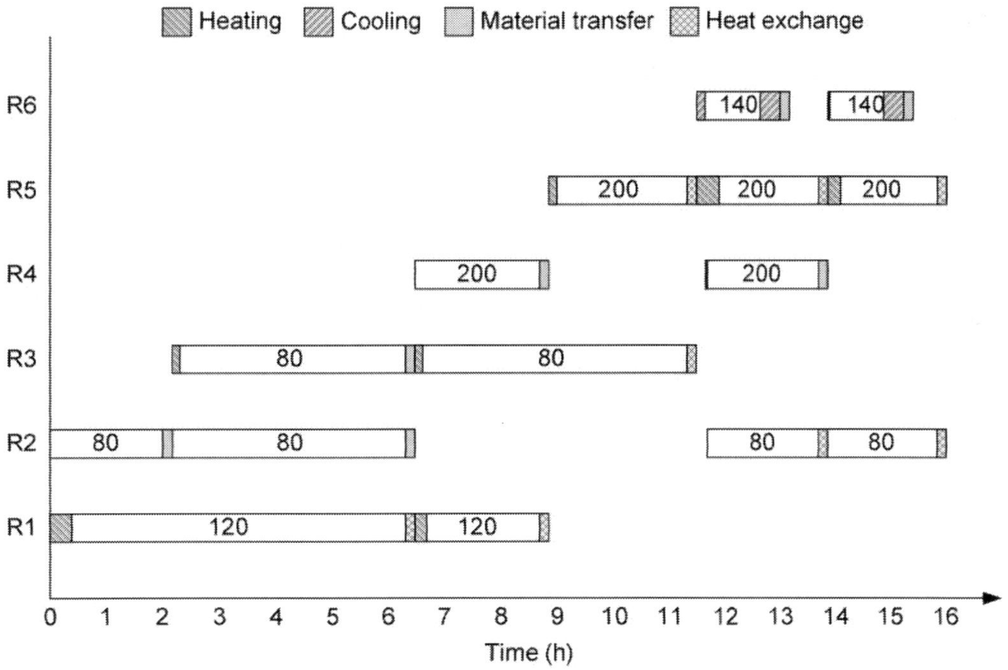

Figure 4. Optimal Schedule with Heat Exchange among Raw Material and Output Streams

Conclusion

Based on a robust scheduling framework adapted from Seid and Majozi,[5] a superstructure-based mathematical model for simultaneous process scheduling and heat integration of batch plants has been developed in this paper. The formulation ensures proper sequencing of tasks over a given time horizon, with the aim of synchronizing the material transfer times so as to maximize heat recovery. An illustrative case study was solved to demonstrate the application of the proposed model. The results suggest that heat integration of process streams enables batch plants to achieve higher production with lower utility consumption. It should be noted that, in the current study, external utilities are only allowed to be used in processing units. This could render utility heating and cooling prior to heat exchange and, consequently, reduce the driving force and potential for heat recovery. In future work, the use of utility heaters and coolers for material transfer streams will be taken into account. Direct and indirect heat integration between processing units may also be considered to provide further energy recovery opportunities.

References

1. Fernández, I., Renedo, C.J., Pérez, S.F., Ortiz, A., Mañana., M., 2012. A review: Energy recovery in batch processes. Renew. sust. energy. rev. 16, 2260–2277.
2. Rosenthal, R.E., 2013. GAMS – A User's Guide, GAMS Development Corporation, Washington, DC.
3. Tawarmalani, M., Sahinidis, N.V., 2005. A polyhedral branch-and-cut approach to global optimization. Math. Program. 103, 225–249.
4. Pinto T, Barbósa-Póvoa APFD, Novais AQ. Design of multipurpose batch plants: a comparative analysis between the STN, m-STN, and RTN representations and formulations. Ind Eng Chem Res 2008;47:6025–44.
5. Seid, R., Majozi, T., 2012. A robust mathematical formulation for multipurpose batch plants. Chem. Eng. Sci. 68, 36–53.

380601 Implementation of an Improved Automatic Operation System for Air Separation Process in Steel Industries

Abstract:

Cryogenic air separation units (ASUs) are widely used for producing gaseous high-purity oxygen in steel industries. As the oxygen demand fluctuates actively in steel manufacturing processes in terms of frequency and amount, the operating condition of ASUs should be accordingly switched rapidly to minimize the waste of oxygen and to save the power consumption in the end. It is thereby financially motivating to develop the automatic operation modification system for ASU. Some studies reported advanced methods such as model predictive control method in the literature. There are many challenges to actually implement such distributed control systems due to the complicating issues of ASUs. They covered the limited snapshot of actual processes. From the hand-on experience, we came to find that it is practically important to introduce simple but straightforward control mechanisms to operate the ASUs in steel industries. This paper proposes a direct approach for the existing control system using parameter tuning based on linear regression of plant data and balanced synchronization between intermediate and final operating target values. Actual case studies of steel industries are conducted in terms of comparing the result according to operator and automatic algorithm for switching operating conditions of varying 15% oxygen production rate. It was found that power consumptions were reduced to 1.4%p for increasing oxygen demand case and 2.4%p for decreasing one.

THE UNIVERSITY of TULSA

Quantifying Model Uncertainty in Scarce Data Regions

TUPSE

Wei Dai, Selen Cremaschi
Department of Chemical Engineering, The University of Tulsa, Tulsa, OK, USA

Background

- Sources of uncertainty in physics-based engineering models and simulations

- Models are validated using a limited set of experimental data
- Data generally lacks in the regions of models predictions

Approach: Gaussian Process Modeling [1]

$y^e \sim y^m + \delta + \epsilon$

- y^e: experimental response
- y^m: model response
- δ: model discrepancy
- ϵ: experimental uncertainty

→ $y^e \sim y^m + \delta$

- assuming Perfect (!) experimental data

→ *From data:* $\delta = y^e - y^m$

- estimate δ distribution
- use estimated δ to predict y^e and its confidence interval

Case study: Erosion Predictions

- Unavoidable extraction of solids from oil and gas bearing reservoirs either onshore or offshore sites
- Solid erosion in the pipelines may result in facility integrity issues
- Erosion is a complex phenomenon depends on many factors
- 1-Dimensional Sand Production Pipe Saver (1-D SPPS)[2] : a widely used model that predicts erosion rate given

➢ fluid properties,
➢ solid characteristics,
➢ the construction material properties and
➢ the geometry of the flow lines

V_L: • Given operating condition, calculate representative impact velocity $V_L = f(geom, \mu_f, \rho_f, \rho_p, d_p, d, v_s)$

ER: • Calculate erosion ratio $ER = AV_L^n$

h: • Calculate erosion rate

Result: Uncertainty predictions for Gas-Solid Flow

- Collect 77 experimental data points for gas-solid flow
- Calculate 1-D SPPS predictions of erosion rate
- Estimate the distribution and confidence intervals of model discrepancy using Gaussian Process Modeling
- Run 4-fold cross-validation 30 times to obtain the average 95% confidence intervals for each discrepancy prediction
- Obtain the probability of under-predicted erosion rate for each test conditions by 95% confidence intervals

Discrepancy = 1-D SPPS prediction of erosion rate – measured erosion rate

Inputs	Ranges/ Instances
Gas Velocity (m/s)	10 ~ 220
Pipe Geometry	(Long) Elbow, Plug Tee
Pipe Material	Carbon Steel
Pipe Diameter (inches)	1 ~ 4
Erosion Rate (mils/lb)	2.4E-4 ~ 4.7

- Average percentage difference between predicted discrepancy and actual discrepancy: 36.2%
- Percentage of under-predicted data points using Gaussian Process Modeling: 13% (10 points out of 77)

Average probability of under-prediction

Particle Size (μm)	No. of Points	Avg. prob. of under-prediction	±95% Confidence
150	31	21.2%	[14.1%, 28.3%]
295	12	31.8%	[7%, 56.5%]
300	8	45.9%	[40.7%, 51.1%]
350	26	15.7%	[2.1%, 29.4%]
Pipe Diameter (in)	**No. of Points**	**Avg. prob. of under-prediction**	**±95% Confidence**
1	14	6.8%	[5.7%, 13.1%]
2	52	24.9%	[15.9%, 33.9%]
3	4	20.3%	[12.2%, 28.4%]
4	7	48.8%	[48.3%, 49.3%]
Geometry	**No. of Points**	**Avg. prob. of under-prediction**	**±95% Confidence**
Elbow	50	21.8%	[14 %, 29.6%]
Long Elbow	19	38.1%	[22.8%, 53.3%]
Plug Tee	8	0.1%	[-0.1, 0.2%]

References

[1] Erosion and Corrosion Research Center, The University of Tulsa, OK, USA
[2] C.E. Rasmussen and C.K. I. Williams, Gaussian Processes for Machine Learning, 2006

Model Discrepancy (δ) and Probability of Under-Prediction Plots

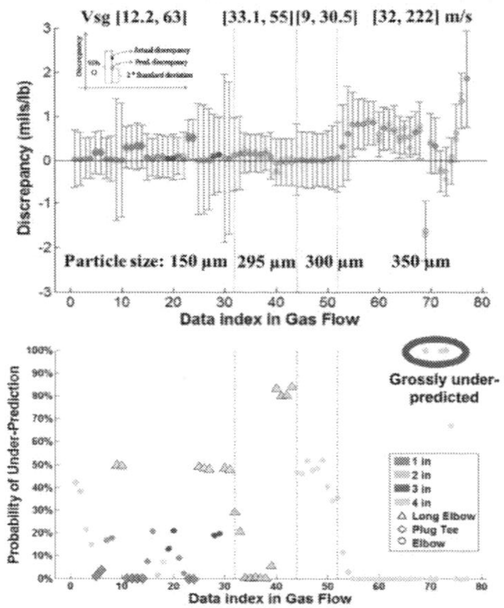

Conclusions and Future Directions

Conclusions

- Gaussian Process Modeling (GPM) has great potential for quantifying model uncertainty.
- Except those data points with very large model discrepancy (>0.5 mils/lb), the 95% confidence intervals of most data points crossed the zero line, showing the potential for under-prediction by 1-D SPPS.
- Three points (350 μm) are grossly under-predicted which is originated from higher experimental uncertainty.
- Four orders of magnitude difference is observed in gas-solid erosion rates.

Future Directions

- Apply data transformation approach to improve GPM predictions.
- Validate model uncertainty in extrapolated input designs.

11/17/2014 2014 AIChE Annual Meeting Atlanta, GA

Financial support from Chevron Energy Technology Company is greatly acknowledged.

A time series Kalman filter(TSKF) for outlier detection in dynamic data sets

Shu Xu[1], Michael Baldea[1], Thomas F. Edgar[*1], Willy Wojsznis[2], Terrence Blevins[2], and Mark Nixon[2]

[1]McKetta Department of Chemical Engineering, The University of Texas at Austin, Austin, TX78712, USA

[2]Process Systems and Solutions, Emerson Process Management, Round Rock, TX78759, USA

October 5, 2014

Abstract

A time series Kalman filter (TSKF) is proposed that successfully handles outlier detection in dynamic systems, where normal process changes often mask the existence of outliers. The TSKF method combines a time series model fitting procedure with a modified Kalman filter to deal with additive outlier (AO) problems in dynamic process data set. Compared with current outlier detection methods, the new method enjoys the following advantages: (a) no prior knowledge of the process model is needed; (b) it is easy to tune; (c) it can be applied to both univariate and multivariate outlier detection; (d) it is applicable to both on-line and off-line operation; (e) it cleans outliers while maintains the integrity of the original data set.

Keywords: outlier detection, Kalman filter, time series modeling, additive outlier, dynamic process modeling.

Introduction

In the process industries, engineers are faced with a new challenge on uncovering useful information from gigabytes of high-frequency sampling data without compression. To overcome such a challenge, we have to go through a knowledge discovery chain[1] consisting of data cleaning and integration, selection and transformation, data mining, evaluation and presentation steps. Data cleaning is critical to knowledge discovery because it removes irregular and noisy information from the original data set which interferes with data analysis and affects model quality.

*Author to whom correspondence should be addressed. E-mail:tfedgar@austin.utexas.edu

Outlier detection is an important task in data cleaning. An outlier is defined as an observation or a subset of observations showing an inconsistent behavior with the remainder of the set of data from statistical perspective[2]. In the process industries, malfunction of sensors, or inappropriate treatment of missing data are typical examples which can generate outliers.

While univariate outliers occur when only a single variable shows abnormal behavior, multivariate outliers describe crossing-boundary behaviors of combinations of variables and they are often treated as univariate ones for simplicity. However, applying univariate procedures on multivariate outliers have several disadvantages, such as over-specification of number of outliers caused by failing to consider interactions between variables[3].

The key assumption underlying many statistical outlier detection methods[2,4–7] that the data are identically and independently distributed (*i.i.d.*) is often compromised by dynamics hidden in process data sets. To overcome such a challenge, we have to separate process dynamics from outliers, and a traditional approach is to apply the moving window technique which assumes that data of a small enough window size follows a normal distribution. However, when the dynamics of data becomes significant, incorporating moving window can no longer obtain satisfactory results. An alternative solution is to approximate dynamics by time-series models such as an autoregressive(AR) model, and separate observations that are inconsistent with the fitted model, as shown in the filter-cleaner example[8]. However, the on-line filter-cleaner[8] is only applicable to univariate additive outlier detection, and it can be improved in several aspects such as the model fitting algorithm, parameter tuning, and an extension to the multivariate cases. Besides, since off-line data analyses are more often encountered in practice, it is worth developing related methods.

Method description

Contamination rate,detection rate,mis-identification rate,and normal data estimation rate

The contamination rate κ is defined as the percentage of outliers in total data, and the detection rate χ is defined as the percentage of outliers being successfully detected. The mis-identification rate β is defined as the percentage of normal data falsely identified as outliers (type I error), and γ is defined as an estimation of the normal data percentage. Generally, γ is set to be larger than 80 %; otherwise, the data will not be used because of a poor quality. Mathematical expressions for κ, χ, β and γ are shown in Eqs. (1) to (4).

$$\kappa = \frac{N_{outliers}}{N_{total\ data}} \tag{1}$$

$$\chi = \frac{N_{successfully\ identified}}{N_{total\ outliers}} \tag{2}$$

$$\beta = \frac{N_{false\ alarm}}{N_{normal\ data}} \tag{3}$$

$$\gamma = \frac{N_{normal\ data}}{N_{total\ data}} \tag{4}$$

Based on the definition, $\kappa = 5\%$ in a data set with 100 observations means the number of actual outliers is 5. In simulation, we are using simulated data sets with outliers added at every 20th sample points.

Reference methods

In univariate cases, the Hampel identifier[9] is a well-known method and its detection condition is shown in Eq. (5):

$$|x_t - med| > g(N, \alpha_N)\, MAD \tag{5}$$

where $g(N, \alpha_N)$ is a function related to observation numbers N and significance level α_N. med and MAD can be calculated based on Eq. (6):

$$med = \frac{x_{[(N+1)/2]:N} + x_{[N/2]+1:N}}{2}, MAD = med\left(|x_1 - med|, ..., |x_N - med|\right) \tag{6}$$

in which $[A]$ rounds A to the nearest integer less than or equal to A, and $x_{1:N}, ..., x_{N:N}$ are the ordered sequence of $\{x_t\}$.

The principal component analysis (PCA) method has been widely used to detect multivariate outliers. First, PCA decomposes the original data set using singular value decomposition shown in Eq. (7). Second, it calculates the Hotelling's T^2 metric defined in Eq. (8), and by monitoring such a metric, disturbances or outliers can be identified if the metrics are larger than the threshold calculated in Eq. (9).

$$\mathbf{X_{PCA}} = \mathbf{U \Sigma V^T} \tag{7}$$

where $\mathbf{X}$ is the training data matrix with n observations and m variables.

$$t^2 = \mathbf{x}^T V \Sigma_a^{-2} V^T \mathbf{x} \tag{8}$$

where V contains the loading vectors associated with the a largest singular values, Σ_a includes the first a rows and columns of Σ, and $\mathbf{x}$ is an observation vector of dimension m.

$$T_\alpha^2 = \frac{(n-1)\, a}{(n-a)} F_\alpha (a, n-a) \tag{9}$$

where $F_\alpha (a, n-a)$ is the critical point of the F-distribution with a and $n - a$ degrees of freedom, and α is the level of significance.

However, the static PCA method does not include the serial correlation for a single variable, and an related upgraded version is the dynamic principal component analysis (DPCA)[10,11], which is proposed for detection and isolation of process disturbances in time series data. The DPCA augments each observation $\mathbf{X}$ matrix at time t with the previous l time instances, as shown in Eq. (10), and similar to PCA, it decomposes $\mathbf{X_{DPCA}}$ and detects outliers by monitoring the Hotelling's T^2.

$$\mathbf{X}_{DPCA}(l) = [\mathbf{X}(t)\, \mathbf{X}(t-1) \cdots \mathbf{X}(t-l)] \tag{10}$$

where $\mathbf{X}(\mathbf{t})$ is the observation matrix at time instance t.

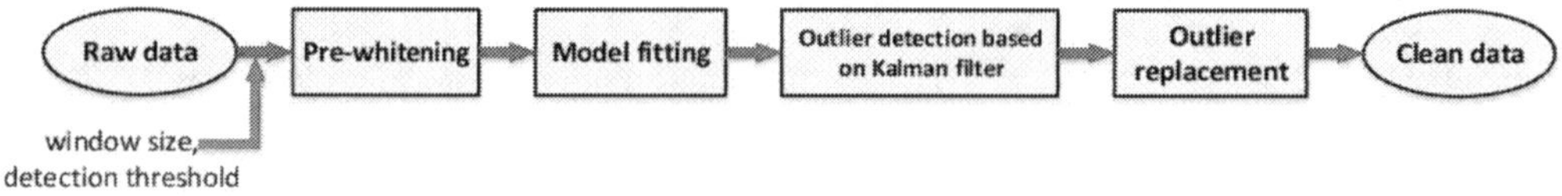

Figure 1: Diagram of time-series Kalman filter

Table 1: Tuning parameters of the TSKF method

Implementation	Tuning parameters
on-line	distance threshold Δ; moving window size(MW)
off-line	γ; partition window size(PW)

Time series Kalman filter(TSKF)

The diagram of TSKF is shown in Fig. 1, and the tuning parameters for on-line and off-line operation are shown in Table 1.

As shown in Fig. 1, the TSKF consists of a series of steps: pre-whitening, model fitting, outlier detection and replacement. For on-line use, since we have no knowledge of how to pre-set the distance threshold Δ, we need to run a training data set.

For off-line use, we can set the γ based on prior knowledge of the raw data set, such as $\gamma = 95\%$ means we estimate that the maximum amount of outliers in the data set should not exceed 5 % of the total number of observations.

Simulation testing results

For illustration, the TSKF method will be applied to both univariate and multivariate outlier detection in the dynamic data sets. For simplicity, only the univariate simulation process is described: following the approach used by Liu et al.[8], we obtain data by simulating the additive outlier (AO) models shown in Eq. (11); v_t has equal probabilities being +*Amp* or -*Amp* (*Amp* is the amplitude of outliers); x_t follows ARMA(1,1) processes for univariate cases. The multivariate simulation process is the similar as the univariate one.

$$y_t = v_t \delta_t^{(T)} + x_t \tag{11}$$

where $\delta_t^{(T)}$ is a pulse function: $\delta_t^{(T)} = 1$ if $t = T$; $\delta_t^{(T)} = 0$ if $t \neq T$. v_t is an outlier. The univariate model was run with 10000 test points for both on-line and off-line cases. For processes with additive outliers(AOs), the data are corrupted with outliers at different contamination rates defined in Eq. (1).

The multivariate models are run with 10000 test points for off-line case and with 1000 test points for on-line case. Similar to the univariate model, data are contaminated with different rates of outliers for the AO process simulation.

ARMA(1,1)model

$$\left(1-\phi z^{-1}\right) x_t=\left(1-\theta z^{-1}\right) \epsilon_t \tag{12}$$

where ϕ, θ are coefficients, ϵ_t is the white noise, $\epsilon_t \sim N(0,1)$ and z^{-1} is a shift operator.

The on-line and off-line additive outlier detection results for data from ARMA(1,1) processes are shown in Tables 2:

Table 2: Additive outlier detection rates for data from ARMA (1, 1) process at $\kappa = 5\%$

Case No.	ϕ	θ	Δ	*Amp*	Liu's filter-cleaner(***)		on-line(*) TSKF		Hampel(***)		off-line(**) TSKF	
					$\beta(/\%)$	$\chi(/\%)$	$\beta(/\%)$	$\chi(/\%)$	$\beta(/\%)$	$\chi(/\%)$	$\beta(/\%)$	$\chi(/\%)$
1	0.0	0.0	2.5	3	0.82	65.13	1.51	70.12	0.72	70.91	1.42	69.03
2	0.0	0.0	2.6	4	0.55	82.83	1.07	91.78	0.63	84.23	0.50	88.17
3	0.5	0.0	2.5	3	1.12	64.38	1.82	67.45	0.94	49.52	1.62	64.65
4	0.5	0.0	2.7	4	0.49	82.44	1.48	89.69	0.71	73.65	0.73	83.92
5	0.9	0.0	3.0	4	0.59	79.84	2.11	82.34	0.47	12.57	1.17	75.47

(*) Simulation condition: N=10000; rep=500; MW=100

(**) Simulation condition: N=10000; rep=500; PW=1000, $\gamma = 95\%$

(***) Some results come from Liu et al.[8].

From Table 2, we can see that for on-line outlier detection rate χ, the results of TSKF method and filter-cleaner[8] are close, and both outperform the Hampel identifier when the process dynamics becomes significant (shown in the first order correlation coefficient ϕ). The result suggests that the Hampel identifier is more sensitive to system dynamics.

Moreover, a larger outlier size *Amp* will make it easier for the TSKF method to detect outliers, as shown in detection rate χ, but it will not necessarily lower the mis-identification rate β.

In addition, though detecting more AOs than the Hampel identifier and on-line filter-cleaner, the TSKF method has a higher mis-identification rate β than the later two, and it increases with a higher correlation ϕ. Such a result is probably caused by a constant Δ which does not show a lot variations as the process evolves. However, the local variance and mean, will be significantly affected by the outliers. Thus, the auto-adjustment of Δ though may lower the mis-identification rate β, will decrease the detection rate χ, as shown by results from several additional simulations.

Figures 2 and 3 show that the TSKF method is able to detect more additive outliers in dynamic process than the Hampel identifier.

VARMA(1,1)model

$$\left(\mathbf{I}-\mathbf{\Phi} \mathbf{z}^{-1}\right) \mathbf{x}_t=\left(\mathbf{I}-\mathbf{\Omega} \mathbf{z}^{-1}\right) \epsilon_t \tag{13}$$

where $\mathbf{\Phi}$,$\mathbf{\Omega}$ are coefficient matrices,ϵ_t is the driving noise of the model, $\boldsymbol{\epsilon}_t \sim N\left(\mathbf{0},\begin{bmatrix}1 & 0.5\\ 0.5 & 1\end{bmatrix}\right)$ and z^{-1} is a shift operator.

The on-line and off-line additive outlier detection results for data from VARMA(1,1) processes are shown in Table 3:

Based on additional simulations, increasing the model complexity by choosing a larger parameter l does not obtain a better performance for PCA and DPCA methods. The parameter

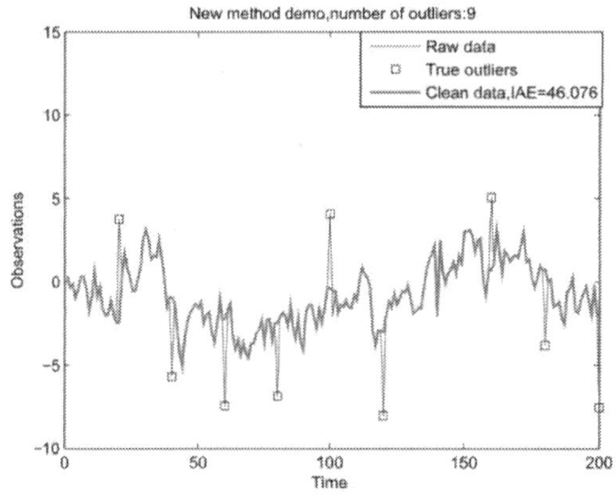

Figure 2: TSKF method for additive outliers

Simulation condition: ARMA (1,1), $\phi = 0.9$; $\theta = 0$; Amp=5; $\kappa = 5\%$; MW=100

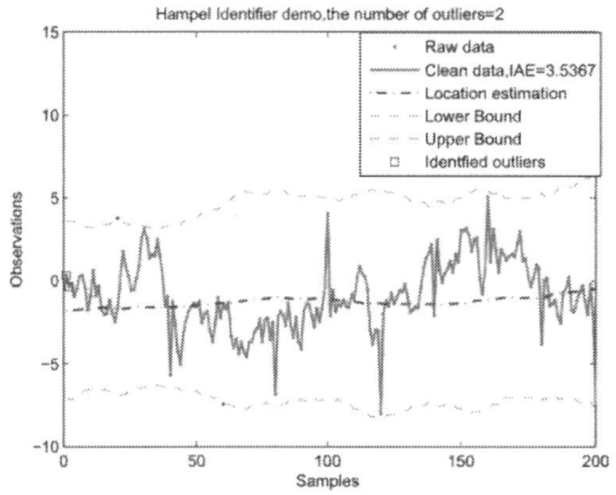

Figure 3: The Hampel identifier for additive outliers

Simulation condition: ARMA (1,1), $\phi = 0.9$; $\theta = 0$; Amp=5; $\kappa = 5\%$; MW=100

Table 3: Additive outlier detection rate $\chi/\%$ for data from VARMA (1,1) process at $\kappa = 5\%$

Case No.	Φ	Ω	Δ	$\begin{bmatrix} Amp \\ Amp \end{bmatrix}$	on-line(*) TSKF	on-line(*) DPCA(I)	on-line(*) PCA(II)	off-line(**) TSKF	off-line(**) DPCA(III)	off-line(**) PCA(IV)
1	**0.0**	**0.0**	3.2	$\begin{bmatrix} 3.0 \\ 3.0 \end{bmatrix}$	90.22	71.82	72.24	88.32	45.16	66.57
2	**0.0**	**0.0**	3.2	$\begin{bmatrix} 4.0 \\ 4.0 \end{bmatrix}$	97.11	86.92	88.72	93.74	66.93	87.11
3	$\begin{bmatrix} 0.9 & 0.1 \\ 0.1 & 0.9 \end{bmatrix}$	$\begin{bmatrix} 0.9 & 0.1 \\ 0.1 & 0.9 \end{bmatrix}$	3.3	$\begin{bmatrix} 4.0 \\ 4.0 \end{bmatrix}$	94.22	87.52	88.66	93.64	66.57	87.38
4	$\begin{bmatrix} 0.9 & 0.1 \\ 0.1 & 0.9 \end{bmatrix}$	$\begin{bmatrix} 0.9 & 0.1 \\ 0.1 & 0.9 \end{bmatrix}$	3.3	$\begin{bmatrix} 5.0 \\ 5.0 \end{bmatrix}$	99.11	95.22	95.78	97.62	81.84	96.35

(*) Simulation condition: N=1000; rep=100; MW=100.
(**) Simulation condition: N=10000; rep=100; PW=1000; $\gamma = 95\%$.
(I) Parameter setting: $\alpha = 0.01$; $l = 1$; $a = 2$.
(II) Parameter setting: $\alpha = 0.01$; $a = 1$.
(III) Parameter setting: $\alpha = 0.01$; $l = 1$; $a = 3$.
(IV) Parameter setting: $\alpha = 0.01$; $a = 1$.

a is chosen to be 2 for on-line and 3 for off-line to ensure the DPCA model captures close to 90% total variance of the data set, and to prevent model over-fitting at the same time. Though a larger significance level α will help raise the detection rate, it will significantly increase the mis-identification rate β (larger than 10%). As a consequence, α was chosen to be 0.01.

By analyzing results shown in Table 3 we can see that the TSKF method generally outperforms the DPCA and PCA method. The DPCA method does not obtain as good detection rates as PCA because without a pre-whitening step, an outlier contamination rate of 5% will severely damage the serial correlation in augmented matrix shown in Eq. (10) and lead to inaccurate t^2 results.

Moreover, a larger outlier size Amp and a lower autocorrelation $\mathbf{\Phi}$ will help the outlier detection by increasing the outlier detection rate χ, which are the same as shown in the ARMA(1,1) case.

Figure 4 demonstrates the on-line multivariate additive outlier detection results of the TSKF method for a VARMA(1,1) process. Figure 5 shows the Hotelling's T^2 record on the first moving window of the DPCA method, it needs to be pointed that though time points 40 and 80 are missed, they can be successfully detected in moving windows afterwards.

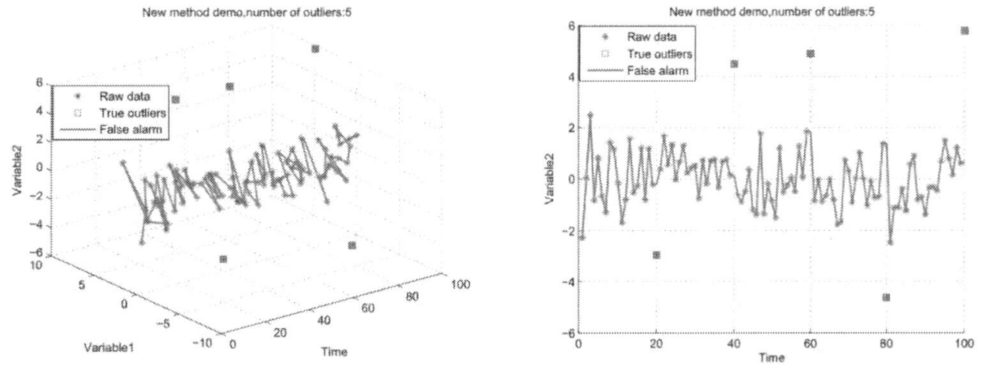

(a) 3D view

(b) Front view

Figure 4: Additive outlier detection results obtained by the TSKF method

Simulation condition: VARMA(1,1); on-line; case No.=3

Summary and discussion of simulation testing results

Based on the simulation results, we can see that a larger autocorrelation contained in the process data will negatively affect the detection results, while a larger outlier size will help the outlier detection. Though only the outlier detection results of ARMA(1,1) and VARMA(1,1) processes are shown, the TSKF method works well for higher order stationary process data, such as ARMA(2,1) and VARMA(2,1). The TSKF method outperforms the Hampel identifier, PCA and DPCA methods in detection of additive outliers.

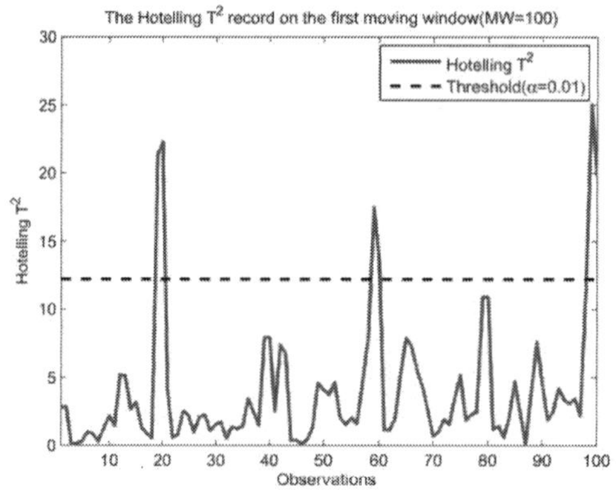

Figure 5: Hotelling's T^2 record on the first moving window of dynamic PCA

Simulation condition: VARMA(1,1); on-line; $\kappa = 5\%$; case No.=3; additive outlier

References

1. Han J, Kamber M, Pei J. *Data mining: concepts and techniques*. San Francisco: Morgan kaufmann, 2006.

2. Barnett V, Lewis T. *Outliers in statistical data*. Wiley Series in Probability and Mathematical Statistics. Applied Probability and Statistics (2nd edition). Chichester: Wiley, 1984.

3. Tsay RS, Peña D, Pankratz AE. Outliers in multivariate time series. *Biometrika*. December 1, 2000; 87(4): 789-804.

4. Hampel FR. A general qualitative definition of robustness. *The Annals of Mathematical Statistics*. 1971; 42(6): 1887-1896.

5. Pearson RK. Outliers in process modeling and identification. *Control Systems Technology, IEEE Transactions on*. 2002; 10(1): 55-63.

6. Rousseeuw PJ. Least median of squares regression. *Journal of the American Statistical Association*. 1984; 79(388): 871-880.

7. Rousseeuw PJ. Multivariate estimation with high breakdown point. *Mathematical Statistics and Applications*. 1985; B: 283-297.

8. Liu H, Shah S, Jiang W. On-line outlier detection and data cleaning. *Computers & chemical engineering*. 2004; 28(9): 1635-1647.

9. Hampel FR. The influence curve and its role in robust estimation. *Journal of the American Statistical Association*. 1974; 69(346): 383-393.

10. Ku W, Storer RH, Georgakis C. Disturbance detection and isolation by dynamic principal component analysis. *Chemometrics and Intelligent Laboratory Systems*. 1995; 30(1): 179-196.

11. Russell EL, Chiang LH, Braatz RD. Fault detection in industrial processes using canonical variate analysis and dynamic principal component analysis. *Chemometrics and Intelligent Laboratory Systems*. 2000; 51(1): 81-93.

367986 Virtual Kinetics Laboratory for Quantitative Model-Based Discrimination of Reaction Mechanisms

Abstract:

The Virtual Kinetics Laboratory (VKL) provides a framework for the automated generation of rate laws, based on proposed reaction mechanism alternatives, and the subsequent automated formulation and numerical solution of the mathematical equations of these rate laws. Subsequently, simulations are performed within a model-based experimental design approach and this can aid to identify the best set of experiments that would yield data that can be used to discriminate between alternative mechanisms. Once the experimental data is collected, VKL can then be used to do parameter estimation in order to derive the rate constants for the various proposed mechanisms. Previous work relied largely on visualization techniques in order to discern between alternative mechanisms. The current work extends the framework, utilizing model-based discrimination techniques including sensitivity and error analyses for the quantitative evaluation of alternative mechanisms.

381715 Ontology Engineering: Support to Industrial Symbiosis Networks

Abstract:

Industrial Symbiosis (IS) is an innovative approach that aims in creating sustainable industrial networks set to process waste into materials, energy and water. Economic benefits are generated by normally low costs of waste or by-products, by using alternative energy sources and by environmental savings. Environmental benefits are inherent in IS and measured by landfill diversion but also by reduction in emissions and by water savings. Operating within confined geographic and administrative boundaries, IS also generates tangible social benefits to local communities, including job generation and retention, as well as new investments.

The key to formation of IS networks is the mediation between participants, the process which requires expertise, hence knowledge in different areas, i.e. waste composition, capability of processing technologies and environmental effect, among the others. The whole process is currently managed by trained practitioners supported by proprietary databases. Limited to the level of expertise and intuition of practitioners and lacking readily available repository of tacit knowledge, the all operation is backward looking and focusing on past successful examples with innovative networks being incidental.

This paper presents design and implementation of a semantic web platform which supports creation and operation of IS networks i) by screening the opportunities based on technological capability and resource availability of registered companies, and ii) by monitoring the IS operation and assessing sustainability using economic, environmental and social parameters. The platform employs ontologies to embed tacit knowledge in the domain of IS, knowledge gained from past experience but also from the latest research and otherwise advances in IS. More specifically, a set of integrated ontologies address off-spec nature of waste, i.e. variability in composition, dynamics in availability and pricing, as well as economic and environmental properties including hazardousness. Similarly, processing technologies are modeled in terms of processing capabilities, which include range of type of inputs, conversion rates, water and energy requirements, range of capacities, emissions as well as fixed and operational costs and environmental effects. Explicit knowledge is collected in the process of ontology instantiation with actual data collected from the IS participants during the registration. The ontologies are designed using ontology web language and hence prepared to grow and to share. In the current implementation more than 1500 different waste types and over 200 different technologies have been included.

Purpose designed matchmaker identify synergies between participants on their semantic and explicit relevance, the process crucial to formation of IS networks. Semantic relevance defines suitability from the type of waste/by-product and range of technology inputs, including complex composites of waste, i.e. biodegradable waste, and it is calculated from distance between the two instances in the respective ontology. Semantic relevance also includes participant general suitability for particular type of IS. Explicit relevance is calculated using vector similarity algorithm for respective properties, such as quantity, availability, geographical location and hazardousness. More intuitive and complex IS networks are proposed by reclusively repeating matches between two participants which in turn gives an opportunity for even better economic and environmental savings and/or targeted production. Both semantic and explicit matching relevance are aggregated in into a numerical value used for match ranking.

The eSymbiosis platform has been implemented as a web service with performance validated verified in the industrial region in Viotia, Greece and with several hundred participating company.

The effort has been funded by the LIFE+ initiative (LIFE 09 ENV/GR/000300), which authors acknowledge.

Markov State Model Construction

Xun Tang, Martha A. Grover

School of Chemical & Biomolecular Engineering, Georgia Institute of Technology, 311 Ferst Dr. NW, Atlanta, GA. 30332-0100.

The Markov state model has been widely used as a tool to model a time-sequential stochastic process in many fields, including biology, engineering, computer science, and economics [1]. The accuracy of a discrete-time Markov state model is mainly affected by the amount of sampling information, the transition time Δt, and the discretization of the state space. Researchers have focused and reported on how to improve the model accuracy by looking into the individual impacts of the aforementioned three factors on a Markov state model [2], [3]. Knowledge of the effects of these factors on model accuracy provides insights for building a reliable model; while an understanding of the relationships among each of these above mentioned factors would significantly improve the efficiency of experimental design to achieve such a reliable model, especially when it is costly to obtain dynamics information. It is particularly important to effectively design the experiments for the purpose of control, which usually requires the model to capture the dynamics in a common state space under each control action.

In this presentation, we investigate the relationships among the state space discretization, the transition time, and the sample size by studying a colloidal self-assembly process simulated with a simplified Langevin dynamics model. We also studied the model accuracy of using two different sampling strategies: many short sets of dynamics data versus one long set of such data, in order to better guide the design of experiments. The results are further applied to guide the construction of an accurate Markov state model with laboratory experimental data for a complex two-dimensional colloidal self-assembly system, for potential use in controlling the assembly process.

References

[1]. Arapostathis, A.; Borkar, V. S.; Fernández-Gaucherand, E.; Ghosh, M. K.; Marcus, S. I., Discrete-time controlled Markov processes with average cost criterion: a survey, SIAM J. Control Optim. 1993, 31, 282-344.

[2]. Prinz, J-H; Wu, H.; Sarich, M. ; Keller, B.; Senne, M.; Held, M.; Chodera, D. J.; Schütte, C.; Noé, F., Markov models of molecular kinetics: Generation and validation. J. of Chem. Phys., 2011, 134, 174105.

[3]. Schütte, C.; Noé, F.; Lu, J; Sarich, M.; Vanden-Eijnden, E., Markov state model based on milestoning. J. of Chem. Phys., 2011, 134, 204105.

Understanding Cognitive Behavior of Process Operators during Abnormal Situations through Eye Tracker Studies

Chandresh Sharma, Punit Kumar, Babji Srinivasan, Rajagopalan Srinivasan
Indian Institute of Technology, Gandhinagar

Abstract:

Ensuring Process safety is critical in process plants. Major accidents such as the ones in Richmond (1993), Longford (1998), and Texas City (2005) highlight that ensuring safety is a continuous battle where the plant operator is at the forefront. Specifically, the human operator in the control room is responsible for effective monitoring and supervision of the process plant. Therefore consideration of human factors especially during abnormal situations is important. To perform his role the control room operator is critically dependent on information conveyed by the Distributed Control System (DCS) through visual displays and alarms [1]. Based on these, during abnormal situations, the operator is expected to quickly and accurately detect when things go wrong, isolate the root cause, and perform corrective / recovery actions. A variety of decision support methods have been proposed to automate monitoring and fault diagnosis [2]. However, to date their deployment and acceptance in actual plant usage is limited. In order to understand why these technologies are unable to penetrate into practice and be deployed for real-time usage, it is necessary to comprehensively understand how the operator uses the control and alarm systems. Specifically, it is critical to understand the cognitive behavior of the operator when he/she is dealing with process upsets and abnormalities. That is the goal of this paper. We seek to develop systematic approaches to understand the cognitive behaviors of process operator during abnormal situations.

For the control room operator, the eyes are a critical source of information. An eye tracker is a device for measuring the point of gaze (where one is looking) in real-time. While studies of eye movement are over 150 years old, it is only in recent years that non-invasive, inexpensive, easy to use eye trackers have become widely available. Eye tracker data is now commonly used for deciding information flow on web pages as per gazing behavior of web page user. Kilingar et al., [3] used eye tracker data to analyze pilot's situation awareness. In this research, Tobii TS300 eye tracker is used to locate the point of gaze of process operators as they use the DCS screens to control a process. We also track the mouse movement and key presses on the DCS computers to fully track the operator's interaction with the DCS (and hence the process). Hence in our study, we have a temporal map of what the operator is looking at, and what actions he performed at a fine granularity. In other words, a spatio-temporal trigger-response chain of the operator's behavior can be developed. Using this, we seek to infer the operator's cognitive behavior especially under abnormal situations.

As a pilot study, the above mentioned methodology has been tested on 68 undergraduate students, who play the role of plant operators seeking to control a simulated ethanol plant. During the course of experiment, we have simulated various abnormal situations. The subjects are then tasked to bring the plant back into normal condition through suitable intervention. In this paper, we will describe the experimental procedure developed to study the cognitive behavior and also report on some of the key insights on the cognitive behaviors observed during normal and abnormal operations.

Reference

1. Aditya A., Chen S.F., Lee Z., and Srinivasan, R. (2014). "Quantifying the Effectiveness of an Alarm Management System through Human Factors Studies". *Computers & chemical engineering* 67(4), 1-12.

2. Venkatasubramanian, V., Rengaswamy., R. Yin, K., and Kavuri, S.N. (2003). "A review of process fault detection and diagnosis: Part I: Quantitative model-based methods." *Computers & chemical engineering* 27 (3), 293-311.

3. Kilingaru, K., Tweedale, J.W., Thatcher, S. and Jain, L.C. (2013). "Monitoring pilot "Situation Awareness"." *Journal of Intelligent and Fuzzy Systems,* 24(3), 457-466.

Process Control Method Based on the Inverse Analysis of Soft Sensors Considering Controllability

THE UNIVERSITY OF TOKYO · Chemical System Engineering

Ippei Kimura, Hiromasa Kaneko, *Kimito Funatsu (*funatsu@chemsys.t.u-tokyo.ac.jp)

Summary

In process industry, effective control methods for set point change and disturbance suppression are required. Optimization of control parameters, however, is difficult when the process is large and complex, resulting in many processes being operated inefficiently. Our goal is to perform more efficient control using information obtained from inefficient daily operation data. In order to achieve this goal, we have been developing a process control method using soft sensor models. The performance was confirmed with a simulated CSTR system.

Objective and Strategy

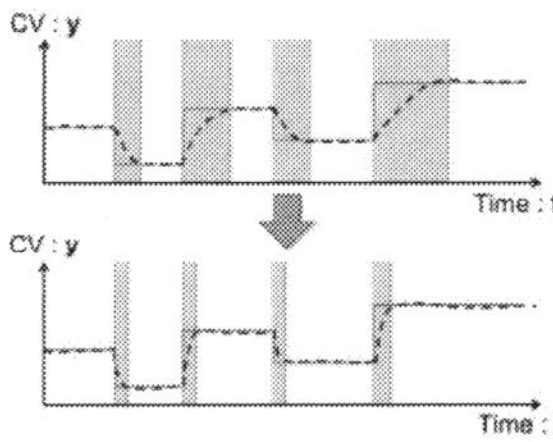

To perform more efficient control using information obtained from inefficient daily operation data

Soft sensors and inverse analysis

Soft Sensor

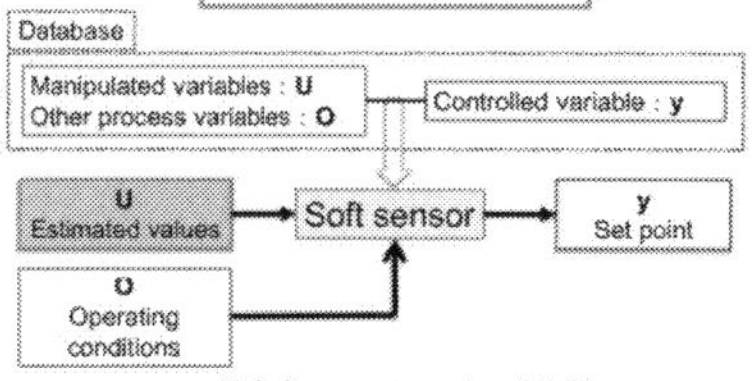

- multiple linear regression (MLR)
- partial least squares (PLS)
- support vector regression (SVR)

Inverse Soft sensor-based Feed Back (ISFB) control

Generation Step

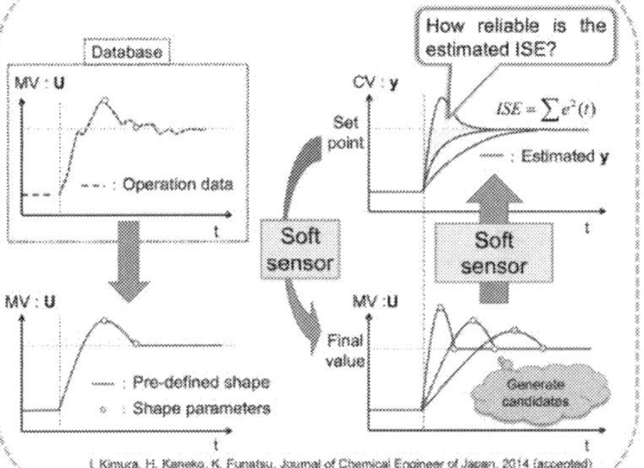

I. Kimura, H. Kaneko, K. Funatsu, Journal of Chemical Engineer of Japan, 2014 (accepted)

Selection Step

Gaussian process

$p(\mathbf{y}) = N(\mathbf{y}|\mathbf{0},\mathbf{C})$

$p(y_{new}|\mathbf{y}) = N(y_{new}|\mathbf{k}^T\mathbf{C}^{-1}\mathbf{y}, c - \mathbf{k}^T\mathbf{C}^{-1}\mathbf{k})$

C : Covariance matrix based on X

k : Covariance vector

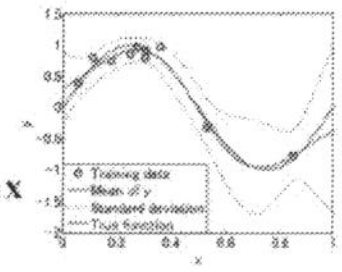

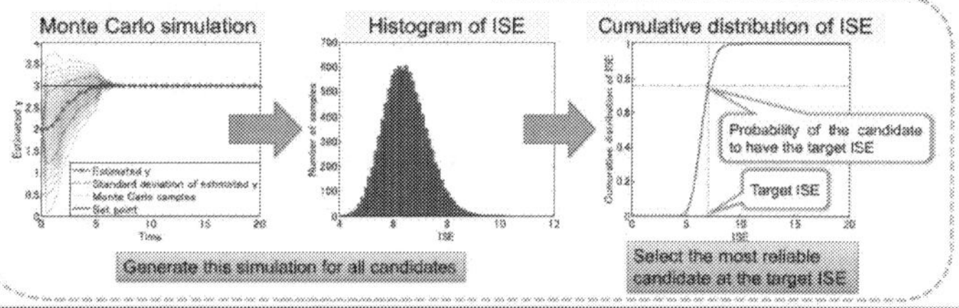

Generate this simulation for all candidates

Select the most reliable candidate at the target ISE

Feed Back Step

Bias correction

- Disturbance suppression
- Modeling error suppression

Model update

- Adaptation to the plant state change

Case Study

CSTR Simulator

Virtual laboratorial plant

- A→B
- 1st order reaction

Variables

- CV(y) : Concentration of B (C_B)
- MV (U) : Concentration of inlet A (C_{A0})
- others (O) : Cooling water temperature (Tc), Inlet temperature (T_0)

Objective

- C_B set point change (10→9 [kmol / m³])

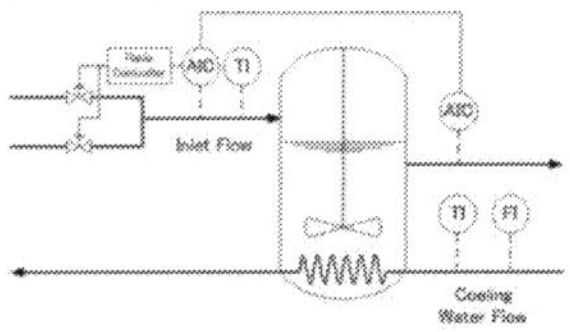

Training Data

- Operation for set point change (28 cases)
- PI control (Tuned by ZN method × 0.2)
- Sampling intervals
  - C_B : 30 [min] (with 30[min] delay)
  - others : 1 [min]
- ISE (10→9 [kmol / m³]) : 8.95 (50~70 [hour])

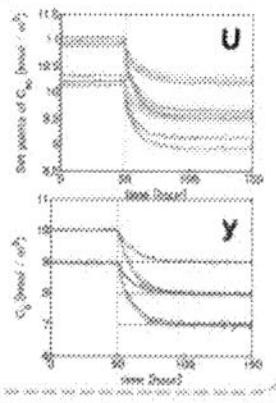

Soft Sensor Modeling

Modeling method : SVR (5 fold cross validation)
Updating method : Online SVR

- Objective variables : C_B
- Explanatory variables : C_{A0} and Tc with time delay (Time delay : 0, 15 , ..., 300 [min])

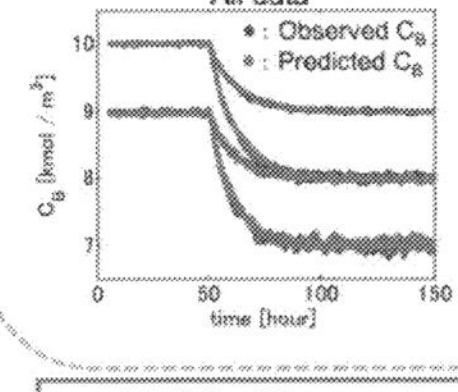

$q^2_{all} = 0.999$
$RMSEcv = 0.021$

$r^2_{all} = 0.999$
$RMSE = 0.023$

Generation Step

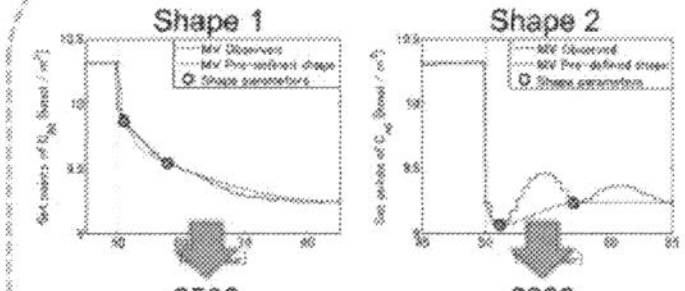

2500 candidates · 6880 candidates

	Shape 1	Shape 2
Similarity to training data	○	△
Control effectiveness	△	○

Selection Step

Target ISE : 7(moderate), 3(challenging)

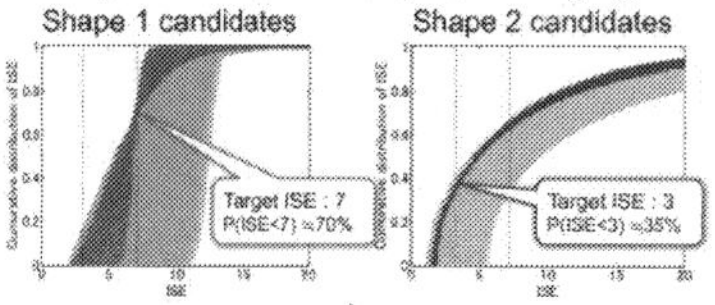

Target ISE : 7

69% < P(ISE < 7) < 71%

\# of candidates : 282 (All shape 1)

Accuracy: $\frac{282}{282} = 100\%$

Target ISE : 3

34% < P(ISE < 3) < 36%

\# of candidates : 975 (All shape 2)

Accuracy: $\frac{22}{975} = 2.26\%$

ISFB Characteristics

- Soft sensors constructed from daily operation data
  → Reducing system identification loads
- Inverse analysis using pre-defined shape
  → Considering plant knowledge

Feed Back Step

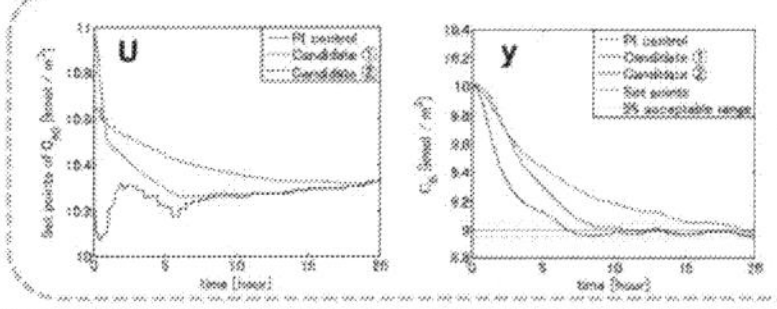

Selected candidates : The highest probability one for each target ISE

	PI control	Target ISE : 7	Target ISE : 3
ISE	8.95	5.96	3.45
5% settling time	15.5	8.07	6.07

383072 Modeling of Alternatives for Recovering the Exhaust Heat from a Gas Turbine Power Plant of Ecopetrol

Abstract:

The current electrical energy demand in GEC-Ecopetrol is 200 MW and an increase of about 240 MW is expected in the next 10 years for new oil wells and enhanced oil recovery. To satisfy part of this new demand, GEC-Ecopetrol is planning to improve the efficiency of the existing power generation facilities and to use a new natural gas source that has become available in Termosuria, one of the main GEC-Ecopetrol power plants. Currently, Termosuria has a gas turbine producing between 34 - 37 MW by a simple Brayton cycle. However, gas turbines in simple cycle are limited by their thermal efficiency, achieving up to 40%. The main energy loss in gas turbines is the heat contained in the exhaust gases, which have a temperature of 840°F in Termosuria. This stream can be used to generate more power by adding a steam Rankine cycle to the process. Generally, the implementation of this combined cycle (CC) results in up to 60% global efficiency increase.

In this paper, two design alternatives of a new combined cycle are modeled for the Termosuria GE LM6000® gas turbine. Since the Heat Recovery Steam Generator (HRSG) is one of the few non-standardized components of the combined cycle that can be customized, both alternatives are focused on the HRSG design, specifically in the implementation, or not, of a post-combustion burner. The influence of environmental conditions (temperature and humidity) on plant performance is studied by applying a typical daily variation curve, obtained from a monitoring meteorological tower installed near the plant site. The thermodynamic behavior is modeled in Gate Cycle® software and the economic analysis is modeled in VBA-Excel®; both models are integrated by CycleLink® utility. For each alternative, the following indicators of plant performance are calculated, hour by hour, following a typical daily curve of environmental conditions: Generated power, thermal efficiency and cost of electricity (*COE*).

364763 Heat Exchanger Network Design with Considering Physical Properties Variation

Abstract:

In this paper energy targeting for an ammonia plant has already been done with and without stream segmentation. Pinch temperature, ΔT_{min}, and the values of the hot and cold utilities in both cases are different and so the HEN has to be changed. The results demonstrates that when splitting streams into two or more segments within the existing ammonia plant, pinch point increases by about 3°C, and hot and cold utilities consumption increases by approximately 23% and 11%, respectively. The area of the heat exchanger network also is increased by about 50%. Finally, an aggregation diagram is introduced for the evaluation of specific heat capacity distribution along the whole streams of the network.

393806 Experimental Stress Testing of Multivariate Data Analysis (MVDA) models in Chromatography

Abstract:

As the biotech industry moves toward implementing proactive practices to ensure product quality and process efficiency, improving data analysis is also becoming necessary. Univariate data analysis can often be inefficient and result in misleading conclusions due to the complexity of biopharmaceutical processes. Multivariate data analysis (MVDA) has several benefits over univariate analysis such as increasing sensitivity to process deviations, reducing the time required to analyze large amounts of data and providing more information to more easily review and identify process deviations. These benefits can standardize monitoring practices, increase process knowledge and potentially save batches. MVDA is proving to be a useful predicting tool for different types of processes. However, all of the current applications are tested based on development or manufacturing data without the freedom to stress test the models for potential future deviations. The models are improved only after specific failures or process deviations are observed. In addition, the model capability to detect certain type of deviation is unknown until the deviation happens. Knowledge of model capability increases the possibility to apply MVDA in manufacturing processes. In this work, we looked at downstream purification as a case study and presented the results from stress testing a cation exchange chromatography step and analyzing the process with an MVDA model. Stress testing was performed in the laboratory setting through controlled experiments. The design of experiments included all the typical deviations in chromatographic processes such as load pool and elution buffer variations, column packing conditions and equipment failures. The effects of different modeling configurations to the MVDA fault detection ability were also studied in a systematic fashion. This work highlights the MVDA model capabilities and limitations in chromatographic process monitoring and demonstrates the methodology to evaluate the MVDA model performance for process monitoring and fault detection.

373518 On Distributed Economic Model Predictive Control of Nonlinear Process Systems

Abstract:

In the context of control of large-scale nonlinear chemical process networks, an attractive control methodology is distributed model predictive control (DMPC) because it has the ability to control multiple-input multiple-output with input and state constraints while remaining computationally feasible to implement on-line through a distributed implementation of the computation (i.e., the computation is distributed to multiple processors) [1]-[2]. Recently, significant effort within the chemical process control community has focused on (centralized) economic model predictive control (e.g., [3]) which combines the advantages of a predictive control methodology with dynamic economic optimization by using a cost function that accounts for the process economics. Since EMPC may use a general (nonlinear) economic cost function and it may dictate a time-varying operation strategy, the on-line computation required to solve EMPC may be significant especially for large-scale process networks. Thus, distributed EMPC (DEMPC) may be one choice to significantly reduce the on-line computational burden. While early work on distributed EMPC (DEMPC) has shown some promising results in this direction [4]-[5], more work in this direction is in order including the development of DEMPC algorithms, rigorous theoretical stability analysis, and novel control loop decompositions methodologies tailored for DEMPC.

In this work, we focus on the development of sequential and iterative DEMPC for nonlinear chemical processes. The theoretical contribution of this work consists of three main parts. First, a novel control loop decomposition methodology on the basis of the process economics is developed. Second, two DEMPC schemes, designed with specific Lyapunov-based constraints [4], are formulated for sequential and iterative implementation, respectively. Third, a rigorous stability analysis of each of the DEMPC architectures is provided. While the stability analysis will be provided for general nonlinear process systems typically arising in chemical process industries, we tailor the formulation of the DEMPC algorithms with a specific focus on chemical processes that achieve better economic performance when operated in a time-varying fashion (e.g., periodic operation). The DEMPC architectures are demonstrated with a chemical process example previously studied in the context of periodic operation.

[1] Christofides PD, Liu J, Munoz de la Pena D. Networked and Distributed Predictive Control: Methods and Nonlinear Process Network Applications. Advances in Industrial Control Series. London, England: Springer-Verlag, 2011.
[2] Christofides PD, Scattolini R, Munoz de la Pena D, Liu J. Distributed model predictive control: A tutorial review and future research directions. Computers and Chemical Engineering. 2013;51:21-41.
[3] Angeli D, Amrit R, Rawlings JB. On average performance and stability of economic model predictive control. IEEE Transactions on Automatic Control. 2012;57:1615-1626.
[4] Chen X, Heidarinejad M, Liu J, and Christofides PD. Distributed economic MPC: Application to a nonlinear chemical process network. Journal of Process Control. 2012;22:689-699.
[5] Lee J and Angeli D. Distributed cooperative nonlinear economic MPC. In: Proceedings of the 20th International Symposium on Mathematical Theory of Networks and Systems. Melbourne, Australia. 2012.

377340 Data-Driven Fault-Tolerant Control of Networked Distributed Processes

Abstract:

The identification and accommodation of faults in process systems are two of the fundamental problems at the intersection between process control and operations. A careful review of the extensive literature on fault detection and fault-tolerant control shows that while the majority of existing results have been developed for spatially homogeneous processes described by lumped parameter systems, methods for the diagnosis and handling of faults in spatially distributed systems described by Partial Differential Equations (PDEs), on the other hand, are few and limited by comparison. This realization, together with the prevalence of spatiotemporal dynamics in such systems, have motivated several efforts in recent years aimed at bridging this gap (see, for example, [1]-[4] and the references therein). In these studies, a framework for the integration of actuator fault detection, isolation and control system reconfiguration for highly-dissipative PDEs was developed on the basis of low-order models that are suitable for the design of practically implementable control and monitoring algorithms, and are amenable to rigorous closed-loop analysis.

Despite the continuing progress in this area, an examination of the above methods reveals that they are developed within the conventional feedback control framework where the sensor-controller communication is assumed to occur continuously and flawlessly. With the advent of networked control systems, however, and the increased reliance on advanced communication technologies in many industrial systems, the fundamental limitations imposed by communication constraints (e.g., in the form of resource constraints, real-time scheduling constraints, delays, etc.) on the closed-loop stability and performance properties can no longer be ignored and need to be explicitly accounted for. Efforts to address sensor-controller communication constraints in the context of spatially distributed control systems were undertaken in a number of recent studies (see, for example, [5]-[7]) which focused on highly-dissipative distributed processes and led to various finite-dimensional networked control and scheduling design methodologies that enforce closed-loop stability with minimal sensor-controller communication requirements. The resulting design methods, however, do not explicitly consider fault diagnosis or handling in the control system design. At this stage, the design of networked fault-tolerant control systems for spatially distributed processes remains an open problem that is in need of investigation and further development.

This work presents a methodology for the integrated identification and accommodation of control actuator faults in a class of spatially distributed systems controlled over a resource-limited communication network. The methodology brings together tools from model reduction, networked control and moving-horizon parameter estimation. Focusing on infinite-dimensional systems described by highly-dissipative parabolic PDEs, modal decomposition is initially used to obtain a finite-dimensional system that captures the slow dynamics of the infinite-dimensional system. The slow system is then used to design a model-based networked controller and explicitly characterize its stability region, in terms of the sensor-controller communication rate, the model uncertainty, the control actuator placement and the controller design parameters. Fault identification is then addressed using a moving-horizon least-squares parameter estimation scheme which is embedded in the sensors to estimate on-line the size of the fault using the sampled state and input data. Once the fault is identified and its magnitude estimated and communicated to the controller, a number of possible fault accommodation strategies are devised to maintain closed-loop stability. These include updating the post-fault control model, adjusting the controller design parameters, or switching to an alternative control configuration. The selection of the appropriate accommodation strategy is made on the basis of the estimated fault magnitude and the characterization of the networked closed-loop stability region. The implementation of the proposed fault identification and accommodation strategies on the infinite-dimensional system is also analyzed. Finally, the proposed methodology is illustrated using a diffusion-reaction process example.

References:

[1] N. H. El-Farra and S. Ghantasala, "Actuator fault isolation and reconfiguration in transport-reaction processes," *AIChE J.*, **53:**1518-1537, 2007.

[2] S. Ghantasala and N. H. El-Farra, "Robust actuator fault isolation and management in constrained uncertain parabolic PDE systems," *Automatica*, **45**:2368-2373, 2009.

[3] Z. Yao and N. H. El-Farra, "Robust fault detection and reconfigurable control of uncertain sampled-data distributed processes," in *Proceedings of 50th IEEE Conference on Decision and Control*, Orlando, FL, 2011, pp. 4925-4930.

[4] S. Ghantasala and N. H. El-Farra, "Active fault-tolerant control of sampled-data nonlinear distributed parameter systems," *Inter. J. Rob. Nonlin. Contr.*, **22**:24-42, 2012.

[5] Y. Sun, S. Ghantasala, and N. H. El-Farra, "Networked control of spatially distributed processes with sensor-controller communication constraints," in *Proceedings of American Control Conference*, St. Louis, MO, 2009, pp. 2489-2494.

[6] Z. Yao, Y. Sun, and N. H. El-Farra, "Resource-aware scheduled control of distributed process systems over wireless sensor networks," in *Proceedings of American Control Conference*, Baltimore, MD, 2010, pp. 4121-4126.

[7] Z. Yao and N. H. El-Farra, "Networked model predictive control of spatially distributed processes," in *Proceedings of American Control Conference*, Washington, DC, 2013, pp. 2062-2067.

Title

Advanced Control Approaches for Combined Cooling and Antisolvent Crystallization (CCAC) in Continuous Mixed Suspension Mixed Product Removal (MSMPR) Cascade Crystallizers

Authors

Yang Yang and Zoltan K. Nagy, Chemical Engineering, Purdue University, West Lafayette, IN

Abstract

In fine chemical and pharmaceutical industry, the control of crystallization process is essential in order to achieve crystals with desired size distribution, shape and crystalline form. Typically these crystal properties could have large influence on the efficiency of downstream operations (e.g. filtration, drying, milling), as well as the physical and chemical properties (e.g. bioavailability, solubility, dissolution rate) of the final active pharmaceutical ingredient (API) [1]. Supersaturation is the driving force for crystallization processes. Cooling and antisolvent addition are commonly used to generate supersaturation. These two methods could either be used separately or in combination. Despite the fact that combined cooling and antisolvent crystallization (CCAC) is a common operation especially in the pharmaceutical industries, a systematic study and design and control procedure was first presented only recently by Nagy et al. [2], who firstly studied the control of CCAC via both model-free and model-based approaches in a batch system. The major advantages of CCAC are the higher yield and the larger control freedom on the multiple crystal properties like crystal size, shape and polymorphic form as compared with cooling or antisolvent addition only crystallization [3]. Additionally, some crystal qualities (e.g. mean crystal size, crystal size distribution (CSD) consistency) could be improved by the CCAC approach [2][4] in batch systems.

As compared with the traditionally used batch crystallization, continuous crystallization has the potential for enhanced reproducibility, reduced operating cost, simplified scale up and improved efficiency of process and equipment [5]. Mixed suspension mixed product removal (MSMPR) crystallizer is an important type of continuous crystallizer. It can be used either as a single stage or as a multistage cascade system. Myerson and Wong used a single stage MSMPR crystallizer with recycle [6], to produce crystals of both high yield and purity. Multistage continuous MSMPR crystallizers were also used in the same group [5][7][8]. These cascade systems were proved experimentally to be able to improve both yield and purity for the crystallization of some organic compounds (e.g. cyclosporine, aliskiren hemifumarate). Vetter et al. modeled the CCAC of aspirin in MSMPR cascade crystallizers and obtained the attainable region of particle sizes [9]. A comparison between MSMPR crystallizer, plug flow crystallizer (PFC) and semi-batch crystallizer is also investigated in the same work.

The control and optimization of a continuous crystallization process is one of the key challenges during its development. This work models a continuous multi-input multi-output (MIMO) system: a two stage MSMPR cascade crystallizers system, with both cooling and antisolvent addition applied at both stages to generate supersaturation. The feed solution is free of crystals. The first stage is operated as a nucleation stage, whereas the second stage is designed for growth and operated at a low supersaturation level. The final mean crystal size and yield are aimed as the two controlled output variables. The temperatures and antisolvent addition rates at both stages are considered as possible manipulated input variables. Two advanced control approaches are proposed and analyzed in this MIMO system: 1) decoupled PID control 2) nonlinear model predictive control (NMPC). Both servo control and regulatory control scenarios are considered to evaluate their overall performances. In addition, a wide operating range is investigated, including both around optimal and non-optimal operating conditions. Additionally, based on different selections of the manipulated input variables, various control methods, including antisolvent control, temperature control, nucleation control, growth control and global control, are discussed and compared. Experimental validations are finally performed to support the simulation results. The results indicated that the MIMO control system is highly complex and nonlinear after some linearization investigations at different operating conditions. The two controlled output variables, mean size and yield are strongly correlated. Thus the decentralized PID control is not applicable. On the other hand, the NMPC shows excellent performance on both the servo and regulatory control. The selection of the manipulated input variables also turns out to be extremely important. Interestingly, the results indicate that the growth control method should be preferred and implemented in practice due to the following three aspects: 1) The growth control method is able to avoid large oscillation and overshoot in output responses. Because it is found that the change of input variables on nucleation stage could result in large oscillation and overshoot. Controlling only on the growth stage could avoid all disturbances from the nucleation stage. 2) The growth control method is able to provide suitable control over a broad operating range hence allowing large changes in target size and yield. This is achieved since this method can easily manipulate the residence time of the growth stage by switching the percentage of cooling and antisolvent addition in the supersaturation generation, while maintaining the best supersaturation level or profile. 3) The growth control method is able to avoid multiplicity of input variables. Because the application of some unnecessary input variables (e.g. inputs on nucleation stage) could result in excess degree of freedom and multiplicity of input variables of the growth stage. The proposed control configuration and nonlinear model predictive control scheme can provide superior control performance for disturbance rejection during regulatory control, for fast target product property change-over, as well as for fast startup operation. The results are demonstrated using simulations and experimental validation.

References:

[1] Nagy, Z. K.; Aamir, E. Systematic design of supersaturation controlled crystallization processes for shaping the crystal size distribution using an analytical estimator. *Chem. Eng. Sci.* **2012**, 84, 656–670.

[2] Nagy, Z. K.; Fujiwara, M.; Braatz, R. D. Modelling and control of combined cooling and antisolvent crystallization processes. *J. Process Control* **2008**, 18, 856–864.

[3] Yang, Y.; Nagy, Z. K. Model-based systematic design and analysis approach for unseeded combined cooling and antisolvent crystallization (CCAC) systems. *Cryst. Growth Des.* **2014**, 14, 687-698.

[4] Hermanto, M. W.; Chow, P. S.; Tan, R. B. H. Operating strategy to produce consistent CSD in combined antisolvent-cooling crystallization using FBRM. *Ind. Eng. Chem. Res.* **2012**, 51, 13773−13783.

[5] Alvarez, A. J.; Singh, A.; Myerson, A. S. Crystallization of cyclosporine in a multistage continuous MSMPR crystallizer. *Cryst. Growth Des.* **2011**, 11, 4392-4400.

[6] Wong, S. Y.; Tatusko, A. P.; Trout, B. L.; Myerson, A. S. Development of continuous crystallization processes using a single-stage mixed-suspension, mixed-product removal crystallizer with recycle. *Cryst. Growth Des.* **2012**, 12, 5701-5707.

[7] Quon, J. L.; Zhang, H.; Alvarez, A.; Evans, J.; Myerson, A. S.; Trout, B. L. Continuous crystallization of aliskiren hemifumarate. *Cryst. Growth Des.* **2012**, 12, 3036-3044.

[8] Zhang, H.; Quon, J.; Alvarez, A. J.; Evans, J.; Myerson, A. S.; Trout, B. Development of continuous anti-solvent/cooling crystallization process using cascaded mixed suspension, mixed product removal crystallizers. *Org. Process Res. Dev.* **2012**, 16, 915-924.

[9] Vetter, T.; Burcham, C. L.; Doherty, M. F. Regions of attainable particle size in continuous and batch crystallization processes. *Chem. Eng. Sci.* **2014**, 106, 167-180.

DESIGN OF SENSOR NETWORKS FOR SIMULTANEOUS FAULTS RESOLUTION AND DETECTION

*Leandro P. F. Rodriguez, Marco V. Cedeño, Mabel C. Sánchez**
Planta Piloto de Ingeniería Química (Universidad Nacional del Sur - CONICET), Bahía Blanca, Argentina

Introduction

Eventhough a great number of works in the literature devoted to the design of sensor networks (SNs) have focused on monitoring the normal process operation, the optimal location of instruments to effectively diagnose plant faults has central importance for safety, environmental protection and process economy.

During the last decade, different sensor network design (SND) approaches for fault diagnosis were presented. Many of them introduced requirements on fault observability (O) and simple resolution (R). In this sense, Bagajewicz et al. (2004) developed a MILP formulation to obtain minimum-cost SNs subject to fault O and simple/multiple fault R, which was stated using matrix algebra concepts. Later on, Bhushan et al. (2008) addressed the robustness of the network to uncertainties/errors in the underlying failure cause-effect model and probability data by modifying a previous design formulation (Bhushan and Rengaswamy, 2002), which was based on quantitative information about the occurrence of faults and sensor failure probabilities. More distributed SNs were preferred to incorporate robustness to modeling errors. Regarding the treatment of inaccurate data, the main idea was to ensure that the constraints involving uncertain data were far from being active at the optimal solution. Recently, Rodriguez et al. (2013) proposed a new strategy to design an optimal SN able to resolve a set of key faults even under the presence of failed sensors or outliers. The procedure deals with key failures isolation when sensor malfunctions occur from a structural point of view. With this purpose key faults Resolution Degree (RD) constraints are incorporated into the minimum-cost design problem, which involves a linear objective function and a set of linear inequality restrictions.

Other contributions analyzed the design of SNs devoted to detect and identify a set of process faults by means of particular monitoring strategies. These works considered the effect of measurement uncertainties, but took no account on the capability of the sensor structure to observe and resolve faults under the presence of sensor failures. In this sense, Musulin et al. (2004) and Gerkens and Heyen (2008) proposed methodologies to design SNs for processes monitored using Principal Component Analysis and model-equation residuals based techniques, respectively. The resulting optimization problems were solved using Genetic Algorithms.

If the strategy used for process monitoring has been selected, it is advisable to design a SN which is able to simultaneously resolve process faults and detect them when that strategy is running on line. With this purpose, a new SND problem is formulated in this work that minimizes the instrumentation cost subject to the resolution of all faults and the detection capability of the monitoring technique. Fault detection (FD) based on Principal Component Analysis (PCA) is selected, and the detection capability of this method is considered using the Fault Size Penalization model introduced by Musulin et al. (2004) which applies the concept of the Minimum Critical Fault Magnitudes (Wang et al., 2002). Furthermore key faults resolution and detectability should be attained even under the presence of sensor malfunctions or outliers. To tackle this issue, the optimization problem is appropriately reformulated.

In this work new formulations of the sensor network design problem for fault diagnosis are presented, and their advantages over other existing procedures are discussed. Application results for a chemical plant with recycle are provided.

Constraints Formulations

The set of faults and variables impacted by their occurrence can be related using the fault-sensor connectivity matrix, **A**, with dimension (I x J), where J is the number of failures (Bagajewicz, et al., 2004). The *j-th* column of the connectivity matrix, $\mathbf{a_j}$, is a binary vector that represents all the variables which reveal the *j-th* failure if they are measured. Hence a SN is represented by a binary vector **q** of dimension I, such that $q_i=1$ if variable i is measured and $q_i=0$ otherwise.

Observability and Resolution

A fault is observable if at least one of the variables affected by its occurrence is measured. Furthermore, given a pair of failures j and k, they are resolvable if the set of variables impacted by the occurrence of the first fault is different in at least one element from the set associated with the second failure. Three different vectors are defined using conjunction operation between the resolution vectors $\mathbf{rv}_j$ and $\mathbf{rv}_k$, where $\mathbf{rv}_j = \mathbf{a_j} \wedge \mathbf{q}$ and $\mathbf{rv}_k = \mathbf{a_k} \wedge \mathbf{q}$. These are:

$$\mathbf{v_1} = \mathbf{rv}_j \wedge \neg \mathbf{rv}_k \tag{1}$$

$$\mathbf{v_2} = \neg \mathbf{rv}_j \wedge \mathbf{rv}_k \tag{2}$$

$$\mathbf{v_3} = \mathbf{rv}_j \wedge \mathbf{rv}_k \tag{3}$$

Observability and R of j-th and k-th faults is achieved if at least two of the three defined vectors are non-null, that is:

$$NNV_{jk}(\mathbf{q}) \geq 2 \quad \forall j \neq k \tag{4}$$

where $NNV_{jk}(\mathbf{x})$ is the amount of non-null vectors obtained as result of the conjunction between $\mathbf{rv}_j$ and $\mathbf{rv}_k$.

General resolution degree

For a given SN, the j-fault has a RD equal to g if the fault remains identifiable eventough g measurements are not available. Then, $t_j = (R_j!)/(g_j!(R_j\text{-}g_j)!)$ sensor configurations exist, with g_j measurements less than those contained in $\mathbf{rv}_j$, which are able to identify the occurrence of the j-th key failure. To verify $RD_j = g$ the following inequalities should be examined:

$$NNV_{j^r k^r}(\mathbf{x}) \geq 2 \quad \forall j^r \neq k^r \quad r = 1..t_j \tag{5}$$

Fault Detectability with Principal Component Analysis

Wang et al. (2002) developed a methodology to determine the detectable fault magnitude using T^2 and SPE statistics. A process failure F_j can be described by the matrix $\boldsymbol{\theta}_j$ of dimension (n x l_j), where l_j is the number of variables affected by j-th failure and are measured by the SN defined by **q**.

By reformulating the expressions of the T^2 and SPE statistics, the fault magnitude that can be detected when PCA is used can be calculated as follows:

$$\|\mathbf{f}_j\| \geq \|\mathbf{f}_{T_j}\| = 2\sigma_{max}^{-1}\left(\mathbf{D}_\lambda^{-1/2}\mathbf{P}^T\boldsymbol{\theta}_j\right)\boldsymbol{\delta}_T \tag{6}$$

$$\|\mathbf{f}_j\| \geq \|\mathbf{f}_{SPE_j}\| = 2\sigma_{max}^{-1}\left(\left(\mathbf{I} - \mathbf{PP}^T\right)\boldsymbol{\theta}_j\right)\boldsymbol{\delta}_{SPE} \tag{7}$$

where $\|\mathbf{f}_{T_j}\|$ is the critical fault magnitude (MCF) of the j-th fault in the subspace of the retained principal components, $\|\mathbf{f}_{SPE_j}\|$ is the MCF of the j-th fault in the residual subspace, and $\boldsymbol{\delta}_T$ and $\boldsymbol{\delta}_{SPE}$ are the confidence limits for T^2 and SPE statistics.

Moreover, Musulin et al. (2004) proposed a methodology for evaluating the FD performance of a given SN. They defined the Minimal Critical Fault Magnitude (*MCFM*) as:

$$MMCF_{p,j} = \min\left\{\|\mathbf{f}_{T_{p,j}}\|, \|\mathbf{f}_{SPE_{p,j}}\|\right\} \tag{8}$$

An instrument configuration is feasible if the following constraint is satisfied:

$$MMCF_{p,j} \leq f_{\sup_{p,j}} \qquad j = 1 \ldots J \tag{9}$$

where $f_{\sup_{p,j}}$ is the norm of the normalized fault vector $\mathbf{f}_j^o$.

Sensor Networks Design Formulations for Fault Diagnosis

The minimum cost Sensor Network Design Problem (SNDP) that satisfies FD constraints using PCA (Musulin et al., 2004) is defined by equation 10:

$$\begin{aligned} &\underset{\mathbf{x}}{Min} \quad \mathbf{c}^T\mathbf{x} \\ &s.t. \\ &\quad MCFM_j \leq f_{\sup_j} \qquad j = 1 \ldots J \end{aligned} \tag{10}$$

where $\mathbf{c}$ is cost vector of instruments.

If O and R restrictions are considered, the SNDP proposed by Rodriguez, et al. (2013) is represented by equation (11):

$$\begin{aligned} &\underset{\mathbf{x}}{Min} \quad \mathbf{c}^T\mathbf{x} \\ &s.t. \\ &\quad \sum_{i=1}^{I}\left(rv_j\right)_i \geq 1 \quad j = 1 \ldots J \\ &\quad \sum_{i=1}^{I}\left(rv_j \wedge \neg rv_k\right)_i + \sum_{i=1}^{I}\left(rv_k \wedge \neg rv_j\right)_i \geq 1 \quad j = 1 \ldots J, j \neq k \end{aligned} \tag{11}$$

This optimization problem involves $J + \dfrac{J(J-1)}{2}$ linear constraints.

By incorporating RD constraints for a set of S key faults to Problem 11 the SN obtained is able to resolve a set of key failures even in the presence of missing measurements or outliers.

$$\begin{aligned}
&\underset{\mathbf{x}}{Min} \quad \mathbf{c}^{\mathrm{T}}\mathbf{x} \\
&s.t. \\
&\qquad \sum_{i=1}^{I}\left(rv_j\right)_i \geq 1 \quad j=1\ldots J \\
&\qquad \sum_{i=1}^{I}\left(rv_j \wedge \neg rv_k\right)_i + \sum_{i=1}^{I}\left(rv_k \wedge \neg rv_j\right)_i \geq 1 \quad j=1\ldots J, j \neq k \\
&\qquad \sum_{i=1}^{I}\left(rv_{s^r}\right)_i \geq 1 \quad s=1\ldots S; \quad r=1\ldots t_s^{'} \\
&\qquad \sum_{i=1}^{I}\left(rv_{s^r} \wedge \neg rv_k^r\right)_i + \sum_{i=1}^{I}\left(rv_k^r \wedge \neg rv_{s^r}\right)_i \geq 1 \quad s=1\ldots S; \quad r=1\ldots t_s^{'}; \quad s \neq k
\end{aligned} \tag{12}$$

where $t_s^{'} = \dfrac{R_s^{'}!}{g_s!(R_s^{'} - g_s)!}$ and $R_s^{'}$ is the cardinality of $\mathbf{rv}_s$. The resulting optimization problem is made up of $\left[J + \dfrac{J(J-1)}{2} + J\sum_{s=1}^{S} t_s^{'}\right]$ linear constraints.

If the J restrictions corresponding to the fault detection capability, which are described by the set of equations 9, are added to the optimization problem 11, problem 13 arises.

$$\begin{aligned}
&\underset{\mathbf{x}}{Min} \quad \mathbf{c}^{\mathrm{T}}\mathbf{x} \\
&s.t. \\
&\qquad \sum_{i=1}^{I}\left(\mathbf{rv}_{\mathrm{j}}\right)_i \geq 1 \quad j=1\ldots J \\
&\qquad \sum_{i=1}^{I}\left(\mathbf{rv}_{\mathrm{j}} \wedge \neg \mathbf{rv}_{\mathrm{k}}\right)_i + \sum_{i=1}^{I}\left(\mathbf{rv}_{\mathrm{k}} \wedge \neg \mathbf{rv}_{\mathrm{j}}\right)_i \geq 1 \quad j=1\ldots J, j \neq k \\
&\qquad MCFM_j \leq f_{\sup_j} \quad j=1\ldots J
\end{aligned} \tag{13}$$

The SN resulting from the resolution of the previous optimization problem satisfies structural determinability and detectability requirements. Similarly, if the J fault detection constraints are added to the optimization problem with RD restrictions (problem 12), problem 14 results.

Problem 13 is subject to two types of inequality constraints. The first two set of inequalities correspond to O and R constraints. The third set is related to *MCFM* restrictions. In Problem 14 the first four sets of linear inequalities include O, R and RD constraints. The fifth set involves the nonlinear restrictions associated with the *MCFM* calculation (equations 6-8). Optimization problems 13 and 14 are solved with an exhaustive search with stopping criterion.

$$
\begin{aligned}
&\underset{\mathbf{x}}{Min} \quad \mathbf{c}^{\mathrm{T}}\mathbf{x} \\
&s.t. \\
&\sum_{i=1}^{I}\left(rv_j\right)_i \geq 1 \quad j=1 \ldots J \\
&\sum_{i=1}^{I}\left(rv_j \wedge \neg rv_k\right)_i+\sum_{i=1}^{I}\left(rv_k \wedge \neg rv_j\right)_i \geq 1 \quad j=1 \ldots J, j \neq k \\
&\sum_{i=1}^{I}\left(rv_{s^r}\right)_i \geq 1 \quad s=1 \ldots S; \quad r=1 \ldots t_s^{'} \\
&\sum_{i=1}^{I}\left(rv_{s^r} \wedge \neg rv_k^r\right)_i+\sum_{i=1}^{I}\left(rv_k^r \wedge \neg rv_{s^r}\right)_i \geq 1 \quad s=1 \ldots S; \quad r=1 \ldots t_s^{'}; \quad s \neq k \\
&MCFM_j \leq f_{\sup_j} \quad j=1 \ldots J
\end{aligned}
\tag{14}
$$

Application Example

The Chemical Plant with Recycle proposed by Belanger and Luyben, (1997) was selected as a case study.

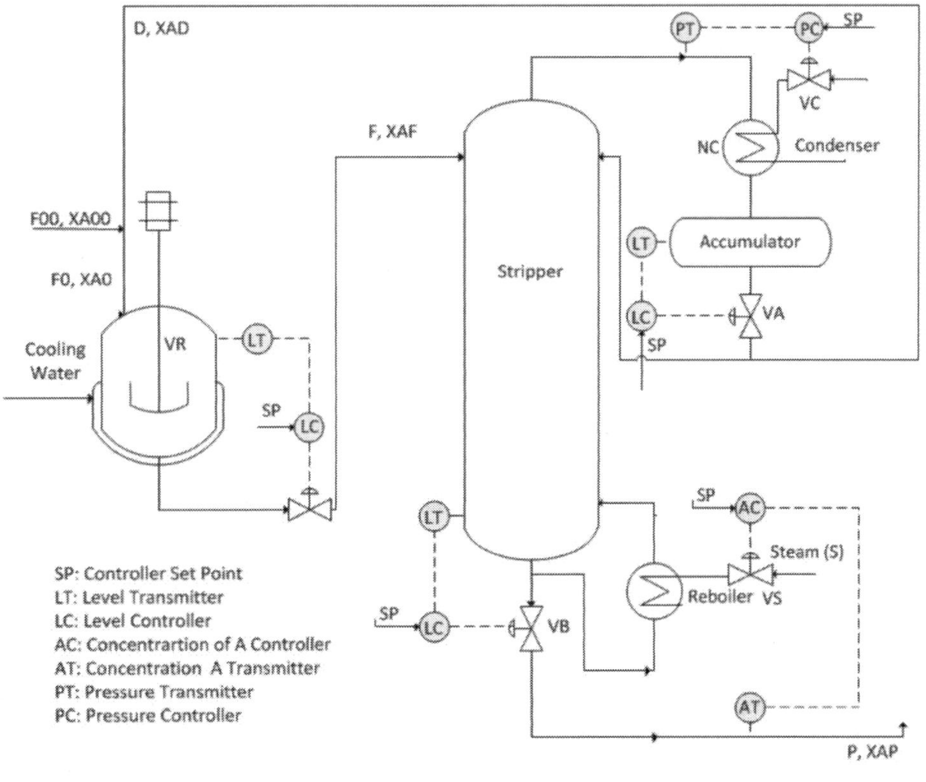

Figure 1. Process and Instrumentation Diagram of Chemical Plant with Recycle

The set of 4 faults, 13 process variables and 4 $\mathbf{f}_j^o$ vectors correspond to those used in Musulin et al., (2004). Instruments costs are introduced in Table 1 and the cause-effect matrix, **A**, takes the absolute values listed in Table 2. The $\boldsymbol{\theta}_j$ matrices of size ($J \times l_j$), used to analyze FD constraints are built with data presented in Table 2, such that the nonzero elements of the j-th failure appear in the intersection between the row associated with the measurement and the column corresponding to that variable.

Table 1. Process Variables.

Notation	Variable	Cost	Notation	Variable	Cost
X_{AF}	Inlet Column Composition	10^5	S	Steam Flow	10^4
D	Distillate Flow	10^4	VS	Vapor Valve	10^3
VC	Valve Condenser	10^3	X_{AP}	Output Composition	10^3
NC	Level Condenser	10^3	VR	Level Reactor	10^3
P	Product Flow	10^4	F	Output Reactor Flow	10^3
VB	Column Base Valve	10^3	F_0	Inlet Reactor Flow	10^4
NB	Column Base Level	10^3			

Solutions obtained for problems 10 and 13 are shown in the first two rows of Table 3. Results obtained for Problem 14 when considering 1 and 4 as key failures are shown in rows 3 to 11.

Table 2. Cause-effect matrix.

Faults	X_{AF}	D	VC	NC	P	VB	NB	S	VS	X_{AP}	VR	F	F_0
F1	-1	-1	-1	-1	0	0	0	-1	-1	0	0	-1	-1
F2	-1	1	1	1	-1	-1	-1	1	1	-1	1	1	1
F3	0	0	0	0	0	0	0	1	1	0	0	0	0
F4	-1	-1	-1	-1	-1	-1	-1	-1	-1	0	1	-1	-1

Table 3. Results.

Restrictions	Solutions	Cost
DM	VC VB VS	3000
O + RS + DM	VC VB VS X_{AP}	4000
O + RS + DM + RD_1=1	VC VB NC VS X_{AP}	5000
O + RS + DM + RD_1=2	VC VB NC VS X_{AP} F	6000
O + RS + DM + RD_1=3	VC VB NC VS X_{AP} F F_0	16000
O + RS + DM + RD_1=4	VC VB NC VS X_{AP} F F_0 D	26000
O + RS + DM + RD_1=5	VC VB NC VS X_{AP} F F_0 D X_{AF}	126000
O + RS + DM + RD_3=1	VC VS X_{AP} P	13000
O + RS + DM + RD_4=1	VC VB NB VS X_{AP}	5000
O + RS + DM + RD_4=2	VC VB NB VS X_{AP} VR	6000
O + RS + DM + RD_4=3	VC VB NB VS X_{AP} VR P	16000

From the results reported in Table 3, it can be observed that O and R restrictions increase the number of instruments in the solution with respect to the case that only considers FD restrictions, because all faults R is not achieved with the first set of restrictions.

Fault 1 admits a RD equal to 5, while the maximum RD of failure 4 is equal to 3. Moreover, it was observed that fault 2 does not accept a RD > 0. When at least one instrument affected by the occurrence of that fault is lost, the resolution vectors of failures 2 and 4 are equal and, therefore, no resolution is achieved. Fault 3 supports a RD = 1, due to its resolution vector has only two elements.

Conclusions

In this work a SND strategy for fault diagnosis that simultaneously satisfies structural criteria and FD constraints, when a monitoring method based on PCA is applied, is presented. Problems involving only structural constraints are MILP, while FD constraints are nonlinear. An exhaustive search with stop criterion is used to find the global optimum. An application example taken from literature is used as case study.

The solution obtained for problems with FD constraints is not able to resolve all failures; therefore it is necessary to incorporate R restrictions to the optimization problem. The RD constraints add robustness to the fault diagnosis system. If one or more measurements are temporarily unavailable, detection and resolution of all faults is possible.

References

1. Bagajewicz, M.; A. Fuxman; A. Uribe. Instrumentation network design and upgrade for process monitoring and fault detection. AIChE Journal, **2004**, 50, 1870.
2. Bhushan, M; R. Rengaswamy. Comprehensive design of a sensor network for chemical plants based on various diagnosability and reliability criteria. 1. Framework. Industrial and Engineering Chemistry Research, **2002**, 41, 1826.
3. Bhushan, M.; S. Narasimhan; R. Rengaswamy. Robust sensor network design for fault diagnosis. Computers and Chemical Engineering, **2008**, 32, 1067.
4. Gerkens, C.; G. Heyen. Sensor placement for fault detection and localization. Computer Aided Chemical Engineering, **2008**, 25, 355.
5. Musulin, E.; Bagajewicz, M.; Nougués, J.M.; Puigjaner, L. Instrumentation Design and Upgrade for Principal Components Analysis Monitoring, Industrial Engineering Chemical Research. **2004**, 43, 2150.
6. Rodríguez, L.; Cedeño, M.; Sánchez, M. A Structural Approach to Design Sensor Networks for Fault Diagnosis. Industrial and Engineering Chemistry Research, **2013**, 52, 17941.
7. Wang, H.; Song, Z.; Li, P. Fault detection behavior and performance analysis of principal component analysis based process monitoring methods. Industrial and Engineering Chemistry Research, **2002**, 41, 2455.
8. Belanger, P.W., Luyben, W.L.: Inventory control in processes with recycle. Industrial and Engineering Chemistry Research, **1997**, 1, 706.

BATCH PROCESSES MONITORING: IDENTIFICATION STAGE

Marco V. Cedeño[1], Leandro P.F. Rodríguez Aguilar[1], Mabel C. Sánchez[1]
[1]Planta Piloto de Ingeniería Química (Universidad Nacional del Sur-CONICET),

Keywords

Multivariable Statistical Process Control, Hotelling Statistic, Fault Identification, Clusters, Fermentation.

Introduction

Batch processes are mainly involved in the production of high-added value products. These processes are characterized by the fact that small variations of operation conditions may have a substantial impact on the final product quality. Therefore a large amount of variables and non-linear relations among them should be quickly and unambiguously taken into account to monitor product quality with the purpose of ensuring a safe and profitable operation.

Many successful applications of Multivariate Statistical Process Control for the monitoring and diagnosis of batch processes have been presented. In general, the batch progress is monitored by exploiting the information contained in a historical database of successful batches using projection techniques.

Monitoring approaches that work in the original measurements space have been successfully applied for batch processes if the number of variables involved is not extremely high and there exist strong non-linear relationships among them that prevent the measurements from being linear combinations. For these cases, original space strategies can perform as well as, or even better than some projection-based techniques (Alvarez et al., 2010). They apply only the Hotelling statistic for detecting the faulty state. If that statistic exceeds its critical value for a given number of consecutive observations, the faulty state is declared and, all the effort must be oriented towards finding what the root cause of the deviation is. The identification of the variables that signal the abnormal behavior is frequently performed by calculating the variable contributions to the inflated statistic.

Recently, a new approach to identify the suspicious measurements when the Hotelling statistic exceeds its critical value was proposed (Cedeño et al., 2012). The methodology consists in finding the Nearest In-Control Neighbour (NICN) of the observation point by solving a minimization problem. The distance between these points is used to evaluate the relative influence of each measured variable on the Hotelling statistic value. Those variables whose distance measures exceed a certain threshold help to isolate the root cause of the fault. To determine the threshold for each variable, its contribution to the inflated statistic for a set of simulated fault cases, in which that variable is not faulty, are evaluated. Then the empirical cumulative distribution function of that variable contribution is obtained. The control limit is selected as the variable-contribution value for which the cumulative probability is (1-α), where α is the probability of wrongly identifying a variable as faulty.

For batch processes, the aforementioned procedure for calculating the variable-contributions control limits requires a large amount of simulations for each observation time interval. A straightforward solution to this problem came from cluster analysis. This is a procedure devoted to explain different behaviors of a data set by grouping them on high-density regions. A clustering technique can be implemented to divide the variable-contributions of batch processes into two groups. One of them corresponds to the measurements that signal the abnormal situation, the other one is associated to the observations which are in control.

In this work the monitoring strategy presented by Cedeño et al. (2012) is extended to tackle batch processes. The technique uses only the Hoteling statistics (T^2) for detection purposes. The identification of the faulty set of observations is performed by applying the NICN concept to calculate the variable contributions to the inflated statistic, and then using cluster theory to point out the suspicious measurements. It avoids the setting of empirical parameters. A fed-batch penicillin fermentation benchmark (Birol et al., 2002) is used to analyze the performance of the presented strategy, which is compared with the most commonly used statistical monitoring strategies.

Nearest In control Neighbourgh for Batch Processes (NICNBP)

The methodology presented by Cedeño et al. (2012) is extended to monitor batch processes as follows. The information of I successful batch runs, which comprises the Normal Operating Condition data set, NOC, is grouped into a three way data matrix $\underline{\mathbf{X}}$ (batch×variables×time). At the beginning of the offline stage, the matrix $\underline{\mathbf{X}}$ is centered and scaled to obtain the matrix $\underline{\mathbf{Z}}$ (I×J×K). Then the relationship among the variables at each time interval is expressed in terms of the correlation matrix $\mathbf{L}_k$, which is calculated using the observations contained in the k-th time slice of matrix $\underline{\mathbf{Z}}$. The correlation matrices are then used to calculate the T^2 critical value for the k-th time interval, $T^2_{C,k}$. Matrices $\mathbf{L}_k$ and $T^2_{c,k}$ values (k=1,…,K) form the local process model that will be used during the on-line monitoring stage.

During the execution of the on-line procedure, the centered and scaled measurement vector for the k-th time interval is used to calculated the statistic T^2_k, which is compared with the corresponding critical value, $T^2_{c,k}$. If $T^2_k > T^2_{c,k}$, the faulty state is declared. Then variable contributions to the inflated T^2 are calculated and analyzed to determine the measurements that signal the out of control state.

Variable contributions are estimated using the knowledge of the nearest neighbour of the observation point that is in statistical control, which is called the NICN, (Cedeño et al., 2012). The information contained in the contribution vector is useful but not enough to determine the set of measurements that reveal the fault. No control limits can be estimated using the reference population because it is associated with successful batches and, the NICN concept used to formulate the variable contributions is related to a batch faulty state. Also the procedure developed by Cedeño et al. (2012) to estimate the control limits for continuous processes is not recommended for the batch ones because it requires many fault simulations.

Several methodologies used the concept of normal measurements to model control limits contributions. It is known that the 99,8 % of normal data are concentrated into a region of three sigma distance around the center. This is called the three sigma rule. Different empirical settings are used to change that traditional value.

Instead of using empirical contribution control limits, in this work it is proposed to concentrate variable contributions on two high-density regions according with their similarities using a hierarchical clustering technique. It is expected that the contributions of the measurements that signal the abnormal situation be in one of the clusters.

Case study

The detection and identification of faults for the bioprocess described by Birol et al. (2002) are analyzed to illustrate the application of the proposed technique. For this fermentation process two sequential stages with different operative modes are analyzed. Ten measurements are taken into consideration for the first stage whereas 13 process observations are used to describe the second one (see Table 1).

The reference population is made up of data provided by the simulation of 187 batches. The code PenSim v.2.0 available at http://simulator.iit.edu/web/pensim/simul.html was used with this purpose.

To evaluate the performance of the proposed strategy, 444 extra runs (Alvarez et al., 2010) were simulated taking into account different fault types (step S; ramp R; initial condition IC), magnitudes, directions, occurrence times and durations, which are summarized in Table 2.

Table 1. Variable tags

Variable	Variable description	Comments
1	Aeration rate	Stages 1 y 2 / Input variable
2	Agitator power input	Stages 1 y 2 / Input variable
3	Substrate feed rate	Stage 2 / input variable
4	Substrate feed temperature	Stages 1 y 2 / Input variable
5	Substrate concentration	Stages 1 y 2
6	Dissolved oxygen concentration	Stages 1 y 2
7	Biomass concentration	Stages 1 y 2
8	Penicillin concentration	Stage 2
9	Volume	Stages 1 y 2
10	Carbon dioxide	Stages 1 y 2
11	pH	Stages 1 y 2
12	Temperature	Stages 1 y 2
13	Cooling water flow rate	Stage 2

Table 2. Fault description

Fault Set	Variable Tag	Fault type	Fault magnitude (% design value)	Starting time (h)	Duration (h)	Number of faults
1	5, 7, 11	IC	±5, ±10, ±15	0	-	18
2	12	IC	±0.5, ±2, ±5	0	-	6
3	1	S	±5, ±10, ±20	0, 100, 200, 300	10, 50, 100	72
4	2	S	±5, ±10, ±20	0, 100, 200, 300	10, 50, 100	72
5	3	S	±5, ±10, ±20	0, 100, 200, 300	10, 50, 100	72
6	1	R	±5, ±10, ±20	0, 100, 200, 300	10, 50, 100	72
7	2	R	±5, ±10, ±20	0, 100, 200, 300	10, 50, 100	72
8	3	R	±5, ±10, ±20	0, 100, 200, 300	10, 50, 100	72

Results and discussion

In this section the application results of the proposed monitoring strategy for the fed-batch penicillin fermentation are presented and compared with those provided by methodologies that work in latent variable (LV) and original variable (OV) spaces. These are the classic MPCA considering the horizontal (Nomikos and MacGregor 1994, 1995) and vertical (Yoo et al., 2004 and Wold et al., 1998) unfoldings, which are referred as N&M and WKFH-AT respectively, and the OSSBP (Alvarez et al. 2010). These methodologies make used of the empirical contribution limits based on the three sigma rule.

To establish a fair comparison basis, it is a common practice that all the techniques involve in the comparison provide similar values for the average number of Type I errors (AVTI) when no faults are present. In this work the approach proposed by Ramaker and Van Sprang (2005) is used to calculate the AVTI. If at least three consecutive observations are outside the control limits, the faulty state is declared.

To assess the identification capabilities of the techniques, the outcome of an identification test is classified as follows:
Precise identification (PI): only the faulty measurements (observations that reveal the fault) are indicated as suspicious,
Ambiguous identification (AI): all the faulty measurements or some of them are indicated as suspicious along with others,
Incorrect identification (II): the variables contained into the faulty variable set are not highlighted as suspicious,
Void identification (VI): No cluster of suspicious measurements is formed.

Table 3 shows the application results for all the strategies. Identification-related indexes highlight NICNBP performance. The proposed strategy shows the highest percentage of PI as well as the lowest values of AI and II.

Table 3. Performance indexes for fault identification

	N&M	WKFH-AT	OSSB	NICNBP
PI(%)	54,1	21,3	70,2	85,5
AI(%)	38,9	72,3	29,7	14,2
II(%)	5,3	5,8	0,0	0,3
VI(%)	1,6	0,6	0,0	0,0

5 Conclusions

A strategy that works in the original variable space is applied to follow the operation of a fed-batch penicillin process. To determine the set of observations that signal the fault, the NICN to the observation point is calculated, and the distance between these two points is used to evaluate the contribution of each variable to the inflated statistic. Then contributions are divided into two groups using a simple hierarchical clustering technique that allows isolating the suspicious measurements in one of the clusters.

In comparison with MPCA-based monitoring strategies and OSSBP, NICNBP provides the best set of identification-related indexes. The on-line identification procedure is very simple, can be run automatically for any process without changes.

Given that NICNBP works in the original variable space, the inversion of the correlation matrix is required to calculate T^2. This condition exists in the case study addressed in this work and in several cases that are analyzed in literature as examples of batch processes.

References

1. Alvarez CR, Brandolin A, Sánchez M. Batch process monitoring in the original measurement's space. J Process Contr 2010; 20: 716-725.

2. Birol G, Ündey C, Çinar A. A modular simulation package for fed-batch fermentation: penicillin production. Comput Chem Eng 2002; 26: 1553-1565.

3. Cedeño MV, Rodriguez LP, Alvarez CR, Sánchez MC. A new approach to estimate variable contributions to Hotelling's statistic. Chemometr Intell Lab 2012; 118: 120-126.

4. Nomikos P, MacGregor JF. Monitoring of batch process using multiway principal component analysis. AICHE J 1994; 40: 1361-1375.

5. Nomikos P, MacGregor JF. Multivariate Spc Charts for Monitoring Batch Processes. Technometrics 1995; 37 (1): 41-59

6. Ramaker HJ, Van Sprang ENM, Westerhuis JA, Smilde AK. Fault Detection Properties of Global, Local and Time Evolving Models for Batch Process Monitoring. J Process Contr 2005; 15 (7): 799-805.

7. Wold S, Kettaneh N, Friden H, Holmberg A. Modelling and diagnostics of batch processes and analogous kinetic experiments. Chemometr Intell Lab 1998; 44: 331-340.

8. Yoo CK, Lee JM, Vanrolleghem PA, Lee IB. On line monitoring of batch process using multiway independient component analysis. Chemometr Intell Lab 2004: 71; 151-163.

PERFOMANCE COMPARISON OF NEW STRATEGIES FOR ROBUST DATA RECONCILIATION

Claudia E. LLanos[1], *Mabel C. Sánchez*[1] *and Ricardo A. Maronna*[2]

[1] *Planta Piloto de Ingeniería Química (UNS-CONICET), Bahía Blanca 8000, Argentina*
[2] *Dpto. de Matemática - Facultad de Ciencias Exactas –Universidad Nacional de La Plata, La Plata 1900, Argentina*

Paper Number N°: **387486**

Introduction

The operation of today's chemical plants is characterized by the stringent need to introduce fast and low-cost changes to improve their performance. The decision-making process about possible modifications in a system requires knowledge about its actual state. This is determined by the values of the process variables contained in the model chosen to represent plant operation. In general, this model is constituted by the mass and energy conservation equations.

The numerical values resulting from the observations do not provide consistent information because they contain some type of error that avoids the conservation equations to be satisfied exactly. Therefore, it is a common practice to incorporate the application of data reconciliation procedures that provide adjusted measured values, which are consistent with the corresponding balance equations (Romagnoli and Sánchez, 2000).

Different approaches are proposed for the simultaneous treatment of random and gross errors in data reconciliation problems. Simultaneous approaches based on the concepts of Robust Statistic have been devised since the last two decades. Robust strategies produce reliable estimates not only when data follows a given distribution exactly, but also when this happens only approximately due to the presence of outliers (Maronna et al., 2006).

Different types of M-estimators, which are generalizations of the Maximum Likelihood Estimator, were used as objective function of the data reconciliation problem instead of the weighted least square estimator. Ozyurt and Pike (2004) presented a comprehensive performance analysis of six robust objective functions and three gross error detection criteria for simulated and industrial case studies operating under steady state condition. Special attention was given to the concept of tuning the loss functions to obtain the same efficiency at the ideal distribution. They concluded that methods based on the Cauchy distribution and Hampel's redescending M-estimator gave promising results with less computation.

A performance analysis of distinct robust M-estimators was also performed by Prata et al. (2008) for dynamic systems who concluded that the Welsch M-estimator showed better performance leading to more robust and reproducible numerical performance.

The Quasi Weighted Least Square (QWLS) estimator was proposed by Zhang et al. (2010), and its behavior was compared with the performances of Least Square (LS), fair function, Hampel M-estimator, linear combination (LCT) and MILP techniques for linear systems. The results of these analysis suggested that QWLS as well as MILP method had the best performance measures. The last method was the most efficient nevertheless it was expensive in computation for large scale process system.

Chen et al. (2013) used the correntropy as an optimality criterion in the estimation problems. By properly adjusting its kernel width, the effectiveness of this estimator was tuned. The optimal kernel width value was chosen by minimizing Aikake information criterion. It was shown that Correntropy estimator was effective even when the random errors did not follow a Gaussian distribution. Its behavior was mainly compared with the quasi-weight least squares estimator for the same type of linear systems, the estimator proposed by Chen gave better results than QWLS.

In the present work a comparative performance analysis is presented regarding the estimation and gross error detection capabilities of the robust data reconciliation strategies that have been presented in the last decade. The comparison is conducted by tuning all loss functions to obtain the same efficiency at the ideal condition. With this purpose, approximate finite sample variances and efficiencies with respect to the normal distribution are calculated using the Jackknife method. A linear example, Steam Metering Network (Serth et al., 1986), extracted from the literature, is used for the analysis. Performance measures are the Mean Square Error, the Average Number of Type I Error and the Overall Performance.

General Formulation

Let us consider a process whose operation is represented by a set of linear algebraic equations which involve only redundant measured variables. It is assumed that measurement vectors $y^T = (y_1, \dots, y_n)$ are available at different time interval. In the absence of gross errors in those vectors, the following measurement model is acceptable

$$y_{ij} = x_i + \varepsilon_{ij}, \tag{1}$$

where y_{ij} represents the measurement of the i-th variable, x_i, at the j-th time period, and e_{ij} stands for the unobservable independent random error, which has a mean value of zero and known standard deviation, σ_i.

The objective of the data reconciliation problem is the minimization of the measurement errors in such a way that the consistency of the balance equations is satisfied. A general formulation of that problem is represented by equation (2), where ρ stands for the loss function, r_{ij} is the normalized residual of the i-th measurement at the j-th time period (equation 3), and constraints represent the process model. The results of the optimization procedure are the adjusted measured variables $\hat{x}$.

$$\begin{aligned} \boldsymbol{Min} \sum_{j=t-N}^{t} \sum_{i=1}^{n} \rho(r_{ij}) \\ st. \;\; Ax = 0 \end{aligned} \tag{2}$$

$$r_{ij} = \frac{y_{ij} - \hat{x}_i}{\sigma_i} \tag{3}$$

In classical data reconciliation problems, ρ corresponds to the least square estimator. Because incorrect estimations arise if there are outliers in the data set, robust estimators were proposed that are not sensitive to the presence of a finite number of gross errors in the measurements. M-estimators, which are generalizations of the maximum likelihood estimator, are applied for solving data reconciliation problems. Recently, the QWLS and the correntropy estimator are presented. In this work their performances are analyzed and compared with those provided by the Welsch estimator and two developed strategies, which combine the strengths of monotone and redescending M-estimators.

Reconciliation strategies based on monotone and redescending estimators

First the Simple (SiM) and Sofisticated (SoM) strategies are presented. These are non-adaptive techniques that combine the strength of monotone and redescending M-estimators. Then, the Welsch estimator that has a redescending behavior and continuous first and second derivatives is introduced. Finally, two adaptive estimators, QWLS and Correntropy, are defined.

1-Simple Method: It involves the following two steps:

Step 1: For the *i*-th measurement at time *t*, a location M-estimator from the bisquare family called $\tilde{y}_{it}$ is computed, which is the solution of:

$$\sum_{j=t-N+1}^{t} \rho_{bis}\left(\frac{y_{ij}-\tilde{y}_i}{\sigma_i}\right) = min \tag{4}$$

$$\rho_{bis} = \begin{cases} 1-[1-(r/k)^2]^3 & if\ |r| \le k \\ 1 & if\ |r| > k \end{cases} \tag{5}$$

The estimator $\tilde{y}_{it}$ is based on the time horizon values $\{y_{ij}, j = t-N+1, \dots, t\}$. Each measurement is dealt separately. The advantage of individual location estimations is the availability of a robust and simple initial estimator, namely the median of $\{y_{t-N+1}, \dots, y_t\}$.

Step 2: The estimator is defined as the solution, $\hat{x}$, satisfying Eq. 6

$$\begin{aligned} &Min \sum_{i=1}^{n} \rho_{Hub}(r_{ij}) \\ &st.\ \ Ax = 0 \end{aligned} \tag{6}$$

where ρ_{Hub} corresponds to the Huber family:

$$\rho_{Hub} = \begin{cases} r^2 & if\ |r| \le k \\ 2k|r| - k^2 & if\ |r| > k \end{cases} \tag{7}$$

The sequential solution of Problems 4 and 6 converges to the solution of Problem 2.

2-Sofisticated Method: It incorporates an extra step to the SiM as follows:

Step 3: Using as initial point the solution of Problem 6. The estimator value $\hat{x}$, is the solution of :

$$\begin{aligned} &Min \sum_{j=t-N+1}^{t} \sum_{i=1}^{n} \rho_{bis}(r_{ij}) \\ &st.\ \ Ax = 0 \end{aligned} \tag{8}$$

In this case all the measurements in the horizon are used in the optimization problem.

The reason to initialize the monotone M-estimator using a location redescending M-estimator arises from the work by Maronna and Arcas(2009). Since for the estimation of x_i the use of redundancy from observations other than *i* is limited, we must take full advantage of the redundancy supplied by the repeated observations y_{ij} in the horizon. The location estimator uses this "self-redundancy" to detect and down weight gross errors.

The values of the adjusted measurements obtained using Welsch, QWLS and Correntropy estimators are based on the measurements vectors corresponding to different time intervals. The least square estimator is used as initial value to solve the equation 2. The parameters of the loss function are adjusted to provide 95% of efficiency relative to the normal distribution using the jackknife method. The equation 9 is solved for each estimator.

$$\mathbf{Min} \sum_{j=t-N}^{t} \sum_{i=1}^{n} \rho_{estimator}(r_{ij}) \qquad (9)$$
$$st. \;\; Ax = 0$$

3-Welsch Estimator

It was introduced by Dennis and Welsch (1976) as a soft redescending estimator, less sensitive to gross errors than the monotone ones. Its influence function asymptotically approaches zero for large values of r_{ij}. The 95 % asymptotic efficiency relative to the standard normal distribution is obtained with a tuning constant C= 2.9846 (W.J.Rey,1983)

$$\cdot \rho_{welsch}(r) = C^2[1 - e^{-(\frac{r}{C_W})^2}] \qquad (10)$$

4- Quasi-Weigthed Least Square

The M-estimator is defined as follows

$$\rho_{QWLS}(r) = \frac{r^2}{1 + \beta|r|} \qquad (11)$$

The addition of the term $\beta|r|$ to the denominator of the least square function reduces the effect of large errors, where β is the tuning constant.

5-Correntropy

This estimator uses the correntropy as an optimal criterion instead of the traditional mean square criterion. The gaussian kernel function depends on its kernel width σ and the normalized error. The value of σ is selected to scale up the errors so that their effects become negligible in the computation of correntropy.

$$\rho_\sigma(r) = \frac{1}{\sigma\sqrt{2\pi}} e^{-(\frac{r^2}{2\sigma^2})} \qquad (12)$$

Comparative Analysis

A comparative analysis is conducted to analyze the performance of the proposed strategies and the three estimators to reconcile the measurements and identify the outliers for a steam metering network of a methanol synthesis plant. This linear example involves 28 streams that interconnect 11 units (see Fig. 1). The flow rates of all streams are measured. Random errors are generated considering that the standard deviations of the observations are 2.5% of their true values.

Three case studies are analyzed with different grade of contamination. The measurement errors for Case Study 1 and 2 are generated using the distributions F = N(0,1) and $F = (1-\varepsilon)N(0,1) + \varepsilon N(0, K^2)$, with ε = 0.1 and $K \in$ [2-20], respectively. Regarding Case Study 3, measurements are generated by randomly adding errors of constant value ($R\sigma_i$) to the true-variable values with a probability ε =0.1. The range of R values is [1 – 10].

Ten thousand simulation trials are performed for each case study, and the length of the data horizon is fixed at H=10. The measures of performance used in this analysis are: MSE, AVTI and OP, which are estimated as follows

$$MSE = \frac{1}{n\,SIM}\sum_{k=1}^{SIM}\sum_{i=1}^{n}\left(\frac{\hat{x}_i - x_i}{\sigma_i}\right)^2 \quad (13)$$

$$AVTI = \frac{\#(gross\ errors\ inconrrectly\ identified)}{SIM} \quad (14)$$

$$OP = \frac{\#(gross\ errors\ correctly\ identified)}{\#(gross\ errors\ simulated)} \quad (15)$$

where SIM is the number of simulation trials.

The efficiencies of the estimators are fixed at 95% by tuning their respective parameters. The cut point of each technique is adjusted by trial and error in such a way that the AVTI is approximately 0.05 for Case Study 1. This provides the same base to compare the performance measure of the different techniques.

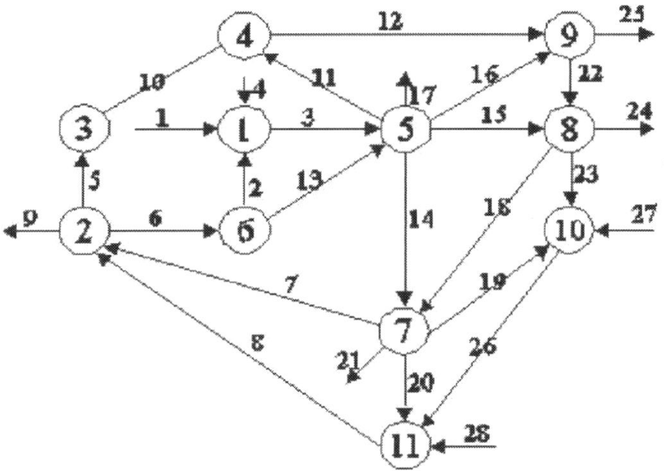

Fig. 1. Steam Metering Network.

RESULTS

The following table and figures present the simulations results for the three cases studies, which are solved using all the methodologies previously mentioned and also the least square technique. Case Study 1 is useful to determine the cut point of the different estimators.

Table 1. Results for Case study 1.

AVTI						MSE*10^2					
LS	SiM	SoM	Correntropy	QWLS	Welsch	LS	SiM	SoM	Correntropy	QWLS	Welsch
0.05	0.0499	0.0499	0.05	0.0499	0.0499	6.074	6.384	6.387	6.4129	6.4109	6.3796

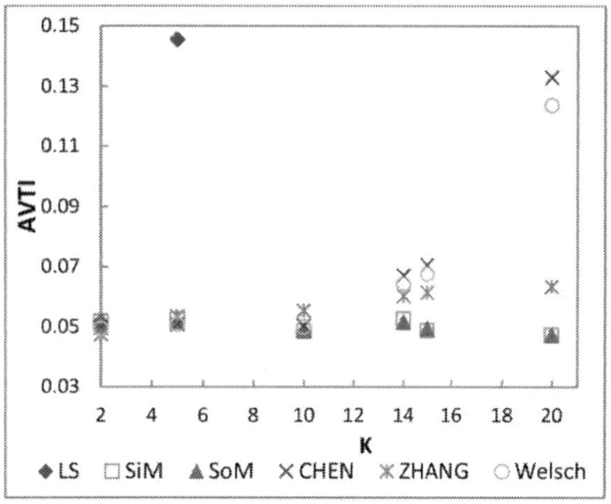

Figure 2 a. AVTI for Case Study 2.

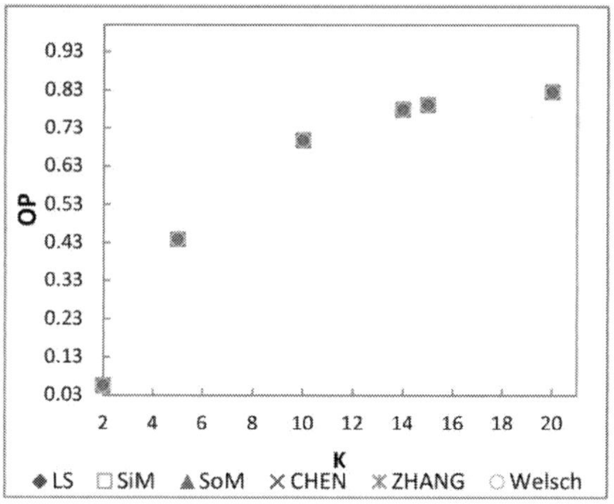

Figure 2 b. OP for Case Study 2.

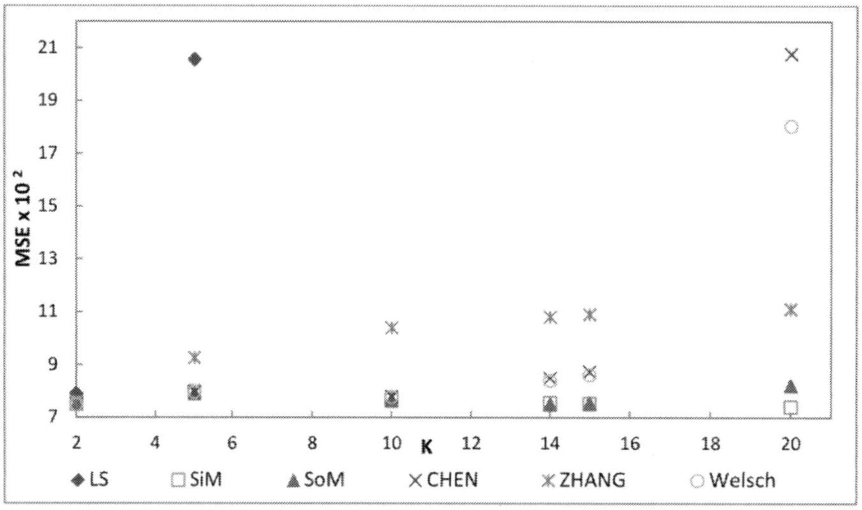

Figure 2 c. MSE x 10^2 for Case Study 2.

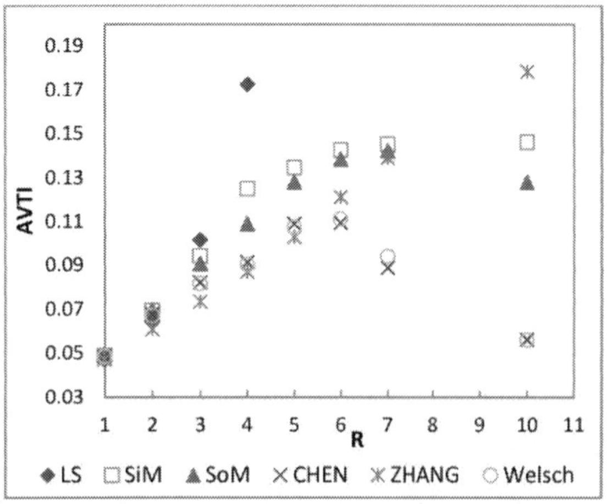

Figure 3 a. AVTI for Case Study 3.

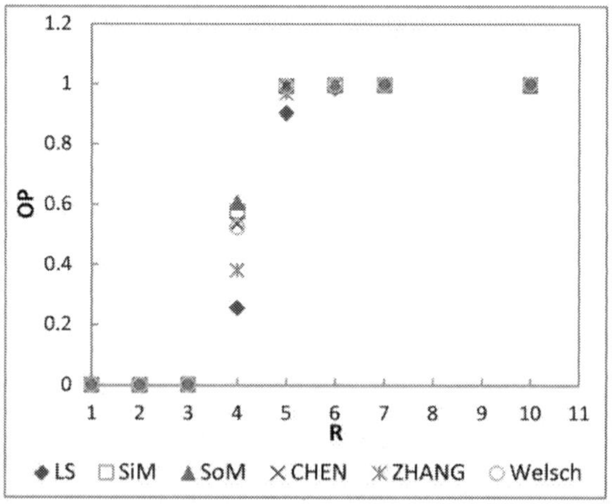

Figure 3 b. OP for Case Study 3.

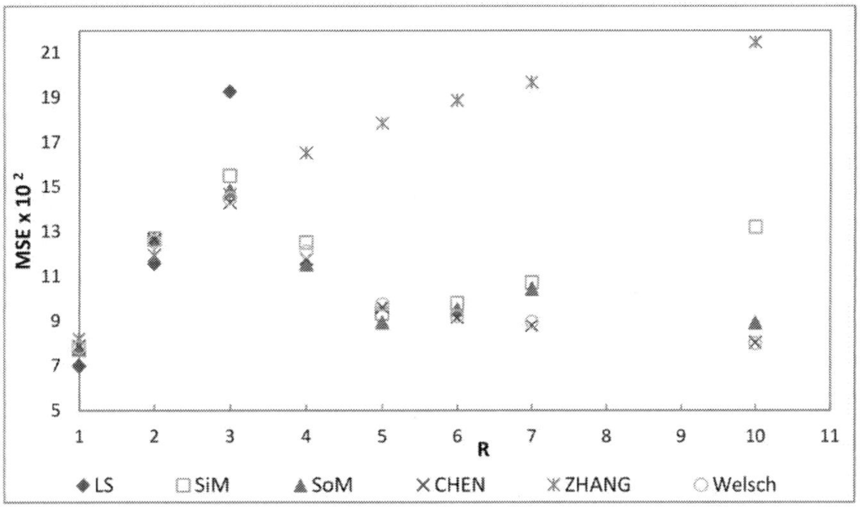

Figure 3c. MSE of SMN for Case Study 3.

As it is expected, when the measurements follow a normal distribution the best results are found with the Least Square estimator (Table 1). However if the distribution is only approximately normal (Case Studies 2 and 3) all robust estimators show a better performance to determine the reconciled values.

When the data sets follow a contaminated normal distribution, SiM and SoM provide the best estimations. The values of AVTI and MSE for Correntropy and Welsch estimators increase for K greater than 15.

The estimation techniques based on Welsch and Correntropy functions and SoM give the best results for Case Study 3. However, SoM is not as good for identifying gross errors (fig 3 a) as the first mentioned strategies. The QWLS M-estimator is the less appropriate for this type of data contamination and, SiM shows an intermediate behavior between all the estimators.

The OP measure for all the estimation methods is similar.. Because SiM and SoM runs require the execution of a sequence of two and three steps, respectively, their computational loads are superior with respect to the other estimation procedures.

Conclusions

In this work five robust estimation strategies are compared for the estimation of adjusted measurements and the identification of outliers for a benchmark problem.

The methodology SoM provides the best results for a contaminated normal distribution, and shows a good behavior when outliers of fixed magnitude are randomly added to the true variable values. For this case, Welsch and Correntropy estimators present a slightly superior performance with respect to SoM.

The computation times of SiM and SoM are superior with respect to those required by Welsch, Correntropy and QWLS methodology, but in the original versions of Correntropy and QWLS techniques parameters should be self-tuning for each process. This extra step increases the computational load.

References

1. Romagnoli, J. and Sánchez, M. (2000), "Data Processing and Reconciliation for Chemical Process Operations", Academic Press, San Diego,.
2. Maronna, R., Martin, R., Yohai, V. (2006), "Robust Statistic: Theory and Methods", John Wiley and Sons Ltd., Chichester,.
3. Özyurt, D. and Pike, R. (2004), "Theory and practice of simultaneous data reconciliation and gross error detection for chemical process," Computers and Chemical Engineering, 28, 381-402.
4. Prata, D., Pinto, J., Lima, E. (2008), "Comparative Analysis of Robust Estimators on Nonlinear Dynamic Data Reconciliation," Computer-Aided Chemical Engineering, 25, 501-506.
5. Zhang, Z., Shao, Z., Chen, X., Wang, K., Qian, J. (2010), "Quasi-weighted least square estimator for data reconciliation" Computers and Chemical Engineering, 34, 154–162.
6. Chen, J., Peng, Y., Munoz, J. (2013), "Correntropy estimator for data reconciliation," Chemical Engineering Science, 104, 1019–1027.
7. Serth, R. and Heenan, W. (1986), "Gross error detection and data reconciliation in steam-metering systems," AIChE J. 32: 733—741.
8. Sánchez,M. and Maronna,R.A (2009), "Simple approaches for robust data reconciliation," AIChE Journal.
9. Maronna, R.A. and Arcas, J.(2009), "Data reconciliation and gross error detection based on regression," Comp. & Chem. Eng. 33, 65—71.
10. Rey W.J.J. (1983), "Introduction to Robust and Quasi-Robust Statistical Methods," Springer-Verlag, Berlin, Heidelberg, New York, Tokyo.

Condition Monitoring of a Gasifier in an Integrated Gasification Combined Cycle Power Plant

Pednekar, P.; Bhattacharyya, D.;
Chemical Engineering Dept., West Virginia University, Morgantown, WV - 26505

Introduction

The IGCC is one of the key technologies for efficient conversion of coal to energy.

- ✓ For high efficiency and availability, early detection of faults is vital
- ✓ Gasifier, a crucial equipment, operates under extreme conditions
- ✓ Placement of sensors to detect faults is not possible at some locations and/or some faults cannot be detected directly

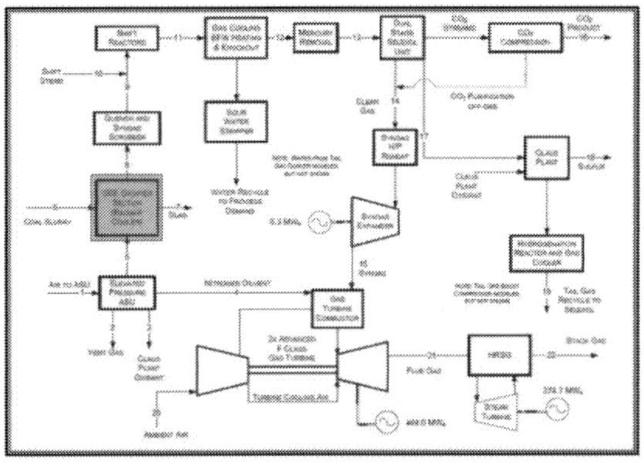

Gasifier

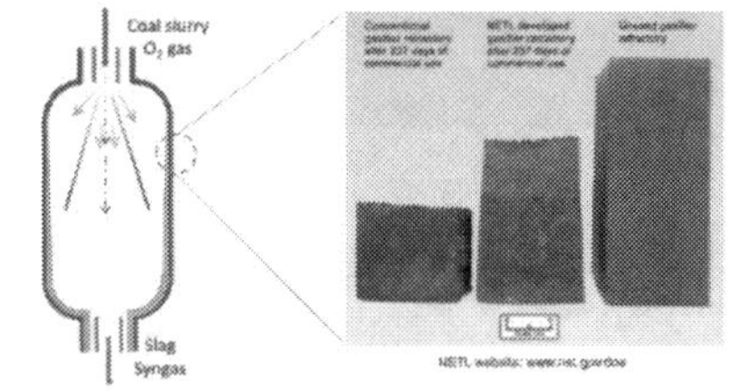

NETL website: www.netl.gov/dw

- ✓ Refractory is expensive to replace
- ✓ Replacement of refractory results in significant downtime for the whole plant
- ✓ In-situ measurement of refractory degradation is not possible
- ✓ Advantageous to monitor the health of the refractory

Summary

Challenge:

Monitoring refractory degradation in a gasifier without being able to measure it directly

Idea:

Use a model-based approach by utilizing suitable measurement(s) at other location(s) to estimate the extent of refractory degradation

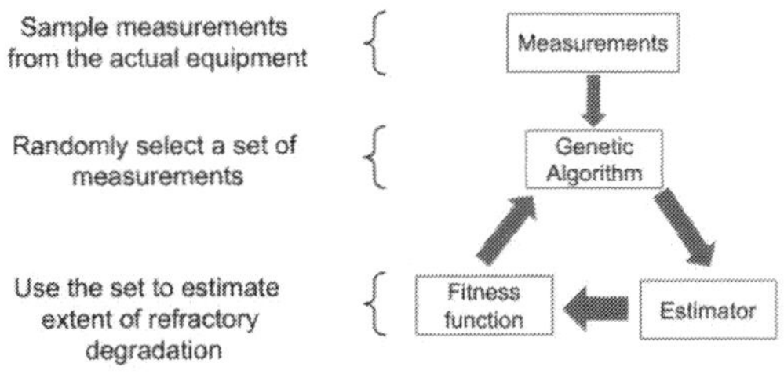

Modelling approach

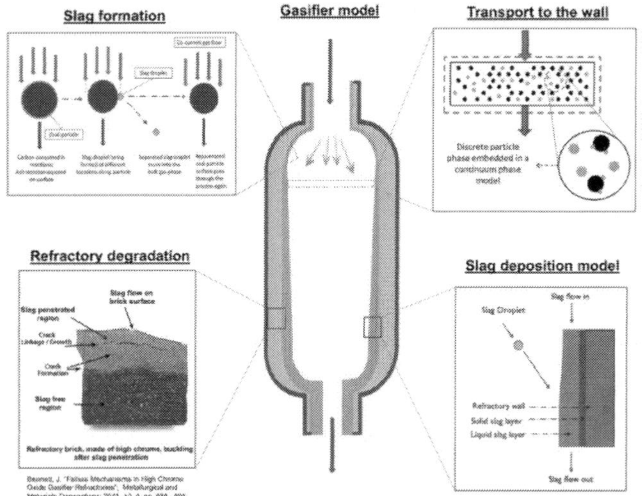

Baumel, J. "Failure Mechanisms in High Chrome Oxide Gasifier Refractories", Metallurgical and Materials Transactions, 2011, 42, 4, pp. 885 - 884

Results

Slag deposition and refractory temperature

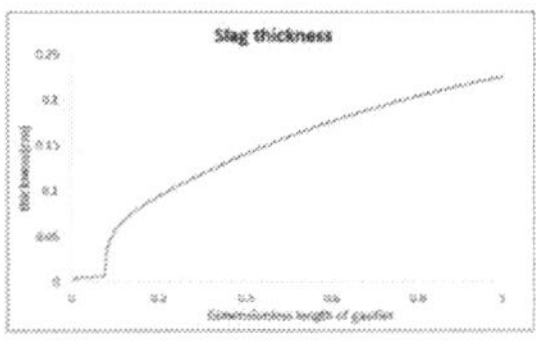

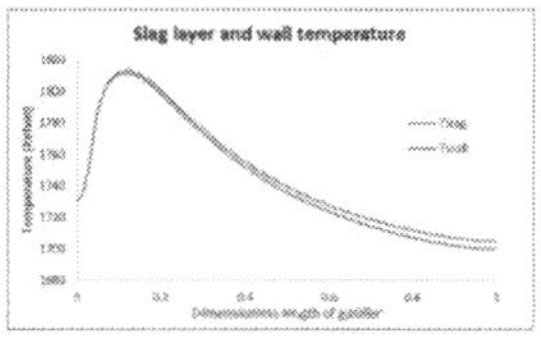

Refractory penetration

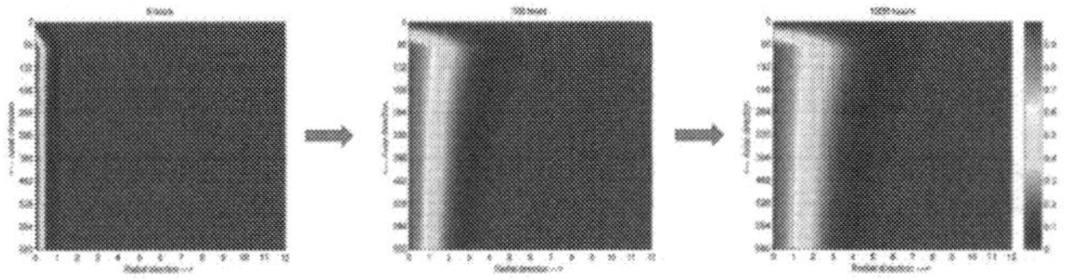

Refractory degradation

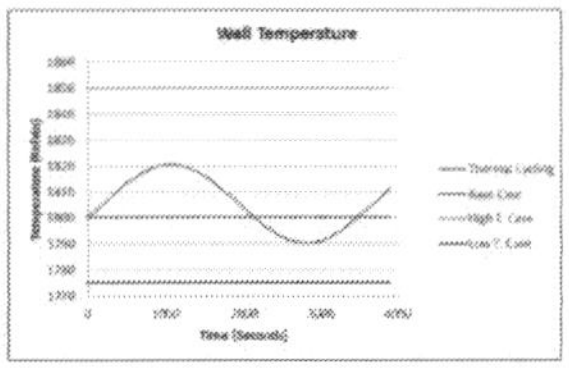

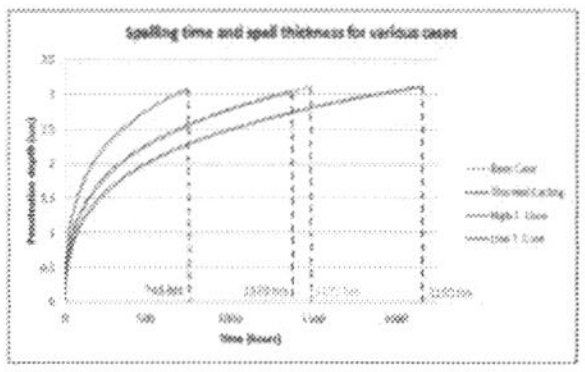

Conclusion

Using a mathematical model of slag penetration and refractory degradation along with an optimally synthesized sensor network, extent of refractory degradation can be estimated satisfactorily.

390772 Dynamic Control of Multiple Water Networks for Agile Manufacturing

Abstract:

Water network design (WND) is of great importance to effective use of valuable water, which plays key role in not only the reduction of freshwater, but also the curbing of wastewater generation. While extensive researches have been conducted for WND, there is a lack of systematic studies on dynamic controls among multiple water networks for producing multiple products from one single production line. This agile manufacturing way shows great potential for future chemicals productions bearing enormous flexibility and profitability. In this paper, a systematic methodology for such a study has been developed. Firstly, a super WND will be optimally obtained, which structurally integrates multiple WNDs for manufacturing multiple products according to their processing recopies. Next, a dynamic control strategy will be developed to optimize the dynamic status transition between any two WNDs. It involves a dynamic optimization problem, which has two objectives: the first is to minimize the dynamic transition time from one WND to another; the second is to minimize the quantity of freshwater and chemical consumption as well as wastewater generation during such transitions. This methodology can be adopted to handle dynamic shift operations among multiple production recipes with the minimum design and operational costs.

Genetic fuzzy decoupling of nonlinear multiple-input multiple-output processes

Sarah Nikbakhsh
John Hedengren

Abstract

There are varying degrees of interaction between inputs and outputs of a multiple-input multiple-output (MIMO) system. In the other words, an input may affect several outputs, and an output may be affected by several inputs. One method to control multivariate systems is to properly pair manipulated and controlled variables and decouple the loops into multiple non-interacting single-input single-output (SISO) systems. Loop interactions can destabilize a control system, and tuning controllers in an interacting MIMO system can be difficult. Therefore, decoupling is one of the important issues in control theory that has been studied in linear and nonlinear system domains. Conventional decoupler design is based on an exact model of the multivariate process. As a less exact model of the process is available, the performance of conventional decoupling approaches degrades. The contribution of this work is to seek one approach that does not require a precise process model. The approach taken is to perform fuzzy decoupling based on a linguistic model of the process. Fuzzy decoupling adopts a series of fuzzy if-then rules instead of a crisp mathematical model of the process. Hence, a set of decoupling rules leads to omit interconnections in a MIMO system. The design of a fuzzy system involves tuning the membership function parameters, fuzzy if-then rules, input scaling factors and output gains. To tackle this problem, either the experts' knowledge or some learning and optimization algorithm is applied. Because these optimization problems are often a large scale mixed-integer problem, evolutionary computations such as genetic algorithms are effective solution methods. The purpose of this work is to design and implement a genetic fuzzy decoupling framework for nonlinear MIMO systems with unknown mathematical models. The genetic algorithm (GA) has been utilized to determine optimal fuzzy rules, output gains and control parameters. The performance of the structure has been assessed for a MIMO nonlinear system via its implementation on chemical benchmark process. The results show the appropriate performance of the genetic fuzzy decoupling method despite the independence from the exact nonlinear model of processes.

Database Management Method Based on Density and Nonlinearity for Locally Weighted Linear Regression Model

Sanghong Kim*, Kazuki Mishima*, Manabu Kano† and Shinji Hasebe*

Abstract

The estimation performance of statistical models significantly depends on quality and quantity of samples in a database, and the limited number of samples can be stored in the database because of the constraint on computational time and the amount of the memory. Thus, a systematic method of selecting samples to be used for modeling is required. In the past research, database management methods based on the density of samples were proposed. This research shows that the strength of the nonlinearity should also be taken into account when locally weighted regression models are used, and proposes a database management method that takes account of both the nonlinearity and the density. In the proposed method, locally weighted linear regression models with different localization parameters are used to evaluate the strength of the nonlinearity. It was confirmed that, by using the proposed method, high prediction accuracy was realized with limited number of samples in a numerical example and an industrial application to a distillation process.

1 Introduction

Soft sensors have been widely used for predicting difficult-to-measure variables by using easy-to-measure variables in many industries, and they have successfully contributed to operation cost reduction in real industrial processes [1, 2, 3]. However, there are still some problems to enhance the estimation performance of soft sensors and make their use more widespread. For example, it is very difficult to select an appropriate set of samples from a given data set although the selection can significantly affect the estimation performance of soft sensors. This problem is related to model maintenance, which is one of the most laborious work in soft sensor design [3], thus, developing a systematic sample selection (database management) method is crucial.

A practically popular database management method is the first-in, first-out method, which uses a certain number of the newest samples. However, this method has the problem that the space covered by the samples tends to be small when the operation condition is not changed. This problem might be solved by using the methods by selecting samples based on the density [4, 5]. Although the sample density is a very important index, it is not sufficient; at least when locally weighted linear regression models are used the strength of the nonlinearity should also be taken into account as explained in Section 3.

This research proposes a database management method which can select samples based on the strength of the nonlinearity as well as the sample density. To evaluate the strength of the nonlinearity, locally weighted linear regression models with different localization parameters are

*Kyoto University, Dept. of Chemical Engineering
†Kyoto University, Dept. of Systems Science

used because the relationship between the estimation error of these models and the localization parameter varies with the strength of the nonlinearity.

The rest of this paper is organized as follows. Section 2 introduces locally weighted linear regression. In Section 3, the algorithm of the proposed method is described after explaining why the strength of the nonlinearity should be taken into account in the sample selection. Section 4 demonstrates two application examples of the proposed method. Conclusions are shown in Section 5.

2 Locally weighted linear regression

The locally weighted linear regression method constructs a local model by prioritizing the similar samples to a query, for which the output estimation is required. The definition of the similarity between samples in the database and the query is usually defined on the basis of the distance from the query to the samples[6, 7, 8]. Other similarity measures proposed so far include the angle [9] and the correlation [10]. In this research, the following definition of the similarity ω is used.

$$\omega_n = \exp\left(-\frac{d_n}{\sigma_d \varphi}\right) \quad (1)$$

$$d_n^2 = (\boldsymbol{x}_n - \boldsymbol{x}_q)^{\mathrm{T}}(\boldsymbol{x}_n - \boldsymbol{x}_q) \quad (2)$$

Here, $\boldsymbol{x}_n$ is the n-th sample in the database, and $\boldsymbol{x}_q$ is the query. σ_d is the standard deviation of d_n $(n = 1, 2, \cdots, N)$, and φ is the localization parameter. N is the number of samples. The similarity decreases steeply when φ is small and gradually when φ is large.

To construct a local linear model, partial least squares (PLS) is adopted. Locally weighted PLS (LW-PLS) calculates the output estimate $\hat{y}_q$ for a given query $\boldsymbol{x}_q$ through the following procedure.

1. Determine the number of latent variables R and set r to 1.

2. Calculate the similarity matrix $\boldsymbol{\Omega} = \mathrm{diag}(\omega_1, \omega_2, \cdots, \omega_N)$.

3. Calculate $\boldsymbol{X}_r$, $\boldsymbol{Y}_r$ and $\boldsymbol{x}_{q,r}$

$$\boldsymbol{X}_r = \boldsymbol{X} - \boldsymbol{1}_N[\bar{x}_1, \bar{x}_2, \cdots, \bar{x}_M] \quad (3)$$

$$\boldsymbol{y}_r = \boldsymbol{y} - \boldsymbol{1}_N \bar{y} \quad (4)$$

$$\boldsymbol{x}_{q,r} = \boldsymbol{x}_q - [\bar{x}_1, \bar{x}_2, \cdots, \bar{x}_M]^{\mathrm{T}} \quad (5)$$

$$\bar{x}_m = \sum_{n=1}^{N} \omega_n x_{nm} \Big/ \sum_{n=1}^{N} \omega_n \quad (6)$$

$$\bar{y} = \sum_{n=1}^{N} \omega_n y_n \Big/ \sum_{n=1}^{N} \omega_n \quad (7)$$

where $\boldsymbol{X}$ is the input variable matrix, $\boldsymbol{y}$ is the output variable matrix and x_{nm} is the m-th component of x_n. In addition, M is the number of input variables and $\boldsymbol{1}_N \in \Re^N$ is a vector of ones.

4. Derive the r-th latent variable of $\boldsymbol{X}$

$$\boldsymbol{t}_r = \boldsymbol{X}_r \boldsymbol{w}_r \quad (8)$$

where $\boldsymbol{w}_r$ is the eigenvector of $\boldsymbol{X}_r^{\mathrm{T}} \boldsymbol{\Omega} \boldsymbol{y}_r \boldsymbol{y}_r^{\mathrm{T}} \boldsymbol{\Omega} \boldsymbol{X}_r$ which corresponds to the maximum eigen value.

5. Derive the r-th loading vector of $\boldsymbol{X}$

$$\boldsymbol{p}_r = \frac{\boldsymbol{X}_r^{\mathrm{T}} \boldsymbol{\Omega} \boldsymbol{t}_r}{\boldsymbol{t}_r^{\mathrm{T}} \boldsymbol{\Omega} \boldsymbol{t}_r} \tag{9}$$

and the regression coefficient

$$q_r = \frac{\boldsymbol{y}_r^{\mathrm{T}} \boldsymbol{\Omega} \boldsymbol{t}_r}{\boldsymbol{t}_r^{\mathrm{T}} \boldsymbol{\Omega} \boldsymbol{t}_r} . \tag{10}$$

6. Derive the r-th latent variable of $\boldsymbol{x}_q$

$$t_{q,r} = \boldsymbol{x}_{q,r}^{\mathrm{T}} \boldsymbol{w}_r . \tag{11}$$

7. If $r = R$, calculate the output estimate

$$\hat{y}_q = \bar{y} + \sum_{r=1}^{R} t_{q,r} q_r \tag{12}$$

and finish estimation. Otherwise, set

$$\boldsymbol{X}_{r+1} = \boldsymbol{X}_r - \boldsymbol{t}_r \boldsymbol{p}_r^{\mathrm{T}} \tag{13}$$
$$\boldsymbol{y}_{r+1} = \boldsymbol{y}_r - \boldsymbol{t}_r q_r \tag{14}$$
$$\boldsymbol{x}_{q,r+1} = \boldsymbol{x}_{q,r} - t_{q,r} \boldsymbol{p}_r . \tag{15}$$

8. Update r to $r + 1$ and go to step 4.

3 Database management method

Before explaining the proposed method, the reason why the strength of the nonlinearity should be taken into account in sample selection is explained through a simple example. In Figure 1, black circles denote samples and the solid lines are the input-output relationship. Suppose that we have to select five samples from ten samples in the initial database. When the conventional method based on the sample density is used, five samples shown in Figure 1 (right top) are selected. If a query is given between samples 1 and 4, a local linear model is constructed by prioritizing them, however, this model is not highly accurate because the input-output relationship has strong nonlinearity between samples 1 and 4. On the other hand, the proposed method selects the samples shown in Figure 1 (right bottom). In this database, samples 1 to 4 are selected to cope with the nonlinearity. Though samples 5 to 9 are not selected, the estimation performance does not deteriorate since the input-output relationship is almost linear between samples 4 and 10.

To evaluate the strength of the nonlinearity, local linear models with the different localization parameter φ are constructed. Figure 2 shows the relationship between the strength of the nonlinearity, the localization parameter and the estimation error. In this figure, deep colored samples have high similarity to the query and local models are constructed by prioritizing them. When the nonlinearity is weak around the query, the estimation error is small even when the localization parameter is large. On the other hand, when the nonlinearity is strong, the estimation error is small only when the localization parameter is small and less samples have high similarity to the query. The proposed method selects the samples based on this difference.

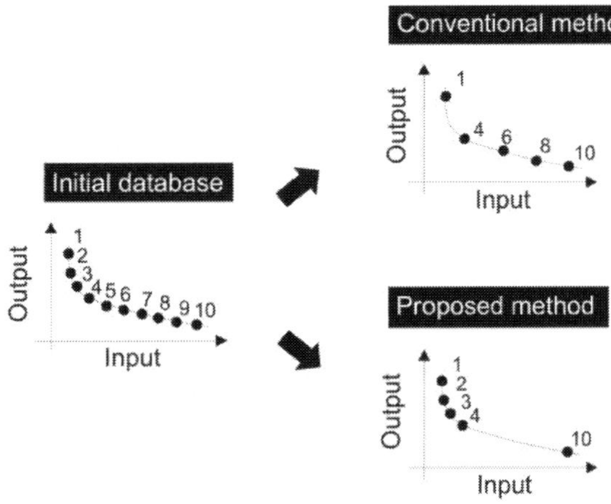

Figure 1: The schematic diagram of the proposed method and the conventional method based on the sample density.

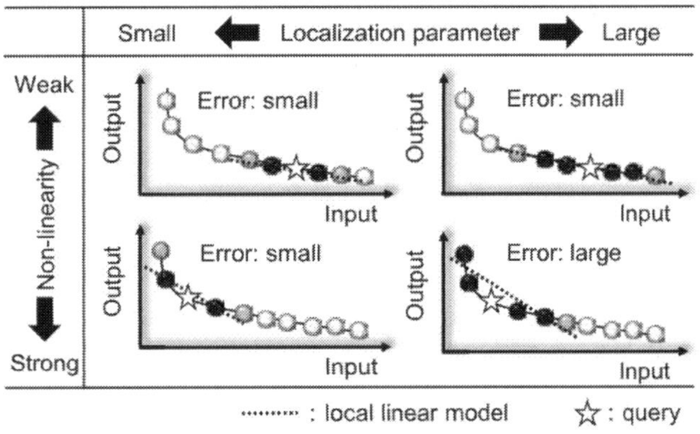

Figure 2: The relationship between the strength of the nonlinearity, the localization parameter and the estimation error.

3.1 Algorithm of the proposed method

The algorithm of the proposed method is as follows.

1. Prepare the initial input variable matrix $\boldsymbol{X}_0$ and the initial output variable matrix $\boldsymbol{y}_0$.
2. Set the threshold of estimation error $e_{\max}$, the number of the latent variables R of LW-PLS models and the number of samples N_{out} to be removed from the initial database.
3. Set i to 1.
4. Set n to 1.
5. Regard the n-th sample $\boldsymbol{x}_n$ as a query.
6. If $\boldsymbol{x}_n$ is in $\boldsymbol{X}_{i-1}$, construct LW-PLS models with different localization parameter φ using $\boldsymbol{X}_{i-1}$ and $\boldsymbol{y}_{i-1}$, and find the minimum value of φ which satisfies the following condition.

$$|\hat{y}_n(\varphi) - y_n| = e_{\max} \tag{16}$$

In addition, calculate the corresponding similarity $\boldsymbol{\omega}_n = [\omega_{1n}, \omega_{2n}, \cdots, \omega_{Nn}]^{\mathrm{T}}$. When Equation (16) is not satisfied with any φ, let $\boldsymbol{\omega}_n = \mathbf{1}_N$. Else if $\boldsymbol{x}_n$ is not in $\boldsymbol{X}_{i-1}$, set $\omega_{n1}, \omega_{n2}, \cdots, \omega_{nN}$ to 0.

7. If $n = N$, go to the next step. Otherwise update n to $n+1$ and go to step 5.

8. Calculate the sum of the contribution of the similarity

$$\omega_{\mathrm{sum},n} = \sum_{n_1=1}^{N} \frac{\omega_{nn_1}}{\sum_{n_2=1}^{N} \omega_{n_2 n_1}} \tag{17}$$

of every sample ($n = 1, 2, \cdots, N$).

9. Remove the sample corresponding to the smallest ω_{sum} from the database. The new input and output matrices are denoted by $\boldsymbol{X}_i$ and $\boldsymbol{y}_i$, respectively.

10. If the number of the samples in the database is $N - N_{\mathrm{out}}$, finish the sample selection. Otherwise, update i to $i+1$ and go to step 4.

In step 6, LW-PLS models are constructed for the queries which are contained in the database to ensure that the estimation error asymptotically becomes 0 when localization parameter φ approaches to 0. In step 8, the sum of the contribution of the similarity is used as the sample selection index since the similarity means the relative importance of every sample.

4 Application examples

This section shows two application examples of the proposed method. One is the numerical example and the other one is the application example in an industrial distillation process.

4.1 Numercal example

In this example, the initial database is prepared by the following equation.

$$y = \exp(x_2) + \exp(x_2) \tag{18}$$

The number of samples N is one hundred and the values of x_1 and x_2 are determined as shown in Figure 3 (left). In this example, the nonlinearity becomes stronger as the value of the input variables become large. Thus, to keep the sample with large value of input variable is important to construct accurate soft sensors. Figures 3 (center) and (right) show the sample selection result of the proposed method and the conventional method based on the density. Here, 58 samples are selected, and the tuning parameters such as $e_{\max}$ and R are determined by trial and error. As shown in Figure 3 (center), the proposed method kept samples with large input values. Consequently, the estimation performance for about 8000 validation samples, which are generated independently from the samples in the initial database and are equally distributed, improved by 32% compared to the conventional method. This result confirmed that the strength of the nonlinearity should be taken into account in sample selection and the proposed method can select the sample based on it.

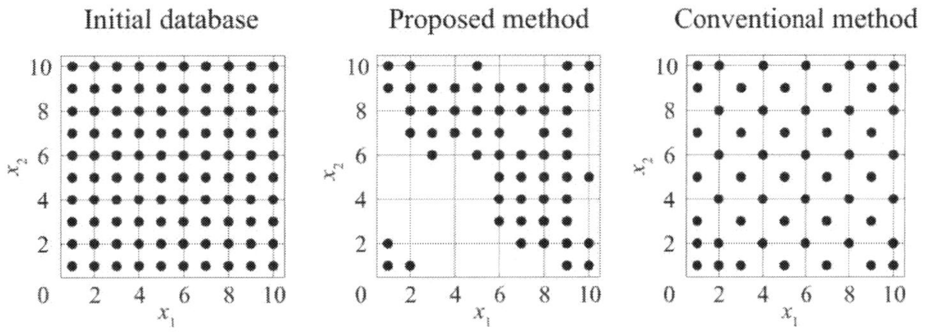

Figure 3: The initial database (left) and selected samples with the proposed (center) and the conventional (right) method.

4.2 Industrial application

This section demonstrates the application result of the proposed method to an industrial distillation process at Showa Denko K.K (SDK) in Japan. The schematic diagram of the target process, which is the cracked gasoline (CGL) fractionator of the ethylene production process, is shown in Figure 4. A soft sensor for estimating the aromatics concentration is developed to realize highly efficient operation of the process.

In this example, one hundred time-series samples are used as the initial database, and the fifty samples are selected by the proposed method. For comparison, two databases (databases 1 and 2) are created: database 1 has fifty newest samples and database 2 has 1st, 3rd, 5th, $\cdots$ 99th samples. The following fifty samples are used for parameter tuning, and further following fifty samples are used for model validation. The tuning parameters such as $e_{\max}$ and R are determined so that the root mean square error (RMSE) for the tuning parameter samples becomes the smallest.

Figure 5 shows the histogram of the estimation error for the model validation samples. RMSE of validation (RMSEV) is the smallest when the proposed method is used. This result also confirms the usefulness of the proposed method.

5 Conclusions

This research showed that the strength of the nonlinearity as well as the sample density is important criterion for selecting samples used for the construction of locally weighted linear regression model. In addition, we developed a data-driven method for the sample selection which can take into account the strength of the nonlinearity. The proposed method improved the estimation performance by 17-26% in an industrial application in the distillation process compared to the conventional methods such as the first-in, first out method. This result clearly shows the usefulness of the proposed method and it can contribute reduce the considerable burden of the model maintenance.

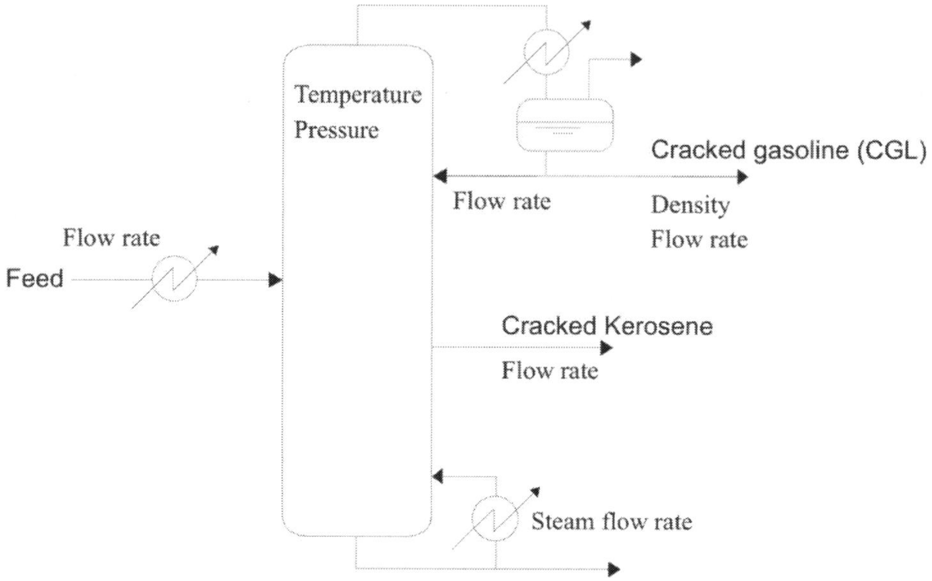

Figure 4: Schematic diagram of the CGL fractionator of the ethylene production process at the SDK.

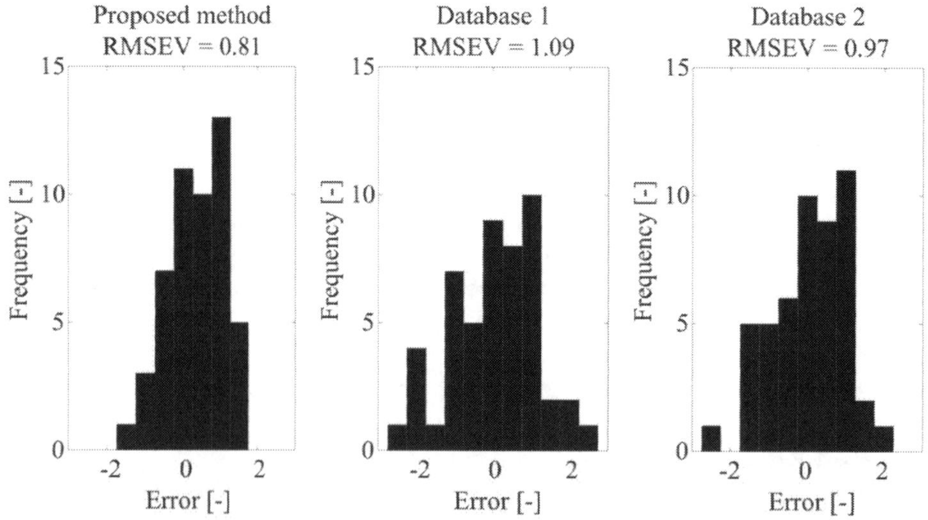

Figure 5: Histogram of the estimation error for the model validation samples.

References

[1] M. Kano and Y. Nakagawa. Data-based process monitoring, process control, and quality improvement: recent developments and applications in steel industry. *Comput. and Chem. Eng.*, 32(1-2):12–24, 2008.

[2] P. Kadlec, B. Gabrys, and S. Strandt. Data-driven soft sensors in the process industry. *Comput. and Chem. Eng.*, 33:795–814, 2009.

[3] M. Kano and K. Fujiwara. Virtual sensing technology in process industries: trends and challenges revealed by recent industrial applications. *J. Chem. Eng. Jpn.*, 46:1–17, 2013.

[4] W.-M. Wu, F.-T. Cheng, and F.-W. Kong. Dynamic-moving-window scheme for virtual-metrology model refreshing. *IEEE Trans. Semicond. Manuf.*, 25(2):238–246, 2012.

[5] H. Kaneko and K. Funatsu. Database monitoring index for adaptive soft sensors and the application to industrial process. *AIChE J.*, 60(1):160–169, 2014.

[6] Z. Ge and Z. Song. A comparative study of just-in-time-learning based methods for online soft sensor modeling. *Chemom. Intell. Lab. Syst.*, 104:306–317, 2010.

[7] S. Kim, M. Kano, H. Nakagawa, and S. Hasebe. Estimation of active pharmaceutical ingredients content using locally weighted partial least squares and statistical wavelength selection. *Int. J. Pharm.*, 421:269–274, 2011.

[8] H. Nakagawa, T. Tajima, M. Kano, S. Kim, S. Hasebe, T. Suzuki, and H. Nakagami. Evaluation of infrared-reflection absorption spectroscopy measurement and locally weighted partial least-squares for rapid analysis of residual drug substances in cleaning processes. *Anal. Chem.*, 84:3820–3826, 2012.

[9] S. Y. Chang, E. H. Baughman, and B. C. McIntosh. Implementation of locally weighted regression to maintain calibrations on FT-NIR analyzers for industrial processes. *Appl. Spectrosc.*, 55:1199–1206, 2001.

[10] K. Fujiwara, M. Kano, S. Hasebe, and A. Takinami. Soft-sensor development using correlation-based just-in-time modeling. *AIChE J.*, 55:1754–1765, 2009.

Development of a Sensor Placement Algorithm for Efficiency Maximization of Processes with Estimator-Based Control Systems

Prokash Paul[1], Debangsu Bhattacharyya[1], Richard Turton[1], and Stephen E. Zitney[2]
[1]West Virginia University, Department of Chemical Engineering, Morgantown, WV, USA; [2]National Energy Technology Laboratory, Morgantown, WV, USA

Motivation

- Measurements are noisy, inaccurate, or unreliable and these can:
  - lead to inefficient/unsafe operation under extreme conditions
  - and damage equipment and reduce their life span.
- In contrast, high accuracy measurement technology is costly or the measurement is available after undesired delay.
- Accurate estimation of the unmeasured /noisy states and disturbances in a process plant can lead to:
  - ✓ better process monitoring
  - ✓ higher performance of the control system
  - ✓ and longer equipment life.
- The location of measured points (i.e., location of sensors) has direct effect on the estimation accuracy of key variables.
- In this work, a Sensor Network Design (SND) algorithm is developed assuming an estimator-based control system to find the location and type of sensors.

Measurement Noise

Effect of measurement noise on Plant Efficiency :

Application: Acid Gas Removal (AGR) Unit of an Integrated Gasification Combined Cycle (IGCC) Power Plant with CO_2 Capture

Operational objective: 90% CO_2 capture

Case 1: The measurement system shows **89.8%** CO_2 capture, even though the actual CO_2 capture is 90%.	*Case 2:* The measurement system shows **90.2%** CO_2 capture, even though the actual CO_2 capture is 90%

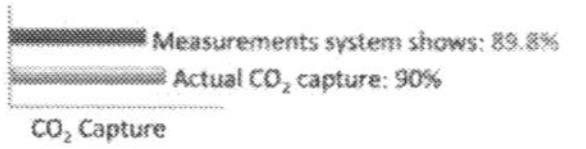

Plant operators/control system will change the operating conditions → increase the CO_2 capture → causing loss in efficiency

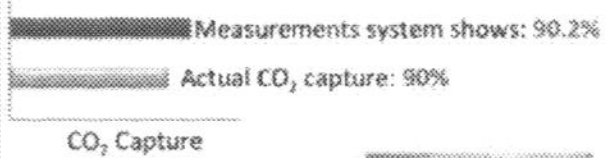

Plant operators/control system will change the operating conditions → decrease the CO_2 capture → penalty from the regulating agencies

Traditional approach to sensor placement

Open-Loop Approach

Input → Process and Control system → Sensor → Measurements → Estimator → Estimated variables

Does the traditional approach to sensor placement have a direct impact on plant efficiency?

Methodology

Estimator-based control

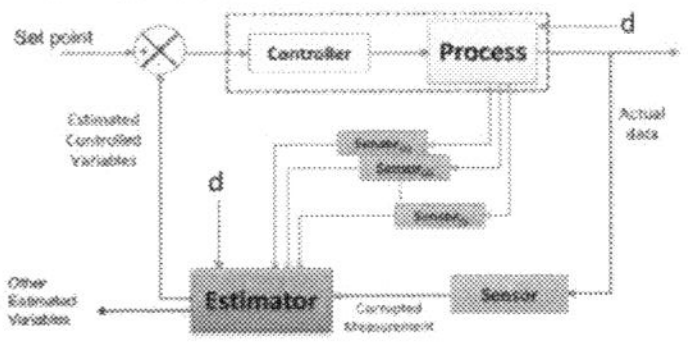

Process Model

$$\frac{dx_{act}}{dt} = Ax_{act} + Bu + w$$
$$y = Cx_{act} + v$$

Estimator (KF)

$$\frac{d\hat{x}}{dt} = A\hat{x} + Bu + K(y - C\hat{x})$$
$$\frac{dP}{dt} = -PC^TR^{-1}CP + PA^T + AP + Q$$
$$K = PC^TR^{-1}$$
$$Q = E[ww^T]$$
$$R = E[vv^T]$$

Comparator

Estimated measurements:

$$\hat{y}_{cont,est} = C_{cont}\hat{x}$$

Error function:

$$\varepsilon(t) = y_{cont,set} - \hat{y}_{cont,est}$$

Controller (PI)

Control action:

$$u_{cont} = K_c\left[\varepsilon(t) + \frac{1}{\tau_I}\int_0^t \varepsilon(t)dt\right]$$

Algorithm to simulate feedback control system with an estimator

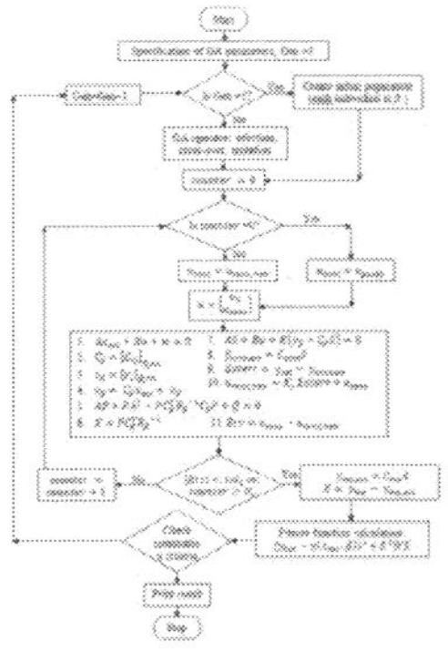

Genetic Algorithm (GA):

GA finds the optimal set of sensors over the successive generations that can maximize objective function subject to the constraints on estimation accuracy and budget (cost of sensors, $)

Objective function:

$$\eta = \frac{F_{in,gas}x_{in,gas} - F_{out,gas}x_{out,gas}}{aF_{in,solvent} + P_1 + P_2 + P_3}$$

IGCC Power Plant

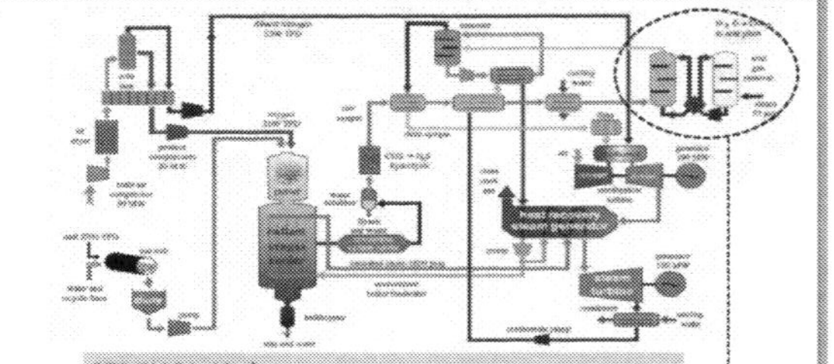

SELEXOL Unit :

Block diagram of AGR process

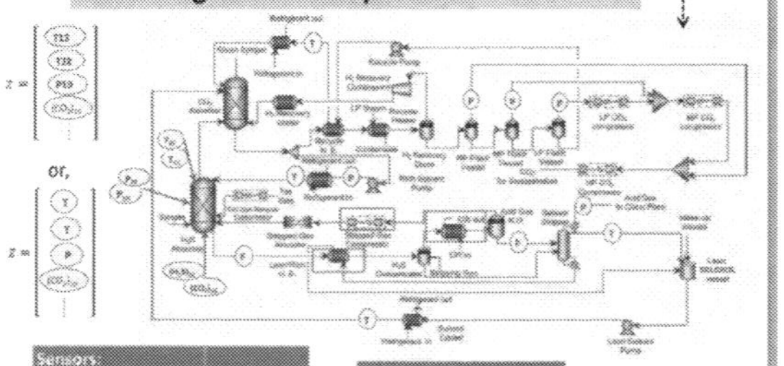

Sensors:	
Flow sensor	H_2S Analyzer
Temperature sensor	CO_2 Analyzer
Pressure measuring device	

AGR process	
Potential no. of sensors	163
Linear model contains	1505 states

Working Approach & Results

Software Packages

- Aspen Plus®
  - Design of the AGR process (Steady-state model)
- Aspen Plus Dynamics®
  - Control system design and simulation of the nonlinear process model
  - Opening the primary controlled variables loop
  - Generation of the linear continuous-time model
- MATLAB®
  - Implementation of the SND algorithm for efficiency maximization

Table: Number of Sensors and the Value of the Objective Function for Different Budgets

Cases	Budget (Cost of sensors, $)	Number of sensors	Efficiency (molCO_2/MWh)
1	431900	75	766.0058
2	322600	66	766.0058
3	229400	64	766.0058
4	187900	62	766.0058
5	149000	50	765.11
6	118700	46	762.6422
7	71200	25	758.2536
8	63700	25	756.7611
9	60200	23	752.0415
10	59700	24	750.0721
11	42500	17	742.8588

Illustration of the underlying reason for the increase in efficiency as the budget for sensors increases.

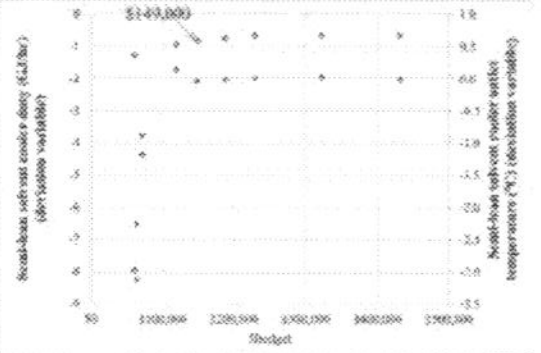

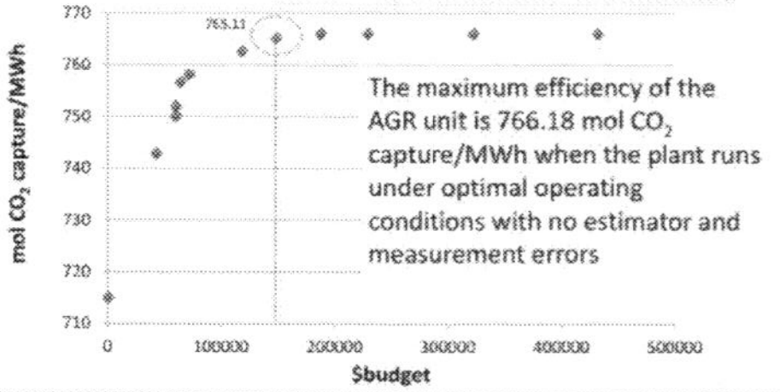

The maximum efficiency of the AGR unit is 766.18 mol CO_2 capture/MWh when the plant runs under optimal operating conditions with no estimator and measurement errors

Future Work

- ➢ Development of dynamic SND algorithm
- ➢ Implementation of the developed SND algorithm on the AGR unit of IGCC power plant

Acknowledgements

As part of the National Energy Technology Laboratory's Regional University Alliance (NETL-RUA), a collaborative initiative of the NETL, this technical effort was performed under the RES contract DE-FE0004000.

388928 A Genetic Algorithm-Based Distributed Sensor Placement for Water Gas Shift Reactor in IGCC Plant Using Kalman Filter

Abstract:

Integrated gasification combined cycle (IGCC) with CO_2 capture is an attracting technology for power generation due to its high efficiency and near-zero emission. A particular unit that is crucial to the performance of IGCC is the water gas shift reactor (WGSR), which is responsible for adjusting the syngas H_2/CO ratio to meet power production and carbon capture targets. The efficacy of the WGSR is greatly affected by number of disturbances, such as upstream feed disturbance, and faults, such as catalyst deactivation due to poisoning and fouling. An early detection of disturbances and faults can facilitate the preventive actions and ensure safe and efficient operation of the reactor. However, faults and disturbances are not essentially observable by measurements and require a state estimation of the internal state of the reactor for them to be detected. This can be done by using a first-principle model of the reactor and series of measurements distributed throughout the reactor. It is therefore necessary to find the optimal number and location of the sensors for efficient estimation of faults and disturbances.

A high-fidelity model of the WGS reactor was developed in our previous work. The model is presented by a system of non-linear differential and algebraic equations (DAE). The discretized equations form a space of grid-points that are available for state measurement. An extended Kalman filter was implemented to estimate the faults using the measurements chosen from the space. A genetic algorithm is then employed to search the space for obtaining an optimal location of the sensors. However, the approach is inefficient due to large search space and computational burden for non-linear state estimation. To address this, a reduced order model of the WGSR is developed by linearization of the non-linear model. The genetic algorithm uses the linearized model and a linear Kalman filter to find the optimal location of the sensors. This approach significantly reduces the computational work compared to our previous method.

A Data Processing Framework for Online Monitoring and Optimization of Emulsion Polymerization Processes

P. Joy (Preet.Joy@avt.rwth-aachen.de), A. Mhamdi (Adel.Mhamdi@avt.rwth-aachen.de), A. Mitsos (Alexander.Mitsos@avt.rwth-aachen.de)

AVT – Process Systems Engineering (SVT), RWTH Aachen University
Turmstr. 46, 52064 Aachen, Germany

Semi-batch emulsion polymerization is used to manufacture several polymer products. Although the technology is fairly mature, most processes in industry are operated according to a fixed recipe with only elementary control loops to maintain the reactor temperature within safety constraints. Also, most processes are not operated at the optimum operating point so as to remain well-within safety limitations. Thus, there exists significant scope for process intensification through online process optimization [1].

An important prerequisite for effective implementation of online process optimization is adequate and accurate process information, which may be gathered by taking measurements of process states. If, however, some process states are not measurable, estimated state values from an estimation algorithm (e.g. Extended Kalman Filter), may be used [1, 2, 6].

In the case of emulsion polymerization processes, the polymer properties are of particular interest and the control scheme needs to ensure the final polymer properties meet customer requirements while the process satisfies necessary safety and environmental constraints. Online measurement of polymer properties is, however, a challenging task and there is no online analysis technique, at present, to achieve this [4, 5]. Also, easy and reliable measurements of temperatures (both reactor and cooling jacket) alone are not sufficient to estimate the polymer properties reliably, because the states which affect the polymer properties are not observable with temperature measurements only [6].

Thus, it is common practice in laboratory processes to use multiple sensors like pH, conductivity, etc. to gain better insight into the process [4, 5]. Recently, there has been significant progress in the use of Raman spectroscopy to determine monomer and polymer concentrations [3]. When multiple sensors are used to gather data from a process, the task of collecting data and establishing a reliable communication between the process, sensors and the estimation/control algorithms becomes a challenging task. One of the main challenges is that each sensor reports data in its own proprietary format. Therefore, in order to effectively use multi-sensor measurements for online estimation and control, it is important to have a common data processing framework which collects multi-sensor data, sends the measured values to the estimation/optimization algorithms and transfers the control actions (e.g. from an optimization algorithm) back to the process.

In this work, an OPC (Open Platform Communications) based data processing framework is proposed as a robust and reliable platform which can be used for multi-sensor data acquisition and online optimization of a semi-batch emulsion polymerization process. The proposed framework is depicted in Figure 1. This framework has been implemented and tested at the University of Hamburg as part of the COOPOL project which is funded by the European Union's Seventh Framework programme for Research.

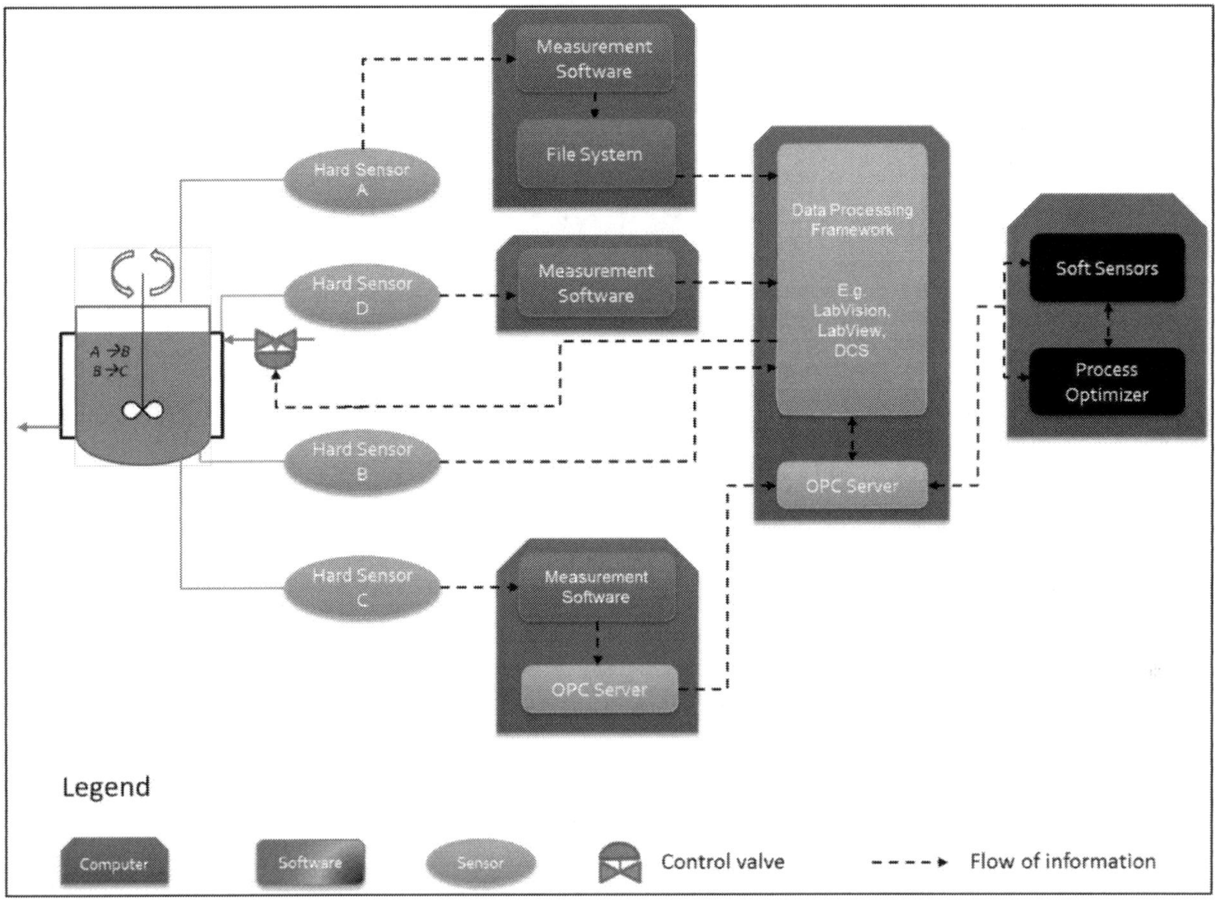

Figure 1: Proposed data processing framework for online monitoring and optimization of emulsion polymerization processes

References

1. Eliçabe, G. E., Meira, G. R., Estimation and Control in Polymerization Reactors. A Review, Polymer Engineering and Science, 1988, 28(3), 121-135.
2. Ellis, M. F., Taylor, T. W., Gonzales, V., Jensen, K. F., Estimation of the Molecular Weight Distribution in Batch Polymerization, AIChE Journal, 1988, 34(8), 1341-1353.
3. Elizalde, O., Azpeitia, M., Reis, M. M., Asua, J. M., Leiza, J.R., Monitoring Emulsion Polymerization Reactors: Calorimetry versus Raman Spectroscopy, Industrial and Engineering Chemistry Research, 2005, 44, 7200-7207.
4. Fonseca, G. E., Dubé, M. A., Penlidis, A., A Critical Overview of Sensors for Monitoring Polymerizations, Macromolecular Reaction Engineering, 2009, 3, 327-373.
5. Frauendorfer, E., Wolf, A., Hergeth, W. D, Polymerization Online Monitoring, Chemical Engineering and Technology, 2010, 33(11), 1767-1778.
6. Kozub, D. J., MacGregor, J. F., State Estimation for Semi-batch Emulsion Polymerization Reactors, Chemical Engineering Sciene, 1992, 47(5), 1047-1062.

Ability to perform efficient fault diagnosis is essential to ensure smooth and safe operation of chemical processes. With increased integration across various units in a modern day chemical process, the need for automated diagnostic techniques is even more acute. Over the last couple of decades, several diagnostic strategies have been proposed in literature. The success of any diagnostic strategy depends on the location of the sensors measuring important variables in the process. Selecting an optimal set of variables to be measured in the process is called as the sensor network design problem. The focus of our current work is on choosing an optimal set of variables that can ensure reliable fault diagnosis. Sensor network design is primarily an activity that is performed at the design stage. Several approaches are available in literature for designing sensor networks. Several of these approaches utilize a digraph (DG) or a signed digraph (SDG) based representation of the process. This is convenient as digraphs are a qualitative representation of the cause-effect relationships amongst process variables. The nodes in a digraph represent process variables and root causes of abnormalities, while the edges represent the cause-effect relationships between adjacent nodes. The edges are unsigned in a DG, while signs (indicating direction of effect) are associated with edges in a SDG. Hence, DGs and SDGs can be generated without having detailed quantitative information (such as values of kinetic parameters) about the process and thus digraph based design approaches can be used at the process design stage. An SDG can be generated based on either on operator experience or from the first principles representation of a process Maurya et al. (2003a,b).

In literature, Raghuraj et al. (1999) used DG representation for sensor network design for fault diagnosis to ensure specific criteria such as observability and resolution of faults. Bhushan and Rengaswamy (2000) extended this work by representing a process as a SDG instead of a DG. Bagajewicz et al. (2004) proposed a Mixed Integer Linear Programming (MILP) based optimization approach for similar problems. Bhushan and Rengaswamy (2002) proposed the concept of unreliability of detection where fault occurrence and sensor failure probabilities are incorporated in the sensor network design procedure. All the above approaches assume that the SDG of the process model is accurate and don't consider errors (model-plant mismatch) in the SDG representation. Such model-plant mismatch can typically manifest in terms of wrong signs being assigned to the edges in the SDG. This can happen, when: (i) some signs on the edges of the SDG are not known and instead represent only an "educated" guess, and (ii) even if the signs on the edges of the SDG are accurate to begin with, they may become erroneous if the true signs change over a period of time due to change in the process conditions. In case the underlying SDG model is not an accurate representation of the process, the resulting sensor network design may just not be suboptimal, but it may also fail to satisfy important reliability or observability requirements. Recognizing this aspect, Bhushan et al. (2008) proposed the idea of distributed sensor network that has a higher chance of ensuring designed performance in the presence of model-plant mismatch. However, the approach considers model-plant mismatch in a generic way and does not take into consideration information about specific uncertain edge-signs that may be available. Recently, Kolluri et al. (2013) proposed set cover problem where sensor networks are designed for ensuring observability and resolution of faults while incorporating information about uncertain edge-signs. Their approach involved designing sensors that could observe and distinguish between various scenarios where faults could occur and the uncertain edges could simultaneously change signs. However, their approach assumed that all scenarios were equally probable. In practice, this will not be the case. Different faults and sensors will have different probabilities of occurrence and failure, as has been considered in the earlier works in literature (Bhushan and Rengaswamy, 2002). Additionally, it may be possible to assign different confidence levels to the uncertain edge signs. For example, it may be known that a particular edge does not change sign even for large variations in the quantitative values of underlying parameters/variables, while some other edge may change sign even for a small variation in the underlying quantitative values. Thus, the sign of the former edge can be specified with a higher confidence than of the latter.

In the current work we present sensor network design approaches that incorporate probabilities of fault occurrence and sensor failures as well as user specified confidence in edge signs. This is done by extending the notion of unreliability of detection presented in literature earlier Bhushan et al. (2008) to the case when some edge signs in the underlying SDG representation are known to be uncertain with varying degrees of confidence. Towards this end, different scenarios namely, of (i) a fault occurring alone with no change in edge signs, and (ii) a fault occurring along with simultaneous change in an uncertain edge sign, are considered. For each of these scenarios, through propagation in the signed digraph, the effect on the process variables is obtained. Under the assumptions of various events (fault occurrences, sensor failures, edge sign changes) happening independently, analytical expression of probability of a scenario occurring and remaining undetected (due to simultaneous failure of sensors) can be obtained. This probability is labeled as the unreliability of detection. The sensor network design problem is then cast as an optimization problem that minimizes the maximum unreliability of detection amongst all scenarios. Incorporating probabilities in the sensor network design approach is specially relevant for cases when there are limitations on the number/type of sensors that can be used in the process due to availability of only limited resources for sensor network design. In such cases, incorporation of probabilities provides a natural way of prioritizing the limited resources towards scenarios that are more likely to occur. The applicability of the proposed sensor network design approach is demonstrated by applying it to various case studies taken from literature. The results indicate the utility of the approach in obtaining a sensor network that can enable simultaneous identification of the fault that has occurred as well as a simultaneous change in edge sign, if any.

References

Bagajewicz, M., Fuxman, A., Uribe, A., 2004. Instrumentation network design and upgrade for process monitoring and fault detection. AIChE J. 50, 1870–1880.

Bhushan, M., Narasimhan, S., Rengaswamy, R., 2008. Robust sensor network design for fault diagnosis. Comput. Chem. Engng. 32, 1067–1084.

Bhushan, M., Rengaswamy, R., 2000. Design of sensor network based on the sdg of the process for efficient fault diagnosis. Ind. Eng. Chem. Res. 39 (4), 999–1019.

Bhushan, M., Rengaswamy, R., 2002. Comprehensive design of a sensor network for chemical plants based on various diagnosability and reliability criteria: I. Framework. Ind. Eng. Chem. Res. 41, 1826–1839.

Kolluri, S., Bajaj, I., Bhushan, M., 2013. Sensor network design for efficient fault diagnosis and signed digraph update. In: 10th International Symposium on Dynamics and Control of Process Systems, Mumbai, India. pp. 821–826.

Maurya, M. R., Rengaswamy, R., Venkatasubramanian, V., 2003a. A systematic framework for the development and analysis of signed digraphs for chemical processes. 1. Algorithms and analysis. Ind. Eng. Chem. Res. 42 (20), 4789–4810.

Maurya, M. R., Rengaswamy, R., Venkatasubramanian, V., 2003b. A systematic framework for the development and analysis of signed digraphs for chemical processes. 2. Control loops and flowsheet analysis. Ind. Eng. Chem. Res. 42 (20), 4811–4827.

Raghuraj, R., Bhushan, M., Rengaswamy, R., 1999. Location of sensors in complex chemical plants based on fault diagnostic observability criteria. AIChE J. 45 (2), 310–322.

Compromise Optimization Tuning Strategy for Model Predictive Controllers

Yamashita, André S [a], Odloak, Darci[a,]*

[a]Department of Chemical Engineering, University of São Paulo, Av. Prof. Luciano Gualberto, trv 3 380, 05424-970 São Paulo, Brazil.
Phone number: +551130912261
e-mail addresses {andre.yamashita,odloak}@usp.br
[*]Corresponding author

Abstract

A novel tuning technique for Mode Predictive Controllers was developed. The tuning goals were defined according to output reference trajectories, which allows for the straightforward inclusion of time-domain metrics. The compromise optimization approach was used to solve the tuning problem. A 3x3 system in closed-loop with a finite horizon MPC was tuned, and the control performance was compared to another multiobjective-based tuning approach from the literature. It was shown that the method developed here is faster and returns a single optimum solution, whereas the method from the literature requires the solution of more than one optimization problem and also, the user must select a single solution from a set of Pareto optima; regarding the closed-loop performance, both techniques yields similar results.

1. Introduction

Model predictive controllers are widely used in the petrochemical industry, mainly because of their ability to incorporate the process constraints directly into the control algorithm formulation and also because it can be straightforwardly extended from single input single output to multiple input multiple output systems. A controller model is used to predict the system outputs future behavior, based on the current input and output values. The MPC is based on the rolling horizon strategy, in which at a given time instant, a vector of optimum control moves is calculated; the first move is injected to the system, and the input and output values are updated using the system model. The process is repeated at the next time instant.

The MPC algorithm is a constrained optimization problem. Its cost function takes into account at least two different terms, related to (i) the differences between the output predictions and the output set points and (ii) the control moves. The constraints take into account the physical limits on input and output values, as well as the upper and lower bounds on the control actions (Qin & Badgwell, 2003).

The cost function weighting matrices, the control and the prediction horizons affect the closed-loop response of the MPC. An appropriate choice of these parameters can improve stability, robustness and performance (Garriga & Soroush, 2010). In this work, we develop a MPC tuning technique based on multiobjective optimization. Section 2 provides the reader a brief background on MPC and multiobjective optimization. Section 3 introduces the Compromise Tuning Technique (CTT), Section 4 shows a case study describing the tuning procedure of a MPC in closed-loop with a 3x3 system; the results are compared to another multiobjective based technique from the literature. Finally, the work closes with some conclusions in Section 5.

2. Methodology

2.1. Model Predictive Control

The MPC used here assumes that the system is represented by state-space model defined in (1).

$$\begin{aligned} x(k+1) &= Ax(k)+B\Delta u(k) \\ y(k) &= Cx(k) \end{aligned} \tag{1}$$

where $x \in \Re^{nx}, u \in \Re^{nu}, y \in \Re^{ny}$. Matrices A, B and C are model matrices that carry all the system information required for future output predictions.

Then, for the system defined in (1), the MPC cost function includes the weighted sum of squared differences between the predicted outputs and set point values over the prediction horizon, and the weighted sum of squared input increments over the control horizon. The control problem can be summarized as follows.

Problem 1

$$\min_{\Delta u_k} V_{1,k} = \sum_{j=0}^{p} \left\| y(k+j\,|\,k) - y_{sp} \right\|_{Q_y}^2 + \sum_{j=0}^{m-1} \left\| \Delta u(k+j\,|\,k) \right\|_R^2 \tag{2}$$

subject to (1) and

$$u_{\min} \le u(k+j) \le u_{\max},\ \ j = 0,\ldots,m-1 \tag{3}$$

$$-\Delta u_{\max} \le \Delta u(k+j) \le \Delta u_{\max},\ \ j = 0,\ldots,m-1 \tag{4}$$

where $y(k+j\,|\,k) \in \Re^{ny}$ is the output prediction calculated at time instant $k+j$ using (1) and information available at time instant k, $y_{sp} \in \Re^{ny}$ is the vector of output set points, p is the output prediction horizon, $\Delta u(k+j\,|\,k) \in \Re^{nu}$ is the control move at time instant $k+j$ calculated using information available at time instant k, m is the control horizon, $Q_y \in \Re^{ny\times ny}$ is a positive semi-definite diagonal matrix, $R \in \Re^{nu\times nu}$ is a positive-definite diagonal matrix, $u_{\min}$, $u_{\max}$, and $\Delta u_{\max}$ are the lower and upper bounds of the system inputs and input increments, respectively.

The tuning algorithms developed here consider the unconstrained version of the MPC, disregarding the constraints defined in (3) and (4), so that the effects of the bounds on the system inputs and input increments do not hinder the tuning results (Shridhar & Cooper, 1998).

2.2. Multiobjective optimization

Multiobjective optimization approaches solve problems with competing goals. There are two main alternatives to deal with the trade-off between competing objectives: either to weight the goals accordingly prior to the problem solution or to chose an optimum solution from a set of 'equally optimum' solutions, based on subjective criteria.

A general multiobjective problem can be posed as follows:

Problem 2

$$\min_x F(x) = \left[F_1(x) \quad F_2(x) \quad \cdots \quad F_w(x) \right]^T \tag{5}$$

subject to

$$g_j(x) \le 0, \ j = 1,\ldots,z \tag{6}$$

$$h_l(x) = 0, \ l = 1,\ldots,e \tag{7}$$

$F(x) \in \Re^w$ is a vector comprised of w objectives $F_i(x)$, $g_j(x)$ and $h_l(x)$ are related with the inequality and equality constraints, respectively, $x \in \Re^{n_{dec}}$ is the vector of decision variables and n_{dec} is the number of decision variables. The feasible design space is defined as $\mathbf{X} = \{x \in \Re^{n_{dec}} \mid g_j(x) \le 0, j = 1,\ldots,z \ and \ h_l(x) = 0, l = 1,\ldots,e\}$, and the feasible criterion space is defined as $\mathbf{Z} = \{z \in \Re^w \mid z = F(x), x \in \mathbf{X}\}$. The following definitions characterize the optimum solutions in multiobjective optimization problems.

Definition 1: A point $x^* \in \mathbf{X}$ is a Pareto optimum *iff* there does not exist another point $x \in \mathbf{X}$, such that $F(x) \le F(x^*)$, and $F_i(x) < F_i(x^*)$ for at least one i.

Definition 2: A point $F^\circ(x) \in \mathbf{Z}$ is an Utopia point *iff* for each i=1,…,w, $F_i^\circ(x) = \min_x \{F_i(x) \mid x \in \boldsymbol{X}\}$.

Compromise optimization

A compromise solution is also a Pareto optimal (Ballestero & Romero, 1991). Moreover, the compromise solution method is attractive because, in real cases, it is virtually impossible to accurately translate the preferences of the decision-maker into utility functions; and it is reasonable to think that the feasible point closest to the Utopia solution (Zelany, 1974) is sought instead.

3. Compromise Tuning Technique

First, the relative importance of the process outputs is defined according to economic, safety or environmental factors. Second, an input-output pairing layout is defined for each process output following their importance order. Either process knowledge, the Relative Gain Array (Bristol, 1966) or other similar techniques are used to select the appropriate pairs. At the third step, inputs, outputs and the system transfer function gains are normalized to achieve better numerical conditioning of the tuning problem. Also, the values of the tuning cost function for different goals can be compared within similar orders of magnitude.

The tuning goals are defined in the fourth step, as the minimization of the quadratic error between the output closed-loop responses and pre-defined reference trajectories. This goal definition strategy has been adopted in other MPC tuning methods (Exadaktylos & Taylor, 2010; Shah & Engell, 2010), as it allows for a simple specification of a desired control performance and time-domain objectives. In this work, an output reference trajectory is defined as the step response of a target transfer function, obtained from the approximation of the open-loop transfer function between the corresponding input-output pair defined in Step 2, by second order transfer functions. The multiobjective goals are defined according to (8):

$$F_i(x) = \sum_{k=1}^{\theta_i} \left(y_i^{ref}(k) - y_i(k) \right)^2 \ \ i = 1,\ldots,w \tag{8}$$

Where θ_t is the tuning horizon, $y_i^{ref}(k) \in \Re^{ny}$ is the discretized reference trajectory of output i, $y_i(k) \in \Re^{ny}$ is the closed-loop trajectory of output i, $k = 1,\ldots,\theta_t$ and x is the vector of decision variables. Observe that if $Q_y = diag(q_1,\ldots,q_{ny})$ and $R = diag(r_1,\ldots,r_{nu})$, then $x = (q_1,\ldots,q_{ny},r_1,\ldots,r_{nu})$. Observe also that since $y_i(k)$ is the response of the system in closed-loop with the MPC, it is a function of the tuning parameters, which are also the decision variables of the tuning optimization problem.
The Utopia solution is obtained solving the following problem.

$$F_i^{\circ}(x) = \min_x F_i(x), \quad i = 1,\ldots,w \tag{9}$$

subject to (8) and

$$LB \le x \le UB \tag{10}$$

Observe that the Utopia point, $F^0(x)$, is unfeasible unless all the tuning goals share the same optimum solution. Once the Utopia point is obtained, the method tries to find the closest feasible solution to it, in terms of the Euclidian distance. This procedure defines the CTT optimization problem.

Problem 3

$$\min_x \left\| F^{\circ} - F(x) \right\|^2 \tag{11}$$

subject to (8) and (10).

Where $F^{\circ} \in \Re^w$ is the vector with components $F_i^{\circ}(x)$.

4. Case study

The transfer function in the Laplace domain used in the tuning technique demonstration is given in (13). It was discretized using a sample time of 1 minute, and converted to matrices A, B and C from the state space model described in (1). The reader is referred to (Porfiirio, Almeida Neto, & Odloak, 2003) for more information about the original system.

$$G(s) = \begin{bmatrix} \dfrac{0.012s - 0.003}{s^2 + 0.988s + 0.065} & \dfrac{-0.004s - 0.004}{s^2 + 0.201s + 0.009} & \dfrac{0.007s + 0.004}{s^2 + 0.511s + 0.027} \\ \dfrac{-0.011s - 0.047}{s^2 + 0.151s + 0.007} & \dfrac{0.012s - 0.009}{s^2 + 0.104s + 0.005} & \dfrac{-0.004s - 0.008}{s^2 + 0.097s + 0.005} \\ \dfrac{-0.06s - 0.094}{s^2 + 0.887s + 0.084} & \dfrac{0.070s - 0.132}{s^2 + 0.860s + 0.039} & \dfrac{-0.043s - 0.010}{s^2 + 1.888s + 0.067} \end{bmatrix} \tag{12}$$

The input-output pair arrangement was y_1-u_1, y_2-u_2 and y_3-u_3, and the transfer function that defines the reference output trajectories is given in (14). The system starts at the origin and the outputs were driven to $y^{sp} = \begin{bmatrix} 1 & 1 & 1 \end{bmatrix}^T$, according to the dynamics defined in (14). The tuning horizon is $\theta_t = 70$ min and the CTT optimization problem (Problem 3) was solved using the *fmincon* (default settings) in MATLAB® 2013. The prediction horizon and the control horizon were chosen as $p = 50$ and $m = 5$, following usual tuning guidelines from literature (Garriga & Soroush, 2010). The Utopia point $F^o = \begin{bmatrix} 0.0587 & 0.0680 & 0.0946 \end{bmatrix}^T$ was obtained, and the optimum tuning parameters were $Q_y = diag\left(\begin{bmatrix} 1.01 & 0.503 & 0.469 \end{bmatrix}\right)$ and $R = diag\left(\begin{bmatrix} 0.214 & 2.004 & 0.01 \end{bmatrix}\right)$, calculated in 34.4 seconds, including the time required to obtain the Utopia solution.

The CTT results were compared to another multiobjective optimization based tuning technique from the literature, which uses the Normal Boundary Intersection (NBI) (Das & Dennis, 1998) method to obtain Pareto optimum solutions. Its advantage is to deliver an evenly-spaced Pareto front and to show a straightforward relationship between the scalarization parameters and the optimum values. The reader is referred to (Vallerio, Van Impe, & Logist, 2014) for more information.

The NBI goals were defined in the same manner as the CTT; a total of 66 Pareto optima were obtained and the set of tuning parameters that yielded the lowest control cost to take the system from the origin to $y^{NBI} = \begin{bmatrix} 1 & -1 & 1 \end{bmatrix}^T$ was selected. The routine *fmincon* was also used to solve the NBI problem, in MATLAB 2013® The optimum set obtained by the NBI method was $Q_y = diag\left(\begin{bmatrix} 1.01 & 0.212 & 0.413 \end{bmatrix}\right)$ and $R = diag\left(\begin{bmatrix} 0.025 & 1.592 & 0.02 \end{bmatrix}\right)$, in 845.1 seconds, including the time needed to obtain the Utopia solution. Observe that except for r_1, the parameter values were numerically close to the ones obtained by the CTT.

$$G(s) = \begin{bmatrix} \dfrac{1}{s^2 + 0.433s + 0.033} & 0 & 0 \\ 0 & \dfrac{1}{s^2 + 0.6s + 0.05} & 0 \\ 0 & 0 & \dfrac{1}{s^2 + 0.40s + 0.01} \end{bmatrix} \quad (13)$$

Figure 1 show the closed-loop responses of the 3x3 system in closed loop with the MPC described in Section 2, using the optimum tuning parameters obtained by the CTT (black line) and by the NBI (blue line) techniques; the dashed red lines indicate the reference trajectories defined in (14). The set points are the same as the ones used in the Pareto optimum decision phase of the NBI technique, $y_{sp} = \begin{bmatrix} 1 & -1 & 1 \end{bmatrix}^T$.

We observe that the strategies perform similarly regarding the response of y_1, the CTT outperforms the NBI regarding y_2 and the opposite is observed regarding y_3. Also, observe that the reference trajectories were conceived faster than the open-loop responses, however, the methodology is flexible enough to accommodate different performance goal definitions, because in real applications, it might be interesting to search for slow responses as well.

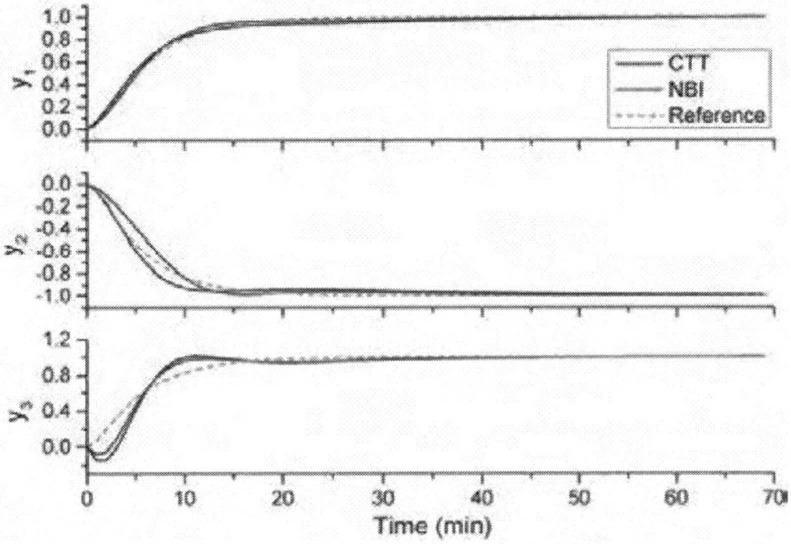

Figure 1: MPC tuning results, system outputs. CTT (black like), NBI (blue line), reference trajectories (red dashed line).

5. Conclusions

A new tuning technique based on compromise optimization was developed. The results showed that both techniques performed similarly, for the first output, while the CTT yielded better tracking performance for y_2 and the NBI prioritized y_3. The main advantages of the technique developed here are twofold: first, it is not necessary to select an solution from a set of Pareto solutions, and second, the CTT technique was 24 times faster than the NBI technique, because the former only solves a single optimization problem, whereas the latter solves as many problems as the size of the scalarization vector dictates, but the lower the number of solutions, the poorer the description of the Pareto frontier..

Acknowledgements

The authors would like to acknowledge the financial support provided by CNPq under grant 140677/2011-9 and FUNDESPA; and Zanin, A. C from CETAI for his support.

Bibliography

Ballestero, E., & Romero, C. (1991). A theorem connecting utility function optimization and compromise programming. *Operations Research Letters*, *10*(7), 421–427.

Bristol, E. (1966). On a new measure of interactions for multivariable process control. *Automatic Control, IEEE Transactions on*, *11*(1), 133–134.

Das, I., & Dennis, J. E. (1998). Normal-Boundary Intersection: A New Method for Generating the Pareto Surface in Nonlinear Multicriteria Optimization Problems. *SIAM Journal on Optimization*, *8*(3), 631–657.

Exadaktylos, V., & Taylor, C. J. (2010). Multi-objective performance optimisation for model predictive control by goal attainment. *International Journal of Control*, *83*(7), 1374–1386.

Garriga, J. L., & Soroush, M. (2010). Model Predictive Control Tuning Methods: A Review. *Industrial & Engineering Chemistry Research*, *49*(8), 3505–3515.

Porfiirio, C. R., Almeida Neto, E., & Odloak, D. (2003). Multi-model predictive control of an industrial C3/C4 splitter. *Control Engineering Practice*, *11*(7), 765–779.

Qin, S. J., & Badgwell, T. A. (2003). A survey of industrial model predictive control technology. *Control Engineering Practice*, *11*(7), 733–764.

Shah, G., & Engell, S. (2010). Tuning MPC for desired closed-loop performance for SISO systems. In *18th Mediterranean Conference on Control and Automation, MED'10* (pp. 628–633).

Shridhar, R., & Cooper, D. J. (1998). A Tuning Strategy for Unconstrained Multivariable Model Predictive Control. *Ind. Eng. Chem. Res.*, *5885*(98), 4003–4016.

Vallerio, M., Van Impe, J., & Logist, F. (2014). Tuning of NMPC controllers via multi-objective optimisation. *Computers & Chemical Engineering*, *61*, 38–50.

Zelany, M. (1974). A concept of compromise solutions and the method of the displaced ideal. *Computers & Operations Research*, *1*(3-4), 479–496.

376160 Run-to-Run Model Predictive Control of Batch Crystallization

Abstract:

The production of crystals with desired size and shape distributions from batch crystallization processes have become of great interest to the pharmaceutical industry, because those characteristics significantly influence the bioavailability of drugs. Recently, model predictive control (MPC) has received a growing attention within the crystallization process community, owing to its unique ability to handle constraints on the inputs, outputs, and the rate of changes of inputs. Another key benefit of using MPC is in its applicability to dynamic and/or multi-input multi-output (MIMO) systems. However, one of the key challenges in the use of MPC to the batch crystallization process is the plant/model mismatch, which persists from run to run and it may deteriorate further due to the lack of *in situ* measurements for the product qualities [1]. Motivated by this, the main ideas of run-to-run (R2R) control are integrated with conventional MPC formulation in an effort to reduce the model mismatch and suppress the uncertainties in the system using the post-batch measurements. The major benefit in the utilization of R2R control is in its unique nature to compensate for drifting processes through correlations developed to model output variations as a function of changes in the process environment [2]-[4]. As a result, small changes have to be made to the optimal jacket temperature computed from the MPC using nominal system parameters in order to produce crystals with desired size and shape distributions from the batch crystallizer.

The present work focuses on the modeling of a batch crystallization process to produce tetragonal HEW lysozyme crystals through kinetic Monte Carlo (kMC) simulation methods in the way described in [5] using the rate equations originally developed by [6]. The kMC simulation is regarded as a representation of a real crystallization process, and the potential disturbances considered in this work include uncertainties in the solubility as well as the nucleation and crystal growth rates. Then, a population balance equation (PBE) is presented to describe the evolution of the crystal volume distribution, and the method of moments (MOM) is applied to derive a reduced-order moment model, which is used for the design of a MPC scheme. Initially, the optimal jacket temperature trajectory is computed by using nominal values for the kinetic parameters and applied to the crystallizer. After the first batch run, a multivariable nonlinear optimization (MNO) problem is solved for the identification of the kinetic parameters while minimizing the difference between the size and shape distributions estimated through model prediction and the post-batch measurements (e.g., measured final crystal size and shape distributions) obtained from the kMC simulation. Furthermore, the constraints on the nucleation and crystal growth rates are imposed in the MNO problem so that the kinetic parameters computed from the MNO result in the generation of physically meaningful rates. Then, the MPC is re-solved to compute a set of new optimal jacket temperatures using the reduced-order model with the new estimated kinetic parameters. This iterative optimization-based two-step control strategy continues until it provides negligible improvement from the previous step. We observe that the result is particularly more sensitive to the solubility than the kinetic parameters for the nucleation and crystal growth rates.

[1] Natarajan S, Lee JH. Repetitive model predictive control applied to a simulated moving bed chromatography system. 2000;24;1127-1133.

[2] Wang Y, Gao F, Doyle FJ. Survey on iterative learning control, repetitive control, and run-to-run control. J. of Process Control. 2009;19:1589-1600.

[3] Campbell WJ, Firth SK, Toprac AJ, Edgar TF. A comparison of run-to-run control algorithms. In Proceedings of the American Control Conference, Anchorage, AK, 2002.

[4] Sachs E, Guo RS, Ha A. On-line process optimization and control using the sequential design of experiments. Symposium on VLSI Technology, Honolulu, HI, 1990.

[5] Kwon JS, Nayhouse M, Christofides PD, Orkoulas G. Modeling and control of crystal shape in continuous protein crystallization. Chem. Eng. Sci. 2014;107:47-57.

[6] Durbin SD, Feher G. Simulation of lysozyme crystal growth by the Monte Carlo method. J. Cryst. Growth. 1991;110:41-51.

370307 A Novel Modeling and Optimization Based Energy Management Solution for Integrated Steel Plants

Abstract:

Energy efficiency and energy cost are of paramount importance for today's world economy, and even more crucial for heavy energy consumers like iron and steel making plants. Coal constitutes major energy input to integrated steel plants, apart from other sources of energy such as electricity, oil and natural gas. At the same time, integrated steel plants produce large amount of byproduct gases such as Blast Furnace Gas (BFG), Coke Oven Gas (COG), Corex Gas (CXG) and others. These are high calorific value (CV) gases hence very rich source of the energy and can be used as an alternative fuels. Thus, steel mill by-product gases can be a clean, oil-replacing energy as fuel, for electric power generation. Byproduct gases are consumed by iron- and steel- making processes and also to generate power in captive power plants. Therefore, the efficient management of byproduct gases plays a central role in the energy cost reduction for the iron- and steel-making industries. Thus, it provides great potential to make self sustaining steel making plants in terms of energy usage.

In recent past, Mixed Integer Linear Programming (MILP) approach for optimal distribution of byproduct gases in an integrated iron and steel making plant have been presented by several researchers (Akimoto et al. 1991; Bemporad and Morari 1999; Kong et al. 2010). Zhang et al. (2011) have presented a prediction method for byproduct gas holder levels using machine learning technique. Jeong et al. (2011) and Zhao et al. (2011) have presented prediction method for various byproduct gas generations based on moving average time-series modeling and echo state network (ESN) modeling, respectively. Thus, there are number of literature available that addresses various aspects of byproduct gas management in an integrated iron and steel plant. However, when it comes to case study, they consider relatively simple byproduct gas network with 2-3 gas generators and 2-3 consumers. On the other hand, in reality, byproduct gas network in a steel plant can be very complex and modeling of this network is a non-trivial task due to existence of different pressure drops in different segment of the gas network and various operational and safety constraints.

This work is geared towards developing simplified modeling strategy for industrial byproduct gas network and followed by formulating MILP optimization problem using this model. Modeling of gas network starts with understanding P&ID of byproduct gas network and then by simplifying the gas network through partitioning the gas network into sub-networks. The partitioning algorithm uses pressure and calorific value of byproduct gases as its criteria.. Further, the various gas generators and consumers in particular sub-network are rearranged in a systematical manner, thereby increasing the modeling ease without losing its mathematical accuracy with respect to optimal scheduling of byproduct gas problem.

It is observed that the proposed modeling strategy is able to reduce significant number of binary as well as continuous variables at modeling stage itself. This model is then used for formulating MILP scheduling problem for optimal distribution of byproduct gases among various external as well as in-house consumers of these gases. Thus, proposed modeling strategy enables one to formulate a tractable MILP based scheduling problem for very large and complex byproduct gas network and consequently achieve an optimal schedule for distribution of byproduct gases such that it

- Reduces flaring of byproduct gases hence improving on economical as well as environmental performance
- Meets the demand of maximum possible consumers of byproduct gases
- Maximizes the usage of byproduct gases in heat and power generation
- Minimizes the use of external fuel such as oil and natural gas

Hence increase overall economic performance of the plant.

References:

Akimoto, K., Sannomiya, N., Nishikawa, Y., and Tsuda, T. (1991). An optimal gas supply for a power plant using a mixed integer programming model. *Automatica*, 27(3), 513-518.

Bemporad, A. and Morari, M. (1999). Control of systems integrating logic, dynamics, and constraints. *Automatica*, 35, 407-427.

Jeong, C., Chu, Y., Han, C. and Yoon, E.,S. (2011). Gasholder level control based on time-series analysis and process heuristics. *Korean J. Chem. Eng.*, 28(1), 16-21.

Kong, H., Qi, E., Li, G. and Zhang, X. (2010). An MILP model for optimization of byproduct gases in the integrated iron and steel plant. *Applied Energy*, 87, 2156-2163.

Zhao, J., Wang, W., Liu, Y. and Pedrycz, W. (2011). A two-stage online prediction method for a blast furnace gas system and its application. *IEEE Trans. on Control Systems*, 19(3), 507-520.

Zhang, X., Zhao, J., Wang, W., Cong, L. and Feng, W. (2011). An optimal method for prediction and adjustment on byproduct gas holder in steel industry. Expert Systems with Applications, 38, 4588-4599.

380262 An Iterative Three-Stage Process for Generating Imidazolium Catalysts from Formaldehyde

Abstract:

Iterative processes that use feedback loops are commonplace in biology. These processes provide a mechanism for organisms to grow and adapt that is important under transient environmental conditions. On the early Earth such processes would have needed to arise spontaneously from available resources; here a model chemical process is shown to produce imidazolium ions that feedback into the process as catalysts for a reaction step. The process can be sub-divided into three stages: two batch reactions (R1 & R2) and one chromatographic separation (S1). Stage R1 converts formaldehyde to a mixture of sugars ($C_nH_{2n}O_n$) via an imidazolium catalyzed polarity reversal. Stage R2 takes the product mixture from R1 and adds aniline to produce new imidazolium catalyst in 0.5% overall yield. Side products generated in R2 decrease reaction efficiency and inhibit catalyst activity. Stage S1 purifies the imidazolium catalysts, allowing them to be recycled back into the process. Over multiple process iterations, the population of imidazolium ions gradually changes with time; however, activity to catalyze sugar formation in stage R1 is retained. This model process based on the formation and sequestration of stable, small molecule organo-catalysts may provide an alternative to template replication in establishing systems chemistry approaches to the origins of life.

Design of T^2 Statistical Control Chart Based on an Economic Criterion

M. A. Bin Shams[1], K. Mahdi[2], A. Elkamel[1], K. Moorthy[1], J. Rafinejad[1], and A. Saxena[1]
[1]Chemical Engineering Department, University of Waterloo, ONT, Canada
[2]Chemical Engineering Department, Kuwait University, Kuwait

Abstract—The objective of this paper is to design the T^2 statistical control chart based on an economical criterion to monitor product quality of the Tennessee Eastman process. An optimization algorithm is proposed to minimize monitoring costs. A sensitivity analysis is conducted on the cost parameters of the economical algorithm to determine their effect on the cost function. The implementation of a multivariate economical criterion on the design of T^2 control chart provides an improved basis for the evaluation and repair of out of control states and result in monitoring cost minimization via optimal sampling schedules. These benefits are in addition to those exist due to the utilization of a statistical based monitoring techniques that are currently used.

Keywords—X-bar control charts, multivariate monitoring, T^2 control charts

I. Introduction

THE aim of this paper is to conduct a preliminary study on the implementation of the economic criteria in the design of T^2 control charts as pioneered by D.C. Montgomery [1]. The algorithm for the economic criteria exists however, is yet to be tested for a real application.

Currently, T^2 control charts are used in industry to provide a compact, but informative graphical representation of process data, obtained through sampling of various process measurements within the plant. They are designed to compromise the inevitable trade-off between type I and type II error often encountered in hypothesis testing based control chart. However, the cost of sampling is not quantified in this design and may be scheduled at suboptimal frequency. [1]

These control charts are based on a multivariate model and are constructed using historical data obtained when the process is operated normal operating condition. The model allows for the examination of the prospect and limitations for improving operations as well as for statistical process monitoring, abnormal situation detection, and diagnosis. In addition, it has been able to generate substantial savings and improved safety conditions in industrial applications that have been operational for several years. This is because a multivariate model is able to integrate the joint effect of the various process parameters that are monitored and provide more accurate control limits. [2]

The implementation of a multivariate economical criterion on the design of T^2 control chart will provide an improved basis for the evaluation and repair of out of control states and result in monitoring cost minimization via optimal sampling schedules. These benefits will be in addition to those that exist in the utilization of a statistical model used at present. [3]

For these reasons, the objectives of this work are three fold:

1. Design the T^2 statistical control chart based on the an economical criterion to monitor product quality of Tennessee Eastman process.
2. Optimize the design of the monitoring chart to minimize the associated monitoring costs.
3. Conduct sensitivity analysis on the cost parameters of the economical algorithm to determine their effect of the cost functions' decision variables.

The paper provides a background on the economical based design of control charts and outlines the potential benefits of implementing this approach. This is followed by the methodology employed to meet the objectives, results, and sensitivity analysis. Finally, the conclusions and recommendations for further study are discussed.

II. Background

A. X-Bar Control Charts

An X-bar control chart is a graphical representation of process data over time against established control limits. The dispersion of data, as shown in Figure 1, is used to monitor the performance of the plant. Figure 1 is representative of an X-bar control chart for a single variable. The red data points illustrate deviation and off-spec conditions. The duration of this condition is known as the out-of-control state. [4]

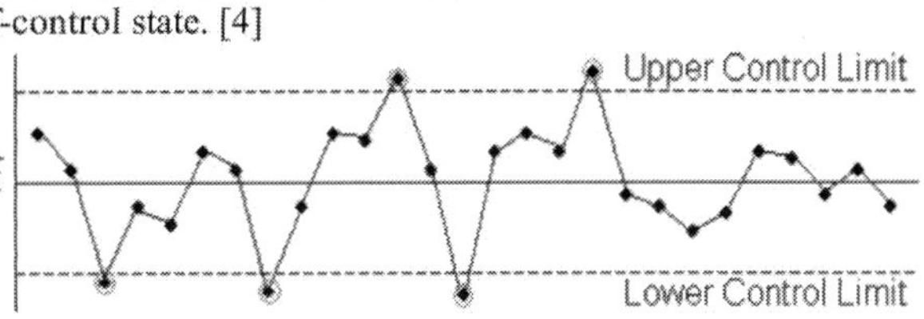

Fig. 1 X-Bar control chart [4]

The X-bar control chart is an example of a univariate model. In other words, it only monitors a single process parameter over time. These process parameters include temperature, pressure, flow, level, etc.

M.A. Bin Shams is with the Department of Chemical Engineering, University of Bahrain, Bahrain (email: mshams@uob.edu.bh)

K. M. Author is with the Chemical Engineering Department, Kuwait University, Kuwait (phone:965.2498.5618;e-mail: khaled.mahdi@gmail.com).

A. E. Author is with The Chemical Engineering department, University of Waterloo, ON, Canada(e-mail: aelkamel@uwaterloo.ca).

L. S. Author is with The Chemical Engineering department, University of Waterloo, ON, Canada(e-mail: lsimon@uwaterloo.ca).

K. M, J. R, and A. S. authors were with The Chemical Engineering department, University of Waterloo, ON, Canada.

B. Multivariate Model

A multivariate model is one that incorporates two or more process parameters. The significance of this model can be illustrated using the graphical representation of Figure 2. Graphs (a) and (b) in Figure 2 are Shewhart control charts based on the univariate model [1]. They illustrate the variance of process parameters x_1 and x_2 over time or sample number. In each case, it can be seen that the process data falls within the control limits as represented by the dotted red lines.

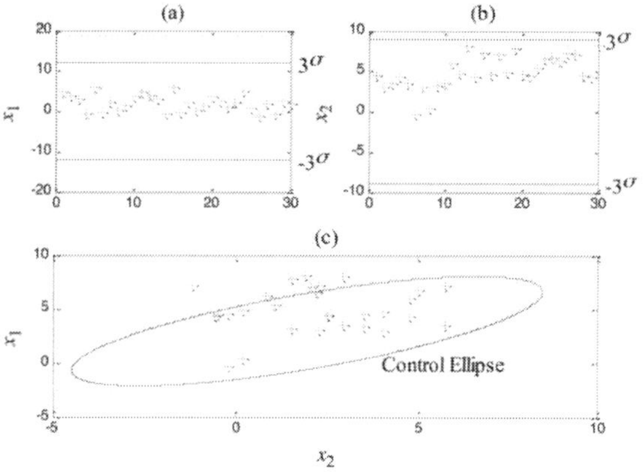

Fig. 2 Significance of multivariate model

However, the calculation of their mean and covariance produced the following results:

$$\mu_1 = \begin{bmatrix} 2 \\ 6 \end{bmatrix}, \quad \sum_1 = \begin{bmatrix} 4 & 0.5 \\ 0.5 & 3 \end{bmatrix}$$

Vector μ_1 states the means of each parameter and the matrix sigma is the covariance matrix of the two process parameters. Matrix sigma 1reflects the fact the covariance structure among the two variables, that is x_1 and x_2 has been changed by a value of 0.5. As a result, plotting the two parameters against each other, as can be seen in Figure 2 (c), shows that there are data points that fall outside of the region of joint control of the two variables. These data points can be considered to be in the out-of-control state and would go undetected in a univariate model. [2]

Therefore, by not utilizing a multivariate design, there is the risk of letting an out-of-control state go undetected. The problem worsens when there are additional variables in a process [2]. Hence, it is important to utilize a multivariate algorithm so as to accurately capture the correlation of various process variables being monitored. This will result in a more accurate determination of the in-control and out-of-control state of the process.

C. T^2 Control Charts

A T^2 control chart is based on a multivariate model and plots the T^2 statistic on the vertical axis against sample number or time, as illustrated in Figure 3. This statistic is based on the covariance of process parameters, process data and the sample size. The chart consists of an upper limit as shown by the blue line and an out-of-control state is determined by the red data points that surpass the upper limit. This upper limit is based on the F-distribution, sample size, and number of parameters. All data points below the blue line are considered to be in an in-control state. [3]

D. Statistical Design

A statistical model aims at ensuring the plant performance is within the established control limits by means of monitoring process parameters such as pressure, temperature, flow, and level. The control limits for this model are based on a standard deviation away from the mean value determined through process design. This standard deviation includes process variables. The T^2 control charts currently used in industry are based on this model. [1]

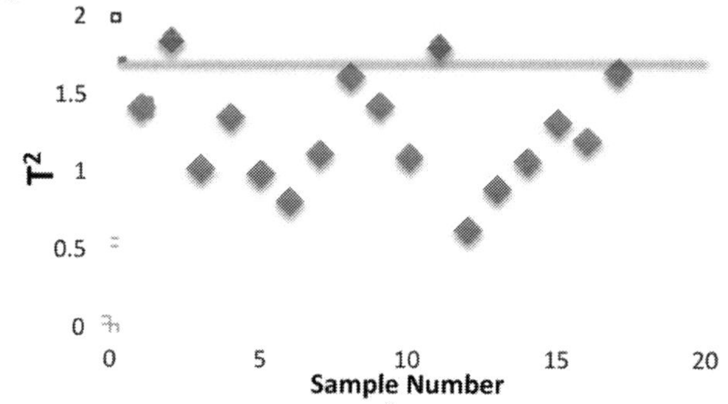

Fig. 3 Sample T^2 control chart

E. Economical Design

Similar to the statistical design, the economical model also aims at ensuring on-spec plant performance. However, it incorporates cost associated variables in addition to process parameters in both the control chart as well the control limits [1]. The algorithm is discussed in more detail in Section 5.

F. Cycle

In this paper, a cycle forms the basis of cost savings. It consists of the in-control state and the out-of-control state. The cycle starts in the in-control state which also includes the last sample before the fault is induced and lasts until the fault starts to take effect. The out of control state includes the first sample after the fault is induced, the detection and diagnosis of the fault, and the re-establishment of the steady state. [5]

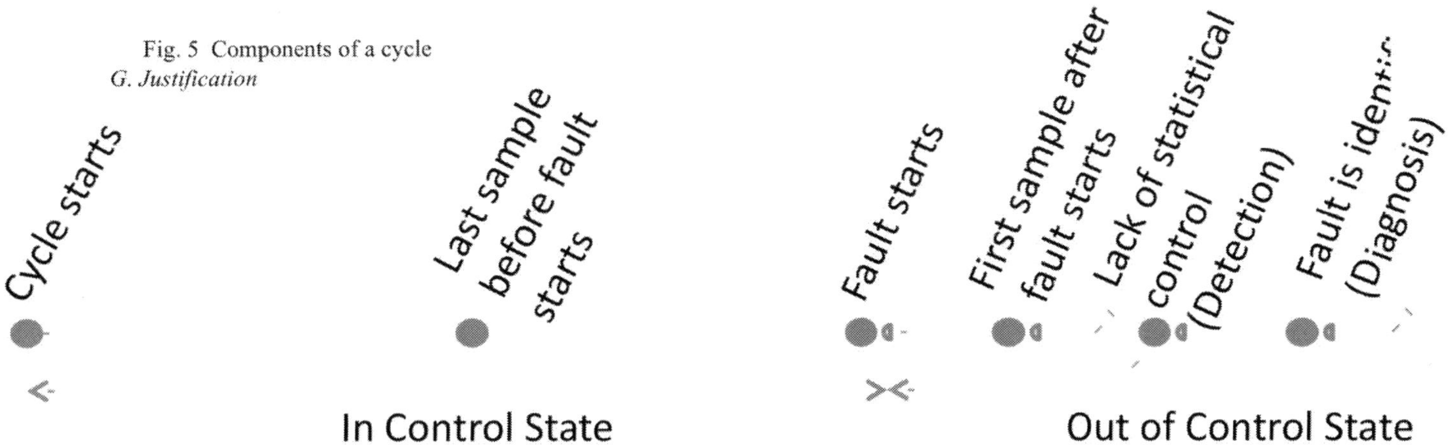

Fig. 5 Components of a cycle

G. Justification

There are two primary justifications for pursuing the design of T^2 control charts based on the economical criterion. The first benefit is the potential minimization of monitoring costs. The economical design algorithm utilizes a cost function that estimates the cost of operation based on costs associated with operations in an out-of-control and in-control states, cost of sampling, and costs of identifying and troubleshooting the assignable cause of out-of-control states. This is further discussed in more detail when we consider implementation on the Tennessee Eastman case study. [1, 2, 4]

Secondly, it may result in the determination of an optimal sampling schedule. This new schedule could provide more opportunities for cost savings and/or better process monitoring. The control chart algorithm requires assigning costs to fixed and variable cost of sampling at different states of the plant operation.

III. Preliminary Test: Application of Economic Criteria on X-Bar Statistical Chart

Prior to the implementation of the economical criterion on the T^2 control chart, initial work was carried out to generate a basic univariate algorithm to design an x-bar control chart. The purpose of developing this algorithm was to create a structural foundation for the more complex multivariate design algorithm required for the T^2 control chart.

The univariate algorithm starts by calculating the mean and deviation for a set of data of a single process parameter when the process is operated under normal condition. The user then specifies the acceptable number of standard deviations that will be used to set the upper and lower control limits for the measured parameter. In addition to these parameters, the number of out-of-control occurrences per hour is also counted. Next, the user is required to input the following process economic information [1, 4]:

1. The time required to take and interpret sample results
2. The time required to find the source of the out-of-control process state after an alarm has been triggered
3. The net income of the process running in an in-control state
4. The net income of the process running in an out-of-control state
5. The fixed and variable samplings costs
6. Cost of finding the assignable cause of process deviance
7. The cost associated with investigating false alarms
8. The hourly production penalty associated with operating in an out-of-control state

The final required inputs are initial guess values for the sample size, sampling time interval, and scaling factor for the control limits. [4]

All the statistical and economic inputs go into an expected cost function that measures the expected hourly losses incurred from the process operation. The cost function implicitly considers statistically criteria including probabilities associated with false alarms (α) and undetected occurrences (β) [1]. The algorithm uses a rigorous minimization function in MATALB® to reduce the expected losses. It accomplishes this by varying three key output parameters: sample size (n), sampling time interval (h) and the control limit scaling factor (k) [4]. All three parameters are optimized by calculating the minimal hourly expected losses and the upper and lower limits of the x-bar control chart as given in [1]:

The formulation of the cost function and the univariate algorithm code is shown in Appendix B. The cost function comprises of four alarm response phases mentioned earlier when an assignable cause is detected and estimates the costs incurred in each phase divided by the expected time of each phase. [4]

The code is tested for a variety of scenarios and produced satisfactory results. For a particular set of inputs including n and varying values of h and k. Table I is generated using an example from a literature source [1]:

TABLE I
Literature values from a univariate economical design model

n	kopt	hopt	Α	1-β	Expected Cost
4	2.84	0.71	0.0045	0.8770	10.51

5	2.99	0.76	0.0028	0.9308	10.38
6	3.13	0.79	0.0017	0.9616	10.39

The optimal cost function value from the example above was 10.38 dollars per hour [1]. Using univariate algorithm developed in MATLAB®, the result shown in TABLE II was produced. The expected cost value is shown to be lower than the literature value indicating that the minimization function in MATALB®was more precise.

TABLE II

Optimal cost obtained using MATALB®

n	k_{opt}	Optimum h_{opt}	A	1-β	Expected Cost ($)
5.3427	3.0310	0.8281	0.0024	0.9443	10.3604

However, the design of control limits and a cost function based on a single measured parameter is not a comprehensive approach to designing monitoring charts [6, 7]. As mentioned earlier, several process parameters tend to co-vary with each other, therefore it is not sufficient to monitor a single variable alone [8]. This is especially true with Chemical Process Industries (CPI), such as the Tennessee Eastman Process, that will be investigated for the design of T^2 control charts [8]. This is further discussed in Section 6.

A multivariate approach is better suited for such process applications [8]. The next step is to develop an algorithm that is able to take multiple data sets to determine multiple means and standard deviations, and also provide a more accurate estimate of the actual number of out-of-control occurrences per hour based on these data sets.

It should be emphasized that x-bar control charts are designed for monitoring purposes [1]. Therefore, the calculation of sampling frequency and number of samples from the design algorithm are set to provide data for the monitoring system.

IV. METHODS

To explain how the economical T^2 control chart was designed, first the basic statistical designed T^2 control chart needs to be explained. The construction of the T^2 control chart consists of two main equations as seen below:

$$T^2 = (X - \mu_0)' S^{-1} (\overline{X} - \mu_0) \tag{1}$$

$$T^2_{\alpha,p,n-p} = \frac{p(n+1)(n-1)}{(n^2 - p)} F_{\alpha,p,n-p} \tag{2}$$

The first equation provides the values of each point on the chart. It consists of X, a vector of the values of the variables that are being monitored, μ, a vector of the means of the variables, and S, the covariance matrix of the variables based on steady state values of the process.

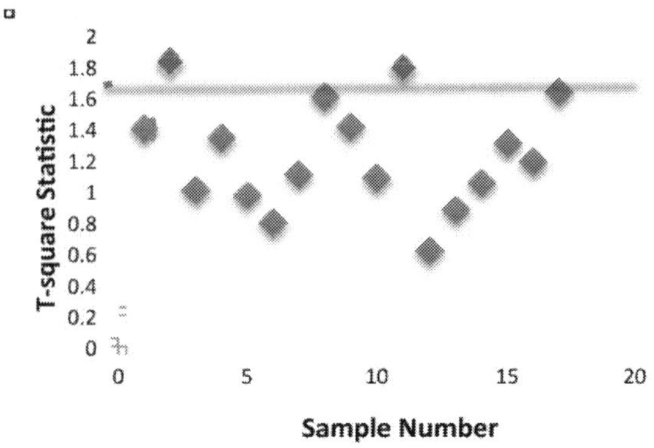

Fig. 4 Sample T^2 control chart

The second equation provides the value of the upper control T^2 Hotelling limit. This equation is based on n, the sample size, p, the number of parameters being monitored, and F, the F-distribution statistic.

Currently the values for the variables that determine the upper control limit are set based on a statistical approach and on past experience in the plant [1].

The economical design however, will also incorporate cost optimization in the calculation of this upper control limit in an attempt to find an optimal value based on economics and statistics.

The approach that is taken is to introduce an objective function, proposed by Montgomery [3].. The objective function consists of four cost variables: fixed cost of sampling (a_1), variable cost of sampling (a_2), cost of investigating and correcting (a_3), and the cost of producing off-spec material (a_4) as seen in the equation below [3].

[illegible] (3)

It also depends on n and k, which is the amount of product produced between successive samples, also related to the sampling interval. In addition, the objective function consists of ρ, β, Φ, and γ, which are probabilities that make the equation a probabilistic cost function.

$$\rho_i = \int_{T^2_{i\ \alpha,p,n-p}}^{\infty} f(T_i^2) dT_i^2 \tag{4}$$

$$\beta_0 = \rho_1 P_0 / (P_1 + \rho_1 P_0) \tag{5}$$

$$\beta_1 = P_1 / (P_1 + \rho_1 P_0) \tag{6}$$

$$\phi_i = 1 - \int_{l_1}^{u_1}\int_{l_2}^{u_2} \cdots \int_{l_p}^{u_p} \left((2\pi)^{p/2} |S|^{1/2}\right)^{-1} \exp\left(1 - \frac{1}{2}(x - \mu_i)' S^{-1} (x - \mu_i)\right) dx_1 dx_2 \cdots dx_p \tag{7}$$

$$\gamma_0 = \beta_0 P_0 + \Delta \beta_0 P_1 \tag{8}$$

$$\gamma_1 = \beta_1 + (1 - \Delta) \beta_0 P_1 \tag{9}$$

The probabilities are based on different states in the process. ρ_i is the conditional probability that the test indicates that the process is out-of-control when it is actually in-control, ρ_0, or out-of-control, ρ_1. β_i is the probability that the process is in-control, β_0, or out-of-control, β_1, when a test is being performed. Φ_i is the conditional probability that off-spec product is being produced when the process is in control, Φ_0, or out-of-control Φ_1. Finally, γi is the probability that the process is in control, γ_0, or out-of-control, γ_1, at any point in time.

The objective function, along with the equations for the probabilities, was coded in MATLAB®. The value of the objective function was minimized by changing three of the variables. These variables were n, k, and the upper control T^2 limit as stated in equations 3 and 4. The initial values of the variables is determined based on plant experience and literature recommendations [3].

The main objective of this paper is to determine whether the economical design has an actual applicability in real case scenarios. Therefore, once this model is attained, it is decided to apply it to a real life case, namely, the Tennessee Eastman Plant.

V. The Tennessee Eastman Plant

The Tennessee East Plant is an industrial chemical plant that has been extensively researched in the process control community due to the many processing units and process control instrumentations that it contains [8]. It has been suggested to be utilized when testing new process control models. Therefore, it was decided to apply the T^2 control chart designed based on economical criteria to this plant.

There are four exothermic reactions that occur in the Tennessee Eastman Plant. The reactions are [8]:

$A_{(g)}+C_{(g)}+D_{(g)} \rightarrow G_{(l)}$ Product 1

$A_{(g)}+C_{(g)}+E_{(g)} \rightarrow H_{(l)}$ Product 2

$A_{(g)}+E_{(g)} \rightarrow F_{(l)}$ Byproduct 1

$3D_{(g)} \rightarrow 2F_{(l)}$ Byproduct 2

As can be seen the reactants are A, C, D, and E and the products are G and H. Due to proprietary reasons some of the processes and units within this plant have been modified. However, it still meets the objective which is real data that can be applied to the model.

Figure 9 is a schematic of the Tennessee Eastman Plant [8]. The reactants in the gas phase enter from the left hand side as Feed A, Feed D, and Feed E. These are mixed together in the reactor where the products are formed in the gas phase [8]. The outlet stream is then passed to the condenser where the products are condensed into the liquid phase [8]. These are then sent to the separator that separates the gas and liquid phases. The reactants in the gas phase go through a compressor and are recycled back into the reactor [8]. The liquid phase enters into the stripper where reactants that have been condensed into the liquid phase are stripped away. At the end of the process, products G and H are obtained.

A FORTRAN code was obtained for this system which simulated data for the plant. When this plant was first introduced, it was an open loop unstable system because of the exothermic reactions occurring in the reactor [8]. However, the FORTRAN code had made this a closed loop process [1].

Data is generated based on steady state values. The next step was to introduce a fault into the process so that the cost savings of switching to economic design criteria over the statistical design criteria used in the development of the upper control limit could be determined.

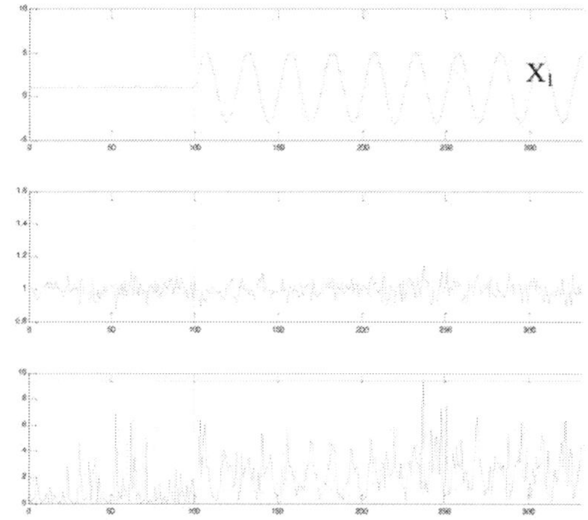

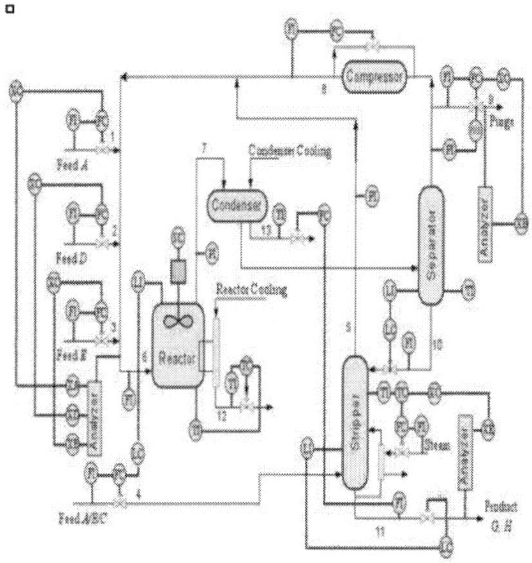

Fig. 5 Tennessee Eastman plant process and instrumentation diagram [8].

VI. Selection of Fault for Cost Savings Analysis

The Tennessee Eastman Plant simulation FORTRAN code had 15 available faults that could be used [8]. The user could control how many faults were initiated at once and when they would occur. Since our basic objective is to perform a preliminary test to investigate potential cost savings of the economic model, it was decided that only one fault would be experimented with. There are two main categories of faults available: step-like faults, and cyclic faults such as control valve stiction faults (see Appendix C for available faults). The step input faults when introduced caused the mean of the process data operating at baseline conditions to shift up or down in a sudden discrete manner for the remainder of the operation cycle. On the other hand, valve stiction faults have a cyclic input effect on process data causing it to fluctuate around the mean. In order to determine which fault would be used, a preliminary test is performed.

The goal of the test is to determine which fault type would be most observable in terms of detection in a T^2 monitoring chart. In order to perform the test, two arbitrary functions were designed. In the first test illustrated in Figure 7, a sinusoid fault input was simulated.

The first two charts represent individual process monitoring charts, and the third is the resulting T^2 monitoring chart developed from the two data sets. The number of samples is shown on x-axis and the process variable measurement on the y-axis. Both charts initially start with a baseline operating value of 1 with a random numbers added to simulating plant noise. The magnitude of this noise is controlled using a specified standard.

After 100 samples, the sinusoid fault was introduced into dataset X_1 alone, while X_2 continued baseline operation.

Fig. 6 Sinusoid fault input simulation

With these two datasets the T^2 monitoring chart is generated by calculating the upper threshold limit and the individual T^2 data points. The results indicate a definite change in the behavior of the T^2 statistic after the fault is introduced indicated by the vertical red dashed line. The statistic appears to be following the sinusoid input. However, there are no points that exceed the upper limit indicated by the horizontal dash line. In this case, the fault would not be observed on the monitoring chart and would continue to affect the process operating in an out-of-control state. The test suggests that a cyclic input would be difficult to track on a T^2 monitoring chart due to the nature of the statistic which is based around tracking shifts in the mean. If a cyclic fault is operating alone without the influence of a step fault, then the mean value of the affected process dataset would not change significantly.

The next test is carried out with a step input function. Results are shown in Figure 8. The step input test followed the same format as the cyclic test, except that after 100 samples a step input fault was introduced increasing the mean by 5 units. The resulting T^2 chart clearly illustrates a shift in the mean of the T^2 statistic above the upper threshold limit. This simulates successful detection of the fault as expected since there was mean shift.

Based on this study it is determined that a step input fault would be selected. Another test are performed with TEP process to determine which step input fault would have the greatest impact on the product stream compositions; primarily looking at the two desired products G and H [1]. The results indicate that fault IDV 6, (Loss of Feed A in Stream 1), is the fault with the greatest effect and is selected. The location of the fault is shown in Figure 9.

An important consideration in the design of the expected cost function is the determination of the initial values that should be used for the decision variables n, k, and T^2; the latter of which is affected primarily by n and α (the probability of type 1 error.). The selected values anchored which local minimum the optimization would approach and hence greatly influenced the cost analysis. Basic sensitivity of the effect of these decision variables was performed and the results can be seen in the sensitivity analysis results section.

The values initially selected were initially based on recommendations from a case study performed by Montgomery for a plant producing discrete units of products [3]. However, since the Tennessee Eastman Plant produces G and H continuously, the values are to be modified. Based on work experiences in various oil and gas companies a value of 200 was selected for n, the number of samples taken per operating cycle. With knowledge of the stream flow rates, k (product produced between successive samples) was determined to be 5.6759 kgmol [9]. The Hotellings' upper threshold limit is calculated knowing that two process parameters, %G and %H, were used with n=200 and α=0.01. This limit is was set at 9.5229 initially.

With the inputs stated above, the minimization of the objective function was performed using the fminsearch function in MATLAB The risk associated with this function is that it does not have a positive constraint on calculated outputs; however this problem is not experienced for the inputs used.

VI. RESULTS

A. Graphical Results

The graphical results are illustrated in Figure 10. The vertical brown dashed line represents the stage at which fault IDV 6 is introduced in the code. This occurs after 160 samples. The first chart of the figure above plots the composition of G versus the sample number for the cycle. There is a noticeable upward shift in the mean after the fault is introduced. The second figure illustrates the composition of product H over the cycle. Unlike the first figure, there is a downward step shift in the composition of H after the fault is introduced.

The third graph illustrates the T^2 monitoring chart that is produced from T^2statistics generated using the two product composition data sets. After the fault is introduced, there appears to be a brief lag phase of about 50 samples after which the effect of the fault is more noticeable. The magnitude of the T^2 statistic increases with larger fluctuations that cause the statistic to surpass both the upper threshold limits represented by the green dashed line for the statistical design, and the red dashed line for the economic design. The economic design upper limit is higher in value than the statistic design limit providing more conservative alarming threshold. Based on the sensitivity analysis on cost parameters that will be described shortly, there is a clear inverse correlation between the number of samples taken per cycle and the magnitude of the T^2 upper limit. Hence, based on the minimization of the expected cost function, it can be seen that although the economically optimized T^2 upper limit is larger, it is less sensitive to the detection of the out-of-control state than the statistically designed upper limit. However, the cost savings from having less sampling outweigh the costs of operating in an out-of-control state slightly more often. These results are shown numerically in Table III.

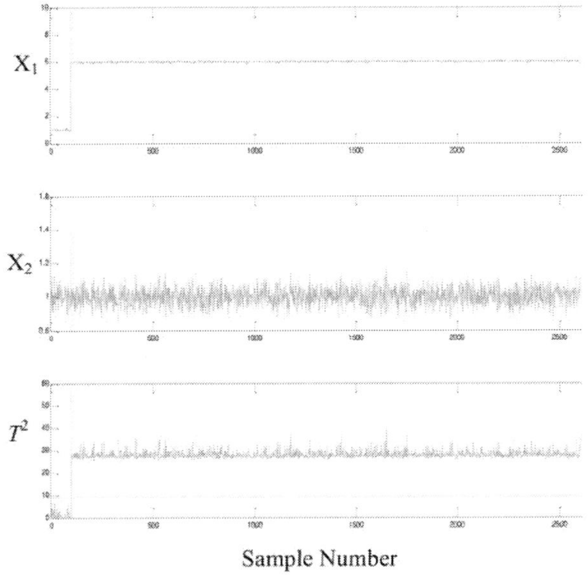

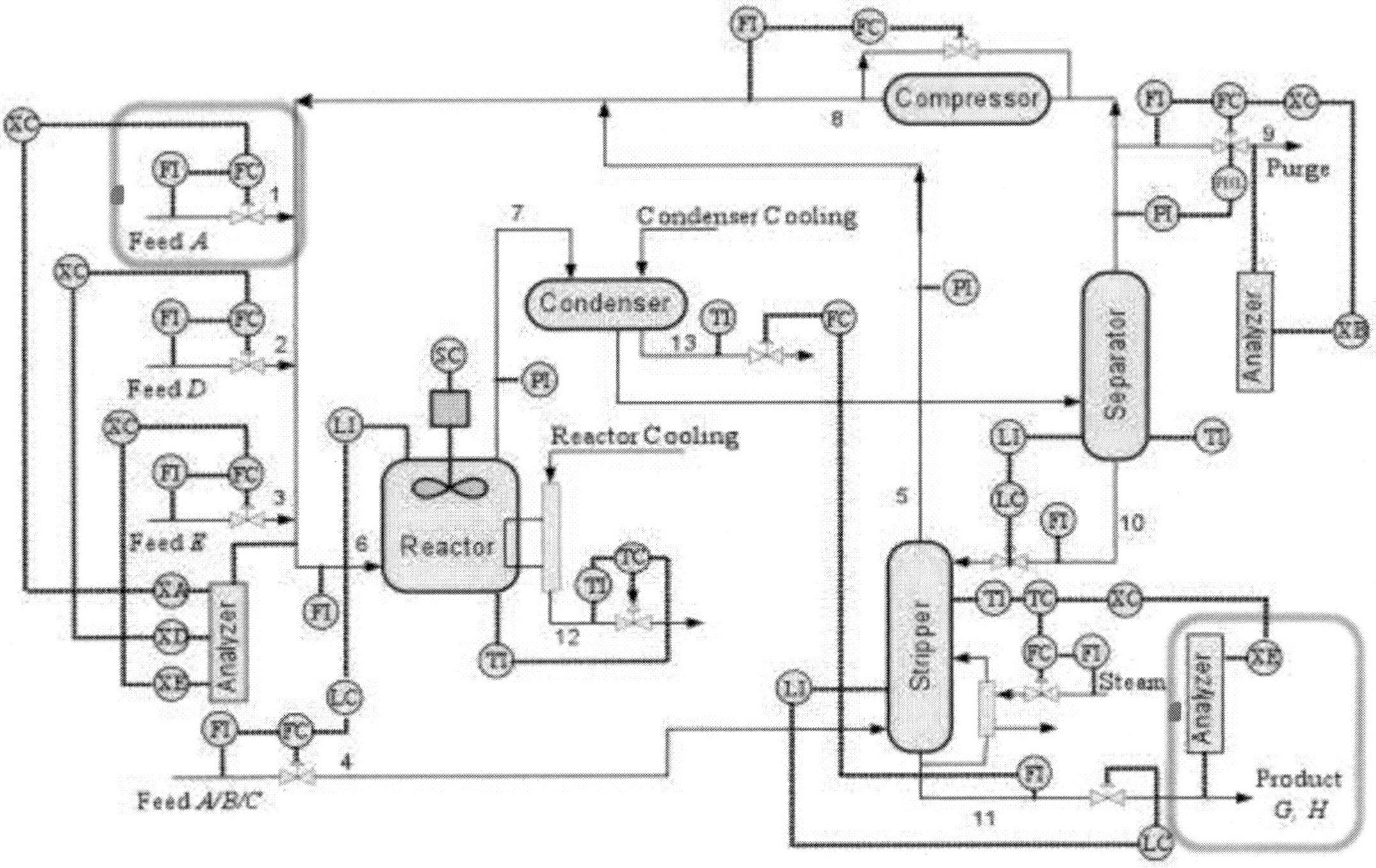

Fig. 8 Fault 6 - Loss of feed A

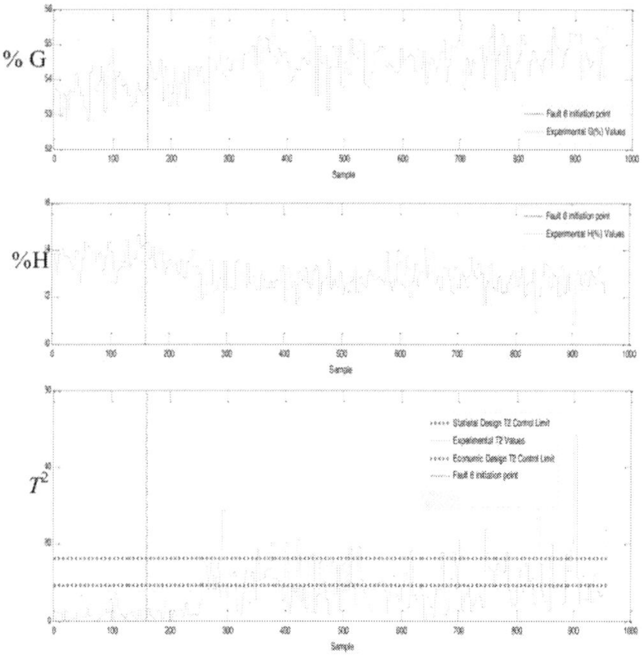

Fig. 9 T^2 control chart based on economical criteria - Tennessee Eastman process (Fault 6)

TABLE III
Fault 6 test results based on economic criterion

Criterion	J ($/kgmol of G and H)	n (# of samples per cycle)	k (product produced per sampling interval) (kgmol)	T^2 (CL)	Sampling interval (min)
Statistical	0.1415	200	5.6759	9.5229	3
Economic	0.0042	8.7559	92.149	16.318	48

The minimization of the cost function dropped the price per kgmole of G and H from $0.1415 to $0.0042 resulting in $290.11 dollars of saving. The number of samples taken per cycle drastically decreases with the economic design from 200 to approximately 9. Consequently, the amount of product produced between sampling intervals and the sampling interval increases. Overall this results in less sampling costs and more savings. The numerical value of the upper T^2 limit increased by almost 70% raising it above the statistical design limit seen in Figure 10.

B. Considerations Associated with Cost Savings

The decrease in price per kgmol of G and H represents a small fraction of a percent of the value of G and H at only 0.46%. However, it should be noted that this is preliminary figure that does not account for other variables that could be affected in the plant. There are 51 potential variables in the Tennessee Eastman plant simulation that can be monitored [8]. In addition, realistic scenarios often incorporate multiple faults occurring simultaneously, however the occurrence of a single fault is only considered here. Finally, the effect the fault can be have more severe consequences than simple loss of product, including damage to equipment that could result in even greater cost savings if addressed properly in the objective function. Nevertheless, the purpose of the study is to determine the potential for cost savings, and it should be noted that any savings gained from the optimization may be fully realized by the plant since the infrastructure required for monitoring is often in place in existing facilities.

C. Sensitivity Analysis of Cost Function Parameters

Initial values for the economic design cost parameters are show in TABLE IV.

TABLE IV
Objective function cost parameter values

Cost Parameter	Value ($)
a_1	1
a_2	0.1
a_3	50
a_4	26.6

The cost a_4 was based on the compositions and costs of various components of the product stream. For sensitivity analysis on the cost parameters, the values were varied individually while keeping the others at their initial values listed above.

Figures 11-14 illustrate the effect of the fixed sampling costs (a_1), the variable sampling costs (a_2), the investigation and troubleshooting costs (a_3), and the penalty cost associated with operation in an out-of-control state (a_4).

The results of the analysis shown below indicate that the sampling costs vary linearly and positively with the cost savings per cycle. The variable sampling costs correlation has a steeper slope indicating a more pronounced effect. As the sampling costs increase, the amount of sampling decreases, hence the savings per cycle increases overall. Figure 13 illustrates the effect of the investigation and troubleshooting costs on the savings. The analysis indicates that this cost had no significant observable effect on the savings per cycle:

Finally, looking at the penalty cost per cycle (Figure 14), there is a clear negative exponential correlation with savings. As the penalty for producing off-spec product increases, the amount of sampling also increases and hence the savings decreases. This is because the process can no longer afford to operate in the out-of-control state for significant amounts of time relative to when this penalty cost is lower. As the penalty cost increases and the savings decrease, the magnitude of the T^2 statistic also decreases to a value similar to the T^2 limit of the statistical design. Overall, the analysis suggests that the troubleshooting costs effect is negated by the greater influence of the sampling costs and penalty operation costs.

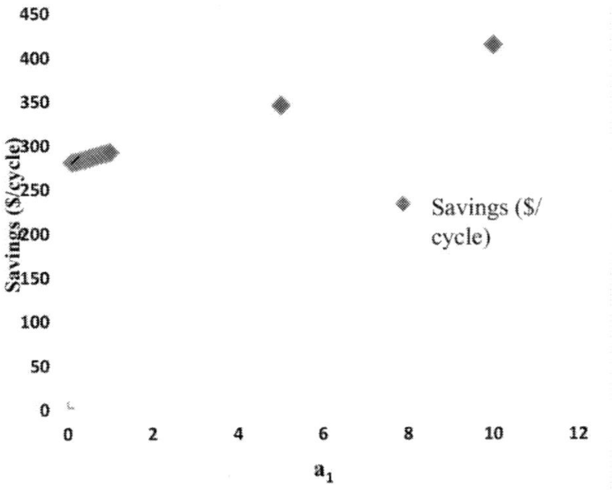

Fig. 10 Effect of a_1 (fixed cost per sample) on cost savings per cycle

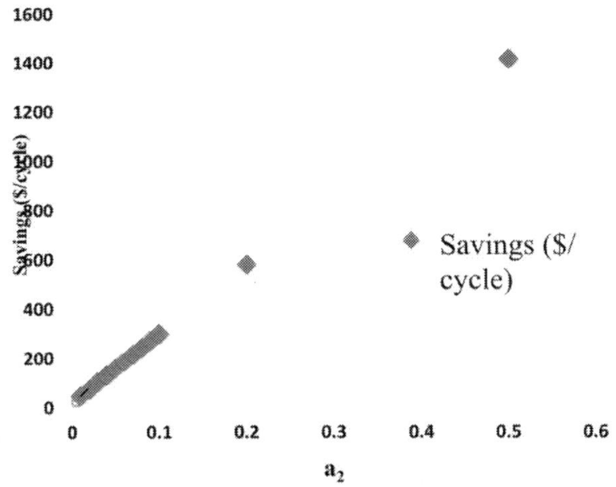

Fig. 11 Effect of a_2 (variable cost per sample) on cost savings per cycle

D. Sensitivity Analysis on Decision Variable n

Preliminary sensitivity analysis was carried out on one of the decision variables, *n*; the number of samples taken per cycle.The purpose of this analysis was firstly to determine whether the minimization of the expected cost function of the economic design would have different local minima depending on the value of the decision variable *n*, and secondly to determine the effect of *n* selected for the statistical design input on the cost savings.

Initially *n* was set at 200, and then adjusted to the following values: 10, 100, 150, 300 and 600. The results indicate that adjusting *n* had no effect on the point at which the optimization converged to, which including the same economical design values shown earlier in Table 3. The results suggest that either a global minimum is achieved for the cost function or that the local minimum spans for a large range of *n* values. The results of the test on the savings per cycle can be illustrated below in Figure 15.

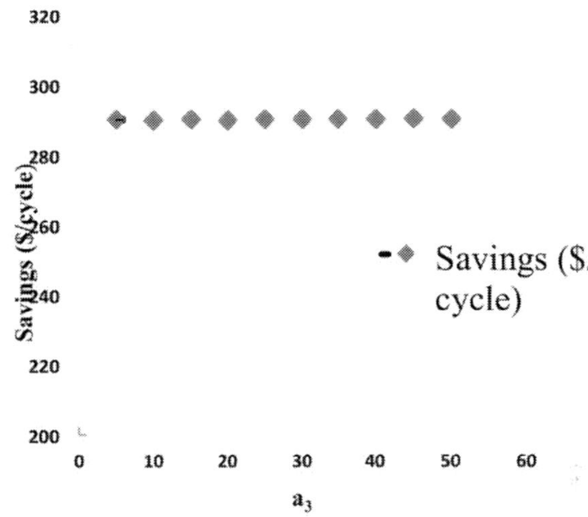

Fig. 12 Effect of a_3 (cost of investigating the alarm and troubleshooting the fault) on cost savings per cycle

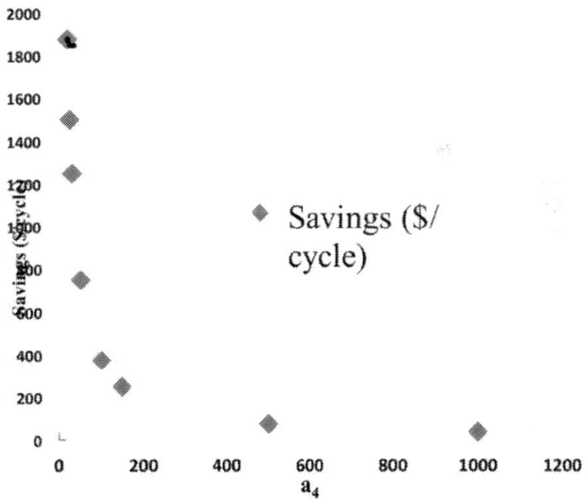

Fig. 13 Effect of a_4 (penalty cost of producing off-spec) on cost savings per cycle

Since the point of convergence is the same, it is logical that as the number of samples used in the statistical design input increases, the savings per cycle that would be earned by switching to an economic design would increase linearly in a positive correlation.

VII. CONCLUSIONS

During this investigation, a T^2 control chart is economically designed and optimized for a closed loop Tennessee Eastman Plant simulation. The economic design is compared to the statistical design to determine potential cost savings associated with the alternative basis for the monitoring design. Changes in parameters are noted to include a reduction in sampling frequency. The economic design showed preliminary opportunities to earn about $290 of savings per cycle. Sensitivity analysis on cost parameters are performed and it is shown that sampling costs varied positively with cost savings, while the troubleshooting and investigation costs have an insignificant effect on savings, and the penalty costs associated with out-of-control operation had a negative effect on savings. Sensitivity analysis on the decision variable n (No. of samples/cycle) is performed and it is observed that the convergence point of the expected cost function does not vary with the range of n values used. Consequently, as the initial value of n used in the statistic design increased, so does the opportunities for saving by switching to the economic design. Further testing of the economical design of T^2 control charts should be performed by introducing alternative faults and looking at the effect of operating the simulation with multiple faults overlapping. In addition, the effect of additional process variables in different units should also be explored to make the study a comprehensive multivariate analysis. It is also recommended to continue researching realistic estimates of statistical model inputs to improve accuracy of cost saving analysis. The values selected in this study were based on initial estimates from a discrete plant production scenario with modifications from personal work experience. Getting more accurate values from the Tennessee Eastman Plant or competitors would help improve the value of the cost analysis.

REFERENCES

[1] Montgomery, D. C,*Introduction to Statistical Quality Control*, 3rd Edition, USA: John Wiley & Sons, Inc., 1996, pp. 349-386.

[2] Kourti, T, "Process Analysis and Abnormal Situation Detection: From Theory to Practice," IEEE Control Systems Magazine, vol. 22, no. 5, pp.10-25, October, 2002.

[3] Montgomery, D. C and Klatt, P.J, "Economic Design of T^2 Control Charts to Maintain Current Control of a Process". Management Science Journal Vol. 19, No. 1 (1972): 76-89.

[4] Montgomery, D. C, "The Economic Design of Control Charts: A Review and Literature Survey". Journal of Quality Technology 12.2 (1980): 75-87.

[5] Lorenzen, T. J, and Vance, L.C, "The Economic Design of Control Charts: A Unified Approach," Technometrics, Vol. 28, No. 1, pp. 3-10, Feb, 1986.

[6] Saniga, E.M, "Economic Statistical Control-Chart Designs with an Application to $\bar{X}$ and R Charts," Technometrics, Vol. 31, No. 3, pp. 313-320, Aug, 1989.

[7] Harris, R.J, *Economic Design of Control Charts for Multivariate, Multistate Processes*, Georgia Institute of Technology: USA, 1977.

[8] Lyman, P.R. and Georgakis, C. "Plant-Wide Control of the Tennessee Eastman Problem," Computers and Chemical Engineering, Vol. 19, No. 3, pp. 321-331, 1995.

[9] Downs, J.J. and Vogel, E.F. "Plant-Wide Industrial Process Control Problem," *Computers and Chemical Engineering*, Vol. 17, No. 3, pp. 245-255, 1993.

[10] Keats, J.B. and Montgomery, D.C., *Statistical Application in Process Control*, Edition 1, Marcel Dekker Inc.: New York, USA, 1996.

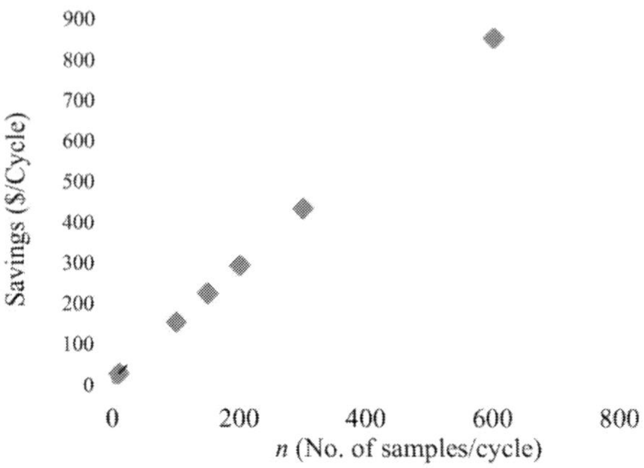

Fig. 14 Effect of decision variable n (# of samples / cycle) on cost savings per cycle

370506 An Efficient Control Formulation for Cyclic Process with Application to an Industrial System

Abstract:

Many industrial processes are cyclic, i.e., they exhibit periodic repetition of certain dynamics. In general, start or stop of individual cycles are subjected to some discrete events such as time event, control event, state events. These events are characterized by opening/closing of solenoid valves, turning on/off the heater, cycle time, etc. Examples of such cyclic processes are found in number of industrial applications such tyre manufacturing [1], upstream oil and gas [2], material processing,etc. In tyre manufacturing, the mixture consisting of latex, cross-linker, filler and other chemicals is set in a mold, which is then heated by a steam jacket to vulcanization temperature and then cooled to achieve desired properties. The discrete events include introduction of the mold in the processing chamber and turning the heating on/off [1]. Similarly, periodic shut-in in upstream oil and gas [3] can also be classified as cyclic processes where system dynamics repeats based on the binary decision of shutting or opening of the production valve. In this contribution, an efficient control methodology will be developed, which is suitable to be implemented in low power controllers such as RTUs/PLCs.

First principles modeling of such process is computationally complex and at times it becomes intractable when used for model based control. On the other hand, available data based modeling, in its current form, is unable to generate control relevant model for such processes. Mainly, because in cyclic process, objective is to control the key performance indices (KPIs) instead of measured variables. These KPIs vary with cycle-to-cycle rather than in time-series fashion. Thus, use of time-series input output data is not required to build control relevant models for such processes. For example, the aim of rubber manufacturing is to ensure tyre with uniform properties, whereas that of periodic well operation is to ensure maximum oil removal from the well. In either case, the key performance indices (KPIs) are not measured and the measured variables e.g. temperature of steam jacket in tyre manufacturing process cannot be used directly to build control relevant model.

In this work, we will develop an efficient data based modeling approach that refrains from using time-series input output data. In particular, time-series input-output data will be mapped into cycle-wise KPIs and all discrete events will be converted into continuous function evaluation in a systematic manner. These batch-wise data will then be used to develop a KPIs-control relevant model and its corresponding control law. Efficacy of the proposed modeling and control scheme will be demonstrated on an industrial relevant application where at least one of the inputs is binary, and at least one of the KPIs need to be calculated from measured data.

References:

[1] A.P. Methal, "Modelling and simulation of degree of crosslinking in tyre manufacturing", MS Thesis, Indian Institute of Technology - Madras (2010)

[2] N.S. Kaisare, A. Gupta, V. Kariwala, N.N. Nandola, J.W. Green, G. Seikel and P.C. Somdecerff, "Control and Optimization Challenges in Liquid-Loaded Shale Gas Wells", in Proceedings of *Dynamics and Control of Process Systems – DYCOPS* (2013), Mumbai, India.

[3] B.R. Knudsen, "Production optimization in shale gas reservoirs", MS Thesis, Norwegian University of Science and Technology (2010).

365925 Design of Industrial Machine Learning System

Abstract:

A large number of real-world machine learning and data mining problems have several crucial issues such as class imbalance and multiple objectives to achieve, which cannot be coped with many learning methods easily. This study proposes an efficient approach to address those issues and applies the approach in design of machine learning system for a real-world problem – inline inspection of surface defects on glass substrates of thin-film transistor liquid crystal display (TFT-LCD). Two major steps of decision support in machine learning system are: (1) use of threshold-moving to handle data imbalance and achieve individual quality objectives, and (2) utilization of Ensemble techniques to handle multiple quality objectives in a classification problem while maintaining low computational complexity. When applied to the industrial case study, the achieved performance shows that utilization of the proposed approach in defect inspection of TFT-LCD glass substrates would be a viable alternative to manual inspections.

386684 Estimation of Distributed Parameter Systems Using Recursively Updated Empirical Basis Functions

Abstract:

While many aspects of state estimation and monitoring have extensively been studied for lumped processes described by ordinary differential equations (ODEs), little work can be found for distributed parameters systems described by partial differential equations (PDEs). The infinite dimensional representation of such processes exemplified by microelectronic fabrications can be approximated using a finite number of ODEs. Such ODE system is constructed by discretization of the PDEs using Galerkin's method [1]. The set of basis functions needed to discretize the system is initially computed and then recursively updated using adaptive proper orthogonal decomposition (APOD) [2], [3]. Thus the estimation problem of distributed parameter system with infinite points inside the domain is reduced to computing the states of reduced order model described by a set of ODE.

Initially, a static observer based on the continuous point measurements is used to estimate the states of ROM. Static observers have two major issues; numerous measurement sensors are required otherwise the static observer gives erroneous estimates, also the static observer existence depends on the location and shape of the measurement sensors. To overcome these issues, we employed a dynamic observer that conceptually needs only one point measurement to predict the dynamic behavior of the modes. Assuming the periodic availability of the process snapshots and continuous availability of point measurements from a restricted number of sensors an APOD-based Luenberger-like dynamic observer is synthesized.

An unavoidable assumption of previous APOD-based observer that employ data driven model reduction technique is the necessity for complete snapshots (in the sense the profiles must span the whole process domain). As expected, even though APOD relaxes the requirement for representative snapshot ensemble, it requires "spatially complete" snapshots. However, obtaining such information might not be feasible owing to high sensor costs and limited availability of sensors [4]. In recent years there has been considerable effort to relax this assumption in the field of model reduction; substituting it with the assumption that periodically all the regions of the process domain are sampled. The basic premise of the research work is that a set of mobile sensors achieve better estimation performance than a set of immobile sensors. To enhance the performance of the state estimator, a network of sensors that are capable of moving within the spatial domain is utilized [5]. Specifically, such an estimation process is achieved by using a set of spatially distributed mobile sensors. The objective is to provide mobile sensor control policies that aim to improve the state estimate. The metric for such an estimate improvement is taken to be the expected state estimation error. The effectiveness of the proposed estimation methods is successfully illustrated on monitoring of spatiotemporal temperature dynamics in a catalytic reactor.

[1] P. D. CHRISTOFIDES, Nonlinear and robust control of PDE systems, Birkhauser, New York, 2000.

[2] D. BABAEI POURKARGAR and A. ARMAOU, Design of APOD-based switching dynamic observers and output feedback control for a class of nonlinear distributed parameter systems, J. Proc. Cont., to appear (2014).

[3] D. BABAEI POURKARGAR and A. ARMAOU, Modification to adaptive model reduction for regulation of distributed parameter systems with fast transients, AIChE J., 59(12) (2013), pp. 4595–4611.

[4] A. A. ALONSO, I. G. KEVREKIDIS, J. R. BANGA, and C. E. FROUZAKIS, Optimal sensor location and reduced order observer design for distributed process systems. Comp. & Chem. Eng., 28:27–45, 2004.

[5] M. A. DEMETRIOU and I. I. HUSSEIN, Estimation of spatially distributed processes using mobile spatially distributed sensor network. SIAM J. Control Optim., 48(1):266-291, 2009.

380137 Integrating PVD Modeling, High-Throughput Experimentation and Big Data into the Discovery of Novel High-Temperature Materials

Abstract:

Integrating PVD Modeling, High-throughput Experimentation and Big Data into the Discovery of New High-Temperature Materials

Jason Hattrick-Simpers, University of South Carolina, Columbia, SC

Jonathan Kenneth Bunn, University of South Carolina, Columbia, SC
Christopher J. Metting, University of South Carolina, Columbia, SC

The identification of new materials and their engineering optimization is a time-intensive process involving the passage of up to 20 years from serendipitous discovery to engineered product. Although the development of high-throughput experimental methods in the early 1990's originally offered the possibility for revolutionizing novel material exploration, researchers encountered several new problems. High-throughput characterization techniques needed to be developed. Such techniques generated terabytes of diffraction and spectroscopy data, which need to be analyzed and then compared systematically. This problem is amplified in the field of high-temperature materials where precise compositional control is required and material properties must be evaluated during prolonged oxidation cycles.

Here we use the Ni-Al binary alloy system to demonstrate a comprehensive modeling-experimental-data minimization methodology for the combinatorial exploration of novel high-temperature alloys. Modeling of the physical vapor deposition process is used to minimize the trial-and-error efforts associated with the identification of composition regions that maximize effective usage of sample real-estate, avoiding low temperature liquidus regions and undesirable phases. A suite of diffraction and spectroscopy techniques are then implemented to monitor the initial phase of the base metal and oxide and their phase evolution in time during oxidation treatments at 1373 K. The resulting data sets are mined individually using a modified version of the CombiView platform and compared to determine prevailing trends. Finally, time dependent phase diagrams of metal oxidation are created, permitting compositional trends to be evaluated and new materials to be identified.

Systematic Chemical Process Design: Sustainable Production of Formaldehyde from Natural Gas

Simon I. Lydom, Søren Mortensen, Patrick N. Pedersen, Emmanouil Papadakis*

Computer Aided Process Engineering Center, Department of Chemical and Biochemical Engineering, Technical University of Denmark, DK-2800 Kongens Lyngby, Denmark
**empap@kt.dtu.dk*

Formaldehyde is an intermediate chemical compound used mostly for polymers – especially glues. Formaldehyde is produced from natural gas using steam reforming and methanol as an intermediate product. In this work, the objective was to design a formaldehyde plant to produce 115,000 ton/year by employing a systematic design procedure consisting of twelve hierarchical tasks. Process optimization, heat integration and environmental impact analysis are also considered in obtaining the final design.

The process itself consists of three parts, the natural gas steam reforming to syngas, the methanol synthesis and the formaldehyde synthesis. After the methanol synthesis, the unreacted syngas is recycled back, and the methanol is separated from the remaining compounds by distillation and it is fed to the formaldehyde reactor. After the formaldehyde synthesis, water is added in an absorber to stabilize the formaldehyde and create formalin. Formaldehyde is present in water as methylene glycol. Methanol is separated from the product by distillation and recycled to the reactor.

Once the base case design is obtained, an economic evaluation is performed in order to identify process bottlenecks. To improve the base case design, optimization and heat integration were performed which resulted in more than 50% reduction in process utilities (energy and water).

To further improve the design, sustainability and LCA analysis were performed to identify the hotspots of the process and subsequently to reduce the environmental effects of the plant. It was found that utilizing purge gas as furnace fuel for the reactors or alternatively if CO_2 is available from a nearby plant, an increased amount of hydrogen can be converted in the methanol synthesis step. The final design had lowered the operational cost and the process became more sustainable. The final design of the production of formaldehyde is an economically viable and a more sustainable process.

Design and Optimization of a Pure Refrigerant System with Subcooling in Liquefied Natural Gas Plant

Inkyu Lee[a], Kyungjae Tak[a], Sunkyu Lee[a], Daeho Ko[b], Il Moon[a,*]

[a]Department of Chemical and Biomolecular Engineering, Yonsei University, 50 Yonsei-ro, Seodaemun-gu, Seoul 120-749, Korea

[b]GS E&C, Gran Seoul 33 Jong-ro, Jongno-gu, Seoul 110-130, Korea

Natural gas liquefaction process is energy intensive because of its cryogenic characteristics.[1] Therefore, minimizing energy consumption is a major concern in design and optimization point of view. [2] This research focused on minimizing the energy consumption of the pure refrigerant system by improving the subcooling system. To minimize energy consumption of the pure refrigerant system, several steps were performed as follows. First, the pure refrigerant system in the propane precooled mixed refrigerant (C3MR) process was simulated and analyzed thermodynamically with and emphasis on subcooling. With the coolant temperature limitation, it was found that high pressure is needed for subcooling. The base case design is shown in Figure 1. Second, new design of subcooling system was proposed for the pure refrigerant system. The proposed design introduces a cooling system using lower pressure pure refrigerant after the cooler. Thus, the temperature is lowered for subcooling instead of the pressure increase which affects to the compression energy increase directly. The energy reduction of the pure refrigerant system is about 16.48% compared with the base case. Figure 2 shows the proposed design. Third, mathematical optimization of the proposed design was performed. The objective function is minimizing total energy consumption of compressors because most of energy in the natural gas liquefaction process is consumed by compression. As the optimization result, an additional 1.26% of energy was reduced and total 17.74% of energy was saved compare to the base case. The results shows the proposed design significantly reduced energy consumption by decreasing the highest pressure of the pure refrigerant.

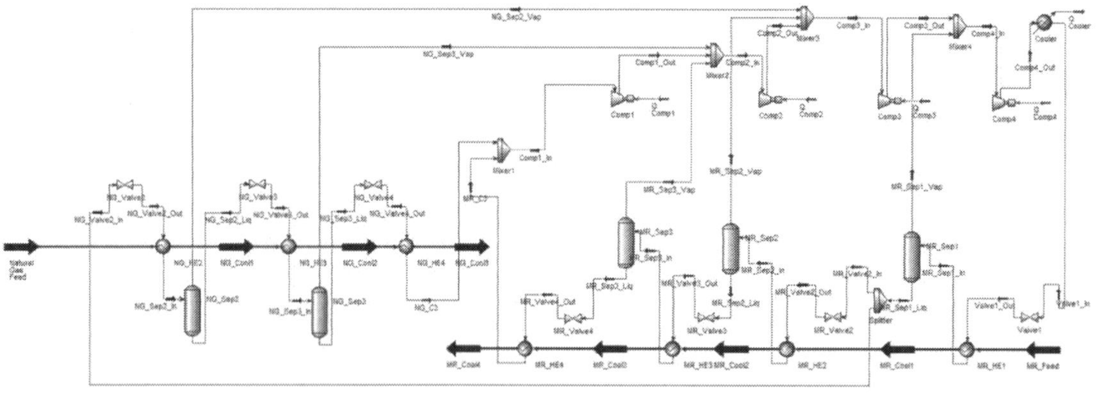

Figure 1. Base case design of a pure refrigerant system

* Corresponding author. Department of Chemical and Biomolecular Engineering, Yonsei University, 50 Yonsei-ro, Seodaemun-gu, Seoul 120-749, Korea

E-Mail address: ilmoon@yonsei.ac.kr

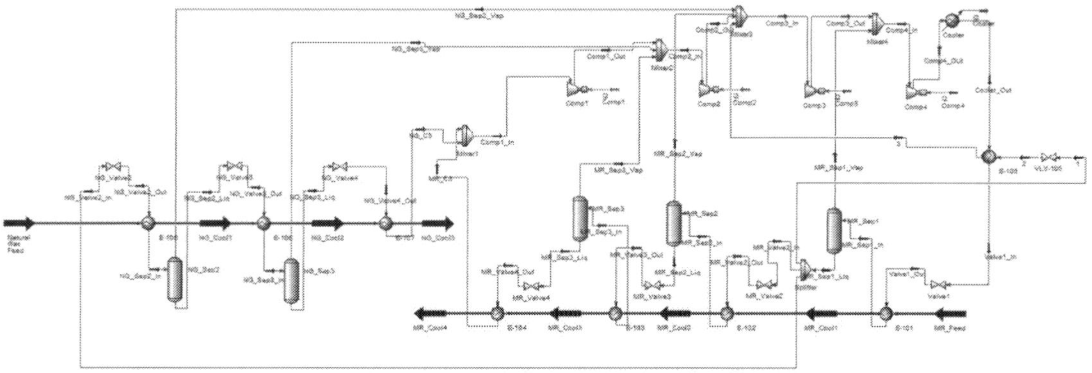

Figure 2. Proposed design of a pure refrigerant system

REFERENCES

[1] Hatcher, P.; Khalilpour, R.; Abbas, A. Optimisation of LNG mixed-refrigerant processes considering operation. Comput. Chem. Eng. 2013, 41, 123-133.

[2] Lim, W.; Choi, K.; Moon, I. Current Status and Perspectives of Liquefied Natural Gas (LNG) Plant Design. Ind. Eng. Chem. Res. 2013, 52, 3065-3088.

* Corresponding author. Department of Chemical and Biomolecular Engineering, Yonsei University, 50 Yonsei-ro, Seodaemun-gu, Seoul 120-749, Korea

E-Mail address: ilmoon@yonsei.ac.kr

Innovative Production of Ethylene Glycol from Ethylene

Yingchao Yue, Emilia Hannele Stenman, Sigyn Björk Sigurdardóttir, Amata Anantpinijwatna

Computer Aided Process Engineering Center, Department of Chemical and Biochemical Engineering, Technical University of Denmark, DK-2800 Lyngby, Denmark
amatana@kt.dtu.dk

Ethylene glycol is a good antifreeze and commercially important raw material for the manufacture of polyester fibers. Global ethylene glycol market is expected to grow, especially in Asia Pacific, which means an increasing demand in the future. Ethylene and oxygen are used as raw material to produce ethylene oxide via direct oxidation process. Then the ethylene glycol is produced by direct hydration of ethylene oxide. After separating from the by-product diethylene glycol and triethylene glycol with distillations, the ethylene glycol (EG) is obtained with 99.9% purity. The annual production is 3.2*10^4 t. The objective of this project is to analyze and optimize the process design, especially with respect to economic potential and sustainability.
The production of ethylene glycol is made up of direct oxidation process and ethylene oxide hydrate process. In the first part, the reactants are introduced into a tubular catalytic reactor and methane is chosen as a diluent to control the temperature. The main product, ethylene oxide, is separated by an ethylene oxide absorber. The recycle is needed to increase the yield of ethylene oxide and a carbon dioxide absorber where quicklime is used to absorb carbon dioxide is also necessary to prevent the accumulation of carbon dioxide. In the second part, ethylene oxide and distilled water are mixed and passed over a non-catalyst tubular reactor. Ethylene oxide is fully converted to ethylene glycol, diethylene glycol and triethylene glycol. Distillation columns are used to separate the product from by-products and water. A recycle is also needed to reduce the consumption of distilled water.
After base case design, economic evaluation was performed which identified the utilities as the variables which could be optimized. In order to improve the base case design, heat integration was performed thereby reducing the utility cost. After optimization with heat integration, sustainability analysis and LCA was done to identify the hotspots. Then an optimized sustainable process were realized which can save 22% cost of utilities and reduce 12% carbon dioxide production, compared with the conventional process.

381632 Systematic Approach for Sustainable Process Design: A Case Study of the Thermal Chlorination of Methane

Abstract:

Systematic Approach for Sustainable Process Design: A Case Study of the Thermal Chlorination of Methane

Kasper Torbensen & Lasse Hansen
empap@kt.dtu.dk

The chlorination of methane is a well-known and studied chemical process, which has supplied chloromethanes for a range of uses since its first industrialization in the first part of the 20th century, and continues to be relevant. The process itself has undergone mechanical and structural changes, as technology and the physical and chemical sciences have progressed over the years. Today, chloromethanes are mainly used as solvents and as precursors for Teflon. This project evaluates this process, with regards to its overall structure, as well as its associated expenses, mechanical and material. In this work, which was part of the MSc course 28350: "Process Design: Principles and Methods" at DTU, a thermal methane chlorination plant was systematically synthesized using the hierarchical decomposition method to produce 90,000 tons chloroform per year.

The overall chlorination process itself is divided into four main parts; reaction, hydrochloric acid (HCl) removal by adsorption, product recovery, and product refining, the latter two by distillation. During the reaction part of the process the four different chloromethanes are produced; methyl chloride (CH_3Cl), methylene chloride (CH_2Cl_2), chloroform ($CHCl_3$), and tetrachloride carbon (CCl_4). The latter three are considered products, while methyl chloride is recycled. The reactor is followed by an HCl-absorption unit, along with a number of preparation steps, before the chloromethanes are recovered in the distillation columns. Recovery of excess, unreacted methane and methyl chloride is recovered by means of distillation, along with trace amounts of chlorine. To avoid accumulation of chlorine in the reactor, a gas-membrane system is implemented as a purging unit.

As proven by a number of studies, the previous use of chloromethanes as refrigerants have been connected to the depletion of the ozone layer, which is why heavy precautions are to be implemented in the plant to prevent these gases to enter the atmosphere, as is common practice when dealing with highly toxic substances. Therefore, high product separations efficiency is necessary and the purge stream cannot contain any chloromethanes, which also is why a gas-membrane is applied in the recycle and purge stream. After obtaining the base case process design, this was optimized, especially with regards to heat integration, which reduces the annual expenses for utilities needed to control the temperature flow through the process. Additionally, to further improve the base case design, sustainability and LCA analysis was performed, to identify the process hotspots.

This environmental aspect, as well as a thorough energy integration of the process, adds a new perspective to this conventional, industrial process. To further elaborate on this, performing hazard analysis of the plant according to the Seveso II directive could provide intel on key aspects of the process, based on the chemical inventory of the plant.

Systematic Synthesis and Optimization of Industrial Styrene Production

Casper Stryhn Svith, Casper Wolf Lundquist, Marina Fedorova

Computer Aided Process Engineering Center, Department of Chemical and Biochemical Engineering, Technical University of Denmark, DK-2800 Lyngby, Denmark
caspersvithi@hotmail.com, casper@lundquist.dk, mfad@kt.dtu.dk

Styrene is an important monomer building block in the synthesis of polymers, including its monopolymer polystyrene and many different co-polymers such as ABS and synthetic rubber. This makes styrene an important product, due to the growth in use of synthetic materials for producing an even bigger range of consumer goods. [1]
Styrene can be produced by various methods, but the dehydrogenation of ethylbenzene is the most common pathway. It is normally coupled with an alkylation process producing ethylbenzene from ethylene and benzene. [2] Combination of these two processes at the same location gives the possibility of heat integration between the exothermic alkylation and the endothermic dehydrogenation processes. Therefore in this work, conducted as part of a MSc course in Process Design at DTU, a styrene chemical plant consisting of both alkylation of benzene with ethylene and dehydrogenation of ethyl benzene to styrene is designed. The design procedure follows the hierarchical decomposition for process design, which is used on the two processes separately. Subsequently the alkylation and dehydrogenation process parts are integrated into a plant design aimed at a total production capacity of 200,000 metric tons of styrene per year. Of special interest in the design is the separation, which makes mass integration possible. This aims to minimize the use of raw materials and water. As a result of this 99 % of the water injected into the dehydrogenation reactor is reused saving 3.34 metric tons of water per ton styrene produced.
An economic evaluation is performed to identify the process bottlenecks, which shows that the cost of raw materials is the largest operational expense. Factors such as plant placement and association with companies working in the petrochemical industry are therefore of much influence on the feasibility of the project.
The base case design shows great potential for heat integration and design optimization for distillation columns, with the biggest operational cost beside raw materials being utility cost for heating. Preliminary heat integration and optimization of distillation columns design shows a reduction in operational cost and the negative environmental impact is lowered, increasing the sustainability of the design.
In order to further improve the design, a software based sustainability and LCA analysis is performed identifying process hotspots and providing insight into further optimization. As the result, more sustainable alternatives to the base case design have been generated and analyzed.

References:

[1] Styrene. (2010). Styrene. ICIS Chemical Business, 278(7), 43
[2] C. Nederlof, 2012, Catalytic dehydrogenation of ethylbenzene to styrene, Doctoral Thesis from The Department of Chemical Engineering at Delft University

Reverse-Circulation Cementing of Offshore Hydrocarbon Wells: Insight from Fundamental and Numerical Analysis

Kyle Macfarlan
Michael Nikolaou
Department of Chemical & Biomolecular Engineering
University of Houston

The conventional cementing process for hydrocarbon wells involves the pumping of fluids (cement and drilling mud slurries) down the metal casing and up through the annulus between the casing and the well. In deepwater, offshore wells, high pressure may complicate the primary cementing process in a number of ways, such as exceeding the formation fracture pressure bound (formation fracturing), causing loss of significant amounts of circulating slurry into the rock formation (lost circulation), and ultimately resulting in cementing of unacceptable quality. While careful design of the cementing process (e.g., pressures, flow rates, cement slurry consistency) may address the problems mentioned, these problems cannot be overcome when cementing deep offshore hydrocarbon wells.

In response, reverse-circulation primary cementing (RCPC) has being explored as a means of overcoming these challenges. In RCPC, fluids are pumped downhole through the annulus, and then up through the casing, in contrast to the conventional cementing. This technique has been used in rare cases onshore and has proven effective at lowering equivalent circulating densities (ECDs). However, technical challenges have prevented it from being used offshore until recent advances in downhole tool technology. For such tools to be effective, quantitative analysis must guide the design and anticipated outcomes. This is the objective of this study.

In any cementing process, it is important to quantify the temperature gradients and ECDs the cement will experience, in order to determine which retarding additives need to be added to the slurry so that the slurry does not set early. The cementing job must be modeled in order to determine these parameters in advance. While many commercially available software packages exist to model conventional cementing jobs, they are unable to properly handle the particular configuration of the RCPC process. To address this need, we used the COMSOL Multiphysics finite-element software package for development of a robust solution.

Results of the detailed numerical simulation will be presented and discussed. An analysis of the fundamental underlying reasons for anticipating the advantages of RCPC will be elaborated on, along with related suggestions for simple calculations that can provide rough quantitative estimates similar to those of the numerical simulation.

DTU Chemical Engineering
Department of Chemical and Biochemical Engineering

Process Design Project

Sustainable Design for Production of Styrene from Ethylene and Benzene

Group 10

Authors:
Magnus Zingler Stummann, s103141
Kasper Hartvig Lejre, s103143
Kasper Fjellerup, s103145

10. June 2014
28350 Process Design: Principles and Methods
Technical University of Denmark
Supervisor: Marina Fedorova

Preface

This project was conducted during the MSc course "28350 Process Design: Principles and Methods" at the Technical University of Denmark where it counts as a part of the final grade in the course.

This report deals with the systematic method for designing a styrene production plant, which produces styrene from benzene and ethylene. One of the main focuses during the design of this process, is to design a process that is sustainable and environmentally friendly and at the same time efficient and commercially feasible. The reason for this focus is that in a world where the amount of fossil fuels are declining, and the global warming is increasing due to emission of carbon dioxide, chemical engineers must be able to find solutions, which are both commercially feasible and has a low impact on the environment.

The main methods used, in order to conduct this project, are described by Biegler et al. [1997].
In order to perform the necessary simulations the software PRO/II is used as a simulation tool for mass and energy balances and the software ECON for the economic evaluation. To ensure that the obtained process is sustainable the software ICAS is used to evaluate the potential environmental impact generated by the process.

It would not have been possible to reach our objectives in this project, if it was not for the teaching assistant Marina Fedorova which was our main supervisor during this project. Due to her great wisdom and knowledge she was able to guide us through the creation of this project. We also want to thank the teaching assistant Emmanouil Papadakis, whose experience was like a rock for us in stormy weather. We are therefore very grateful for all the help we have received from these great people. There was always an open door for us when frustrations arose.

Kongens Lyngby, Technical University of Denmark, June 10, 2014

Magnus Zingler Stummann

Kasper Hartvig Lejre

Kasper Fjellerup

Abstract

The usage of styrene monomer has been increasing for several decades, and there is no sign of the consumption slowing down. Therefore increasing production facilities are needed to satisfy the global demand.

There are several pathways to produce styrene. This project focuses on the production of styrene from ethylene and benzene, where the intermediate product, ethylbenzene, is produced during a high pressure alkylation of ethylene and benzene. Styrene is afterwards formed by a dehydrogenation of ethylbenzene.

The process design is made using a systematic method for chemical process design. A base case design is achieved from the hierarchical decomposition approach. According to the design, the process will produce, on an annual basis, 168,000 metric tons of styrene with a usage of 1.10 kg raw material per kg product. This gave a base case design with a Net Present Value (NPV) of 71,732,000 $, a rate of return of 9 % and a payback time of 7.6 years.

By performing an economical evaluation of the plant it is possible to identify the major expenses and perform optimization of those. Optimization of the heat exchangers are archived by the use of the heat exchanger network synthesis method. Due to these optimizations the energy consumption decreased by 43 %. The potential environmental impact due to this process is calculated by the Waste Reduction Algorithm (WAR). This algorithm shows that one of the waste streams should be optimized. Due to this, further optimization of the plant is performed, which gives a higher recycle rate of benzene and ethylbenzene, thereby lowering the consumption of raw materials. Thereby, a production stream of 4,118 tonnes/year of 99.7 % pure toluene (molar basis) is produced. This transforms most of a waste stream into a saleable product.

The final process design from this project has an NPV of 258,931,000 $, which is 261 % higher than the NPV for the base case design, a rate of return on 41 % and a payback time of 2.4 years.

Nomenclature

Symbol	Units	Description
A	m^2	Area
C	\$	Price
C_p	kJ/(kg K)	Specific heat capacity
F	kg/h	Mass flow rate
f	mol/h	Molar flow rate
I	-	Potential Environmental Impact
M	g/mol	Molar weight
m	kg	Mass
n	mol	Molar quantity
P	atm	Pressure
Q	kJ/h	Heat duty
T	K	Temperature
T_b	K	Boiling point
T_m	K	Melting point
U	W/(m^2 K)	Overall heat transfer coefficient
V	m^3	Volume
W	kW	Work
W_b	kW	Brake horsepower

Abbreviations

Name	Description
BZ	Benzene
C	Compressor
CI	Cost Index
CN	Controller
DEB	Di-ethylbenzene
Dev	Deviation
DFH	Direct Fired Heater
E	Heat Exchanger
EB	Ethylbenzene
EP	Economic Potential
EPS	Expanded Polystyrene
Et	Ethylene
F	Flash
FCI	Fixed Capital Investment
H	Direct Fired Heater
HIE	Heat Integration Exchanger
M	Methane
MPF	Material and Pressure correction Factor
NC	Number of Streams

NE	Number of Equations
NM	Number of streams into mixer
NPV	Net Present Value
NR	Number of Parallel Reactions
NRTL	Non-Random Two-Liquid model
NS	Number of outlet Streams
NV	Number of Variables
P	Pump
PEB	Poly-ethylbenzene
PEI	Potential Environmental Impact
PS	Polystyrene
R	Reactor
S	Stream
SC	Stream Calculator
SP	Separator
ST	Styrene
T	Distillation tower
T-c	Condenser at distillation tower
T-r	Reboiler at distillation tower
TEB	Tri-ethylbenzene
Tol	Toluene
UF	Update Factor
WAR	Waste Reduction
WCI	Working Capital Investment

Greek letters

Symbol	Unit	Description
γ	-	Stoichiometric coefficient/ratio between C_p and C_v
Δ	-	Change
ϵ	-	Void fraction
η	-	Conversion
η_c	-	Compression efficiency
η_m	-	Motor efficiency
η_p	-	Pump efficiency
ξ	-	Splitter Fraction/Recovery
ρ	kg/m^3	Density
τ	s	Residence time

Subscripts

Symbol	Description
$catalyst$	Catalyst
con	Condensation
D	Divider

F	Flash
gen	Generated
i	Component number
in	In
j	Stream number
k	Component number
L	Liquid
lm	Logarithmic
out	Out
r	Reaction number
S	Stream Calculator
$syst$	System
vap	Evaporation

Contents

List of Figures

List of Tables

Key numbers: NPV, Rate of Return and Payback Time of the process for the final design.
Overview of economical key numbers for the plant design in the different steps of the process.
Overview of the overall economical effect of the optimizations.
Key numbers for the final design.
Data from the calculation of EP.

Chapter 1: Introduction

The scope of this project is to construct a preliminary design for a styrene monomer (ST) production plant. It is desired that the plant should both be sustainable and able to generate profit for the investors.
In order to achieve this objective a systematic method for chemical process design is applied. This implies that the design problem is broken down into 12 sub-problems or tasks, which then will be solved one by one.
The first 3 tasks are to find information about the product and the process, and from this information, generate a simple process flowsheet. Task 4 is to perform a simple mass balance and make design decisions related to reactions and separations. In task 5 and 6 the process conditions, such as temperature and pressure, are decided, and mass and energy balances are performed in order to define heat duties for the different unit operations. In task 7 the simulation advances from the simple flowsheet to the flowsheet where rigorous models are used in order to simulate separations etc. For the different units heat duties are then calculated on the basis of the energy balance. On the basis of the rigorous models sizing and costing of the used equipment are performed in task 8. In task 9 an economic evaluation of the base case is performed, in order to find equipment and parts of the flowsheet, which needs to be optimized. To optimize the flowsheet further, heat integration is performed in task 10. On the basis of the optimized flowsheet an environmental analysis is performed in task 11, to investigate if the process is harmful to the environment and to identify which parts of the flowsheet that should be optimized. In task 12 additional optimization is performed and/or further optimization problems are defined.
During the execution of the above mentioned tasks, optimization of the plant design is performed, both before the heat integration and after the environmental analysis. This is done in order to perform heat integration on the plant design, which is as close as possible to the final preliminary design.

1.1 Product Information

In this chapter, the chosen product is investigated and described. This is done in order to give an idea of what kind of product it is, and to find relevant physical parameters needed for designing the process.
The chosen product for this project is ST (Figure 1), which is an organic compound with the chemical formula $C_6H_5CH{=}CH_2$.
ST is the second most important monomer in the chemical industry [Tamsilian et al., 2012], because it is a basic building block in the onrushing plastic industry. It is used in large quantities for production of a large range of polymers and co-polymers, including polystyrene (PS) which is used in CD cases, packaging and disposable razors etc. Other major products from ST are expanded polystyrene (EPS), SB Latex, SB Rubber, styrene block co-polymers (e.g. ABS, MBS, SBS) and unsaturated polyester resins [Sanfilippo et al., 2007]. ST owes its popularity

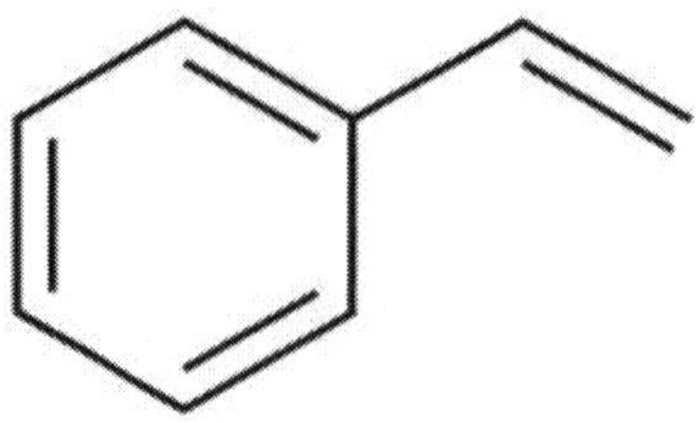

Figure 1: Styrene monomer

to the easily reacting double bond, which can undergo polymerization reactions - both with itself or with other monomers.
The process for production of ST was developed by BASF and DOW Chemical in the 1930s, and they are still the major licensors of production together with Badger, Lummus, Shell and LyondellBasell [Nederlof, 2012].

Naturally it occurs in cinnamon and strawberries, but industrially it is produced by dehydrogenation of ethylbenzene (EB). This process accounts for about 85 % of the total produced ST [Nederlof, 2012]. It means that ST is always produced in elongation of an EB production process. EB is commonly produced by the alkylation of benzene (BZ) with ethylene (Et).
The remaining fraction of 15 % is produced by a peroxidation process together with propene oxide [Nederlof, 2012]. It is chosen to produce ST from the dehydrogenation of EB, since it is the most used production path in the industry, and it only yields ST as major product. Nederlof [2012] reports that both processes have high energy demands, low equilibrium conversions and high capital costs. Other production processes are also being developed, which have lower energy consumption. However, these processes have not been commercialised yet and are still on a research scale.
In Table 1 different physical parameters of ST can be seen.

Table 1: Physical parameters for styrene [Lide, 2009].

Physical parameter	**Styrene**
Molar weight $\left[\frac{kg}{kmol}\right]$	104.15
Density $\left[\frac{kg}{m^3}\right]$	901.6
Melting point [°C]	-30.65
Boiling point [°C]	145

At standard conditions ST is an organic, colorless liquid. Additionally, ST is highly flammable, easily ignited and is under suspicion for being carcinogenic for humans.

It is however not labelled as carcinogenic at the moment. It irritates skin and eyes, and is toxic to inhale [CLH, 2011].

1.2 Market Demand

The demand of ST has increased over the last couple of years, and it is estimated, that it will increase even further in the future, as shown in Figure 2. This increase is due to a significantly increasing demand from especially the Asian market.

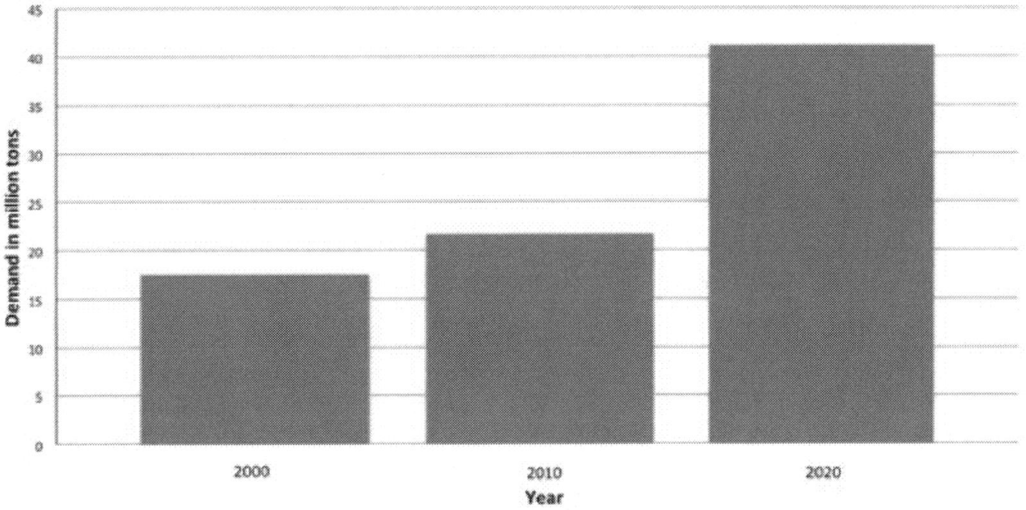

Figure 2: Global styrene demand in million tonnes as a function of year [PR Newswire, 2011].

From Figure 2 it is shown, that the demand in 2010 was approximately 21 million tonnes of ST, and the demand is increasing even further in the future roughly following the global economic growth. It is estimated that the demand will exceed 41 million tonnes of ST in 2020. From this demand analysis, it is chosen, that the process should produce at least 150,000 tonnes of ST yearly, resulting in 0.7 % of the global market demand in 2010. It should be possible to reach this production rate, since a plant in Texas, running with multiple reactors, had a total annual capacity of about 1.1 million tonnes of ST in 1993 [EPA, 1994]. It is further reported by Haslego [2010] that the typical capacity of one reactor lies about 100,000 tonnes per year. Therefore it is concluded that it is possible to reach 150,000 tonnes of ST annually for this project. From the demand analysis it is further concluded that it is rather safe to enter the ST market - even though there are a high amount of competitors, and the price of ST has a tendency to fluctuate due to overcapacity in the market, compared to the demand, because of newly built plants every couple of years. But it is a fact that the demand of ST is booming, which allows new plants to be built.

Chapter 2: Process Description

This chapter concerns the available information in the literature of the production of EB from BZ and Et, and the production of ST from EB. Physical parameters for the different components, which are used in the process, are also discussed in this chapter.

2.1 Reactions

2.1.1 Ethylbenzene Production by Alkylation of Benzene by Ethylene

ST is produced from EB, which therefore must be produced first. EB is produced by alkylation of BZ by Et as shown in reaction scheme (2.1) [Ebrahimi et al., 2011].

$$C_6H_6(BZ) + C_2H_4(Et) \leftrightarrow C_6H_5C_2H_5(EB) \qquad (\Delta H_{298} = -114\text{kJ/mol}) \qquad (2.1)$$

During the formation of EB, the undesired by-products diethylbenzene (DEB) and tri-ethylbenzene (TEB) are formed [Liu et al., 2009, Ebrahimi et al., 2011]. The selectivity for TEB is so low that the formation of TEB can be neglected [Ebrahimi et al., 2011].

$$C_6H_5C_2H_5(EB) + C_2H_4(Et) \leftrightarrow C_6H_4(C_2H_5)_2(DEB) \qquad (2.2)$$
$$C_6H_4(C_2H_5)_2(DEB) + C_2H_4(Et) \leftrightarrow C_6H_3(C_2H_5)_3(TEB) \qquad (2.3)$$

A complete conversion of BZ is possible at 500 °C and atmospheric pressure, but under this condition polyethylbenzene (PEB) is produced and the reaction must therefore take place in the presence of a catalyst and at low temperature [Ebrahimi et al., 2011]. Ebrahimi et al. [2011] found that in a two serial reactor process the EB selectivity was close to 90 %, and a selectivity of DEB on 0.69 %.
An example of a flowsheet for the production of EB is shown in Figure 3 [Ebrahimi et al., 2011]. The first two reactors (R1, R2) on the flowsheet performs alkylation of BZ, while the last reactor (R3) performs transalkylation. In the first separation tower, BZ is separated from EB, whereafter the lighter products from the BZ mixture are removed in separation tower T2, and the BZ is recycled to the reactors. BZ is recycled so that the ratio between the recycled BZ and the stoichiometric required BZ for the alkylation is 6:1, while keeping the amount of raw materials as low as possible [Ebrahimi et al., 2011]. The excess amount of BZ increases the selectivity of EB, and a BZ/Et ratio between 3 and 6 is common [Ebrahimi et al., 2011, Liu et al., 2009].
EB is separated in the third separation tower T3, from PEB and heavier products. In the final tower T4, DEB and TEB are separated from undesired heavier products. In the third reactor EB is produced by transalkylation of DEB in the following reaction [Ebrahimi et al., 2011]:

$$C_6H_4(C_2H_5)_2(DEB) + C_6H_6(BZ) \rightarrow 2\,C_6H_5C_2H_5(EB) \qquad (2.4)$$

The EB produced in reactor R3 is sent to separation tower T1 [Ebrahimi et al., 2011].

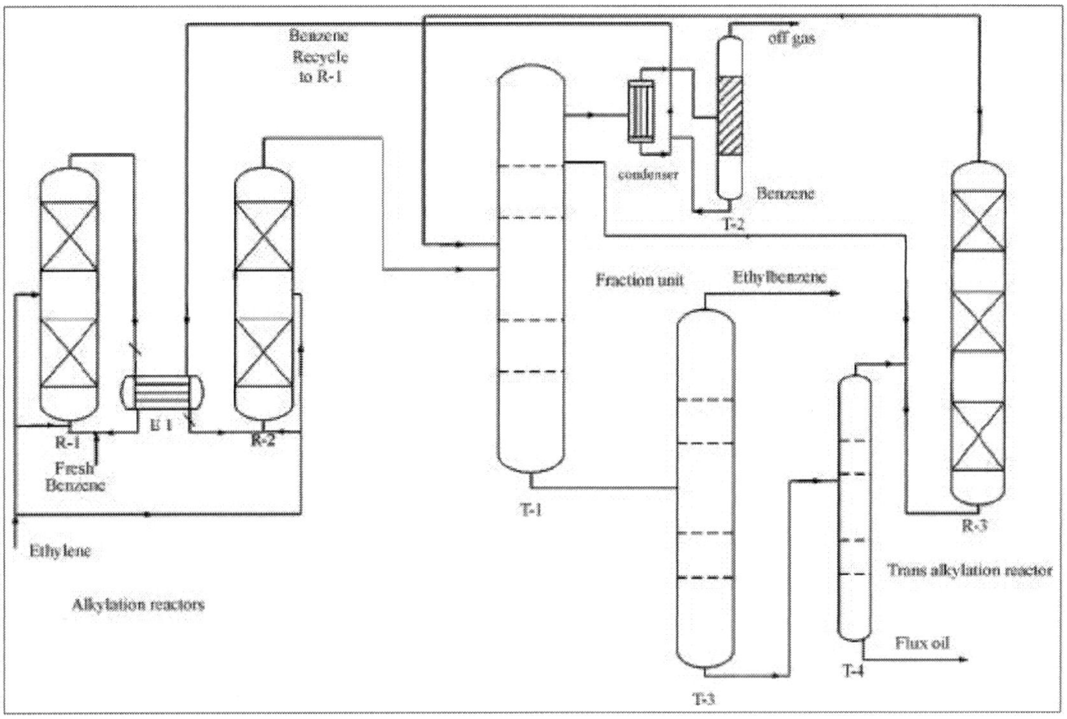

Figure 3: Flowsheet for production of ethylbenzene from the alkylation of benzene with ethylene [Ebrahimi et al., 2011].

Table 2 shows different process conditions for the alkylation of BZ with Et. The table shows that a high BZ/Et ratio is common, and that the EB selectivity generally is close to 90 %. The temperature, pressure and catalyst have a significant influence on the selectivity for DEB. The temperatures shown for the catalysts are the optimum temperatures. When increasing or decreasing the operating temperature in the reactor, the selectivity of EB is decreasing heavily [Ebrahimi et al., 2011].

Table 2: Process conditions for production of ethylbenzene from benzene and ethylene. Where T is the temperature and P is the inlet pressure.

Catalyst	Process conditions			Selectivity		Reference
	T [°C]	P [bar]	$\frac{n(C_6H_6)}{n(C_2H_4)}$	EB	DEB	
Zeolite-Y	260	37.5	6	89.87	0.758	[Ebrahimi et al., 2011]
3998	160	16-21	3	88.1	-	[Liu et al., 2009]
MCM-22	220	34.4	4	89.83	9.470	[Cheng et al., 1998]

When achieving the described temperatures, pressures and ratios of BZ and Et, it is possible to achieve complete conversion of Et. Therefore, it is only the selectivity, which is considered in the production of EB. It is chosen to use the Zeolite-Y catalyst, because it reduces the selectivity of DEB. When the molar ratio between BZ and Et is 6, a 100 % conversion of Et is achieved.

2.1.2 Transalkylation

Some DEB is formed during the process, which has to be transalkylated back into EB using BZ in excess to ensure almost complete conversion of DEB. Here Zeolite-Y can also be used as catalyst. Data for the transalkylation reaction with Zeolite-Y can be seen in Table 3.

Table 3: Data for the transalkylation reactor [Gerzeliev et al., 2010].

Catalyst	P [atm]	T [°C]	Conv. DEB [%]	BZ/DEB mass ratio
Zeolite-Y	24.67	290	99	>8

From Gerzeliev et al. [2010] it is determined that the minimum mass ratio between BZ and DEB is 8 in order to achieve a conversion degree of 99 % of DEB. By approaching these data, it is possible to transalkylate almost all the DEB back into EB.

2.1.3 Styrene Production from Ethylbenzene

For production of ST by dehydrogenation of EB, adiabatic reactors are commonly used. Most of the commercial adiabatic dehydrogenations have two catalytic beds. This normally gives a conversion between 60 % and 75 % and a selectivity between 87 % and 96 % [Tamsilian et al., 2012]. Three stage reactors have also been used to increase the conversion. A high temperature of 900 K and an inlet pressure of 0.67 bar is used to achieve a high conversion of EB [Tamsilian et al., 2012]. Since the reaction is endothermic, the conversion decreases during the reaction due to the decrease in temperature. Heating between reactor beds can therefore be used to maintain a high conversion [Tamsilian et al., 2012].

Figure 4 shows a typical flowsheet for a three stage reactor network to dehydrogenation of EB. The flowsheet shows that O_2 is added to the streams between reactor 2 and 3. It is added to oxidize the hydrogen, which is formed during the dehydrogenation. A fraction unit is used to separate the formed ST from the by-products: Toluene (Tol), BZ and the unreacted EB. The unreacted EB is removed by distillation from Tol and BZ, and is recycled back to the reactors. Tol and BZ are afterwards separated by distillation.

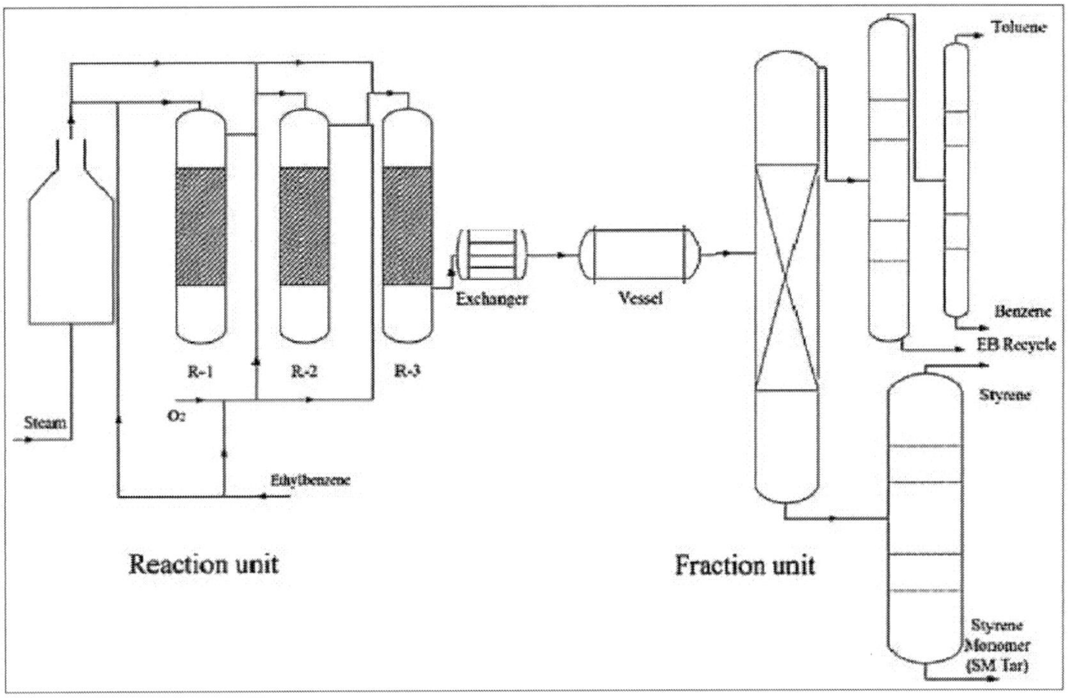

Figure 4: Styrene production process [Tamsilian et al., 2012].

The three main reactions during the dehydrogenation of EB are shown in reactions (2.5) to (2.7), where reaction (2.6) and (2.7) are the formation of the two main by-products BZ and Tol [Tamsilian et al., 2012, Cavani and Trifiro, 1995].

$$C_6H_5C_2H_5(EB) \leftrightarrow C_6H_5C_2H_3(ST) + H_2 \quad (2.5)$$
$$C_6H_5C_2H_5(EB) \rightarrow C_6H_6(BZ) + C_2H_4(Et) \quad (2.6)$$
$$C_6H_5C_2H_5(EB) + H_2 \rightarrow C_6H_5CH_3(Tol) + CH_4 \quad (2.7)$$

There is further four notable reactions [Tamsilian et al., 2012]:

$$H_2O + \tfrac{1}{2}\,C_2H_4(Et) \rightarrow CO + 2\,H_2 \quad (2.8)$$
$$H_2O + CH_4(M) \rightarrow CO + 3\,H_2 \quad (2.9)$$
$$H_2O + CO \rightarrow CO_2 + H_2 \quad (2.10)$$
$$H_2 + \tfrac{1}{2}O_2 \rightarrow H_2O \quad (2.11)$$

The last of these reactions is taken into account in this process.
Tamsilian et al. [2012] and Nederlof [2012] reports the conversion of EB together with the selectivity of the different compounds, which can be seen in Table 4.

Table 4: Inlet concentrations, conversion of ethylbenzene and selectivity for styrene, toluene and benzene. The inlet conditions are 0.67 bar, 625 °C, 150.74 kmol/h of EB in feed.

EB(wt. %) - inlet	ST(wt. %)-inlet	Conv. EB (%)	ST (%)	BZ (%)	Tol (%)
98.28	0.836	69	97.97	1.27	2.95

The table shows that the conversion of EB generally is quiet low (69 %), and recycling of unreacted EB is therefore necessary to ensure a healthy business, and almost complete conversion. The conversion found is an average of different values found in the literature. Tamsilian et al. [2012] observed a high selectivity for ST, and a generally low selectivity for BZ and Tol. The produced BZ could, for the case for production of ST from BZ and Et, be recycled to the reactor where BZ is alkylated again.

2.2 Raw Materials and By-products

In Table 5 physical parameters for the raw materials, the intermediates and the by-products are given.

Table 5: Physical parameters for ethylbenzene, diethylbenzene, benzene, ethylene and toluene [Lide, 2009].

Physical parameter	**Ethyl-benzene**	**para-Diethyl-benzene**	**Benzene**	**Toluene**	**Ethylene**
Molar weight $\left[\frac{kg}{kmol}\right]$	106.165	134.218	78.112	92.139	28.05
Density $\left[\frac{kg}{m^3}\right]$	862.6	862.0	876.5	862.3	1.18
Melting point [°C]	-94.96	-43.3	5.49	-94.95	-169.2
Boiling point [°C]	136.16	184	80.09	110.63	-103.7

It is important to establish the purities of the raw materials and product, since it determines, how the reactions proceeds. The Et and BZ are bought from another company with purities of respectively 99.5 % and 99 % [Sigma-Aldrich, 2013], whereby assuming that the raw materials are 100 % pure is acceptable. It is also found that ST should be sold as at least 99.5 % pure. Tol is a by-product, which is collected and sold, as at least 99.5 % pure [Sigma-Aldrich, 2013]. It could also be a possibility to buy Et from another company, where it is collected as by-product, but with a lower purity (and to a lower price).

2.3 Safety

When dealing with a chemical process, it is very important to make sure that the safety of the workers on the plant is satisfactory. In this process, the workers are working with toxic and carcinogenic compounds, including BZ and ST. It is essential to construct the plant in a way, so the risks for the workers are minimized. It can be achieved by suiting the workers with safety rules, safety equipment, safety training and so on. But it is also important to ensure that the plant is constructed from quality materials. A hazard and operability analysis study of the plant should also be performed before construction. However, this matter is beyond the scope of this project.

2.4 Location of the Plant

It is assumed that the plant is located in Texas, USA, because it has an enormous number of chemical plants located already. Thereby, it is easy to acquire raw materials and sell the product, because they can be transferred through pipes or by train. It means that e.g. the salaries and taxes, used in this project, corresponds to how they are in Texas.

2.5 Economic Potential

The informations about raw materials and products are now mapped, including the reactions and temporary flowsheet diagram. In order to move further in the design process, it is important to establish, whether the process has a reasonable and satisfactory economic potential, so there is a chance that it generates profit. Otherwise, it is not desirable to move further with the design process.
To determine the economic potential, the demand analysis from section 1.2 is used. The demand is 150,000 tonnes of ST per year. The prices of ST, Et and BZ are respectively 1.60 $/kg, 1.00 $/kg and 1.02 $/kg [ICIS, 2014]. By using the molar mass of ST from Table 1, the production of ST in kmol/y is calculated.
Since it is known how much ST the process should produce every year, it is possible to go back in the process and calculate, how much Et and BZ is needed. To perform these calculations, the first step is to define the selectivity and conversion from EB to ST. The next step is then to consider the selectivity and conversion from Et and BZ to EB. By this approach it is possible to calculate the amount of Et and BZ needed to produce 150,000 tonnes of ST.
The selectivity and conversion of EB to ST is taken from Table 4 given in section 2.1.3, which gives respectively 97.97 % and 69 %. Recycling is included, which means that unreacted EB is put back into the reactor. For the calculation of EP, the conversion from EB to ST is therefore assumed to be 95 %.

In section 2.1.1 it was stated that most of the BZ is recirculated and therefore, the EP should only consider the reaction ratio of 1:1 with respect to Et and BZ. The selectivity and conversion from Et and BZ to EB is given as 89.97 % and 100 %. The production of ST in kmol/h is then given as:

$$n_{ST} = \frac{150 \cdot 10^6 \frac{\text{kg}}{\text{y}}}{104.15 \frac{\text{kg}}{\text{kmol}}} = 1.44 \cdot 10^6 \frac{\text{kmol}}{\text{y}} \tag{2.12}$$

From using the selectivity and conversion of EB to ST, it is possible to calculate the consumption of EB. Further, by using the selectivity and conversion of Et and

BZ to EB, it is then possible to calculate the consumption of Et and BZ in kg, whereby the economic potential is calculated:

$$
\begin{aligned}
EP &= F_{ST}C_{ST} - F_{BZ}C_{BZ} - F_{Et}C_{Et} \\
&= 150 \cdot 10^6 \frac{\text{kg}}{\text{y}} \cdot 1.60 \frac{\$}{\text{kg}} - 134.35 \cdot 10^6 \frac{\text{kg}}{\text{y}} \cdot 1.02 \frac{\$}{\text{kg}} - 48.24 \cdot 10^6 \frac{\text{kg}}{\text{y}} \cdot 1.00 \frac{\$}{\text{kg}} \\
&= 54.72 \text{ million } \frac{\$}{\text{y}} \\
&= 0.36 \frac{\$}{\text{kg}}
\end{aligned}
$$

The EP generates profit and it is decided to work further with the production of ST. From the EP it was assumed that EB is recycled, which gives a conversion of 95 % of EB. This has to be achieved, in order to generate profit.
The intermediate calculations can be seen in Appendix A. It has now been established that it is possible to produce ST (from the literature) and it is also economically favourable.

Chapter 3: Process Flowsheet

In this chapter, the process flowsheet is constructed through 4 levels given by the hierarchical decomposition approach.

3.1 Level 1

Level 1 is about deciding whether the process should be run as a batch or a continuous process. This is done on the basis of different considerations.
The first consideration is whether any of the needed apparatus runs in batch mode. As can be seen in section 2.1, the process involves two main reactors and a couple of separation steps that are presumably consisting of distillations. Therefore no apparatus needs to be operated as batch equipment.
The second consideration concerns the production rate. A production of 150,000 tonnes of ST per year is chosen, which is a high rate that points towards continuous operation.
The third consideration is about the product lifetime. ST has been produced for a couple of decades, and nothing suggest that the demand would decrease, as discussed in section 1.2. Therefore the product should have a long lifetime. This means that a plant, running for many years before breaking even, would be possible.
Taking all these considerations into account, a continuous process would be the best choice since this would be able to have a high production rate, while producing ST for many years.

3.2 Level 2

This level in the hierarchical decomposition approach concerns the input-output structure.
As raw materials Et and BZ are used. It is assumed that these does not contain any notable impurities and therefore no initial separation is needed. The raw materials are therefore ready to be used as they are delivered.
The reactions yield ST, by-products and unreacted reactants. The conversion is 100 % with respect to Et, but BZ is present in excess. Therefore this should be recycled. The reaction of EB is not complete, thus excess amounts should be recycled. From the formation of ST, BZ is also produced, which is also recycled. A by-product from the reaction from EB to ST is Tol, which is recovered and sold if possible. Other by-products are sent for waste-treatment. The off gas from the BZ recovery is flared and the heat can possibly be used by the heat exchangers. These considerations can be summed up to give the flowsheet that can be seen in Figure 5.

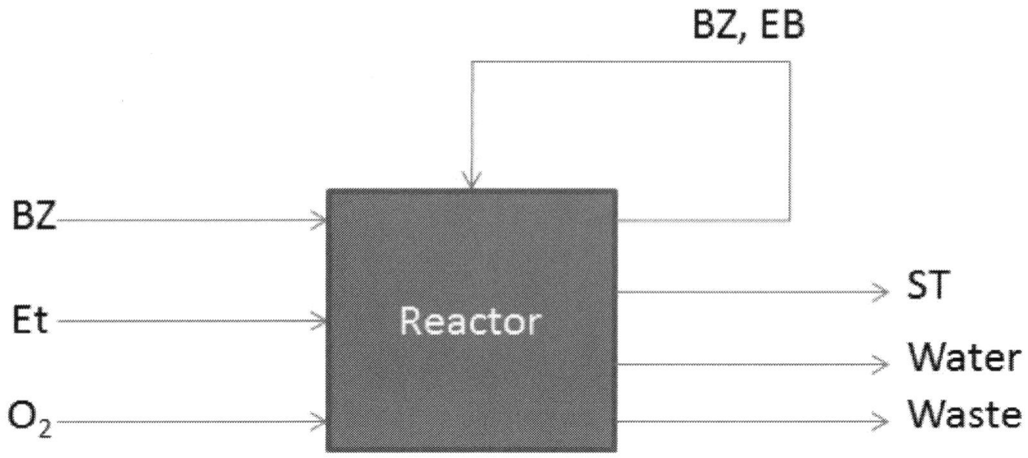

Figure 5: Flowsheet after level 2 of the hierarchical decomposition approach.

3.3 Level 3

This level concerns the recycle structure and reactors. Since two main reactions are involved in the production of ST from Et and BZ, two reactors are needed. In the literature these reactors might consist of multiple units, but for now they will simply be noted as two reactors. Between these two reactors and after the second, separation steps are needed in order to recycle un-reacted reactants and remove by-products.
In the first reactor, BZ is present in excess amount. Therefore the left-over BZ needs to be separated from the EB and recycled to the reactor. The conversion in Reactor 1 is 100 % with respect to Et, thus the exit stream consists only of EB and BZ. As shown in Figure 3 more by-products are present after Reactor 1. These could be treated by separating them, converting them back, and recycling them. For now that is a part of the Separation 1 unit.
After the second reactor a stream containing ST, EB, BZ, water and Tol are present together with some minor by-products. Here BZ should be separated and re-cycled to the first reactor. EB should be separated and recycled to the second reactor. Tol should be separated and sold as a by-product. Then only ST is left which is the final product.
All these considerations leads to the flowsheet that can be seen in Figure 6. In order to pump the liquids around in the system, pumps are needed and likewise with compressors for the gas transportation. Furthermore, the process is kinetically controlled which means that catalysts are required in order to obtain a suitable reaction rate [Tamsilian et al., 2012, Ebrahimi et al., 2011].

When considering whether the reactors are adiabatic or isothermal, Figure 3 shows that when splitting up reactor 1, the reactor where the alkylation of BZ takes place, into two serial connected reactors it can be an advantage to use the intermediate stream to heat the inlet feed, thereby having a more optimal use of the energy. It can therefore be an advantage to run the alkylation reaction in an isothermal reactor. Ebrahimi et al. [2011] further reports that the optimum selectivity of EB is achieved at a specific temperature, meaning that the temperature should be kept constant. This is the case for the first reactor - the alkylation reactor - and the transalkylation reactor.

For the dehydrogenation of EB to ST, both isothermal and adiabatic dehydrogenation is used in the industry [Behr, 1999, Nederlof, 2012, Society, 2006]. For the isothermal case, the reactor temperature is commonly maintained at 580 - 610 °C. For the adiabatic reactor the inlet temperature is instead higher than for the isothermal reactor and is between 610 - 660 °C [Society, 2006].

The most widely used reactor type is the adiabatic, stated by Tamsilian et al. [2012], which also investigates the temperature effect upon the EB conversion. The conversion of EB is increasing linearly for increasing temperature, whereby it is decided to construct the dehydrogenation reactor as an adiabatic reactor, in order to achieve as high conversion as possible.

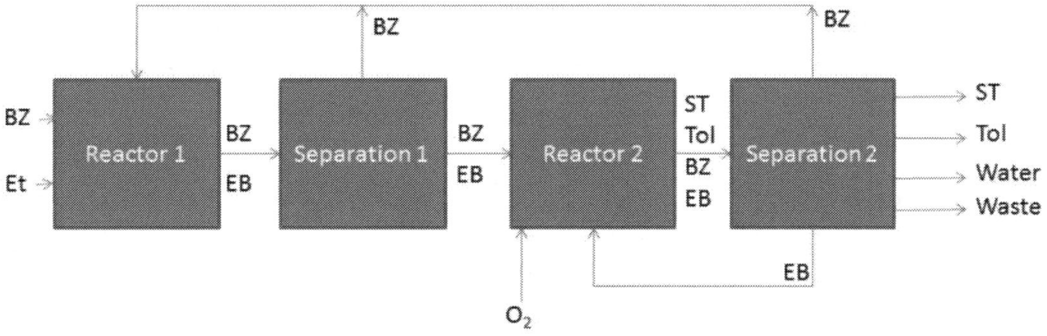

Figure 6: Flowsheet after level 3 of the hierarchical decomposition approach.

3.4 Level 4

Level 4 of the hierarchical decomposition approach concerns the decisions on the separating system.

After Reactor 1 the feed is fed into a separation tower. Here BZ is separated and recycled to Reactor 1, where the heavy part is EB and DEB. These are sent into another distillation tower. Here the heavy part, consisting mostly of DEB, is sent into a transalkylation tower where it is transformed into EB by a reaction with BZ, and sent back into the first separation tower.

The light part of the second distillation tower is EB and this is sent into Reactor 2. To this reactor O_2 is also added in order to remove the H_2 that is formed in the reactor by a reaction forming water.

In the flow exiting reactor 2 ST, Tol, BZ, H_2O and EB are the major compounds. First ST is removed from the rest to give the product stream and a stream of by-products and left overs.
The light compound from Reactor 2, consisting mainly of Tol, BZ and EB is sent into a series of distillation towers. The first tower separates EB as the heavy part and this is recycled into Reactor 2. The light part of the distillation is sent into another distillation tower, where BZ is separated as the light part. The heavy part is sent into another tower separating the two major components H_2O and Tol.
Thereby the two flowsheets, presented in section 2.1, have been validated using the hierarchical decomposition approach.

The final flowsheet can be seen in Figure 7.

It is noteworthy that the assumptions about how the compounds separates are based on Table 6.

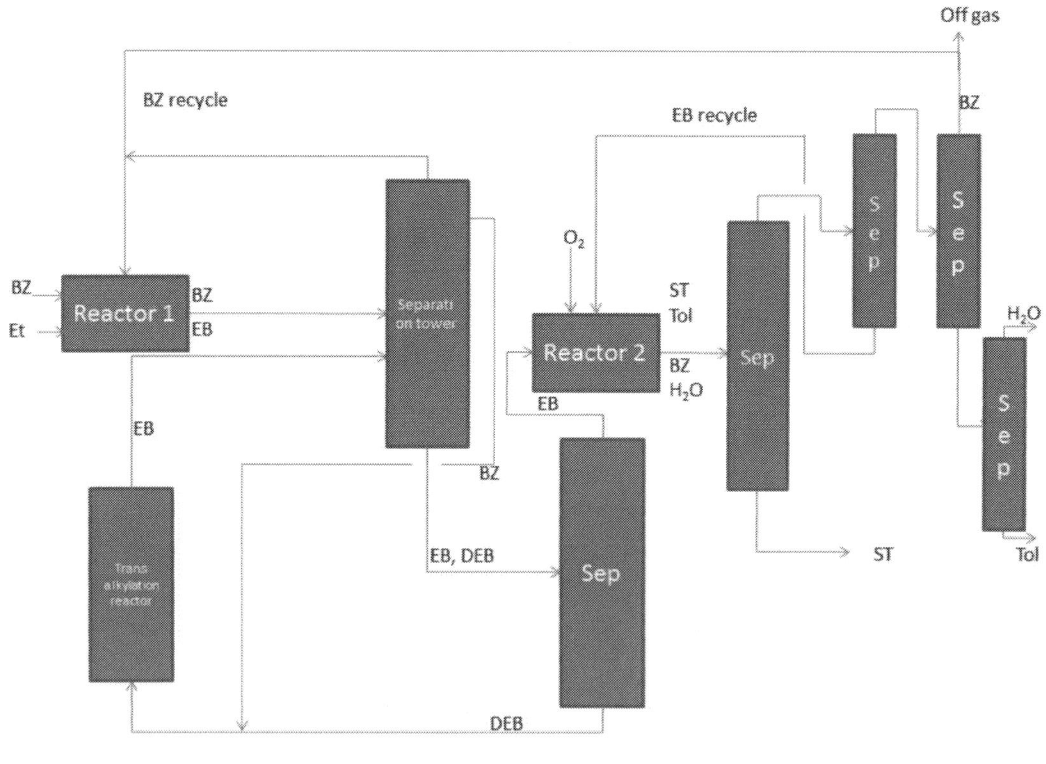

Figure 7: Flowsheet after level 4 of the hierarchical decomposition approach.

Here, the ratios of the boiling points for the different compounds have been calculated, by acting that a compound is first present in the lower phase and then in the upper phase.
As an example of the calculation and investigation of these ratios the separation between BZ and ST it presented. If ST is in the heavy phase the ratio equals 1.81

Table 6: The ratio of boiling point temperature between the lower and upper phase.

		Lower				
		Styrene	Ethyl-benzene	Diethyl-benzene	Benzene	Toluene
Upper	Styrene	-	0.94	1.27	0.55	0.76
	Ethylbenzene	1.06	-	1.35	0.59	0.81
	Diethylbenzene	0.79	0.74	-	0.44	0.60
	Benzene	1.81	1.70	2.30	-	1.38
	Toluene	1.31	1.23	1.66	0.72	-

while ST in the light phase would give a ratio of 0.55. Thereby it can be concluded, that ST is present in the heavy phase and BZ is in the upper (lighter) phase since the ratio value for this situation is significantly larger than 1. Because of the significantly larger value it can be concluded, that the separation between ST and BZ is easy. If the ratio lies around 1 the separation requires a significant effort, which results in a lower degree of separation or a larger use of energy to achieve the desired degree. This would be the case for the separation between ST and EB, since the ratio is equal to 1.06 for ST in the heavy phase and EB in the light phase.

When considering at the separation of the compounds from Reactor 2, it is seen that the ratio of the boiling temperatures of ST and EB is around 1 for ST in the lower phase and EB in the upper phase. Thereby it could be an advantage to make the separation in a way, a small amount of ST is separated into the upper phase, whereby ensuring that as pure ST as possible is present in the lower phase. Furthermore, the small amount of ST in the upper phase is then separated with the EB into the lower phase in the next separator and then recycled back to reactor 2.

When comparing the final flowsheet given by the hierarchical decomposition approach to the flowsheet given in Figure 4 it might be noted that the separation of SM Tar is not present in the final flowsheet. This is due to lacking physical data about the substance and the amount is presumably neglectable. Therefore it had to be assumed that it was not present in this process. In order to make this process flowsheet perfect, experiments will have to be done in order to establish the physical behaviour of SM Tar and the compositions.

Chapter 4: Mass Balance

In this task, the mass balance of the system is produced and investigated. The variables in the process are investigated by a degree of freedom analysis, and the calculation order is found. This information is used in order to implement the process in PRO/II, to produce the mass balance of the process. Before analysing the process by a degree of freedom analysis, the flowsheet from Figure 7 is extended by introducing mixers and a splitter, which can be seen in Figure 8. Here separation step number 7 is a flash.

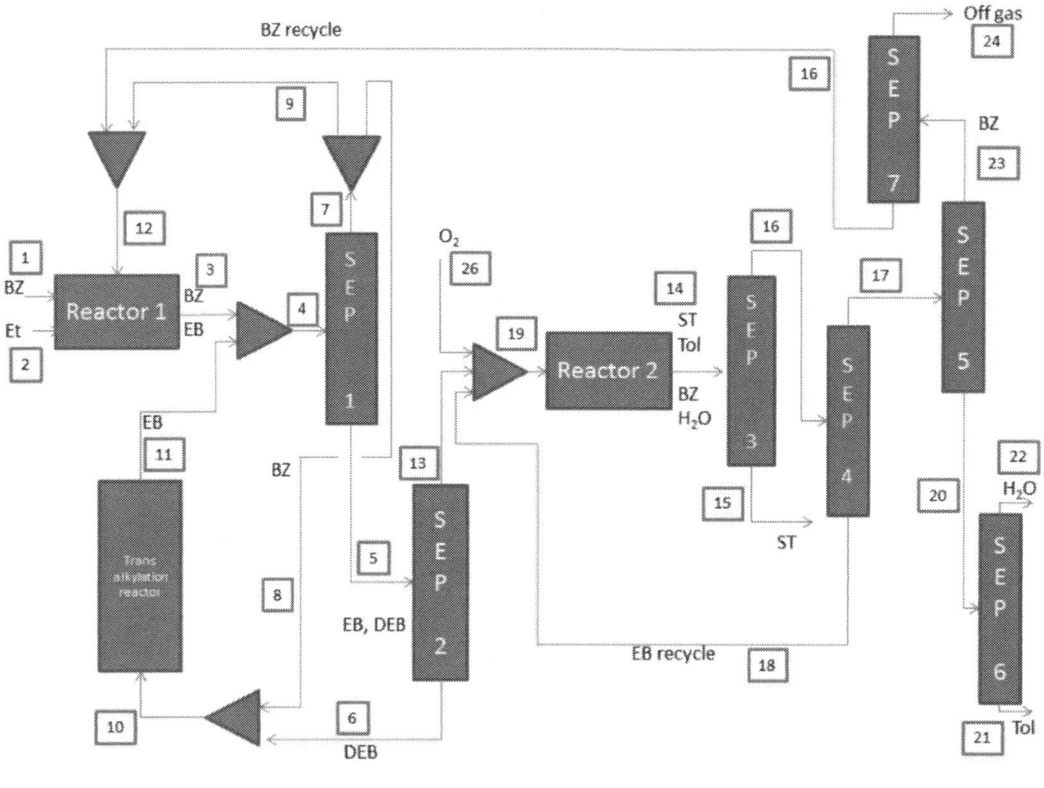

Figure 8: Extended flowsheet by introducing mixers and a divider.

4.1 Degree of Freedom Analysis

In order to investigate which parameters that are required to describe the system fully, and to find out which parameters have to be specified, a degree of freedom analysis is performed. The scope of the analysis is to define which parameters in the stream summary table that has to be calculated and from which parameters they should be calculated. The degree of freedom is found by subtracting the number of equations, NE, from the number of variables, NV. Thereby, it is calculated, how many variables that have to be specified.

The first step is to establish the total number of streams; NC. The total number of streams is found, from Figure 8, to be 26. The number of components is 10; BZ, Et, EB, O_2, H_2O, ST, Tol, DEB, M and H_2.

It is then possible to set up mass balances which are used to determine the mass flows in the system. The first step is to establish the mass balances for the different units.

Eq. (4.1) gives the mass balance for a reactor:

$$f_{i,j+1} = f_{i,j} + \sum_{r=1}^{NR} \gamma_{r,i} \eta_{r,k} f_{k,j} \tag{4.1}$$

Where subscript j is the stream number, i is the component, NR is the number of parallel reactions, r is the reaction number and k is the limiting component number. $\gamma_{r,i}$ is the normalized stoichiometric coefficient for each component, k, and for each reaction, r, normalized based on the limiting component for the desired reaction. $\eta_{r,k}$ is the conversion for component, k, for each r. Since the conversions are based on the limiting compound, the flowrate of the limiting compound for the specific stream is multiplied by the conversion and coefficient, γ. The coefficient is negative for reactants, positive for products and zero for inerts. Thereby, it is possible to find the flowrate, f, for each component and stream by approaching Eq. (4.1).

For a mixer the mass balance equation is:

$$f_{i,\mathrm{NM}+1} = \sum_{j=1}^{\mathrm{NM}} f_{i,j} \tag{4.2}$$

Here NM is the number of streams into the mixer. Eq. (4.2) gives the outlet flowrate from a mixer, which is just a summation of the ingoing flowrates of component i.

The splitter and divider units have the following mass balance equations:

$$f_{i,j+1} = \xi_{i,U} f_{i,j} \tag{4.3}$$

$$f_{i,\mathrm{NS}} = \left(1 - \sum_{j=1}^{\mathrm{NS}-1} \xi_{i,U}\right) f_{i,j} \tag{4.4}$$

A stream is divided into NS streams by a splitter or divider. The mass balances above are also used for describing a flash. The outlet streams from j to NC - 1 is described by Eq. (4.3), and are determined by a value of the splitter fraction/recovery, $\xi_{i,U}$, which is specified for each compound, i, for each of the streams. The flowrate of the last outlet stream (NS) is determined from Eq. (4.4), which is the remaining flowrate of compound i from the inlet stream. It means, that for NS outlets it is necessary to specify NS - 1 numbers of $\xi_{i,U}$.

Now it has been established which equations are used to calculate the mass balance of the process. The next step of the analysis is to determine which equation describes the different units, which can be seen in Table 7.

Table 7: Analysis results of the number of equations.

Unit	**Equation**	**Number of units**	**Number of Eq.**
Mixer	Eq. (4.2)	4	4·NC
Stream Calculator	Eq. (4.3)	6	6·NC
	Eq. (4.4)	6	6·NC
Divider	Eq. (4.3)	1	NC
	Eq. (4.4)	1	NC
Flash	Eq. (4.3)	1	NC
	Eq. (4.4)	1	NC
Reactor	Eq. (4.1)	3	3·NC
		Total NE	**23·NC**

Here each of the mass balance equations has been attached to the corresponding unit. The equations are then used to determine the outlet streams for each of the units. It gives a total number of equations (NE) of 23·NC. Described more deeply; for a divider, Eq. (4.3) and Eq. (4.4) are used to determine all the outlet flowrates for each of the components. Here, the divider is only dividing one stream into two streams. This means that only one equation of (4.3) is used, for each component, to calculate the flowrate for the first outlet stream. The second outlet stream is then determined from Eq. (4.4). This gives a total of 2·NC equations as seen in Table 7.

The number of variables (NV) is then found. The results can be seen in Table 8.

Here the number of variables in the system are presented. There are NC variables for each stream, resulting in 26·NC variables for the flowrates. For each reaction in the reactors, it is necessary to have a value of γ for each component, and a conversion, η. Thus, for each reaction it is necessary to establish NC + 1 variables. There are seven reactions in the three reactors. There are two reactions in reactor 1; the formation of EB and the formation of DEB. In the transalkylation reactor the only reaction is the conversion of DEB to EB. In the dehydrogenation reactor, four reactions are occurring. The formation of ST, Tol, H_2O and Et + BZ. A total of seven reactions, yields 7·NC + 7 variables for the reactors. The divider is only used to divide a stream into two, whereby NC variables occurs. It is also the case for the stream calculators and the flash, giving 6·NC and NC variables, respectively.
It gives a total of 41·NC + 7 variables. Subtraction of NE from NV gives the

Table 8: The results of the total number of variables analysis with the resulting degree of freedom.

Number of Variables (NV)	
Component flow rates, f_1-f_{26}	26·NC
Reactor parameters: γ, η_k	7·NC+7
Stream Calculators: ξ_S	6·NC
Divider: ξ_D	NC
Flash, ξ_F	NC
Total NV	**41·NC+7**
Degrees of freedom, NV-NE	**18·NC+7**

degree of freedom, which is then 18·NC + 7, meaning that 18·NC + 7 variables have to be specified in order to achieve a uniquely specified process.

The variables required to be specified are the three known input streams ($3 \cdot$ NC) and the equipment parameters, γ, η_k, ξ_S and ξ_D (11·NC+3), giving a total of 15·NC+7, which then gives a uniquely specified process, because the 23·NC process variables are found from the 23·NC equations. Table 9 shows the variables, which have to be specified.

Table 9: Variables that have to be specified.

Variables to specify (decisions)	
f_1, f_2, f_{26}	3·NC
Equipment parameters:	
Reactor: γ, η_k	7·NC + 7
Divider ξ_D	NC
Flash: ξ_F	NC
Stream Calculator ξ_S	6·NC
Total Variables	**18·NC + 7**
Unknown process variables f_3-f_{25}	**23·NC**

The conclusion from the analysis is that 18·NC + 7 variables needs to be specified from the three known process streams and the equipment parameters, and 23·NC variables needs to be calculated from 23·NC numbers of equations. First, the variables to specify should be presented, before calculating the mass balance.

The conversions in the three different reactors are shown in Table 10. The conversions are based on the conversions found in the literature [Ebrahimi et al., 2011,

Liu et al., 2009, Tamsilian et al., 2012] as described in section 2.1. The selectivity for the different compounds are also taken into account in the table, so that the conversion in the table is the conversion of the reactant to the specified compound.

Table 10: The conversions and stoichiometric coefficients for the different reactions in the three reactors.

Reactor	Reaction	Conversion [%]	Based on
R1	BZ + Et → EB	99	Et
	EB + Et → DEB	1	Et
R2	DEB + BZ → 2 EB	99	DEB
R3	EB → ST + H_2	66.13	EB
	H_2 + 1/2 O_2 → H_2O	100	H_2
	EB → BZ + Et	0.86	EB
	EB + H_2 → Tol + M	1.99	EB

It is seen that there are seven specified conversions for the key compounds. The stoichiometric coefficients are given from the reactions in the table. The flowrates, f_1, f_2 and f_{26} are pure BZ, Et and O_2 respectively. The input of Et is calculated to be 216 kmol/h by the EP in order to achieve the desired value of ST, and the oxygen added to the process is determined from how much H_2 that is present, thus no H_2 is present after R3. The BZ added to the system is also dependent of how much BZ is recycled and the excess amount of BZ compared to Et. The exact values are determined from convergence calculations, because it is known that the BZ in stream S12 and S1 (Figure 8) and Et in S2 should have a ratio between Et and BZ of 6.

The divider parameter is also dependent of other streams. In order to achieve high conversion in the transalkylation reactor it is desirable to have a ratio between BZ and DEB of at least 8. The divider parameter is therefore dependent of the needed amount of BZ compared to the DEB. Here convergence calculations are also used to determine the exact value of the divider parameter - a value is guessed, and further iterations are performed. It is then checked, if the demand of a ratio of at least 8 is reached and if the system converges. It is done until convergence, and thereby the parameter is found. It is possible to calculate a preliminary value. It is known, that the flowrate of Et is 216 kmol/h. Et and BZ are reacting 1:1, so at least 216 kmol/h BZ is also needed, if assuming that 100 % of BZ is converted. BZ is recycled in excess of 8, thereby feeding 1728 kmol/h into reactor 1. The excess 1512 kmol/h is then approaching separation tower 1, where 99 % is recovered. The amount of DEB formed is 2.16 kmol/h, with a conversion of Et to DEB of 1 %. It means that at least 17.28 kmol/h of BZ should be divided from separation tower 1. The excess amount of BZ is the light product from separation tower 1, 1512 kmol/h if assuming 100 % recovery of BZ. It means that the divider parameter is 17.28/1512 which gives around 0.011, meaning that 1.1 % should be divided into the transalkylation reactor.

In Table 11 the recovery in the different stream calculators (SC) are shown with the recoveries in molar fractions.

Table 11: The stream calculators with the recoveries of the top and bottom in molar fractions.

Stream calculator	**Top fract. [%]**	**Bottom fract. [%]**	**Separation Between**
SC1	99	99	BZ/H_2O
SC2	98	98	EB/ST
SC3	99	99	EB/ST
SC4	95	95	Tol/EB
SC5	95	99	BZ/H_2O
SC6	95	95	H_2O/Tol
SC7	100	95	Et/BZ

When dealing with compounds where the difference in boiling points is large, the recovery is set to 99 %, while when the difference is low the recovery is set to 95 %. In SC7 (the flash) the recovery of the top product is set to 100 %. This is because the top product is a gas and the bottom product is a liquid. The recovery of the bottom product is set to 95 %, since it is desired to purge a part of the bottom product before recycling. The purge is therefore combined with the stream calculator. In SC5 the bottom recovery is set to 99 % and the top is set to 95 %. This is because it is desired to remove as much water from the top fraction as possible, while it is less important to decrease the amount of BZ in the bottom product. The next step is then to derive the solution strategy. All the equations are established for the mass balance.

4.2 Flowsheet Decomposition

This analysis is done in order to establish which streams should be the tear streams in the iterations, and based on this determining a calculation order for the units. Only the internal streams are investigated thus, leaving out the input and exit streams, giving the flowsheet that can be seen in Figure 9.
The analysis is shown in Table 12. The first task is to identify all the internal streams and assign numbers to these. These numbers can be seen on the flowsheet in Figure 9, and are written in the first column of the table.

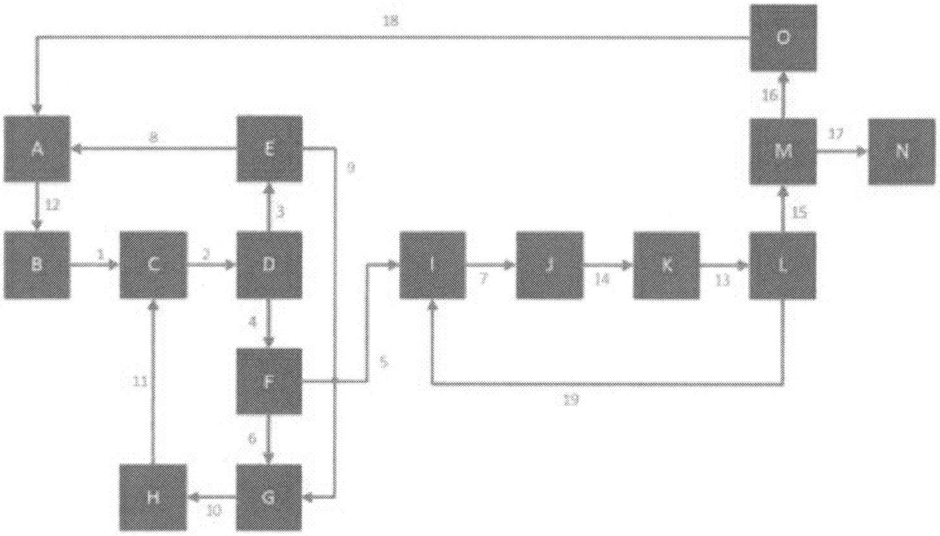

Figure 9: The decomposed flowsheet.

Table 12: Table for the flowsheet decomposition leading to the finding of the tear streams.

Streams	**Precursors**	**Precursors** (2)	**Precursors** (3)
1	12		
2	1 , 11	12 , 10	2 , 2 cut (2)
3	2		
4	2		
5	4		
6	4		
7	5 , 19	2 , 7 cut (7)	
8	3		
9	3		
10	6 , 9	2 , 2	2
11	10		
12	8 , 18	2 , 7	2
13	14		
14	7		
15	13		
16	15		
17	15		
18	16		
19	13		

In the second column, the streams that are precursers to each stream is denoted. In column 3 all the streams only dependent on one precurser are eliminated, and the precursers are changed to the ones that are remaining. Hereby it can be seen that stream 7 is a precurser to itself, and thereby it can be concluded that stream 7 should be cut and used as tear stream.
In column 4 stream 7 is eliminated since it has been cut, and it can be seen that all the streams left, including stream 2, has stream 2 as precurser, and thereby also stream 2 should be cut and used as tear stream.
To summarize, it is concluded, during the flowsheet decomposition that stream 2 and 7 should be used as tear streams for the iterations.
A proposal for the unit calculation order, based on the determined tear streams, could be J-K-L-M-N-O-A-B-D-F-E-G-H-C-I.

4.3 Flowsheet and Mass Balance

The system is simulated in PRO/II, and the flowsheet for this simulation is shown in Figure 10. In the flowsheet there are three controllers. The first controller ensures that the molar fraction of BZ in stream S12, the stream to the transalkylation reactor, is 0.889, since a significant excess of BZ is necessary to ensure a high conversion by regulating the separator SP1. The second controller CN2 regulates the amount of BZ fed to the system, by measuring the fraction of BZ in the inlet stream (S14) to R1. The last controller measures the amount of H_2 formed, and ensures that a sufficient amount of oxygen is fed to R3 in order to transform all the H_2 to H_2O.
The reactor, where ST is formed, is in the flowsheet represented by two reactors R3 and R4. In R3 ST is formed, while in R4 the H_2, which is formed during the production of ST, is converted into water. The reactor is separated into two reactors to ensure that the controller CN3 works. These two reactors will in reality be one unit.
Molar flow rate of the different components are shown in Table 13. The table shows that the production of ST is 202 kmol/h.

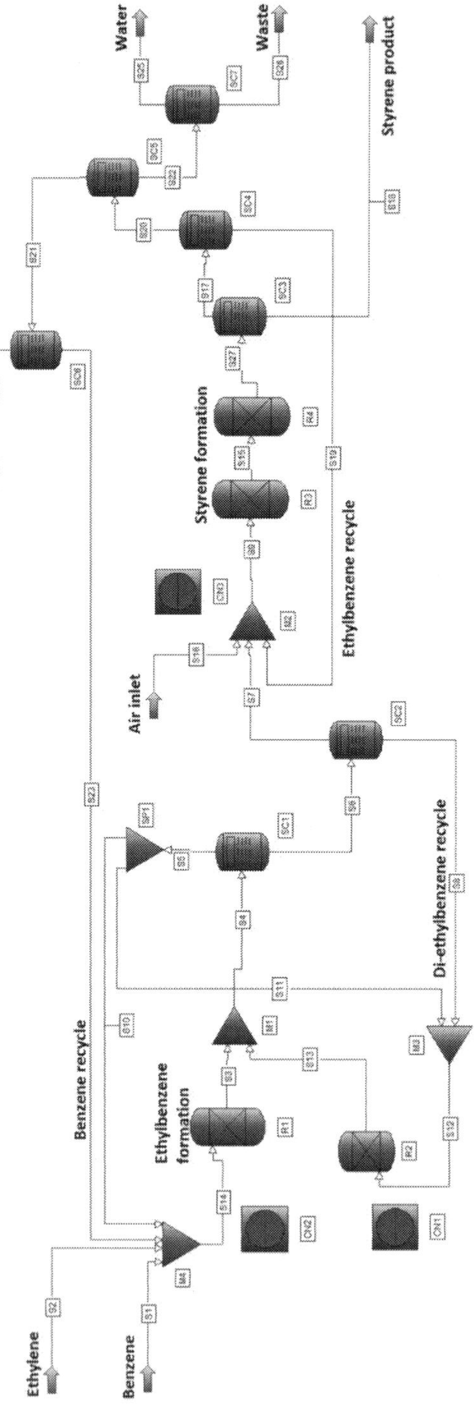

Figure 10: Simple flowsheet from PRO/II.

Stream Name	S1	S2	S3	S4	S5	S6	S7	S8	S9	S10	S11	S12	S13	S14
Total flow [kmol/h]	213.59	216.00	1739.43	1798.28	1559.72	240.24	233.33	6.91	438.89	1507.78	51.94	58.85	58.85	1955.43
H_2 [kmol/h]	0.00	0.00	0.00	0.00	0.00	0.00	0.00	0.00	0.00	0.00	0.00	0.00	0.00	0.00
O_2 [kmol/h]	0.00	0.00	0.00	0.00	0.00	0.00	0.00	0.00	102.67	0.00	0.00	0.00	0.00	0.00
Methane [kmol/h]	0.00	0.00	0.00	0.00	0.00	0.00	0.00	0.00	0.32	0.00	0.00	0.00	0.00	0.00
Ethylene [kmol/h]	0.00	216.00	0.00	0.00	0.00	0.00	0.00	0.00	0.14	0.00	0.00	0.00	0.00	216.00
Benzene [kmol/h]	213.59	0.00	1521.44	1571.50	1557.45	15.73	15.42	0.31	16.36	1505.59	51.86	52.18	50.06	1735.28
H_2O [kmol/h]	0.00	0.00	1.87	1.91	0.02	1.89	1.86	0.04	12.14	0.02	0.00	0.04	0.04	1.87
Toluene [kmol/h]	0.00	0.00	0.06	0.06	0.00	0.06	0.06	0.00	0.38	0.00	0.00	0.00	0.00	0.06
Ethylbenzene [kmol/h]	0.00	0.00	213.88	222.59	2.23	220.36	215.96	4.41	304.92	2.15	0.07	4.48	8.71	2.20
Styrene [kmol/h]	0.00	0.00	0.00	0.01	0.00	0.01	0.00	0.01	1.93	0.00	0.00	0.01	0.01	0.00
Di-ethylbenzene [kmol/h]	0.00	0.00	2.18	2.20	0.02	2.18	0.04	2.14	0.04	0.02	0.00	2.14	0.02	0.02

Stream Name	S15	S16	S17	S18	S19	S20	S21	S22	S23	S24	S25	S26	S27
Total flow [kmol/h]	643.15	102.43	340.36	205.00	103.13	237.23	31.54	205.69	18.06	13.48	185.74	19.95	545.36
H_2 [kmol/h]	195.57	0.00	0.00	0.00	0.00	0.00	0.00	0.00	0.00	0.00	0.00	0.00	0.00
O_2 [kmol/h]	102.67	102.43	4.83	0.05	0.24	4.59	4.36	0.23	0.00	4.36	0.22	0.01	4.88
Methane [kmol/h]	6.39	0.00	6.32	0.06	0.32	6.01	5.71	0.30	0.00	5.71	0.29	0.02	6.39
Ethylene [kmol/h]	2.75	0.00	2.72	0.03	0.14	2.59	2.46	0.13	0.00	2.46	0.12	0.01	2.75
Benzene [kmol/h]	18.97	0.00	18.78	0.19	0.94	17.84	16.95	0.89	16.10	0.85	0.85	0.04	18.97
H_2O [kmol/h]	12.14	0.00	205.63	2.08	10.28	195.35	1.95	193.40	1.86	0.10	183.73	9.67	207.71
Toluene [kmol/h]	6.45	0.00	6.38	0.06	0.32	6.06	0.06	6.00	0.06	0.00	0.30	5.70	6.45
Ethylbenzene [kmol/h]	94.59	0.00	93.65	0.95	88.96	4.68	0.05	4.64	0.04	0.00	0.23	4.40	94.59
Styrene [kmol/h]	203.58	0.00	2.04	201.54	1.93	0.10	0.00	0.10	0.00	0.00	0.01	0.10	203.58
Di-ethylbenzene [kmol/h]	0.04	0.00	0.00	0.04	0.00	0.00	0.00	0.00	0.00	0.00	0.00	0.00	0.04

Table 13: Molar flow rate of the different components in the different streams.

In Table 14 the mass balance for the simulated system is shown. The deviation between inlet and outlet is 0.1 kg/h or less than 0.01 %. The small deviation makes the simulation reliable, and a heat balance can therefore be produced for the system. This is done in Chapter 6.

Table 14: Mass balance for the production of styrene from ethylene and benzene with the deviations (Dev.).

Comp.	**Feed [kg/h]**	**+Reaction[kg/h]**	**-Product[kg/h]**	**=Dev.[kg/h]**	**Dev. [%]**
H_2	0	0.00	0.00	0.00	0.00
O_2	3285.3	-3129.03	156.26	0.00	0.00
CH_4	5.07	97.41	102.48	0.00	0.00
C_2H_4	6063.43	-5986.28	77.15	0.00	0.00
C_6H_6	140902.7	-16664.96	124237.77	-0.07	0.00
H_2O	253.12	3523.28	3776.41	-0.01	0.00
C_7H_8	40.19	559.45	599.65	0.00	0.00
EB	33081.61	592.79	33674.42	-0.02	0.00
ST	202.31	21001.48	21203.79	0.00	0.00
DEB	295.7	5.85	301.56	0.00	0.00
Total	184129.40	0.00	184129.49	-0.10	0.00

Chapter 5: Process Conditions

This chapter concerns defining the conditions for the reactors and the separation columns. Thereby the temperatures and pressures of all the streams will also be found. The stream numbers correspond to the ones given in the simple flowsheet in Figure 10.

Firstly, the temperature of stream S1, S2 and S16 are chosen. These are the feed streams of reactants into the system (Et, BZ and O_2). The temperatures are chosen to be 300 K for S1 and S2 and 898 K for S16, because it is the desired temperature of the inlet stream to the reactor. A Direct Fired Heater (DFH) is used to heat the stream in the rigorous modelling. The pressures are standard atmospheric pressure of 1 atmosphere.

Reactor 1, where EB are formed from the reaction between Et and BZ, the temperature and pressure are found from Ebrahimi et al. [2011] to be 533 K and 37.5 atmosphere as discussed in section 2.1.1. This temperature gives the conversion and selectivity chosen in Chapter 4.

Reactor 2, which is the transalkylation of DEB to EB should, according to Gerzeliev et al. [2010], be 563 K and 24.67 atmosphere to give the highest yield.

Reactor 3 and 4 are identical. They are only split into two reactors for calculation purposes, and therefore the process conditions for these two reactors are identical. According to Tamsilian et al. [2012] the optimal inlet temperature is 898 K and the pressure should be 0.667 atmosphere, as described in section 2.1.3.

The process conditions for the four reactors are defined in Table 15.

Table 15: Table of the conditions for the reactors.

Reactor number	**Temperature [K]**	**Pressure [atm]**
1	533	37.5
2	563	24.67
3	898	0.667
4	898	0.667

For the stream calculators the output temperatures are chosen to be the dew point in the overhead product stream and the bubble point temperature in the bottom product stream.

Chapter 6: Energy Balance

In order to perform an accurate energy balance that accounts for the vapor-liquid equilibrium, it is necessary to choose a more accurate thermodynamic model than the ideal model - the equation of state. Since BZ, EB and ST all are aromatic compounds it is decided to use Wilson, when only the organic phase is present, this is due to Wilsons ability to describe aromatic compounds and their temperature dependency. Due to the formation of water, during the production of ST, a two phase system appear; it is therefore necessary to use the Non-Random Two-Liquid model (NRTL), which in contrary to the Wilson model, is able to handle liquid-liquid systems [Suppes, 2004]. NRTL is therefore used during simulations of the unit operations where both a water and an organic phase are present. For the part of the plant where high pressure (>12 atm) is applied, Wilson is used, due to the fact that only one liquid phase is present and that Wilson is able to handle high pressures.
With all the temperatures and pressures known, an energy balance for the simple flowsheet is now constructed. Pumps, relief valves, heat exchangers and compressors are added to the flowsheet where they are needed in order to make the stream temperatures and pressures correspond to the ones chosen in Chapter 5. The flowsheet can be seen in Figure 12. A summation of the flows, the components, temperatures and pressures can be seen in Table 16 and 17.
A change that can not be seen from the flowsheet is that nitrogen has been added to the component list. This is done because of considerations about the source of oxygen. If oxygen is to be bought from an external provider it would be a significant expense and therefore it is chosen to use atmospheric air as inlet, because it is free of charge. The air is assumed to consist of 79% nitrogen and 21% oxygen, and for now, it is assumed to have an inlet temperature of 898 K.
Comparing the flowsheet in Figure 12 with the flowsheet in Figure 10, shows that two stream calculators and one recycle loop are removed, while a flash is installed. This is due to the discovery of azeotropes between water and ST, EB, Tol and BZ. The azeotrope between BZ and water is shown in Figure 11. The figure shows that it is not possible to get a recovery above approximately 20 % of BZ at 1 atm, and not higher than 40 % at 10 atm. Due to the azeotrope between ST and water it is not possible to get a solution with a higher molar fraction of ST than 30 %. Therefore the BZ separation and recycle loop is dropped for now. XY-diagrams for water and ST, EB and BZ are shown in Appendix B.
Due to the fact that Tol forms an azeotrope with water and the molar fraction of Tol is only 3 % in stream S20 in Figure 12, it is decided to drop the separation of Tol from water. This means that two separation towers are removed from the flowsheet.

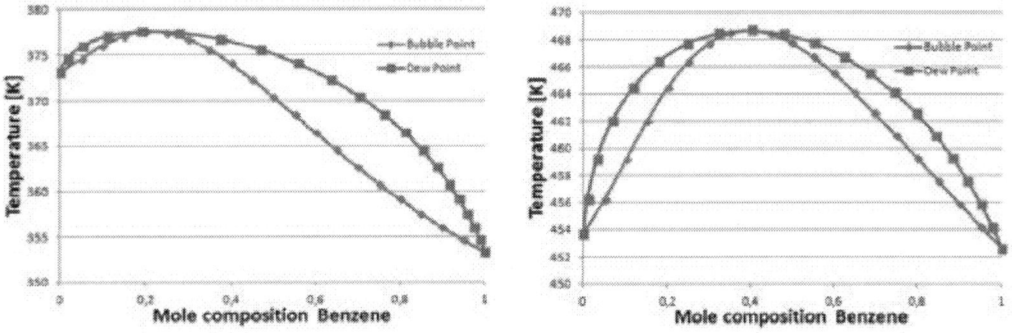

(a) At 1 atm

(b) At 10 atm

Figure 11: T,X,Y-diagram for a mixture of benzene and water at different pressures.

Since the azeotropes would make the distillation very expensive it was decided to remove the water by decantation in the flash. Due to this it is still possible to get a high recovery of ST and EB. It was decided still not to recycle the BZ and recover the Tol since it was assumed that it would be too expensive to perform the necessary distillations compared to a possible revenue.
This was a poor assumption, and as it is shown in Chapter 13 and 14 recycling of BZ and recovery of Tol both made the process better from an economic and an environmental point of view.

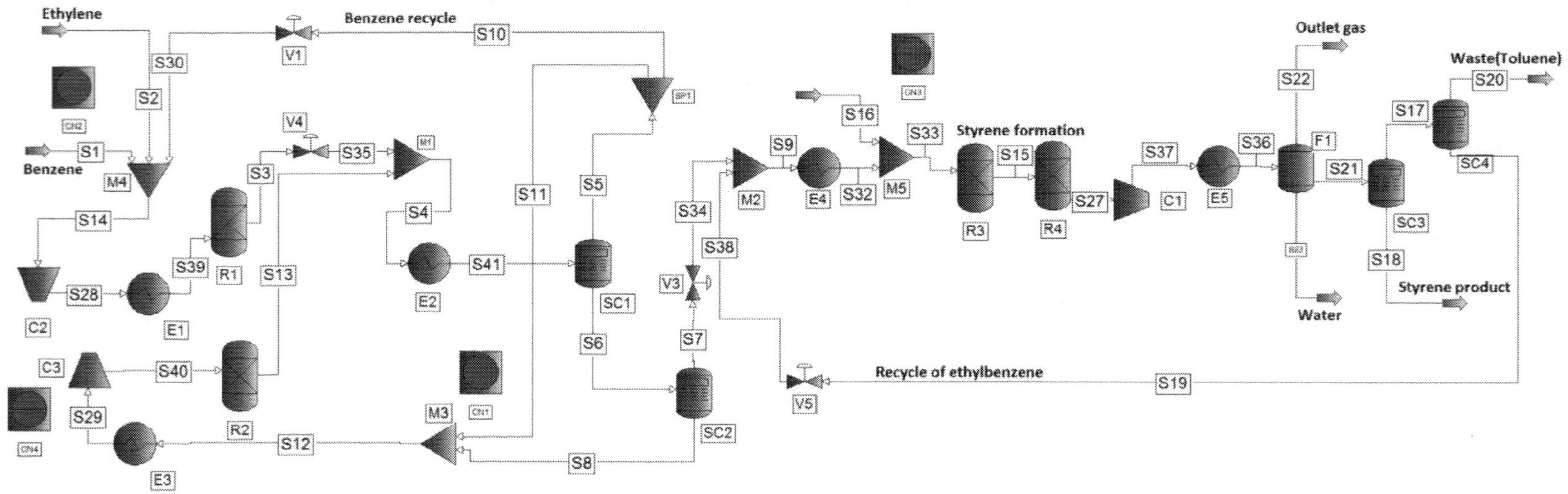

Figure 12: The flowsheet after task 6.

Table 16: Stream summary for streams S1-S19.

Stream Name	S1	S2	S3	S4	S5	S6	S7	S8	S9	S10	S11	S12	S13	S14	S15	S16	S17	S18	S19
T. [K]	300.0	300.0	533.0	510.4	501.9	563.3	565.2	571.8	466.4	501.9	501.9	502.7	563.2	423.1	638.0	898.0	404.1	421.6	399.3
P. [atm]	1.0	1.0	33.5	22.2	22.2	19.8	19.8	17.6	0.667	22.2	22.2	17.6	22.2	1.0	0.6	1.0	1.0	1.1	0.9
Enthalpy [GJ/hr]	0.8	0.4	76.7	80.7	87.2	14.2	20.2	0.5	22.4	84.2	2.9	3.4	4.1	85.5	35.9	16.5	7.0	6.0	2.2
Rate [kmol/h]	228.1	216.0	1718.6	1776.9	1542.0	238.1	231.2	6.9	322.3	1490.5	51.5	58.4	58.4	1934.6	997.0	471.1	118.8	201.1	91.1
H_2 [kmol/h]	0.00	0.00	0.00	0.00	0.00	0.00	0.00	0.00	0.00	0.00	0.00	0.00	0.00	0.00	194.91	0.00	0.00	0.00	0.00
N_2 [kmol/h]	0.00	0.00	0.00	0.00	0.00	0.00	0.00	0.00	0.01	0.00	0.00	0.00	0.00	0.00	372.18	372.17	0.19	0.00	0.01
O_2 [kmol/h]	0.00	0.00	0.00	0.00	0.00	0.00	0.00	0.00	0.00	0.00	0.00	0.00	0.00	0.00	98.93	98.93	0.00	0.00	0.00
Methane [kmol/h]	0.00	0.00	0.00	0.00	0.00	0.00	0.00	0.00	0.00	0.00	0.00	0.00	0.00	0.00	6.05	0.00	0.02	0.00	0.00
Ethylene [kmol/h]	0.00	216.00	0.00	0.00	0.00	0.00	0.00	0.00	0.00	0.00	0.00	0.00	0.00	216.00	2.61	0.00	0.04	0.00	0.00
Benzene [kmol/h]	228.08	0.00	1502.57	1552.19	1539.76	15.55	15.24	0.31	16.09	1488.33	51.43	51.74	49.62	1716.41	18.70	0.00	16.98	0.17	0.85
H_2O [kmol/h]	0.00	0.00	0.00	0.00	0.00	0.00	0.00	0.00	0.04	0.00	0.00	0.00	0.00	0.00	0.04	0.00	0.79	0.01	0.04
Toluene [kmol/h]	0.00	0.00	0.00	0.00	0.00	0.00	0.00	0.00	0.31	0.00	0.00	0.00	0.00	0.00	6.36	0.00	6.15	0.06	0.31
Ethylbenzene [kmol/h]	0.00	0.00	213.83	222.54	2.23	220.32	215.91	4.41	303.89	2.15	0.07	4.48	8.71	2.15	94.27	0.00	92.61	0.94	87.97
Styrene [kmol/h]	0.00	0.00	0.00	0.00	0.00	0.00	0.00	0.00	1.92	0.00	0.00	0.00	0.00	0.00	202.88	0.00	2.02	199.84	1.92
Di-ethylbenzene [kmol/h]	0.00	0.00	2.18	2.20	0.02	2.18	0.04	2.14	0.04	0.02	0.00	2.14	0.02	0.02	0.04	0.00	0.00	0.04	0.00

Table 17: Stream summary for streams S20-S41.

Stream Name	S20	S21	S22	S23	S27	S28	S29	S30	S32	S33	S34	S35	S36	S37	S38	S39	S40	S41
T. [K]	375.7	288.2	288.2	288.2	1034.4	559.8	548.7	474.9	898.0	898.0	694.6	517.7	300.0	1070.4	393.8	533.0	563.2	320.0
P. [atm]	0.9	1.1	1.1	1.1	0.667	37.5	17.6	1.0	0.667	0.667	0.667	24.7	1.1	1.1	0.667	37.5	24.7	22.2
Enthalpy [GJ/h]	1.3	0.8	0.3	0.2	91.8	114.9	3.6	84.2	42.3	58.8	20.2	76.7	2.7	107.8	2.2	102.8	4.1	11.8
Rate [kmol/h]	27.7	319.9	391.5	188.2	899.5	1934.6	58.4	1490.5	322.3	793.4	231.2	1718.6	899.5	899.5	91.1	1934.6	58.4	1776.9
H_2 [kmol/h]	0.00	0.00	0.00	0.00	0.00	0.00	0.00	0.00	0.00	0.00	0.00	0.00	0.00	0.00	0.00	0.00	0.00	0.00
N_2 [kmol/h]	0.18	0.19	371.99	0.00	372.18	0.00	0.00	0.00	0.01	372.18	0.00	0.00	372.18	372.18	0.01	0.00	0.00	0.00
O_2 [kmol/h]	0.00	0.00	1.48	0.00	1.48	0.00	0.00	0.00	0.00	98.93	0.00	0.00	1.48	1.48	0.00	0.00	0.00	0.00
Methane [kmol/h]	0.02	0.02	6.03	0.00	6.05	0.00	0.00	0.00	0.00	0.00	0.00	0.00	6.05	6.05	0.00	0.00	0.00	0.00
Ethylene [kmol/h]	0.04	0.04	2.57	0.00	2.61	216.00	0.00	0.00	0.00	0.00	0.00	0.00	2.61	2.61	0.00	216.00	0.00	0.00
Benzene [kmol/h]	16.13	17.15	1.55	0.00	18.70	1716.41	51.74	1488.33	16.09	16.09	15.24	1502.57	18.70	18.70	0.85	1716.41	51.74	1552.19
H_2O [kmol/h]	0.75	0.79	5.99	188.16	194.95	0.00	0.00	0.00	0.04	0.04	0.00	0.00	194.95	194.95	0.04	0.00	0.00	0.00
Toluene [kmol/h]	5.84	6.21	0.15	0.00	6.36	0.00	0.00	0.00	0.31	0.31	0.00	0.00	6.36	6.36	0.31	0.00	0.00	0.00
Ethylbenzene [kmol/h]	4.63	93.54	0.73	0.00	94.27	2.15	4.48	2.15	303.89	303.89	215.91	213.83	94.27	94.27	87.97	2.15	4.48	222.54
Styrene [kmol/h]	0.10	201.86	1.01	0.00	202.88	0.00	0.00	0.00	1.92	1.92	0.00	0.00	202.88	202.88	1.92	0.00	0.00	0.00
Di-ethylbenzene [kmol/h]	0.00	0.04	0.00	0.00	0.04	0.02	2.14	0.02	0.04	0.04	0.04	2.18	0.04	0.04	0.00	0.02	2.14	2.20

The heat duties for all the units are now calculated. These are calculated from the enthalpies for the streams that can be seen in Table 16 and 17. However, for the reactors, the heat of reaction also plays a role which is taken into account. In Table 18 the heat duties for all the units can be seen. The heat duty for a single unit is then calculated by subtracting the enthalpy of the ingoing stream from the enthalpy of the outgoing stream.

Table 18: Table of heat duties for the units.

Unit	Stream In	Stream Out	Enthalpy In [GJ/h]	Enthalpy Out [GJ/h]	Heat Duty [GJ/h]
Reactor					
R1	S39	S3	102.66	76.53	-48.64
R2	S40	S13	4.06	4.07	$2.41 \cdot 10^{-2}$
R3	S33	S15	58.79	35.94	0.00
R4	S15	S27	35.94	91.84	0.00
Heat Exchanger					
E1	S28	S39	115.17	102.52	-12.237
E2	S4	S41	80.87	11.82	-69.04
E3	S12	S29	3.39	3.57	0.186
E4	S9	S32	22.40	42.20	19.79
E5	S37	S36	95.59	2.70	-92.90
Stream calculator					
SC1	S41	S5+S6	11.83	101.53	89.70
SC2	S6	S7+S8	14.20	20.69	6.49
SC3	S21	S17+S18	1.49	12.79	11.30
SC4	S17	S19+S20	6.88	3.47	-3.41
Compressor					
C1	S27	S37	91.66	95.59	3.93
C2	S14	S28	85.64	115.17	29.53
C3	S29	S40	3.57	4.06	0.48

Here a negative heat duty corresponds to withdrawal of energy from the unit, and a positive value corresponds to energy added to the unit.
It should be noted that R3 and R4 in reality is one reactor, which is operated adiabatic, and the heat duty is therefore equal to zero.

Chapter 7: Rigorous Process Simulation

For this process it is only necessary to replace stream calculators with distillation columns in order to obtain a rigorous process simulation.
In the simple model the stream calculators are only used to define recovery for the heavy and the light phases. In order to obtain a rigorous model, shortcut distillation calculations are performed for each distillation. In the method an average volatility is assumed for each tower. For the shortcut distillation the Fenske method is used to estimate the separations and the necessary number of trays. The minimum reflux is calculated from the Underwood method, while the Gilliland method is applied for the calculation of the number of theoretical trays necessary in order to perform the distillation and the actual reflux ratio, and the appertaining reboiler and condenser duties. Lastly the Kirkbride method is used to calculate the optimum feed tray location. These calculations are performed by use of PRO/II's shortcut model using the inside out algorithm. This algorithm is not able to take two liquid phases in one tray into account, but since only one liquid phase is present this algorithm is usable. If the water phase was removed by distillation this algorithm would not have been suitable [Bagajewicz et al., 1994]. The shortcut calculations give more than one possible combination of number of trays and feed tray location with the corresponding reflux ratios and reboiler and condenser duties. In order to determine which combination is the most optimal combination, the total energy need, which is the sum of the reboiler and the condenser duty, is plotted against the number of theoretical trays as shown for the first tower in Figure 13. The feed tray location is then selected as the one giving the lowest total duty for the chosen number of trays.

From this figure it can be observed that for the first tower 13 trays give the lowest energy consumption. 13 trays are therefore applied to the first distillation tower. The shortcut calculations showed that for 13 trays the most optimal feed location is tray 10.
Based on this method the number of theoretical trays and the feed location for each tower are estimated as shown in Table 19.

Table 19: Results from replacing the stream calculators with shortcuts.

Tower number	**Number of trays**	**Feed tray**
1	13	10
2	14	12
3	48	26
4	18	9

The shortcuts are then replaced by distillation tower, where the number of trays and the feed positions are the ones found by the shortcut method. Furthermore

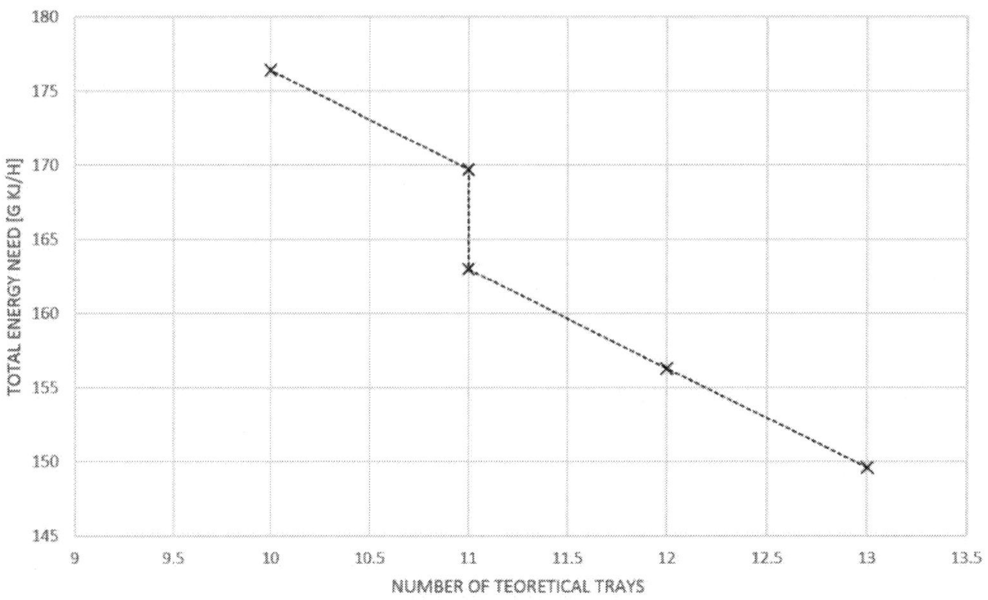

Figure 13: Plot of the duty as function of number of trays for the first distillation tower.

the pressure drop over the towers is chosen to be 0.01 atm for all of the towers. The thermodynamic model for the distillation towers are chosen to be the Wilson model. For R4 and the subsequent heat exchanger and flash the thermodynamic model is chosen to be NRTL. The reason for this is that two liquid phases are present, and the Wilson model is not able take the liquid-liquid equilibrium into account.
In Table 20 all the parameters are defined for the distillation towers in the base case design.

Table 20: Table of the specified parameters for each distillation tower.

	Tower 1	Tower 2	Tower 3	Tower 4
Nr. of trays	13	14	48	18
Feed tray	10	12	26	9
Recovery heavy key	0.99	0.985	0.99	0.99
Recovery light key	0.99	0.985	0.99	0.95

For the simulations performed using PRO/II, the heat duties for the reboiler and the condenser are defined as variables that are changed in order to reach the defined recovery rates.

During the rigorous simulations it is discovered that ST has a tendency to polymerize at high temperature and pressure. Vacuum distillation (0.2 atm) is therefore applied for the separation of ST and EB in T3 [Behr, 1999]. Due to this discovery it is decided not to increase the pressure after the dehydrogenation reactor (R4). This would have lowered the utility, however, since a pressure increase also increases the temperature, the ST might polymerize. The product from R4 is instead cooled immediately after the reactor. It is furthermore decided to operate the flash drum, F1, which removes water and gas from the organic phase, at the same pressure as the dehydrogenation reactors (0.667 atm). This gives a small, but acceptable, decrease in the recovery of the organic phase. Due to these changes and considerations it is assumed that ST does not polymerize in the plant. To prevent the polymerization, inhibitor can be added to the distillations [Behr, 1999].

In Figure 14 the detailed flowsheet for the base case design is shown. Due to the vacuum distillation in T3 and implementation of rigorous unit simulations the flowsheet is changed compared to the previous version in Chapter 6. After T3 the pressure is increased to 1 atm. Since the top product from T3 is a gas, the pressure is increased by first using a compressor (C4). After the compression the temperature of the gas is decreased, due to interstage cooling, which decreases the compressor duty. When the gas is cooled, some of it condensates. Since a compressor is only suitable for gases, a flash drum (F2) is installed, which separates the liquid and the gas. The gas is sent to a compressor (C5), the liquid to a pump (P1), and the pressure is increased to 1 atm. The mixture is then sent to distillation tower 4 (T4).
The stream summery appertaining the flowsheet is shown in Table 21 and 22.

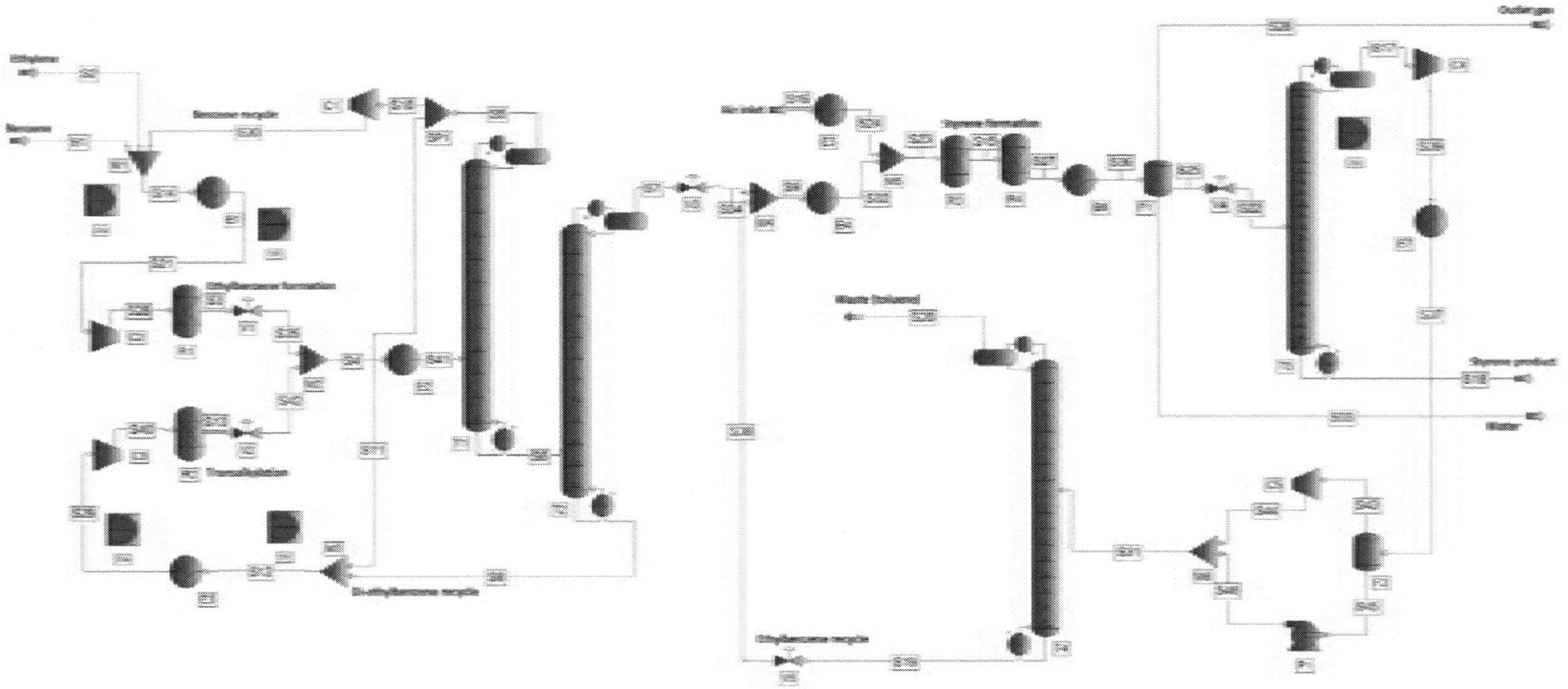

Figure 14: The flowsheet of the base case design.

Table 21: Stream summary for streams S1-S23 for the base case design.

Stream Name	S1	S2	S3	S4	S5	S6	S7	S8	S9	S10	S11	S12	S13	S14	S15	S16	S17	S18	S19	S20	S21	S22	S23
Phase	Liquid	Vapor	Liquid	Vapor	Vapor	Liquid	Vapor	Liquid	Mixed	Vapor	Vapor	Mixed	Vapor	Mixed	Vapor	Vapor	Vapor	Liquid	Liquid	Vapor	Vapor	Mixed	Vapor
T. [K]	300.0	300.0	533.0	369.4	353.2	402.3	407.0	422.4	393.3	353.2	353.2	368.7	563.2	349.1	573.6	293.0	354.5	367.8	409.0	368.4	391.5	300.0	898.0
P. [atm]	1.0	1.0	33.5	1.0	1.0	1.0	1.0	1.0	0.667	1.0	1.0	1.0	22.2	1.0	0.6	1.0	0.2	0.2	1.0	1.0	1.0	0.2	0.667
Enthalpy [GJ/h]	0.8	0.4	79.0	82.4	66.3	5.9	14.1	0.2	16.6	64.4	1.8	2.0	3.4	66.0	36.0	0.3	6.0	3.6	2.5	1.0	82.2	1.5	59.1
Rate [kmol/h]	235.3	216.0	1772.1	1821.2	1580.5	237.3	231.9	5.4	326.1	1536.8	43.7	49.2	49.2	1988.1	999.9	468.4	115.5	200.2	94.2	21.3	1988.1	315.6	794.4
H_2 [kmol/h]	0.00	0.00	0.00	0.00	0.00	0.00	0.00	0.00	0.00	0.00	0.00	0.00	0.00	0.00	196.72	0.00	0.00	0.00	0.00	0.00	0.00	0.00	0.00
N_2 [kmol/h]	0.00	0.00	0.00	0.00	0.00	0.00	0.00	0.00	0.00	0.00	0.00	0.00	0.00	0.00	370.03	370.03	0.11	0.00	0.00	0.11	0.00	0.11	370.03
O_2 [kmol/h]	0.00	0.00	0.00	0.00	0.00	0.00	0.00	0.00	0.00	0.00	0.00	0.00	0.00	0.00	98.36	98.36	0.00	0.00	0.00	0.00	0.00	0.00	98.36
Methane [kmol/h]	0.00	0.00	0.00	0.00	0.00	0.00	0.00	0.00	0.00	0.00	0.00	0.00	0.00	0.00	6.11	0.00	0.01	0.00	0.00	0.01	0.00	0.01	0.00
Ethylene [kmol/h]	0.00	216.00	0.00	0.00	0.00	0.00	0.00	0.00	0.00	0.00	0.00	0.00	0.00	216.00	2.63	0.00	0.02	0.00	0.00	0.02	216.00	0.02	0.00
Benzene [kmol/h]	235.30	0.00	1556.09	1597.63	1578.30	15.94	15.94	0.00	15.94	1534.63	43.67	43.67	41.54	1769.93	18.54	0.00	14.56	0.00	0.00	14.56	1769.93	14.56	15.91
H_2O	0.00	0.00	0.00	0.00	0.00	0.00	0.00	0.00	1.08	0.00	0.00	0.00	0.00	0.00	1.08	0.00	1.12	0.00	1.08	0.04	0.00	1.13	1.08
Toluene [kmol/h]	0.00	0.00	0.00	0.00	0.00	0.00	0.00	0.00	0.30	0.00	0.00	0.00	0.00	0.00	6.40	0.00	5.93	0.00	0.30	5.64	0.00	5.93	0.30
Ethylbenzene [kmol/h]	0.00	0.00	213.83	221.44	2.22	219.22	215.93	3.29	306.71	2.15	0.06	3.35	7.60	2.15	95.15	0.00	91.69	0.93	90.78	0.92	2.15	92.62	306.71
Styrene [kmol/h]	0.00	0.00	0.00	0.00	0.00	0.00	0.00	0.00	2.01	0.00	0.00	0.00	0.00	0.00	204.84	0.00	2.01	199.21	2.01	0.00	0.00	201.22	2.01
Di-ethylbenzene [kmol/h]	0.00	0.00	2.16	2.18	0.00	2.18	0.03	2.15	0.03	0.00	0.00	2.15	0.02	0.00	0.03	0.00	0.00	0.03	0.00	0.00	0.00	0.03	0.03

Table 22: Stream summary for streams S24-S46 for the base case design.

Stream Name	S24	S25	S26	S27	S28	S29	S30	S31	S32	S33	S34	S35	S36	S37	S38	S39	S40	S41	S42	S43	S44	S45	S46
Phase	Vapor	Liquid	Vapor	Vapor	Vapor	Vapor	Mixed	Mixed	Vapor	Water	Vapor	Mixed	Mixed	Mixed	Mixed	Mixed	Vapor	Liquid	Vapor	Vapor	Vapor	Liquid	Liquid
T. [K]	898.0	300.0	300.0	1034.2	533.0	458.5	353.6	354.4	898.0	300.0	407.0	367.9	300.0	354.2	394.9	381.0	563.2	320.0	544.4	354.2	378.4	354.2	354.2
P. [atm]	1.0	0.667	0.667	0.667	37.5	1.0	1.0	1.0	0.667	0.667	0.667	1.0	0.667	0.5	0.667	0.5	24.7	1.0	1.0	0.5	1.0	0.5	1.0
Enthalpy [GJ/h]	16.4	1.5	1.4	119.2	109.0	2.7	64.8	1.7	42.7	0.4	14.1	79.0	3.3	1.7	2.5	6.3	3.4	12.1	3.4	0.0	0.0	1.7	1.7
Rate [kmol/h]	468.4	315.6	410.7	901.5	1988.1	49.2	1536.8	115.5	326.1	175.2	231.9	1772.1	901.5	115.5	94.2	115.5	49.2	1821.2	49.2	0.2	0.2	115.2	115.2
H_2 [kmol/h]	0.00	0.00	0.00	0.00	0.00	0.00	0.00	0.00	0.00	0.00	0.00	0.00	0.00	0.00	0.00	0.00	0.00	0.00	0.00	0.00	0.00	0.00	0.00
N_2 [kmol/h]	370.03	0.11	369.91	370.03	0.00	0.00	0.00	0.11	0.00	0.00	0.00	0.00	370.03	0.11	0.00	0.11	0.00	0.00	0.00	0.09	0.09	0.02	0.02
O_2 [kmol/h]	98.36	0.00	0.00	0.00	0.00	0.00	0.00	0.00	0.00	0.00	0.00	0.00	0.00	0.00	0.00	0.00	0.00	0.00	0.00	0.00	0.00	0.00	0.00
Methane [kmol/h]	0.00	0.01	6.10	6.11	0.00	0.00	0.00	0.01	0.00	0.00	0.00	0.00	6.11	0.01	0.00	0.01	0.00	0.00	0.00	0.01	0.01	0.00	0.00
Ethylene [kmol/h]	0.00	0.02	2.61	2.63	216.00	0.00	0.00	0.02	0.00	0.00	0.00	0.00	2.63	0.02	0.00	0.02	0.00	0.00	0.00	0.01	0.01	0.01	0.01
Benzene [kmol/h]	0.00	14.56	3.97	18.54	1769.93	43.67	1534.63	14.56	15.94	0.00	15.94	1556.09	18.54	14.56	0.00	14.56	43.67	1597.63	41.54	0.06	0.06	14.51	14.51
H_2O [kmol/h]	0.00	1.13	21.49	197.80	0.00	0.00	0.00	1.12	1.08	175.18	0.00	0.00	197.80	1.12	1.08	1.12	0.00	0.00	0.00	0.00	0.00	1.12	1.12
Toluene [kmol/h]	0.00	5.93	0.47	6.40	0.00	0.00	0.00	5.93	0.30	0.00	0.00	0.00	6.40	5.93	0.30	5.93	0.00	0.00	0.00	0.01	0.01	5.92	5.92
Ethylbenzene [kmol/h]	0.00	92.62	2.53	95.15	2.15	3.35	2.15	91.69	306.71	0.00	215.93	213.83	95.15	91.69	90.78	91.69	3.35	221.44	7.60	0.06	0.06	91.63	91.63
Styrene [kmol/h]	0.00	201.22	3.62	204.84	0.00	0.00	0.00	2.01	2.01	0.00	0.00	0.00	204.84	2.01	2.01	2.01	0.00	0.00	0.00	0.00	0.00	2.01	2.01
Di-ethylbenzene [kmol/h]	0.00	0.03	0.00	0.03	0.00	2.15	0.00	0.00	0.03	0.00	0.03	2.16	0.03	0.00	0.00	0.00	2.15	2.18	0.02	0.00	0.00	0.00	0.00

Chapter 8: Sizing of the Equipment

The next step is then to size the equipment, so the evaluation of the costing of the process can be performed. The methods used for sizing are from Biegler et al. [1997].

8.1 Reactor Sizing

Since catalysts are used in the three reactors of the process (when considering R3 and R4 as one reactor), the reactor sizes can be determined based on the void fraction of the catalyst, ϵ, by:

$$V_{reactor} = \frac{V_{catalyst}}{1-\epsilon} \tag{8.1}$$

Where the volume of the catalyst is determined by:

$$\frac{1}{\tau} = \frac{f}{\rho \cdot V_{catalyst}} \tag{8.2}$$

Where τ is the residence time, ρ is the molar density and f is the molar flow rate. All the reactors in the system are fixed bed catalytic reactors.

By using data from the literature the following sizing data have been found; amount of catalyst needed, height/diameter of reactor and material, which can be seen in Table 23. The following subsections describes the method and data used to size the reactors.

Table 23: Overview over the reaction sizes for the base case design.

	R1	**R2**	**R3+R4**
Type of catalyst	Zeolite-Y	Zeolite-Y	Iron Oxide based
Amount of catalyst [kg]	18354	610	56710
Diameter [m]	4	1	4
Height [m]	5.85	3.11	5.24
Material	Stainless Steel	Stainless Steel	Stainless Steel

8.1.1 Alkylation Reactor

The first reactor to size is R1, where the conversion of Et and BZ to EB occurs. From Ganji et al. [2004] it is reported, that 8,740 kg of catalyst is used for the reaction, for an ingoing flow of Et and BZ of respectively 3,564 kg/h and 63,743 kg/h. The equivalent stream for this process yields an ingoing flow of Et and BZ of respectively 6,060 kg/h and 137,462 kg/h, which means, that the ingoing flow in this process is 2.1 times greater than the process reported by Ganji et al. [2004].

It is assumed, that there is a linear relationship between the flow rate into the reactor and the catalyst needed for the reaction. By approaching this, the needed amount of catalyst is:

$$m_{catalyst} = 2.1 \cdot 8,740\text{kg} = 18,354\text{kg} \tag{8.3}$$

In the reactor, Zeolite-Y is used as the catalyst. Ngandjui et al. [1997] reports that the density is equal to 480 kg/m^3, whereby the volume of the catalyst is calculated to 38.2 m^3. The porosity of Zeolite-Y is reported by Price [2014] to be 0.48. The volume of the reactor is then calculated to 73.5 m^3. If choosing a diameter of 4 meter, the corresponding height of the reactor becomes 5.85 meters. The temperature of the reactor is 533 K, and since no corrosive compounds are present, the chosen material is stainless steel.

8.1.2 Transalkylation Reactor

For the transalkylation reactor, the catalyst Zeolite Y is also used. Ganji et al. [2004] reports that they use 6,042 kg of catalyst in the transalkylation reactor, with a total inflow of reactants of 39,777 kg/h, which mainly consist of BZ. This is also the case in this report where 84 % of the inlet flow consists of BZ. In this process, the total inflow is only 4,036 kg/h, which gives that 9.9 times less catalyst has to be used for the reported process, resulting in a total mass of catalyst of 610 kg. Approaching the density of Zeolite Y, the volume of the catalyst is calculated and thereafter the volume of the reactor, resulting in a volume of the reactor of 2.4 m^3. By choosing a diameter of 1 meter, the resulting height of the reactor is 3.11 meter. The temperature is 563 K, whereby stainless steel is chosen once again as the material for the reactor.

8.1.3 Dehydrogenation Reactor

The last reactor of the system is converting EB into ST. As described earlier, the two remaining reactors, in Figure 14, represents, in reality, only one reactor, where all the reactions occurs. Therefore, R3 and R4 are calculated as one reactor in the following calculations. Mousavi et al. [2012] reports that the catalyst bed depth of their process is 1.7 m and the diameter is 1.95 m. It gives a total volume of the catalyst bed of 5.1 m^3, where an EB flow of 36.87 kmol/h is flowing through. In this report, the flow of EB into the reactor is 309 kmol/h, i.e. 8.4 times more flow for this process. It gives a volume of the catalyst of 42.8 m^3. Tamsilian et al. [2012] reports that the potassium-promoted iron oxide catalyst has been extensively used for the reaction, which has a porosity of 0.35 and a density of 1325 kg/m^3, resulting in 56,710 kg of catalyst. The reactor volume is then calculated to 85.6 m^3. By choosing a diameter of 4 m, the resulting height is 5.24 m. Since the reactor is operated adiabaticly, the inlet temperature is 898 K and the outlet temperature is about 1034 K. According to Biegler et al. [1997], a special type of stainless steel (type 446) is chosen, which can withstand up to about 1400 K.

8.2 Flash Drum Sizing

The flash drums are then estimated by the following design equation for a vessel:

$$V = 2 \cdot \left(\frac{F_L \cdot \tau}{\rho_L} \right) \tag{8.4}$$

Where F_L is the liquid flow rate leaving the flash drum, ρ_L is the liquid density and τ is the residence time, which is typically set to five minutes [Biegler et al., 1997]. From Table 22 it can be found that the liquid flow rate is 315.6 kmol/h, the decanted water flow rate is 175.2 kmol/h, and the molecular weight is 103 kg/kmol and 18 kg/kmol respectively. By weighting the compounds with the fraction present, the liquid density is calculated to 887.9 kg/m^3. By approaching these values, the volume of the flash drum is calculated by summing the two volumes:

$$V = 2 \cdot \left(\frac{315.6\frac{\text{kmol}}{\text{h}} \cdot 103\frac{\text{kg}}{\text{kmol}} \cdot \frac{5\text{min}}{60\frac{\text{min}}{\text{h}}}}{887.9\frac{\text{kg}}{\text{m}^3}} \right) + 2 \cdot \left(\frac{175.2\frac{\text{kmol}}{\text{h}} \cdot 18\frac{\text{kg}}{\text{kmol}} \cdot \frac{5\text{min}}{60\frac{\text{min}}{\text{h}}}}{1000\frac{\text{kg}}{\text{m}^3}} \right) = 6.63\text{m}^3 \tag{8.5}$$

Biegler et al. [1997] states that the optimal value of L/D, due to cost, is 4, whereby the resulting calculated diameter is 1.3 m with a corresponding length of 5 m. If the diameter is larger than 1.2 m the vessel should be sized as a horizontal vessel. This requires more space, but less cost for structural support. By designing a horizontal vessel, L/D is not longer 4. It is chosen, that the diameter of the horizontal flash drum is 1 meter, which gives a length equal to 8.4 meter. The temperature is 300 K during the separation, which means that carbon steel is used as material.

The above method is also used to size the other flash drum. The flow is 115.2 kmol/h, which then gives a total volume of 2.24 m^3. In order to reach a L/D value of 4, the resulting diameter is 0.9 meter and the corresponding length is then 3.52 meter. The temperature is about 350 K, which means that the material is also carbon steel.

8.3 Storage Tanks

It is assumed that the raw materials, waste and products are transferred with pipes to and from the facility, whereby no storage tanks for the substances are needed.

8.4 Distillation Tower Sizing

Based on the reflux ratio and the state of the feed, the flow rates in the towers are determined. Based on these flow rates, and the flooding velocity, the minimum diameter for each tray can be determined [Biegler et al., 1997]. It is assumed that valve trays are used for all the towers. The calculations are carried out by use of PRO/II. The largest minimum tray diameter for a tower is then set to be the

diameter of the entire tower. If the diameter of the tower is larger than 4 meters, the distillation tower is divided into several towers, so the diameter of each tower becomes lower than 4 m.
The height of the distillation towers are found from Eq. (8.6) [Biegler et al., 1997].

$$H = \text{Tray stack} + \text{Extra feed space} + \text{Disengagement space} + \text{skirt height} \tag{8.6}$$

Here the tray stack height is the number of trays, excluding the boiler and condenser, multiplied with the tray height, which is set to 0.6096 m. The extra feed space is set to 1.5 m. The disengagement space is set to 1.5 m for both the top and the bottom. The skirt height is set to 1.5 m. It is assumed that the maximum height of the distillation towers is 60 m. The values for the extra feed space, disengagement space and skirt height followed the recommendations by Biegler et al. [1997].
The calculated size of the different distillation towers are shown in Table 24. The table shows that distillation tower 1 and 3 must be divided into several towers in order to ensure that the towers are not too wide.

Table 24: Sizes of the distillation towers from the flowsheet in Figure 14.

Distillation tower	**T1**	**T2**	**T3**	**T4**
Separation between	B/EB	EB/DEB	S/EB	EB/T
Condenser [M kJ/h]	-15.61	-4.34	-55.98	-4.74
Reboiler [M kJ/h]	74.92	12.76	64.10	6.49
Nr. Trays	11	12	48	16
Tray stack [m]	6.71	7.32	29.26	9.75
Total height [m]	12.71	13.32	35.26	15.75
Diameter [m]	4.267	1.829	5.639	1.372

The number of necessary towers in order to keep the diameter under 4 m, can be calculated on the basis of cross section area, as shown in Eq. (8.7), where the necessary numbers of towers is rounded up from the calculation to the next integer.

$$\text{Nr. of towers} = \frac{(\text{Calculated diameter})^2}{(\text{Maximum diameter})^2} \tag{8.7}$$

Table 25 shows the size of the towers after T1 and T3 are divided into two towers each. Since all towers are significantly lower than 60 m it is assumed that none of them needs to be placed in a serial connection.

Table 25: The updated sizes of the distillation towers from the flowsheet in Figure 14.

Distillation tower	**T1**	**T2**	**T3**	**T4**
Separation between	B/EB	EB/DEB	S/EB	EB/T
Nr. of towers	2	1	2	1
Condenser [M kJ/h]	-7.80	-4.34	-27.99	-4.74
Reboiler [M kJ/h]	37.46	12.76	32.05	6.49
Nr. Trays	11	12	48	16
Tray stack [m]	6.71	7.32	29.26	9.75
Total height	12.71	13.32	35.26	15.75
Diameter [m]	3.017	1.829	3.987	1.372

The size of the condensers and reboilers for the distillation towers are calculated in section 8.8.

8.5 Compressor Sizing

The five compressors are then sized. It is assumed that the compressors are adiabatic, whereby the theoretical work can be calculated from the enthalpies:

$$W = f \cdot [H_v(P_2, T_2) - H_v(P_1, T_1)] \tag{8.8}$$

Where f is the molar gas flow rate through the compressor. This is the theoretical work, which is required for the compression of the flow rate. By considering the efficiencies of the compressors, the brake horsepower can be calculated from:

$$W_b = \frac{W}{\eta_m \cdot \eta_c} \tag{8.9}$$

It is assumed that the compressors are carried out by a shaft-driven electric motor, whereby the efficiency of the motor (η_m) is assumed to be 0.9. For compression, the efficiency (η_c) is assumed to equal 0.8. Thereby, it is possible to calculate the actual power in order to perform the desired compression, by approaching Table 21 and Table 22 for the concerned values of flow rates and enthalpies.

For two of the compressors (C2 and C3), there are pressure build-ups to respectively 37.5 atm and 24.67 atm. Biegler et al. [1997] invokes a compression ratio guideline of 2.5, which means that C2 and C3 have to be divided into several compressors in series in order to achieve the desired pressures. To do this, the following equations are used:

$$W = f \cdot \left(\frac{\gamma}{(\gamma - 1)}\right) \cdot R \cdot T_i \cdot \left((\mathrm{P_{i+1}/P_i})^{(\gamma - 1)/\gamma} - 1\right) \tag{8.10}$$

$$T_{outlet} = (P_{i+1}/P_i)^{(\gamma-1)/\gamma} \cdot T_i \tag{8.11}$$

Where γ is the ratio between C_p and C_v. These equations assumes an ideal system, which is not completely precise at these pressures, but for preliminary designs, it is suitable. The resulting works gets slightly higher for Eq. (8.10) than by using Eq. (8.8), but this difference is assumed to be adequate.
The compressors in series are decided to be operated adiabatically, which means, that the temperature between the compressors are increasing, and more and more work is required for the compressors.
The number of stages, N, is found by using Eq. (8.12), where P_0 is the pressure in the inlet of the first compressor and P_N is the outlet pressure.

$$(P_N/P_0)^{1/N} = 2.5 \tag{8.12}$$

Which means that four compressors in series are required for both C2 and C3, with a compression ratio of 2.47 and 2.23 respectively. The data for the staged relation are found from the mass balance, which yields Table 26, showing the works for each of the compressors. γ is found by dividing the C_p by C_v for the specific stream. The outlet temperature is calculated for each of the compressors, whereby it is used to determine the work of the next compressor.

Table 26: The work of the staged, adiabatic compressors in series for C2 and C3.

	W [kW]	W_b **[kW]**	W_b **[bhp]**	**P outlet [atm]**
C2-1	1676.93	2329.06	3123.33	2.47
C2-2	1896.77	2634.40	3532.79	6.12
C2-3	2145.43	2979.76	3995.93	15.15
C2-4	2426.69	3370.41	4519.79	37.50
Sum	**8145.82**	**11313.64**	**15171.84**	
C3-1	44.11	61.27	82.16	2.23
C3-2	49.20	68.34	91.64	4.97
C3-3	54.88	76.22	102.21	11.07
C3-4	61.21	85.01	114.00	24.67
Sum	**209.40**	**290.83**	**390.01**	

The remaining three compressors have been sized after Eq. (8.8), resulting in Table 27, which presents the theoretical work and the actual work for the three different compressors (C1, C4 and C5).

Table 27: The calculated theoretical and actual work of the three compressors. The actual work is given in the units of kW and horsepower.

Compressor	W [kW]	W_b [kW]	W_b [bhp]
C1	93.47	129.82	174.09
C4	75.75	105.21	141.09
C5	0.08	0.12	0.16

Biegler et al. [1997] reports that the basic configuration of a compressor is a centrifugal compressor with a carbon steel circuit and a maximum pressure of 1000 psig (about 68 atmosphere gauge), which is greater than the operating pressures of the compressors in this plant, and therefore the compressors are constructed from the just described guideline.

8.6 Pump Sizing

At the entry to tower 4, a pump is used to pump the liquid from the flash drum into the tower. The pressure increases from 0.5 atmosphere to 1 atmosphere. The brake horsepower is calculated by assuming that the liquid is incompressible:

$$W_b = f \cdot \frac{P_{i+1} - P_i}{\rho \cdot \eta_p \cdot \eta_m} \tag{8.13}$$

Where η_p is the pump efficiency, which is assumed to be 0.5 and η_m is the motor efficiency, assumed to be 0.9. The liquid flow (S45) into the pump is 115.2 kmol/h, with a corresponding density of 865.2 kg/m^3.
The brake horsepower of P1 is then calculated to 0.42 kW. Biegler et al. [1997] recommends a centrifugal pump with cast iron, which is operating below a temperature of 400 K and a maximum suction pressure of 11 atmosphere. This configuration includes the driver, coupling and base plate, which are taken into account in the costing of the pump.

8.7 Direct Fired Heater Sizing

In order to provide the required reaction heat for the conversion of EB to S, DFH are used to provide the reactor with steam, because of the high temperature (898 K). The DFHs are symbolized as simple heat exchangers, whereby it is possible to determine the heat duties. The DFHs are symbolized by E4 and E5. E4 is heating the EB stream into the reactor and E5 is supplying heat to the atmospheric air stream.
The heat duties are calculated by approaching:

$$Q = f \cdot C_p \cdot \Delta T \tag{8.14}$$

Whereby the heat duties for E4 and E5 are calculated to 26,370 MJ/h and 8,750 MJ/h, respectively. By approaching these values, the cost of the DFH can be established by using a process heater with cylindrical construction and the material is stainless steel [Biegler et al., 1997]. The type of stainless steel should withstand a temperature of above 900 K.

8.8 Heat Exchanger Sizing

All heat exchangers have been sized under the assumption that they are counter-current and shell and tube heat exchangers. Figure 15 shows a schematic drawing of the used heat exchangers, where T_1 and T_2 are respectively the temperature of the inlet and outlet high temperature stream, while t_1 and t_2 are the inlet and outlet temperature of the low temperature stream.

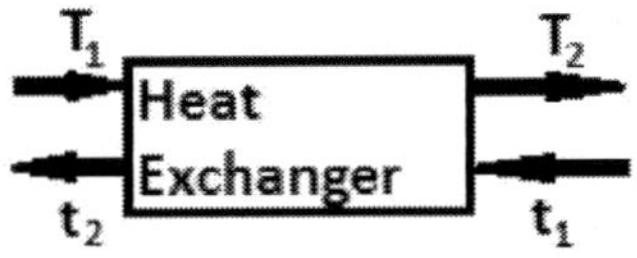

Figure 15: Schematic drawing of a heat exchanger.

The area of a heat exchanger (A) can be calculated from Eq. (8.15).

$$A = \frac{Q}{U \cdot \Delta T_{lm}} \tag{8.15}$$

Where Q is the heat duty, which was found in the energy balance and U is the overall heat transfer coefficient, which is estimated from Biegler et al. [1997]. ΔT_{lm} is the log mean temperature, which is given in Eq. (8.16).

$$\Delta T_{lm} = \frac{(T_1 - t_2) - (T_2 - t_1)}{\ln\left(\dfrac{T_1 - t_2}{T_2 - t_1}\right)} \tag{8.16}$$

When phase changes occurs in the heat exchangers, the U value changes. For these cases, the heat exchanger is divided into serial units, and U values are estimated for all the different phases. In the case of a media, which is first cooled and then condensed, the U-value is determined for both the vapour media and the condensing media. An example of the temperature profile for this case is shown in Figure 16. The total area of the heat exchanger is then calculated from Eq. (8.17), Eq. (8.18) and Eq. (8.19).

$$A_{vap} = \frac{Q_{vap}}{U_{vap} \cdot \Delta T_{lm-vap}} \tag{8.17}$$

$$A_{con} = \frac{Q_{con}}{U_{con} \cdot \Delta T_{lm-con}} \tag{8.18}$$

$$A_{total} = A_{vap} + A_{con} \tag{8.19}$$

Here A_{vap} and A_{con} are respectively the area for the part of the heat exchanger where the media is a vapour and the part of the heat exchanger where the media is condensing, and A_{total} is the total area of the heat exchanger. U_{vap} and U_{con} are respectively the overall heat transfer coefficient for the part of the heat exchanger where the media is vapour and the part where the heat exchanger is condensing. ΔT_{lm-vap} and ΔT_{lm-con} are respectively the log mean temperature for the part of the heat exchanger where the media is vapour, and the part where the media is condensing and cooling, given as:

$$\Delta T_{lm-vap} = \frac{(T_1 - t_2) - (T_c - t_3)}{\ln\left(\dfrac{T_1 - t_2}{T_c - t_3}\right)} \tag{8.20}$$

$$\Delta T_{lm-con} = \frac{(T_c - t_3) - (T_2 - t_1)}{\ln\left(\dfrac{T_c - t_3}{T_2 - t_1}\right)} \tag{8.21}$$

Here T_c is the temperature of the hot media when it begins to condense and t_3 is the temperature of the cold media when the hot media begins to condense. Generally, it is decided that the area of the heat exchangers must not be larger than 1000 m^2 [Biegler et al., 1997], heat exchangers larger than this value is divided into several heat exchangers in parallel connection.

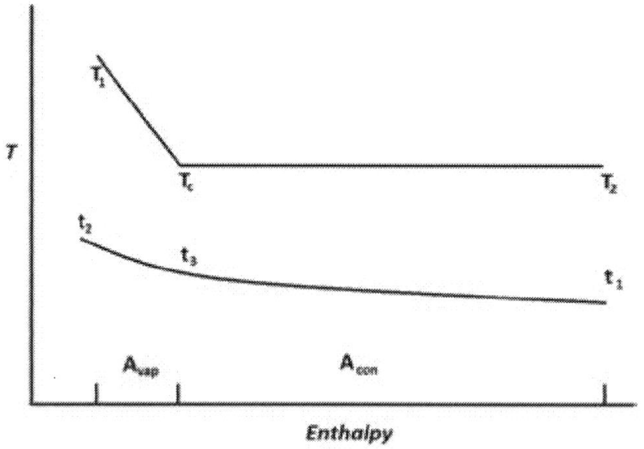

Figure 16: A typical temperature profile of condensation of a gas.

Table 28 shows total duty and area of the different heat exchangers. For heating and evaporation, hot steam is used, while water is used for the cooling and condensation.

The size of the condenser and reboiler for the different distillation towers are shown in Table 29. For the calculation of the condensers it is assumed that water is used for the cooling media, while for the calculations of the reboilers, hot steam is assumed to be the heating media. It is also assumed that the energy is only used

Table 28: Sizes of heat exchanges from the flowsheet in Figure 14. Here "cool", "heat", "con" and "vap" stands for respectively cooling, heating, condensation and evaporation.

Heat Exchanger	**E1**	**E2**	**E3**	**E6**	**E7**
Operation	Heat+vap	Cool+con	Heat+vap	Cool + con	Cool+con
Q_{total} [MJ/h]	16329	68952	680	89204	4631
A_{total} [m^2]	49.0	659.8	7.7	324.5	30.0
T_{feed} [K]	349.1	369.4	369.1	1034.1	381.1
$T_{Product}$ [K]	391.5	320	458.8	300	354.2
Cooling/Heating media infomation					
Media	Steam	Water	Steam	Water	Water
F_{media} [ton/h]	70.2	1649	2.92	2133	110.7
T_{in} [K]	450	288.15	470	288.15	288.15
T_{out} [K]	449	298.15	469	298.15	298.15

to make phase shifts. Table 29 shows that none of the condensers or boilers needs to be divided into two, since they all have an area below 1000 m^2 [Biegler et al., 1997]. It is important to stress that the used method for calculating the area of the heat exchanger, is a simple method, and it is therefore rather inaccurate. An example of the calculation for a heat exchanger is shown in Appendix C.

Table 29: Areas and duties for the condensers and reboilers from the flowsheet in Figure 14. It is assumed, that water is used as the cooling media for the condensers and condensing steam as the hot media for the reboilers. The utilities are given per heat exchanger.

Distillation tower	**T1**	**T2**	**T3**	**T4**
Separation between	BZ/EB	EB/DEB	ST/EB	EB/Tol
Nr. of towers	2	1	2	1
Condenser				
T_{product}	354	407	355	369
Q_{total} [GJ/h]	-7.80	-4.34	-27.99	-4.74
A [m^2]	63.4	9.07	223.4	30.8
$F_{\text{cooling water}}$ [ton/h]	186.6	103.8	669.4	113.3
$T_{in,\text{ cooling water}}$ [K]	288.2	288.2	288.2	288.2
$T_{out,\text{ cooling water}}$ [K]	298.2	298.2	298.2	298.2
Reboiler				
$T_{product}$ [K]	403	422	368	409
Q_{total} [GJ/h]	37.5	12.8	32.0	6.49
A [m^2]	197	115	95.9	39.2
$F_{\text{heating steam}}$ [ton/h]	188	64	322	33
$T_{in,\text{ heating steam}}$ [K]	450	450	450	450
$T_{out,\text{ heating steam}}$ [K]	449	449	449	449

Chapter 9: Costing of the Equipments and Utilities

The costing calculations of all the above described equipments are then performed. The method is based on scaling, where the cost of an equipment is estimated by using data from a similar piece of equipment but with a different capacity, by:

$$\text{Cost of equipment a (C)} = (\text{Cost of equipment b}) \cdot S^{\beta} = \alpha \cdot S^{\beta} \quad (9.1)$$

Where S is the the ratio between the capacity of equipment a and b, and β is a cost capacity exponent which varies dependently of the type of equipment.
It is then possible to determine the cost of the equipments, by collecting data of the base equipments, i.e. capacity of equipment b, α and β for each of the equipment types. Both Biegler et al. [1997] and Timmerhaus et al. [2003] gives data for these parameters for various types of equipment. It is decided to use the data from Timmerhaus et al. [2003], since this data is from January 2002, and the data from Biegler et al. [1997] is from the research of Guthrie, which is from 1969. However, it is possible to update the costs by using an update factor (UF), but since prices are changing considerably with time, due to changes in economic conditions, it is chosen to use the newest data. The method of updating the prices becomes more inaccurate, when the time period increases. The update method, to account for the inflation, uses UF which is defined as:

$$\text{UF} = \frac{\text{Present Cost Index}}{\text{Base Cost Index}} \quad (9.2)$$

The Chemical Engineering Plant Cost Index is used as an index for the UF. Timmerhaus et al. [2003] reports that the index is 390.4 in 2002. By using data from 1957 to 2002 from both Biegler et al. [1997] and Timmerhaus et al. [2003], it is possible to extrapolate to the value in 2014 by constructing a trendline, which is shown in Figure 17.

It is worth to note, that the index for 1979 is omitted, because it reduces the R^2-value significantly. It can be discussed, whether the trendline only should describe the years from about 1990 and forward, since the course is flattening out. However, it is decided to use all the data points - except from the index from 1979.

By approaching the trendline for Figure 17, the index for year 2014 is calculated to 521, which then gives a UF of:

$$\text{UF} = \frac{521}{390.4} = 1.33 \quad (9.3)$$

In order to get an updated base cost, the equipment cost (C) is multiplied with UF and furthermore MPF (Material and Pressure correction Factor):

$$\text{Updated Base Cost} = \text{C} \cdot \text{UF} \cdot \text{MPF} \quad (9.4)$$

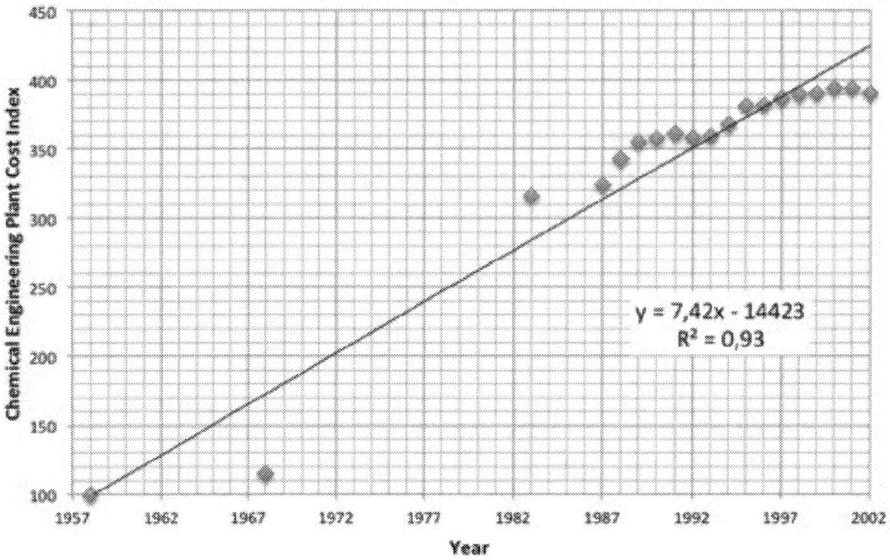

Figure 17: Chemical Engineering Plant Cost Index as a function of year from 1957 to 2002, with the respective linear trendline.

MPF corrects for special materials and high pressures. Thereby, the updated base cost of an equipment can be found for all the equipment by using the above equations.

Timmerhaus et al. [2003] uses log-log plots of capacity versus cost for each of the equipment types (plot of Eq. (9.1) with the base equipment data entered). It means that it is possible to read off the base cost of an equipment, when having the capacity.

A calculation example of a compressor is in Appendix D by using a log-log plot for compressors. Here it is found that the specified compressor (C4) has a base cost of 93,100 $ and a utility cost of 50,228 $ per year.

The equipment costings are performed using the software ECON, which contains the cost model, data and method by Timmerhaus et al. [2003]. All the log-log plots are therefore incorporated in the software which makes it easier to get a more precise purchase cost of the equipment. The utilities of the equipments are calculated based on the utility prices defined by ECON. They are calculated by multiplying the size of the utility by the corresponding price.
All the equipment defined in Chapter 8 are then costed by the described method, giving Table 30-33.

Table 30 shows the purchase cost and utility cost for compressors and pump. It is worth noting that C2 and C3 consists of four compressors each, but it is only the

Table 30: Purchase cost and utility cost for the material handling equipment, including compressors and pump.

Unit	**Purchase Cost [$]**	**Utility Cost [$]**
	Compressor	
C1	68,801	35,925
C2	8,449,822	5,419,247
C3	313,281	164,920
C4	94,389	50,228
C5	114,767	61,791
Sum	9,041,060	5,732,111
	Pump	
P1	11,632	201
Sum	11,632	201

total unit that is shown. It is also observed that both equipment cost and utility cost of C2 is very high, which is linked to the high amount of materials that have to be compressed every hour because of the huge amount of recycled BZ - and a pressure increase from 1 atmosphere to 37.5 atmosphere.

Table 31: Purchase cost and utility cost for the heat transfer equipment, including heat exchangers, direct fired heaters and cooling/ heating jackets for reactors.

Unit	Purchase Cost [$]	Utility Cost [$]
Heat Exchanger		
E2	38,506	250,088
E3	10,144	1,061,169
E4	5,328	44,270
E5	63,289	193,454
E8	6,852	12,897
Sum	124,119	1,561,878
Direct Fired Heaters		
H1 (E4)	839,072	223,470
H2 (E5)	394,716	74,151
Sum	1,233,788	297,621
Cooling/ heating jacket for reactor		
U-R1-Cooling	-	147,126
U-R2-Heat	-	5,605
Sum	-	152,731

Table 31 shows the purchase and utility cost of the heat exchangers and cooling/heating jackets for the reactors, where the reboilers and condensers are not included. Especially E3 is very expensive regarding the utility cost, because a relatively large amount of steam required. It is assumed that the purchase cost of the jackets are zero, thus it is only the steam utility and cooling water utility that are included. H1 and H2 are DFHs. The used utility of those are coal.

Table 32: Purchase cost for the reactors and flashes.

Unit	Purchase Cost [$]
Flash	
F1	11,065
F2	12,228
Sum	23,293
Reactor	
R1	758,299
R2	106,254
R3	236,968
Sum	1,101,521

Table 32 provides the purchase costs of the three reactors and the two flashes. R1 is significantly larger than the two others, again because of the large amount of BZ that has to be present in the reactor in order to maintain a high conversion. Lastly, the purchase cost and utility cost of the tower units, which includes column, trays, condenser and reboiler, are presented in Table 33.

The units T1 and T3 are split into two towers each, but are given as a total price in Table 33, which is also valid for the corresponding condensers and reboilers. The equipment price for the towers and trays for unit T3 are highest due to the height. It is seen that the utility costs of the reboilers of T1 and T3 constitutes a major part of the total utility cost. T1 has a large reboiler duty due to the large amount of BZ that has to be distilled off. T3 has a high duty because of the binary VLE diagram between ST and EB for the given conditions. The gap between the diagonal and the XY-composition plot is relatively small, giving a more difficult separation (higher tower, higher duty) - especially when approaching a separation degree of 99 %.

The total purchase cost of all the equipment is then summed to 12,937,717 $ and the utility cost becomes 18,191,035 $. Thereby, the updated base costs of the equipment have been found, and the corresponding utilities for 330 operation days per year.
The next step is then to perform the economic evaluation which determines, if the constructed base case has the potential to generate enough money for the investors.

Table 33: Purchase cost and utility cost for the tower units including column, valve trays, condenser and reboiler.

Unit	Purchase Cost [$]	Utility Cost [$]
	Tower Including Trays	
T1	275,376	-
T2	88,804	-
T3	766,708	-
T4	62,604	-
Sum	1,193,492	-
	Condenser	
T1-c	24,212	43,176
T2-c	5,328	12,056
T3-c	58,750	157,006
T4-c	8,134	13,177
Sum	96,424	225,415
	Reboiler	
T1-r	53,064	4,804,558
T2-r	18,167	826,801
T3-r	32,012	4,166,554
T4-r	9,145	423,166
Sum	112,388	10,221,079

Chapter 10: Economic Evaluation

The economic evaluation begins by calculating the capital cost and operating cost associated with the process. The capital cost are the one-time investments that the investors have to invest in order to build and start the plant. The operating costs are continuous expenses, given on an annual basis, including raw materials, utilities, labour, R&D, administration etc.

10.1 Capital Cost

The capital cost is divided into three major parts - the direct costs, the indirect costs and the working capital investment (WCI). The direct costs include the equipment cost, piping, buildings, installation etc. The indirect costs include the engineering, supervision, contingency, contractor's fees etc. The WCI represents funds required to fill the raw material tanks, goods in process, e.g. catalysts, and one month of production costs, hence the initial payroll and expenses from the production can be paid from the start.
The different parts of the capital cost are found by a percentage of the cost of the purchased equipment including the delivery. Since the plant is a fluid processing plant, Timmerhaus et al. [2003] gives percentages for the different expenses. The percentages are based on additions to an existing site, which means that transportation facilities, supplying service facilities, loading terminals and so on are present at the land purchased. In Table 34, the different expenses are shown with the corresponding percentages of the delivered equipment cost and the resulting expense of each of the costs.

Since the percentages are based on the equipment cost including delivery, the delivery cost should be estimated. Timmerhaus et al. [2003] recommends that the cost of the total delivery should be 10 % of the base equipment cost. The base cost of the equipment is calculated to 12,937,717 \$, which means that the delivery cost is 1,293,771.7 \$. The purchased equipment cost including delivery is then 14,231,488.7 \$, which is then the base for costing the capital cost.

Table 34 yields a total capital investment of 84,392,727.99 \$, which is the amount that the investors have to invest in order to construct the plant and start the operation of it. It is worth to note that the WCI is calculated as 89 % of the purchased equipment cost delivered.

Table 34: The total capital investment cost including direct cost, indirect cost and working capital investment.

Direct Cost	**Percent of delivered equipment**	**Result [$]**
Purchased Equipment Delivered		$ 14,231,488.70
Purchased Equipment Installation	47%	$ 6,688,799.69
Instrumentation and Controls (installed)	36%	$ 5,123,335.93
Piping (Installed)	68%	$ 9,677,412.32
Electrical Systems (Installed)	11%	$ 1,565,463.76
Buildings (Including Services)	18%	$ 2,561,667.97
Yard Improvement	10%	$ 1,423,148.87
Service Facilities (Installed)	70%	$ 9,962,042.09
	Total Direct Cost	$ 51,233,359.32
Indirect Cost		
Engineering and Supervision	33%	$ 4,696,391.27
Construction Expenses	41%	$ 5,834,910.37
Legal Expenses	4%	$ 569,259.55
Contractor's Fees	22%	$ 3,130,927.51
Contingency	44%	$ 6,261,855.03
	Total indirect cost	$ 20,493,343.73
Fixed-capital Investment		
Fixed-capital Investment		$ 71,726,703.05
Working Capital Investment		
Working Capital Investment	89%	$ 12,666,024.94
	Total Capital Investment (TCI)	**$ 84,392,727.99**

10.2 Operating Cost

The next step is then to determine the operating cost, which includes a large range of different subjects. Operating cost is divided into two categories; manufacturing costs and general expenses. The manufacturing costs are divided into the categories; variable production cost, fixed charges and plant overhead costs. The variable cost includes raw materials, operating labour, utilities, maintenance, repairs, operating supplies, laboratory charges and royalties. The fixed charges include taxes, interest, insurance and rent. Depreciation are included later in the evaluation. Plant overhead includes every thing from plant superintendence over safety to restaurant visits. The general expenses covers the costs of administration, distribution, selling and R&D. Timmerhaus et al. [2003] suggest factors that can determine a major part of the just described factors, but some of the expenses must be determined by considering the plant in greater detail, including raw materials, operating labour and utilities.

First, the annual cost of the raw materials are determined, which is done from the stream summaries in Table 21 and Table 22. The amount of Et and BZ added to the process every hour are found, and the catalyst prices are added, which yields Table 35. Behr [1999] reports that the specified catalysts have a lifetime of approximately two years after which the catalyst beds are replaced. The deactivation of catalyst is due to loss of promoters and poisoning by impurities. The WCI takes the purchase of the catalysts into account, but because of the relatively short lifetime, it is decided to include the price of the catalysts as a raw material as well. The price of the catalysts are therefore presented as an annual price. The compound that can be added to the process, in order to prevent and inhibit polymerization of ST and EB, are not taken into account in this pricing, since more research is needed to account for the effectiveness etc.

Table 35: Total raw material prices. The raw materials are ethylene and benzene, and the catalysts are Zeolite-Y (for R1 and R2) and an iron oxide catalyst for R3. The prices of the catalysts are found from Alibaba Group [2014].

Compound	**Amount [kg/y]**	**Price [$/kg]**	**Annual Price [$/y]**
Benzene	145,567,666	1.02	148,479,018
Ethylene	47,992,129	1	47,992,129
R1-Catalyst	9,177	6	55,062
R2-Catalyst	305	6	1,830
R3-Catalyst	28,355	6.8	192,814
Total raw material price including catalysts			**196,720,853**

The amount of operating labour is estimated to be 20 workers, 2 managers and 2 lab technicians. By estimating the different salaries to respectively 55,000 $/y,

93,000 $/y and 37,000 $/y [Indeed, 2014, Find The Data, 2014], the resulting salary consists of around 2 % of the total fixed capital investment (FCI).

The utilities are estimated in Chapter 9 to be 18,191,035 $ in total. It is then possible to perform the operating cost calculations from the guidelines given by Timmerhaus et al. [2003]. This is seen in Table 36. Here, the factors which determines the costs of a various number of items, are given together with the basis. As an example, the operating labour cost is 2 % of the FCI, and operating supervision is given as 15 % of the cost of the operating labour cost.

Table 36: The total product cost (operating cost), which includes the manufacturing costs and the general expenses. The items of operating cost are given as a percentage of a basis.

Items of Operating Cost	Factor [%]	Basis	Cost, [$/year]
Raw Material			196,720,853
Operating Labour	2	Fixed Capital Investment	1,434,534
Operating Supervision	15	Operating Labour	215,180
Utilities			18,191,035
Maintenance and Repairs	6	Fixed Capital Investment	4,303,602
Operating Supplies	15	Maintenance and Supplies	645,540
Laboratory Charges	15	Operating Labour	215,180
Royalties	1	Total Product Cost	2,512,532
		Variable Cost	**224,238,456**
Property Taxes	2	Fixed Capital Investment	1,434,534
Financing (interest)	0	Fixed Capital Investment	-
Insurance	1	Fixed Capital Investment	717,267
Rent	0	Fixed Capital Investment	-
		Fixed Charges	**2,151,801**
Plant Overhead	60	Labour + Supervision + Maintenance	3,571,990
		Plant Overhead Costs	**3,571,990**
		Manufacturing Cost	**229,962,247**
Administration	20	Labour + Supervision + Maintenance	1,190,663
Distribution & selling	4	Total Product Cost	10,050,126
Research & Development	4	Total Product Cost	10,050,126
		General Expense	**21,290,915**
Total Product Cost without Depreciation			**251,253,152**

The total product cost without depreciation is then determined to 251,253,152 \$ from Table 36 by using the guidelines provided by Timmerhaus et al. [2003].

10.3 Evaluation

The next step is then to perform an evaluation in order to evaluate the profitability of the process by introducing some key measures.

It is assumed that it takes three years to construct the plant. In these years, the expenses are the total capital investment and land cost. It is assumed, that the land is bought for 2,000,000 \$. The construction inflation rate is assumed to be 2 % per year, which is also the case for the inflation rates of the sales price of the product and the total product manufacturing cost.

When the plant has been constructed, it is ready for operation. The profit calculations are based on an equally distributed production capacity every year, i.e. the same amount of ST is produced every year. The income tax rate is set to 35 %, which is the federal corporate tax rate in USA [Landel, 2008], and the type of depreciation used is MACRS, with a recovery period of 5 years, since it is a chemical manufacturing plant [Timmerhaus et al., 2003]. It gives depreciation rates of various sizes over the depreciation period. It is then possible to calculate the gross profit as the gap between the total annual sales and the total annual costs. The net profit is defined as the income tax subtracted from the gross profit. The net profit is increasing as the years are going because of the inflation rates of 2 % of both product price and the product cost. From these considerations and by assuming a project lifetime of 20 years, it is possible to construct the cumulative cash flow, which can be seen in Figure 18 as a function of the project lifetime.
Here it is observed that the three first years, from year -2 to 0, is used to construct the plant, whereby the cumulative cash flow is decreasing heavily. From the first year, the plant begins to operate, and generates a profit. It is shown that the break-even point is reached after about 7.6 years, which means that the plant will generate money for the investors from this point. After 20 years of operation, the cumulative cash flow ends at about 143 MM\$. In order to evaluate the profitability of the investment, the measure *Net Present Value* (NPV) is used. It is described as the difference between the cash inflow and cash outflow, when comparing the present value of money to the value of money in the future (the future money are discounted back to its present value), taking the inflation and the interest rate into account. From this, the capital investment is subtracted. If the NPV is positive it means that the investment in the plant is a good idea, compared to investing the money e.g. in a bank. In Table 37 the key measures, including NPV are found.
It means that the investors will gain an NPV of about 72 MM \$ if investing in the process, with a rate of return of 9 %, and it will take 7.6 years to earn the invested money back. In reality the investors will have earned 143 MM\$, but the present value is only 72 MM\$, because those cash flows will be spread out over 23

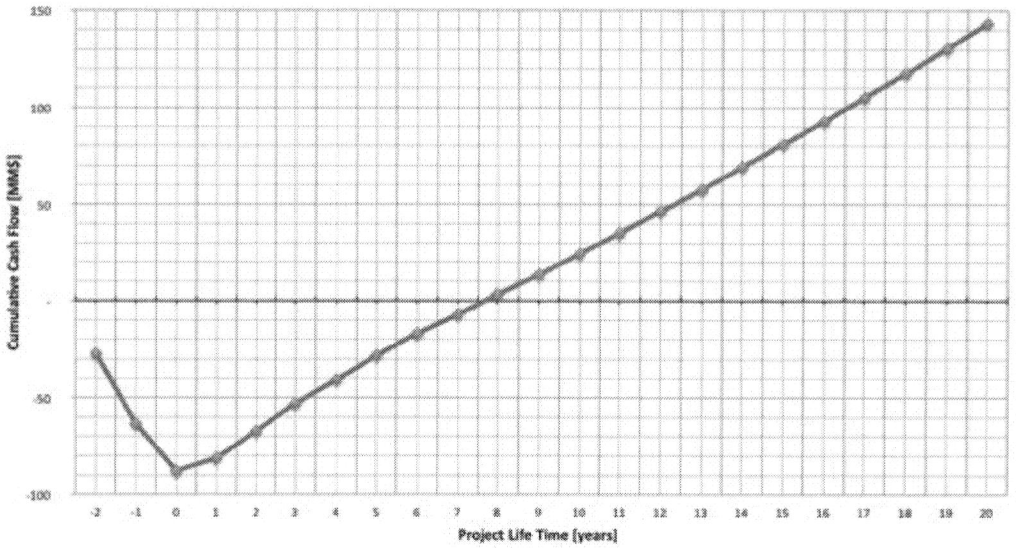

Figure 18: Cumulative cash flow in MM$ as a function of the project lifetime in years.

Table 37: Key numbers: NPV, Rate of Return and Payback Time of the process for the base case.

Measure	Value
NPV	71,731,904 $
Rate of Return	9 %/y
Payback Time	7.6 years

years and could have earned interests during those years. The rate of return is a measure for the actual earning rate of the investment, so it is known what rate of interest the investors should obtain elsewhere in order to decide to invest in the process or not. The key numbers; NPV, rate of return and payback time are then used to determine, if a change in the process is favourable or not.

It is concluded that the base case design is acceptable, because of the positive NPV and the payback time is within the foreseeable future. But, in order to attract investors, it is necessary to optimize the process even further. This is done in the next section, where the bottlenecks of the process are analysed, which determines, where optimizations are necessary in order to increase the profit and the load of the environment (minimizing the utilities, especially). Furthermore the measures given in Table 38 are also used to account for the performance of the optimization.

Here it is observed that there is used a considerable amount of utility per kg of ST. This has to be reduced. The gab between the operating cost and the sales

Table 38: Key numbers for the base case.

Measure	Value	Unit
Total Production per year	165,106,620	kg/y
Energy needed/kg of product	26.3	MJ/kg
Water needed/kg of product	381.1	kg/kg
Operating cost/ kg of product	1.52	$/kg
Raw material/ kg of product	1.17	kg/kg
Utility cost/kg of product	0.11	$/kg

price of the product is also too small, and has to be increased. The raw material consumption degree is satisfying - almost a conversion degree from raw materials to product of 0.85 is achieved.

10.4 Analysis of Purchase Cost and Utility Cost

In order to analyse the bottlenecks, the equipment and utility costs are investigated by the pie charts in Figure 19.

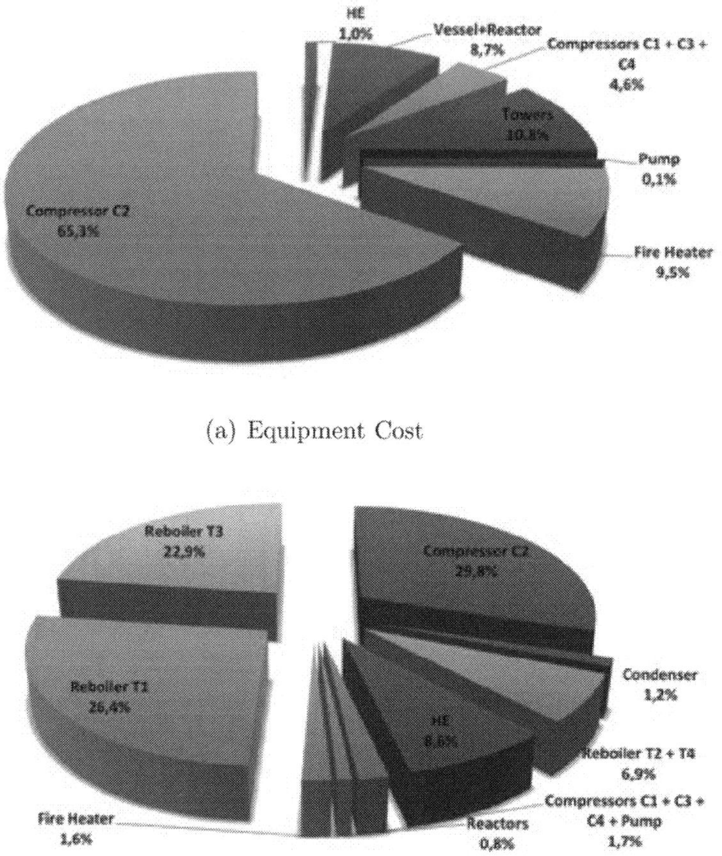

(a) Equipment Cost

(b) Utility Cost

Figure 19: Pie charts for categorizing the (a) equipment costs and (b) utility costs. HE stands for heat exchanger.

For the equipment cost C2 is categorized by itself, because of the enormous cost, which constitutes 65 % of the total equipment cost. For the utility costs, the reboilers for tower 1 and 3 as well as C2 are also categorized by itself. Reboilers for tower 1 are responsible for 26 % of the total utility cost, and 23 % for reboiler T3. The electricity cost for C2 is 30 % of the total utility cost.

This shows that it is desirable to decrease the cost for C2 by optimization. Especially the utility cost for this compressor should be optimized. It is also important to reduce the steam use of the reboilers belonging to towers T1 and T3. In order to evaluate the effect of a reduction of the equipment cost and utility costs, a sensitivity analysis is performed, Figure 20.

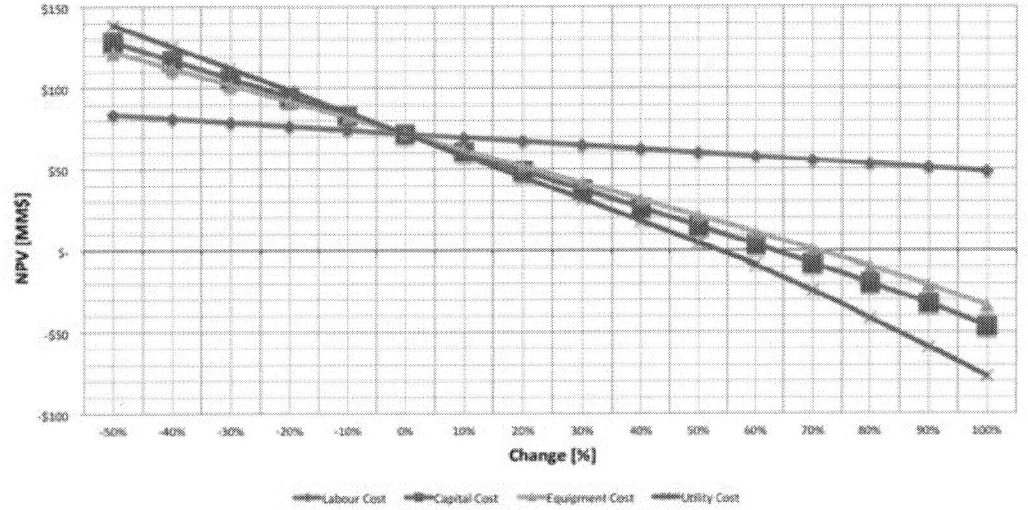

Figure 20: Sensitivity analysis. The effect on the NPV by percentage change of the labour cost, capital cost, equipment cost and utility cost.

The sensitivity analysis investigates the effect on the NPV by changing prices for different items. For Figure 20 it is seen that especially the utility price has a significant effect on the NPV. If the utility cost is reduced by 50 % the NPV increases about 100 %. A change in equipment cost affects the NPV also, but not as much as the utility cost. It can be seen that an increase in equipment cost, which results in a decrease in utility cost, could be an optimization, because the utility cost is more dependent of the change. It is also seen that the capital cost and equipment cost are almost equally dependent of a change. From the plot it can be concluded that a decrease in equipment cost and utility cost are preferred, in order to optimize the NPV, since they are most dependent of changes. But since the utility cost most sensitive, it is more optimal to decrease this cost.

Figure 21 presents the NPV as a function of the percentage change of respectively raw material prices and product price.
It is seen that just a slight change in one of those prices affects the NPV significantly - they are much more affected by a change than the case before. Therefore, it is important to make sure that the product can be sold to the specified price and that the raw materials can be bought to the specified price, since a minor change in the market can have negative consequences for the plant. As described earlier in the report, it is assumed that the prices only follows the inflation rate, but since the market is growing with a different ratio between demand and supply, the price of especially ST can fluctuate. If the price of ST is reduced by about 7 %, the NPV reaches zero. It is therefore necessary to increase the NPV, in order to make

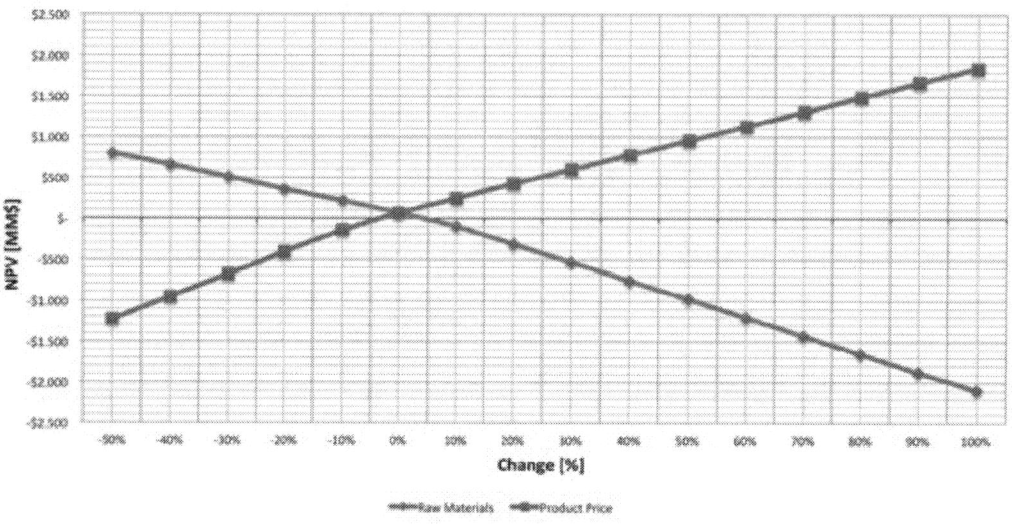

Figure 21: Sensitivity analysis. The effect on the NPV by percentage change of the raw material prices and product price.

sure that the process generates profit even though the prices are fluctuating.

10.5 Subconclusion: Economical Evaluation

The conclusion of the above analysis is that the base case design of the process is acceptable, but further optimization is required in order to increase the NPV and rate of return and decrease the payback time to below 3 years. It is particularly the utility cost that has to be decreased - and it is especially the utilities of compressor C2 and the steam to the reboilers of respectively tower unit T1 and T3 that should be decreased in order to increase the profitability of the process. By optimizing these utilities the energy need per kg of product will decrease together with the operating cost. These will affect the economic measures in a positive way.

In the next chapters, possibilities of optimizing the process and heat integration are investigated. Firstly it is analysed, whether a change in the process, by using the equipment in a different way, could have a positive effect. Hereafter, the heating/cooling sources are investigated in order to integrate the heat, and thereby decrease the utility need. Figure 22 shows a pie chart of the energy consumption. It is seen that heating utilities accounts for 38 % of the total energy consumption, and cooling accounts for 45 % of the total energy consumption.
If it is possible to integrate some of the energy used for cooling and heating the utility demand will decrease heavily, which is a gain for the environmental impact and from an economical point of view. These considerations are investigated and analysed in the following chapters.

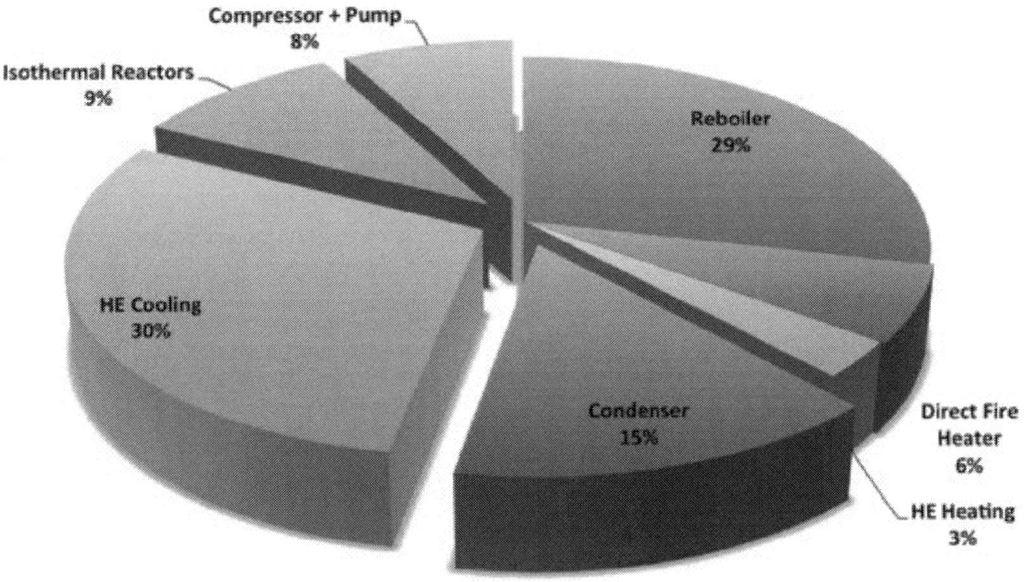

Figure 22: Pie chart of the total energy consumption, distributed by categories. The total amount of energy used is $4.34 \cdot 10^6$ GJ/y. HE stands for heat exchangers.

The waste stream (Figure 14, stream S20) consists of 69 % of BZ, 26% Tol and 4 % of EB. The residual 1 % consists mostly of H_2O and N_2, but a minor part is consisting of M and Et. In the following economic evaluation, no actions due to the waste stream are described. When having a waste stream with the desired composition it has to be cleaned at a treatment facility, so that it will not constitute a danger for the labour or the environment. In this process it is assumed for now that the waste stream can be used as utility for the heating processes. Thereby, it is possible to use the waste stream to save money on the utilities. As described before, it is also a possibility to send the waste stream to a treatment plant, but this solution is very expensive. These considerations are not taken into account in the economic evaluation.

The outlet gas stream (S26) is also a waste stream, where the majority is N_2 (90 %). The remaining 10 % consists of M, Et, BZ, H_2O (5.2 %), Tol, EB and ST. Here it is also possible to use the gas stream as a utility for the heating, thereby lowering the utility need. Again, these considerations are not taken into account when performing the economic evaluation. However, it is cheapest to use the waste streams as utility instead of sending them to a treatment plant.
Pure water is decanted from the organic phase, however, traces of e.g. organic compounds could be present, whereby cleaning of the water stream is required to meet the limits defined by the authorities. This cleaning is not taken into account in this costing.

Chapter 11: Optimization

The economic evaluation shows that heat transfer operations are the most expensive part of the utilities. This utility can be reduced by heat integration as shown in Chapter 12. In order to ensure that it is not necessary to perform the heat integration twice, the rest of the plant is first optimized. The reboiler utilities account for 56 % of the total utility cost and therefore, large savings could be achieved by optimizing the reboiler duties. The operating cost of the compressors should also, if possible, be minimized since they also account for a significant part of the utility cost, about 31 %.

11.1 Optimization of Distillation Towers

To cope with the large expense of the utility for the reboilers, optimization is investigated in this section. It is done by finding the most optimal feed tray location in order to minimize the heat duties of the reboilers. By plotting the duty of the reboiler as a function of the feed tray location, it is possible to find the minimum heat duty needed. It is the reboiler duties that are optimized, because steam is much more expensive than cooling water. It is also considered to optimize the towers, by making high pressure distillation, however, this would increase the risk of polymerization of ST and EB, and it is therefore decided not to increase the pressures in the towers.

11.1.1 Separation Between Ethylbenzene and Benzene

In order to minimize the utility cost for the separation between EB and BZ in the first distillation tower, the reboiler duty is plotted against the feed tray location as shown in Figure 23. The figure shows that the feed should be fed to tray 9, since the graph has its minimum at feed tray 8.7.

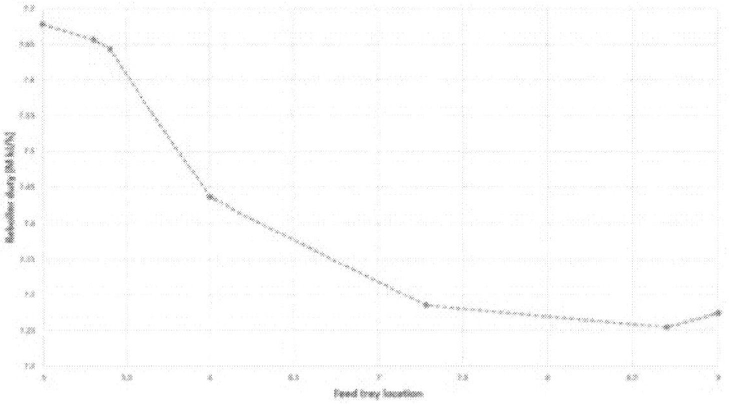

Figure 23: Reboiler duty as a function of feed tray location for separation between EB and BZ.

11.1.2 Separation Between Ethylbenzene and Diethylbenzene.

The exact same principle is carried out for the separation between EB and DEB which yields Figure 24. Based on this figure it is found that the most optimal feed tray location is tray 9.

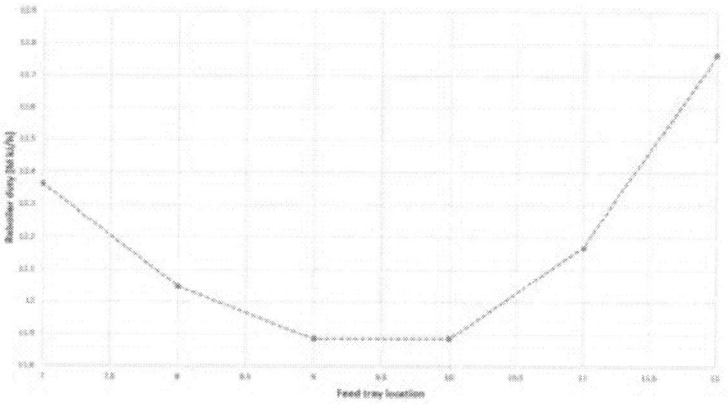

Figure 24: Reboiler duty as a function of feed tray location for separation between EB and DEB.

11.1.3 Separation Between Styrene and Ethylbenzene

The most expensive distillation tower regarding utility is tower 3, where ST and EB is separated. The utility for the reboilers to these towers accounted for 23 % of the total utility cost. In the base case distillation tower 3 had 48 theoretical trays. The literature showed that it is common to have 70 trays for this separation [Behr, 1999]. 70 stages are therefore applied for this distillation tower. This increase in number of trays decreases the reboiler utility from 64 GJ/h to approximately 43 GJ/h, depending on the new feed tray location.
The reboiler duty is again plotted as a function of the feed tray location in Figure 25. The figure shows that the feed should be introduced to the tower though tray 28 in order to achieve a minimum of utility for the reboiler.

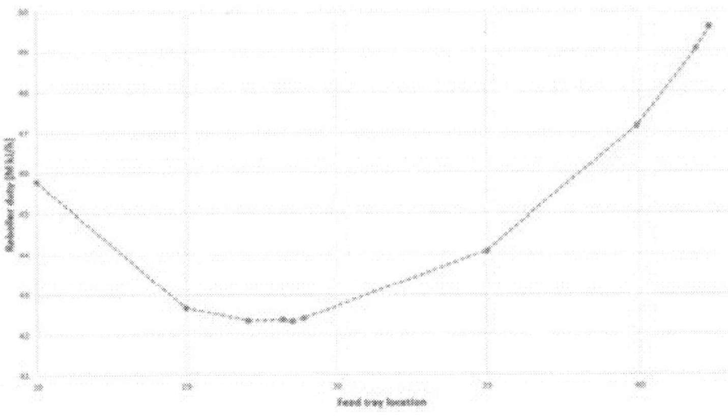

Figure 25: Reboiler duty as a function of the feed tray location for the distillation tower 3.

11.1.4 Separation Between Ethylbenzene and Toluene

In order to minimize the utility cost for the separation between EB and Tol the reboiler duty is plotted as a function of the feed tray location, which is shown in Figure 26. Based on this figure it is found that the most optimal feed tray is tray 8.

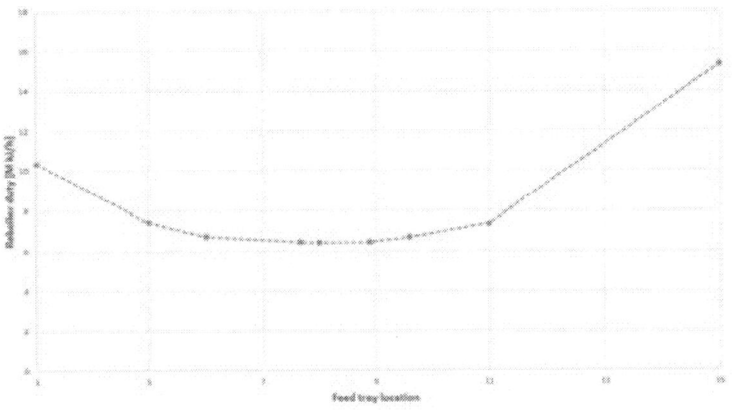

Figure 26: Reboiler duty as a function of the feed tray location for the distillation tower 4.

11.2 Optimization of Compressors

The compressors accounts for about 31 % of the total utility cost in the base case. Decreasing the compressor utility is therefore important to ensure a shorter payback time for the plant.
The most expensive compressor (C2) is located just before the feed enters reactor 1, where EB is formed. The set up for the base case is shown in Figure 27. In the base case, the feed is first heated to ensure that only vapour is present before the

pressure is build-up by use of compressors. The compressors are run adiabatically, which means that during the pressure build-up the temperature is increasing. Due to this the heat exchanger only needs to increase the temperature with 42 K, while the temperature increases with 141 K during the compression. The high temperature in the compressors, together with the large flow and pressure build-up, increases the utility cost for the compressors.

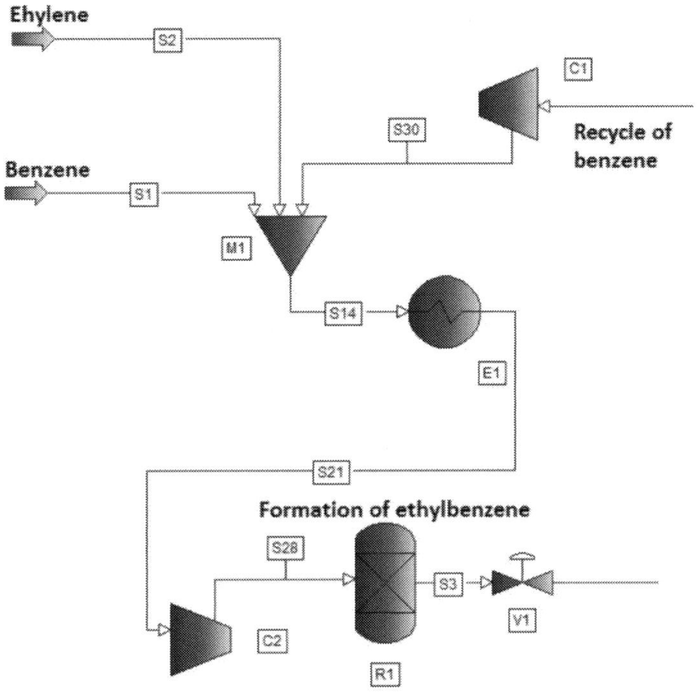

Figure 27: Base case set up for the pressure build-up before reactor 1.

In order to decrease the utility cost for this part of the plant, two different proposals for lowering the utility cost are tested.
The first proposal for an optimized case is shown in Figure 28. In this optimization the pressure is separately increased for the three streams; Et feed, BZ feed and recycle of BZ, instead of first connecting them and then increase the pressure as done in the base case. For the Et stream the pressure build-up is done by use of compressors with interstage cooling. C2 on Figure 28 represents four compressors, where each increases the pressure with a ratio of about 2.5. Interstage cooling is applied in order to reduce the utility cost for the compressors. The BZ feed stream is liquid, thus a pump is used to build-up the pressure. The recycle stream with BZ is vapour and a compressor is therefore used to increase the pressure. After the first interstage cooling the stream is only liquid and a pump is therefore used to increase the pressure to the desired 37.5 atm. The heat exchanger before the reactor, E3, should in this case increase the temperature from 300 K to 533 K.

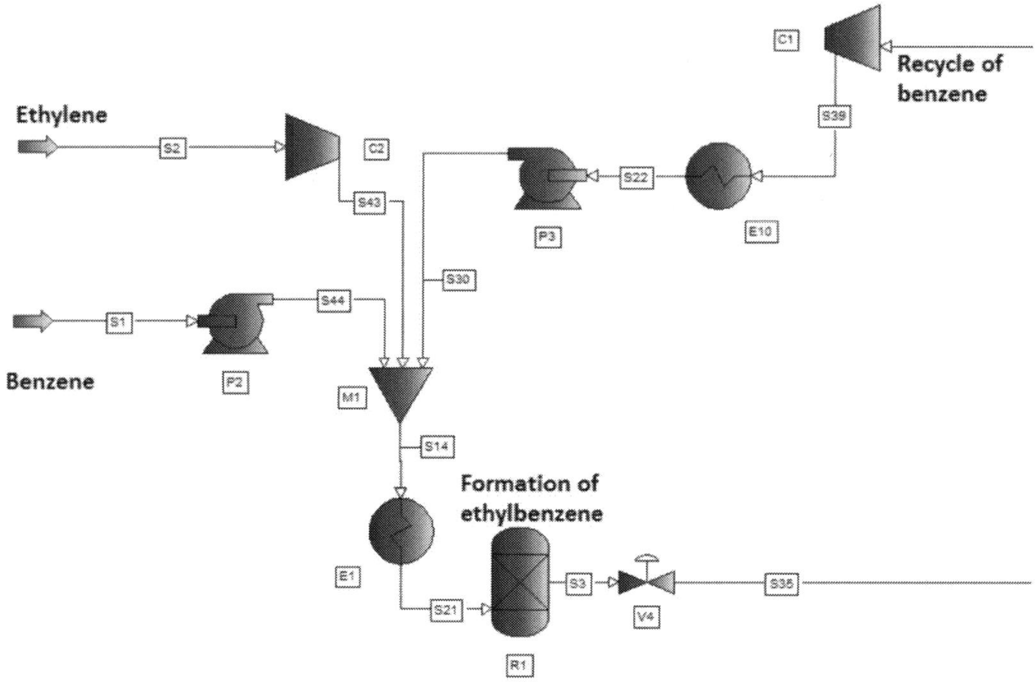

Figure 28: First proposal for optimization of the compressor utility.

Implementation of this design to the plant decreased, the utility cost for pressure increase in this part of the plant with 53 %, and decreased the equipment cost to pumps and compressors with 70 %. The new design appeared as an excellent optimization of the plant, but the utility cost for preheating of the stream in E1 increased with 417 %. Due to this increase in the utility cost for heating, the overall utility cost for this part of the plant increased with 12 %, and the plant operation would therefore cost 800,000 $ more per year. This design is therefore not applied.

The second suggestion for optimization of the compressor before the first reactor is shown in Figure 29. In this suggestion the feed streams and the recycle stream is first mixed and then send to a flash, where the vapour and gas phase is separated. The pressure is then increased to the desired pressure, with an adiabatic compressor for the vapour phase and with a pump for the liquid phase. The two streams (S30 and S22) are mixed after they have reached the desired pressure and then heated to the desired inlet temperature by heat exchanger E1. In this case the temperature increases with 170 K during the adiabatic compression. The mixed feed has therefore a temperature of 518 K when it reaches E1, which increases the temperature to 533 K.

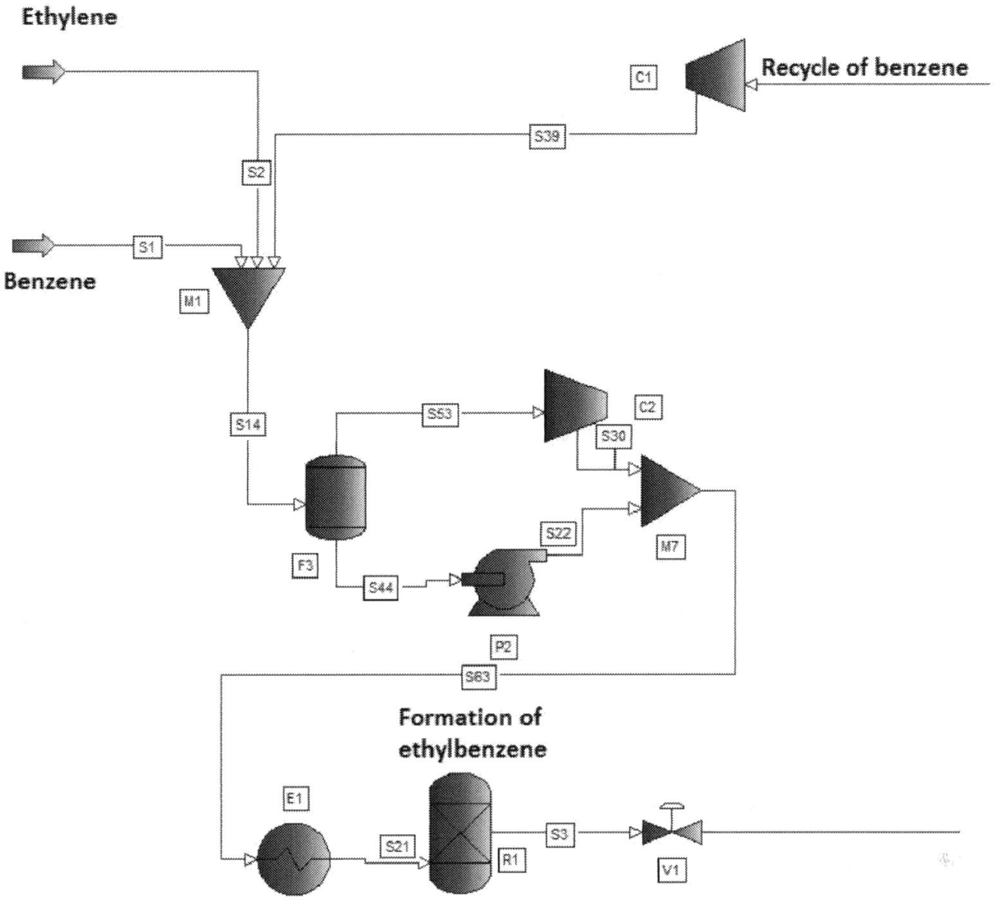

Figure 29: The second proposal for optimization of compressor utility.

This design decreases the equipment cost with 13 % and the utility cost with 7 % for this part of the plant. Under the assumption that the plant will have a life time of 20 years this improvement will save 10,000,950 $ in equipment and utility cost.

It was also investigated if the utility cost for the compressor, before the transalkylation reactor, could be reduced. This was done even though this compressor utility only accounted for 1 % of the total utility cost, because it was assumed that the base case design, shown in Figure 30, could be improved.
An improved design for this part of the plant is shown in Figure 31. Here the pressure is increased in the two streams before mixing. The BZ stream condensates after the stage cooling in E8, and a pump is therefore used to increase the pressure further.

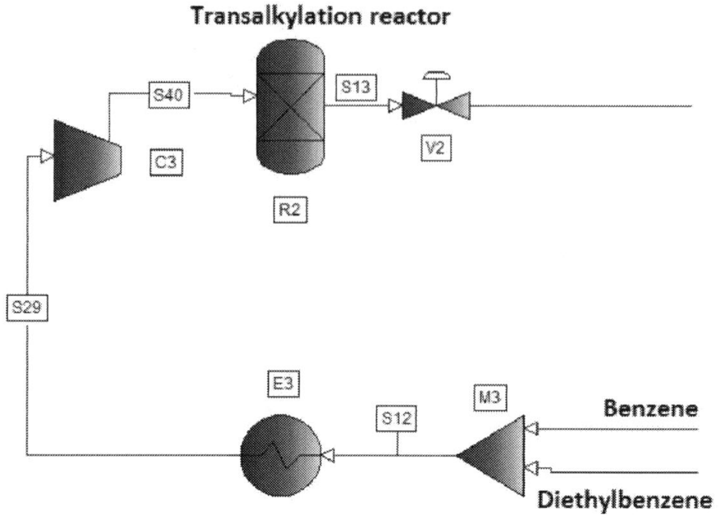

Figure 30: Base case set up for the pressure build-up before the transalkylation reactor.

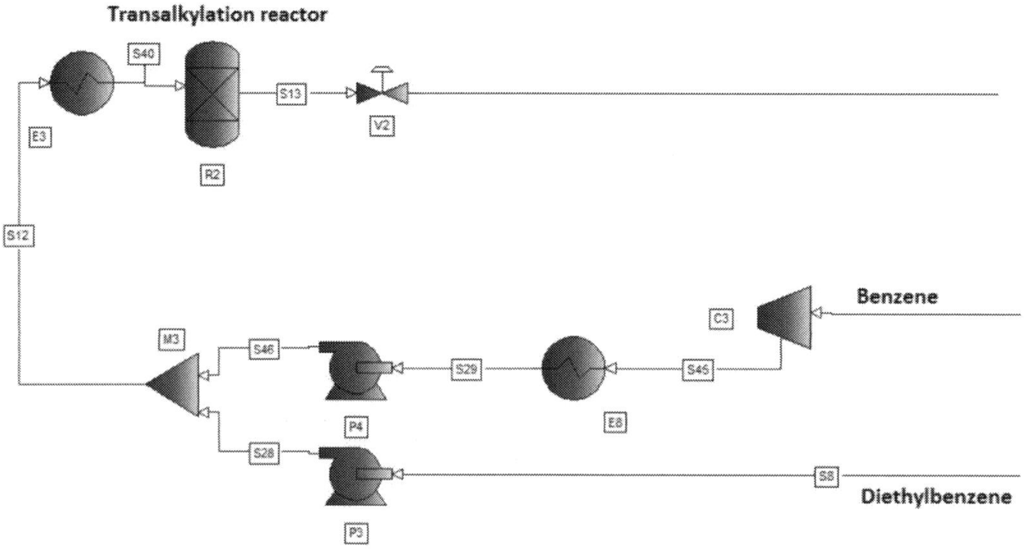

Figure 31: Set up for the optimized pressure build-up before the transalkylation reactor.

Due to this optimization the equipment cost is reduced with 68 % and the utility cost is reduced with 17 % for this part of the plant. Under the assumption that the plant has a life time of 20 years this optimization will increase the earnings with 930,000 $ and this optimization is therefore implemented.

11.3 Optimized Flowsheet

The flowsheet for the optimized case is given in Figure 32 and the stream summary is given in Table 39 and Table 40. The flowsheet shows that there is added a heat exchanger (E9) after T4, the purpose for this heat exchanger is to cool down the vapour phase from T4. This is done in order to make it easier to store the waste stream but also to give the possibility of save energy by heat integration.

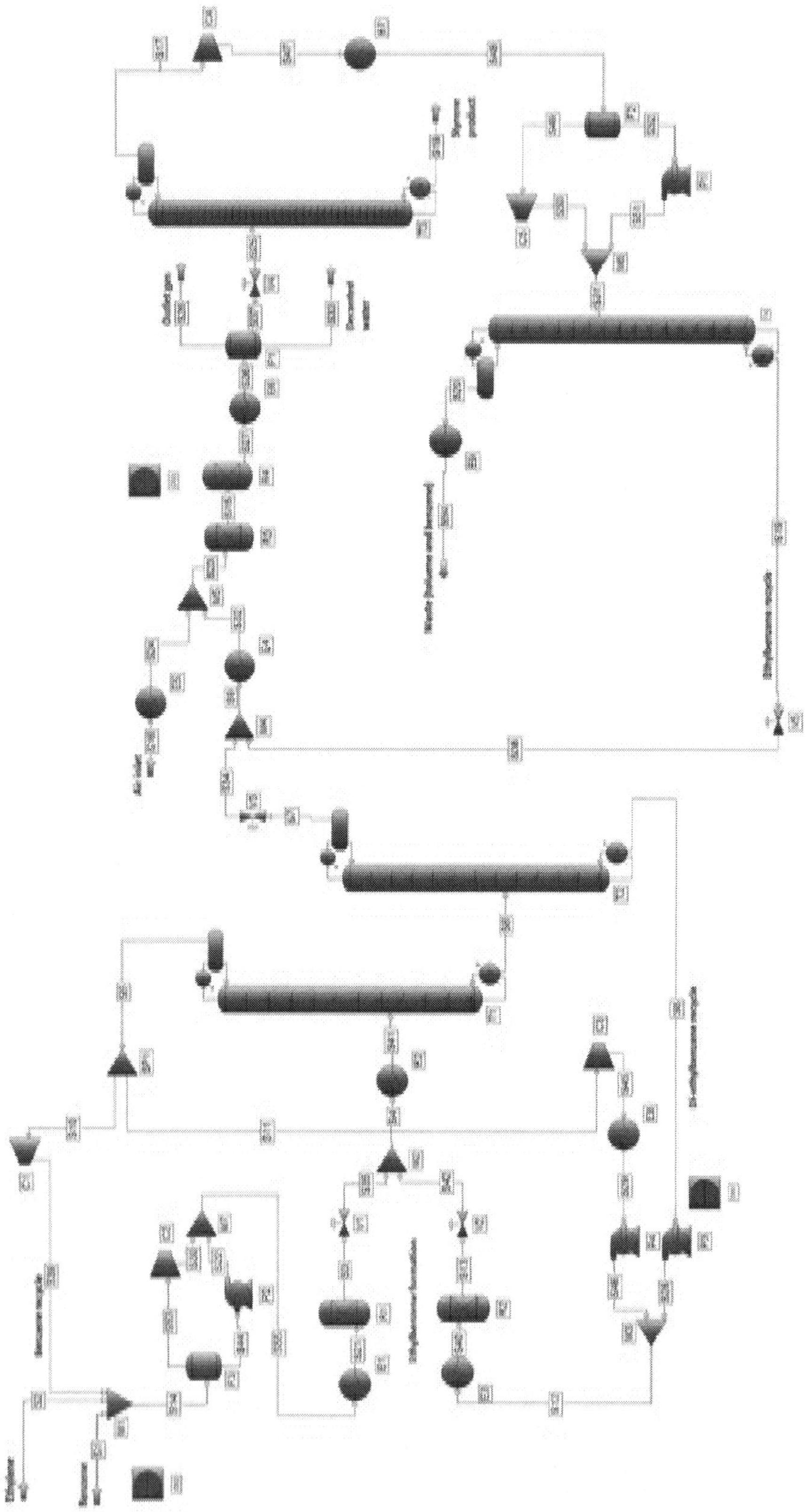

Figure 32: Optimized flowsheet.

Table 39: Stream summary of streams S1-S27 for the optimized base case design.

Stream Name	S1	S2	S3	S4	S5	S6	S7	S8	S9	S10	S11	S12	S13
Phase	Liquid	Vapor	Liquid	Vapor	Vapor	Liquid	Vapor	Liquid	Mixed	Vapor	Vapor	Liquid	Vapor
T. [K]	300.0	300.0	533.0	369.4	353.2	402.4	407.0	422.5	393.3	353.2	353.2	366.6	563.2
P. [atm]	1.0	1.0	33.5	1.0	1.0	1.0	1.0	1.0	0.667	1.0	1.0	24.7	22.2
Enthalpy [GJ/hr]	0.8	0.4	77.5	80.9	65.0	5.9	14.1	0.2	16.6	63.2	1.8	0.7	3.4
Rate [kmol/h]	230.68	216.00	1737.72	1786.92	1550.80	237.05	231.61	5.44	325.78	1507.04	43.76	49.19	49.19
H_2 [kmol/h]	0.00	0.00	0.00	0.00	0.00	0.00	0.00	0.00	0.00	0.00	0.00	0.00	0.00
N_2 [kmol/h]	0.00	0.00	0.00	0.00	0.00	0.00	0.00	0.00	0.00	0.00	0.00	0.00	0.00
O_2 [kmol/h]	0.00	0.00	0.00	0.00	0.00	0.00	0.00	0.00	0.00	0.00	0.00	0.00	0.00
Methane [kmol/h]	0.00	0.00	0.00	0.00	0.00	0.00	0.00	0.00	0.00	0.00	0.00	0.00	0.00
Ethylene [kmol/h]	0.00	216.00	0.00	0.00	0.00	0.00	0.00	0.00	0.00	0.00	0.00	0.00	0.00
Benzene [kmol/h]	230.68	0.00	1521.73	1563.30	1548.58	15.64	15.64	0.00	15.64	1504.89	43.69	43.69	41.57
H_2O [kmol/h]	0.00	0.00	0.00	0.00	0.00	0.00	0.00	0.00	1.09	0.00	0.00	0.00	0.00
Toluene [kmol/h]	0.00	0.00	0.00	0.00	0.00	0.00	0.00	0.00	0.30	0.00	0.00	0.00	0.00
Ethylbenzene [kmol/h]	0.00	0.00	213.83	221.44	2.21	219.22	215.93	3.29	306.71	2.15	0.06	3.35	7.61
Styrene [kmol/h]	0.00	0.00	0.00	0.00	0.00	0.00	0.00	0.00	2.01	0.00	0.00	0.00	0.00
Di-ethylbenzene [kmol/h]	0.00	0.00	2.16	2.18	0.00	2.18	0.03	2.15	0.03	0.00	0.00	2.15	0.02

Stream Name	S14	S15	S16	S17	S18	S19	S20	S21	S22	S23	S24	S25	S26	S27
Phase	Mixed	Vapor	Vapor	Vapor	Liquid	Liquid	Vapor	Mixed	Liquid	Vapor	Vapor	Mixed	Vapor	Vapor
T. [K]	349.0	573.2	293.0	353.2	366.4	409.3	368.8	533.0	351.4	898.0	898.0	300.0	300.0	1034.2
P. [atm]	1.0	0.6	1.0	0.2	0.2	1.0	1.0	37.5	37.5	0.667	1.0	0.2	0.667	0.667
Enthalpy [GJ/hr]	64.7	36.0	0.3	6.0	3.5	2.5	1.0	103.6	3.1	59.1	16.4	1.5	1.4	92.4
Rate [kmol/h]	1953.72	999.62	468.38	115.26	200.16	94.17	21.09	1953.72	284.55	794.16	468.38	315.42	410.65	901.26
H_2 [kmol/h]	0.00	196.72	0.00	0.00	0.00	0.00	0.00	0.00	0.00	0.00	0.00	0.00	0.00	0.00
N_2 [kmol/h]	0.00	370.02	370.02	0.11	0.00	0.00	0.11	0.00	0.00	370.02	370.02	0.11	369.91	370.02
O_2 [kmol/h]	0.00	98.36	98.36	0.00	0.00	0.00	0.00	0.00	0.00	98.36	98.36	0.00	0.00	0.00
Methane [kmol/h]	0.00	6.11	0.00	0.01	0.00	0.00	0.01	0.00	0.00	0.00	0.00	0.01	6.10	6.11
Ethylene [kmol/h]	216.00	2.63	0.00	0.02	0.00	0.00	0.02	216.00	0.27	0.00	0.00	0.02	2.61	2.63
Benzene [kmol/h]	1735.57	18.28	0.00	14.36	0.00	0.00	14.36	1735.57	283.07	15.65	0.00	14.36	3.92	18.28
H_2O [kmol/h]	0.00	1.09	0.00	1.12	0.00	1.09	0.04	0.00	0.00	1.09	0.00	1.13	21.49	197.81
Toluene [kmol/h]	0.00	6.40	0.00	5.93	0.00	0.30	5.63	0.00	0.00	0.30	0.00	5.93	0.47	6.40
Ethylbenzene [kmol/h]	2.15	95.15	0.00	91.69	0.93	90.77	0.92	2.15	1.21	306.71	0.00	92.62	2.53	95.15
Styrene [kmol/h]	0.00	204.84	0.00	2.01	199.20	2.01	0.00	0.00	0.00	2.01	0.00	201.21	3.62	204.84
Di-ethylbenzene [kmol/h]	0.00	0.03	0.00	0.00	0.03	0.00	0.00	0.00	0.00	0.03	0.00	0.03	0.00	0.03

Table 40: Stream summary of streams S28-S55 for the optimized base case design.

Stream Name	S28	S29	S30	S31	S32	S33	S34	S35	S36	S37	S38	S39
Phase	Liquid	Liquid	Mixed	Mixed	Vapor	Water	Vapor	Mixed	Mixed	Liquid	Mixed	Mixed
T. [K]	423.9	353.5	519.7	353.0	898.0	300.0	407.0	368.1	300.0	300.0	394.9	353.6
P. [atm]	24.7	2.5	37.5	1.0	0.667	0.667	0.667	1.0	0.667	0.667	0.667	1.0
Enthalpy [GJ/h]	0.2	0.5	80.0	1.6	42.7	0.4	14.1	77.5	3.3	1.5	2.5	63.5
Rate [kmol/h]	5.44	43.76	1669.18	115.26	325.78	175.19	231.61	1737.72	901.26	315.42	94.17	1507.04
H_2 [kmol/h]	0.00	0.00	0.00	0.00	0.00	0.00	0.00	0.00	0.00	0.00	0.00	0.00
N_2 [kmol/h]	0.00	0.00	0.00	0.11	0.00	0.00	0.00	0.00	370.02	0.11	0.00	0.00
O_2 [kmol/h]	0.00	0.00	0.00	0.00	0.00	0.00	0.00	0.00	0.00	0.00	0.00	0.00
Methane [kmol/h]	0.00	0.00	0.00	0.01	0.00	0.00	0.00	0.00	6.11	0.01	0.00	0.00
Ethylene [kmol/h]	0.00	0.00	215.73	0.02	0.00	0.00	0.00	0.00	2.63	0.02	0.00	0.00
Benzene [kmol/h]	0.00	43.69	1452.50	14.36	15.64	0.00	15.64	1521.73	18.28	14.36	0.00	1504.89
H_2O [kmol/h]	0.00	0.00	0.00	1.12	1.09	175.19	0.00	0.00	197.81	1.13	1.09	0.00
Toluene [kmol/h]	0.00	0.00	0.00	5.93	0.30	0.00	0.00	0.00	6.40	5.93	0.30	0.00
Ethylbenzene [kmol/h]	3.29	0.06	0.95	91.69	306.71	0.00	215.93	213.83	95.15	92.62	90.77	2.15
Styrene [kmol/h]	0.00	0.00	0.00	2.01	2.01	0.00	0.00	0.00	204.84	201.21	2.01	0.00
Di-ethylbenzene [kmol/h]	2.15	0.00	0.00	0.00	0.03	0.00	0.03	2.16	0.03	0.03	0.00	0.00

Stream Name	S40	S41	S42	S44	S45	S46	S47	S48	S49	S50	S51	S52	S53	S54	S55
Phase	Vapor	Liquid	Vapor	Liquid	Mixed	Liquid	Mixed	Mixed	Vapor	Vapor	Liquid	Liquid	Vapor	Mixed	Mixed
T. [K]	563.2	320.0	544.4	349.0	386.6	355.0	381.0	352.7	352.7	377.7	352.8	352.7	349.0	320.0	519.1
P. [atm]	24.7	1.0	1.0	1.0	2.5	24.7	0.5	0.5	0.5	1.0	1.0	0.5	1.0	1.0	37.5
Enthalpy [GJ/h]	2.7	11.9	3.4	3.0	2.0	0.5	6.3	1.6	0.0	0.0	1.6	1.6	61.5	0.1	83.1
Rate [kmol/h]	49.19	1787.84	49.19	284.55	43.76	43.76	115.26	115.26	0.21	0.21	115.05	115.05	1669.18	21.09	1953.72
H_2 [kmol/h]	0.00	0.00	0.00	0.00	0.00	0.00	0.00	0.00	0.00	0.00	0.00	0.00	0.00	0.00	0.00
N_2 [kmol/h]	0.00	0.00	0.00	0.00	0.00	0.00	0.11	0.11	0.08	0.08	0.03	0.03	0.00	0.11	0.00
O_2 [kmol/h]	0.00	0.00	0.00	0.00	0.00	0.00	0.00	0.00	0.00	0.00	0.00	0.00	0.00	0.00	0.00
Methane [kmol/h]	0.00	0.00	0.00	0.00	0.00	0.00	0.01	0.01	0.01	0.01	0.00	0.00	0.00	0.01	0.00
Ethylene [kmol/h]	0.00	0.00	0.00	0.27	0.00	0.00	0.02	0.02	0.01	0.01	0.01	0.01	215.73	0.02	216.00
Benzene [kmol/h]	43.69	1564.23	41.57	283.07	43.69	43.69	14.36	14.36	0.05	0.05	14.31	14.31	1452.50	14.36	1735.57
H_2O [kmol/h]	0.00	0.00	0.00	0.00	0.00	0.00	1.12	1.12	0.00	0.00	1.12	1.12	0.00	0.04	0.00
Toluene [kmol/h]	0.00	0.00	0.00	0.00	0.00	0.00	5.93	5.93	0.01	0.01	5.92	5.92	0.00	5.63	0.00
Ethylbenzene [kmol/h]	3.35	221.44	7.61	1.21	0.06	0.06	91.69	91.69	0.05	0.05	91.64	91.64	0.95	0.92	2.15
Styrene [kmol/h]	0.00	0.00	0.00	0.00	0.00	0.00	2.01	2.01	0.00	0.00	2.01	2.01	0.00	0.00	0.00
Di-ethylbenzene [kmol/h]	2.15	2.18	0.02	0.00	0.00	0.00	0.00	0.00	0.00	0.00	0.00	0.00	0.00	0.00	0.00

11.4 Sizing of Optimized Case

In order to investigate the effect of the optimizations, the units, affected by the changes, are re-sized. The new dimensions for the distillation towers are shown in Table 41. The table shows that distillation tower 1 and 3 still needs to be divided into two towers each, as in the base case. The distillation towers and their reboilers and condensers are sized from the principles explained in Chapter 8. The cooling streams and the hot steam streams, used to design the condenser and reboiler, are calculated with the same temperatures as in Table 29, section 8.8.

Table 41: Dimensions and duties for the distillation towers in the optimized case.

Distillation tower	**T1**	**T2**	**T3**	**T4**
Sep. Between	BZ/EB	EB/DEB	ST/EB	EB/Tol
Nr. of towers	2	1	2	1
Tower dimensions				
Nr. Trays	11	12	68	16
Feed tray location	9	9	28	8
Diameter [m]	2.910	1.829	3.449	1.372
Total height [m]	12.7	13.3	47.5	15.8
Condenser				
Duty [M kJ/h]	-6.38	-3.45	-17.23	-4.65
A [m^2]	52.1	7.22	141	30.1
Reboiler				
Duty [M kJ/h]	36.3	11.9	21.9	6.5
A [m^2]	188	107	64.6	39.3

The sizes of the heat exchangers are shown in Table 42. The table shows that in the optimized case E3 needs to heat 1.99 G/h, where it in the base case only needs to heat 680 MJ/h.

Table 42: Sizes of the heat exchangers used in the optimized the base case. Here "cool", "heat", "con", and "vap" stands for cooling, heating, condensation and evaporation, respectively.

Heat Exchanger	**E1**	**E3**	**E8**	**E9**
Operation	Heat+vap	Heat+vap	Cool+con	Cool+con
Q_{total} [GJ/h]	21.2	1.99	1.47	0.82
A_{total} [m^2]	321	14.8	9.24	7.27
T_{feed} [K]	518	367	387	369
$T_{Product}$	533	563	354	320
Cooling/Heating media info				
Media	Steam	Steam	Water	Water
Flow [ton/h]	91.2	8.57	35.2	19.6
T_{in} [K]	288	600	288	288
T_{out} [K]	549	599	298	298

The work for the optimized pumps and compressors are shown in Table 43. The table shows that the compressor before the reactor where EB is formed, still is the most energy consuming pressure build-up unit. C2 is in reality four compressors in series, but the total sum of the work is shown.

Table 43: Work for optimized pumps (P) and compressors (C).

Unit	C2	C3	P2	P3	P4
Work [kW]	9656	49	62	1.1	5.4

The flash shown on Figure 29 (F2) is sized to have a diameter of 1 m and a height of 5.8 meters.

11.5 Subconclusion Effect of Optimization

The effect of the optimizations is shown in Table 44. In the table; T1, T2, T3 and T4 stands for distillation tower 1 to 4, while PB1 and PB2 stands for respectively pressure build-up to reactor 1 and reactor 2, which includes all the equipment to increase the pressure. The table shows that the largest optimization is the utility cost for distillation tower 3, where ST and EB are separated from each other. The increase in number of trays, and optimized feed tray locations, increase the cumulative cash flow over 20 years with 27,414,000 $. The second largest optimization was in the pressure build-up to reactor 1, which increased the earnings with 10,001,000 $.

Table 44: Overview of effect of the different optimizations. Here T refers to distillation towers, and PB1 is pressure build-up to the first reactor where EB is formed, and PB2 is the pressure build-up to the transalkylation reactor.

	T1	**T2**	**T3**	**T4**	**PB1**	**PB2**
			Utility			
Base [$]	4,848,000	839,000	4,324,000	436,000	6,480,00	209,190
Optimized [$]	4,749,000	778,000	2,948,000	430,000	6,036,000	173,365
Saved [%]	2.0	7.3	32	1.6	6.9	17
			Equipment			
Base [$]	353,000	112,000	857,000	80,000	8,460,000	318,609
Optimized [$]	349,000	111,000	956,000	80,000	7,346,000	103,000
Saved [%]	1.1	0.8	-11	0.2	13	68
Total saved [$]	1,984,000	1,223,000	27,414,000	136,000	10,001,000	932,000

The overall effect of the optimization is shown in Table 45. The table shows that due to these optimizations the utility cost is reduced with 11 %, and the equipment cost is reduced with 10 %. This increases the NPV with 36 % or about 26 MM$. From an economic point of view these changes should therefore be applied to the plant. The optimization also decreases the energy consumption with 8.4 %, and the optimized plant is therefore more climate friendly.

Table 45: Effect of optimization.

	Base case	**Optimized case**	**Change [%]**
Utility cost [$]	18,191,035	16,170,652	11.1%
Equipment cost [$]	12,937,717	11,707,184	9,5%
NPV [$]	71,731,904	97,487,064	35.9%
Total energy consumption [GJ/y]	$4.337 \cdot 10^6$	$3.971 \cdot 10^6$	8.44%
Rate of return [%]	9	14	55.6%
Payback Time [y]	7.6	5.4	28.9%

Chapter 12: Heat Integration

12.1 Constructing the Heat Exchanger Network

During the economical analysis it is obvious that the utility cost for the reboilers and the heat exchangers are a quite significant part of the expenses. This implies that there is a significant room for optimization. Therefore heat integration is performed for the optimized base case.
Initially an overview of the heat exchangers, the reboilers and the condensers are performed, which can be found in Table 46.

Table 46: Overview of the heat exchangers, the reboilers and the condenser.

Unit	Streams	Flow size $\left[\frac{kg}{h}\right]$	C_p In $\left[\frac{kJ}{kg\,K}\right]$	C_p Out $\left[\frac{kJ}{kg\,K}\right]$	Heat Flow Out $\left[\frac{GJ}{h}\right]$	Temp. In [K]	Temp. Out [K]
				Cold streams			
E1	S14-S21	143,565	2.09	0.43	21.21	518.4	533
E3	S12-S40	4,055	1.95	1.75	1.99	367	563
E4	S9-S32	34,047	1.52	2.05	26.11	393	898
E7	S16-S24	13,513	1.01	1.13	8.68	293	898
T1-r		24,805			72.74	389	402.3
T2-r		638			11.88	414	422.5
T3-r		20,851			42.45	366.4	367.4
T4-r		9,894			6.46	409.2	409.3
				Hot streams			
E2	S20-S54	1.76	1.21	1.78	-0.82	369	320
E6	S27-S36	47,575	2.27	1.74	-89.76	1034	300
E7	S47-S48	11,649	1.54	1.87	-4.67	381	353
E8	S45-S29	3.42	1.45	1.89	-1.47	387	354
E9	S4-S41	147,618	1.38	1.79	-69.87	369	320
T1-c		122,814			-13.01	353.5	353.2
T2-c		24,168			-3.45	407.4	407
T3-c		11,649			-34.39	355.8	353.2
T4-c		1,755			-4.65	379.4	368.7

For the reboilers and the condensers the input temperature and the output temperature correspond to the temperature found for the unit and the temperature in the preceding tray. The "Heat flow out" is the duties in the unit.
The heat integration is investigated using the tabular method. Therefore an $F \cdot C_p$ needs to be determined for each stream. However, as it can be seen in Table 46, for most of the heat exchangers C_p is not constant across the heat exchanger. This is both due to the heat capacities temperature dependency, but also phase changes in the heat exchangers. In the previous analysis and sizing of the heat exchangers it is established that for E3, E6 and E9 the phase changes in the stream would be most precisely described as happening at a specific temperature. The energies

needed for the phase changes is calculated during the heat exchanger sizing. $F \cdot C_p$ is calculated as the rest of the "Heat Flow Out" distributed uniformly over the entire temperature change.
For the rest of the heat exchangers with phase changes it is most precisely described as a phase change happening across the entire heat exchanger. Therefore $F \cdot C_p$ is calculated as the entire "Heat Flow Out" uniformly distributed on the entire temperature change. This simplified the calculations, but also decreased the accuracy of the calculation. In spite of this the assumptions are still considered as fairly good approximations.
This gives the values that can be seen in Table 47.

Table 47: Overview of the heat exchangers with $F \cdot C_p$ and the heat duties for the phase changes.

Unit	$F \cdot C_p$ $\left[\frac{kJ}{K\,h}\right]$	**Phase Change 1** $\left[\frac{kJ}{h}\right]$	**at temp.** [K]	**Phase Change 2** $\left[\frac{kJ}{h}\right]$	**at temp.** [K]
E1	1,453,014	-	-	-	-
E2	16,755	-	-	-	-
E3	7,551	521,000	525	-	-
E4	51,703	-	-	-	-
E5	14,347	-	-	-	-
E6	121,087	49,800	380	832,000	350
E7	166,786	-	-	-	-
E8	44,545	-	-	-	-
E9	255,102	57,400,000	360	-	-

The reboilers and the condensers also need a value for $F \cdot C_p$. These are found by uniformly distributing the heat duties across a temperature interval representative for each unit.
Now ΔT for the heat exchanger network is to be chosen. A value of 10 K is chosen in order to ensure a relatively high driving force for the heat exchangers as a minimum.
Thereby all the preparations for the tabular method are finished. The next step is to set up the temperature intervals and showing the units with arrows on respectively the hot and the cold side. This can be seen in Figure 33.

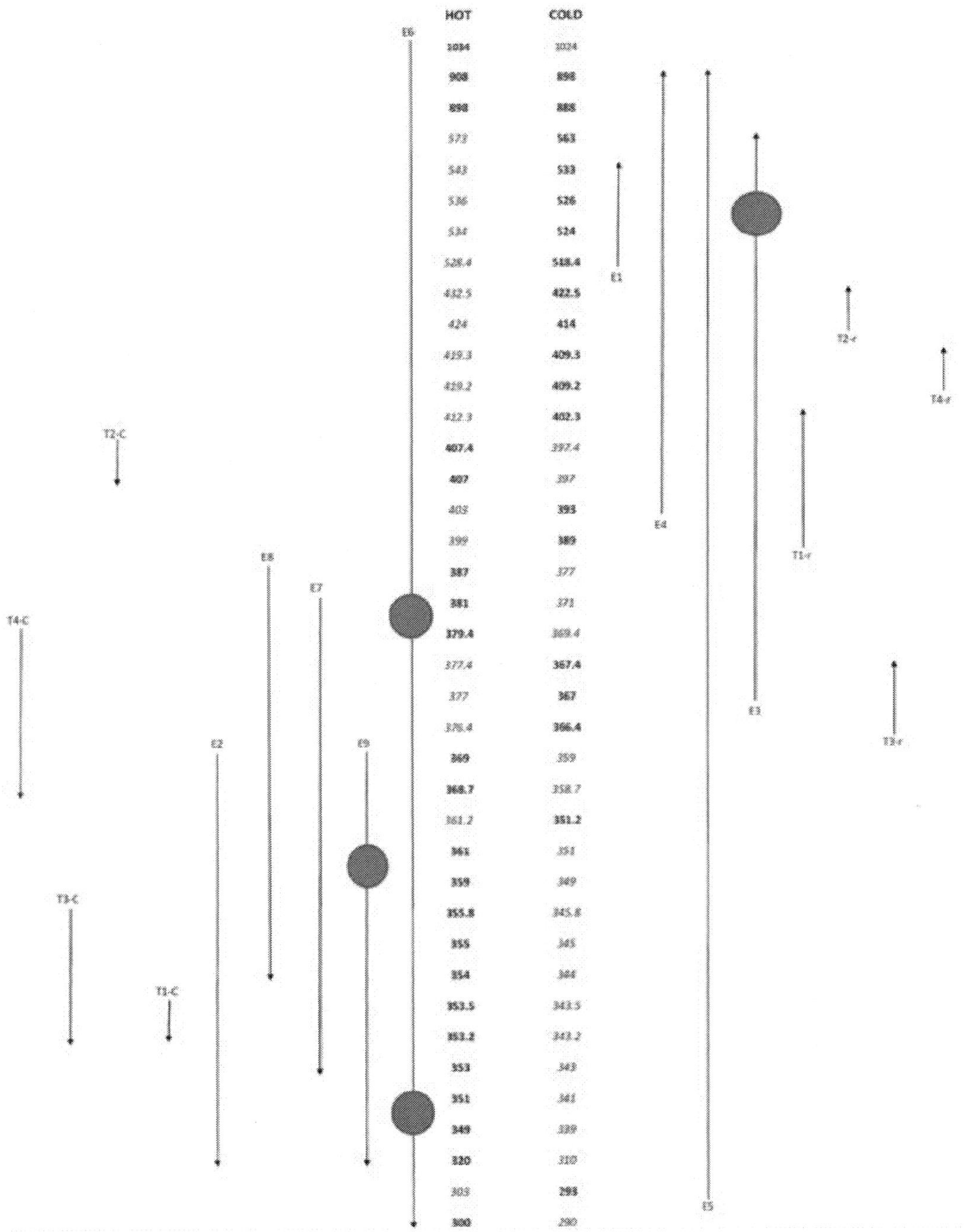

Figure 33: The tabular heat exchanger network problem. The temperature changes for each unit is represented by arrows. The phase changes happening at a constant temperature are represented by a blue circle. The temperatures written with a bold font in the middle are the ones given in Table 46 and 47. The ones in italic font are the corresponding temperatures on the opposite stream with regards to ΔT.

In this figure it can be seen that on the hot side, for the highest temperatures only, E2 can exchange energy with the cold streams, while a significant amount of the cold streams needs to reach temperatures in this region. However, since E6, according to Table 47 has a high $F \cdot C_p$, it will probably be able to exchange a significant amount of energy.
With this figure constructed it is now possible to perform the actual tabular method. The first calculations performed is the determination of the required heat and the available heat for each interval. This is determined by:

$$q = F \cdot Cp \cdot (T_{upper} - T_{lower}) + q_{\text{phase change}} \tag{12.1}$$

Then the cascaded heat for respectively the cold and the hot site can be found. This is found from:

$$Q_i = Q_{i-1} + q_i \tag{12.2}$$

Where Q is the cascaded heat. Thereby the cascaded heat is the cumulated heat. The next thing calculated is the net heat. This is calculated as the difference between the available and the required heat for each interval:

$$q_{net} = q_{available} - q_{required} \tag{12.3}$$

Using this, the cascaded net heat for each temperature interval could be calculated by:

$$Q_{net,i} = Q_{net,i-1} + q_{net,i} \tag{12.4}$$

Examining the column with the values for the cascaded net heat the lowest value is determined. Another column is then constructed by adjusting the cascaded net heat with this value. This gives a new column with the adjusted cascaded net heat where the lowest value is 0, which represents the pinch point. This is found to be the interval 376.4 K - 366.4 K.
All the values for the different calculations are found in Appendix E. The first value in the adjusted cascaded net heat column represents the minimum amount of energy needed to be added to the system. This is 120 GJ/h, while the last value in the column represents the minimum amount of cooling needed which is 138 GJ/h. Using only the minimum amounts means a 37.4% saving on heating and 38.2% saving on the cooling.
Using the results from the tabular method the heat exchanger network is now constructed.
At first the network above the pinch point is constructed. Here it is shown that right at the pinch point, four hot streams are present, but only two cold streams. This means that in order to take advantage of all the available energy from the hot streams, one of the cold streams needs to be split up. From Table 47 it is shown that the reboiler in tower 3, T3-r, needs a significant amount of energy. Therefore this reboiler is split up into four streams which harvests energy from E6, E7, E8 and T4-c. T3-r consumes all of the energy in T4-c, E8 and E7, but not all of E6. The rest of E6 is used to heat up E4 from 393 K to the desired 898 K, and E1

from 527.9 K to the desired 533 K.
The only hot stream not accounted for is then T2-c. This stream also contains a significant amount of energy, and in order to exchange all of it, it needs to be divided into three streams exchanging energy with E3, E5 and T1-r. By setting up four coupled heat duty equations with four unknown variables it is established that T2-c is capable of heating the three cold streams up to 395.2 K. The proposed heat network above the pinch point can be seen in Figure 34.

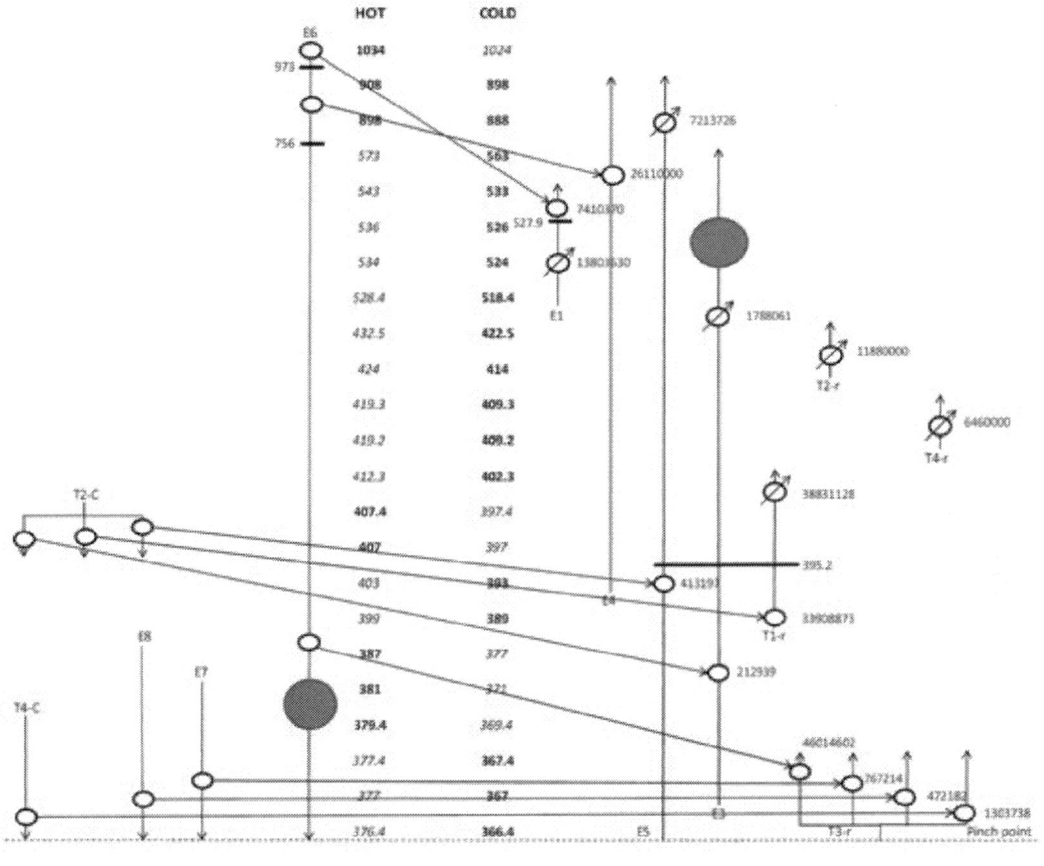

Figure 34: Heat exchanger network above the pinch point. The values next to the heat exchangers represent the duties given in kJ/hr. Blue circles represent phase changes.

For the heat streams below the pinch point only one cold stream (E5) which needs heating is present. This can be achieved by a heat exchange with E6 which will be cooled from the pinch point of 376.4 K and down to 366.4 K. The rest of the hot streams will need to be cooled with utility - cooling water. This gives the heat exchanger network below the pinch point that is shown in Figure 35.

By investigation of the proposed heat exchanger network it can be seen that all the

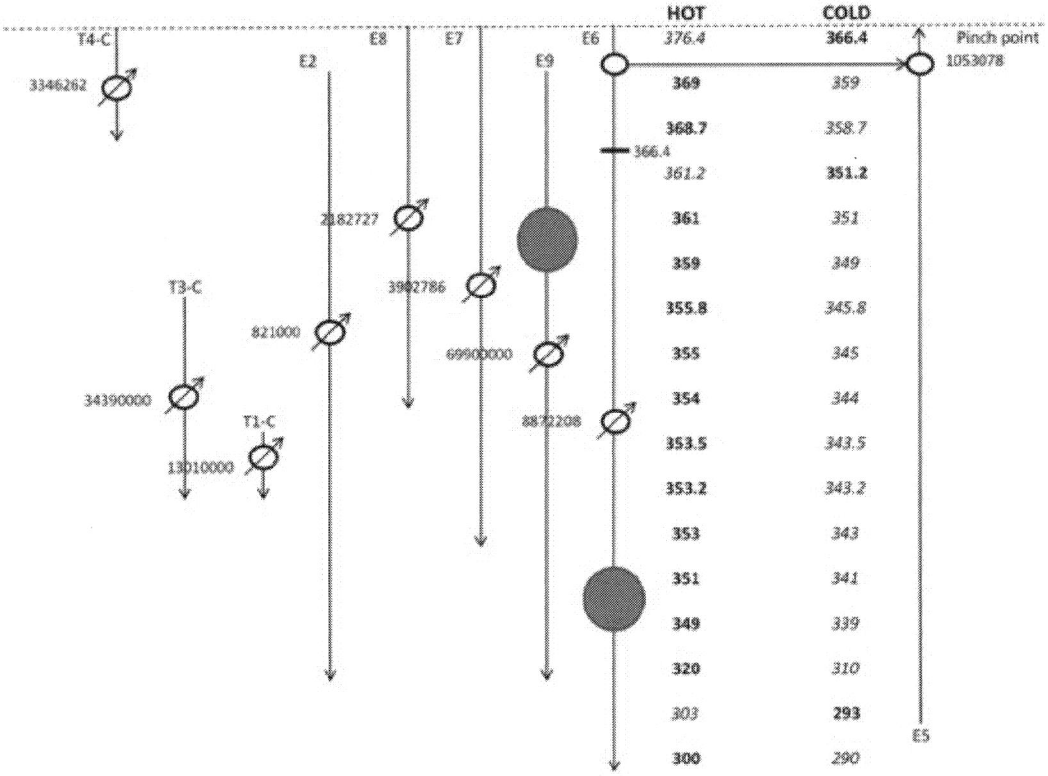

Figure 35: Heat exchanger network below the pinch point. The values next to the heat exchangers represent the duties given in kJ/hr. Blue circles represent phase changes.

available energy has been used above the pinch point, and below the pinch point all the required energy is exchanged. This means that only the minimum amount of heating and cooling, found from the tabular method, is necessary, leading to the utility savings established previously of 37.4% on the heating and 38.2% on the cooling, and thereby lowering the utility cost for the optimized base case.

12.2 Sizing of the New Heat Exchangers

Due to the implementation of heat integration, the heat exchangers needs to be resized. It is done from the approach described in section 8.8. The resulting heat exchangers, after the sizing work, are shown in Table 48 with specifications and costs. The table yields the integrated equipment, the temperature difference between the two streams at the inlet and outlet, the heat duty, resulting heat exchanger area, the equipment cost and the utility cost. For the DFH, some energy still needs to be added, which is illustrated by the heat duty. When two heat exchangers are integrated, no utility is used. The abbreviation 'cool' and 'steam' means, that cooling water or hot steam is used as utility source to heat/cool the stream. An asterix is used to illustrate that the condensers are divided up into two, because some of the distillation towers needs to be divided into two towers according to section 8.4. It is assumed that the two distillation towers which T1 consist of shares their reboiler, and two separate pumps ensure that each tower receives the necessary reflux ratio. This is also the case for the reboiler in T3. A flowsheet for the heat integrated system is shown in Figure 36. The figure shows which streams are exchanging heat. The figure does not describe which stream is heated and which stream is cooled, but this is shown in Table 48.

Table 48: The heat exchangers after heat integration with costs and specifications. Here "cool" stands for cooling water and steam strands for stands for hot condensing steam.

Heat Exchanger	Cool stream/ hot stream	ΔT at inlet [K]	ΔT at outlet [K]	Q_{total} [GJ/h]	A [m^2]	Equipment Cost [$]	Utility Cost [$/y]
HIE-1	E4/E6	363	75	61.1	492	51,512	-
HIE-2	E1/E6	454.6	510.5	7.41	30.1	8,006	-
HIE-3	T3r/E6	10	388.6	39.9	763	70,536	-
HIE-4	E5/E6	10	66.4	1.05	173	24,390	-
HIE-5	cool/E6	11.85	68.25	7.79	179	25,016	21,869
HIE-6	E1/steam	30.56	17	21.2	321	38,210	1,381,147
HIE-7	E3/T2-c	11.9	40.42	0.213	10.3	5,460	-
HIE-8	E3/steam	203.8	37	1.78	6.58	5,328	115,882
HIE-9	cool/E2	31.85	71.23	69.9	655	63,289	193,454
HIE-10	E5/T2-c	40.6	12.2	0.41	49.5	10,453	-
HIE-11	E5/DFH	-	-	11.4	-	472,964	96,608
HIE-12	T3-r/E7	14.6	13.6	0.767	107	17,265	-
HIE-13	cool/E7	66.05	82.85	3.86	27.1	7,636	10,654
HIE-14	T3-r/E8	20.6	19.6	0.47	46.0	10,019	-
HIE-15	cool/E8	65.85	88.85	1.00	7.04	5,328	4,121
HIE-16	T1-r/T2-c	18	12.2	33.9	456	48,874	-
HIE-17	T1-r /steam	103.8	47.7	38.8	124	19,169	2,525,973
HIE-18	cool/T1-c*	65.05	55.85	12.8	104	42,052	96,446
HIE-19	T2-r /steam	35	27.5	11.9	107	17,265	768,208
HIE-20	T3-r/T4-c	10	12	1.30	233	30,329	-
HIE-21	cool/T3-c*	64.85	57.65	34.5	281	21,524	35,326
HIE-22	T4-r /steam	121.05	111.15	6.46	30.1	9,145	416,655
HIE-23	cool/T4-c	66.05	82.85	3.35	22.1	6,994	9,252
					Total Cost	1,010,764	5,675,595

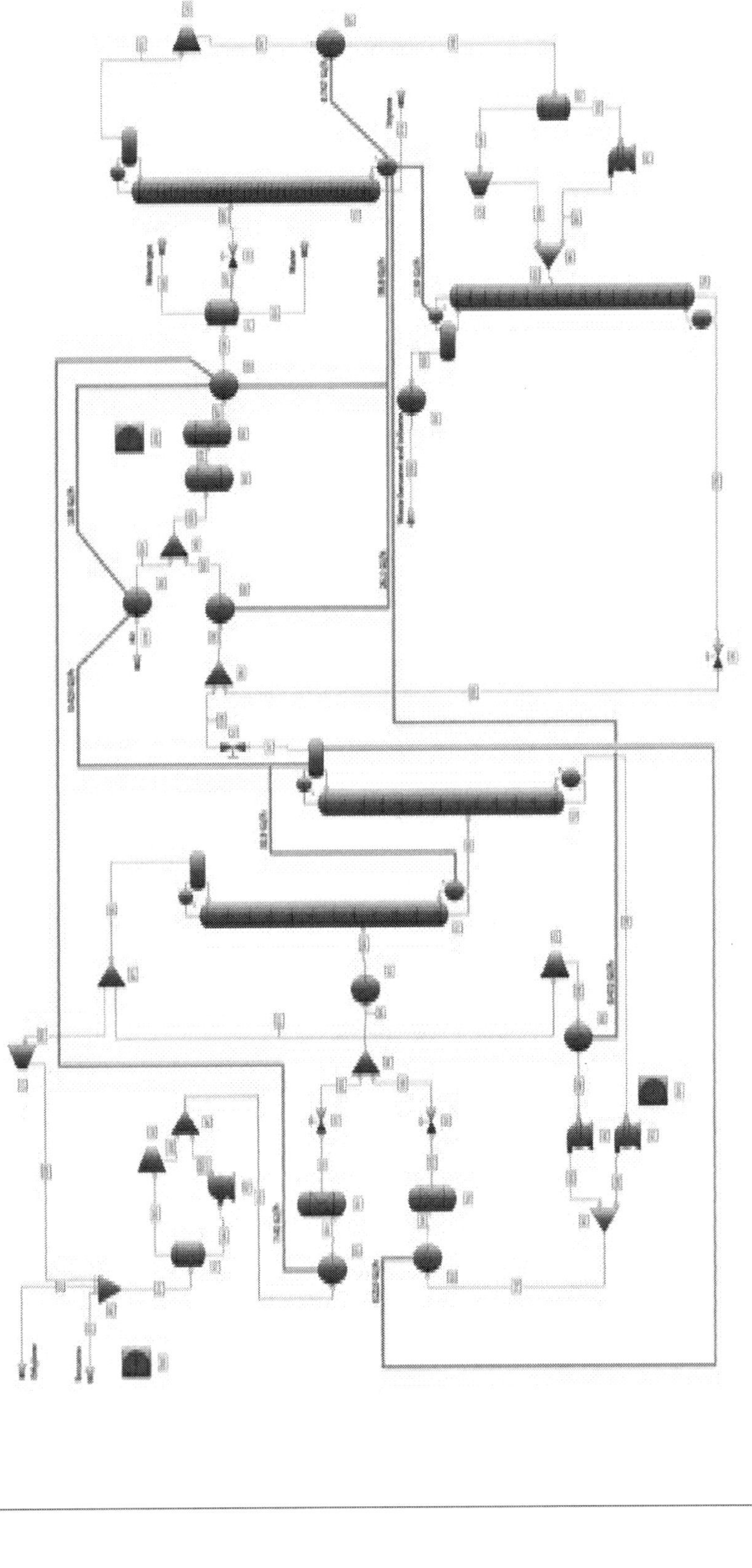

Figure 36: The flowsheet after heat integration, the red lines indicate which heat exchangers there is connected. The number on the red line is the amount of energy, which is exchanged between the different streams.

12.3 Heat Integration Effect Analysis

The effect of the heat integration is now to be analysed. The costing of the heat exchangers are calculated in Table 48. It is then possible to sum up all the equipment costs and utility costs to evaluate the resulting effect of the heat integration. Furthermore, the economic evaluation is performed once again with the new values of equipment and utility cost, as described in section 10. The resulting key measures after the heat integration are shown in Table 49 and Table 50, where it is compared to the optimized case and the base case.

Table 49: Key numbers of the heat integration compared to the optimized case.

	Base Case	**Optimized case**	**After HI**
Utility cost [$]	18,191,035	16,170,652	10,670,447
Equipment cost [$]	12,937,717	11,707,184	11,140,699
Total energy consumption [GJ/y]	$4.337 \cdot 10^6$	$3.971 \cdot 10^6$	$2.471 \cdot 10^6$
NPV [$]	71,731,904	97,487,064	143,081,264
Rate of return [%]	9	14	23
Payback Time [y]	7.6	5.4	3.7

As expected the utility cost is decreased significantly - 35 % - about 5.5 MM$ per year. It is achieved by a slight increase of the equipment cost. The energy consumption has also decreased significantly (38 %). This results in an increase in the NPV of 48 %, giving a rate of return of 23 % and a payback time of 3.7 years, which is within an acceptable range and foreseeable future. The rate of return is also quiet high, which makes the process more favourable for the investors.

Table 50: The percentage difference between the base case (BC) and the case after the optimization and the percentage change between the BC and the case after Heat Integration (HI) for the key measurements.

	Change optimized to HI [%]	**Change BC to HI** [%]
Utility cost	-34.0	-41.3
Equipment cost	4.84	-13.9
Total energy consumption	-37.8	-43.0
NPV	46.8	99.5
Rate of return	64.3	155.6
Payback Time	-31.5	-51.3

In Table 50 the percentage change from the base case to the case after heat integration is compared. Data from Table 45 and Table 49 are used to the comparison.

Since the heat exchangers and reboilers/condensers are intercorrelated, it makes the heating and cooling in the different parts more complicated, because they are dependent of each other. But, this is not taken into account, and it is just assumed that the heat integration network is not interrupted.

From Table 50 it is concluded that the costs are reduced considerably from the base case to the case after the heat integration. Especially the utility cost is reduced significantly, which affects the other measures heavily. The NPV is increased from about 72 MM$ to about 143 MM$. The total energy consumption has also decreased, which makes the process more favourable from an environmental point of view.

12.4 Subconclusion: Heat Integration

From the heat integration it is concluded that a significant amount of utility is saved by integrating the reboilers, condensers, direct fired heaters and heat exchangers. This give a total decrease of the energy consumption of 38 %. This makes the process more favorable for the investors, because the NPV has increased considerably (to 143 MM$ from 72 MM$). The payback time after heat integration is 3.7 years, which is within a foreseeable future, with a rate of return of 23 %.

The heat integration is therefore applied on the process plant, and is considered as a part of the process when further analysis and optimization is performed.

Even though the costs have decreased, increasing the earning of the investment, it is also important to look at the environmental impact of the process. It is important to find the bottlenecks of the process from an environmental point of view in order to make the process more environmental friendly and sustainable.

Chapter 13: Sustainability Analysis

A sustainability analysis is performed on the optimized base case design in order to make sure that the pollution is kept at a minimum. This is done by the use of the Waste Reduction Algorithm (WAR).
From WAR the Potential Environmental Impact (PEI) is obtained, which is a quantitative measure of the potential effect that the materials would have on the environment if they were emitted [Young and Cabezas, 1999]. PEI for a system is found from the general PEI balance that is based on the principle that PEI can enter the system, leave the system and be generated in the system [Cabezas et al., 1999]:

$$\frac{\partial I_{syst}}{\partial t} = I_{in} - I_{out} + I_{gen} \tag{13.1}$$

Where I is the amount of PEI. The subscript in means amount going into the system, out means the amount leaving the system and gen is the amount generated in the system. This balance only takes the chemicals into account and does not deal with the potential environmental impact from the energy going in and out of the system. A more thorough balance was proposed by Young and Cabezas [1999] which does take the energy into account, but for this project the balance in Eq. (13.1) is used. By assuming that the system is in steady state, the balance can be further simplified into:

$$I_{gen} = I_{out} - I_{in} \tag{13.2}$$

For the optimized base case the system boundaries are placed so that only the streams feeding raw materials and streams exiting the process is treated. This means that the process is simplified into what can be seen in Figure 37.

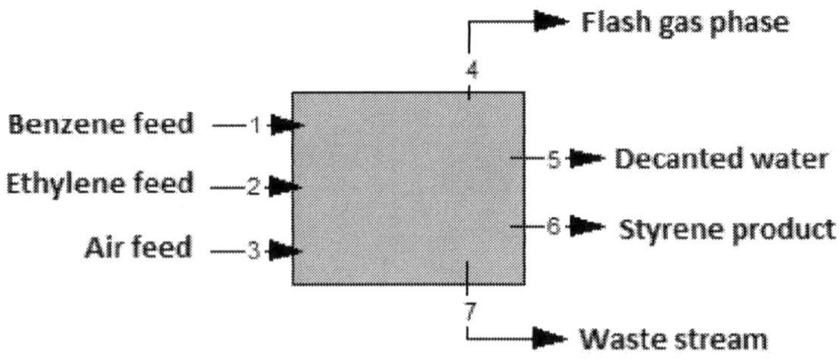

Figure 37: The simplified optimized base case system.

The PEI is therefore calculated by summing the PEI amount in the inputs and summing the PEI amount in the outlets. The PEI amount for each compound can be seen in Table 51.

Table 51: Potential Environmental Impact for the compounds.

Compound	PEI/kmol
H_2O	0.00
Nitrogen	0.00
Oxygen	0.00
Methane	0.66
Benzene	57.03
Ethylene	58.08
Di-ethylbenzene	61.07
Toluene	122.03
Styrene	151.69
Ethylbenzene	152.90

In this table it can be seen that ST and EB gives the highest PEI/kmol.
The PEI balance calculations are performed using ICAS 15.0, where the WAR algorithm is implemented. The results are presented in Table 52 for the optimized base case design.

Table 52: PEI for input and output streams as well as total impact generated. Stream names correspond to Figure 32.

Stream	Stream Name	Stream Description	Total PEI
Input 1	S1	Benzene feed	13,221.25
Input 2	S2	Ethylene feed	12,545.18
Input 3	S16	Air feed	0.00
Output 4	S26	Flash gas phase	1,374.19
Output 5	S33	Decanted water	0.00
Output 6	S18	Styrene product stream	30,363.29
Output 7	S54	Waste stream (i.e. Toluene)	1,627.73
Impact generated			7,598.79

Here it can be seen that the total PEI generated is 7,598.79. More interesting is it to note, that the by far biggest contributor to the total impact generated is output stream 6 which has a PEI of 30,363.29, this is a logical consequence of the high PEI/kmol of ST. With the goal of lowering the PEI but still keeping a high production rate in mind, it will be more or less impossible to change the high potential environmental impact from the ST product stream. This leaves output stream 4 and 7 to be optimised. Output stream 4, consisting mostly of nitrogen and water, is sent into a flare where the harmful components are combusted. However output stream 7 is mostly consisting of BZ and Tol which gives the possibility of recycling

the BZ, but also the smaller amount of EB could be recycled. This could also have economical benefits since the purer Tol could be sold as a by-product.
The recycling of BZ was a part of the initial flowsheets but was dropped since it was considered to be too expensive compared to the relatively low amount of BZ. But now, due to environmental considerations it is brought back into the process. Implementing it in the optimized base case design means that three distillation towers and a flash needs to be added, as shown in the new flowsheet in Figure 38. The first added distillation tower (T5) separates BZ from Tol and EB. The BZ is sent to a flash where the light substances are removed in order to avoid accumulation in the process. The BZ is afterwards recycled to the inlet stream of BZ to the process, while EB and Tol are sent to the next distillation tower (T6). Here EB and Tol are separated, and EB is recycled to the reactor where ST is formed. In order to achieve a sufficient quality of Tol, making it saleable, remains of BZ in the stream must be removed by a last distillation (T7). After this distillation Tol has a purity of 99.7 % (molar based), and BZ can be recycled to the reactor where EB is formed. These optimizations results in an increased ST production, decreased BZ consumption and an additional saleable product stream with Tol.
Even though these alterations of the design adds more equipment to the process flowsheet, the NPV is increased with 76 % compared to the optimized base case with heat integration. The final design is presented in Chapter 14.

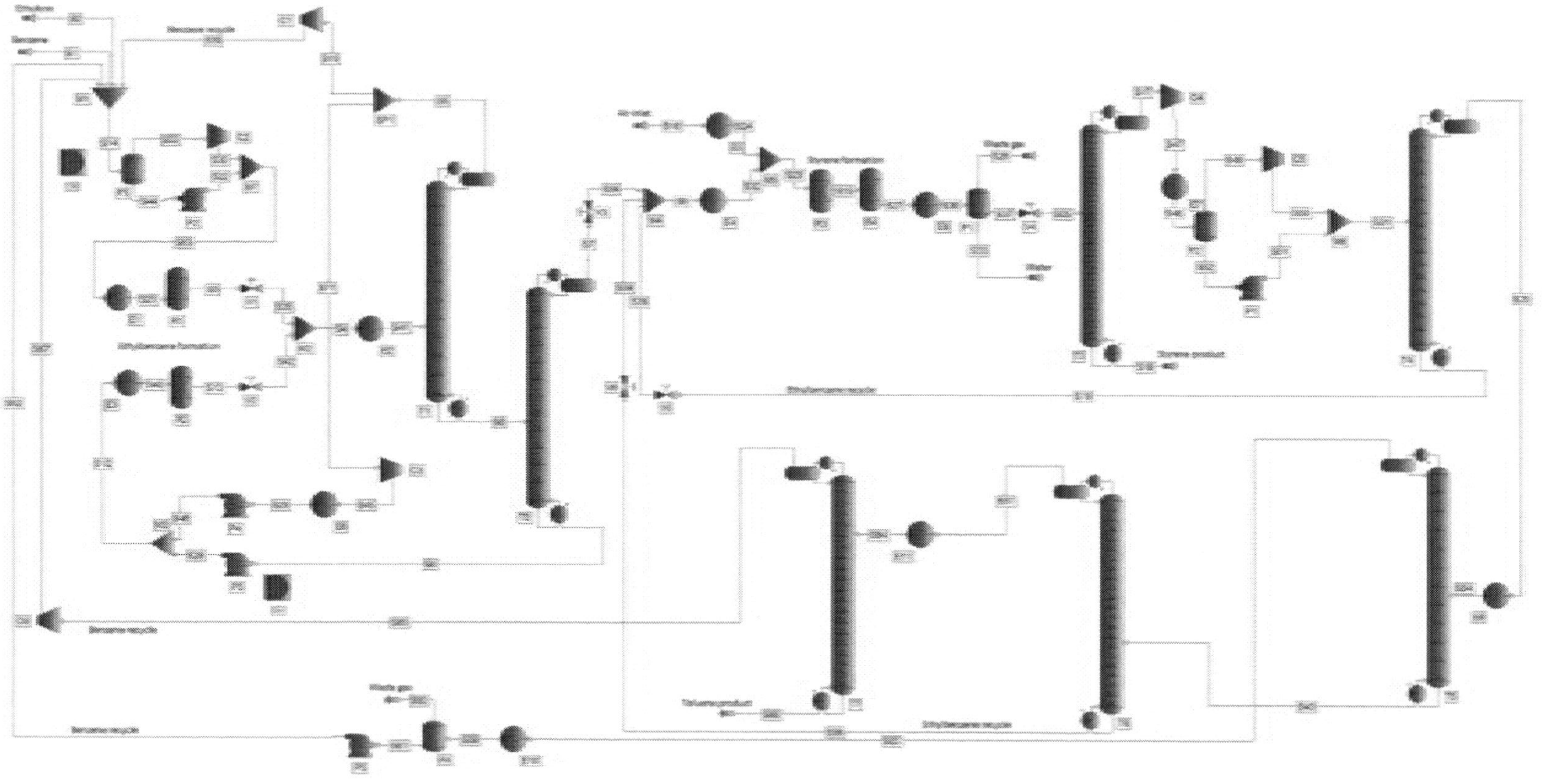

Figure 38: Optimized sustainable flowsheet. The stream summary table is found in Chapter 14.

Performing the WAR algorithm on this design gives the results that can be seen in Table 53.

Table 53: PEI for input and output streams as well as the total impact generated by the new optimized base case design. Stream names correspond to Figure 38.

Stream	Stream Name	Stream Description	Total PEI
Input 1	S2	Ethylene feed	12,545.18
Input 2	S1	Benzene feed	12,392.92
Input 3	S16	Air feed	0.00
Output 4	S26	Flash gas phase	1,395.94
Output 5	S33	Decanted water	0.00
Output 6	S18	Styrene product stream	30,480.81
Output 7	S57	Recycle purge	1.44
Output 8	S60	Toluene product stream	687.74
Impact generated			7,627.83

It appears that the total PEI generated has increased compared to the first case. However, this increase comes from the increased ST production rate. Since this increased production rate is desired from a process design point of view, a non-product analysis is performed on the process. This means that the product stream is excluded when performing the WAR algorithm. Only the ST product is excluded though since this is the main product, even though the Tol stream becomes a product stream after the sustainable optimization. This gives the new total generated PEI given in Table 54.

Table 54: PEI generated for standard balance and non-product balance.

Case	PEI generated	Non-product PEI generated
Optimized base case	7,598.79	-22,764.50
New optimized base case	7,627.83	-22,852.98

Here it can be seen that the new sustainable design decreases the PEI generated by the process when not taking the product into consideration.
It is important to note that the PEI balance used, does not take the energy into account. The addition of three extra distillation towers means that extra utility will be needed. This increased utility need, will lead to more pollution which could possibly mean that the new and apparently more sustainable process design could actually be worse. In order to take this into account the extended balance proposed by Young and Cabezas [1999] could be used, but since the total distillation reboiler and condenser duties only increases by 2.7 %, the increase in PEI generated due to increased duty is assumed to be negligible.

Chapter 14: The Final Process Design

Based on optimizations carried out in Chapter 13 the final flowsheet is designed. The flowsheet is shown in Figure 39. Here the blue "S" streams indicates flows of mass and the embedded heat, while the red "H" streams indicates heat transfer through heat integration.

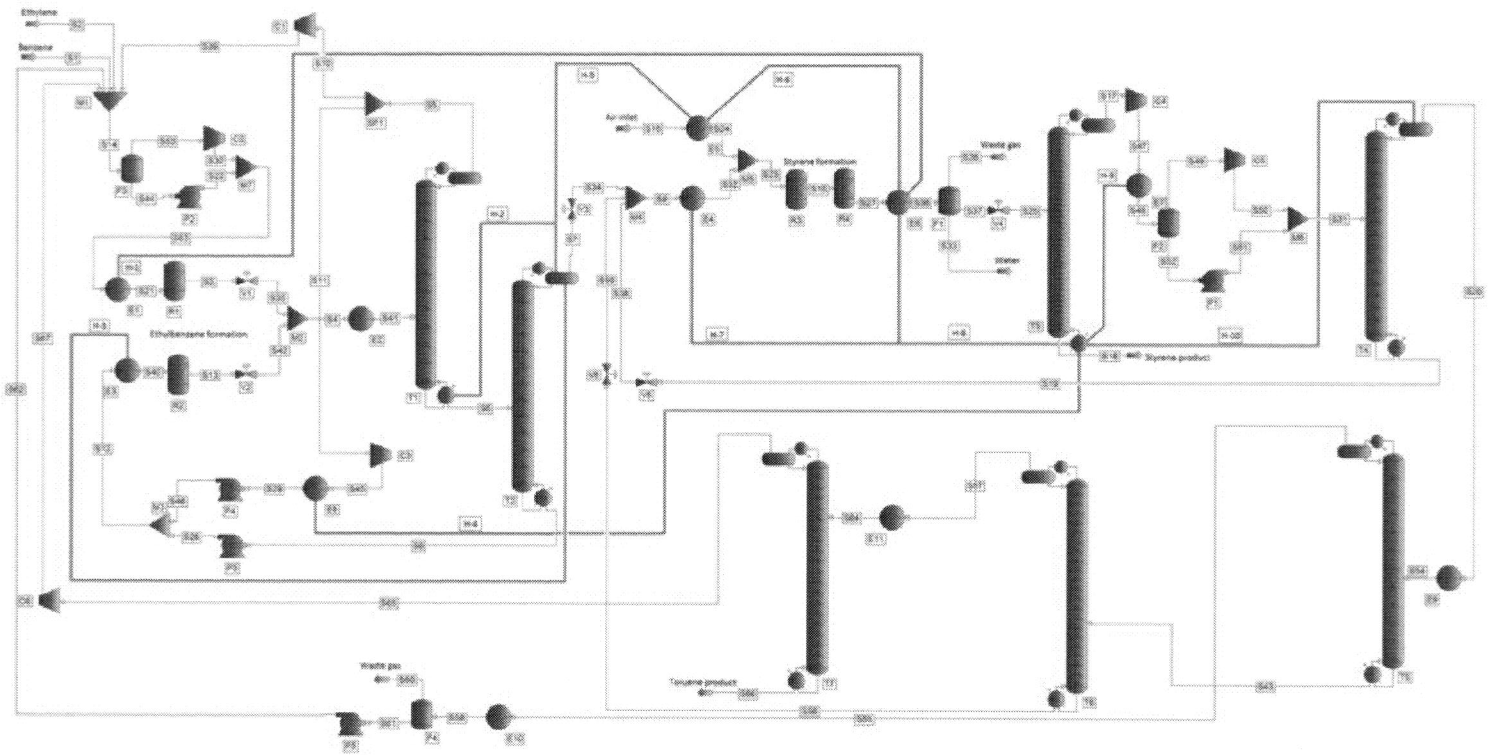

Figure 39: The final flowsheat. Here the blue "S" streams indicates flow of mass and energy, while the red "H" streams indicates flow energy though heat integration.

Table 55: Stream summary for streams S1-S34 for the final process design.

Stream Name	S1	S2	S3	S4	S5	S6	S7	S8	S9	S10	S11	S12
Phase	Liquid	Vapor	Liquid	Vapor	Vapor	Liquid	Vapor	Liquid	Mixed	Vapor	Vapor	Liquid
T. [K]	300.0	300.0	533.0	369.6	353.2	402.3	407.0	422.5	369.6	353.2	353.2	366.4
P. [atm]	1.0	1.0	33.5	1.0	1.0	1.0	1.0	1.0	0.3	1.0	1.0	24.7
Enthalpy [G/h]	0.8	0.4	78.0	81.5	65.5	5.9	14.1	0.2	16.6	63.6	1.9	0.7
Rate [kmol/hr]	217.3	216.0	1750.2	1800.3	1562.8	237.5	232.1	5.4	327.6	1518.1	44.6	50.1
H_2 [kmol/hr]	0.00	0.00	0.00	0.00	0.00	0.00	0.00	0.00	0.00	0.00	0.00	0.00
N_2 [kmol/hr]	0.00	0.00	0.63	0.65	0.65	0.00	0.00	0.00	0.00	0.63	0.02	0.02
O_2 [kmol/hr]	0.00	0.00	0.00	0.00	0.00	0.00	0.00	0.00	0.00	0.00	0.00	0.00
M [kmol/hr]	0.00	0.00	0.12	0.13	0.13	0.00	0.00	0.00	0.00	0.12	0.00	0.00
E [kmol/hr]	0.00	216.00	0.00	0.00	0.00	0.00	0.00	0.00	0.00	0.00	0.00	0.00
B [kmol/hr]	217.30	0.00	1532.73	1575.13	1559.38	15.75	15.75	0.00	15.75	1514.85	44.54	44.54
H_2O [kmol/hr]	0.00	0.00	0.00	0.00	0.00	0.00	0.00	0.00	1.13	0.00	0.00	0.00
T [kmol/hr]	0.00	0.00	0.70	0.71	0.38	0.33	0.33	0.00	0.70	0.37	0.01	0.01
EB [kmol/hr]	0.00	0.00	213.84	221.45	2.21	219.24	215.95	3.29	307.98	2.15	0.06	3.35
S [kmol/hr]	0.00	0.00	0.00	0.00	0.00	0.00	0.00	0.00	2.02	0.00	0.00	0.00
DEB [kmol/hr]	0.00	0.00	2.16	2.18	0.00	2.18	0.03	2.15	0.03	0.00	0.00	2.15

Stream Name	S13	S14	S15	S16	S17	S18	S19	S20	S21	S22	S23
Phase	Vapor	Mixed	Vapor	Vapor	Vapor	Liquid	Liquid	Vapor	Mixed	Liquid	Vapor
T. [K]	563.2	349.1	637.8	293.0	353.1	366.4	409.3	369.1	533.0	351.4	898.0
P. [atm]	22.2	1.0	0.2	1.0	0.2	0.2	1.0	1.0	37.5	37.5	0.3
Enthalpy [G/h]	3.5	65.2	36.4	0.3	6.1	3.5	2.5	1.0	104.2	3.2	59.5
Rate [kmol/hr]	50.1	1966.2	1008.9	475.0	116.1	201.0	94.6	21.5	1966.2	296.1	802.6
H_2 [kmol/hr]	0.00	0.00	197.53	0.00	0.00	0.00	0.00	0.00	0.00	0.00	0.00
N_2 [kmol/hr]	0.02	0.63	375.25	375.25	0.11	0.00	0.00	0.11	0.63	0.00	375.25
O_2 [kmol/hr]	0.00	0.00	99.75	99.75	0.00	0.00	0.00	0.00	0.00	0.00	99.75
M [kmol/hr]	0.00	0.12	6.13	0.00	0.01	0.00	0.00	0.01	0.12	0.00	0.00
E [kmol/hr]	0.00	216.01	2.64	0.00	0.02	0.00	0.00	0.02	216.01	0.28	0.00
B [kmol/hr]	42.41	1746.58	18.39	0.00	14.41	0.00	0.00	14.41	1746.58	294.33	15.75
H_2O [kmol/hr]	0.00	0.00	1.13	0.00	1.13	0.00	1.09	0.04	0.00	0.00	1.13
T [kmol/hr]	0.01	0.70	6.83	0.00	6.33	0.00	0.32	6.01	0.70	0.25	0.70
EB [kmol/hr]	7.61	2.15	95.54	0.00	92.04	0.98	91.12	0.92	2.15	1.23	307.98
S [kmol/hr]	0.00	0.00	205.69	0.00	2.02	199.99	2.02	0.00	0.00	0.00	2.02
DEB [kmol/hr]	0.02	0.00	0.03	0.00	0.00	0.03	0.00	0.00	0.00	0.00	0.03

Stream Name	S24	S25	S26	S27	S28	S29	S30	S31	S32	S33	S34
Phase	Vapor	Mixed	Vapor	Vapor	Liquid	Liquid	Mixed	Mixed	Vapor	Water	Vapor
T. [K]	898.0	300.0	300.0	1034.4	423.9	353.5	519.7	353.0	898.0	300.0	407.0
P. [atm]	1.0	0.2	0.667	0.667	24.7	2.5	37.5	1.0	0.3	0.667	0.667
Enthalpy [G/h]	16.6	1.5	1.4	93.0	0.2	0.5	80.0	1.6	42.9	0.4	14.1
Rate [kmol/hr]	475.0	317.0	417.4	910.2	5.4	44.6	1670.1	116.1	327.6	175.7	232.1
H_2 [kmol/hr]	0.00	0.00	0.00	0.00	0.00	0.00	0.00	0.00	0.00	0.00	0.00
N_2 [kmol/hr]	375.25	0.11	375.14	375.25	0.00	0.02	0.63	0.11	0.00	0.00	0.00
O_2 [kmol/hr]	99.75	0.00	0.98	0.98	0.00	0.00	0.00	0.00	0.00	0.00	0.00
M [kmol/hr]	0.00	0.01	6.12	6.13	0.00	0.00	0.12	0.01	0.00	0.00	0.00
E [kmol/hr]	0.00	0.02	2.62	2.64	0.00	0.00	215.73	0.02	0.00	0.00	0.00
B [kmol/hr]	0.00	14.41	3.98	18.39	0.00	44.54	1452.25	14.41	15.75	0.00	15.75
H_2O [kmol/hr]	0.00	1.13	21.84	198.66	0.00	0.00	0.00	1.13	1.13	175.68	0.00
T [kmol/hr]	0.00	6.33	0.51	6.83	0.00	0.01	0.45	6.33	0.70	0.00	0.33
EB [kmol/hr]	0.00	92.97	2.57	95.54	3.29	0.06	0.92	92.04	307.98	0.00	215.95
S [kmol/hr]	0.00	202.01	3.68	205.69	0.00	0.00	0.00	2.02	2.02	0.00	0.00
DEB [kmol/hr]	0.00	0.03	0.00	0.03	2.15	0.00	0.00	0.00	0.03	0.00	0.03

Table 56: Stream summary for streams S35-S67 for the final process design.

Stream Name	S35	S36	S37	S38	S39	S40	S41	S42	S43	S44	S45
Phase	Mixed	Mixed	Liquid	Mixed	Mixed	Vapor	Liquid	Vapor	Liquid	Liquid	Mixed
T. [K]	368.1	300.0	300.0	394.9	353.6	563.2	326.0	544.4	381.3	349.0	386.6
P. [atm]	1.0	0.667	0.667	0.667	1.0	24.7	1.0	1.0	1.0	1.0	2.5
Enthalpy [GJ/h]	78.0	3.3	1.5	2.5	63.9	2.7	12.0	3.5	0.1	3.1	2.0
Rate [kmol/hr]	1750.2	910.2	317.0	94.6	1518.1	50.1	1800.3	50.1	7.4	296.1	44.6
H_2 [kmol/hr]	0.00	0.00	0.00	0.00	0.00	0.00	0.00	0.00	0.00	0.00	0.00
N_2 [kmol/hr]	0.63	375.25	0.11	0.00	0.63	0.02	0.65	0.02	0.00	0.00	0.02
O_2 [kmol/hr]	0.00	0.98	0.00	0.00	0.00	0.00	0.00	0.00	0.00	0.00	0.00
M [kmol/hr]	0.12	6.13	0.01	0.00	0.12	0.00	0.13	0.00	0.00	0.00	0.00
E [kmol/hr]	0.00	2.64	0.02	0.00	0.00	0.00	0.00	0.00	0.00	0.28	0.00
B [kmol/hr]	1532.73	18.39	14.41	0.00	1514.85	44.54	1575.13	42.41	0.72	294.33	44.54
H_2O [kmol/hr]	0.00	198.66	1.13	1.09	0.00	0.00	0.00	0.00	0.04	0.00	0.00
T [kmol/hr]	0.70	6.83	6.33	0.32	0.37	0.01	0.71	0.01	5.71	0.25	0.01
EB [kmol/hr]	213.84	95.54	92.97	91.12	2.15	3.35	221.45	7.61	0.92	1.23	0.06
S [kmol/hr]	0.00	205.69	202.01	2.02	0.00	0.00	0.00	0.00	0.00	0.00	0.00
DEB [kmol/hr]	2.16	0.03	0.03	0.00	0.00	2.15	2.18	0.02	0.00	0.00	0.00

Stream Name	S46	S47	S48	S49	S50	S51	S52	S53	S54	S55	S56
Phase	Liquid	Mixed	Mixed	Vapor	Vapor	Liquid	Liquid	Vapor	Mixed	Vapor	Liquid
T. [K]	355.0	381.0	352.7	352.7	377.7	352.8	352.7	349.0	320.0	353.4	406.5
P. [atm]	24.7	0.5	0.5	0.5	1.0	1.0	0.5	1.0	1.0	1.0	1.0
Enthalpy [GJ/h]	0.5	6.3	1.6	0.0	0.0	1.6	1.6	61.5	0.1	0.6	0.0
Rate [kmol/hr]	44.6	116.1	116.1	0.2	0.2	115.9	115.9	1670.1	21.5	14.1	1.0
H_2 [kmol/hr]	0.00	0.00	0.00	0.00	0.00	0.00	0.00	0.00	0.00	0.00	0.00
N_2 [kmol/hr]	0.02	0.11	0.11	0.08	0.08	0.03	0.03	0.63	0.11	0.11	0.00
O_2 [kmol/hr]	0.00	0.00	0.00	0.00	0.00	0.00	0.00	0.00	0.00	0.00	0.00
M [kmol/hr]	0.00	0.01	0.01	0.01	0.01	0.00	0.00	0.12	0.01	0.01	0.00
E [kmol/hr]	0.00	0.02	0.02	0.01	0.01	0.01	0.01	215.73	0.02	0.02	0.00
B [kmol/hr]	44.54	14.41	14.41	0.05	0.05	14.36	14.36	1452.25	14.41	13.69	0.00
H_2O [kmol/hr]	0.00	1.13	1.13	0.00	0.00	1.13	1.13	0.00	0.04	0.00	0.03
T [kmol/hr]	0.01	6.33	6.33	0.01	0.01	6.32	6.32	0.45	6.01	0.30	0.06
EB [kmol/hr]	0.06	92.04	92.04	0.05	0.05	91.99	91.99	0.92	0.92	0.00	0.91
S [kmol/hr]	0.00	2.02	2.02	0.00	0.00	2.02	2.02	0.00	0.00	0.00	0.00
DEB [kmol/hr]	0.00	0.00	0.00	0.00	0.00	0.00	0.00	0.00	0.00	0.00	0.00

Stream Name	S57	S58	S59	S60	S61	S62	S63	S64	S65	S66	S67
Phase	Vapor	Mixed	Mixed	Vapor	Liquid	Liquid	Mixed	Liquid	Vapor	Liquid	Mixed
T. [K]	380.4	300.0	369.2	300.0	300.0	300.0	514.0	370.0	354.2	383.0	355.3
P. [atm]	1.0	1.0	0.3	1.0	1.0	1.0	37.5	1.0	1.0	1.0	1.0
Enthalpy [GJ/h]	0.3	0.1	0.0	0.0	0.1	0.1	83.2	0.1	0.0	0.1	0.0
Rate [kmol/hr]	6.4	14.1	1.0	0.1	14.0	14.0	1966.2	6.4	0.7	5.6	0.7
H_2 [kmol/hr]	0.00	0.00	0.00	0.00	0.00	0.00	0.00	0.00	0.00	0.00	0.00
N_2 [kmol/hr]	0.00	0.11	0.00	0.10	0.01	0.01	0.63	0.00	0.00	0.00	0.00
O_2 [kmol/hr]	0.00	0.00	0.00	0.00	0.00	0.00	0.00	0.00	0.00	0.00	0.00
M [kmol/hr]	0.00	0.01	0.00	0.01	0.00	0.00	0.12	0.00	0.00	0.00	0.00
E [kmol/hr]	0.00	0.02	0.00	0.01	0.01	0.01	216.01	0.00	0.00	0.00	0.00
B [kmol/hr]	0.72	13.69	0.00	0.02	13.68	13.68	1746.58	0.72	0.72	0.00	0.72
H_2O [kmol/hr]	0.00	0.00	0.03	0.00	0.00	0.00	0.00	0.00	0.00	0.00	0.00
T [kmol/hr]	5.65	0.30	0.06	0.00	0.30	0.30	0.70	5.65	0.03	5.62	0.03
EB [kmol/hr]	0.01	0.00	0.91	0.00	0.00	0.00	2.15	0.01	0.00	0.01	0.00
S [kmol/hr]	0.00	0.00	0.00	0.00	0.00	0.00	0.00	0.00	0.00	0.00	0.00
DEB [kmol/hr]	0.00	0.00	0.00	0.00	0.00	0.00	0.00	0.00	0.00	0.00	0.00

Table 57: Stream summary for the heat flows in the final process design.

Stream	Cool/Hot	Q [GJ/h]	A[m²]	$T_{C,in}$	$T_{C,out}$	$T_{H,in}$	$T_{H,out}$
H-1	E1/E6	7.60	30.8	518	524	1034	973
H-2	T1-r/T2-c	33.9	456	389	395	407	407
H-3	E3/T2-c	0.213	10.3	367	396	407	407
H-4	T3-r/E8	0.998	7.04	366	367	387	387
H-5	E5/T2-c	0.413	49.5	366	395	407	407
H-6	E5/E6	1.05	173	300	366	376	366
H-7	E4/E6	61.1	492	393	898	973	756
H-8	T3-r/E6	39.9	763	366	367	756	376
H-9	T3-r/E7	0.767	107	366	367	381	381
H-10	T3-r/T4-c	1.30	233	366	367	379	376

14.1 Equipment size

In this section the sizing for the different equipments in the final design for the plant are presented.
In the final design of the plant, seven distillation towers are needed. The dimensions and duties for the different towers are shown in Table 58. The table shows that two towers are needed for the separations in T1 and T3. The duties, area and utility for these towers are therefore given for each of the towers. The table shows that the condenser in T2 and the reboiler in T3 do not need any utility, due to heat integration. The towers are calculated based on the method described in Chapter 8

Table 58: Size and utility for the distillation towers in the final design. Sep. between stands for separation between.

Tower	**T1**	**T2**	**T3**	**T4**	**T5**	**T6**	**T7**
Sep. between	BZ/EB	EB/DEB	EB/ST	Tol/EB	BZ/Tol	Tol/EB	BZ/Tol
Nr. of towers	2	1	2	1	1	1	1
Recovery							
Light key [%]	99	98.5	99	95	95	99	99.5
Heavy key [%]	99	98.5	99	99	95	99	99.5
Condenser							
Duty [GJ/h]	-6.52	-3.45	-17.3	-4.65	-0.439	-0.318	-0.283
A [m^2]	53.2	516	141	255	3.57	1.79	2.56
F_{water}[ton/h]	183	-	107	80.0	10.5	7.61	6.76
Reboiler							
Duty [GJ/h]	36.4	11.9	21.3	6.51	1.02	0.534	0.317
A [m^2]	290	107	573.9	39.6	3.66	3.04	1.96
F_{steam}[ton/h]	97.4	59.5	-	32.7	5.12	2.68	1.59
Dimensions							
Diameter [m]	3.017	1.829	3.449	1.372	0.762	0.610	0.610
Nr. Trays	11	12	68	16	11	20	19
Total height [m]	12.7	13.3	47.5	15.8	12.7	18.2	17.6

For this design, eleven heat exchangers are needed. The sizes and utilities needed for these heat exchangers are shown in Table 59. The table only shows the total energy need and the total needed area. Many of the heat exchangers are connected to each other, which is shown in Figure 39 and Table 57. Some of the heat exchangers are therefore in reality divided into several parts. E1 and E4 does not need any utility in order to achieve the desired outlet temperature due to the heat integration network covering the entire energy need. E5 only needs heating by a DFH due to the high outlet temperature which would have made it impossible to heat the stream by use of condensing steam. The area and amount of utility needed are calculated based on the method described in section 8.8.

Table 59: Overview over sizes and utility needed for the heat exchangers in the final design. DFH stands for Direct Fired Heater. Cooling and heating stands for respectively cooling by cooling water and heating by condensing steam.

	Q_{total}[GJ/h]	A_{total} [m^2]	T_{in} [K]	T_{out} [K]	Utility need	$F_{utility}$ [ton/h]
E1	21.2	354	518	533	-	-
E2	69.9	655	369	320	Cooling	1671
E3	1.99	16.8	367	563	Heating	8.94
E4	61.1	492	393	898	-	-
E5	12.9	222	300	898	DFH	-
E6	118	1640	1034	300	Cooling	196
E7	4.67	134	381	354	Cooling	93.3
E8	1.47	53.0	387	354	Cooling	23.9
E9	0.846	7.43	369	320	Cooling	20.2
E10	0.544	6.95	353	300	Cooling	13.0
E11	0.224	1.33	380	370	Cooling	5.35

In the final design, nine compressors and five pumps are needed. The sizes of the different pumps and compressors are shown in Table 60. In the table, C2 is divided into four compressors due to the high increase in pressure, which is too large for a single compressor to handle. Compressors and pumps are calculated by the approach described in section 8.5 and 8.6.

Table 60: Overview of pumps and compressors needed for the final design.

	ΔP [atm]	Work [kW]
	Compressors	
C1	0.01	129
C2-1	1.47	1983
C2-2	3.65	2252
C2-3	9.03	2544
C2-4	22.35	2877
C3	1.51	49
C4	0.31	104.86
C5	0.5	0.12
C6	0.03	0.08
	Pumps	
P1	0.5	0.38
P2	36.50	29
P3	23.68	1
P4	22.17	3
P5	0.02	0.04

In order to produce ST, three reactors are needed; a high pressure alkylation reactor between Et and BZ, which occurs in R1, a transalkylation of DEB and a dehydrogenation of EB, which generates ST. The last reactor is shown as two reactors in the flowsheet in Figure 39. The sizes and amount of catalysts needed for the three reactors are shown in Table 61.

Table 61: Overview over reactor sizes in the final plant design.

	R1	**R2**	**R3+R4**
Type of catalyst	Zeolite-Y	Zeolite-Y	Iron Oxide based
Amount of catalyst [kg]	18354	610	56710
Diameter [m]	4	1	4
Height [m]	5.85	3.11	5.24

Four flashes are needed in this process. Their sizes are shown in Table 62.

Table 62: Overview of the sizes for flash drums in the final design.

	F1	**F2**	**F3**	**F4**
Position	Horizontal	Vertical	Horizontal	Vertical
Diameter [m]	1	1	1	5
Height [m]	8.7	3.52	5.8	1.1

14.2 Economic Evaluation of Final Design

This section deals with the economics for the final plant design; if it is profitable to invest the money in this plant rather than investing the money other places - e.g. in the stock market. The costing of the equipment from the previous section, and the economic evaluation have been performed in the same way as in section 10.

The key costs are summarized in Table 63.

Table 63: Key costs for the final design: Raw materials (including catalysts), product sales (ST), By-product sale (Tol), equipment cost, utility cost, capital cost and operating cost.

Cost	Value
Raw Materials [$/y]	185,365,348
Product Cost (ST) [$/y]	265,253,858
By-product Cost (Tol) [$/y]	4,278,650
Equipment Cost [$]	11,459,633
Utility Cost [$/y]	10,831,709
Capital Costs	
Fixed-Capital Investment [$]	63,532,205
Working Capital Investment [$]	11,218,981
Total Capital Investment [$]	74,751,186
Operating Costs	
Manufacturing Cost [$/y]	206,428,221
Total Operating Cost [$/y]	228,963,872

Here it is shown ST is sold for 265 MM$ per year, and it costs 229 MM$ per year to produce it. In order to start up the plant, it is necessary to invest a total of almost 75 MM$.

Figure 40 presents pie charts of the distributions of the cost of the different parts of the equipment and utilities.
Heat transfer equipment for Figure 40 (a) covers all the equipment costs of the reboilers, condensers, DFH, heating/cooling jackets and heat exchangers. Especially the compressor C2 network is very expensive - both with respect to equipment purchase cost, but also for the utility cost. It accounts for 63.5 % of the total equipment cost and 42.7 % of the utility cost. The reboilers are responsible for 35.4 % of the total utility cost, which makes them almost as expensive. The compressor cost is this high, due to the high pressure increase and large amount of recycled BZ. The reboiler cost is high, since high separation degrees are approached and hot steam is expensive.

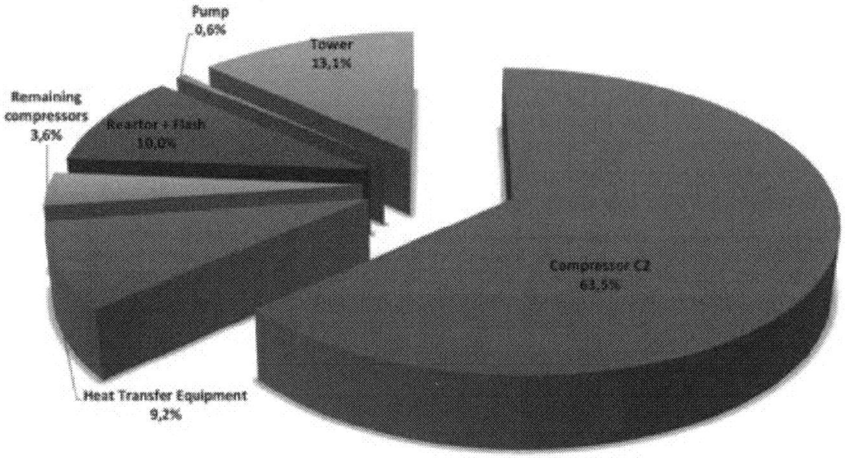

(a) Equipment Cost.

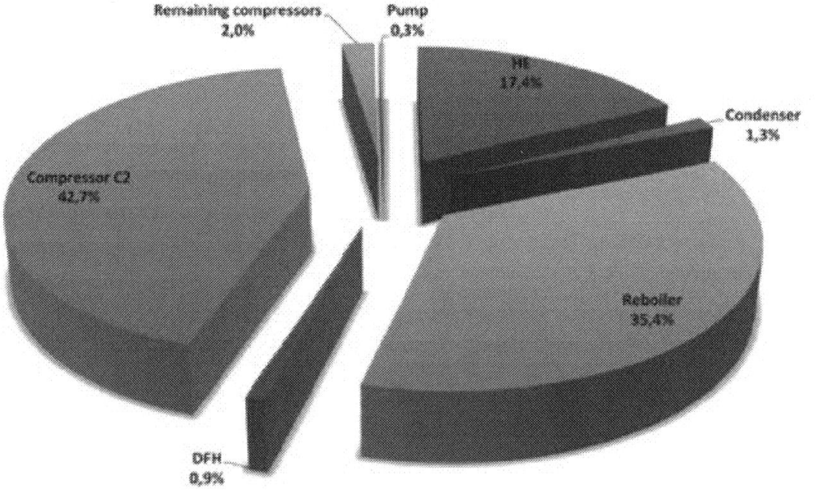

(b) Utility Cost.

Figure 40: Pie charts for categorizing the (a) equipment costs and (b) utility costs for the final process. Here HE stands for heat exchangers.

By evaluating the costs in Table 63, the economic evaluation can be performed as described in section 10, which yields the cumulative cash flow curve, Figure 41.

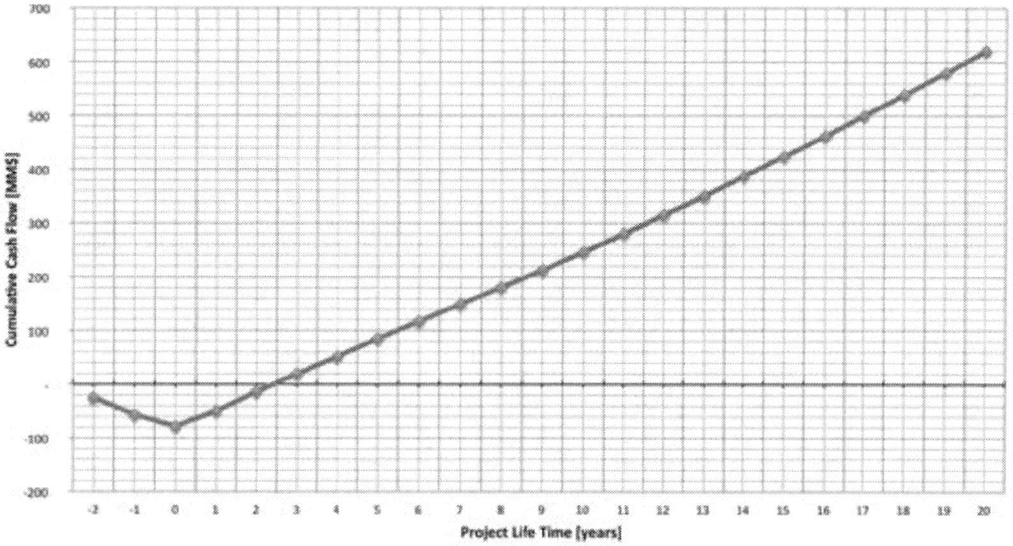

Figure 41: Cumulative cash flow in MM$ as a function of the project life time in years.

Here it is shown that it costs around 80 MM$ to construct the plant. From the point where the process is producing ST, the process is generating profit. It is symbolized by the linear line, which reaches the break-even point after 2.4 years. This means that the investors will break-even after only 2.4 years of operation, and the process is hereafter generating money to the investors. In order to evaluate the economics of the process even further, two additional key measures are presented, Table 64.

Table 64: Key numbers: NPV, Rate of Return and Payback Time of the process for the final design.

Measure	Value
NPV	258,930,848 $
Rate of Return	41 %/y
Payback Time	2.4 years

The NPV is established to be 259 MM$ with a corresponding rate of return of 41 %. It means that the investors will gain 259 MM$ by investing in this process, instead of investing the money elsewhere.

It is highly profitable to invest in this process based on these key numbers, since the investment is earned back after 2.4 years of operation with a rate of return of

41 % of the investment and 259 MM$ in NPV.

The key numbers assume that no fluctuations of e.g. the product price are happening, and that it just follows the inflation rate of 2 %. Figure 42 investigates the robustness and sensitivity of the process by changing different costs by a certain percentage and then observe the effect in NPV.

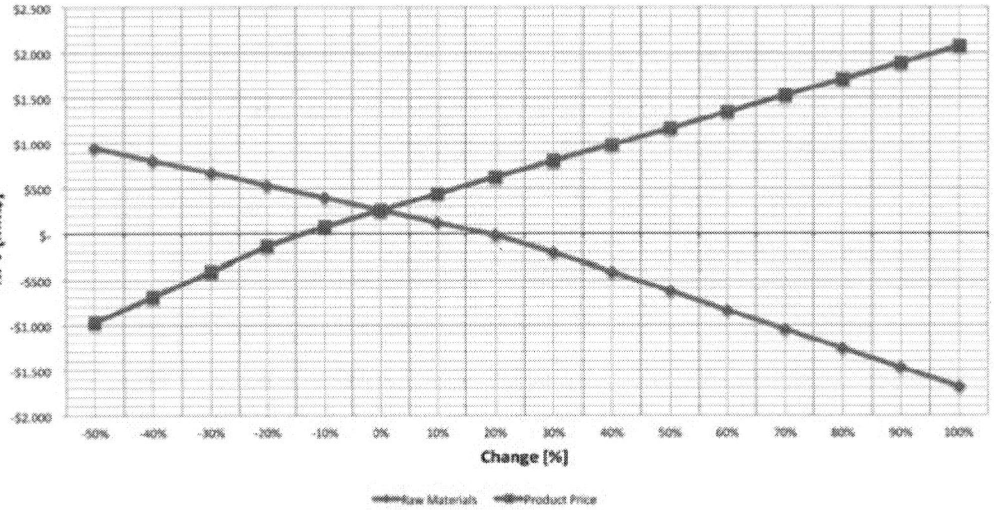

(a) Sensitivity analysis for changes in raw material price and product price.

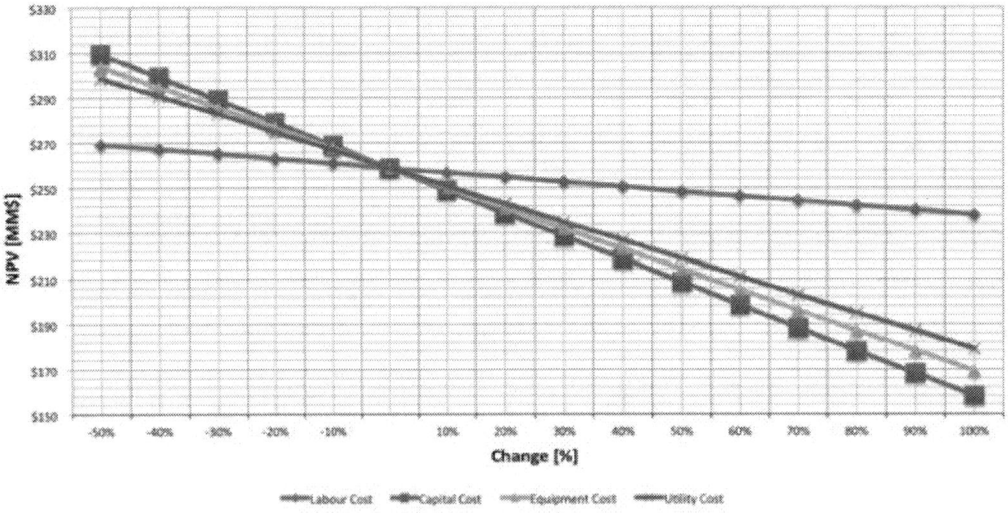

(b) Sensitivity analysis for changes in labour cost, capital cost, equipment cost and utility cost.

Figure 42: Sensitivity analysis investigating the effect on NPV by changing: (a) Raw material and product price and (b) Labour, capital, equipment and utility cost.

In Figure 42(a) the effect on NPV is investigated by changing the raw material price and the product price. If the raw material price increases by 19 %, the NPV reaches zero. If the product price decreases by 14 % the NPV reaches zero as well. It means that there is not much room for price fluctuations on the market. On the contrary, if the product price is increasing or the raw material price is decreasing it leads to a large increase in the NPV.

Figure 42(b) shows the effect on NPV by changing the labour cost, capital cost, equipment cost and utility cost. It is worth to note that the y-axis is starting in the point 150 MM$. It means that even though one of the costs are increasing by 100 %, the NPV is still positive. An increase in capital cost of 100 % decreases the NPV from 259 MM$ to just above 150 MM$. Therefore it is especially the change of raw material price and product price which have the largest impact on the NPV.

When elaborating on the above evaluation it can be concluded that investing in this process is a profitable investment.
In order to make sure that the process will gain money, it is important to establish, in which degree the raw material and product prices fluctuate. Assuming that the prices only follows the inflation rate, it is a very good investment. The investment is already paid back after 2.4 years of operation with a rate of return of 41 % and an NPV of 259 MM$.

Chapter 15: Discussion

This project deals with the design of a styrene plant, where styrene is produced from ethylene and benzene through the intermediate compound ethylbenzene.

The scope of this project was to design a profitable and sustainable styrene plant with an annual production capacity of 150,000 tonnes. In order to achieve this goal a systematic method for chemical process design was applied.

The first step in designing the process was to design a flowsheet through the hierarchical decomposition approach. This approach gave a flowsheet that has similarities to flowsheets found in the literature. Mass and energy balance was made on the generated flowsheet, and on this basis rigorous simulations were performed, and sizing and costing of the used equipment was estimated. Thereby was a base case design for the process constructed.
An economical analysis showed that the base case had an NPV of 71,732,000 $, with a payback time of 7.6 years. The plant had an energy consumption of $4.34 \cdot 10^6$ GJ/y.
Since it is desired to design a plant with a maximum payback time of 3 years, further optimization was necessary. By analysing the utility costs for the base case it was discovered that a single compressor was responsible for approximately 30 % of the total utility cost. A part of the reason for the high utility cost was the preheating of the feed to the compressor. This was done in order to ensure that the entire stream was vapour when it reached the compressor. Instead, by separating the feed stream into a liquid and a gas stream, and increasing the pressure by a pump and a compressor, it was possible to decrease the utility cost for this pressure build-up by approximately 7 %.
The economical analysis also showed that the utilities for the reboilers, connected to the distillation towers, were responsible for 56 % of the total utility cost, and that the plant was not energy efficient due to lack of interconnection between heating and cooling parts of the plant. It was therefore decided to investigate the possibility for heat integration. The heat exchanger network synthesis was therefore applied on the plant. Due to this heat integration and the optimization of the compressors, the total energy consumption decreased by 43 % and the utility cost decreased by 41 %. This decreased the payback time to 3.7 years.
In order to evaluate the sustainability of the process design the Waste Reduction Algorithm (WAR) was used. This gave a total generated Potential Environmental Impact (PEI) on the optimized case of -22,764.50, if the product stream was not taken into account.
By analysing the results obtained from WAR it was discovered that two waste streams had a high environmental impact. One of the streams was a waste stream containing toluene, benzene and ethylbenzene. In order to reduce the PEI for this waste stream, three distillation towers were added to the process. Due to these distillation towers, a significantly larger amount of the ethylbenzene and benzene

could be recycled, and thereby reducing the PEI. At the same time this involved an increase in the amount of product and a decrease in the benzene consumption. Due to these distillations, a 99.7 % pure toluene stream (molar based) was achieved. This means that the quality of the toluene was sufficiently high to make it saleable. This lead to an annual production of 4,118 tonnes of toluene. This new sustainable design decreased the PEI to -22,852,98, and at the same time increased the profit from the plant by making the by-product saleable, reducing the raw material cost and yielding more product. The other outlet stream from the plant was a gas waste stream containing mainly N_2, light organic compounds and small amounts of styrene. In order to reduce the PEI from this stream, it would be necessary to either increase the pressure or decrease the temperature below 300 K. Reducing the temperature was considered too expensive compared to the possible decrease in PEI, since this would also require a considerable amount of energy to reduce the temperature further. Increasing the pressure would increase the risk of polymerization of styrene and would also be expensive in compressor duty. It was therefore decided not to make any sustainability optimization on this stream.

The decanter is used to decant water from the organic phase and separation of the vapour. It is possible that the water phase will contain traces of e.g. benzene or styrene, which then has to be removed, by sending the water stream to a treatment plant, if the emission limit exceeds the one defined by the authorities. These considerations are not taken into account in the report, but have to be taken into consideration before constructing the plant. The separated gas phase consists mainly of nitrogen, but also a small fraction of styrene. If using pure oxygen, instead of air, it is possible to gain a higher yield of styrene, but the cost of oxygen is expensive. It is therefore decided to use the free air, which is more economical efficient, and also environmental friendly since it is an energy highly consuming process to produce pure oxygen.

In Table 65 the economical key numbers for the plant, in different parts of the design process, are shown. Due to the described optimizations the NPV is increased with 261 % to 258,930,848 \$, while the rate of return is increased from 9 % to 41 %, and the payback time is reduced from 7.6 years to 2.4 years. The table also shows that during the decrease of the utility cost, the equipment cost is also decreased. The table shows that the sustainable process design did make the utility cost increase, compared to the case after heat integration, but due to the amount of raw material saved, the NPV increases with more than 100,000,000 \$. Table 66 shows that due to the optimizations the energy consumption is decreased with 42.2 % compared to the base case design. The final design is therefore significantly more environmentally friendly than the base case design. The table also shows that due to the optimizations, the rate of return has increased with 355.6 %.

Table 65: Overview of economical key numbers for the plant design in the different steps of the process.

Item	Base Case	Optimized	After HI	Final design
Utility cost [$]	18,191,035	16,170,652	10,670,447	10,831,709
Equipment cost [$]	12,937,717	11,707,184	11,140,699	11,459,633
NPV [$]	71,731,904	97,487,064	143,081,264	258,930,848
Total energy consumption [GJ/y]	$4.337 \cdot 10^6$	$3.971 \cdot 10^6$	$2.470 \cdot 10^6$	$2.505 \cdot 10^6$
Rate of return [%]	9	14	23	41
Payback Time [y]	7.6	5.4	3.7	2.4

Table 66: Overview of the overall economical effect of the optimizations.

Item	Change BC to final
Utility cost	-40.5%
Equipment cost	-11.4%
NPV	261.0%
Total energy consumption	-42.2%
Rate of return	355.6%
Payback Time	-68.4%

The design of this process is based on a preliminary design approach. Simplified sizing and costing correlations are used to rapidly estimate the cost of the process. Biegler et al. [1997] reports a 25 % to 40 % level of accuracy for the sizing and costing correlations. Therefore, the final economics are not detailed estimates, and therefore acceptable deviations appear. Instead, the method is used as a screening method to establish qualitative trends among different process designs. As discussed in Chapter 13, the final, sustainable design is the most profitable design, when comparing to the other design alternatives. Therefore, this design is the process design that should be investigated further, if choosing to invest in this process.

This plant design is based on a series of assumptions:

- It is assumed that it is possible to buy high quality benzene and ethylene for the plant. Based on this assumption it is assumed that the feed stream for benzene and ethylene are 100 % pure. Buying this pure is expensive and not profitable.
  Due to this assumption it is not necessary to purge the recycled benzene stream, which otherwise would have contained small amounts of light hydrocarbons. If the feed stream contains impurities, these would have made different by-products during the reaction where ethylbenzene was formed. In a real plant the raw material cost would therefore have been higher due to benzene lost during the purge of the recycle stream.

- The price for ethylene and benzene were set to a relatively low value compared to the assumption that the purity was close to 100 %. A potential buyer of this plant design should therefore first investigate if they can buy ethylene and benzene for the assumed prices.

- During the economic analysis it was assumed that the price of styrene would be constant during the entire lifetime of the plant. The price for styrene is unfortunately known for fluctuation. This can, in spite of the increasing demand for styrene, make the market more insecure to enter as a producer.

- During the sizing of the heat exchangers it was not possible to find the exact overall heat transfer coefficients for the different shell and tube side systems. This was in particular a problem during the heat integration where in some cases a condensing stream at one side and an evaporating stream at the other side of the shell wall is handled. Due to these sizing problems it is expected that some uncertainties are present in the heat exchanger costing.

- During the heat integration the location of the different heat exchangers was not taken into account. Heat was for instance moved from some condensers to reboilers. This would, at a minimum, require an extra utility cost due to the need for pumps or compressors, which was not taken into account. It is therefore possible that some of the heat integration decisions should be reconsidered when making the final P&ID.

- During the economic analysis it was assumed that the plant would have a constant production capacity of 100 %. This is probably unlikely because it is common that there are some production problems during the first start ups of a chemical plant. The salary for the workers on the plant was calculated on basis of the capital cost of the plant and not on the amount of unit operations and their possible need for maintenance. This estimate is therefore relatively rough, but since the worker salary is only a small fraction of the operating cost this is minor uncertainty.

- During the design of the process, some reasonable assumptions have been taken, when lack in information from the literature was present. Therefore, some engineering reflections were used to take a reasonable assumption. Since a theoretical approach is used for the designing, estimates and assumptions are used to account for data, that are unknown. Further research have to be carried out in order to establish the exact physical parameters.

- For the PEI calculations the energy need was not taken into account. Considering that nowadays it is necessary to reduce the emission of carbon dioxide, the energy consumption should have been taken into account, since it is very likely that the energy would come from combustion of coal.

Before commercializing the plant, heat integration should be applied on the part of the plant which was introduced in order to make the plant more sustainable.

This would possibly make a small decrease in the utility cost for the final process design.
During the design and economical analysis of the plant, a possible combustion of the gas streams was not considered. A possible energy saving due to this combustion, and the equipment cost for this combustion, should be calculated in order to achieve a more accurate capital cost and utility cost for the plant.
Since high pressure of up to 37.5 atmosphere is applied in some parts of the plant, it should be considered to implement turbines when the stream goes from this pressure to atmospheric pressure. This would generate power, which makes the plant more environmentally friendly. Before applying turbines to the system an economic analysis should indicate that the payback time for the turbines would not be much larger than 3 years in order to ensure that the investment would pay off.
Before building the plant it should be considered to find a location with a production plant for benzene and ethylene nearby, in order to limit the transportation expenses. It should therefore also be considered to be situated close to a proportional buyer of styrene.

There is one direct fired heater in the process which currently uses coal as utility. In order to make the process more sustainable, it is a possibility to use another utility source for this heater. Natural gas could be a possible candidate since it releases fewer greenhouse gases and pollutants. By replacing coal with natural gas, the utility cost of the heater increases from 96,608 \$/y to 541,728 \$/y, which decreases the NPV with 1.3 %. Therefore the investors should decide whether this should be implemented or not; it is a balance between the economics and the environmental considerations.

Chapter 16: Conclusion

This report describes the design of a chemical plant, which produces styrene from ethylene and benzene through the intermediate compound ethylbenzene. A market analysis showed that it is possible to enter the market with a production plant capacity of 150,000 ton styrene per year. In order to make a profitable and sustainable design a systematic method for chemical process design is applied.

The first step in designing the plant is to make a base case design. This is done through the hierarchical decomposition approach, giving a plant design with similarities to styrene production plants described in the literature. The base case design plant design has an NPV of 71,732,000 $, a payback time of 7.6 years and a rate of return of 9 %. In order to investigate where to optimize the production plant, an economical analysis is performed. This analysis shows that the utility cost for a compressor series, which is used to build-up pressure to the reactor where ethylbenzene is formed, is responsible for approximately 30 % of the utility cost. It also shows that the reboilers, connected to the distillation towers, are responsible for 56 % of the total utility cost for the plant. The compressor duty is decreased by relocation of a heat exchanger and installation of a flash drum and a pump. The utility cost for heating and cooling is decreased by use of heat integration. The heat exchanger network is made on the basis of the heat exchanger network synthesis. These optimizations decreases the utility cost with 41 % and the energy consumption with 43 %.

In order to evaluate the sustainability of the designed plant the Waste Reduction Algorithm (WAR) is applied. This algorithm gives a generated Potential Environmental Impact (PEI) for the plant of -22,764.50. This value is negative due to the assumption that the product stream is not taken into account. The PEI-analysis shows that a waste stream containing toluene, benzene and ethylbenzene should be optimized. In order to further decrease the PEI, three extra distillation towers are added to the process. This made it possible to recycle a larger amount of benzene and ethylbenzene, produce 4,118 ton/year of 99.7 % pure toluene (molar basis) which is sold, and reduce the waste stream to a small waste gas. This decreases the PEI generated by the plant further to -22,852,98, while decreasing the amount of raw material needed.

Table 67 shows the key numbers for the final process design. The table shows that the total production of styrene per year is higher than the goal of 150,000 tonnes/year. This is due to a higher conversion than initially assumed. The final process design needs 15.1 MJ/kg produced styrene, which is high, but this is partly due to a large recycle of benzene, used in significant excess during the formation of the intermediate, ethylbenzene. 203.9 kg water is needed per kg produced styrene, which is due to cooling and heating during the production. It is unfortunately not possible to make integration between all the heat exchangers, and a large amount of cooling water and hot steam is therefore still necessary during the production. The operating cost for the process is 1.38 $/kg, covering both the utility cost, raw material cost and work salary etc. Since the product is sold for 1.6 $/kg, the earning of the process is 0.22 $/kg. The utility cost is 0.07 $/kg which is only 5 % of the total operating cost. This shows that the optimization of the plant has successfully decreased the utility cost to a very satisfactory level. The produced styrene has a purity of 99.5 %, which is in line with the scope of the report. The by-product toluene is produced as 99.7 % pure, which makes it saleable. It is important to achieve these purities in order to maintain the quality of the products and the market share. The process uses 1.1 kg of raw material per kg product which gives a conversion from raw materials to product of 99 %. It is therefore concluded that a satisfactory usage of raw materials are achieved.

Table 67: Key numbers for the final design.

Measure	**Value**	**Unit**
Total Production per year	165,783,661	kg/y
Energy needed/kg of product	15.1	MJ/kg
Water needed/kg of product	203.9	kg/kg
Operating cost/ kg of product	1.38	$/kg
Raw material/ kg of product	1.10	kg/kg
Utility cost/kg of product	0.07	$/kg

For future work it is recommended that the following is investigated before commercializing the plant.

- For the three distillation towers where toluene is concentrated and benzene and ethylbenzene is recycled, heat integration should be investigated.
- It should be investigated how much energy is saved by combustion of the waste gas from the plant. The price of a potential combustion plant should also be investigated.
- Since a part of the plant is operated at high pressure, it should be investigated if it is economically viable to implement turbines where the pressure is decreasing to atmospheric pressure.
- When considering the environmental impact of the process, it is important to involve a considerable amount of different parameters. In the PEI analysis, the energy consumption has not been taken into consideration, which needs to be done in the further work. Thereby, it is possible to identify the bottlenecks more precisely.

These key numbers, including the key numbers of the economics, gives an idea of, if further work of the process should be done. The final design has an NPV of 259 MM$, with a payback time of 2.4 years, and a rate of return of 41 %. This indicates, that there is a potential of reaching a large profit by commercialising the plant. It is therefore recommended that further work on the final design should be carried out, into the next phase of the design process.

References

Alibaba Group [2014], 'Product search of zeolite-y and shift catalyst', URL: `http://www.alibaba.com/`. Last Visited: 25.05.2014.

Bagajewicz, M., Bondy, R., Cathcart, B., Champagnie, A., Kovach, J., Leung, G., Parikh, R., Schmid, C., Shah, V. and Yu, R. [1994], *PRO/II Unit Operations Reference Manual*, Simulation Sciences Inc.

Behr, P. D. A. [1999], *Experiment Styrene Production from Ethylbenzene*, Faculty of Biochemical and Chemical Engineering - University of Dortmund.

Biegler, L. T., Grossmann, I. E. and Westerberg, A. W. [1997], *Systematic Methods of Chemical Process Design*, Prentice Hall PTR, Upper Saddle River, New Jersey.

Cabezas, H., Bare, J. C. and Mallick, S. K. [1999], 'Pollution prevention with chemical process simulators: The generalized waste reduction (war) algorithm - full version', *Computers and Chemical Engineering* **23**, 623–634.

Cavani, F. and Trifiro, F. [1995], 'Alternative processes for the production of styrene', *Applied Catalysis A* **133**, 213–239.

Cheng, J. C., T. G. Degnan, J. B., Huang, Y. Y., Kalyanaraman, M., Kowalski, J. A., Loehr, C. A. and Mazzone, D. N. [1998], 'A comparison of zeolites mcm-22, beta, and usy for liquid phase alkylation of benzene with ethylene', *Science and Technology in Catalysis* **121**, 53–60.

CLH [2011], 'Proposal for harmonised classification and labelling', URL: `http://www.compositesworld.com/cdn/cms/uploadedFiles/danish_epa_styrene_review(2).pdf`. Last Visited: 11.02.2014.

Ebrahimi, A. N., Zolfaghari, A., Mousavi, S. A., Aghazadeh, F. and Soltani, A. [2011], 'Modification and optimization of bezene alkylation process for production of ethylbenzene', *Chemical Engineering and Processing* **50**, 31–36.

EPA [1994], 'Oppt chemical facts sheet, styrene'.

Find The Data [2014], 'Salaries by city', URL: `salaries-by-city.findthedata.org/`. Last Visited: 26.05.2014.

Ganji, H., Ahari, J. S., Farshi, A. and Kakavand, M. [2004], 'Modelling and simulation of benzene alkylation process reactors for production of ethylbenzene', *Petroleum & Coal* **46**, 55–63.

Gerzeliev, I. M., Khadzhiev, S. N. and Sakharova, I. E. [2010], 'Ethylbenzene synthesis and benzene transalkylation with diethylbenzenes on zeolite catalysts', *Petroleum Chemistry* **51**(1), 39–48.

Haslego, C. [2010], 'Basics of polystyrene production', URL: http://www.cheresources.com/content/articles/processes/basics-of-polystyrene-production. Last Visited: 17.02.2014.

ICIS [2014], 'Indicative chemical prices a-z', URL: http://www.icis.com/chemicals/channel-info-chemicals-a-z/. Last Visited: 18.03.2014.

Indeed [2014], 'Average salary of jobs by city', URL: http://www.indeed.com/salary. Last Visited: 26.05.2014.

Landel, M. [2008], 'Federal corporate tax rates', URL: http://www.marielandel.com/federal-corporate-taxe-rates.html. Last Visited: 07.06.2014.

Lide, D. R. [2009], *Handbook of Chemistry and Physics*, CRC Press.

Liu, S., Chen, F., Xie, S., Zeng, P., Du, X. and XU, L. [2009], 'Highly selective ethylbenzene production through alkylation of dilute ethylene with gas phase-liquid phase benzene and transalkylation feed', *Journal of Natural Gas Chemistry* **18**, 21–24.

Mousavi, S. M., Panahi, P. N., Niaei, A., Farzi, A. and Salari, D. [2012], 'Modeling and simulation of styrene monomer reactor: Mathematical and artificial neural network model', *International Journal of Scientific & Engineering Research* **3**, 1–7.

Nederlof, C. [2012], Catalytic dehydrogenations of ethylbenzene to styrene, PhD thesis, Delft University of Technology.

Ngandjui, L. T., Louhibi, D. and Thyrion, F. [1997], 'Kinetic analysis of diethylbenzene-benzene transalkylation over faujasite y', *Chemical Engineering and Processing* **36**, 133–141.

PR Newswire [2011], 'Sytrene market to exceed demand of 41 million tons in 2020, asia pacific will drive growth', URL: http://www.prnewswire.com/news-releases/sytrene-market-to-exceed-demand-of-41-million-tons-in-2020-...asia-pacific-will-drive-growth-134865908.html. Last Visited: 11.02.2014.

Price, G. L. [2014], 'Zeolite y - structure and characterization', URL: http://www.personal.utulsa.edu/~geoffrey-price/zeolite/fau.htm. Last Visited: 31.05.2014.

Sanfilippo, D., Capone, G., Cipelli, A., Pierce, R., Clark, H. and Pretz, M. [2007], 'Snow: Styrene from ethane and benzene', *Natural Gas Conversion VIII* pp. 505–510.

Sigma-Aldrich [2013], 'Information about products', URL: http://www.sigmaaldrich.com/. Last Visited: 17.02.2014.

Society, S. E. [2006], 'Manufacturing proceses and selection', URL: `http://www.sbioinformatics.com/design_thesis/Styrene/Styrene_Methods-2520of-2520Production.pdf`. Last Visited: 19.02.2014.

Suppes, G. J. [2004], *Selecting Thermodynamyc Models for Process Simulation of Organic VLE and LLE Systems*, Department of Chemical Engineering The University of Missouri-Columbia.

Tamsilian, Y., Ebrahimi, A. N., S.A., A. R. and Abdollahzadeh, H. [2012], 'Modeling and sensitivity analysis of styrene monomer production process and investigation of catalyst behavior', *Computers and Chemical Engineering* **40**, 1–11.

Timmerhaus, K. D., Peters, M. S. and West, R. E. [2003], *Plant Design and Economics for Chemical Engineers, Fifth Edition*, McGraw-Hill Companies, Inc., 1221 Avenue of the Americas, New York.

Young, D. M. and Cabezas, H. [1999], 'Designing sustainable processes with simulation: The waste reduction (war) algorithm', *Computers and Chemical Engineering* **23**, 1477–1491.

Appendix

A EP Calculation

Table 68: Data from the calculation of EP.

Production Styrene [kmol/y]	1,440,230.44
Selectivitet Styrene	0.98
Conversion to Styrene	0.95
Comsuption of ethylbenzene [kmol/y]	1,547,445.18
Selectivitet EB	0.90
Conversion to EB	1.00
Comsuption of ethylene [kmol/y]	1,719,956.85
Comsuption of B [kmol/y]	1,719,956.85
Comsuption of ethylene [kg/y]	48,244,789.58
Comsuption of B [kg/y]	134,345,829.37
EP [$/y]	54,722,464.46
EP [$/kg]	0.36

B Diagrams for Azeotropes

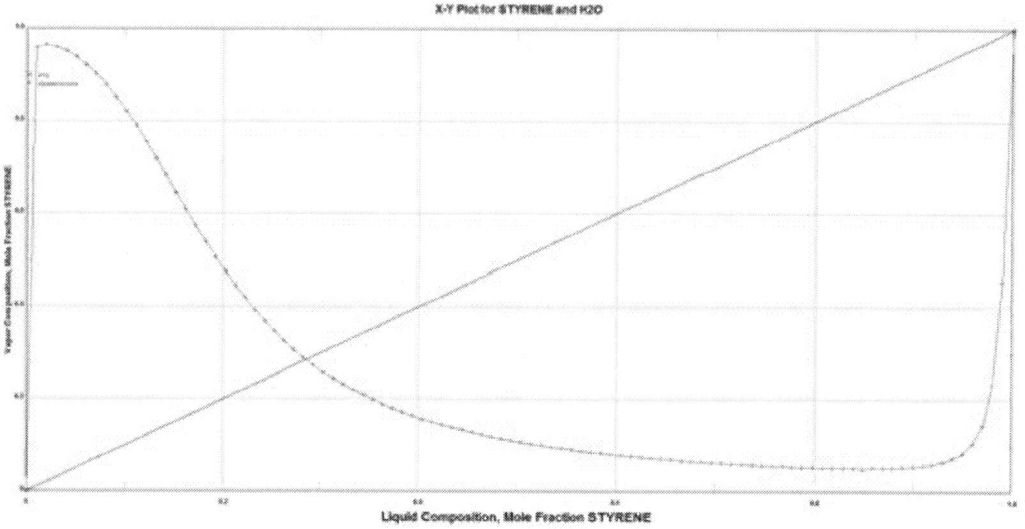

Figure 43: XY-diagram for styrene and water at 1 atm. Generated by use of UNIQUAC as thermodynamic model.

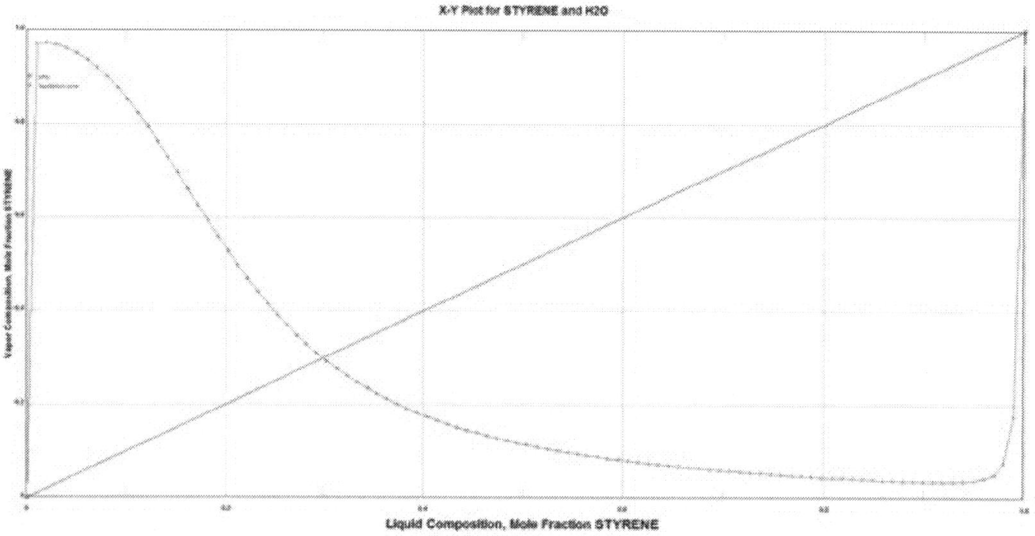

Figure 44: XY-diagram for styrene and water at 0.2 atm. Generated by use of UNIQUAC as thermodynamic model.

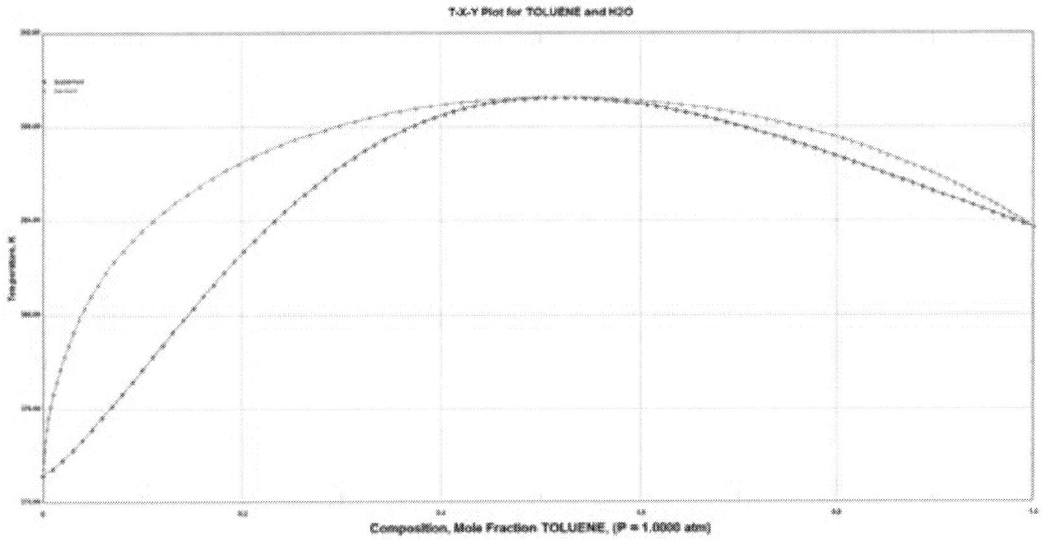

Figure 45: TXY-diagram for toluene and water at 1 atm. Generated by use of Wilson as thermodynamic model.

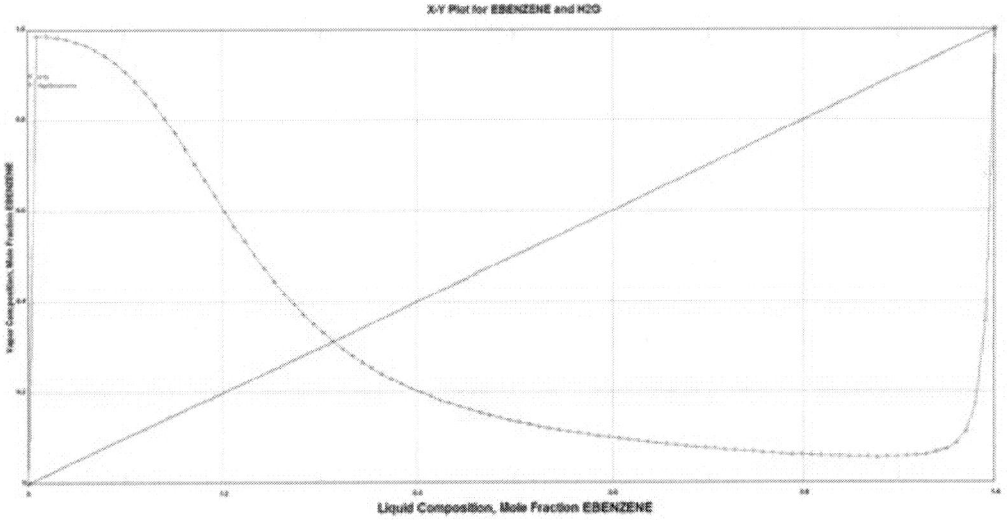

Figure 46: XY-diagram for ethylbenzene and water at 1 atm. Generated by use of UNIQUAC as thermodynamic model.

C Calculation for Area of Heat Exchanger E7

In this example the calculation for the area of the heat exchanger E7 will be reviewed. All the heat exchangers have been calculated from the same principles as this heat exchanger. The temperature profile for the hot side of heat exchanger E7 is shown in Figure 47. Based on this figure, it is assumed, that the stream is first condensed and then cooled. It is assumed, that the condensation of all the compounds takes place at one time.

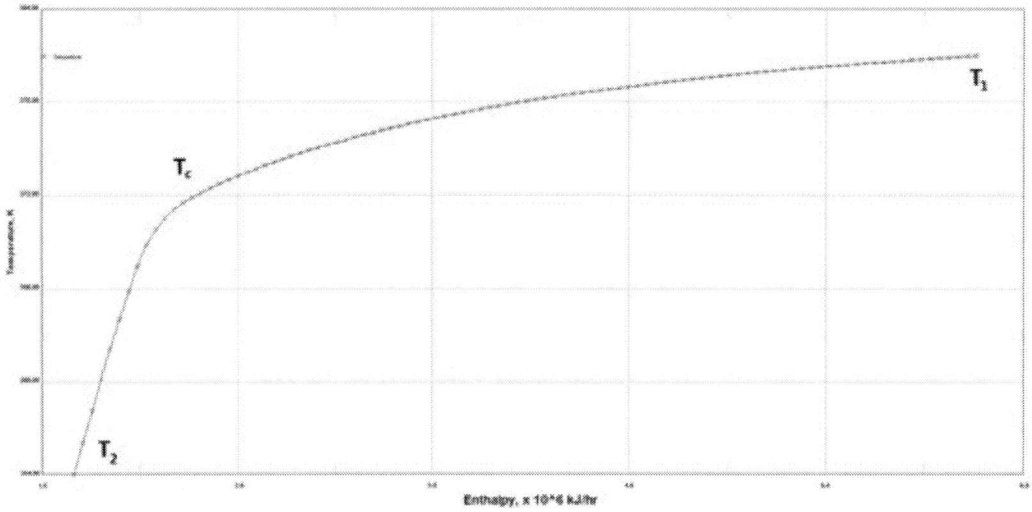

Figure 47: Temperature profile for the hot side of heat exchanger E7.

The total duty of the heat exchanger (Q_{Total}) is found from the energy balance:

$$\begin{aligned} Q_{Total} &= Q_{In} - Q_{Out} = (H_{in} - H_{out}) \cdot F \\ &= (541.9\text{kJ/kg} - 143.3\text{kJ/kg}) \cdot 1.162 \cdot 10^4\text{kg/h} = 4.631 \cdot 10^6\text{kJ/h} \end{aligned}$$

The change in enthalpy due to cooling is found by:

$$Q_{cool} = C_p \cdot \Delta T \cdot F = 1.879 \frac{\text{kJ}}{\text{kg} \cdot \text{K}} \cdot (381.1\text{K} - 354.2\text{K}) \cdot 1.162 \cdot 10^4\text{kg/h} = 5.862 \cdot 10^5\text{kJ/h}$$

Here it is assumed that the heat capacity of the media is constant during the cooling. Since the vapour is first condensed and then cooled, it is assumed, that the heat capacity of the media is equal to the heat capacity of the outlet stream from the heat exchanger.

The change in enthalpy due to condensation (Q_{con}) is found from the different between the total change in enthalpy and the change in enthalpy due to cooling of the media:

$$\begin{aligned} Q_{con} &= -\Delta H_{vap} \cdot f_{\text{feed in vapor phase}} = Q_{Total} - Q_{cool} \\ &= 4.631 \cdot 10^6\text{kJ/h} - 5.862 \cdot 10^5\text{kJ/h} = 4.045 \cdot 10^6\text{kJ/h} \end{aligned}$$

Due to the assumption that the stream condenses immediately when it enters the heat exchanger, $T_c = T_1$, is a rough assumption, but as mention in section 8.8, this is only a rough calculation, which main purpose is to give an estimate of the cost. Water is used to cool the hot stream, and it is assumed, that the water stream has an inlet temperature of 288.15 K and an outlet temperature of 398.15 K.
The needed flow of cooling water can be found from:

$$F_{water} = \frac{Q_{Total}}{C_{p-water} \cdot \Delta T} = \frac{4.631 \cdot 10^6 \text{kJ/h}}{4.181 \frac{\text{kJ}}{\text{kg} \cdot \text{K}} \cdot (298.15\text{K} - 288.15\text{K})} = 1.107 \cdot 10^5 \text{kg/h}$$

The cooling streams temperature at the point, when the hot side begins to condensate (t_3), can be found from:

$$t_3 = t_2 - \frac{Q_{con}}{m_{water} \cdot c_{p-water}} = 298.15 - \frac{4.045 \cdot 10^6 \text{kJ/h}}{1.107 \cdot 10^5 \text{kg/h} \cdot 4.181 \frac{\text{kJ}}{\text{kg} \cdot \text{K}}} = 289.4\text{K}$$

The log mean temperature for the condensing part of the heat exchanger can be found as:

$$\begin{aligned} \Delta T_{lm-con} &= \frac{(T_1 - t_2) - (T_c - t_3)}{\ln\left(\dfrac{T_1 - t_2}{T_c - t_3}\right)} \\ &= \frac{(381.1\text{K} - 298.15\text{K}) - (381.1\text{K} - 289.4\text{K})}{\ln\left(\dfrac{381.1\text{K} - 298.15\text{K}}{381.1\text{K} - 289.4\text{K}}\right)} = 87.21\text{K} \end{aligned}$$

The log mean temperature for the cooling part of the heat exchanger can be found:

$$\begin{aligned} \Delta T_{lm-cool} &= \frac{(T_c - t_3) - (T_2 - t_1)}{\ln\left(\dfrac{T_c - t_3}{T_2 - t_1}\right)} \\ &= \frac{(381.1\text{K} - 289.5\text{K}) - (354.11\text{K} - 288.15\text{K})}{\ln\left(\dfrac{381.1\text{K} - 289.5\text{K}}{354.2\text{K} - 288.15\text{K}}\right)} = 78.15\text{K} \end{aligned}$$

The overall heat transfer coefficient for the condensing part (U_{con}) and cooling part (U_{cool}) of the heat exchanger are estimated to respectively $2044 \dfrac{\text{kJ}}{\text{h} \cdot \text{m}^2 \cdot \text{K}}$ and $1022 \dfrac{\text{kJ}}{\text{h} \cdot \text{m}^2 \cdot \text{K}}$. The area of the condensing part and cooling part of the heat exchanger can be found by:

$$A_{con} = \frac{Q_{con}}{U_{con} \cdot \Delta T_{lm-con}} = \frac{4.045 \cdot 10^6 \text{kJ/h}}{2044 \dfrac{\text{kJ}}{\text{h} \cdot \text{m}^2 \cdot \text{K}} \cdot 87.21\text{K}} = 22.7\text{m}^2$$

$$A_{cool} = \frac{Q_{cool}}{U_{cool} \cdot \Delta T_{lm-cool}} = \frac{5.862 \cdot 10^5 \text{kJ/h}}{1022 \dfrac{\text{kJ}}{\text{h} \cdot \text{m}^2 \cdot \text{K}} \cdot 78.15\text{K}} = 7.33\text{m}^2$$

The total area of the heat exchanger (A_{Total}) is the sum of the different area of the different parts of the heat exchanger and can therefore be calculated as:

$$A_{Total} = A_{con} + A_{cool} = 22.7\text{m}^2 + 7.33\text{m}^2 = \underline{30.0\text{m}^2}$$

D Costing Example of Compressor C4

In this appendix a calculation example is provided for compressor C4 by using the method and data by Timmerhaus et al. [2003].

The first step is to determine the type of equipment, so the correct plot can be found. C4 is just a normal compressor, giving Figure 48.

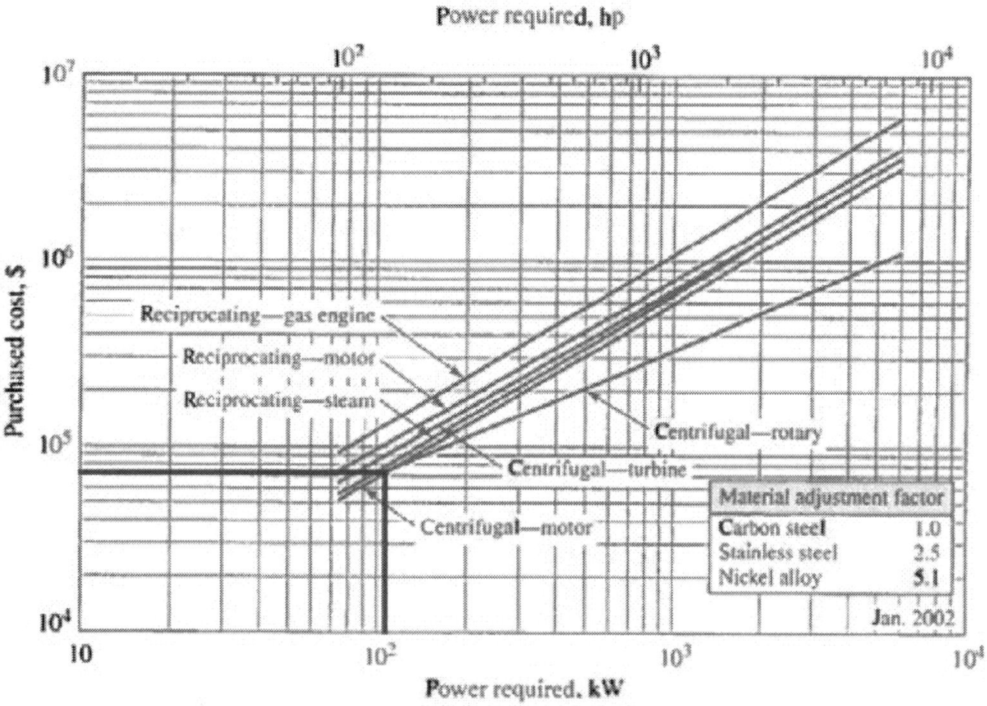

Figure 48: Purchased cost of compressors. Price includes drive, gear mounting, baseplate and normal auxiliary equipment; operating pressure up to 69 atmosphere.

The capacity of compressor C4 is calculated to 104.86 kW (according to Table 27). Thereby, it is possible to find the purchase cost by approaching Figure 48. C4 is a centrifugal-motor compressor. The base purchase cost is then estimated to be around 70,000 \$. For C4, MPF is equal to 1, because it is constructed of carbon steel. The updated base cost is then calculated, by multiplying with UF, so the 2002 price is translated into a 2014 price:

$$\text{UBC} = \text{C} \cdot \text{UF} \cdot \text{MPF} = 70,000\$ \cdot 1.33 \cdot 1 = 93,100\$$$

It means that the specified compressor has a price of 93,100 \$. All the other equipment in the process are costed in the same way by using log-log plots from Timmerhaus et al. [2003].

Furthermore the electricity of the compressor has to be calculated, which is done by multiplying the work with the price of electricity. From ECON it is found, that the price of electricity is 16.8 \$/GJ. 104.86 kW is then equal to a total utility cost of 50,228 \$ per year by assuming 330 operating days per year.

E Results from the tabular method

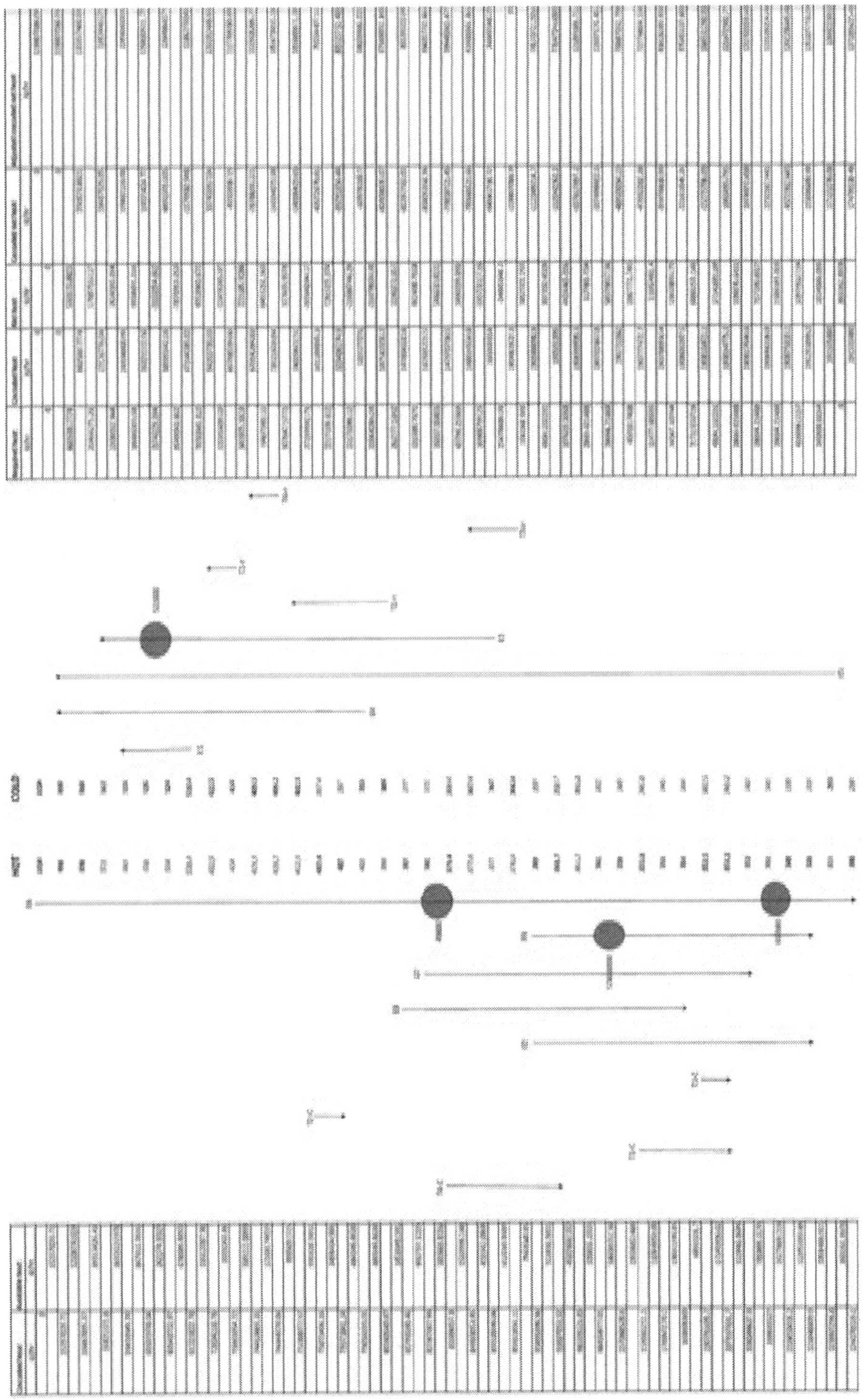

381852 Systematic Design and Optimization of Acrylonitrile Production

Abstract:

Systematic Design and Optimization of Acrylonitrile Production

Steen M. Christensen, Anna Leth-Espensen, Marina Fedorova*

Computer Aided Process Engineering Center, Department of Chemical and Biochemical Engineering, Technical University of Denmark, DK-2800 Lyngby, Denmark

*mfad@kt.dtu.dk

Acrylonitrile is an important starting material for a wide range of polymers and copolymers and is found in numerous everyday products like conveyer belts, Lego blocks, and fabric material. In this project, which is part of a MSc Process Design course at the Technical University of Denmark, a production facility of acrylonitrile is designed. The objective is to design a process for the production of 100 000 tons of acrylonitrile per year. The designed process must be environmentally acceptable, must apply heat integration and be optimal with respect to cost.

The production of acrylonitrile is nowadays performed through the Sohio process. This process is a reaction between ammonia, air, and propene that react through the catalytic mechanism of ammoxidation. Other production methods for acrylonitrile including the ethylene cyanohydrine process and the ammoxidation of propane are also possible. A new process that could be interesting in the future is the ammoxidation of propane, because of the lower cost of propane compared to propylene. However, the selectivity for product in this process is too low. Therefore this project focuses mainly on designing and optimizing the known Sohio process.

The design is done systematically from generating a simple flow sheet to a rigorous model of the plant. The simple flow sheet is created by the Hierarchical Decomposition Douglas method. The design has then been optimized to find a more sustainable solution, both with respect to economics and the environment. Also, heat integration has been performed to reduce the energy consumption.

With respect to the economical optimization especially the by-products have been taken into account. In the production of acrylonitrile two other saleable by-products are produced though in considerably smaller amounts than acrylonitrile. One is hydrogen cyanide which is converted into acetone cyanohydrin for easy shipping and another is acetonitrile. The purification of these two by-products and the acrylonitrile is a difficult process, which has been considered in detail in this project. The separation methods include neutralization, absorption by water, stripping, decanting, distillation, reactive distillation and a pressure swing distillation column of which especially the last two are interesting and crucial in order to obtain the required purity of the acetonitrile.

Several alternatives to the base case design were obtained after performing optimization and sustainability and LCA analyses. With respect to the environmental optimization especially the recycle of process water and definition of the optimum pressure and temperature settings for the distillation columns save energy and resources.

386703 Optimal Design of an Integrated Algae-Based Biorefinery for Biodiesel, Astaxanthin and PHB Production

Abstract:

In this work, we propose an MINLP model for the design of an integrated microalgae-based biorefinery for the simultaneous production of biodiesel, astaxanthin and polyhydroxybutyrate (PHB), as well as a combined heat and power cycle. The biodiesel production process under study consists of four main stages namely: microalgae cultivation, harvesting and dewatering, lipid extraction and oil upgrading. Astaxanthin, a natural ketocarotenoid that is a secondary metabolite in several microalgae species, is a powerful antioxidant with application in nutraceuticals, pharmaceuticals, cosmetics and food industries. We analyze the possibility to maximize both astaxanthin and oil production from microalgae species like *Haematococcus pluvialis*, as their accumulation takes place under the same culture conditions (Huntley and Redalje, 2007). Astaxanthin can be produced after harvesting and dewatering, employing a drying and pulverization step (Li et al., 2011). On the other hand, PHB is a biopolymer that constitutes an alternative to fossil fuel based polymers, with similar properties. It also has promising uses as in drug delivery and food packaging and it is produced by microorganisms that can uptake glycerol as substrate. Glycerol, in turn, is a by-product in the biodiesel production process. In this way, the PHB production process is composed of three main steps: glycerol purification, fermentation, and PHB extraction. Residues from these processes, as well as glycerol, can be considered as an option to feed a power generation step. The produced energy is a significant alternative that can be used in the processing sections described before. In this work, we consider the anaerobic digestion of the algal biomass cake (after lipid extraction), using glycerol and waste paper as additional substrates. The addition of glycerol to the anaerobic digester, decreases PHB productivity but enhances methane yield, so a lower bound is set to the flowrate of glycerol submitted to the anaerobic digestion step (Garcia Prieto et al., 2014).

The proposed MINLP model is implemented in GAMS (Brooke et al., 2013) for net present value maximization. Model equality constraints include mass balances, yield equations and correlations, while inequality constraints include process specifications such as a requirement on produced CO_2 to be less than the required amount for biomass generation in the photobioreactors. In this way, a global reduction of the CO_2 emissions to the atmosphere is ensured, as the rest of the required amount of CO2 is provided by a flue gas stream from a thermoelectrical plant. The upper bound of CO_2 is limited by the CO_2that is produced in this plant, located next to Bahia Blanca city. Furthermore, the carbon to nitrogen ratio in the anaerobic digester must be between specific values for optimal operation. An upper bound for the waste paper amount fed to the anaerobic digester is needed. The constraint of this stream, which ensures the substrate requirements in the digester, is assumed to be the total amount of the domestic recycled paper in Bahia Blanca's city. A lower bound constraint on the glycerol flowrate stream to the biodigester is required to ensure the amount of carbon necessary for biogas production.

Numerical results show that an important reduction in biodiesel production cost can be obtained for the integrated biorefinery scheme as compared to the biodiesel process from microalgae with power and heat generation unit by anaerobic digestion (0.98 $/kg biodiesel against 1.26 $/kg biodiesel). Even though the contribution of capital, raw material, utilities and operating and maintenance cost are quite similar, the production cost is lower, mainly due to revenues from PHB (6.25 $/kg), which is a value-added product with a selling price much higher than glycerol (0.26 $/kg), its main raw material. The integrated scheme including astaxanthin production is the optimal alternative as it includes an extra value-added product such as astaxanthin (7.78 $/kg).

References

Brooke, A.; Kendrick, D.; Meeraus, A.; Raman, R.; GAMS- A User's Guide, GAMS Development Corporation: Washington, 2013.

Huntley, M. E.; Redalje, D.G. CO2 Mitigation and Renewable Oil from Photosynthetic Microbes: A new Appraisal. Mitigation and Adaptation Strategies for Global Change, 2007, 12, 573-608.

Garcia Prieto, C., F Ramos, V. Estrada, M. S. Diaz, Optimal Design of an Integrated Microalgae Biorefinery for the Production of Biodiesel and PHBs, Chemical Engineering Transactions, 2014

Li, J.; Zhu, D.; Niu, J.; Shen, S.; Wang, G. An Economic Assessment of Astaxanthin Production by large Scale Cultivation of Haematococcus pluvialis. Biotech. Adv. 2011, 29, 568-574.

388364 Optimal Design of Thermal Membrane Distillation Systems and Heat Integration with Process Plants

Abstract:

This work presents a mathematical programing model for the optimal design of thermal membrane distillation (TMD) systems that are thermally coupled with processing facilities. TMD is an emerging technology that lies at the interface between thermal and membrane technologies. It is a non-isothermal separation process that is based on heating up the feed solution to be distilled to some moderate temperature to create partial evaporation of water, the vapor that permeates through the membrane is condensed and collected as a highly pure liquid on the permeate side. A superstructure representation and an optimization formulation are introduced to synthesize the TMD network and the heat-exchange network that integrates heating and cooling in the process facility. The superstructure and associated optimization formulation seek to identify the system configuration along with design and operating variables such as heat-exchanger areas, membrane area, extent of thermal coupling between the process and TMD, and the TMD feed-preheating temperature. The objective function maximizes the net annual profit which accounts for the revenues from the sales of purified water, the avoided cost of the treated wastewater, and the total annualized costs involving the capital investment of the added heat transfer units and the TMD network, the operating costs for the heating and cooling utilities and the operating expenses for the TMD system. Three scenarios are analyzed. The first one considers that there is not energy interaction between the TMD unit and the process facility. The second scenario considers only the heat integration among the process streams within the process facility (no heat integration with TMD). The third scenario involves the heat integration between the process facility and the TMD unit. The proposed optimization formulation is applied to a case study where a TMD system is integrated with a methanol plant and the results show significant economic benefits for the implementation of the proposed methodology.

390628 Globally Optimal Sensor Network Design Revisited: Application of the Generalized Bender Decomposition

Abstract:

In a series of papers [1, 2, 3] the problem of optimal sensor selection for linear systems has been shown to be of the Mixed Integer Convex Programming (MICP) class. While the use of such formulations has opened the door to a guarantee of global optimality, the use of a branch and bound search procedure has limited application of this approach to fairly small systems. The particular bottleneck can be attributed to the fact that during each iteration of the branch and bound search a fairly slow Semi-Definite Programming (SDP) problem needed to be solved to it global optimum.

In this work, we illustrate that a simple reformulation of the MICP and subsequent application the Generalized Benders Decomposition (GBD) will result in massive reductions in computational effort. While the resulting algorithm must solve multiple Mixed Integer Linear Programs (MILPs), this increase in computational effort is significantly outweighed by the reduction in the number of SDP problems one must solve. The approach will be illustrated using steady-state type processes (that use data reconciliation based estimation) as well as closed-loop dynamic process (that Kalman filtering based estimation).

[1] Chmielewski D. J., T. Palmer and V. Manousiouthakis, "On the Theory of Optimal Sensor Placement," AIChE J., **48**(5), pp 1001-1012, (2002).

[2] Peng, J.K., and D.J. Chmielewski, "Covariance Based Hardware Selection-Part II: Equivalence Results for the Sensor, Actuator and Simultaneous Selection Problems," IEEE Trans. Cont. Sys. Tech., **14**(2), pp 362-368, (2006).

[3] Ahmed S.K., J.-K. Peng, and D.J. Chmielewski, "Covariance-Based Hardware Selection-Part III: Distributed Parameters Systems" AIChE J.,**58**(9), pp 2705-2713, (2012).

388662 Development of an Energy Management System with Energy Storage System and Its Operation Optimization Using Time Series Analysis

Abstract:

Many issues still need to be tackled for the wide introduction of renewable energies. In addition to economic technological maturity, unpredictable nature makes it more challenging to employ renewable energy sources in practice. At the same time, energy demand continuously increases and is subject to vary without considering its generation. The corresponding energy system is under heavy challenges in terms of providing sustainable energy generation methodologies as well as energy demand management.

This paper is concerned with an energy system in terms of both of supply and demand. There are two important features that a new energy system should address. At first, increasing portion of energy supply should be obtained from the renewable energy sources to reduce the negative impact of currently dominating fossil fuels. In evaluating the performance of an energy system, it is thereby important how to hand the resulting irregularity of the renewable energy supply. The estimation of output with regard to external environment should be developed and implemented with the existing energy supply management framework. Secondly, the capricious energy demand should be addressed in the energy system. It is fully motivating to develop a methodology to provide how much energy should be required at the process or a complex of processes in a foreseeable near future. It is of course very difficult to forecast the demand but the previous operating data and the external situation may be of some help.

The corresponding energy system for next generation should incorporate the varying energy supply and demand at the same time. We need a kind of buffer between them and energy storage system can play such a role. To predict the energy generation output of the renewable energies based on the previously obtained data. In this paper time series analysis methods is used to model and forecast the photovoltaic electricity generation and electricity demand. The necessary size of buffering energy storage system is determined by a new framework using Monte Carlo simulation.

373767 Integrated Solvent and Process Design Exemplified for a Diels-Alder Reaction

Abstract:

The purpose of this work is to develop practical and systematic methods for (1) reaction solvent design and (2) integrated reaction solvent and process design. Firstly, a new kind of solvent theoretical descriptor that is generated from quantum chemical COSMO calculations is introduced. Group contributions to each solvent descriptor are regressed for 71 UNIFAC groups. A reaction model is built by correlating a small set of experimental reaction rate constants in various solvents with the corresponding solvent theoretical descriptors. Based on the reaction kinetic model and the developed group contribution method, a computer-aided molecular design (CAMD) problem is formulated. Optimal reaction solvents with highest reaction rates are identified from the solution of the CAMD problem. On the other hand, considering the strong interdependence between solvent molecular characteristics and the process performance, an integrated reaction solvent and process design is performed. Solvent molecular structures and process operations are simultaneously optimized by the formulation and solution of an MINLP-based (mixed integer nonlinear program) optimization problem where process-wide objectives instead of a single reaction performance criterion are used. The implementation of the proposed design methodologies is exemplified, step by step, through an application to a Diels-Alder reaction.

389838 Signature Descriptors for Structure Generation of Optimal Reactants and Products

Abstract:

Selection of optimal chemicals and maximization of process efficiency has successfully been carried out by utilizing process design techniques that integrate product design considerations in a reverse problem formulation. The combined process-product design problem is solved for design targets in terms of properties that are later used to design optimal products. However, most of these techniques have been implemented in non-reactive systems. Different efforts to address the paucity of techniques for reactive systems have been restricted to single unknown reactants. Also, utilization of linear property models and solution schemes that do not have a provision for treating property constraints are among some of the other restrictions. In this work an algorithm has been developed for the design of optimal reactants and products that is not restricted by the number of unknown reactants and products. Nonlinear property models can be incorporated in the algorithm and the solution scheme considers property constraints. The products of the reaction(s) are constrained by property bounds arising from the process to ensure the optimality of the process. An optimization problem is formulated for each of the products in order to select the best chemical structures that satisfy the property bounds and have the best dominant property. The recently developed molecular signature descriptors have been utilized to perform molecular design. The signature descriptor is a systematic codification system over an alphabet of atoms describing the extended valence of the atoms of a molecule. The signature of a molecule can be expressed as a linear combination of its atomic signatures. Topological indices of molecules that form various property models can be linearly represented in terms of molecular signature descriptors [1]. Different quantitative structure activity/property relationships (QSARs/QSPRs) have been applied in the developed algorithm to estimate properties from molecular structures. QSARs/QSPRs can be expressed in terms of different topological indices. Thus the target properties can be expressed in terms of signatures. In the developed method, the problem is formulated in terms of the number of occurrences of each signature, i.e. the number of times each signature appears in the molecular structure. The dominant property to be optimized and the property constraints can thus be ultimately expressed in terms of the number of occurrences of the signatures in the product molecule. After identifying the occurrence values from each of the formulated problems, the molecular structure of products is generated by using an algorithm developed by Chemmangattuvalappil et al. [2]. Next, using the chemical equation to track the migration of groups from reactants to products, the structure of the reactants is determined by considering the appropriate addition and/or removal and/or rearrangement of groups. Thus, candidate reactant structures can be generated irrespective of the type and number of reactions as the introduced algorithm focuses on the design of products first and then uses the chemical equations to identify the reactant structures. Graph theory principles have been utilized to track the signatures and to avoid the generation of infeasible molecular structures. To solve the resulting nonlinear integer optimization problem(s), MI-LXPM, a real coded genetic algorithm (RCGA) developed by Deep et al. [3] has been incorporated in the solution scheme. RCGAs are general purpose population based search techniques that mimic the principles of natural selection and natural genetics laid down by Charles Darwin and encode the solution (represented as chromosomes) in real numbers. Property functions that may be discontinuous, non-differentiable and/or highly non-linear can be handled by genetic algorithms (GAs). Real coding is generally more efficent than binary and gray coding and it helps avoid the "Hamming-Cliff" problem. MI-LXPM is a robust and efficient GA for solving nonlinear integer programming problems. It utilizes the recently developed and self adaptive Laplace crossover (LX) operator and a tunable power mutation (PM) operator [4]. A special truncation procedure is utilized to handle integer restrictions on variables instead of rounding off carried out in previous algorithms. One may not begin with an array of feasible points since a parameter free constraint handling procedure dependent on tournament selection is used. Thus search of feasible points is not a constraining factor. This contribution will illustrate the developed methods and highlight their use through a case study.

[1] J.L. Faulon, D.P. Visco Jr., R.S. Pophale, 2003, Journal of Chemical Information and Computer Sciences, 43, 707-720

[2] N.G. Chemmangattuvalappil, M.R. Eden, 2013, Industrial & Engineering Chemistry Research, 52, 7090-7103

[3] K. Deep, K.P. Singh, M.L. Kansal, C. Mohan, 2009, Applied Mathematics and Computation, 212, 505-518

[4] K. Deep, M. Thakur, 2007, Applied Mathematics and Computing, 193, 211-230

372705 Adaptive Energy Management in Renewable Energy Smart Grids Using the Power Grand Composite Curves Approach

Abstract:

To address the intermittent nature of largely unpredictable environmental phenomena, smart grids transform renewable energy sources (RES) into dependable power flows by simultaneous utilisation of different types of conversion and storage equipment (e.g. fuel cells, electrolysers, accumulators, etc.). The resulting infrastructures combine multiple subsystems that need to operate efficiently while satisfying power demands with minimum contribution of non-RES components. This is generally approached through the development of power management strategies (PMS) [1] typically accounting for decisions regarding the appropriate instance to activate/ deactivate different subsystems, the duration and conditions of their operation, the amount/type of energy carrier to use (e.g. power or hydrogen) and so forth. The participation in grids of numerous subsystems of heterogeneous technical features results in a large number of potential PMS. The selection of the appropriate PMS is a key requirement to enable optimum grid operation. This is a non-trivial task that requires intense effort to review, analyse and eventually exploit the complex interactions observed among the subsystems. Due to this complexity conventional systems mainly utilize a pre-specified PMS which is repeated throughout the cyclic system operation. This is clearly inefficient and results in energy wasting or increased non-RES power utilization because the system remains unresponsive to variability.

The recently proposed power pinch analysis [2] represents a method which allows the investigation of complex energy systems based on the systemic identification of insights pointing towards optimum operating decisions. A major advantage of the method is its implementation in the form of intuitive and easy to develop graphical interfaces (e.g. grand composite curves vs. time and so forth), while the underlying principles may also be combined with rigorous mathematical tools (e.g. flexible models combined with optimization algorithms). The current work discusses the implementation of the power grand composite curves method (PGCC) in the analysis of smart grids and addresses several issues including the simultaneous appearance of multiple operating limits, pinch points and operating targets that need to be satisfied. Furthermore, a novel approach to identify the optimum PMS in RES-based smart-grids using the PGCC to adaptively adjust the system operation in short-term power requirements is proposed. The main aim is to identify the optimum PMS within recurrent subsequent time intervals in order to minimize external non-RES utilization. This is approached by a) using an estimated PGCC to target the optimum power requirements for a predefined future prediction time horizon and b) then adjusting the system PMS in the current time-interval in order to meet the identified target within the prediction horizon. First the system PGCC is estimated for a future time horizon, indicating the external non-RES input required to satisfy the expected power demands. An appropriate shift in the PGCC sets a target of the minimum power inventory needed by the end of the current interval to completely avoid the use of non-RES power input during the desired time horizon. The development of this power inventory is then achieved through the selection of the most suitable PMS that best matches this target. The proposed method is presented within a formal mathematical framework supporting a systematic algorithm to select the optimum PMS within different time intervals in year-round operation. Pinch analysis is also formally associated with a generic model considering numerous potential structural and temporal interactions observed in smart grids and transforming them into a pool of reasonable PMS from where the optimum is selected.

The proposed method is illustrated through implementation on a hybrid smart grid considering multiple RES power generation and storage options (i.e. photovoltaics, wind generators, batteries, electrolyzer, fuel cell, intermediate and final hydrogen storage tanks, a compressor and a diesel generator) that is currently in operation in Greece. Numerous PMS [1] are considered through a generic and inclusive model able to capture different operating realizations in view of external environmental variability. Results indicate considerable performance improvements due to adaptive PMS utilization.

[1] D. Giaouris, A. I. Papadopoulos, C. Ziogou, D. Ipsakis, S. Voutetakis, S. Papadopoulou, P. Seferlis, F. Stergiopoulos, C. Elmasides, 2013, Performance investigation of a hybrid renewable power generation and storage system using systemic power management models, Energy, 61, 621-635

[2] Wan Alwi S.R., Rozali N.E.M., Manan Z.A., Klemeš J.J., 2012, A process integration targeting method for hybrid power systems, Energy, 44, 6-10.

389910 3D Molecular Descriptors Relating Solvent Structure to Crystal Morphology: An Efficiency Study of PCA, GA and Ann Algorithms in QSPR Model Development

Abstract:

The paper outlines the development of a quantitative structure-property relationship (QSPR) that relates solvent structure to the morphology of ibuprofen crystals grown within that solvent. Morphology can be quantified by aspect ratio, and ibuprofen aspect ratio data was obtained for crystals grown in 16 different organic solvents. To provide a quantitative representation of the geometry optimized solvent molecules, 3D molecular descriptors were generated. Three different empirical force fields were used to estimate the three-dimensional structure of the solvent molecules while their effect on the developed models is analyzed. The descriptor data matrix, containing a multitude of descriptor types, is reduced in size, using Principal Component Analysis (PCA), Genetic algorithm (GA), Artificial Neural Network (ANN), and also combination of such algorithms for regression into linear models. The Singular Value Decomposition (SVD) procedure was used for PCA due to its increased stability in primary component prediction. The predictive capabilities of these models were also analyzed through means of internal and external validation methods.

373752 Comparison of Two Different Liquefaction Models for Cryogenic Carbon Capture Process

Abstract:

An excellent understanding of design and operating requirements is of importance in a natural gas liquefaction plant to improve the plant energy efficiency. This paper presents and analyzes steady state and dynamic models of two different natural gas liquefaction processes. Both models were generated using Aspen HYSYS and were optimized to reduce energy consumption and capital costs while maintaining an optimum liquefaction rate. The overall energy requirements, capital costs, and operating costs are compared between the two natural gas liquefaction processes and the advantages and disadvantages of each are outlined.

384637 The Concept of Entropic Machines for Water Desalination

Abstract:

Scarcity of water resources is becoming an increasing challenging problem in the world. This calls for new means of water utilization, including seawater. Entropy generation has been implemented in designing a new device for desalination process based on nonequilibrium thermodynamics and interfacial/electro chemistry concepts. The efficiency of the new proposed device is almost twice of the previous apparatus proposed in the literature and therefore it is classified as a new generation membrane-free machines for the desalination process required for purification of the natural seawater used as drinking water and in related chemical and biological applications.

391795 Model-Based Prediction of Drop-Drop Coalescence in an Electric Field

Abstract:

Electrocoalescence describes the phenomenon by which coalescence of an electrically conducting phase, dispersed in an insulating phase, is achieved by applying an electric field between a high voltage and a ground electrode. In separation technology, electro-coalescers help to enhance the performance of common sedimentation units. Since larger particles sink faster, coalescence of small droplets is desired so as to decrease the residence time, and thus, separator sizes. Although electro-coalescers are broadly applied in industry, i.e. for water-crude oil separation, the system behaviour is not yet fully understood.

Coalescence enhancement of water droplets in oil emulsions is commonly contemplated for the separation of an aqueous phase dispersed in a dielectric oil phase with a considerably lower dielectric constant than that of the dispersed phase. The characteristics and geometry of the electrode system have a large impact on the performance of an electrostatic coalescer and are actually strictly linked to the type of the applied electric field and the emulsion used. Furthermore, addition of chemicals and heating has also been revealed to further enhance the electrocoalescence of water droplets. In this work, the coalescence of two water drops sinking in a dielectric oil phase at an applied high voltage, pulsed dc electric field, in particular with regards to the effects of pressure and temperature on coalescence performance is investigated. The developed model helps to recognise and prove approaches to electrocoalescence mechanisms, the dispersion flow direction with respect to the applied electric field, as well as the electric field configuration. A model development and experimental verification will be presented in detail.

389485 CFD-Based Modeling and Simulation of the Sloshing Effect to the Gtl Reactor on the FPSO

Abstract:

The design and operation of the reactor almost decides the economics of a chemical process and it is no exception to the GTL process operated on the FPSO off-shore environment. This reserarch develops CFD-based models for the evaluation of the sloshing effect coming from the unsteady waves the FPSO endures until it reaches the shutdown and evacuation stage. Previously developed models of the reactor, reaction mechanism are combined together with wave equations

390173 De Novo Molecular Design with Graph-Based Evolutionary Algorithm Considering Multi-Dimensional Constraints

Abstract:

The area of computer-aided molecular design has greatly influenced the rate and cost at which novel chemicals with desired attributes have been identified. As such, great effort has been invested in new methodologies which allow for the solution of larger and more complex problems of this nature. The application of evolutionary algorithms is one such technique which has shown promise in the solution of large combinatorial, and highly non-linear molecular design problems. The typical approach begins with a population of randomly generated individuals and the more fit members are stochastically selected to undergo computational analogues of natural recombination and mutation. This process is iterative until the resultant population possesses the desired attributes, which could be evaluated with existing quantitative structure property (activity) relationships (QSPR's) or other types of property models.

It has been shown that many molecular properties or attributes are often best characterized by a combination of descriptors with varying dimensionality. This could, for example, include a combination of graph theoretical two-dimensional descriptors, such as the chi connectivity index, along with three-dimensional descriptors, which capture important spatial characteristics. The inverse solution to property models of this nature, which entails identifying candidate molecular structures with the desired properties as defined by the given model, is often highly non-linear in nature. In addition, the use of molecular fragments, as often practiced in the de novo design of novel structures, can lead to a combinatorially large search space which becomes intractable for exhaustive solution techniques. The application of an evolutionary algorithm provides a powerful method for the solution of these types of molecular design problems in which there are often multiple objective constraints with high computational complexity and a large search space.

This approach utilizes a fragment based descriptor known as the Signature descriptor, which is represented as a molecular graph, as building blocks to generate candidate solutions. The nature of signature descriptors allows for control over the desired chemical search space as well as convenient reconstruction of the resultant molecular graphs. This descriptor has proven useful in the solution of problems with topological constraints and has recently been extended to tackle problems of higher dimensionality, including three dimensional as well as four dimensional (conformational ensemble) constraints. The graph-based operators necessary for such an approach will be outlined as well as exemplified through a case study which will highlight the advantages of this algorithm.

364758 Different Pump-Around in Transient Modeling of Atmospheric Distillation Unit

Abstract:

In this research, steady state and dynamic simulation of atmospheric distillation unit of Shiraz oil refinery have been done. To verify obtained data by simulation, experimental ASTMD86 curves can be used.

Pump-around simulation is of great importance in simulation. Atmospheric unit of Shiraz refinery has two pump-arounds. One of which is used to preheat crude oil entering into the Tower. Since Feed temperature has a considerable effect on both product composition and heat into the tower, three different simulations were presented and the best practical simulation was chosen.

386783 Water Recycle and Reuse Software Design

Abstract:

Recycling industrial process water increases the efficiency of water use, resulting in both freshwater conservation and raw material cost efficiency. Thus, designing networks for allocating and recycling process water is a key competency for chemical engineering students. Software was designed with MATLAB® and Simulink® as a tool to aid in solving industrial-scale problems. Generating visual tools for designing networks, the program includes a graphical user interface, graphical stream matching, a safety analysis module, and an economic analysis module. For a given set of process water sources and sinks, the program identifies the thermodynamic maximum for water recycle with a material recycle pinch diagram. The program also generates a mass-mapping diagram that presents the relative flow rates and concentrations of all streams. Within an interactive block diagram interface, the user matches sources of process water for recycle to sinks while considering safety and operations constraints. The program indicates progress towards maximum recycle and provides residual diagrams to guide further network design. The economics module enables comparison of networks, and the safety analysis module guides the user through a hazard and operability study.

387156 Economic Viability of Consolidated Bioprocessing As Compared to Separate Hydrolysis and Fermentation for the Production of Biochemically Derived Ethanol from Biomass

Abstract:

As oil prices continue to increase worldwide, the current research has targeted various alternative energy sources. One such energy source is ethanol, an interesting renewable liquid fuel option, which can be produced via biochemical means from many different types of biomass. This biochemical pathway involves two steps, the first of which involves the conversion of the sugar polymers found in plant cells to simple sugars via a chemical pretreatment and enzyme hydrolysis. The second step involves the mixed fermentation of these sugars using a recombinant organism. The resulting ethanol must them be separated from a very dilute water mixture into a pure product.

Our previous research in biochemically derived ethanol production using separate hydrolysis and fermentation (SHF) processes has shown that this technology is economically feasible for large scale process development. The mixing of multiple feedstock options with corresponding pretreatment methods were considered, and an optimization of the resulting mixed-integer nonlinear program (MINLP) showed that a viable process design producing approximately 62 million gallons of ethanol per year calls for the use of sugarcane bagasse and sodium hydroxide pretreatment with a minimum ethanol selling price (MESP) of $2.42 per gallon of ethanol. However, this research also showed that a major cost factor was the purchase of enzymes. The removal of enzymes in the production process, thereby eliminating the hydrolysis step, can be a means to further reduce the MESP.

The development of an enzyme-free bioethanol production process has already been considered by many researchers (1-4). Mascoma Corporation, based out of New Hampshire, USA, has developed a consolidated bioprocessing (CBP) design that uses microorganisms capable of performing the hydrolysis and fermentation in a single step. In this research we look to analyze the consolidated bioprocessing method to determine if the use of costly enzymes is a necessary evil for the biological method bioethanol production. The optimization of a consolidated bioprocessing plant will be done, allowing for various biomass feedstocks and pretreatment methods as in previous studies, using an MINLP model to find the MESP associated with this conversion method. The results will then be compared with the previously studied SHF design to determine the most cost effective large scale method of bioethanol production.

1. Lynd, L. R., Weimer, P. J., van Zyl, W. H. & Pretorius, I. S. Microbial Cellulose Utilization: Fundamentals and Biotechnology. Microbiol Mol Biol Rev 66, 506–577 (2002).
2. Lynd, L. R., Zyl, W. H. van, McBride, J. E. & Laser, M. Consolidated bioprocessing of cellulosic biomass: an update. Current Opinion in Biotechnology 16, 577–583 (2005).
3. Mascoma Corporation. http://www.mascoma.com/index.php
4. Ryu, S. & Karim, M. N. A whole cell biocatalyst for cellulosic ethanol production from dilute acid-pretreated corn stover hydrolyzates. Appl Microbiol Biotechnol 91, 529–542 (2011).

A Comparison of Neural Networks, Extreme Learning Machines, and Support Vector Regression for Surrogate Models of Chemical Processes

Brianna Christian, Selen Cremaschi
Department of Chemical Engineering, University of Tulsa, 800 South Tucker Drive, Tulsa, Ok, 74104, USA

Complex mathematical models have become an integral part of chemical process simulation. Academically and industrially these simulators have been used for optimum design of chemical process equipment. Often solutions to the mathematical models used to build the simulator are found using numerical techniques. Numerical solutions of these models are generally highly accurate but come with a high computational cost especially as the complexity of the model increases. When considering optimal design of process equipment, the iterative nature of process simulators can cause the resulting optimization problem to become computationally intractable. One approach to solving this problem is to generate a surrogate model using sampled data points from the original model. In this work, we compare the quality of solution, the regression or training time, and the time to calculate an output for three different surrogate modeling methods. The three methods studied in this work are Artificial Neural Neural Networks, Extreme Learning Machines, and Support Vector Regression. The surrogate models in this work have been used to predict the simulation outputs of six different unit operations from a biodiesel production plant.

Artificial Neural Networks (ANNs) are a type of surrogate model that attempts to mirror the way the brain works. The neural network mimics the brain by using a set of artificial neurons or nodes. The connections between the nodes represent the axons and dendrites and the weights represent the synapses (Basheer 2000). The values for the synaptic strength (weights) are found through a learning process, which incrementally adjusts the network parameters. The learning process usually uses a gradient-based algorithm that leads to slow convergence and/or convergence to local minima. The result, then, may be a neural network that poorly predicts the original response.

Huang et al. (2006) proposed a quick learning, single-layer feed forward network called Extreme Learning Machines (ELMs). The authors show that as long as the node activation function is infinitely continuous, the input weights and biases can be randomly set. Once the weights and biases are randomly set, the only unknowns, the output weights, are determined analytically by solving the linear system of equations. Huang et al. (2006) concluded that ELMs can be used efficiently in many applications, however the authors mention that the universal prediction capability and sparse high dimensional applications need further investigation.

Support Vector Machines (SVMs) are a statistical regression technique that uses a fixed non-linear mapping to map inputs into an m-dimensional feature space. Once inputs are mapped to feature space, a linear model is created. The linear model predicts the target value by using a linear combination of the transformed inputs. In order to calculate the weights of the transformed inputs, linear regression is conducted. The linear regression is formulated as a constrained quadratic program. The objective function is a combination of a loss function with a regularization term. The regularization term is defined as the norm of the weights. Often the loss function is chosen heuristically.

The performances of these three surrogate modeling methods were compared for predicting the outputs for five different common unit operations: heat exchanger, liquid-liquid extraction, reactor, distillation column, and mixer. Performance of each method was measured using five different measurements. Models for each unit operation are fit using varying sizes of training data sets. The sets vary in size from 50 to 5000 points. The time needed to train these models has been recorded for comparison. The quality of prediction measures used were the RMSE, Bayesian Information Criteria, and Akaike Information Criteria. The quality of prediction measures were calculated using a 500 point testing set. The time taken to predict the entire set was recorded. For statistical significance each of the models was fit ten times. The results of our study find that for a NaOH biodiesel reactor, a neural network achieves the best fit (minimum Root Mean Squared Error) at all training set sizes, however the neural network has the highest training time required for each training set. The average training time for the neural network on the 2500 point training set is approximately 3.5 times higher than that of the support vector regression and the extreme learning machine. The predication time for this case did not yield a statistically conclusive result.

Sources

Haykin, Simon. *Neural networks: a comprehensive foundation*. Prentice Hall PTR, 1994.

Huang, Guang-Bin, Qin-Yu Zhu, and Chee-Kheong Siew. "Extreme learning machine: theory and applications." *Neurocomputing* 70.1 (2006): 489-501.

Basak, Debasish, Srimanta Pal, and Dipak Chandra Patranabis. "Support vector regression." *Neural Information Processing-Letters and Reviews* 11.10 (2007): 203-224.

387355 Economics of a Macroalgae Biorefinery: (How) Can It be Viable?

Abstract:

Macroalgal biomass is generally considered as a prospective feedstock for biofuel production as it offers a number of benefits [1-4] including: 1) it is an abundant renewable carbon source, 2) its growth does not require arable land, fertilizer or fresh water, 3) it is not a primary food crop, 4) it grows rapidly (e.g., 2,960 dry MT/km^2/yr [2]). The presence of significant quantities of polysaccharides (i.e., laminarin, alginate, and small amounts of other sugars) and sugar alcohols (i.e., mannitol), the absence of a secondary cell wall (i.e., no lignin) and the low levels of hemicellulose make certain species of brown macroalgae (e.g., *Saccharina Latissima*) particularly attractive for the production of biofuels [5]. Successful deployment of macroalgae-based biorefineries, however, depends on their economic viability at industrial scale. With that in mind, the key objective of this study was to carry out a detailed technoeoconomic analysis (TEA) of this process to understand the economic potential and cost drivers of macroalgal biorefineries.

In this work, a detailed process model for a macroalage-to-ethanol biorefinery was built. Given the current status and possible technological advances, several scenarios were constructed by varying key process and economic parameters. These include: 1) macroalgae price ($50 to 200/dry MT), 2) overall yield (50 to 80%), 3) solids loading in hydrolysis (5 to 20%), and 4) enzyme loading (10 to 20mg protein/g polysaccharide). With a delivered macroalge price of $100/MT, depending on the maturity of the other process parameters (i.e., yield, solids, and enzyme loading), the Minimum Ethanol Selling Price (MESP) was observed to be in the range of $3.6-$8.9/gal. Overall, the price of macroalgae and the yield remain key parameters in determining the economic feasibility of the biorefinery. For instance, the feedstock price needs to be less than $50/MT to ensure MESP is lower than $3/gal (at 80% yield) and with every $50/MT increase in the price of macroalgae, the MESP would increase by $0.6-1/gal (for the yield range studied). Solids loading was observed to be as significant as the yield as it affects both the capital and operating expenses. In addition to the macroalgae-to-ethanol biorefinery configuration, we have studied other configurations that could possibly improve the economics of biorefinery – these additional configurations include the possibility to co-produce alginate and ethanol.

References

1. Jung KA, Lim S-R, Kim Y, Park JM: Potentials of macroalgae as feedstocks for biorefinery. *Bioresource technology* 2013, 135:182-190.

2. Roesijadi G, Jones S, Snowden-Swan L, Zhu Y: Macroalgae as a biomass feedstock: a preliminary analysis, PNNL 19944. *Pacific Northwest National Laboratory, Richland, WA* 2010.

3. Wargacki AJ, Leonard E, Win MN, Regitsky DD, Santos CNS, Kim PB, Cooper SR, Raisner RM, Herman A, Sivitz AB: An engineered microbial platform for direct biofuel production from brown macroalgae. *Science* 2012, 335:308-313.

4. Adams JM, Gallagher JA, Donnison IS: Fermentation study on Saccharina latissima for bioethanol production considering variable pre-treatments. *Journal of applied Phycology* 2009, 21:569-574.

5. Li K, Liu S, Liu X: An overview of algae bioethanol production. *International Journal of Energy Research* 2014.

372406 Optimal Integration of Industrial Wastewater Discharges through Dynamic Models

Abstract:

This work presents a dynamic model to determine the optimal location of a new industrial plant considering simultaneously the integration of the water network system inside of the new industrial facilities and its effect on the surrounding watershed to satisfy environmental regulations and constraints over the water quality. The model was formulated to predict the behavior for the watershed system impacted for the new polluted discharges. In this sense, the material flow analysis (*MFA*) technique is useful to properly track the pollutants throughout the watershed system. This model considers the variation of the existing discharges, as well as the fluctuation respect to the time of the extractions carried out over the watersheds and their surroundings. In this regard, the model accounts for all the types of inlet streams in the watershed (i.e., pluvial, residential, sanitary and industrial discharges, etc.); in addition to the natural extractions (i.e., evaporation, filtration) or not natural (i.e., irrigation). The mathematical formulation includes a disjunctive model for the optimal selection of new sites to install new facilities; each possible alternative is available to discharge new wastewater streams to the section nearest of the river satisfying the environmental regulations and all the environmental restrictions imposed over the water quality along the watershed. Thus, the new wastewater discharge is integrated to the existing processes and its compositions are tracked throughout the watershed. In this way, it can be determined the required treatment for the new wastewater stream, which operates dynamically considering the temporal variations, as well as the chemical and biochemical reactions carried out in the water bodies for the pollutants degradation.

Finally, the objective function consists in minimizing the total annual cost that is constituted by the cost of installation of the new plant (including the transportation for raw materials, products and services, as well as the land cost) and the wastewater treatment costs. The proposed model is a Mixed Integer Non Linear Programming problem and it was implemented in the software GAMS, where the solver DICOPT was used to solve a case study from Mexico. Results show that there is very important to consider the dynamic behavior for the system to track the pollutants through the watershed; this way, the model allows to yield sustainable solutions during the installation of a new industrial plant.

OPTIMIZATION ANALYSIS OF A BIOREFINERY BASED ON OLIVE STONE

V. Hernández[a], .M. Romero- García[b], E. Castro[b], C.A. Cardona[a]

[a] *Instituto de Biotecnología y Agroindustria, Departamento de Ingeniería Química. Universidad Nacional de Colombia Sede Manizales*

[b] *Department of Chemical, Environmental and Materials Engineering, Agrifood Campus of International Excellence (ceiA3), University of Jaén, Spain*

14AIChE Annual Meeting, Atlanta, GA

Abstract

Currently, most of the energy and chemicals are derived from fossil raw materials, although there are several environmental, economic and social concerns related to its use which have increased the interest on renewable raw materials. Replacing either fully or partially the fossil raw materials is an objective in many countries, including Spain. Since the fossil supply in the country is almost entirely of foreign origin, is of special interest the use of local biomass such as agricultural, forest, agro-industrial and industrial wastes, due to its low cost and large availability. Olive oil and pitted table olive production are the most important agrifood industries in the Mediterranean countries, being Spain the largest producer in the world. One important by-product generated in the olive oil extraction and pitted table olive production is the olive stone. The olive stone represents 10-30 % wt of the fruit, which implies an annual production of some three million tonnes, approximately. Currently, the main use of this by-product is the direct combustion to produce energy as electricity or heat. In this work, a biorefinery integrating the production of xylitol, ethanol, furfural, poly-3-hydroxybutyrate (PHB) and energy through a cogeneration system was proposed. Then an optimization based on technologies and raw material distribution was carried out. The approach used in this work allowed finding the best scheme for producing value-added products from a feedstock which currently does not have and industrial application.

Introduction

Olive oil and pitted table olive production are the most important agrifood industries in the Mediterranean countries, being Spain the largest producer in the world. Currently, this crop presents a global cultivated area of almost 10 million hectares with an annual production of olives over 18 million tonnes. One important by-product generated in the olive oil extraction and pitted table olive production is the olive stone. The olive stone represents 10-30 % wt of the fruit, which implies an annual production of three million tonnes, approximately. For example, in the 2009/2010 season, approximately $1.7 \cdot 10^5$ tons and 2.1 million tons of stone from table olives and olive oil industries, respectively, were produced. Currently, the main use of this by-product is the direct combustion to produce energy as electricity or heat.

Methodology

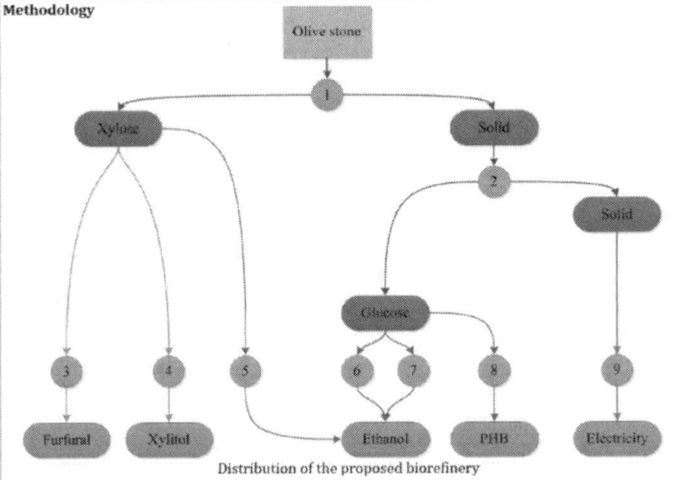

Distribution of the proposed biorefinery

(1) Acid pretreatment using sulfuric acid at 2% w/v (Experimental stage)
(2) Dilute acid hydrolysis using sulfuric acid at 0.4% (Rinaldi and Schüth 2009).
(3) Furfural production via xylose cyclodehydration (Agirrezabal-Telleria, Gandarias et al. 2013)
(4) Xylitol production by *Candida moggi* (Tochampa, Sirisansaneeyakul et al. 2005)
(5) Ethanol production from xylose using *E. coli* MM160 (Experimental stage)
(6) Ethanol production from glucose using recombinant *Zymmomonas mobilis*
(7) Ethanol production from glucose using recombinant *Saccharomyces serevisiae*
(8) PHB production from glucose using *Ralstonia eutropha* (Shahhosseini 2004)
(9) Electricity production by gasification from lignin (Rincón, Becerra et al. 2013)

Results

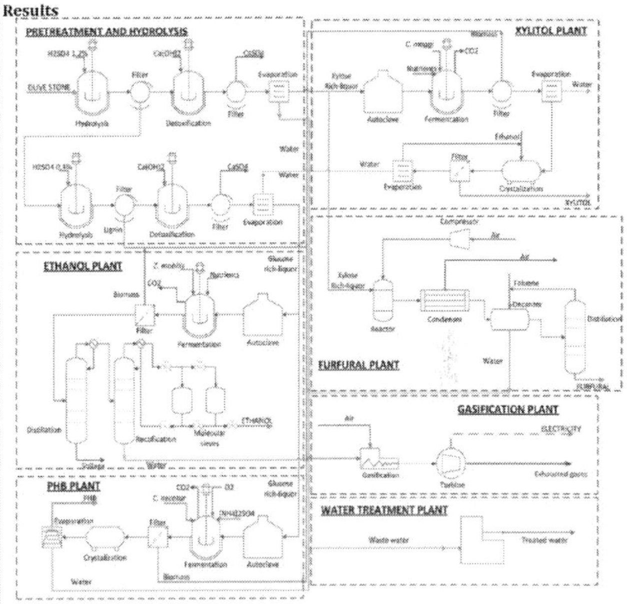

Biorefinery scheme from olive stone

Optimization model

$$\max(Z) = \sum_i m_i p_i + \sum_k w_k p_k - m_{OS} p_{OS} - \sum_j m_j p_j$$

$m_{Cellulose} = 0{,}201 m_{OS}$

$m_{Hemicellulose} = 0{,}2992 m_{OS}$

$m_{Lignin} = 0{,}3887 m_{OS}$

$m_{Xylose,1} = y_1 m_{Hemicellulose}$

$m_{Glucose,2} = y_2 m_{Cellulose}$

$W_{Electricity} = y_9 m_{Lignin}$

$m_{Xylose} = m_{Xylose,3} + m_{Xylose,4} + m_{Xylose,5}$

$m_{Xylose,3} \le M_3 X_3 \qquad m_{Xylose,4} \le M_4 X_4 \qquad m_{Xylose,5} \le M_5 X_5$

$X_3 + X_4 + X_5 = 1$

Technology	Yield	Unit
1	0.74	g xylose/g hemicellulose
2	0.89	g glucose/g cellulose
3	0.41	g furfural/g xylose
4	0.26	g xylitol/g xylose
5	0.48	g ethanol/g xylose
6	0.47	g ethanol/g glucose
7	0.49	g ethanol/g glucose
8	0.31	g PHB/g glucose
9	5108.78	MJ/ton lignin

$m_{Glucose} = m_{Glucose,6} + m_{Glucose,7} + m_{Glucose,8}$

$m_{Glucose,6} \le M_6 X_6 \qquad m_{Glucose,7} \le M_7 X_7 \qquad m_{Glucose,8} \le M_8 X_8$

$X_6 + X_7 + X_8 = 1$

$m_{Furfural} = y_3 m_{Xylose,3}$

$m_{Xylitol} = y_4 m_{Xylose,4}$

$m_{Ethanol,5} = y_5 m_{Xylose,5}$

$m_{Ethanol,6} = y_6 m_{Glucose,6}$

$m_{Ethanol,7} = y_7 m_{Glucose,7}$

$m_{Ethanol} = m_{Ethanol,5} + m_{Ethanol,6} + m_{Ethanol,7}$

$m_{PHB} = y_8 m_{Glucose,8}$

Results

Share of production costs

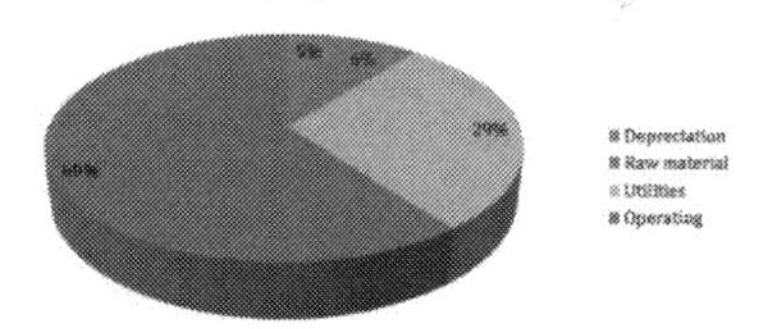

Environmental assessment

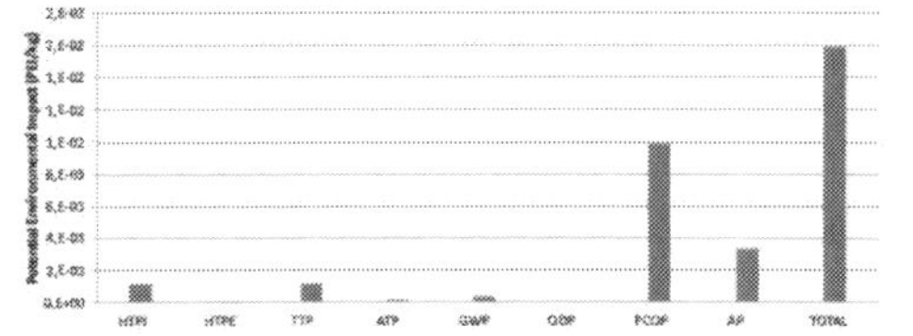

HTPI, Human Toxicity by ingestion; HTPE, Human Toxicity by Dermal Exposition or inhalation; TTP, Tettrestrial Toxicity Potential; ATP, Aquatic Toxicity Potential; GWP, Global Warming (GWP); ODP, Ozone Depletion Potential; PCOP, Photochemical Oxidation Potential; AP, Acidification Potential

Conclusions

According to the results, the olive stone is an interesting material to be used as feedstock under a biorefinery concept to obtain value-added products.

The formulation of optimization problems in biorefinery design plays an important role, especially for decision support in distribution within the selection of the most promising process pathway in terms of technology.

Acknowledgements

The authors express their gratitude to the Universidad Nacional de Colombia at Manizales, to the Investigation and extension direction, from the Engineering and Architecture Faculty from Universidad de Colombia Sede Manizales, to the Design of Sustainable Biorefineries (code 157) and Chemical, catalytic and biotechnological Processes (code 16081) projects from the Investigation Direction at Manizales (DIMA), to the Administrative Department of Science, Technology and Innovation (COLCIENCIAS), to the Agrifood Excellence Campus (ceiA3) in University of Jaén and to the Agrifood International Doctorate School (eidA3) for funding this work. Financial support from Junta de Andalucía (Proyecto de Excelencia AGR-6103) is also gratefully acknowledged. JMRG thanks his scholarship to Junta de Andalucía.

References

Hernández V, Romero-García JM, Dávila JA, Castro E, Cardona CA. Techno-economic and environmental assessment of an olive stone based biorefinery. Resources, Conservation and Recycling. 2014;92:145-50.
Romero-García JM, Niño L, Martínez-Patiño C, Álvarez C, Castro E, Negro MJ. Biorefinery based on olive biomass. State of the art and future trends. Bioresource Technology. 2014;159:421-32.
Rincón LE, Becerra LA, Moncada J, Cardona CA. Techno-Economic Analysis of the Use of Fired Cogeneration Systems Based on Sugar Cane Bagasse in South Eastern and Mid-Western Regions of Mexico. Waste and Biomass Valorization. 2013.
Cerón IX, Hernández V, Cardona CA. Biorefinery Design for Value Added Products from Amazon Fruits. 2nd Iberoamerican Congress on Biorefineries. Jaen, Spain 2013.
Garrido-Fernández, A., M.J. Fernández-Diez, and M.R. Adamos, *Physical and Chemical Characteristics of the Olive Fruit*, in *Table Olives*, C. Hall, Editor 1997.

391453 Computer Aided Design of Ionic Liquids for Thermal Energy Storage Applications

Abstract:

Ionic liquids (ILs) and salts have properties that make them attractive for thermal energy storage (TES). In particular, ILs have high heat capacity, density, and thermal decomposition temperatures which are key properties for TES. Further, ILs and salts can be tuned to optimize these properties. Since very large numbers of IL structural variations are possible we utilize a computer-aided ionic liquid design (CAILD) method to select optimal IL structures for this application. The design problem is posed as an MINLP model where the objective is to maximize energy density and constraints relate to structural feasibility, and bounds on viscosity, melting point (M.Pt.) and thermal decomposition temperature. This MINLP model is solved to design optimal IL structures. Three specific problems related to design of ionic salts (M.Pt. < 140 C), ionic liquids (M.Pt. < 100 C), and room temperature ionic liquids (M.Pt. < 25 C) are presented. In depth analysis on structural features and functional groups that contribute to optimized properties are discussed. Comparison between designed ionic liquids and current industrial thermal storage fluids is also presented.

372418 Optimal Planning for Ambrox® Production Based on Ageratina Jocotepecana

Abstract:

All industry throughout its development has different types of problems to hold on in the market, most of these problems are tackled by creating new alternatives for carrying out the process making it more efficient, less costly and environmentally friendly. In this context, the perfume industry is dedicated to sell fragrances, and the main problem is how make that this fragrance remains the longest possible time; for many years this problem has been solved using the ambergris as fixative, which is obtained from the cachalot. However, during the last years, the social pressure associated to the killing of this cetaceous has promoted the development of new fixatives in the perfume industry.

Until today, the chemical compounds developed to replace the ambergris represent complex chemical synthesis with high costs. Recently, in the Chemical and Biochemical Institute from the Universidad Michoacana has been synthesized a fixative named *Ambrox®* from the endemic plant *Ageratina Jocotepeca*; this new chemical route is very simple and may represent an attractive solution to replace the ambergris. However, there are several challenges for the industrial application of this plant, this is because this grows just in specific areas due to the ambient conditions. Therefore, there is required an optimization study for the entire supply chain associated to the new process.

Therefore, this paper proposes an optimization formulation for the entire supply chain networks associated to the production and distribution of *Ambrox®* from *Ageratina Jocotepeca*. The proposed optimization model accounts for the cultivation, harvesting, processing and distribution of products and raw materials. This supply chain network has special characteristics such as the high value of products and the low volume of products required, as well as the long distance for distribution due to the markets are located mainly in Europe, whereas the cultivation areas are in Mexico. In the optimization model were considered simultaneously economic, environmental and social objectives. This way, the economic objective function accounts for the maximization for the total profit, whereas the environmental impact was measured through the Life Cycle Assessment for the entire supply chain, and social aspects like the generated jobs and land conversion were also considered. Results show that there is possible to replace the current processes for the *Ambrox®* production with the one proposed and simultaneously can be obtained economic, environmental and social benefits.

FAILURE ANALYSIS OF A SECOND GENERATION BIOREFINERY FROM OIL PALM EMPTY FRUIT BUNCHES USING GRAPH THEORY

14AIChE Annual Meeting, Atlanta, GA

V. Hernández[a], E. Castro[b], J.B. Restrepo[a], A.A. González[a], R.A. Tolosa[a], C.A. Cardona[a]

[a] *Instituto de Biotecnología y Agroindustria, Departamento de Ingeniería Química. Universidad Nacional de Colombia Sede Manizales*

[b] *Department of Chemical, Environmental and Materials Engineering, Agrifood Campus of International Excellence (ceiA3), University of Jaén*

Abstract

The oleochemical sector in Colombia has been growing during the last decade as effect of the development of the biodiesel industry and the increase on demand of petrochemical substitutes and specialty fats. Currently, the oil palm is the most important oleochemical feedstock by yield, harvested area and production in the country. This crop reached a cultivated area of 477,000 hectares in Colombia in 2013, while the production of crude palm oil exceeded the one million tonnes for the same year. The palm oil extraction leads to obtain different by-products and residues with economical potential. Among these, the empty fruit bunches (EFB) are the solid residue that is produced in the highest amount from the fresh fruit bunches (FFB) of the oil palm. Usually the EFB represent 12 to 26% of the FFB in dry basis. This means that approximately 438 tonne of EFB are produced per km^2. Due to its high moisture content, this residue is not appropriate as fuel and currently it does not have any industrial application. In this work, experimental and simulation procedures were performed in order to propose a biorefinery scheme to take full advantage of this waste. In order to complement the simulation and quantitative analysis of the proposed process flowsheet, a qualitative analysis is performed in order to assess security issues. With the aid of graph theory most sensitive units to cascade failures are identified.

Introduction

The empty fruit bunches (EFB) are the solid residue that is produced in the highest amount from the fresh fruit bunches (FFB) of oil palm. Colombia is the fourth world producer of oil palm due to its favorable agro-ecological conditions. In 2013, the oil palm reached a cultivated area of 477,000 hectares. Considering that that 4,380 kg of EFB can be produced per hectare per year, more than 1 million tons of EFB were produced in 2013.

In this work the techno-economic assessment of an integrated biorefinery to obtain xylitol, ethanol, furfural, poly-3-hydroxybutyrate (PHB) and energy was carried out. Besides, in order to complement the simulation and quantitative analysis of the proposed process flowsheet, a qualitative analysis was performed to assess security issues. With the aid of graph theory most sensitive units to cascade failures were identified.

Methodology

Characterization → Biorefinery simulation → Economic assessment → Failure analysis

Process description

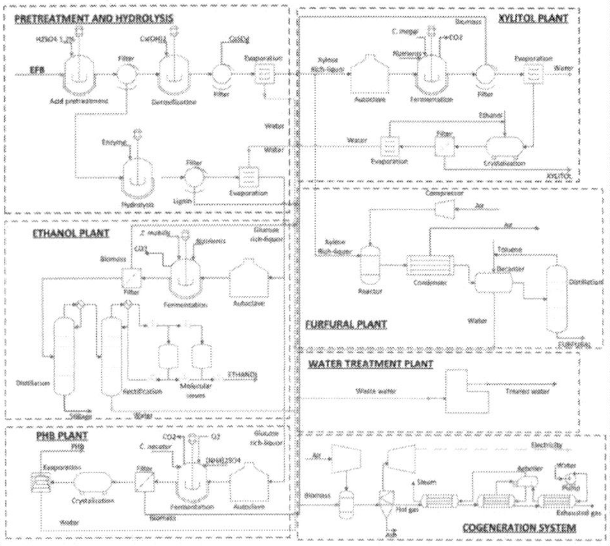

Figure 1. Biorefinery scheme from oil palm empty fruit bunches

Results

Table 1. Characterization of the raw material

Characteristic	% dry basis
Cellulose	36.63
Hemicellulose	22.36
Lignin	19.28
Extractives	12.01
Ash	0.97

Moisture content: 4.33%

Table 2. Overall material balance of the system

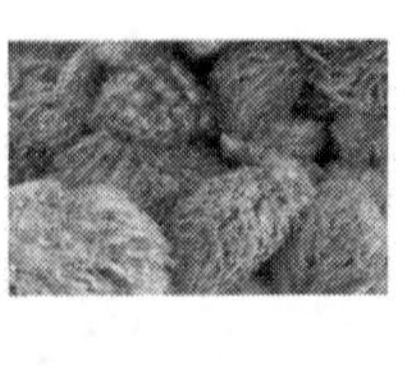

Stream	Production capacities
Empty fruit bunches	10,000 Kg/h
Xylitol	387 Kg/h
Ethanol	757 l/h
Furfural	371 Kg/h
PHB	192 Kg/h
Electricity	1,481 kW

Economic assessment

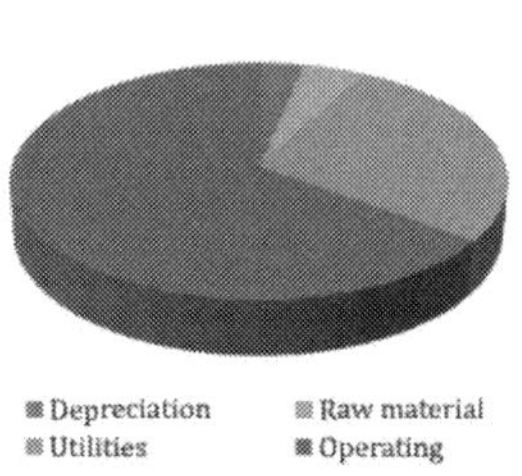

Figure 2. Share of annualized cost

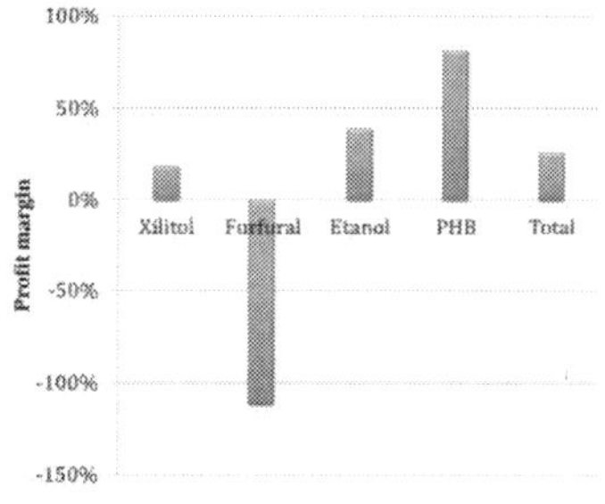

Figure 3. Unit and net profit margin

Graph analysis

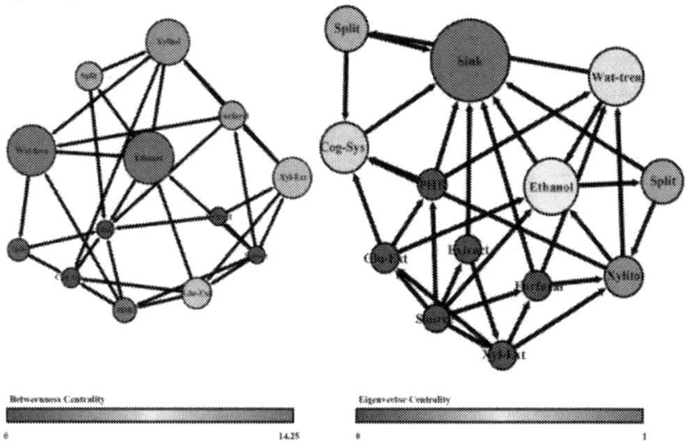

Figure 4. Betweenness and Eigenvector centrality analysis of the resulting graphs.

Conclusions

According to the results, the oil palm is an interesting material to be used as feedstock under a biorefinery concept to obtain value-added products.

Both analyses eigenvector centrality and betweenness centrality agree that the most relevant processes in the plant in the case of failures are the ethanol production and the water treatment process. Nevertheless, in the case of cascade failures that propagates through the networks describing shortest paths xylitol process is also sensitive, and in the case that the failure do not propagate by shortest paths the cogeneration system is more prone to be affected by failures.

Acknowledgements

The authors express their gratitude to the Universidad Nacional de Colombia at Manizales, to the Investigation and extension direction, from the Engineering and Architecture Faculty from Universidad de Colombia Sede Manizales, to the Design of Sustainable Biorefineries (code 157) and Chemical, catalytic and biotechnological Processes (code 16081) projects from the Investigation Direction at Manizales (DIMA), to the Administrative Department of Science, Technology and Innovation (COLCIENCIAS), to the Agrifood Excellence Campus (ceiA3) in University of Jaén and to the Agrifood International Doctorate School (eidA3) for funding this work. Financial support from Junta de Andalucía (Proyecto de Excelencia AGR-6103) is gratefully acknowledged.

References

Rincón LE, Hernández V, Cardona CA. Process Biochemistry. 2014;49:474-89.
Hernández V, Castro E, Restrepo JB, Cardona CA. 21st International Congress of Chemical and Process Engineering CHISA 2014, Prague, Check Republic 2014.
Rincón LE, Becerra LA, Moncada J, Cardona CA. Waste and Biomass Valorization. 2013.
Cerón IX, Hernández V, Cardona CA. 2nd Iberoamerican Congress on Biorefineries. Jaen, Spain 2013.
Albert, R., Jeong, H., & Barabasi, A.-L. (2000). *Nature, 406*(6794), 378–382.
Boccaletti, S., Latora, V., Moreno, Y., Chavez, M., & Hwang, D. U. (2006). *Physics Reports, 424*(4-5), 175–308. doi:10.1016/j.physrep.2005.10.009
Bonacich, P. (1972). *Journal of Mathematical Sociology, 2*(1), 113–120.
Borgatti, S. P. (1995). *Connections, 18*(1), 112–114.
Borgatti, S. P. (2005). *Social Networks, 27*(1), 55–71.
Cohen, R., & Havlin, S. (2010). Cambridge University Press.
Dehmer, M. (Ed.). (2011). Birkh user.
Freeman, L. C. (1979). *Social Networks, 1*(3), 215–239.
Koschützki, D., Lehmann, K., Peeters, L., Richter, S., Tenfelde-Podehl, D., & Zlotowski, O. (2005). *Network Analysis* (Vol. 3418, pp. 16–61). Springer Berlin Heidelberg.
Newman, M. E. J. (2005). *Social Networks, 27*(1), 39–54.

Spatial Modeling of a Tubular Reactor for the High Pressure Synthesis of LDPE

TECHNISCHE UNIVERSITÄT DARMSTADT

Sebastian Fries, Markus Busch
Technische Universität Darmstadt, Ernst-Berl-Institut, Alarich-Weiss-Str. 8, 64287 Darmstadt, Germany, email: markus.busch@pre.tu-darmstadt.de

Abstract

Low-density polyethylene (LDPE) is produced at high temperatures (140 °C – 330 °C) and high pressures (1000 bar – 3500 bar) under free radical conditions. The heat of polymerization (3.5 kJ/g) must be continuously removed to maintain a steady reaction. However, in continuous operation often a reduction in heat exchange can be observed. This undesired process known as fouling can affect the quality of the product as well as the productivity and operating safety of a tubular reactor. Although LDPE has been produced for more than 70 years the cause of fouling is still not completely revealed. In literature it is discussed as an insulating polymer film on the reactor wall or as a viscous polymer rich hydrodynamic boundary layer.

Deeper understanding could be achieved by a model capturing the reaction environment and the microstructure of the polymer at the reactor wall. For this the solution of a system of partial differential equations in radial and axial position along the reactor axis is required. Due to the complexity associated with solving coupled equations for heat and mass balance and differential equations needed to determine radial dependend quantities, a three-stage model family has been developed:

Generic Process

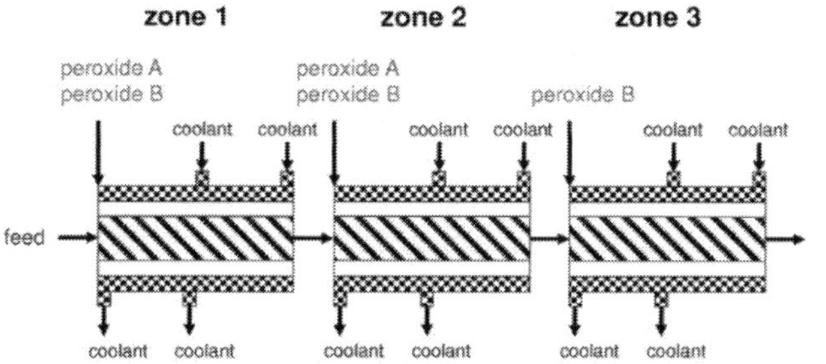

Model Family

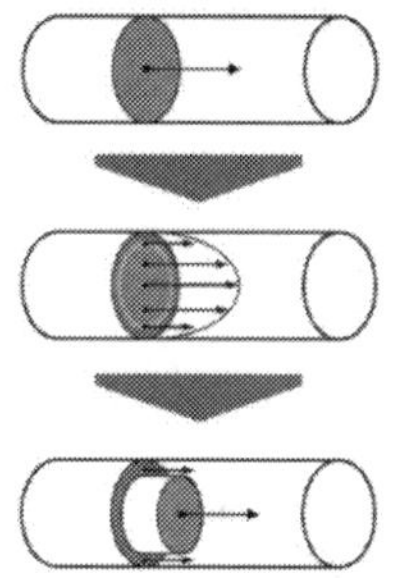

one dimensional module
- plug flow model based on ordinay differential equations
- incorporation of a complex reaction network with primary and secondary radicals
- calculation of rigorous molecular weight distribution

radial module
- radial profiles (partial differential equations)
- wall temperature obtained by one dimensionale module used as boundary condition
- appliance of empirical power-law turbulent velocity profiles
- mean polymer properties by method of moments

compartment module
- representing center and wall flow layer by two ideally mixed compartments (ordinary differential equations)
- compartment temperature and velocity according to radial module
- calculation of rigorous molecular weight distribution

One Dimensional Module

- iterative approach (shooting method) for solving the heat balance with counter-current cooling
- empirical correlation predicts increasing fouling layer thickness with lower Reynolds number
- LCB density grows due to higher polymer content SCB density rises due to elevated reactor temperature
- formation of new macromolecules and β-scission leads to decreasing M_n and M_w

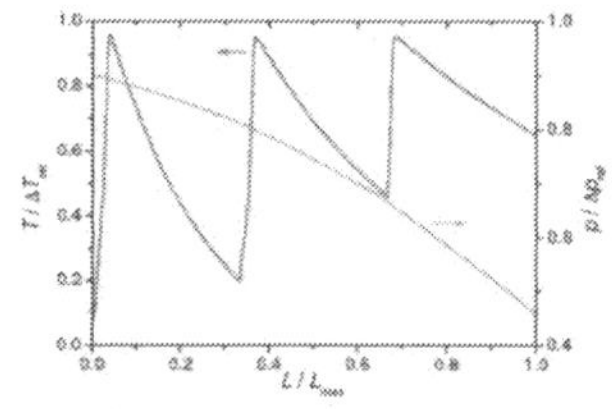

Figure 1: Axial temperature- and pressure profiles in the generic tubular reactor.

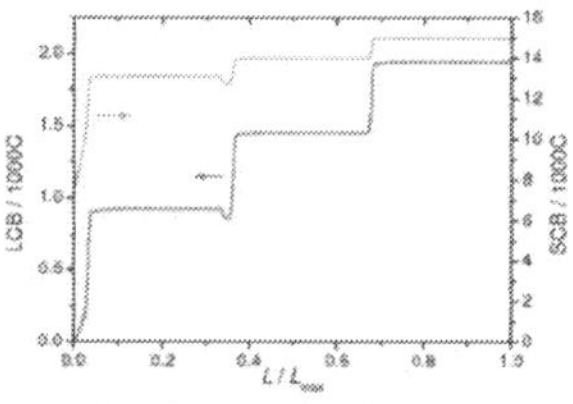

Figure 2: Axial short chain (SCB) and long chain branching density (LCB) in the generic tubular reactor.

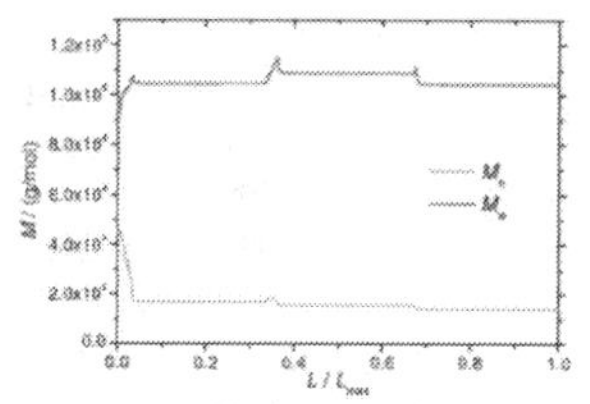

Figure 3: Axial number average (M_n) and mass average molar mass (M_w) of the polymer in the generic tubular reactor.

Radial Module

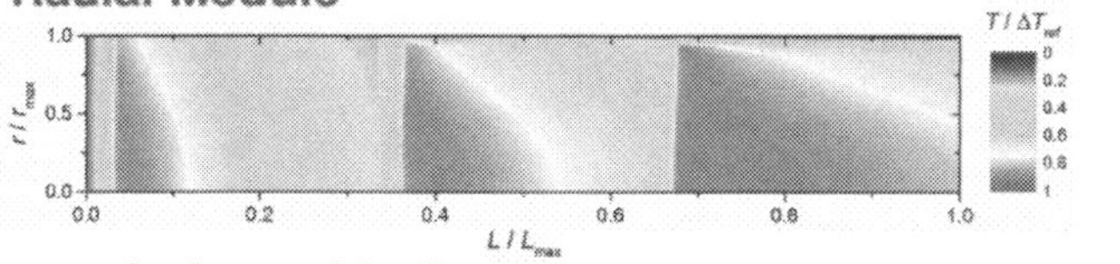

Figure 4: Temperature contours for all zones in the generic tubular reactor.

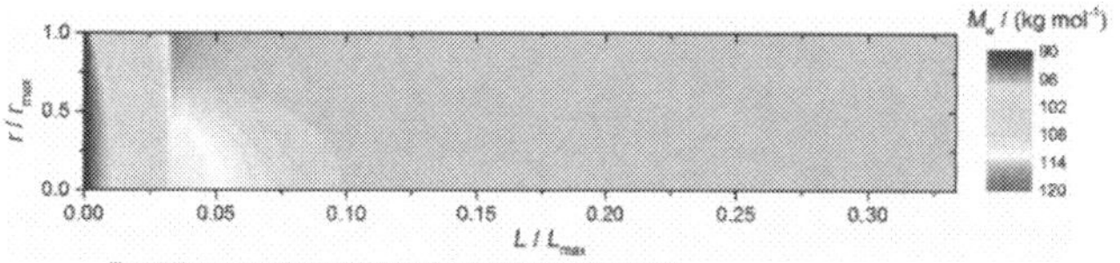

Figure 5: Mass average molar mass (M_w) contours for zone 1 in the generic tubular reactor.

- solving of partial differential equations is computationally demanding and therefore time consuming
- due to decreasing turbulent diffusion coefficients at the end of the reactor heat exchange is reduced
- less transfer reactions to small species lead to higher polymer molar masses at the wall

Compartment Module

- delivers same predictions as radial module regarding the mean polymer properties but in much shorter time
- molecular weight distribution of one dimensional and compartment module virtually the same ⟶ plug flow is a good approximation for turbulent flow
- slightly elevated values for conversion and M_w at the wall cannot explain formation of fouling

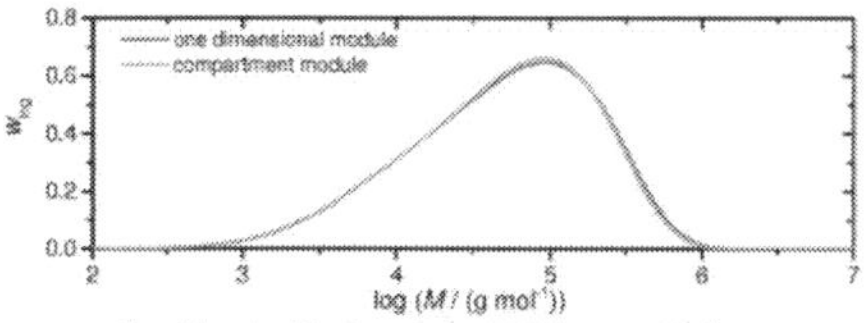

Figure 6: Comparison of the polymer molecular weight distribution generated by the one dimensional and compartment module at the end of the generic tubular reactor.

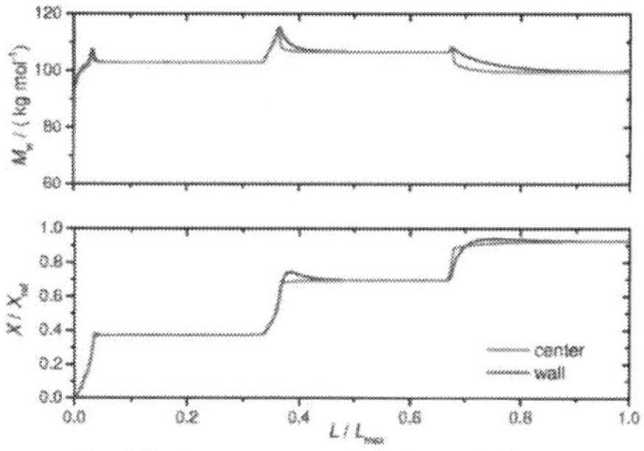

Figure 7: Axial polymer mass averaged molar mass (M_w) and conversion (X) of the center and wall compartment in the generic tubular reactor.

Outlook

- further validation of model family with data of a lab scale tubular reactor
- further work at compartment module to find origins of fouling

Acknowledgment

The authors greatly acknowledge Saudi Basic Industries Corporation for funding and support of this work.

Furthermore the authors thank Michael Wulkow, CIT, Rastede, Germany for support in model development with the software Predici.

378570 Fast Tracking Process Development Life Cycle Based on State-of-the-Art Process Simulation Models

Abstract:

Fast tracking process development life cycle based on state-of-the-art process simulation models

Abstract

The ability to fast track the process of the development life cycle by effectively utilizing state-of-the-art process simulation models is described in this article. The process development life cycle, encompassing various stages, such as research stage, process development stage and technology deployment stage, depends heavily on validated process models to enable researchers to reduce excessive and unnecessary research spending by efficiently optimizing the process schemes and process parameters. Recently, this novel approach of integrating process simulation efforts into process technology development throughout its life cycle has been successfully implemented for the development of Butenes to Butanols process technology at Saudi Aramco. Motivated by the opportunity to develop technologies to valorize undervalued streams and toward capturing refinery petrochemical integration benefits, researchers at Saudi Aramco invented a distinctive one-step process to convert mixture of Butenes into mixed Butanols liquid without the need for separation of the isomers. The mixed Butanols with high octane rating and energy content can be used as valuable oxygenates for clean fuel combustion and octane enhancers.

Preliminary simulation models developed during the research stage revealed significant issues associated with the Azeotropic nature of the mixed Butanol product, which aided researchers during the scale-up process to incorporate necessary process changes to the pilot plant. Validated simulation models based on pilot plant data were key in developing a comprehensive large scale integrated pilot plant with associated recycle loops and distillation columns that not only predicted optimized process schemes but also aided the researchers to foresee potential technical difficulties associated with product extraction and separation and to devise an action plan to successfully demonstrate the technology at a 60 kg/day capacity unit. In addition, process models are further modified and utilized to depict an 80 KTA commercial Butanol plant towards the development of a comprehensive feasibility study that enables the modelers and researchers to conduct rigorous process economics for potential technology deployment and licensing.

Authors:

Kadhim A MOHAMMED, kadhim.mohammed@aramco.com

Engineering Specialist

Process & Control Systems Department

Saudi Aramco

Dhahran 31311 SAUDI ARABIA

Kareemuddin M SHAIK, kareemuddin.shaik@aramco.com

Research Specialist

Research & Development Center

Saudi Aramco

Dhahran 31311 SAUDI ARABIA

Hiren SHETHNA, hiren.shethna@aramco.com

Engineering Specialist

Process & Control Systems Department

Saudi Aramco

Dhahran 31311 SAUDI ARABIA

In the context of radical polymerizations the intermolecular transfer to polymer commonly denotes a reaction, in which the radical functionality of a growing chain is passed on to a dead polymer. This requires the abstraction of a methylene hydrogen atom and thus leads to the formation of a secondary macroradical. If a sequence of propagation steps follows such a transfer reaction, a so called long chain branch (LCB) will be created. [1] Together with the short chain branches (SCB), which only contain a few monomer units, the LCBs determine the overall topology of the polymer. The latter plays an important role in many respects, e.g. for radii of gyration as a microscopic quantity or rheological properties of the material in a macroscopic sense. [2,3] Hence it is desirable for polymer engineers to gain control over the mechanism that generates LCBs. This would be a progress towards the long-term goal of influencing the polymeric properties just by sophisticated design of the production process.

However, controlling the formation of LCBs first requires knowledge of the rate for chain transfer to polymer. Either it must be assessed as an absolute value at defined conditions or as a ratio with respect to any of the well-characterized elementary reactions in the polymerization such as the propagation. Moreover a supposed non-trivial dependency on different reaction parameters should be considered for the rate coefficient. Not only temperature and pressure but also viscosity of the reaction mixture, final degree of conversion, degree of polymerization or initial monomer concentration could have an impact on the rate of intermolecular transfer to polymer. [4]

Some of the aforementioned correlations for the chain transfer to polymer have already been successfully tested in special systems. Ahmad and coworkers for example have examined the variation of LCB density with the initial monomer concentration for the radical solution polymerization of n-butyl acrylate. [5] Nevertheless, there is no comparable study concerned with the polymerization of ethylene yet. This lack may be due to the extraordinary process parameters under which the branched, low-density type of polyethylene is typically being produced.

The polymerization of ethylene is done in mass, involves high pressures of up to 300 MPa and high temperatures in the range of 450-550 K. This prevents a straightforward analysis of reactions such as the chain transfer to polymer which is part of a complex reaction network. In fact kinetic studies under high pressure require material-consuming and also otherwise costly experiments. For this reason a careful preparation, which allows to perform only a minimum of experiments for the investigation of the chain transfer to polymer, is attractive.

One possibility to designate the most appropriate approach is to simulate experiments of different design and analyze their outcome with respect to the variation of detectable properties. Therefore we use an established model that takes into account the complete reaction network of the polymerization, the physical properties of the reaction mixture, the reactor geometry and the residence time. It employs differential equations which are solved numerically. We can evaluate the quality of different sensors for the reaction rate in focus and differentiate between more and less favorable experimental parameters.

[1] M. Busch, *Macromol. Theory Simul.* **2001**, *10*, 262-274.
[2] T. B. McLeish, *Adv. Phys.* **2002**, *51*, 1379-1527.
[3] F. Stadler, J. Kaschta, H.Münstedt, F. Becker, M. Buback, *Rheol. Acta.* **2009**, *48*, 479-490.
[4] T. Herrmann, "Modellierung technischer Hochdruck-LDPE-Reaktoren", Der Andere Verlag, Uelvesbüll, **2011**.
[5] N. M. Ahmad, F. Heatley, P.A. Lovell, *Macromolecules* **1998**, *31*, 2822-2827.

380917 Process Design and Simulation for the Conversion of Coal to Methanol

Abstract:

Abundantly available Low Rank Coal (LRC) in Indonesia is one of promising resources to be converted to valuable intermediate chemicals such as Methanol which can be further converted to useful derivatives such as Gasoline, Olefin (Ethylene and Polypropylene), Ethanol, or Dimethyl-ether (DME). The conversion of Coal to Methanol (CtM) involves coal gasification to produce syngas (CO & H_2) and then followed by syngas conditioning and cleaning for methanol synthesis. In an effort to meet energy and chemical demand in Indonesia, the coal based methanol (CtM) project is developed, aiming to produce 800.000 tones/year AA Grade Methanol from LRC with moisture content of 33% planned to be used as a feedstock to Ethanol Plant.

In order to provide suitable and better process design during detailed feasibility study, one of the challenges within this framework is to carry out rigorous and robust modelling for the coal based methanol plant which integrating the coal/solid processing with gaseous and liquid process modelling.

Aspen Plus^{TM} (Version 8.4) process simulation software was used for facilitating the process modelling and simulation since it can provide such an integrated solid and gas/liquid modelling approach. A hierarchy model in Aspen Plus was applied for each process section to build an integrated coal based methanol Plant with process model involving solid, gas, and liquid processing. This approach provides opportunity to specify different suitable property packages. For instance, PR-BM *(modified Peng Robinson equation for solid processing)* property package was used in the Coal Pretreatment and *Gasification and Peng Robinson* property estimation method was applied in the Gas Conditioning and Cleaning Unit. As Aspen Plus process simulator facilitates process simulation for rigorous solid modelling, coal compositions and its properties (proximate and ultimate analyses including their heating value and moisture content) are possible to be used as basic input for the simulation. Furthermore, Aspen Plus^{TM} also allows for specifying coal particle size distribution (PSD) and grind-ability quality (HGI-Index) within process simulation.

Prior to gasification process, the process model is well established for drying of coal feedstock from moisture content of 33% to 5% in combining of coal grinding to form fine powder coal (average PSD of 90 μm) using available drying and grinding unit model in Aspen Plus. The power requirement for this process as a simulation result is comparable with vendor proposition. Furthermore, impact of particle size and coal quality (particularly grind-ability, HGI) on the grinding performance can be examined.

As the heart of the process, the coal gasification process is well modelled to incorporate coal decomposition mechanism and syngas gas formation using an equilibrium/thermodynamic approach. The process conditions and requirement of the gasification (pressure of 50 bar, temperature of 1500 °C, O_2/Coal ratio about 0.5) are established accordingly to simulate syngas production (CO & H_2). Literatures data was used to verify the simulation results. Although the coal processing simulation is well established, a rigorous model involving detailed kinetics on the coal decomposition and syngas formation for coal gasification is still needed to be developed for the selected gasification technology *(an entrained flow gasification technology)*. Combining with dynamic simulation approach and detailed computational fluid dynamics, this process model would not only suit for process performance evaluation but also provides insight on the reliability and safety aspect of the gasification process. This simulation result is then used for challenging and justifying vendor proposal regarding to the coal consumption, utilities requirement, and waste estimation during the project development.

After gasification process, further process model is developed to integrate this coal solid modelling with Gas Conditioning (Water Gas Shift Reaction), Gas Cleaning Unit (Acid Gas Removal Unit using Rectisol® Process and Sulfur Recovery Unit using Clauss Process) and Methanol Synthesis to produce Methanol and Syngas suitable for feedstock of Ethanol Plant. This approach resulted logic Heat & Material Balances, Utilities Requirement (water, steam, and power), and Equipment Sizing. Furthermore, capital expenditure of CtM Plant is well estimated using Aspen Economic Evaluation for further economic evaluation.

Population Balance Equation Models for Emulsified Product Design: Achieving Target Drop Size Distributions in High Pressure Homogenization

Shashank N. Maindarkar[1], Hans Hoogland[2] and Michael A. Henson[1]

[1]Department of Chemical Engineering, University of Massachusetts, Amherst, MA 01003
[2]Unilever R & D, Vlaardingen, Netherlands

Abstract

Oil-in-water emulsions are ubiquitous dispersed phase systems with diverse applications that include consumer products, processed foods, pharmaceutical delivery and petroleum recovery. Emulsified products are typically formed by first preparing a coarse premix using a low shear stator-rotor type device that mixes the various formulation ingredients into a stable form. This premix is then processed with a high shear device such as a high pressure homogenizer in which relatively large drops are broken into much smaller drops. Emulsion formulation variables and process operating conditions both impact the drop size distribution, a key property that influences emulsion rheology, stability, texture and appearance. A typical emulsified product requires the drop size distribution to be maintained within acceptable limits, which includes achieving a prescribed mean drop size, maintaining small variations about the mean and avoiding very small or very large drops that may adversely affect product properties such as texture and appearance. Due to a lack of quantitative understanding, emulsified products are currently manufactured by combining a broad knowledge of previous product formulations with empirical scientific experimentation.

An alternative to trial-and-error experimentation is to utilize a suitable mathematical model to predict the drop size distribution for different emulsion formulations and processing conditions. The population balance equation (PBE) modeling framework is particularly well suited for this problem as size distribution dynamics can be captured using mechanistic functions for drop breakage and coalescence phenomena which occur during emulsification. In this study, we developed a PBE model for prediction of homogenized drop size distributions and utilized the model to achieve target distributions by changing formulation and homogenization variables. An experimentally validated model (Maindarkar et al., 2013) was used to optimally determine the amount of surfactant, the number of homogenization passes and the pressure of each pass such that specified emulsion drop size properties were achieved. Two optimization objectives that differed with respect to the distribution specifications were formulated and solved using nonlinear optimization techniques. The model predictions were successfully validated by performing homogenization experiments using the optimal formulation and homogenization variables.

Maindarkar, S. N., Bongers, P. and Henson, M. A. (2013). Predicting the effects of surfactant coverage on drop size distributions of homogenized emulsions. Chemical Engineering Science, 89, 102-114.

Medium-Term Scheduling of a Multiproduct Industrial Batch Plant

Esmael Seid, Thokozani Majozi, Jui-Yuan Lee

NRF-DST Chair Sustainable Process System Engineering
School of Chemical and Metallurgical Engineering
University of the Witwatersrand, Johannesburg

November 16-21, 2014

AICHE Annual Meeting

www.wits.ac.za

Medium-term scheduling of a multiproduct batch plant

- A medium-term scheduling technique for a multiproduct batch has been developed
- The intractability of the short-term scheduling models when directly applied to the medium-term scheduling problems was solved by applying the decomposition method

First level mathematical model

Period 1	Period 2	Period 3	Period N-1	period N
P1 = X1kg	P1 = X3Kg	P1 = 0	○ ○ ○ ○ ○ ○	P1 = 0
P2 = X2kg	P2 = 0	P2 = X6Kg	○ ○ ○ ○ ○ ○	P2 = 0
P3 = 0	P3 = X4Kg	P3 = X7Kg	○ ○ ○ ○ ○ ○	P3 = Xn-1Kg
P4 = 0	P4 = X5Kg	P4 = X8Kg	○ ○ ○ ○ ○ ○	P4 = XnKg

- Objective function minimize the model size of each subproblem

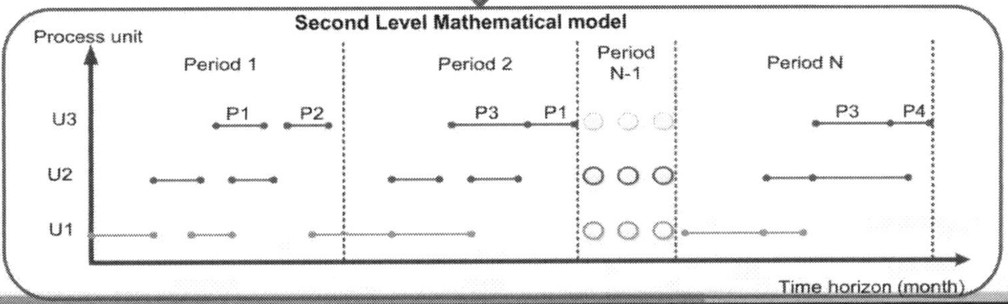

2

Medium-term scheduling of a multiproduct batch plant

Case study

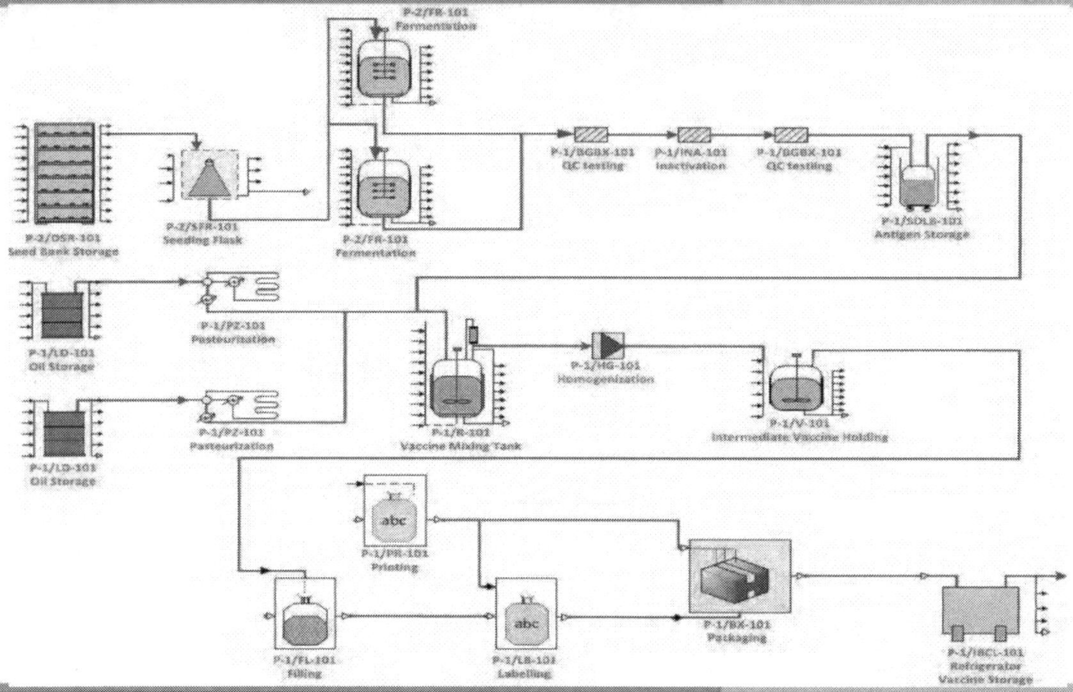

Medium-term scheduling of a multiproduct batch plant

Results

Computational results for the first level decomposition model

Products	Time period N1	Time period N2	Time period N3	Time period N4	Time period N5
product 1	905.2	0	0	0	0
product 2	2722	0	0	2579.5	0
product 3	0	0	2610	0	0
product 4	566.6	1469	0	0	0
product 5	0	2722	1863	0	0
product 6	0	2722	1075.3	4890	0
product 7	0	0	0	0	5129

Computational results for the second level decomposition model

Time periods	Time points	Binary variables	Continuous variables	Constraints	Time horizon
N1	10	950	5022	5045	16.12
N2	12	876	4586	5011	30.27
N3	10	840	4422	4618	56.03
N4	13	767	5280	4827	82.47
N5	13	299	1458	3006	95.7

4

360261 Process Simulation-Based Optimization of a Commercial Amine Gas Sweetening Unit

Abstract:

Amine sweetening units are integral parts of any natural gas plant with the objective of recovering mainly H_2S and CO_2 from gas streams before the latter are transported or used to produce LNG. Safety, environmental, operational and economic considerations govern the essence of removing these acid gases: H_2S is toxic and corrosive while CO_2 reduces the heating value besides being corrosive. In most cases, sales gas revenues far outweigh the capital and operating costs of these gas plants. Nevertheless, Emphasis in recent years has been on increasing gas plants efficiency due to exploration of sour gas fields with high H_2S and CO_2 concentrations which pose challenge to their profitability. Indeed, Abu Dhabi has significant developed and undeveloped sour gas reserves with high H_2S and CO_2 content (5% to 30%). This indicates the necessity of investigating several optimization schemes in order to maintain the economics of Abu Dhabi different gas plants.

This paper discusses the prominence of optimum operating conditions in amine gas units to best improve Abu Dhabi gas plants' efficiency without additional capital expenditure. Effect of lean amine temperature and amine strength on plant performance in several aspects is explained through sensitivity analysis. To this end, a kinetics based process simulation model has been developed for a commercial gas sweetening unit of a gas plant located in Abu Dhabi. Trends of various process variables such as sweet gas H_2S, CO_2 & H_2O, steam consumption rate and solvent circulation rate as function of the two abovementioned parameters are explained in detail. It is demonstrated through this study that lower amine temperature (47 ^{O}C) provides significant reduction in solvent circulation rate, steam consumption rate, pumping duty, dehydration unit load while posing no risk of hydrocarbon condensation and hydrate formation. Also, it is found that amine strength of 50 % is best suited for amine-sweetening unit's operating conditions as it resulted in potential savings in operating costs without increasing the risk of corrosion and fouling. To provide guidelines for optimum plant operations, DCS (Distributed Control System) lookup tables for the commercial amine-sweetening unit have also been generated and successfully tested.

383651 Integrated Simulation and Optimization Platform for System Design and Operation

Abstract:

Integrated simulation and optimization platform for system design and operation

Rui Huang

United Technologies Research Center

East Hartford, CT 06108

Computer-aided system simulation has been increasingly important in industrial environments particularly for the smart manufacturing in the digital age. System modeling and simulation can contribute insights to engineers with more information about the process and the consequences of the design and control decisions. There are many commercial modeling and simulation packages on the market, such as Aspen Suite for chemical processes simulation, Dymola (such as Aspen for the simulation of chemical processes or the general-purpose object-oriented modeling tool Dymola) for computational fluid simulation, etc. In addition, many companies and institutes have custom-built in-house simulation tools tailored to particular processes.

On the other hand, more and more system design and operation tasks involve solving optimization problems. Although significant efforts are devoted in system modeling and simulation in the simulation platforms, optimization engineers usually find it hard to extract the system equations to perform equation-oriented optimization tasks. They are often left without choices to run exhaustive search or simulation-based optimization such as genetic algorithms.
In this presentation, an integrated framework for system modeling, simulation and optimization is described. This framework allows engineers to model the systems by compiling different components as the state-of-the-art practice used in Aspen and Dymola. Then a set of equations of the system is assembled in the background, and derivative information is calculated based on finite differences or automatic differentiation. In addition, different simulation solvers and optimization solvers are integrated to the backend of the platform. As a result, engineers are able to simulate the system by solving the set of system equations or optimize the system with user-defined objectives constrained by the set of system equations. A vapor compression cycle is presented as an example.

The benefit of the framework is to integrate the system modeling and simulation with the optimization. Given the system is properly modeled, systematic equation-based optimization can be performed without significant additional efforts.

Efficient Ant Colony Optimization (EACO) Algorithm for Deterministic Optimization

Ant colony optimization (ACO) is a new metaheuristic optimization algorithm for tackling large set of optimization problems. Initially, ACO has been practiced with great success for combinatorial optimization problems. In recent years, the research focus has been on extending the combinatorial ACO algorithm to continuous and mixed variable nonlinear programming problems. In this work, a new variant of ACO algorithm to solve deterministic optimization problems is proposed. In the proposed algorithm, a new and efficient sampling technique, Hammersley sequence sampling (HSS), is implemented to initialize the solution archive for the continuous and mixed variable domains and to generate multidimensional random number for the implementation of the probabilistic rule of the combinatorial ACO algorithm. The better uniformity properties of HSS are exploited in developing the efficient ACO algorithm (EACO).

Nine distinctive benchmark problems are considered for the algorithm. We compare the results from the EACO algorithm of these benchmark problems with those obtained by the conventional ACO algorithms. The results show that the EACO algorithm improves the computational efficiency by a range from 3% to 71%. Thus, this new method can be a useful tool for large-scale and wide range of optimization problems. Finally, we investigate how the EACO algorithm performs on a real-world solvent selection optimization problem.

Solvent selection is an important step in process synthesis, design, or process modification. The computer-aided molecular design (CAMD) approach, based on the reverse use of group contribution methods, provides a promising tool for solvent selection. The existence of a combinatorial search space in molecular design poses a challenge for traditional deterministic optimization methods. In this work, we present a new approach to solvent selection using the EACO algorithm. A real world case study of acetic acid extraction from water, based on the infinite dilution activity coefficient model is examined. The UNIFAC model is used to estimate the mixture properties. As deterministic optimization technique, the EACO algorithm outperforms its counterparts, the conventional ACO, simulated annealing and genetic algorithm optimization techniques. Using the proposed algorithm method, new solvents with better targeted properties are found.

Ant Colony Optimization

The Ant Colony Optimization (ACO) metaheuristic defines a class of optimization algorithms inspired by the foraging behavior of real ants Dorigo and Blum (2005). Natural ants randomly search food by exploring the area around their nest. If an ant locate a food source, while returning back to the nest, it lay down a chemical pheromone trail that marks its path. This pheromone trail will indirectly communicate with other members of the ant colony to follow the path. Over time the pheromones will start to evaporate and therefore reduce the attraction of the path. The routes that are used frequently will have higher concentration of the pheromone trial and remain attractive. That means, the shorter the path between the nest and food source imply short cycle time for the ants and these paths will have higher concentration of pheromones than the longer routes. As a result more ants are attracted by the shorter paths for the future. Finally, the shortest path will be discovered by the ant colony.

In ACO algorithms, artificial ants are stochastic candidate solution construction procedures that exploit a pheromone model and possibly available heuristic information of the mathematical problem (E.g. traveling sales man problem). In a similar fashion, an artificial ants walk on a connected graph $G(C, L)$, where C represents the vertices and L is the set of connections. The set C of vertices represent solution components, which every ant chooses incrementally to create a full route. The pheromone model consists of a set of numerical values, called pheromones that are updated at each iteration in order to bias ants toward the most promising regions of the search space. Like the natural ants, by updating the pheromone values according to the information gained on the search domain, the algorithmic procedure leads to very good and hopefully, a global optimal solution. The main algorithmic components of the ACO metaheuristic are the ants' solution construction and the update of the pheromone information.

In combinatorial optimization, the problem is partitioned into a finite set of components, and the combinatorial optimization algorithm attempts to find the optimal combination or permutation.

Continuous variable optimization problems could be tackled with a combinatorial optimization algorithm only once the continuous *ranges* of the decision variables values are converted into finite sets. If the initial possible range is wide, this approach is difficult. At each construction step, ants make a probabilistic choice of the solution component from the set of available components according to the transition probability function. The probabilities associated with the elements of the set of solution components make up a *discrete probability distribution* that an ant samples in order to choose a component to be added to the current partial solution. The ACO algorithm for continuous domain is based the extension of the original ACO algorithm developed for combinatorial optimization problems without any major conceptual change, see Socha and Dorigo (2008). The fundamental idea underlying ACO algorithm for continuous domain is the shift from using a discrete probability distribution to using a continuous one, that is, a probability density function (PDF). In ACO continuous domain, instead of choosing a solution component based the *discrete probability distribution*, an ant samples a PDF.

The outline of ACO algorithm for the mixed variable optimization problem is given in table 1. Liao *et. al*. (2013).

Table.1 Algorithm for mixed variable ant colony optimization

- *Set K, nAnts, NCONV, NORDV, NCATV,* ρ, q, and termination criterion
- Initialize the solution archive *(T(K,NDIM))* and evaluate the K solutions
- Rank solutions according to the quality of the objective function $(T = Rank(S_1 \cdots S_k))$

While Termination criterion is not satisfied

- Generate *nAnts* new solutions

For all nAnts

- Incremental Solution Construction

For all NDIM

- o Probabilistically construct continuous decision variables
- o Probabilistically construct ordinal decision variables
- o Probabilistically construct categorical decision variables

End for #NDIM

- Store and evaluate the objective function value of newly generated solution

End for #nAnts

- Rank solutions and select the best K solutions, $T = Best(rank(S_1 \cdots S_{k+nAnt}), K)$
- Update Solution archives

End while

---

where T is the solution achieve and K is size of the solution archive. nAnts is the number of ants. *NCONV, NORDV, NCATV* and *NDIM* are the number of continuous, ordinal, categorical and the total number of variables, respectively.

References

1. Diwekar, U.M., W. Xu. 2005. Improved genetic algorithms for deterministic optimization and optimization under uncertainty. part i. algorithms development. *Industrial and Engineering Chemistry Research* 44(18) 7132-7137.
2. Dorigo, M., C. Blum. 2005. Ant colony optimization theory: A survey. *Theoretical Computer Science* 344(2-3) 243-278.
3. Kim, K.-J., U.M. Diwekar. 2002a. Efficient combinatorial optimization under uncertainty. 1. Algorithmic development. *Industrial and Engineering Chemistry Research* 41(5) 1276-1284.
4. Kim, K.-J., U.M. Diwekar. 2002b. Hammersley stochastic annealing: Efficiency improvement for combinatorial optimization under uncertainty. *IIE Transactions (Institute of Industrial Engineers)* 34(9) 761-777.
5. Schlter, M., J.A. Egea, L.T. Antelo, A.A. Alonso, J.R. Banga. 2009a. An extended ant colony optimization algorithm for integrated process and control system design. *Industrial and Engineering Chemistry Research* 48(14) 6723-6738.
6. Schlter, M., J.A. Egea, J.R. Banga. 2009b. Extended ant colony optimization for non-convex mixed integer nonlinear programming. *Computers and Operations Research* 36(7) 2217-2229.
7. Xu, W., U.M. Diwekar. 2005. Improved genetic algorithms for deterministic optimization and optimization under uncertainty. part ii. solvent selection under uncertainty. *Industrial and Engineering Chemistry Research* 44(18) 7138-7146.
8. Xu, W., U.M. Diwekar. 2007. Multi-objective integrated solvent selection and solvent recycling under uncertainty using a new genetic algorithm. *International Journal of Environment and Pollution* 29(1-3) 70-89.

9. Socha, K., M. Dorigo. 2008. Ant colony optimization for continuous domains. European Journal of Operational Research 185(3) 1155 -1173.
10. Tianjun Liao, Krzysztof Socha, Marco A. Montes deOca, Thomas Stutzle and Marco Dorigo 2013. Ant Colony Optimization for Mixed-Variable Optimization Problems

Application of better optimization of nonlinear uncertain systems (BONUS) algorithm for adaptive sensor placement

AIChE Annual Meeting, Atlanta, GA, November 16-21, 2014
For submission to "Design and Operations Under Uncertainty"

Rajib Mukherjee, Urmila Diwekar
Vishwamitra Research Institute, Center for Uncertain Systems: Tools for Optimization and Management (CUSTOM), 368,
56th Street, Clarendon Hills, IL-60514

Abstract:

Sensors are generally used for monitoring large systems. They are generally strategically located so that they can achieve maximum information. They can be placed in the most effective locations according to specified objectives and constraints such as cost, performance, sensitivity of the variation in uncertain environment etc. In many cases, the subject of interest can be a large number of objects. The objects generally forms clusters around point of interest. The point of interest can move and the number of point of interest can also change. In those cases, the object can distribute dynamically in a given region where the sensor is monitoring. The distribution of the objects in a cluster around the point of interest can be Gaussian in nature. In those cases the sensor can perform effectively if it can dynamically locate the mean of the clusters and adapt its position.

In the present work we have used the example of grazing animals as the object of interest and develop an algorithm for adaptive sensor placement based on BONUS algorithm. The area where the animal can graze was evenly divided into certain number of regions based on the number of sensors. The number of regions is constrained by the number of sensors. Every sensor will scan the region around it, make local estimate of the objects and find the mean and variance of the clusters. The sensors talk to each other in a series and form the cumulative estimate of the clusters through cumulative probability density. Once a sensor identifies a cluster in its region, it adaptively optimizes its position by finding the minimum Mahalanobis distance between the cluster and its location. At different timestep, the sensors will scan its region and update the information about the clusters and adapt its position accordingly.

In our problem we have looked into the problem of adoptive sensor placement from uncertainty perspective. The sensor can cover a certain region of interest but its location at a given time step is uncertain. The locations of the object clusters are also uncertain. We have used an effective sampling technique with Hammersely sequence sampling (HSS) for the distribution of the clusters and location of the sensors. The sampling along with Better Optimization for Nonlinear Uncertain Systems (BONUS) algorithm is used for solving the problem. The BONUS algorithm avoids calculation of the objective functions in subsequent iterations at different timesteps. The simulation result shows the effectiveness of the algorithm for adoptive sensor placement.

DEVELOPMENT OF THE START-UP PROCEDURE FOR A THERMALLY COUPLED DISTILLATION COLUMNS SYSTEM THROUGH DYNAMIC SIMULATION

Angela L. U. Vasconcelos, Federal University of Campina Grande, Campina Grande, PB, Brasil
Luis G. S. Vasconcelos, Federal University of Campina Grande, Campina Grande, PB, Brasil
Romildo P. Brito, Federal University of Campina Grande, PB, Brasil

Abstract

The operation of a continuous column distillation is periodically interrupted in order to perform maintenance procedures. After the maintenance, it is required perform the start-up operation. The start-up procedure of a distillation column can take several hours to days to reach the steady state. The start-up operation of a chemical process is one of the most complicated processes observed in industrial practice. The complication of the start-up operation becomes larger for a thermally coupled distillation columns system. The system studied is two thermally coupled columns where one is an extraction column used to separate the ethanol-water azeotropic mixture with ethylene glycol as solvent. The work aims to study the dynamic behavior of a thermally coupled distillation columns system during the start-up operation and based on this study propose a procedure suitable to start-up the system. Dynamic simulation and the start-up procedure were developed in Aspen Tech software. From the model developed in Aspen Tech was possible to analyze the dynamic behavior of the start-up operation of thermally coupled distillation columns system and propose an appropriate procedure satisfactorily.

Introduction

With the globalization of the economy, industrial processes have increasingly sought total quality in services, products and environmental management. For these objectives to be achieved it is necessary to carry out preventive maintenance of equipment constituting the industrial processes. The maintenance process aims to generate operational conditions for equipment, facilities and services function properly. After maintenance, the start-up operation is performed until the process reaches the steady state.

One of the most complicated processes observed in industrial practice is to start operation of a distillation column due to simultaneous changes of several variables relevant to the process (Kruse et al. 1996).

During the startup operation, the dynamic behavior of a distillation column presents a trend that can be represented by three characteristic stages. The first stage, called the discontinuous phase, is characterized by its short time period and the discontinuous nature of all variables. The second stage, the semi-continuous stage is characterized by the non-linear and transient behavior of the variables and the approximation of hydraulic variables from their steady state values. The time period of the second stage is greater than that of the first stage. The third stage, called the continuous phase, is characterized by linear and transient responses of all variables. At this stage, all variables reach their steady state values (Ruiz et al. 1988).

In order to develop a suitable startup procedure is necessary to know and understand the dynamic behavior of a distillation column during startup operation.

Thermally Coupled Distillation Columns

The system of thermally coupled distillation columns studied consists of two distillation columns connected by two streams, one stream in the vapor phase (VAP) and a stream in the liquid phase (LIQ) as shown in Figure 1.

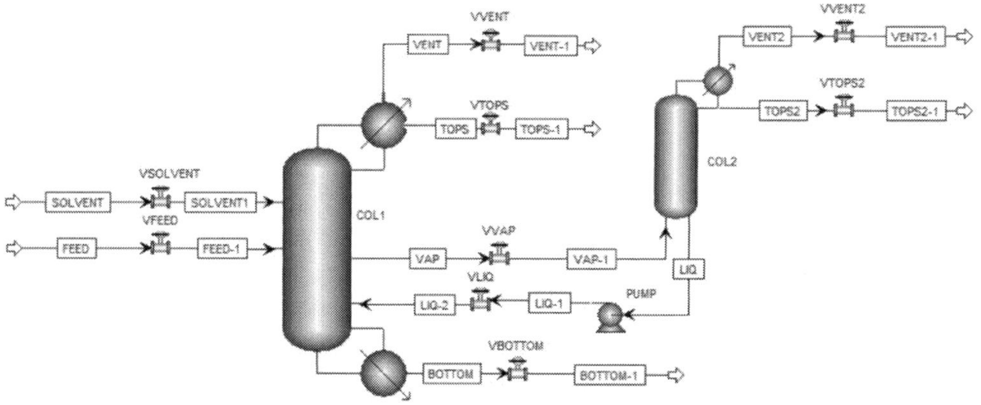

Figure 1. Flowsheet of thermally coupled distillation columns system.

The reboiler was removed from the second distillation column due to the link between the columns. The first column (COL1) is an extractive distillation column used to separate the azeotropic mixture ethanol-water and has as solvent ethylene glycol. The first column contains 29 stages, including reboiler and condenser.

The second column (COL2) is a distillation column whose top product is water. The second column contains five stages, including condenser, and hasn't reboiler. This column is fed by the VAP stream that exiting the first column from stage 24. The product stream LIQ returns the first column in stage 25. The characteristics of the streams SOLVENT and FEED can be seen in Table 1.

Methodology

Initially the model of the studied system at steady state, was built in the Aspen Plus. After achieved the convergence, this model was transferred to Aspen Plus Dynamics. In Aspen Plus Dynamics all controllers were added and the adjusted parameters. Then ran the pre-built script "Empty" that leaves the column in empty and cold state which corresponds to its initial state. In reality, the column is not empty but full of nitrogen at 20 ° C temperature and pressure of 1.01 bar.

During the startup operation the inert gas, nitrogen, must be purged from the interior of the distillation column and thus should have two product output currents at top column in the model built in Aspen Plus, one for the output of desired product and one for nitrogen purging. After running the script "Empty", all controllers were placed in manual mode and the output signals of the controllers adjusted their minimum values.

Then was executed the initialization mode of the system to fit the initial state with all flows equal to 0 kg/hr. After running the initialization, added a "Task" to the model, which is the procedure adopted for the startup of the column. Then changed the mode to run for dynamic and ran the simulation.

Results

Three start-up procedures were simulated varying the steam heating flow rate in the reboiler and their results were compared. In the first case, the flow rates of the heating steam were adjusted to lower values. In the second case the flow rates of the heating steam were adjusted median values. In the third case the flow rates of the heating steam were adjusted to higher values.

Start-up procedures

At the beginning of the start-up operation all controllers are in manual mode.

At t = 0, the VFEED and VSOLVENT valves which allow the passage of the azeotropic mixture and of the solvent respectively, begin to be opened. Then, in the first moments the valves undergo a gradual opening of the position 0% to 8%.

It is expected until the moment that the level in the base of the first column reaches the value of 0.34 m so that the column pressure controller 1 is put in automatic mode by specifying a set point of 1 bar in the first stage.

After 6 minutes, it starts the flow of heating vapor in the reboiler, at a flow rate 500 kg/hr. At the same instant of time, the nitrogen purge valve of the column 1 (VVENT) is fully open.

While the composition of nitrogen in a purge stream (VENT-1) is greater than 0.5 the heating steam flow rate increases to 700 kg/hr.

At instant that the composition of nitrogen in stream VENT-1 becomes less than 0.5 is expected 3 minutes for the position of the purge valve (VVENT) assume the position of 50% and 9 minutes for the steam heating flow rate increase to 1000 kg/hr. In the next moments, the VVENT valve is gradually closed up to 0.1% position.

Then waits until the moment when the level in the reflux vessel of the column 1 reached the value of 0.5 m to begin reflux in column 1 at a flow rate of 1000 kg/hr. After three minutes, the reflux flow rate increases to 1500 kg/hr. In the next moments, the VVENT valve is gradually closed until the 0% position. After nine minutes, the steam heating flow rate is increased to 1300 kg/hr.

It is expected until the moment that the level in the reflux vessel reaches 2 m so that the level controller, COL1_DrumLC, is placed in automatic mode with the set point of 2 m and so that the heating steam flow rate is adjusted to 1500 kg/hr.

After 24 minutes, VFEED valve assumes the position of 9% and the VSOLVENT valve assumes the position of 11%.

After 15 minutes, the sump level control of the column 1, COL1_SumpLC is placed in automatic mode with set point of 1.7 m.

Over the next 117 minutes, the steam heating flow rate undergoes a gradual increase of 1500 kg/hr to 2400 kg/hr, the reflux flow rate increases from 1500 kg/hr to 2500 kg/hr, the open position of the VFEED valve increases from 9% to 18% and the open position of the VSOLVENT valve increases from 11% to 20%. It is expected until the moment when the water composition in the VAP stream is greater than 0.8 so that the VVAP valve is opened by 2%.

After 9 minutes, the pressure controller in column 2, COL2_CondPC, is placed in automatic mode with set point of 0.6 bar in stage 1 and the nitrogen purge valve, VVENT2, is fully open.

After 3 minutes, the position of the VVAP valve increases to 5%.

It is expected until the moment that the nitrogen composition in a purge stream from column 2, VENT2-1, is less than 0.01 so that the VVENT2 purge valve be fully closed. After 15 minutes, the position of VVAP valve increases to 8%.

After 9 minutes, the position of the VFEED valve increases to 20% and the position of the VSOLVENT valve increases to 22%. After 3 minutes, the set point level controller is set to 1.5 m.

Over the next 180 minutes, the steam heating flow rate undergoes a gradual increase of 2400 kg/hr to 3999 kg/hr (in the first case), 2400 kg/hr to 4300 kg/hr (in the second case) and 2400 kg / hr to 4629 kg/hr (in the third case), the open position of the VFFED valve increases from 20% to 45%, the open position of the VSOLVENT valve increases from 22% to 47% and the position of the VVAP valve goes 8% to 25%. Then, the position of the VVAP valve will to 30%.

After 9 minutes, the steam heating flow rate in the reboiler increases to 4278 kg/hr (in the first case), to 4600 kg/hr (in the second case) and to 4838 kg/hr (in the third case).

After 3 minutes, the feed flow rate control of the azeotropic mixture is placed in automatic mode with set point of 4186 kg/hr. At the same instant, the feed flow rate control of the solvent is too placed in automatic mode with set point of 5586 kg/hr and the steam heating flow rate increases to 4371 kg/hr (in the first case), to 4700 kg/hr (in the second case) and to 4841 kg/hr (in the third case).

After 9 minutes, the steam heating flow rate undergoes a further increase and goes to 4464 kg/hr (in the first case), to 4800 kg/hr (in the second case) and to 4944 kg/hr (in the third case). After 6 minutes, the temperature control is placed in automatic mode with set point at 170.49 ° C.

When the level of liquid in the reflux vessel of the column 2 attains 0.7 m, starts the reflux column 2 at a flow rate of 50 kg/hr, which increases within 15 minutes to 150 kg/hr. After 9 minutes, the position of the VVAP valve goes to 35% and the reflux flow rate increases up to 200 kg/hr. In the following 33 minutes, the reflux flow rate increases of 200 kg/hr to 400 kg/hr.

Then, it waits till the moment that the level in the sump of column 2 is larger than 0.8 m so that the position of VLIQ valve, which allows the return of liquid to the column 1, is be set to 10%.

After 9 minutes, the position of the valve VLIQ goes to 15% and the set point of the level control of the reflux vessel of the column 1, COL1_DrumLC, is set at 1.25 m.

During the following 36 minutes, VLIQ valve position increases from 10% to 35%.

After 9 minutes, the reflux flow rate of the column 2 increases to 500 kg/hr.

Over the next 18 minutes, the position of the VLIQ and VVAP valves goes from 35% to 45%.

After 9 minutes, the flow rate control of the VAP stream is placed in automatic mode with the set point of 572.4 kg/hr and the position of VLIQ valve is adjusted by 50%.

It is expected until the moment that the level in the reflux vessel of column 2 is larger than 0.9 m so that the vessel level control is placed in automatic mode with a set point of 0.5 m.

It is expected until the moment that the level in the sump of column 2 is larger than 2.5 m so that the reflux flow rate of column 2 is set at 350 kg/hr.

Graphical results

Figure 2 shows the dynamic behavior of the composition of VAP stream during the startup operation of the first case.

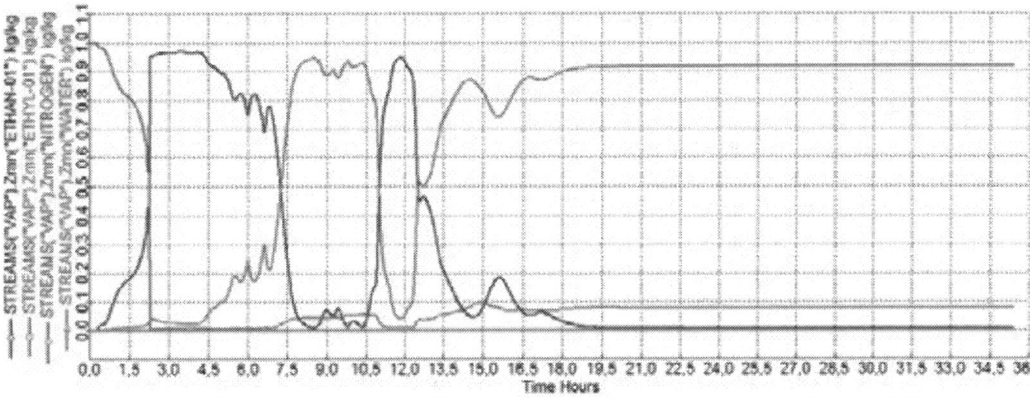

Figure 2. Composition of VAP stream of the first case.

Figure 3 shows the dynamic behavior of the heating steam flow rate in the reboiler of column 1 during the startup operation of the first case.

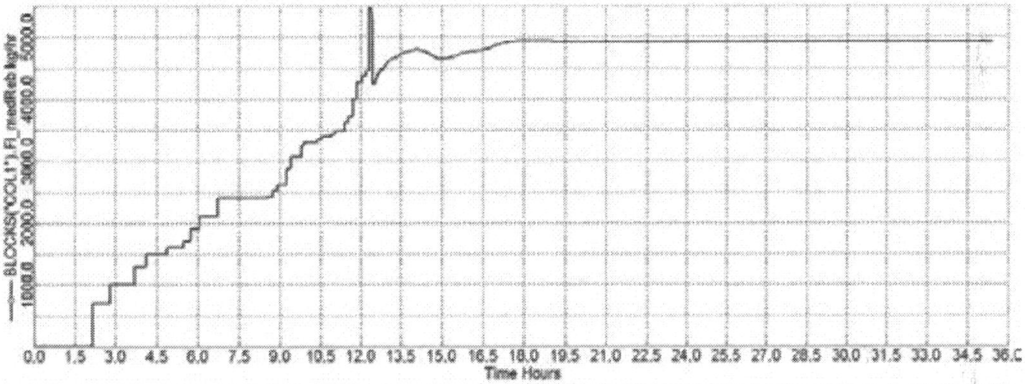

Figure 3. Heating steam flow rate in the reboiler of column 1 of the first case.

Figure 4 shows the dynamic behavior of the distillate stream composition of the column 2, TOPS2-1, during the startup operation of the first case.

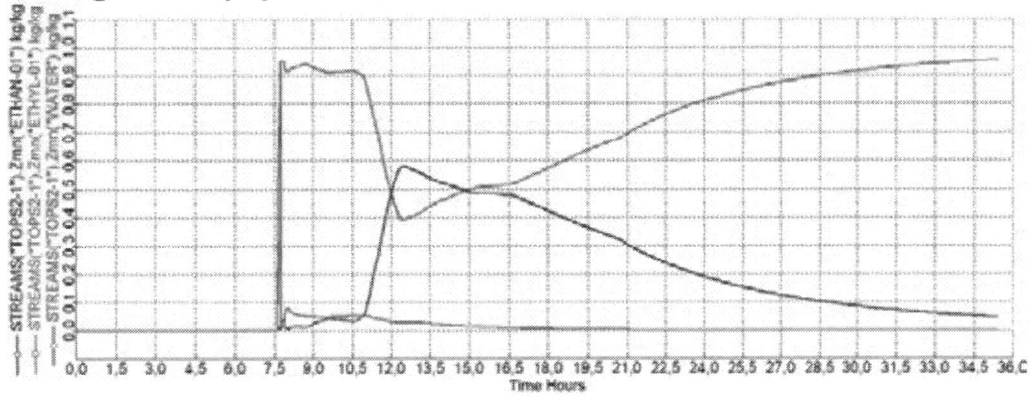

Figure 4. Distillate stream composition of the column 2 in the first case.

Figure 5 shows the dynamic behavior of the composition of VAP stream during the startup operation of the second case.

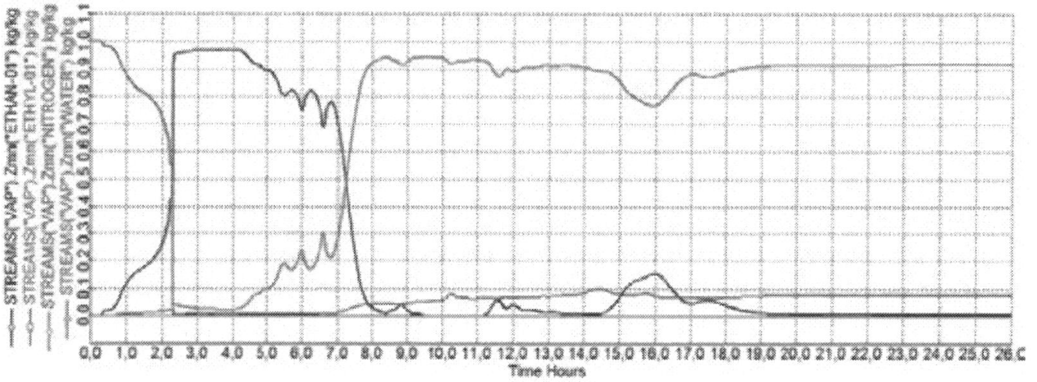

Figure 5. Composition of VAP stream of the second case.

Figure 6 shows the dynamic behavior of the heating steam flow rate in the reboiler of column 1 during the startup operation of the second case.

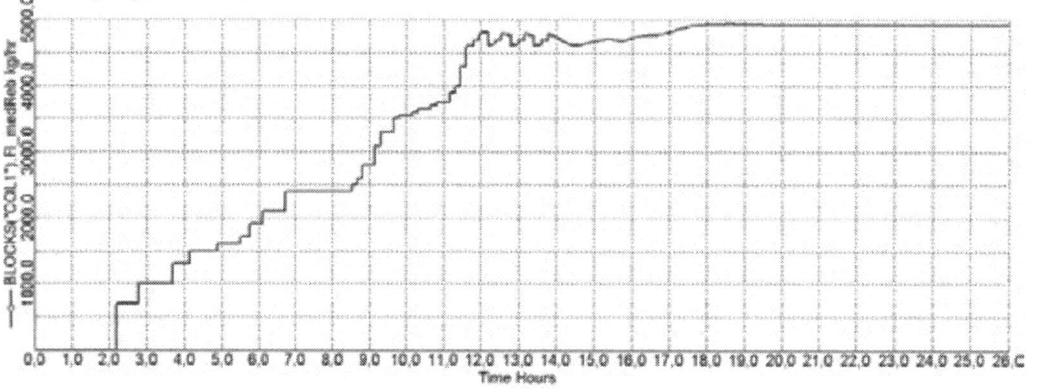

Figure 6. Heating steam flow rate in the reboiler of column 1 of the second case.

Figure 7 shows the dynamic behavior of the distillate stream composition of the column 2, TOPS2-1, during the startup operation of the second case.

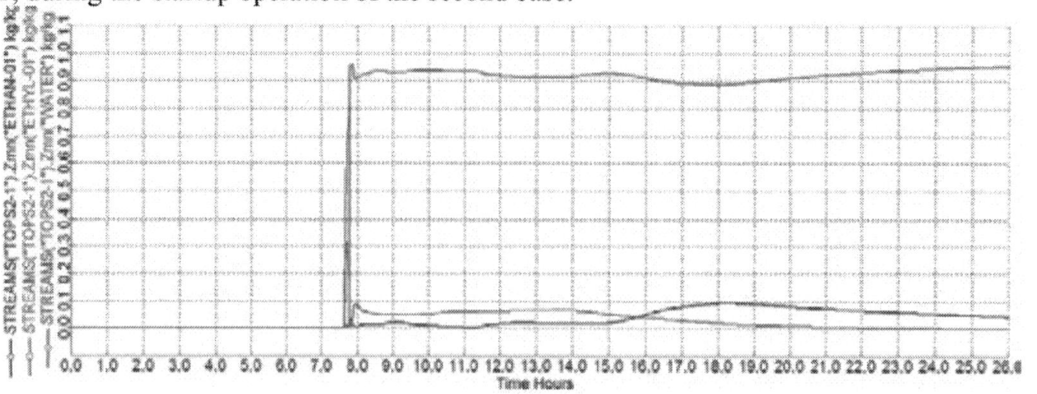

Figure 7. Distillate stream composition of the column 2 in the second case.

Figure 8 shows the dynamic behavior of the composition of VAP stream during the startup operation of the third case.

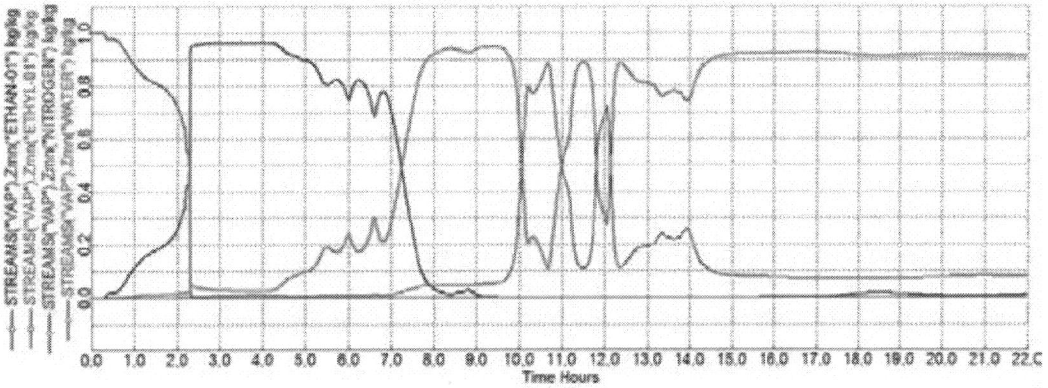

Figure 8. Composition of VAP stream of the third case.

Figure 9 shows the dynamic behavior of the heating steam flow rate in the reboiler of column 1 during the startup operation of the third case.

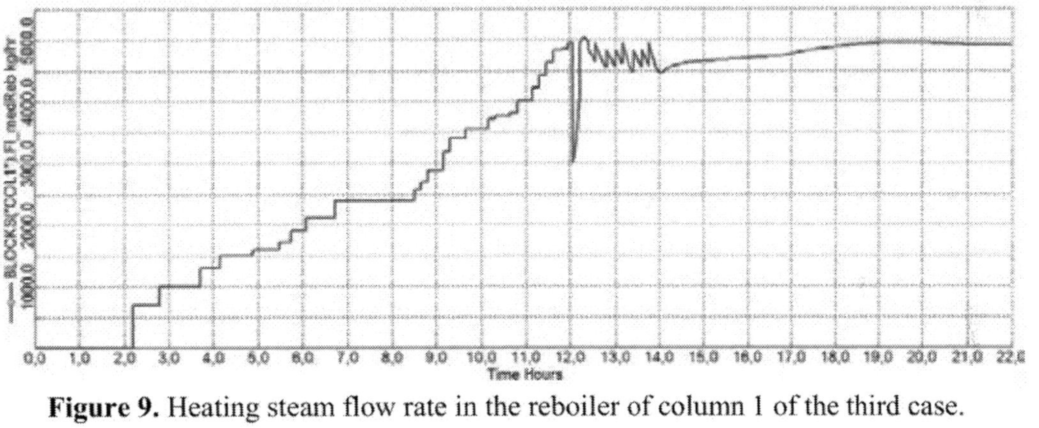

Figure 9. Heating steam flow rate in the reboiler of column 1 of the third case.

Figure 10 shows the dynamic behavior of the distillate stream composition of the column 2, TOPS2-1, during the startup operation of the third case.

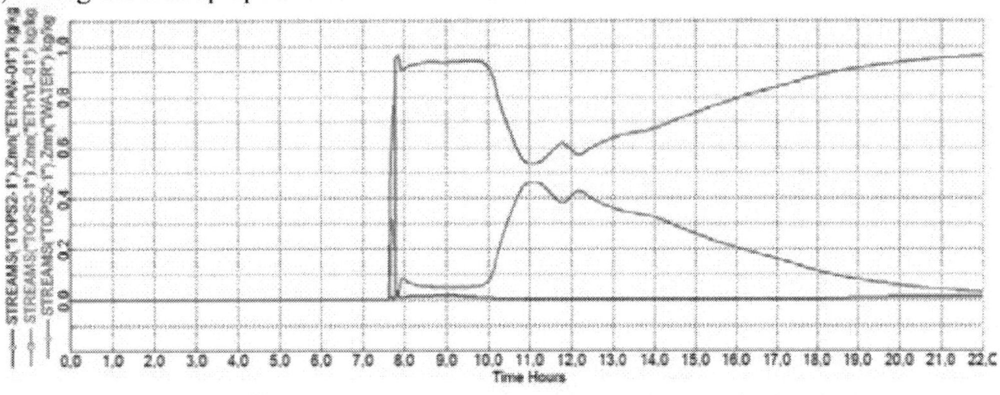

Figure 10. Distillate stream composition of the column 2 in the third case.

From the results it was observed that the start-up procedure for the third case had the lowest start-up time, 22 hours, followed by the second case, 26 hours, and the first case had the largest start-up time of 35,5 hours.

As the heating steam flow rate in reboiler is increased the temperature at each stage of the column increase too. Thus, the dynamic behavior of the components composition in the stages of the column and consequently in the VAP stream is influenced by the dynamic behaviour of the heating steam flow rate in reboiler.

From figures 2 and 3 it can be seen that in the first case the flow rate of steam heating was not sufficient to maintain the water composition above 0.9 so that the ethanol composition in a few moments become higher than the water composition.

Observing figures 5 and 6 it follows that in the second case the flow rate of heating steam was manipulated to maintain the water composition above 0.9 until the instant at which the temperature control was placed in automatic mode and the water composition had a slight decrease and the ethanol composition had a slight increase until it reaches the steady state.

By analyzing Figures 8 and 9 is seen that in the third case, the flow rate of heating steam was adjusted to higher values, so that changes were observed in the water and ethylene glycol compositions in the VAP stream until the instant the temperature control was placed in automatic mode and then the water composition increased while the ethylene glycol composition decreased and then rapidly reached a steady state value.

Conclusions

From the results it was observed that the start-up procedure for the third case had the lowest start-up time, 22 hours, followed by the second case, 26 hours, and the first case had the largest start-up time of 35,5 hours.

Hypothetically, the presence of ethanol in the second column raises the start-up time of the system. The higher the flow rate of heating steam in the reboiler the lower the ethanol composition in the stage 24 of the column 1, therefore the lower the amount of ethanol that enters the column 2 by the VAP stream. Therefore, it is necessary to keep the composition of ethanol in stage 24 of the column 1 at minimum value by manipulating the heating steam flow rate in reboiler of the column 1.

From the model developed in Aspen Tech was observed as relevant variables in the startup operation dynamically behave and how they are influenced by changes in other variables. Furthermore, it was possible to propose a proper startup procedure for system for thermally coupled distillation columns studied based on the analysis of the dynamic behavior of the startup operation.

References

1. Kruse, C., Fieg, G., Wozny, G. (1996), "A new time-optimal strategy for column startup and product changeover," *Journal of Process Control*, pp. 187-193.
2. Ruiz, C. A., Cameron, I. T., Gani, R. (1988), "A generalized dynamic model for distillation column – III. Study of startup operations," *Computers & Chemical Engineering*, pp. 1-14.

Optimization of Process Parameters in Bioremediation of Cr(VI) Contaminated Aqueous Solution through Response Surface Methodology

Subhajit Majumder[1], Pratyusha Naini[2], Smita Raghuvanshi[3], Suresh Gupta[4*]

[1]*Lecturer, Department of Chemical Engineering*
Birla Institute of Technology and Science (BITS), Pilani- 333031, Rajasthan, India
Email: subhajit@pilani.bits-pilani.ac.in; subh.jit@gmail.com

[2]*Post-Graduate Student, Department of Chemical Engineering*
Birla Institute of Technology and Science (BITS), Pilani- 333031, Rajasthan, India
Email: nainiprathyusha@gmail.com

[3]*Assistant Professor, Department of Chemical Engineering*
Birla Institute of Technology and Science (BITS), Pilani- 333031, Rajasthan, India
Email: smita@pilani.bits-pilani.ac.in

[4]*Assistant Professor and Head, Department of Chemical Engineering,*
Birla Institute of Technology and Science (BITS) Pilani, -333031, Rajasthan, India
Email: sureshg@pilani.bits-pilani.ac.in

EXTENDED ABSTRACT

Bioremediation of metals wastes has gained significant attention in recent years. It is used to degrade toxic organic pollutants, heavy metals using living microorganisms (bacteria, fungi, algae etc.). Microorganisms show moderate to high uptake of heavy metals via different processes such as transport through cell-membrane, biosorption on cell walls and subsequent entrapment in extracellular materials, redox reaction, precipitation, complexion etc. (Rai et al., 1981; Macaskie & Dean, 1989; Avery & Tobin, 1993, Brady et al., 1994, Malik, 2004). For the last few decades, separate batch studies were carried out to optimize various process parameters such as pH, initial concentrations, contact time. But, this approach could not determine the interaction between the various process parameters. Combined effect of all the process parameters on bioremediation of heavy metals needs to be analyzed which helps during the scale-up studies. Conventional batch process is time consuming and requires a large number of experiments to be carried out. Concept of statistical methods can be exploited to mitigate the problems. Response surface methodology (RSM) is used to optimize all the process parameters collectively which reduces number of experiments and thereby reducing the overall cost of the

*Corresponding author. Tel.: +91-1596-515224; Fax: +91-01596-244183
E-mail address: sureshg@pilani.bits-pilani.ac.in, sureshg25@gmail.com

process. RSM is a collection of mathematical and statistical methods used for modeling and optimizing the process parameters by considering the complex combined effect. The present study is aimed to optimize the bioremediation of Cr(VI) by using indigenous chromium resistant bacteria isolated from activated sludge of sewage treatment plant (BITS, Pilani, Rajasthan, India). The influence of initial Cr(VI) concentration, pH and contact time of the solution on bioremediation is studied. RSM is applied for the optimization of these process parameters to evaluate their interaction effect on bioremediation process. The optimization procedure through RSM shows a close interaction between the experimental and simulated values of Cr(VI) removal.

Keywords: *Bioremediation, Activated sludge, Indigenous bacteria, Process parameters, Response surface methodology.*

References

Avery, S. V., Tobin, J. M. (1993). Mechanism of Adsorption of Hard and Soft Metal Ions to Saccharomyces cerevisiae and Influence of Hard and Soft Anions. Appl. Environ. Microbiol. 59: 2851-2856.

Brady, D., Glaum, D., Duncan, J. R. (1994). Copper tolerance in Saccharomyces cerevisiae. Letters in Appl. Microbiol. 18: 245-250.

Macaskie, LE, Dean, AC. (1989). Microbial metabolism, desolubilization, and deposition of heavy metals: metal uptake by immobilized cells and application to the detoxification of liquid wastes. Adv Biotechnol Processes. 12: 159-201.

Malik A. (2004). Metal bioremediation through growing cells. Environ Int. 30: 261-278.

Rai, L. C., Gaur, J. P., Jumar, H. D. (1981). Phycology and heavy-metal pollution. Biol. Rev. 56: 99-151.

STRATEGIC DESIGN OF BIOSNG SUPPLY CHAINS

Andres J. Calderon[1], Paolo Agnolucci[2], Lazaros G. Papageorgiou[1,*]

1. Centre for Process Systems Engineering, Department of Chemical Engineering, UCL (University College London), London WC1E 7JE, UK
2. UCL Institute for Sustainable Resources, UCL (University College London), London WC1H 0NN, UK

* Corresponding author. E-mail address: l.papageorgiou@ucl.ac.uk (L.G. Papageorgiou)

Key words
Synthetic Natural Gas (SNG), Forestry waste, Solid Recovered Fuels (SRF), energy crops, gasification, torrefaction, pyrolysis, supply chain optimisation.

Abstract

Natural gas could play an important role in the transition between fossil and renewable fuels as a potential substitute for fuels with higher CO2-intensity like coal. Although unconventional fuels like tight gas sands and shale gas are growing in popularity, the debate about their environmental impact is still open (EA, 2013; EPA, 2011). In this context, the development of new technologies for the production of synthetic natural gas (BioSNG) from renewable resources emerges as an important alternative to be considered.

While extensive research has been devoted to develop efficient and scalable technologies for the production of BioSNG from different sources (e.g. wood, energy crops and municipal solid waste), most of the efforts have been focused on thermochemical routes. Gasification, whose applicability has been extensively demonstrated for coal, represents one of the emerging commercial technologies for thermal conversion of biomass. Biomass gasification is a very efficient technology with the potential of becoming a critical stage in the production of synthetic fuels (NREL, 2012; van der Meijden, Veringa, & Rabou, 2010).

Despite all best efforts, production of BioSNG is rarely regarded as profitable, with high investment costs being a key obstacle in the economy of biofuels production. Several authors are working on the development of pre-treatment technologies as a previous step of gasification to obtain intermediate products with higher energy density. The implementation of these technologies may improve the economics of the global production chain by reducing both transport costs and the size of installed capacities required for gasification (Magalhães, Petrovic, Rodriguez, Putra, & Thielemans, 2009; Uslu, Faaij, & Bergman, 2008).

Unfortunately, there are still numerous aspects to be considered to achieve a better understanding of the system and identify synergies inherent to the nature of the problem. In order to address the problem, this paper presents a spatially explicit, multi-period mixed integer linear programming (MILP) to address the strategic design of a second generation supply chain for BioSNG production. The decisions include feedstock logistic (transport modes, capacities, destinations); technology selection, locations and scales of BioSNG production facilities; economies of scale; location and cultivation rate of new energy crops; BioSNG production rates and transportation logistics as well as optional generation of heat and power.

A case study of the UK is proposed to demonstrate the applicability of the model. The relevance of pre-treatment technologies on the production of BioSNG is studied in a national context. In addition, it is also evaluated the impact of power generation on the economics of BioSNG production as well as the mechanisms implemented by the government to support the sustainable generation of synthetic fuels.

References

EA. (2013). *An Environmental Risk Assessment for shale gas exploratory operations in England* (pp. 1–32). Retrieved from https://www.gov.uk/government/uploads/system/uploads/attachment_data/file/296949/LIT_8474_fbb1d4.pdf

EPA. (2011). *Plan to Study the Potential Impacts of Hydraulic Fracturing.* Washington, D.C. Retrieved from http://water.epa.gov/type/groundwater/uic/class2/hydraulicfracturing/upload/FINAL-STUDY-PLAN-HF_Web_2.pdf

Magalhães, A. I. P., Petrovic, D., Rodriguez, A. L., Putra, Z. A., & Thielemans, G. (2009). Techno-economic assessment of biomass pre-conversion of biomass-to-liquids line-up. *Biofuels, Bioproducts & Biorefining, 3*, 584–600. doi:10.1002/bbb

NREL. (2012). *Biomass Gasification Technology Assessment Consolidated Report.* Retrieved from http://www.nrel.gov/docs/fy13osti/57085.pdf

Uslu, A., Faaij, A. P. C., & Bergman, P. C. a. (2008). Pre-treatment technologies, and their effect on international bioenergy supply chain logistics. Techno-economic evaluation of torrefaction, fast pyrolysis and pelletisation. *Energy, 33*(8), 1206–1223. doi:10.1016/j.energy.2008.03.007

Van der Meijden, C. M., Veringa, H. J., & Rabou, L. P. L. M. (2010). The production of synthetic natural gas (SNG): A comparison of three wood gasification systems for energy balance and overall efficiency. *Biomass and Bioenergy, 34*(3), 302–311. doi:10.1016/j.biombioe.2009.11.001

385037 Homotopy Continuation Solution Method with Advanced Step Implementation for Nonlinear Model Predictive Control

Abstract:

Nonlinear model predictive control (NMPC) is a powerful control algorithm that enables the satisfaction of complex control objectives, and the systematic consideration of all process dynamic interactions for multivariable control systems. The implementation of model predictive control in highly nonlinear dynamic systems that exhibit fast dynamics may become prohibitive if the solution time of the associated dynamic optimization problem is comparable to the duration of the control interval suitable for adequate control performance. Therefore, either a delayed control action is applied to the system or a simplified model usually a linear realization of the nonlinear dynamic model is used hence compromising the achieved control performance.

Over the years a large number of efficient real-time strategies have been proposed such as explicit NMPC, Newton-type controllers and NLP sensitivity-based controllers. In the present work, the NMPC problem is formulated as a homotopy between the calculated known optimal solution at the previous control interval and the seeking optimal solution at the current control interval. Specifically, the nonlinear dynamic optimization problem is discretized using orthogonal collocation on finite elements and transformed into a large-scale nonlinear program. The Karush-Kuhn-Tucker optimality conditions parameterized with respect to the initial conditions are then solved within a homotopy continuation framework. Starting from a known optimal control sequence that corresponds to the initial states of the previous control interval the optimal solution is pathfollowed using a sparse arc-length continuation solution technique to the new set of initial conditions that reflect the current plant state after the latest measurement information has been taken into consideration. Solution of the sparse Newton step is performed using UMFPACK /1, 2/ whereas the implementation of the continuation solution method is performed using PITCON /3/. Pathfollowing the optimal solution is appropriately modified in order to identify active set changes imposed by either the violation of model constraints or the complementarity slackness condition along the solution trajectory. Furthermore, other singular points due to linear independence constraint qualification or second-order optimality condition violations are effectively treated along the continuation path.

The implementation of the homotopy NMPC exploits the advanced-step algorithm /4/ by utilizing the control interval for the pathfollowing procedure to target the predicted state for the initial state vector of the system during the current control interval for further reduction in the computational time. The correction to the state values imposed after the incorporation of the new measurement set starts from the predicted initial state values. The method appears to provide acceptable speed of solution with enhanced robustness in terms of convergence for a highly nonlinear system with considerable fast dynamics (control interval equal to 50 ms) consisted of a cart with a double pendulum. A second example considers the control of the molecular weight distribution for a bulk polymerization reactor.

1. A column pre-ordering strategy for the unsymmetric-pattern multifrontal method, T. A. Davis, ACM Transactions on Mathematical Software, vol 30, no. 2, June 2004, pp. 165-195.
2. Algorithm 832: UMFPACK, an unsymmetric-pattern multifrontal method, T. A. Davis, ACM Transactions on Mathematical Software, vol 30, no. 2, June 2004, pp. 196-199.
3. Rheinboldt W.C., and J.V. Burkardt (1983). A locally parameterized continuation process. ACM Trans on Mathematical Software, 9, 215-235.
4. Zavala V. M., Biegler L. T., (2009). The advanced-step NMPC controller: Optimality, stability and robustness, Automatica, 45, 86-93.

CHALLENGES IN THE DEVELOPMENT OF THE MATHEMATICAL MODELING OF A DELAYED COKING PROCESS

Cláudio N. Borges[1, 2], *Maria Anita Mendes*[2] *and Rita M. B. Alves*[2]

[(1)] *Polytechnic School - Department of Chemical Engineering, University of São Paulo, São Paulo, Brazil,*
[(2)] *Presidente Bernardes Refinery - RPBC, Petrobras, Cubatão, Brazil.*
claudio.neves@petrobras.com.br; rmbalves@usp.br

Introduction

The Delayed Coking Unit is a thermal conversion process used in many refineries to convert vacuum residue into lower molecular weight high-value products (gases, naphtha and gasoil) and petroleum coke. A slight increasing of liquid yield of delayed coking unit could bring very considerable economic benefits, especially in the liquid distillate. Market competition, restrictions on product specifications and bottleneck condition require better control of production planning. Therefore, developing new strategies and mathematical models focused on optimal operating condition of industrial processes and formulations of products is essential to achieve better yields and a more accurate product quality monitoring.

Constraints in carry out modeling and simulation in good adherence to the industrial plant, which is empirical, are the data quality and degree of detail of the models used. This work is based on the experimental data analysis and operating conditions from an industrial delayed coking unit. Therefore, the aim is to discuss the main challenges in the development of mathematical model of the furnace and coke drum such as rigorous molecular characterization of the vacuum residue and products, to find out parameters that affect coke morphology, to detail reaction zones inside the coke drum, to select correlations for critical properties and formation enthalpy.

Delayed Coking Process Description

Usually, this process is composed by three main unit operations: the main fractionator, the coking furnace and at least two batch coke drums. The main fractionator has as main purposes: fractionating the effluents from the coke drums and the fresh feed from the vacuum tower distillation. In the coking furnace, heating and thermal cracking reactions of the feed are carried out.

As shown in **Figure 1**, the hot vacuum residue feed flows directly into the bottom of the main fractionator.The bottom section acts as an accumulator drum from the coking furnace and also as a heat exchange section between the coke drum effluent and the vacuum residue. The effluent heavy fraction is condensed and incorporated into the vacuum residue, generating a combined feed coking furnace. The condensed fraction is called the *effluent recycle.*

The furnace raises the combined feed temperature to 940 °F, considering the desired heat flux and the minimum residence time to minimize coking in furnace itself. The furnace effluent flows to the bottom of one of the pair of coke drums, where it thermally cracks to gas, naphtha, light gasoil, heavy gasoil and solid coke. The coke gradually fills the drum, usually over a period of 21 hours. Meanwhile, the effluent from the coke drum enter the lower part of the main fractionator to be separated into gas naphtha, light and heavy gasoils, and recycled oil. Since the coking reaction is endothermic, the vapors leaving the top of the coke drum are roughly 110 °F colder than the furnace coil outlet temperature.

While coke bed is being formed in one active drum, the other one is being removed by high pressure water jets.

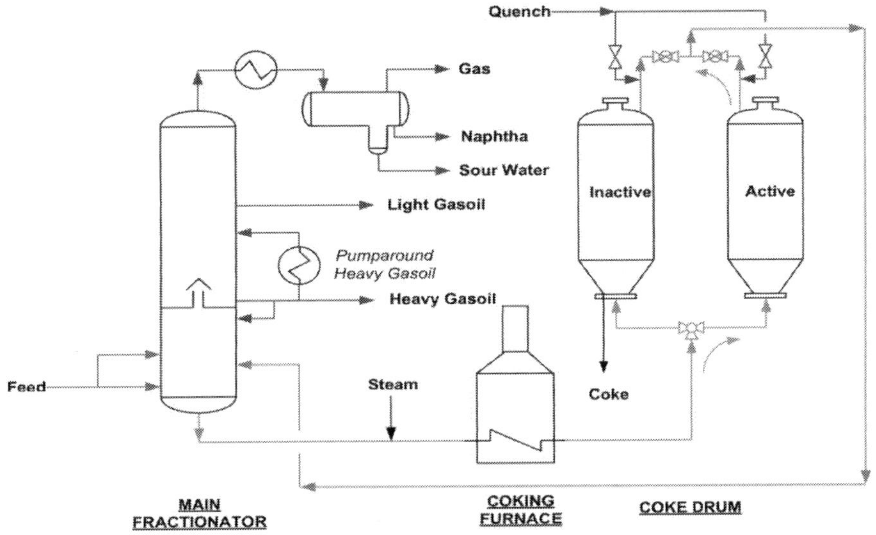

Figure 1. Simplified process flow diagram.

Development Mathematical Model of Thermal Cracking

Mass and energy balance of the system furnace-coke drum

Following are the main related mass and energy balance equations:

Boundary N. 1:

$$X_1 + S = X_0 + X_2 + X_3 + X_4 + X_5 + X_6 \quad \text{Equation (1)}$$

Boundary N. 2:

$$X_1 + S = Y_0 + Y_2 + Y_3 + Y_4 + Y_5 + Y_7 \quad \text{Equation (2)}$$

Outlet coil:

$$Y_i = \alpha_i \cdot X_i \quad \text{Equation (3)}$$

Boundary N. 3:

$$\Delta X_i = (1 - \alpha_i) \cdot X_i \quad \text{Equation (4)}$$

Boundary N. 1 and N. 2:

$$(\Delta H)_{IN} + Q_0 = (\Delta H)_{OUT} + (\Delta H)_{REACTION} + \sum(\Delta H)_{LOSS} \quad \text{Equation (5)}$$

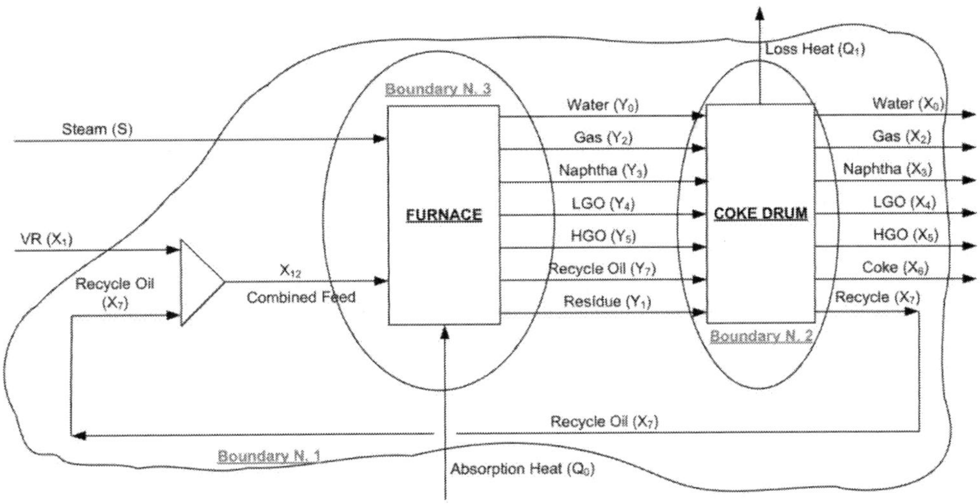

Figure 2. The boundaries for the mass and energy balance in the system.

Coking furnace

The furnace can be divided into two sections: pre-heating and reaction. The furnace reaction section can be considered as a tubular reactor steady-state with two-phase flow. The simultaneous resolution of the mass and energy balance, heat transfer and liquid-vapor equilibrium predicts the lumps composition through the coil.

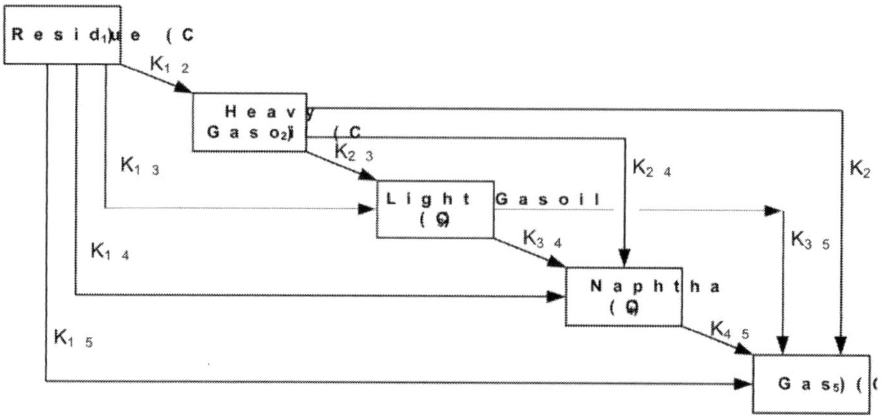

Figure 3. Proposed five lumped kinetic models with 10 rate parameters.

Given the lumps composition from mass and energy balance, it is possible to predict the furnace coil kinetics. The kinetic constants of the Arrhenius equation are calculated as function of the feed physical-chemical properties (1st order).

$$ln(K_{ij}) = ln(A_{ij}) - \frac{E_{ij}}{R \cdot T} \qquad \text{Equation (6)}$$

Coke drum

The coke drum is considered as a perfect continuous stirred tank reactor in pseudo-steady state. The kinetics is based on a thermal cracking model and it is divided into three phases: vapor, liquid and solid.

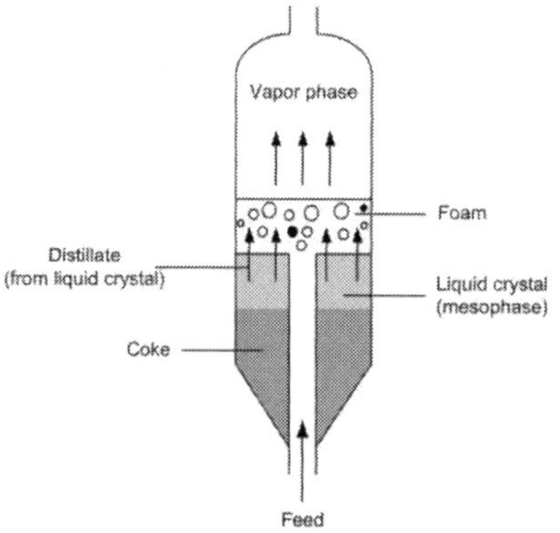

Figure 4. The coke drum filling system.

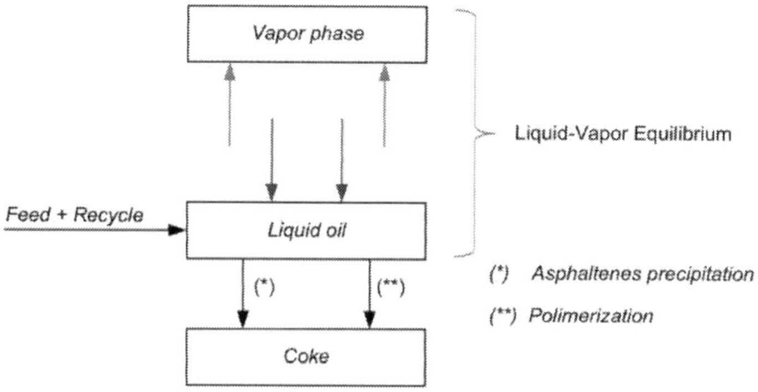

Figure 5. The coke drum modeling strategy

The coke formed from asphaltenes precipitation tends to be amorphous and it is a poor quality sponge coke, while the coke derived from polymerization and polycondensation tends to be more crystalline.

Results

Vacuum residue characterization

The vacuum residue, consisting of four fractions: saturates (*Sa*), aromatics (*Ar*), resins (*Re*) and asphaltenes (As), may be defined by the overall mass balance equation as:

$$VR = Sa + Ar + Re + As \qquad \text{Equation (7)}$$

Based on the results of SARA analysis, the Conradson carbon residue (CCR) was determined by the contribution of each fraction.

$$CCR = 0.1423 \cdot Ar + 0.2325 \cdot Re + 0.6774 \cdot As \qquad \text{Equation (8)}$$

Table 1 shows the H/C atomic ratio for the four fractions present in the vacuum residue (saturates, aromatics, resins and asphaltenes), obtained from various literature (Chen, et al., 2012; Zhao, et al., 2010). The H/C atomic ratio decreases in the order: saturates > aromatic > resins > asphaltenes.

Table 1. H/C atomic ratio.

	Saturates	Aromatics	Resins	Asphaltenes
Average	1.8657	1.4939	1.3844	1.1451
Maximum	1.9735	1.5950	1.4352	1.2343
Minimum	1.7349	1.3678	1.3249	1.0873

To determine the average H/C atomic ratio is applied a simple rule of mixture as shown in equation (9).

$$\left(\frac{H}{C}\right)_{VR} = \left(\frac{H}{C}\right)_{Sa} \cdot X_{Sa} + \left(\frac{H}{C}\right)_{Ar} \cdot X_{Ar} + \left(\frac{H}{C}\right)_{Re} \cdot X_{Re} + \left(\frac{H}{C}\right)_{As} \cdot X_{As} \qquad (9)$$

The hydrogen content and H/C atomic ratio are very important parameters for the heavy fractions processing. A higher hydrogen content or higher H/C atomic ratio is typically correlated with a higher liquid yield upgrading (Xu, et al., 2014). A simple way to determine the aromatic character of an oil fraction is to use the H/C atomic (Wiehe, 1992). This H/C atomic has a strong correlation with the aromaticity factor fa but has relatively weak correlation with molecular structure. The aromaticity factor was defined by Corbett (1964). Based on experimental data from various publications (Guo, et al., 2008; Zhao, et al., 2005; Yang, et al., 2005; Rogel, et al., 2003) was determined a relationship between aromaticity factor and the H/C atomic as shown in equation (10)

$$f_a = 1{,}1810 - 0{,}5911 \cdot \left(\frac{H}{C}\right) \qquad (R = -0{,}9930) \qquad \text{Equation (10)}$$

Effluent of the coke drum – True boiling point curve

Figure 6 shows the strategy to determine the recycled charactheristics while the Figure 7 shows the result from the mass and energy balance in the main fractionator bottom section and the overhead coke drum based on this strategy.

During the coke formation, in the coke drum a carbon concentration occurs, causing a decrease in the H/C atomic ratio, while there is an increase in the H/C ratio, from heavy gasolil to light end (Table 2).

Table 2. Lumps yields from effluent coke drum.

Lump	**Yield, LV%**	**H/C atomic ratio**
Gas	14.36	↑
LPG	6.48	
Naphtha	17.87	
Light Gasoil	27.61	
Heavy Gasoil + Recycled	33.68	

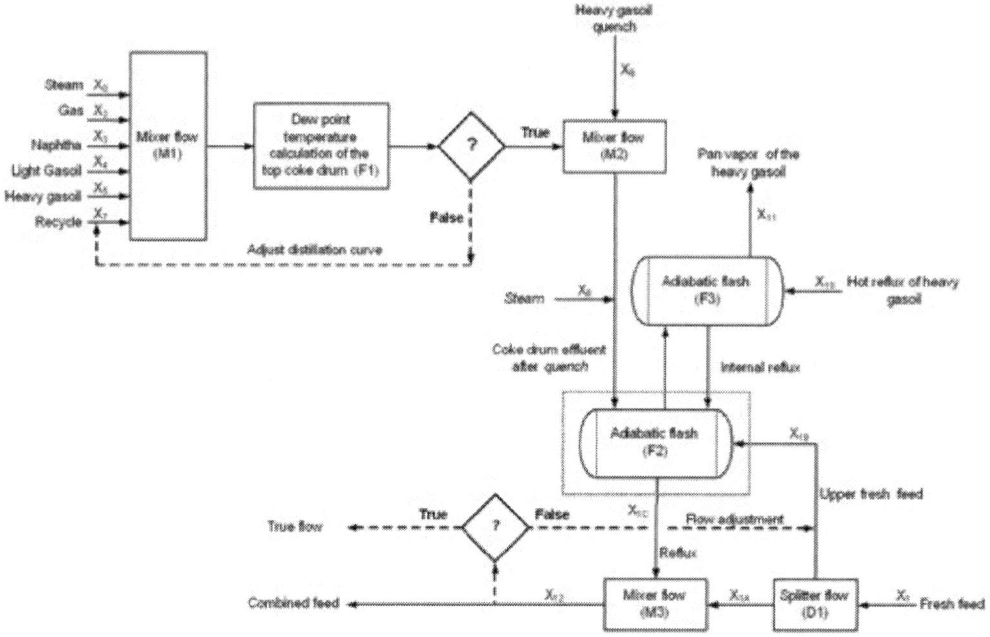

Figure 6. Strategy to determine the recycled characteristics.

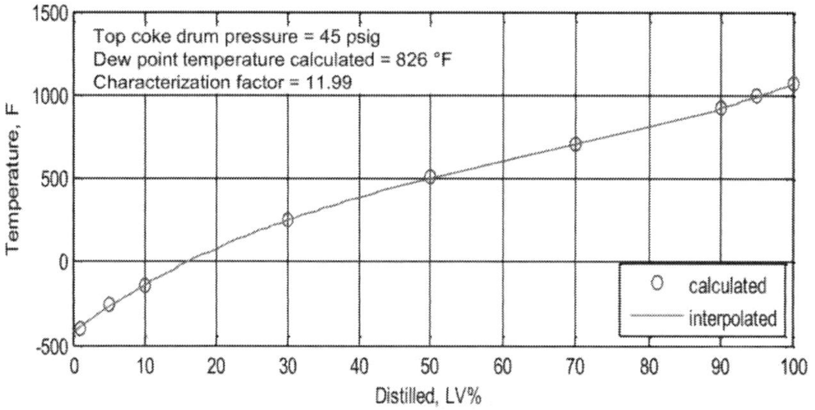

Figure 7. Effluent true boiling point distillation curve from the coke drum.

Coke bed

Petroleum coke contains a small amount of liquid hydrocarbons trapped in the pores, known as volatile combustible material (*VCM*), which is related to the severity degree of the process and feed characteristics. Typically, the volatile matter content varies from 8.0 to 14.0 weight%.

To determine estimate the total volatile matter content are proposed three distinct types: light (VCM_1), intermediate (VCM_2) and heavy (VCM_3).

$$VCM_{TOTAL} = VCM_1 + VCM_2 + VCM_3 \qquad \text{Equation (11)}$$

The Hardgrove Grindability Index (*HGI*) is influenced by the morphology and volatile matter of the coke bed, as well as, carbon, hydrogen and impurities content. This affects the hardness, where a high value of HGI is normally an indication of a soft coke while a hard coke usually scores a low value of HGI (Joshi, 1979). The value of HGI can be estimated from asphaltenes and Conradson Carbon Residue contents weight ratio fresh feed using equation (12).

$$\left(\frac{HGI}{VCM}\right) = a - b \cdot \left(\frac{As}{CCR}\right) \qquad \text{Equation (12)}$$

Moreover, the HGI can be influenced by turbulence and viscosity of the liquid phase.

Discussion

✓ Using molecular information allows to build a rigorous mathematical model consistent with the phenomena observed in the actual industrial unit.

✓ Only aromatics, resins and asphaltenes have significant contribution for the Conradson carbon residue, being asphaltenes the most significant.

✓ When coke is formed from low asphaltene content vacuum residue, the value of the Hardgrove Grindability Index (HGI) tends to increase due to the influence of turbulence and low viscosity of the reaction system.

✓ The H/C atomic ratio is important when processing heavy fractions. Decreasing this ratio causes increasing of the coke and gas production. The H/C of the naphtha, light gasoil, heavy gasoil and coke are 2.10, 1.66, 1.51, and 0.53, respectively.

✓ The correlations to determine the molecular weight, viscosity and critical properties of heavy gasoil, recycle, and vacuum residue present divergent results (N.B.P. > 1,000 °F).

References

Chen, K. et ali, Sutdy of the thermal performance and interaction of petroleum residue fractions during the coking process, Energy & Fuels 2012, 26, 6343 - 6341.

Corbett, L. W., Densimetric method for characterizing asphalt, Analytical Chemistry 1964, 36 (10), 1967 - 1971.

Ellis, P. J. et. ali. Tutorial: Delayed Coking Fundamentals, AIChE 1998 Spring National Meeting, International Conference on Refinery Processes Topical Conference Preprints, Session 29, Paper 29a, 1998, New Orleans, p. 151 – 169.

Guo, A. et. ali., Simulated delayed coking characteristics of petroleum residues and fractions by thermogravimetry, Fuel Processing Technology 2008, 89, 643 - 650.

Joshi, N. R., Relative grindability of bituminous coals on volume basis, Fuel 1979, 58, pp. 477 - 478.

León, A. Y. et. ali. Estimation of Critical Properties of Typical Colombian Vacuum Residue SARA Fractions, Ciencia, Tecnologia y Futuro 2008, 3(4), 129 – 142.

Riazi, M. R., Characterization and properties of petroleum fractions. 1st ed. Philadelphia: ASTM manual series: MNL50, p. 407, 2005.

Rogel E. and Carbognani, L., Density estimation of asphaltenes using molecular dynamics simulations, Energy & Fuels 2003, 17, 378 - 386.

Wiehe, I. A., A solvent-resid phase diagram for tracking resid conversion, Ind. Eng. Chem. Res. 1992, 31, 530 - 536.

Xu, Z. et ali, Detailed characterization of virgin heavy oil resid and its thermally cracked resid, Energy & Fuels 2014, 28, 1664 - 1673.

Yang, C. et ali., Hydroconversion characteristics and kinetics of residue narrow fractions, Fuel 2005, 84, 675 - 684.

Zhao, S. et ali, China refinery tests asphaltenes extraction process, Oil & Gas Journal, April 5, 2010.

Zhao, S. et ali., Systematic characterization of petroleum residua based on SFEF, Fuel 2005, 84, 635 - 645.

Acknowledgements

Authors would like to acknowledge CAPES/PROEX and PETROBRAS for financial and technical support.

A Framework for Stochastic Modelling and Optimisation of Chemical Engineering Processes

Usman Abubakar, Srinivas Sriramula and Neill C. Renton

Lloyd's Register Foundation (LRF) Centre for Safety & Reliability Engineering, University of Aberdeen, UK.

This paper presents a new framework, termed "Stochastic Process Performance Modelling Framework (SPPMF)", which combines traditional deterministic process simulation, response surface modelling techniques and advanced structural reliability analysis methods to facilitate efficient performance modelling and optimisation of chemical process systems under uncertainties. Stochastic constraints have been added to the conventional process optimisation formulation as indicated below:

$$\boldsymbol{Min/max}\ f^{\circ}\left(\underline{d}, \underline{c}, \underline{\psi}, \underline{x}\right)$$

$$\boldsymbol{subject\ to}\text{: } \underline{h}\left(\underline{d}, \underline{c}, \underline{\psi}, \underline{x}\right) = 0\ ;\ \underline{g}\left(\underline{d}, \underline{c}, \psi, \underline{x}\right) \leq 0$$

$$\boldsymbol{Min/max}\ P = \int_{g(\underline{d}, \underline{c}, \underline{\psi}, \underline{x}) \leq 0} f\left(\underline{d}, \underline{c}, \underline{\psi}, \underline{x}\right).d\underline{x}$$

$$x_i \in R_i$$

where $\underline{\psi}$, $\underline{d}$, $\underline{c}$ and $\underline{x}$ are the vectors of state variables, equipment sizes/dimension specifications, control variables and uncertain parameters respectively. While the joint probability density function (pdf) is represented by $f(.)$; $f^{\circ}(.)$ is the objective function, $\underline{h}$ and $\underline{g}$ are the vectors of equality and inequality constraints respectively. Each of the parametric ranges R_i is characterised by a pdf D_i while P is the probability value.

Both First Order Reliability Method (FORM) and Monte Carlo Simulation (MCS) are then applied to gain a wide range of performance measures. A number of other Structural Reliability Analysis (SRA) concepts have been adopted and cross-applied to chemical engineering processes. For example, while SRA is mainly concerned with the effects of random forces and mechanical properties on structural performance, Process Reliability Analysis is focused on random process conditions (e.g. changes in pH, reaction rates, etc) and their effects on both product quantity and quality. Finally, SPPMF has been successfully applied to model stochastic properties of a range of typical process systems. The results show that the new framework can be efficiently implemented in process engineering with significant benefits over the traditional methods.

387274 The Integration of Data and Models for the Optimization of Integrated Biorefineries

Abstract:

The great advancements in the field of biorefineries have resulted to a vast amount of models of different scale. The models include conceptual, detailed, experimental, pilot, cost analysis, short-cuts, supply chain, LCA creating a huge database that remains unstructured and disjoint. Additionally, the various models are based on heterogenous data sources and softwares. The paper addresses the challenge of collocating all the models and available knowledge in a common platform with perspective of model integration. The approach proposes a Java based platform which manages decision-making, extracts and distributes data selectively to systematize the development and execution of optimization models for integrated biorefineries. Automation of the models development using natural language is important to render the platform accessible also by non-experts of the field hiding the complex mathematical formulas on the background. The combination and comparing of inputs-outputs of different models (GAMS, MATLABL, Aspen) to serve different purposes is coordinated through the platform providing a more accurate and complete viewing of the integrated biorefineries. Ontologies are employed to succeed the selective extraction of data sets which fill the models context. Furthermore, the dynamic framework established translates the user input through successive queries and allocation algorithms directly to models ready to provide the targetted outcome without further modifications. The features of dynamic configuration of data (ontologies) and the visualization of the results conclude the objectives of the approach. Implementations keep extending and currently include the automated development of models (GAMS, MATLAB) that optimize chemical superstructures and the environmental assessment of waste for integrated biorefineries.

ANALYSIS OF THE ANAEROBIC DIGESTION PROCESS USING COMPUTATIONAL FLUID DYNAMICS

V. Hernández[a], E. Castro[b], C.A. Cardona[a]

[a] *Instituto de Biotecnología y Agroindustria, Departamento de Ingeniería Química. Universidad Nacional de Colombia Sede Manizales*

[b] *Department of Chemical, Environmental and Materials Engineering, Agrifood Campus of International Excellence (ceiA3), University of Jaén, Spain*

14AIChE Annual Meeting, Atlanta, GA

Abstract

Anaerobic digestion is a complex multi-stage process involving physical, biochemical and fluid dynamic processes. It is a biological conversion process in which microorganisms transform the organic matter into a gaseous mixture mainly composed of methane and carbon dioxide and a digestate or slurry which is rich in ammonium and other nutrients through the concerted action of a close-knit community of bacteria. The primary objective of anaerobic digestion is the stabilization of organic matter, with a concurrent reduction in odors, pathogen concentration, and mass of solid organic material. It includes a series of biochemical processes by different microorganisms to degrade organic matter under anaerobic conditions. One of the main variables in anaerobic digestion for providing an optimum performance is a proper mixing. It provides intimate contact between the feed sludge and active biomass, yielding uniformity of temperature and substrate concentration throughout the digester, and preventing both the formation of surface scum layers and the deposition of sludge on the bottom of the tank. The most efficient mixing device in terms of power consumed per mixed volume is the mechanical mixer. In this work, the anaerobic digestion process was modeled and simulated creating a 2-D model in the computational tool COMSOL Multiphysics.

Introduction

According to the United Nations the world population could reach 9.15 billion persons by 2050. Despite of the benefits of using agrochemicals in agriculture practices, the excessive use and poor application methods of chemical fertilizers can cause unwanted side effects, such as the environmental depletion as well as its volatile price.

As an alternative to counteract these environmental and economic problems, different efforts have been developed in order to obtain biofertilizers form lignocellulosic residues from agroindustrial processes. The biofertilizers are defined as substances obtained from organic matter that when applied to seed, plant surfaces or soil promotes growth by supplying primary nutrients to the plant. In this work the case of the production of two types of biofertilizers: a liquid fertilizer of foliar application and a solid fertilizer to be applied on the field.

One of the main variables in anaerobic digestion for providing an optimum performance is a proper mixing. It provides intimate contact between the feed sludge and active biomass, yielding uniformity of temperature and substrate concentration throughout the digester, and preventing both the formation of surface scum layers and the deposition of sludge on the bottom of the tank. The most efficient mixing device in terms of power consumed per mixed volume is the mechanical mixer.

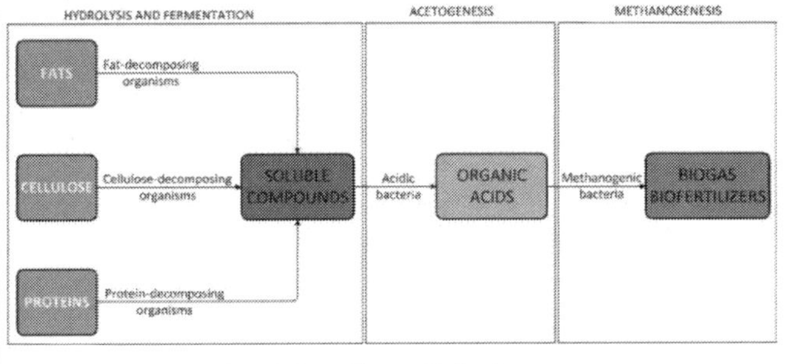

Figure 1. Stages of the anaerobic digestion process

Methodology

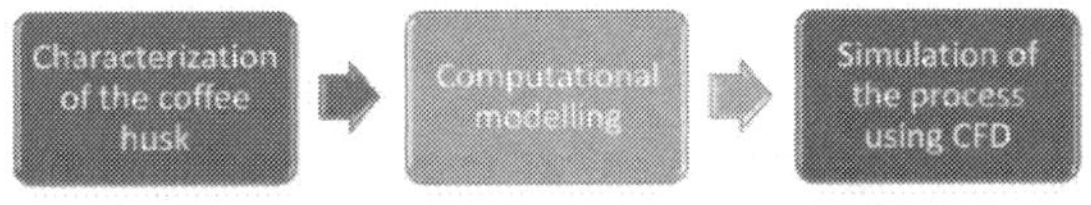

Results

Characterization of the raw material

Parameter	Percentage
Cellulose	34.38
Hemicellulose	25.55
Lignin	24.38
Extractives	11.51
Ash	0.65
O-acilo group	3.52
Nutrients	
Nitrogen (%)	0.933
Phosphorous (ppb)	1.043
Potash (ppb)	2,334.45

Moisture content 10.5%

Modelling

Computational modelling

Continuity equation

$$\frac{\partial\rho}{\partial t}+\frac{\partial(\rho u)}{\partial x}+\frac{\partial(\rho v)}{\partial y}=0$$

Momentum equations

$$\rho\left(\frac{\partial u}{\partial t}+u\frac{\partial u}{\partial x}+v\frac{\partial u}{\partial y}\right)=-\frac{\partial p}{\partial x}+\frac{\partial}{\partial x}\left(\mu\frac{\partial u}{\partial x}\right)+\frac{\partial}{\partial y}\left(\mu\frac{\partial u}{\partial y}\right)+F$$

$$\rho\left(\frac{\partial v}{\partial t}+u\frac{\partial v}{\partial x}+v\frac{\partial v}{\partial y}\right)=-\frac{\partial p}{\partial y}+\frac{\partial}{\partial x}\left(\mu\frac{\partial v}{\partial x}\right)+\frac{\partial}{\partial y}\left(\mu\frac{\partial v}{\partial y}\right)+\rho g+F$$

Energy equation

$$\frac{\partial T}{\partial t}+u\frac{\partial T}{\partial x}+v\frac{\partial T}{\partial y}=\frac{k}{\rho c}\left[\frac{\partial}{\partial x}\left(\frac{\partial T}{\partial x}\right)+\frac{\partial}{\partial y}\left(\frac{\partial T}{\partial y}\right)\right]$$

Kinetic model of the anaerobic digestion

$$-\frac{dS_{VSS}}{dt}=k_1S_{VSS}$$

$$\frac{dS_{VDS}}{dt}=k_1S_{VSS}-\frac{k_2S_{VDS}}{k_3+S_{VDS}}$$

$$\frac{dS_{TVA}}{dt}=\frac{k_4S_{VDS}}{k_5+S_{VDS}}-\frac{k_6S_{TVA}}{k_7+S_{TVA}}$$

$$\frac{dS_M}{dt}=\frac{k_6S_{TVA}}{k_7+S_{TVA}}$$

k_1=0.054 day^{-1}
k_2=4.2 g/l
k_3=9.8 g/l
k_4=3.6 g/l
k_5=10.2 g/l
k_6=4.3 g/l
k_7=3.1

VSS	Volatile suspendedsolids
VDS	Volatile disolved solids
TVA	Total volatile acids
M	Methane

Results

Temperature profiles

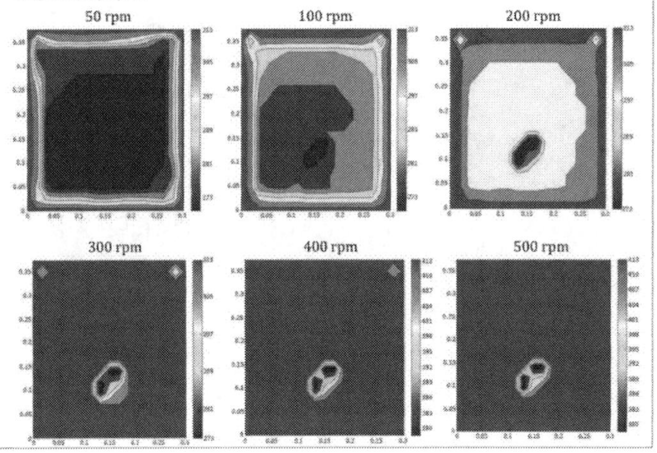

Conclusions

In this work the mixing performance of a stirrer biodigester was simulated using a 2-D model. According to the results, to increase the impeller rotational speed improves the heat distribution increasing the homogeneity into the digester.

Acknowledgements

The authors express their gratitude to the Universidad Nacional de Colombia at Manizales, to the Investigation and extension direction, from the Engineering and Architecture Faculty from Universidad de Colombia Sede Manizales, to the Design of Sustainable Biorefineries (code 157) and Chemical, catalytic and biotechnological Processes (code 16081) projects from the Investigation Direction at Manizales (DIMA), to the Administrative Department of Science, Technology and Innovation (COLCIENCIAS), to the Agrifood Excellence Campus (ceiA3) in University of Jaén and to the Agrifood International Doctorate School (eidA3) for funding this work. Financial support from Junta de Andalucía (Proyecto de Excelencia AGR-6103) is also gratefully acknowledged. JMRG thanks his scholarship to Junta de Andalucía.

References

Borja, R., A. Martín, et al. (2005). "Kinetic modelling of the hydrolysis, acidogenic and methanogenic steps in the anaerobic digestion of two-phase olive pomace (TPOP)." Process Biochemistry **40**(5): 1841-1847.
Latha, S., D. Borman, et al. (2009). CFD Multiphase Modelling for Evaluation of Gas Mixing in an Anaerobic Digester. 14th European Biosolids and Organic Conference and Exhibition. Leeds, UK, .
Li, Y., U. Merrettig-Bruns, et al. (2014). "Optimization of ammonia pretreatment of wheat straw for biogas production." Journal of Chemical Technology & Biotechnology: n/a-n/a.
Wu, B. (2012). "Integration of mixing, heat transfer, and biochemical reaction kinetics in anaerobic methane fermentation." Biotechnology and Bioengineering **109**(11): 2864-2874.
Wu, B. (2012). "Large eddy simulation of mechanical mixing in anaerobic digesters." Biotechnology and Bioengineering **109**(3): 804-812.

Simulation Based Optimization of Stirred Tank Reactors

Extended Abstract for the 2014 Annual AIChE Conference

Yuvraj Dewan[*], Ashish Newale[†], Ravindra Aglave[‡] and Thomas Eppinger[§]

CD-adapco

May 12, 2014

The design of a stirred tank reactor is a multi-input and multi-output problem. Design parameters may include geometric variables (impeller geometry, baffle geometry etc.) and operating conditions (impeller RPM, fill level etc.). Some common design objectives are minimizing mixing time, maximizing mixing quality, achieving a specific scale of agitation etc. Additionally, various constraints can be imposed on the performance of a reactor, such as avoiding overflow, avoiding mixing 'dead zones', imposing a maximum shear stress etc. Designing to these multiple objectives and constraints by modifying a large number of design parameters is a challenging task since the relationships of these multiple inputs and outputs can be highly non-linear. Therefore, the design of stirred tank reactors has traditionally been a very repetitive and time-consuming task. This is the primary motivation for developing a design methodology that is both autonomous, and efficient in targeting specific reactor performance.

Optimization algorithms lend themselves well to solve such multi-input, multi-output problems. An optimization algorithm iteratively runs evaluations and understands the sensitivities between all inputs and combination of inputs to all outputs. The algorithm successively guesses a better design until a converged design is reached that optimizes all the required objectives while satisfying all constraints. This study details the implementation of this optimization methodology to design optimal stirred tank reactors for various design objectives and constraints.

The use of an optimization algorithm for design studies requires two components: an analysis tool and an optimization algorithm. The optimization algorithm uses the analysis tool to generate the relationship between a set of inputs and outputs. Figure 1 graphically represents the implementation of the analysis tool within the optimization process.

The analysis tool for this study is STAR-CCM+, which is a multi-physics simulation platform provided by CD-adapco. The reactor is analyzed using a Computational Fluid Dynamics (CFD) approach. The use of STAR-CCM+ for mixing tank performance prediction has been validated by the work of Maaß et al [1]. The stirred tank reactor geometry was created to be fully parametric with 22 different geometric parameters. These parameters deal with overall tank geometry, impeller

[*]Application Engineer; yuvraj.dewan@cd-adapco.com
[†]Application Support Engineer; ashish.newale@cd-adapco.com
[‡]Director: Chemical Processing Industry; ravindra.aglave@cd-adapco.com
[§]Technical Specialist: Chemical Processing Industry; thomas.eppinger@cd-adapco.com

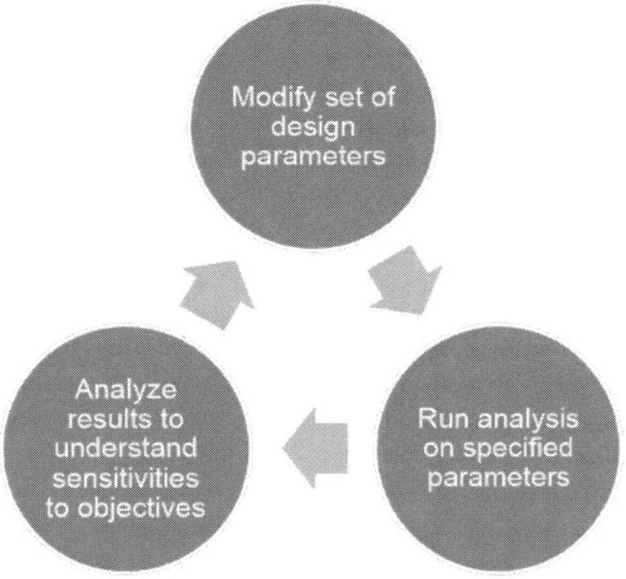

Figure 1: Schematic of optimization process

geometry, impeller blade geometry and baffle geometry. A polyhedral computational mesh with prism layers on all walls was used. Depending on the design objectives, two types of simulations were run, steady and transient. The steady state (time averaged) cases use a Moving Reference Frame (MRF) approach, which solves the fluid equations in a rotating reference frame around the impeller. The transient (time accurate) cases use a Rigid Body Motion (RBM) approach, which physically rotates computational mesh for the spinning region. For all of the simulations, a passive scalar approach is taken to track the quantity of mixing throughout the reactor.

The highly efficient SHERPA algorithm was used to drive the optimization study [2]. The acronym SHERPA stands for Simultaneous Hybrid Exploration that is Robust, Progressive and Adaptive. During a single search, SHERPA uses multiple search methods simultaneously (not sequentially). This approach takes advantage of the best attributes of each method and reduces a method's participation in the search if/when it is determined to be ineffective. From a usability standpoint, the SHERPA algorithm is extremely easy to use since it only has a single tuning parameter (total number of evaluations), as compared to traditional algorithms which require a high level of expertise to apply.

The authors have previously used the same methodology and tools to generate some preliminary results [3]. This study will extend the previous work by running optimization cases with more rigid requirements and constraints. This will serve to highlight the applicability of this method for cases in which manual designer intuition can be extremely difficult. Finally, some data will be reduced about the trends observed during the optimization process which will help gain further insight into the behavior of the generic mixing tank reactor system.

References

[1] **Maaß, S., Eppinger, T., Altwasser, S., Rehm, T., and Kraume, M.**, "Flow Field Analysis of Stirred Liquid-Liquid Systems in Slim Reactors", *Chemical Engineering and Technology*, **34**, pp. 1215-1227 (2011)

[2] **Red Cedar Technology**, "SHERPA - An Efficient and Robust Optimization/Search Algorithm", *http://www.redcedartech.com/pdfs/SHERPA.pdf*

[3] **Newale, A., Dewan, Y., Aglave, R., Eppinger, R.**, "Simulation Based Optimization of Stirred Tank Reactors", presented at the 2014 AIChE Spring National Meeting, New Orleans, LA (March 30-April 3, 2014)

Dynamic Analysis of Offshore Natural Gas Conditioning: TEG Dehydration, Joule-Thomson Expansion and Membrane Separation

Lara de O. Arinelli, Federal University of Rio de Janeiro, RJ, Brazil
Ofélia Q. F. Araújo, Federal University of Rio de Janeiro, RJ, Brazil
José L. de Medeiros, Federal University of Rio de Janeiro, RJ, Brazil

Abstract

Natural gas extracted from reservoirs may contain a significant amount of contaminants that have to be removed before the gas transportation and/or use in order to prevent damage to pipeline and equipment. Water and hydrocarbon dew point adjustments are the main processes on offshore gas conditioning, but depending on the gas composition, other treatments may be used. The Brazilian Pre-Salt reservoirs are known to have a high CO_2 content, which is a challenge for the oil and gas industry. Offshore processing is inherently related to flow and pressure oscillations from production risers that may cause serious operational problems, losses and environmental and safety issues. Therefore, the control system must be sufficiently robust to ensure the stability of the process. This paper approaches the dynamic behavior of an offshore natural gas treatment plant with the following processes: dehydration by TEG absorption, Joule-Thomson expansion for hydrocarbon dew point adjustment and membrane permeation for CO_2 separation. The transient responses of this plant were analyzed under three dynamic flow patterns in the production riser – pulse, sine wave and slug flow – and the overall robustness is tested. The results confirm a satisfactory control of gas production and specifications. As expected, the slug flow perturbation has more impact on the process due to its high frequency severe oscillations.

Introduction

Natural gas (NG) is composed mainly by methane and is considered a cleaner fuel than oil because it emits lower rates of greenhouse gases. It has an increasing participation in worldwide energy supply and in Brazil it represented 12.8% in 2013, against 11.5% in the previous year (9). The 2008 Pre-Salt discoveries introduced a new horizon and challenges for the brazilian oil and gas industry. The expectation back then was to double the national production capacity by 2020 and until now it has been proving its potential (6). The typical Pre-Salt production profile exhibits high gas-oil ratio (GOR) with high CO_2 content (5). In order to ensure secure flow, handling and storage conditions, NG is treated offshore before it can be used as fuel or transported to onshore facilities. Therefore, there are many studies focusing on NG conditioning processes applicable to offshore platforms with CO_2 capture technologies. Among them, membrane permeation (MP) has been chosen as adequate for NG with high CO_2 content. MP is simple and easy to operate and has low footprint and weight when compared to other technologies. MP was selected for CO_2 separation on Pre-Salt platforms operating on Lula field.

Offshore processes are constantly susceptible to pressure and flow oscillations in the riser systems. Such disturbances are associated with slug flow regimes in risers and can cause operational problems on the platforms, eventually leading to non-programmed stops. If a shutdown occurs, startup procedures are initiated and they are generally associated with emission of greenhouse gases on flares, affecting the environment. The platform overall safety may also be compromised due to instabilities and risk scenarios. In addition, maintenance and production losses reflect on high costs for the oil and gas industry.

Slug flow is an intermittent multiphase flow pattern that is very common in production risers, where gas and liquid flow simultaneously. In offshore processes, the slug flow can cause a lot of trouble and damage due to its severe flow and pressure oscillations (12).

Dynamic simulation allows the analysis of the process behavior under system perturbations. It also enables elaboration of control strategies, commissioning of controllers, simulation of startup and shutdown procedures, support for safety studies, operators training, besides other applications. Therefore, it is an essential tool for both conceptual project and operation steps of offshore processes.

The objective of this project is to develop and analyze the dynamic behavior of a conventional NG conditioning plant with the following operations: initial three-phase separation, triethyleneglycol (TEG) absorption for NG dehydration, Joule-Thomson expansion (JT) for hydrocarbon dew point adjustment (HCDPA) and membrane permeation (MP) for CO_2 removal.

Methodologies

The simulation software used for both steady and dynamic states is Aspentech HYSYS V8.4 with Peng-Robinson Equation of State Package, which is appropriate for NG systems. For MP response, a HYSYS dynamic extension was developed using Visual Basic 6.0, based on a steady state HYSYS User Unit Operation (HUO) code developed by H2CIN Laboratory at UFRJ (14).

A base case was first simulated in steady state. Its parameters were analyzed and selected using the HYSYS Case Study tool. Before changing to dynamic mode, equipment sizing was accomplished by literature correlations and HYSYS sizing tools. In addition, the control system was proposed and tuned in accordance with literature data. In dynamic mode, three scenarios of feed flow oscillation were simulated: pulse perturbation, sine wave and slug flow.

Conceptual Project

Assumptions

- ✓ Inlet multiphase stream at 960 t/h, 75 bar and 40°C. Its composition (Table 1) was adapted from Vaz, 2009 (13) work with the purpose of representing a Pre-Salt feed with a moderately high CO_2 content (~20% molar base on gas phase – similar to Lula field, Santos Basin) (5).

Table 1. Inlet Stream Composition (% mol.)

Inlet Stream Composition (% mol.)					
Component	**Composition**	**Component**	**Composition**	**Component**	**Composition**
Methane	33.45%	i-Pentane	0.43%	n-C13	0.43%
Ethane	4.28%	n-Pentane	0.09%	n-C14	0.09%
Propane	2.95%	n-Hexane	0.17%	n-C15	0.17%
CO_2	10.32%	n-Heptane	1.12%	n-C16	1.12%
Nitrogen	0.22%	n-Octane	1.54%	n-C17	1.54%
Water	11.92%	n-Nonane	1.36%	n-C18	1.36%
H_2S	0.01%	n-Decane	1.36%	n-C19	1.36%
i-Butane	1.44%	n-C11	1.19%	C20+	1.19%
n-Butane	0.21%	n-C12	1.18%	TEG	1.18%

- ✓ The regulatory specifications considered for the final NG are defined by ANP (Brazil National Agency of Oil, Gas and Biofuels) in the 2008 n. 16 resolution (1) (Table 3).

Table 2. Brazilian regulatory specifications for natural gas

Property	Unit	Value
max. WDP, at 1.01 bar	°C	-45
max. HCDP, at 45 bar	°C	0
max. CO_2 content	% mol.	3.0

✓ Export gas pressure of 200 bar (8).
✓ CO_2 rich stream final destination: Enhanced Oil Recovery (EOR) at 250 bar.

Base Case

The overall process, divided by operational areas, can be seen on Figure 1.

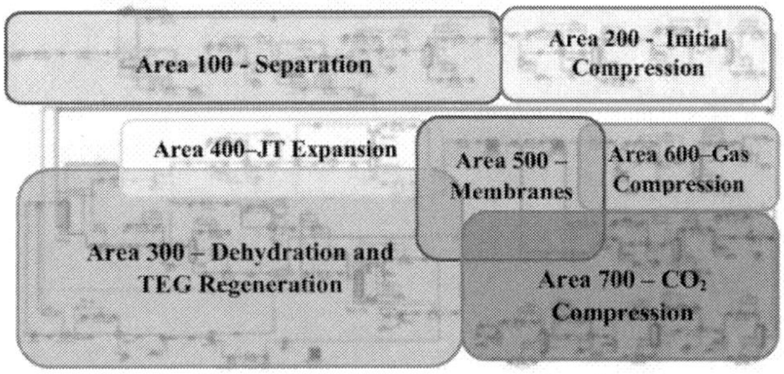

Figure 1. Natural gas conditioning process flowsheet

- Separation (Area 100): For the inlet stream phase separation, a three-phase and a two-phase horizontal vessels were used (Figure 2). The stream is first expanded to reach the V-100 operating pressure (10.5 bar). The outlet oil stream has its pressure reduced to 4.5 bar and is then cooled to meet the two-phase separator conditions. The gas outlet from V-101 is compressed and mixed with the V-100 gas stream. Oil and water separated in this area are designated to their treating processes.

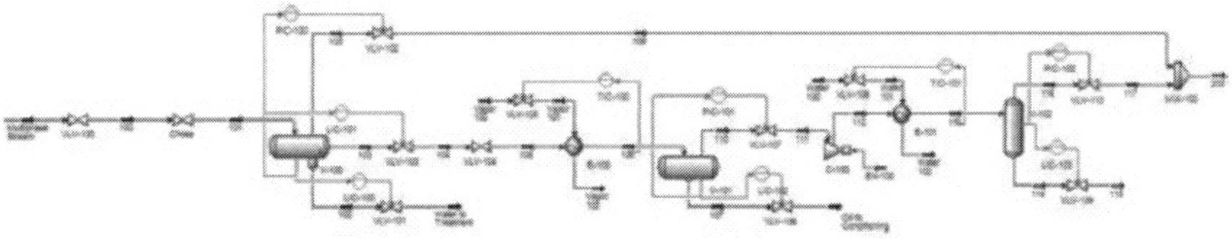
Figure 2. Separation (Area 100)

- Compression (Area 200): Compressors are common equipment on gas systems. Multistage centrifugal compressors are used to meet high pressure targets with all stages using the same compression ratio for better energy usage. An intercooler and a knockout vessel are prescribed after each compressor stage to prevent damage on downstream equipment. In Area 200, NG passes through a 2-stage compressor to reach the absorber column operating pressure (Figure 3).

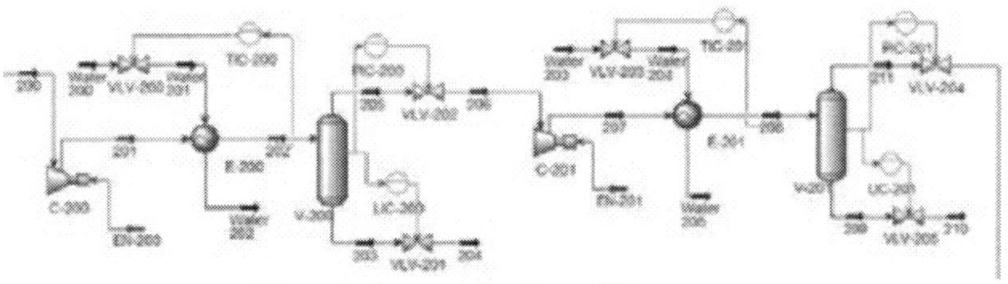
Figure 3. Compression (Area 200)

- Dehydration and TEG Regeneration (Area 300): Even after the phase separation area, NG still contains saturation water. Under high pressure and/or low temperature, hydrates can be formed creating troubles to pipeline and equipments. Hence, NG must be dehydrated before transportation or use. TEG absorption process is used because it presents high efficiencies and easy solvent regeneration. As can be seen in Figure 4, NG is admitted at the bottom of the absorber while countercurrent lean TEG is admitted at the top. The absorber is filled with structured packing and operates at approximately 60 bar. Rich TEG stream (301) is expanded and pre-heated before entering the flash tank for removal of residual gas. It is then heated in a glycol-glycol heat exchanger (E-301) with the regenerated TEG for energy recovery. Rich TEG proceeds to the top of the regeneration column, while a stripping gas stream is added at the reboiler to help the regeneration process. The regenerator is an atmospheric random packed column with refluxed condenser and reboiler. The condenser is used to minimize TEG loss at the top of the column. The reboiler temperature must be controlled at 204°C in order to prevent TEG degradation above 206°C. A TEG make up stream is added to compensate TEG losses. The dehydrated NG leaving the absorber has a water dew point (WDP) of -45.27°C at 1.01 bar in accordance with the specifications.

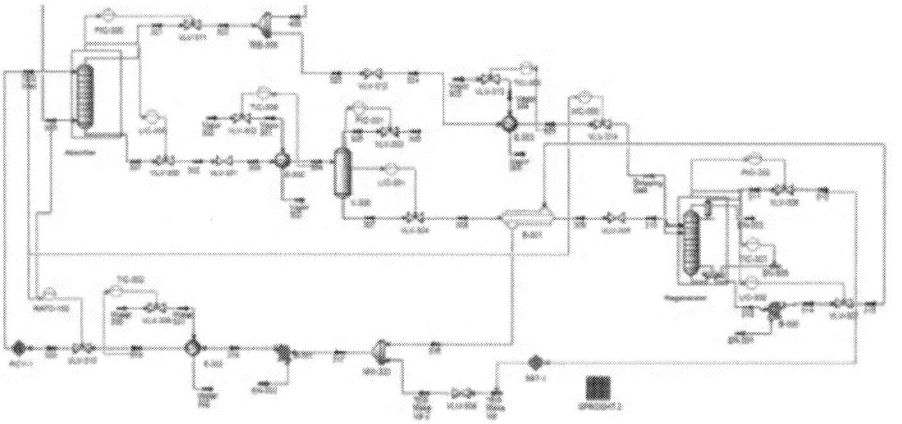

Figure 4. Dehydration and TEG Regeneration (Area 300)

- Joule-Thomson Expansion (Area 400): This area promotes HCDPA by removing condensables, also called natural gas liquids (NGL). The JT process was selected due to its simplicity and easy operation. As shown in Figure 5, the feed stream is first compressed, then pre-cooled on E-400, and cooled in a gas-gas countercurrent exchanger with the treated gas (E-401). The gas is then expanded in an isenthalpic valve to about 52 bar, reaching approximately -10°C and enters the low temperature separation vessel, where NGL is removed at the bottom. The HCDP of stream 409 is -10.94°C at 45 bar. This over specification is necessary because MP increases HCDP.

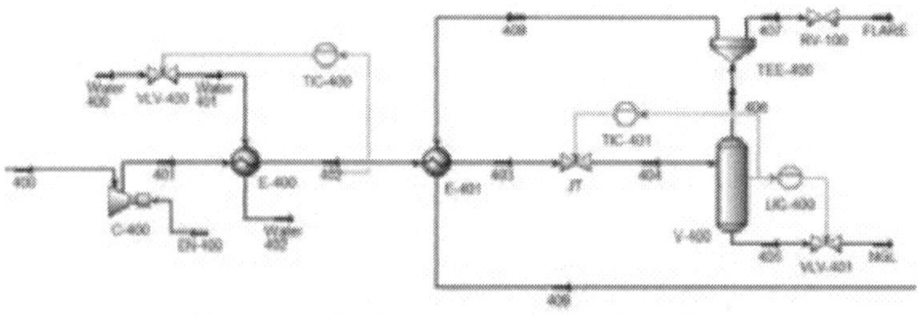

Figure 5. Joule-Thomson Expansion (Area 400)

- Membrane Separation (Area 500): When combined with water, CO_2 is highly corrosive, so its presence in NG may cause problems during gas processing and handling. It also reduces the NG calorific value and the transportation capacity of pipelines (4). A two stage MP system was used with specification of 187.000 m² of area each (Figure 6). Each MP stage produces a retentate (with lower CO_2 content) and a CO_2 rich permeate. The retentate from stage 1 (503) proceeds to stage 2, whose retentate (507) attains 2.97% mol of CO_2 – within the required specification. The two permeate streams (501 and 505) are mixed and sent to Area 700.

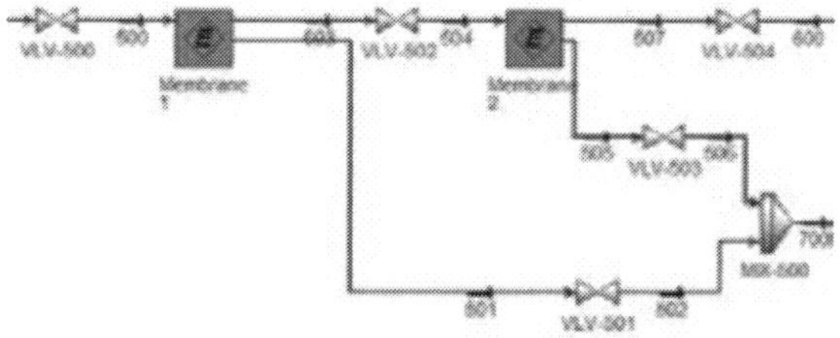

Figure 6. Membrane Separation (Area 500)

- Final Compressions (Areas 600 and 700): In Area 600 the treated NG is compressed in two stages to the export pressure of about 200 bar (Figure 7a). Area 700 compresses the final permeate stream from 1 bar to near 250 bar through five stages and sent it to EOR application (Figure 7b). Tables 3 and 4 show the final gas stream condition and composition.

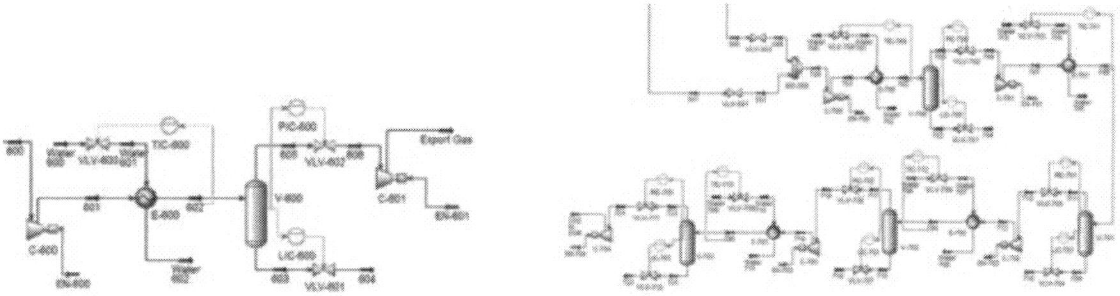

(a) (b)

Figure 7. Final Compressions: Gas - Area 600 (a) - and CO_2 - Area 700 (b)

Table 3. Export Gas Stream Composition (% mol.)

Export Gas Composition (% mol.)					
Component	**Composition**	**Component**	**Composition**	**Component**	**Composition**
Methane	79.19%	i-Pentane	0.15%	n-C13	0.00%
Ethane	10.29%	n-Pentane	0.02%	n-C14	0.00%
Propane	5.12%	n-Hexane	0.01%	n-C15	0.00%
CO_2	2.97%	n-Heptane	0.01%	n-C16	0.00%
Nitrogen	0.52%	n-Octane	0.01%	n-C17	0.00%
Water	0.00%	n-Nonane	0.00%	n-C18	0.00%
H_2S	0.02%	n-Decane	0.00%	n-C19	0.00%
i-Butane	1.53%	n-C11	0.00%	C20+	0.00%
n-Butane	0.17%	n-C12	0.00%	TEG	0.00%

Table 4. Export Gas Stream Conditions

Condition	Value
Mass Flow (t/h)	49.71
Temperature (°C)	94.29
Pressure (bar)	204.07
WDP (°C), at 1.01 bar	-54.48
HCDP (°C), at 45 bar	-4.80

Dynamic Analysis

Control System

Process control on offshore platforms is usually classic with a large number of simple loops with PID controllers in SISO (Single Input Single Output) configuration. The controllers for the present process were selected in accordance with literature data (3, 10, 11, 13).

Control parameters suggested by HYSYS Dynamic Modeling Manual (2) and by Luyben, 2002 (7) were used as base values for the controllers initial tuning (Table 5). The process was able to initialize in dynamic state with the defined parameters with no major problems. Some instabilities had to pass through a manual fine-tuning in order to reduce response oscillations.

Table 5. Control parameters for initial tuning

Controlled Variable	Kc	Ti	Td
Level	2.0	1.0	-
Pressure	2.0	2.0	-
Temperature	2.0	2.0	-
Composition	0.4	0.3	-

Dynamic Scenarios

Pulse perturbation: positive and negative pulses with 20% of amplitude simulated by varying the inlet stream mass flow rate.

Sine wave flow: a sine wave generator was used, which is a tool available on HYSYS. The oscillations were soft, with 10% of amplitude and frequencies of 1 and 0.5 cycles/min.

Slug flow: an emulator was developed to perform the intermittent flow oscillation caused by slug flow. A vessel and two valves were used (Figure 8) such that slug cycles were created by coordinating valves opening and closing. The HYSYS tool Event Scheduler was responsible for managing actions at each period of time. The slug flow oscillation in gas and in liquid streams can be seen in Figure 9 (a) and (b). It performed for about 150 min with 9 min/cycle and then the emulator was "turned off" to analyze the process stability after the end of the perturbation.

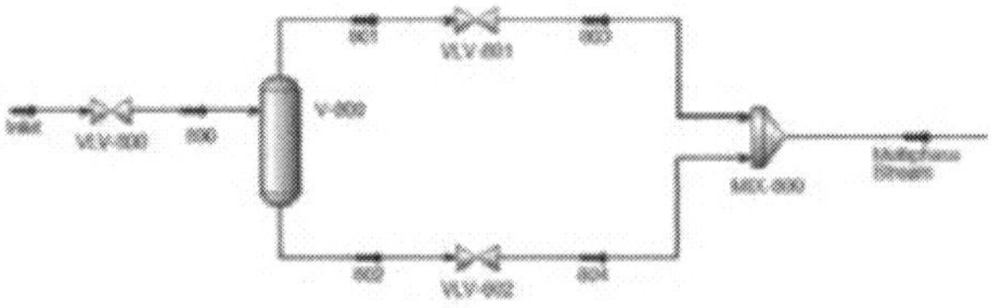

Figure 8. Slug flow generator scheme

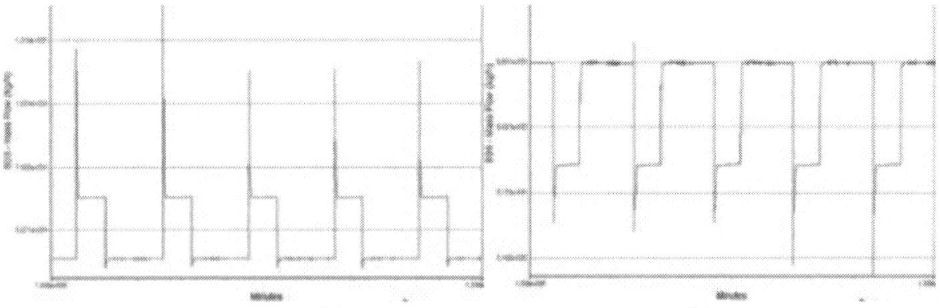

(a) (b)

Figure 9. Slug flow on gas (a) and liquid (b) streams

Results

The process responses were similar in each dynamic scenario. As expected, the slug flow presented the most severe impacts. The negative pulse perturbation was more impacting than the positive, displaying higher oscillations, despite having the same amplitude. For the sine wave flow, the 0.5 cycle/min case was harsher to the process than the case with doubled frequency.

The separation area is responsible for absorbing the first perturbation impacts. The vessels pressure and level were extremely affected by inlet oscillations. Under slug flow the two-phase separator reached levels near 100%, but it was still able to operate without any trouble (Figure 10b).

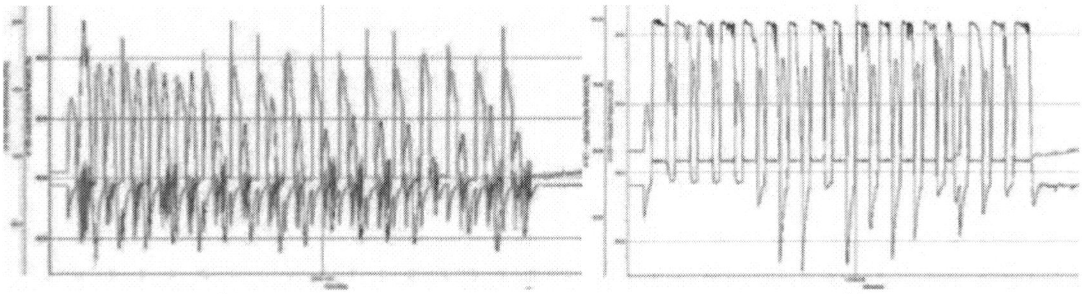

(a) (b)

Figure 10. Three-phase (a) and two-phase (b) separators pressures (red) and levels (blue): Slug flow case

At the absorber, TEG mass flow was the most affected variable for all cases. As soon as the perturbation occurs, TEG flow rate is affected. However, after a short period of time, the ratio controller manipulates the TEG stream valve opening to try to maintain an optimum proportion between gas and TEG flow rates and minimize WDP oscillations. Figure 11 illustrates the positive pulse case.

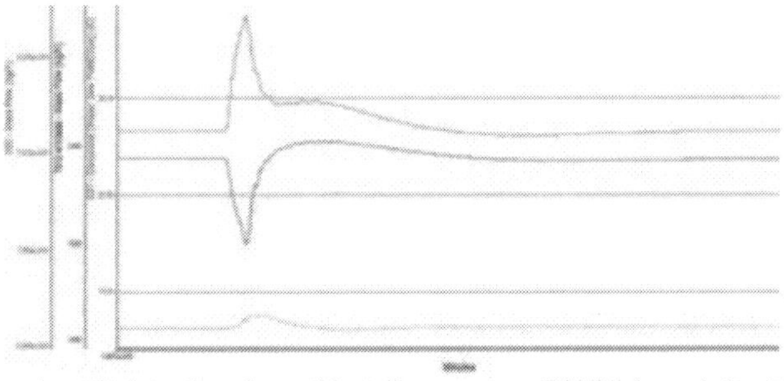

Figure 11. Absorber TEG (red) and gas (blue) flow rates and WDP (green): Positive pulse case

In the JT process, it was possible to notice the strict relation between HCDP and separator temperature in all cases, which was already expected. Figure 12 displays the sine wave case.

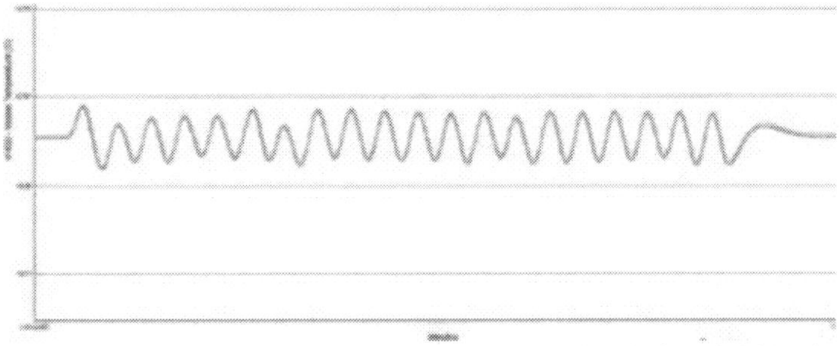

Figure 12. JT separation vessel temperature (blue) and adjusted HCDP (dotted pink): Sine wave case

At the regenerator, stripping gas (SG) mass flow rate and condenser and reboiler pressures show similar profiles (Figure 13). When TEG mass fraction varies, the controller manipulates the SG flow rate directly affecting the pressure of the regeneration system, though not in a severe way.

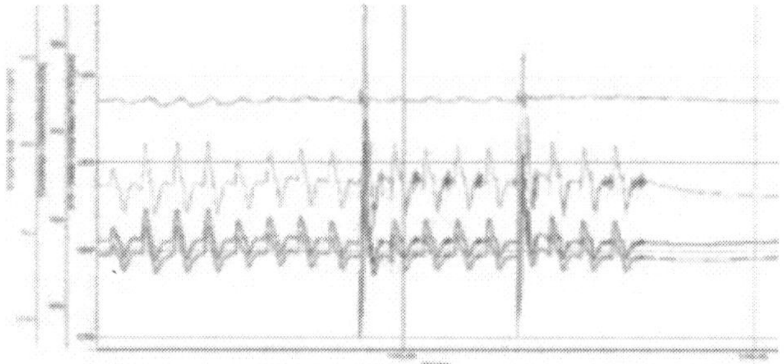

Figure 13. Regenerator TEG w/w (dotted pink), SG flow rate (yellow), condenser and reboiler pressures (pink, red): Slug flow case

Figure 14 depicts the evolution of methane and CO_2 contents of the processed gas through the MP stages. The difference between feed and 1st retentate curves is greater than between 1st and final retentate curves. It means that it is easier to remove CO_2 from NG in the first MP stage than in the second, which is expected, since the latter CO_2 content is lower than the former.

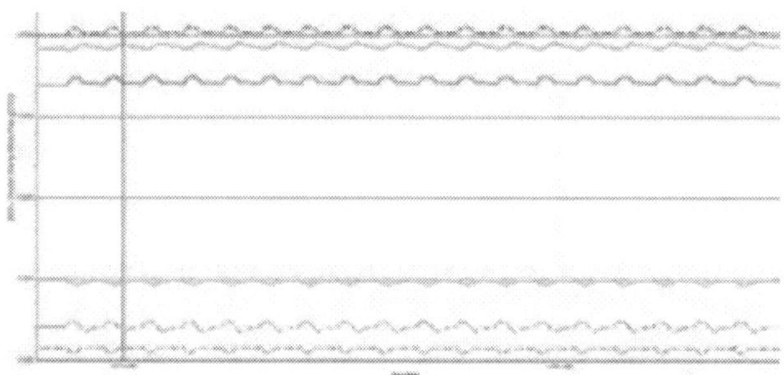

Figure 14. Methane (pink) and CO_2 (blue) contents - MP feed (solid line), 1st (dotted) and final (dashed) retentates: Slug flow case

The export NG properties and conditions presented oscillations under all the proposed scenarios. However, they are much softener than the actual inlet perturbations. Figure 15 shows flow rate, WDP, HCDP and methane and CO_2 contents of the final NG for the slug flow case. It is important to state that the export gas did not present any harsh peaks or change of levels that could take it out of specification, even in the most severe case.

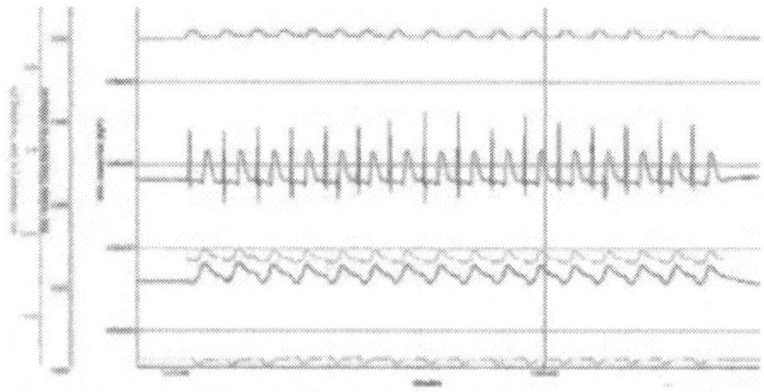

Figure 15. Export gas flow rate (red), methane (pink) and CO_2 (light blue) contents, WDP (blue) and HCDP (green): Slug flow case

Conclusions

The base case was able to plausibly represent a NG conditioning plant operating on steady state. The final gas stream was within the ANP regulatory specifications for WDP, HCDP and CO_2 content. In addition, the equipment sizing results were compatible with an offshore plant project.

It is possible to conclude that the process responses to the pulse and sine wave perturbations indicated that in general the proposed control system was properly tuned. The sine wave oscillations had little impact on the operation and the lower frequency perturbation (0.5 cycles/min) presented more disturbance in the responses. It is believed that the inlet flow rate oscillation at 1 cycle/min happens to be very fast, so the process does not "have time" to respond to it, noticing it less intensely than the sine wave with lower frequency.

The process response to the pulse perturbation was fast and returned to stability after a period of time with no major disturbances. However, the negative pulse was harsher than the positive, so the process displayed directionality in this case – distinct gains depending if the perturbation is negative or positive – due to the non-linearity of the system.

As expected, the operational impacts caused by slug flow were more severe. Despite this, the control system still performed well and in most cases it was possible to notice the softening of the oscillations over time. The slug flow emulator is not a completely accurate representation of the phenomenon but it is considered that it successfully approached its inherent intermittent profile.

The MP dynamic extension presented fast transient responses under system disturbances according to the expected for this unit operation. It was possible to conclude that the 1st MP stage removes more CO_2 from the gas stream than the 2nd. The first retentate stream has already a lower CO_2 content, which makes it more difficult to separate and meet the specification requirement.

It was possible to realize that the impact on the gas dew points must be monitored during all operation. In some instants they increased, which may cause loss of specification and hence potential risk situations, although it has not reached that point in any of the simulated scenarios.

It is considered that the present study can be used to predict dynamic behaviors in the NG processing, allowing to analyze and test control strategies, besides other applications like operator training and HSE (Health, Safety & Environment) assessments.

Acknowledgements

To the financial support of PETROBRAS S.A., "Agência Nacional do Petróleo, Gás Natural e Biocombustíveis" - ANP - and "Financiadora de Estudos e Projetos" - FINEP - through the ANP Human Resources Program for Oil & Gas - PRH-ANP / MCT, in particular to the PRH-13 from "Escola de Química/UFRJ" - Processing, Management and Environment in Oil and Natural Gas Industry.

References

1. ANP – National Agency of Oil, Gas and Biofuels. "Resolução ANP nº 16." Brasília, 2008.
2. ASPEN TECHNOLOGY INC. Hysys Dynamic Modeling. Cambridge: Aspentech, 2005.
3. CAMPBELL, John M. Gas Conditioning and Processing - Volume 2: The Equipment Modules. Norman, Oklahoma: Campbell Petroleum Series, 2004.
4. DORTMUNDT, D., Dosh, K.. Recent developments in CO_2 removal membrane technology. Des Plaines, IL: UOP LLC, 1999.
5. FORMIGLI, J. CO_2: Challenges & Opportunities. 2nd Brazilian Congress of CO_2 in Oil, Gas and Biofuels Industry, April 8-9, 2013.
6. ERNST & YOUNG TERCO. "Perspectivas para a indústria de Petróleo e Gás no Brasil: Uma análise das lições aprendidas no Mar do Norte e da retomada de licitações em 2013". Ernst & Young Terco Brasil, 2014.
7. LUYBEN, W. L. Plantwide Dynamic Simulators in Chemical Processing and Control. CRC Press, 2002

8. MACHADO, P.B. "Análise comparativa de tecnologia de separação supersônica para o condicionamento de gás natural". MSc Thesis. Rio de Janeiro: UFRJ, 2012.
9. MME. "Ministério Minas e Energia" – "Balanço Energético Nacional". Brasília: EPE, 2014.
10. MOKHATAB, S., POE, W.A., SPEIGHT, J.G. Handbook of Natural Gas Transmission and Processing. Burlington: Elsevier, 2006.
11. NUNES, G.C., MEDEIROS, J.L., ARAÚJO, O.Q.F. "Modelagem e Controle na Produção de Petróleo - Aplicações em Matlab". São Paulo: Blucher, 2010.
12. SAUSEN, A., SAUSEN, P., CAMPOS, M. The Slug Flow Problem in Oil Industry and Pi Level Control. In: New Technologies in the Oil and Gas Industry, 103-118. Intech, 2012.
13. VAZ, J.C. "Síntese de Controle e Análise de Cenários em Planta Offshore de Óleo e Gás". MSc Thesis. Rio de Janeiro: UFRJ, 2009.
14. de Medeiros, J.L., Araújo, O.Q.F. "MP HUO", H2CIN-UFRJ, 2010.

389064 Oil Scheduling Incluing Smoothening Transitions in Distillation Units

Abstract:

Oil blending is part of the normal scheduling operations in oil refineries. Mix of different petroleum types are prepared in charging tanks to meet a variety of quality specifications normally imposed by required production profiles and/or equipment design and defined on the planning level. Changing from one mix to another results in changeover costs because distillation units have to dynamically go from one operating condition to a new one. Oil companies may adopt the strategy of simultaneously feeding a distillation unit with the tank leaving the feeding operation and the tank taking over for a short period of time with the purpose of smoothening transition. The temporary simultaneous feeding operation has not been addressed in past literature. Most of the works reported in the literature disregard changeover times. This work conducts a study of the effects of incorporating such a smoothening transition step as part of the decisions involved in the scheduling of oil supply. The **MOS** (Multi-Operation Sequencing) model proposed by Mouret *et al.* [*Comp & Chem Eng*, **35**, 1038 (2011)], was taken as the starting development point since it is considered one of the state-of-the-art models for oil scheduling. A real-world case has been used as a testing problem. Results show that the produced scheduling can be quite different from that in which transition is considered abrupt.

390486 Optimization of Water Supply Chain Design and Long-Term Planning for Shale Gas Production: Milfp Model and Algorithms

Abstract:

Natural gas is playing a significant role in meeting the global energy demand as well as behaving as a bridge fuel as the U.S. develops more sustainable and renewable fuel options [1]. Shale gas, as the unconventional natural gas extracted from shale rock, has emerged as one of the most promising energy source within the past decade [2]. The large-scale production of shale gas is not possible until the hydraulic fracturing and horizontal drilling technologies are developed [3, 4], and the combination of these two technologies resulted in exponential increase in gas production [5]. However, accompanying the benefits of the shale gas production are environmental concerns. Issues regarding water usage and environmental impacts have brought the shale gas extraction into a highly controversial situation [6, 7]. Many works have been done on the optimization problem regarding shale gas and water [8, 9]. In this work, we propose an optimization model for the design and operation of the water management systems and supply chain in shale gas production process. Distinct from most of the existing works in this field, to take both economic and environmental impacts into account, we present a novel fractional objective function defined as the net profit per unit freshwater consumption. In this way, we are able to maximize the profit by consuming the least freshwater, thus leading to a more profitable and sustainable water management system.

Due to the fractional form objective function, the resulting problem is a mixed-integer linear fractional programming (MILFP), which can be computationally intractable for large-scale problems due to its combinatorial nature and pseudo-convexity. Therefore, we presented tailored algorithms to facilitate the solution process: parametric algorithm based on exact Newton's method and reformulation-linearization method to globally optimize the MILFP problem [10-12].

To illustrate the application of the proposed supply chain model and solution approaches, we investigate two case studies with different scales, a smaller case study to demonstrate the optimal water management strategy and a larger case to compare the computation performance of proposed algorithms. In this model we consider a set of freshwater sources, shale sites with multiple wells in each one and three different water treatment options, including disposal wells, centralized wastewater treatment (CWT) facility and onsite treatment methods. Water is classified into three levels with different total dissolved solids (TDS) concentration. The total time span is set to 10 years, divided into 520 periods. As the optimization result shows, almost all the freshwater are transported by pipelines. For level 1 water, CWT facilities and onsite treatment methods are applied; for level 2 water, all the water is treated by CWT facilities; for level 3 water, CWT facilities and disposal wells are applied to treat the produced water. In all, CWT facilities treat more than half of the produced water, and the rest is treated by disposal well s and onsite treatment. The optimal objective value is $36.4 per thousand barrels freshwater consumption for the small case, and for both cases, the proposed parametric algorithm based on exact Newton's method is proven the most efficient solution approach in solving such MILFP problem.

References

[1] D. M. Kargbo, R. G. Wilhelm, and D. J. Campbell, "Natural Gas Plays in the Marcellus Shale: Challenges and Potential Opportunities," *Environmental Science & Technology*, vol. 44, pp. 5679-5684, 2010/08/01 2010.

[2] U. S. E. I. Administration, "World Shale Gas Resources: An Initial Assessment of 14 Regions Outside the United States " April 2011

[3] K. B. Gregory, R. D. Vidic, and D. A. Dzombak, "Water Management Challenges Associated with the Production of Shale Gas by Hydraulic Fracturing," *Elements*, vol. 7, pp. 181-186, Jun 2011.

[4] A. Vengosh, N. Warner, R. Jackson, and T. Darrah, "The effects of shale gas exploration and hydraulic fracturing on the quality of water resources in the United States," *Proceedings of the Fourteenth International Symposium on Water-Rock Interaction, Wri 14*, vol. 7, pp. 863-866, 2013.

[5] J. P. Nicot, B. R. Scanlon, R. C. Reedy, and R. A. Costley, "Source and Fate of Hydraulic Fracturing Water in the Barnett Shale: A Historical Perspective," *Environmental Science & Technology*, vol. 48, pp. 2464-2471, Feb 2014.

[6] S. G. Osborn, A. Vengosh, N. R. Warner, and R. B. Jackson, "Methane contamination of drinking water accompanying gas-well drilling and hydraulic fracturing," *proceedings of the National Academy of Sciences*, vol. 108, pp. 8172-8176, 2011.

[7] S. C. Schon, "Hydraulic fracturing not responsible for methane migration," *Proceedings of the National Academy of Sciences*, vol. 108, pp. E664-E664, 2011.

[8] D. C. Cafaro and I. E. Grossmann, "Strategic planning, design, and development of the shale gas supply chain network," *AIChE Journal*, pp. n/a-n/a, 2014.

[9] Y. Linlin and I. E. Grossmann, "Superstructure-Based Shale Play Water Management Optimization," in *AIChE Annual Meeting*, San Francisco, CA, 2013.

[10] F. You, P. M. Castro, and I. E. Grossmann, "Dinkelbach's algorithm as an efficient method to solve a class of MINLP models for large-scale cyclic scheduling problems," *Computers and Chemical Engineering*, vol. 33, pp. 1879-1889, 2009.

[11] D. J. Yue, G. Guillen-Gosalbez, and F. Q. You, "Global Optimization of Large-Scale Mixed-Integer Linear Fractional Programming Problems: A Reformulation-Linearization Method and Process Scheduling Applications," *AIChE Journal*, vol. 59, pp. 4255-4272, Nov 2013.

[12] Z. X. Zhong and F. Q. You, "Globally convergent exact and inexact parametric algorithms for solving large-scale mixed-integer fractional programs and applications in process systems engineering," *Computers & Chemical Engineering*, vol. 61, pp. 90-101, Feb 2014.

377206 Rigorous Dynamic Simulation Study on the Ethylene Refinement Process within the Bio-Ethanol to Ethylene Section

Abstract:

Rigorous dynamic simulation study on the ethylene refinement process within the bio-ethanol to ethylene section

Yingzhe Yu, Lu Kong, Zhongfeng Geng, Xiuqin Dong, Minhua Zhang*

Key Laboratory for Green Chemical Technology of Ministry of Education, R&D Center for Petrochemical Technology, Tianjin University, Tianjin 300072, P. R. China

*Corresponding author Tel. & fax: +86-22-27406119, E-mail address: mhzhang@tju.edu.cn

In this work, two types of ethylene rectification simulation process are developed to refine the raw ethylene flow (ethylene content > 98 wt %). The normal simulation method is that two *Radfrac* columns with the condenser and reboiler inside the model. The more rigorous one is a two-rectification columns process with a propylene refrigerant cycle, which provides the heat and cooling energy. This Aspen simulation utilizes explicit heat-exchanger for the condensers and the reboilers, which means the column itself is a *Radfrac* model without any reboiler or condenser. These are added externally as Aspen *HeatX* models, as are separate reflux drums and circulation from the columns through the reboilers. Design and control of the processes are investigated using Aspen Plus and Aspen Dynamics.

When the steady-state simulation is completed and proper geometric parameters of the equipment are added, it is just the time to do the dynamic works. As the feed flow consists of high concentration ethylene, it is quite difficult to choose a sensitive plate when it comes to lay out a temperature control structure on the column. However, a combination of the traditional method in steady simulation and the dynamic simulation could provide an effective way to deal with this situation. Then with the control structures for the two processes, the dynamic simulation results are achieved using Aspen Dynamics. The feed flow rate and feed composition disturbances are taken using dynamic simulation to evaluate the control structure. With the proper complete process in Aspen Dynamics, the safety analysis, such as condenser failure and heat-input surge, is explored. It is indicated that the traditional distillation simulation with a *Radfrac*column alone could not accurately predict the process variables changes during the disturbances.

Keywords: Dynamic simulation, ethylene distillation, rigorous distillation process, sensitive plate selection, safety analysis

391769 Multi-Criteria Optimization of Vaccine Supply Chain in Low-and Middle- Income Countries

Abstract:

With increasing population in developing countries and introduction of new vaccines (such as Rotavirus and Pneumococcal vaccines) that are more expensive, designing reliable and green vaccine supply chain is vital. This is especially important because the new vaccines are of greater volume and require significantly more packaging and cold storage in comparison to traditional vaccines. The existing cold storages capacity and transportation plans are not sufficient in many low- and middle- income countries for meeting their immunization demands. This study suggests economic and ecofriendly alternatives for vaccine supply chain through multi-criteria optimization using Non-Dominated Sorted Genetic Algorithm. The modeled supply chain includes manufacturing of vaccine, transportation, cold storage and disposal. We consider different technology options for each section with varying environmental emissions and costs. The proposed method evaluates the optimal combination of technology options. The model considers the demand for immunization as a function of time and this demand is based on infectious disease spread in developing countries. The results presented would be helpful for governments, policy makers and donors (e.g. UNICEF, WHO, GAVI Alliance) to design future immunization supply chains reducing both environmental burden and cost.

367461 Saudi Aramco Enterprise-Wide Hydrocarbon Chain Optimization Initiative

Abstract:

In an environment of very volatile and soaring demand for refined products in the Kingdom of Saudi Arabia, there is an increasing need for optimizing the entire hydrocarbon supply chain. . As the major stakeholder of most of the components of this supply chain, Saudi Aramco has recognized this need and embarked on a strategic initiative for optimizing and integrating the various components of its vast hydrocarbon operations covering oil, gas and NGL production, in addition to refinery production and product distribution network via different transportation modes. In order to build such an optimized and integrated supply chain network, it is essential to develop a suite of best-in-class software that is capable of handling the complexity of this system. The corporate initiative entails the development of advanced model-based optimization applications, high-fidelity process models, and planning & scheduling solutions. Full implementation of this strategic initiative will enhance the utilization of corporate assets, capture revenue enhancement opportunities and ensure proper, timely, and proactive decision making process while managing the company's complex hydrocarbon supply chain. This paper will focus on describing the business problem involved in the distribution of refined products by various transportation modes in addition to the integration of refinery, pipeline and ship scheduling solutions. The paper will highlight the benefits that can be achieved from using various optimization techniques coupled with enhanced visualization to implement an optimized solution which facilitates streamlining the process of generating schedules. The paper will illustrate how Key Performance Indicator (KPI) driven solutions are structured, integrated and mapped with planning decisions. It will further highlight the integration scheme between the planning and scheduling of various components of the supply chain.

379006 Dynamic Simulation Model Based Analysis for Process Development Window of Operability and Control

Abstract:

The ability to use dynamic simulation models for the process life cycle development by effectively determining the process window of operability and select the appropriate control strategy is described in this paper. The process development life cycle, encompassing various stages, such as research stage, process development stage and technology deployment stage, depends heavily on the validated process models to enable researchers to reduce excessive and unnecessary research spending by efficiently optimizing the process schemes and process parameters. Researchers at Saudi Aramco invented a distinctive one-step process to convert mixture of Butenes into mixed Butanols liquid without the need for separation of the isomers. Preliminary simulation models developed during the research stage revealed significant issues associated with the Azeotropic nature of the mixed Butanol product, which aided researchers during the scale-up process to incorporate necessary process changes to the pilot plant. Validated simulation models based on pilot plant data were key in developing a comprehensive large scale integrated pilot plant.

The objective of this study is to understand the scaled up commercial unit operation during its startup, analyze appropriate control strategy and optimize process basis design parameters. The process constitutes of multiple reactions of isomers of Butenes to isomers of Butanols with various reaction rates. The mixed Butanols product has different ternary separation regions with different Azeotropic points. In addition, the impact of varying isomer conversion, overall mixed Butenes conversion requiring substantial recycle rates, water specification in the mixed Butanols product substantially impact the overall process design mandates a thorough understanding of the process through dynamic modelling of the Butenes to Butanols process to provide insight into the process window of operability during startup and normal operations. Through this study the process constraints were investigated and optimal control layer configurations was established. The work is part of a comprehensive feasibility study that enables the modelers and researchers to conduct rigorous process economics for potential technology deployment and licensing.

389189 Global Optimization of a General Class of Nonconvex Optimization Problems

Abstract:

Title: Global Optimization of a General Class of Nonconvex Optimization Problems

In this work, a general class of nonconvex optimization problems is considered that possesses: a separable objective function consisting of general nonconvex terms; a single equality constraint requiring that a product of general nonconvex single variable functions is equal to a constant parameter, and upper and lower bounds on all optimization variables. This class of problems is shown to arise frequently in practice, especially during the optimization of serial systems.

The proposed solution method employs optimality properties of the considered optimization problem, in combination with interval analysis, to guarantee that upper and lower bounds converging to the problem's global optimum are identified for all optimization problem instances created by varying the equality constraint's constant parameter.

The proposed method is illustrated on a number of case studies.

390077 A Decomposition Method Via Surrogate Modeling for Integrated Planning, Scheduling, and Dynamic Optimization of Sequential Batch Processes

Abstract:

Today's process manufacturing enterprises call for new solutions so as to survive and grow in this highly competitive international marketplace [1]. To address the emerging challenges, enterprise-wide optimization (EWO) has attracted much interest [2-4]. An integrated optimization method proves to achieve a better performance than a conventional method which optimizes the subproblems independently in a sequential way [5, 6]. In recent decade, many integrated methods are presented, e.g. integrated scheduling and dynamic optimization [6-11], scheduling and control [12-14], etc. However, the integrated method usually suffers from severe computational complexity. Consequently, most existing integrated methods only focus on incorporating two decision layers. However, to exploit the full potential of EWO, more decision layers should be integrated and optimized collaboratively. Our goal is to devise efficient solution method, which can solve a complex problem integrating more than two decision layers.

In this work, we propose a novel framework to solve the integrated planning, scheduling, and dynamic optimization problem for batch processes. The integrated problem is formulated into a mixed-integer dynamic optimization problem (MIDO) [15], which contains binary decision variables associated with planning and scheduling decisions, and constraints of differential equations representing the dynamic models. The differential equations are then discretized by the collocation method [16], and they are transformed into a large set of nonlinear algebraic equations. The discretization procedure reformulates the integrated problem into a large-scale mixed-integer nonlinear program (MINLP). However, the formulated MINLP is too complex to be solved by simply invoking a commercial general-purpose MINLP solver [12].

To conquer the computational complexity, we develop efficient decomposition methods based on surrogate modeling. The decomposition methods exploit the special structure of the integrated problem to reduce the computational complexity. The integrated problem has a planning problem in the top level, a number of scheduling problems in the middle level, and a larger number of dynamic optimization problems in the bottom level. The problems in two adjacent levels are coupled by linking functions. Representing these linking functions by surrogate models breaks the coupling between the upper level problems and the lower level problems. Specifically, the lower level problems are solved to generate data points for building the surrogate models. The surrogate models are then substituted into the upper level problems replacing the detailed lower level problems. Next, the upper level problems are solved with the surrogate models. The determined variables of the upper level problems are passed to the lower level problems which are solved accordingly.

We develop two decomposition methods. The first method decomposes the dynamic optimization problems from the planning and scheduling problem. The large set of nonlinear constraints is replaced by the simple surrogate models representing the task operational recipes. The integrated MINLP is decomposed into a mixed-integer linear program (MILP) for the planning and scheduling problem and a set of separable dynamic optimization problems. The resulting MILP is much easier to solve than the complicated MINLP. Based on the first method, the second decomposition method further decomposes the planning and scheduling problem. The scheduling problem in a planning period is solved first to generate data points for building the surrogate models, which represent the production costs as a function of the production quantities. The generated surrogate models are used to solve the planning problem replacing the detailed scheduling problems.

The decomposition methods significantly reduce the computational complexity of the integrated problem. The computational efficiency is demonstrated in two case studies, including a large problem which the direct solution method fails to find a feasible solution after a long time period. The case studies also demonstrate the advantage of the integrated optimization methods in comparison to the conventional sequential method which neglects the interaction between the dynamic optimization problems and the scheduling problems.

References

[1] P. D. Christofides, J. F. Davis, N. H. El-Farra, D. Clark, K. R. D. Harris, and J. N. Gipson, "Smart plant operations: Vision, progress and challenges," *Aiche Journal,* vol. 53, pp. 2734-2741, Nov 2007.

[2] A. Quaglia, B. Sarup, G. Sin, and R. Gani, "A systematic framework for enterprise-wide optimization: Synthesis and design of processing networks under uncertainty," *Computers & Chemical Engineering,* vol. 59, pp. 47-62, Dec 2013.

[3] J. M. Wassick, "Enterprise-wide optimization in an integrated chemical complex," *Computers & Chemical Engineering,* vol. 33, pp. 1950-1963, Dec 2009.

[4] V. A. Varma, G. V. Reklaitis, G. E. Blau, and J. F. Pekny, "Enterprise-wide modeling & optimization - An overview of emerging research challenges and opportunities," *Computers & Chemical Engineering,* vol. 31, pp. 692-711, May-Jun 2007.

[5] S. Engell and I. Harjunkoski, "Optimal operation: Scheduling, advanced control and their integration," *Computers & Chemical Engineering,* vol. 47, pp. 121-133, Dec 20 2012.

[6] Y. Chu and F. You, "Integrated Scheduling and Dynamic Optimization of Sequential Batch Processes with Online Implementation," *AIChE Journal,* vol. 59, pp. 2379-2406, 2013.

[7] Y. Chu and F. You, "Integrated scheduling and dynamic optimization by stackelberg game: Bilevel model formulation and efficient solution algorithm," *Industrial & Engineering Chemistry Research,* vol. 53, pp. 5564-5581, 2014/04/02 2014.

[8] Y. F. Chu and F. Q. You, "Integrated Scheduling and Dynamic Optimization of Complex Batch Processes with General Network Structure Using a Generalized Benders Decomposition Approach," *Industrial & Engineering*

Chemistry Research, vol. 52, pp. 7867-7885, Jun 2013.

[9] Y. Chu and F. You, "Integration of Scheduling and Dynamic Optimization of Batch Processes under Uncertainty: Two-Stage Stochastic Programming Approach and Enhanced Generalized Benders Decomposition Algorithm," *Industrial & Engineering Chemistry Research,* vol. 52, pp. 16851-16869, 2013.

[10] Y. Chu and F. You, "Integration of production scheduling and dynamic optimization for multi-product CSTRs: Generalized Benders decomposition coupled with global mixed-integer fractional programming," *Computers & Chemical Engineering,* vol. 58, pp. 315-333, 11/11/ 2013.

[11] A. Flores-Tlacuahuac and I. E. Grossmann, "Simultaneous cyclic scheduling and control of a multiproduct CSTR," *Industrial & Engineering Chemistry Research,* vol. 45, pp. 6698-6712, Sep 27 2006.

[12] Y. Chu and F. You, "Moving horizon approach of integrating scheduling and control for sequential batch processes," *AIChE Journal,* vol. 60, pp. 1654-1671, 2014.

[13] Y. Chu and F. You, "Integration of scheduling and control with online closed-loop implementation: Fast computational strategy and large-scale global optimization algorithm," *Computers & Chemical Engineering,* vol. 47, pp. 248-268, Dec 20 2012.

[14] J. Zhuge and M. G. Ierapetritou, "Integration of scheduling and control with closed loop implementation," *Industrial & Engineering Chemistry Research,* vol. 51, pp. 8550-8565, Jun 27 2012.

[15] P. I. Barton, R. J. Allgor, W. F. Feehery, and S. Galan, "Dynamic optimization in a discontinuous world," *Industrial & Engineering Chemistry Research,* vol. 37, pp. 966-981, Mar 1998.

[16] J. E. Cuthrell and L. T. Biegler, "On the optimization of differential algebraic process systems," *AIChE Journal,* vol. 33, pp. 1257-1270, Aug 1987.

370436 Coopetitive Supply Chain Relationship Model: Application to the Smartphone Manufacturing Network

Abstract:

In the era of turbulent global economy, "supply chain 2.0" has to be lean, agile and structurally flexible, thus underscoring the need of strong supply chain relationship. Most of the relationship literature focused on vertical collaboration between cooperative buyers and suppliers. However, there are some cases of stable and volatile cooperation amongst competitors such as airline alliances and Apple-Samsung product manufacturer-component supplier relationship, respectively, which is known as coopetition. Despite several qualitative studies on coopetition in general, only a handful of attempts, mainly based on game theory, have been made for quantitative characterization of such coopetitive dynamic behavior in supply chain relationship. Thus, in this regard, we developed a mixed-integer non-linear programming model describing manufacturer-supplier relationship for a cost efficient supply chain coopetition network. Notably, the coopetition factor was newly introduced to quantify the degree of coopetition taken in the supply chains and to help the end-product manufacturer determine the optimal level of coopetition to engage in. The applicability and utility of the model were successfully demonstrated through a case study of a hypothetical smartphone supply chain network. The coopetition factors of end-manufacturing firms were measured under different scenarios, as such suggesting their optimal supply chain interactions for strategic viability. Hence, the current exploratory research can pave the way in enabling manufacturers to drive coopetitive business globally.

An Efficient Adaptive Sampling Approach for Optimization with Surrogate Models

Justin Smith, Daniel Crunkleton, Selen Cremaschi
The University of Tulsa
800 South Tucker Drive, Tulsa, OK, 74104

Fueled by the increase in computational power, more and more complex models, such as CFD, and process simulators such as Aspen Hysys, are routinely used for design of chemical processes. When these models are to be used for optimization, the difficulty of estimating the derivatives and long run-times required to evaluate these models prohibits the use of traditional derivative-based optimization solvers. It is also often not possible to directly study the process's entire design space due to issues of run times or cost. In recent years, surrogate-based optimization has been suggested and successfully used for optimization of such models [1,2]. In this approach, a simpler surrogate model is constructed to accurately replicate the input-output relationship of the original complex model, and derivative-based optimization algorithms are used to optimize the design of the process using this surrogate rather than the original model. Surrogate models, such as Kriging[3], and artificial neural networks[4] (ANNS), have been shown to significantly reduce the time required to reach a numerical solution for a process model from several hours or days to one which is negligible [5].

The main computational cost of surrogate-based optimization is the construction of an accurate surrogate, which may require many evaluations of the original rigorous model. Therefore, the location of these rigorous model evaluations, i.e., sample points, should be selected to minimize the overall computational cost. Several methods have been proposed for better choosing the location of these sample points, ranging from using a Latin hypercube sampling design to using incremental methods such as adaptive sequential sampling [6], incremental Latin hypercube sampling (iLHS)[7], and local linear approximations[8]. Most of these approaches, however, focus on generating samples for constructing a global surrogate model with certain accuracy throughout the whole search space. When the goal of surrogate model construction is optimization, these approaches may generate more sample points than necessary in suboptimal regions.

In this presentation, we introduce and test a method, Optimization Directed INcremental sampling (ODIN), which has the potential to avoid this issue. ODIN is a sequential sampling algorithm that relies on both exploration and exploitation of the search space to locate the optimum. The first step of ODIN is to construct an initial surrogate using sample points generated according to a space-filling design. For this work, we used a Latin hypercube as the initial space-filling design. ODIN, then, determines the "global optimum" along with up to 30 of the best branches using BARON's option to find k-best solutions. A large number of new sample points is then proposed, and a score, which falls between zero and two, is calculated for each of these points. The score combines the exploration and exploitation potential of each proposed point. The exploitation potential is estimated based on its distance from any of the k-solutions obtained from the optimization of the surrogate model. Here, a score of zero is assigned to the points that are furthest away, and one to the points that are closest. Each proposed point is also assigned a score between zero and one based on its distance from the existing sampled points, i.e., the ones that were used to construct the surrogate. For this score, which is an estimate of the exploration potential of the proposed point, a score of zero means the proposed point is the closest to an existing point and a score of one means the proposed point is the furthest away from all existing points. The sum of these two scores is taken to obtain the overall score for the proposed points, and the points with the highest overall scores are chosen as new sample points for the next iteration of the algorithm. The algorithm is terminated once the change in the optimum solution obtained from the surrogate, ε, is less than a user defined value for successive iterations or when the maximum number of iterations (specified by the user) is reached.

The performance of ODIN is tested to solve optimization test problems from the Virtual Library of Simulation Experiments [10], and compared to a previously proposed mixed-adaptive sampling algorithm [6] in terms of solution quality and number of required function evaluations. Artificial neural networks were used as the surrogate model. The sample sizes were incremented geometrically, increasing by one-third of the previous sample size in each iteration. The maximum number of iterations was set to 10. ODIN was able to locate the global optimum for 13 of the 21 problems. For problems where the global optimum was not located, this failure seemed to be due to rapid gradient changes around local minimas, causing the global minima not to be discovered within the 10 iterations used. ODIN performed at a level equivalent or better than the mixed adaptive sampling algorithm for 12 out of 21 test problems. This talk will discuss the details of these tests along with the algorithm.

References:

[1]Biegler, L.; Lang, Y.; Lin,W. *Multi-Scale Optimization for Process Systems Engineering*. Computers and Chemical Engineering. 2014. 60,10, 17-30

[2]Cozad, A.; Sahinidis, N.; Miller, D.; Learning Surrogate Models for Simulation-Based Optimization. AIChE Journal. 2014. 60,6, 2211-2227

[3] Park, K.; Oh, P.; Lim, H. *The application of the CFD and Kriging method to an optimization of heat sink* Int. J. Heat Mass Transfer 2006, 49, 34-39

[4] Lang, Y.; Malacina, A.; Biegler, L.; Munteanu, S.; Madsen, J.; Zitney, S. *Reduced Order Model Based on Principal Component Analysis for Process Simulation and Optimization* Energy Fuels 2009, 23, 1695

[5]Smith, J.; Neto,A; Cremaschi,S.; Crunkleton,D. *CFD-Based Optimization of a Flooded Bed Algae Reactor.* Ind. Eng. Chem. Res. 2013. 52(22) 7181-7188

[6]Eason,J; Cremaschi,S. *Efficient surrogate model generation with adaptive sequential sampling* . Presented at the 2012 AIChE Annual Meeting , Pittsburgh, PA. 2012.

[7]Nuchitprasittichai, A.; Cremaschi,S. *An Algorithm to Determine Sample Sizes for Optimization With Artificial Neural Networks*. AIChE Journal. 2013. 59 (3) 805-812

[8] Crombecq,K.; Surrogate Modeling of Computer Experiments With Sequential Experimental Design. Doctoral dissertation, Ghent University. 2011

[9] Tawarmalani, M.; Sahinidis N. , *A polyhedral branch-and-cut approach to global optimization.*Mathematical Programming, 2005. 103 (2), 225-249

[10] Surjanovic, S. ;Bingham, D. (2013). Virtual Library of Simulation Experiments: Test Functions and Datasets. Retrieved May 12, 2014, from http://www.sfu.ca/~ssurjano.

374580 Value Chain Optimization Model for a Softwood Biorefinery Based in Scotland

Abstract:

There is a push towards sourcing chemicals and materials from renewable feedstock such as lignocellulosic biomass. Value chain assessment is used to evaluate the feasibility of the use of a certain technology and feedstock to produce various chemical sources in a given location. In this work an optimisation model for the value chain assessment of a lignocellulosic biorefinery was developed using mixed integer linear programing. The model allows for a comparison of two product sources which undergo mechanical and/or chemical pretreatment prior to processing by the biorefinery into three product streams, delivered to the customer. Optimisation identifies the optimal source or sources of feedstock and the locations of intermediate storages, pretreatments, biorefinery(ies) and customers with respect to maximising profit. The model was verified based on a case study detailed in Scotland. The case study evaluates the use of felled softwood and/or to the use of sawmill by-products with the production of hemicellulose, lignin and cellulose. The results and implications of the optimisation of the scenario are discussed with respect to costs of transport, processing and product values.

DISTILLATION CURVE OPTIMIZATION USING MONOTONIC INTERPOLATION

Brenno C Menezes, PETROBRAS, Rio de Janeiro, RJ, Brazil
Jeffrey D Kelly, Industrial Algorithms, Toronto, ON, Canada
Ignacio E Grossmann, Carnegie Mellon University, Pittsburgh, PA, United States

Introduction

Oil refinement involves a series of complex manufacturing processes in which final products such as fuels, lubricants and petrochemical feedstocks are produced from crude-oil feedstocks by separation and conversion unit-operations in coordination with tanks, blenders and transportation vessels. To manage the processing of the hydrocarbon streams, well-known distillation curves or assays of both the crude-oil and its derivatives are decomposed or characterized into several temperature "cuts" based on what are known as the True Boiling Point (TBP) temperature distribution or distillation curves.[1,2] These curves are relatively simple and one-dimensional representations of how a complex hydrocarbon material's yield and quality data such as density, sulfur and pour-point are distributed or profiled over its TBP temperatures where each cut is also referred to as a component, pseudocomponent or hypothetical in process simulation and optimization technology. Throughout the oil-refinery process the full range of hydrocarbon components is transformed (blended, reacted, separated) to smaller boiling-point temperature ranges resulting in intermediate and final products in which planning and scheduling optimization using TBP curves of the various streams can be used to effectively model these process unit-operations and predict the macroscopic (i.e., cold flow) properties of their outlet streams.[3] The entire refining process can be categorized into three distinct areas: crude-oil blending, refinery unit-operation processing and product blending[4]. Our focus is related to the last two areas.

Our proposed new technique is to integrate both the optimization of blending several streams' distillation curves together with also shifting or adjusting the cutpoints of one or more of the stream's initial boiling point (IBP) and/or final boiling point (FBP) in order to manipulate its TBP curve in an either offline or online environment. This shifting or adjusting of the TBP curve's IBP and FBP (front and back-end respectively) ultimately requires that the upstream unit-operation has sufficient handles or controls to allow this type of cutpoint variation where the solution from this higher-level optimization would provide set points or targets to a lower-level advanced process control systems, which are now commonplace in oil refineries. By shifting or adjusting the frontend and backends of the TBP curve for one or more distillate blending streams, it allows for improved control and optimization of the final product demand quantity and quality, affording better maneuvering closer and around downstream bottlenecks such as tight property specifications and volatile demand flow and timing constrictions.

Specifically, we propose a novel technique using monotonic interpolation to blend and cut distillation temperatures and evaporations for petroleum fuels in an optimization environment is proposed. Blending distillation temperatures is well known in simulation whereby cumulative evaporations at specific temperatures are mixed together then these data points are used in piece-wise cubic spline interpolations to revert back to the distillation temperatures. Our method replaces the splines with monotonic splines to eliminate Runge's phenomenon and to allow the distillation curve itself to be adjusted by optimizing its initial and final boiling points known as cutpoints. By optimizing both the recipes of the blended material and its blending component distillation curves, very significant benefits can be achieved especially given the global push towards ultralow sulfur fuels (ULSF) due to the increase in natural gas plays reducing the demand for other oil distillates. One example is provided to highlight and demonstrate the technique.

Distillation Curve Overview

Hydrocarbon streams (crude-oils, fuels, petrochemical feeds) are identified or characterized by distillation curves in terms of their quantity (yield) and quality (properties) and how they vary with respect to temperature. These data are determined by experimental methods using small-scale laboratory distillation columns in which hydrocarbon fractions are collected at certain boiling-point temperatures to define their quantity and quality data or assay at each boiling temperature. These data can then be used in mathematical models, either in rigorous engineering or simplified empirical simulation and optimization environments, to determine the yields and properties of distillation, separation or fractionation unit-operation outputs.[2] Integrated with this can be the final product recipes necessary to match demand quantity and quality specifications such as specific gravity, sulfur, pour-point, and evaporation temperatures of the fuels so that simultaneous distillation blending and cutpoint temperature shifting or adjusting can be simulated and optimized together as is proposed in this paper.

Since obtaining TBP curve data directly from a laboratory distillation run is expensive and time-consuming, the normal procedure is to determine distillation curves by applying American Society for Testing and Material (ASTM) methods such as D86 (atmospheric) and D1160 (vacuum) or simulated distillation (SD) methods, using gas chromatography (GC) such as D2887 and D7169. These are based on reduced numbers of separation stages and can fortunately be easily interconverted to TBP data with sufficient accuracy and precision.[1] Figure 1 shows both the ASTM D86 and TBP data from the Colorado School of Mines (2012) for a straight-run gasoline stream from a crude-oil distillation unit (CDU), plotted with volume yield percent as the ordinate and temperature as the abscissa.

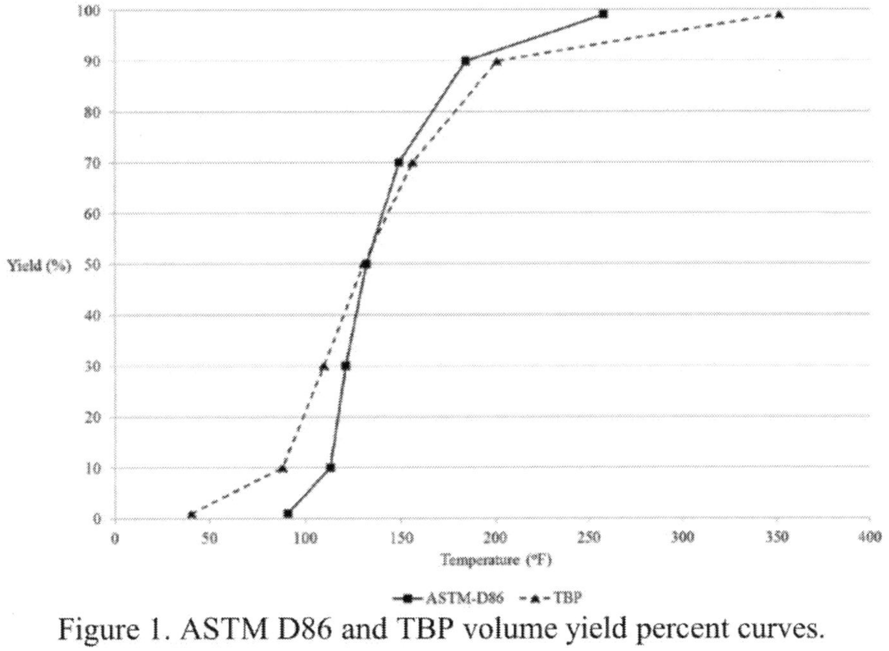

Figure 1. ASTM D86 and TBP volume yield percent curves.

A well-known observation, with regard to Figure 1, is that the ASTM D86 and TBP temperatures are near equal at their 50% yield point, whereas D86 yields are underestimated below the 50% yield and D86 yields are overestimated above the 50% yield. Riazi correlations used to convert ASTM D86 (D01 to D99) to TBP (T01 to T99) temperatures at conventional yield percentages of 1%, 10%, 30%, 50%, 70%, 90%, and 99%, where the same equations can be simply inverted in order to convert from TBP to ASTM D86 can be found in Colorado of Mines reference.[5]

Note that for the IBP and FBP we use the percent evaporated at 1% and 99% instead of 0% and 100%, and this is a well-known recommendation found in several sources. Other interconversion correlations can be found in Riazi and Daubert[6] and Daubert.[7] Also note that for temperatures greater than ca. 650 degrees F and at atmospheric pressure, thermal cracking may occur due to pyrolysis reactions. In these situations, ASTM D1160 should be use, which is conducted at a lower pressure (near vacuum), where its temperatures need to be corrected back to atmospheric pressure.

For product streams separated from unit-operations such as distillation columns and fractionation towers, the experimental TBP and/or ASTM distillation curves reflect the nonideal separation inside the vessels. For example, if we compare a CDU's crude-oil feed distillation curve with all of its product distillation curves as shown in Figure 2, it is interesting to see what are known as overlaps between the light and heavy adjacent products; this represents what is referred to as nonsharp or nonideal fractionation.[8]

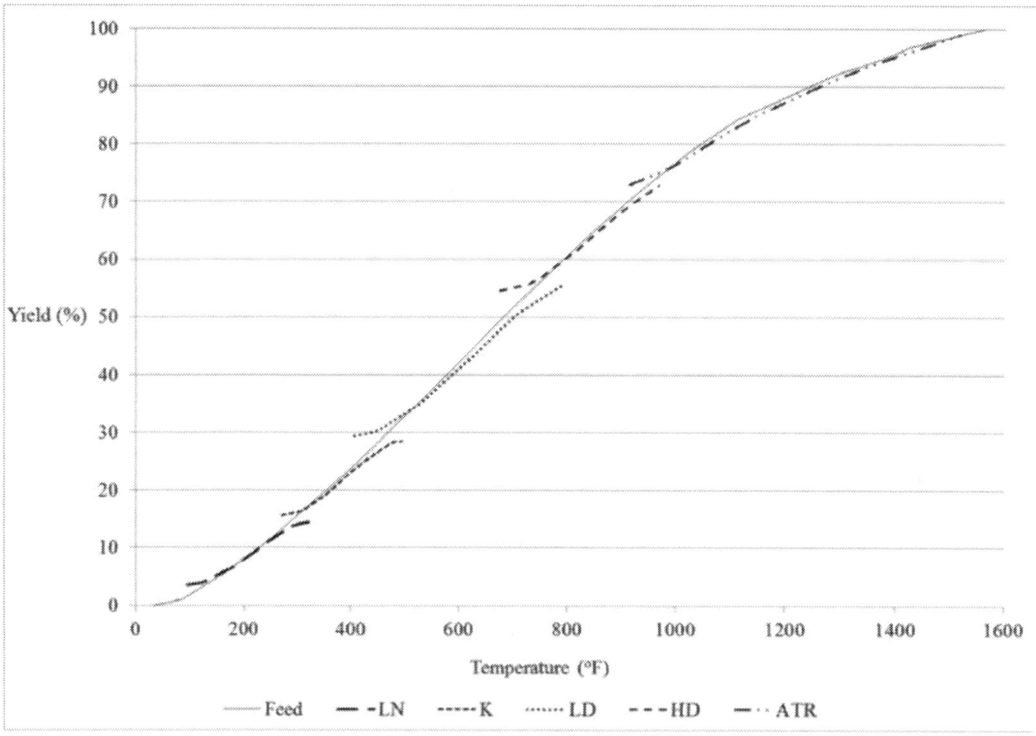

Figure 2. Feed and product yield curves.

Distillation Blending Using Monotonic Interpolation

The difficulty with blending the distillation temperatures from several component materials is that these TBP distillation temperatures do not blend linearly on a mass or volume basis. Although there is some use of what are commonly referred to as blending indices or property transforms for distillation temperatures (and used effectively for other properties such as cloud-point, pour-point, viscosity, etc.) which then allow the indexed or transformed properties to be blended linearly, literature on their mathematical structure and coefficients is scarce. Instead, an acceptable approach is to convert the TBP temperatures to evaporation cumulative compositions at several pre-defined temperatures using monotonic interpolation of which linear interpolation is inherently monotonic but may not be as accurate as higher-order monotonic interpolations. Unfortunately if non-monotonic interpolation is used such as cubic splines, then oscillating behavior around the breakpoints will occur and is well known as Runge's phenomenon. The temperature range should be defined to span all of the expected TBP temperatures for all of the distillates included in the blend at a suitable level of resolution or temperature discretization. After the set of evaporation components have been blended together using the recipes fractions of each component blended, the blended product's distillation temperatures at 1%, 10%, 30%, 50%, 70%, 90% and 99% evaporated, distilled or yielded are computed by again using monotonic interpolation where the evaporation components constitute the abscissa and the distillation temperatures the ordinate.

Monotonic interpolation is defined as a variant of cubic interpolation that preserves monotonicity of the data set being interpolated. Monotonicity is maintained by linear interpolation but not guaranteed by cubic interpolation. As is observed in both Figure 1 and 2, distillation curves are inherently monotonic, and therefore not using monotonic interpolation will result in unrealistic and unexpected prediction variability. Although linear interpolation will produce monotonic interpolations it does not preserve the shape of the curves. Shape-preserving cubic splines are available known as Piecewise Cubic Hermite Interpolation Polynomials (PCHIP). More recently, another shape-preserving or what has been termed "constrained" cubic spline interpolation method has appeared and was developed based on practical arguments by Kruger[9] and employed in the financial sector for bond yields for example.

To summarize, the overall calculation process of distillation blending is shown in Figure 3. We first inter-convert from ASTM D86 to TBP temperatures, and using the TBP temperatures at their defined yields of 1%, 10%, 30%, etc., evaporations are determined by interpolating the TBP distillation curve at selected predefined TBP temperature increments. These evaporation cumulative compositions for each blending component are mixed using the ideal blending law i.e., no excess properties and no nonlinear behavior where if the amounts or flows of components are fixed then this is also called linear blending.

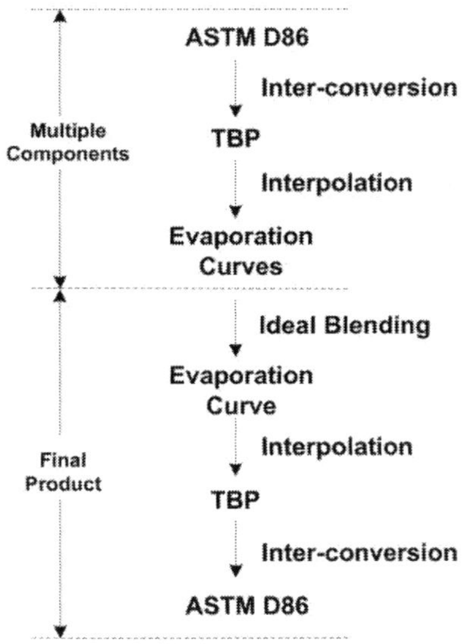

Figure 3. Flowchart of distillation blending calculation process.

After the blending of all of the components' evaporations, another interpolation step is used to compute the TBP temperatures of the final blended product at the defined yield values. Then, inter-conversion is again used to calculate the ASTM D86 temperatures which can be downloaded to various control strategies as mentioned.

Cutpoint Temperature Optimization

Distillation curves of petroleum products are typically plotted as TBP temperature versus relative yield percent, where we have chosen to represent the curves in the opposite way, indicating that the TBP temperature is now our independent variable or degree-of-freedom in the optimization. When yields, evaporations, or relative amounts distilled of the distillates are used as a function of temperature, the IBP (T01) and FBP (T99) can be manipulated for example to adjust or shift the distillation curve by interpolation and/or extrapolation in order to increase or decrease the frontend and backend of the curves, as graphically depicted in Figure 4. The incremental or marginal adjustment or shifting can also be likened to fine-tuning the distillation curve, and this will ultimately translate to changing all of the relative yield values along the total or overall TBP temperature range of the distillate stream. This will have the desired effect that we can alter the evaporation amounts in the downstream blending to better control and optimize the blend quantity and quality specifications to meet market demands. However, when we manipulate the T01 and/or T99 for any distillate component stream, these values (interconverted to D01 and D99 if required) must be sent down as targets or setpoints to an appropriate advanced process control strategy on the respective upstream unit-operation such as an atmospheric crude-oil distillation column or the main fractionator tower of a fluidized bed catalytic cracking unit (FCCU). Fortunately, such strategies are available whereby reliable online analyzers measure in real-tim or near-time (including sampling and analysis delays) ASTM D86 temperatures, which can be linked to pump-around heat exchanger duty and draw tray temperature controls to achieve the desired increase and/or decrease in initial and final cutpoint temperatures.

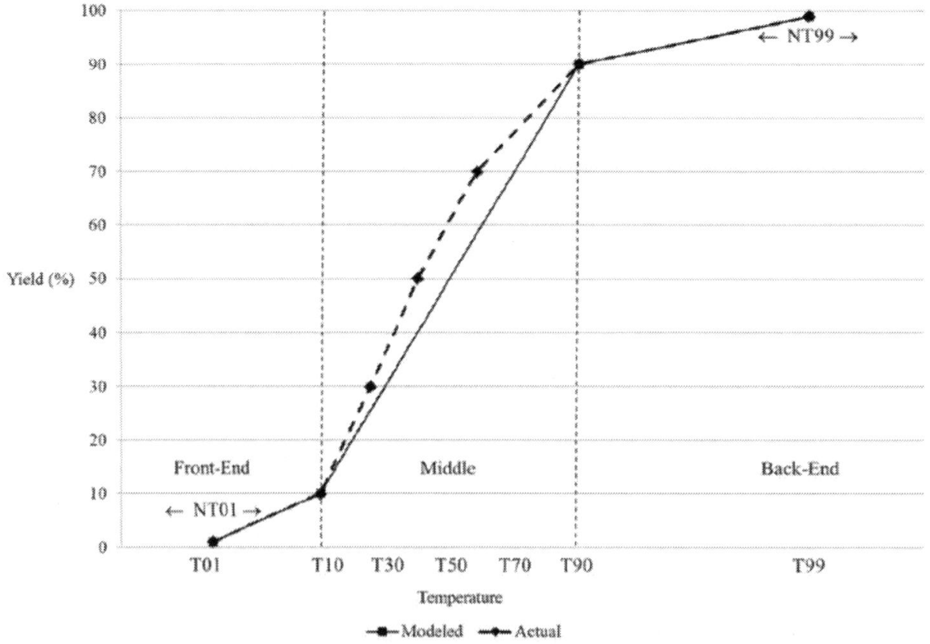

Figure 4. Distillation curve adjustment or shifting, as a function of TBP temperature.

Our new distillation curve adjustment or shifting technique to perform temperature cutpoint optimization is straightforward to implement mathematically. The concept is to assume that a typical distillation curve can be reasonably decomposed or partitioned into three distinct regions or parts i.e., a frontend, middle and backend. The different sections can be sufficiently approximated as straight-lines or piece-wise continuous from OT01 to OT10 for the frontend, OT10 to OT90 for the middle and OT90 to OT99 for the back-end where "O" stands for "old" and of course "N" stands for "new" for the new adjusted or shifted 1% and 99% temperatures. The complete formulation is found in Kelly et al.[10]

The idea for this formulation is straightforward, in the sense that if we increase or decrease the frontend and backend, then given their delta yield amounts we can easily adjust the new yields given the old yields, where the old yields are, of course, the well-defined 1%, 10%, 30%, etc. values. However, it is re-emphasized that our approach is to incrementally or marginally shift or adjust the NT01 and NT99 so as not to significantly disrupt the adjacent distillation streams, and can be controlled by specifying acceptable bounds on NT01 and NT99 in the optimization.

Also, it should be pointed out that our use of the term cutpoint is somewhat different than is found, for example in th work of Mahalec and Sanchez[11] and Menezes et al.,[2] in which a cutpoint is defined as the lower/initial and upper/final temperatures on the TBP of th entire crude oil, which hypothetically represents the crude-oil fraction distilled or evaporated from the distillation tower such as a CDU. Instead given that the TBP curve of a CDU's crude-oil mixture is not normally available in practice, and especially if the feed is to an FCCU main fractionator, which is rarely known, we use the definition commonly found in oil-refinery operations of defining cutpoint temperatures as the smallest and/or largest controllable separation or fractionation temperature (i.e., IBP, 5% or 10% and 90%, 95% or FBP) between adjacent cuts, fractions or streams from a distillation tower.

Example: Diesel Blending and Cutpoint Temperature Optimization

This case is taken, in part, from Erwin[12] in terms of using his ASTM D86 (1 atm) temperatures for his four experimental diesel components (DC1 to DC4), as shown in Table 1 and Figure 5, where the dotted line represents the final product blend distillation curve.

Table 1. Example 2's InterConverted TBP (ASTM D86) Temperatures in Degrees F.

	DC1	DC2	DC3	DC4
1%	305.2 (353)	322.2 (367)	327.0 (385)	302.4 (368)
10%	432.9 (466)	447.1 (476)	405.2 (435)	369.7 (407)
30%	521.6 (523)	507.1 (509)	457.1 (462)	441.0 (449)
50%	565.3 (551)	549.5 (536)	503.3 (492)	513.8 (502)
70%	606.4 (581)	598.4 (573)	551.1 (528)	574.3 (550)
90%	668.3 (635)	666.1 (634)	605.8 (574)	625.4 (592)
99%	715.7 (672)	757.7 (689)	647.0 (608)	655.2 (620)

Our blending and cutpoint temperature optimization objective function is to maximize the flow of DC1 and DC2, subject to their relative and arbitrary pricing of 0.9 for DC1 and 1.0 DC2, with lower and upper bounds of 0.0 and 100.0 each. For simplicity, we have fixed each of the flows for DC3 and DC4 to a marginal and arbitrary value of 1.0, and the total blend flow cannot exceed 100.0. The typical

ASTM D86 temperature specification for international diesel sales is D10 ≤ 480, 540 ≤ D90 ≤ 640, and D99 ≤ 690 degrees F. To make this problem more interesting, we have set the bounds to D10 ≤ 470, 540 ≤ D90 ≤ 630 and D99 ≤ 680. This ensures that it is not possible to satisfy the diesel blend with the most valuable DC2 component as its D10, D90, and D99 temperatures are all greater than our blended diesel's distillation temperature specification. Furthermore, it is also not possible to fill the blend with all DC1, because its D90 does not comply with the specification. As such, this forces a mixture of DC1 and DC2 and as we shall see, this also requires an adjustment or shifting to the distillation curve for DC1 where only the DC1 material is allowed to have manipulated cutpoint temperatures. At the optimized solution NT01 = 312.8 and NT99 = 689.3, which are both different than their original values of OT01 = 305.2 and OT99 = 715.7. Since both their frontend and backend temperatures have been reduced we expect that the new flow for DC1 is less than the old flow and it is NF = 37.06 and OF = 39.24; consequently, the flow for DC2 is 60.94, which is consistent with the total flow of 100.0. The new and optimized TBP curve for DC1 given its front- and back-end shifts is now [(1.053%,312.8), (10.015%,432.9), (31.188%,521.6), (52.361%,565.3), (73.534%,606.4), (94.707%,668.3), (98.995%,689.3)], where the new yields are computed.

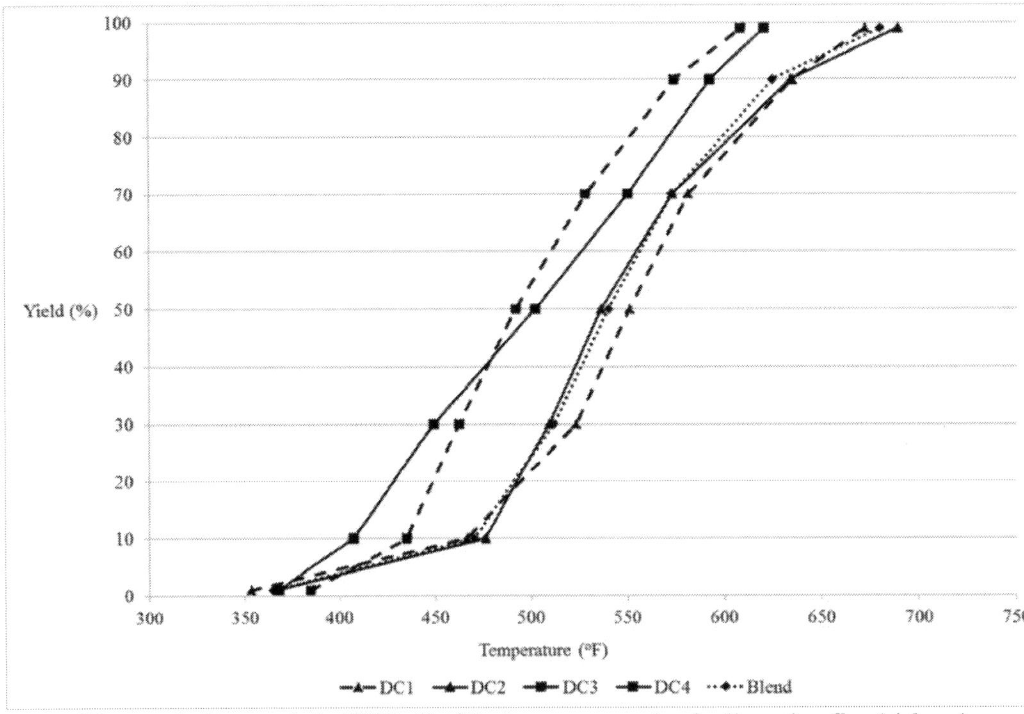

Figure 5. Example's TBP distillation curves, including the final blend.

Conclusions

In this paper, we have presented the fine points of performing distillate blending and cutpoint temperature optimization using monotonic interpolation. The first step of the procedure is to convert from faster and less-expensive experimental methods, such as ASTM D86, to the slower and more expensive TBP temperatures, using well-established interconversion analytical expressions. These interconverted TBP temperatures are converted to cumulative evaporations similar to pseudocomponents in process design simulators using monotonic interpolation, blended linearly by mass or volume, then converted back to TBP temperatures using another monotonic interpolation. If required, these values are interconverted back to ASTM D86, and this defines the second step. The third step is to adjust or shift the frontend and/or backend of one or more of the component TBP distillation curves by optimizing the T01 (1%) and/or T99 (99%) cutpoint temperatures, while respecting the relative yields of the adjusted or shifted distillation curves. It is the third step that, to our knowledge, has not been discussed in the literature prior to this work and if distillation blending indices are used instead of mixing, cutting, and interpolating with evaporation components then this type of temperature cutpoint optimization is not possible. As highlighted in the Introduction, significant economic benefits can be achieved by implementing such an algorithm, especially for ultralow sulfur fuels which always require accurate prediction of key evaporation points on the final product distillation curves.

1. Riazi, M. R. (2005), "Characterization and Properties of Petroleum Fractions. American Society for Testing and Materials".
2. Menezes, B.C.; Kelly, J. D. and Grossmann, I. E. (2013), "Improved Swing-Cut Modeling for Planning and Scheduling of Oil-Refinery Process. *Ind. Eng. Chem. Res.*, 52, pp. 18324-18333.
3. Kelly, J. D. (2004), "Formulating Production Planning Models". *Chem. Eng. Prog.*, 100, pp. 43-50.
4. Jia, Z.; Ierapetritou, M. and Kelly, J. D. (2004, "Refinery Short Term Scheduling Using Continuous Time Formulation: Crude-Oil Operations". *Ind. Chem. Eng. Res.*, 42, pp. 3087-3097.
5. Colorado School of Mines. Refinery Feedstocks and Products - Properties & Specifications. http://home.comcast.net/~jjechura/CHEN409/02_Feedstocks_&_Products.pdf (accessed Jan 25, 2014).

6 Riazi, M.R.; Daubert, T. E. (1986), "Analytical Correlations Interconvert Distillation Curve Types". *Oil & Gas Journal*. 84, pp.50-57.

7 Daubert, T. E. (1994), "Petroleum Fraction Distillation Interconversion". *Hydrocarbon Process*, 73(9), pp. 75-78.

8 Li, W.; Hui, C.; Li, A. (2005), "Integrating CDU, FCC and Blending Models into Refinery Planning". *Comput. Chem. Eng*. 29, pp. 2010-2028.

9. Kruger, C. J. C. (2002), "Constrained Cubic Spline Interpolation for Chemical Engineering Applications". www.korf.co.uk./spline.pdf (accessed Jan 15, 2014).

10 Kelly, J. D.; Menezes, B.C. and Grossmann, I. E., (2014), "Distillation Blending and Cutpoint Temperature Optimization using Monotonic Interpolation". Ind. Eng. Chem. Res., 53, pp. 15146-15156.

11 Mahalec, V.; Sanchez, Y. (2012), "Inferential Monitoring and Optimization of Crude Separation Units via Hybrid Models". *Comput. Chem. Eng.*, 45, pp. 15-26.

12 Erwin, J. (1992). "Assay of Diesel Fuel Components Properties and Performance". Symposium of Processing and Product Selectivity of Synthetic Fuels. American Chemical Society: Washington, DC, August

Uncertainty Quantification in Stochastic Multiscale Models of Heterogeneous Catalysis

Marcel Núñez and Dionisios G. Vlachos

In the design of novel catalysts, first-principles based microkinetic modeling plays a key role in the identification of material properties that affect catalytic activity. Spatial kinetic Monte Carlo (KMC) in particular is an integral tool in the study of complex physical phenomena present in heterogeneous catalytic systems. Although KMC has been successfully applied in several systems, the accuracy of its predictions is not well understood. Disparity in length and time scales as well as the stochastic nature of KMC pose challenges for the development of tools for meaningful uncertainty quantification. A framework is needed in which model reliability can be assessed with respect to two requirements. First, model structure must account for all of the relevant physics. Second, parameters must be sufficiently accurate.

Sensitivity analysis addresses the latter. It identifies which parameters most heavily impact model predictions and provides physical insights by identifying the rate-determining steps of a reaction network. The usual finite-difference application of sensitivity analysis is computationally intensive for stochastic models. Reweighting techniques reduce the computational cost by obtaining all gradient information in a single run. However, disparity in reaction time scales introduces statistical noise into reweighting derivative estimates that can lead to physically wrong conclusions. In this work, sensitivity analysis through reweighting is extended to systems with multiple time scales. We show that through this approach, the noise in derivative estimates for stiff systems can be reduced considerably. The approach is demonstrated in an analytically solvable linear system.

Eigenvalue Analysis of Stochastic Non-linear Chemical Reaction Networks using Moment Closure

Patrick Smadbeck, Yiannis Kaznessis

Random noise has long been known to contribute significantly to the dynamics of chemical reaction networks in systems that contain species with low copy numbers (e.g. DNA in a bacterium). These systems are common in biology and are typically referred to as being far from the thermodynamic limit. The fact that random noise or fluctuations are vital to understanding the behavior of such systems necessitates the use of stochastic simulation algorithms. While it is getting progressively easier to run the large kinetic Monte Carlo (kMC) simulations necessary for the modeling of biological systems, many issues remain. When compared to deterministic models, for example, it is not easy to obtain steady-state results or perform prior analysis for stochastic models using kMC simulations.

Decades ago it was recognized that while individual kMC trajectories are non-deterministic, the underlying distribution and its statistics are indeed deterministic. The equations that govern the dynamics of these statistics are rather simple to obtain, but the set of equations is often under-defined. In such cases a *moment-closure* technique is needed. This closure scheme is an additional statistical relationship that allows for the determination of moment dynamics. We have recently developed a numerical closure scheme called Zero-Information Closure[1] that shows promise in its versatility and accuracy. This research is focused on extending our moment closure research to the deterministic analysis of stochastic systems.

Using several examples we show that moment-closure techniques may offer a novel deterministic option for the analysis of stochastic chemical reaction network simulations. In particular, steady-state distributions are obtained quickly, the resulting linearized Jacobian is determined, and an eigenproblem is solved to determine both eigenvalues and moment vectors. We were also able to show, using comparisons to kinetic Monte Carlo correlation functions, that these eigenvalues and eigenvectors provide an accurate representation of the steady-state dynamics of these systems. The possibility of using moment-closure techniques to quickly identify interesting steady-state and dynamic behavior in stochastic chemical reaction simulation is a landmark achievement for master equation closure methods. By changing the focus of stochastic simulation from the non-deterministic trajectories of kMC algorithms to the deterministic statistics of the ensemble distribution a whole panel of new analytical techniques may become available to biological modeling in the future.

[1] P Smadbeck and YN Kaznessis. "A closure scheme for chemical master equations." Proceedings of the National Academy of Sciences (2013)

The role of time-scale analysis in simulation of ALD and CVD surface reaction kinetics

Elizabeth Remmers, Curtisha D. Travis, David Arana-Chavez, and Raymond A. Adomaitis*

Department of Chemical & Biomolecular Engineering, Institute for Systems Research, University of Maryland, College Park, MD

Atomic layer deposition (ALD) and chemical vapor deposition (CVD) are thin-film manufacturing techniques that share a number of common characteristics. From a process perspective, each is used to deposit thin films from gas-phase reactants in a controlled and conformal manner. Likewise, both deposition processes are inherently nonlinear and time-dependent, and mathematical model elements describing the deposition surface reaction and gas-phase precursor transport are strongly coupled. The transport and reaction models for both must describe chemical and physical phenomena over multiple time and length scales. Because of this overlap, some well-established modeling concepts developed by the CVD community (Jones and Hitchman, 2009; Middleman and Hochberg, 1993; Smith, 1995) can be used directly in the analysis of ALD systems.

One important distinction, however, can be made between the two deposition processes: the notion of the steady-state deposition *rate* in CVD does not exist for ALD. The *instantaneous* rate of ALD depends strongly on the state of the growth surface, and this state changes continuously through each exposure and purge period until saturation is reached. The completely dynamic nature of the ALD process adds considerably to the difficulty of developing simulators, because the entire process cycle must be modeled, i.e., the extent of reactions taking place during half-cycle A influence those taking place during half-cycle B, and vice versa. Due to the heterogeneous nature of these ALD reactions, adsorption, desorption and surface reaction kinetics play a crucial role in the deposition process (Elliott and Greer, 2004; George, 2010; Puurunen, 2005). Careful chemical and mathematical analysis is required to resolve the multiple time scales present in this process to identify the rate-limiting steps.

A range of ALD models has been developed at various levels of detail and theory (Elliott, 2007; Holmqvist *et al.*, 2012) to accommodate specific modeling objectives. Of these, atomic-scale models have produced insightful results by employing *ab initio* methods, such as density functional theory (DFT), to investigate reaction mechanisms, energetically favored reaction pathways, proton transfer mechanisms, structures of surface reaction intermediate species and reaction energetics corresponding to reactions along each stage of ALD growth (see Elliott (2012) for a comprehensive review of atomic-scale ALD models). Results of such studies have been combined with kinetic Monte Carlo (kMC) simulations to quantify nucleation (Dkhissi *et al.*, 2009) and growth kinetics (Dkhissi *et al.*, 2008; Mazaleyrat *et al.*, 2005). Empirical growth models have also been developed to describe ALD kinetics during the initial stages of nucleation (Alam and Green, 2003; Yim *et al.*, 2008) and island growth (Nilsen *et al.*, 2007; Puurunen *et al.*, 2004; Puurunen and Vandervorst, 2004). Additionally, surface reaction models have been developed to predict growth kinetics and the effects of operating parameters (e.g., temperature, pressure, precursor exposure time) during undersaturating conditions and leading to self-limiting ALD growth (Holmqvist, 2013; Lim, Park, and Wang, 2000a,b, 2001; Park *et al.*, 2000; Travis and Adomaitis, 2013a).

The ALD modeling efforts described above are revealing an increasingly complex picture of the surface processes taking place even in those systems once thought to represent "ideal ALD," such as alumina ALD using trimethylaluminum and water as precursors (Ott, *et al.*, 1996; Puurunen, 2005; Travis and Adomaitis, 2013c). When broken down into elementary reaction steps with reaction rates determined

*Corresponding author: adomaiti@umd.edu

using Absolute-Rate or Transition-State Theory (Glasstone, Laidler, and Eyring, 1941), many of the ALD surface processes can be written as

$$A \overset{K}{\rightleftharpoons} A^{\ddagger} \overset{k}{\rightarrow} B$$

with equilibrium and rate constants K and k, respectively, and transition state or critical complex $A^{\ddagger}$. Even in this simple monomolecular reaction we observe the chemical kinetics will ultimately be modeled by a set of differential equations describing the irreversible and rate-limiting steps, and many of the surface states will be governed by equilibrium processes and so will be described by algebraic models.

In this paper, we will describe our work in developing a rational framework for the analysis and simulation of the increasingly complex ALD and CVD surface process models and to make a connection between the model dynamic reduction work carried out, for example, in the combustion (Lam and Goussis, 1994; Maas and Pope, 1992) and chemical process systems (Vlachos, 1996; Vora and Daoutidis, 2001) communities and model reduction for *deposition system dynamics*. To put the need in perspective, consider modeling the HfO_2 ALD process used in current semiconductor manufacturing, where thirteen classes of chemical reactions (each with numerous sub-cases) are modeled by Shirazi and Elliott (2014). Many CVD models are equally complex; for example, Coltrin, Kee, and Miller (1986) studied reaction and transport modeling of gas-phase species for Si CVD with 27 reactions and 11 species, and Norinaga and Deutschmann (2007) modeled pyrolysis of light hydrocarbons for CVD of carbon using 227 species and 827 reactions.

Our primary objective is to investigate the mathematical structure of the differential-algebraic (DAE) systems of equations describing surface reaction species dynamics during thin-film deposition processes. Our research is motivated by the challenges presented by writing a well-posed DAE model for these systems as well as the difficulties encountered when numerically solving these systems. Using a perturbation analysis approach (Baldea, Daoutidis and Kumar, 2006; Kumar, Christofides, and Daoutidis, 1998; Nayfeh, 1981), we demonstrate that the deposition kinetics decomposes naturally into slow (deposition reactions), fast (equilibrium reactions), and instantaneous (conserved quantities) time scales. A key contribution of our work is the development of a reaction network factorization procedure that partitions the surface reaction and deposition species *dynamic* balances into the distinct time scale ranges described. Under what conditions this procedure works, understanding the implications of fixed points for dynamic ALD processes, interpreting reaction fluxes, and extending the methods to spatially distributed processes in the context of representative thin-film application domains will be discussed. Physical interpretation of DAE system initial conditions (Biegler, 2000; Pantelides, 1988) for these surface processes constitutes another important research direction in this project; results will be presented which illuminate ALD growth surface dynamics at the start of each exposure and purge period.

References

Alam, M. A., and M. L. Green, Mathematical description of atomic layer deposition and its application to the nucleation and growth of HfO_2 gate dielectric layers, *J. Appl. Phys.* **94** 3403-3413 (2003).

Baldea, M., P. Daoutidis, and A. Kumar, Dynamics and control of integrated networks with purge streams, *AIChE J.* **52** 1460-1472 (2006).

Biegler, L. T., *Differential-Algebraic Equations (DAEs)*, Lecture notes (2000).

Coltrin M. E., R. J. Kee, and J. A. Miller, A mathematical model of silicon chemical vapor deposition, *J. Electrochem. Soc.* **133** 1206-1213 (1986).

Dkhissi, A., A. Estéve, C. Mastail, S. Olivier, G. Mazaleyrat, L. Jeloaica, and M. D. Rouhani, Multiscale modeling of the atomic layer deposition of HfO_2 thin film grown on silicon: How to deal with a kinetic Monte Carlo procedure, *J. Chem. Theory Comput.* **4** 1915-1927 (2008).

Dkhissi, A., G. Mazaleyrat, L. Jeloaica, and M. D. Rouhani, Nucleation and growth of atomic layer deposition of HfO_2 gate dielectric layers on silicon oxide: A multiscale modelling investigation, *Phys. Chem. Chem. Phys.* **11** 3701-3709 (2009).

Elliott, S. D., Models for ALD and MOCVD growth of rare earth oxides, *Topics Appl. Phys.* **106** 73-86 (2007).

Elliott, S. D., Atomic-scale simulation of ALD chemistry, *Semicond. Sci. Technol.* **27** 074008 (2012).

Elliott, S. D. and J. C. Greer, Simulating the atomic layer deposition of alumina from first principles, *J. Mater. Chem.* **14** 3246-3250 (2004).

George, S. M., Atomic layer deposition: An overview, *Chem. Rev.* **110** 111-131 (2010).

Glasstone, S., K. J. Laidler, and H. Eyring, *The Theory of Rate Processes*, First edition McGraw-Hill Book Company, New York, (1941).

Holmqvist, A., *Model-based Analysis and design of Atomic Layer Deposition Processes*, PhD Thesis, Lund University (2013).

Holmqvist, A., T. Törndahl, and S. Stenström, A model-based methodology for the analysis and design of atomic layer deposition processes–Part I: Mechanistic modelling of continuous flow reactors, *Chem. Eng. Sci.* **81** 260-272 (2012).

Jones, A. C. and M. L. Hitchman, *Chemical Vapour Deposition*, The Royal Society of Chemistry, London (2009).

Kumar, A., P. D. Christofides, P. Daoutidis, Singular perturbation modeling of nonlinear processes with nonexplicit time-scale multiplicity, *Chem Engng Sci.* **53** 1491-1504 (1998).

Lam, S. H. and D. A. Goussis, The CSP method for simplifying kinetics, *Int. J. Chem. Kinetics* **26** 461-486 (1994).

Lim, J., J. Park, and S. Wang, Kinetic modeling of film growth rates of TiN films in atomic layer deposition, *J. Appl. Phys.* **87** 4632-4634 (2000).

Lim, J., J. Park, and S. Wang, Analysis of a transient region during the initial stage of atomic layer deposition, *J. Appl. Phys.* **88** 6327-6331 (2000).

Lim, J., J. Park, and S. Wang, Kinetic modeling of film growth rate in atomic layer deposition, *J. Electrochem. Soc.* **148** C403-C408 (2001).

Maas, U. and S. B. Pope, Simplifying chemical kinetics: Intrinsic low-dimensional manifolds in composition space, *Combustion and Flame* **88** 239-264 (1992).

Mazaleyrat, G., A. Estéve, L. Jeloaica, and M. Djafari-Rouhani, A methodology for the kinetic Monte Carlo simulation of alumina atomic layer deposition onto silicon, *Comput. Mater. Sci.* **33** 74-82 (2005).

Middleman, S. and A. K. Hochberg, *Process Engineering Analysis in Semiconductor Device Fabrication*, McGraw-Hill, Inc., (1993).

Nayfeh, A. H., *Introduction to Perturbation Techniques*, J. Wiley, New York, (1981).

Nilsen, O., C. E. Mohn, A. Kjekshus, and H. Fjellvåg, Analytical model for island growth in atomic layer deposition using geometrical principles, *J. Appl. Phys.* **102** 024906:1-024906:13 (2007).

Norinaga, K. and O. Deutschmann, Detailed kinetic modeling of gas-phase reactions in the chemical vapor deposition of carbon from light hydrocarbons, *Ind. Eng. Chem. Res.* **46** 3547-3557 (2007).

Ott, A. W., K. C. McCarthy, J. W. Klaus, J. D. Way, and S. M. George, Atomic layered controlled deposition of Al_2O_3 films using a binary reaction sequence chemistry, *Appl. Surf. Sci.* **107** 128-136 (1996).

Pantelides, C. C., The consistent initialization of differential-algebraic systems, *SIAM J. Sci, Stat. Comput.* **9** 213-231 (1988).

Park, H., J. Min, J. Lim, and S. Kang, Theoretical evaluation of film growth rate during atomic layer epitaxy, *Appl. Surf. Sci.* **158** 81-91 (2000).

Puurunen, R. L., W. Vandervorst, W.F. A. Besling, O. Richard, H. Bender, T. Conard, C. Zhao, A. Delabie, M. Caymax, and S. de Gendt, Island growth in the atomic layer deposition of zirconium oxide and aluminum oxide on hydrogen-terminated silicon: Growth mode modeling and transmission electron microscopy, *J. Appl. Phys.* **96** 4878-4889 (2004).

Puurunen, R. L. and W. Vandervorst, Island growth as a growth mode in atomic layer deposition: A phenomenological model, *J. Appl. Phys.* **96** 7686-7695 (2004).

Puurunen, R. L., Surface chemistry of atomic layer deposition: A case study for the trimethylaluminum/water process, *Appl. Phys. Rev.* **97** 121301:52 (2005).

Shirazi, M. and S. D. Elliott, Atomistic kinetic Monte Carlo study of atomic layer deposition derived from density functional theory, *J. Comput. Chem* **35** 244-259 (2014).

Smith, D. L., *Thin-Film Deposition; Principles and Practice* McGraw-Hill, Inc. New York (1995).

Travis, C., D. and R. A. Adomaitis, Modeling ALD surface reaction and process dynamics using absolute reaction rate theory, *Chem. Vapor Depos.* **19** 4-14 (2013).

Travis, C., D. and R. A. Adomaitis, Modeling alumina atomic layer deposition reaction kinetics during the trimethylaluminum exposure, *Theo. Chem. Accounts* **133** 1414 (2013).

Vlachos, D. G., Reduction of detailed kinetic mechanisms for ignition and extinction of premixed hydrogen/air flames, *Chem. Engng Sci.* **51** 3979-3993 (1996).

Vora, N. and P. Daoutidis, Nonlinear model reduction of chemical reaction systems, *AIChE J.* **47** 2320-2332 (2001).

Yim, S., D. Lee, K. Kim, S. Kim, T. Yoon, and K. Kim, Nucleation kinetics of Ru on silicon oxide and silicon nitride surfaces deposited by atomic layer deposition, *J. Appl. Phys.* **103** 113509:1:9 (2008).

MODEL PREDICTIVE CONTROL OF THERMAL STRESSES IN CZOCHRALSKI CRYSTAL GROWTH PROCESS

JAVAD ABDOLLAHI, MOJTABA IZADI, STEVAN DUBLJEVIC

The evolution of thermal stresses in Czochralski crystal growth process has an important role in dislocation density creation and crystal grown quality [1]. The thermal stresses are direct result of developed thermal gradients within the crystal domain and successful regulation of the CZ crystal growth process implies tight manipulation of the temperature distribution evolution in order to minimize the effect of thermal stresses on crystal quality.

In this work, an adequate axisymmetric crystal growth model is developed and the heat transfer dynamics modelled by the two-dimensional (2D) parabolic PDEs and thermal stress model are considered within the optimal regulator setting. By assuming that the temperature evolution within the grown crystal can be reconstructed form the top surface of grown crystal measurement [2], a model predictive control (MPC) is proposed in order to minimize the thermal stresses over the entire crystal domain during the growth process, see Figure below. The crystal radius and length evolution control is realized by the feedback linearization and yields an ODE which generates the evolution of the conduction-convection PDE model with time-varying domain. A mesh moving scheme is utilized to discretize the time-varying geometry of the domain in developing the finite-element model of the process [3]. Due to the fact that the evolution of the crystal domain is known from the crystal growth dynamics, the Arbitrary Lagrangian-Eulerian method is used to spatially discretize the domain. Thermal stresses are obtained through the elastic thermal model and calculated assuming quasi steady state. Finite element method is utilized for order reduction and solution of the coupled temperature and thermal stress distribution evolution over the crystal domain. Model predictive control strategy is realized to maximize crystal cooling rate and satisfy the constraints on thermal stresses within the crystal. Finally, the proposed method is implemented on a high fidelity finite element model of the process and the simulation results are presented to validate the proposed method.

(1) Sinno T., Brown R. Modelling microdefect formation in Czochralski silicon, J. Electrochemical Soc. 1999;146(6):230–2312.

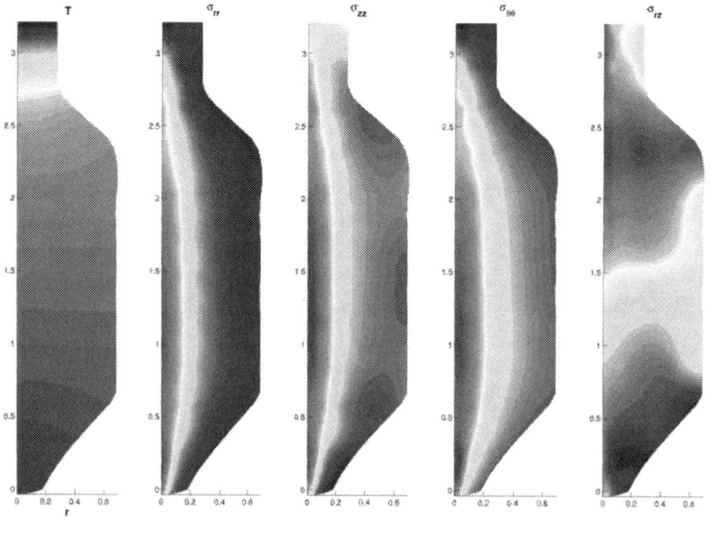

FIGURE 1. Stress distribution in the grown CZ crystal.

(2) Javad Abdollahi, Mojtaba Izadi, and Stevan Dubljevic, Temperature Distribution Reconstruction in Czochralski Crystal Growth Process, AIChE Journal, accepted, 2014.
(3) Reddy JN., An Introduction to Nonlinear Finite Element Analysis, Oxford University Press, 2004.

379245 Multi-Length Interfacial Dynamics in Microchannels Via Spectral Boundary Element Algorithms

Abstract:

The dynamics of multi-phase soft particle, such as droplets, capsules and erythrocytes, in confined solid geometries under Stokes flow conditions is a problem of great technological and fundamental interest since it is encountered in a broad range of industrial, biomedical and physiological processes. The last few decades, interfacial dynamics in Stokes flow via the solution of boundary integral equations has developed considerably. The main benefits of this approach are the reduction of the problem dimensionality by one and the great parallel scalability. Thus, considerable progress has been made for the study of droplets and bubbles in viscous flows, as well as membrane-like interfaces such as those in artificial capsules and erythrocytes, in microfluidic channels and the microcirculation.

In this talk we will present our efforts to study efficiently a family of particularly challenging interfacial problems; namely, the case of multi-length interfacial dynamics in Stokes flow. In many interfacial processes, multiple length scales appear and affect the interfacial dynamics. Common examples constitute the drop coalescence process (which, beyond the drop size, also includes the small gap between the interfaces), droplets/cells in close proximity to solid surfaces (as in microfluidic channels, porous media, and the microcirculation) as well as the appearance of tips and necks during large interfacial deformation in strong flows.

Currently the study of multi-length interfacial dynamics in three dimensions constitutes a computational challenge owing to the requirement of extreme numerical accuracy for the accurate determination of the dynamics between the regular (particle-size) scale and the small length scale present in these problems. To overcome this difficulty, we have developed a series of efficient and highly-accurate interfacial algorithms based on our Spectral Boundary Element implementation for Stokes flow. As applications for multi-length interfacial systems, we will present our investigation of large deformation of soft particles, involving pointed tips and tails in micro-channels. We will also present our investigation of die particles, which can be orders of magnitude smaller than the channel size.

388832 Interval-Analysis-Based Determination of Globally Optimal Total Annualized Cost for Compressor Sequences

Abstract:

In this work, the minimum total annualized cost (TAC) problem is studied for a series of non-isentropic compressors and coolers that brings a gas with constant compressibility factor from a specified initial pressure and temperature to a specified final pressure and the same temperature. For constant heat capacity, constant compressibility factor gases, additional properties of the globally optimal compressor sequence are analytically established. These properties permit development of an interval-analysis-based solution strategy that identifies compressor configurations yielding the globally-minimum TAC for varying network outlet pressures. Several case studies are presented to illustrate the developed theorems and solution strategy.

Parameter estimation in cellular systems modeled as stochastic differential equations

Kirti M. Yenkie, [a,b] Urmila M. Diwekar [a,b] and Andreas A. Linninger[a]
[a]*Department of Bio Engineering, University of Illinois, Chicago, IL 60607 - USA*
[b]*Center for uncertain Systems: Tools for Optimization & Management (CUSTOM), Vishwamitra Research Institute, Clarendon Hills, IL 60514 – USA*

Abstract

In cellular systems the factors like low-dimensionality, high macromolecular content, spatial heterogeneity render the classical laws of mass action invalid. Stochastic models such as the popular Gillespie algorithm are more suitable for modeling cellular events. However, the determination of rate laws from data in stochastic cellular environments is a difficult task, because the parameter estimation involves non-differentiable functions. Moreover, reaction propensities vary at different times, so their expected values cannot be computed with classical inversion methods.

To overcome these problems, we introduce a novel parameter estimation technique using stochastic differential equations. Stochastic differential equations(SDEs) are widely used in disciplines like finance, engineering, environment, physics, population dynamics and medicine. The predictive power of SDEs lies in the choice of the parameter values which can describe the real noisy data. Our research shows that SDEs can be used for finding an accurate and computationally feasible method for estimation of unknown reaction rates in cellular events. Even though parameter estimation in SDEs is a well studied problem in literature, most methods are time consuming, involve computational difficulties and at times result in objective functions which are non-differentiable.

Our aim is to model and estimate parameters in cellular systems using SDEs with a sutiable Ito representation for the stochastic behavior. The SDE consist of two parts; the drift term or the deterministic part and the diffusion term or the stochastic component. We decompose the parameter estimation problem in two levels; the inner level is the estimation of parameter (σ) due to randomness, usually addressed as the 'standard deviation' in the diffusion part and the outer level involves determination of parameters (θ) in the drift part (see eq.1).

$$dx_t = a_1(x_t, \boldsymbol{\theta}, t)dt + \sigma b_1(x_t, t)dz_t \tag{1}$$

In the novel parameter estimation approach, σ is estimated by analyzing the available data and, thus, the parameter in the diffusion term is evaluated without any optimization or probabilistic methods. The parameters in the drift term (θ) are estimated by using kinetic inversion methods available for deterministic systems. However, instead of using the least square error minimization as the objective, the objective function is modified to include the parameter due to randomness. The results of the SDE are evaluated by using the estimated parameters in the model and solving the system for the expected value. The expected value is then compared with the actual data and the results from the deterministic model.

The goal is to apply the proposed method to case studies like cellular pathways and metabolic models of different species and compare the SDE model projections and parameter estimates with deterministic models, stochastic simulation algorithms and conventional parameter estimation techniques to pave a new path in cellular systems modeling.

Keywords: stochastic, randomness, Ito, inversion

386984 Well Placement Optimization Under Geological Uncertainty with Optimally Selected Realizations

Abstract:

Uncertainty in the geological properties of a reservoir has a major impact in determining well locations and well production parameters [1]. To quantify reservoir production performance, flow simulations are performed on multiple geological model realizations generated by geostatistical tools. Since reservoir flow simulation is a computationally intensive task due to complex geological heterogeneities and numerical thermal modeling, generally only few selected realizations are chosen for flow simulations. In our previous work [2], an optimization based geological realization selection method has been proposed, which minimize the probability distance between the discrete distribution represented by the superset of realizations and the reduced discrete distribution represented by the selected realizations.

In this work, the reservoir well placement optimization problem is studied with the consideration of geological uncertainty. To account for the geological uncertainty, the well placement problem is formulated as a mean-variance optimization problem by maximizing the risk averted expected cumulative oil production. The objective function evaluation is based on time-consuming reservoir flow simulation, so derivative-free optimization techniques have been adopted to solve the optimization problem. Specifically, we have implemented a Gaussian surrogate model based optimization algorithm and have also adopted a mesh adaptive direct search algorithm NOMAD [3] to generate the well placement plan. To compare the solution quality, different well placement plans are generated by using: 1) the superset of realizations; 2) the subset of realizations determined by the optimal realization selection method [2]; and 3) the subset of realizations selected through traditional ranking method [4].

Case study results demonstrate the benefits of optimal realization selection. The well placement plan obtained using optimally selected realizations has higher cumulative oil production rates than the well placement plan obtained based on realizations generated through ranking method. The optimally selected realizations lead to a distribution of cumulative oil production that is close to the cumulative oil production distribution obtained using the superset of realizations. Those results suggest that optimal realization selection algorithm is promising in selecting the geological realizations and reducing the flow simulation efforts, and can lead to good well placement plans that capture the distribution of the geological uncertainty well.

References:

[1] Bangerth, W., Klie, H., Wheeler, M. F., Stoffa, P. L., & Sen, M. K. On optimization algorithms for the reservoir oil well placement problem. Computational Geosciences, 2006, 10, 303.

[2] Rahim, Shahed, Li, Zukui, Trivedi, Japan. Reservoir Geological Uncertainty Reduction: an Optimization based Method using Multiple Static Measures. Mathematical Geosciences, 2014, under review.

[3] S. Le Digabel. Algorithm 909: NOMAD: Nonlinear optimization with the MADS algorithm. ACM Transactions on Mathematical Software, 2011, 37, 44.

[4] Fenik D.R., Nouri A., Deutsch C.V. Criteria for ranking realizations in the investigation of SAGD reservoir performance. Canadian International Petroleum Conference. 2009, 191.

390540 Multi-Objective Optimal Sensor Deployment Under Uncertainty for Advanced Power Systems

Abstract:

Advanced power plants using an integrated gasification combined cycle (IGCC) offer a competitive and economical means to produce electricity and thermal energy with significantly reduced emission levels. An efficient, safe, and reliable operation of an IGCC plant requires effective strategies for monitoring and control. Sensors play a key role in monitoring and control of the IGCC plant. The sensors in an IGCC plant operate under harsh environments leading to failures. Sensor failure can not only create operational difficulties but can also lead to poor monitoring of the process equipment and inability to meet environmental emission limits. Sensors for IGCC application vary widely in cost, accuracy, and reliability. The objectives of the proposed research is to develop a fundamental understanding of the relationships among the sensor placement, interaction with the process, and hierarchal interactions of the sensor intelligence that can help in identifying the type, number, and location of sensors for maximum effectiveness and efficiency of the measurement technology and the process. The optimization of the location, number, and type of sensors will contribute to enhanced control of a process, and to this end, new fundamental algorithms and hybrid hardware- virtual sensor architectures are proposed to be developed. In order to synthesize a sensor network that can maximize the observability, and efficiency, a multi-objective optimization framework is necessary. Uncertainties are inherent in such a problem. Therefore, the problem of sensor placement is formulated as a multi-objective stochastic programming problem, the solution of which is the optimal sensor network for an IGCC plant in the face of uncertainties in performance, reliability, efficiency, and power demand. In order to solve this large scale multi-objective optimization under uncertainty, a new algorithmic framework is developed. The framework have components such as (1) an efficient sampling technique for uncertainty analysis, (2) the decomposition strategies used in L-shaped method for stochastic programming, (3) a better optimization of nonlinear system (BONUS) algorithm to circumvent multiple model runs for sampling at each optimization iteration and (4) an efficient multi-objective optimization algorithm that evaluates trade-offs in sensor network designs for various objectives such as efficiency, observability, and cost.

Formatting Issue - Best Available Paper/Graphic

Optimal design of chemical processes with chance constraints

We consider a problem of a chemical process (CP) design in the case of the presence of uncertainty in the process models. One of two following formulations of a problem of the optimal design of chemical processes (CP) under uncertainty is usually used: a) the formulation of *the two-stage optimization problem* (TSOP) takes into account the possibility of the change of control variables at the operation stage b) the formulation of *the one-stage optimization problem* (OSOP) implies that the control variables are constant at the operation stage.

Methods of solving the TSOP with hard constraints (TSOPHC) and the OSOP with chance constraints (OSOPCC) have been developed extensively. Considerable contribution to the development of the methods of solving TSOPHC was made by Grossmann, Halemane (1983), Floudas, Pistikopoulos, Ierapetritou et al. Methods of solving the OSOPCC were developed by Diwekar, Kalagnanam (1997). Significantly less attention was given to formulation and solving of the TSOP with soft constraints (Bernardo, Pistikopoulos, Saraiva, 1999). One of the main issues in solving the TSOPCC is a calculation of multiple integrals to determine the expected value of the objective function and the probability of constraints satisfaction. This operation is very intensive computationally. The use of the standard Gaussian quadrature for the calculation of the multiple integrals is very intensive computationally even for small dimensionality of vector θ of the uncertain parameters. In connection with this the problem of simplification of a multidimensional integrals calculation is very important. One can note a series of papers which consider this problem (Pintaric, Kravanja, 2004), (Li et al., 2008). We consider a method of solving TSOPCC based on the reduction of chance constraints to deterministic ones.

The two-stage optimization problem with chance constraints (TSOPCC) has the following form

$$f_1 = \min_{d,z(\theta)} \int_T f(d,z(\theta),\theta)\rho(\theta)d\theta \qquad (1)$$

$$\Pr\{g_j(d,z(\theta),\theta) \le 0\} \ge \alpha_j \quad j=1,...,m \qquad (2)$$

where $0 \le \alpha_j \le 1, \quad j=1,...,m$.

We showed that constraint (2) can be substituted with the following two constraints

$$\Pr\{\theta \in T_{\alpha_j}\} \ge \alpha_j \qquad (4)$$

$$\max_{\theta \in T_{\alpha_j}} g_j(d,z(\theta),\theta) \le 0, \qquad (5)$$

where T_{α_j} is a region satisfying condition (4). Hence, problem (1) can be rewritten in the following form

$$f^* = \min_{d,z(\theta),T_{\alpha_j}} E[f(d,z(\theta),\theta)] \qquad (6)$$

$$\max_{\theta \in T_{\alpha_j}} g_j(d,z(\theta),\theta) \le 0, \; j=1,\mathrm{K},m \qquad (6)$$

$$\Pr\{\theta \in T_{\alpha_j}\} \ge \alpha_j, \qquad (7)$$

The peculiarity of problem (6) is that

a) we should look for optimal forms and positions of the regions T_{α_j} and

b) the search variables $z(\theta)$ are multivariate functions

We will develop an iteration method of solving problem (6) which will be based on a partition of the uncertainty region T into subregions (multidimensional rectangles $R_l^{(k)}$). Let at the k-th iteration the region T be partitioned on a set $R^{(k)}$ of the subregions $R_l^{(k)}$

$$R_r^{(k)} = \{\theta_i : \theta_i^{(k),L,r} \le \theta_i \le \theta_i^{(k),U,r}, r=1,...,N_k\}, \qquad (8)$$

$$R_1^{(k)} \cup \ldots \cup R_{N_k}^{(k)} = T \tag{9}$$

$$(R_l^{(k)}) \cap (R_q^{(k)}) = \varnothing, \quad \forall l, q \; 1 \le q \le N_k, \quad 1 \le q \le N_k \tag{10}$$

where $(R_l^{(k)})$ is a set of interior points of the region $R_l^{(k)}$, N_k is a number of regions $R_l^{(k)}$ at the k-th iteration and l is a number of a subregion in the set $R^{(k)}$. Unfortunately it is very difficult to look for the optimal form and location of the region T_{α_j}. Therefore, we will restrict a class of possible regions T_{α_j} and look for some approximation $\tilde{T}_{\alpha_j}^{(k)}$ of the region T_{α_j}. At the k-th iteration of this procedure we will solve the following problem

$$\min_{d, \tilde{z}^{(k)}(\theta), \tilde{T}_{\alpha_j}^{(k)}} E_{ap}^{(k)}[f(d, \tilde{z}^{(k)}(\theta), \theta)] \tag{11}$$

$$\max_{\theta \in \tilde{T}_{\alpha_j}^{(k)}} g_j(d, \tilde{z}^{(k)}(\theta), \theta) \le 0 \qquad j = 1, \ldots, m \tag{12}$$

$$\Pr\{\theta \in \tilde{T}_{\alpha_j}^{(k)}\} \ge \alpha_j. \qquad j = 1, \ldots, m, \tag{13}$$

where $\tilde{z}^{(k)}$ is an approximation of multivariate function $z(\theta)$, $E_{ap}^{(k)}[f(d, \tilde{z}^{(k)}(\theta), \theta)]$ is an approximation of $E[f(d, z(\theta), \theta)]$.

Approximation of the region T_{α_j}. We will look for the approximation $\tilde{T}_{\alpha_j}^{(k)}$ in the form of some combination of multidimensional rectangles $T_{\alpha_j}^{l,(k)} = \{\theta_i : \theta_i^{L,j,l,(k)} \le \theta_i \le \theta_i^{U,j,l,(k)}, i = 1, ..., p\}$

$$\tilde{T}_{\alpha_j}^{(k)} = T_{\alpha_j}^{1,(k)} \cup T_{\alpha_j}^{2,(k)} \cup \ldots \cup T_{\alpha_j}^{N_k,(k)} \tag{14}$$

where $T_{\alpha_j}^{i,(k)}$ is the l-th subregion in the set $\tilde{T}_{\alpha_j}^{(k)}$ and N_k is the number of subregions $T_{\alpha_j}^{j,l,(k)}$ at the k-th iteration and

$$(T_{\alpha_j}^{s,(k)}) \cap (T_{\alpha_j}^{t,(k)}) = \varnothing, \quad \forall s, t = 1, ..., N_k. \tag{15}$$

Note that the number of regions $T_{\alpha_j}^{l,(k)}$ is equal to the number of the regions $R_l^{(k)}$. We will require satisfaction of the following conditions

$$T_{\alpha_j}^{l,(k)} \in R_l^{(k)}, \quad l = 1, ..., N_k \quad j = 1, ..., m. \tag{16}$$

Since conditions (10) are met then it follows from (16) that conditions (15) are met. Taking into account (14), (15) we obtain

$$\Pr\{\tilde{T}_{\alpha_j}^{(k)}\} = \Pr\{T_{\alpha_j}^{1,(k)}\} + \Pr\{T_{\alpha_j}^{2,(k)}\} + \ldots + \Pr\{T_{\alpha_j}^{N_k,(k)}\} \tag{17}$$

In this case the search of the optimal forms and locations of the regions T_{α_j} is reduced to the search of the optimal upper and lower bounds $\theta_i^{L,j,l,(k)}$, $\theta_i^{U,j,l,(k)}$ of the sides of the multidimensional rectangles $T_{\alpha_j}^{l,(k)}$. Since all the parameters θ_i are independent and have the normal distribution then the probability measure of the multidimensional rectangle $T_{\alpha_j}^{l,(k)}$ is equal to multiplication of the probability measures of the intervals $I_i^{j,l,(k)} = [\theta_i^{L,j,l,(k)} \le \theta_i \le \theta_i^{U,j,l,(k)}]$. Thus, we have

$$\Pr\{\theta \in T_{\alpha_j}^{l,(k)}\} = \prod_{i=1}^{p} [\Phi((\theta_i^{U,j,l} - E[\theta_i])\sigma_i^{-1}) - \Phi((\theta_i^{L,j,l} - E[\theta_i])\sigma_i^{-1})], \tag{18}$$

where $\Phi(\eta)$ is the standard normal distribution function

Substituting the expressions for $\Pr\{T_{\alpha_j}^{l,(k)}\}$ from (18) in (17) we obtain

$$\Pr\{\tilde{T}_{\alpha_j}^{(k)}\} = \sum_{l=1}^{N_k} \prod_{i=1}^{p} [\Phi(\hat{\theta}_i^{U,j,l,(k)}) - \Phi(\hat{\theta}_i^{L,j,l,(k)})]. \tag{19}$$

Formatting Issue - Best Available Paper/Graphic

Approximation of multivariate functions. We will look for an approximate solution of problem (11) in the supposition that the control variables $z(\theta)$ will be piece-wise constant functions $\tilde{z}^{(k)}(\theta)$ of the following form

$$\tilde{z}^{(k)}(\theta) = z^{l,(k)} \text{ if } \theta \in R_l^{(k)},\ \ l = 1,\dots,N_k \tag{20}$$

It is easy seen that if at each iteration $N_k = 1$ then TSOPCC is transformed into OSOPCC. For the approximate calculation of the expected value of the function $f(d,\tilde{z}^{(k)},\theta)$ we will use the piecewise linear approximation $\tilde{f}(d,\tilde{z}^{(k)},\theta)$ of the function $f(d,\tilde{z}^{(k)},\theta)$ of the following form

$$\tilde{f}(d,\tilde{z}^{(k)},\theta) = \tilde{f}(d,\tilde{z}^{l,(k)},\theta,\theta_l^{(k)}) \text{ if } \theta \in R_l^{(k)}\ l = 1,\dots,N_k$$

$$\tilde{f}(d,z^{r,(k)},\theta,\theta_r^{(k)}) = f(d,z^{r,(k)},\theta_r^{(k)}) + \sum_{i=1}^{N_k} \frac{\partial f(d,z^{r,(k)},\theta_r^{(k)})}{\partial \theta_i}(\theta_i - \theta_{ri}^{q}) \tag{21}$$

and $\theta_{ri}^{(k)}$ is the *i*-th component of the vector $\theta_r^{(k)}$. For the expected value approximation we use the same partition of the uncertainty region *as* for approximation of the regions T_{α_j} and multivariable functions. The right-hand side of formula (21) is the linear part of the Taylor's expansion of the function $f(d,\tilde{z}^{(k)},\theta)$ at the point $\theta_r^{(k)}$. The points $\theta_r^{(k)}$, $r = 1,\dots,N_k$ are the *linearization points*. They will coincide with the middle points of the subregions $T_r^{(k)}$. It is easy to obtain the following expression

$$\int_{T_q^{(k)}} \tilde{f}(d,\tilde{z}^{(k)},\theta,\theta_r^{(k)})\rho(\theta)d\theta = a_r f(d,\tilde{z}^{(k)},\theta_r^{(k)}) + \sum_{i=1}^{p} \frac{\partial f(d,\tilde{z}^{(k)},\theta_r^{(k)})}{\partial \theta_i}(E[\theta_i;T_r^{(k)}] - a_r\theta_r^{(k)}) \tag{22}$$

where $a_r = \int_{T_r^{(k)}} \rho(\theta)d\theta$, $E[\theta_i;T_r^{(k)}] = \int_{T_r^{(k)}} \theta_i\rho(\theta)d\theta$. One can show that calculation of the values a_r and $E[\theta_i;T_r^{(k)}]$ is reduced to calculation of some one-dimensional integrals.

As an approximation of the expected value $E[f(d,\tilde{z}^{(k)}(\theta),\theta);T]$ we will use the following expression

$$E_{ap}^{(k)}[f(d,\tilde{z}^{(k)}(\theta),\theta)] = \sum_{r=1}^{N_k} \int_{T_q^{(k)}} \tilde{f}(d,z^{r,(k)},\theta,\theta_r^{(k)})\rho(\theta)d\theta \tag{23}$$

Using (23), (19), (22) it is easy to transform problem (11) into the following optimization problem which will be solved at *k*-th iteration

$$f^{(k)} = \min_{d,\tilde{z}^{(k)},\theta_i^{L,j,l,(k)},\ \theta_i^{U,j,l,(k)}} E_{ap}^{(k)}[f(d,\tilde{z}^{(k)}(\theta),\theta);T] \tag{24}$$

$$\max_{\theta \in T_{\alpha_j}^{l,(k)}} g_j(d,z^{l,(k)},\theta) \le 0,\ \ j = 1,\dots,m, \qquad l = 1,\dots,N_k$$

$$\sum_{l=1}^{N_k}\prod_{i=1}^{p} [\Phi(\hat{\theta}_i^{U,j,l,(k)}) - \Phi(\hat{\theta}_i^{L,j,l,(k)})] \ge \alpha_j,\ j = 1,\dots,m$$

$$T_{\alpha_j}^{l,(k)} \in R_l^{(k)},\ \ l = 1,\dots,N_k\ \ j = 1,\dots,m$$

Problem (24) is a semi-infinite programming problem. For solving problem (24) one can use the outer approximations method (Hettich, Kortanek, 1993). We will partition one or several subregions $R_l^{(k)}$ at each iteration for improvement of the approximations of the regions T_{α_j}, the piece-wise constant approximation of multivariate functions $z(\theta)$ functions and the approximation of the expected value of the goal function. Thus, the solution of problem (24) does not require the numerical calculation of multiple integrals.

As an illustration we consider a problem of a design of a chemical process consisting of a continuous stirred tank reactor and a heat exchanger.

Formatting Issue - Best Available Paper/Graphic

References

Bernardo, F.P., Pistikopoulos E.N., Saraiva, P,M., 1999. Robustness criteria in process design optimization under uncertainty. Comput.Chem. Eng., S459-S462

Diwekar U.M. and Kalagnanam J.R., 1997, An efficient sampling technique for optimization under uncertainty, AIChE Journal, 43, 440-447.

Halemane K.P. and Grossmann I.E., 1983, Optimal Process Design under Uncertainty, AIChE Journal, 29, 425-433.

Hettich R. and Kortanek K.O., 1993, Semi-infinite programming: Theory, methods and applications, SIAM Review, 35, 380-429.

Li P., Arellano-Garcia H. and Wozny G., 2008, Chance constrained programming approach to process optimization under uncertainty, Computers and Chemical Engineering, 32, 25-45.

Pintaric Z.N. and Kravanja Z., 2004, A strategy for MINLP synthesis of flexible and operable processes, Computers and Chemical Engineering, 28, 6-7, 1105-1119.

Data-Driven Individual and Joint Chance-Constrained Optimization via Kernel Density Estimation

Bruno A. Calfa[1], Ignacio E. Grossmann[1],
Anshul Agarwal[2], Scott Bury[2], John M. Wassick[2]

Optimization under uncertainty is an important and challenging topic in the Process Systems Engineering (PSE) community. Its importance stems from the fact that the decision-making capability of optimization models is subject to the uncertainty of the parameters used. For instance, forecast product demand is not free of uncertainty since one cannot exactly predict future events. Consequently, optimization models under uncertainty are generally more difficult to model and solve than their deterministic counterpart. A few frameworks to account for uncertainty in optimization have been proposed, such as Stochastic Programming (Birge & Louveaux, 2011), Robust Optimization (Ben-Tal, Ghaoui, & Nemirovski, 2009), Approximate Dynamic Programming (Powell, 2011), and Chance-Constrained Optimization, which is the focus of this talk.

Chance- or probabilistic-constrained optimization was introduced in the seminal paper by Charnes & Cooper (1959). Briefly, the use of chance constraints (CCs) allows the modeler to specify a *confidence level* or *reliability level* of satisfying some of the constraints in an optimization model. In other words, the solution to the model satisfies those constraints at least a percentage of all possible realizations of the uncertain parameter present in the respective constraints. The literature considers two types of CCs: *individual* and *joint*.

Classical CCs (when strong assumptions about parametric models for probability distributions are made) have been extensively studied and applied in the literature. In this talk, we focus on data-driven rather than axiomatic or model-driven approaches for reformulating CCs into algebraic constraints. Jiang & Guan (2013) proposed φ-divergences to construct data-driven, density-based confidence sets for CCs. φ-divergences (Pardo, 2006) model the distance of probability distributions. This distance measure can be used to define a confidence set for the estimated distribution that contains the unknown "true" distribution within some specified φ-divergence tolerance. The authors showed that, when using this confidence set, the data-driven CCs are equivalent to classical CCs, but with a reduced risk level.

We propose a data-driven, nonparametric approach to reformulate (conditional) individual and joint chance constraints with right-hand side uncertainty into algebraic constraints. The approach consists of using Kernel Density Estimation (KDE) to approximate unknown "true" continuous probability density/distribution

[1] Department of Chemical Engineering. Carnegie Mellon University. Pittsburgh, PA. 15213

[2] The Dow Chemical Company. Midland, MI. 48674.

functions. The generality and flexibility of the approach renders typically-made strong and parametric assumptions about the distributions unnecessary in order to model chance constraints. Given historical data values for continuous univariate or multivariate random variables (uncertain parameters in an optimization model), the inverse cumulative distribution function (quantile function) and the joint cumulative distribution function are estimated for the univariate and multivariate cases, respectively. The approach uses φ-divergences, which model the distance between distributions, to construct confidence sets that contain the true distribution within a specified φ-divergence tolerance. We propose a new way of specifying the φ-divergence tolerance based on point-wise standard errors of the smoothing estimates with respect to the asymptotic distribution.

The approach is illustrated with a production planning problem with uncertain plant capacity. The system consists of a network of three plants, one raw material, one intermediate product, and two finished products. The deterministic model is a Linear Programming (LP) formulation with 68 constraints and 76 variables. We model the maximum capacity of two of the plants in the network as individual and joint chance constraints. In this test case, historical data for the actual production rates of the two plants are artificially generated in order to assess the efficiency of the proposed KDE-based approach. The reformulated individual chance-constrained model is an LP with 68 constraints and 68 variables, whereas the reformulated joint chance-constrained model is an NLP with 64 constraints and 68 variables. All models are solved in less than one second. Results show that the solution obtained by the proposed approach is very similar to the solution obtained when exact distributions are used. Moreover, the more the sample size of the data the smaller the gap between exact and kernel-based approaches.

References

Ben-Tal, A.; Ghaoui, L. E.; and Nemirovski, A. 2009. *Robust Optimization.* Princeton Series in Applied Mathematics.

Birge, J. R., and Louveaux, F. 2011. *Introduction to Stochastic Programming*. Springer, second edition.

Charnes, A., and Cooper, W. W. 1959. Chance-Constrained Programming. *Management Science.* 6(1):73–79.

Jiang, R., and Guan, Y. 2013. Data-driven Chance Constrained Stochastic Program. *Optimization Online.* Retrieved from http://www.optimization-online.org/DB_FILE/2013/09/4044.pdf on Feb 24, 2014.

Pardo, L. 2006. *Statistical Inference Based on Divergence Measures.* Statistics: A Series of Textbooks and Monographs. Chapman & Hall/CRC. Boca Raton, FL. USA.

Powell, W. B. 2011. *Approximate Dynamic Programming: Solving the Curses of Dimensionality*. Wiley Series in Probability and Statistics. John Wiley & Sons, Inc., second edition. Hoboken, NJ. USA.

Models and Computational Strategies for Multistage Stochastic Programming under Endogenous and Exogenous Uncertainty

Robert Apap & Ignacio Grossmann

In multiperiod planning models, the input data often contains some level of uncertainty. This uncertainty can be classified into two categories depending upon the way the uncertainty is resolved: exogenous uncertainty, where realizations occur independently of the decisions; and endogenous uncertainty, where the realizations are affected by the decisions. In the endogenous case, the decisions may affect the timing of the realizations, or their underlying probability distribution[1]. This work considers the class of endogenous uncertainty where decisions affect the timing of realizations. In the context of process systems engineering, exogenous uncertainty often corresponds to a market uncertainty, such as crude-oil prices. These realizations occur in each period of the time horizon, independently of any decisions. Endogenous uncertainty is a technical uncertainty, such as oilfield size or quality, that is only resolved after a decision is made at a given point in time (e.g., to drill a given oilfield). It is very common for both types of uncertainty to be present in the same problem; however, to the best of our knowledge, there are no published papers that present numerical results for problems containing both endogenous and exogenous parameters.

A common approach for optimization under exogenous uncertainty is two-stage stochastic programming. In this scenario-based approach, first-stage decisions are made 'here and now' at the beginning of the first time period, and are then followed by the resolution of uncertainty. Second-stage, or recourse, decisions are taken as corrective action at the end of the period(s). For problems spanning multiple stages, this becomes a multistage stochastic program, and the decision maker is given the additional freedom to make new 'here-and-now' decisions at the beginning of each stage. Decisions, realizations, and recourse actions occur sequentially, allowing for a more accurate description of the decision-making process for long-term planning projects. Multistage stochastic programming also provides a more suitable framework for endogenous uncertainties, as these realizations can occur at any point in the time horizon.

In this work, we address the modeling and solution of multistage MILP stochastic programming problems involving both endogenous and exogenous uncertain parameters. The general model for this class of problems was presented by Goel and Grossmann[2] as a mixed-integer disjunctive program. Because the number of non-anticipativity constraints (NACs) grows exponentially as the number of time periods, uncertain parameters, or realizations increases, most problems of practical interest are too large to be solved directly with commercial MILP solvers. To address this issue, we first propose the idea of a composite scenario tree for stochastic processes that contain both endogenous and exogenous parameters. The structure of the scenario tree naturally leads to two new theoretical model-reduction properties based on adjacency and scenario grouping, respectively. These new properties drastically reduce the number of NACs and the computational time required to solve the model, and complement the properties previously proposed by Goel and Grossmann[2] and Gupta and Grossmann[3]. Even with these reduction properties, however, the reduced model is often still intractable, so we also propose two special solution approaches. The first approach is Lagrangean decomposition, in which the model is decomposed into independent scenario subproblems that can be solved in parallel. The second is a sequential-decomposition heuristic which involves sequentially solving endogenous MILP subproblems to determine the binary investment decisions, fixing these decisions to satisfy the first-period and exogenous NACs, and then solving the original problem as an LP. This procedure quickly provides a high-quality feasible solution to the original problem. We present numerical results for a process network example and an oilfield development planning problem. These results clearly demonstrate the efficiency of these special solution methods over solving the reduced model directly.

References:

[1] Jonsbraten, T. W. (1998). Optimization Models for Petroleum Field Exploitation. Ph.D. thesis, Norwegian School of Economics and Business Administration.

[2] Goel, V. & Grossmann, I. E. (2006). A class of stochastic programs with decision dependent uncertainty. *Mathematical Programming*, 108 (2-3, Ser. B), 355-394.

[3] Gupta, V. & Grossmann, I. E. (2011). Solution strategies for multistage stochastic programming with endogenous uncertainties. *Computers and Chemical Engineering*, 35, 2235–2247.

377971 Optimal Placement of Imperfect Detectors in the Design of Mitigation Systems: A Non-Uniform Unavailability P-Median Formulation

Abstract:

Mitigation systems are designed to act as the last layer of protection against health, personnel, and environmental hazards. Several of these mitigation systems rely on properly placed detectors to efficiently acknowledge the hazards before issuing executive actions. One such example is the placement of gas detectors within chemical process facilities to provide rapid detection of flammable or toxic gas releases. Most strategies in use for design of mitigation systems like this are based on qualitative or semi-quantitative approaches rather than in the rigorous quantitative consideration of the hazardous scenarios. To address this gap, while considering the inherent uncertainty associated with mitigation systems, the use of a stochastic programming formulation (SP) equivalent to the P-median Problem (PMP) was proposed and validated in the mitigation of contamination scenarios within water distribution systems. This formulation has also been validated for optimal design of other types of mitigation systems. The use of an extended version of formulation SP resulted in an order of magnitude improvement in the performance of gas detection systems when compared to some of the prescriptive detector placement approaches used in the industry. These validation and comparison studies were performed using real sets of CFD generated data provided by GexCon. The data corresponds to FLACS dispersion simulations including the full geometric features of an offshore facility and capturing the uncertainty in the leak characteristics.

Previous work by the authors extended formulation SP into formulation SPU to allow the consideration of imperfect detection (detector unavailability). Unavailability represents the probability of a false negative, i.e., the probability that the detector will not be able to perform its intended function when required. Detector unavailability impacts the system performance, and explicit consideration of detector unavailability affects the optimal placement of detectors. Moreover, neglecting detector unavailability when optimizing the layout will result in a significant deterioration of the mitigation system effectiveness in realistic situations. Formulation SPU assumes that all the detectors have a uniform unavailability, that is, an identical unavailability is assumed for all detectors. In this work, we present formulation SPq, an extended version of formulation SPU that relaxes this assumption allowing for non-uniform detector unavailabilities. This leads to a mixed-integer nonlinear programming formulation due to the need to model products of non-uniform detector unavailabilities. However, formulation SPq explicitly considers individual detection levels, allowing the modeler to specify the number of detection levels, dramatically reduce the computational complexity of the problem. Furthermore, the effect of truncating the number of detection levels for the real gas detector mitigation system problem was analyzed. Based on this analysis, a quadratic formulation that provides a good compromise between accuracy and complexity is presented. Three solution strategies were proposed for this quadratic problem: SPq-Q, SPq-L1, and SPq-L2. The quadratic formulation performance was tested on the same real sets of CFD generated data. From a solution quality perspective, the proposed quadratic formulation outperforms previously proposed approximate solution strategies in the literature.

Decomposition strategy for a class of nonconvex two-stage stochastic programs

Rohit Kannan and Paul I. Barton

Stochastic programming with recourse is a natural way of addressing uncertainties in a problem formulation, and has a wide range of applications in areas such as production planning, capacity expansion, environment and water management, scheduling, natural gas production network design, design and optimization of chemical processing systems, etc. [1, 2, 3].

This work deals with nonconvex two-stage stochastic programs with recourse, whose deterministic equivalent form can be written as:

$$\begin{aligned} \min_{x_1,\cdots,x_s,y,z} \quad & \sum_{h=1}^{s} p_h \left(c_{y,h}^{\mathrm{T}} y + c_{z,h}^{\mathrm{T}} z + f_h(x_h)\right) && \text{(P)} \\ \text{s.t.} \quad & g_h(x_h) + B_{y,h} y + B_{z,h} z \le 0, \quad \forall h \in \{1,\cdots,s\}, \\ & x_h \in X_h, \quad \forall h \in \{1,\cdots,s\}, \\ & y \in Y, \ z \in Z \end{aligned}$$

where $X_h = \{x_h \in \Pi_h \subset \{0,1\}^{n_{x_b}} \times \mathbb{R}^{n_{x_c}} \,|\, r_h(x_h) \le 0\}$, $Y = \{y \in \{0,1\}^{n_y} \,|\, A_y y \le d_y\}$, $Z = \{z \in \mathbb{R}^{n_z} \,|\, A_z z \le d_z\}$, functions $f_h : \Pi_h \to \mathbb{R}$, $g_h : \Pi_h \to \mathbb{R}^m$, $r_h : \Pi_h \to \mathbb{R}^{m_p}$ are continuous, and it is assumed that at least one function in problem (P) is nonconvex. The uncertainties are modeled using a finite number, s, of realizations/scenarios indexed by h, each with a positive probability of occurrence $p_h > 0$. The integer and continuous first-stage decisions y and z, respectively, are made before the realization of the uncertainties, while the mixed-integer second-stage decisions, x_h, are made after the realization of scenario h.

Due to their decomposable structures, two-stage stochastic programs with recourse have traditionally been solved using duality-based decomposition techniques. Lagrangian relaxation is one such popular decomposition technique for solving problem (P). The conventional Lagrangian relaxation approach [4, 5] involves creating copies of all the first-stage variables for each scenario, and adding the so-called non-anticipativity constraints that restrict these copies to take the same value across all scenarios. The non-anticipativity constraints, which couple the scenario problems, are then dualized to obtain the following decomposable lower bounding problem:

$$\max_{\substack{\alpha_1,\cdots,\alpha_{s-1},\\ \beta_1,\cdots,\beta_{s-1}}} \ \min_{\substack{x_1,\cdots,x_s,y_1,\cdots,y_s,\\ z_1,\cdots,z_s}} \ \sum_{h=1}^{s} p_h \left(c_{y,h}^{\mathrm{T}} y_h + c_{z,h}^{\mathrm{T}} z_h + f_h(x_h)\right) + \sum_{h=1}^{s-1} \alpha_h^{\mathrm{T}}(y_h - y_{h+1}) + \sum_{h=1}^{s-1} \beta_h^{\mathrm{T}}(z_h - z_{h+1}) \qquad \text{(LRP)}$$

$$\begin{aligned} \text{s.t.} \quad & g_h(x_h) + B_{y,h} y_h + B_{z,h} z_h \le 0, \quad \forall h \in \{1,\cdots,s\}, \\ & x_h \in X_h, \ y_h \in Y, \ z_h \in Z, \quad \forall h \in \{1,\cdots,s\}, \end{aligned}$$

where α_h, β_h are Lagrange multipliers (of appropriate dimensions) for the non-anticipativity constraints. The solution of problem (P) via the relaxation of the non-anticipativity constraints of both the continuous (z) and integer (y) first-stage variables, *i.e.* problem (LRP), may not be possible in reasonable computational times since the number of Lagrangian dual iterations required for convergence is usually large, each Lagrangian dual is a nonconvex MINLP, and a branch-and-bound framework is required to guarantee convergence.

In this work, an improved decomposition approach, which relaxes the non-anticipativity constraints of only the z variables is proposed, and the resulting lower bounding problem is solved using nonconvex generalized Benders decomposition. Nonconvex generalized Benders decomposition (NGBD) [6, 7] is an extension of generalized Benders decomposition [8] when problem (P) is separable in the first- and second-stage variables and involves only integer first-stage variables. Since strong duality does not necessarily hold for problem (P), the above solution strategy is employed in a branch-and-bound framework to guarantee convergence. By relaxing (and consequently dualizing) the non-anticipativity constraints of only the continuous first-stage variables, the following stronger lower bounding problem, which can be solved using NGBD, is obtained:

$$\begin{aligned} \max_{\beta_1,\cdots,\beta_{s-1}} \quad \min_{\substack{x_1,\cdots,x_s,y,\\ z_1,\cdots,z_s}} \quad & \sum_{h=1}^{s} p_h \left(c_{y,h}^{\mathrm{T}} y + c_{z,h}^{\mathrm{T}} z_h + f_h(x_h)\right) + \sum_{h=1}^{s-1} \beta_h^{\mathrm{T}} (z_h - z_{h+1}) \qquad \text{(LRP-NGBD)} \\ \text{s.t.} \quad & g_h(x_h) + B_{y,h} y + B_{z,h} z_h \leq 0, \quad \forall h \in \{1,\cdots,s\}, \\ & x_h \in X_h, \ z_h \in Z, \quad \forall h \in \{1,\cdots,s\}, \\ & y \in Y \end{aligned}$$

Because problem (LRP-NGBD) provides tighter lower bounds than problem (LRP), a smaller number of branch-and-bound nodes will be required for convergence in general. Further, the first-stage integer variables are more efficiently handled using NGBD than in the conventional Lagrangian relaxation approach. Other numerical benefits such as fathoming a node by value dominance using the subproblems of NGBD [which are much easier to solve than problem (LRP-NGBD)] without fully solving problem (LRP-NGBD) at that node can also be exploited in the improved Lagrangian relaxation approach. Computational experience, which demonstrates the effectiveness of the proposed decomposition approach, is provided through case studies.

References

[1] Birge, J. R., & Louveaux, F. (2011). Introduction to stochastic programming. Springer.

[2] Sahinidis, N. V. (2004). Optimization under uncertainty: state-of-the-art and opportunities. Computers & Chemical Engineering, 28(6), 971-983.

[3] Li, X., Armagan, E., Tomasgard, A., & Barton, P. I. (2011). Stochastic pooling problem for natural gas production network design and operation under uncertainty. AIChE Journal, 57(8), 2120-2135.

[4] Guignard, M., & Kim, S. (1987). Lagrangean decomposition: a model yielding stronger Lagrangean bounds. Mathematical Programming, 39(2), 215-228.

[5] Carøe, C. C., & Schultz, R. (1999). Dual decomposition in stochastic integer programming. Operations Research Letters, 24(1), 37-45.

[6] Li, X., Tomasgard, A., & Barton, P. I. (2011). Nonconvex generalized Benders decomposition for stochastic separable mixed-integer nonlinear programs. Journal of Optimization Theory and Applications, 151(3), 425-454.

[7] Li, X., Tomasgard, A., & Barton, P. I. (2012). Decomposition strategy for the stochastic pooling problem. Journal of Global Optimization, 54(4), 765-790.

[8] Geoffrion, A. M. (1972). Generalized Benders decomposition. Journal Of Optimization Theory and Applications, 10(4), 237-260.

382235 Coal and Biomass to Liquid Transportation Fuels: Process Synthesis and Global Optimization Strategies

Abstract:

The volatility of the global oil market, coupled with increasing levels of greenhouse gas (GHG) emissions, has sparked the search for alternative methods for the production of liquid transportation fuels from domestically available carbon-based resources. The United States has three feedstocks available to gradually replace petroleum as its primary energy source, namely: coal, biomass, and natural gas. The major contributions to the production of liquid fuels from these three feedstocks have been highlighted in a recent review [1]. Hybrid energy systems utilizing coal and biomass have been the center of interest in recent years. Coal has a lower delivered cost than biomass, but conventional coal-to-liquids (CTL) refineries result in almost twice the life-cycle GHG emissions of a petroleum-based plant. Biomass, however, has the ability to reduce greenhouse gas emissions through the capture of CO2 during photosynthesis. The economic and environmental benefits that both these feedstocks offer can be exploited by investigating coal-and-biomass-to-liquid fuels (CBTL) systems.

Using an optimization-based process synthesis framework, the thermochemical conversion of coal and biomass will be presented [2]. Synthesis gas (syngas) will be generated via gasification of the biomass and coal feedstocks through separate gasification units. The syngas can then be directed to either a dedicated water-gas-shift reactor or passed directly to the scrubbing system. Hydrocarbon production will proceed through either Fischer-Tropsch synthesis or methanol formation. The Fischer-Tropsch hydrocarbons can be upgraded over a ZSM-5 reactor or through a series of treatment units that will produce distillate and gasoline-range products. The methanol can be passed to the methanol-to-gasoline (MTG) process or to the methanol-to-olefins/olefins-to-gasoline-and-distillate (MTO/MOGD) processes. Simultaneous heat and power integration is included in the model to convert waste heat into electricity. The optimal process topology that can produce gasoline, diesel, and kerosene at the lowest cost is determined using rigorous deterministic global optimization and process synthesis strategies [3-13].

This study presents a baseline so that a fair comparison between different process alternatives can be made. Multiple case studies are presented to investigate the effect of plant capacity and product distribution on the overall cost of the refinery, the process topology, the total plant cost, and the break-even oil price. The results demonstrate that coal and biomass refineries can become economically competitive with petroleum-based processes while simultaneously reducing life-cycle greenhouse gas emissions.

[1] Floudas, C. A.; Elia, J. A.; Baliban, R. C. Hybrid and single feedstock energy processes for liquid transportation fuels: A critical review. Computers & Chemical Engineering 2012, 41 (6), 24-51.
[2] Niziolek, A. M.; Onel, O.; Elia, J. A.; Baliban, R. C.; Xiao, X.; Floudas, C. A. Coal and Biomass to Liquid Transportation Fuels: Process Synthesis and Global Optimization Strategies. Industrial & Engineering Chemistry Research 2014 DOI: 10.1021/ie500505h.
[3] Baliban, R. C., Elia, J. A., Floudas, C. A. Toward novel biomass, coal, and natural gas processes for satisfying current transportation fuel demands, 1: Process alternatives, gasification modeling, process simulation, and economic analysis. Industrial & Engineering Chemistry Research 2010, 49, 7343-7370.
[4] Elia, J. A., Baliban, R. C., Floudas, C. A. Toward novel biomass, coal, and natural gas processes for satisfying current transportation fuel demands, 2: Simultaneous heat and power integration. Industrial & Engineering Chemistry Research 2010, 49, 7371-7388.
[5] Baliban, R. C., Elia, J. A., Floudas, C. A. Optimization framework for the simultaneous process synthesis, heat and power integration of a thermochemical hybrid biomass, coal, and natural gas facility. Comp. Chem. Eng. 2011, 35, 1647-1690.
[6] Baliban, R. C., Elia, J. A., Floudas, C. A. Simultaneous process synthesis, heat, power, and water integration of thermochemical hybrid biomass, coal, and natural gas facilities. Comp. Chem. Eng. 2012, 37, 297-327.
[7] Baliban, R. C., Elia, J. A., Misener, R., Floudas, C. A. Global Optimization of a MINLP Process Synthesis Model for Thermochemical Based Conversion of Hybrid Coal, Biomass, and Natural Gas to Liquid Fuels. Comp. Chem. Eng. 2012, 42, 64-86.
[8] Baliban, R. C., Elia, J. A., Weekman, V., Floudas, C. A. Process Synthesis of Hybrid Coal, Biomass, and Natural Gas to Liquids via Fischer-Tropsch Synthesis, ZSM-5 Catalytic Conversion, Methanol Synthesis, Methanol-to-Gasoline, and Methanol-to-Olefins/Distillate Technologies. Comp. Chem. Eng. 2012, 47 (12), 29-56.
[9] Baliban, R. C.; Elia, J. A.; Floudas, C. A. Biomass to liquid transportation fuels (BTL) systems: process synthesis and global optimization framework. Energy Environ. Sci. 2013, 6 (1), 267-287.
[10] Baliban, R. C.; Elia, J. A.; Floudas, C. A. Novel Natural Gas to Liquids Processes: Process Synthesis and Global Optimization Strategies. AIChE Journal 2013, 59 (2), 505-531.
[11] Baliban, R. C.; Elia, J. A.; Floudas, C. A. Biomass and Natural Gas to Liquid Transportation Fuels: Process Synthesis, Global Optimization, and Topology Analysis. Industrial & Engineering Chemistry Research 2013, 52 (9), 3381-3406.
[12] Baliban, R. C.; Elia, J. A.; Floudas, C. A.; Gurau, B.; Weingarten, M. B.; Klotz, S. D. Hardwood Biomass to Gasoline, Diesel, and Jet Fuel: 1. Process Synthesis and Global Optimization of a Thermochemical Refinery. Energy & Fuels 2013, 27 (8), 4302-4324.
[13] Baliban, R. C.; Elia, J. A.; Floudas, C. A.; Xiao, X.; Zhang, Z.; Li, J.; Cao, H.; Ma, J.; Qiao, Y.; Hu, X. Thermochemical Conversion of Duckweed Biomass to Gasoline, Diesel, and Jet Fuel: Process Synthesis and Global Optimization. Industrial & Engineering Chemistry Research 2013, 52 (33), 11436-11450.

385237 Modeling of an Indirect Coal-Biomass to Liquids (CBTL) Plant with CO2 Capture and Storage (CCS)

Abstract:

Coal to liquids plants produce substantial amount of CO_2 in comparison to the petroleum –based fuel production processes. However, if moderate amount of biomass is co-fed with the coal, the life cycle CO_2 emissions can be substantially reduced[1]. Further reduction in CO_2 emissions is possible by optimally designing these plants while ensuring that they remain cost-competitive with conventional petroleum-based processes. With this motivation, a high-fidelity model of an indirect coal-biomass to liquids (CBTL) plant with CO_2capture and storage (CCS) has been developed. The plant is integrated with a combined cycle plant that produces steam and electric power by utilizing light hydrocarbons and excess heat that is available from the CBTL plant.

In the indirect CBTL plant, syngas is produced from a coal-biomass feed in a fluidized-bed gasifier. A water gas shift (WGS) reactor is used to obtain the desired H_2/CO ratio in the syngas for maximizing the amount of liquid produced in the slurry-phase Fischer-Tropsch (FT) reactor while satisfying the desired extent of CO_2 capture in the overall process. A significant portion of the light gases produced in the FT reactor is recycled back via an autothermal reformer (ATR) in order to increase the fuel yield with low utility penalty[2]. The remaining portion is sent to the pressure swing adsorption (PSA) H_2 unit to satisfy the H_2 requirement in the product upgrading section. The liquid products are sent to the product upgrading section where a novel integrated hydrotreater is used to achieve higher thermodynamic efficiency and smaller footprint than the traditional configuration where multiple hydrotreaters are used. The straight-run gasoline obtained from the FT crude is processed in an isomerization unit in combination with a catalytic reforming unit (CRU) to obtain the gasoline product with desired specifications. The off-gas from the entire process is used in a combined cycle plant that uses a gas turbine integrated with a heat recovery steam generator (HRSG) coupled with a steam turbine to produce electric power. The steam required in the CO_2 capture units is extracted from the steam turbine. Substantial amount of CO_2 is captured before the FT reactor by using a dual-stage, selective physical solvent-based process. For selection of the optimal post-FT CO_2capture technology, three candidate technologies, namely Selexol, MEA and MDEA/PZ, are evaluated.

The plant-wide model of the CBTL plant including the combined cycle plant has been developed in Aspen Plus integrated with additional models developed in Excel. The steam turbine model involves stage-by-stage calculations and has been developed in Matlab. In this presentation, we will discuss details of the models for all key unit operations. In addition, we will present results from the optimization studies that show that the energy penalty due to CO_2capture can be substantially reduced while having negligible effect on the fuel yield.

References

1. Affordable, Low Carbon Diesel Fuel from Domestic Coal and Biomass, DOE/NETL 2009/1349, January 2009

2. Steynberg, A.P.; Dry, M.E. Fischer-Tropsch Technology, Elsevier Science: San Diego, 2004.

386766 A Systems Approach for the Integration of Biorefineries with Conventional Refineries

Abstract:

A systems approach for the integration of biorefineries with conventional refineries

Eleanna Andritsou, Marinella Tsakalova, Antonis Kokossis

National Technical University of Athens, Greece

Biobased chemical production (biofuels, polyesters coating, glycerol, resins, etc.) would account for 10 percent of the 1.5 trillion worldwide annual chemical market (McKenzie and Co), shaping up a new biobased industry. Within the frames of redefining the chemical industries a great challenge is to examine the integration, redesign and retrofit of conventional refineries with bio-refineries. Biorefineries bare similarities and analogies with the conventional refineries making the synergies of mass and energy flows an interesting and unexploited area.

The paper outlines a methodology for the integration of biorefineries with conventional refineries. The process invites a conceptual design of processes considering the processes as black boxes with input and output streams. The implementation takes the form of a synthesis model that is generalized with process and intermediates as appropriate to the case. The implementation features four (4) biobased (woodchips, corn stover, municipal waste, plant oi) and three (3) conventional (crude oil, natural gas and coal) feedstocks, fifty (50) different bio and conventional processes (fermentation, gasification, anaerobic digestion, transesterification, reforming, refining, hydrocracking, catalytic cracking, dehydrogenation, synthesis, etc.) , sixteen (16) intermediate chemicals (ethanol, methanol, naptha, glycerol, syngas etc.) and eight (8) final products (ethylene, propylene, propylene glycol, acetone, butanol etc.) The model identifies the optimal integrated flows and synergies in terms of energy production and consumptions. The synthesis model performs screening over multiple production routes subjected to economic and energy restrictions including the impact of LCA aspects.

Keywords: integration, biorefineries, conventional refineries, optimization, synthesis

389717 Superstructure Optimization for Catalytic Conversion of Lignocellulosic Biomass to Fermentable Sugars

Abstract:

The major challenge in utilizing the lignocellulosic biomass for production of chemicals and fuels is the extraction of sugars trapped inside the biomass effectively. Recently, a novel nonenzymatic technology[1] has been reported for production of sugars *via* single-stage conversion of lignocellulosic biomass using biomass-derived gamma-valerolactone (GVL) as a solvent. Although this single-stage conversion technology is very promising since it leads to a high sugar yield while eliminating costly pretreatment steps, obtaining fermentable sugars with high purity and developing a separation system for effective GVL recovery and recycling are critical for economic feasibility of the process. Therefore, a series of effective separation technologies should be developed and integrated with the upstream conversion technology for a cost-effective production of fermentable sugars.

There are various alternatives for separating biomass-derived sugars from the GVL solvent, lignin and biomass residues. The biomass-derived sugars can be separated from GVL, lignin and residues by supercritical CO_2 extraction, phase separation upon salt addition, or multi-stage extraction using an alkylphenol solvent. Following the separation of sugars, GVL solvent can be recovered from lignin and residues using precipitation of solids, evaporation of GVL, or adsorption of lignin. Later, GVL can be recycled back to the sugar production step, while lignin and biomass residues can be utilized for steam and power generation in a utility plant. However, it is not clear that what combination of separation technologies will make the production of fermentable sugars more cost-effective, which requires a systematic evaluation of all the alternatives. To this end, we generated a superstructure that integrates the upstream sugar production technology with the separation alternatives to obtain fermentable sugars with high purity, while achieving high recovery and recycling of the GVL solvent, and utilizing the biomass residues in a utility plant for steam and power generation.

In the utility plant, the energy generated from combustion of biomass residues is used to produce high-pressure (HP) steam in a boiler. In addition, HP steam can be purchased to satisfy the utility demand. HP steam can be used for heating, electricity generation, and/or converted to medium-pressure (MP) or low-pressure (LP) steam to be used for electricity generation and heating. Electricity can be generated in steam turbines and/or purchased to meet the electricity demand. In case of electricity surplus, excess electricity can be sold to the grid to generate extra revenue. The utility demands are determined by heat integration of process streams. To minimize the requirement for utilities, the maximum heat recovery should be obtained in a heat exchange operation, which is referred as a *heat cascade*. The heat cascade can be modeled as a *transshipment* problem[2], but with variable heat duties (to be determined by the superstructure model). After the heat integration, remaining heating requirement can be satisfied using steam produced in the utility plant, which is also included in the superstructure model. Selection of how much electricity to produce and/or purchase, selection of the type and amount of steam (HP, MP, LP) to produce and its conditions (temperature and pressure; superheated, saturated or wet steam) are the decision variables for the utility plant.

In summary, our superstructure model includes: (1) sugar production alternatives, (2) the heat recovery network, and (3) the utility system. The modeling of the major unit operations is based on experimental parameters (e.g. conversions, partition coefficients). The proposed optimization formulation is a mixed-integer nonlinear programming (MINLP) model with detailed equipment models, constraints imposing experimental inlet/outlet conditions and purity constraints. The objective function is the maximization of the net revenue obtained from the sugar product and excess electricity sales minus the costs of raw materials and utilities as well as operating and annualized capital costs of selected process equipment units. Using this model, we identified various process alternatives which we discuss in detail.

References

1. J. S. Luterbacher, J. M. Rand, D. M. Alonso, J. Han, J. T. Youngquist, C. T. Maravelias, B. F. Pfleger, and J. A. Dumesic, *Science*, 2014, 343, 277-280.
2. S. A. Papoulias and I. E. Grossmann, *Computers and Chemical Engineering*, 1983, 7, 707-721.

Systematic development of fermentation and downstream processing concepts for an efficient production of itaconic acid

Ulonska, K., Viell, J., Mitsos, A., Marquardt, W.,
Process Systems Engineering, RWTH Aachen University, Aachen

Limited fossil sources and an increasing awareness of sustainability problems pave the way for integrated process concepts for the conversion of renewable raw materials. The efficient production of platform molecules represents the key for a sustainable and economic competitive process concept of future biorefineries.
Itaconic acid was identified as a promising platform molecule from lignocellulosic biomass [1] for the production of fuels [2], polymers and a variety of additional products [3], but it has not yet reached industrial scale production. While the global production in 2009 was only 80 kt/a [4], the potential non-fuel market volume for 2020 is envisioned to be 200 to 6000 kt/a [5]. For the envisioned market volume and a successful replacement of petrochemicals in the named applications, the process efficiency in terms of economy, energy efficiency and productivity needs to be improved. These open tasks can only be successfully addressed by a systematic analysis of existing and development of alternative processing strategies. This work shows the improvement potential for the production of itaconic acid by analyzing and developing alternative fermentation and downstream processing concepts.

The fermentation is conducted using the industrial strain *Aspergillus terreus*, which offers the highest yield and titer to date [6]. However, the productivity of the batch fermentation is low compared to state-of-the-art fermentations. A few downstream processing strategies for the recovery of itaconic acid from the fermentation broth were proposed in the literature. At small scale, precipitation using calcium hydroxide and sulfuric acid is often used. Due to impurities from the fermentation broth, the resulting calcium sulfate cannot be sold and the scale-up to industrial-scale production is not favorable [7]. The industrial process, on the other hand, uses a combination of evaporation, centrifugation and crystallization to recover solid itaconic acid [8]. A promising alternative might be liquid-liquid extraction, if the extracting agent can be utilized as solvent in subsequent reaction steps. The potential of this concept is shown with the in-situ conversion of itaconic acid to methyltetrahydrofuran

(MTHF) [2], using MTHF as extracting agent and reaction solvent. In addition a cooling crystallization concept is analyzed taking advantage of the strongly temperature-dependent solubility of itaconic acid. The solubility limit is increased in a preheating step preventing crystallization in the subsequent reverse osmosis, where the acid concentration is increased by removing water. In the final step, cooling crystallization is applied and solid itaconic acid is recovered, while the dissolved acid is recycled. The fermentation conditions of Kuenz et al. [6] form the basis of the calculations, which were executed in Matlab and Aspen Plus.

For downstream processing, evaporation represents the benchmark because of its straightforward implementation. Due to the low concentration of itaconic acid in water, the total energy demand for product recovery is very high. The total energy demand for agitation and aeration in the fermenter and evaporation account for about 39.2 MJ/kg itaconic acid, assuming conversion efficiency from steam to electricity of 40% and a residual moisture content of 10% in the final product.
The energy demand for the production of itaconic acid using cooling crystallization as downstream processing concept is reduced by 58% compared to the benchmark, with the electricity for agitation and aeration being the highest contribution. A fair comparison of the two alternative downstream processing concepts, extraction and cooling crystallization, is only possible for the conversion of sugar to the final product MTHF, as itaconic acid is not recovered in its solid form in the extraction process. Due to a large reduction of the energy demand compared to the evaporation, both downstream processing concepts are promising alternatives, if the extracting agent can be directly utilized in the subsequent reactor as solvent. For the production of solid itaconic acid cooling crystallization is preferred.
The energy demand for the fermentation and downstream processing can be further reduced by shortening the fermentation time and hence increasing the productivity. The lower titer will on the other hand increase the purification effort. The integrated consideration of fermentation and downstream processing results in an increase of the overall productivity and a reduction of the energy requirement. Hence, the energy requirement for the cooling crystallization concept is successfully reduced by 64% and the productivity increased by 24%, compared to the benchmark, showing the high potential of a systematic improvement of downstream processing concepts as well as fermentation strategies. Although all figures are subject to model uncertainty

they provide an evidence of the potential of process integration in biocatalytic processes.

Acknowledgement

This work was performed as part of the Cluster of Excellence "Tailor-Made Fuels from Biomass", which is funded by the Excellence Initiative by the German federal and state governments to promote science and research at German universities.

Literature

[1] Werpy, T. and Petersen, G., U.S. Department of energy: Top value added chemicals from biomass (2004).

[2] Geilen, F., Engendahl, B., Harwardt, A., Marquardt, W., Klankermeyer, J. and Leitner, W., Selective and Flexible Transformation of Biomass-Derived Platform Chemicals by a Multifunctional Catalytic System, Angew Chem 49:5510-5514 (2010).

[3] Weastra, BioConSepT, Determination of market potential for selected platform chemicals acid, available online: http://www.bioconsept.eu/ (2013).

[4] Willke, T., Vorlop, K.-D., Biotechnological production of itaconic acid, Appl Microbiol Biotechnol, 56:289–295 (2001).

[5] Medway, A. and Sperry, J., Heterocycle construction using the biomass-derived building block itaconic acid, Green Chem 16, 2084–2101 (2014).

[6] Kuenz, A., Gallenmüller, Y., Willke, T., Vorlop, K.-D., Microbial production of itaconic acid: developing a stable platform for high product concentrations, Appl Microbiol Biotechnol 96:1209-1216 (2012).

[7] Kurzrock, T., Weuster-Botz, D., Recovery of succinic acid from fermentation broth, Biotechnol Lett 32:331–339 (2010).

[8] Okabe, M., Lies, D., Kanamasa, S.m Park, E. Y., Biotechnological production of itaconic acid and its biosynthesis in Aspergillus terreus, Appl Microbiol Biotechnol 84:597–606 (2009).

A superstructure approach for the application of waste treatment management in biorefineries

Aikaterini D. Mountraki, Marinella L. Tsakalova, Antonis C. Kokossis
School of Chemical Engineering, National Technical University of Athens, Greece

Abstract

This paper outlines a synthesis approach for the integrated management of waste in multi-product biorefineries. Even though targeted at specific components, the treatment of individual biorefinery waste streams can be costly as the by-product volumes are often small in comparison. Moreover, the heavy organic content provides incentives to valorize the biomass content rather than to treat it as waste. An integrated approach can reduce treatment costs and add value to those by-products that are impossible to include in the value chain.

The work introduces a systematic methodology based on synthesis and optimization technologies. A syperstructure approach is set up to screen the waste management technologies and to select treatment policies either as single or hybrid solutions. The systematization is based on a preliminary stream classification [1,2] that is used to allocate by-product streams to potential treatment units. The number, type, capacity and the particular waste treatment technology are addressed as degrees of freedom to optimize. The allocation of streams to technologies and processing units yields a synthesis model that is formulated using cost models for CAPEX and OPEX to trade against benefits produced by the use of waste (cogeneration, biogas production etc).

The synthesis model is optimized using discrete-continuous optimization techniques and can be extended with additional objectives that account LCA metrics. The method has been applied and illustrated in an organosolv biorefinery that produces sugars (C6 and C5) and lignin as intermediates. The biorefinery features fifteen (15) production processes with a possibility to produce a list of twelve (12) products. The analysis considers six (6) different treatment technologies for liquid stream wastes, six (6) for solid stream wastes, seven (7) for gas stream wastes, two (2) for streams that can be re-used and two (2) different approaches for streams containing catalysts that may be regenerated. The application of the method produced a model with 45 binary variables and 26 continuous variables. The optimization has been applied to assess the benefits in using the integrated approach, sensitivities associated with technology uncertainties and the impact of product portfolios in the analysis.

Keywords: Waste handling, Synthesis, Biorefinery, Lignocellulosic, Mixed Integer.

REFERENCES

[1] Woodard, F. (2001), Industrial Waste Treatment Handbook, Butterworth-Heinemann

[2] Chakraborty A., Linninger A.A, (2002). Plant-Wide Waste Management. 1. Synthesis and Multiobjective Design, Ind. Eng. Chem. Res. (41) 4591-4604

®

Nonlinear Model Predictive Control of an Industrial Polymerization Process

Rahul Bindlish
Engineering Solutions Technology Center
Dow Chemical Company

AIChE Annual Meeting 2014

Outline

- Literature
- NMPC highlights
- NMPC development
- NMPC application
- Results
- Conclusions

Literature

- Survey of Industrial MPC:

  Qin and Badgwell (2003)
- Industrial NMPC Simulation cases:

  Congalidis et al. (1989), Bindlish and Rawlings (2003), Benamor et al. (2004), Negrete et al. (2013)
- Industrial NMPC Plant implementations:

  Ogunnaike (1994), Seki et al. (2001)

AIChE Annual Meeting 2014

DOW

NMPC Highlights

Application of a commercial NMPC technology that uses the laboratory quality measures in the feedback control loop:

1. Long sampling times for polymer quality attributes (0.5 -1 day)
2. Varying dead times (2 - 7 days) and gains (multiplier of 1 - 20) for product quality with respect to reactor temperatures
3. Process models need to extend for extremely low feed rates (approximately 35 percent of normal rates)
4. Process also occasionally operates with one of the seven reactors bypassed for maintenance

AIChE Annual Meeting 2014

Viscosity response

Interacting nonlinearities with feed and temperature

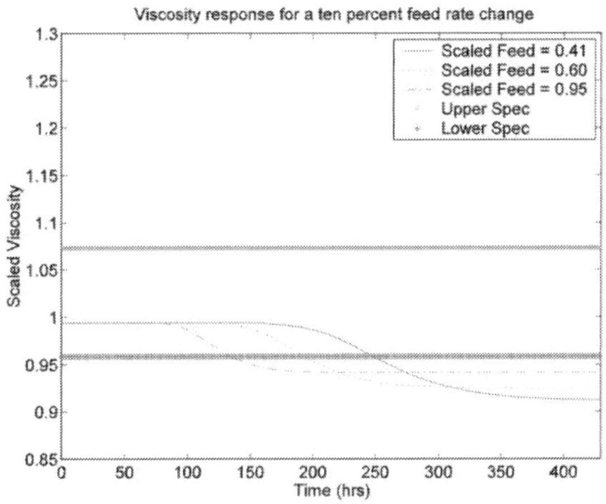

AIChE Annual Meeting 2014

Polymerization Process

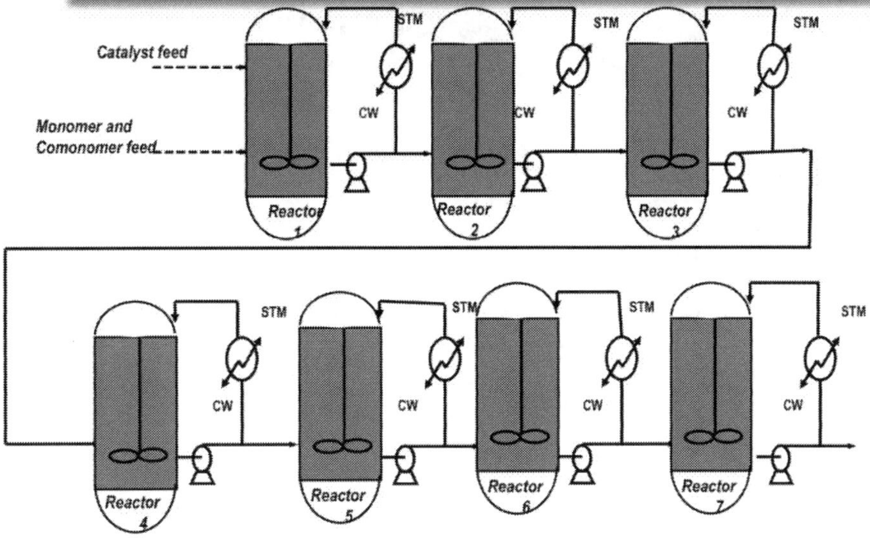

AIChE Annual Meeting 2014

NMPC model development

- Using Aspen Technology, Inc.'s Aspen Non-linear controller based on bounded-derivative-neural network (BDN) models
  - Turner and Guiver (2005), Naidoo et al. (2007)
- Initial BDN model based on simulations of the fundamental kinetic model (approximately 50,000 cases)

$$x_{k+1} = f(x_k, u_k, d_k)$$
$$y_k = h(x_k, u_k, d_k)$$

DOW

NMPC design

State estimation:

Output disturbance model with state estimates using an Extended Kalman filter (Lee and Ricker (1994))

Steady-state target calculation:

Nonlinear objective function to calculate steady-state targets for the outputs, inputs and states

$$\min_{y_{ss}, x_{ss}, u_{ss}} \Phi = (y_{ss} + p - \bar{y})^T Q_s (y_{ss} + p - \bar{y}) + (u_{ss} - \bar{u})^T S_s (u_{ss} - \bar{u})$$

$$f(x_{ss}, u_{ss}, d) = 0, y_{ss} = h(x_{ss}, u_{ss}, d), u_{\min} \le u_{ss} \le u_{\max}, y_{\min} \le y_{ss} \le y_{\max}$$

NMPC design

Dynamic regulation:

Open-loop, nonlinear objective function to calculate future control moves

$$\min_{u^k..u^{k+m}} \Phi_k = \sum_{j=0}^{N} (y_{k+j} - y_{ss})^T Q (y_{k+j} - y_{ss}) + \sum_{j=0}^{m} (\Delta u_{k+j}^T S \Delta u_{k+j})$$

$$\hat{x}_{k+j+1,k} = f(\hat{x}_{k+j,k}, u_{k+j}, d_k)$$

$$\hat{p}_{k+j+1,k} = \hat{p}_{k+j,k}$$

$$\hat{y}_{k+j} = h(\hat{x}_{k+j}, u_{k+j}, d_k) + \hat{p}_{k+j}$$

$$u_{\min} \le u_{k+j} \le u_{\max}, y_{\min} \le y_{k+j} \le y_{\max}, \Delta u_{\min} \le \Delta u_{k+j} \le \Delta u_{\max}$$

Original Control Design

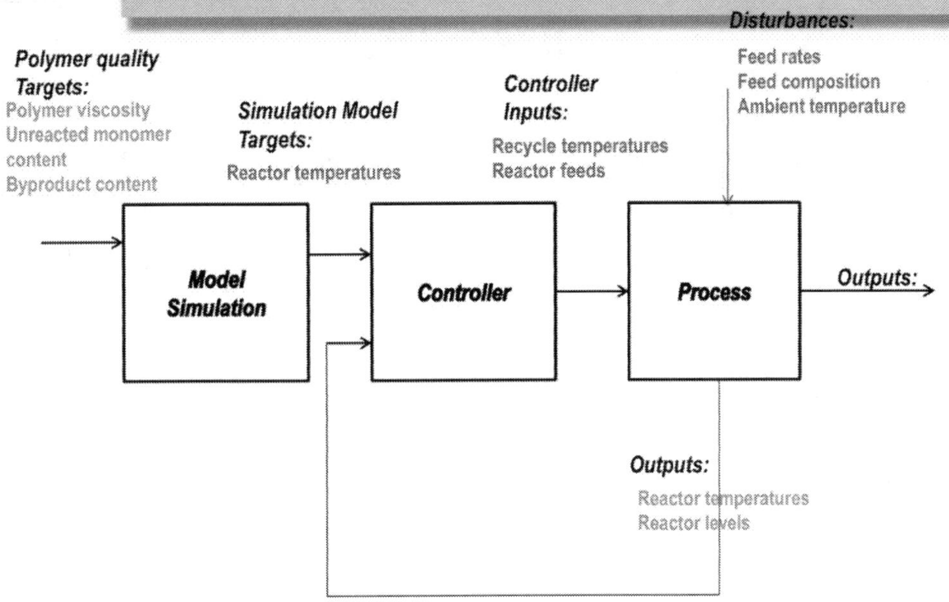

NMPC Goal

Implement inputs that minimize expected deviation of outputs from control objectives over a future horizon

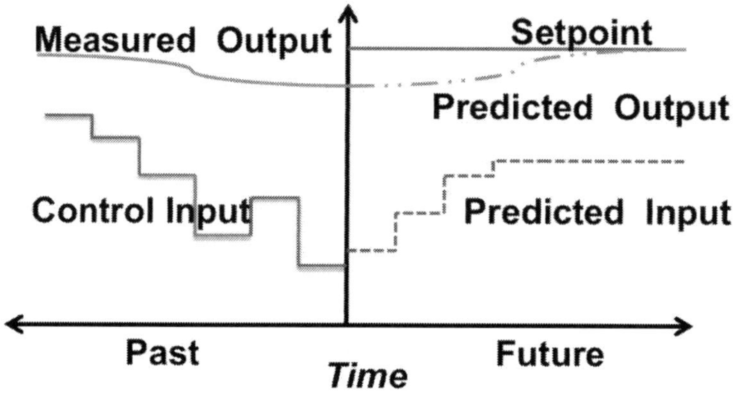

May 2013 – R. Bindlish

AIChe Annual Meeting 2014

NMPC Objectives

Controller Objective	Relative Ranking (Priority)
Valve limits for recycle streams	1
Polymer spec limits	2
Polymer viscosity target	3
Limits for reactor temperature differences	4
Minimize byproduct content	5

AIChE Annual Meeting 2014

Cascade NMPC

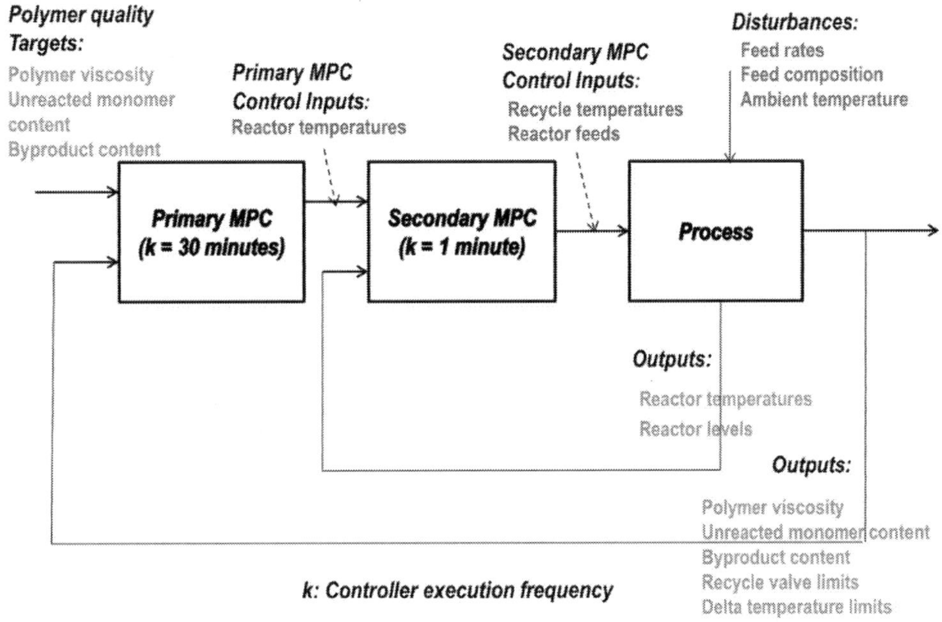

AIChE Annual Meeting 2014

Feed rate variation

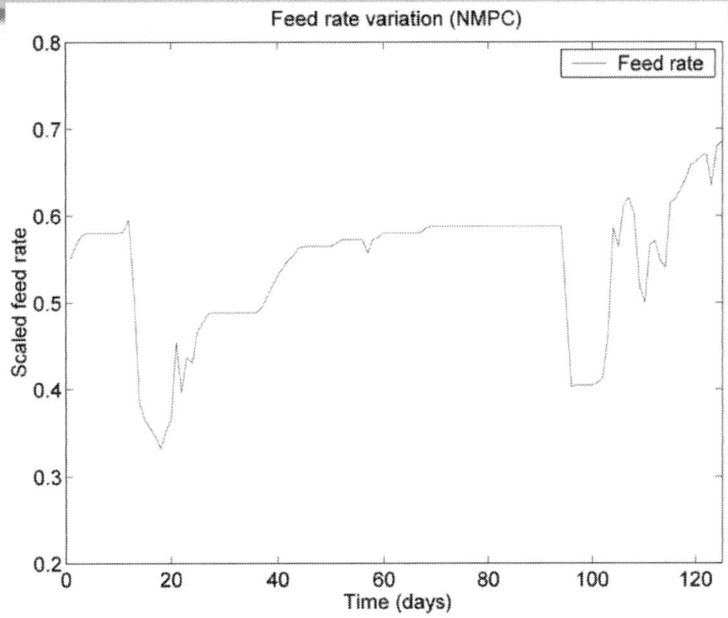

AIChE Annual Meeting 2014

Polymer Viscosity

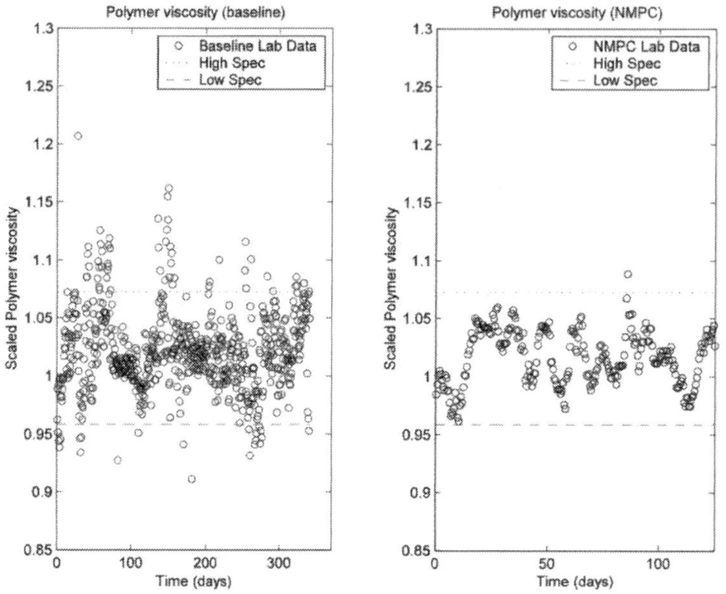

AIChE Annual Meeting 2014

Unreacted monomer content

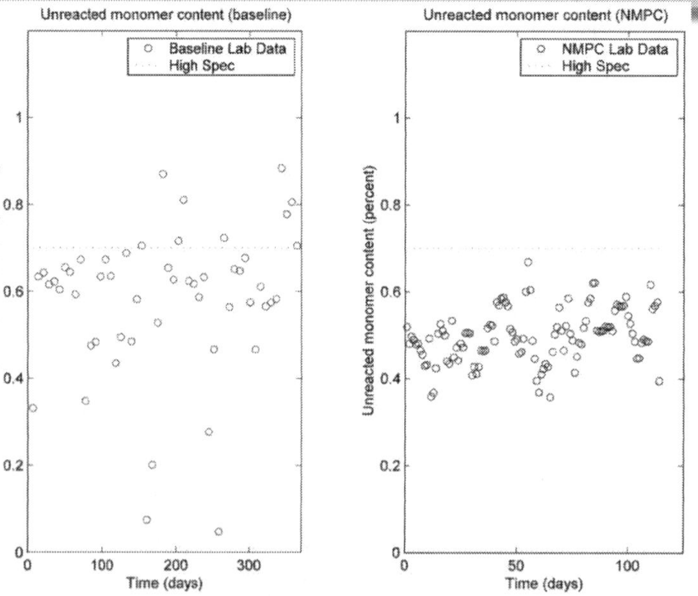

AIChE Annual Meeting 2014

Byproduct content

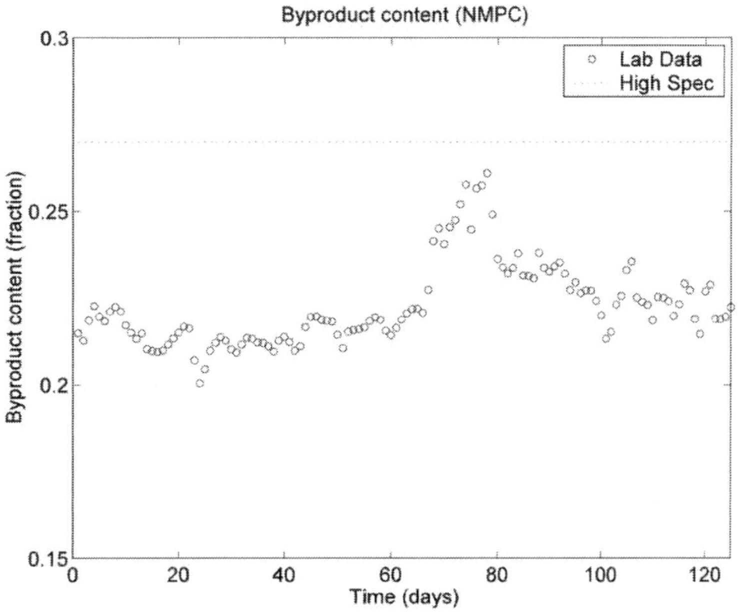

AIChE Annual Meeting 2014

Results

- Better control of polymer quality measures
- 34% reduction in standard deviation of polymer viscosity
- Enabled minimization of byproduct content

AIChE Annual Meeting 2014

Conclusions

- Nonlinear model predictive control (NMPC) has been in continuous use since October 2012 to maintain
  - Polymer quality at specified production rate requirements
- Cascade NMPC is successful for
  - Process objectives with different settling times
  - Computational needs of the associated dynamic optimization problem

AIChE Annual Meeting 2014

Questions

AIChE Annual Meeting 2014

AUTHOR INDEX

AUTHOR INDEX

AUTHOR INDEX

AUTHOR INDEX

AUTHOR INDEX

AUTHOR INDEX

AUTHOR INDEX

AUTHOR INDEX

AUTHOR INDEX

Computing and Systems Technology Division 2014

Core Programming Area at the 2014 AIChE Annual Meeting

Atlanta, Georgia, USA
16-21 November 2014

Volume 2 of 2

ISBN: 978-1-5108-1255-0

Printed from e-media with permission by:

Curran Associates, Inc.
57 Morehouse Lane
Red Hook, NY 12571

Some format issues inherent in the e-media version may also appear in this print version.

Printed by Curran Associates, Inc. (2015)

For permission requests, please contact AIChE
at the address below.

AIChE
120 Wall Street, FL 23
New York, NY 10005-4020

Phone: (800) 242-4363
Fax: (203) 775-5177

www.aiche.org

Additional copies of this publication are available from:

Curran Associates, Inc.
57 Morehouse Lane
Red Hook, NY 12571 USA
Phone: 845-758-0400
Fax: 845-758-2634
Email: curran@proceedings.com
Web: www.proceedings.com

TABLE OF CONTENTS

VOLUME 1

VOLUME 2

385148 Model Predictive Control for Multiplexed Microfluidic Particle Manipulation

Abstract:

A major challenge in nanoparticle design and fabrication is the ability to build complex structures in controlled and directed processes. In the past, self-assembly of particles has been pursued to generate 2D and 3D ordered structures of nanoparticles and colloids; however, development of new methods for the controlled fluidic-directed assembly of anisotropic and/or highly ordered structures would enable new routes for the manufacturing of advanced materials. In this work, we report a fundamentally new technology that will allow for multiplexed particle trapping and directed assembly of small particles using a combination of microfluidics and feedback control. We used nonlinear model-predictive control to manipulate an arbitrary number of small particles in arbitrary directions using microfluidic-based control. In particular, we utilize a Hele-Shaw microfluidic cell to generate hydrodynamic forces on particles under viscous-dominated flow conditions, and we model the flow field inside the microfluidic device as a function of the device geometry and flow rates at the periphery of the cell. In this talk, we will discuss the numerical formulation of this flow model along with the solution of the associated constrained optimal control problem and its implementation as a feedback controller. Finally, we will discuss the experimental implementation of the multiplexed microfluidic trap.

Centre for Process Systems Engineering

Imperial College London

ETH Eidgenössische Technische Hochschule Zürich
Swiss Federal Institute of Technology Zurich

ChromaCon
A new dimension in purification

On the Development of Multi-Parametric Controllers for the Twin-Column MCSGP

Maria M. Papathanasiou, Fabian Steinebach,
Nikolaos A. Diangelakis, Guido Stroehlein,
Massimo Morbidelli, Athanasios Mantalaris, Efstratios N. Pistikopoulos

Centre for Process Systems Engineering — Imperial College London

Intensification of Monoclonal Antibody Production

- ✓ Sustainability
- ✓ Cost reduction
- ✓ Maximization of productivity

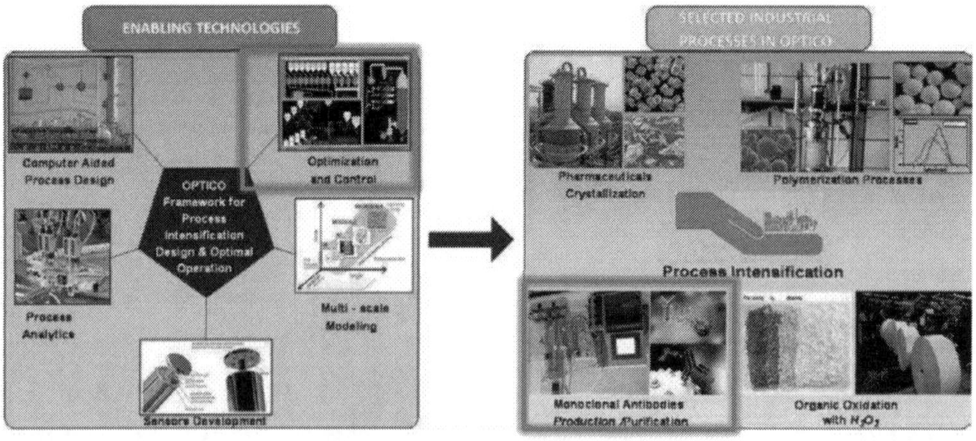

(http://www.opticoproject.eu/)

Intensification of Monoclonal Antibody Production

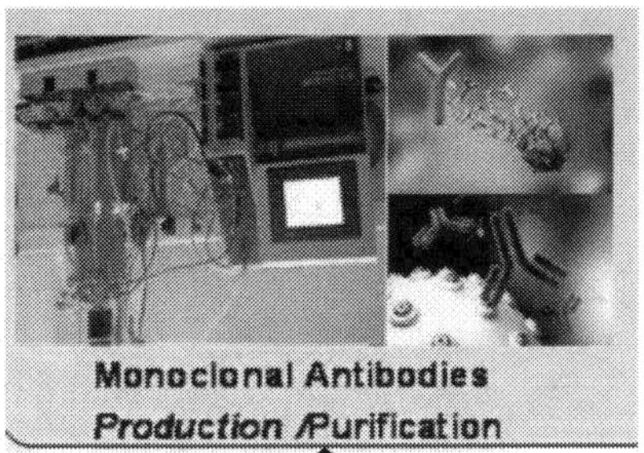

Upstream
- Cell Culture
- mAb mixture

Downstream
- Separation
- Purification of mAb

→ Purity
→ Yield

- Process Optimization
- Control strategy for process monitoring

(http://www.opticoproject.eu/)

Centre for Process Systems Engineering
Imperial College London

The twin-column MCSGP – *The Process*

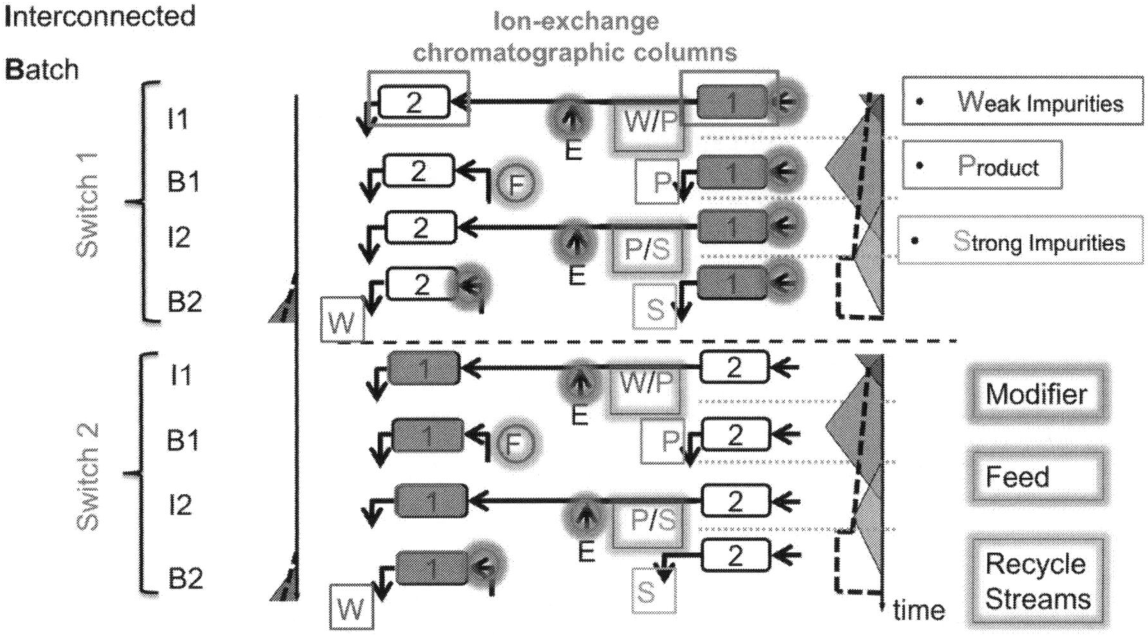

→ **High product purity & recovery yield**

(Aumann and Morbidelli, 2007, Krättli et al., 2013)

Centre for Process Systems Engineering

Imperial College London

The twin-column MCSGP – *The Model*

$$\frac{\partial c(z,t)_{i,h}}{\partial t} = D_{ax}\frac{\partial^2 c(z,t)_{i,h}}{\partial z^2} - \frac{Q_h}{A_{col}\varepsilon_t}\frac{\partial c(z,t)_{i,h}}{\partial z} - \frac{(1-\varepsilon_t)}{\varepsilon_t}\frac{\partial q(z,t)_{i,h}}{\partial t}$$

➡ **Mass balance equation of chromatography**

$$\frac{\partial q(z,t)_{i,h}}{\partial t} = k_i\left(q^*\left(c(z,t)_{i,h}\right) - q(z,t)_{i,h}\right)$$

➡ **Kinetic equation**

$$c(z,t=0)_{i,h} = c^0_{i,h} \quad \text{I.C.}$$
$$c(z=0,t)_{i,h} = c^{in}_{i,h} \quad \text{B.C.}$$
$$\frac{\partial c}{\partial z}(z=L_{col},t)_{i,h} = 0 \quad \text{B.C.}$$

Initial & boundary conditions

$$t\in[0,t_{end}],\ z\in[0,L_{col}],\ i=1\ldots n_{comp},\ h=1,2$$

$$q^*\left(c(z,t)_{1,h}\right) = H_{mod}\cdot c(z,t)_{1,h}$$

Solid phase concentration at equilibrium

$$H_{i,h} = \alpha_{i,1}\left(c(z,t)_{1,h}\right)^{\alpha_{i,2}}$$
$$H^{II}_{i,h} = \alpha_{i,3}\left(H_{i,h}\right)^{\alpha_{i,4}}$$
$$H^{I}_{i,h} = H - H^{II}_{i,h}$$

Henry constants

$$q^{I}_{i,h} = \alpha_{i,5}\left(H^{I}_{i,h}\right)^{\alpha_{i,6}}$$
$$q^{II}_{i,h} = \alpha_{i,7}\left(H^{II}_{i,h}\right)^{\alpha_{i,8}}$$

Saturation capacities for the adsorption sites

$$q^*\left(c(z,t)_{i,h}\right) = \frac{c_{i,h}H^{I}_{i,h}}{1+\sum_{j=2}^{n_{comp}}\frac{c_{j,h}H^{I}_{j,h}}{q^{I}_{j,h}}} + \frac{c_{i,h}H^{II}_{i,h}}{1+\sum_{j=2}^{n_{comp}}\frac{c_{j,h}H^{II}_{j,h}}{q^{II}_{j,h}}},\quad i=2\ldots n_{comp}$$

Solid phase concentration at equilibrium

(Müller-Späth et al., 2008)

Governing Equations

Isotherms

✓ **Mixing Nodes & Feeding** (depending on configuration)

Centre for Process Systems Engineering

Imperial College London

The twin-column MCSGP – *The Model in numbers…*

$$\frac{\partial c(z,t)_{i,h}}{\partial t} = D_{ax}\frac{\partial^2 c(z,t)_{i,h}}{\partial z^2} - \frac{Q_h}{A_{col}\varepsilon_t}\frac{\partial c(z,t)_{i,h}}{\partial z} - \frac{(1-\varepsilon_t)}{\varepsilon_t}\frac{\partial q(z,t)_{i,h}}{\partial t}$$

$$\frac{\partial q(z,t)_{i,h}}{\partial t} = k_i\left(q^*\left(c(z,t)_{i,h}\right) - q(z,t)_{i,h}\right)$$

$$c(z,t=0)_{i,h} = c^0_{i,h} \quad \text{I.C.}$$
$$c(z=0,t)_{i,h} = c^{in}_{i,h} \quad \text{B.C.}$$
$$\frac{\partial c}{\partial z}(z=L_{col},t)_{i,h} = 0 \quad \text{B.C.}$$

$$t\in[0,t_{end}],\ z\in[0,L_{col}],\ i=1\ldots n_{comp},\ h=1,2$$

$$q^*\left(c(z,t)_{1,h}\right) = H_{mod}\cdot c(z,t)_{1,h}$$

$$H_{i,h} = \alpha_{i,1}\left(c(z,t)_{1,h}\right)^{\alpha_{i,2}}$$
$$H^{II}_{i,h} = \alpha_{i,3}\left(H_{i,h}\right)^{\alpha_{i,4}}$$
$$H^{I}_{i,h} = H - H^{II}_{i,h}$$
$$q^{I}_{i,h} = \alpha_{i,5}\left(H^{I}_{i,h}\right)^{\alpha_{i,6}}$$
$$q^{II}_{i,h} = \alpha_{i,7}\left(H^{II}_{i,h}\right)^{\alpha_{i,8}}$$

$$q^*\left(c(z,t)_{i,h}\right) = \frac{c_{i,h}H^{I}_{i,h}}{1+\sum_{j=2}^{n_{comp}}\frac{c_{j,h}H^{I}_{j,h}}{q^{I}_{j,h}}} + \frac{c_{i,h}H^{II}_{i,h}}{1+\sum_{j=2}^{n_{comp}}\frac{c_{j,h}H^{II}_{j,h}}{q^{II}_{j,h}}},\quad i=2\ldots n_{comp}$$

(Müller-Späth et al., 2008)

35 parameters (2 columns, 4 components)

79 variables (2 columns, 4 components)

1 distribution domain

DOF:

- 2 flow rates and their composition
- Buffer composition
- Step length of I- and B- phases

Centre for Process Systems Engineering

Imperial College London

The twin-column MCSGP – *The Model in numbers…*

$$
\begin{aligned}
\frac{\partial c(z,t)_{i,h}}{\partial t} &= D_m \frac{\partial^2 c(z,t)_{i,h}}{\partial z^2} - \frac{Q_h}{A_{col}\varepsilon_t}\frac{\partial c(z,t)_{i,h}}{\partial z} - \frac{(1-\varepsilon_t)}{\varepsilon_t}\frac{\partial q(z,t)_{i,h}}{\partial t} \\
\frac{\partial q(z,t)_{i,h}}{\partial t} &= k_i\left(q^*\left(c(z,t)_{i,h}\right) - q(z,t)_{i,h}\right) \\
c(z,t=0)_{i,h} &= c^0_{i,h} \quad \text{I.C.} \\
c(z=0,t)_{i,h} &= c^{in}_{i,h} \quad \text{B.C.} \\
\frac{\partial c}{\partial z}(z=L_{col},t)_{i,h} &= 0 \quad \text{B.C.}
\end{aligned}
$$

$$t \in [0,t_{end}], z \in [0,L_{col}], i = 1 \ldots n_{comp}, h = 1,2$$

$$\boxed{q^*\left(c(z,t)_{i,h}\right) = H_{mod} \cdot c(z,t)_{i,h}}$$

$$H_{i,h} = \alpha_{1,i}\left(c(z,t)_{i,h}\right)^{\alpha_{2,i}}$$

$$H^B_{i,h} = \alpha_{3,i}\left(H_{i,h}\right)^{\alpha_{4,i}}$$

$$H^A_{i,h} = H - H^B_{i,h}$$

$$q^A_{i,h} = \alpha_{5,i}\left(H^A_{i,h}\right)^{\alpha_{6,i}}$$

$$q^B_{i,h} = \alpha_{7,i}\left(H^B_{i,h}\right)^{\alpha_{8,i}}$$

$$q^*\left(c(z,t)_{i,h}\right) = \frac{c_{i,h}H^A_{i,h}}{1+\sum_{j=2}^{n_{comp}}\frac{c_{j,h}H^A_{j,h}}{q^A_{j,h}}} + \frac{c_{i,h}H^B_{i,h}}{1+\sum_{j=2}^{n_{comp}}\frac{c_{j,h}H^B_{j,h}}{q^B_{j,h}}}, \quad i = 2 \ldots n_{comp}$$

(Müller-Späth et al., 2008)

Typical Simulation…

- **2** columns
- **4** components
- **50** grid points

- **4119** DAEs
- **3302** *(algebraic & differential)*

Centre for Process Systems Engineering

Imperial College London

The twin-column MCSGP – *Input/Output*

$$
\begin{aligned}
\frac{\partial c(z,t)_{i,h}}{\partial t} &= D_m \frac{\partial^2 c(z,t)_{i,h}}{\partial z^2} - \frac{Q_h}{A_{col}\varepsilon_t}\frac{\partial c(z,t)_{i,h}}{\partial z} - \frac{(1-\varepsilon_t)}{\varepsilon_t}\frac{\partial q(z,t)_{i,h}}{\partial t} \\
\frac{\partial q(z,t)_{i,h}}{\partial t} &= k_i\left(q^*\left(c(z,t)_{i,h}\right) - q(z,t)_{i,h}\right) \\
c(z,t=0)_{i,h} &= c^0_{i,h} \quad \text{I.C.} \\
c(z=0,t)_{i,h} &= c^{in}_{i,h} \quad \text{B.C.} \\
\frac{\partial c}{\partial z}(z=L_{col},t)_{i,h} &= 0 \quad \text{B.C.}
\end{aligned}
$$

$$t \in [0,t_{end}], z \in [0,L_{col}], i = 1 \ldots n_{comp}, h = 1,2$$

$$\boxed{q^*\left(c(z,t)_{i,h}\right) = H_{mod} \cdot c(z,t)_{i,h}}$$

$$\boxed{H_{i,h} = \alpha_{1,i}\left(c(z,t)_{i,h}\right)^{\alpha_{2,i}}}$$

$$\boxed{H^B_{i,h} = \alpha_{3,i}\left(H_{i,h}\right)^{\alpha_{4,i}}}$$

$$H^A_{i,h} = H - H^B_{i,h}$$

$$\boxed{q^A_{i,h} = \alpha_{5,i}\left(H^A_{i,h}\right)^{\alpha_{6,i}}}$$

$$\boxed{q^B_{i,h} = \alpha_{7,i}\left(H^B_{i,h}\right)^{\alpha_{8,i}}}$$

$$q^*\left(c(z,t)_{i,h}\right) = \frac{c_{i,h}H^A_{i,h}}{1+\sum_{j=2}^{n_{comp}}\frac{c_{j,h}H^A_{j,h}}{q^A_{j,h}}} + \frac{c_{i,h}H^B_{i,h}}{1+\sum_{j=2}^{n_{comp}}\frac{c_{j,h}H^B_{j,h}}{q^B_{j,h}}}, \quad i = 2 \ldots n_{comp}$$

(Müller-Späth et al., 2008)

- Identical technical characteristics
- Same physicochemical properties

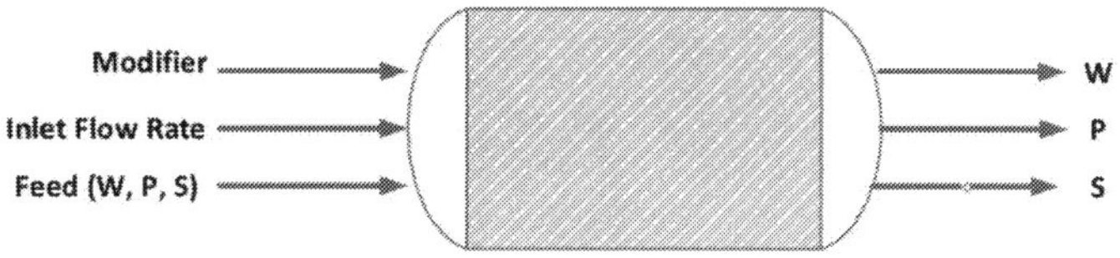

- 5 inputs (per column)
- 3 outputs (per column)
- Highly nonlinear terms
- Periodic operation profile

*Control development becomes challenging!!!

Centre for **Process Systems Engineering**

Imperial College London

General Control Strategy

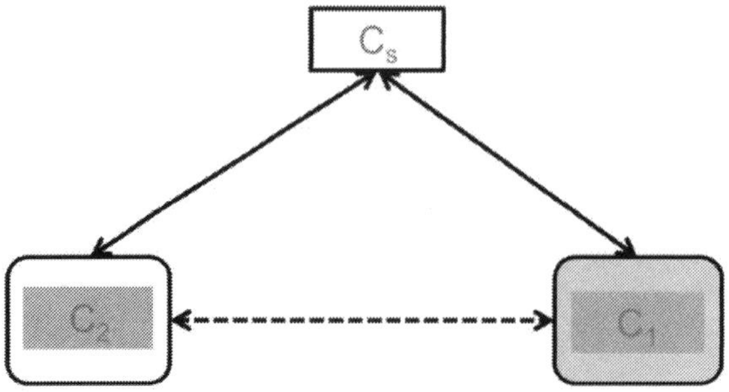

- Identical control design for C_1 and C_2
- Supervisory controller to coordinate C_1/C_2 actions
- Model-based PAROC framework & software platform

Centre for **Process Systems Engineering**

Imperial College London

PAROC Framework – *Theoretical Background*

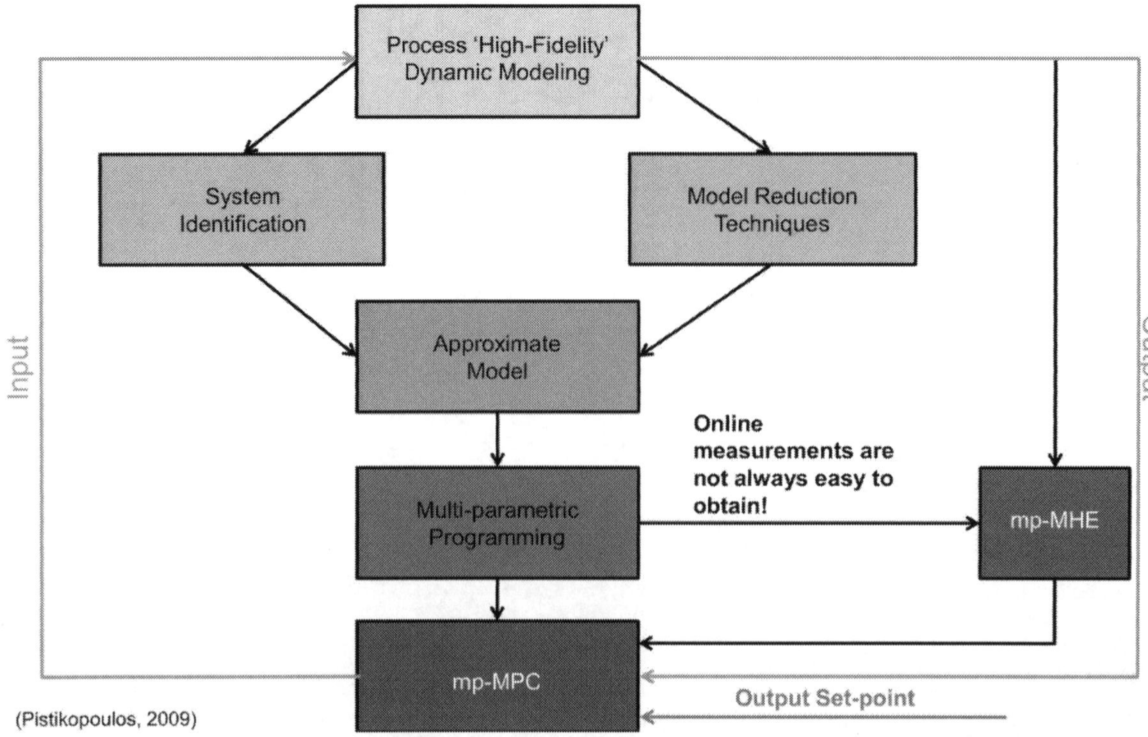

(Pistikopoulos, 2009)

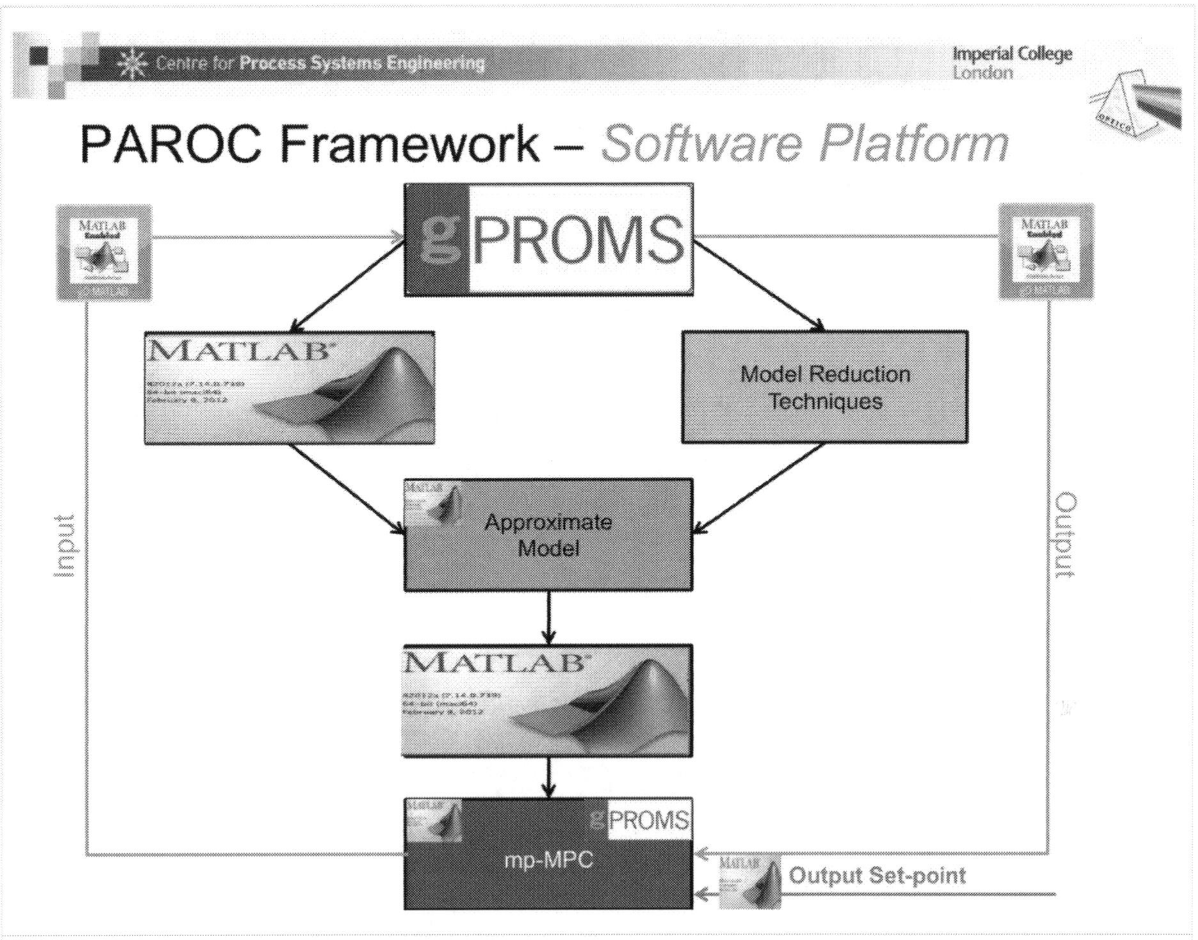
Centre for Process Systems Engineering
Imperial College London
PAROC Framework – Software Platform
gPROMS
MATLAB
Model Reduction Techniques
Approximate Model
MATLAB
gPROMS
mp-MPC
Input
Output
Output Set-point

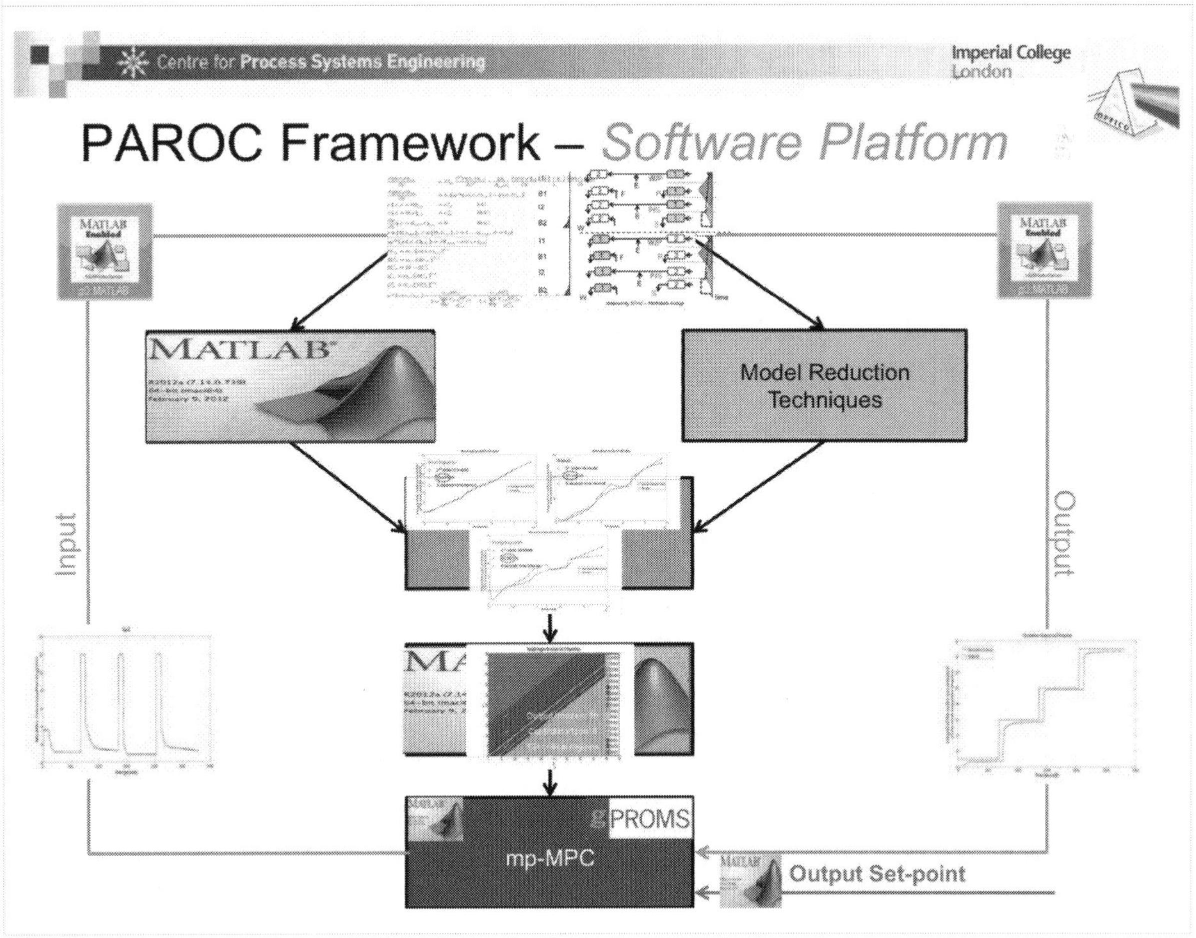
Centre for Process Systems Engineering
Imperial College London
PAROC Framework – Software Platform
MATLAB
Model Reduction Techniques
gPROMS
mp-MPC
Input
Output
Output Set-point

Centre for Process Systems Engineering

Imperial College London

Application of PAROC to MCSGP

Process 'High-Fidelity' Dynamic Modeling

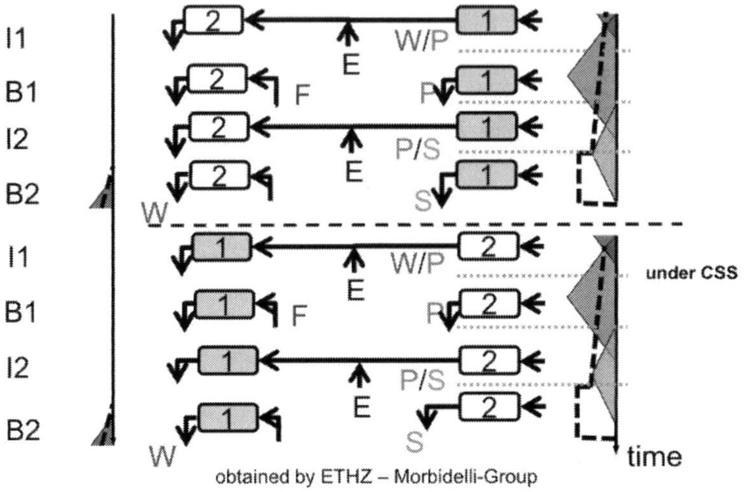

obtained by ETHZ – Morbidelli-Group

Centre for Process Systems Engineering

Imperial College London

Application of PAROC to MCSGP

Process 'High-Fidelity' Dynamic Modeling

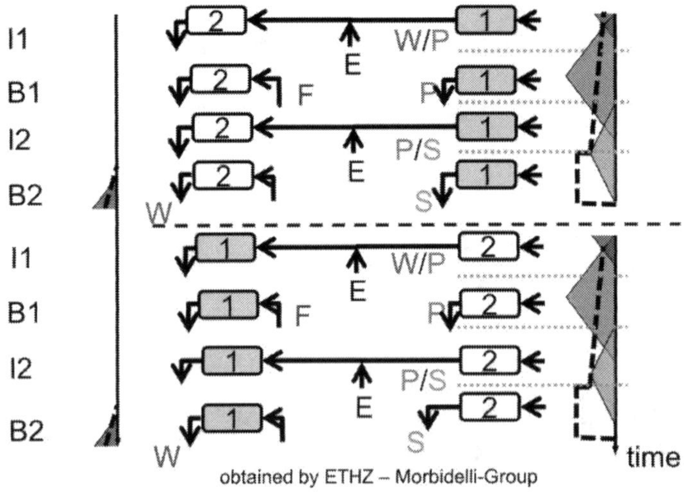

obtained by ETHZ – Morbidelli-Group

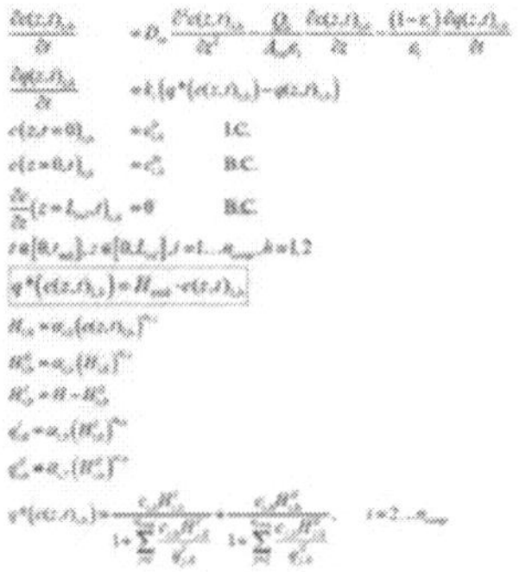

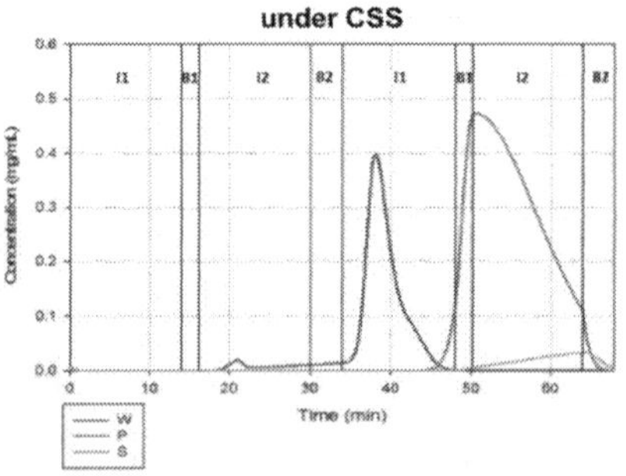

Application of PAROC to MCSGP

Process 'High-Fidelity' Dynamic Modeling

Centre for Process Systems Engineering
Imperial College London

Application of PAROC to MCSGP

Process 'High-Fidelity' Dynamic Modeling

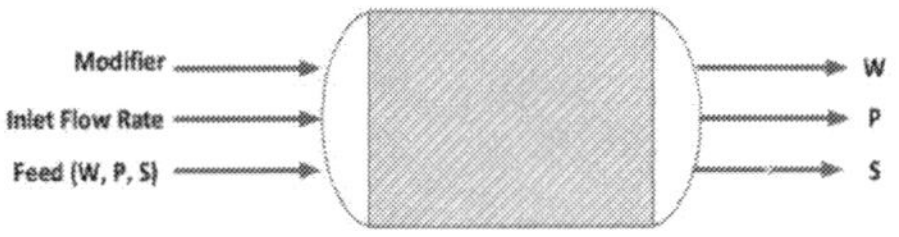

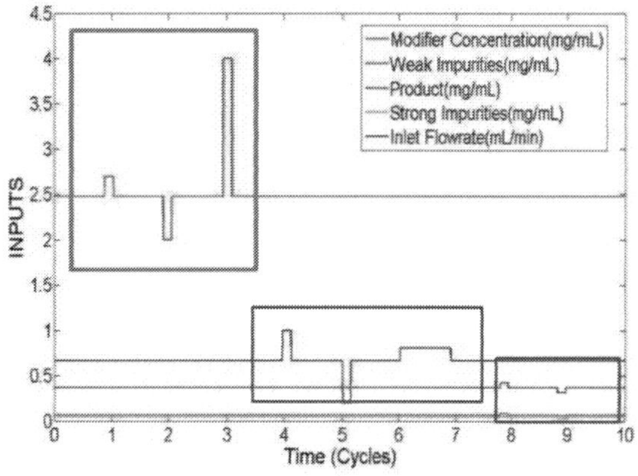

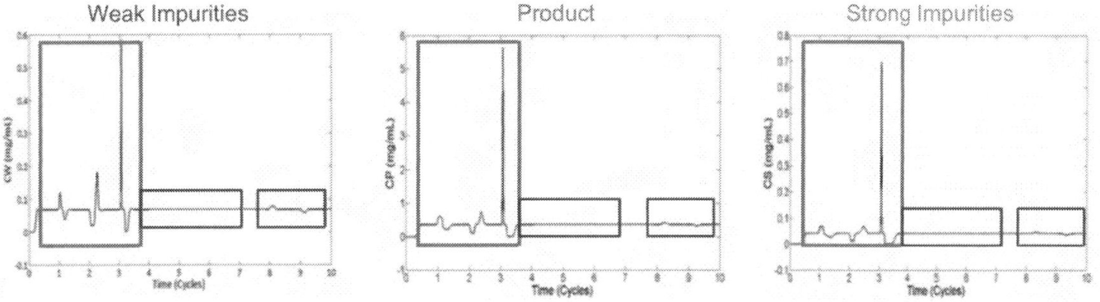

Application of PAROC to MCSGP

Process 'High-Fidelity' Dynamic Modeling

- Modifier Concentration
- Feed concentration
- Flow rate

$$\text{Yield} = \frac{C^{av}_{\text{Product}}}{C^{av}_{\text{Feed}}}$$

$$\text{Purity} = \frac{C^{av}_{\text{Product}}}{C^{av}_{\text{Weak}} + C^{av}_{\text{Product}} + C^{av}_{\text{Strong}}}$$

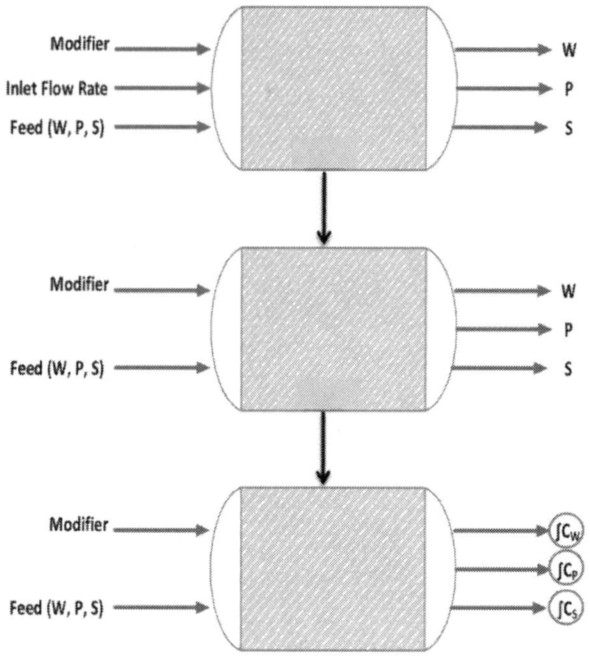

Centre for Process Systems Engineering
Imperial College London

Application of PAROC to MCSGP

Process 'High-Fidelity' Dynamic Modeling

System Identification

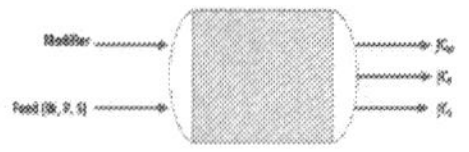

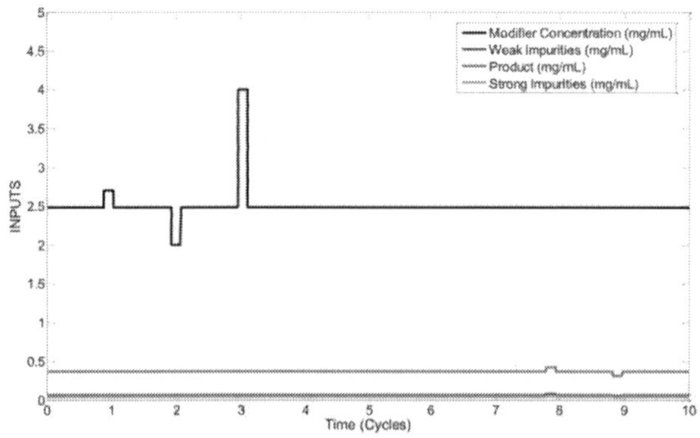

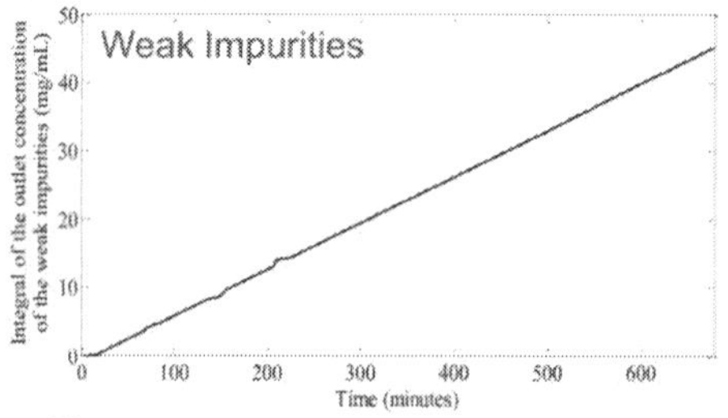

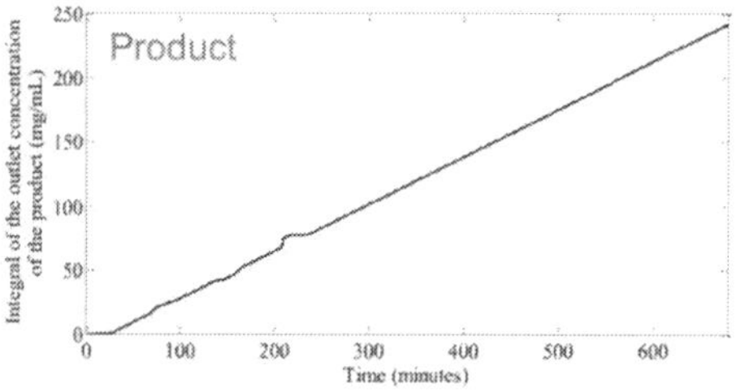

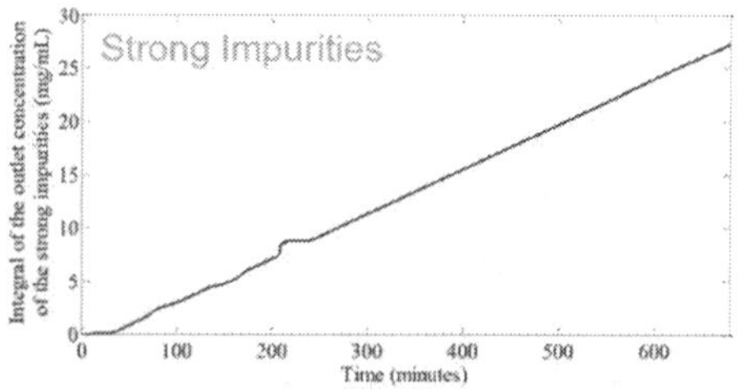

Application of PAROC to MCSGP

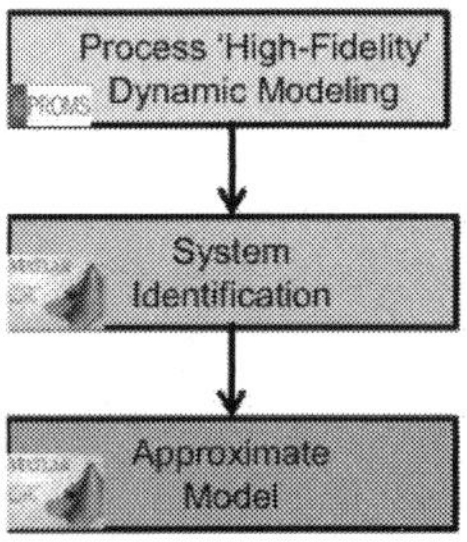

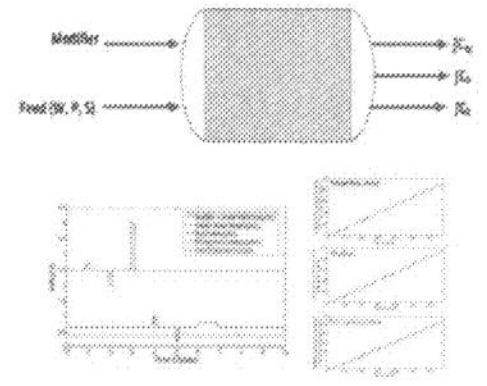

- **4** states
- **1** input (modifier concentration)
- **3** outputs ($\int C_W$, $\int C_P$, $\int C_S$)
- Measured disturbances (feed composition)
- Fit: **94.88%, 94.93% , 93.06%** for the 3 outputs respectively

$$x(t+Ts)=\begin{bmatrix}0.9998 & 0.0003667 & 0.0004885 & -9.137e-05\\ 0.0009448 & 0.9971 & -0.003154 & -0.001599\\ -0.001027 & 0.0003536 & 0.999 & 0.001881\\ 0.000969 & 0.001559 & 0.0005033 & 0.9976\end{bmatrix}\cdot x(t)+\begin{bmatrix}-6.252e{-}06\\ 4.514e{-}05\\ 6.337e{-}05\\ -3.396e{-}05\end{bmatrix}\cdot u(t)$$

$$+\begin{bmatrix}1.557e-05 & 8.449e-05 & 9.616e-06\\ -9.018e-05 & -0.0004893 & -5.57e-05\\ -3.487e-05 & -0.0001892 & -2.154e-05\\ 3.698e-05 & 0.0002007 & 2.284e-05\end{bmatrix}\cdot d(t)$$

$$\mathrm{y}(t)=\begin{bmatrix}545.6 & 101.1 & 17.69 & 7.422\\ 2925 & 660.2 & 102.6 & -67.49\\ 331.9 & 78.42 & 17.12 & -11.23\end{bmatrix}\cdot x(t)$$

Application of PAROC to MCSGP

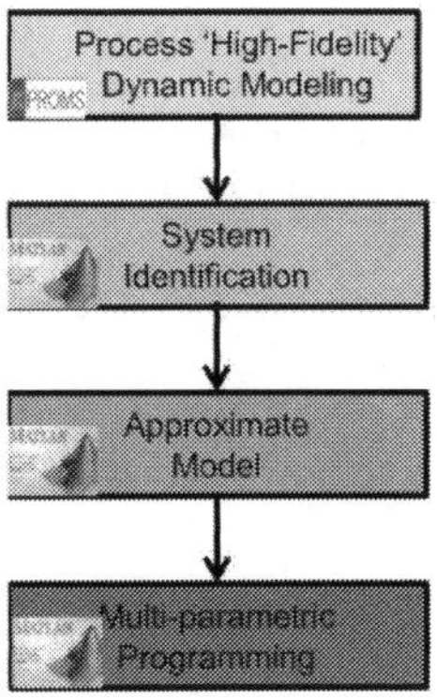

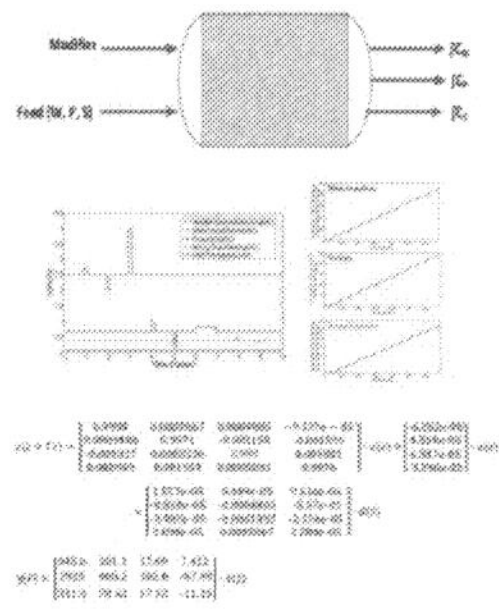

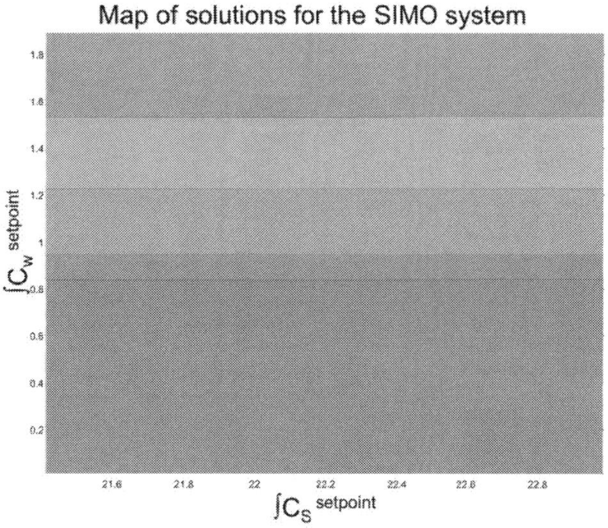

- ➢ Output Horizon: **4**
- ➢ Control Horizon: **2**
- ➢ Total number of parameters: **22**
- ➢ Total number of Critical Regions: **9**

Centre for Process Systems Engineering

Imperial College London

Application of PAROC to MCSGP

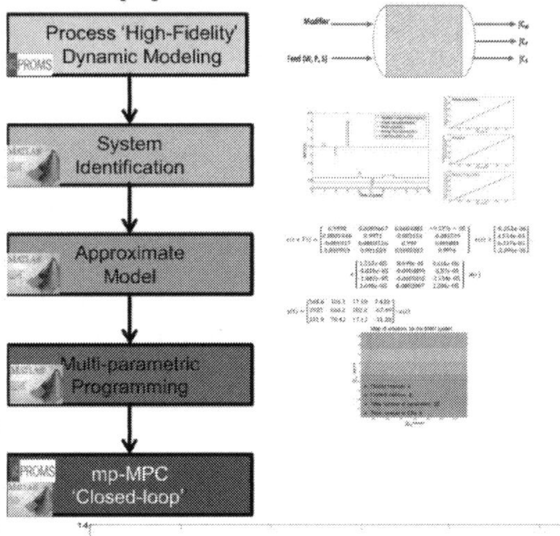

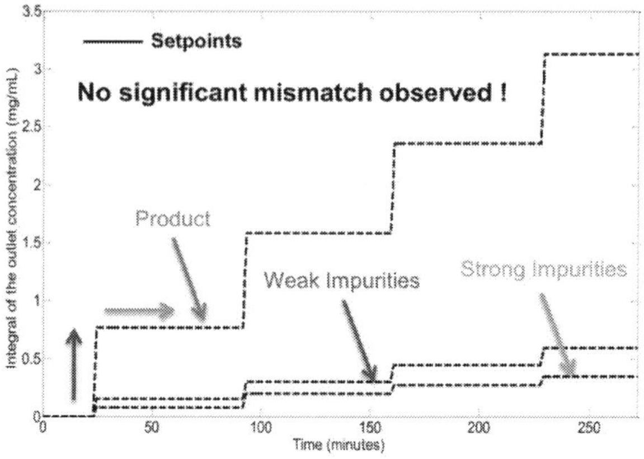

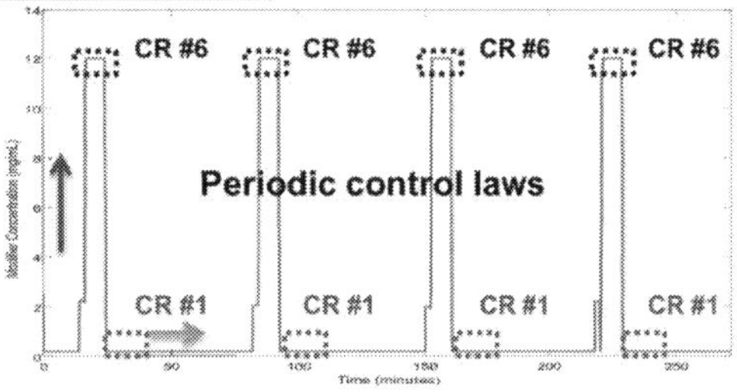

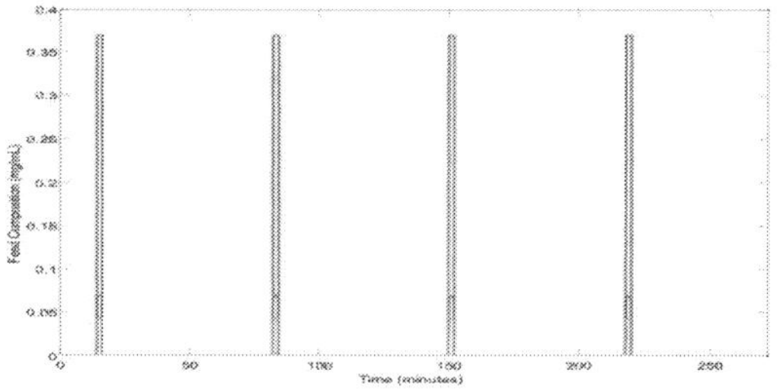

Centre for Process Systems Engineering

Imperial College London

Concluding Remarks & Future Developments

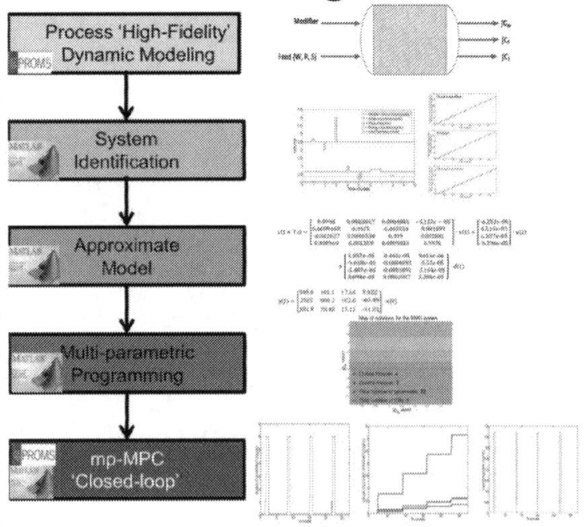

✓ Application of the generic framework & software platform **PAROC to periodic systems**

✓ SIMO controller under measured disturbances for a chromatographic separation

- **Periodic** profile captured
- Tracking of the **integral** of the outlet concentration allows **continuous, tailor-made** process monitoring

➢ Robustification of the controller to account for various flow rates

➢ Design of supervisory controller linking the 2 columns

➢ Application of Moving Horizon Estimation to overcome difficulties with online measurements. *(Miss Ioana Nascu, Thursday, November 20, Advances in Process Control I)*

Centre for
Process
Systems
Engineering
Imperial College
London
ETH
Eidgenössische Technische Hochschule Zürich
Swiss Federal Institute of Technology Zurich
ChromaCon
A new dimension in purification
On the Development of
Multi-Parametric Controllers for
the Twin-Column MCSGP
Maria M. Papathanasiou, Fabian Steinebach,
Nikolaos A. Diangelakis, Guido Stroehlein,
Massimo Morbidelli, Athanasios Mantalaris, Efstratios N. Pistikopoulos
Acknowledgements
OPTICO project (G.A. No 280813) for "Model-based Optimization and Control for Process-Intensification in Chemical and Biopharmaceutical Processes"
Mr. Richard Oberdieck

382586 Quality Control of Penicillin Production Based on Multiple Data-Driven Models

Abstract:

Penicillin production plays an important role in pharmaceuticals and bio-chemicals industries for the production of low-volume but high-value-added secondary metabolites for antibiotics. In order to form the target product, it is common to grow the cells in a batch culture followed by a fed-batch operation to facilitate the synthesis of the antibiotic. The main advantage of batch operation in Penicillin production is consistently producing products of the desired quality.

Multiway analysis based inferential models are developed through batch-wise unfolding of process measurement trajectories in order to control the end-of-batch product quality [1]. The multiway analysis based inferential model, such as multiway partial least squares (MPLS), calls for the entire batch trajectory to predict the final quality in batch processes. Nevertheless, the process measurements beyond current sampling instance are not available, which are required for quality prediction. The unavailability of future trajectories is treated as a missing data problem in the framework of latent variable methods, which ignores the causal relationship between control action and process outputs [2]. Therefore, a data-driven dynamic model is recently developed to characterize the input-output dynamics for quality control in batch processes [3].

However, in the Penicillin process, the quality variables are also available intermittently, instead of only at the end, and existing approaches, including those that use a causal model do not utilize the improved information available through intermittent quality measurements. In addition, one global data-driven dynamic process model may not well characterize the essentially nonlinear dynamics in batch processes. Motivated by these considerations, this work addresses the problem of quality control for Penicillin production by developing a framework that utilizes the increased availability of quality measurements. To this end, the whole batch process is divided into several modeling phases based on the available intermediate and final product qualities. Then, each modeling phase is characterized by a localized multiple linear discrete time model, with a corresponding MPLS quality inferential model to predict the end-of-batch product quality. The developed multiple inferential and dynamic models are integrated with MPC framework so that causality and nonlinear relationships between the future inputs and outputs are accounted for in predicting the intermediate/final qualities and computing the control inputs.

The proposed multiple models based MPC results in significant improvement on final product quality in application to a highly nonlinear batch fermentation process, with comparison to trajectory tacking control and single model based MPC.

[1] Golshan, M., MacGregor, J. F., Bruwer, M. J., & Mhaskar, P. (2010). Latent Variable Model Predictive Control (LV-MPC) for trajectory tracking in batch processes. Journal of Process Control, 20(4), 538-550.

[2] Flores-Cerrillo, J., & MacGregor, J. F. (2004). Control of batch product quality by trajectory manipulation using latent variable models. Journal of Process Control, 14(5), 539-553.

[3] Aumi, S., Corbett, B., Clarke Pringle, T., & Mhaskar, P. (2013). Data driven model predictive quality control of batch processes. AIChE Journal, 59(8), 2852-2861.

385349 Hierarchical Control of a Drop-on-Demand Manufacturing System for the Production of Pharmaceuticals

Abstract:

The support of the FDA in the development of more efficient production technologies, with frequent use of on-line measurement and process control creates new possibilities for innovation in pharmaceutical production processes. As part of this renewed emphasis on improvement of manufacturing, the pharmaceutical industry has begun the selective transition from traditional batch processing to continuous manufacturing. Continuous production processes offer the potential for reduced production costs, faster product release, reduced variability, increased flexibility and efficiency, and improved product quality [1].

As a part of the research agenda of the Engineering Research Center for Structured Organic Particulate Systems, a mini-manufacturing process for drug formation has been under development. The process utilizes the drop-on-demand (DoD) printing technology for predictable and highly controllable deposition of active pharmaceutical ingredients (API) onto an edible substrate, using a semi-continuous operation suitable for low volume production of personalized dosage forms [2]. The DoD mini-manufacturing system consists of a precision positive displacement pump, xy-staging, a hot air based heating system, online imaging and sensing, and temperature, pump and stage controllers. While developed for small scale production, the drop on demand concept is readily scalable to intermediate and event large scale production levels.

Here, we present a hierarchical control framework for the formation of individual dosage forms with precise control of formulation composition, dosage size, and deposit morphology. Implementation of a supervisory control system on the mini-manufacturing process, including on-line monitoring, automation and closed-loop control, is essential for producing individual dosage forms with the desired properties. The drop volume is monitored using the imaging system, to ensure consistent drop size and hence deposit amount. The xy-staging and synchronization logic allows precise drop positioning on the substrate while printing and enables layering of different drugs. The temperature control system allows maintaining printable material rheological properties for the production of melt-based dosage forms, i.e. co-melts of polymer-API systems. Since crystallization temperature has an effect on product morphology, influencing the dissolution properties and hence the bioavailability of the drug, precise control of the drop solidification process once the drop is deposited on the substrate is important [3]. This is achieved indirectly by manipulating the substrate temperature profile using varying temperature gradients. As in the case of bulk crystallization processes, cycling of temperature also can be an effective mechanism for control of crystal size – in this case feature granularity [4].

A polynomial chaos expansion (PCE) based nonlinear surrogate model family has been developed to predict the dissolution profile of the solidified drug deposition given the temperature profile applied on the substrate for several drop sizes. Using the proposed substrate temperature control strategy, the crystallization behavior can be tailored and consistent drug morphology achieved. A hierarchical control system is implemented by monitoring the drop size online and controlling the temperature profile to achieve the desired dissolution profile for the dosage forms created by the deposited drop sizes. The proposed control strategy can eliminate variations in the dissolution profiles due to variable dosage amounts, hence enabling the application of the DoD system for the production of individualized dosage regimens for adaptive clinical trials and personalized treatments.

1. K. V. Gernaey, A. E. Cervera-Padrell, and J. M. Woodley, "A perspective on PSE in pharmaceutical process development and innovation," *Computers & Chemical Engineering*, vol. 42, pp. 15–29, Jul. 2012.
2. L. Hirshfield, A. Giridhar, L. S. Taylor, M. T. Harris, G. V. Reklaitis, "Dropwise Additive Manufacturing of Pharmaceutical Products for Solvent-based Dosage Forms," *Journal of Pharmaceutical Sciences*, vol. 103, issue 2, pages 496-506, February 2014.
3. E. Icten, Z. K. Nagy, G.V. Reklaitis, "Supervisory Control of a Drop-on-Demand Mini-manufacturing System for the Production of Pharmaceuticals," *Proc. of 24th Eur. Symp. Comput. Aided Process Eng. ESCAPE 24*, vol. 24, June 2014.
4. M. Fujiwara, Z. K. Nagy, J. W. Chew, and R. D. Braatz, "First-principles and direct design approaches for the control of pharmaceutical crystallization," *Journal of Process Control*, vol. 15, no. 5, pp. 493–504, Aug. 2005.

Economic Model Predictive Control of an Industrial Fluid Catalytic Cracking Plant

Hasan Sildir[a], Yaman Arkun[a,*], Asli Ari[b], Ibrahim Dogan[c], Murat Harmankaya[c]

[a]*Department of Chemical and Biological Engineering, Koc University, Rumeli Feneri Yolu, 34450 Sariyer, Istanbul, TURKEY*

[b]*TUPRAS R&D Center, Korfez, Kocaeli, TURKEY*

[c]*TUPRAS İzmir Refinery, Aliaga, İzmir, TURKEY*

*Corresponding author: yarkun@ku.edu.tr; Tel: +90-212 338 1313, fax: +90-212 338 1548

Fluid catalytic cracking units (FCCU) are designed to crack heavy hydrocarbons over catalyst particles to form lighter products. These units consist of the riser (reactor) and the regenerator which are followed by the fractionator that separates the riser effluent into valuable products. In our previous work, we have developed a nonlinear dynamic model and validated with industrial data. In this study, our focus involves the development of a real-time hierarchical optimizing control structure which accounts for plant profit directly as well as satisfying safety and stability.

Proposed hierarchical control structure is based on temporal decomposition of the control tasks depending on their function and frequency (see Fig.1). At the upper control layer, an economic model predictive controller (EMPC) provides the riser and the regenerator temperature set points which are determined through solving a dynamic optimization problem involving the plant profit explicitly. EMPC is stabilized by terminal region constraint which is imposed by the steady state optimization. In the lower level, a regulatory model predictive controller (RMPC) adjusts the catalyst circulation rate ($\dot{M}_{cat}$) and the air flow rate ($\dot{M}_{air}$) to regulate the plant at the temperature set points imposed by EMPC. In this cascaded structure EMPC drives the RMPC along an economically optimal path.

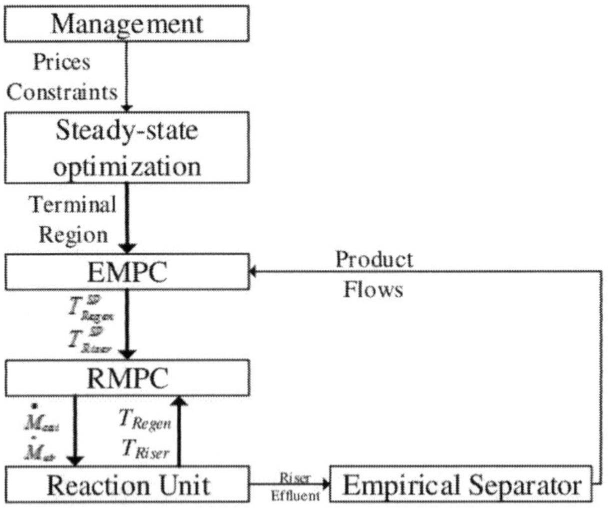

Fig. 1: Hierarchical control structure implemented in the study.

Once the product distribution of the riser effluent is predicted by the model, this distribution is mapped to the final products separated in the fractionator. The empirical separator shown in Fig.1 provides the mapping between the riser effluent and final products. The empirical separator uses historical plant data to calculate the temperature cut points which calculates the amounts of final products without developing a rigorous fractionation model.

The management layer determines and updates the prices of products regularly considering the market demand and interactions between different sub-plants in the refinery. The proposed control approach is proven to provide fast transition which is triggered by changing optimal operating conditions due to shifts in product prices. The detailed formulations of individual elements of the current approach and its impact on overall plant performance will be presented.

Model Predictive Control of Reactive Distillation for Benzene Hydrogenation

The U.S. Environmental Protection Agency (EPA) classifies benzene as a Mobile Source Air Toxic (MSAT) [1]. MSATs are known carcinogens and/or cause other serious health or environmental effects. On road vehicles have been a significant contributor to benzene emissions [2]. As such, the EPA requires all refiners to limit the amount of benzene in gasoline to 0.62 vol% [1]. The main source for producing benzene in a gasoline pool is the reformer unit and thus benzene is present in significant amounts in the reformate stream. Benzene can be removed by hydrogenation in the presence of a catalyst. However, the catalyst used for the reaction is not selective for benzene, and toluene, which is present in the reformate stream in considerable quantities, will also be hydrogenated. Toluene hydrogenation is undesirable as toluene has a high octane rating (RON) and, should be retained in the final product.

$$benzene\ (100\ RON) + 3H_2 \rightarrow\ cyclohexane\ (83RON)$$
$$toluene\ (120\ RON) + 3H_2 \rightarrow\ methylcyclohexane\ (75\ RON)$$

One option to overcome this problem is to use a reactive distillation column. By combining reaction with separation it is possible to selectively react one component in a specified region of the column while suppressing unwanted reactions of other components. Use of reactive distillation columns can result in savings in terms of capital investment and utilities [3, 4]. While reactive distillation can have significant advantages over traditional designs, there are also downsides that need to be considered. For example, the simultaneous presence of reaction and separation phenomena can result in complex dynamic behavior; this is especially so as the feed benzene concentrations can regularly vary between 3 vol% to 11 vol%. In order to investigate the use of reactive distillation for benzene hydrogenation, a detailed first-principle based dynamic model has been developed in gPROMS which includes multi-loop PI controllers [5]. The results from this model indicate that the response of the column, when subjected to feed concentration disturbances, is sluggish and results in significant settling times [5].

In order to overcome the limitations of the multi-loop control system, a model predictive control (MPC) strategy has been developed in this work. The MPC controller is applied to the rigorous dynamic model in order to investigate the closed-loop performance of the process. Responses have been evaluated for different sets of manipulated and controlled variables. A number of simulation studies have been performed where the controller was used to reject step disturbances, such as changes in the feed composition. It has been observed that the settling time of the column has been significantly reduced and the closed loop dynamics have improved for the system under MPC control compared to the multi-loop control scheme.

References:

[1] Control of Hazardous Air Pollutants from Mobile Sources: Final Rule to Reduce Mobile Source Air Toxics, *United States Environmental Protection Agency,* EPA420-F-07-017, Feb 2007.

[2] Mobile Source Air Toxics: Control of Hazardous Air Pollutants from Mobile Sources, *United States Environmental Protection Agency,* EPA420-R-05-901, Nov 2005.

[3] Harmsen, G. J. (2007). Reactive distillation: The front-runner of industrial process intensification: A full review of commercial applications, research, scale-up, design and operation. *Chemical Engineering and Processing: Process Intensification*, *46*(9), 774-780.

[4] Taylor, R., & Krishna, R. (2000). Modelling reactive distillation. *Chemical Engineering Science*, *55*(22), 5183-5229.

[5] Mahindrakar, V., & Hahn, J. (2014). Dynamics and control of benzene hydrogenation via reactive distillation. *Journal of Process Control*, *24*(3), 113-124.

MODEL PREDICTIVE CONTROL OF POST-COMBUSTION CO_2 CAPTURE PROCESS INTEGRATED WITH SUPERCRITICAL COAL-FIRED PLANT

14AIChE
Annual Meeting, Atlanta, GA

Qiang Zhang (Speaker)
Dr. Debangsu Bhattacharyya
Dr. Richard Turton

2014 American Institute of Chemical Engineering Annual Meeting , 18 Nov 2014

1. Motivation
2. Dynamic Modeling
3. Application of PID and Linear Model Predictive Control
4. Conclusion & Future Work

Motivation

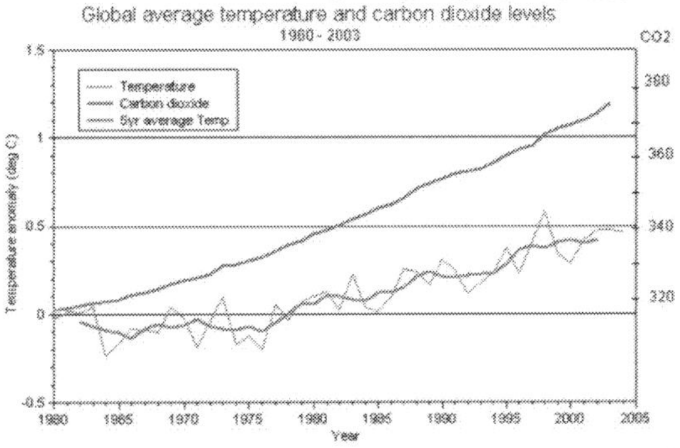

Source: the UK's Climate Research Unit (CRU)

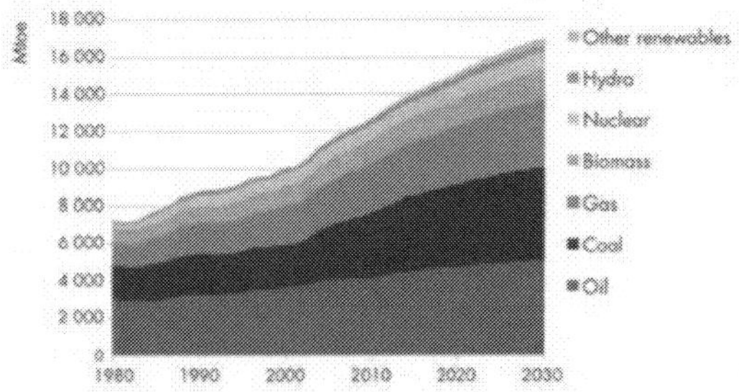

Source: IEA World Energy Outlook 2008 Reference Case

- Increasing man-made CO_2 emission contributes to global warming.
- Fossils are dominant in global energy sources.
- A major source of CO_2 is from stationary energy generation (power plants).

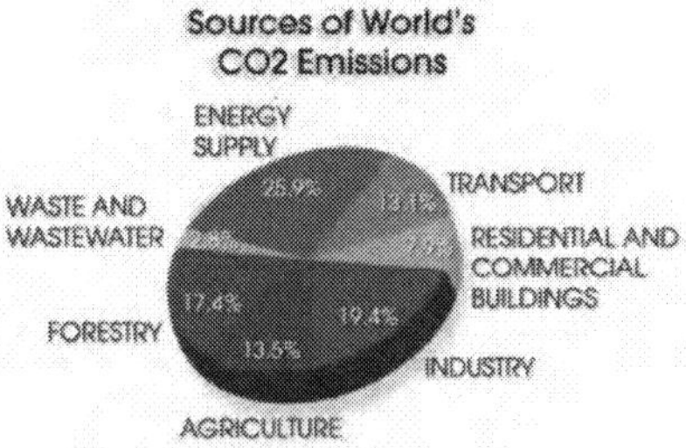

Source: http://www.pbs.org/wgbh/pages/frontline/heat/etc/worldco2.html

Motivation

Postcombustion capture (absorption process)

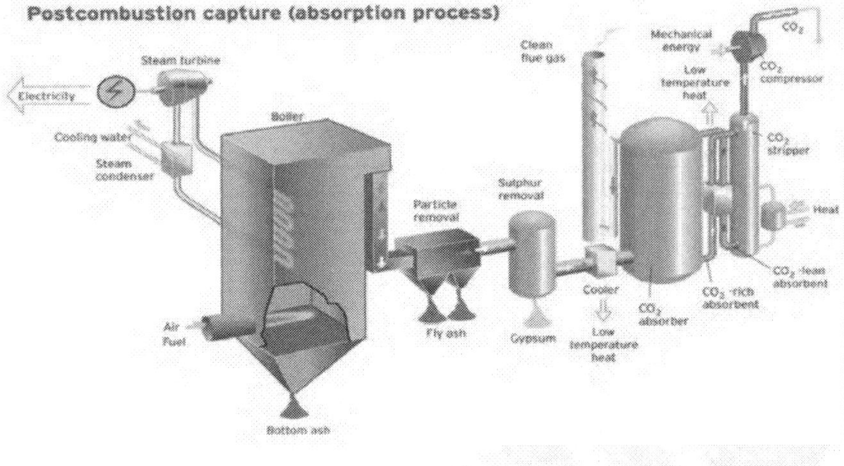

Advantages

- Can be easily retrofitted to the existing conventional pulverized-coal power plant.
- Has been used for commercial flue gas treating for over 20 years.
- Operates under normal/ambient conditions.
- Experiment and pilot plant data are available.

Disadvantages

- Large energy penalty and high capital cost.
- Forms carbamate that is corrosive at high temperature.
- MEA thermal degradation requires significant make-up solvent.

WEST VIRGINIA UNIVERSITY
CHEMICAL ENGINEERING

Dynamic Modeling

Net Power Output (without CO_2 capture): 550 MWe
Net Power Output (with CO_2 capture): 421 MWe

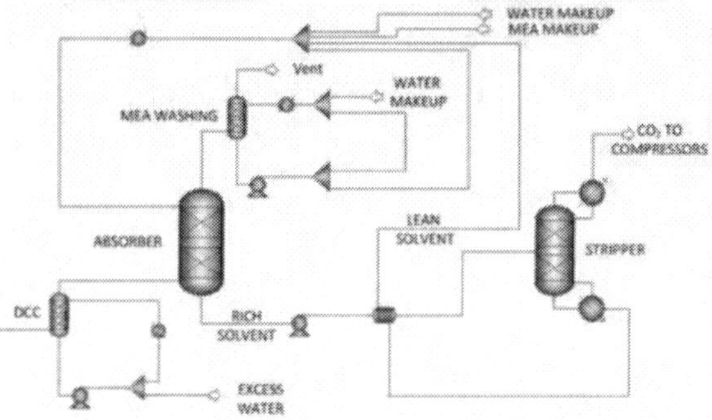

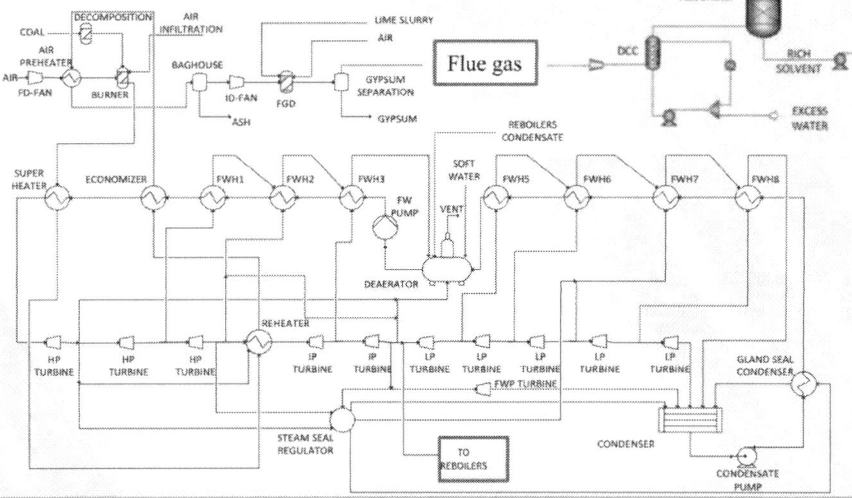

No. of Parallel CO_2 Capture Process: 6

*Cost and Performance Baseline for Fossil Energy Plants Volume 1: Bituminous Coal and Natural Gas to Electricity

WEST VIRGINIA UNIVERSITY
CHEMICAL ENGINEERING

Dynamic Modeling

Modeling of CO_2 Capture Process in Aspen Plus & Aspen Plus Dynamics

Steady State

Steady state equilibrium integrated system model
(lack of accuracy)

Steady state rate-based integrated system model
(Rigorous and good prediction on performance)

Simulated in Aspen Plus

Dynamic

Dynamic equilibrium integrated system model
(lack of accuracy but full process modeled)

Dynamic rate-based Absorber & Stripper model
(Not supported by Aspen Plus Dynamics)

Simulated in Aspen Plus Dynamics

WEST VIRGINIA UNIVERSITY
CHEMICAL ENGINEERING

Dynamic Modeling

Model Validation

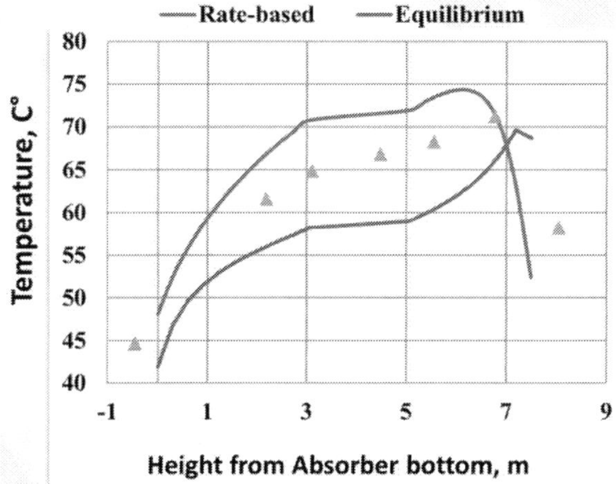

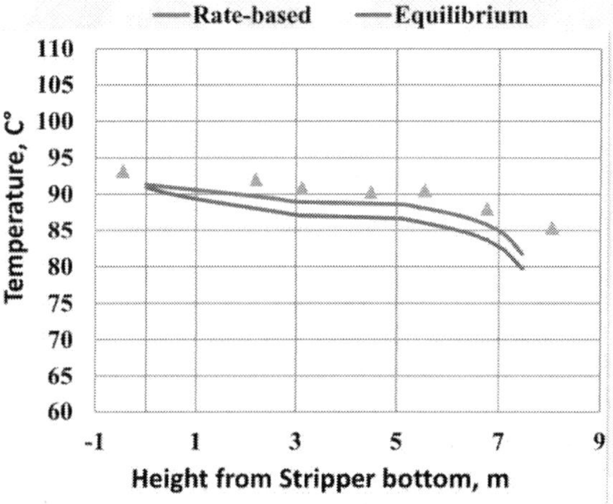

	Pilot Plant	Rate-based model	Equilibrium model
Capture rate	69%	69%	89%

*Dugas; Case 47, Pilot Plant Study of Carbon Dioxide Capture by Aqueous Monoethanolamine; 2006

WEST VIRGINIA UNIVERSITY
CHEMICAL ENGINEERING

Dynamic Modeling

Approach to get reliable results from dynamic equilibrium Aspen Plus Dynamic model

- Due to the differences in calculation methods and reaction specifications, the equilibrium-based simulation can not match the rate-based steady state results.
- Component Murphree efficiency is very difficult to predict mathematically as the separation on each stage is a result of both physical and chemical equilibrium under different conditions.
- Based on the efficiency given by Aspen Plus at rate-based steady state design point, a correlation of component efficiency with key variables as parameters was developed.

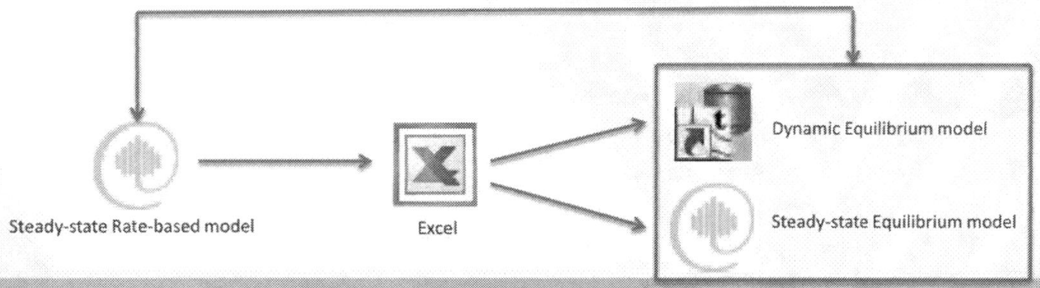

WEST VIRGINIA UNIVERSITY
CHEMICAL ENGINEERING

Dynamic Modeling

Efficiency Model

Correlated component efficiency

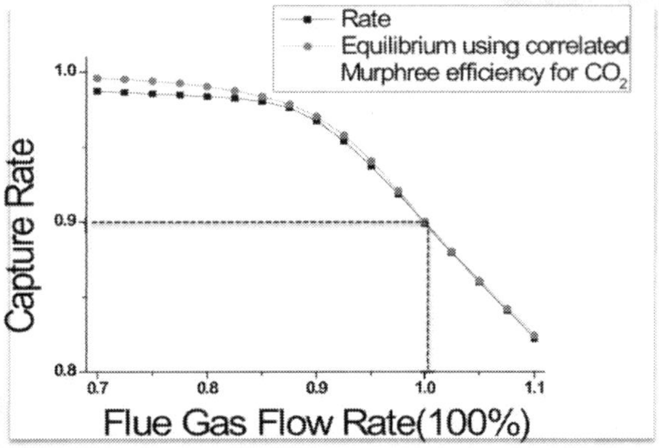

$$\varepsilon_{CO2,i} = \frac{1}{1.3}\varepsilon_{CO2,i}^{D}\left(\frac{F_{FG}^{D}}{F_{FG}}\right)^{a}\left(\frac{F_{CO2}^{D}}{F_{CO2}}\right)^{b}\left(\frac{F_{Sol}}{F_{Sol}^{D}}\right)^{c}$$

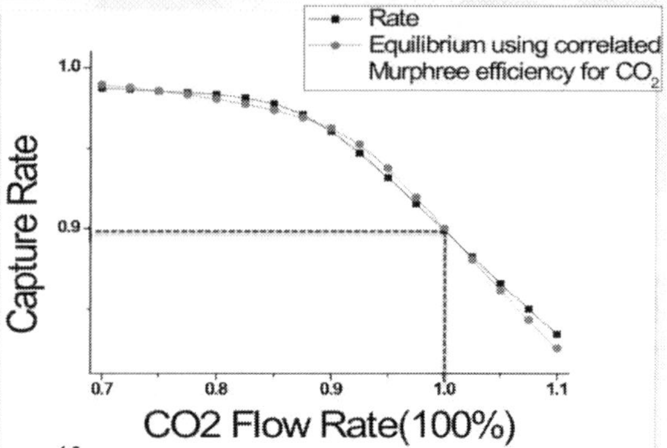

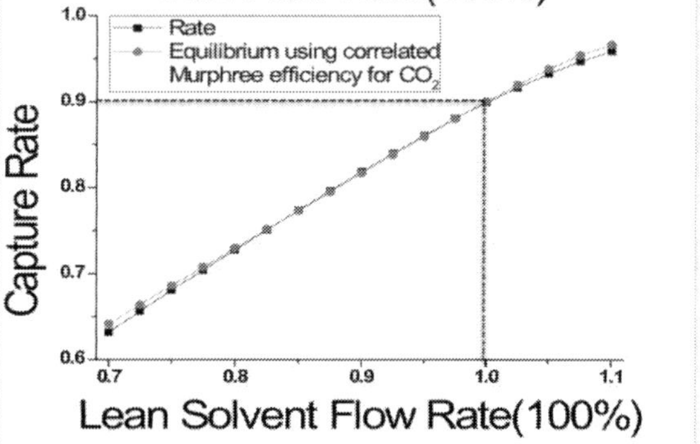

Dynamic Modeling

Packing Pressure Drop Model

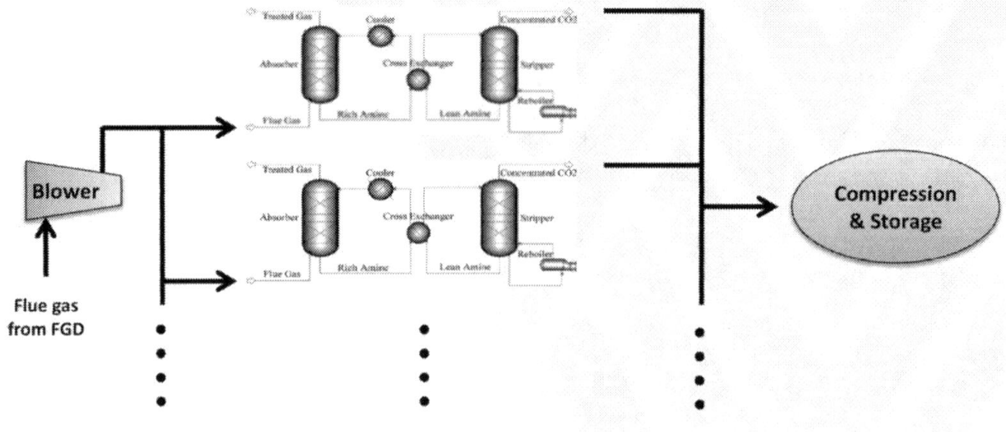

- Nonlinear process and highly interactive multivariable system.
- Variability in flue gas flow due to load-following by the power plant.
- Achieve regulation of flue gas flow through each train is difficult due to small pressure drop that is affordable.

Dynamic Modeling

IMTP Packing Pressure Drop Model

$$\Delta P_o = \frac{2a\left(\frac{s}{d_n}\right)^b H\rho_G\left(\frac{V}{\varepsilon}\right)^2}{d_h} \qquad \Delta P = \frac{\Delta P_o}{(1-A_o-\Delta A)^3} \qquad A_o = cRe_L^d Fr_L^e Eo^f \qquad \Delta A = gFr_L^h\left(\frac{L\rho_L}{V\rho_G}\right)^i Eo^j$$

Dry Bed *Wet Bed*

Packing Pressure Drop Correlation in Aspen Plus Dynamics

$$\Delta P = K_o\rho_G V^2$$

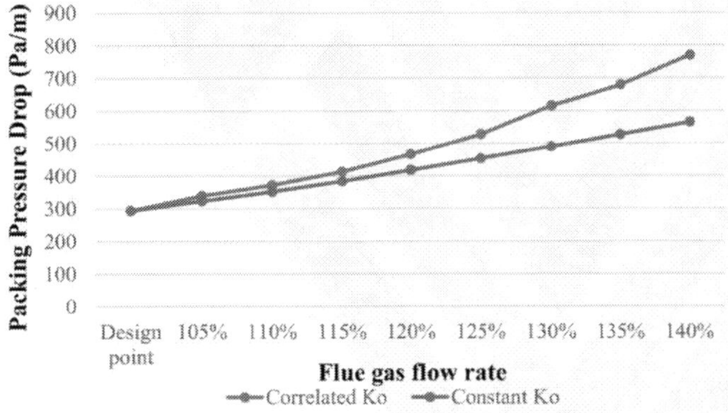

WEST VIRGINIA UNIVERSITY
CHEMICAL ENGINEERING

Dynamic Modeling

Pressure Drop Model – Absorber

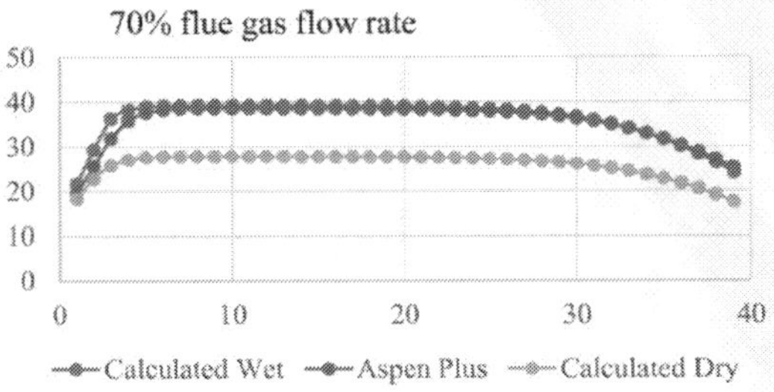

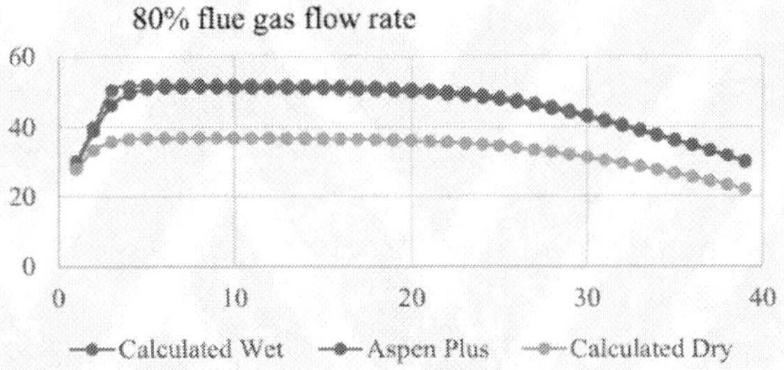

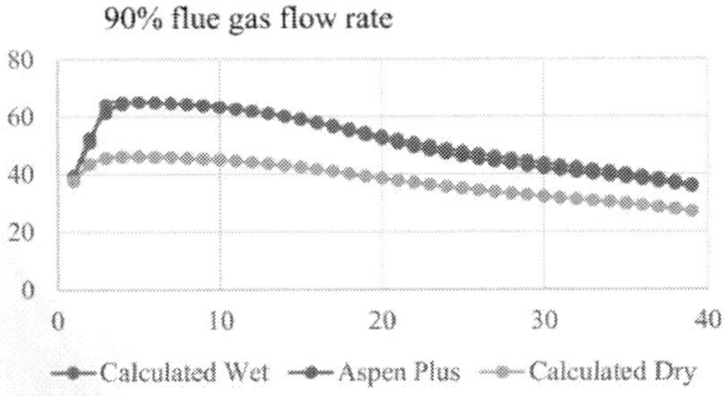

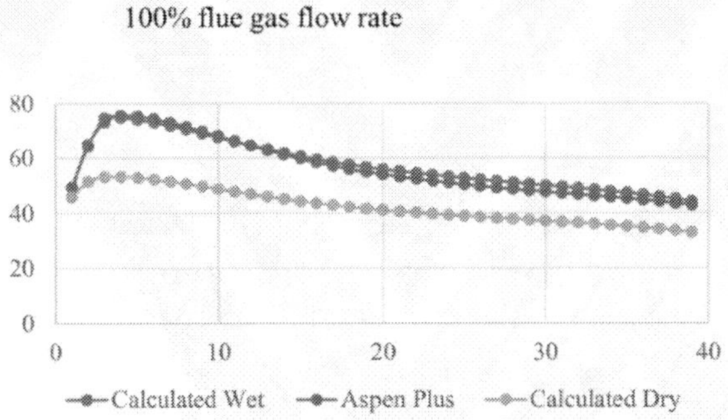

X-stage numbers; Y-Section Pressure Drop (Pa)

WEST VIRGINIA UNIVERSITY
CHEMICAL ENGINEERING

Dynamic Modeling

Pressure Drop Model – Absorber

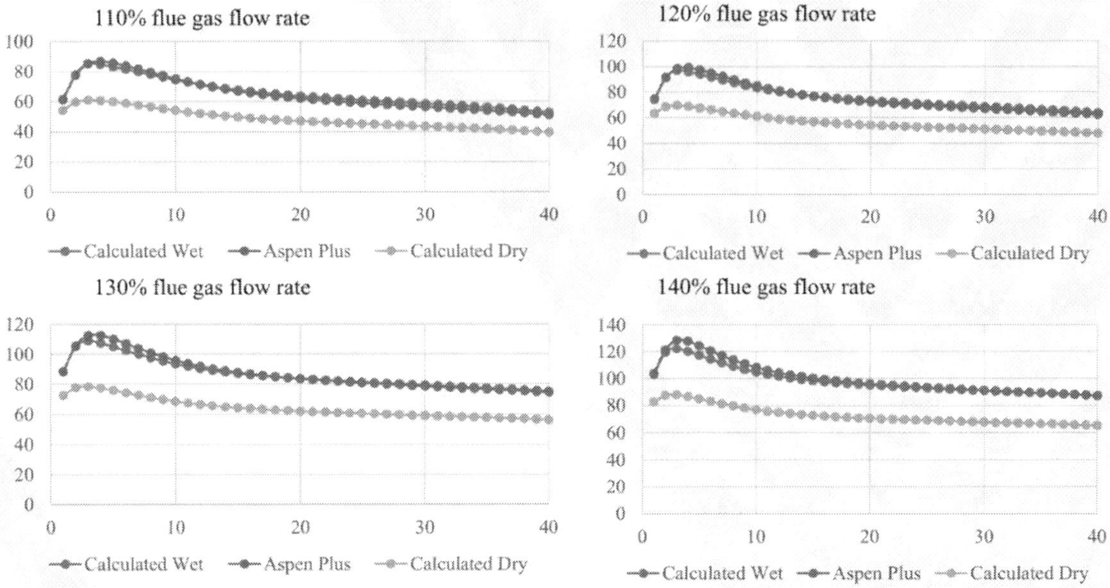

X-stage numbers; Y-Section Pressure Drop (Pa)

WEST VIRGINIA UNIVERSITY
CHEMICAL ENGINEERING

Application of PID and Linear Model Predictive Control

PI & PID Controllers

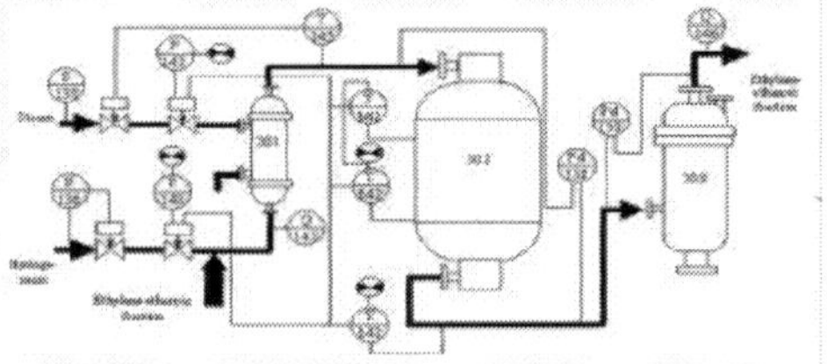

- Most widely used controller using a generic control loop feedback mechanism for industrial control systems.
- Satisfactory performance without the need for process model.
- Inner and outer loop PID controller provide cascade control.
- Usually can not provide optimal control.

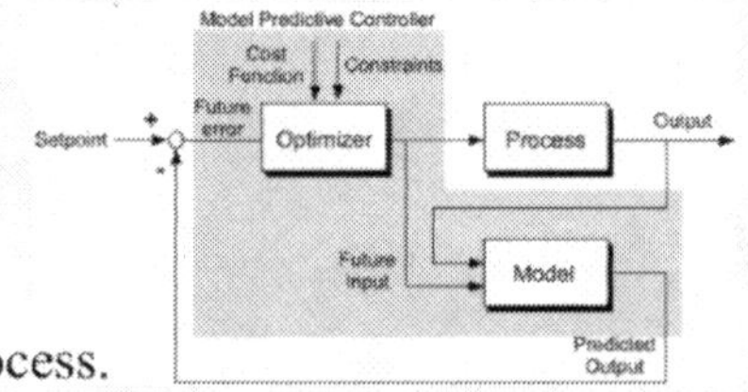

Model Predictive Controllers (MPC)

- An advanced control method applicable for chemical process.
- Predict the change in dependent variables based on the changes in independent variables.
- Perform optimal control by considering both current and future timeslot.
- Rely on dynamic models of process which require system identification.

WEST VIRGINIA UNIVERSITY
CHEMICAL ENGINEERING

Application of PID and Linear Model Predictive Control

System Identification

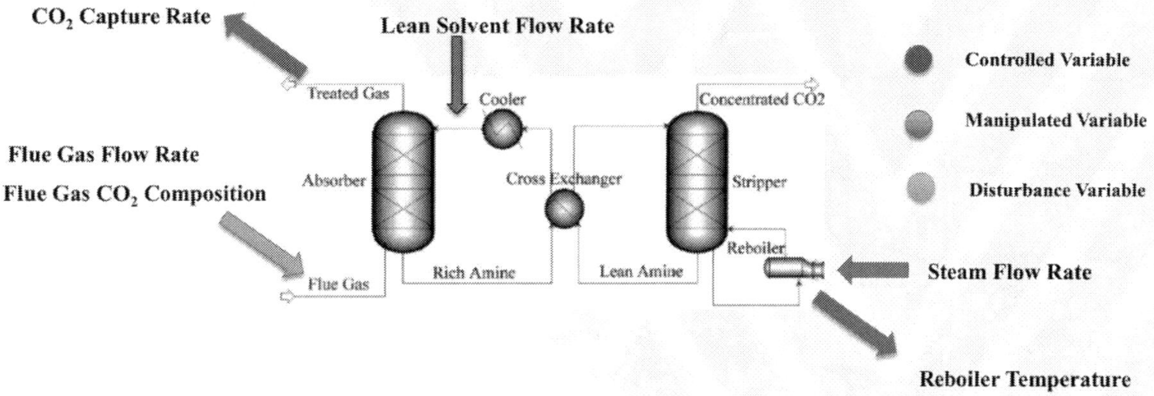

$$G(s) = \frac{K(1 + T_z s)}{1 + T_p s}$$

$1st$ order process model with process zero

- Data was generated by applying Pseudo Random Binary Signal to input variables.
- Max deviation amplitude of PRBS is 5% of the value at designed steady-state.

WEST VIRGINIA UNIVERSITY
CHEMICAL ENGINEERING

Application of PID and Linear Model Predictive Control

System Identification

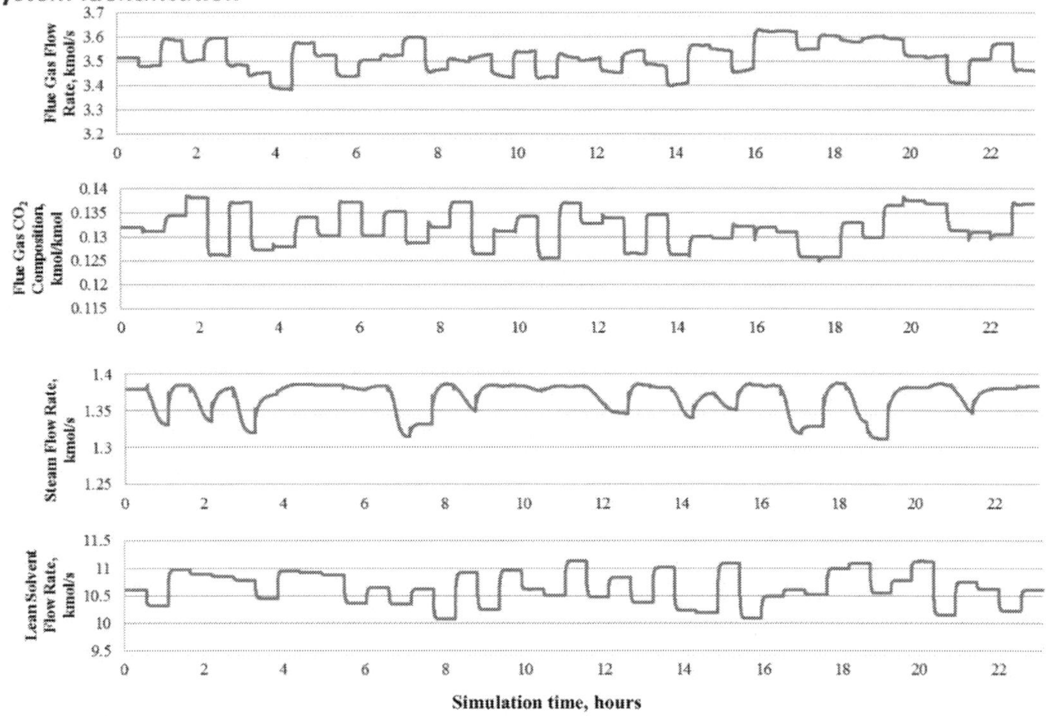

Application of PID and Linear Model Predictive Control

System Identification

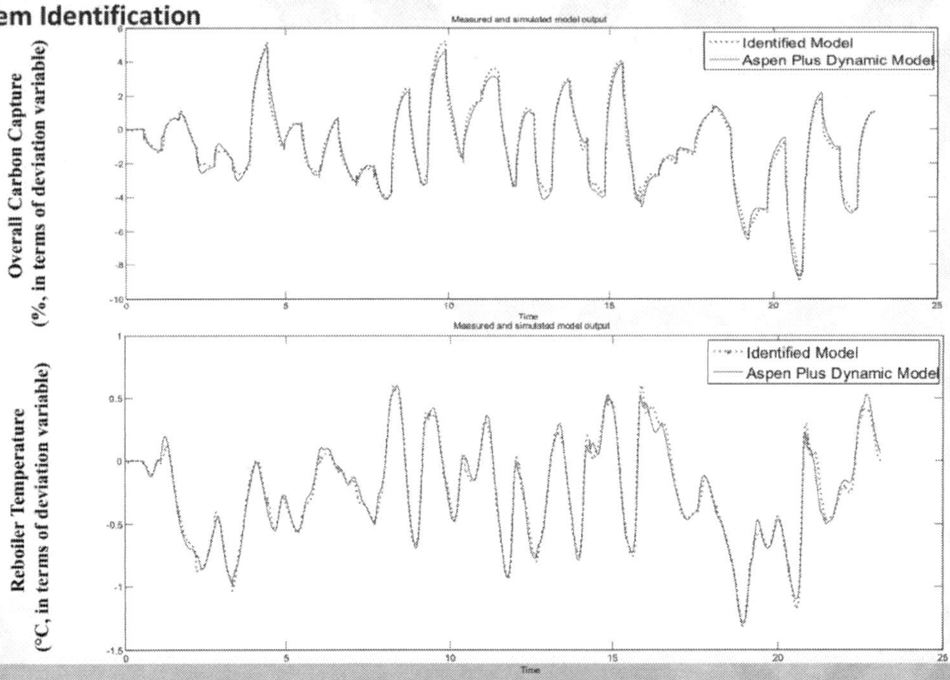

WEST VIRGINIA UNIVERSITY
CHEMICAL ENGINEERING

Application of PID and Linear Model Predictive Control

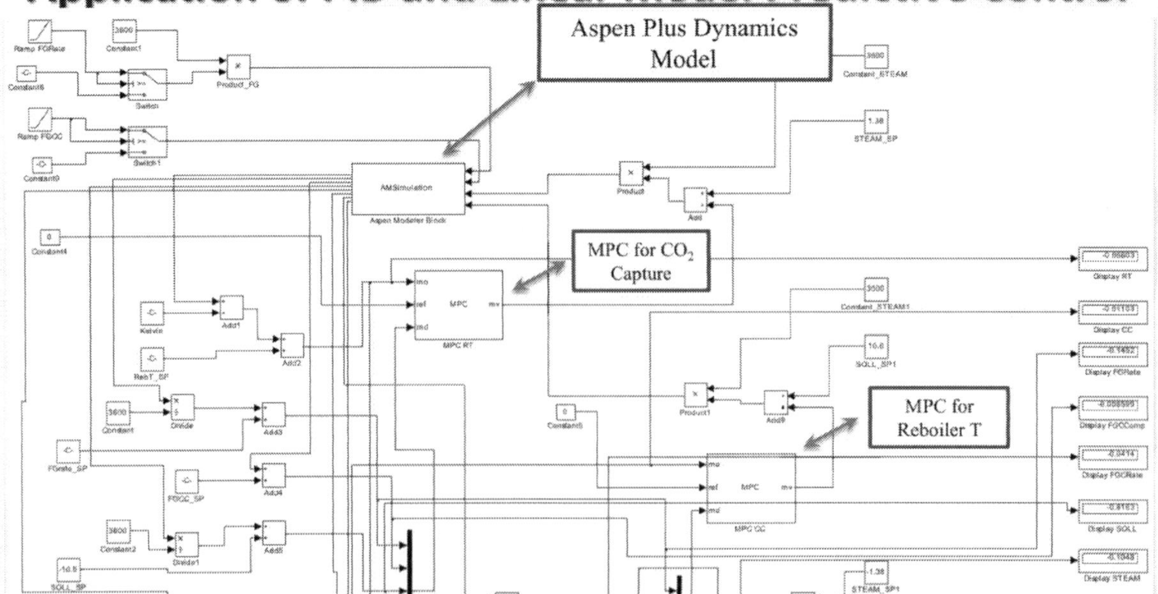

Application of PID and Linear Model Predictive Control

Dynamic Response to 5% Ramp down in both Flue Gas Flow Rate and CO_2 Composition

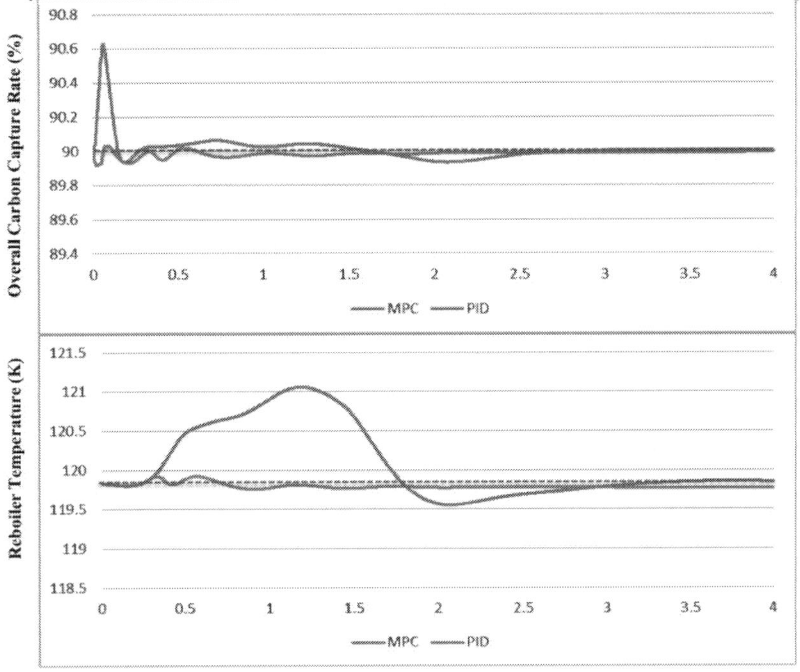

Simulation time, hours

Integral Square Error (CO_2 Capture Rate)

PID	0.02469159
MPC	0.0017583
Percent Improvement	92.88%

Application of PID and Linear Model Predictive Control

Dynamic Response to 10% Ramp down in Flue Gas Flow Rate and 5% in Flue Gas CO_2 Composition

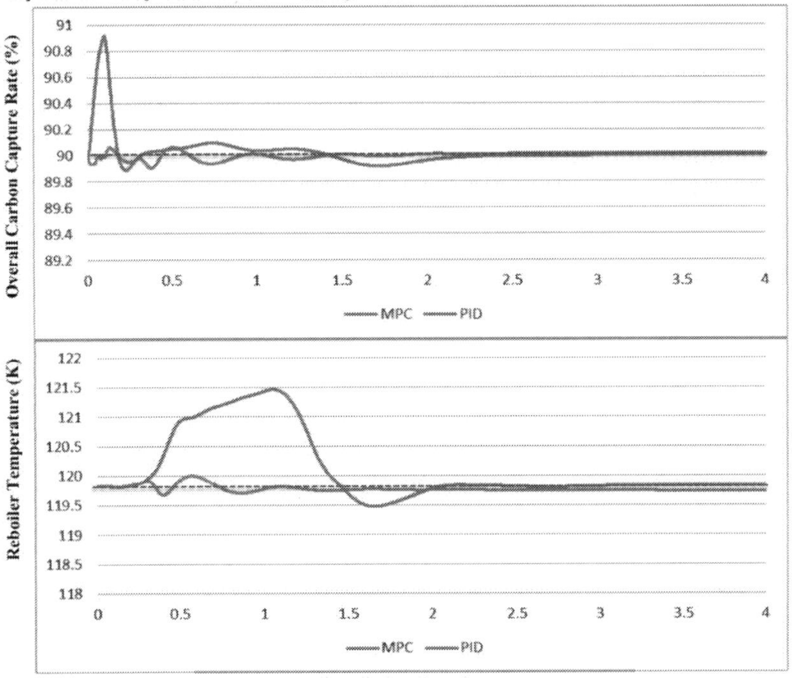

Simulation time, hours

Integral Square Error (CO_2 Capture Rate)

PID	0.06796127
MPC	0.00239925
Percent Improvement	96.15%

Application of PID and Linear Model Predictive Control

Dynamic Response to 15% Ramp down in Flue Gas Flow Rate and 5% in Flue Gas CO_2 Composition

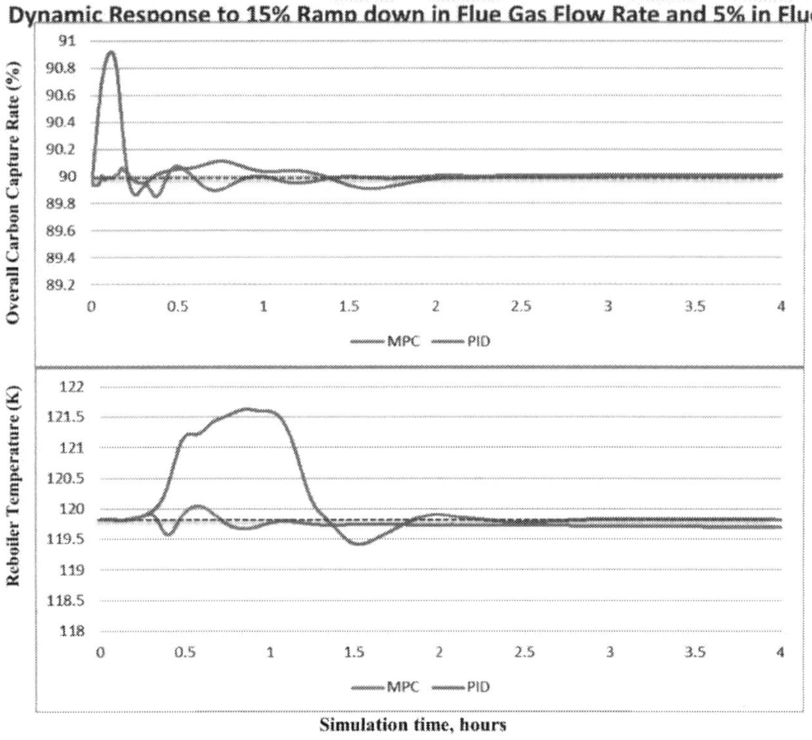

Integral Square Error (CO_2 Capture Rate)

PID	0.09364864
MPC	0.00444926
Percent Improvement	95.25%

Application of PID and Linear Model Predictive Control

Dynamic Response to 2.5% Random Step Change in both Flue Gas Flow Rate and CO_2 Composition

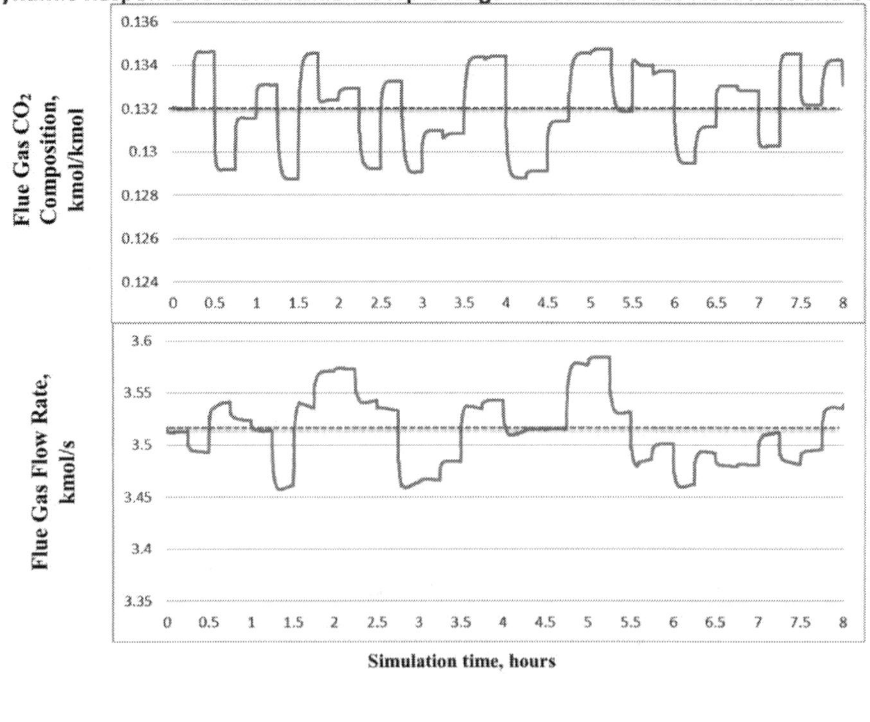

Application of PID and Linear Model Predictive Control

Dynamic Response to 2.5% Random Step Change in both Flue Gas Flow Rate and CO_2 Composition

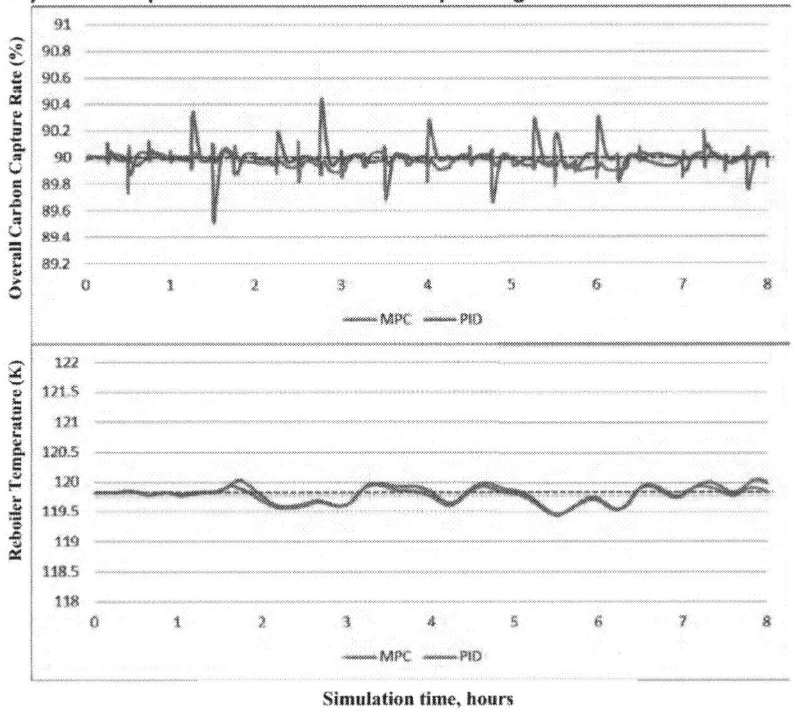

Integral Square Error (CO_2 Capture Rate)

PID	0.05151658
MPC	0.01795874
Percent Improvement	65.14%

Application of PID and Linear Model Predictive Control

Dynamic Response to 5% Ramp down in both Flue Gas Flow Rate and CO_2 Composition (90% Absorber Efficiency)

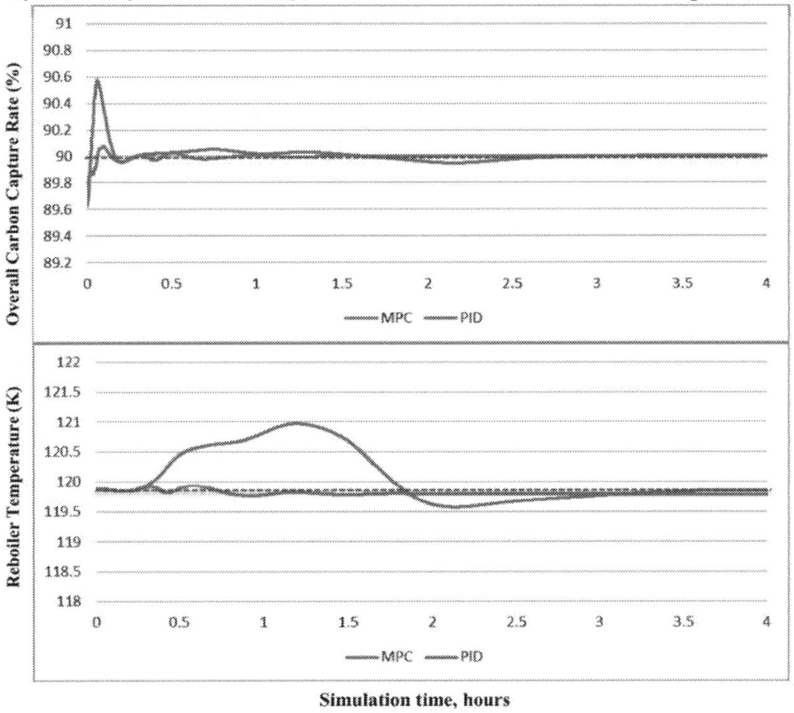

Integral Square Error (CO_2 Capture Rate)

PID	0.0216522
MPC	0.00160997
Percent Improvement	92.56%

Application of PID and Linear Model Predictive Control

Dynamic Response to 5% Ramp down in both Flue Gas Flow Rate and CO_2 Composition (100% Absorber Efficiency for Train 1 & 90% Absorber Efficiency for Train 2)

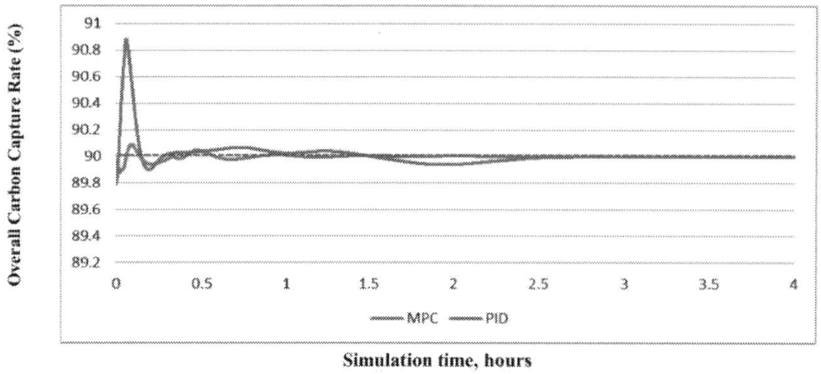

Integral Square Error (Overall CO_2 Capture Rate)

PID	0.04210578
MPC	0.00149806
Percent Improvement	96.44%

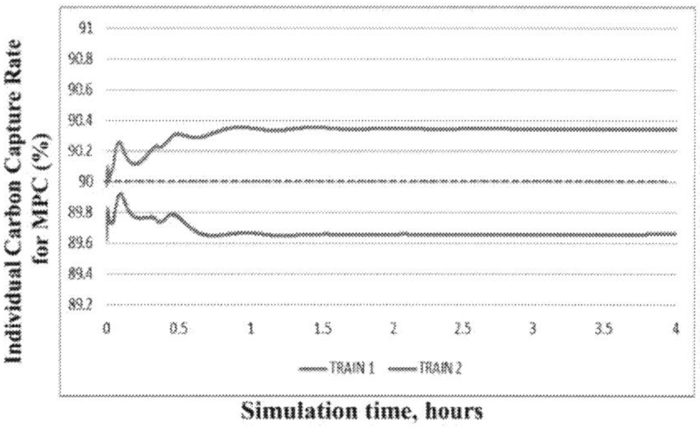

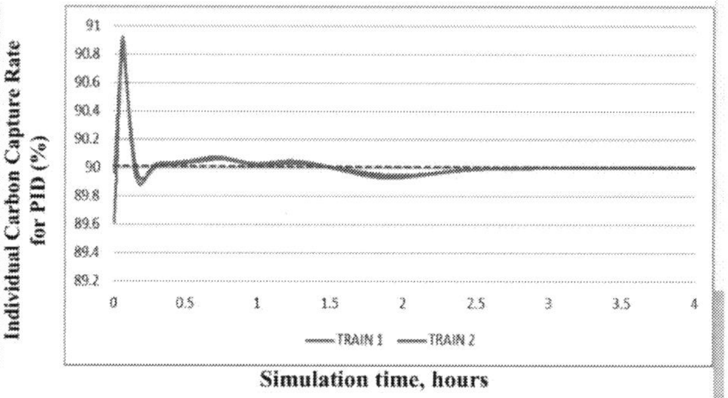

Conclusions

- By applying the correlations of CO_2 Murphree efficiency and packing pressure drop, the performance of equilibrium model can satisfy with the rate-based model under both design and off-design points.
- Model predictive controller shows a superiority over traditional PID controller in controlling both CO_2 capture rate and stripper reboiler temperature in face of a random or ramp disturbance in flue gas.
- For multiple trains carbon capture process, model predictive controller can provide an optimal control to maintain the overall carbon capture rate instead of individual capture rate for each train.

WEST VIRGINIA UNIVERSITY
CHEMICAL ENGINEERING

Future Work

- Applying different type of disturbance to this CO_2 capture process, such as a mixed ramp and random disturbance, to investigate the performance of model predictive controller in reference tracking.
- Since chemical process is inherently nonlinear, the application of nonlinear model predictive control will be studied.
- Since the dynamic behavior of chemical process can never be modeled perfectly in face of system uncertainties, a robust controller will be designed to maintain the stability and robust performance.

WEST VIRGINIA UNIVERSITY
CHEMICAL ENGINEERING

Acknowledgment

❖ This project is a subtask from the collaborative activities of the US Advanced Coal Technology Consortium (US ACTC) as part of the US-China Clean Energy Research Center (CERC) supported by the US Department of Energy (DOE) under Cooperative Agreement DEPI0000017.

U.S. - China Clean Energy Research Center
中美清洁能源研究中心

U.S. DEPARTMENT OF ENERGY

WEST VIRGINIA UNIVERSITY
CHEMICAL ENGINEERING

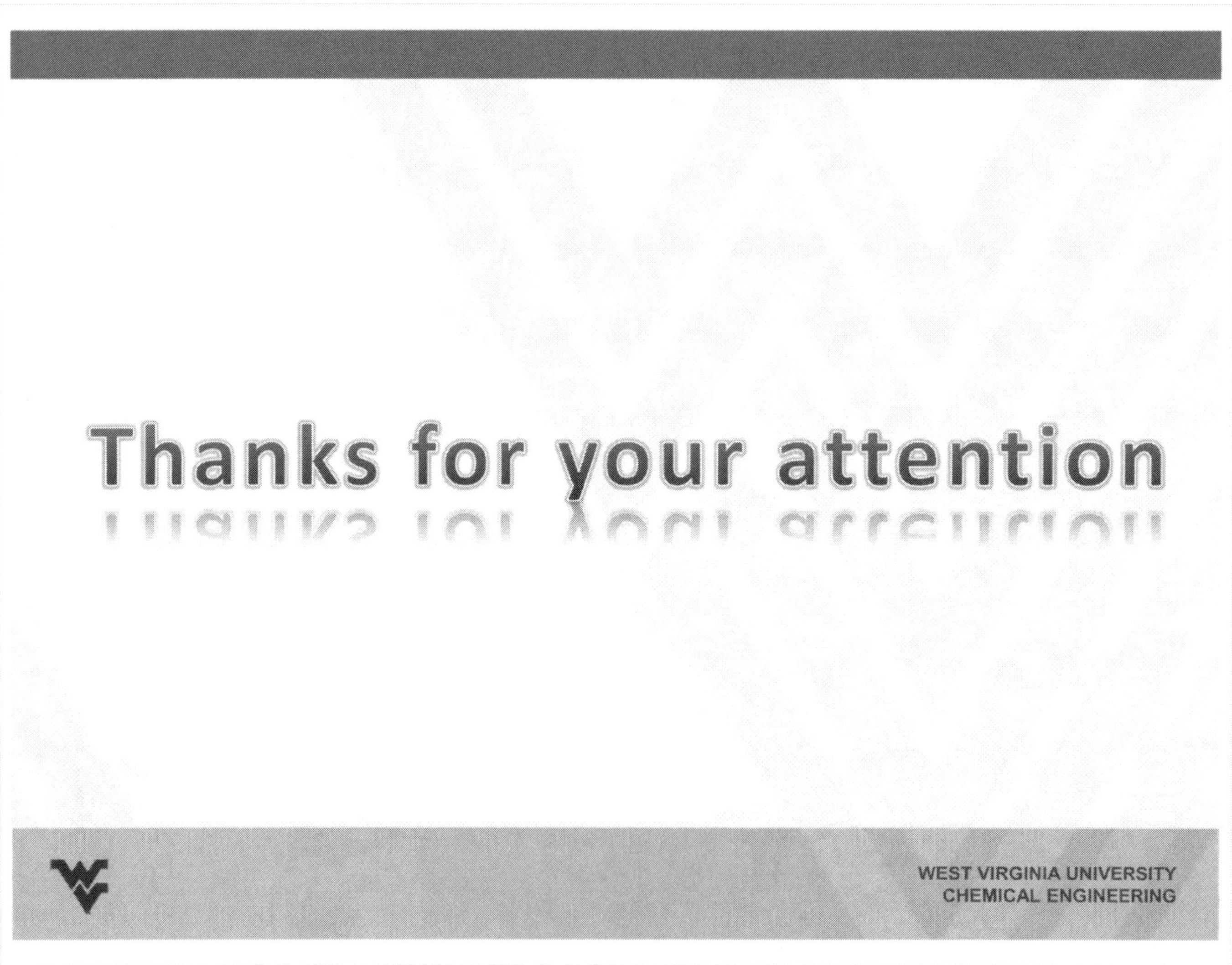
Thanks for your attention
WEST VIRGINIA UNIVERSITY
CHEMICAL ENGINEERING

13C Metabolic Flux Analysis of Metabolic Reprogramming Induced by AKT Signaling in Cancer Cells

At a molecular level, cancer is a disease of deregulated cell growth arising from genetic mutations that result in improper cell signaling. A key enzyme in mediating cellular growth response is the Akt serine/threonine kinase enzyme. This enzyme mediates a host of proliferative and survival responses including advancing cell cycle progression, increasing rate of biosynthesis, and increasing apoptotic resistance. Hyperactivation of the Akt signaling axis is implicated in a number of cancers including prostate, breast, lung, and colon cancers. Akt is also known to exhibit profound regulation of central carbon metabolism at the transcriptional, translational, post-translational, and allosteric levels. Prior biochemical studies of Akt have shown that its activation leads to increased expression of glycolytic enzymes including the GLUT1 glucose transporter and hexokinase (HK), two rate-limiting enzymes in glycolysis.

Our group has employed metabolic flux analysis using both isotopically labeled glucose and glutamine to map the flow of carbon through central metabolic processes in a wild-type pro-B-cell line, a derivative line expressing constitutively active myr-Akt, and a third line exhibiting hyperactive glucose metabolism via simultaneous overexpression of GLUT1 and HK. Through comparison of flux maps generated for these three cell lines, we have discovered significant metabolic reprogramming in multiple central carbon pathways as a result of Akt signaling that is not recapitulated in our simple model of GLUT1 and HK overexpression.

AKT activation leads to a 2.5x increase in glucose consumption and a 5.8x increase in lactate secretion compared to the wild type cells. Additionally, AKT activation results in 16.5% of incoming glucose to be sent down the pentose phosphate pathway, compared to <1% of incoming glucose in wild type cells. This result has interesting consequences on cellular NADPH production and redox balance in these cells. Perhaps most interesting, AKT activation reduces mitochondrial pyruvate carbon uptake from 48.7% in wild type cells to 3.9% in AKT activated cells, relative to incoming glucose flux. Additionally, all mitochondrial pyruvate uptake is seemingly linked to fatty acid synthesis, as evidenced by labeling data and TCA cycle fluxes. Full characterization of the metabolic changes induced by AKT signaling to support proliferation and survival of cancer cells has identified several candidate targets for inhibition that will result in better therapeutic and adjuvant treatments for the host of cancer subtypes with aberrant Akt signaling.

386069 A Generalizable Optimization Algorithm for Clinically-Relevant Patient-Specific Cancer Chemotherapy Scheduling

Abstract:

Chemotherapy is the most commonly employed method for systemic cancer treatment of solid tumors and their metastases. The balance between successful killing of cancer cells and minimizing the toxicity to the host remains a challenge for clinicians in deploying chemotherapy treatments. With toxicity as the primary reason a patient is either dose-reduced or dose-delayed in the clinic, our approach explicitly incorporates treatment-induced toxicities into the schedule design. As a case study, we synthesize administration schedules for docetaxel, a widely used chemotherapeutic ($765 million in sales in 2012) employed as a monoagent or in combination for the treatment of breast, non-small-cell lung, prostate, and head and neck cancers. The primary adverse effect of docetaxel treatment is myelosuppression, characterized by neutropenia, a low plasma absolute neutrophil count (ANC). Through the use of model-based systems engineering tools, we aim to provide treatment schedules for docetaxel and its combination therapies that reduce toxic side effects and improve patient treatment outcome.

Mirroring our past work, the current algorithm employs models of tumor growth, drug pharmacokinetics, and pharmacodynamics – both anticancer effect and toxicity, as characterized by ANC. Also included is a toxicity-rescue therapy, in granulocyte colony stimulating factor (G-CSF), that serves to elevate ANC. The single-agent docetaxel chemotherapy schedule minimizes tumor volume (the performance objective), over a multi-cycle horizon, subject to toxicity and logistical constraints imposed by clinical practice. In docetaxel treatment, the solving horizon length is 8 cycles(24 treatment weeks) where the tumor size is minimized every week along the treatment horizon. The numbers of variables and constraints in the docetaxel scheduling problem are 408,000 and 408,200 respectively. This problem is solved through a dynamic optimization step following an enumeration step. The enumeration step is carried to solve for all of the discrete variables (when to dose) and provide an initial guest for the dynamic optimization. Pyomo with Ipopt as the solver is then used to optimize the dose magnitude. The optimal solution provides a clinically-feasible treatment schedule for docetaxel chemotherapy within an order of 1-2 hours of computational time.

This single-agent chemotherapy scheduling problem formulation is extended to combination chemotherapy using docetaxel-cisplatin or docetaxel-carboplatin drug pairs. The two platinum agents display different toxicities, with cisplatin exhibiting kidney function damage (dynamically modelled through creatinine clearance changes) and carboplatin demonstrating the same myelosuppression effects as docetaxel. These case studies provide two different challenges to the algorithm: (i) cisplatin scheduling significantly increases in the number of variables and constraints (585,000 and 885,000 respectively with the same treatment horizon of 8 cycles and same objective function of minimizing tumor size weekly), thereby challenging the computational engine and formulation; (ii) carboplatin's overlapping toxicity tests the ability of the algorithm to schedule drugs with different mechanisms of action (they act in different phases of the cellular growth cycle) that have the same toxic side effect. Oftentimes in the clinic, the use of agents with overlapping toxicities is avoided, for fear of severe toxic side effects. The simulated results demonstrate the algorithm's flexibility, with docetaxel and cisplatin or carboplatin scheduled to provide tumor kill and toxicity within clinically acceptable limits. With an increase in number of variables and constraints in the combination chemotherapy, the algorithm is able to provide an optimal treatment schedule within 24 hours of computational time.

Overall, a clinically-relevant chemotherapy scheduling optimization algorithm is provided for designing single agent and combination chemotherapy, when toxicity and pharmacokinetic/pharmacodynamic information for the agents in question is available. Furthermore, the algorithm can be extended to patient-specific treatment via updating of the pharmacokinetic/pharmacodynamic models as data (e.g., tumor volume, toxicity measurements) is collected during treatment.

Design of a Digital Controller for a Fully Implantable Closed-Loop Artificial Pancreas

Lauren M. Huyett, Dale E. Seborg, Francis J. Doyle III

UCSB Department of Chemical Engineering, Santa Barbara, CA
Sansum Diabetes Research Institute, Santa Barbara, CA

In type 1 diabetes mellitus (T1DM), the body loses the ability to regulate glucose metabolism via the production of insulin. Current treatment involves manual administration of exogenous insulin based on frequent blood glucose (BG) measurements; however, it is difficult to achieve optimal results with this method[1]. A closed-loop artificial pancreas (AP) that automatically regulates BG has the potential to improve health outcomes in T1DM. This work details the design and *in silico* testing of an AP control algorithm for use with a pump and sensor implanted in the intraperitoneal (IP) space, eliminating problematic delays observed in current subcutaneous-only systems.

An autoregressive model with exogenous input (ARX) estimation procedure was used to perform control-relevant order reduction[2] of a personalized discrete model of IP insulin pharmacodynamics[3]. The starting model was

$$p\left(z^{-1}\right)=\frac{G(z^{-1})}{U(z^{-1})}=\frac{15(TDI^{-1})z^{-3}}{(1-0.98z^{-1})(1-0.75z^{-1})^{2}}$$

where G is the glucose concentration (mg/dL), U is the insulin delivery (units/h), z^{-1} is the backshift operator ($u(k-1)=z^{-1}u(k)$), TDI is the personalized total daily insulin dose (units), and the sampling period, Δt, is 5 min. The lower-order control-relevant model

$$\tilde{p}\left(z^{-1}\right)=\frac{K(TDI^{-1})z^{-1}\left(1-\beta z^{-1}\right)}{(1-a_1z^{-1})(1-a_2z^{-1})}$$

was determined by iteratively passing the full-order model step response through the prefilter

$$L(z)=B'(z)^{-1}\tilde{\varepsilon}(z)\tilde{\eta}(z)\left[r(z)-d(z)\right]$$

and fitting the filtered data to the model $\tilde{p}$. Here, B' is the numerator of $\tilde{p}$, $\tilde{\varepsilon}$ and $\tilde{\eta}$ are the sensitivity and complementary sensitivity operators based on $\tilde{p}$, and r and d are the reference and disturbance variables. The prefilter was designed to retain characteristics of the full-order model that are important during closed-loop operation[2]. The model $\tilde{p}$ was used to determine the parameters for a digital proportional-integral-derivative (PID) controller

$$u(k)=u(k-1)+K_c\left[e(k)-e(k-1)+\Delta t e(k)\tau_I^{-1}-\tau_D\Delta t^{-1}(e(k)-2e(k-1)+e(k-2))\right]$$

based on internal model control (IMC) tuning rules:

$$K_c=\frac{\delta(1-\delta)}{K(TDI^{-1})(\delta-\beta)}\left[(a_1+a_2)-2a_1a_2\right],$$

$$\tau_I=\frac{\left[(a_1+a_2)-2a_1a_2\right]\Delta t}{\left[1-(a_1+a_2)+a_1a_2\right]},$$

$$\tau_D = \frac{a_1 a_2 \Delta t}{\left[(a_1 + a_2) - 2a_1 a_2\right]}.$$

The output $u(k)$ is the calculated insulin dose (units/h), $e(k)$ is the difference between the measured and desired BG, and k is the time step. The single tuning parameter δ was used to specify the controller aggressiveness. The controller was tested *in silico* in the 10-subject UVA/Padova metabolic simulator[4] using a protocol with three meal disturbances: dinner at 19:00 (70g carbohydrates (CHO)), breakfast at 08:00 (40g CHO), and lunch at 13:00 (70g CHO). During the testing, an aggressive (δ = 0.87) or conservative (δ=0.92) tuning was used. In addition, the robustness of the controller to a 50% over- or under-estimation of the model gain was determined.

The results of the study are summarized in Table 1.

Table 1: Summary of results from *in silico* controller testing (mean ± standard deviation).

Controller	% Time 80-140 mg/dL	% Time 70-180 mg/dL	Maximum BG (mg/dL)	Minimum BG (mg/dL)
$\delta = 0.92$	74 ± 10	94 ± 4	200 ± 13	81 ± 7
$\delta = 0.87$	79 ± 8	97 ± 3	190 ± 10	84 ± 5
$\delta = 0.87$, 150% Gain	73 ± 10	93 ± 5	200 ± 13	81 ± 7
$\delta = 0.87$, 50% Gain	88 ± 4	100 ± 1	175 ± 6	91 ± 4

For all cases, nearly the entire 27 hour study period was spent within the clinically acceptable BG range of 70-180 mg/dL, with a large portion of the time spent within the tight control range of 70-140 mg/dL. The mean maximum BG values show that forays above 180 mg/dL were relatively small. The mean minimum BG value was above the 70 mg/dL safety threshold even when the gain was over- or underestimated. Neither model estimation error resulted in a dangerous condition for the subjects, demonstrating that the controller was robust to plant/model gain mismatch. The results showed a marked improvement over a previous *in silico* study using subcutaneous insulin, which achieved 58% and 83% of time within the tight control and clinically acceptable glucose ranges, respectively[3].

A fully implantable AP using an IMC-PID controller has shown great promise for improved regulation of blood glucose concentration in T1DM. The next step will be to evaluate this algorithm in an animal model, before moving on to clinical testing to demonstrate that a fully-implantable AP can achieve better results than conventional T1DM treatment.

1. The effect of intensive treatment of diabetes on the development and progression of long-term complications in insulin-dependent diabetes mellitus. The Diabetes Control and Complications Trial Research Group. *N Engl J Med,* 329(14):977-86, 1993.
2. Rivera D. E., and S. V. Gaikwad. Digital PID controller design using ARX estimation. *Computers & Chemical Engineering,* 20(11):1317-1334, 1996.
3. Lee J. J., E. Dassau, H. Zisser, and F. J. Doyle III. Design and in silico evaluation of an intraperitoneal-subcutaneous (IP-SC) artificial pancreas. *Computers & Chemical Engineering*, 2014 (in press).
4. Kovatchev B. P., M. Breton, C. D. Man, and C. Cobelli. In silico preclinical trials: a proof of concept in closed-loop control of type 1 diabetes. *J Diabetes Sci Technol,* 3(1):44-55, 2009.

Approximately 90% of cancer-associated mortality is a consequence of distant metastasis, a multistep process whereby cells from the primary tumor invade or migrate from a primary site to secondary organs. These cancerous cells may either proliferate immediately or lay dormant (*e.g.* as pre-malignant micrometastases) for years prior to detection. Clinically, undetected metastases have serious implications for cancer patients; exemplified by ~ 33% of women whom suffer a metastatic relapse within 5 years following removal of breast cancer. Distant metastases are generally more resistant to treatments than the primary tumor, hence facilitates the need to develop improved therapeutic approaches rationally designed based on the molecular pathophysiology within the metastatic microenvironment. As such, the liver represents an ideal system to evaluate both micrometastasis and the efficacy of cancer therapeutics as: (i) the preferred site of metastasis for many carcinomas, *e.g.* breast, lung, colon, prostate, brain; (ii) a major organ for drug metabolism, *i.e.* activation and detoxification, to assess efficacy and limiting toxicities of cancer therapies, and (iii) one of the primary locations of systemic regulation of circadian rhythms via modulation of hormones, nutrients, and inflammatory cytokines that significantly impact liver metabolism.

In vitro human disease models that capture the complexity of *in vivo* tissue behaviors are crucial to gain mechanistic insights of disease progression and enable the development of therapies that are effective across a broad range of patient populations [1]. Here, the integration of tissue engineering, primary cell sources, computational and systems biology approaches, as well as emerging biomaterials strategies and microfabrication processes has enabled novel experiments and analyses to investigate breast cancer metastasis. To capture the complexity of human physiology *in vitro* demands the evaluation of interacting cell types (*i.e.* hepatic tissues and tumor cells) coexisting in a hierarchical 3-dimensional structure influenced by gradients in nutrients, mechanics, and cell composition. We recapitulated the fundamental physiologic functions and conditions of the liver hepatic niche, including multi-cellular composition (*e.g.* tumor, stromal, and parenchymal cells), metabolism, protein production and evaluated circadian cycles, as reviewed [2]. Ultimately, to computationally model integrated responses of multiple cell types (*e.g.* primary human hepatocytes and a full complement of non-parenchymal cells, at physiological-relevant ratios) requires the need to assess systems under both homeostatic and pathologic conditions. In these studies, we aim to understand how the deregulation of cellular networks underlies complex disease states, such as cancer, using systems-based approaches to elucidate cellular responses to external stimuli. Specifically, we investigated how changes in the populations of multiple cell types, in the presence or absence of MDA-MB-231 cancer cells, regulate the response to inflammatory signals, such as cytokines, chemokines, and growth factors while corroborating the effects of cell function and viability across a broad spectrum of metrics (*e.g.* markers of tissue health: AST, ALT, function: cytochrome P450 activity, glucose consumption, urea catabolism, acute phase protein production, as well as tissue formation and structural integrity via imaging techniques). An all-human cellular composition of our experimental system led to the investigation of human-specific cellular crosstalk between tumor and hepatic tissues. Multivariate analyses of resultant phenotypic responses to external cues were assessed in a dynamic fashion. In this study, we systematically evaluated crosstalk in the presence of multiple factors, including stimuli that mimic bacteria (*e.g.* LPS) and/or combinations of chemotherapeutic agents (*e.g.* doxorubicin, cisplatin, camptothecin) [3]. An experimental system composed of primary human cells facilitated the evaluation of our results amongst donors to assess the robustness of our computational models.

Due to the intrinsic complexity of biological systems, integration of experimental and computational approaches is crucial to transform the way we study human tissue physiology and pathophysiology *in vitro*. Here, a systems biology perspective – including the use of statistical techniques such as PCA and PLSDA – led to novel insights attributed to cytokine profiles and signaling networks across a range of physiological perturbations, with emphasis on development and validation of predictive models for cell and tissue behavior relevant in drug discovery. Collectively, our approach has provided biological insights of plausible signatures of early metastatic disease and response to therapies.

[1] M. R. Ebrahimkhani, C. L. Young, D. A. Lauffenburger, L. G. Griffith, J. T. Borenstein *Approaches to in vitro tissue regeneration with application for human disease modeling and drug development.* 2014, **Drug Discovery Today** (in press)
[2] S. E. Wheeler, *et al. All-human microphysical model of metastasis therapy.* **Stem Cell Research & Therapy**, 2013, 4(Suppl 1):S11.
[3] A. M. Clark, *et al. A Microphysiological System Model of Therapy for Liver Micrometastases.* 2014, **Experimental Biology & Medicine** (in press)

Research was sponsored by the U.S. Army Research Office and the Defense Advanced Research Project Agency and was accomplished under Cooperative Agreement Number W911NF-12-2-0039 and funded by NIH 1UH2TR000496-01: All-Human Microphysical Model of Metastasis Therapy. The views and conclusions contained in this document are those of authors and should not be interpreted as representing the official policies, either expressed or implied, of the Army Research Office, DARPA, or U.S. Government. The U.S. Government is authorized to reproduce and distribute reprints for Government purposes notwithstanding any copyright notation hereon.

380212 Tumor Vascular Status Regulates Molecular Dissemination into Regional and Systemic Tissues

Abstract:

Tumors secrete biomolecules (proteins, exosomes) of a broad range of physiochemical properties (chain-like biopolymers, membrane-enclosed particulates) and sizes (1-5 nanometer (nm) in hydrodynamic radius, 30-100 nm in diameter) that are hypothesized to precondition the metastatic niche. However, it is not well understood by what route these biomolecules escape from the primary tumor to accumulate within their site of action. Since normal blood capillaries are permeable only to low molecular weight (MW) solutes (<5 nm in diameter), lymphatic drainage supports the clearance of intermediate MW solutes (>40kDa or >5nm diameter but <100 nm) from the intercapillary tissue space, and cell-mediated trafficking clears large particulates (>100 nm), molecular size characteristics are hypothesized to crucially influence their clearance profile and capacity to disseminate systemically. However, though the tumor vascular plexus is well described to be irregular and highly permeable, how it influences the extent and rate of molecular clearance as well as blood versus lymphatic partitioning of low versus high MW tumor-derived molecules remains ill-defined. To address this issue, the contribution of the developing tumor vasculature on molecular clearance and dissemination patterns was investigated by comparing the transport patterns of fluorescently labeled molecules from healthy versus malignant tissues that display well-documented changes in their complex neo-vascular plexus with tumor progression. We found that progression of B16F10 melanoma in C57Bl6 animals for seven days or more resulted in dramatically reduced (>75%) lymphatic transport relative to that from normal (naïve) skin tissue, which resulted in reduced sentinel (tumor-draining) lymph node molecular accumulation. Advanced tumors also demonstrated disrupted permeability profiles that resulted in altered systemic accumulation of injected molecular beacons relative to healthy tissues. Experimental results from these temporal biodistribution studies and pharmacodynamic modeling will be discussed in depth. In summary, this work suggests that altered vascular profiles within malignant skin results in molecular clearance and dissemination patterns that are absent from healthy animals and may reveal the utility of angioplastic therapeutics for mitigation of the role of molecular transport in metastasis.

Impact of Necrotizing Enterocolitis on the Barrier Properties of Intestinal Mucus

Carlson, T.L.[1], Yildiz, H.[1], Claud, E.[2], Carrier, R. L.[1]

[1]Department of Chemical Engineering, Northeastern University, Boston, Massachusetts 02115

[2]Department of Pediatrics, University of Chicago Medicine, Chicago, IL 60637

Introduction: Mucus functions as a protective layer to selectively control transport to the underlying epithelia. Different disease states have been associated with changes in mucus production, quantity, and composition, which may lead to altered barrier properties and pathogen invasion. Necrotizing enterocolitis (NEC) is a devastating disease occurring primarily in preterm and low birth weight infants. NEC involves infection of intestinal epithelium and is associated with 15-30% mortality [1]. Multiple causes for NEC are currently being investigated, including different infective agents and changes in microbiome, innate immunity, and epithelial barrier properties; however, the changes in the mucosal barrier are not well understood. We hypothesize that the barrier properties of the mucus lining the GI tract are altered in NEC. If this hypothesis is correct, it may be possible to modulate mucus barrier properties to protect preterm infants against invading microbes and development of NEC.

Materials and Methods: Intestinal segments were excised from 3 day-old Sprague Dawley "NEC induced" and healthy breastfed rat pups. NEC inducing stress was applied with hypoxia, cold stress, and formula feeding. This study included pups in stages 0-1 which have potential to develop NEC. Two sections (~1 cm) of ileum were extracted, cut lengthwise, and placed on a glass slide. Particle suspensions (5 μl, carboxylate-modified yellow-green fluorescent microspheres, 200 nm, 2% solids in maleate buffer) were carefully pipetted onto the mucus surface. Samples were incubated in a dark, humid chamber for 90 min. Twenty second videos were captured using Olympus DP imaging software, with $n \geq 100$ particles analyzed in each excised segment. Particle trajectories were analyzed to calculate their time dependent ensemble averaged mean-squared displacements (<MSD>) and effective diffusion coefficients (D_{eff}).

Results and Discussion: Diffusive transport in intestinal mucus of NEC induced rat pups was significantly higher (2-fold) at a time scale of 10 seconds than in intestinal mucus of healthy control pups (Fig 1). Altered mucus barrier properties in NEC induced pups could be due to lower Muc 2 concentration, low hydrophobicity, lowered phospholipid concentrations, loss of epithelial growth factor or high bile acid accumulation [2-5].

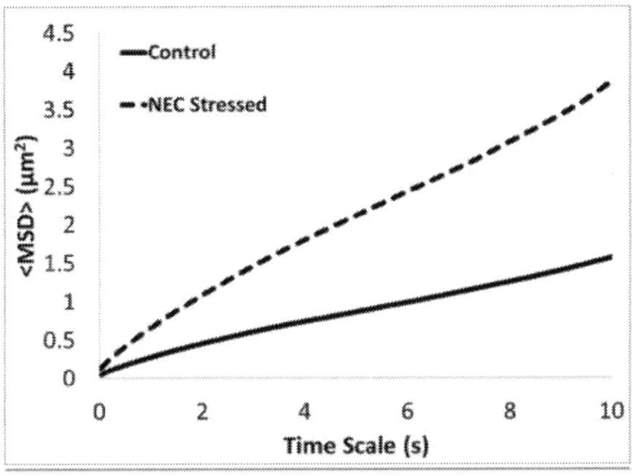

Fig 1. Transport of carboxyl fluospheres in rat intestinal mucus. Average <MSD> of particles diffusing in intestinal mucus of control vs NEC induced rat pups.

Conclusions: NEC was shown to have an impact on diffusive transport of sub-micron particles through intestinal mucus, indicating the mucus barrier is changed in this disease state which may increase NEC susceptibility. These results support exploration of modulation of the mucus barrier to help prevent the onset of NEC in premature infants.

References:

1. Gregory, K.E., *Clinical Predictors of Necrotizing Enterocolitis in Premature Infants.* Nursing Research, 2008. **57**(4): p. 260-270 10.1097/01.NNR.0000313488.72035.a9.
2. Martin, N.A., et al., *Active Transport of Bile Acids Decreases Mucin 2 in Neonatal Ileum: Implications for Development of Necrotizing Enterocolitis.* PLoS ONE, 2011. **6**(12): p. e27191.
3. Dvorak, B., et al., *Comparison of Epidermal Growth Factor and Heparin-binding Epidermal Growth Factor–like Growth Factor for Prevention of Experimental Necrotizing Enterocolitis.* Journal of Pediatric Gastroenterology and Nutrition, 2008. **47**(1): p. 11-18 10.1097/MPG.0b013e3181788618.
4. *Intestinal barrier failure during experimental necrotizing enterocolitis: protective effect of EGF treatment*, ed. J.A. Clark, et al. Vol. 291. 2006. G938-G949.
5. Cetinkaya, M., et al., *CDP-choline reduces severity of intestinal injury in a neonatal rat model of necrotizing enterocolitis.* J. Surg. Res., 2013. **183**(1): p. 119-128.

Elective administration of bacterial endotoxin (lipopolysaccharide, LPS) has been used as a reproducible experimental model of systemic inflammatory response syndrome (SIRS); allowing the analysis of response to infectious stress at multiple physiological levels. In this model, administration of low dose LPS to healthy human subjects transiently alters many physiologic and metabolic processes in a qualitatively similar manner to those observed after acute injury and systemic inflammation (1, 2). Recently, we performed the first complete metabonomic analysis of experimental endotoxemia in humans by describing the temporal alterations of plasma metabolites within the first day of LPS administration (3). This analysis uncovered the temporal patterns in the host metabolism reflecting collective impacts of regulations at various organs and at multiple levels of cellular processes including transcription, translation and signal transduction. In the current study, we sought to assess the concordance of metabolic response to endotoxemia and clinical sepsis cases of various origins. Our hypothesis was that the metabolic response patterns elicited by endotoxemia would agree with those clinically observed in sepsis similar to the previously defined agreement between leukocyte transcriptional profiles of complementary experimental and clinical cases of systemic inflammation (4).

Significant alterations in the plasma metabolome of the subjects challenged with endotoxin were statistically compared and contrasted with that of septic patients (5) who had been stratified into three distinct groups: Sepsis survivors with uncomplicated recovery courses, sepsis non-survivors and non-infected patients who exhibited SIRS criteria. Common metabolites between endotoxemia and each of these three groups were individually identified together with their direction of change and functional classifications. Furthermore, a core group of metabolites which displayed coherent changes in all three comparisons were determined and their relevance to metabolic response to systemic inflammation is discussed.

In summary, this study benchmarks an instrumental experimental model against actual clinical data from three distinct classes of patients with systemic inflammation. Although at moderate levels, observed concordance between metabolomic responses warrants that this experimental model effectively induces the molecular cues that convey complementary mechanisms at the systemic level which are clinically observed in septic patients and therefore, can facilitate further research on the development and assessment of rational clinical therapies to prevent or attenuate systemic inflammatory response syndrome (SIRS) as well as providing data for novel *–omics* studies to elucidate intricate regulatory schemes governing the human inflammatory response.

References

1. Lowry SF. Human endotoxemia: a model for mechanistic insight and therapeutic targeting. Shock. 2005;24 Suppl 1:94-100. Epub 2005/12/24. PubMed PMID: 16374380.

2. Calvano SE, Coyle SM. Experimental human endotoxemia: a model of the systemic inflammatory response syndrome? Surg Infect (Larchmt). 2012;13(5):293-9. Epub 2012/10/18. doi: 10.1089/sur.2012.155. PubMed PMID: 23072275; PubMed Central PMCID: PMC3503465.

3. Kamisoglu K, Sleight KE, Calvano SE, Coyle SM, Corbett SA, Androulakis IP. Temporal metabolic profiling of plasma during endotoxemia in humans. Shock. 2013;40(6):519-26. Epub 2013/10/04. doi: 10.1097/shk.0000000000000063. PubMed PMID: 24089011; PubMed Central PMCID: PMCPMC3970546.

4. Haimovich B, Reddell MT, Calvano JE, Calvano SE, Macor MA, Coyle SM, et al. A novel model of common Toll-like receptor 4- and injury-induced transcriptional themes in human leukocytes. Crit Care. 2010;14(5):R177. Epub 2010/10/12. doi: 10.1186/cc9283. PubMed PMID: 20929567; PubMed Central PMCID: PMC3219281.

5. Langley RJ, Tsalik EL, van Velkinburgh JC, Glickman SW, Rice BJ, Wang C, et al. An integrated clinico-metabolomic model improves prediction of death in sepsis. Science translational medicine. 2013;5(195):195ra95. Epub 2013/07/26. doi: 10.1126/scitranslmed.3005893. PubMed PMID: 23884467; PubMed Central PMCID: PMCPMC3924586.

376319 The Effects of Fluid Shear Stress on Cancer Stem Cells

Abstract:

Cancer is the second leading cause of death in the United States. Despite improvement in treatment options, understanding more about the disease mechanisms is an important step in cancer treatment. It has been theorized that cancer stem cells play an important role in patients undergoing relapse. As a growing theory, it is important to understand more about these cancer stem cells and how they are affected by different environment. The physical environment that cancer stem cells experience can lead to various cell signaling. Conventional concepts from chemical engineering such as fluid flow can be used to model the external effect that cancer cells experience. Fluid flow and fluid shear stress results within the body especially within the blood stream of patients undergoing metastasis where cancer cells detach from the primary tumor and enter the blood stream as circulating tumor cells to form the secondary tumor. This project simulates the fluid shear stress experienced by both circulating tumor cells and individual cancer cells in the vascular system of the human body. We developed a cost-effective microfluidic system to test the effects of fluid shear stress on breast cancer cells (MCF7 and MDA-MB-231), glioblastoma cells (U87-MG), and leukemia cells (KG-1α). Physiological fluid shear stresses of 5, 20, and 60 dyne/cm^2 were tested. Cancer stem cell traits were observed under shear stress by testing the expression of cancer stem cell markers. Results on cancer stem cell viability and cancer stem cell marker expression under fluid shear stress conditions will be presented, and the implications of these results on circulating tumor cells and metastasis will be discussed.

ID# 359244

Identification and Estimation of the Influential Parameters in Bioreaction Systems

Mordechai Shacham, Chem. Eng. Dept., Ben-Gurion University, Beer-Sheva, Israel
Neima Brauner, School of Engineering, Tel-Aviv University, Tel-Aviv, Israel

Introduction

In recent years there has been a rising interest in mathematical modeling in systems biology in particular in dynamic modeling of bioreaction systems. The goal of the mathematical modeling is to obtain expressions that describe the dynamic behavior of the systems under consideration. The resultant models can be applied for analysis, design optimization and control of the biological systems. The mathematical models are often described by systems of ordinary differential equations (ODE's). The models usually contain unknown parameters that need to be determined using a set of measurements (experimental data). Typically, the estimation of the parameter values is performed using a maximum likelihood approach, where the objective is to minimize the weighted squared error between the set of the measured data and the corresponding model predictions. Recently, Sun et al. (2012) reviewed the methods used for parameter estimation, and classify the optimization algorithms used as deterministic global, metaheuristic global (such as evolutionary algorithms, EA) and deterministic local (gradient-based) algorithms. They point out that deterministic global algorithms are very time consuming and usually cannot obtain satisfactory solutions in a reasonable time frame when applied to practical problems of realistic size. Consequently, the use of metaheuristics for identifying near optimal parameter values is recommended, followed by gradient based methods in order to refine these values.

Kravaris et al. (2013) point out that parameter estimation for practical, complex systems is an ill-conditioned problem. The main reasons mentioned for the ill-conditioning are excessive number of unknown parameters in the fundamental models, uncertainty regarding the suitability of the mathematical model to represent the data and insufficient amount and/or low precision, noisy experimental data. In order to overcome these limitations Kravaris et al. (2013) recommend that the parameter estimation problem of complex systems should not deal solely with the determination of the optimal solution, but concentrate on computing a solution which is robust to variations in the experimental data. McLean and McAuley (2012) emphasize the point that in realistic parameter estimation problems it is impossible to identify all the parameters. McLean et al. (2012) review some of the methods for selecting subsets of parameters for estimation. Liu and Wang (2009) present a kinetic model of the ethanol fermentation process, where the presumed model contains 19 parameters to be calculated by regression of experimental data. Due to several orders of difference between the estimated values of the parameters (between 10^{-1} to 10^{12}) they eventually recommend the use of a "pruned model" with only 13 parameters.

Shacham and Brauner (2014) presented a stepwise regression procedure which starts with a minimal set of parameters that can yield a feasible solution (satisfying the state variable and their derivative values at time = 0) and continues with addition of parameters until reaching an unstable set (as measured by confidence interval to parameter value ratios) or obtaining a physically infeasible parameter value. The method of Shacham and Brauner (2014) is applicable to parameter estimation problems where no constraints are imposed on the parameter values. In this paper a new stepwise regression algorithm, which is applicable to constrained dynamic parameter estimation, is presented. The proposed method is demonstrated by identifying the maximal set of influential parameters and determination of their optimal values for the model of the "ethanol fermentation" process presented by Liu and Wang (2009) using the data provided by Gennemark and Wedelin (2009).

Basic Concepts

Adopting the terminology used in Floudas et al., 1999, the parameter estimation problem involving an ODE model can be formulated as follows:

$$\min_{\theta} \Phi(\theta) = \sum_{\mu=1}^{n} \sum_{i=1}^{m} \left(x_{\mu,i} - x_{\mu,i}^{c}\right)^2 * W_i \tag{1}$$

s.t.

$$\begin{aligned} &\frac{d\mathbf{x}}{dt} = \mathbf{F}(\mathbf{x}, \boldsymbol{\theta}, t) \\ &\mathbf{x}(0) = \mathbf{x}_0 \\ &\boldsymbol{\theta}_{\min} \le \boldsymbol{\theta} \le \boldsymbol{\theta}_{\max} \end{aligned} \tag{2}$$

where **F** is a system of *m* ordinary differential equations, **x** is a vector of *m* dynamic variables, $\boldsymbol{\theta}$ is a vector of *p* parameters, $\mathbf{x}_{\mu}$ is a vector of *m* observed values at the μth data point, $\mathbf{x}_{\mu}^{c}$ is a vector of *m* calculated variable values at the μth data point, *n* is the total number of the data points, $\boldsymbol{\theta}_{min}$ and $\boldsymbol{\theta}_{max}$ are vectors of lower and upper bounds, respectively, on the parameter values, and **W** is a vector of *m* weighting factors.

This formulation enables solution of the problem using the, so called, sequential (or feasible path) approach. Alternative formulations, which enable converting the differential parameter estimation problems into an algebraic parameter estimation problems (simultaneous approach) are presented for example by Shacham and Brauner (2014). Only the sequential approach is discussed in this paper.

Using the sequential approach, the minimization defined in Eq (1) is carried out in an outer loop, while in the inner loop an integration routine is used to determine the state variable values at time intervals where experimental data are available. In this study, the integration is usually carried out by the MATLAB *ode45* function, which is based on an explicit Runge-Kutta (4,5) formula, the Dormand-Prince (1980) pair. This algorithm monitors the estimate of the integration error, and controls the step size of the integration in order to keep the error below a specified threshold. The accuracy requested is that both the relative and absolute (maximal) errors be less than the truncation error tolerance. The default value of this tolerance is 1.0E-6. The integration is carried out in a piecewise manner from the point t_{μ} to $t_{\mu+1}$. At this point the μth component of the objective function is calculated and added to the previous components of the objective function. This process is continued until reaching the last set of data points where $t_{\mu+1} = t_n$. For integration of stiff systems the MATLAB library function *ode15s* is used, which is based on the backward difference formulas (BDF) method of Gear (1971).

In the outer loop minimization subject to constraints is carried out using the *fmincon* function. A sequential quadratic programming method is combined with the "active set" strategy (Powel, 1978) for optimization. A quasi-Newton approximation to the Hessian of the

Lagrangian is calculated at each iteration. This method belongs to the gradient-based category, where derivatives of the objective function and the constraints need to be evaluated. Regarding the weighting factor W_i: in the statistical literature the reciprocal of the variance of the $\mathbf{x}_i$ data is used as weighting factor. As the variance is proportional to the experimental error in the data, such scaling reduces the impact of the residual of noisy data on the value of objective function, $\Phi(\boldsymbol{\theta})$. The estimation of the variance requires performance of repetitive experiments, while in most of the practical dynamic parameter estimation problems, only one set of state variable values is available for every t_μ point. Obviously, scaling must be introduced in problems where the values of the state variables are associated with different units, or different magnitude. Therefore, the scaling we introduce is:

$$W_i = (\max_{\mu} |x_{\mu,i}|)^{-1} \tag{3}$$

This formulation actually scales all the variables in the objective function to the [-1,1] interval.

The robustness of the solution is measured by the relative magnitude of the change of the parameter values when the model is fitted to different sets of experimental data. Small change of the parameter values indicates robust solution. The objective of the stepwise regression process is to identify the set of parameters whose inclusion in the model yields a robust solution with minimal objective function value. The proposed stepwise regression procedure is describe in the next section.

Stepwise regression for dynamic parameter estimation problems with parameters subject to constrains

Stepwise regression has been extensively used for determining the most influential terms and parameters in linear regression model. Various aspects of the stepwise regression procedure for identifying the most influential terms and parameters and establishing stopping criterion for addition of more terms and parameters to a linear regression model were discussed in detail, for example, by Brauner and Shacham (1999) and Shacham and Brauner (2002). In a typical linear regression model there is one dependent variable and the parameters (except the "free" parameter) are associated with "explanatory" (independent) variables. The parameters that are included in the regression model (and the associated terms of the explanatory variables) at any stage are denoted as basic parameters, and the remaining parameters as non-basic parameters. At every step the objective function calculated while adding each of the remaining non-basic parameter (and the associated variable) into the basic set, in turn. The parameter whose addition causes the greatest reduction of the objective function value is selected for inclusion in the basic parameter set at this particular stage. The addition of the parameters is continued as long as the parameter confidence intervals indicate that all the basic parameters are significantly different from zero. In the linear case the stepwise regression can be initiated with a null set of basic parameters.

In the nonlinear case (characteristic of most dynamic parameter estimation problems) the stepwise regression must be preceded with an initial stage of model analysis and creation of the initial basic set. This stage is described by Shacham and Brauner (2014) for the case where none of the parameters is subject to constrains. Here the algorithm is extended where some or all the parameters are subject to constraints. The various steps of the initial analysis and set up are the following.

1. Definition of new parameters to enable addition of one parameter at a time to the basis and setting non-basic parameters at zero

Modification of the definitions of the parameters is often required in order to avoid singularity when non-basic variables are set at zero value. Furthermore, if a term contains a product of several parameters this product must be replaced by a new parameter in order to enable consideration of a single parameter at a time.

2. Classification of parameters' hierarchy within the model various expressions

The model equations can usually be divided to groups of expressions linked by multiplication and division operators. Zeroing one (or more) of the parameter values may cause removal of the whole group from the model. Such parameters are categorized as of first hierarchy (H = 1), as they must be included in the basis prior to considering the inclusion of additional parameters of lower hierarchy of the same expression in the basis.

3. Bringing parameters constrained to nonzero values into explicit forms

Parameters that are constrained to nonzero values must be included in the basis. The modification of the parameters (step 1 above) may cause some of the nonzero constrained parameters to appear in implicit form. Yet, these parameters must appear in the parameter list in an explicit form.

4. Selection of an initial basic set from amongst the H = 1 and nonzero constrained parameters, which is sufficient for representing the initial trend of the state variables and ensures that none of the state variables will remain constant throughout the integration.

The various stages of the proposed stepwise regression algorithm will be demonstrated by solving the example problem Wang et. al., 2001.

Description of the example problem: Kinetics Model of ethanol fermentation (Wang et. al. 2001, Liu and Wang, 2009).

The fermentation process, described by Wang et. al. 2001 and Liu and Wang, 2009, uses high ethanol tolerance yeast, Saccharomyces diastaticus LORRE 316, to produce ethanol. The mathematical model proposed by these authors includes the material balance equations that describe the growth of microorganisms, consumption of glucose and formation of products and reads:

$$\begin{aligned}
\frac{dx}{dt} &= \mu x \\
\frac{ds}{dt} &= -\frac{q_{p1}}{Y_{p1/s}} - \frac{q_{p2}}{Y_{p2/s}} \\
\frac{dp_1}{dt} &= q_{p1}x \\
\frac{dp_2}{dt} &= q_{p2}x
\end{aligned} \tag{4}$$

where x is the concentration of cell mass, s is the concentration of glucose, p_1 is the concentration of ethanol, p_2 is the concentration of glycerol, $Y_{p1/s}$ is the ethanol yield factor, and $Y_{p2/s}$ is the glycerol yield factor. The unstructured kinetic model for the specific cell growth and product formation is expressed as (Liu and Wang, 2009):

$$\begin{aligned}
\mu &= \frac{\mu_m s}{K_s + s + s^2/K_{SI}} \frac{K_{p1}}{K_{p1} + p_1 + p_1^2/K_{p1I}} \frac{K_{p2}}{K_{p2} + p_2 + p_2^2/K_{p2I}} \\
q_{p1} &= \frac{v_{p1} s}{K_s^{'} + s + s^2/K_{SI}^{'}} \frac{K_{p1}^{'}}{K_{p1}^{'} + p_1 + p_1^2/K_{p1I}^{'}} \\
q_{p2} &= \frac{v_{p2} s}{K_s^{''} + s + s^2/K_{SI}^{''}} \frac{K_{p2}^{'}}{K_{p2}^{'} + p_1 + p_1^2/K_{p2I}^{'}}
\end{aligned} \tag{5}$$

Wang et al. 2001 pointed out that ethanol/glycerol fermentation is a typical example of a process in which utilization of the carbon source cannot be split into two processes, namely, production of biomass and formation of product. Ethanol and glycerol are the end products of energy metabolism associated directly with biomass growth. These products serve, however, as inhibitors capable of retarding metabolic activity. In the expression for μ in Eq. (5) the glucose, ethanol, and glycerol inhibition on biomass growth are considered. The formation of ethanol and glycerol is associated with the microbial growth (Eq. 4), but also with the inhibition of the glucose and themselves (Eq. 5).

There are 19 parameters that have to be estimated from experimental observations. Wang et al. (2001) carried out several batch experiments to provide experimental data to be used for estimating the parameters of the kinetic model. Details of the experimental materials and methods are available in the same reference. Gennemark and Wedelin (2009) provided the results of 3 of the experiments (that were carried out by Wang et al. 2001) in their website of benchmark problems. The experimental data for the 3 experiments that were started at approximate glucose concentrations of 100, 150 and 200 g/L are shown in Tables 1, 2 and 3, respectively. The data of the 3 experiments were used in the objective function for determination of the model parameters.

Liu and Wang (2009) attempted to find the parameter values using the search space of $[10^{-12}, 10^{15}]$, except for the yield coefficient for ethanol, $Y_{p1/s}$ = [0.1, 0.51], and the yield coefficient for glycerol $Y_{p2/s}$ = [0.1, 0.2]. We will be using the values shown for $Y_{p1/s}$ and $Y_{p2/s}$ as lower and upper limits on the values of these parameters, while the suggested search space of the other parameters is constrained to positive values (≥ 0).

Application of the stepwise regression algorithm to the example problem – Initial stage

The first step if the initial stage involves modifications of some of the parameters, to enable setting them at zero value, if necessary. Equations (4) and (5) contain many expressions which involve division by parameters, and thus singularity is introduced when those parameters are set at zero value (i.e., not included in the basis). To avoid this, the model equations are rearranged to following form:.

$$\begin{aligned}
\frac{dx}{dt} &= \mu x \\
\frac{ds}{dt} &= -q_{p1}\pi_7 x - q_{p2}\pi_8 x \\
\frac{dp_1}{dt} &= q_{p1}\pi_2 x \\
\frac{dp_2}{dt} &= q_{p2}\pi_3 x
\end{aligned} \qquad (6)$$

where:

$$\begin{aligned}
\mu &= \frac{\pi_1 s}{1+\pi_4 s+\pi_9 s^2}\,\frac{1}{1+\pi_{10}p_1+\pi_{11}p_1^2}\,\frac{1}{1+\pi_{12}p_2+\pi_{13}p_2^2} \\
q_{p1} &= \frac{s}{1+\pi_5 s+\pi_{14}s^2}\,\frac{1}{1+\pi_{15}p_1+\pi_{19}p_1^2} \\
q_{p2} &= \frac{s}{1+\pi_6 s+\pi_{16}s^2}\,\frac{1}{1+\pi_{17}p_2+\pi_{18}p_2^2}
\end{aligned} \qquad (7)$$

The 19 original parameters were replaced by new parameters $\pi_1, \pi_2, \dots \pi_{19}$. Observe that in this formulation in principle each of the parameters can be set at zero. The definitions of these parameters in terms of the original parameters are shown in Table 4.

The second step involves separation of the parameters into groups and determination of the parameter hierarchy within the groups. Observe, for example, that the term associated with π_1 contains the parameters π_4, π_9, π_{10}, π_{11}, π_{12} and π_{13}. Thus, in order to consider the inclusion of the latter six parameters in the basic parameter set, π_1 must be already included in this set. Accordingly we denote the expression associated with the π_1 parameter as "Group 1", where the parameter π_1 is at the highest hierarchy (H = 1), and the other parameters in the same group (*i.e.*, π_4, π_9, π_{10}, π_{11}, π_{12}, π_{13}) are at lower hierarchy level (H = 2). Table 4 lists the four groups (associated with the four differential equations) and the initial parameter hierarchy (H) within these groups.

The original parameters $Y_{p1/s}$ and $Y_{p2/s}$ are constrained to nonzero values and as such they must be included in the basis. Therefore, they are introduced in an explicit form in the revised parameter list. Nonzero values for the associated parameters ($\pi_7 = \pi_2/Y_{p1/s}$ and $\pi_8 = \pi_3/Y_{p2/s}$) can be obtained upon inclusion of π_2 and π_3 in the basis.

According to the criteria stated in the 4th step of the "initial setup" section, the initial basic parameter set must include at least 4 parameters of $H = 1$ from among: π_1, π_2, π_3, π_7 and π_8. However, as $Y_{p1/s}$ and $Y_{p2/s}$ are constrained to nonzero values, (these parameters replace π_7 and π_8), the initial selected basic parameter set includes 5 parameters π_1, π_2, π_3, $Y_{p1/s}$ and $Y_{p2/s}$. The upper limits ($Y_{p1/s} = 0.51$ and $Y_{p2/s} = 0.2$) were used as initial estimates on the constrained parameter values. For the rest of the parameters, linear regression was applied to the estimated data of state variable derivative values to provide the following initial estimates for the parameter values: $\pi_1 = 0.002$, $\pi_2 = 0.01$ and $\pi_3 = 0.0008$. The objective function value at the initial was $\phi = 0.107$. Using *fmincon* to find the optimal values of the 5 parameters of H=1 (see Table 5) yielded moderate changes of the parameter values and reduced the objective function value by almost threefold to $\phi = 0.0388$.

Stepwise regression for the example problem

The stepwise regression was carried while including the data of all 3 experiments (Tables 1, 2 and 3) in the objective function. As the numbers of data points differ in the various data sets a revised form of the scaling factor:

$$W_i = (\max_{\mu} |x_{\mu,i}| * n)^{-1} \tag{3A}$$

was employed. This type of scaling was used in an attempt to equalize the contributions of the errors in the various sets to the objective function.

To ensure consistency of the ODE solver algorithm used in the various steps of the stepwise regression process only the stiff method (ode15s) was used in all the stages.

The optimal solution for the 5 parameter's model was used as initial estimate for the 1st stage of the stepwise regression process. All the remaining parameters of $H = 2$ were tested, in turn, for entering the basis. A small positive value (10^{-12}) was used as initial estimate for the newly entering parameters. The *fmincon* function uses the upper and lower limits on the variable to restrict the search range. For $Y_{p1/s}$ and $Y_{p2/s}$ these limits were specified already. As for the rest of the parameters the lower limit was set at zero (only positive values are allowed). After some experimentation to ensure that the upper limit constraint is inactive in all the cases this limit was set at 20. The π_9 parameter, whose addition reduced the objective function value to $\phi = 0.00769$ was selected to enter the basis in the 1st stage of the stepwise regression process.

The results of the stepwise regression are summarized in Table 5 and Figure 1. Addition of more parameters to the basis enables reducing the objective function value up to the 7th stage of the stepwise regression process. Observe (Fig. 4) that the reduction is sharp at the beginning, but becomes much more moderate toward the final stages. Starting at the 8th stage, no (positive) value of the remaining parameters can further reduce the objective function value.

Thus, the minimal objective function value obtained is: ϕ = 0.00105. Setting the remaining non-basic parameters at zero value, Eq. 7 reduces to the following simplified form:

$$
\begin{aligned}
\mu &= \frac{\pi_1 s}{1+\pi_4 s+\pi_9 s^2}\frac{1}{1+\pi_{12} p_2} \\
q_{p1} &= \frac{s}{1+\pi_5 s}\frac{1}{1+\pi_{19} p_1^2} \\
q_{p2} &= \frac{s}{1+\pi_{16} s^2}\frac{1}{1+\pi_{18} p_2^2}
\end{aligned} \tag{8}
$$

Introducing the optimal parameter values obtained in the 7th stage of the stepwise regression process (Table 5) yields the calculated state variable values shown in Figures 2, 3 and 4 for initial glucose concentrations of 100, 150 and 200 g/L, respectively. Observe that the fit between the experimental and calculated values is very satisfactory.

To assess the stability of the solution obtained, the parameter values of the optimal 12 parameters model were calculated using four different experimental data sets: a. a set the includes the data of the 3 experiments (Tables 1, 2 and 3); b. a set that includes the data from experiments 1 and 2; c. a set that includes the data from experiments 1 and 3, and d. a set that includes the data from experiments 2 and 3. The parameter values for the different cases together with the respective ϕ values are shown in Table 6. The percent differences between the smallest to largest individual parameter values $\Delta\theta_i$ were calculated for the 12 parameters:

$$\Delta\theta_i = 100(\theta_{iL} - \theta_{iS})/\theta_{iL} \tag{9}$$

where θ_{iL} is the largest value of the particular parameter and θ_{iS} its smallest value. The $\Delta\theta_i$ values for the optimal stepwise regression model are plotted in Fig. 5. Observe that for 8 parameters the $\Delta\theta_i$ values are very small (2% or less) or small (20 % or less), for 3 parameters medium values of 50% or less, and only for the 12th parameter (π_{18}) the $\Delta\theta_i$ value reaches 100%. Thus, most of the parameter values are stable and depend on the particular data used only to a small extent.

Liu and Wang (2009) used a hybrid differential evolution algorithm to identify the 19 parameters of the model in its original form (Eqs. (4) and (5)). Based on the magnitude of the calculated parameter values they suggested that some of them may be insignificant, and proposed a "pruned" kinetic model with only 13 parameters. Their pruned model is similar to the model shown in equations (6) and (7), except that the terms associated with the parameters π_{12}, π_{19}, π_{16} and π_{18} are removed and the terms associated with the parameters π_{10}, π_{11}, π_{15}, π_6 and π_{17} are included. The best solution we have found using this reduced parameter set yields and objective function value of ϕ = 0.00146, which is 40% larger than the value obtained using the stepwise regression process. The parameter values for the four sets of the experimental data were calculated for this "pruned" model, as well, and the resultant parameter and objective function values are shown in Table 7. In this case the $\Delta\theta_i$ *values* (Figure 6) exceed 80% for 6 out of the 13 parameters. Thus, there is much higher uncertainty regarding the parameters of that model compared to the model obtained by the proposed stepwise regression algorithm.

Conclusions

A new stepwise regression algorithm has been developed for efficient solution of parameter estimation problems, where some or all of the parameters are subject to constraints, and it may be impossible to evaluate all the parameter values. The objective is to obtain reliable estimates to the maximal number of physical parameters in a stable regression model, where the modeling of the noise in the data is avoided. The algorithm requires redefinition of some of the parameters to enable addition of one parameter at a time to the model and setting the remaining parameters at zero. With the reformulated model equations, the algorithm considers the hierarchy among the parameters, presence of parameters constrained to non-zero values, the initial trends of the experimental data curves and the initial values of the state variables in order to establish a minimal initial set of parameters to be included in the model. More parameters are then added in a stepwise manner, while considering the hierarchy of the parameters and the associated reduction of the objective function value. The stepwise procedure is halted when addition of more parameters cannot further reduce the objective function value.

In the example solved the proposed method managed to identify 12 parameters (out of the 19 proposed). Comparison of this solution with an alternative solution proposed by original authors of this problem (Liu and Wang 2009) shows that the stepwise regression yields a solution of lower objective function value and more stable parameter values than alternative methods. As multiple experimental data sets were available, the stability of the parameter values was examined by the spread of the parameter values obtained using different sets of experimental data for regression. However, if multiple experimental data sets are not available, they can be generated by perturbing the available data set with a random noise at the experimental uncertainty level.

We believe that the proposed stepwise regression algorithm represents a systematic method for identifying the most influential parameters in nonlinear dynamic parameter estimation problems, where some or all the parameter values are subject to constraints. The method yields parameter values consistent with the minimal objective function value, and enables also assessment of the stability of the parameter values obtained.

References

1. Biegler, L. T., Damiano, J. J. and Blau, G. E. (1986) Nonlinear parameter estimation: a case study comparison. AIChE Journal 32(1), 29-45.
2. Brauner, N. and M. Shacham., " Regression Diagnostic Using an Orthogonalized Variable Based Stepwise Regression Procedure", Computers chem. Engng., 23(suppl.), S327-S331 (1999)
3. Dormand, J. R. and P. J. Prince, "A family of embedded Runge-Kutta formulae," J. Comp. Appl. Math., Vol. 6, 1980, pp 19-26.
4. Floudas, C. A., Pardalos, P. M., Adjiman, C. S., Esposito, W. R., Gümüs, Z. H., Harding, S.T., Klepeis, J. L., Clifford A. Meyer, C. A. and Schweiger, C. A. (1999). Handbook of Test Problems in Local and Global Optimization, Kluwer, Dordrecht, The Netherlands
5. Gear, C. W., Numerical Initial Value Problems in Ordinary Differential Equations, Prentice Hall, Englewood Cliffs, 1971
6. Gennemark, P. and Wedelin, D., Benchmarks for identification of ordinary differential equations from time series data, Bioinformatics, 25, 780-786, 2009.
7. Kravaris, C., Hahn, J. and Chu, Y., 2013, Advances and selected recent developments in state and parameter estimation, Computers and Chemical Engineering, 51, 111-123, 2013.
8. Liu, P.-K. and Wang, F.-S., Hybrid Differential Evolution with Geometric Mean Mutation in Parameter Estimation of Bioreaction Systems with Large Parameter Search Space, Computers and Chemical Engineering, 33, 1851-1860, 2009.
9. McLean, K. A. P., Wu, S., & McAuley, K. B. (2012). Mean-squared-error methods for selecting optimal parameter subsets for estimation, Industrial and Engineering Chemistry Research, 51, 6105-6115.
10. McLean, K. A. P. & McAuley, K. B. (2012). Mathematical modelling of chemical processes - obtaining the best model predictions and parameter estimates using identifiability and estimability procedures, Canadian Journal of Chemical Engineering, 90, 351-366.
11. M.J.D. Powell, 1978, A Fast Algorithm for Nonlinearly Constrained Optimization Calculations, Numer. Analysis, ed. G.A. Watson, Lecture Notes in Mathematics, Springer Verlag, Vol. 630
12. Shacham, M. and N. Brauner, "The SROV Program for Data Analysis and Regression Model Identification", Computers chem. Engng. 27(5), 701-714(2003)
13. Shacham, M. and Brauner, N., Application of stepwise regression for dynamic parameter estimation, Computers and Chemical Engineering. 69, 26-38, 2014
14. Sun, J., Jonathan M. Garibaldi, J. M.and Hodgman, C., Parameter Estimation Using Metaheuristics in Systems Biology: A Comprehensive Review, IEEE/ACM Transactions on Computational Biology and Bioinformatics, 9, 185-202, 2012
15. Wang, F. S., Su, T.L.and Jang, H. J. Hybrid differential evolution for problems of kinetic parameter estimation and dynamic optimization of ethanol fermentation process. Industrial and Engineering Chemistry Research, 40, 2876-2885, 2001.

Table 1. Experimental data (Wang et al. 2001, Gennemark and Wedelin, 2009) for ethanol fermentation with initial glucose concentration of 100 g/L

Point No.: i	t (h)	x_i(g/L)	s_i(g/L)	p_{1i}(g/L)	p_{2i}(g/L)
1	0	0.32138	98.4237	0	0
2	1	0.55582	96.1842	0	0
3	2	1.1656	93.1433	0.8453	0
4	3	1.9665	89.4814	3.9807	0.2561
5	4	3.1822	81.6356	6.9194	0.3673
6	5	5.4744	69.6836	11.5906	1.0449
7	6	7.594	50.4962	18.4895	1.4041
8	7	10.2692	31.2942	26.995	1.897
9	8	11.1342	15.0058	35.0239	2.4328
10	9	12.3686	3.8	40.5563	2.7454
11	10	12.6412	0	42.9921	2.8563

Table 2. Experimental data (Wang et al. 2001, Gennemark and Wedelin, 2009) for ethanol fermentation with initial glucose concentration of 150 g/L

Point No.: i	t (h)	x_i(g/L)	s_i(g/L)	p_{1i}(g/L)	p_{2i}(g/L)
1	0	0.2607	151.241	0	0
2	1	0.3449	150.2995	0	0
3	2	0.64775	149.389	0	0
4	3	1.2406	147.1655	1.9422	0.23365
5	4	1.9698	142.819	4.5141	0.3827
6	5	3.0478	132.7825	7.4811	0.99425
7	6	5.0755	116.6395	13.6387	1.495
8	7	7.5743	94.7079	19.0816	1.9185
9	8	8.8361	77.12	26.5594	2.41
10	9	9.7001	61.6384	34.4949	3.0079
11	10	10.444	45.3708	42.5252	3.6632
12	11	11.4506	25.0062	51.3984	4.2621
13	12	12.1132	10.8632	57.676	4.5932
14	13	12.659	2.0748	63.2968	4.9134
15	14	12.9694	0.0528	65.8565	5.072

Table 3. Experimental data (Wang et al. 2001, Gennemark and Wedelin, 2009) for ethanol fermentation with initial glucose concentration of 200 g/L

Point No.: i	t (h)	x_i(g/L)	s_i(g/L)	p_{1i}(g/L)	p_{2i}(g/L)
1	0	0.25794	197.938	0	0
2	1	0.35163	196.3455	0	0
3	2	0.50495	195.443	0	0
4	3	0.91191	192.0625	3.0597	0.2169
5	4	1.512	187.3675	4.8155	0.83731
6	5	2.8031	178.5835	7.5538	1.1603
7	6	4.4029	166.403	13.0554	1.6974
8	7	6.2635	144.1565	19.5166	2.4533
9	8	8.6109	124.5265	25.3135	3.0111
10	9	9.6147	102.8158	35.0076	3.8146
11	10	11.2518	85.5552	43.1744	4.5128
12	11	12.1692	68.7319	50.5293	5.1625
13	12	13.1241	51.4266	59.1247	5.8394
14	13	14.2433	36.2531	66.1442	6.2358
15	14	15.0539	24.549	71.8272	6.574
16	15	16.1507	8.4547	78.4983	6.8424
17	16	16.5711	2.1814	81.2294	6.9718
18	17	16.9961	0.52	85.8299	7.142
19	18	17.2255	0	85.9216	7.2436

Table 4. List of the original and definition of the revised parameters for the ethanol fermentation problem

No.	Original parameter	Revised parameter	Definition	Group	Initial hierarchy
1	μ_m	π_1	μ_m/K_s	1	1
2	v_{p1}	π_2	v_{p1}/K'_s	2	1
3	v_{p2}	π_3	v_{p2}/K''_s	3	1
4	$Y_{p1/s}$	π_4	$1/K_s$	1	2
5	$Y_{p2/s}$	π_5	$1/K'_s$	2	2
6	K_s	π_6	$1/K''_s$	3	2
7	K_{s1}	π_7	$v_{p1}/(Y_{p1/s}K'_s)$	4	1
8	K_{p1}	π_8	$v_{p2}/(Y_{p2/s}K''_s)$	5	1
9	K_{p1I}	π_9	$1/(K_{SI}K_s)$	1	2
10	K_{p2}	π_{10}	$1/K_{pI}$	1	2
11	K_{p2I}	π_{11}	$1/(K_{p1I}K_{pI})$	1	2
12	K'_s	π_{12}	$1/K_{p2}$	1	2
13	K'_{sI}	π_{13}	$1/(K_{p2I}K_{p2})$	1	2
14	K'_{p1}	π_{14}	$1/(K'_{SI}K'_s)$	2	2
15	K'_{p1I}	π_{15}	$1/K'_{pI}$	2	2
16	K''_s	π_{16}	$1/(K''_{SI}K''_{s})$	3	2
17	K''_{SI}	π_{17}	$1/K'_{p2}$	3	2
18	K'_{p2}	π_{18}	$1/(K'_{p2}K'_{p2I})$	3	2
19	K'_{p2I}	π_{19}	$1/(K'_{p1}K'_{p1I})$	2	2

Table 5. Results of the stepwise regression process for the ethanol fermentation problem

No.	Parameter	Initial basis		1st stage	2nd stage	3rd stage	4th stage	5th stage	6th stage	7th stage
		Estimate	Optimal							
1	$Y_{p\ 1/s}$	0.51	0.41295	0.48023	0.51	0.51	0.51	0.51	0.51	0.50797
2	$Y_{p\ 2/s}$	0.2	0.194	0.16435	0.2	0.18102	0.18565	0.19092	0.18958	0.19845
3	π_1	0.002	0.0043824	0.014124	0.014223	0.013354	0.015737	0.015539	0.053186	0.068368
4	π_2	0.01	0.020197	0.027201	0.015755	0.0343	0.051212	0.055077	0.092276	0.14571
5	π_3	0.0008	0.001717	0.0021648	0.0012555	0.0011674	0.0013901	0.0014115	0.001499	0.0015456
6	π_9			0.0002359	0.0001288	0.0001113	0.0001243	0.000123	0.000187	0.0001897
7	π_{12}				0.50679	0.50754	0.86131	0.86309	1.2941	1.2813
8	π_5					0.016797	0.037006	0.035587	0.070942	0.11794
9	π_{16}						3.61E-05	3.885E-05	5.696E-05	5.181E-05
10	π_{19}							0.0001275	0.0001606	0.0002239
11	π_4								0.043109	0.062764
12	π_{18}									0.011142
Objective func.		0.10696	0.038847	0.0076891	0.0040976	0.0028047	0.0019276	0.0016473	0.0011797	0.0010517

Table 6. Parameter values obtained using four different data sets, for the stepwise regression optimal model

No.	Parameters	Optimal values			
		3 Sets	Set 1 and 2	Set 1 and 3	Set 2 and 3
1	$Y_{p1/s}$	0.5080	0.5049	0.5075	0.5078
2	$Y_{p2/s}$	0.1985	0.1964	0.1978	0.1979
3	π_1	0.0684	0.0746	0.0683	0.0649
4	π_2	0.1457	0.1454	0.1440	0.1466
5	π_3	1.546E-03	1.594E-03	1.511E-03	1.919E-03
6	π_9	1.897E-04	3.394E-04	1.681E-04	1.720E-04
7	π_{12}	1.2813	1.2905	1.2898	1.2901
8	π_5	0.1179	0.1179	0.1200	0.1174
9	π_{16}	5.181E-05	5.958E-05	5.092E-05	8.993E-05
10	π_{19}	2.239E-04	1.328E-04	2.413E-04	1.965E-04
11	π_4	6.276E-02	5.776E-02	6.280E-02	6.578E-02
12	π_{18}	1.114E-02	1.000E-12	1.199E-02	1.205E-02
	ϕ	**1.052E-03**	**5.442E-04**	**4.655E-04**	**6.522E-04**

Table 7. Parameter values obtained using four different data sets, for the "pruned" model presented by Liu and Wang (2009)

Parameters	**Optimal values**			
	3 Sets	**Sets 1 and 2**	**Sets 1 and 3**	**Sets 2 and 3**
$Y_{p1/s}$	0.5100	0.5090	0.5100	0.5100
$Y_{p2/s}$	0.1985	0.1947	0.1989	0.2000
π_1	2.821E-02	4.896E-02	2.542E-02	9.803E-03
π_2	6.364E-02	6.955E-02	8.627E-02	7.444E-02
π_3	2.005E-03	2.346E-03	2.415E-03	1.851E-03
π_9	1.576E-04	2.850E-04	1.477E-04	4.962E-05
π_{10}	5.836E-02	4.351E-02	1.212E-02	6.528E-02
π_5	3.663E-02	4.508E-02	4.245E-02	4.407E-02
π_6	8.957E-03	1.444E-02	9.100E-03	6.375E-03
π_4	1.580E-02	3.533E-02	1.783E-02	1.903E-03
π_{15}	1.099E-02	6.783E-03	2.203E-02	1.621E-02
π_{17}	4.811E-02	1.375E-02	1.324E-01	1.115E-01
π_{11}	7.168E-04	1.961E-03	1.300E-03	1.000E-12
ϕ	1.408E-03	3.341E-04	4.059E-04	3.355E-04

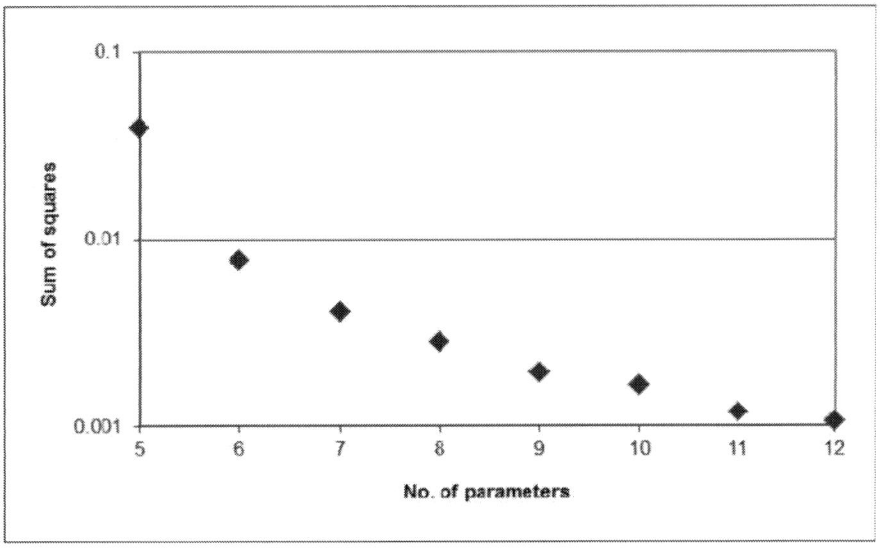

Figure 1. Reduction of the objective function value throughout the stepwise regression process

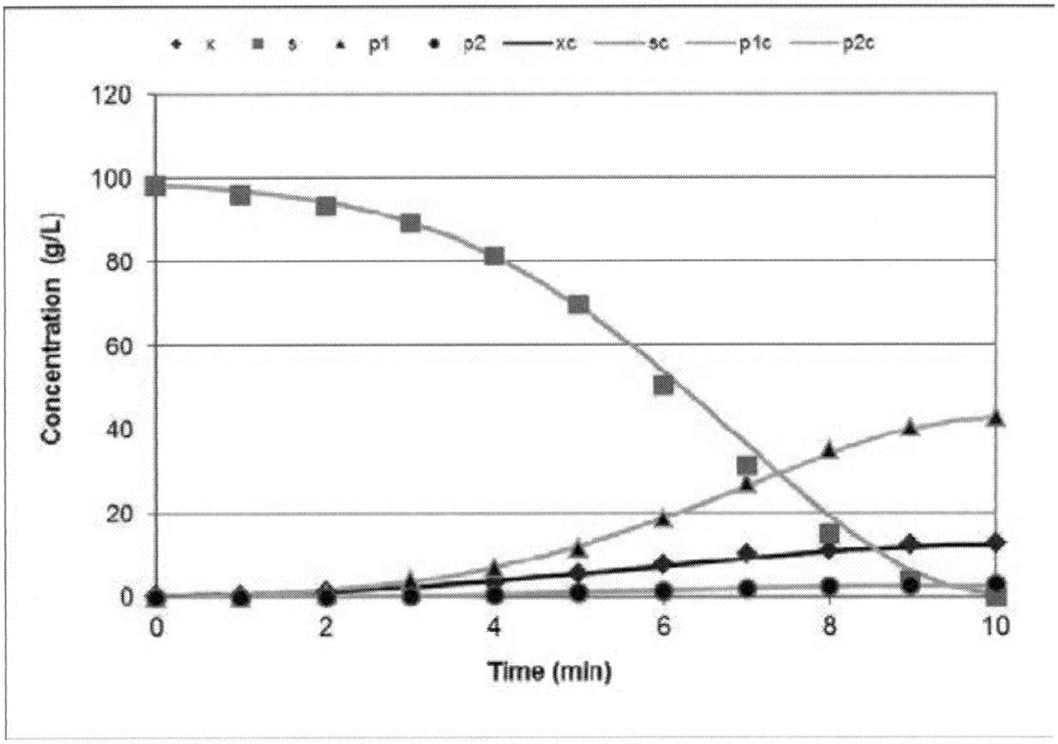

Figure 2. Plot of experimental data (*x*, s, p1 and p2) and calculated values (*x*c, sc, p1c and p2c) using the optimal set of parameter values for ethanol fermentation with initial glucose concentration of 100 g/L

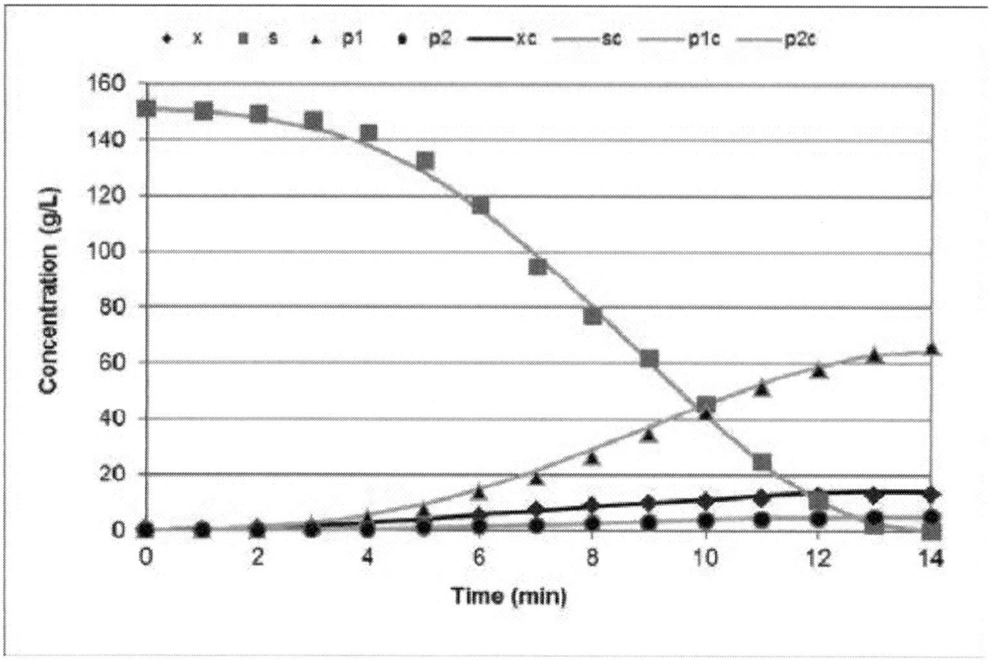

Figure 3. Plot of experimental data (*x*, s, p1 and p2) and calculated values (*x*c, sc, p1c and p2c) using the optimal set of parameter values for ethanol fermentation with initial glucose concentration of 150 g/L

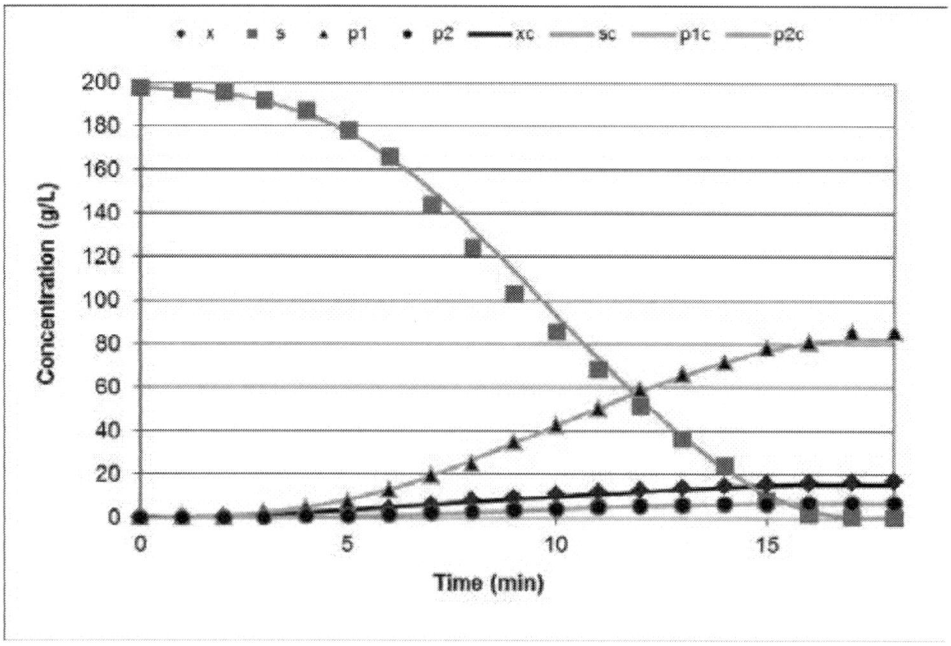

Figure 4. Plot of experimental data (*x*, s, p1 and p2) and calculated values (*x*c, sc, p1c and p2c) using the optimal set of parameter values for ethanol fermentation with initial glucose concentration of 200 g/L

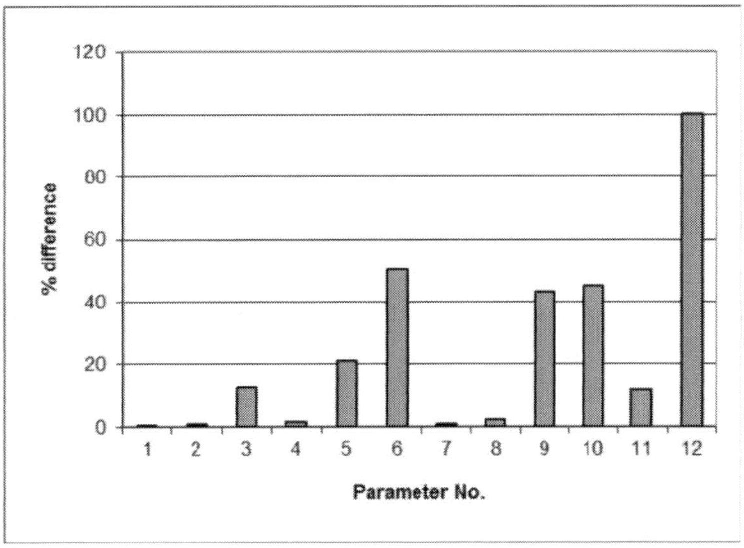

Figure 5. Percent difference between smallest and largest parameter values for the stepwise regression optimal model

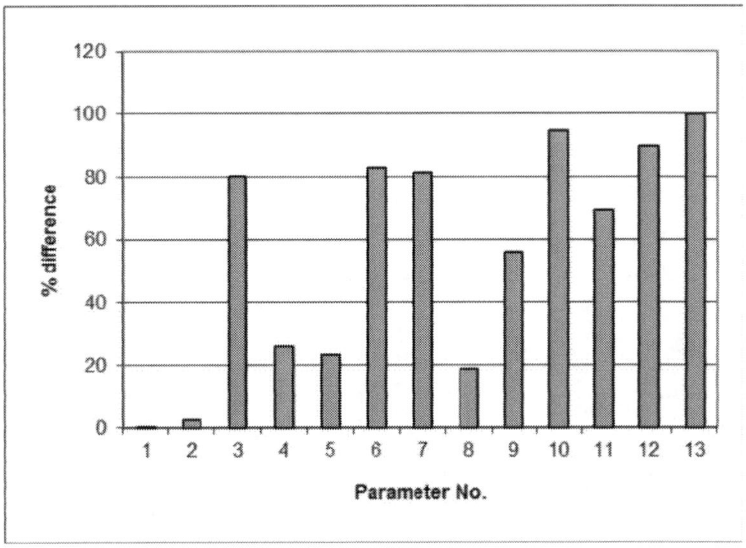

Figure 6. Percent difference between smallest and largest parameter values for the "pruned" model presented by Liu and Wang (2009)

Simulation and Analysis of Biologically Inspired Chemical Reaction Networks using Moment Closure

Patrick Smadbeck, Yiannis Kaznessis

When considering simulation options biological systems are often thought of as a large set of chemical reactions. The simulation of such chemical networks is often performed using ordinary differential equations (ODEs), and the subsequent analysis of such systems is straightforward. Simulations using ODEs, however, are deterministic and assume that there is little internal or external noise affecting the system. Biological systems are markedly different in that noise can have a profound impact on system dynamics. This random or probabilistic element becomes important for systems where components (e.g. DNA promoter regions, mRNA) exist at low copy number. In such systems stochastic simulations become necessary to accurately capture system dynamics.

Stochastic simulation algorithms for large reaction networks have become very efficient over the past decade using fast hybrid and distributed kinetic Monte Carlo (kMC) algorithms. While dynamic modeling of biologically-relevant stochastic simulations is now attainable, analysis using kMC methods are still wanting with regards to analysis. In particular, deterministic analytical methods well-known to biological researchers are not easily applied to kMC results. One option for avoiding such an issue is to, instead, focus on the statistics of the underlying probability distributions for these stochastic systems. The statistics are deterministic, and are in theory subject to deterministic analysis techniques as well. The issue is that the differential equations governing the statistical dynamics of reaction networks are often under-defined, and require what is called a moment-closure scheme.

We have recently developed a numerical moment closure technique called Zero-Information Closure.[1] We have been able to show that the method is both accurate and versatile, even for complex underlying chemical networks. The potential of such a system is in its ability to apply deterministic analytical techniques to stochastic systems. In systems biology non-linear analysis of reaction networks is commonly used to identify areas of interesting or complex behavior (e.g. oscillations or limit cycles). The approaches are well-established and well-known to both computational and experimental researchers. Through moment closure it may be possible to apply non-linear analysis, model reduction and other deterministic analysis to stochastic biological system in a mathematically rigorous, but readily understandable manner. This presentation will go through initial non-linear analysis results with a focus on biologically relevant networks.

---

[1] P Smadbeck and YN Kaznessis. "A closure scheme for chemical master equations." Proceedings of the National Academy of Sciences (2013)

372701 De Novo Multifunctional Peptide Design Using Sequence Similarity Metrics

Abstract:

Sequence similarity is a design principle for *de novo* peptide development. Given peptides with known functions, sequence similarity may be used to create new sequences with the same functions or a combination of those functions. In previous work[1], a method for finding scoring matrices in pairwise global sequence alignments between peptide groups of varied level of function was used to maximize the level of that function with *de novo* peptides. This work proposes a way of designing multifunctional peptides. Peptides with known therapeutic or other property values are evaluated in terms of sequence similarity, with the goal of creating structure-activity relationships for the novel design of peptides within a molecular design framework. The linked-antimicrobial peptide database (LAMP)[2] provides a set of known sequences which serve as a base set to create the property prediction model. The peptides in this database either have shown antibacterial or antifungal properties, or both. A paired-alignment tree for peptides in each of these classes was constructed to extract common sequence features. The molecular design problem was then formulated as a graph maximization, and a Tabu search algorithm was then applied to find peptides that align well with multiple sequence features from the trees of subsequences representing antimicrobial properties. These novel peptides were then compared quantitatively with those peptides known to be both antibiotic and antifungal, to validate the design methodology.

1. Oren, E.E., Tamerler, C., Sahin, D., Hnilova, M., Seker, U.O.S., Sarikaya, M. Samudrala, R., 2007. A novel knowledge-based approach to design inorganic-binding peptides. *Bioinformatics*, 23(21).

2. Zhao, X., Wu H., Lu H., Li G., Huang Q., 2013. A database linking antimicrobial peptides. *PLoS ONE*, 8(6).

390033 Inference of Putative Metabolite-Dependent Regulatory Interactions from Metabolomics Data

Abstract:

Despite providing one of the most direct readouts of cell phenotype available, to date there has been surprisingly little application of metabolomics (the systems-level measurement and analysis of metabolites, which are the small molecules that supply and drive cellular processes) to metabolic engineering strain design. Similarly, very few of the computational methods that have been commonly used for strain design, such as constraint-based modeling strategies, take advantage of metabolomics. In order to address this gap, we are developing new computational strategies that use dynamic metabolomics data to improve the accuracy of strains designed *in silico*.

Metabolite-dependent regulatory interactions, such as allosteric regulation, are not as well- studied as other types of regulation (e.g., transcriptional regulation). Here, we use metabolomics data and existing knowledge of metabolic network structures to infer putative regulatory interactions via structure learning algorithms including Bayesian Networks. To infer these interactions more directly, we use dynamic flux estimation and metabolomics data to generate a flux distribution complementary to the metabolite abundances; the structure learning is performed on this augmented dataset. The resulting network interactions can be sorted by likelihood and used as a list of putative interactions which can then be tested by adding them to a metabolic model and assessing whether their presence is beneficial or detrimental to model accuracy.

384846 Theoretical Prediction of Organic Reaction Yields and Predicting Retrosynthetic Pathways with Network Chemistry

Abstract:

Free energy and extent of the reaction (yield) are predicted by a thermodynamically guided approach based upon estimation of free energies of compounds. The non-ideality of the system and the effect of solvent are taken into account through the fugacity coefficient calculated by Perturbed-Chain Induced-Polar Statistical Associating Fluid Theory (PCIP-SAFT) molecular theory. A structure-based formulation is defined for the Gibbs free energy and the yield is obtained from the relationship between the free energy of reaction and equilibrium constant considering temperature effects. The model is trained using a database containing 10000 cited reactions. Regression of computed Gibbs free energy and yield of reaction versus experimental values resulted in a coefficient of determination (R2) of about 0.95. The absolute average deviation of the predicted yield is ~0.15 that has been validated against sets of 2000 reactions. Finally, the proposed theory is implemented to predict retrosynthetic pathways of chemical reactions with Chematica which is the most advanced network chemistry package.

389719 Multiple-Resource Multiple-Species Autocatalysis

Abstract:

Study of autocatalytic reactions has been a popular practice in complex systems science, as a result of the generalizable results of such analysis, which has implications reaching out to broad fields such as population dynamics [1, 2], RNA polymerization, DNA replication [3-5] and stock market dynamics [6-8]. In the past, several studies were carried out using models of cubic autocatalysis in CSTRs. In the case of multiple-species autocatalysis carried out in a CSTR, where at least two autocatalytic species compete for a single resource and none of them are present in the feed stream, it was earlier shown that the coexistence of autocatalysts were globally unstable at steady state [1, 2].

The availability of multiple resources opens up various scenarios that can lead to coexistence of the competing species and has not been studied so far. In this study, we have added another resource to the inlet stream, which can be utilized by either or both autocatalytic species in the CSTR. Bifurcation analyses were carried out using the residence time of the reactor, so as to explore the effect on the stability of coexistence steady states. Three types of scenarios were employed in the analysis, where both species utilize both resources, only one species utilizes both resources, or each species utilizes a single resource.

The results of the bifurcation analyses indicate that the system exhibits a variable multiplicity of 7 to 9 steady states, depending on the residence time of the reactor. It was shown that, while the coexistence steady states are unstable when there are 7 steady states, an additional interaction isola that has a stable zone forms at certain values of the reactor residence time. The robustness of the stable coexistence steady state depends on the specific scenario and rate constants of autocatalysis. This shows that, through careful manipulation of reactor residence time and addition of a secondary resource, stable coexistence of multiple species can be established.

References

[1] İ. Birol, F. Teymour, Phys. D Nonlinear Phenom. 144 (2000), 279-297.

[2] W. Chaivorapoj, İ. Birol, F. Teymour, Ind. Eng. Chem. Res. 41 (2002), 3630-3641.

[3] L. da Silva, K. C. Mundim, C. Tsallis, Physica A 259 (1998) 415–429.

[4] P. R. Wills, S. Kauffman, B. M. R. Stadler, P. F. Stadler, Bull. Math. Biol. 60 (1998) 1073–1098.

[5] P. Schuster, W. Fontana, Physica D 133 (1999) 427–452.

[6] S. Kauffman, McKinsey Quart. (1995) 118–129.

[7] C. Blomberg, J. Theoret. Biol. 187 (1997) 541–554.

[8] J. Hofbauer, J. W. Weibull, J. Econ. Theory 71 (1996) 558–573.

380535 On Integrating the Droop Model with the Flux Balance Analysis Method to Predict Metabolic Shifts in Microalgae Growth

Abstract:

Identifying the mechanism for and predicting the metabolic shift between lipid accumulation and cell growth is a key research issue for microalgal biodiesel production. In this research, we propose a novel way to integrate a metabolic network model with a semi-empirical model called "Droop model" to predict the lipid accumulation and cell growth simultaneously. At each time instant of mass balance model integration, the Droop model is used to predict the cell growth rate and the nutrient consumption. Then, the Flux Balance Analysis (FBA) model is used to predict the rate of lipid accumulation, which is biochemically consistent with the predicted growth rate. In order to test the validity of the proposed approach, experiments are conducted for growing microalgae specie *C. reinhardtii* in a batch photo-bioreactor. Parameters of the Droop model are estimated from the experimental data and model predictions for the lipid contents are verified with the measured data. Parameter sensitivity analysis is conducted to investigate how the various parameters affect the cell growth and lipid accumulation. Finally, changes in the intracellular flux distribution are analyzed and predicted over the course of microalgae growth and nitrogen starvation with the integrated model.

366773 Redemption: Reduced Dimensional Ensemble Modeling and Parameter Estimation

Abstract:

The creation of kinetic ordinary differential equation (ODE) models of biological networks is often hampered by imprecise knowledge of reaction rate equations and their associated parameters. The model is generally written as:

dX/dt = Sv(x,p) (1)

where X is the vector of species concentrations, S is the stoichiometric matrix, v(X,p) is the vector of rate equations, and p is the vector of kinetic parameters. Here, the estimation of unknown kinetic parameters from time-series concentration data frequently becomes the bottlenecking step in the model building, motivating the development of a large number of methods [1]. Existing methods can generally be divided into two groups: integral and differential approach, based on whether or not the ODE model is integrated during the estimation. In the integral approach, the parameter estimation is formulated as a constrained optimization to minimize model prediction error. Meanwhile, the differential approach involves a pre-processing step, in which the time-series data are smoothen and differentiated to give estimates of dX/dt and Eq (1) is then used to compute dynamic reaction flux estimates v. Afterwards, the parameters are estimated by minimizing flux prediction errors. The main advantage of the differential over the integral approach is computational speed, as the ODEs need not be integrated and the parameter estimation can be done one flux at a time. However, the differential approach is known to give biased parameter estimates, and the parameter accuracy is sensitive to the data-smoothing step.

In biological networks, the number of species often exceeds that of reactions, and thus the matrix S does not have a full column rank. Consequently, the differential approach cannot be directly applied to metabolic networks, as the dynamic flux estimation becomes underdetermined. Despite this restriction, it may be possible to estimate the dynamic fluxes when the data are sufficiently dense or by taking minimum norm solution [2, 3]. In a new class of methods, we have used the relationship between dX/dt and v in Eq. (1) to reduce the dimensionality of the parameter estimation [4]. Specifically, we first divided the fluxes into independent and dependent subsets, vi and vd, such that the dependent fluxes could be uniquely determined from the independent fluxes according to Eq. (1). Given the values of the (independent) parameters of vi(X,p) and data of X, the flux vi could be calculated according to the rate equations vi(X,p). Subsequently, the flux vd were obtained by solving Eq. (1) using the estimates of dX/dt from data and the vi values above. The (dependent) parameters of vd(X,p) could then be estimated by least square optimization of the differences between the flux vd and the flux function vd(X,p), performed one flux at a time. Therefore, the parameter estimation was done only over the independent parameters of vi(X,p). We have shown using case studies that such a strategy, called incremental parameter estimation (IPE), could provide more than two orders of magnitude improvement in the computational speed over the traditional parameter estimation using the same error function [4].

However, the strategy above still depends on time-series data preprocessing and thus suffers from the same issues affecting any differential estimation method. Bhatt and coworkers [5-7] recently proposed differential estimation methods that avoided time-series data smoothing and differentiation, by estimating the extents of reactions, not the reaction fluxes. Nevertheless, the estimation of reaction extents requires the S matrix to have a full column rank. In this work, we developed a new reduced dimensional parameter estimation, motivated by the extents of reactions. The new method involved the calculation of integrated fluxes V, based on rewriting Eq. (1) in the integral form:

$X(t) - X(0) = SV(X,p)$ (2)

We divided the integrated fluxes V into independent and dependent components, and performed the parameter search only for those that appeared in the independent subset. We demonstrated the performance of the integrated flux estimation method using two case studies: an ODE model of branched metabolic network and a lin-log model of *L. lactis* metabolic pathways. The case studies showed that the integrated flux strategy could provide parameter estimates with better accuracy than the IPE method without the need to perform data smoothing. For the lin-log model, the integrated flux could provide a solution while previous methods failed to converge.

The difficulty in the parameter estimation above has been attributed to the parameter identifiability, or the lack thereof. In other words, there can exist multiple parameter combinations that are indistinguishable according to the data. This issue has motivated the application of ensemble modeling through the creation of a family of models. Within the umbrella of ensemble modeling, we have created a method for the generation of an ensemble of parameters that can provide statistically equivalent fit to the experimental data [8]. The method was based on the parameter estimation approach above. Here, instead of finding the global minimum solution of the error function, we used a combination of an out-of-equilibrium Adaptive Monte Carlo method and a multiple ellipsoids-based sampling method [9] to identify a set of parameters whose error functions were within a specified threshold. The threshold error value could be set using statistical score (e.g. F-distribution) or estimated by bootstrapping the original data using Monte Carlo approach. Our ensemble modeling method has the advantage that the ensemble could be compactly defined over only the space of the independent parameters.

The methods mentioned above are available as a MATLAB based toolbox, called REDEMPTION (REduced Dimensional Ensemble Modeling and Parameter estimaTION). Here, we have used eSS (Enhanced Scatter Search) algorithm from SSmGO (Scatter Search for Matlab Global Optimization) Toolbox for the global optimization [10, 11], and CVODE from SUNDIALS (SUite of Nonlinear and DIfferential/ALgebraic equation Solvers) for solving ODEs [12]. For the exploration of parameters for ensemble modeling, we have used the HYPERSPACE toolbox [9].

1. Chou, I.C. and E.O. Voit, *Recent developments in parameter estimation and structure identification of biochemical and genomic systems.* Mathematical Biosciences, 2009. **219**(2): p. 57-83.

2. Chou, I.C. and E.O. Voit, *Estimation of dynamic flux profiles from metabolic time series data.* BMC Systems Biology, 2012. **6**.

3. Voit, E.O., *Characterizability of metabolic pathway systems from time series data.*Mathematical Biosciences, 2013.

4. Jia, G.J., G. Stephanopoulos, and R. Gunawan, *Incremental parameter estimation of kinetic metabolic network models.* BMC Systems Biology, 2012. **6**.

5. Amrhein, M., et al., *Extents of Reaction and Flow for Homogeneous Reaction Systems with Inlet and Outlet Streams.* Aiche Journal, 2010. **56**(11): p. 2873-2886.

6. Bhatt, N., M. Amrhein, and D. Bonvin, *Incremental Identification of Reaction and Mass-Transfer Kinetics Using the Concept of Extents.* Industrial & Engineering Chemistry Research, 2011. **50**(23): p. 12960-12974.

7. Bhatt, N., et al., *Incremental identification of reaction systems-A comparison between rate-based and extent-based approaches.* Chemical Engineering Science, 2012. **83**: p. 24-38.

8. Jia, G., G. Stephanopoulos, and R. Gunawan, *Ensemble Kinetic Modeling of Metabolic Networks from Dynamic Metabolic Profiles.* Metabolites, 2012. **2**(4): p. 891-912.

9. Zamora-Sillero, E., et al., *Efficient characterization of high-dimensional parameter spaces for systems biology.* BMC Systems Biology, 2011. **5**.

10. Egea, J.A., et al., *Scatter search for chemical and bio-process optimization.* Journal of Global Optimization, 2007. **37**(3): p. 481-503.

11. Rodriguez-Fernandez, M., J.A. Egea, and J.R. Banga, *Novel metaheuristic for parameter estimation in nonlinear dynamic biological systems.* BMC Bioinformatics, 2006. **7**.

12. Hindmarsh, A.C., et al., *SUNDIALS: Suite of nonlinear and differential/algebraic equation solvers.* Acm Transactions on Mathematical Software, 2005. **31**(3): p. 363-396.

PROCESS SIMULATION IN THE 1960'S AND 1970'S

ABSTRACT

Highlights of the early work in process simulation, including the Process Assembly Case Evaluator Routine (PACER) and the Chemical Engineering Simulation System (CHESS) are recalled – focusing on designs of the first thermophysical property systems, automated translators, automatic program generation routines to model process units, and dynamic process simulators. These led to NSF funding of faculty to collaborate in seeking more effective ways to use computers in chemical engineering education – resulting in the Computer Aids for Chemical Engineering Education (CACHE) Committee. Due to the advanced architecture of FLOWTRAN, the Monsanto process simulator, a CACHE Task Force wrote a monograph and made FLOWTRAN available to universities worldwide. This led to the development of a prototype simulator, PLEXSYS, and the U.S. DOE support of the Advanced System for Process Engineering (ASPEN) project. This talk traces the role of Lawrence B. Evans and others in many of these and other early developments.

371407 A Tribute to Lawrence B. Evans: History of Aspen Technology from the Aspen Project to the Year 2000

Abstract:

MIT Chemical Engineering Professor Larry Evans was Principal Investigator of the ASPEN Project, a major research and development effort funded in 1976 by the U.S. Department of Energy and 65 companies in the process industries worldwide. The purpose of the project was to develop a "'third generation" process modeling and simulation system that could be used to evaluate proposed synthetic fuel processes both technically and economically.

When the project was completed in 1981 Larry, along with seven key members of the project staff, founded Aspen Technology, Inc. (AspenTech) to license the technology from MIT and to further develop, support and commercialize it. With Larry as CEO AspenTech greatly expanded the breadth and depth of the technology over the ensuing years, and brought on board a wide range of complementary products. The company grew from a 10-person startup to a public company with a capitalization value of $3.9B as of the end of 2013. This presentation is intended to chronicle that journey as it unfolded through the year 2000.

388751 CO2 Capture, Energy Savings and Heat Pump Integration

Abstract:

We begin by describing the impact of Professor Larry Evans' leadership in promoting the technology training and industrial applications of process simulation, advanced process control and supply chain optimization in the Asia-Pacific, particularly in China. We then focus our discussion on CO_2 capture from power plant flue gas. While amine-based CO_2 capture systems have been designed since 1927, concern over anthropomorphic climate change sparked renewed interest in the 1980's. Subsequent research has led to significant advances such as well-fit thermodynamic models, rigorous rate-based models for distillation, and improved process designs. The aims of this presentation are to (1) summarize proposed process improvements; (2) demonstrate a model integrating all of these improvements into one process in Aspen Plus; and (3) report an optimized design. Additionally, we report the use of a chemical heat pump for waste heat recovery with predicted amine-based solvent regeneration energy well below 2 GJ/tonne CO2 captured.

385493 Impact of Process Simulation Tools on Process Synthesis, Analysis and Design

Abstract:

The combination of advances in numerical methods and the ever-decreasing cost of computational power have led to almost universal adoption of process simulation software for the design of chemical processes. Chemical engineers have overcome the challenges of adapting and customizing commercial process simulation programs to be able to address synthesis, analysis and design problems across the whole spectrum of process industries. The software vendors have worked alongside their customers to address not only customization, but also usability, speed and workflow, providing skilled process engineers with a toolkit that enables better optimization and faster development of design alternatives. Despite the great accomplishments in the field over the past forty years, there remain some significant challenges still to be addressed, and a few of these challenges will be discussed in the paper.

Process and Product Engineering in the Watson Era: Challenges and Opportunities in Data Science

Venkat Venkatasubramanian
Complex Resilient Intelligent Systems Laboratory
Department of Chemical Engineering,
Columbia University
New York, NY 10027

Abstract

"Who is Bram Stoker?" – With this $1 million prize winning final question in the game show *Jeopardy,* IBM's Watson supercomputer using DeepQA technology ushered in a new era in artificial intelligence, cyberinfrastructure and informatics. This has far reaching implications for knowledge management in a number of fields including process and product systems engineering.

Chemical engineering, too, has entered an important era. Driven by a convergence of powerful forces such as the great progress in molecular sciences and computer/communications technologies, ever increasing automation of globally integrated operations of our enterprises, tightening regulatory constraints, and competitive business pressures demanding speed to market for products and services, our discipline is in an unprecedented transition. One important common outcome from this convergence is the generation, use, and management of massive amounts of diverse data, information, and knowledge. We are moving from an era of limited data obtained through time consuming experiments to one of a tsunami— it's a dramatic transition from a "data poor" to a "data rich" paradigm. However, a systematic way to convert the raw data into information and first-principles knowledge that can be used for real-time decision making is very much lacking. A new paradigm is needed that increases the idea flow, broadens the search horizon, and archives the knowledge from today's successes to accelerate those of tomorrow. This is where the recent advances in the field of data science can help.

Data science, loosely defined as a body of knowledge comprising of machine learning, natural language processing, databases and informatics, will play a crucial role in this endeavor by streamlining information gathering, data integration, model development, and managing all these for easy and timely access and reuse. In this talk, I will discuss the challenges, opportunities and recent trends using a number of case studies.

371088 Application of Computer-Aided Process Optimization Technologies – Progress, Opportunities & Challenges

Abstract:

Title: Application of Computer-Aided Process Optimization Technologies – Progress, Opportunities & Challenges

Author: Manolis Kotzabasakis

The impact of computer-aided process optimization technologies has grown exponentially over the past 35 years. These technologies have evolved from in-house main-frame programs developed and run by a few select experts to user-friendly products with ever-increasing intelligence and adaptive capabilities that are accessible anytime and anywhere. They are routinely used every day throughout the lifecycle of a process plant and in day-to-day operations, including conceptual and detailed design, production planning and scheduling, process monitoring and control, and supply chain management and optimization. The ready availability and increased power of computer-aided process optimization technologies have resulted in better process designs and more efficient plant operations, and has accelerated the rate of new process technology innovation. But what does the future hold?

This presentation will examine trends that are impacting the application of computing-aided process optimization technologies in the process industries, including the need to respond rapidly to changing business conditions, more stringent operating requirements, increased need for collaboration across different business functions, advances in computing and real-time systems, new developments in modeling, control and optimization algorithms, the impact of mobile computing and social media, and the changing user demographic. It will discuss our perspectives on how computer-aided process optimization technologies will need to adapt to this changing landscape, and explore the potential to provide benefits to the process industries beyond current practices.

387454 Energy-Efficient Smart Manufacturing: Application to the Control of Austenitizing Furnaces

Abstract:

Energy constitutes the second largest operating cost in manufacturing industries within the United States [1]. In many manufacturing facilities, significant energy efficiency improvements could be gained from advanced sensing, monitoring, control and optimization techniques. However, the expertise required to implement them is not available in-house, and the associated capital expenditures often exceed the resources available to small and medium-sized manufacturers. The scope of our smart manufacturing work is to democratize access to model-based optimization for a broad class of manufacturing processes by developing a generic modeling and optimization platform, which can be modularized, configured and deployed with minimal expertise in a largely existing communication and control infrastructure.

Within this broader scope, this paper describes new developments concerning the operation and control of an austenitizing furnace. Austenitizing furnaces are high-temperature-high-throughput processes in which steel billets are heated in a semi-batch manner. The billets travel on rollers through the furnace, which is heated by ceiling and floor radiation tubes spanning its entire length. The overall heat input from the burners is on the order of 10 million BTU per hour to achieve furnace temperatures of 1400F.

Many such furnaces are still operated based on heuristics focused on the material properties, rather than overall process operation and efficiency, making this class of systems a prime candidate for optimization-based control. Furthermore, advanced sensing and online process monitoring are required to detect the potential for temperature maldistribution, which is a significant contributor to quality control issues and heat losses.

In our paper, we develop a mathematical model of an industrial austenitizing furnace, which encompasses the burner, hearth and billet domains. We validate the model based on data collected from furnace operations at an industrial partner. Based on this model, we develop a model predictive control scheme aimed at minimizing energy use while following the austenitization program prescribed for the metal products. The control scheme contains an additional feedback element based on the quality control checks that the parts are subjected to after they leave the furnace. Simulation results emphasize the energy savings that can be derived from model-based, smart manufacturing. Finally, we present incipient results concerning implementing these tools and algorithms in a smart manufacturing platform that will facilitate their wide deployment across manufacturing facilities in the United States.

References:

[1] Energy Use, Loss and Opportunities Analysis, US Manufacturing and Mining, US Department of Energy Report, 2004.

390667 Optimal Closed-Loop Design of Grid Scale Energy Storage Systems

Abstract:

It is widely recognized that a major concern with renewable energy is the fact that wind and solar sources are non-dispatchable. That is, the power produced from renewable sources is dependent on environmental conditions and is likely uncorrelated with the power demand from load centers. While fossil based sources are dispatchable and currently have the ability to respond to the full range of consumer loads, the additional range imposed by renewable sources is expected to exceed the dispatch capability of these fossil plants at the point of 20% renewable power. Thus, many have advocated the use of massive energy storage systems to provide the additional level of dispatch capability required to maintain grid solvency.

Due to the uncertainty of consumer demand as well as that of renewable generation, the problem of optimal placement of these storage units within the grid along with the selection of equipment sizes must be formulated as a stochastic program. However, rather than being a fairly simple two-stage stochastic program, the dynamics imposed by the storage devices requires the formulation to be of the far more challenging multistage class.

In this work, we propose a novel approximate solution procedure for this class of multistage stochastic programming problems. The approach utilizes the computational efficiency of the recently developed method of Economic Linear Optimal Control (ELOC) and its extension Constrained ELOC. The proposed method occurs in two stages. The first is a global search over the here-and-now variables as well as the parameters of the ELOC policy. However, this first search assumes a statistical constraint enforcement mechanism (similar to chance constrained optimization). In the second step is a gradient based search over the here-and-now variables, but used the Constrained ELOC policy to enforce point-wise-in-time constraints. To address the possible occurrence of an infeasible solution from the Constrained ELOC policy a fairly simple barrier type approach is used to improve convergence properties.

388658 Multivariate Nonlinear Model Predictive Controller for Managed Drilling Processes

Abstract:

Multivariate nonlinear model predictive controller for managed drilling processes

Reza Asgharzadeh Shishavan, Hector Perez and John D. Hedengren, Chemical Engineering, Brigham Young University, Provo, UT

This paper details a comprehensive control structure and the application in managed pressure drilling (MPD). The goal of this controller is to regulate the downhole pressure and rate of penetration (ROP) at desired values and attenuate any unwanted gas influx. For this purpose, a moving horizon estimator (MHE), an Extended Kalman Filter (EKF) and a nonlinear model predictive controller (NMPC) are designed to estimate and regulate the process. MHE and EKF provide the estimated unmeasured variables values to the controller to update the NMPC model and enable the controller to capture varying dynamics of the drilling process.

During MPD, the main pump sends the drilling mud into the middle of the drill string pipe. This mud travels downward through the drill bit and into the annulus and upward to the choke valve. Mud is used to remove the drilling cuttings and also to maintain a hydrostatic pressure at the drill bit against reservoir pore pressure. Rotation speed of the drill string (RPM) and weight on the bit (WOB), the amount of downward force placed on the drill bit, are two of the main factors that affect the drilling rate. Drill string rotation is adjusted with a top drive and WOB is adjusted by actuation above ground. As a result of hydraulic forces, torsional spring characteristics of the drill string, frictional losses, non-Newtonian mud flow, and other effects, the downhole WOB and RPM may be dynamically different than those applied topside.

While process automation and control has found many applications in the downstream industry, it is still immature in the upstream processes. The prior incidents while drilling have motivated research in drilling automation [1, 2]. One of the principle reasons that automation has not been progressed in drilling industry is lack of a high speed data communication network. Mud pulse telemetry has been the dominant technology available to send the data from sensors to topside equipment. However, mud pulse telemetry has limited bandwidth that is not capable, in many circumstances, for high-speed automated control systems.

Recently, a new high speed technology system has been developed and introduced that sends the data with a high bandwidth (57,600 bits/sec) to the surface. In this technology, electrical wires and conductors are used to transfer signals through drill pipe and across each threaded pipe to the next segment. Therefore the controller designed in this work is based on the wired pipe technology [3].

Two main goals in most drilling operations are safety and economics. In relation to safety, kick attenuation is vital to minimize operational risks. Kicks are the unexpected intrusion of formation gas into the drilling annulus due to a higher formation pressure. Kick attenuation can be achieved through annulus pressure regulation [2].In relation to economics, a higher ROP will lower the drilling cost. Therefore, ROP maximization is a motivating factor [4].

It is known that ROP depends on many factors including RPM, WOB, and downhole pressure. Previous research has attempted control and real-time optimization of ROP by only changing RPM and WOB, choosing to approach downhole pressure as a disturbance variable [4-6]. On the other hand, downhole pressure depends on the mud pump flow rate, choke valve opening, ROP and RPM (through an effect on the friction factor). Among all prior work, only mud pump flow rate and choke valve opening have been considered as manipulated variables in designing the pressure controllers [1, 7-10] for kick attenuation, while ROP and RPM have been considered as disturbance variables.

Prior research has focused on designing separate independent controllers and optimizers to regulate either pressure or ROP and excluded the effect of the multivariate nature of the control. This research offers a comprehensive controller that considers the interactions between ROP and pressure, regulating both in a single application. This multivariate controller uses pump flow rate, choke valve opening, rotation speed and WOB as manipulated variables and seeks to maintain ROP and downhole pressure.

Annulus density, friction factor and unwanted gas influx flow rate are not directly measured during drilling and therefore are considered as estimated variables. For this purpose, a MHE was designed to estimate annulus friction factor and density, and an EKF was designed to estimate unwanted gas influx flow rate. The nonlinear model predictive controller (NMPC) designed for this application uses a novel ℓ_1-norm with a dead band structure in the objective function. This is known to improve robustness of the controller while rejecting data outlier, noise and drift [11]. The NMPC uses both interior point and active set point optimization methods.

The control methods during normal drilling or kick conditions are detailed as follows: During normal drilling conditions, the goal is to maximize the ROP while maintaining the downhole pressure within a specified range. Therefore the controller calculates the required downhole RPM, WOB and downhole pressure values that lead to highest possible ROP. Next, it manipulates the pump flow rate and choke valve opening to regulate the downhole pressure to specified set points and changes topside WOB and RPM until the required WOB and RPM are achieved at the Bottom Hole Assembly (BHA).

During a kick event, the primary goal is to attenuate the kick while maintaining the ROP constant. First, the controller detects the kick condition through a rise in the downhole pressure and estimated unwanted gas influx value. Second, the application switches the controlled variable from downhole pressure to choke valve pressure and sets a new set point as the current value plus the desired change observed in the downhole pressure. Third, the controller calculates the optimum pump flow rate, choke valve opening and downhole RPM that lead to the new choke valve pressure set point. Finally, the controller adjusts the choke valve opening and pump flow rate and topside rotation speed to target values. This increases the pressure of the upper parts of the annulus that have not been affected by the kick and results in kick attenuation. WOB only affects ROP. Therefore, the controller also changes the topside WOB to compensate for the effect of the change in RPM to maintain a constant ROP. While attenuating the kick, the new reservoir pore pressure is estimated using the available topside and along-string measurements. When the unexpected gas influx is suppressed, the controlled variable is switched back to downhole pressure with the new set point calculated based on the new reservoir pore pressure.

Case study results show that integrated multivariate ROP and downhole pressure have superior performance compared to the case where each ROP and pressure have separate independent controllers. The benefits are faster and more effective kick attenuation along with a higher ROP leading to more efficient drilling operations.

References:

1. Hauge, E., et al., *A novel model-based scheme for kick and loss mitigation during drilling.* Journal of Process Control, 2013. **23**(4): p. 463-472.

2. Godhavn, J.-M., et al. *Drilling seeking automatic control solutions.* in *18th IFAC World Congress, August 28, 2011 - September 2, 2011.* 2011. Milano, Italy: IFAC Secretariat.

3. Reeves, M., et al. *High-speed drillstring telemetry network enables new real-time drilling and measurement technologies.* in *2006 IADC/SPE Drilling Conference - Achieving our Goals: People, Planning and Performance, February 21, 2006 - February 23, 2006.* 2006. Miami Beach, FL, United states: Society of Petroleum Engineers (SPE).

4. Arabjamaloei, R. and B. Karimi Dehkordi, *Investigation of the most efficient approach of the prediction of the rate of penetration.* Energy Sources, Part A: Recovery, Utilization and Environmental Effects, 2012. **34**(7): p. 581-590.

5. Eren, T., *Real-Time-Optimization of drilling parameters during drilling operations*, in *Petroleum and natural gas engineering,* 2010, Middle east technical university.

6. Irawan, S., A.M. Abd Rahman, and S.Q. Tunio, *Optimization of weight on bit during drilling operation based on rate of penetration model.* Research Journal of Applied Sciences, Engineering and Technology, 2012. **4**(12): p. 1690-1695.

7. Carlsen, L.A., G. Nygaard, and M. Nikolaou, *Evaluation of control methods for drilling operations with unexpected gas influx.* Journal of Process Control, 2013. **23**(3): p. 306-316.

8. Gravdal, J.E., et al., *Improved Kick Management During MPD by Real-Time Pore-Pressure Estimation.* SPE Drilling & Completion, 2010. **25**(4): p. 577-584.

9. Zhou, J., et al. *Adaptive observer for kick detection and switched control for bottomhole pressure regulation and kick attenuation during managed pressure drilling.* in *2010 American Control Conference, ACC 2010, June 30, 2010 - July 2, 2010.* 2010. Baltimore, MD, United states: IEEE Computer Society.

10. Zhou, J., et al., *Switched control for pressure regulation and kick attenuation in a managed pressure drilling system.* IEEE Transactions on Control Systems Technology, 2011. **19**(2): p. 337-350.

11. Hedengren, J., et al., *Nonlinear Modeling, Estimation and Predictive Control in APMonitor.* Computers & Chemical Engineering, 2014. In Press, Accepted Manuscript.

Advanced Process Controls across the hydrocarbon value chain

Opportunities & Challenges

(Paper# 345e)

Hamza Al-Hamadah, APC Engineer, *Saudi Aramco*
Co-Author: Dr. Rohit Patwardhan, Engineering Specialist, *Saudi Aramco*
2014 AIChE Annual Meeting, Atlanta, GA

Outline

* Overview
* Opportunities
* Technology Management
* Challenges & Trends
* Conclusions

Overview

ارامكو السعودية
Saudi Aramco

* Saudi Aramco is a fully integrated, global petroleum and chemicals enterprise based in Dhahran, Saudi Arabia
* Over 80 years in the oil business
* A world leader in hydrocarbons exploration, production, refining, distribution, shipping and marketing
* World's top Oil & NGL exporter, supplying ~10% of global demand of oil
* Over 56,000 employees, 80+ nationalities

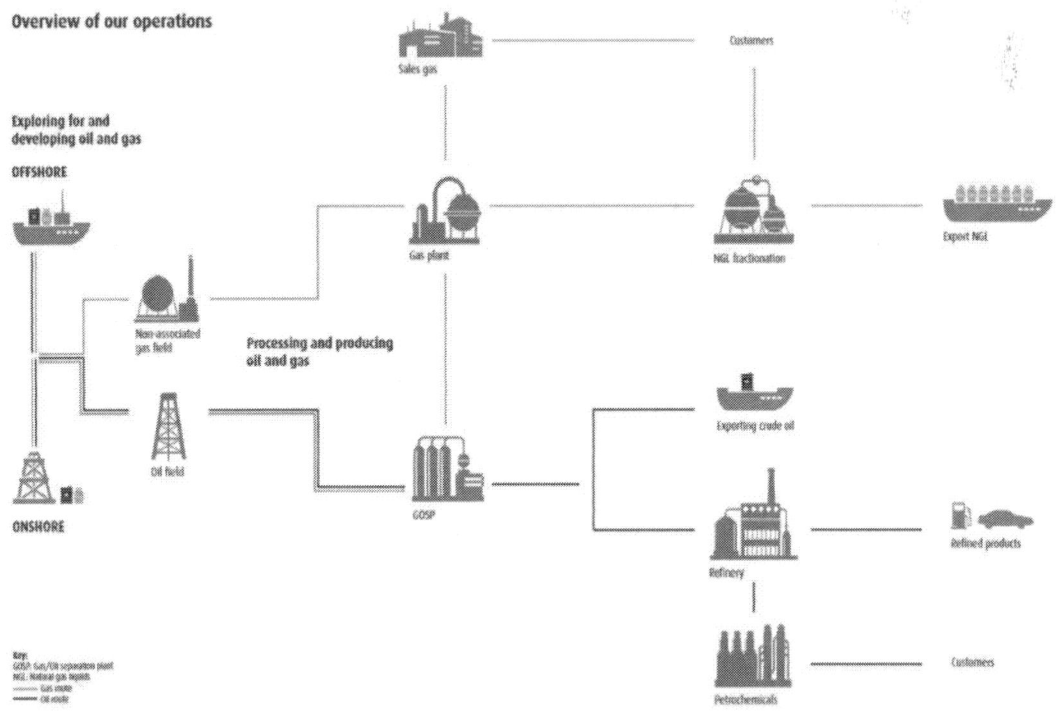

APC Experience

* Over 20 years of experience in APC (Linear MPC, RTO)
* Breadth of applications: Over 100 controllers (Oil, Gas & NGL, Refining)
* Technology diversity: Aspen DMCplus®, Honeywell®, Yokogawa®
* Central engineering with well over 100 man-years of collective APC expertise
* Critical component of ERP

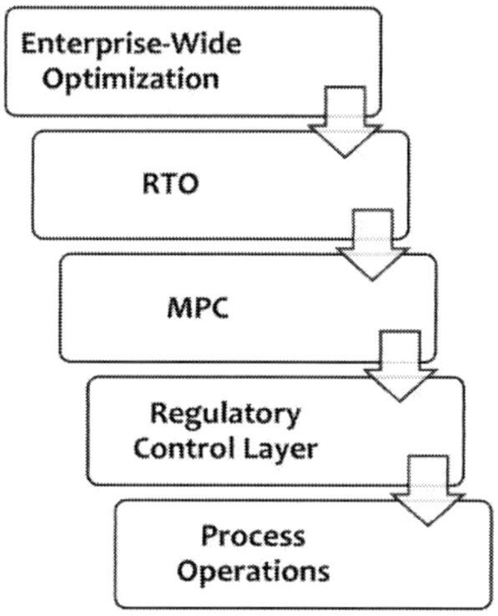

Opportunities

* APC maximizes value-added, higher margins
* Reducing product giveaway, and maximizing throughput
* Over 90% coverage in downstream

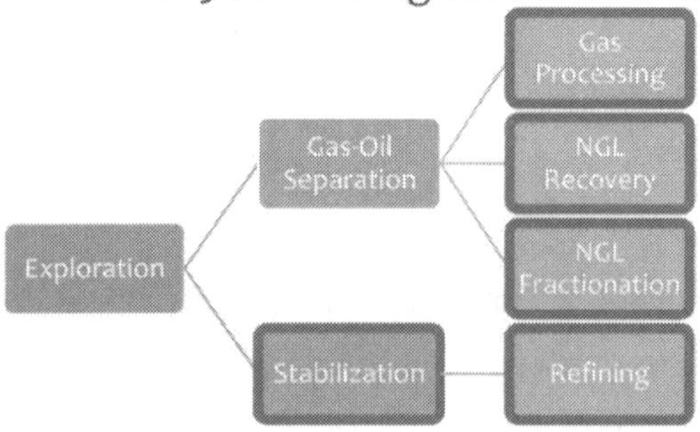

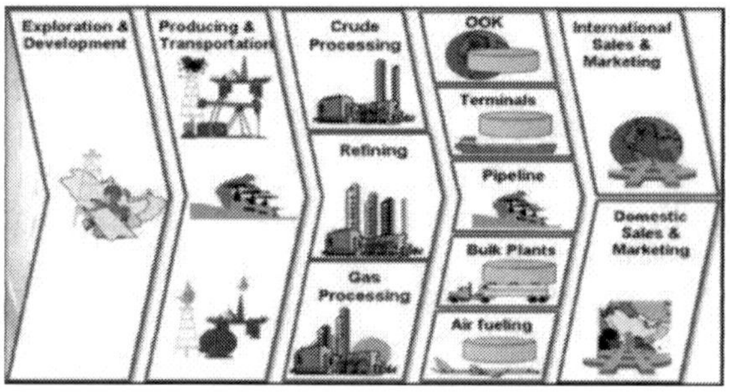

Technology investments demands rigorous feasibility analysis
APC typically, but not always, high ROI

Opportunities

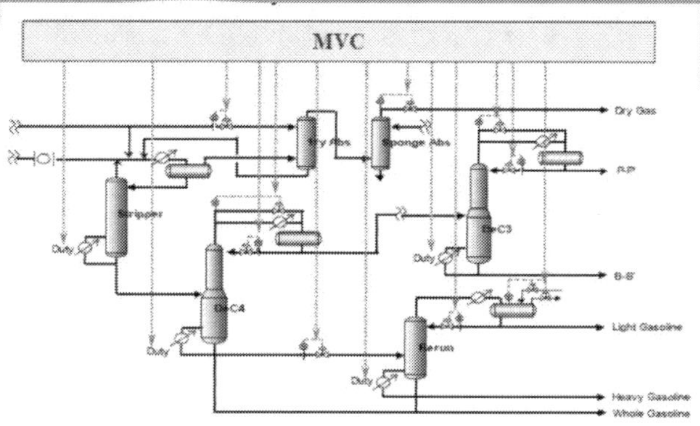

Maturity: level of experience acquired within industry for particular process application

Complexity: Level of controller complexity (i.e., matrix size, (non)linearity, LP equation complexity, disturbances)

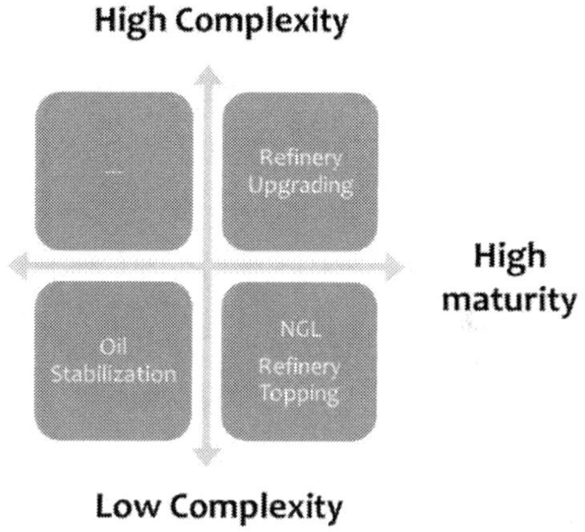

Technology Management

* Sustainability of benefits is biggest challenge
* Changes of feed, equipment degradation, unmeasured disturbances, skilled-support, etc.
* Progressive area of research, increasing industry attentions

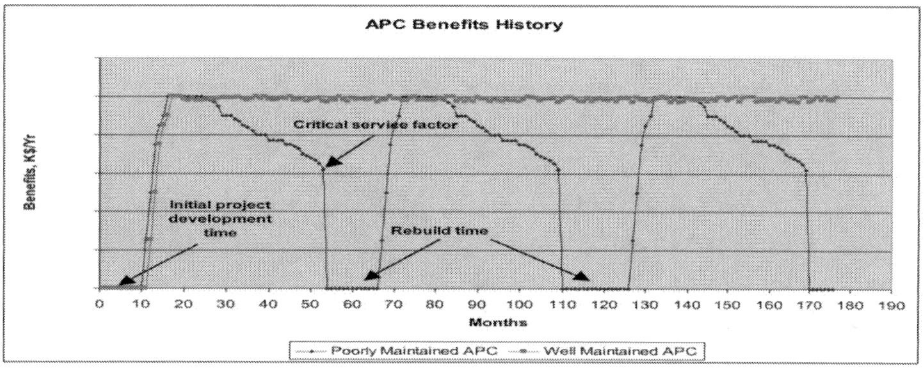

Technology Management

- Multiple stakeholders
- Central Engineering → Technology Leadership
- Work processes:
  - Standardization
  - Engineering Solutions
  - Technology Monitoring & Benchmarking

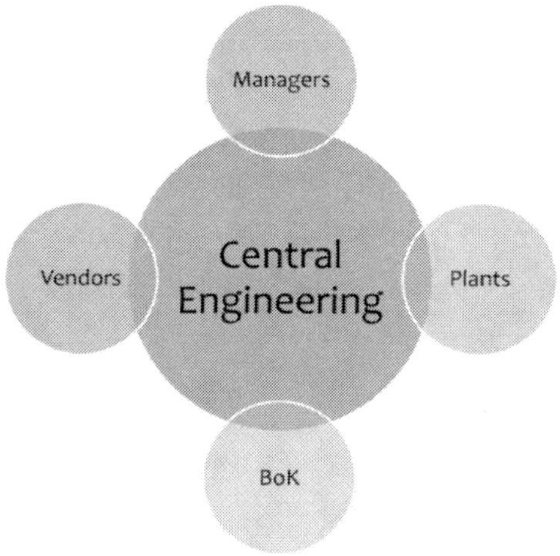

Technology Management

- What to monitor?
- Progressive research in monitoring, but industry standards are non-existent
- Saudi Aramco Monitoring Pyramid:
  - Profitability-driven
  - Base-layer control performance is strongest influencer
  - Engineering monitoring layers are symptomatically-driven with emphasis on control engineering

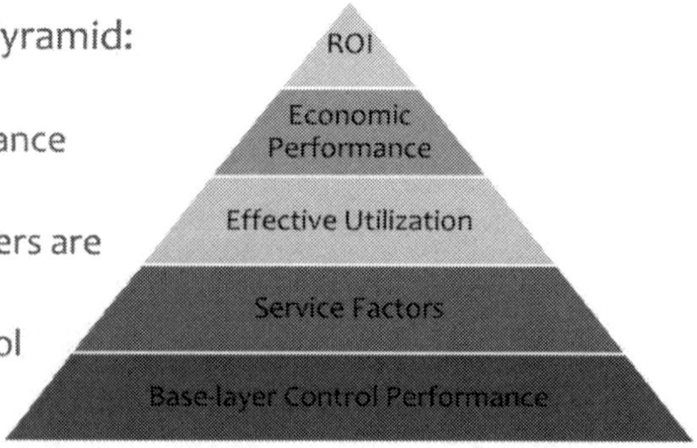

MPC Monitoring

Monitoring Hierarchy	Level of Complexity	Emphasis	Frequency*
Management	Low	Application Utilization & Financial Impact	M/Q
Engineering	High	Control Engineering	M/Q
Operations	Medium	Service factors, Base-layer control	W

*(Q=Quarterly, M=Monthly, W: Weekly)

- Monitoring needs to be systematic, methodical, hierarchal
- Communicate the right information to the right people

MPC Monitoring

- Management Monitoring through corporate-wide benchmarking
  Monitor → Identify performance gaps →Avail resources →Improve
- Engineering & Operations Monitoring
  - Assessing Model qualities → tuning/re-vamping
  - Inferential models
  - PID loops performance
  - Constraint-analysis
  - Controller behavior
- Economic Performance:
  - Base-lining approach
  - ON/OFF approach
- Performance Benchmarking: over 15% improvement in technology utilization

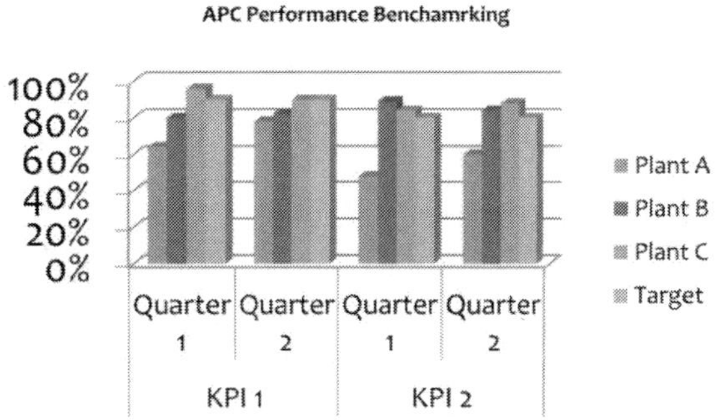

Challenges & Trends

* System Engineering (connectivity, configurations, virtualization, etc.)
* Control Engineering Issues
  * Disturbance rejection vs. set-point tracking
  * Nonlinearities
  * Closed-loop testing
  * Decentralized MPC vs. Distributed MPC
  * Base layer controls performance
* MPC Algorithms (S-MPC, NLMPC, etc)
* Experience Retention, scarcity of competent APC engineers
* Academic research focuses on MPC algorithms, Industry demands more focus on Control Engineering (Qin et al)

Conclusions

* MPC is an indispensable profit-maximizing technology
* Effective technology monitoring & servicing is biggest industry challenge
* Industry: exploring for new MPC applications to maximize profits
* Academia: More effective MPC algorithms

Acknowledgments

AIChE®

ارامكو السعودية
Saudi Aramco

References

* Canney, W. (2003). *The future of advanced process control promises more benefits and sustained value.* Oil & Gas Journal, April 21, 2003.
* Froisy, J. B. (2006). *MPC: Building a bridge between theory and practice.* Computers and Chemical Engineering 30 (1426-1435)
* Jokinen, P. (1996). *A life-cycle approach to automation information management in process industries.* ISA Transactions Vol. 35, pp. 285-296.
* Kano, M., and Ogawa, M. (2010). *The state of the art in chemical process control in Japan: Good practice and questionnaire survey.* Journal of Process Control Vol. 20, pp. 969-982.
* Lee, J. (2011). *Model Predictive Control: Review of the Three Decades of Development.* International Journal of Control, Automation, and Systems. Vol. 9(3), pp. 415-424.
* Lodolo, S., Vederinkov, O. (2014) *Revamping Advanced Process Control.* Digital Refining. Retrieved at :http://www.digitalrefining.com/data/articles/file/42214904.pdf
* Paulonis, M., and Cox, J. (2003). *A practical approach for large-scale controller performance assessment, diagnosis, and improvement.* Journal of Process Control Vol. 13, pp. 155-168.
* Qin, S. J., and Badgwell T. A. *A survey of industrial model predictive control technology.* Control Engineering Practice Vol. 11, pp. 733-764

Thank you..

Q&A

CFD-SIMULATION OF AN INDSTRIAL FURNACE IN THE HOT-DIP GALVANIZATION PROCESS

Christoph Triebl [a], Christoph Spijker [a], Harald Raupenstrauch [a], Alexander Jarosik [b], Gerhard Angeli [b]

[a] *Chair of Thermal Processing Technology, Montanuniversitaet Leoben, Leoben, Austria*
[b] *voestalpine Stahl GmbH, Linz, Austria*

Abstract

In a hot-dip galvanization line, the direct-fired furnace used for pre-treatment of the steel strip is simulated to investigate the operating conditions in the furnace and the heating effects on the strip. A computational fluid dynamics (CFD) simulation of the furnace is set up to model the heating of the steel strip by non-premixed burners. For simulating turbulence and combustion, the realizable-k-ε-model and the Equilibrium combustion model are used. The discrete-ordinate method is chosen for modeling radiation in the furnace. The results show the influence of the geometry on the conditions above and below the steel strip in the direct-fired furnace. The convective and the radiant heat flux through the strip surface are investigated as well as the impact of the flames on the heat treatment of the steel strip. The temperature profile of the strip is analyzed in different sections of the furnace. Regarding emission values, the reburning zone for minimizing the carbon monoxide content is studied.

Introduction

Zinc coatings are widely spread for protecting steel against corrosion in many different applications. The commercially most important method for producing zinc coatings is hot-dip galvanization. State of the art in continuous processes is invented by Sendzimir, at which pre-heating of the strip is implemented prior dipping into the liquid zinc bath. The surface of the steel strip is first heat-treated and cleaned by removing oil residues in a direct-fired furnace by flash evaporation due to the fast temperature increasing. Installed swirl burners in the furnace are providing the heat input to heat up the strip by radiation and thermal convection. To regulate the heat treatment in the furnace, the operating conditions of the burners are adjusted. [1] Carbon of the steel strip surface is reacting with the combustion gas, which leads to a carbon-free strip surface [2]. To minimize carbon monoxide emission, air is injected downstream the burner zones. After the direct-fired furnace, the steel strip is passed through an annealing furnace for reducing the surface to eliminate oxides and improve the surface quality for the galvanization. Although the Sendzimir-process is more than 70 years old, detailed conditions in the pre-heating furnaces are not determined and potential for optimization still exists. [3, 4]

To investigate the operation conditions of a direct-fired furnace in a hot-dip galvanizing process, the furnace was modeled by using computational fluid dynamics (CFD). The commercial software ANSYS Fluent 13.0 was chosen to model the turbulent combustion in the direct-fired furnace. The flow conditions were investigated as well as the heat treatment of the steel strip. The reburning zone was analyzed downstream the burner zones to ensure low carbon monoxide at the exit of the direct-fired furnace.

Model Description

The modeled furnace is a horizontal direct-fired furnace with asymmetrical arranged swirl burners. Shieldgas is streaming countercurrent to the treated steel strip to regulate the oxygen content and dew point in the furnace for a controlled oxidation of the strip surface. The shieldgas is consisting of nitrogen with small fractions of hydrogen. Since free jets have to be simulated, the realizable-k-ε-model is used for modeling turbulence. This model solves the transport equations for the turbulent kinetic energy k and the turbulent dissipation rate ε, which are derived from the dynamic equation of the mean square vorticity fluctuation [5]. In the furnace, non-premixed burners are installed, which are fed by pre-heated air and natural gas. In modeling non-premixed combustions, convective and diffusive time scales are of the same magnitude, but the chemical time scale is smaller by several orders of magnitude. For this reason, fast chemistry and subsequently local chemical equilibrium are assumed [6, 7]. Simplifying the model by introducing fast chemistry, equal diffusivities, isobaric and adiabatic conditions, the Equilibrium model is chosen for simulating combustion in the direct-fired furnace [8]. Due to the fast chemistry, reaction takes place in asymptotically thin layers, which leads to a one-dimensional combustion normal to these layers, at which only the transport equation for the mixture fraction Z has to be solved [6, 7]. Species mass fractions and temperature are depending on the mixture fraction, which is a conserved scalar and characterizes the local mixture of fuel and oxidizer [9, 10]. The fluctuation of Z caused by turbulence is taken into account by a probability density function (PDF) [8]. Therefore, chemistry is solved prior the simulation and implemented into the calculation [10]. For simulating radiation, the discrete ordinate method (DOM) is used, which models scattering, diffusive reflection, as well as refraction and can be adopted in complex geometries, such as furnaces [11]. The radiative transfer equation is calculated for a finite number of ordinate directions that are fixed in the coordinate system by solving the transport equations for the radiation intensity. The resolution of the discrete-ordinate method is set by selecting the number of solid angles. [5, 12]

Since the mesh at the burners has to be in high resolution, the mesh of the total furnace would be large including all burners. Hence, for every different operation mode of the burners, one single burner is simulated and cut at a specified section. The conditions at this section are stored and used for the simulation of the furnace. The mesh of the total direct-fired furnace is cut at the burners at the same section and the prior stored profiles are patched on these sections to provide the input data of the burners for the simulation of the furnace.

Results and Discussion

In the first step, single burners for each operation mode are simulated in a separate model. The geometry includes the burner port and a small building block for forming a flame. An appropriate section plane is chosen for saving the required profiles for the simulation of the direct-fired furnace. Figure 1 shows the temperature profile in the section plane of the simulated single burner. Due to the adjustment of the nozzles for air injection, a stellar reaction zone is resulting. Data of the velocities, temperature, pressure, turbulence, mean mixture fraction and its variance are stored and implemented to the furnace simulation. Figure 2 and Figure 3 show the profiles of the temperature and the velocity, respectively, in the cross section of two burners in the furnace. In the temperature profile in Figure 2, the formation of the flames is apparent. Due to radiation, the temperature is, except at the flames, quite constant in the cross sections of the furnace. The space above the steel strip is larger than under the strip, which results in marginal higher profiles of the velocity and the temperature below the steel strip.

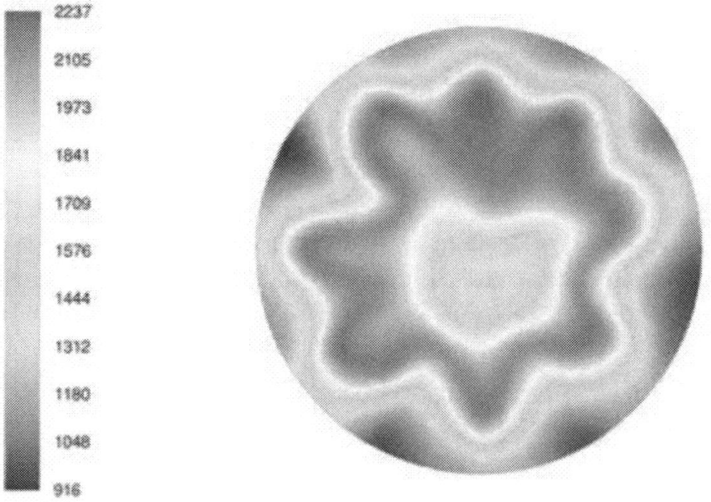

Figure 1: Temperature profile [K] in the section plane of the burner

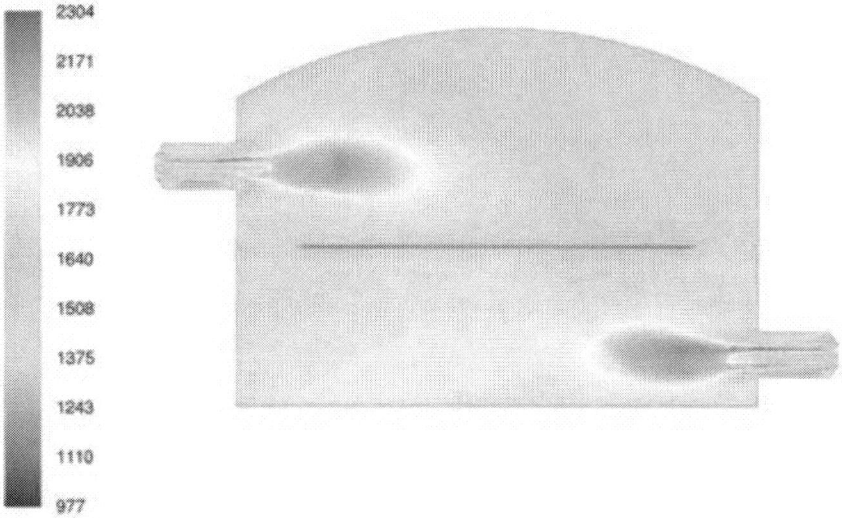

Figure 2: Temperature profile [K] in the cross section of two burners.

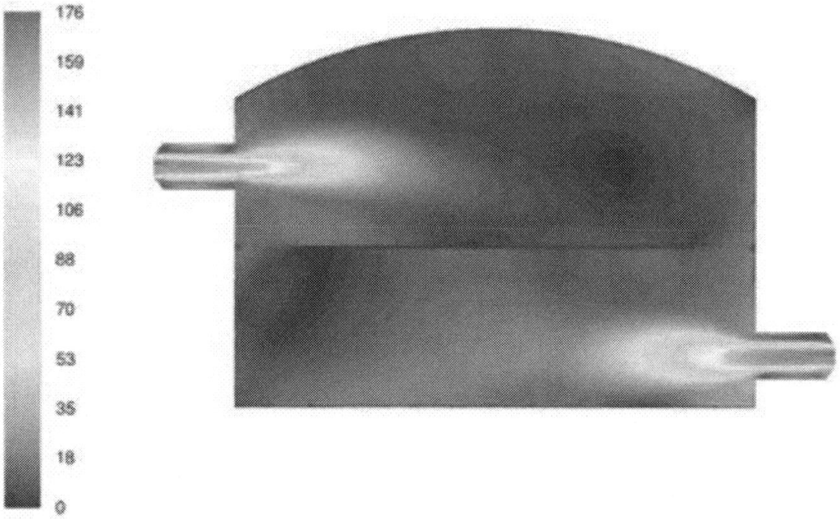

Figure 3: Velocity profile [m/s] in the cross section of two burners.

In Figure 4 and Figure 5, the profiles of temperature and velocity are shown in the cross sections of the upper and the lower burners, respectively. The shieldgas is streaming from the bottom to the top of the figures and the steel strip is moving in the opposite direction. The inflowing shieldgas is mixed with the flue gases and heated up by the first burners from 960 K to approximately 1700 K. As mentioned before, the temperature is under the steel strip a bit higher than above the strip. Due to the increasing gas flow in consequence of the adding flue gases, the flames are deflecting downstream the furnace. The asymmetric arrangement of the burners leads to an interaction of the flames, which causes slight transient effects in the furnace.

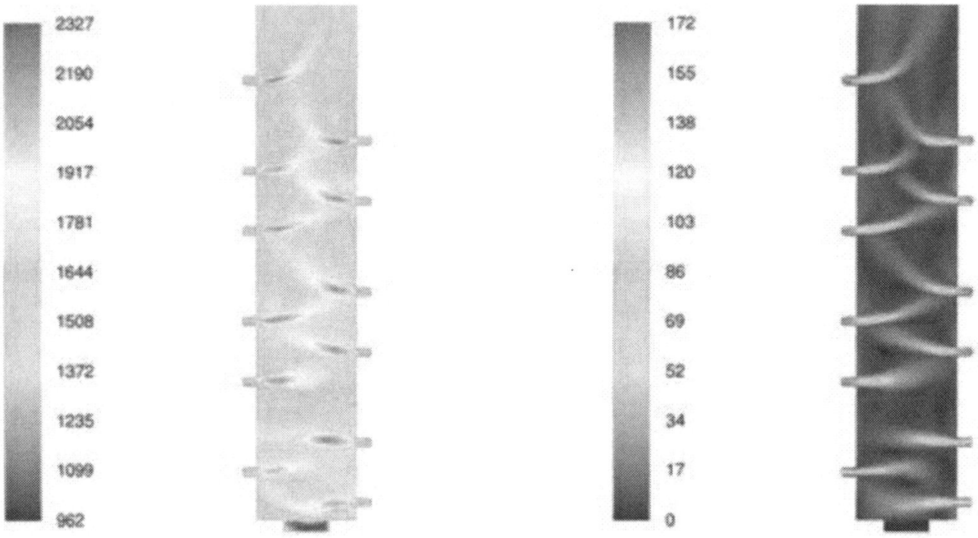

Figure 4: Profiles of temperature [K] (left) and velocity [m/s] (right) in the cross section of the upper burners.

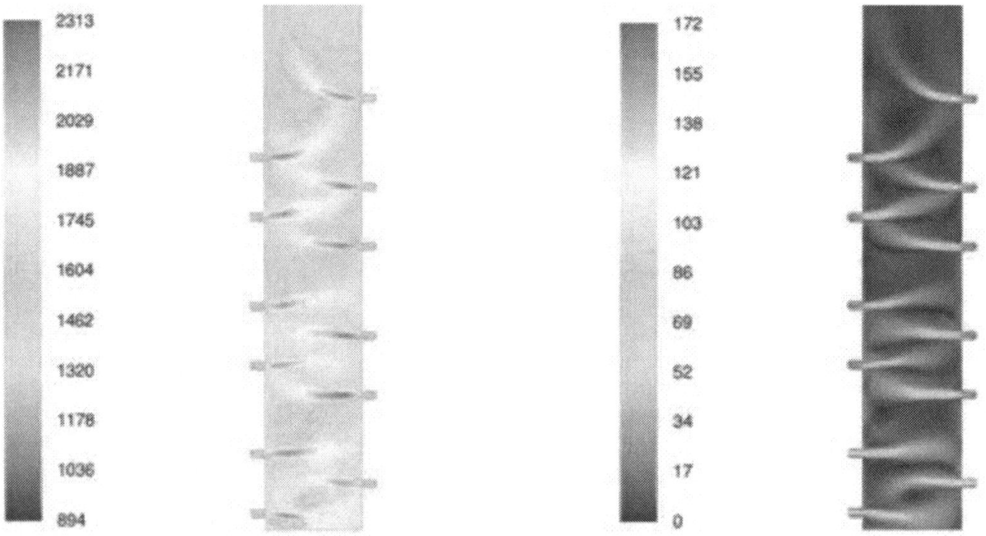

Figure 5: Profiles of temperature [K] (left) and velocity [m/s] (right) in the cross section of the lower burners.

Figure 6 shows the convective heat flux on the left side and the radiation heat flux on the right side through the strip surface in the burner zones. Positive values stand for heat release and negative values for heat absorption of the steel strip. The maximum convective heat flux through the strip surface is given when the steel strip enters the burner zones. The reason for this is the highest temperature difference between the strip and the gas in this section. The contact of the flames with the surface of the steel strip is seen in Figure 6 by the zones with the higher heat input. In the exit section of the strip in the direct-fired furnace, which is in the bottom in Figure 6, the steel strip releases heat by thermal convection due to the relative cold shieldgas. On the profile of the radiation heat flux in Figure 6, the contours of the flames are seen. The arrangement of the burners and the deflection of the flames lead to a Z-shaped pattern of the maximum radiation heat flux through the strip surface. In contrast to the thermal convection, there is no heat release of the steel strip by thermal radiation.

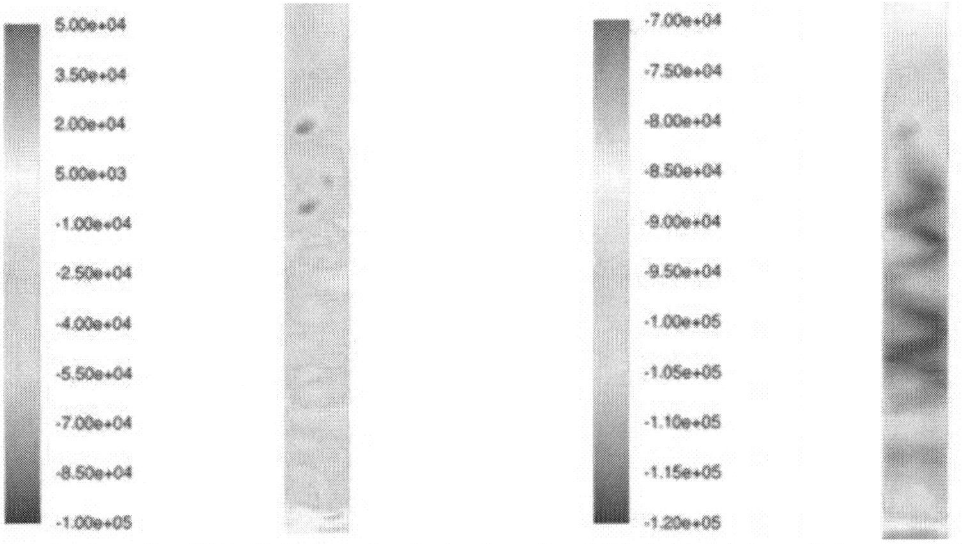

Figure 6: Convective heat flux [W/m²] (left) and radiation heat flux [W/m²] (right) through the strip surface in the burner zones.

Since the direct-fired furnace has an oxidizing atmosphere, the oxygen distribution plays an important role for the oxidation of the steel strip. Figure 7 shows the profile of the oxygen mass fraction of the furnace gas on the strip surface in the burner zones. Due to the asymmetric arrangement of the burners, the oxygen distribution on the surface of the steel strip is not symmetrical. This has an impact on the uniformity of the oxide formation on the strip surface and should be noted for further treatment of the steel strip.

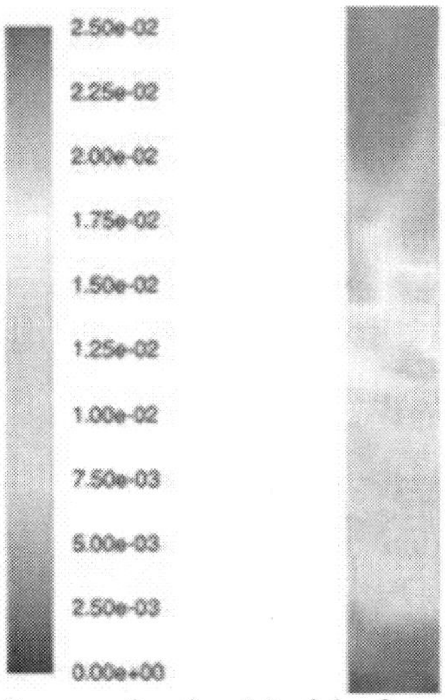

Figure 7: Profile of the O_2 mass fraction [-] of the furnace gas on the strip surface.

In Figure 8, the temperature profile of the strip surface is shown in the total furnace (left), the beginning (middle) and the end section (right) of the furnace. The constant heating of the steel strip is possible due to the countercurrent flow of the shieldgas and the radiation from the furnace wall. In the first section of the direct-fired furnace, the edges of the strip are heated up faster due to the view factor (Figure 8, middle). In the burner zones, the flames contact the steel strip in the central zone, which leads to a faster heating in the central section compared to the edges of the strip (Figure 8, right).

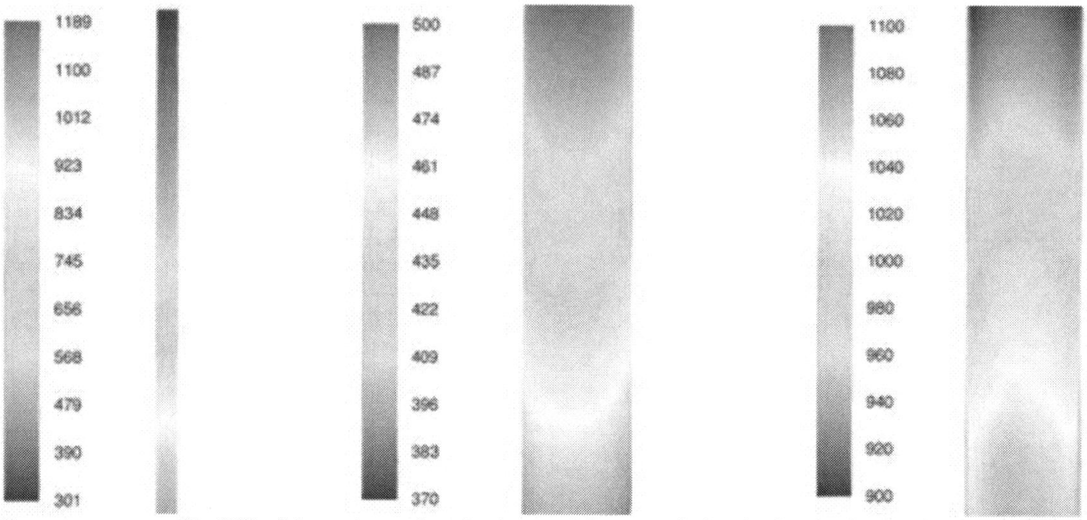

Figure 8: Temperature profile [K] of the strip surface in the total furnace (left), the beginning (middle) and the end section (right) of the furnace.

Figure 9 shows the temperature profile of the steel strip along the standardized furnace length. The strip enters the furnace at ambient temperature of approximately 300 K and is heated up in the first 20 % of the furnace to 400 K. In the middle of the furnace, the temperature of the steel strip reaches 600 K and is then increasing nearly constant to approximately 1100 K at the exit of the direct-fired furnace. The dwell time of the strip in the furnace is about half a minute.

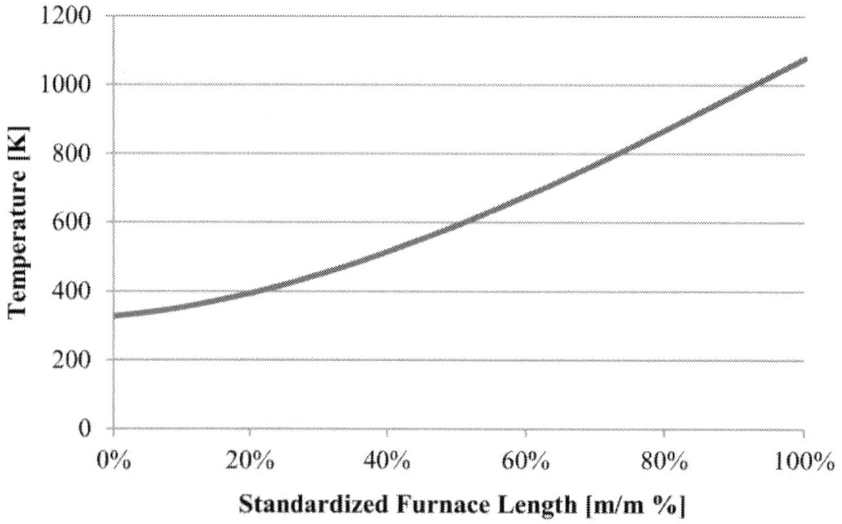

Figure 9: Temperature of the steel strip in relation to the standardized furnace length [m/m %]

In Figure 10, the profile of the mass fraction of carbon monoxide in the section of the reburning zone is shown. The positions of the upper and lower air injections are marked. The CO content is already quite low before the reburning section, and is above the steel strip approximately twice as high as below the strip, namely 0.17 % and 0.07 %, respectively. This is caused by a slight difference of the mean mixture fraction above and below the steel strip due to the furnace geometry. The fraction of carbon monoxide after the reburning zone is under 10 ppm and hence negligible.

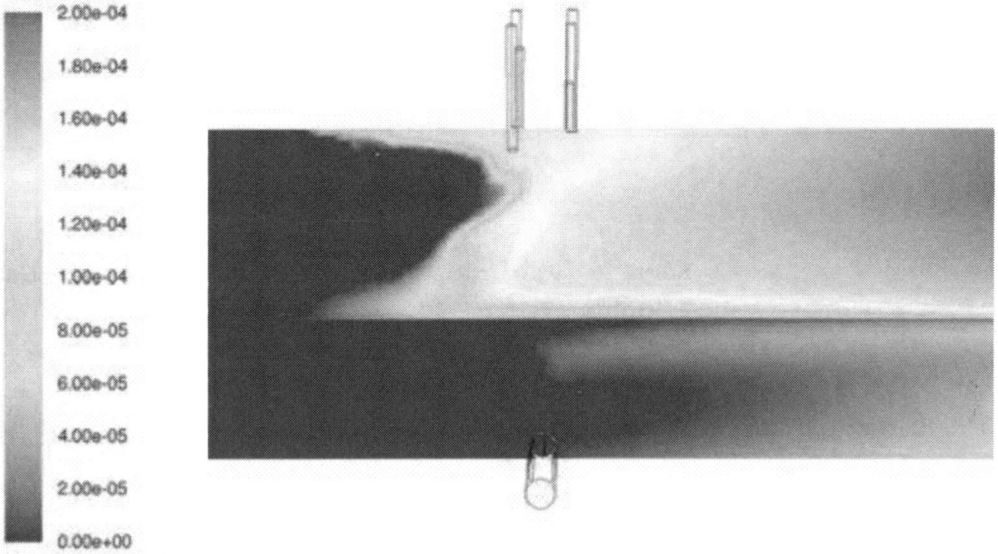

Figure 10: Profile of the CO mass fraction [-] in the reburning zone. The positions of the air injections are shown.

Figure 11 shows the pathlines of the air injections for reburning, colored by the O_2 mass fraction. The oxygen is reacting in large part when entering the furnace, which leads to a decrease of the O_2 mass fraction from 23 % to 5 %. Due to the lower carbon dioxide content below the steel strip, the oxygen fraction is slightly higher in this area. Since no CO is remaining in the exhaust gas, the oxygen content is kept constant after the reburning zone.

Figure 11: Pathlines of the air injections for reburning colored by the O_2 mass fraction [-].

Conclusion

The results of the simulation show the slightly different conditions above and below the steel strip that is caused by the geometry of the direct-fired furnace. The shieldgas deflects the flames of the burners, which has an impact on the profiles of the convective and radiant heat flux through the strip surface. Due to the high temperature of more than 1500 K, the steel strip heats up immediately after entering the furnace, which plays a major role in the heat treatment of the strip. Due to the flames of the burners, the strip is heated in the central zone faster than at the edges. The relative low carbon monoxide content before the reburning section results in a complete oxidation of the carbon monoxide to carbon dioxide, which plays a major role in the compliance of the emission standard.

Acknowledgement

Financial support by the Austrian Federal Government (in particular from Bundesministerium für Verkehr, Innovation und Technologie and Bundesministerium für Wissenschaft, Forschung und Wirtschaft) represented by Österreichische Forschungsförderungsgesellschaft mbH and the Styrian and the Tyrolean Provincial Government, represented by Steirische Wirtschaftsförderungsgesellschaft mbH and Standortagentur Tirol, within the framework of the COMET Funding Programme is gratefully acknowledged. The author is thankful for the collaboration with voestalpine Stahl GmbH.

References

1. Jenkins, D. R., et al. (1990), "Transient model of a gas-fired furnace", in *Proceedings of the 1990 Mathematics-in-Industry Study Group*, N. G. Barton, ed. CSIRO Australia, pp. 1-14.
2. Wang, R. (1991), "Direct-fired heating in continuous hot-dip galvanizing lines", *MPT Metallurgical Plant and Technology International*, Volume 14, no. 4, pp. 126-130.
3. Marder, A. R. (2000), "The metallurgy of zinc-coated steel", *Progress in Materials Science*, Volume 45, Issue 3, pp. 191-271.
4. Sendzimir, T. (1938), "Process for coating metallic objects with layers of other metals", U.S. Patent No. 2,110,893.
5. Fluent Inc. (2006), "FLUENT 6.3 User's Guide".
6. Peters, N. (1984), "Laminar diffusion flamelet models in non-premixed turbulent combustion", *Progress in Energy and Combustion Science*, Volume 10, Issue 3, pp. 319-339.
7. Peters, N. (1988), "Laminar flamelet concepts in turbulent combustion", *Symposium (International) on Combustion*, Volume 21, Issue 1, pp. 1231-1250.
8. Goldin, G. M. (2005), "A priori investigation of the constructed PDF model", *Proceedings of the Combustion Institute*, Volume 30, Issue 1, pp. 785-792.
9. Claramunt, K., et al. (2006), "Analysis of the laminar flamelet concept for non-premixed laminar flames", *Combustion and Flame*, Volume 145, No. 4, pp. 845-862.
10. Pitsch H., et al. (1998), "Unsteady flamelet modeling of turbulent hydrogen-air diffusion flames", *Symposium (International) on Combustion*, Volume 27, Issue 1, pp. 1057-1064.
11. Thynell, S. T. (1998), "Discrete-ordinates method in radiative heat transfer", *International Journal of Engineering Science*, Volume 36, Issues 12–14, pp. 1651-1675.
12. Siegel, R., Howell, J. R. (1992), "Thermal Radiation Heat Transfer", Third Edition, *Hemisphere Publishing Corporation*, Washington, DC.

379094 Simulation and Optimal Operation Strategy for Continuous Biobutanol Production

Abstract:

Biobutanol is considered to be a promising renewable energy source, especially as a gasoline substitute, owing to its high energy content, low volatility, and low water solubility. However, the volumetric productivity of biobutanol in batch fermentation is limited by the toxicity of butanol: the microorganism's cell growth gets inhibited when its concentration in the fermentation broth reaches a certain critical level. For this reason, produced biobutanol needs to be removed from the fermentation broth during the operation, in order to achieve a high level of volumetric productivity needed for commercialization. In this study, a continuous extractive fermentation process integrated with an ex-situ adsorption recovery process is investigated as a way to overcome this limitation. The spatial segregation of the adsorption system and the fermentation process enables continuous biobutanol production without the need to stop the fermentation, and can maintain the concentration of butanol in the fermentor below the threshold of toxicity thanks to removal of biobutanol in the adsorption column. Since the adsorption column has to be switched periodically due to the limited capacity of the adsorbent, the overall operation follows a cyclic pattern and the proposed continuous extractive fermentation process shows a cyclic steady state behavior. A dynamic model for the integrated process is developed and used for simulating the concentration profiles at cyclic steady state. Finally, major operating variables are optimized based on the cyclic steady state behavior in order to design an optimal operation strategy that satisfies given requirements.

385773 MPC of a Heat Pump: Achieving Energy Efficiency through Improved Tuning

Abstract:

The comfort level within the indoor environment of a building is maintained by mechanical and electrical systems classified as heating, ventilating and air-conditioning (HVAC) systems. In many residential and industrial settings this conditioning is provided through the heating and cooling mechanisms of a heat pump. The most common realization of a heat pump consists of four components: a compressor, outdoor heat exchanger, expansion valve, and an indoor heat exchanger. In the cooling mode, conditioning performance is tied directly to the tracking performance of the indoor coil's supply air temperature , while safety-related issues center upon the above-zero regulation of the refrigerant superheat temperature at the outlet of the indoor coil. Multiple actuators are available for manipulation to satisfy these control objectives, however one caveat in their operation is the vast range which exists amongst each of their individual energy demands. This includes the most energy intensive actuator, the compressor, drawing electricity to regulate it's motor frequency which directly affects the inlet/outlet refrigerant compression ratio. Less energy intensive in nature are the fan drives which are used in setting the air flow rate over both the indoor and outdoor coils. The final actuator as well as the least energy intensive one, is the expansion valve. Through manipulating the valve flow area, conditioning ability of the indoor coil is directly affected (variation to refrigerant vapor/liquid fraction, refrigerant mass flow rate). On top of the large energy discrepancy amongst actuators, a high level of interaction and subsequent nonlinearities persist within the heat pump dynamics, causing single-input-single-output (SISO) control approaches (classical PI/PID or decoupled versions of it) to be far from superior when one must consider satisfying the interconnected control objectives described previously.

A more suitable control approach, in terms of effectively accounting for such system dynamics and control objectives, is that of model predictive control (MPC). Given the known system nonlinearities, a traditional linear MPC framework must be extended to include an augmented offset-free mechanism if any potential plant-model mismatch is to be captured. Such offset-free MPC structures have been seen to be an effective means at regulating both local HVAC systems (see [1] and [2]) as well as in building-based set-point management [3]. Though each have been shown effective in mitigating disturbances and conserving energy usage within the system, a general tuning framework for tuning the offset-free MPC structures remain unavailable. In a recent work [1], a sequential tuning approach has been shown to aid in tuning. In [2], the tuning approach involved first tuning the state-estimated and augmented model parameters through open-loop step changes and then moving on to tune the MPC parameters. There is still a significant energy gap that can be bridged by designing and implementing a comprehensive tuning framework.

Motivated by these considerations, a framework is proposed predicated on the idea of internal model control, where the `desired' response is specified through appropriate first-order time constants. Preserving the strength of the MPC in handling constraints, limits as to how fast a desired response could be achieved without causing chattering are specified. Subsequently, the tuning of the observer and control parameters are done together within an integrated offset-free MPC framework. The efficacy of the proposed tuning framework is demonstrated on the experimentally-validated heat pump model coupled with realistic variations to both ambient and internal conditions, as well as measurement noise.

[1] M. Wallace, B. Das, P. Mhaskar, J. House, T. Salsbury, Offset-free model predictive control of a vapor compression cycle, Journal of Process Control 22 (2012) 1374-1386

[2] M. Wallace, P. Mhaskar, J. House, T. Salsbury, Offset-free model predictive control of a heat pump, Ind. & Eng. Chem. Res. (submitted),

[3] C. Touretzky, M. Baldea, Nonlinear model reduction and model predictive control of residential buildings with energy recovery, in: Journal of Process Control (invited paper for Special Issue on Control of Buildings), 2013.

379755 Model Predictive Control of Dual Mixed Refrigerant Liquefaction Process for Stable and Economic Operation in Offshore FLNG

Abstract:

Dual mixed refrigerant (DMR) liquefaction cycle is one of the promising processes to produce LNG in offshore FLNG because of its high process efficiency and compactness. However, stable and economic operation of DMR liquefaction process is a key to harness its advantages in the offshore or marine operation environments. For example, roll and pitch and heave movements are major disturbances in offshore FLNG process. Furthermore, frequent changes in LNG shipping schedule due to the uncertainty in offshore weather condition and the limited LNG storage volume can also lead to frequent LNG production load variations compared to the onshore LNG production. This study proposes a model predictive control (MPC) strategy for the DMR process in order to reject such process disturbances and minimize the energy consumption, production loss, subcooling or insufficient cooling of LNG under the load variation conditions.

A rigorous process dynamic model was first constructed in Aspen HYSYS Dynamics^{TM}and MATLAB/Simulink environments that can simulate process disturbances including steep load variations. An energy optimized steady-state model was first developed to identify an appropriate nominal operating condition. The corresponding pressure and temperature conditions were used as a design basis for dynamic model. Using the step response data of from dynamic simulation software, multivariate transfer function matrix was identified for the purpose of control system design. In this study, LNG flowrate, cold mixed refrigerant liquid (CMRL) and cold mixed refrigerant vapor (CMRV) were manipulated to regulate LNG temperature, LNG flowrate, and warm side temperature difference of multi-stream heat exchanger and optimize accumulated energy consumption and production loss.

The dynamic behaviors and the features of process with MPC are compared with a conventional PI control scheme with loop pairing and show the applicability and advantage of implementing multivariate control for the offshore plant.

372984 Process Synthesis without Integer Variables: Using Complementarity Conditions for Thermodynamic and Distillation Models

Abstract:

Discrete decisions are essential to process synthesis; they allow for units to be enabled and disabled in an optimization formulation. Unfortunately, the resulting mixed integer optimization problems may be difficult to solve, especially with highly nonlinear, non-convex detailed process equipment models (e.g. non-ideal thermodynamics). Typically researchers are forced to simplify models (e.g. linearization to MILP, simplifying assumptions regarding physics, etc.), develop complex decomposition strategies using generalized disjunctive programming and/or reduce the scope of the problem to be computationally tractable. All three of these options are not ideal, as they may produce unrealistic results, require substantial effort to implement and may not answer the actual design question.

As an alternative to mixed integer programming, researchers are exploring the application of complementarity conditions (and related reformulations) that allow (some) process synthesis problems to be reformulated as nonlinear programs (optimization problems). This approach is adventitious, as it enables optimization of the detailed, typically non-convex, process models along with some discrete decisions using generic, large-scale NLP algorithms. In this paper, we focus on two applications of complementarity conditions in process synthesis problems: (1) embedded cubic equation of state (e.g. Peng-Robinson) models and (2) optimization of distillation cascade size with tray-by-tray models.

Regarding embedded thermodynamic models, Kamath et al (2010) proposed an equation-based method for distinguish between liquid, vapor and spurious roots in a cubic equations of state. Phase disappearance is accommodated using complementarity conditions. We will summarize this approach, and propose an extension to prevent spurious vapor-liquid equilibrium solutions that occur when K = 1 (see Gundersen, 1982).

Furthermore, we will present a novel model for distillation optimization that replaces integer variables in the Viswanathan-Grossmann (1990) formulation with tray bypasses. This is essentially a relaxation of the integer problem, except that thermodynamics drives the continuous bypass fractions/efficiency to binary values, because mixing is inefficient. Care is taken in the model formulation to avoid degeneracies when distillation trays are bypassed. The efficacy of this approach and the tendency towards integer solutions will be demonstrated using a cryogenic air separation unit synthesis case study.

References:

Kamath, R. S., Biegler, L. T., & Grossmann, I. E. (2010). An equation-oriented approach for handling thermodynamics based on cubic equation of state in process optimization. Computers & Chemical Engineering, 34(12), 2085–2096.

Gundersen, T. (1982). Numerical aspects of the implementation of cubic equations of state in flash calculation routines. Computers & Chemical Engineering, 6(3), 245–255.

Viswanathan, J., & Grossmann, I. E. (1990). A Combined Penalty Function and Outer-Approximation Method for MINLP Optimization. Computers & Chemical Engineering, 14(7), 769–782.

379762 Integrated Environmental, health and safety index (IEHS): An index's evolution into a flowsheet design tool

Abstract:

A great wealth of knowledge about environmental impacts, employee health and process safety is available from databases, simple formulas, or first principle knowledge. A designer charged with developing the process flowsheet for a new plant often cannot access this information due to time restraints, market pressures or difficult interpretation. This is heightened in operations that are small, produce specialty chemicals or use novel processes. Ruling the design decisions during this stage are quick economic analyses and rules of thumb. This stage is often overlooked by many who want to focus on the initial reaction selection or the general input output structure of the plant. In terms of design decisions, EHS mistakes are often made in this stage or can only start being examined here due to lack of specificity.

The proposed index will adapt the EHS index developed by Koller et. al. [1] specifically for this process flowsheet development phase. It will add information to design decisions that is easy to compute and interpret, while replacing rules of thumb for safety. From start to finish, the procedure will be built around the lack of information from new chemicals and novel units. This will work for designs ranging from an incomplete flowsheet to a rigorous technoeconomic study. The index will provide information on the effects of design decisions on the entire plant, as well as their environment, health and safety consequences.

The main addition to the index will be a set of factors quantifying different interaction effects. These factors can be seen as a different perspective on the same design. The chemical mixing interactions will be assessed using first principle equations and predicative variables. The perspective of the maintenance engineer changes the potential failure rate of the unit through unit-process interactions. A simplified and automated HAZOP team will look at the unit-unit interactions to understand the operability of the design. These new variables will be available for optimization and gives the designer a sense of how well the pieces "fit" together. The inclusion of EHS aspects in design choices will better represent the decision to management as well as the designer. This is in addition to preventing unsafe plants from advancing past this phase by working with the designer to pinpoint the problem in a transparent manner.

References

[1] Koller, G., Fischer, U., and Hungerbühler, K. (2000) Assessing Safety, Health, and Environmental Impact Early during Process Development. *Ind. Eng. Chem. Res. 39*, 960–972.

Innovation and Sustainability in Process design by including Ecosystems as Unit Operations

Varsha Gopalakrishnan, Bhavik R. Bakshi, Guy Ziv

Despite the critical importance of ecological systems, engineering, like most other disciplines, takes nature for granted. For over a century, processes have been designed without considering the role of nature in supporting them by providing a large diversity of essential goods and services. These benefits from nature, called ecosystem services, include goods such as fossil resources, minerals, water, air and biodiesel; and services such as cycling of materials, photosynthesis, regulation of air and water quality, and soil formation. Such ignorance about nature results in atleast two major shortcomings: 1) Processes designed by engineers may demand more from ecosystems than can be provided without ecological degradation, and 2) Engineering misses opportunities of working with nature to address many industrial needs in an economically and environmentally superior manner. The first reason is a root cause for the inherent unsustainability of many engineering activities, while the second reason reflects the engineering attitude of dominating nature. Recent methods for Sustainable Engineering and life cycle considerations in process design also fail to address these shortcomings. [1]

This presentation will describe progress toward addressing the shortcomings listed above by developing methods for including ecosystems in process design in a manner analogous to unit operations. The ecological components included in this work are a forest ecosystem and treatment wetlands. Services which the trees provide to the manufacturing site include regulating air quality by taking up pollutants, reducing energy consumption in buildings, and reducing water runoff. They could also be a source of fuel for the process, and provide other benefits to the society such as climate regulation, recharging aquifers, and supporting biodiversity. Wetlands can treat a large variety of pollutants, including aliphatic and aromatic compounds, various pesticides, heavy metals and excess nutrients. In addition, they also support biodiversity, help maintain the water table and provide biomass.

In this work, these ecosystems are included in the flowsheet along with conventional unit operations to design a process that develops synergies between technological and ecological systems. The forest ecosystem is modeled using the i-Tree software and the wetland ecosystem by using appropriate non-mechanistic first-order rate equations. Thus, the technological and ecological systems are designed in a mutually beneficial manner. For example, if the wetland cannot treat certain molecules or if some emissions exceed the capacity of the forest to capture them, the technological system may be modified to overcome these constraints. Similarly, the ecological systems are tailored to the types of emissions from the technological equipment and goods and services such as wood from the trees and water from wetlands can be recycled for process use. Designing such synergies results in innovative designs which cannot be found by the techno-centric approach of conventional engineering.

Sustainability metrics for this integrated system are quantified by the demand and supply of ecosystem services. The demand for such services is created by the emissions and resource use by processes, including wastewater emissions. The supply of ecosystem services is estimated by the capacity of an ecosystem process to capture completely or support and emissions and nutrient that are present in the environmental on a local scale. For designing a sustainable system with effective utilization of natural capital, the demand for ecosystem services should not exceed their supply. The overshoot in demand for multiple services at a particular scale indicates that the process currently does not function within the regenerative capacity of the local ecosystems. To map the demand for an ecosystem service with its supply, details about spatial characterization of functions and its effect on different scales become relevant. Such an approach encourages not just technological improvements, but also restoration and protection of ecosystems.

This approach for achieving Techno-Ecological Synergy is applied to a case study of a soybean – biodiesel production process for the hexane-oil extraction and methanol transesterification phases. The process is designed with different utility generating systems, including a traditional coal combined heat and power, and a biomass co-firing unit where the biomass fuel is derived by harvesting 15% of the trees, resulting in a positive feedback from the forest ecosystem.

Cost analysis for this integrated design indicates that utilization of natural systems and green infrastructure for reducing environmental impacts instead of conventional end-of-pipe control technologies are economically superior and it can provide a multitude of benefits when compared to traditional gray infrastructure. Thus, by designing integrated networks of technological and ecological systems, insight into the trade-offs between the two systems can be obtained, which may encourage industries to reduce the overshoot in demand due to their activities. The ultimate outcome of such efforts is expected to be "engineering within ecological constraints" [2] – a required step toward sustainability. Challenges in the practical implementation of such ideas will also be discussed.

[1] Urban, Robert A., et al. "Towards sustainability of engineered processes: Designing self-reliant networks of technological–ecological systems." Computers & chemical engineering 34.9 (2010): 1413-1420.

[2] National Academy of Engineering. *Engineering Within Ecological Constraints*. Washington, DC: The National Academies Press, 1996

Bridging the gap between LCA and PSE via a framework for multi-scale sustainable process design

Rebecca J. Hanes and Bhavik R. Bakshi

William G. Lowrie Department of Chemical and Biomolecular Engineering
The Ohio State University

November 18, 2014

The Ohio State University

14AIChE Annual Meeting, Atlanta, GA

OSU Process Systems Engineering Group

Process systems engineering (PSE)

Process systems engineering relies on a **broad analysis boundary**.

- Consider systems at regional, national spatial scales
- Account for external activities and other relevant factors
- Mitigate risks of narrow analysis boundary, including
  - Issues shifting outside boundary instead of being solved
  - Decisions being infeasible or sub-optimal due to incomplete information

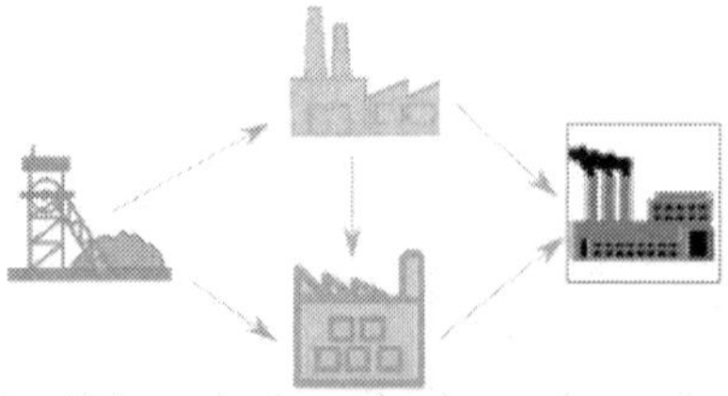

Traditional chemical engineering

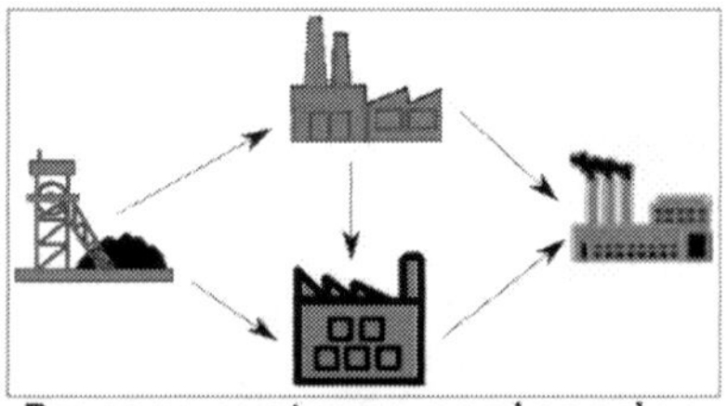

Process systems engineering

Grossmann, I.E. and A.W. Westerberg, *AIChE J.*, 46 (9), 2000.

PSE and sustainability

Sustainability concerns are accounted for at the **life cycle** scale.

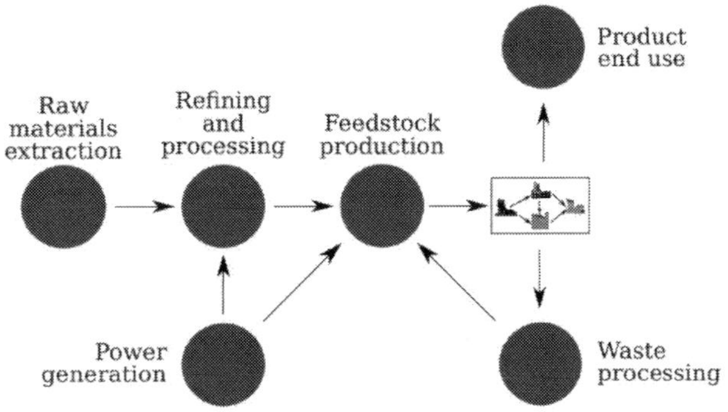

- Compromise between *process* (blue) economics and *life cycle* (red) environmental impacts
- Also possible to account for social criteria, such as jobs created

Kniel, G.E., K. Delmarco and J.G. Petrie, *Env. Prog.*, 15 (4), 1996.
Yue, D., M.R. Slivinsky, J. Sumpter and F. You, *Ind. Eng. Chem. Res.*, 53 (10), 2014.

PSE and sustainability

Why the expanded analysis boundary?

- Considering a single process does not guarantee a net reduction in environmental impacts
- Impact generation is shifted outside the boundary, is no longer captured

PSE and sustainability

Why the expanded analysis boundary?

- Considering a single process does not guarantee a net reduction in environmental impacts
- Impact generation is shifted outside the boundary, is no longer captured
- **Examples**
  - Purchasing ready-to-use materials instead of raw materials
  - Sending wastewater to an off-site treatment center

Sustainable process design: Current approach

Sustainable process design problem as a multi-objective optimization:

$$\text{maximize} \quad Z_\$; \qquad \text{minimize} \quad Z_\text{e}$$
$$\text{subject to} \quad f(x) \geq 0; \qquad Z_\$ = Pg(x); \qquad Z_\text{e} = Bh(x)$$

where P is price data on *process* inputs and outputs and B is data on *life cycle* environmental impacts

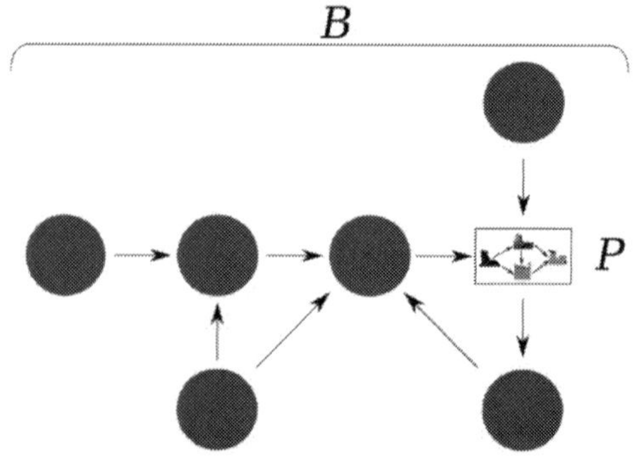

Cano-Ruiz, J.A. and G.J. McRae, *Annu. Rev. Energy Environ.*, 23 (1), 1998.

Sustainable process design: Current approach

B is quantified with **process analysis** life cycle assessment.

- Consider only significant life cycle processes (dark red)
- Boundary is usually arbitrary
- 20 - 50% of life cycle is neglected (light red)

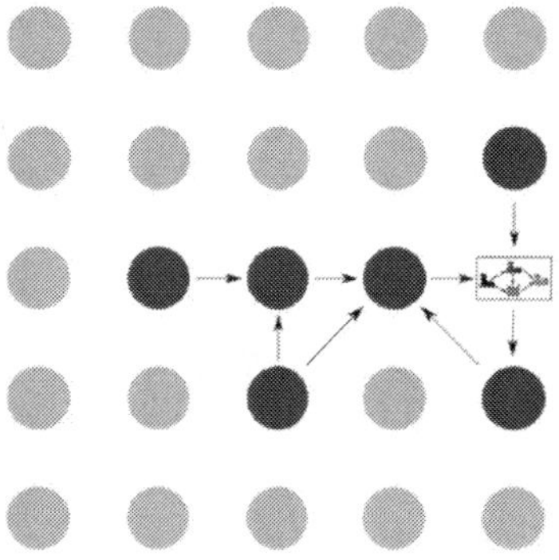

Lenzen, M., *J. Ind. Ecol.*, 4 (4), 2000.

Sustainable process design: Current approach

B is quantified with **process analysis** life cycle assessment.

- Consider only significant life cycle processes (dark red)
- Boundary is usually arbitrary
- 20 - 50% of life cycle is neglected (light red)

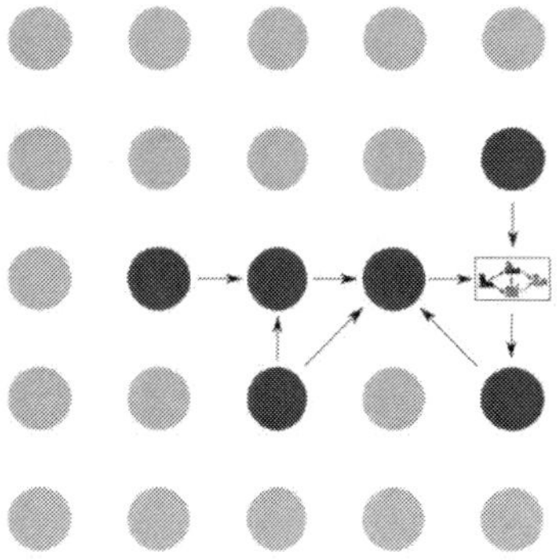

⇒ Impact generation is still shifted outside the system

⇒ Decisions are still based on incomplete information, may still be sub-optimal or infeasible

⇒ *Computationally intractable* to model the full life cycle using process analysis

Lenzen, M., *J. Ind. Ecol.*, 4 (4), 2000.

Life cycle assessment: State-of-the-art

Hybrid life cycle assessment expands the boundary with a coarse, large scale economic model

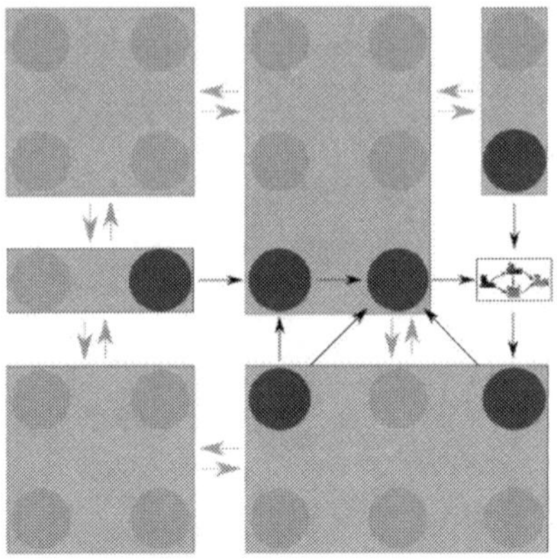

- Accounts for indirect contributions to the life cycle
- Captures flows lacking life cycle scale models as inputs from economy
- Lacks detailed process scale models
- Optimization methods are not applied

Not currently utilized in sustainable process design applications

Joshi, S. *J. Ind. Ecol.*, 3 (2-3), 1999.

Sustainable process design: Proposed approach

Proposed **multi-scale modeling framework** integrates:

- An *economic input-output* (EIO) model
- A *value chain* model analogous to a process analysis life cycle
- Theoretical *engineering process* models of the system of interest

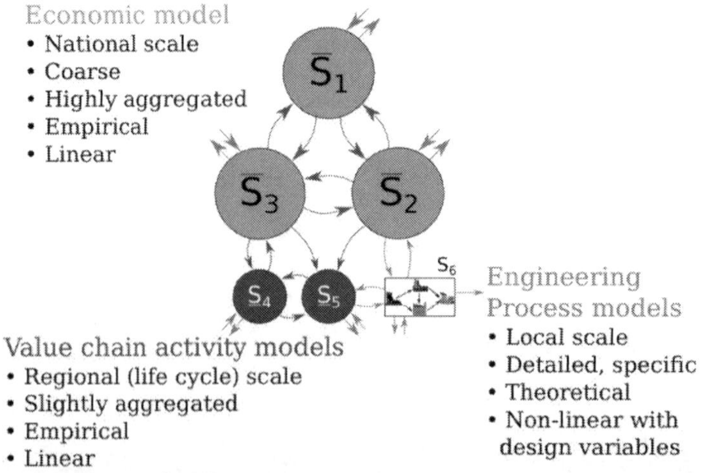

Sustainable process design: Proposed approach

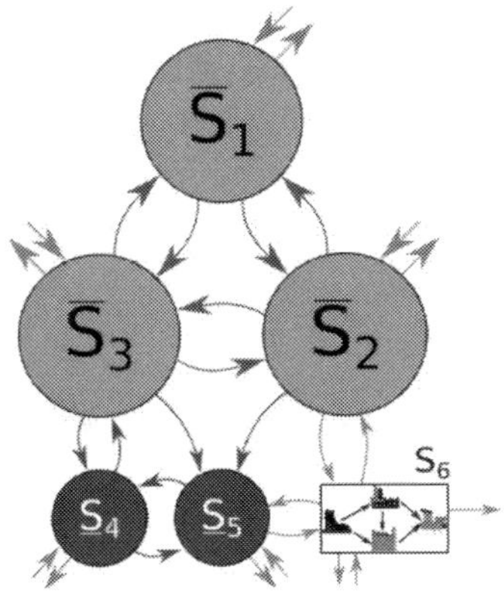

- Coarser models at large scales allow for a national or global analysis boundary
- Decisions can now be made based on the full life cycle rather than a portion
- Reduced or eliminated risk of infeasible, sub-optimal decisions
- Economic input-output model captures infinite order contributions to the life cycle
- Inputs to process system that lack life cycle data are modeled as inputs from the economy

Input-output representation

Represent process n as an **input-output (IO) vector**:

$$\begin{bmatrix} -y_{1n} \\ \vdots \\ -y_{mn} \\ +y_{m+1,n} \\ \vdots \\ +y_{Mn} \end{bmatrix} \quad \text{Technological Interactions}$$

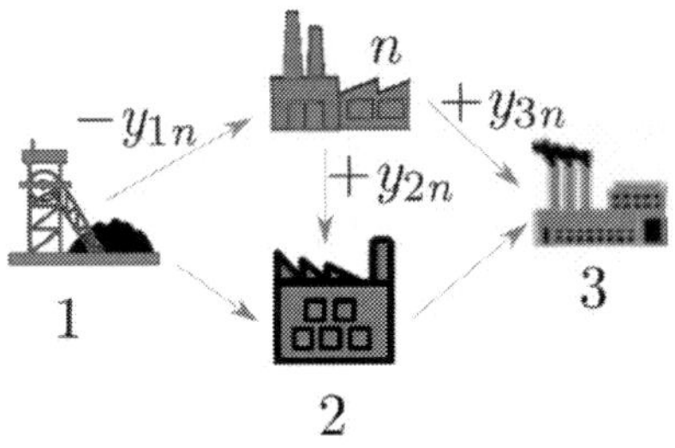

$$\begin{bmatrix} -b_{1n} \\ \vdots \\ -b_{rn} \\ +b_{r+1,n} \\ \vdots \\ +b_{Rn} \end{bmatrix} \quad \text{Environmental Interactions}$$

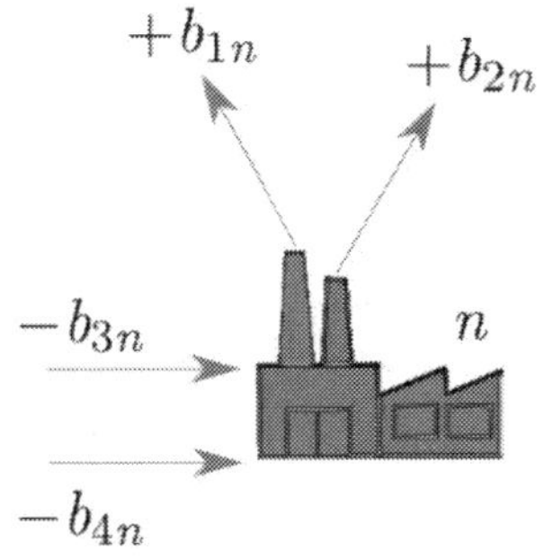

Input-output representation

Incorporate engineering process models into process n IO vector:

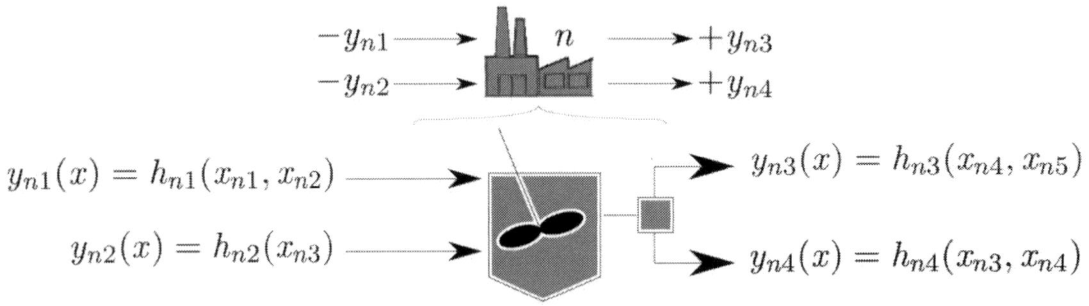

– Write IO vector elements as *functions* of unit operation design variables $\mathbf{x}_n = \{x_{n1}, x_{n2}, x_{n3}, x_{n4}, x_{n5}\}$

Input-output representation

Incorporate engineering process models into process n IO vector:

$$\begin{bmatrix} -y_{1n}(\mathbf{x}_n) \\ \vdots \\ -y_{mn}(\mathbf{x}_n) \\ +y_{m+1,n}(\mathbf{x}_n) \\ \vdots \\ +y_{Mn}(\mathbf{x}_n) \end{bmatrix}$$

$$\begin{bmatrix} -b_{1n}(\mathbf{x}_n) \\ \vdots \\ -b_{rn}(\mathbf{x}_n) \\ +b_{r+1,n}(\mathbf{x}_n) \\ \vdots \\ +b_{Rn}(\mathbf{x}_n) \end{bmatrix}$$

– Unit operation models $\{h_n(\mathbf{x}_n)\}$ impose constraints on design variables $\mathbf{x}_n$:

$$h_{n1}(x_{n1}, x_{n2}) \geq 0$$
$$h_{n2}(x_{n3}) \geq 0$$
$$h_{n3}(x_{n4}, x_{n5}) \geq 0$$
$$h_{n4}(x_{n3}, x_{n4}) \geq 0$$

– Each process n has a unique set of $\{h_n(\mathbf{x}_n) \geq 0\}$ and unique $\mathbf{x}_n$

Input-output representation

Represent process n as a **multi-scale IO vector**:

$$\left[\begin{array}{c} -\mathbf{y}^{E}_{u\,n}(\mathbf{x}_n) \\ \hline -\mathbf{y}^{V}_{u\,n}(\mathbf{x}_n) \\ \hline -y_{1n}(\mathbf{x}_n) \\ \vdots \\ +y_{Mn}(\mathbf{x}_n) \end{array}\right]$$

$$\left[\mathbf{b}_n(\mathbf{x}_n)\right]$$

$$\{h_n(\mathbf{x}_n) \geq 0\}$$

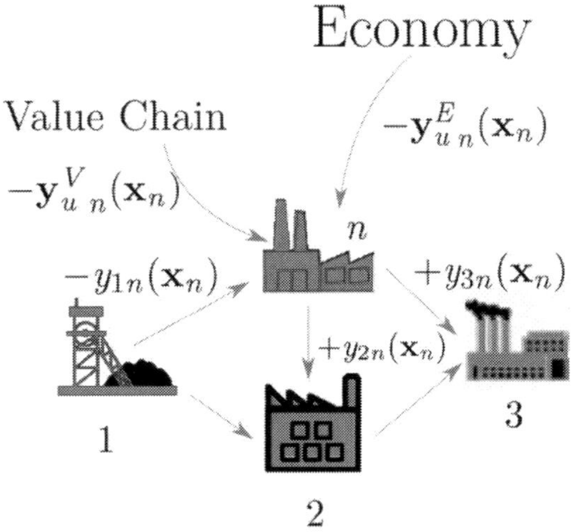

Vector elements are all derived from unit operation models $\{h_n(\mathbf{x}_n)\}$.

Input-output representation

Represent **value chain activity** l as a multi-scale IO vector:

$$\left[\begin{array}{c} -\underline{\mathbf{y}}_{ul} \\ \hline \underline{\mathbf{y}}_{l} \\ \hline -\mathbf{y}^{V}_{d\,l}(\mathbf{x}_n) \end{array}\right]$$

$$\left[\underline{\mathbf{b}}_l\right]$$

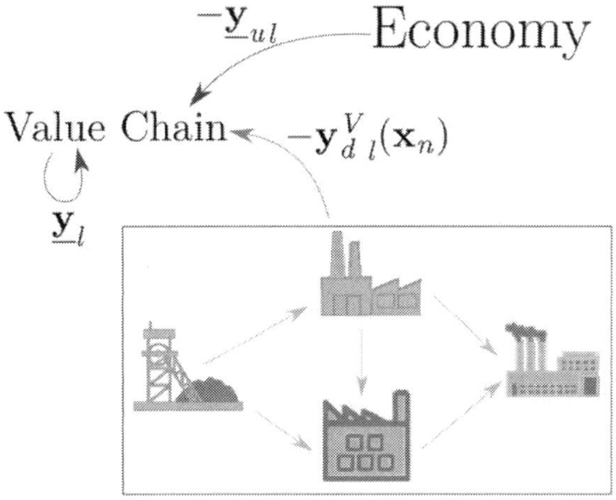

Input-output representation

Represent **industrial sector** j as a multi-scale IO vector:

$$\left[\begin{array}{c} \bar{\mathbf{i}}_j - \bar{\mathbf{a}}_j \\ \hline -\underline{\mathbf{y}}_{dj} \\ \hline -\mathbf{y}^{E}_{d\,j}(x) \end{array}\right]$$

$$\left[\overline{\mathbf{b}}_j\right]$$

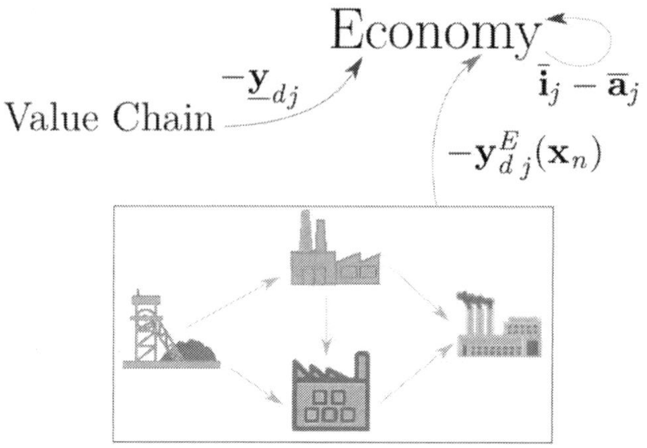

Multi-scale modeling framework

Multi-scale transactions matrix:

$$\begin{array}{ccc} \text{Industrial sectors} & \text{Value chain activities} & \text{Processes} \end{array}$$

$$\left[\begin{array}{c|c|c} \cdots \quad \bar{\mathbf{i}}_j - \bar{\mathbf{a}}_j \quad \cdots & \cdots \quad -\underline{\mathbf{y}}_{ul} \quad \cdots & \cdots \quad -\mathbf{y}^{E}_{u\,n}(\mathbf{x}_n) \quad \cdots \\ \hline \cdots \quad -\underline{\mathbf{y}}_{dj} \quad \cdots & \cdots \quad \underline{\mathbf{y}}_{l} \quad \cdots & \cdots \quad -\mathbf{y}^{V}_{u\,n}(\mathbf{x}_n) \quad \cdots \\ \hline \cdots \quad -\mathbf{y}^{E}_{d\,j}(\mathbf{x}_n) \quad \cdots & \cdots \quad -\mathbf{y}^{V}_{d\,l}(\mathbf{x}_n) \quad \cdots & \cdots \quad \mathbf{y}_{n}(\mathbf{x}_n) \quad \cdots \end{array}\right]$$

Multi-scale modeling framework

Multi-scale transactions matrix:

$$\begin{array}{ccc} \text{Industrial sectors} & \text{Value chain activities} & \text{Processes} \end{array}$$

$$\left[\begin{array}{ccc|ccc|ccc} \cdots & \bar{\mathbf{i}}_j - \bar{\mathbf{a}}_j & \cdots & \cdots & -\underline{\mathbf{y}}_{ul} & \cdots & \cdots & -\mathbf{y}^E_{u\,n}(\mathbf{x}_n) & \cdots \\ \hline \cdots & -\underline{\mathbf{y}}_{dj} & \cdots & \cdots & \underline{\mathbf{y}}_l & \cdots & \cdots & -\mathbf{y}^V_{u\,n}(\mathbf{x}_n) & \cdots \\ \hline \cdots & -\mathbf{y}^E_{d\,j}(\mathbf{x}_n) & \cdots & \cdots & -\mathbf{y}^V_{d\,l}(\mathbf{x}_n) & \cdots & \cdots & \mathbf{y}_n(\mathbf{x}_n) & \cdots \end{array}\right]$$

Multi-scale environmental interactions matrix:

$$\left[\begin{array}{ccc|ccc|ccc} \bar{\mathbf{b}}_1 & \cdots & \bar{\mathbf{b}}_J & \underline{\mathbf{b}}_1 & \cdots & \underline{\mathbf{b}}_L & \mathbf{b}_1(\mathbf{x}_1) & \cdots & \mathbf{b}_N(\mathbf{x}_N) \end{array}\right]$$

Multi-scale modeling framework

Multi-scale transactions matrix:

$$\begin{array}{ccc} \text{Industrial sectors} & \text{Value chain activities} & \text{Processes} \end{array}$$

$$\left[\begin{array}{ccc|ccc|ccc} \cdots & \bar{\mathbf{i}}_j - \bar{\mathbf{a}}_j & \cdots & \cdots & -\underline{\mathbf{y}}_{ul} & \cdots & \cdots & -\mathbf{y}^E_{u\,n}(\mathbf{x}_n) & \cdots \\ \hline \cdots & -\underline{\mathbf{y}}_{dj} & \cdots & \cdots & \underline{\mathbf{y}}_l & \cdots & \cdots & -\mathbf{y}^V_{u\,n}(\mathbf{x}_n) & \cdots \\ \hline \cdots & -\mathbf{y}^E_{d\,j}(\mathbf{x}_n) & \cdots & \cdots & -\mathbf{y}^V_{d\,l}(\mathbf{x}_n) & \cdots & \cdots & \mathbf{y}_n(\mathbf{x}_n) & \cdots \end{array}\right]$$

Multi-scale environmental interactions matrix:

$$\left[\begin{array}{ccc|ccc|ccc} \bar{\mathbf{b}}_1 & \cdots & \bar{\mathbf{b}}_J & \underline{\mathbf{b}}_1 & \cdots & \underline{\mathbf{b}}_L & \mathbf{b}_1(\mathbf{x}_1) & \cdots & \mathbf{b}_N(\mathbf{x}_N) \end{array}\right]$$

Unit operation models:

$$\left\{\begin{array}{c} \{h_1(\mathbf{x}_1) \geq 0\} \\ \vdots \\ \{h_N(\mathbf{x}_N) \geq 0\} \end{array}\right\}$$

Multi-scale modeling framework

Smaller scale models are captured implicitly in larger scale models.

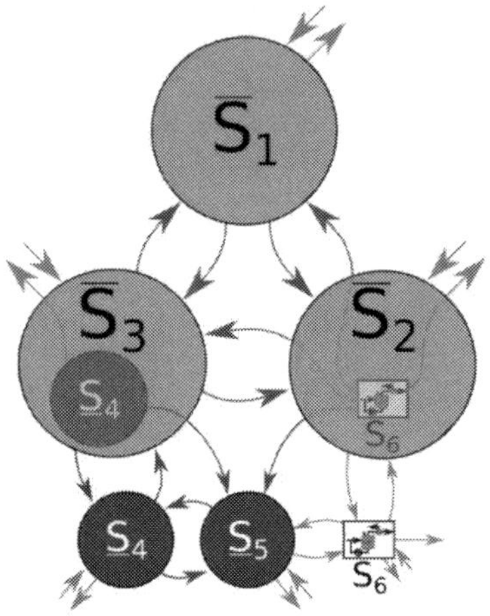

Multi-scale modeling framework

Smaller scale models are captured implicitly in larger scale models.

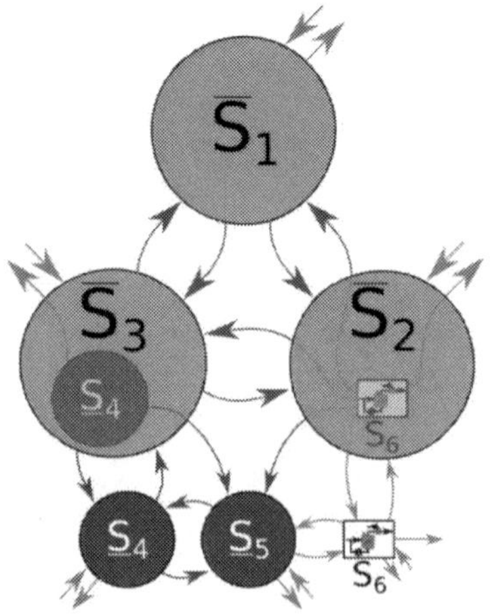

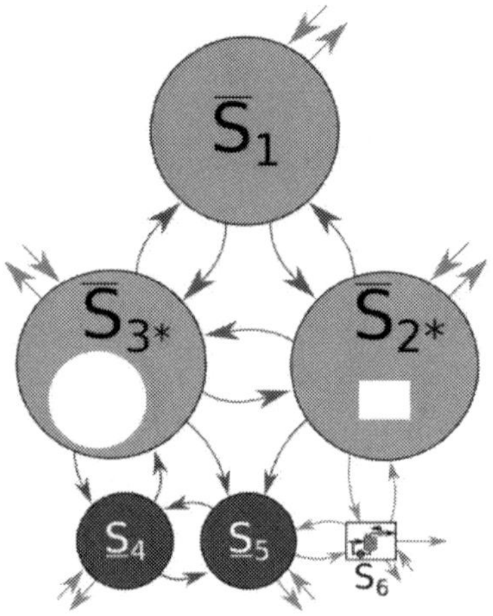

Disaggregation of the economy and value chain scale avoids double-counting interactions and increases modeling accuracy.

Sustainable process design: Optimization

Economic objective

$$\text{maximize} \quad Z_\$ = \sum_{n=1}^{N} Z_{1n}(\mathbf{x}_n)$$

Environmental objective

$$\text{minimize} \quad \underline{\overline{Z}}_s = \sum_{j=1}^{J} \overline{Z}_{2j}(\{\mathbf{x}\}) + \sum_{l=1}^{L} \underline{Z}_{2l}(\{\mathbf{x}\}) + \sum_{n=1}^{N} Z_{2n}(\mathbf{x}_n)$$

subject to

$\underline{\overline{\mathbf{A}}}^*(\{\mathbf{x}\})\underline{\overline{\mathbf{s}}} \geq \underline{\overline{\mathbf{f}}}$		System-wide product balance
$\{h_n(\mathbf{x}_n) \geq \mathbf{0}\}$,	$\forall n \in N$	Unit operation models
$Z_{1n} = \mathbf{p}_n(\mathbf{x}_n) \cdot \mathbf{s}$,	$\forall n \in N$	Process economics
$\overline{Z}_{2j} = \overline{s}_j \overline{\mathbf{b}}_j(\{\mathbf{x}\})$,	$\forall j \in J$	Economy environmental impacts
$\underline{Z}_{2l} = \underline{s}_l \underline{\mathbf{b}}_l(\{\mathbf{x}\})$,	$\forall l \in L$	Value chain environmental impacts
$Z_{2n} = s_n \mathbf{b}_n(\mathbf{x}_n)$,	$\forall n \in N$	Process environmental impacts

Application: Ethanol production system

Optimize process for **minimum CO_2 emissions** at two scales:

1. Value chain scale: Value chain and process emissions
   - *Current approach* to sustainable process design

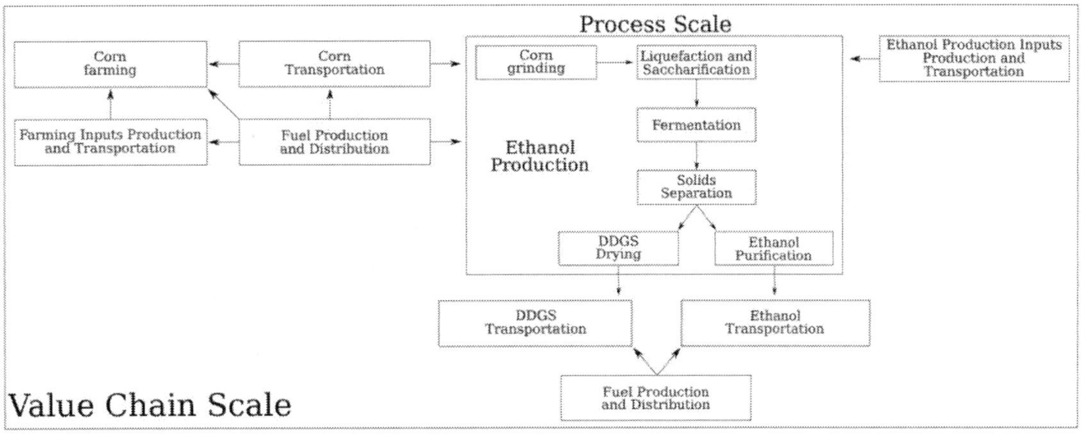

Karuppiah, R., A. Peschel, I.E. Grossmann, M. Martín, W. Martinson and L. Zullo, *AIChE J.*, 54 (6), 2008.

Application: Ethanol production system

Optimize process for **minimum CO_2 emissions** at two scales:

1. Value chain scale: Value chain and process emissions
2. Economy scale: Total system-wide emissions
   - *Proposed multi-scale approach* to sustainable process design

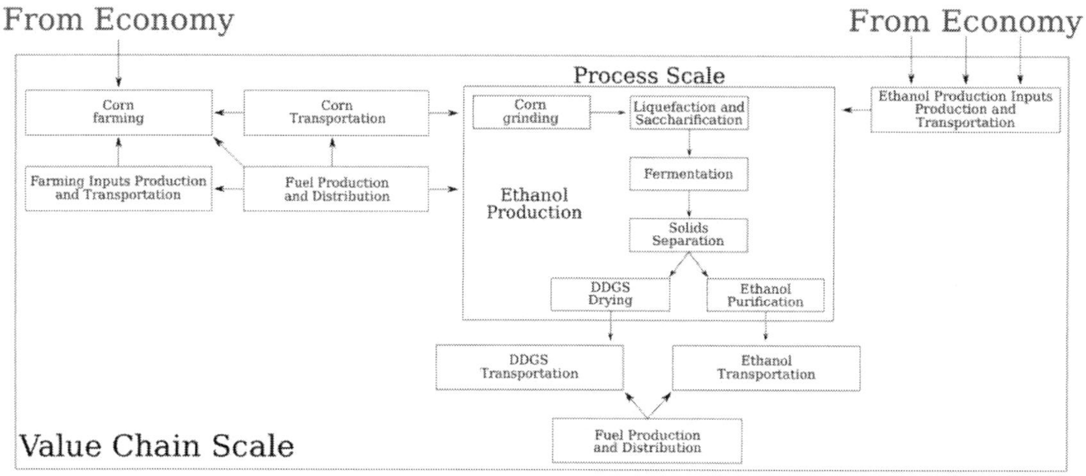

Application: Ethanol production system

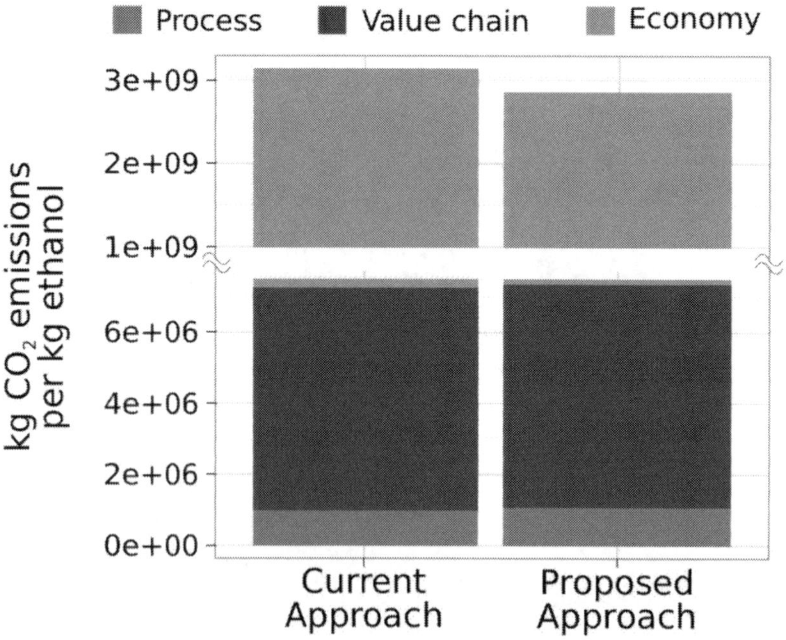

Perez R.E., Jansen P.W., and Martins J.R.R.A. (2012) pyOpt: A Python-Based Object-Oriented Framework for Nonlinear Constrained Optimization, Structures and Multidisciplinary Optimization, 45(1):101-118.

Application: Ethanol production system

- The current and proposed approaches resulted in *different* optimal process designs
- Design found by proposed approach is *environmentally superior* with a *net decrease* in emissions relative to current approach
  - Decrease = **8.94%** relative to current approach design
- Design found by current approach is *sub-optimal* when evaluated using the multi-scale system boundary

Conclusions

Multi-scale modeling framework expands analysis boundary of sustainable process design by integrating **fundamental** process models with **empirical** life cycle and economy models.

- Accounts for previously neglected portions of the life cycle
- Can lead to environmentally superior designs
- Potentially results in innovative win-win designs
- Advantages of proposed approach are expected to be more significant with larger scale design problems
  - Example: Bio-fuel and bio-product supply chain design
- Framework can potentially be used for comprehensive *uncertainty analysis*

Future Work

1. Additional case study: Life cycle design
   - Design a life cycle by selecting an optimal combination of production or conversion pathways
   - Earlier talk: "Embedding Fundamental Models of Technological Alternatives within a Hybrid Inventory for Multi-Scale Life Cycle Design"

2. Extend modeling framework to account for *ecological* system components
   - Design for synergy between industrial activities and ecosystems

Acknowledgements

Partial funding provided by USDA 2012-38202-19288.

Many thanks to Dr. Mariano Martín for furnishing the original GAMS model of the ethanol process.

Questions?

Computer-Aided framework for Sustainable Process Design - targeting conceptual and detailed engineering phases

Carina L. Gargalo[1], John P. Duque[2], Gürkan Sin*[1]

[1] Computer Aided Process Engineering Center, Department of Chemical and Biochemical Engineering, Technical University of Denmark, DK-2800 Lyngby, Denmark
[2] Department of Chemical Engineering, Utrecht University, Budapestlaan 6, 3524CD Utrecht, The Netherlands
*gsi@kt.dtu.dk

Designing sustainable chemical and biochemical processes requires systems approach by which sustainability is integrated systematically into design. Sustainability is commonly defined as finding solutions that avoids compromising the needs of future generations while meeting the needs of the present generation. The implications for process design is that design decisions must be expanded to consider simultaneously not only the economic and technical but also environmental, safety and social aspects of process and products (Azapagic 2010, Shokrian & High, 2014).
Along the years, several studies have developed methods to create new designs or improve an already existing design by retrofit techniques with the objective of decreasing the operating cost. These studies included but not restricted to Rapoport et al.(1994), Uerdingen et al. (2003) and Carvalho et al. (2008), among others. Carvalho et al. (2008) extended the Uerdingen et al. (2003) methodology by adding the IChemE (Azapagic et al., 2002) metrics calculation and identifying the process critical points regarding the consumption of resources. Several other metrics and methodologies have also been suggested over the years to define a systematic and holistic way to evaluate the overall chemical/biochemical process design performance, such as, Azapagic et al. (2002) and Sikdar (2009), among others.
Despite the fact that several aspects of sustainability are well described, there is still a need to accurately analyze and incorporate sustainability into process synthesis of new processes. Consequently, complex decision-making scenarios may appear from a substantial number of design objectives and constraints that must be concurrently and efficiently satisfied. Additionally, data and models used as input for sustainability analysis may be incomplete or inaccurate. The current methodologies mostly target a specific aspect of sustainability for a given process design, separately focusing only on the optimization of the economic or the environmental performance, hence there is a need for a comprehensive and generally applicable methodology for chemical engineers to use in design and retrofitting stages of chemical processes in practice (Shokrian & High, 2014). To address these challenges and to facilitate the sustainability analysis in industrial practice, the present work aims to propose a systematic computer-aided framework for sustainable process design that accommodates and quantitatively integrates the sustainability aspects into the early stage of process design, through the integration of different methods, tools, algorithms and databases. It intends to guide the user through the mathematical complexity of the problems and, by enabling the user the link between sustainability potential, process debottlenecking, improvements and possible trade-offs though multi-criteria decision analysis, to identify a suitable process subject to design constraints and a set of performance criteria. The framework also integrates uncertainty and sensitivity analysis to tackle any gap in process knowledge and incomplete/inaccurate data and models on the sustainable process design.
The framework workflow is divided into four parts. Part 1 is where the user describes the design problem, identifying the needs and constraints, along with the objective function. Part 2 comprises the generation of a superstructure following the method proposed by Quaglia et al., (2013). It has as particular objective to create a comprehensive database on the case study and to, based on the objective function, pre-select the

base-case(s) that will be analyzed in the next stage using a two-stage mathematical programming (with deterministic and stochastic parts to tackle uncertainties and in accuracies in data and models and thereby generate robust solutions and ranking against uncertainties). In Part 3, the pre-selected base-case(s) will be evaluated, firstly by rigorous process simulation (e.g. Aspen/ProII) followed by a complete sustainability analysis based on the whole process life-cycle (economic, environmental and social sustainability). From Part 3, the user is able to identify the process hot spots, which lead to the definition of the related design targets and debottlenecking with the generation of new design alternatives. Part 4 is characterized by a multi-criteria decision making procedure that will lead the user to the sustainable process design that better responds to the design problem.

To highlight the framework features, a comprehensive study of the possible options to convert glycerol to a platform of value-added chemicals (usually obtained from the petro-chemical industry) is analyzed by superstructure optimization and the optimal options for different scenarios are identified. The biodiesel production from vegetable oil has glycerol as main by-product, where for every 10 kg of biodiesel produced, 1 kg of glycerol is produced as a by-product (Shams et al., 2008) thus, converting glycerol into value-added chemicals, would allow higher flexibility and higher profits in the biodiesel industry, potentially increasing the overall process sustainability. Over the years, it has been defended that the concept of integrated biorefinery is a more sustainable way of producing energy and value-added chemicals than the petro-chemical industry producing the same chemical or range of chemicals. Hence, this contribution will analyze quantitatively and elucidate the factors contributing to sustainable design of integrated biorefinery.

References:

Azapagic, A., Howard, A., Parfitt, A., Tallis, B., Duff, C., & Hadfield, C. (2002). Sustainable Development Progress Metrics. *IChemE*. Retrieved from http://nbis.org/nbisresources/metrics/triple_bottom_line_indicators_process_industries.pdf

Carvalho, A., Gani, R., & Matos, H. (2008). Design of sustainable chemical processes: Systematic retrofit analysis generation and evaluation of alternatives. *Process Safety and Environmental Protection*, *86*(5), 328–346.

Quaglia, A., Sarup, B., Sin, G., & Gani, R. (2013). A systematic framework for enterprise-wide optimization: Synthesis and design of processing networks under uncertainty. *Computers & Chemical Engineering*, *59*, 47–62.

Rapoport, H., & R. Lavie. (1994). Retrofit Design of New Units into an Existing Plant. *Computers & Chemical Engineering*, *18*, 743–753.

Shams Yazdani, S., & Gonzalez, R. (2008, November). Engineering Escherichia coli for the efficient conversion of glycerol to ethanol and co-products. *Metabolic Engineering*. doi:10.1016/j.ymben.2008.08.005

Shokrian, M., & High, K. A. (2014). An efficient multi criteria process optimization framework: Sustainable improvement of the Dimethyl Ether Process. *Computers and Chemical Engineering*, *60*, 213–230.

Sikdar, S. K. (2009). On aggregating multiple indicators into a single metric for sustainability. *Clean Technologies and Environmental Policy*, *11*(2), 157–161. doi:10.1007/s10098-009-0225-4

Uerdingen, E., Fischer, U., Hungerbühler, K., & Gani, R. (2003). Screening for Profitable Retrofit Options of Chemical Processes□: A New Method. *AIChE Journal*, *49*(9), 2400 – 2418. doi:10.1002/aic.690490915

378797 Refinery Optimization Under Uncertainties

Abstract:

In the refining industry, crude oil procurement is the largest expenditure and these purchases have a large impact on refinery profitability. Moreover, once a specific crude oil is selected and purchased, the refining process must be operated to extract the most value possible from the crude based on the characteristics of the crude and the constraints of the refinery, while ensuring that all refined products meet quality specifications. However, a major challenge in optimizing this system is that the crude qualities are not known accurately at the time of purchase and the refined product market prices may change markedly between the date of crude purchase and the date when that crude is refined. Hence, the goal of this research is to provide a systematic approach to maximize refining profitability in the presence of uncertainties.

Towards this end, a simplified refinery scheme [1] including crude distillation unit, reformer, cracker, desulfurization and isomerization is studied and modeled as a scenario-based two-stage stochastic programming formulation. In stage I, the best crude oil combination is selected among several candidates and their purchase amounts are determined to maximize the expected gross margin across all scenarios. In stage II, the uncertainties are realized and the optimal operations for the plant are determined according to the real quality of crude, associated with each scenario, such that both the market demand and quality specifications are satisfied. The proposed scheme takes into account uncertainties in the crude quality, nonlinearity from the pooling and unit operation modes, and logic decisions into account, resulting in a large-scale mixed-integer non-linear programming (MINLP) problem. In order to obtain the global optimal solution for this problem efficiently, the variable discretization, feasibility and optimality-based domain reduction techniques are integrated into the non-convex generalized Benders decomposition (NGBD) methodology [2,3], which takes full advantage of the decomposable structure of the scenario-based stochastic programming problem. The results of the stochastic scheme are compared with a deterministic approach to demonstrate the benefits of expected profit maximization and variance reduction. Even though the complexity of this approach is considerably increased due to the multi-scenario and non-convex formulation, the proposed enhanced NGBD approach is still able to find and verify the global optimal solution within a couple of hours, whereas the state-of-the-art commercial software cannot find the global optimal solution in several days. Moreover, the NGBD method exhibits a linear growth in solution time with respect to the number of scenarios.

Reference

[1] J. P. Favennec, Refinery operation and management, Editions TECHNIP, Paris, 2001.

[2] X. Li and P. I. Barton, Decomposition strategy for the stochastic pooling problem, Journal of Global Optimization, vol. 54, 2012, pp. 765-790.

[3] X. Li, A. Tomasgard and P. I. Barton, Nonconvex generalized Benders decomposition for the stochastic separable mixed-integer nonlinear problem, Journal of Optimization Theory and Applications, vol. 151, 2011, pp. 425-454.

A mathematical model for the electrochemical conversion of gaseous CO_2 to liquid formic acid

Georgopoulou C.[1], Jain S.[2], Agarwal A.[2], Rode E.[2], Dimopoulos G.[1], Sridhar N.[2], Kakalis N.M.[1]
DNV GL Strategic Research and Innovation, Piraeus, Greece[1]
DNV GL Strategic Research and Innovation, Dublin, Ohio, USA[2]

In recent years, CO_2 capture and utilization has gained attention in the industry, due to the increased concern in the environmental impact from anthropogenic greenhouse gas emissions. One possible pathway for CO_2 utilization is the conversion of CO_2 into chemical feedstock. There is a wide range of conversion processes that can use CO_2 as a source material for producing valuable products, like polymers and inorganic minerals; however, their technical feasibility depends on process economics and scale-up capabilities.

In 2011, a novel concept was introduced by DNV for the Electrochemical Reduction of CO_2 to Formate/Formic Acid (ECFORM) [1-5]. A schematic diagram of the ECFORM process is shown in Figure 1. The gaseous CO_2 is inserted into an electrochemical cell through the porous cathode electrode, where it is absorbed in the liquid electrolyte solution and chemically converted to formic acid, at the presence of electric voltage difference between the electrodes. The cell chambers are separated from each other via an ion-exchange membrane, which allows for electrical continuity but prevents the mixing of the catholyte and anolyte streams. The process has been demonstrated on experimental scale by DNV GL Strategic Research and Innovation in Ohio, US.

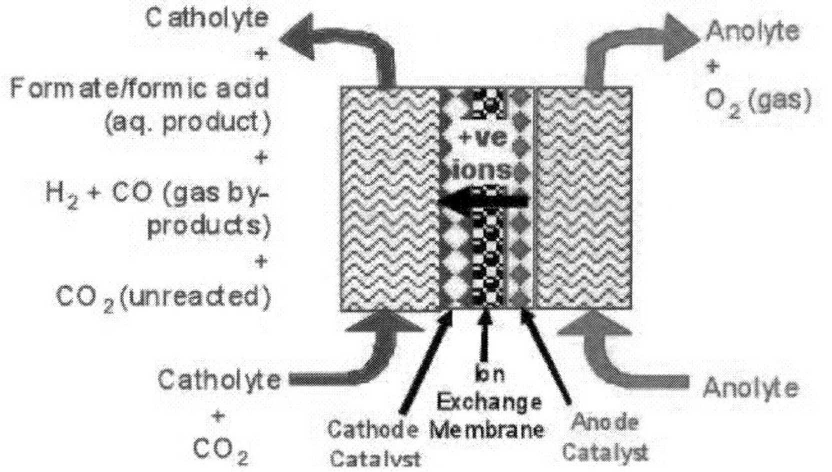

Figure 1. Schematic representation of the ECFORM process.

A cost-effective way to estimate the technology capabilities at different system sizes is via the use of process modelling and simulation techniques. Computer-aided methodologies can successfully address the complexity of processes governed by interdisciplinary phenomena and provide insight on the interactions of the key process parameters, how they are interconnected, and what is their influence on the overall system performance. This approach offers the advantage to design experiments and reduce the laboratory costs, as well as to conduct studies on steady-state and transient operation modes at different system sizes, using the same modelling resources.

In this paper, we present a novel phenomenological mathematical model for the ECFORM process. The model consists of a system of Partial Differential and Algebraic Equations (PDAEs) that describe the gas and liquid mass transport, reaction kinetics and electrochemistry. The model captures the steady-state and dynamic behavior of the gas and liquid flows in the cell. In addition, the PDAEs are discretized in the horizontal and vertical directions of each component, providing 2-dimensional profiles of the physical and electrochemical properties within the cell geometrical boundaries. Electrochemical and liquid bulk chemical reactions are also considered by applying rates for the forward and backward reactions [6]. The cell potential is modelled by considering that the total electric current, caused by the voltage differential and ion migration, remains unchanged.

A generic model structure is employed (Figure 2), by decomposing the ECFORM process into individual components that correspond to the physical modules of the system, including the gas and liquid flows, the anolyte, the catholyte, the membrane and the electrodes. The set of PDAEs is solved simultaneously using advanced numerical solvers [7].

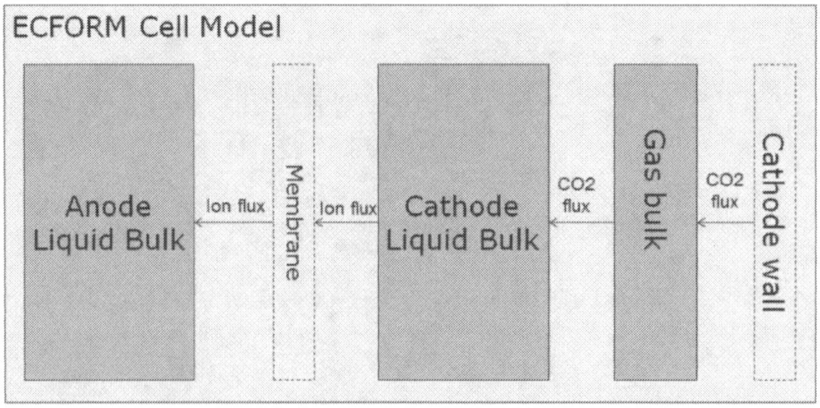

Figure 2. Schematic diagram of the ECFORM process model structure.

Model calibration with respect to experimental results is illustrated and the key parameters that affect system performance are discussed. Challenges with regards to calibrating and fitting the model to the experimental data are analyzed. The model captures the behavior of the ECFORM process and provides insights on system scale-up capabilities. Figure 3 presents model results for the p_H profile in the cell, at normal operating conditions. Figure 4 presents the calculated potential profiles in the anolyte, showing that the highest potential reduction is close to the electrodes. Figure 5, finally presents the concentrations of the key process products, namely formic acid (HCOOH) and potassium formate (HCOOK).

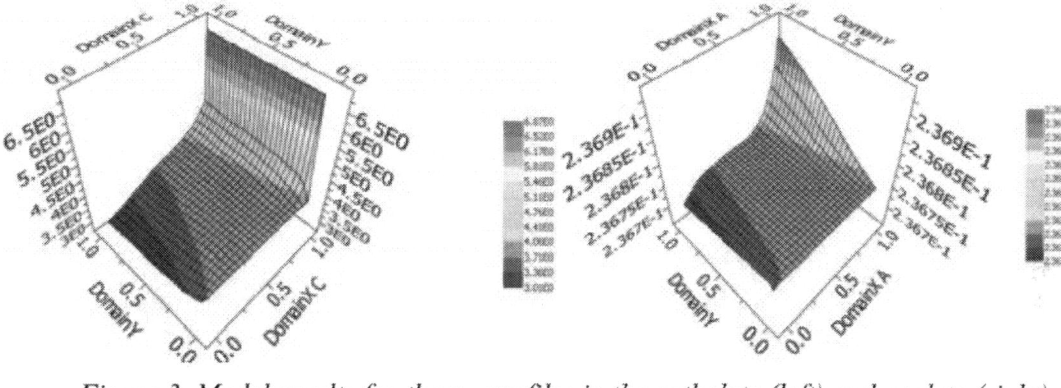

Figure 3. Model results for the p_H profiles in the catholyte (left) and anolyte (right).

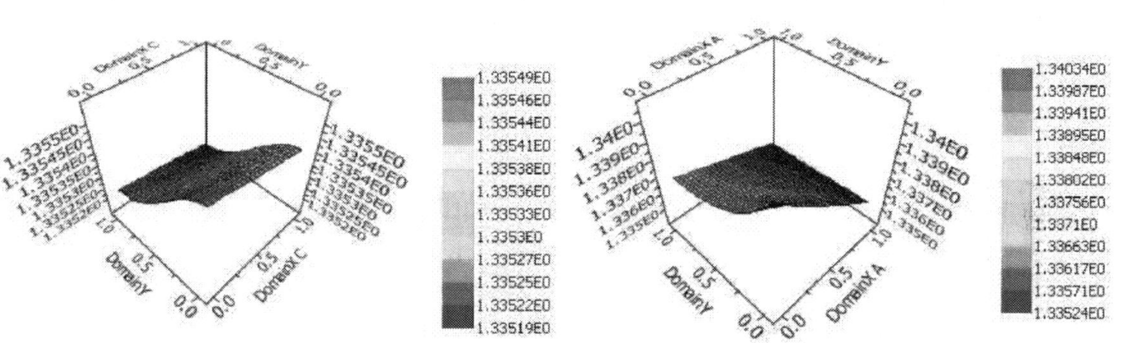

Figure 4. Model results for the potential (in V) profiles in the catholyte (left) and anolyte (right).

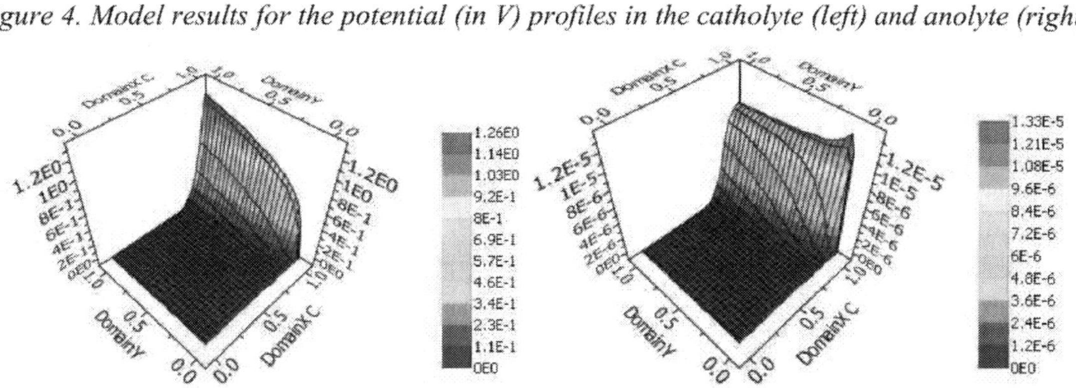

Figure 5. Model results for the HCOOH and HCOOK concentration (in mol/m^3) profiles in the catholyte.

References

[1] Sridhar N., Zhai Y., Agarwal A., Chiachiarelli L., Hill D. *Longterm demonstration of the electrochemical reduction of CO_2 to formic acid.* The CO_2 Challenge Forum, September 27-28, 2010, CPE Lyon, Lyon, France.

[2] Zhai Y., Agarwal A., Chiacchiarelli L., Hill D., Sridhar N. *Evaluation of tin electrocatalyst for conversion of CO_2 to formate salt via long term cathodic half-cell and continuous full cell testing.* Future Directions in CO_2 Conversion Chemistry Workshop, October 21, 2010. Department of Chemistry at Princeton University, Princeton.

[3] Hill D., Zhai Y., Agarwal A., Rode E., Ayello F., Sridhar N. *Energy storage via electrochemical conversion of CO_2 into specialty chemicals.* Proceedings of the ASME 2011 5th International Conference on Energy Sustainability ES2011 August 7-10, 2011, Washington, DC, USA. Paper N^{o} ES2011-540.

[4] Sridhar N., Hill D. *Carbon dioxide utilization; Electrochemical conversion of CO_2 – Opportunities and challenges.* DNV Research and Innovation, Position Paper 07, 2011.

[5] Agarwal A., Zhai Y., Hill D., Sridhar N. *The Electrochemical Reduction of Carbon Dioxide to Formate/Formic Acid: Engineering and Economic Feasibility.* ChemSusChem (2011), 4, 1301 – 1310.

[6] Gupta N., Gattrell M., Macdougall B. *Calculation for the cathode surface concentrations in the electrochemical reduction of CO_2 in $KHCO_3$ solutions.* Journal of Applied Electrochemistry (2006) 36:161–172.

[7] Process systems enterprise. gPROMS, http://www.psenterprise.com/gproms/ , 2014-05-01.

The increase in the demand for energy and current concerns on sustainability are supporting the development of technologies that use solar radiance and wind and biomass as sources of energy and, in the case of biomass, also as a source of chemicals, since it is a carbon source. One of the main features of renewable sources of energy is their variability across regions and with time. In particular, renewable sources constitute a major challenge due to their non-consistent daylong and yearlong availability. Therefore, storage systems or supplementary sources of energy or a combination of them are needed in facilities that are based on renewables in order to operate under a steady state, Weekman (2010) and Yaun and Chen (2012). However, it is also possible to produce chemicals, e.g. methane, via water electrolysis (Levene et al., 2005) and using the CO_2 present in flue gases instead of dumping it to the atmosphere. This means that we are not only capturing the CO_2 to further produce chemicals, but we are also storing energy for a continuous operation of the system downstream.

We present the optimal year round production of methane from water electrolysis and CO_2 comparing the use of solar PV systems and wind turbines on a monthly basis. The process starts by obtaining electricity in a wind farm or a solar field, which is used to produce oxygen and hydrogen from water in a system of electrolyzers. The oxygen is then purified and stored using a series of equipment, while the hydrogen is purified using a deoxygenating reactor and is then used to react with CO_2 to synthesize methane. We operate the plant for a year considering the wind or solar availability. We evaluate several alternatives: the use of wind alone, constant and variable methane production, the optimal renewable source selecting between solar and wind, and we finally screen Europe and the US for the use of both sources of energy. The problem is formulated as a multiperiod MINLP optimization.

For the use of wind alone, the investment of a plant devoted to synthetic natural gas production is 375 M€ for a production cost of synthetic methane of 13.1 €/MMBTU (0.48€/Nm^3), a price that is currently over that of the selling price of natural gas. If the plant operates at a constant methane production rate, the investment and production costs are almost double, 546 M€, but we can obtain credit from the excess of electricity produced, 375 GWh/yr, during some months. Furthermore, 2.9 kg of CO_2 can be reused per kg of CH_4 produced in this process.

However, for a case of study in a particular location, the use of solar energy is cheaper than the use of wind energy in terms of inversion (240 M€) and production costs (0.25 €/Nm^3 vs. 0.48 €/Nm^3), while the monthly operation of the process differs due to the availability of renewable energy.

The use of wind and/or solar energy depends on the region and the area available to install either the wind turbines of the PV panels. We screen the US and the EU for the preferable renewable source of energy as function of the wind velocity and solar incidence. It turns out that solar is the most preferable in most regions. However, the limitation comes from the usage of the ground. While a solar field requires certain preparation in order to install the panels and once the solar field is in place the ground cannot be used for other activities, the wind farm allows further use of the land, such as agricultural activities, since the actual area that the turbines use is significantly lower than that of the panels.

References

Levene, J.I., Mann, M.K. , Margolis, R. , Milbrandt, A., 2005. An Analysis of Hydrogen Production from Renewable Electricity Sources Conference Paper . NREL/CP-560-37612 September 2005

Weekman, V.W., 2010. Gazing into an energy crystal ball. CEP June 2010; 23-27

Yuan, Z., Chen, B., 2012. Process Synthesis for Addressing the Sustainable Energy Systems and Environmental Issues. AIChE J. 58 (11), 3370-3389

374168 Mechanistic Insights into Early Endoderm Differentiation of Human Embryonic Stem Cells Using Systems Level Analysis of Signaling Interactions

Abstract:

Introduction: Human Embryonic Stem Cells (hESCs) are an attractive raw material for regenerative medicine due to their unique properties of self-renewal and lineage specific differentiation. hESCs are directed to mature cell types using various chemical cocktails that replicate signaling programs active during *in vivo* development. These signaling pathways employ large number of mediators that are controlled by several regulatory mechanisms. Current efforts to develop chemical modulators for controlling differentiation are purely experimental and they have successfully identified the key players. However, systems level analysis of the dynamics of this process is still under-developed. We recently demonstrated the strength of quantitative systems level approaches to uncover the role of regulatory mechanisms in the self-renewal state of hESCs [1]. **Here, we present a complete systems level analysis of the dynamics of regulatory interactions in the TGF-β/SMAD and PI3K/AKT pathways, directing the early endoderm differentiation of hESCs, by integrating efficient computational tools and targeted experimental perturbations.**

Materials and Methods:

Experimental analysis: For self-renewal, H1 hESCs were maintained on matrigel-coated plates in mTeSR1. Endoderm differentiation was performed using 100 ng/ml Activin A (to activate TGF-β/SMAD2,3) with or without modulation of PI3K/AKT pathway using PI3K inhibitor, Wortmannin. The phosphorylation dynamics of participating signaling molecules were measured using MagPix Luminex xMAP technology. The initial selection of key molecules was based on the study by Singh *et al.* [2]. Nucleo-cytoplasmic shuttling rates of molecules were measured using

Fluorescence Recovery After Photobleaching (FRAP) analysis. Key transcription factors characterizing endoderm were measured using qRT-PCR. Mathematical and computational tools described in the next section were used to guide generation of data by in-house experiments.

Mathematical analysis: We first employed data-driven modeling tools like Partial Least Squares Regression (PLSR) and Dynamic Bayesian Network Analysis (DBN) to identify key molecules, hypothesize interactions and identify most informative time points. Detailed mechanistic Ordinary Differential Equation (ODE) model for the TGF-β/SMAD2,3 pathway with crosstalk interactions of PI3K/AKT was developed for a systems level analysis. The model was calibrated using Replica Exchange Ensemble Modeling (EM) and sensitive reactions were identified using computationally efficient Global Sensitivity Approach (GSA).

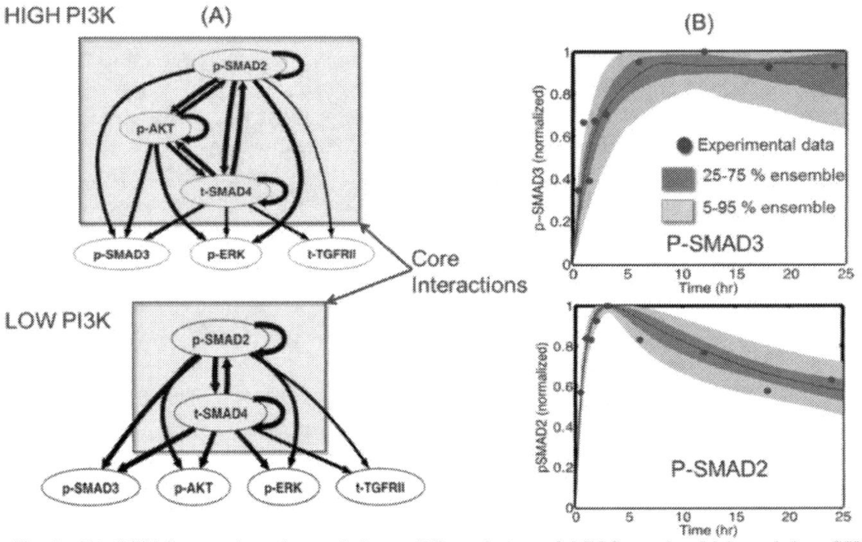

Fig 1. (A) DBN interactions for endoderm differentiation of hESCs under high and low PI3K conditions. (B) Ensembles of dynamic profiles for p-SMAD2 and p-SMAD3 under endoderm differentiation predicted by detailed ODE model compared to experimental observations.

Results and Discussion: PLSR results indicated that endoderm markers of hESCs correlate well with the early but not the late signaling events. Application of DBN on the early signaling dynamics showed that the molecules, p-SMAD2 and SMAD4 form the core interactions (± p-AKT) and, p-SMAD3 and p-ERK were only influenced by the core interactions (Fig. 1A). Interestingly, our experiments showed that the dynamics and fold-change of p-SMAD2 and p-SMAD3 diverge, a novel result seen only in hESCs. The reason for the divergence was investigated using a detailed ODE model with crosstalk interactions with AKT. We generated hESC specific model with SMAD2 and SMAD3 interactions modeled separately. EM on the detailed ODE model captured the parameter ensembles that explained the differentiating hESC system (Fig. 1B). The nucleo-cytoplasmic shuttling rates of SMAD2 and SMAD3 (separately) measured through FRAP analysis was used to constrain the parameter ensembles. From among the various hypotheses, AKT was found to primarily influence the phosphorylation rates of SMAD molecules. GSA results indicated that the sensitive parameters for p-SMAD2 and p-SMAD3 were of similar ranking but of varying strengths resulting in the divergent dynamics. Phosphorylation and de-phosphorylation of SMADs were the most sensitive reactions, while SMAD nucleo-cytoplasmic shuttling rates were associated with modulation of the aforementioned reactions. Further, negative feedback via SMAD7 constrained the propagation of parameter uncertainty as well as promoted a robust signaling response.

Conclusions: Our results show that early signaling dynamics of p-SMAD, p-AKT and p-ERK encode the long-term endoderm differentiation response of hESCs. We provide mechanistic explanations for the divergence of p-SMAD2 and p-SMAD3 dynamics in hESCs. Further, our results present the importance of negative feedback via SMAD7 in controlling the long-term signal propagation and population variability in hESCs. Application of mechanistic information revealed by such an analysis will direct precise perturbations through designed small molecules, hence offering an avenue to remove xenogenic factors in current culture conditions.

References:

[1] Mathew *et al*. Bioinformatics 2014, Article in Press, ID: bioinf-2013-2200.r1 (btu209).

[2] Singh *et al*. Cell Stem Cell 2012, 10(3): 312-26

380636 Multi-Agent Model of Vascularized Bone Tissue Growth within Porous Biomaterials

Abstract:

Engineering bone tissue from stem cell seeded biomaterials has been recognized as a promising technique on the repair of bone tissue defects above the critical size. Rapid and stable vascularization of scaffolds is required to supply nutrients and oxygen that stem cells need to survive as well as to go through osteogenic differentiation. The process of bone formation from stem cells and simultaneous vascularization involve the combination of many complex interacting factors and provide challenges to tissue engineers. Experiments alone are not sufficient to study the effects and interactions of various components involved in tissue formation such as biodegradable scaffolds, signaling molecules and different types of cells. A computational model that can realistically simulate the behavior of such systems over time would be beneficial for many different theoretical and practical purposes. These purposes include rapid testing of alternative hypotheses, better understanding the mechanisms involved, estimation of the important variables of the system and prediction of their effects on system level behaviors. A simulation based on the computational model will provide a tool for rapidly screening alternatives, determining the most promising experimental search space and investigating ways of intervening in such systems for functional tissue formation.

Agent-based modeling techniques can be used to study complex systems with many interacting elements. The central idea in agent-based models (ABM) is to define agents that represent the building blocks of a system and to develop rules that regulate their interactions. The rules originate from the vast qualitative and quantitative knowledge gained through many years of studying the individual components of these biological systems. Agent-based systems are naturally suitable for modeling biological systems as they are comprised of individual constituents like cells that interact with each other to form macro-scale bodies like tissue.

In this study, a multi-layer ABM is presented to simulate tissue growth and vascular network formation within porous biodegradable scaffold. In our model, one layer describes the event occurring within the scaffold, the second layer describes the angiogenesis, focusing the individual behavior of endothelial cells, and finally the third layer simulates the behavior of stem cells and their osteogenic differentiation. This computational framework, coupling two different interacting cell types, is used to investigate cellular response to structural changes of biomaterials and different strategies to maximize vascularization and bone regeneration.

A SIMPLE MODEL OF THE ANTHOCYANINS EFFECT ON CARCINOMA GROWTH

2014 AIChe Annual Meeting

Javier A. Dávila, Juan C. Higuita, Moshe Rosenberg, Carlos A. Cardona
Universidad Nacional de Colombia, Manizales. University of California, Davis

Research group in Chemical, Catalytic and Biotechnological Processes

OUTLINE

1. Introduction
2. Methodology
3. Results and discussion
4. Conclusions

1. INTRODUCTION

CANCER. IMPORTANCE

Cancer is based on the uncontrollable growth of abnormal cells that form a tissue named tumor

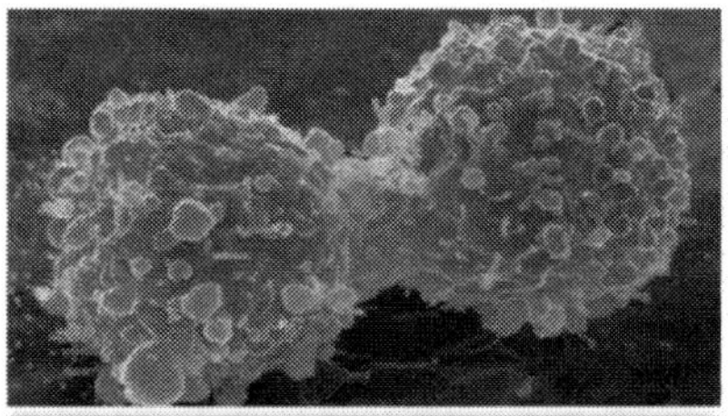

8.2 million Deaths/ 2012

- The main disease in the world
- Several types of cancer
- For the next 2 decades, cancer will increase until 22 million of deaths around the globe

1. INTRODUCTION

CANCER GROWTH

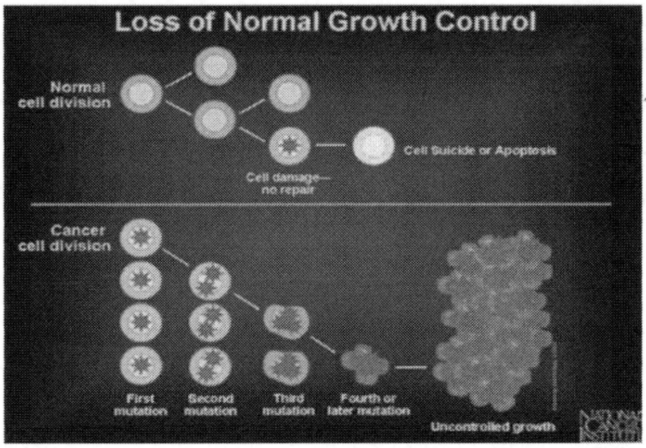

Natural process

- Cell damage
- Apoptosis does not occur
- Uncontrolled cell growth

1. INTRODUCTION

ANTHOCYANINS

Anthocyanins

- Induce Apoptosis
- Antioxidant Activity
- Anti-proliferative activity towards multiple cancer cell types

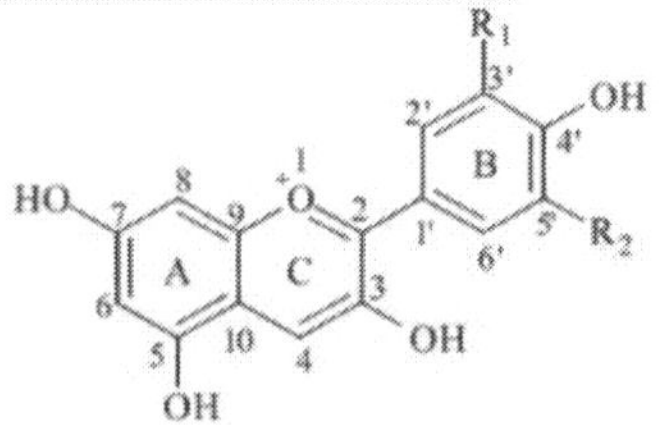

Name	R1	R2
Delphinidin	OH	OH
Petunidin	OCH_3	H
Cyanidin	OH	H
Pelargonidin	H	H
Peonidin	OCH_3	H
Malvidin	OCH_3	OCH_3

1. INTRODUCTION

CARCINOMA GROWTH

- Avascular growth
- Angiogenesis
- Vascular growth

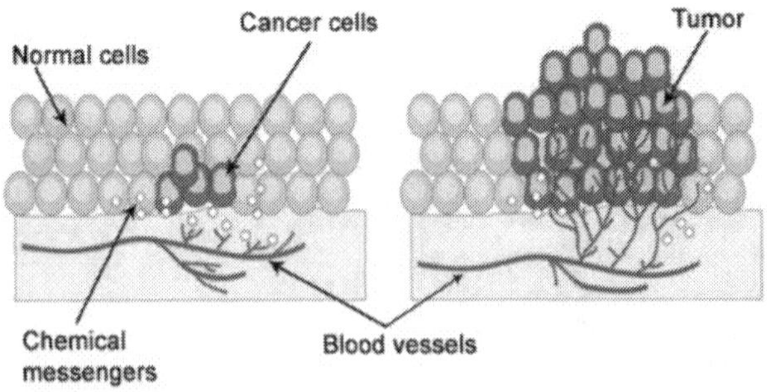

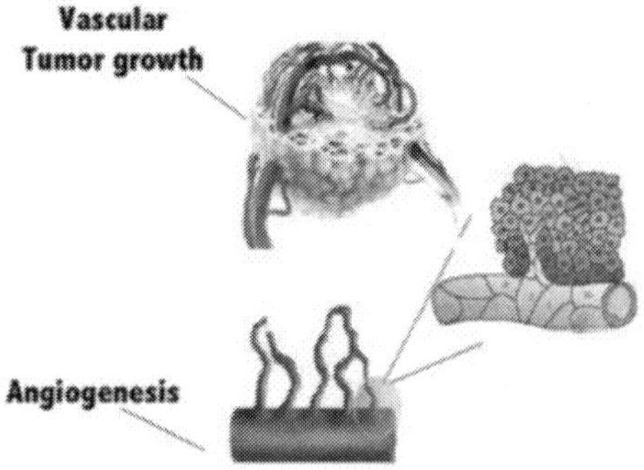

1. INTRODUCTION

CARCINOMA GROWTH IN THE AVASCULAR STEP

All carcinomas in the avascular step evolve according to a universal scheme of progression.

- Net interaction: (Cells and extracelular matrix)
- Initial promotion of cancer
- Mechanims of nutrients supply

Allows the modeling and simulation of carcinoma growth

Interactions expressed as energies

$$E = E_{Adhesion} + E_{Area} + E_{Chemical} + \dots + E_{Other}$$

1. INTRODUCTION

AIM OF THIS RESEARCH

To modeling and simulate a carcinoma growth by a simple model taken into account the anthocyanin effect over the number of cancer cells

2. METHODOLOGY

MODELING OF CARCINOMA GROWTH

A modified Cellular Potts Model (CPM) and Metropolis Monte Carlo (MMC).

- CPM uses a lattice to describe N cells.
- Cells are composed by a number of pixels with a position (i,j).
- Each pixel is associated to a spin (σ) to identify the space occupied for a cell.
- Cell is taken as a group of subcomponents that can be rearranged.
- Movement of cells is due to a mutation of the spins in a random manner (MMC).
- Cellular interactions and changes are described in terms of energies.

1	1	1	1	1	2	2	2	2
1	1	1	1	1	2	2	2	2
3	1	1	4	4	4	2	2	2
3	3	4	4	4	4	4	2	2
3	3	4	4	4	4	4	2	2
3	3	4	4	4	4	4	6	2
3	5	5	4	4	4	6	6	6
5	5	5	5	4	6	6	6	6
5	5	5	5	5	6	6	6	6

(Glazier et al., 2007; Foty et al., 1994)

2. METHODOLOGY

MODELING OF CARCINOMA GROWTH

Cellular interactions

$$E = \underbrace{\sum_{(i,j)(i',j')} J_{\tau(i,j)\tau(i',j')}(1-\delta_{\sigma(i,j)\sigma(i',j')})}_{\text{Adhesion energy}} + \underbrace{\sum_{Cells} \lambda_a (a(\sigma,\tau) - A(\sigma,\tau)_T)^2}_{\text{Area of the cell}} + \underbrace{AF}_{\text{Anthocyanin effect}}$$

- J is the adhesion energy
- δ is Kronecker function (=1 for the same cell)
- λ a is the elasticity parameter
- a(σ, τ)) and (A(σ, τ)T) are actual and permitted areas
- AF is the anthocyanin effect

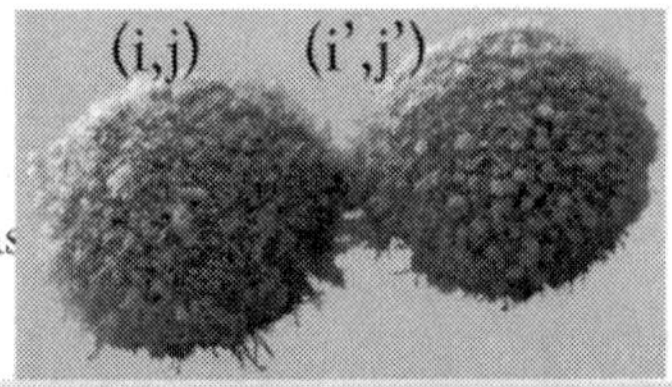

2. METHODOLOGY

MODELING OF CARCINOMA GROWTH

Cellular interactions

Parameter	Value
J11	2
J22	14
J12=J21	7
λ	0.1
a(σ,τ)	8
A(σ,τ)T	16
AF	0.125, 0.28, 0.4

(Glazier et al., 2007; Foty et al., 1994)

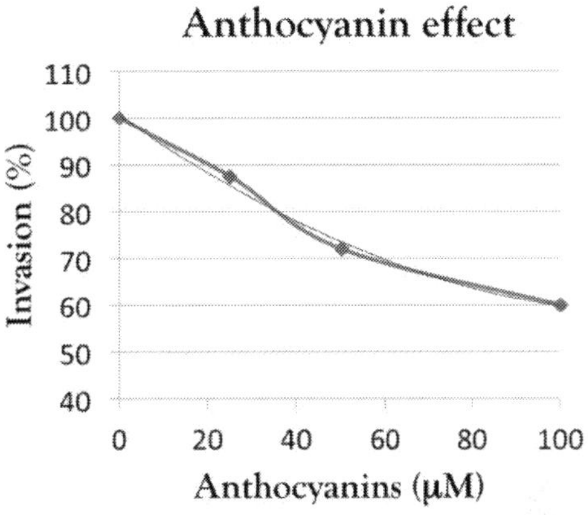

11 Research group in Chemical, Catalytic and Biotechnological Processes

2. METHODOLOGY

MODELING OF CARCINOMA GROWTH

A Lattice of 150x150

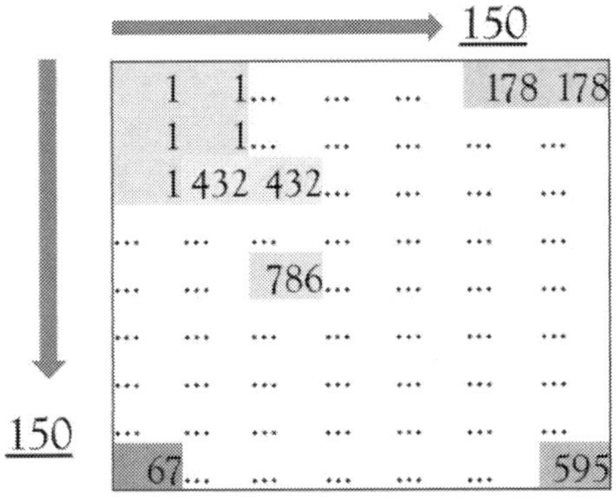

Anthocyanin (μM)	Reduction (%)
25 μM	12.5
50 μM	28
100 μM	40

(Chen et al., 2006)

12 Research group in Chemical, Catalytic and Biotechnological Processes

2. METHODOLOGY

MODELING OF CARCINOMA GROWTH

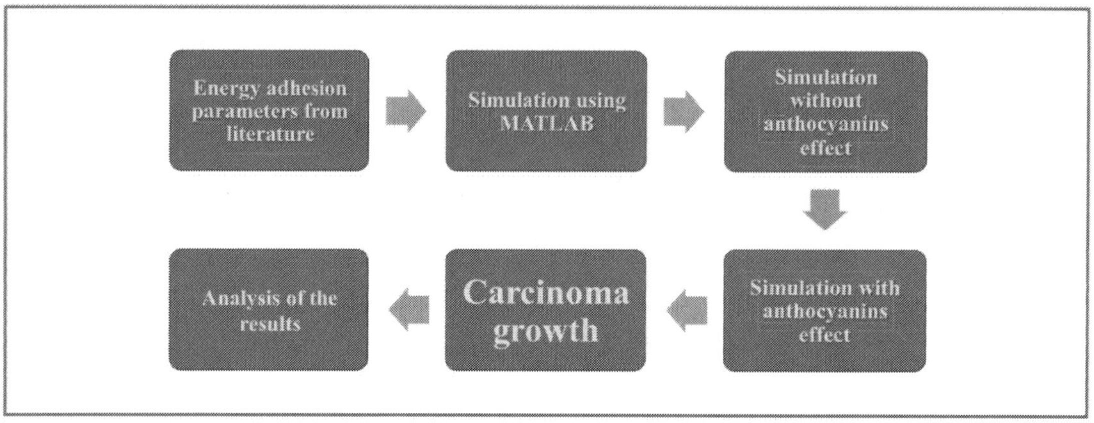

13 Research group in Chemical, Catalytic and Biotechnological Processes

2. METHODOLOGY

MODELING OF CARCINOMA GROWTH

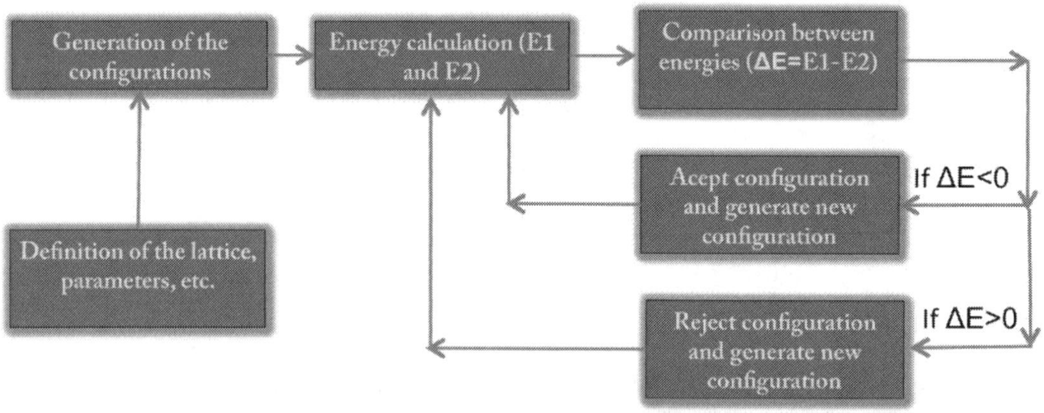

14 Research group in Chemical, Catalytic and Biotechnological Processes

3. RESULTS AND DISCUSSION

Without Anthocyanin effect

2231 cancer cells

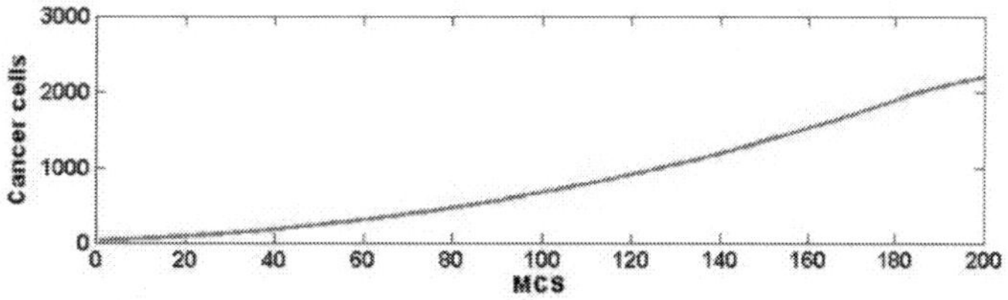

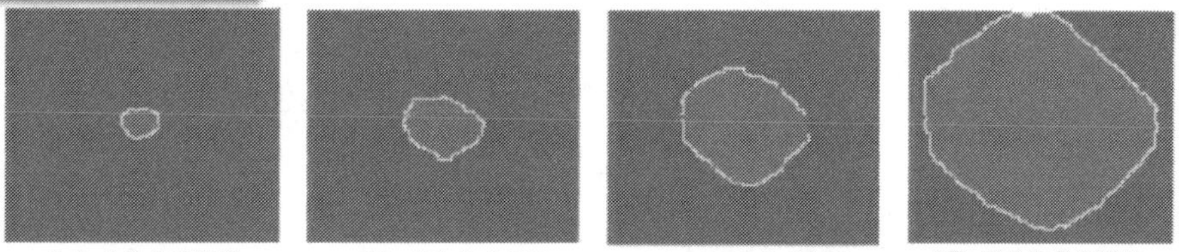

(a) 10 MCS (b) 50 MCS (c) 100 MCS (d) 200 MCS

3. RESULTS AND DISCUSSION

CARCINOMA GROWTH MODELING

With Anthocyanin effect

2046 cancer cells

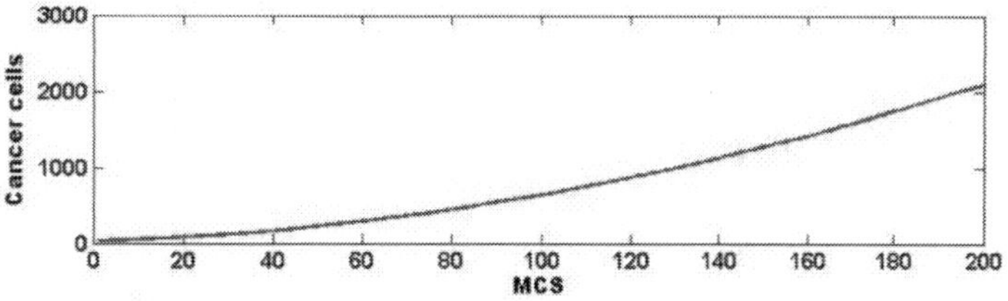

(a) 10 MCS (b) 50 MCS (c) 100 MCS (d) 200 MCS

3. RESULTS AND DISCUSSION

CARCINOMA GROWTH MODELING

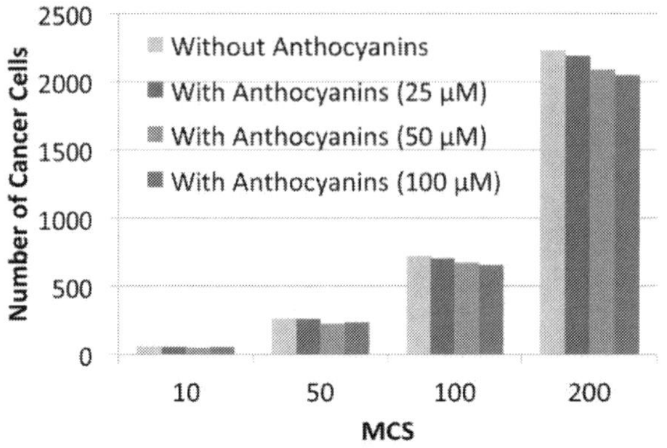

3. RESULTS AND DISCUSSION

CARCINOMA GROWTH MODELING

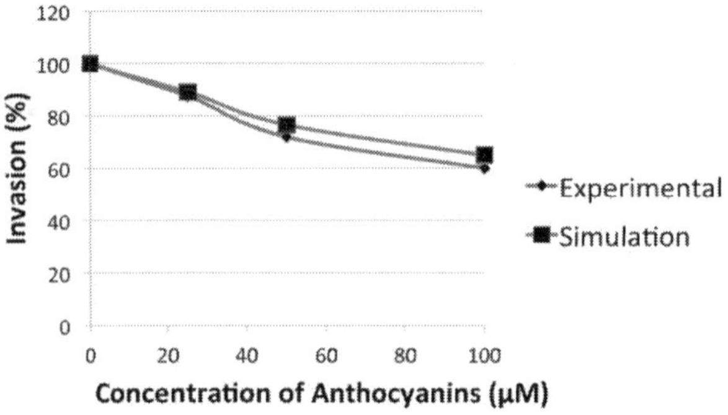

(Chen et al., 2006)

CONCLUSIONS

- A simple model taken into account area and adhesion energies offers an approximation of anthocyanin effect on carcinoma growth
- Modeling and simulation become in important tools for provide a framework to investigate and make predictions for carcinoma growth
- The anthocyanins has an important effect on carcinoma growth

AKNOWNLEDMENT

The authors express their gratitude to Universidad Nacional de Colombia, at Manizales for your economical support as well as to University of California in Davis for your contributions

380233 Registration and Dimensionality Reduction Algorithms for the Analysis of Drosophila Embryogenesis Images

Abstract:

Over the past several years, there have been significant algorithmic developments in image analysis. These include new algorithms for image registration [1], as well as methods to compute image features which are both informative and invariant to symmetries such as translations and rotations [2, 3]. Simultaneously, dimensionality reduction techniques, such as principal component analysis and diffusion maps [4], have proven useful in uncovering patterns and structure in large, complex data sets. Vector diffusion maps [5] is a recently developed algorithm that combines diffusion maps and the factoring out of symmetries in a single algorithmic step, which simultaneously registers and extracts low-dimensional structure from imaging data sets. We discuss these algorithms and apply them to imaging data collected from studies of *Drosophila* embryogenesis to extract parsimonious descriptions of developmental dynamic pattern formation processes.

[1] Singer, A. "Angular synchronization by eigenvectors and semidefinite programming." *Applied and computational harmonic analysis* 30.1 (2011): 20-36.

[2] Mallat, Stéphane. "Group invariant scattering." *Communications on Pure and Applied Mathematics* 65.10 (2012): 1331-1398.

[3] Zhao, Zhizhen, and Amit Singer. "Fourier–Bessel rotational invariant eigenimages." *JOSA A* 30.5 (2013): 871-877.

[4] Coifman, Ronald R., et al. "Geometric diffusions as a tool for harmonic analysis and structure definition of data: Diffusion maps." *Proceedings of the National Academy of Sciences of the United States of America* 102.21 (2005): 7426-7431.

[5] Singer, Amit, and H T. Wu. "Vector diffusion maps and the connection Laplacian." *Communications on pure and applied mathematics* 65.8 (2012): 1067-1144.

365335 Effects of Thermal Stress on the Mechanical Deformation of Cardiac Tissue

Abstract:

The electrophysiological changes initiate mechanical contraction in the cardiac tissue via excitation contraction coupling (ECC) while changes in tissue length affect the electrical activity, via stretch-activated channels (SACs) which is known as mechanoelectrical feedback (MEF) (Kiseleva I et al., 2000). Under certain conditions of abnormal mechanical deformations of tissue, MEF induce local electrical depolarizations that can lead to cardiac arrhythmias (Nazir S A, 1996b; Kuijpers et al., 2007), which may lead to sudden cardiac death and ventricular fibrillation.

Nash-Panfilov (NP) model (Nash and Panfilov, 2004) is mostly used to represent electromechanical coupling and takes into account both electrical and mechanical properties of cardiac tissue to link the excitation with contraction. This model includes an additional variable to represent an instance of active stress responsible for mechanical deformation and is coupled to the mathematical equation describing the tissue's mechanics model. A Mooney–Rivlin type model is used to describe the passive mechanical properties of the cardiac tissue. In addition to electromechanical features within cardiac tissue one needs to consider thermal effects associated with cardiac system. In particular, the effect of temperature in electrophysiology has been studied using experimental settings (Collins and Rojas, 1982; Sitsapesan et al., 1991). However, the effect of temperature on cardiac electrocardiology has been limited to the cell mathematical models. Recent works have included the effect of temperature in FitzHugh-Nagumo (FHN) and Hodgkin–Huxley mathematical model representing the propagation of neural signal taking into account dynamical heat transfer within biological tissue.

In this study, we illustrate the role of thermal stress, induced by the effects of the temperature variations, on the mechanical deformation of the cardiac tissue, adopting a model containing all the key ingredients that account for heat, ECC and MEF coupling. Numerical examples will be provided to illustrate the effects of the thermal stress on mechanical deformation of the cardiac tissue. This may also serve to demonstrate the influence of thermal stress on the cardiac alternans, which have been shown to be a precursor to arrhythmias.

References:

Collins C A, and Rojas E. Temperature dependence of the sodium channel gating kinetics in the node of Ranvier. Q. J. Exp. Physiol., 67, 41–55, 1982.

Sitsapesan R, Montgomery RA, MacLeod K T, and Williams A J. Sheep cardiac sarcoplasmic reticulum calcium-release channels: modification of conductance and gating by temperature. J. Physiol., 434, 469–488, 1991.

Kiseleva I,Kamkin A, Wagner KD, Theres H, Ladhoff A, Scholz H, Gunther J, Lab MJ,

Echebarria B. Mechano-electric feedback after left ventricular infarction in rats. Cardiovascular research: 45,370 – 378, 2000.

Nazir S A, and Lab M J. Mechanoelectric feedback in the atrium of the isolated guinea-pig heart. Cardiovasc. Res., 32, 112–119, 1996b.

Kuijpers N H, ten Eikelder H M, Bovendeerd, P H, Verheule S, Arts T, and Hilbers P A. Mechanoelectric feedback leads to conduction slowing and block in acutely dilated atria: a modeling study of cardiac electromechanics. Am. J. Physiol. Heart Circ. Physiol., 292, H2832–H2853, 2007.

Nash M P, Panfilov A V. Electromechanical model of excitable tissue to study re-entrant cardiac arrhythmias. Prog. Biophys. Mol. Biol. 85, 501–522, 2004.

Uncertainty in clinical data and stochastic model for in-vitro fertilization.

Kirti M. Yenkie[a,b] and Urmila Diwekar[a,b]

[a]*Department of Bioengineering, University of Illinois, Chicago, IL 60607 - USA*
[b]*Center for Uncertain Systems: Tools for Optimization & Management (CUSTOM), Vishwamitra Research Institute, Clarendon Hills, IL 60514 – USA*

Abstract

In-vitro Fertilization (IVF) is the most common technique in Assisted Reproductive Technology (ART). It has been divided into four stages; (i) superovulation, (ii) egg retrieval, (iii) insemination/fertilization and (iv) embryo transfer. The first stage of superovulation is a drug induced method to enable multiple ovulation, *i.e.*, multiple follicle growth to oocytes or matured follicles in a single menstrual cycle. IVF being a medical procedure that aims at manipulating the biological functions in the human body is subjected to inherent sources of uncertainty and variability. Also, the interplay of the hormones with the natural functioning of the ovaries to stimulate multiple ovulation as against a single ovulation in a normal menstrual cycle makes the procedure dependent on several factors like the patient's condition in terms of cause of infertility, actual ovarian function, responsiveness to the medication. The treatment requires continuous monitoring and testing and this can give rise to errors in observations and reports. These uncertainties can be observed in the form of measurement noise in the available data. Thus, it becomes essential to look at the process noise and think of a way to account for it and build better representative models for follicle growth. The purpose of this work is to come up with a robust model which can project the superovulation cycle outcome based on the hormonal doses and patient response and hence provide a treatment guideline to enhance the success rate of the procedure.

Keywords-infertility, assisted reproduction, multiple ovulation, uncertainty, stochastic differential equations

1. Introduction

Infertility is the inability of a couple to achieve conception or to bring a pregnancy to term after a year or more of regular, unprotected intercourse. The World Health Organization has estimated that about 8-10% couples experience some kind of infertility problems. Medical science has come up with ARTs like IVF to treat these problems. IVF involves fertilization of oocytes by a sperm in an IVF laboratory simulating similar conditions as in the human body. It is a four stage medical procedure and the success of superovulation, its first stage, is critical to proceed with the next stages in the cycle. Also, superovulation requires maximum medical attention, investment of time and money as compared to the other stages.

The major hormonal drug involved in superovulation is the follicle stimulating hormone (FSH), which is responsible for follicle growth and development. In our previous work (Yenkie et. al. (2013)), a deterministic model for prediction of multiple follicle growth, dependent on the amount of FSH dosage was built by drawing similarities with the particulate process of batch crystallization. This model was validated with clinical data from 50 IVF cycles, available from our collaborating hospital in India. The model fitted very well to most of the data and almost 85% predictions were within 0 to 30% range of the modeling error. However, some of the projected results had deviations from the expected behavior and hence could be used to model the uncertainties in the process.

2. The Determinstic Model

Thus, the concept of the moment model in batch crystallization was used for modeling superovulation in IVF (Yenkie et. al. (2013)). The growth term in batch crystallization is temperature dependent and hence it becomes the decision variable for controlling the process. On similar lines in IVF the follicle growth is dependent upon the doses of hormones injected to the patient.

The follicle size is converted into mathematical moments by assuming them to be spherical in shape. The eq. (1) is used for converting follicle size to moments.

$$\mu_i = \sum n_i(r,t) r_i^i \Delta r_i \qquad (1)$$

Here, μ_i - ith moment, $n_j(r, t)$ - number of follicles in j^{th} bin with mean radius as r at time t, r_j - mean radius of jth bin and Δr - range of radii variation in bins. Each moment corresponds to a feature of the follicles, like the zeroth moment represents the number, first moment represents the size, etc.

The follicle growth term (G) is dependent on the amount of follicle stimulating hormone (FSH) injected (C_{fsh}) to the patient at the particular time (t) in the cycle and is represented in eq. (2). Here, k and α are kinetic constants.

$$G(t) = k \Delta C_{fsh}(t)^{\alpha} \qquad (2)$$

In the literature by Baird (1997) it was suggested that the number of follicles activated for growth during that particular superovulation cycle are always constant for a patient, hence the zeroth moment has a constant value. The moment equations for the follicle dynamics can be written as in eqs. (3) and (4).

$$\mu_0 = constant \qquad (3)$$

$$\frac{d\mu_i}{dt} = iG(t)\mu_{i-1}(t);\ (i = 1,2,\dots 6) \qquad (4)$$

The data obtained from our collaborating hospital is converted to moments as shown earlier in eq. (1) and used for model fitting and validation. The validation method requires the conversion of predicted moments back to the follicle size and number and this was done by adapting the methodology proposed by Flood (2002).

3. The Stochastic Model

The aim is to develop a stochastic model for the superovulation stage in terms of Ito processes. Initially, these Ito processes were mostly applied in financial field and stock price modeling to model uncertainties. Since they can characterize time-dependent uncertainties and can be integrated and differentiated using the rules of Ito's stochastic calculus, they can prove beneficial in modeling biological processes. The simplest example of an Ito process is the Brownian motion or Wiener process. For any stochastic process to be characterized as the Wiener process it must follow three important properties (Diwekar (2008)).

1. It should follow the Markov property: probability distribution for all future values of the process depend only on its current value.
2. It should have independent increments in time. probability distribution for changes in the process over any time interval is independent of any other time interval.
3. Changes in the process over a finite time interval should be normally distributed, with variance linearly dependent on the length of time interval (i.e., N (0, $\sqrt{dt}$) for all t > 0).

We check whether superovulation classifies as a Wiener process. As per the current protocols, hormonal dosage is decided based on daily monitoring of patient's response and the follicle size observed. Thus, the next follicle size distribution (FSD) after medication is dependent on FSD at that time, hence it is reasonable to categorize it as a Markov process.

The follicle growth term developed in the determinstic model and which satisfies the data is dependent on the amount of FSH injected (C_{fsh}) to the patient at the particular day (t) in the cycle (eq. (2)). Thus it satisfies the second property, where FSH dosage have independent increments.

The follicle growth and number is represented as the FSD, where the follicle size is divided into equispaced bins and number of follicles are represented as the discrete markers in each of the size bins as shown in Fig. 1. When these markers are connected with a smooth curve, the data tends to follow a normal distribution.

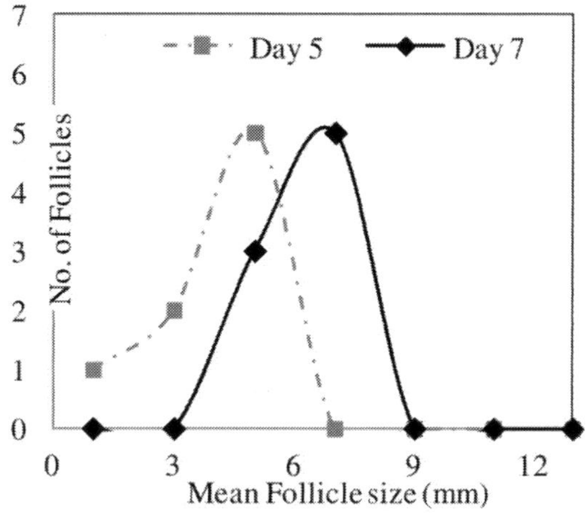

Figure 1 Follicle size showing normal distribution on Day 5 and Day 7.

The Ito processes are of various types whose basic building block is a Wiener process. Depending upon the behavior of the system under random influences it can modeled as one of the suitable Ito processes. It was found that the uncertainties could be represented using the simple brownian motion type of Ito processes, resulting in a (eq. (7) to eq. (12)) set of stochastic differential equations (SDEs).

$$d\mu_1(t) = G(t)\mu_0 dt + \sigma_1 \epsilon_t \sqrt{dt} \quad (7)$$

$$d\mu_2(t) = 2G(t)\mu_1(t) + \sigma_2 \epsilon_t \sqrt{dt} \quad (8)$$

$$d\mu_3(t) = 3G(t)\mu_2(t) + \sigma_3 \epsilon_t \sqrt{dt} \quad (9)$$

$$d\mu_4(t) = 4G(t)\mu_3(t) + \sigma_4 \epsilon_t \sqrt{dt} \quad (10)$$

$$d\mu_5(t) = 5G(t)\mu_4(t) + \sigma_5 \epsilon_t \sqrt{dt} \quad (11)$$

$$d\mu_6(t) = 6G(t)\mu_5(t) + \sigma_6 \epsilon_t \sqrt{dt} \quad (12)$$

Here, μ_i - ith moment, σ_i - standard deviation terms for each moment equation corresponding to noise in the clinical data, ϵ_t - set of random numbers derived from unit normal distribution and dt - time interval.

4. Results and Discussions

The results for the FSDs are shown in Fig 2A and 2B for day 5 and day 9 respectively. The plots show observed clinical data as discrete data points (O), deterministic model (D) projections as dotted profiles and stochastic model (S) projections as continuous profiles. The results projected by the stochastic model match the data better when compared to deterministic results.

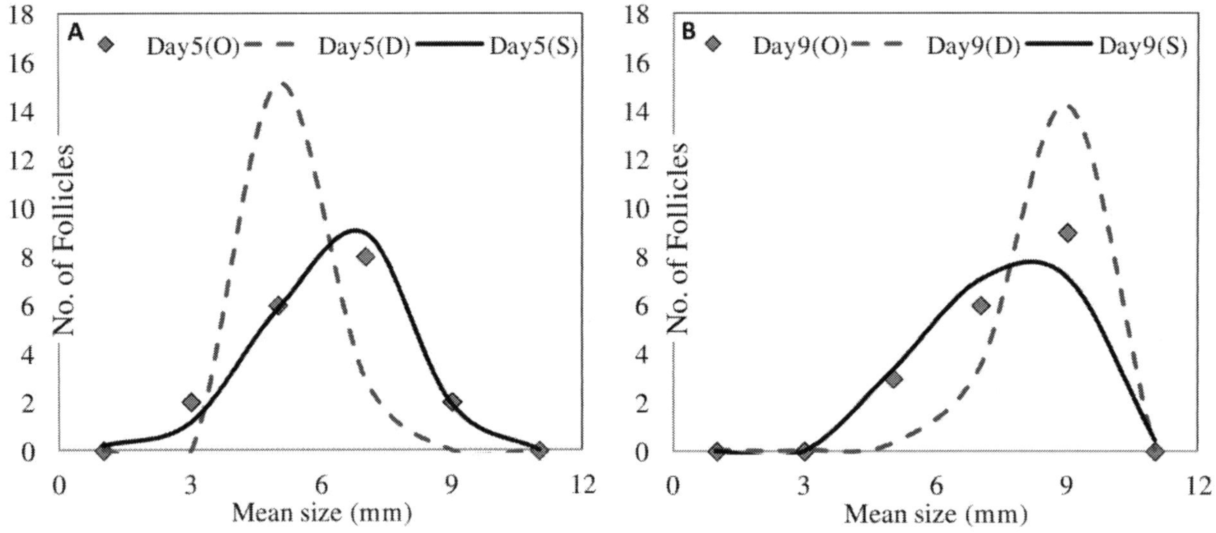

Figure 2. Comparison of FSDs from observed clinical data (O), deterministc (D) and stochastic (S) model predictions for Patient-I.

5. Conclusions

The approach to develop a stochastic model for the superovulation stage in IVF in terms of Ito form of stochastic differential equations proves promising. The approach used for stochastic model development involving a novel parameter estimation procedure is an added contribution in this work. The model predictions obtained from the stochastic model are inclusive of all possible noise in the data and hence match better as compared to deterministc results. The robust predictions from the SDE model can be used for predicting robust drug dosing policies using stochastic control methods for achieving the desired outcome in superovulation for enhancing the success rate of the overall IVF cycle.

386746 Optimization of HIV Treatment Schedules Based on the Mitochondrial Toxicity of Nucleoside-Analogue Antiretroviral Agents

Abstract:

Nucleoside-analogue Reverse Transcriptase Inhibitors (NRTIs) are an important class of drugs used for the treatment of HIV infection. Although these drugs are effective in maintaining the viral load below detectable limits, their prolonged use leads to several medical complications. Clinical and molecular biological evidences suggest that NRTIs cause toxic side-effects by interfering with normal mitochondrial function. Since NRTIs are structurally similar to the natural deoxynucleotides (dNTPs), they competitively inhibit the mitochondrial DNA (mtDNA) replication and transcription processes. These negative effects lead to a depletion of mtDNA and mRNA transcripts and therefore a reduced expression of proteins for the electron transport chain.

To minimize drug toxicity and drug expenses, structured interruptions in treatment are widely practiced. Mathematical models describing the disease dynamics have been employed to minimize the intake of drug by optimizing these interruptions. However, it is of utmost importance to ensure that the drug intake is within the toxic threshold. To achieve this, a mathematical model that provides mapping between the NRTI drug concentration and mitochondrial protein levels is necessary. We developed a hybrid model of the mitochondrial gene expression process that accounts for the effects of NRTIs. The copy numbers of mtDNA and mRNA and the protein concentrations are tracked using a deterministic system of ordinary differential equations, while the parameters that depend on the drug concentration are computed from repeated event-based stochastic simulations. Since NRTIs are known to inhibit the rate of mtDNA replication and transcription, the corresponding parameters are obtained through stochastic simulations of the respective processes. These processes are polymerase chain reactions and they are modeled and simulated using Gillespie's algorithm, a kinetic Monte Carlo method. At each location on the mtDNA strand, the polymerases can perform one of these actions: (1) addition of a nucleotide/NRTI (2) excision of the previously added block (3) Disassociation of the polymerase from the strand (4) Re-association of the polymerase to the strand. The simulation stops either when a strand is completely replicated or transcribed or when the polymerase disassociates and fails to re-associate. The occurrence of the latter is more probable when a NRTI or AVRN was added in the previous step and was failed to be removed. The training and validation datasets for the model are obtained experimentally.

An optimization problem is formulated to minimize the average maximal intracellular drug concentration with an inequality constraint on the mitochondrial protein concentration. This constraint is the protein threshold required for normal cellular function. Different treatment schedules that achieve similar reductions in average drug concentration are investigated.

DEVELOPMENT OF A GENERIC PHYSIOLOGY BASED BIOKINETIC MODEL FOR PREDICTING INTERNAL DOSE AND ASSIMILATION OF BIOMONITORING DATA FOR INDUSTRIAL CHEMICALS

D.A. Sarigiannis[1], *S.P. Karakitsios*[1], *A. Gotti*[1], *E. Handakas*[1], *K. Papadaki*[1]

[1]*Department of Chemical Engineering, Aristotle University of Thessaloniki, 54124, Thessaloniki, Greece*

Introduction

Physiology Based BioKinetic (PBBK) models are continuously gaining ground in regulatory toxicology, describing in quantitative terms the absorption, metabolism, distribution and elimination (ADME) processes in the human body, with a focus on the effective dose at the expected target site. This trend is further amplified by the continuously increasing scientific and regulatory interest about aggregate and cumulative exposure. PBBK models translate external exposures from multiple routes into internal exposure metrics, addressing the effects of exposure route in the overall bioavailability, or the dependence on critical developmental windows characterized by enhanced susceptibility, such as pregnancy, lactation and infancy. Recently, efforts have shifted also towards the integration of whole-body physiology, disease biology, and molecular reaction networks, as well as integration of cellular metabolism into multi-scale whole-body models.

The current study aims at the development of (a) a lifetime PBBK mode that takes stock of gestation and breastfeeding while covering a large chemical space and (b) a framework for biomonitoring data assimilation. The model captures multi-route exposure (including inhalation, skin absorption and gastro-intestinal absorption), and incorporates detailed compartmental description, plasma protein/red cell binding and estimates the concentration of parent compounds and metabolites in several biological fluids including breast milk and the contribution of lactation to breast feeding exposure, as well as the potential of bioaccumulation. The latter is very important when dealing with persistent/bioaccumulative compounds. Assessment of internal dose through PBBK modeling allows the estimation of internal doses of xenobiotics that exceed levels associated with biological pathway alterations and, eventually, induction of biological perturbations that may lead to health risk. The biological pathway altering dose (BPAD) outperforms current metrics of risk in that it combines dose–response data with analysis of uncertainty and population variability so as to derive exposure limits. BPADs, which can be derived from high-throughput screening (HTS) in vitro data, are more pertinent in the case of pathways the perturbation of which is a key event in the mode of action (MOA) leading to an apical of systemic adverse outcome.

Methodology

Conceptualization of the generic PBPK model

The model is designed to describe as much as possible the actual ADME processes occurring in human body, so as to be easily applicable for a broad variety of chemicals under proper parameterization. The model includes the parent compound and a number of three potential metabolites. For each compound/metabolite all major organs is included and the link among the compounds and the metabolites is through the metabolizing tissues. This is mainly the liver, but also other sites of metabolism might be considered based on the presence or not of the enzymes involved in the metabolism of the compound of interest. Both phase I and phase II metabolism are described. Both membrane and blood flow limited processes are included. In order to capture the in-utero exposure, the model is also replicated in order to describe the functional interaction of the mother and the developing fetus through the placenta. The anthropometric parameters of the models are age dependent covering the whole lifespan from the moment of conception, so as to provide a lifetime internal dose assessment. Details on the mathematical framework so as to describe the structure of the generic-lifetime PBTK model are given elsewhere [1].

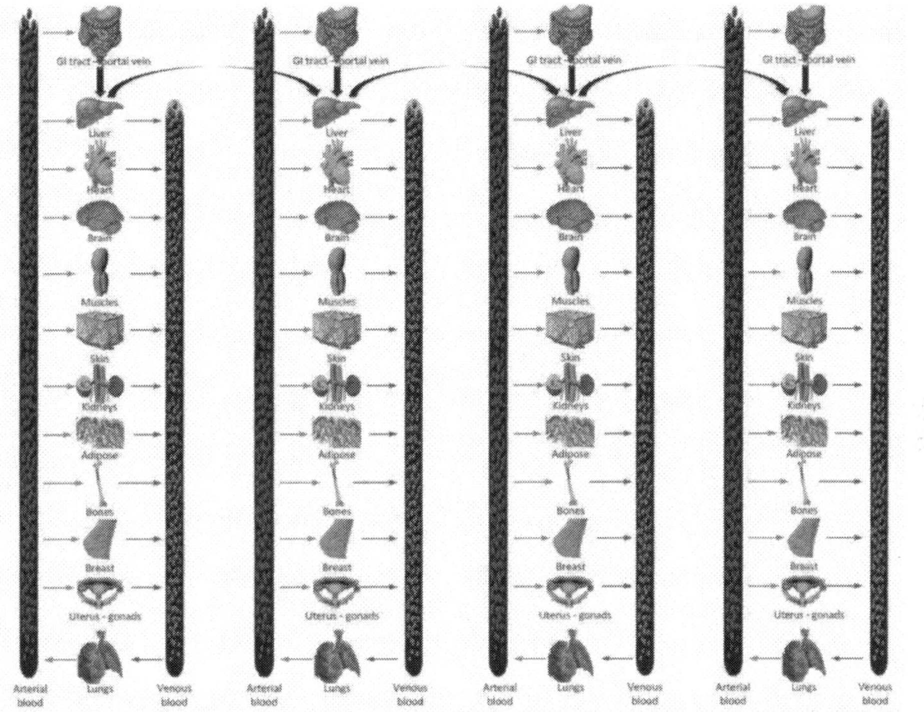

Figure 1. Conceptual representation of the Mother-Fetus PBTK model

QSAR Model

In order to expand the applicability of the generic PBBK model to cover as wide as possible a chemical space, parameterization of the model for data poor chemicals is done using advanced quantitative structure-activity relationships (QSARs). This pertains to all necessary parameters such as blood-tissue partition coefficients, clearance and elimination kinetics. Several QSAR modeling approaches have been investigated, including (a) the Peyret, Poulin and Krishnan algorithm, which is based on the fractional content of cells, interstitial fluid in tissue, plasma in blood, erythrocyte in blood, tissue lipids and the lipophilicity of the compound of interest; (b) the molecular fractions algorithm proposed by Béliveau et al. that takes into account the frequency of occurrence of the several molecular fragments of the compounds and (c) Abraham's solvation equation for estimating biological properties, which takes into account the excess molar refraction that can be determined simply from knowledge of the compound refractive index, the compound dipolarity/polarizability, the solute effective or summation hydrogen-bond acidity, the solute effective or summation hydrogen-bond basicity and the McGowan characteristic volume that can trivially be calculated for any solute simply from a knowledge of its molecular structure. The most efficient of all the examined methods was Abraham's solvation equation combined with Artificial Neural Networks (ANN). The whole data set was divided into training and test sets (85% and 15% respectively of the total number of compounds) and the model developed from training set is externally validated using the test set. The data sets of 33 and 29 industrial chemicals are used in order to predict the partition coefficients of seven tissues and the metabolic constants, respectively. The examined QSAR model was analyzed with Multi-Layer Perceptron (MLP) followed by BFGS Quasi-Newton. The Multi-layer network contain a single input layer, which consists of the values of molecular descriptors, one or more hidden layers, which process the descriptors into internal representations and an output layer utilizing the internal representation to produce the final prediction [2].

Reverse Dosimetry Model

The PBBK model is geared with reverse modeling algorithms in order to reconstruct exposure from human biomonitoring (HBM) data. A tiered approach is followed as a function of data availability (periodicity and size of sampling, specimen type) and requirements of the exposure reconstruction analysis (temporal analysis of exposure, contribution from different routes), ranging from Exposure Conversion Factors (ECFs), up to Markov Chain Monte Carlo analysis. Probabilistic techniques are used to validate the exposure outcome on the basis of actual environmental and population biomonitoring data. Assimilation of human biomonitoring data and their translation into intake distribution amounts to a computational inversion problem, where the objective is to identify the specific input distributions that best explain the observed outputs while minimizing the residual error. Inputs involve spatial and temporal information on micro-environmental media concentrations of xenobiotics and corresponding information on human activities, food intake patterns or consumer product use that result in intakes; outputs are the observed biomarkers. The error metric can be defined in terms of population variation (the latter has to be lower than the intra-individual variation, which may be associated to measurement or other random error source). At the individual level, the PBBK model is combined with multimedia models and survey questionnaires to identify exposure sources, used as ancillary information, aiming to predict exposure magnitude and eventually the timing of exposure events.

More in detail, a computational framework was developed based on Bayesian Markov Chain Monte Carlo (MCMC) combined with the generic Physiological Based Pharmacokinetic (PBPK) model to perform accurate exposure reconstruction (ER). The ER framework developed consists of 3 basic steps:

At first the prior parameter distribution, the joint probability distribution, the population model and the determination of the measurement model have to be specified. At the next step exposure is calculated using MCMC simulation considering the observed biomonitoring data. Finally, the evaluation of the results is realized using MC simulation, with emphasis to the comparison of prior and posterior distribution as well as parameter independence. MCMC simulation refers to a class of iterative simulations in which the random variables of interest are drawn from a sequence, or chain, of distributions that eventually converge to a stable posterior distribution. Moreover, Differential Evolution (DE) and MCMC algorithms have been combined to this problem for the first time. Differential Evolution Markov Chain is a population MCMC algorithm, in which multiple chains run in parallel. In fact DE is a simple genetic algorithm for numerical optimization in real parameter spaces. As a result, this combined computational framework speeds up the calculation and convergence, even for nearly collinear parameters and multimodal densities. The analysis of exposure reconstruction problems is based on the MCMC and DEMC technique and it has been carried out according to the following steps: The exposure route models are posed to the PBBK models that they simulate the exposure of human to chemicals via inhalation, skin and oral route. Then, the probability model is specified according the prior parameter distributions as well as the calculated and/or fixed distribution of the pharmacokinetic mode. The PBBK model has been combined with the Bayesian MCMC [3, 4] and DEMC [5] techniques in order to simulate and calculate the expected value of the exposure. The model has been developed and applied to software acslX version 3.0. The distribution plots and histograms are created using Matlab® 2014. The sampling is set appropriate according to the problem and the available data for the proposal function.

The computer used in this study was a Intel(R) Core(TM) i7- 3537U CPU 2.00 GHz 4.00 GB RAM. The flowchart diagram of the procedure is demonstrated in Figure 2.

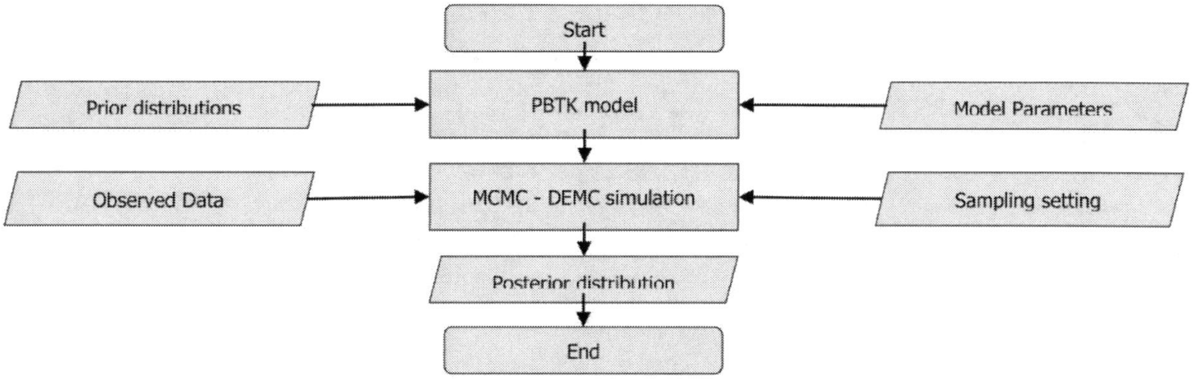

Figure 2. Exposure reconstruction flowchart procedure

The uncertainty of the results depends on the uncertainty of the prior distributions as they have influence to the algorithms. Also, although that the estimation of the parameters' distribution provides a mathematical and physiological fit of the model, it can lead to a non-negligible error and especially when it is assumed to be representative for all individuals' parameters. It has to be noted that DEMC algorithm has a better approached to the actual dose as well as and the new distribution density has an obvious and strong trend to move toward the correct direction. Moreover, comparing the computation needs of the algorithms, the DEMC is 3 to 4 time slower than MCMC but the predictions are closer to the actual value because of parallel running of multiple chains. Also, the model increases the computation needs in case of the existence of multi metabolic pathway(s) and mechanisms as well as of high frequency exposures. The need for better awareness of biomarker levels is obvious in case of exposure to chemical substances and this becomes more important when the xenobiotics are rapidly conjugated into metabolites. Hence, knowledge of the metabolite prior distribution could decrease the uncertainty of the results.

Results and discussion

QSAR Modelling results

The statistical qualities of the QSAR model are assessed by the square of the correlation coefficient (R^2). The predicted and experimental values of liver:blood and kidney:blood partition coefficient are presented in the Figure 3. Michaelis-Menten constant (K_m) and normalized maximal velocity (V_{max}) are presented in Figure 4. As shown in the figures below, Abraham's equation, combined with ANN, gives high values of R^2 for both physicochemical and kinetic parameters of the QSAR model. To better highlight the difference in performance of this new proposed QSAR, **Error! Reference source not found.** presents the experimental and predicted values of the normalized maximal velocity obtained by another QSAR model retrieved from Price and Krishnan [6].

The significantly improved performance of Abraham's equation and ANN combination can be ascribed to its capacity to represent mathematically the complex interactions of biochemical micro-processes, which are lumped into the metabolic and physicochemical parameters. Generally, the predictive capacity of the Abraham's model for new chemical entities is influenced by chemical nature of the training set molecules used for development of the model. In actual case, the test set molecules can be predicted well when these molecules are structurally very similar to the training set molecules. The reason is that the model has captured all features common to the training set molecules. Nonetheless not even robust and validated QSARs can be expected to reliably predict the response for the entire chemical space. Only the predictions for chemicals falling within their applicability domain can be considered reliable.

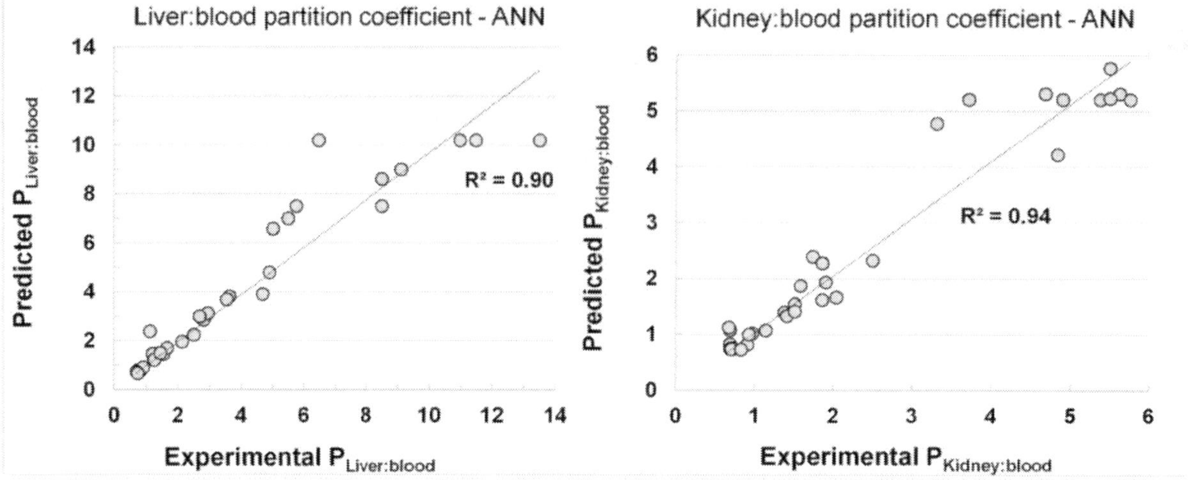

Figure 3. Experimental vs predicted liver:blood and kidney:blood partition coefficients

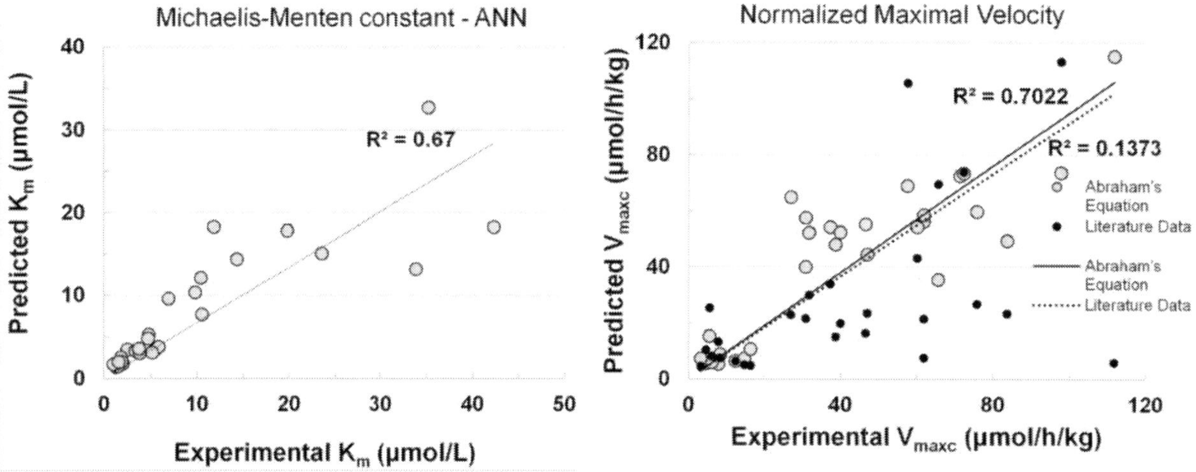

Figure 4. Experimental vs predicted Michaelis-Menten constant and Normalized maximal velocity

Test case of the generic PBPK model to bisphenol-A (BPA)

The generic model was applied for the assessment of a highly controversial industrial chemical with widespread applicability in consumer goods, namely bisphenol-A (BPA). Exposure scenarios were built based on an extensive literature review of BPA exposure data.

In order to associate the risk of the several exposure scenarios based on the Biologically Effective Dose (BED) derived by the PBBK model, two different exposure metrics were used:

- The Tolerable Daily Intake (TDI) of 50 μg/kg_bw/d proposed by the European Food Safety Authority [7], was translated into internal exposure, found to correspond to a concentration of 0.16 μg/L of free plasma BPA.
- In vitro, the ToxCast assays provided six ER agonist or binding AC50 values for BPA, ranging from 0.6 to 1.7 μM. To calculate a conservative Biological Pathway Altering Dose (BPAD), the lowest ToxCast AC50 was selected (0.64 μM for Attagene Factorial cis ERE assay) [8, 9]. Based on our refined PBBK model, the estimated concentration (145 μg/l), is 3 orders of magnitude higher than the equivalent derived from the EFSA TDI (0.16 μg/l).

The results of plasma free-BPA under the the respective exposure scenarios, in contrast to the respective internal exposure reference values are illustrated in Figure 5.

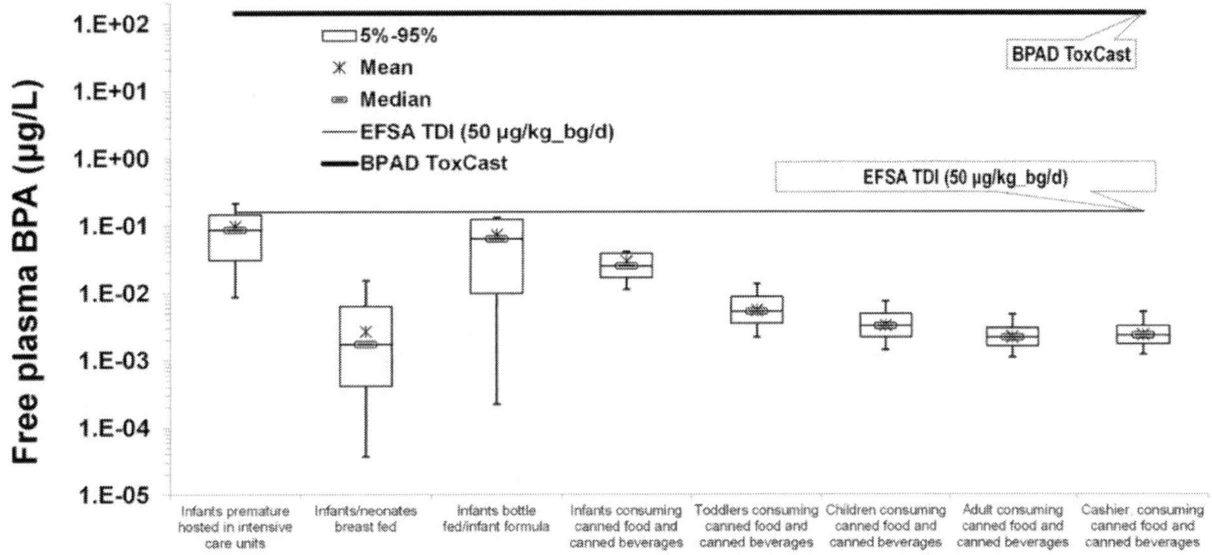

Figure 5. Internal exposure to free plasma BPA under several major exposure scenarios

For the majority of scenarios, the estimated internal dose was close to 0.002 µg/L and only in the case of bottle fed infants, internal exposure concentrations were up to 0.023 µg/L. This is partially explained by the neonates immaturity of the detoxification pathway, resulting to higher internal doses for the same bodyweight normalized dose compared to children older than 1 year old or adults. The biologically effective dose of the developing fetus during gestation is highly linked to the one in maternal blood. According to our model and based on a conservative exposure scenario for the mother (e.g. 5 µg/kg_bw/d), free plasma BPA in maternal blood is almost 0.006-0.007 µg/L, which is slightly higher than what expected for a non-pregnant woman (0.005 µg/L). Placental concentration is 0.0013 µg/L and the corresponding fetal concentration is 0.004-0.005 µg/L. Maternal BPA-Glu bioavailability is also very important in the case of breast-fed infants. Transfer of BPA through milk is not sufficient enough to explain exposure of breast-fed infants; the overall BPA exposure through breast-feeding can only be explained by BPA-Glu cleavage in the gastrointestinal tract. Even when the worst-case scenario is taken into account, breast fed infants seem to be significantly less exposed compared to the bottle fed infants and neonates.

In addition, exposure to BPA was reconstructed based on real-life HBM data. Average urine BPA-Glu was 2.8 µg/L across Europe, covering different age groups, using an average urine BPA-Glu equal to 2.8 µg/L. In particular the algorithm has been tested under the assumption that the average amount of 2.8 µg/L BPA-Glu to human' s urine is the results of an ordinary adult dietary schedule that includes 3 different meals: i) breakfast at 7:00 am (dose 1), ii) lunch at 2:00 pm (dose 2) and iii) dinner at 7:00 pm (dose 3). The results of the exposure reconstruction algorithms after 1000 iterations converge to the available biomonitoring data. All the prior distribution have been moved through to the direction of the actual exposure dose (Figure 6). The results indicated that the overall daily intake is very low, approximately 1 µg/kg_bw/d, which is far below the Tolerable Daily Intake (TDI) of 50 µg/kg_bw/d proposed by the European Food Safety Authority (EFSA). Reconstruction of dose from the actual biomonitoring data, allows as to run the model in forward mode and to estimate the biologically effective dose through the target tissue. In that case the estimated internal dose was close to 0.003 µg/l, far below any internal dose derived reference value, corresponding to the lower estimates of the already considered exposure scenarios.

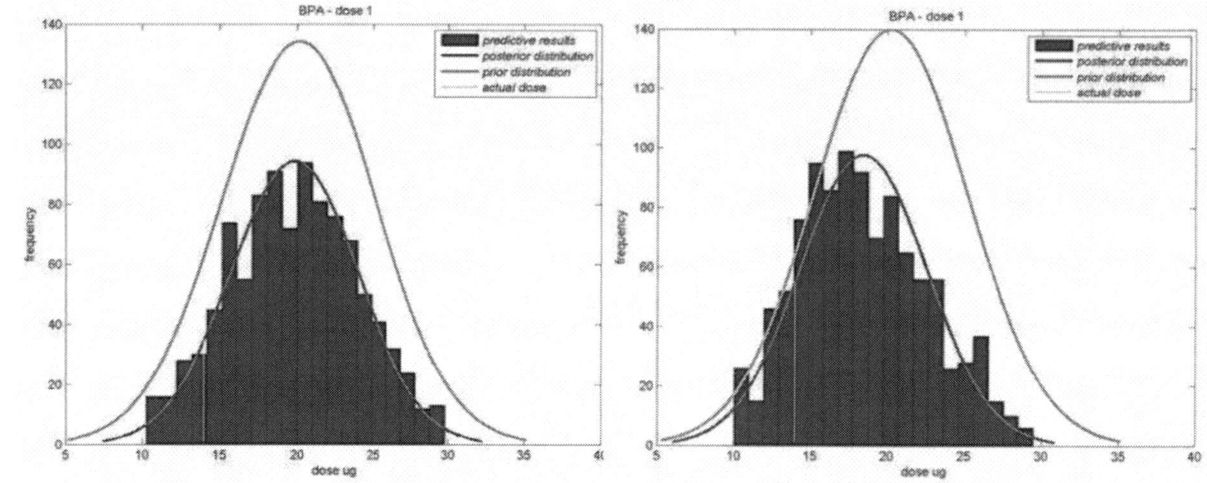

Figure 6. Reconstruction of oral exposure to BPA – dose 1 using two different algorithms (left side MCMC – right side DEMC)

Conclusions

The aim of the study is the development of a generic PBBK model, aiming at associating the different types of environmental, exposure, biomonitoring as well as multiple type toxicity data. The need to associate these types of data is increased by the world-wide interest for "exposure based" risk assessment. A current limitation to further introducing PBBK models in the risk assessment arena is the lack of generic character of these models. By introducing a multi-route, multi-compartmental, life-time PBBK model that describes both phase I and phase II metabolism, perfusion as well as membrane limited blood flow, the limiting step of describing the ADME process of a large chemical space is the proper parameterization for "data poor" compounds. Using QSARs for predicting key model parameters for a broad chemical space seems to be a promising way around this problem. Towards this aim, an original QSARs methodology was developed, coupling Abraham's solvation equation to Artificial Neural Networks. Our method yielded significantly improved results compared to the existing methodologies. Biomonitoring data assimilation through exposure reconstruction, allows the exploitation of the vast amount of biomonitoring data collected worldwide. Exposure reconstruction allows not only the evaluation of given HBM data against external exposure reference values, but also to estimate the actual biologically effective dose (by re-running forward the estimated external dose) and to compare this to internal exposure metrics. This concept is in perfect agreement with the rapidly produced high-throughput screening (HTS) *in vitro* data. Overall, our comprehensive and generic life stage PBBK model supports the association of a variety of environmental, exposure and biomonitoring data, as well as the incorporation of recent advances of in vitro toxicology using high-throughput systems in the risk assessment process enhancing thus significantly the artillery of environmental health science and chemical safety regulators.

References

1. Sarigiannis D.A., Karakitsios S.P. (2012) A dynamic physiology based pharmacokinetic model for assessing lifelong internal dose. in AIChE 2012. Pittsburgh, PA.
2. Dudek A.Z., Arodz T., Gálvez J. (2006). Computational methods in developing quantitative structure-activity relationships (QSAR): A review. Combinatorial Chemistry and High Throughput Screening. 9(3), pp. 213-228.

3. Gilks W.R., Roberts G.O. (1996) Strategies for improving MCMC, in Markov chain Monte Carlo in practice. 1996, Springer. p. 89-114.
4. Haario H., Laine M., Mira A., Saksman E. (2006). DRAM: efficient adaptive MCMC. Statistics and Computing. 16(4), pp. 339-354.
5. Ter Braak C.J. (2006). A Markov Chain Monte Carlo version of the genetic algorithm Differential Evolution: easy Bayesian computing for real parameter spaces. Statistics and Computing. 16(3), pp. 239-249.
6. Price K., Krishnan K. (2011). An integrated QSAR-PBPK modelling approach for predicting the inhalation toxicokinetics of mixtures of volatile organic chemicals in the rat. SAR and QSAR in Environmental Research. 22(1-2), pp. 107-128.
7. EFSA (2008) 2008. Toxicokinetics of Bisphenol A - Scientific Opinion of the Panel on Food additives, Flavourings, Processing aids and Materials in Contact with Food European Food Safety Agency.
8. Judson R.S., Kavlock R.J., Setzer R.W., Cohen Hubal E.A., Martin M.T., Knudsen T.B., Houck K.A., Thomas R.S., Wetmore B.A., Dix D.J. (2011). Estimating toxicity-related biological pathway altering doses for high-throughput chemical risk assessment. Chemical Research in Toxicology. 24(4), pp. 451-462.
9. Judson R.S., Houck K.A., Kavlock R.J., Knudsen T.B., Martin M.T., Mortensen H.M., Reif D.M., Rotroff D.M., Shah I., Richard A.M., Dix D.J. (2010). In vitro screening of environmental chemicals for targeted testing prioritization: The toxcast project. Environmental Health Perspectives. 118(4), pp. 485-492.

368818 Evolution of Mixed-Integer Programming and Its Applicationss to Process Systems Engineering

Abstract:

In this presentation, which is in honor of Larry Evans for his 80th birthday, we review the evolution and great progress that has been achieved in mathematical programming, particularly mixed-integer programming, over the last 50 years. We first focus on mixed-integer linear programming (MILP), and its great progress that has been achieved in the development of powerful branch-and-cut methods and its implementation in commercial software. We show the impact that MILP has had in solving large-scale supply chain, process planning and scheduling problems. We then review developments in mixed-integer nonlinear programming (MINLP), where there has been an increased availability of software packages implementing methods ranging from branch and bound to decomposition methods. Furthermore, there has been significant progress in deterministic global optimization algorithms for solving MINLP problems involving nonconvex constraints. We show the impact that MINLP is having in solving increasingly larger and more complex nonlinear planning and scheduling problems, as well as certain classes of process synthesis problems. We next provide next an overview of Generalized Disjunctive Programming (GDP), a logic-based modeling framework which is an alternative to the fully algebraic mixed-integer programming approach. Throughout the presentation we illustrate the application of MILP, MINLP and GDP techniques with a number of examples arising in the synthesis, design and enterprise-wide optimization of process systems. We close by highlighting some of the outstanding challenges in advancing the area of discrete-continuous optimization

371653 Optimizing the Refinery – Advances in Planning and Scheduling Decision-Making Strategies

Abstract:

Despite significant progress in the last decade, refining planning and scheduling processes continue to present very challenging modeling and optimization problems. In this work we discuss recent advances made in the area of refinery optimization. In particular, we cover three prominent areas that significantly affect the overall refinery economics: (i) crude assay characterization, (ii) optimization and analysis of crude purchasing decisions and (iii) optimal blending strategies for refinery products. The resulting problems represent some of the most challenging large-scale, mixed integer, non-convex optimization problems.

For the crude acquisition problem, we present advances in two key areas: (a) the creation of a novel assay modeling framework based on molecular characterization that describes crude assays at the detailed molecular level; (b) the creation of a robust optimization framework based on a combination of Pareto-type analysis, parametric optimization, and goal programming. This approach allows users to accurately represent the key refinery raw materials and their critical properties that affect the crude selection process. Given this assay representation, we evaluate the range of actionable decision-making options around the optimal solution that simultaneously preserve the economic optimization while accounting for important strategic and operational goals not explicit in the economic objective function (such as supply uncertainty, scheduling and maintenance).

For the product blending problem, we present a novel modeling and optimization methodology that determines the optimal sequence and timing of blend events, as well as rundown component tank switches in order to handle the blending of "hot" streams into a finished product tank. This solution incorporates multiple blend headers and multiple blends in a multi-period, event-driven campaign, using open-equation based optimization and modeling technology. This development is comprehensive, practical, and superior to current practices.

371336 The Role of Molecular Thermodynamics in Process Modeling

Abstract:

Molecular thermodynamics is the scientific foundation for process modeling and simulation, a universal engineering technology used in support of chemical product research and development, process development, design and debottlenecking, and process optimization and control. As chemical engineers strive to address critical societal challenges such as sustainable energy, global warming, and the aging population, molecular thermodynamics and process modeling technology must advance. This presentation reviews major developments in molecular thermodynamics and process modeling technology over the past thirty years. In addition, we showcase several recent high-impact advances in molecular thermodynamics relevant to the environment, energy, and medicine: 1) thermodynamic modeling of hydraulic fracturing fluids and produced water, 2) molecular characterization of crude oils, and 3) predictive solubility modeling of drug molecules.

379039 The Challenges of Product Development in High-Pressure Polymerization Technology - How Simulations Can Help

Abstract:

Product development in high-pressure polymerization technology is a specifically demanding task. First, operation procedures from large-scale technical practice cannot be scaled down to typical lab application one to one. Second, competitive high-pressure polymerization plants operate on a scale of significant more than 100 kt/á production capacity. Both, the significant scale up as well as limitations in miniaturizing laboratory equipment are still a remarkable challenge in designing strategies for product development in high-pressure polymerization technology. The contribution will take a look on how simulation technologies can help to overcome these challenges and were actual the limitations occur.

The first target is the prediction of microstructural polymer properties from process conditions and strategy. During the recent years there has been remarkable progress. Specifically, the combination of deterministic modeling with Monte-Carlo techniques, so-called as hybrid technology, brought significant improvement with respect to detail of predicted molecular structure and limitation of computational efforts. By this the detail is sufficient for feeding rheological models and demands on computational equipment together with the speed of computation is sufficient to establish such technologies in standard research. The contribution review the current status present examples of such investigations and discuss the portability of such technologies.

With increasing insight into the polymer structure aspects of reactor operation come into focus again that have been accepted for a long time as non ideal behavior to be adapted as the impact of the correlation of process conditions in polymer microstructure had been the dominant problem. Now, as a second aspect the modeling of species none uniformly distributed along the reactor axis come into focus again and the resulting consequences. The contribution will have a look on radial distribution in high-pressure polymerization reactors together with origin and consequences of varying heat transfer in reactor tubes.

Another issue are the accessible thermo physical properties of supercritical media. For a long time their precision seemed to be sufficient as the dominant field of uncertainty was the reaction mechanism together with its kinetics. Again, progress in this field reopened the question of precise information in this field. The contribution will evaluate that current status and discuss consequences and perspectives.

379095 On the Solution of Large-Scale Chemical Production Scheduling Problems

Abstract:

In this talk, we first present an overview of the different types of chemical production scheduling problems and the mixed-integer programming (MIP) models that have been proposed to address them. We then discuss four solution methods we have recently developed to enhance the solution of MIP scheduling models: (1) algorithms for the generation of unit-, task-, material-, and resource-specific nonuniform grids used to formulate multi-grid discrete-time models; (2) four classes of tightening inequalities based on a demand propagation algorithms; (3) a branch-and-bound algorithm designed to harness parallel computational resources; and (4) a number of reformulations and branching strategies. We also present computational results showing that the proposed methods lead to computational improvements of two to four orders of magnitude, thereby enabling us to address problems that were thought to be intractable.

Integrated Planning and Scheduling for Large Scale Semi-Continuous Process Plants

Vladimir Mahalec
Department of Chemical Engineering, McMaster University, Hamilton, ON, Canada

Large scale semi-continuous process plants (e.g. refineries, petrochemical sites) produce many products by transforming its feedstocks into pure products or components which are blended into final products. Processing steps are carried out in separation unit, reactors, or blending units; accurate prediction of behavior of all of these units as a rule requires nonlinear models.

Optimal long term product of such a plant requires that the materials be processed in such a manner that the value of the products minus (the cost of the feedstocks plus operating costs) is maximized. Operating costs include cost utility costs (energy, water, etc.) and the costs of switching operation of various units from producing one slate of products to producing another slate of products.

For large scale semi-continuous plants, the costs of feedstocks and utilities are two or more orders of magnitude higher than the costs of switching process units from one mode of operation to another. Long term production plan for semi-continuous multi-product plant optimizes use of raw materials and utilities, while disregarding the costs of switching. Such plans are most often computed via multi-period month by month models over the planning horizon (which usually is twelve months). Requirement that the production planning model contains monthly models leads to large, multi-period models; if process units are represented by nonlinear models, then the planning model becomes difficult to converge. Hence, the planning models comprise typically of linearized models or simplified nonlinear models.

Recently introduced inventory pinch based planning models enable computation of production plans by a two level decomposition. The first level consists of a nonlinear model which has only a small number of periods delineated by the products cumulative demand/production pinch point. The second level uses a fine grid multiperiod linear model (corresponding to the optimal states from the first level) to compute (i) short term production plan – e.g. first 30 days, and (ii) long term production plan – e.g. 12 months. Production plan computed via decomposition is at the same optimum as the plan that would be computed via fine-grid nonlinear model, but it takes a lot less computational effort.

Short term production plans includes mixed integer constraints that govern allocation of swing storage and threshold production constraints. Results of the short term production plan become constraints for computing production schedules. Since scheduling is in general a NP-hard problem, one can not guarantee computation of the optimal schedule in polynomial time. We compare various methods (decomposition, heuristics) for computing the schedules and discuss their applicability in industrial practice.

Ontology design for physical chemical systems

Heinz A Preisig NTNU - Trondheim, Norway heinz.preisig@chemeng.ntnu.no

Background

Computational engineering is getting more and more a central activity in all engineering disciplines. It certainly will also increasingly affect chemical engineering as the trend is towards multi-scale computations in particular simulation covering the full range: quantum and nano scale ab initio, micro scale and meso-scale structures, as well as macroscopic design and operation-related computations. This requires a shift in education and specifically a shift towards more computation related education as well as a widening of the knowledge base that chemical engineers take along from their education.

On the research side we increasingly see an integration of material-related computational activities with traditional chemical engineering computation tools. Specifically large-scale optimisation has been applied to the particulate domain, which intrinsically is a very large-scale energy minimisation problem. Otherwise, many of the associated techniques, such as computational fluid mechanics, ab initio computations, micro and meso-structure computations have not really made the impact on the profession than one would expect.

Whilst one will focus on continuously improving the computational capabilities by designing new and improved algorithms, programs and work on implementing workflows across the tools and scales, one should also observe a refocus of a large group of users towards the structures of the problem they work on in contrast to developing algorithms. This will increase the productivity of the engineers in particular the quality of the work will increase whilst the required time will decrease. The change will really only going to happen in a larger scale if one succeeds to have more people have a broad knowledge and if the providers construct a computational frame work, that is easy to use and constructed around consistenat basic information. To mention an example it should not be the case that one constructs and performs simulations that do not close the basic conservation principles.

Presentation

Besides the background, I will discuss a structure for an ontology suitable for the represenation of physical-chemical-biological systems. The structure is buidling systematically from a starting piece of information, namely a directed graph consisting of nodes and arcs. Information is added each time defining a new piece based on what has been defined before. Thus the information is build in a lower triangular form, or what in compiler construction is called a context-free definition.

One or two example will be used to explain the use of the ontology: a reactive flash and a stirred tank reactor with feed, some outlets and a breathing pipe, with a three phase system in the contents, an extractive reactor.

These two systems have most of the components one would normally find in the processing industry.

Comment Could also fit into 10E00 or 10E04

386771 Ontology Engineering and Its Applications to the Development of Biorefineries

Abstract:

The paper addresses the deployment of ontology engineering in modelling, scheduling and planning aspects of the biorefineries. A vast amount of data, models and designs are available in the area yet scattered and heterogeneous. Models and tools are designed requiring experts to handle codes and databases to further proceed studies and analysis. The scattered parts of the models unknown to the non-experts require modification selectively depending on the scenario and criteria examined. These parts are consisted of complex networks of qualitative data (chemicals, technologies, products etc), quantitative data (cost, enery consumption, environmental impact etc) and scheduling problems for the field. A key challenge in the development of biorefining is to establish connections among the complex networks in a dynamic framework easily used by non-experts combining all the scattered parts of the data sources. Through ontology engineering the available data and knowledge are concentrated in a structural and flexible form to contribute directly to the development of biorefinery-based models (GAMS, MATLAB, Aspen). The changing parts synthesizing the models are organized in the ontology and correlated to each other to serve all possible combinations. A methodology to automatically complete the context of any scenario of the biorefining models depending on selective criteria is illustrated by this paper.

363444 The Ontology System for Easy and Reusable Model Knowledge Representation

Abstract:

The development and application of process models are usually not trivial due to the complexity of the system being studied and/or scales being considered. Therefore, it is useful to store the modelling results in some knowledge base with the possibility to update, access and extract information about the model, in order not to repeat the same model development steps, and also to reuse or improve existing know-how. Hence, there is a need for a simple but efficient system for knowledge representation and management of models and their connections to other tools for different model-based applications.

This work presents design and implementation of an ontology for knowledge representation related to computer-aided modelling. This ontology is built on the basis of the modelling language, which includes classes, layers, and blocks (these classification names are related to the graphical representation of the model in the modelling-tool software). Class is a pattern with specific parameters or characteristics that is used as template for creation of modelling objects. These objects are then called the instances of the class. Classes interrelate between each other; upper layers represent classes and contain sub-layers; sub-layers contain blocks. The collection of instances of the classes represents the model in the system.

The structure of the knowledge ontology is based on a model decomposition technique [1]. Every model representation includes four upper layers, which are *System information, Balance equations, Constitutive equations* and *Connection equations*. Every process model could be decomposed to the equations and relations, which can be linked to one of the upper layers. Each upper layer includes a number of sub-layers, representing specific phenomena occurring in the system. In the upper layer there are predefined sub-layers (giving the basic and essential description of the system) as well as user-defined sub- layers, which might be unique for the specific case. The predefined sub-layers are, for example, number of balance-volumes, number of phases, presence of the reaction etc., given in the *System information* upper layer; similarly, mass balance, energy balance, momentum balance equations belong to the *Balance equations* upper layer; kinetic, thermodynamic, diffusion models belong to the *Constitutive equations* upper layer; relations between volumes inside the system and between system and outside world, closure equations, initial conditions belong to the *Connection equations* upper layer. Each sub-layer contains one or more blocks, representing various options or scenarios, which are possible in relation to the phenomena, linked to the corresponding sub-layer. Model equations are included in the corresponding blocks and this gives the model developer the possibility to quickly generate-create the required type of model by choosing one of the options available in the sub-layers. Also, the system allows updating of existing models, therefore, storing knowledge for further reuse.

On top of simple representation of the model structure this system is linked to a tool for model solution and analysis. Also, a model transfer via XML option for the created or existing model is available. Therefore, any tool, which accepts this type of XML file as an input, could be linked to the modelling tool-box.

This modelling ontology can be applied to a wide range of modelling applications in chemical engineering, as it is not associated to any specific modelling tools. The software implementation of the ontology makes it also easy to use even for the model developer with limited knowledge about modelling. The ontology also represents the knowledge in a way that allows model reuse and transfer to other software applications.

[1] I. Cameron, R. Gani, 2011, Product and Process Modelling. A Case Study Approach., Elsivier.

Rapid ontology alignment in large metabolic information databases

Akhil Kumar[2] and Costas D. Maranas[1]

[1]Department of Chemical Engineering, Pennsylvania State University, University Park, PA; [2] The Huck Institutes of the Life Sciences, Pennsylvania State University, University Park, PA;

A large number of resources are focused on metabolic information such as metabolites, enzymatic reactions, metabolic pathways and metabolic network models. MetRxn is a freely available searchable web-resource that integrates heterogeneous information from 8 metabolic databases and 112 metabolic models. The metabolic information resources include databases such as BRENDA, KEGG, MetaCyc, ChEBI, HMDB, Rhea, Reactome, BKM and metabolic network models from all taxa of life. Information heterogeneity arises due to disparity in semantics and incompleteness of annotation. MetRxn uses a suite of custom algorithms to automatically curate heterogeneous metabolic information using metabolite atom and bond connectivity details as inputs. The size of the curated dataset in MetRxn is currently at 115512 metabolites, 42965 reactions, 455 metabolic pathways and 112 metabolic models (May 2014). We leverage this curated dataset with complete atom and bond connectivity details to automatically categorize metabolites with semantic annotations. Semantic annotations include metabolite functional group and reaction mechanism identification. Semantic annotations are identified using a novel polynomial time substructure search algorithm CLCA (Canonical Labelling for Clique Approximation). CLCA applied to a single molecular graph identifies labels similar to chemical fingerprints, and when applied to two or more molecular graphs, identifies conserved moieties. Functional groups (e.g. ester, carboxyl groups, etc.) are identified by common labels and reaction mechanisms are identified by the moiety conserved between the substrate to product transformations. Ontology classifiers inspired from existing efforts such as CHEBI, KEGG and IUBMB are applied by aligning metabolic entities for common semantic annotations. Ontologies represent knowledge in a hierarchical form that is amenable to manual as well computational analysis. A hierarchical representation allows users to trace parent, child and siblings across a cascade of semantic relationships for each metabolic entity in MetRxn. Ontology information for metabolites, reactions, pathways, metabolic network and additional annotations such as atom maps, UniProt and NCBI gene id cross references for all reactions in MetRxn can be accessed at http://www.metrxn.che.psu.edu/.

14AIChE Annual Meeting, Atlanta, GA

American Institute of Chemical Engineering
14 AIChE Annual Meting
November 16-21, 2014 – Atlanta, GA. USA.

Operations Research Ontology for Improving Optimization Frameworks

Edrisi Muñoz, Elisabet Capón, José M. Laínez, Antonio Espuña, Luis Puigjaner

ETH Eidgenössische Technische Hochschule Zürich Swiss Federal Institute of Technology Zurich

UNIVERSITAT POLITÈCNICA DE CATALUNYA BARCELONATECH

University at Buffalo The State University of New York

CIMAT RESEARCH CENTER FOR MATHEMATICS

Decision-making on Enterprise Wide Optimization (EWO)

- Processes at industry are highly complex systems with different space and time scales.
- EWO is a new emerging area that lies at the interface of chemical engineering and operations research.
- Relevant challenges in EWO are the integrated and coordinated decision-making across:
  - Various functions in a company
  - Various geographically distributed organizations
  - Various levels of decision-making

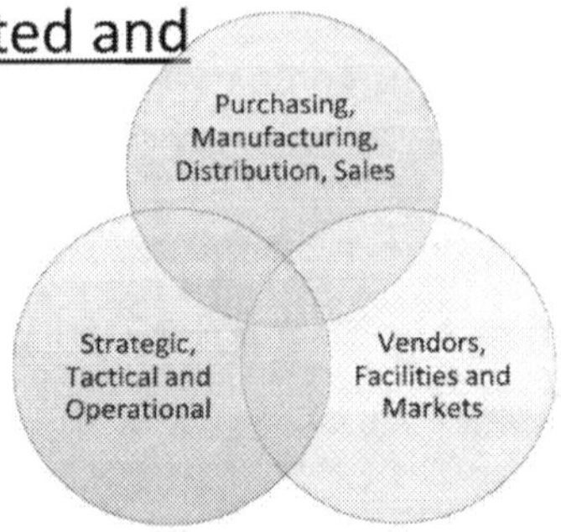

Analytical models in EWO

- Many models are in the scope of **Operations Research**
- The <u>more the detail</u> in analytical models, the <u>higher</u> the <u>complexity</u>
- Different models could represent the <u>same given problem</u>
- Models depend on <u>system conception and modeler skills</u>
  - Different system requirements
  - Different system constraints
  - Different mathematical notation

$$\min \sum_{t \in T} f_t(x, w_t, w_{t-1}, \theta_t)$$

$$s.t.\ g_t(x, w_t, w_{t-1}, \theta_t) \leq 0 \quad t \in T$$

$$x \in R^{n_x}, x_i = 0,1 \quad i = 1, m_x$$

$$w_t \in R^{m_{wt}}, w_{tj} = 0,1 \quad j = 1, m_w, t \in T$$

Enterprise Ontology Project (EOP)

- An ontological model that <u>represents the enterprise</u> operational, tactical and strategic <u>functions</u>.
- It allows a <u>better understanding</u> of the real and assumed <u>system</u>.
- It <u>integrates and standardizes</u> data/data/analytical/analytical models.
- It is based on *ISA 88 & 95* automation and control <u>standards</u>.
- Define **Process, Procedure, Physical, and Recipe Models**.

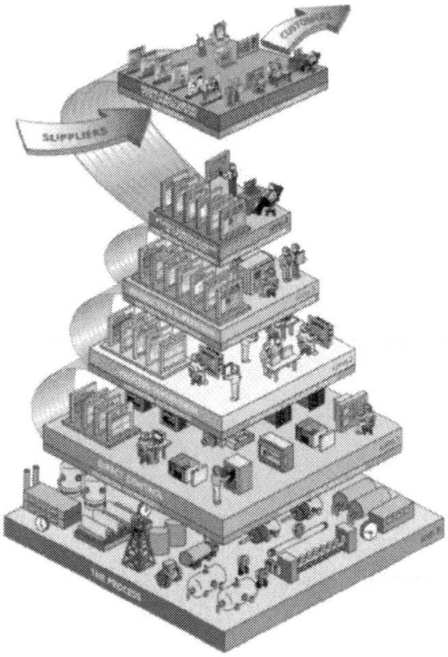

Mathematical Modeling Ontology (MMO)

- An ontological model that represents the <u>mathematical modeling bases</u>.
- It allows the <u>integration</u> of mathematical models from <u>different hierarchical levels</u>.
- It <u>relates (V&V)</u> mathematical elements and real system elements.
- Introduces the **Conceptual Mathematical Modeling** approach for real system conception.

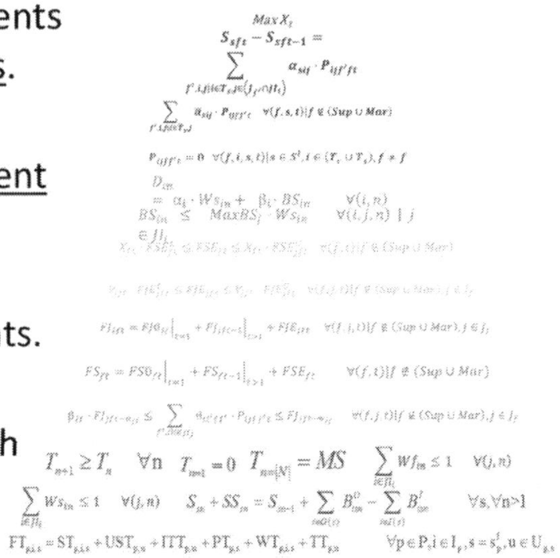

Outline

Proposal

Previous work

Case study

Conclusions

Operations Research Ontology (ORO)

- Operations Research Ontology (ORO) aims to **support problem design and mathematical formulations** (modeling & programming) for decision making in order to reach EWO.
  - It standardizes and facilitates the decision making process based on operations research.
  - It creates and classifies systems of mathematical equations according to the system's specification and decision type.
  - It re-uses knowledge from Enterprise Ontology Project and Ontological Math Representation.

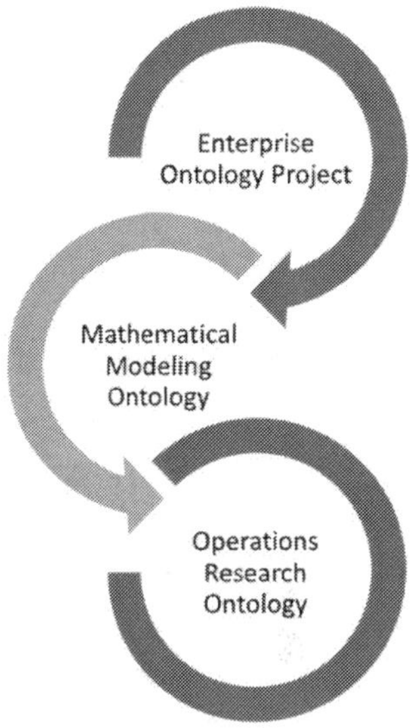

Operations research ontology

- MMO metrics
  - 52 classes
  - 64 properties
  - 9 axioms
- Classes:
  - algorithm, decision-making, feasible solution, model construction, model, model validation, optimal solution, optimization, problem definition, problem recognition, system
- Properties:
  - appliesAlgorithm, testsModel, hasPhase, derivesInOptimalSolution, examinesSystem, establishesGoals, hasMathematicalConstraint

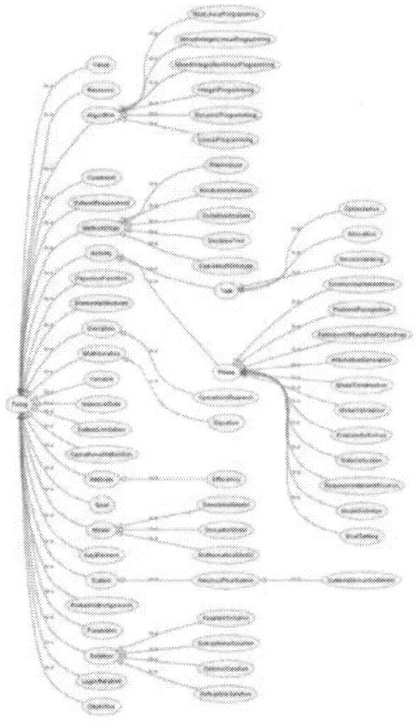

ORO main features

- The <u>functionality</u> of the ontology applies theory, methods, and techniques related to **Operations Research**.
- It <u>verifies and validates</u> OR models (structure and solution) applied at EWO.
- It captures <u>key elements</u> of operations research useful for <u>decision-making task</u>.

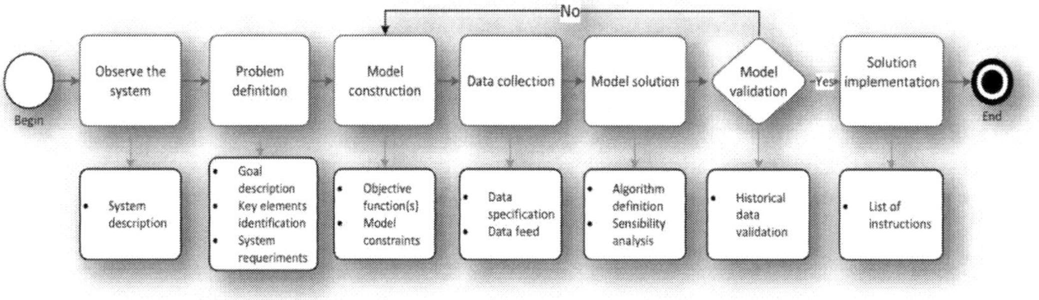

Knowledge of optimization models

Extract of the correlation matrix of scheduling models and system & mathematical elements **data base**

Concept	Symbol	Notation - Modeling aspects (Review on scheduling by Mendez et al. (2006)								
		STN-discrete	RTN-discrete	STN-continuous	RTN-continuous	Unit specific time event	Time slots	Unit specific immediate precedence	Immediate precedence	General precedence
Indices										
Task		i	i	i	i	i	J_l			
Equipment/Unit		j	R_i^J	j	R^J	j	j	j	j	j
TimePeriod/EventPoint/Slot		t	t	n	n	n	k			
State		s		s						
Family		f								
Resource		r	r	r	r					r
Resource unit										z
Batch							i	i	i	i
Stage							l			l
Variables										
Allocation	W	W_{ijt}	W_{iRJ},Jt	Ws_{in};Wf_{in};V_j sn	$W_{inn'}$	Ws_{in};W_{in};Wf_{in}	W_{ilk}	WF_{ij};$X_{ii'}$	WF_{ij};W_{ij}	W_{ij}; Y_{iz}
Batch size	B	B_{ijt}	B_{iRJ},Jt	Bs_{in};Bp_{in};Bf_{in}	$B_{inn'}$	B_{in};Bs_{in};Bf_{in}				
Storage capacity	S	S_{st}	S_1	S_{sn};S_{sin}		S_{sn}				
Minimum unit capacity	Vmin	$Vmin_{ij}$	$Vmin_{iRJ}$J	V_j^{min};V_j^{max};C_j	V_i^{min},V_i^{max}	V_i^{min};V_i^{max}				
Maximum unit capacity	Vmax	$Vmax_{ij}$	$Vmax_{iRJ}$J							
Minimum inventory requirement	Cmin	$Cmin_s$	Cmin	C_s^{max}		C_s^{max}				

* Mendez et. al 2006, *State of the art review of optimization methods for short-term scheduling of batch processes.*
* Harjunkoski et. al 2014, *Scope for industrial applications of production scheduling models and solutions methods.*

Operations Research Model

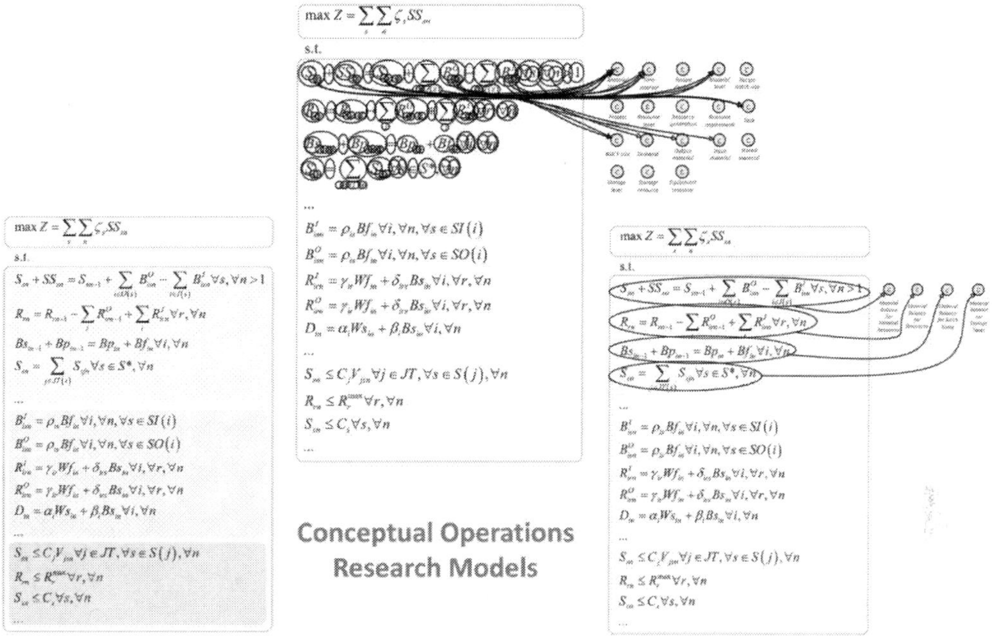

Proposed framework workflow

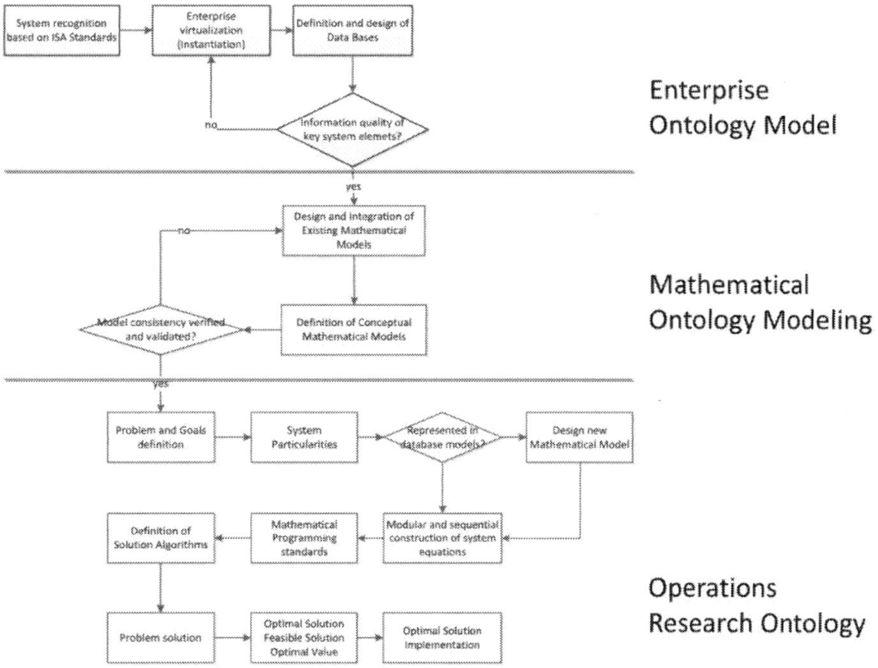

Case Study

- Three suppliers, Four potential locations (Distribution Center/Production site. Each site is represented by a STN production processing and six markets.

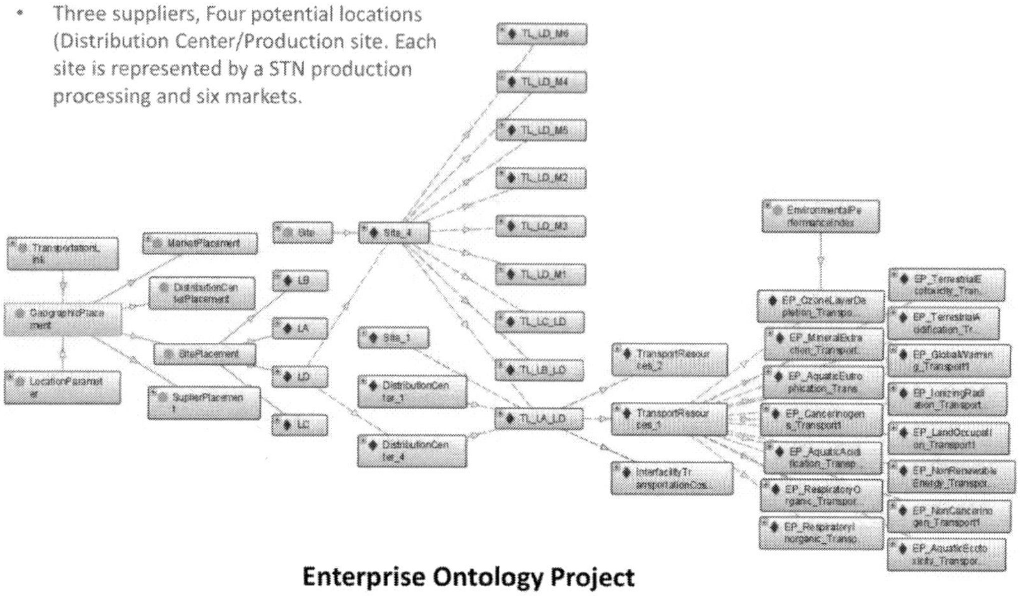

Enterprise Ontology Project

Case Study

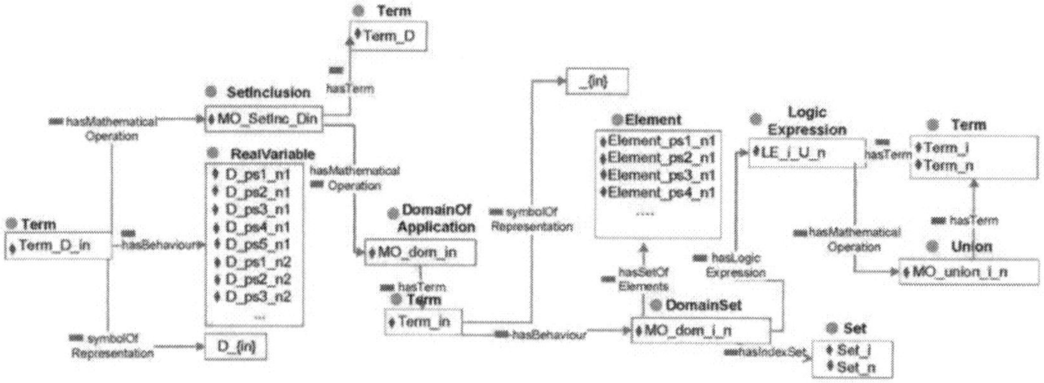

Mathematical Modeling Ontology

Case study

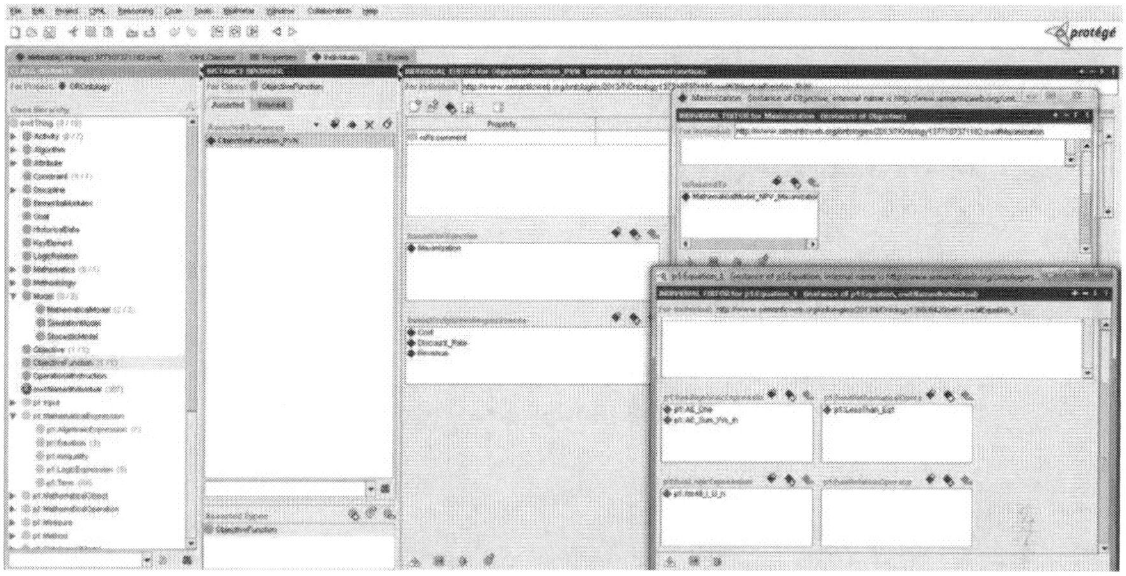

Operation Research Ontology

Summary

- System to support EWO based on knowledge technologies
- Facilitates the tasks of systems modeling and analytics based on OR

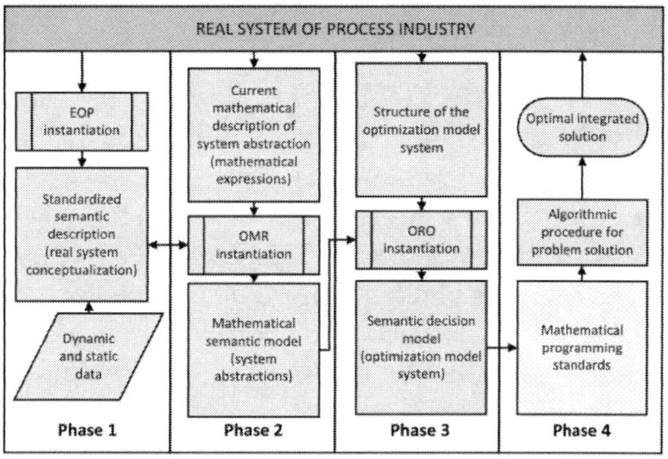

Conclusions

- Informal representation
  - Different human readable representations with real meaning of mathematical expressions, such as, graphics, natural languages and symbols.
- Formal representation
  - Machine reading and understanding mathematical expressions, such as, Content MathML, Markup MathML, OWL, Mathematical programming languages, UML.
- Current approach
  - A framework capable of delivering quality information, standardizing and integrate analytical models.

Acknowledgments

We want to fully thanks to the financial support received from the Spanish Ministerio de Economía y Competitividad and the European Regional Development Fund are acknowledged for supporting the present research with projects EHMAN (DPI2009-09386) and SIGERA (DPI2012-37154-C02-01), and the "Agència de Gestió d'Ajuts Universitaris i de Recerca (2014SGR1092)". In addition also Centro de Investigacion en Matematicas A. C. (CIMAT), from México is fully acknowledged.

Automated Learning of a Materials Science Ontology using Natural Language Processing of Abstracts

Matthew Conway, Jonathan Packer, Daniel Esposito and Venkat Venkatasubramanian

Complex Resilient Intelligent Systems Laboratory
Department of Chemical Engineering
Columbia University
New York, NY 10025

The scientific literature is vast, growing rapidly, even in narrow topical domains, making it challenging for both novice researchers and seasoned experts, who frequently find themselves having to explore new disciplines or sub-disciplines outside their area of expertise. Typically researchers proceed through the literature in an ad-hoc exploratory fashion and may end up drawing wrong conclusions about relevance of concepts in the papers they read [Jardine, 2014]. A properly comprehensive view of a discipline can be provided by an expert review, but for fast-emerging fields one might not yet be available. Also, some fields may not be well defined enough to have experts. While there has been plenty of work in the information retrieval field on many-document summarization, some domain-specific customization is necessary to get these methods to work in the scientific fields, where non-standard abbreviations, measurement units and other complex terms abound [Harmston, Filsell, and Stumpf, 2010]. In practice, this means that any attempt to do text mining on the literature quickly becomes a data-cleanup operation.

In order to make this process transparent, explicit and repeatable, we have developed a performant, extensible, term extractor for text-mining of materials science and engineering literature. It uses the information-theoretic approach of Bouma [2009] to identify collocations (roughly equivalent to noun phrases) and also resolves acronyms and alternative uses of hyphenation and pluralization. The program takes its input as a simple list of plain-text abstracts and provides hooks for user-defined domain-specific preprocessing rules. It retains positional information so each disambiguated term it identifies can be mapped back to the section of the original text from which it came. The results of using this approach to investigate recent trends in the thin-film solar cells will be presented and compared with a recent overview of the field [Polizzotti et al., 2013] to identify strategies towards automatic review generation. We will also present results of running topic modeling [Blei and Lafferty, 2007] on the "bag of terms" the model outputs and first steps toward a tool that can help a user organize the domain knowledge into a comprehensive ontology.

References:

1. Bouma, Gerlof. 2009. "Normalized (Pointwise) Mutual Information In Collocation Extraction." Proceedings of GSCL.

2. Blei, David M., and John D. Lafferty. 2007. "A Correlated Topic Model of Science." *The Annals of Applied Statistics* 1 (1) (June): 17–35. doi:10.1214/07-AOAS114. http://projecteuclid.org/euclid.aoas/1183143727.

3. Harmston, Nathan, Wendy Filsell, and Michael P H Stumpf. 2010. "What the Papers Say: text Mining for Genomics and Systems Biology." *Human Genomics* 5 (1) (October): 17–29. http://www.pubmedcentral.nih.gov/articlerender.fcgi?artid=3500154\&tool=pmcentrez\&rendertype=abstract.

4. Jardine, James G. 2014. "Automatically Generating Reading Lists" (848).

5. Polizzotti, Alex, Ingrid L. Repins, Rommel Noufi, Su-Huai Wei, and David B. Mitzi. 2013. "The State and Future Prospects of Kesterite Photovoltaics." *Energy & Environmental Science* 6 (11): 3171. doi:10.1039/c3ee41781f. http://xlink.rsc.org/?DOI=c3ee41781f.

388945 Semantically Enabled Technology for Port Symbiosis

Abstract:

Ontology Engineering is proved to be a leading technology in the field of Industrial Symbiosis. Classification of processes, raw materials, products, byproducts, emissions and energy as well as several relationships among them are already presented in several studies. Primary data are used to produce new metadata, where numerous connections are uncovered. On top of this, enablers can serve as imaginary industrial units, to reveal initially impossible connections. Enablers can be mechanical, chemical, biochemical, thermochemical, electrochemical units etc. Additionally, it is apparent that for every new type of classification, such as EWC, a number of new possibilities is added to the system.

Although ontology engineering is a reliable way to produce superstructures of material valorization routes, manual work is required in order to refine and quantify results. For this reason an expansion to the conventional practice is proposed, where tacit knowledge is valorized in a mathematical way. Semantic distance should not be limited to the "number of production steps" and "geospatial distance". The user needs to optimize the superstructure using the criteria an Engineer would use on "ITSELF" (Infrastructure, Technical, Social, Environmental, Legal, Financial) methodology.

During the EPIC2020 project, workshops and interviews took place as communication channels for both collecting data and disseminating the project. The workshops were attended by stakeholders, academics, policy makers, SME's and individuals. They all brought up significant views concerning new bioenergy production activities. These should be modeled, investigated and evaluated for maximizing the project's impact.

The above input is grouped in six categories. Infrastructure: information dealing with the existance, size and development status of Natural Gas Network, District Heating Network, Steam Network, car biofuels and current symbiotic solutions. Technical: Technology compatibility based on factors such as moisture content, technology maturity, chemical resemblance (i.e. Hydrogen to Carbon ratio and Oxygen to Carbon ratio of the functional units' inputs and outputs for estimation of entropy difference). Social: these models evaluate mainly the number of new jobs, both construction workers and operators. Environmental: Pollution levels, noise levels, greenhouse gas savings. Legal issues: biogas feed mixture, maximum capacity eligible for national financing, bioenergy production subsidy. Financial information concerns the supply and demand of the candidate products as well as the capital investments and their payback time.

381733 Reusing Ontologies in the Domain of Process Systems Engineering

Abstract:

The number of ontologies available today, more than 10,000 as identified by dedicated search engines[1], is a good indicator of the progress made so far. Ontologies are developed for and used in several applications ranging from knowledge representation and semantic search up to data integration and web service discovery (Cheung, Cheung & Kwok 2012)(Trokanas, Cecelja & Raafat 2013, Raafat et al. 2013b)(Raafat et al. 2013a). Much of recent ontology development efforts have been reported in the domain of Process Systems Engineering.(Trokanas et al. , Cecelja et al. 2013, Muñoz et al. 2012).

It is the fact that ontologies are by definition conceptualisations that are supposed to be shared and reused (Gruber 1993, Gruber 1995). Ontology reuse is a practical and useful approach for knowledge engineers. It has the potential to reduce the cost of developing ontology from scratch, as well as promote interoperability among applications. In addition, many of the existing ontologies cover similar or overlapping domains. Ontology reuse is also part of many of existing frameworks and established methodologies (Noy, McGuinness 2001, López et al. 1999) for ontology development. However, by reviewing the literature (Bock et al. 2010, Brandt et al. 2008, Fernandes et al. 2011, Giménez et al. 2008), it became apparent that ontology reuse is a seldom occurring task. Researchers have identified a lack in robust and pragmatic methods for evaluating and identifying ontologies for reuse.

This paper presents a metric for the evaluation of ontologies for reuse. The metric account for ontology metadata such as the datatypes, external resources, terminology and languages used for modelling. It also accounts for structural characteristics of ontologies such as size, breadth and width of the evaluated ontologies. All the available information is translated into vectors which are consequently compared. The resulting similarity scores for each aspect are aggregated and normalised into a single metric ranging between [0,1]. The metric has been tested and verified using experiments for the development of an ontology for the domain of Industrial Symbiosis.

The effort has been partly funded by the LIFE+ initiative (LIFE 09 ENV/GR/000300) and EPSRC, which authors acknowledge.

Bock, C., Zha, X., Suh, H. & Lee, J. 2010, "Ontological product modeling for collaborative design", *Advanced Engineering Informatics,* vol. 24, no. 4, pp. 510-524.

Brandt, S.C., Morbach, J., Miatidis, M., Theißen, M., Jarke, M. & Marquardt, W. 2008, "An ontology-based approach to knowledge management in design processes", *Computers & Chemical Engineering,* vol. 32, no. 1, pp. 320-342.

Cecelja, F., Trokanas, N., Raafat, T. & Kokossis, A. 2013, "Ontology Engineering for the Development of Industrial Symbiosis Networks", *2013 AIChE Annual Meeting Global Challenges for Engineering a Sustainable Future*.

Cheung, C.F., Cheung, C. & Kwok, S. 2012, "A knowledge-based customization system for supply chain integration", *Expert Systems with Applications,* vol. 39, no. 4, pp. 3906-3924.

Fernandes, R.P., Grosse, I.R., Krishnamurty, S., Witherell, P. & Wileden, J.C. 2011, "Semantic methods supporting engineering design innovation", *Advanced Engineering Informatics,* vol. 25, no. 2, pp. 185-192.

Giménez, D.M., Vegetti, M., Leone, H.P. & Henning, G.P. 2008, "PRoduct ONTOlogy: Defining product-related concepts for logistics planning activities", *Computers in Industry,* vol. 59, no. 2, pp. 231-241.

Gruber, T.R. 1995, "Toward principles for the design of ontologies used for knowledge sharing", *International journal of human computer studies,* vol. 43, no. 5, pp. 907-928.

Gruber, T.R. 1993, "A translation approach to portable ontology specifications", *Knowledge acquisition,* vol. 5, no. 2, pp. 199-220.

López, M.F., Gómez-Pérez, A., Sierra, J.P. & Sierra, A.P. 1999, "Building a chemical ontology using methontology and the ontology design environment", *Intelligent Systems and their Applications, IEEE,* vol. 14, no. 1, pp. 37-46.

Muñoz, E., Capón-García, E., Espuña, A. & Puigjaner, L. 2012, "Ontological framework for enterprise-wide integrated decision-making at operational level", *Computers & Chemical Engineering,* vol. 42, no. 0, pp. 217-234.

Noy, N.F. & McGuinness, D.L. 2001, "Ontology development 101: A guide to creating your first ontology", .

Raafat, T., Trokanas, N., Cecelja, F. & Bimi, X. 2013a, "An Ontological Approach Towards Enabling Processing Technologies Participation in Industrial Symbiosis", *Computers & Chemical Engineering,* .

Raafat, T., Trokanas, N., Cecelja, F. & Bimi, X. 2013b, "An ontological approach towards enabling processing technologies participation in industrial symbiosis", *Computers & Chemical Engineering,* vol. 59, no. 0, pp. 33-46.

Trokanas, N., Cecelja, F. & Raafat, T. 2013, "Semantic approach for pre-assessment of environmental indicators in Industrial Symbiosis", *Journal of Cleaner Production,* .

Trokanas, N., Raafat, T., Cecelja, F., Kokossis, A. & Yang, A. "Semantic Formalism for Waste and Processing Technology Classifications Using Ontology Models" in *Computer Aided Chemical Engineering* Elsevier, , pp. 167-171.

---

[1] http://swoogle.umbc.edu/

Robust Design and Operation of Red Blood Cell Production in a Parallelised Hollow Fibre Bioreactor

R. Misener[1], M. Fuentes-Garí[1], E. G. Velliou[1], N. Panoskaltsis[2], E.N. Pistikopoulos[1], A. Mantalaris[1]

1 Department of Chemical Engineering, Imperial College London, South Kensington Campus, London SW7 2AZ, UK

2 Department of Hematology, Imperial College London, Northwick Park & St. Mark's Campus, London HA1 3UJ, UK

Recent work developed a novel, biomimetic, cost-effective three-dimensional hollow fiber bioreactor for growing healthy red blood cells *ex vivo* (Panoskaltsis et al., 2012). This promising bioreactor recapitulates the architectural and functional properties of erythrocyte formation and thereby reduces the need for expensive growth factors by more than an order of magnitude. Individual experiments to empirically improve the bioreactor are cost- and labor-intensive, so we propose global superstructure optimization for bioreactor design. Our approach integrates topological design choices with operating conditions. Design choices include: number of parallelized bioreactors; number and type of hollow fibers in each reactor; size and aspect ratio of each bioreactor. Operating conditions are: feed concentrations; flow rate through the reactor. This presentation quantifies the potential for *ex vivo* red blood cell production to compete openly against the transfusion market for rare blood. We discuss the potential of our superstructure design strategy not only on this individual bioreactor but also more generally on bioprocess optimization.

The bioreactor is modeled using the Krogh cylinder approximation (Krogh, 1919). Five species represent mass exchange in the bioreactor: glucose corresponds to cellular nutrients; lactate models waste; oxygen stands in for cellular metabolism; stem cell factor (SCF) represents cellular expansion; erythropoietin (EPO) is mapped to cellular differentiation. Mass transfer parameters are derived from prior work (Macedo, 2011; Panoskaltsis et al., 2012). We cannot assume steady state operation; representing cellular growth, proliferation, and differentiation requires modeling multiple operation phases (including start-up) with a dynamic model. To create an MINLP, we discretize the ordinary differential equation (ODE) model of hematopoiesis (*i.e.*, blood production) introduced by Colijn and Mackey (2005) and extended by Ma et al. (2007, 2012). We address the MINLP using the solver software ANTIGONE (Misener and Floudas, 2013, 2014). We consider parameter uncertainty based on measured ranges (Macedo, 2011; Panoskaltsis et al., 2012) and use robust optimization strategies to inoculate bioreactor performance against uncertainty.

Recall that optimizing a bioreactor by discretizing a dynamic optimization problem and solving the resulting MINLP has been previously proposed by Banga et al. (1998, 2005). The mathematical novelty of our approach stems from explicitly considering simultaneous bioreactor design and operation and incorporating mathematical models of hematopoiesis into the optimization problem. The engineering contribution is showing how this superstructure design strategy can generally impact bioprocess optimization and studying the impact of robust strategies on the design choices. The industrial contribution is demonstrating the potential for *ex vivo* red blood cell production to compete openly against the transfusion market for rare blood. This effort belongs to the Imperial College *Framework for the Design, Modelling and Optimization of Biomedical Systems* (Velliou et al., 2014).

References

J. R. Banga, R. Irizarry-Rivera, and W. D. Seider. Stochastic optimization for optimal and model-predictive control. *Comput. Chem. Eng.*, 22(4-5):603–612, **1998**.

J. R. Banga, E. Balsa-Canto, C. G. Moles, and A. A. Alonso. Dynamic optimization of bioprocesses: Efficient and robust numerical strategies. *J. Biotechnol.*, 117(4):407–419, **2005**.

C. Colijn and M. C. Mackey. A mathematical model of hematopoiesis - I. Periodic chronic myelogenous leukemia. *Journal of Theoretical Biology*, 237(2):117–132, **2005**.

C. Y. J. Ma, R. Kumar, X. Y. Xu, and A. Mantalaris. A combined fluid dynamics, mass transport and cell growth model for a three-dimensional perfused biorector for tissue engineering of haematopoietic cells. *Biochem. Eng. J.*, 35(1):1–11, **2007**.

C. Y. J. Ma, N. Panoskaltsis, R. Kumar, X. Y. Xu, and A. Mantalaris. Simulation of ex vivo bone marrow culture: Application to chronic myeloid leukaemia growth model. *Biochem. Eng. J.*, 61:66–77, **2012**.

H. M. Macedo. A Novel 3D Dual Hollow Fibre Bioreactor for the Production of Human Red Blood Cells. *PhD thesis*, Imperial College London, **2011**.

R. Misener and C. A. Floudas. GloMIQO: Global Mixed-Integer Quadratic Optimizer. *Journal of Global Optimization*, 57:3–30, **2013**.

R. Misener and C. A. Floudas. ANTIGONE: Algorithms for coNTinuous / Integer Global Optimization of Nonlinear Equations. *Journal of Global Optimization*; In Press, **2014** (DOI: 10.1007/s10898-014-0166-2).

N. Panoskaltsis, H. M. M. Macedo, M. T. M. Blanco, A. Mantalaris, and A. G. Livingston. 3-dimensional hollow fibre bioreactor systems for the maintenance, expansion, differentiation and harvesting of human stem cells and their progeny, **2012**. Patent. WO 2012/069841 A1.

E. Velliou, M. Fuentes-Garí, R. Misener, E. Pefani, M. Rende, N. Panoskaltsis, A. Mantalaris, E. N. Pistikopoulos. A framework for the design, modeling and optimization of biomedical systems. In J. D. Siirola, M. Eden and G. P. Towler, editors, Proceedings of the 8th International Conference on Foundations of Computer-Aided Process Design - FOCAPD, Computer-Aided Chemical Engineering, **2014**. Accepted.

Acknowledgements

This work is supported by ERC-BioBlood (no. 340719), ERC-Mobile Project (no. 226462), by the EU 7th Framework Programme [MULTIMOD Project FP7/2007-2013, no 238013] and by the Richard Thomas Leukaemia Research Fund. R.M. is further thankful for a Royal Academy of Engineering Research Fellowship.

363026 A Simultaneous Optimization Approach for Synthesis and Design of Process and Water Networks

Abstract:

Recently, significant efforts are being made to reduce freshwater consumption and wastewater generation in process industries. These efforts are regarded as critical concerns among industrial practitioners due to the rise of freshwater and effluent treatment costs and stringent regulations. One of the more efficient ways to reduce freshwater consumption in the process is by reusing wastewater that is generated by the process or utility after being treated in the wastewater treatment plant to acceptable limits. This can be done by considering various treatment technologies and exploring numbers of alternatives. Furthermore, different process technologies and design alternatives are also being considered in the processing network to transform raw materials into products while reducing freshwater consumption and wastewater generation. In order to accomplish this goal, many design alternatives can be considered. One of the techniques used to represent the design space and identify best technology network is superstructure-based optimization techniques [1, 2].

Usually, in process synthesis and design, subsystems (e.g. water network and heat exchanger network) are dealt and solved sequentially or separately after process synthesis and design is performed and an optimal process flow sheet is identified. In this work, a simultaneous synthesis and design of process and wastewater networks or also known as a multi-network problem is presented. The integration of water consumption early at process synthesis and design stage is expected to reduce the consumption of freshwater and reusing more processed water. In this end, we address the challenge to manage the complexity of early-stage decision making in synthesizing and designing process and wastewater networks.

In order to address these challenges, a systematic framework recently developed by Quaglia et al. [1] is used as a basis and expanded use of the superstructure-based optimization approach for the optimal synthesis and design of processing plant network that is connected with a wastewater treatment network. A solution strategy to solve the multi-network problem accounts explicitly the interactions between the process and wastewater networks by selecting appropriate technologies and alternatives in order to convert raw materials into products and produce cleaned water to be reused in the process. The strategy accounts explicitly for the interactions between the process and water networks (so called as *a network within a network*) via selection of appropriate raw materials, technologies and alternatives for process and water treatment as well as products. The systematic approach is used to manage the difficulty and solving simultaneously process synthesis and water synthesis network problems with respect to economics, resources consumption and sustainability.

A new superstructure is formulated for the simultaneous synthesis of the process and water networks in order to achieve this task optimally and efficiently. All possible alternatives with respect to the topology of the process and wastewater network are represented in the superstructure at different process and treatment tasks. The resulting synthesis-design problem size is large and complex. The optimization problem is formulated as a Mixed Integer Non Linear Programming (MINLP) and solved in GAMS under different objective function scenarios (e.g. minimize total cost, minimize waste etc.) and constraints (e.g. effluent discharge limits, process pollutant limit etc.). The applicability of the systematic approach is demonstrated using a conceptual case study, especially developed to test all the features of the solution approach.

Reference:

[1] A. Quaglia, B. Sarup, G. Sin, R. Gani, 2012. Computers and Chemical Engineering, 38, 213.

[2] H. Yeomans, I.E. Grossmann, 1999. Computers and Chemical Engineering, 23, 709.

386467 Process Modeling and Kinetic Parameter Estimation for the Design of Photocatalytic Wastewater Treatment Systems

Abstract:

Semiconductor photocatalysis in principle relies on the use of irradiation for catalyst activation and has received considerable attention over the past decades toward achieving efficient removal of organic toxic substances from wastewater effluents. Moreover, solar photocatalysis is considered as a cost-effective and sustainable water treatment technique, due to the utilization of the abundant solar energy and the guarantee of no toxic emissions. Several heterogeneous photocatalysts have been investigated under ultraviolet and/or visible light irradiation conditions. However, the most acute problem of slurry photocatalysis is the recognized need for a post-treatment catalyst recovery step. Therefore, research has focused on supported catalysts that can prevent particle migration into the treated water and reduce photocatalytic process costs even more, by eliminating post-treatment.

The efficiency of solar photocatalytic removal of Bisphenol A (BPA) from aqueous samples has been previously studied by immobilizing two different photocatalytic materials (ZnO and TiO_2) via heat attachment onto a glass substrate and subsequently investigating the effect of several operating parameters known to affect this environmentally promising process: these include the immobilized catalyst amount, the initial BPA concentration, the treatment time, the water matrix, the addition of hydrogen peroxide and the simultaneous presence (or absence) of other organic substances.

A series of kinetic data (time-dependent concentration vectors) from sixteen (16) experimental campaigns has been employed in order to systematically extract a multiparametric kinetic model which can be efficiently used for photocatalytic wastewater purification design. The identification of the optimal model that can be formulated relies on the combined employment of (a) global optimization for kinetic parameter estimation, and (b) quantitative criteria for optimal model selection. The latter are namely the Akaike Information Criterion (AIC) and the Bayesian Information Criterion (BIC), which are measures of the relative quality of a statistical model (for a given set of experimental data) that rely on the likelihood function and can be used for model selection among a finite set of postulated models. As ad hoc parameter set expansion can easily increase the likelihood metric but also result in overfitting, both AIC and BIC resolve this problem by introducing a penalty term for the number of parameters in the model. The identification of the optimal multiparametric model is then implemented toward preliminary design of a photocatalytic wastewater purification system and subsequent sensitivity analysis.

REFERENCES

Donoso-Bravoa, A., Mailier, J., Martin, C., Rodríguez, J., Aceves-Larae, C.A., Vande Wouwer, A. (2011). Model selection, identification and validation in anaerobic digestion: A review, *Water Research* **45**: 5347–5364.

Flores-Sánchez, A., Flores-Tlacuahuac, A., Pedraza-Segura, L.L. (2013), Model-based experimental design to estimate kinetic parameters of the enzymatic hydrolysis of lignocellulose, *Ind. Eng. Chem. Res.* **52**(13): 4834–4850.

Miró A., Pozo C., Guillén-Gosálbez, G., Egea, J.A., Jiménez, L. (2012). Deterministic global optimization algorithm based on outer approximation for the parameter estimation of nonlinear dynamic biological systems, *BMC Bioinformatics* **13**(90): 1–12.

Singer A.B., Taylor J.W., Green W.H., Barton, P.I. (2006). Global dynamic optimization for parameter estimation in chemical kinetics, *J. Phys. Chem. A* **110**(3): 971–976.

Zacharakis, A., Chatzisymeon, E., Binas, V., Frontistis, Z., Venieri, D., Mantzavinos D. (2013). Solar photocatalytic degradation of Bisphenol A on immobilized ZnO or TiO_2 *Int. J. Photoenergy* **570587**: 1–9.

Zhao, Y., Stadtherr M.A. (2011). Rigorous global optimization for dynamic systems subject to inequality path constraints, *Ind. Eng. Chem. Res.* **50**(22): 12678–12693.

382505 Overall Process Synthesis of Seawater Desalination

Abstract:

Population and economic growth, along with climate change, exert an increasing pressure on the availability of adequate water supplies. Today 36% of the global population live under severe water stress with the prediction of reaching 52% by 2050. This rising issue can be alleviated and resolved by undertaking responsible measures such as developing efficient water treatment and management systems [1,2]. Early stage design in water treatment processes, in particular, can provide safe and affordable water by offering maximum design flexibility at lowest system change costs. One promising water treatment process is desalination due to the abundance of the saline water resources on Earth and their sustainable utilization [3]. However, no work has been done so far on the systematic synthesis and optimization, neither in desalination, nor in water and wastewater treatment.

The present work proposes a mathematical framework based on mixed integer non-linear optimization techniques for the systematic synthesis of seawater desalination processes. In the proposed model, two major indicative groups of contaminants are studied, i.e. total suspended and dissolved solids, along with eight candidate technologies targeting the removal of the contaminants. A pre-treatment system of the potential conventional and non-conventional steps is considered [4] before the desalination process which is modelled to accommodate for non-conventional and emerging technology candidates. Optimization of the flowsheet is achieved through increasing the separation efficiencies of the candidate technologies by manipulating their operating conditions within an allowable range [5]. The economic aspects cover operating costs for electricity, chemicals, cleaning, equipment replacement and maintenance. The capital costs account for the capacity of the facility to be designed [6,7,8]. The model is solved for desired water purity to meet drinking water standards with the objective of minimizing the investment and operating costs of the overall process.

The results generally agree with the current world practices in desalination design as a preference towards non-conventional methods is observed, producing water at the lowest price limit. Hence, the developed tool has a significant benefit in reducing the design time, and simultaneously, increasing the cost effectiveness of seawater treatment processes.

Key References

1. WWAP (United Nations World Water Assessment Programme), (2014), The United Nations *World Water Development Report 2014: Water and Energy*, Paris, UNESCO, vol. 1, pp.1 - 299

2. Veolia Water, (2010), *Finding the Blue Path for Sustainable Economy. A White Paper*, North America, pp.1-12

3. National Centre of Excellence in Desalination Australia, (2011), *Australian Desalination Research Roadmap*, Murdoch University, Western Australia, pp.1 - 78

4. Voutchkov, N., (2009), *SWRO Pre-treatment Systems: Choosing Between Conventional and Membrane Filtration*, Filtration & Separation, vol. 46, pp.5 - 9

5. Koleva, M.N., Polykarpou, E.M. and Papageorgiou, L.G., (2013), *An MILP Model for Cost Effective Water Treatment Synthesis*, Proceedings from the 20th International Congress on Modelling and Simulation, Adelaide, pp. 2716 - 2722

6. Pickering, K.D. and Wiesner, M.R., (1993), *Cost Model for Low-pressure Membrane Filtration*. Journal of Environmental Engineering, vol. 119, pp.772 - 797

7. Lu, Y., Hu, Y., Xu, D., and Wu, L., (2006), *Optimum Design of Seawater Desalination System Considering Membrane Cleaning and Replacing*, Journal of Membrane Science, vol. 282, pp.7 - 13

8. Skiborowski, M., Mhamdi, A., Kraemer, K. and Marquardt, W., (2012), *Model-based Structural Optimization of Seawater Desalination Plants*, Desalination, vol. 292, pp.30 - 44

Solution of Mixed-Integer Nonlinear Programming Problems in Hydraulics of Water Distribution Networks

Arpan Seth[1] Carl D. Laird[1]

[1] School of Chemical Engineering, Purdue University, West Lafayette, IN

Optimization provides a great tool for many applications related to public water distribution networks. These include, but are not limited to, optimal operation of the distribution networks by pump scheduling for minimizing energy costs, estimation of model parameters including network demands, real-time data reconciliation and gross error detection. However, the majority of techniques proposed in the literature solve these problems using a simulation-optimization framework with modeling software (e.g. EPANET) as a black-box. One reason for the preference for these techniques is the complex nature of the water distribution network hydraulic model. This model is a DAE system with non-linear pressure drop relationships and differential equations describing storage tank level dynamics. Additionally, along with discrete pump schedules, control rules require models with discrete variables. Therefore, simultaneous optimization techniques where the hydraulic model equations are embedded directly into the optimization formulation can result in large-scale Non-Linear Programming (NLP) or Mixed Integer Non-Linear Programming (MINLP) formulations that can be very challenging to solve.

However, recent developments in both NLP and MINLP algorithms, and parallel computing architectures have shown good promise in solving such large-scale problems. In this work we propose a water distribution system hydraulic modeling framework for generating explicit set of model equations that can be embedded directly into an optimization formulations.

First, we validate the proposed model against EPANET, the established water distribution system simulator. Next, we present the application of this modeling framework to solve continuous nonlinear optimization problems (e.g. demand estimation, real-time parameter estimation), and mixed-integer nonlinear programming problems (e.g. structural error detection, pump scheduling). We discuss a large-scale algorithm for parallel decomposition of the continuous problem, and provide a solution technique for the MINLP that is tailored to handle the non-linearities arising from the pressure drop equations.

CCSI

Carbon Capture Simulation Initiative

Advanced Heat Integration Tool for Simulation-based Optimization Framework

Yang Chen [a,b], **John Eslick** [a,b], **Ignacio Grossmann** [a], **David Miller** [b]

a. Dept. of Chemical Engineering, Carnegie Mellon University

b. National Energy Technology Laboratory

November 18, 2014

Simulation-Based Optimization

+ Treats simulation as black box (does not require mathematical details of model)
  - ⟶ **Easy to implement**
+ Does not require simplification of the process model
  - ⟶ **High-fidelity models applied**
+ Readily adapted for parallel computing
  - ⟶ **Computational time reduced**

− Not well suited for problems with many variables such as heat integration, and superstructure optimization
  - ⟶ **Heat integration is a separate module linked to simulation-based optimization algorithm**

Goal: Develop a simulation-based optimization framework with heat integration for large-scale high-fidelity process models.

Simulation-Based Optimization with Heat Integration

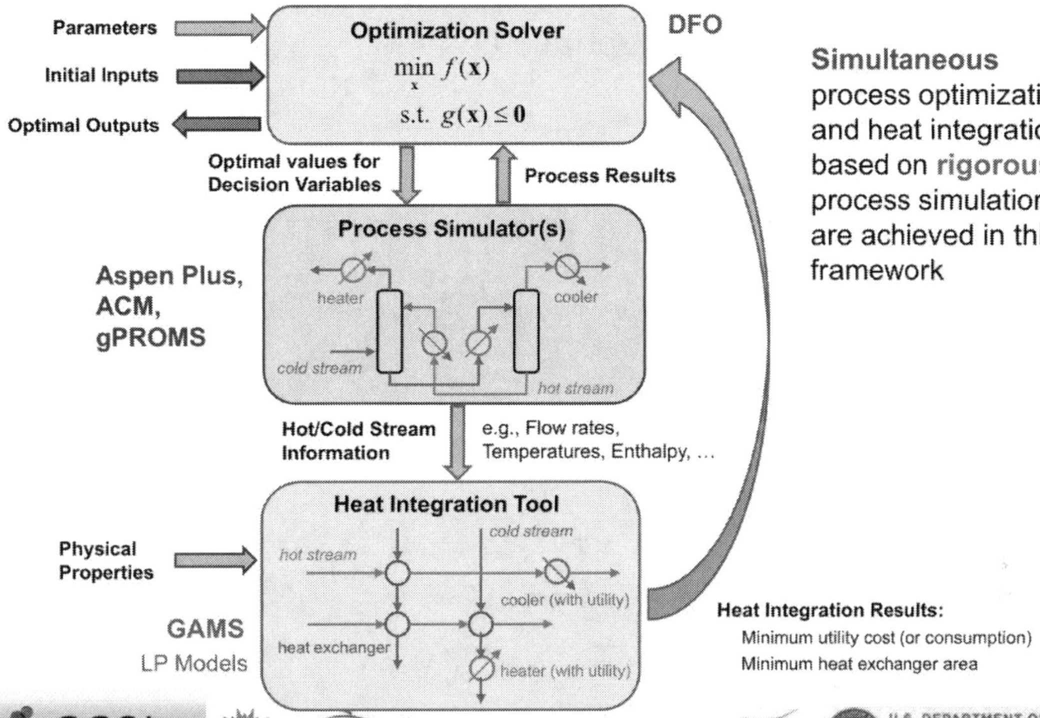

Simultaneous process optimization and heat integration based on **rigorous** process simulations are achieved in this framework

CCSI Carbon Capture Simulation Initiative · NETL · Berkeley Lab · Lawrence Livermore National Laboratory · Los Alamos · Pacific Northwest · U.S. DEPARTMENT OF ENERGY

3

Minimum Utility Cost (Consumption)

- LP Transshipment Model

$$\min\ Z = \sum_{m\in S} c_m Q_m^S + \sum_{n\in W} c_n Q_n^W$$

$$\text{s.t.}\quad R_{ik} - R_{i,k-1} + \sum_{j\in C_k} Q_{ijk} + \sum_{n\in W_k} Q_{ink} = Q_{ik}^H \qquad i \in H_k'$$

$$R_{mk} - R_{m,k-1} + \sum_{j\in C_k} Q_{mjk} - Q_m^S = 0 \qquad m \in S_k'$$

$$\sum_{i\in H_k} Q_{ijk} + \sum_{m\in S_k} Q_{mjk} = Q_{jk}^C \qquad j \in C_k$$

$$\sum_{i\in H_k} Q_{ink} - Q_n^W = 0 \qquad n \in W_k \quad k = 1,\dots,K$$

$$R_{ik},\ R_{mk},\ Q_{ijk},\ Q_{mjk},\ Q_{ink},\ Q_m^S,\ Q_n^W \ge 0 \qquad R_{i0} = R_{iK} = 0$$

Q^S	heat load of hot utility
Q^W	heat load of cold utility
Q^H	heat load of hot process stream
Q^C	heat load of cold process stream
Q	exchange of heat
R	heat residual
c	unit cost of utility
k	temperature interval
i	hot process stream
j	cold process stream
m	hot utility
n	cold utility

- Heat loads of the streams are calculated directly from the total change of enthalpy from the simulation results.
- Assumption: **Constant** heat capacity flowrates (FCps) for streams.

Papoulias SA, Grossmann IE. *Comput. & Chem. Eng.* 1983;7(6):707-721.

CCSI Carbon Capture Simulation Initiative · NETL · Berkeley Lab · Lawrence Livermore National Laboratory · Los Alamos · Pacific Northwest · U.S. DEPARTMENT OF ENERGY

4

Stream with Variable FCp

- A process stream with phase change

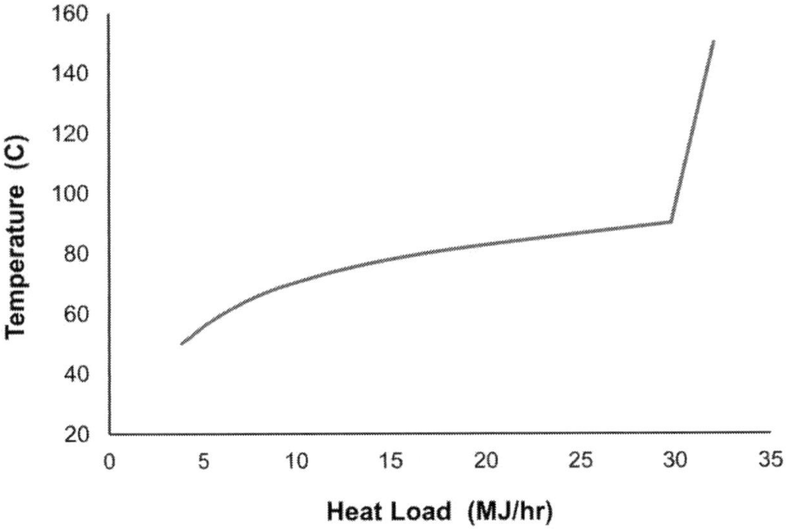

A mixture stream of CO_2 and H_2O (CO_2: 40%, H_2O: 60%; 1kmol/hr; 1 bar)

Problems with Constant FCps

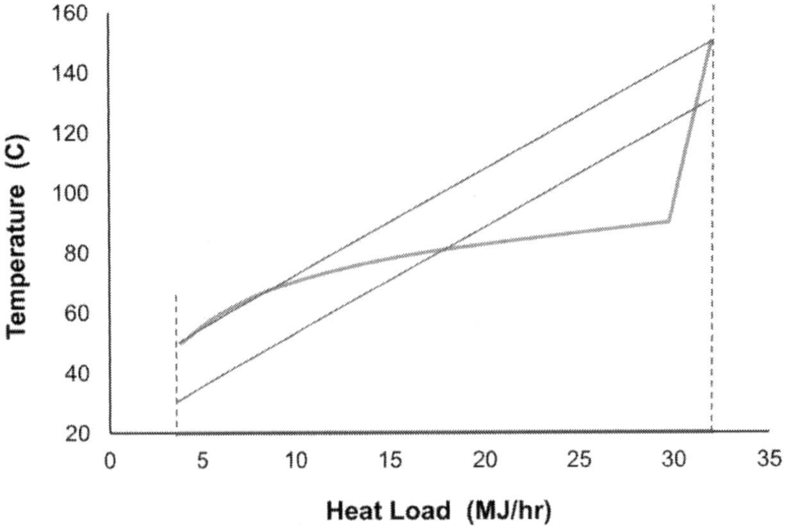

- Overestimate the heat recovery
- Infeasible heat exchanger network design

Piecewise Linear Approximation

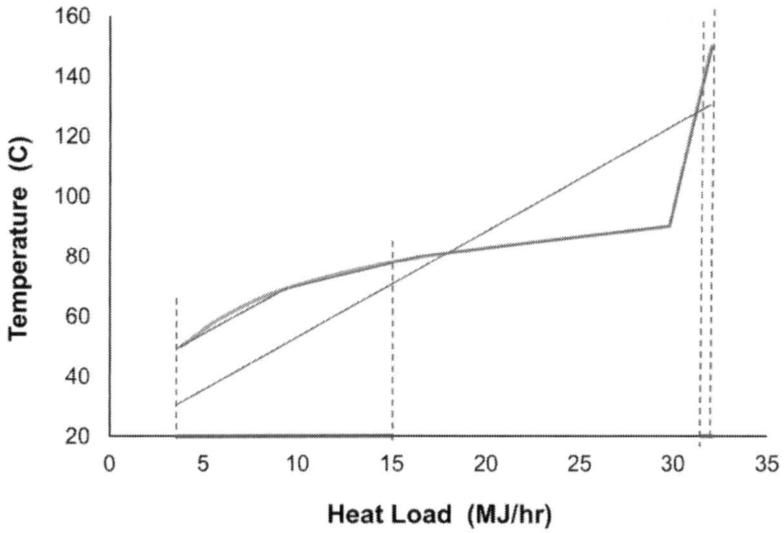

- More accurate heat integration results
- Assume constant FCps in each small temperature interval
- Build a series of sub-streams with identical temperature change or heat load in process models

Minimum Heat Exchanger Area

- LP Area Targeting Model (Modified from LP Transportation Model)

$$\min \frac{1}{\text{Ft}} \sum_{k=1}^{K} \sum_{l=1}^{K} \frac{1}{\text{LMTD}_{k,l}} \sum_{i \in H_k} \sum_{j \in C_l} \frac{q_{ik,jl}}{h_i + h_j}$$

$$\text{s.t.} \quad \sum_{l=k}^{K} \sum_{j \in C_l} q_{ik,jl} = Q_{ik}^{H} \qquad i \in H_k \quad k = 1,\dots,K$$

$$\sum_{k=1}^{l} \sum_{i \in H_k} q_{ik,jl} = Q_{jl}^{C} \qquad j \in C_l \quad l = 1,\dots,K$$

Q^H	heat load of hot stream
Q^C	heat load of cold stream
q	exchange of heat
Ft	correction factor for a non-countercurrent flow
h	stream film heat transfer coefficient
LMTD	logarithmic-mean temperature difference
k	temperature interval
l	temperature interval
i	hot stream
j	cold stream

- Temperature interval should be smaller than the minimum utility problem for accurate area targets.
- Number of temperature intervals: accurate results vs. CPU times.
- Double-temperature approach: HRAT & EMAT.

Jezowski JM, Shethna HK, Castillo FJL. *Ind. Eng. Chem. Res.* 2003;42(8):1723-1730.

Implementation - Graphical User Interface

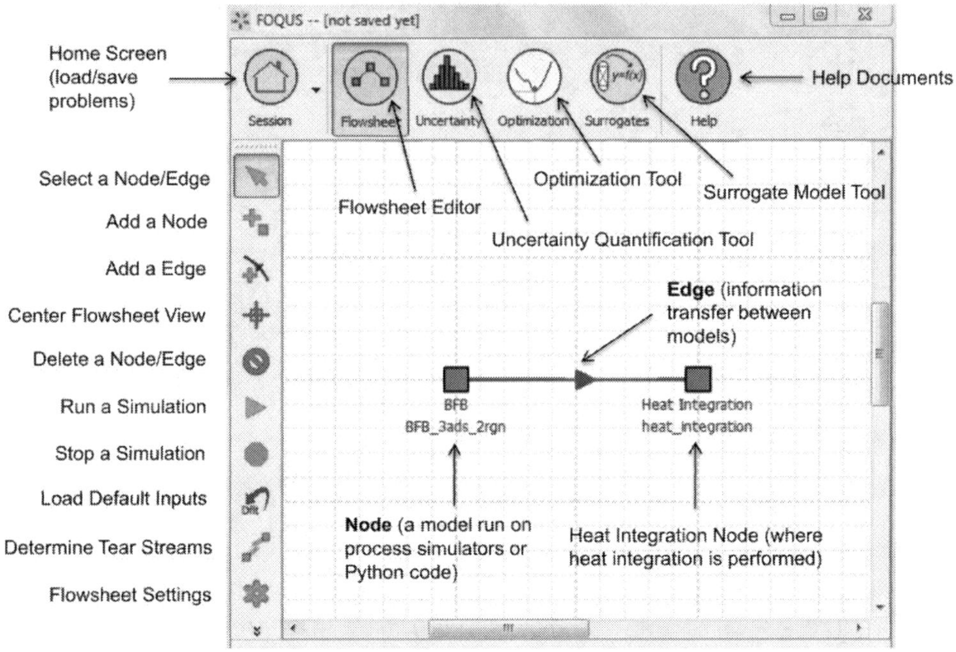

Framework for **O**ptimization and **Q**uantification of **U**ncertainty and **S**ensitivity (FOQUS)

9

Simulation Model (1)

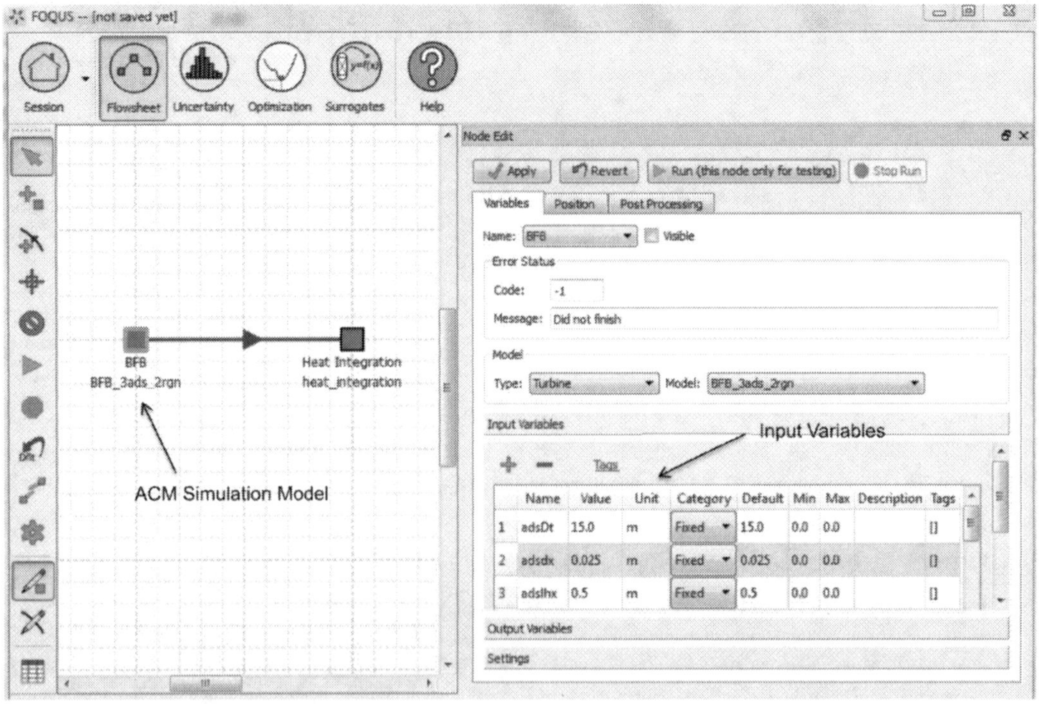

10

Simulation Model (2)

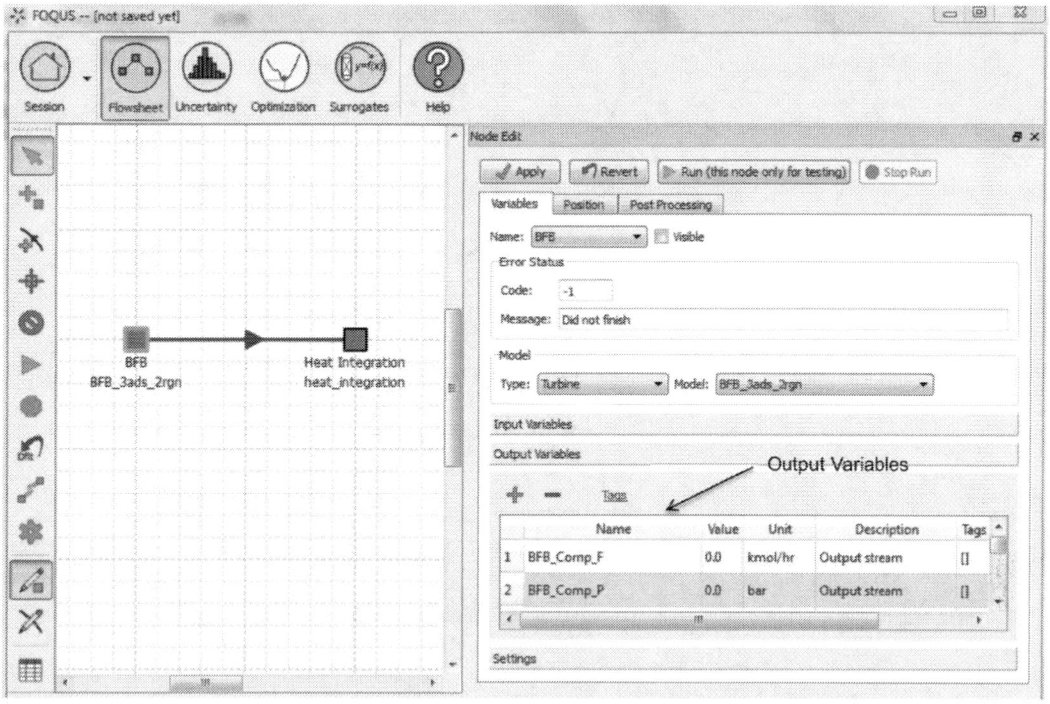

11

Heat Integration Tool (1)

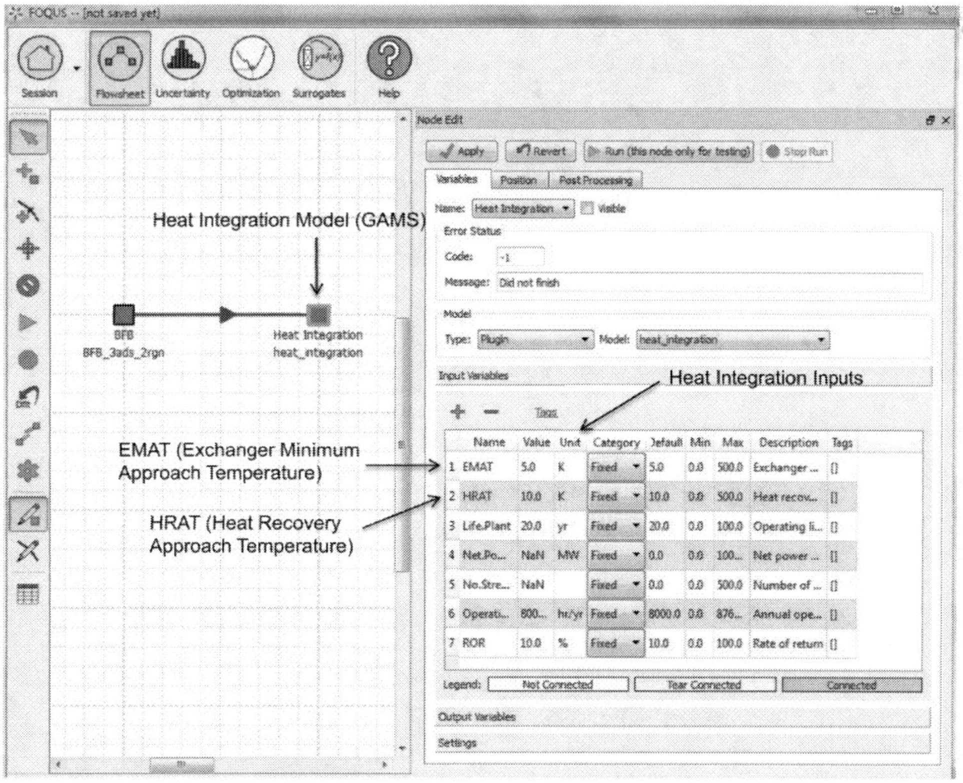

12

Heat Integration Tool (2)

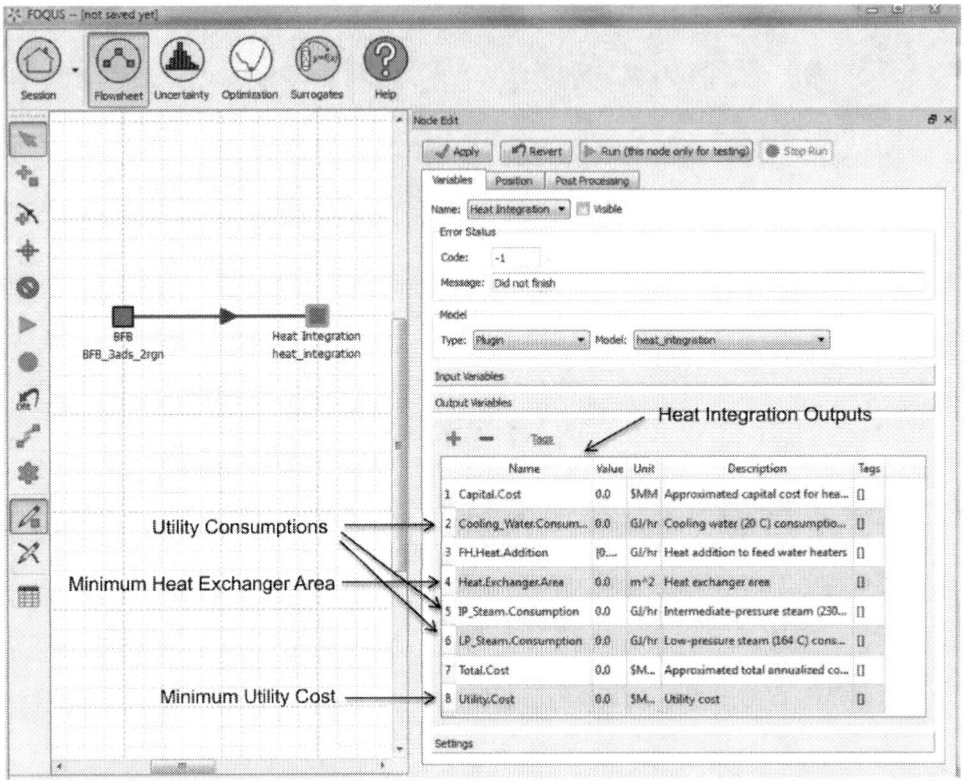

13

Optimization Solver

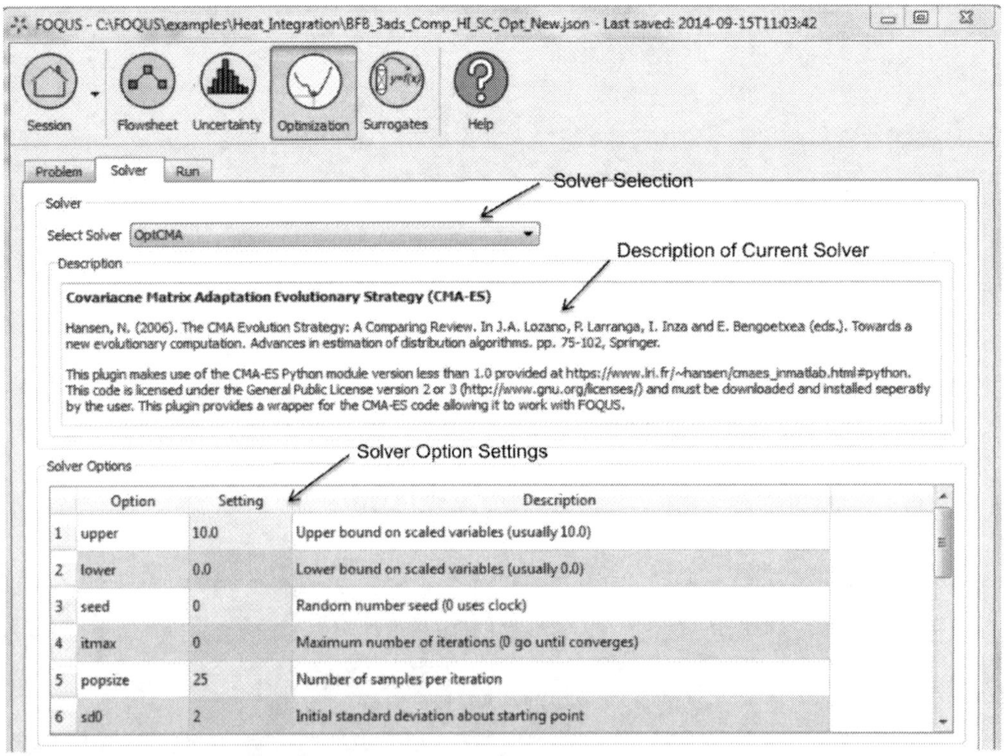

14

Optimization Problem Setting

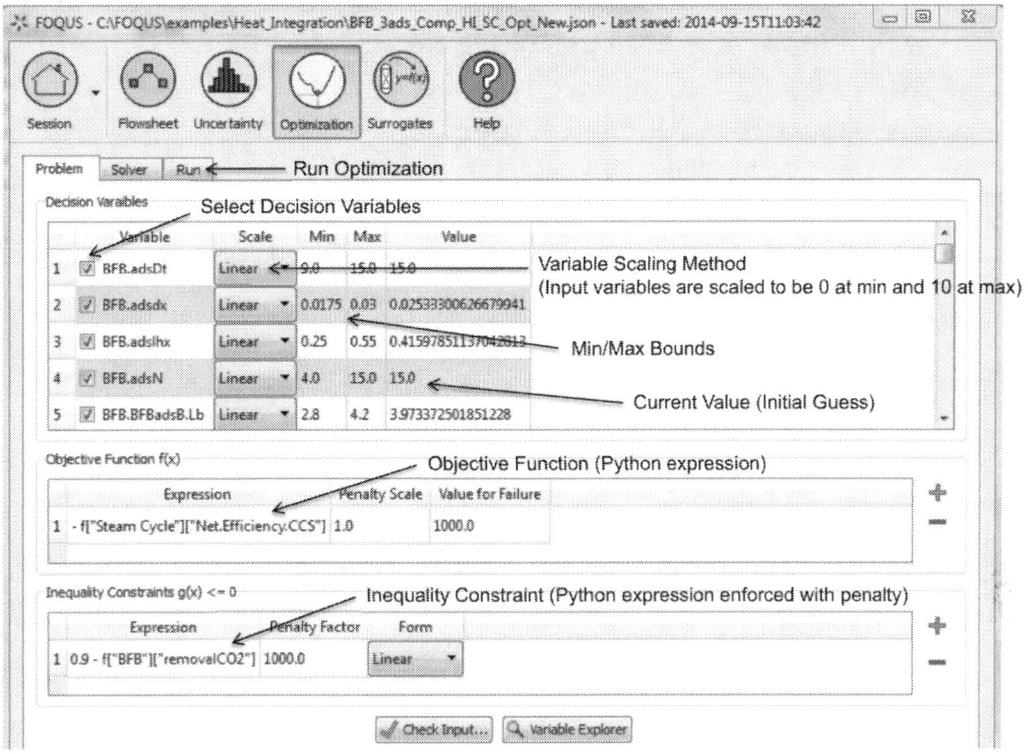

15

Case Study – A Power Plant with CO_2 Capture

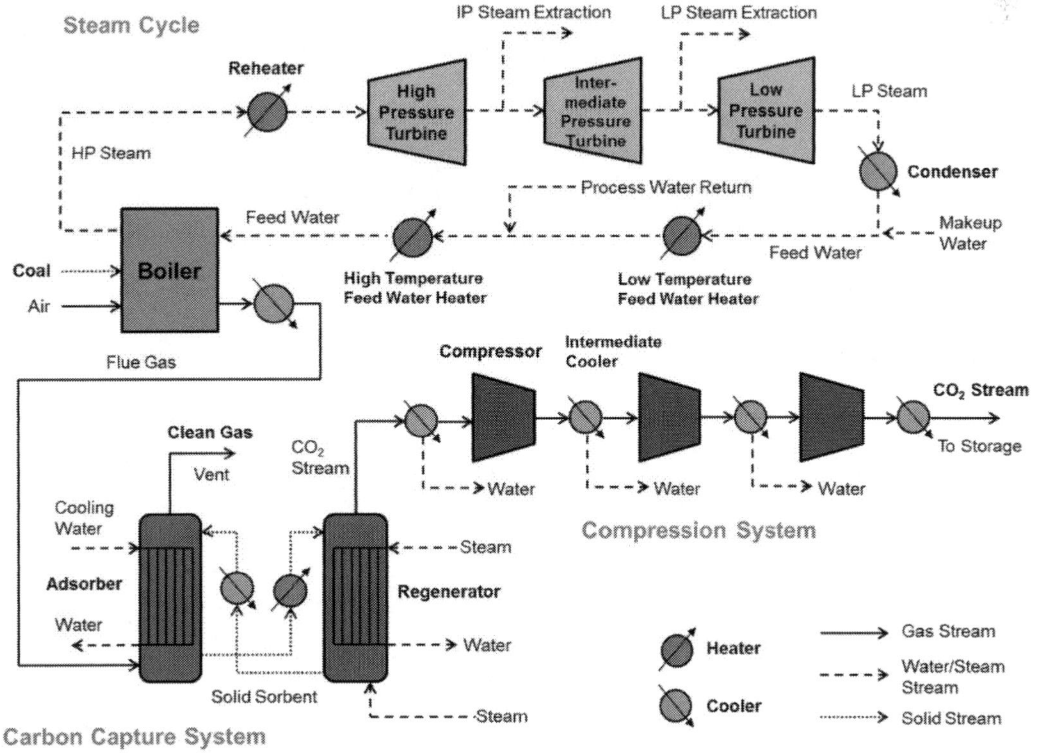

16

Problem Statement

Objective Function: Maximizing **Net efficiency**

Constraint: **CO_2 removal** ratio ≥ 90%
Flowsheet evaluation (via process simulators)
Minimum utility and area target (via heat integration tool)

Decision Variables (23): Bed length, diameter, sorbent and steam feed rates, temperatures

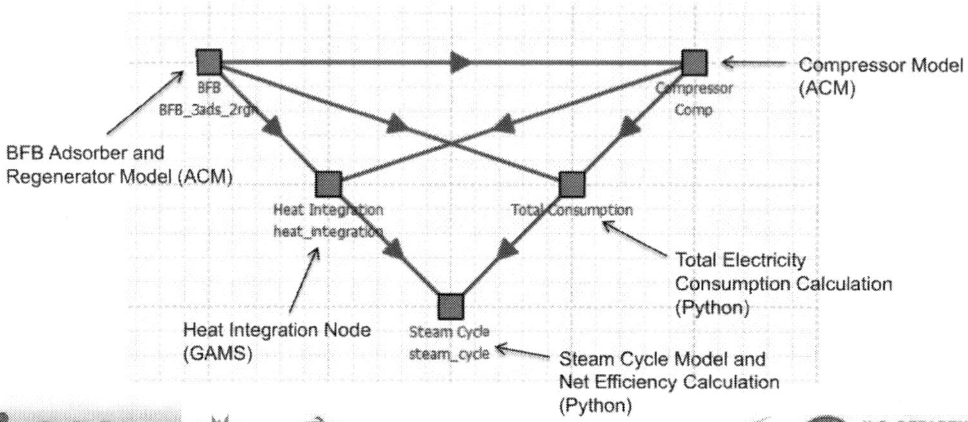

Case Study Results (1)

Base case w/o CCS: 650 MW_e, **42.1 %** with CCS: 419.6 MW_e, **27.2 %**	**Simultaneous** optimization and heat integration approach	**Sequential** optimization and heat integration approach	Optimization **w/o heat** integration
Net power efficiency (%)	**33.8**	**32.2**	**30.5**
Net power output (MW_e)	522.2	497.9	471.1
CO_2 removal ratio (%)	**90.2**	**90.1**	**90.1**
Electricity consumption (MW_e)	85.2	73.8	73.8
IP steam withdrawn (GJ/hr)	0	0	0
LP steam withdrawn (GJ/hr)	768.5	1113.7	1231.9
Cooling water consumption (GJ/hr)	1820.3	1594.2	3333.6
Heat addition to feed water (GJ/hr)	562.9	467.4	0
Heat exchanger area (million m^2)	0.751	1.125	

Note: Constant FCps are assumed here and piecewise linear approximation is not used.

Optimization and heat integration significantly increased the net efficiency of the power plant with CCS.

Case Study Results (2)

Base case w/o CCS: 650 MW_e, **42.1 %** with CCS: 419.6 MW_e, **27.2 %**	Heat integration with **constant** FCps	Heat integration with **variable** FCps (5 segments)	w/o heat integration
Net power efficiency (%)	**33.8**	**31.9**	**30.5**
Net power output (MW_e)	522.2	493.4	471.1
CO_2 removal ratio (%)	**90.2**	**90.0**	**90.1**
Electricity consumption (MW_e)	85.2	72.0	73.8
IP steam withdrawn (GJ/hr)	0	0	0
LP steam withdrawn (GJ/hr)	768.5	1089.7	1231.9
Cooling water consumption (GJ/hr)	1820.3	1700.8	3333.6
Heat addition to feed water (MW_{th})	562.9	313.9	0
Heat exchanger area (million m^2)	0.751	0.923	

After considering variable FCps and using piecewise linear approximation of the composite curve, the net efficiency is somewhat decreased but the obtained results become much more **realistic**.

CCSI Carbon Capture Simulation Initiative · NETL · Lawrence Livermore National Laboratory · Los Alamos · Pacific Northwest · U.S. DEPARTMENT OF ENERGY 19

Conclusions

- Simulation-based optimization framework with heat integration is a suitable tool for optimization of large-scale high-fidelity process models.
- This framework can be easily implemented in the software FOQUS.
- Performance of power plant with CCS can be significantly increased by simultaneous optimization and heat integration.
- More accurate heat integration results are obtained by using piecewise linear approximation for the composite curve of process streams.

CCSI Carbon Capture Simulation Initiative · NETL · Lawrence Livermore National Laboratory · Los Alamos · Pacific Northwest · U.S. DEPARTMENT OF ENERGY 20

Acknowledgement

Jim Leek (LLNL)
Josh Boverhof (LBNL)
Juan Morinelly, Melissa Daly (CMU)

DOE: Carbon Capture Simulation Initiative (CCSI)

Disclaimer

CCSI Carbon Capture Simulation Initiative · NETL · Berkeley Lab · Lawrence Livermore National Laboratory · Los Alamos National Laboratory · Pacific Northwest National Laboratory · U.S. Department of Energy

21

Simultaneous Synthesis of Heat Exchanger Networks and Work Exchange Networks

Kefeng Huang and I A Karimi*
Department of Chemical & Biomolecular Engineering
National University of Singapore
4 Engineering Drive 4, Singapore 117585

Abstract

In the last several decades, energy has increasingly become an immediate global concern. High energy cost due to depleting fossil energy resources, increasing global energy demands, stringent environmental regulations on CO_2 emissions, alarming climate changes, growing international competition, and market globalization highlight the importance of energy utilization. As the largest industrial energy consumer over the past decade, the chemical industry accounted for nearly 20% of all energy used in the industrial sector. With affordable renewable energy still not within reach, utilizing the limited fossil energy efficiently remains crucial for the chemical industry in the near future. That is why improving energy efficiency has been a major concern in the chemical industry over the several decades.

The varying temperature and pressure requirements in a chemical process demand the use of energy in two common forms, heat and work. Besides consuming various utilities for heating and cooling, many chemical plants operate at high pressures and need compression work extensively. The compression section is a major consumer of energy and critical to the overall energy efficiency and cost effectiveness. In such plants, some streams can produce work through expansion turbines. Thus, it would be possible to conserve energy via integrating compressors with turbines to exchange work by using simple single-shaft-turbine-compressor (SSTC) units. The key question is how to optimally integrate both heat and work simultaneously to improve energy efficiency further in the same network. The chemical processes are still struggling to utilize energy optimally and often use enumerative, try-and-see heuristic procedures based on experience. The complex interactions between heat exchanger network (HEN) and work exchange networks (WEN) involve highly nonlinear dynamics. While process integration literature has addressed some of these issues in the form of unintegrated design and synthesis of HENs and WENs, most effort has been devoted to the heat exchanger network synthesis (HENS) or work exchange network synthesis (WENS) rather than overall heat and work integration and synthesis of WEN and HEN.

HENs are crucial for energy integration in chemical processes, and recent progress have been made on non-isothermal mixing (Huang et al., 2012), phase changes (Hasan et al., 2009), multiple utilities placement

*Corresponding author: cheiak@nus.edu.sg, +65 6516-6359 (V), +65 6779-1936 (F)

(Huang and Karimi, 2013) and efficient algorithm (Huang and Karimi, 2014) for HENS problems. But surprisingly, limited work has so far addressed exchanging work by matching appropriate high-pressure (HP) and low-pressure (LP) steams in a network. Razib et al. (2012) introduced WENS problem and developed a mixed-integer nonlinear programming (MINLP) model to minimize the total annualized cost (TAC) for a constant speed for the single shaft WEN without considering heat integration of process streams. Onishi et al. (2014) combined the possible expansions and compressions route of Wechsung et al. (2011) and the HENS model of Yee and Grossmann (1990) for the integration between heat and work with the objective of minimizing TAC. However, the predetermined pressure-manipulating route, which is proposed based on reducing energy consumption, can lead to sub-optimal solutions. Furthermore, they just considered the match between one compressor and one turbine on a single common shaft. Later, Onishi et al. (2014) modified their approach by including the work of Razib et al. (2012) on WEN and SSTC. Again, the predetermined heaters (coolers) at the end of HP (LP) streams may give sub-optimal solutions. Moreover, due to the economic metrics they selected, their approach gives final solution networks with quite complicated configurations that seem not realistic. Thus, a systematic synthesis approach simultaneously considering the interactions between HEN and WEN must be exploited in order to achieve full and efficient energy and cost savings.

In this work, we first define the problem of simultaneous synthesis for HENs and WENs. Then, we propose a superstructure and the corresponding MINLP formulation with an objective of TAC on appropriate economic metrics. We allow the flexible choice between a heater or cooler at the end of each stream to ensure that the final temperature can always be reached. Finally, we use several examples to demonstrate the effectiveness of our proposed approach We show that the existing approaches in literature even fail to obtain optimal networks in case of some small problems.

References

1. Hasan, M.M.F., Jayaraman, G., Karimi, I.A., Alfadala, H.E., 2009. Synthesis of heat exchanger networks with nonisothermal phase changes. AIChE J. 56, 930–945.
2. Huang, K.F., Al-mutairi, E.M., Karimi, I.A., 2012. Heat exchanger network synthesis using a stagewise superstructure with non-isothermal mixing. Chemical Engineering Science 73, 30-43.
3. Huang, K.F., Karimi, I.A., 2013. Simultaneous synthesis approaches for cost-effective heat exchanger networks. Chemical Engineering Science-Under Review.
4. Huang, K.F., Karimi, I.A., 2014. Efficient algorithm for simultaneous synthesis of heat exchanger networks. Chemical Engineering Science 105, 53-68.
5. Onishi, V.C., Ravagnani, M.A.S.S., Caballero, J.A., 2014. Simultaneous synthesis of heat exchanger networks with pressure recovery: Optimal integration between heat and work. AIChE J. 60, 893-908.
6. Onishi, V.C., Ravagnani, M.A.S.S., Caballero, J.A., 2014. Simultaneous synthesis of work exchange networks with heat integration. Chemical Engineering Science 112, 87-107.
7. Razib, M.S., Hasan, M.M.F., Karimi, I.A., 2012. Preliminary synthesis of work exchange networks. Computers and Chemical Engineering 37, 262– 277.
8. Wechsung, A., Aspelund, A., Gundersen, T., Barton, P.I., 2011. Synthesis of heat exchanger networks at subambient conditions with compression and expansion of process streams. AIChE J. 57, 2090–2108.
9. Yee, T.F., Grossmann, I.E., 1990. Simultaneous optimization models for heat integration—II. Heat exchanger network synthesis. Computers & Chemical Engineering 14, 1165–1184.

Integrating Compressors into Heat Exchanger Networks

Chao Fu, Truls Gundersen

Department of Energy and Process Engineering, Norwegian University of Science and Technology, Kolbjoern Hejes vei 1.A, NO-7491 Trondheim, Norway

The concept of Appropriate Placement is fundamental in Pinch Analysis and addresses how different pieces of equipment can be integrated with heat recovery processes to ensure it results in energy savings, also referred to as Correct Integration. The concept of Appropriate Placement is also closely linked to the so-called Plus/Minus principle that is used to suggest process modifications that can improve the level of heat recovery.

A qualitative approach to Appropriate Placement is based on Pinch Decomposition and the identification of heat sources and heat sinks. A quantitative approach is based on the Grand Composite Curve (GCC) and allows the calculation of the amount of heat that can be correctly integrated. While this type of analysis is simple for distillation columns, evaporators, heat pumps and heat engines, it is considerably more complicated for pressure changing equipment such as compressors and expanders. The main reason is that the shape of the GCC will change. Another complicating factor is that new pinch points may occur.

The placement of compressors and expanders refers to the inlet temperature of these units. For larger pressure ratios, the division into multiple stages is also part of the Appropriate Placement issue. In this work, focus has been on heat recovery problems where compression is needed, however, the results are applicable also to expansion by using similar arguments. The insight obtained from various case studies has been generalized and formulated as a set of theorems with corresponding assumptions.

Since both heat and power are involved, the objective has been to minimize exergy consumption for the integrated processes. Actually, the work results in a procedure for targeting minimum exergy consumption. While establishing these targets, the optimal use of Pinch Compression (compressor inlet is at pinch temperature) and/or Ambient Compression (compressor inlet is at ambient temperature) will be identified and the corresponding possible split of streams into sub-streams (or branches) make the design task easy and straightforward.

Ambient Compression is normally used to minimize work consumption. The compression heat is then difficult to utilize due to its low temperature and is thus normally wasted. Recently the appropriate placement of compressors has been discussed in literature [1-3] and stated as follows: compression adds heat to the system and should thus preferably be done at the pinch point so that compression heat can be utilized in the heat deficit region above pinch. The topic of operation of compressors was also briefly discussed in previous works [4-5] focusing on reactor systems, however, without the use of exergy.

Pinch Compression reduces the heat demand, however, the compression work increases. The target is then to minimize exergy consumption. Several challenges exist when applying the rule about "Pinch Compression" for integrating compressors into a Heat Exchanger Network (HEN): (1) the streams to be compressed are included in the HEN, thus the heat demand for streams is changing; (2) the role of a stream (as a hot or cold stream) to be compressed may change when the compression heat is utilized; (3) Pinch Compression may produce more heat than required, thus stream splitting is required; and (4) the heating or cooling of a stream with pressure manipulation may create new pinch points. The integration of compressors into HENs is thus a rather complex problem.

The objective of this paper is to develop design guidelines for integrating compressors into HENs with a target of minimizing exergy consumption. Analytical methods are used to prove why Pinch Compression is preferred with respect to exergy consumption. In addition, graphical methods are used to illustrate how Pinch Compression can be implemented. A design procedure that achieves the minimum exergy target is presented. Several examples are used to test the procedure. Industrial applications are also included to show the importance and advantages of Pinch Compression.

References

[1] A. Aspelund, D. O. Berstad, T. Gundersen. An Extended Pinch Analysis and Design procedure utilizing pressure based exergy for subambient cooling. Appl. Therm. Eng. 2007; 27(16): 2633-2649.
[2] A. Wechsung, A. Aspelund, T. Gundersen, P.I. Barton. Synthesis of heat exchanger networks at subambient conditions with compression and expansion of process streams. AIChE J. 2011;57(8): 2090–2108.
[3] T. Gundersen, D. Marmolejo-Correa. Appropriate placement of compressors and expanders. AIChE Spring Meeting 2012, April 1-4, Houston, USA.
[4] Z. Kravanja, P. Glavic. Heat integration of reactors – II. Total flowsheet integration. Chem. Eng. Sci. 1989; 44(11): 2667-2682.
[5] M. Homsak, P. Glavic. Appropriate placement of Energors against local pinch temperatures. Trans. IchemE. 1995; 73(A): 871-879.

A new rigorous decomposition method for mixed-integer linear programs with both linking variables and linking constraints

Emmanuel Ogbe, Xiang Li*

Department of Chemical Engineering, Queen's University,
19 Division Street, Kingston, ON, K7L 3N6, Canada

Many process systems engineering problems can be cast as a large-scale mixed-integer linear programming (MILP) problem, such as supply chain optimization, process design, production planning and schedule. The large size of the problem typically comes from the need to characterize uncertainties with a scenario approach, e.g., using a scenario-based two-stage stochastic programming formulation [1]. In this case, the problem exhibits the structure illustrated in Figure 1(a). This problem can be decomposed into a set of subproblems with much smaller sizes if the values of a set of linking variables are known. Benders decomposition (BD) [2] is the classical approach for efficient solution of this problem.

A large-scale linear or mixed-integer programming problem may also come from a planning/scheduling problem over a large number of time periods, where there exist some linking constraints coupling decision variables over different time periods. In this case, the problem exhibits the structure illustrated in Figure 1(b). This problem can be decomposed into a set of subproblems with much smaller sizes if the linking constraints are not present. Dantzig-Wolfe decomposition (DWD) [3] can be employed for efficient solution, if the problem is a linear programming (LP) problem. But DWD is not rigorous for MILP problems.

However, neither BD nor DWD can be employed for solving problems with the structure illustrated in Figure 1(c). This problem structure can come from a scenario-based stochastic programming with a risk measure, e.g., conditional value-at-risk [4][5]. One approach to exploit this structure is Lagrangian decomposition (LR) [6][7], but LR is not rigorous when integer variables are present, and it may converge slowly due to the difficulty in generating valid Lagrange multipliers.

This presentation introduces a new decomposition method that can solve problems with the third structure rigorously and efficiently, if integer variables are not present or integer variables are all linking variables. In this decomposition method, a set of restricted master problems and pricing problems are solved to yield a sequence of upper bounds for the problem, and a set of relaxed master problems are solved to yield a sequence of lower bounds for the problems. An optimal solution can be obtained in a finite number of steps when the upper and lower bounds converge. The computational advantage of the decomposition method is demonstrated through an energy and bioproduct supply chain optimization problem [8], in comparison to a state-of-the-art optimization solver.

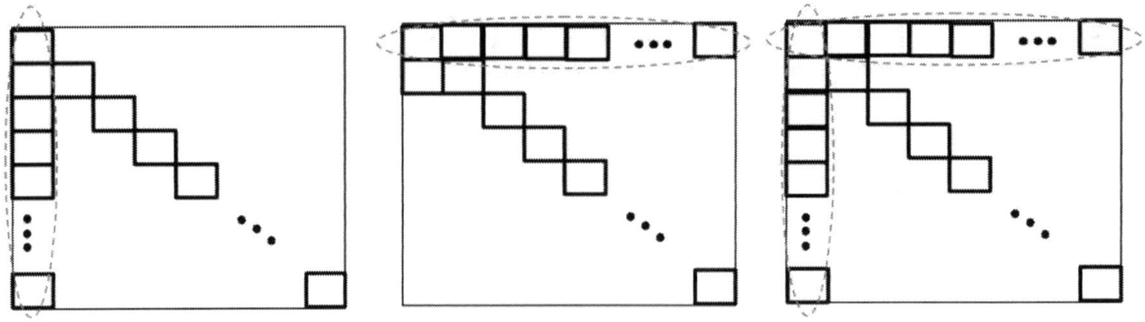

(a) Structure 1 – Only linking variables

(b) Structure 2 – Only linking constraints

(c) Structure 3 - Both linking variables and liking constraints

Figure 1. Different decomposable problem structures.

References

[1] Birge, J. R.; Louveaux, F. *Introduction to stochastic programming*, Second Edition; Springer: New York, 2011.

[2] Benders, J. F. Partitioning procedures for solving mixed-variables programming problems. *Numer. Math.* **1962**, 4, 238-252.

[3] Dantzig, G. B.; Wolfe, P. Decomposition principle for linear programs. *Operations Research* **1960**, 8, 101-111.

[4] Rockafellar, R. T.; Urysaev, S. Conditional value-at-risk for general loss distributions. Journal of Banking & Finance **2002**, 26, 1443-1471.

[5] Bruno, S. V. B.; Sagastizabal, C. Optimization of real asset portfolio using a coherent risk measure: applications to oil and energy industries. *Optimization and Engineering* **2011**, 12, 257-275.

[6] Fisher, M. L. The Lagrangian relaxation method for solving integer programming problems. *Management Science* **1981**, 27, 1–18.

[7] Carøe, C. C.; Schultz, R. Dual decomposition in stochastic integer programming. Operations Research Letters **1999**, 24, 37–45.

[8] Čuček, L.; Lam, H.; Klemes, J.; Varbanov, P.; Kravanja, Z. Synthesis of regional networks for the supply of energy and bioproducts. *Clean Tech. and Environ. Pol.* **2010**, 12, 635-645.

384004 Structured Regularization for Degenerate Models in Primal-Dual Interior Point Method

Abstract:

Nonlinear programing (NLP) is an essential tool for formulating large-scale mathematical models of chemical engineering processes. Active set methods and interior point methods are both popular strategies for solving NLP problems. The active set method involves the combinatorial complexity of identifying the active constraints and therefore avoids the influence of inactive dependent constraints. However, identifying the active constraints slows down the algorithm for large-scale problems especially when there are many degrees of freedom. Alternatively, interior point methods introduce a barrier parameter and add a penalty term to the objective function for the inequality constraints resulting in cheaper iterations with newton steps. The high performance of the interior point method makes it widely used for NLP problems. However, interior point methods require certain regularity conditions for the solution path to exist. Dependent constraints, both equalities and inequalities, are common in chemical processes and violate the regularity conditions which may cause the interior point method to fail.

Material flow equations for multicomponent streams exist widely in chemical process models. When composition is unknown, the mass balance equations involve bilinear terms. In addition, synthesis or design MINLP problems derive NLP subproblem relaxations with the input flow rate vanishing when the corresponding unit is unselected. In that case, bilinear terms lead to local dependence in the equations at optima, which violate Linear Independent Constraints Qualification (LICQ) and make Lagrange multipliers undefined in the Karush-Kuhn-Tucker (KKT) matrix. The interior point solver IPOPT introduces regularization terms to fix the KKT matrix, however in some cases the required regularization terms are too large and the solver will fail. Moreover, inequality constrained problems in process optimization frequently violate constraint qualifications (such as LICQ and the Mangasarian-Fromovitz Constraint Qualification (MFCQ)), and therefore lead to unbounded multipliers. For instance, semi-definite programming problems that are reformulated as NLPs lead to inequalities that can violate the MFCQ. In the NLP solver, this leads to unbounded multipliers, which causes the Lagrange Hessian to blow up.

To deal with these challenges, we present a new regularization algorithm, which uses ideas from active set methods, and identify an independent subset of active constraints. For interior point methods, this is achieved by adding big-M terms in dependent rows that essentially eliminate dependent constraints. This results in more accurate Newton steps and faster convergence to a solution. Numerical experiments on hundreds of modified examples from comprehensive test sets are implemented in the C++ version of IPOPT with linear solver HSL_MA57, HSL_MA97 and MUMPS. The test results show the effectiveness of the new algorithm with average reduction of more than 50% of the iterations. Moreover, several large-scale nonlinear blending problems and semi-definite optimization problems are solved with the proposed algorithm and the improvements over existing regularization are demonstrated.

386109 Clustering-Based Interior-Point Strategies for Stochastic Programs

Abstract:

We present a clustering-based interior-point strategy for two-stage stochastic programs. This problem class arises in stochastic optimal control, robust design, and parameter estimation and has the property that an arrowhead block representation of the KKT system can be obtained. Each block corresponds to a scenario, which is typically obtained by sampling a probability distribution. Our key observation is that scenarios can be clustered on the fly inside the linear solver based on the influence of the associated block on the Schur complement system (first-stage decision). This results in a much smaller compressed KKT system compared to scenario aggregation typically performed prior to the solution of the problem. We show how to use the compressed system as a pre-conditioner for the full space KKT system. We also describe the features of our implementation in C++, demonstrate that high compression rates of 50-90% are possible, and that speedups of an order of magnitude are achievable.

371248 A Systematic Approach for Infeasibility Diagnosis in NLP and MINLP

Abstract:

With increase in the computing power available, optimization models used to make business and engineering decisions are becoming more and more complex. As models continue to grow larger, analyzing models giving erroneous results and correcting them becomes increasingly difficult. Identification of Irreducible Inconsistent Sets (IIS) in a model can help speed up the process of correcting infeasible models [1]. An Irreducible Inconsistent Set is defined as an infeasible set of constraints with every proper subset being feasible. Identifying an IIS provides the modeller with a set of mutual inconsistencies that need to be eliminated. However, efficient implementations for IIS isolation are only available for linear programs (LPs).

We propose a novel approach for IIS identification that is applicable to NLPs and MINLPs. This approach makes use of feasibility-based reduction techniques in a computationally inexpensive preprocessing stage to test for infeasibility in subparts of the infeasible model. This stage allows for rapid elimination of a large number of constraints in the model. Further, this preprocessing step itself could be sufficient to eliminate all constraints not part of an IIS for a large number of problems. The reduced model obtained can be filtered with any standard IIS isolation algorithm to obtain an IIS. The benefits of the approach lie in the efficient reduction in the model obtained by the preprocessing stage which leads to speedups in IIS identification.

We implement our proposed preprocessing algorithm in the global solver BARON along with four different filtering algorithms: deletion filter, addition filter, addition-deletion filter and depth-first binary search filter [2]. We have compiled a test library of more than a 1000 infeasible models derived from problems submitted to BARON over the NEOS server in 2012 and 2013. We exhaustively test our implementation on this library and present computational results on the efficacy of the proposed preprocessing algorithm, as well as comparisons on the performance of different filtering algorithms.

This work was inspired by interactions with Air Liquide.

[1] H. J. Greenberg. An empirical analysis of infeasibility diagnosis for instances of linear programming blending models. IMA Journal of Mathematics in Business and Industry, 3:163-210, 1992.

[2] J. W. Chinneck. Feasibility and infeasibility in optimization. Springer, 2008.

385139 Solving Joint Chance Constrained Problem Via Robust Optimization Approximation

Abstract:

Chance/probabilistic constraint enforces probabilistic guarantee on constraint satisfaction and provides an important way of quantifying solution reliability for optimization under parameter uncertainty. Chance constrained optimization problem is generally very challenging due to the difficulties in checking the feasibility of the probabilistic constraint and the nonconvexity of the feasible region. As a result, chance constrained problems are normally solved through approximation except for the cases that a deterministic equivalent model can be obtained (e.g., normal distribution) [1][2].

As an important methodology for dealing with optimization problems with data uncertainty, uncertainty set induced robust optimization has been successfully used to safely approximate individual chance constraint in the past [3][4][5]. In set induced robust optimization, the uncertain data is assumed to be varying in an adjustable uncertainty set, and the aim is to choose the best solution among those "immunized" against data uncertainty.

While individual chance constraint focus on probabilistic guarantee on individual constraint satisfaction, joint chance constraint enforces that several constraints are satisfied simultaneously under parameter uncertainty. Thus, it is useful in modeling uncertainty correlation in different constraints. However, it becomes more complicated to handle joint chance constrained optimization problem, and its solution is not as satisfactory as individual chance constraint so far [6]. In this work, we extend the robust optimization approximation framework to the solution of joint chance constrained problem. Specifically, joint chance constraint is converted to single chance constraint problem through the introduction of inner maximization problem, and then safely approximated with the aid of probability inequalities. Different robust optimization formulations are then derived for joint chance constraint. To illustrate the proposed robust optimization approximation framework, numerical examples are studied and the solution is compared to the traditional decomposition method for approximating joint chance constraint.

References:

[1] A. Nemirovski, A. Shapiro. Convex approximations of chance constrained programs. SIAM J. Optim. 2006, 17(4), 969

[2] S. Ahmed and A. Shapiro. "Solving chance-constrained stochastic programs via sampling and integer programming," in Tutorials in Operations Research, Z.-L. Chen and S. Raghavan (eds.), INFORMS, 2008

[3] Z. Li, R. Ding, C.A. Floudas. A comparative theoretical and computational study on robust counterpart optimization: I. Robust linear optimization and robust mixed integer linear optimization. Industrial & Engineering Chemistry Research, 2011, 50, 10567-10603.

[4] Z. Li, Q. Tang, C.A. Floudas. A comparative theoretical and computational study on robust counterpart optimization: II. Probabilistic guarantees on constraint satisfaction. Industrial & Engineering Chemistry Research. 2012, 51, 6769-6788.

[5] Z. Li, C.A. Floudas. A Comparative Theoretical and Computational Study on Robust Counterpart Optimization: III. Improving the Quality of Robust Solutions. Industrial & Engineering Chemistry Research. Under review. 2014

[6] Chen, W., Sim, M., Sun, J., & Teo, C. P. From CVaR to uncertainty set: Implications in joint chance-constrained optimization. Operations research, 2010, 58, 470-485.

377851 Towards Gradient-Based Algorithms for the Optimal Design of Hybrid Renewable Energy Systems with Stochastic Generators, Storage, and Discrete Control Logic

Abstract:

Reliable power generation from renewable resources such as wind and solar is difficult because these resources are intermittent and unpredictable. A potential solution to this problem is to combine multiple renewable resources with an array of energy storage options that are complementary in their time-scales of operation, capacities, peak power outputs, and costs. Clearly, the design of such a system (e.g., technology selection, component sizing, etc.) involves a number of complex trade-offs and is highly dependent on the uncertain resource availability and demand. Moreover, with multiple generation and storage units present, operating such a system requires complex control logic that determines when and to what extent each component is used. This is typically called the energy management policy (EMP).

It is well known that designing renewable energy systems according to average resource availabilities leads to highly suboptimal or infeasible designs. In particular, stochasticity during real operation may lead to the violation of operating constraints and the frequent inability to meet demand. Thus, true optimal design of renewable energy systems requires an optimization subject to a detailed stochastic simulation, potentially with discrete events arising from the action of the EMP. Due to the complexity of this problem, it has thus far been treated using heuristic algorithms that view the stochastic simulation as a black-box (e.g., they do no require differentiability or even continuity per se). Unfortunately, these algorithms tend to exhibit slow convergence and are not guaranteed to furnish a local solution or even a KKT point.

In this talk, we consider 'sequential' NLP formulations for the above problem and analyze their regularity. For important classes, it is shown that the problem is indeed differentiable. We then undertake the computation of derivatives and assess the utility of applying efficient gradient-based NLP solvers to this class of problems. This approach has the potential for significant efficiency gains over heuristic approaches, and furthermore furnishes true local optima. Since renewable generation and storage technologies remain expensive, overdesigning systems according to suboptimal solutions is very costly and can quickly lead to economic infeasibility of the system. Thus, algorithms that provide higher quality solutions are of great interest for more accurately assessing the value of intermittent renewable resources.

360529 On the Implementation of a Structured-Exploiting Interior-Point Solver for Structured Nonlinear Programs

Abstract:

We describe the implementation of a fillter line-search interior point solver capable of exploiting multiple embedded structures at the linear algebra level (PIPS-NLP). We place special emphasis on the issue of inertia detection because we argue that this prevents the use of modular linear algebra implementations and thus hinders scalability. We prove that a simple test for curvature along the computed direction is sucient to guarantee global convergence. In addition, using CUTEr and energy optimization problems, we demonstrate that the strategy is as eective as inertia detection obtained through symmetric indenite factorizations in converging to second order points. We provide scalability results for our linear algebra implementation using stochastic optimal control problems of natural gas networks in which we exploit stochastic and reduced space structure.

392084 Thermodynamics, Chemical Process Stability and Control

Abstract:

The basic geometric structure of the thermodynamic phase space is based on the idea that the second law of thermodynamics characterizes concave (differentiable) equilibrium manifold. The primary coordinates in phase space are determined by extensive variables, such as energy, volume and mol-numbers. The secondary variables are given by the tangent space and these correspond to intensive variables such as temperature, pressure and chemical potentials. In the past decade this geometric structure has led to different developments that relate to modeling control and optimization of chemical process systems. These include nonlinear geometric control, passive based adaptive control, energy shaping control and control of process networks using circuit theory or port-Hamiltonian descriptions. These theories all describe how systems evolve on equilibrium manifold and the theory thereby provides a generalization of Gibbs stability theory for isolated to thermodynamically open systems. The theory has been used to show that multi-component, multi-phase and reactive equilibrium systems have stable dynamics provided suitable control strategies are applied. It has been applied to non-equilibrium reactive systems, networked systems and more generally a class of distributed parameter systems satisfying the hypothesis of local equilibrium. Such systems include, distillation and absorption columns, tubular reactors, particulate systems and even chemical plants. The purpose of the paper is to review the basics of the theory, provide an overview of potential application domains and list outstanding problems.

391974 Recent Advances in Accelerated Molecular Dynamics Methods

Abstract:

Many important materials processes take place on time scales that vastly exceed the nanoseconds accessible to molecular dynamics simulation. Typically, this long-time dynamical evolution is characterized by a succession of thermally activated infrequent events involving defects in the material. Over the fifteen years, we have been developing a new class of methods, accelerated molecular dynamics, in which the known characteristics of infrequent-event systems are exploited to make reactive events take place more frequently, in a dynamically correct way. For certain processes, this approach has been remarkably successful, offering a view of complex dynamical evolution on time scales of microseconds, milliseconds, and sometimes beyond. Examples include metallic surface diffusion and growth, radiation damage annealing in ceramics, and carbon nanotube dynamics. After an introduction to these methods, I will present some recent advances and results, and then describe the major ongoing challenges and our current thinking on how to overcome them.

392330 Lattice-Based Kinetic Monte Carlo Approaches for Mesoscale Phenomena

Abstract:

Lattice-based, or "on-lattice", kinetic Monte Carlo simulations are attractive because of their relative computational simplicity and efficiency, and have been employed to simulate an enormous range of non-equilibrium physical, chemical and biological phenomena. In this talk, the on-lattice kinetic Monte Carlo (LKMC) method is analyzed in the context of several distinct applications, ranging from microstructure evolution in semiconductor crystals, to cellular aggregate formation in blood flow, to coarse-grained simulations of phase evolution in liquid-vapor systems. These diverse examples are used to illustrate the flexibility of the general LKMC framework and also the potential pitfalls related to its application in some cases. In particular, special emphasis is placed on the discussion of (1) reduced degree-of-freedom representations (via coarse-graining) and the concomitant loss of entropy, and (2) LKMC simulation of particulate systems subject to external advective fields such as fluid flow.

392461 Analysis of External-Field-Driven Surface Morphological Evolution: Stabilization and Nanopatterning

Abstract:

Directed assembly on material surfaces and surface morphological evolution are governed by strongly nonlinear dynamics and lead to intriguing pattern formation. Externally applied fields, such as electric fields and temperature gradients, can be used to drive such surface morphological evolution and directed assembly in crystalline solids. We have studied the use of external fields to prevent surface morphological instabilities in stressed solids and to drive the assembly of single-layer epitaxial islands. Toward this end, in this presentation, we examine the surface morphological stability of electrically and thermally conducting crystalline elastic solids in uniaxial tension and of biaxially coherently strained epitaxial films on substrates under the combined action of an electric field and a temperature gradient. We use linear stability theory and self-consistent dynamical simulations based on a surface mass transport model that accounts for surface electromigration and thermomigration induced by the applied electric field and thermal gradient, respectively, surface diffusional anisotropy, and the temperature dependence of surface diffusivity. We find that properly directed external fields of magnitude higher than a critical value can stabilize the planar surface morphology and derive the conditions for synergistic action of the external fields, as well as the criticality conditions for surface stabilization.

We have also developed and validated a nonlinear model for the current-driven dynamics of single-layer epitaxial islands on crystalline substrates. Simulations based on the model show that the dependence of the stable steady island migration speed on the inverse of the island size is not linear for larger-than-critical island sizes. In this nonlinear regime, we report morphological transitions, Hopf bifurcations, and necking instabilities for various surface crystallographic orientations and island misfit strains. Moreover, we have studied the evolution of pairs of different-size islands driven to coalescence and explored the effects of three key geometrical parameters: the sizes of the two islands of the pair and their center-to-center line misalignment with respect to the electric-field direction. We discovered various patterns ranging from equal- and different-size stable steady island-pair configurations to many-island patterns that can be tailored by controlling the initial-pair geometrical parameters and the duration of application of the electric field.

392086 A Process Systems Engineering Perspective on Health and Disease

Abstract:

With our increasing understanding of life's multi-scale complexity, we suggest that living systems share characteristics common to engineering, and chemical engineering in particular. A chemical plant can be perceived as a complex network of interacting sub-systems, with their dynamics constrained to tight requirements of robustness (to maintain safe operation) on one hand, while maintaining a certain degree of flexibility to accommodate changeover on the other. In a manner similar to the nested hierarchy of physiological systems (cell ó organ ó organism), physico-chemical processes in a chemical plant take place at the at the molecular level within the confines of a process unit, which is part of an integrated plant, which in turn is a component of a world-wide supply chain. The aim of analysis, synthesis, and design of complex supply chains is to identify the laws governing optimally integrated systems. Optimality of operations is not a uniquely defined property. Optimality usually expresses the decision maker's balance between alternative, often conflicting, objectives. Both living and engineered complex constructs have evolved through multiple iterations, the former by natural processes and the latter by design, to optimize survival in a dynamically changing environment by maintaining systemic responses within acceptable ranges. A deviation from these limits leads to possibly irreversible damage. Stability and resiliency of these constructs is the result of dynamic interactions among constitutive elements.

Accordingly, in this presentation we attempt to draw analogies between fundamental concepts pervasive in chemical engineering theory and practice and conceptual/theoretical models of health and disease. We opt to establish connections between these concepts in the context of the analysis, synthesis, design, control and operation of complex chemical plants (and supply chains) and how these concepts mirror themselves onto critical aspects of notional physiological counterparts of engineered systems.

386353 Framework for the Design and Optimization of Biorefinery Networks

Abstract:

Advances in the processing of biomass feedstocks has resulted in the development of a large number of potential bio-based fuels and specialty chemicals, each of which can be produced using many different conversion technologies. To address questions such as; which is the most attractive set of products that can be produced from a particular feedstock, which are the best processing pathways, and which are the costs and environmental benefits associated with these conversions, researchers have performed detailed techno-economic analyses of particular technologies and developed biomass conversion superstructures to link feedstocks to products through all the possible technologies. In all cases, the set of feedstocks and potential products was considered to be fixed and the biorefinery was optimized to find which technologies should be used and which is the optimal set of process design parameters.

In this presentation, we will propose an alternative framework for the design and optimization of biorefineries in which we aim to introduce the flexibility required for the identification of new bio-based "products". We propose to do so by establishing a natural decomposition of the biorefinery network in a supply and a demand problem that are linked through a set of fixed intermediates. These intermediates are the result of the fractionation of biomass through a sequence of pre-treatment technologies that create streams of different compositions, purities and costs. The structure of the supply problem is optimized to establish the pathways that lead to the minimum cost of production of the different intermediate compound streams. The demand side problem focuses on the fast screening of new technologies and products that use these intermediate chemicals at the prices determined by the solution of the supply problem. In here, simple evaluations are used to asses the potential of a particular process or product and advantage in the use of a less pure but less costly intermediate stream is taken whenever possible.

This presentation depicts joint research activities at Masdar Institute and MIT and the proposed framework is demonstrated by analysing a biorefinery that conforms to the needs of the United Arab Emirates.

This work was funded by the Cooperative Agreement between the Masdar Institute of Science and Technology (Masdar Institute), Abu Dhabi, UAE and the Massachusetts Institute of Technology (MIT), Cambridge, MA, USA - Reference 02/MI/MI/CP/11/07633/GEN/G/00 for work under the Second Five Year Agreement.

A Mathematical Programming Approach to Apply Total Site Analysis in Integrated Biorefineries

Konstantinos A. Pyrgakis and Antonis C. Kokossis
School of Chemical Engineering, National Technical University of Athens, Athens, Greece

The development of biorefineries constitutes a really complex and challenging problem since a multitude of candidate products and processes are possible to enter into an overall plant. Moreover, the high majority of bioprocesses appear really energy intensive dealing with numerous separation systems for the purification and dehydration of final products making them even unprofitable. Process integration has a critical role to secure sustainability and profitability of processes and consequently constitutes indispensable need for the design of upcoming biorefineries. Energy integration is able to detect the feasible margins of energy efficiency while Total Site Analysis - a powerful process integration tool - is capable to manage energy interactions among different processes.

Applying Total Site Analysis for a cluster of given processes can secure maximum energy efficiency for the overall plant. However, since the processes of upcoming biorefineries are not yet known, integration should be applied in another way to assess and design understudied biorefineries. Process synthesis tools appear indispensable to manage multiplicities and complexities of large problems. Under these circumstances, candidate processes can be treated as degrees of freedom and the designing problem can be structured in a systematic way in order to assess successfully complex chemical production schemes including numerous products. As a result, there are strong incentives to investigate possibilities for better integration screening combinations among the set of the energy intensive processing routes. Following the systematic approach for the selection of candidate processes it is able to develop a mathematical programming approach applying Total Site Analysis to preview multiple options and select the appropriate product portfolios that offer the highest energy benefits.

Synthesizing different processes together in a systematic way and applying Total Site Analysis in the same time enables the investigation and selection of the best biorefinery structure that offers the maximum energy integration potential and consequently the minimum utility cost for the overall plant. The proposed systematic approach is further complemented by economical factors like annualized investment costs and selling product prices establishing a novel, complete and pivotal mathematical programming tool to coordinate strategic decisions for the design of future biorefineries. The mathematical approach enables the selection of the appropriate product portfolios that secure optimal energy integration and maximum profitability for the overall biorefinery, using an MILP model that guarantees the optimality of the solution.

Integrating and collocating processes together appear to offer higher integration potentials for the overall plant instead of integrating processes individually. This happens because high available energy of one process is able to cover energy needs of other incoming processes. Through the proposed approach it is able to detect those promising processes that exchange energy as much as possible and benefit at the utmost from Total Site Integration resulting optimally integrated biorefineries.

The proposed mathematical formulation has been applied to assess a lignocellulosic biorefinery including a set of 8 real pilot plants and 10 candidate products. The mathematical approach reduced the utility costs up to 83%, while 6 of the processes raised their profit in the range of 1,3 - 3,3 times and the other 2 transferred from the unprofitable zone to high levels of profitability.

Cost Estimation for Early-Stage Synthesis and Design of Biorefinery Networks

Peam Cheali, Krist V. Gernaey and Gürkan Sin

Department of Chemical and Biochemical Engineering, Søltofts Plads, Building 229, Technical University of Denmark, DK-2800 Lyngby, Denmark

Developing a new and sustainable (bio) chemical process is resource and time-consuming, and thus a costly investment. After the screening of initial ideas, laboratory experiments and pilot scale plant tests are normally performed. As a result, the capital and operating cost of the full-scale plant are determined on the basis of such experiments in order to evaluate economic feasibility, which is one of the important criteria to decide on whether or not to move ahead with a biorefinery project. However, determination of capital or operation cost can be quite challenging at the early stage of idea generation and screening for biorefinery concepts, due to inadequate cost data or a lack of complete design information. This is in particular relevant for biorefinery systems research, since potentially many new ideas are generated and suggested that originate from laboratory or theoretical holistic studies. In the early design state, most of the ideas have not as such been subjected yet to further scrutiny and testing at pilot, demonstration or full-scale. This is of course quite logic, since before submitting to field testing and validation, one needs to identify the most feasible idea against the background of many potentially other plausible ideas, and finding a good solution to this problem is quite a challenge. This sets the motivation for this contribution, where the aim is to systematically study the uncertainties involved in cost data and to apply a Monte Carlo based flexible yet rigorous basis to factor in the uncertainties in cost data to support risk based and reliable decision making for early-stage biorefinery concept screening and identification of an optimal solution.

The Association for Advancement of Cost Engineering International (AACE International) developed recommended practices for cost estimation in the process industries (Christensen and Dysert, 2011). This reference document provides details on classifications and principles for cost estimation during the life cycle of the project, including the quality of the estimation as well. At the early-stage design, order-of-magnitude estimate ("Class 5 estimate", "ratio estimate", "rule of thumb") is assigned, that is the expected accuracy range is -20% to -50% and +30% to +100% for low and high ranges, respectively. This confirms that the quality of the estimation is low during the early-stage where the availability and accuracy of the data is low as well. This is in general similar to early-stage design of biorefinery networks in which the design information (i.e. the equipment list, operating conditions) is incomplete and significant uncertainties (i.e. market prices of feedstock and products, product yield, and cost estimation due to incomplete design information) exist. Therefore, reasonably accurate cost estimation is required to make an effective decision on identifying an optimal biorefinery concept.

In our earlier work (Cheali et al., 2014), a systematic methodology to generate and identify the optimal processing networks under uncertainties was applied to a biorefinery system. A superstructure was formulated consisting of corn stover and wood feedstock, biochemical and thermochemical conversion platforms, Fischer-Tropsch (FT) products, bioethanol, lignin (as a product) and electricity. The optimization problem, considering a generic model for each unit operation in combination with collected data, was then formulated and solved using GAMS. The identification of the optimal and robust biorefinery networks producing transportation biofuels (FT products, bioethanol and bioethanol derivatives) under market price uncertainties was then performed with respect to techno-economic criteria.

In this study, we extend the models and database by incorporating the early-stage cost estimation with its uncertainty analysis. Methods for estimating capital investment that require less information during the early-stage of process development (order-of-magnitude) were used with Monte Carlo simulation. Appropriate ranges for uncertain parameters (i.e. capital investment, plant capacity and market prices of feedstock and products) are identified with their correlations, and Latin Hypercube Sampling (LHS) is then used to sample parameters from their respective domain of uncertainty. The samplings are therefore used as input data for solving deterministic and stochastic optimization problems to identify the optimal solution under uncertainties. The application of the methodology is highlighted on a transformation of lignocellulosic feedstock to FT products, bioethanol, and bioethanol derivatives as a case study. The results from the methodology and the impact of uncertainties in cost estimation on the selection of the optimal biorefinery network at early-stage design are discussed.

The expanded database and models integrating the early-stage cost estimation with uncertainty analysis provide a highly effective decision-support toolbox for early-stage synthesis and design of future biorefineries.

Reference:

Cheali, P.; Quaglia, A.; Gernaey, K. V.; Sin, G. Effect of Market Price Uncertainties on the Design of Optimal Biorefinery Systems—A Systematic Approach. *Ind. Eng. Chem. Res.*, **2014**, 53 (14), 6021–6032.

Christensen, P.; Dysert, L. R. Cost Estimate Classification System – as Applied in Engineering, Procurement, and Construction for the Process Industries. *AACE International Recommend Practice No. 18R-97*, **2011**.

The transportation sector is highly dependent on petrol based fuels such as gasoline and diesel. The need for alternatives has focused the research on the ones capable of substituting crude derivatives directly, particularly bioethanol and biodiesel. Biofuels have become a short and medium term alternative to reduce our dependency on fossil fuels. Most biorefineries are devoted for the production of one biofuel alone, bioethanol or biodiesel. So far, the most flexible biorefinery in terms of the range of products produced has been based on the Fischer-Tropsch technology. It can use biomass, coal, and/or natural gas as raw materials for the production of light hydrocarbons, synthetic gasoline and diesel as well as heavier products (Baliban et al., 2013, Martín & Grossmann, 2011).

In a recent paper, Martín & Grossmann (2013) presented the possibility of obtaining ethanol and biodiesel (FAEE) from algae simultaneously due to the composition of the algae. The lipids in the alge can be used to obtain oil which is used for biodiesel production, while the starch is a source of glucose and thus ethanol can also be produced by fermentation. Finally, the protein is an additional asset to the process. Due to the lower cost associated to the production of biodiesel, the plant was shown to be self-sufficient, using the ethanol to transesterify the oil, but the main product was FAEE with only 10% of the total production going to bioethanol. Recently, some experimental studies have proved the possibility of obtaining FAEE from lignocellulosic raw materials by aerobic fermentation of glucose (Paap et al., 2013; Valle-Rodríguez, 2014). Therefore, it would seem to be natural to evaluate the competitiveness of the simultaneous production of bioethanol and biodiesel from lignocelulosic switchgrass and compare it to the one that uses algae as raw materials, and both alternatives to the stand-alone processes.

In this work, we evaluate the optimal operation for the simultaneous production of biodiesel (FAEE) and ethanol from switchgrass. A superstructure embedding a number of alternatives is proposed. Two technologies are considered for switchgrass pretreatment, dilute acid and ammonia fibre explosion (AFEX) so that the structure of the grass is broken down. Next, enzymatic hydrolysis follows any of the pretreaments to obtain fermentable sugars, mainly xylose and glucose. We separate the lignin and with the sugars we consider the production of bioethanol and/or FAEE. The problem is formulated as an MINLP with simultaneous optimization and heat integration. Finally, an economic evaluation is performed and water consumption is minimized.

It is shown that although the investment cost of the plant decreases with the fraction of FAEE produced, the biofuels cost per gallon increases from $0.8/gal for ethanol production alone to values beyond $1.5/gal even for high conversions of sugars to FAEE. Therefore, the simultaneous production of bioethanol and FAEE from switchgrass is not competitive with using algae as raw material (Martín & Grossmann, 2013) at the current state of development of the synthesis. However, it might become promising if the conversion from sugars to biodiesel exceeds 0.5. Although the facility is feasible and more flexible as a biorefinery, the optimal operating conditions reveal that the preferred product is bioethanol due to the higher yield.

References

Baliban, R.C., Elia, J.A., Floudas, C.A.. (2013) Biomass and natural gas to liquid transportation fuels: Process synthesis, global optimization, and topology analysis. Ind. Eng. Chem. Res. 52, 9(6), 3381-3406

Martín, M., Grossmann, I.E. (2011) Process optimization of FT- Diesel production from biomass. Ind. Eng. Chem Res, *50* (23),13485–13499

Martín, M., Grossmann, I.E. (2013) Optimal engineered algae composition for the integrated simultaneous production of bioethanol and biodiesel AIChE J. 59 (8) 2872–2883

Valle-Rodríguez, J.O.; Shi, S.; Siewers, V., Nielsem J (2014) Metabolic engineering of Saccharomyces cerevisiae for production of fatty acid ethyl esters, an advanced biofuel, by eliminating non-essential fatty acid utilization pathways. Applied Energy 115, 226–232

Paap, S.M.; West, T.H.; Manley, D.K.; Steen, E.J.; Beller, H.R.; Keasling, J.D.; Dibble, D.C.; Chang , S.; Simmons, B.A. (2013) Biochemical production of ethanol and fatty acid ethyl esters from switchgrass: A comparative analysis of environmental and economic performance. Biomass and Bioenergy 49, 49-62

383870 Optimal Design and Synthesis of Integrated Algal Biorefinery for Biofuel and Chemical Production Under Economic and Environmental Criteria

Abstract:

Recent progress in developing sustainable products is largely spurred by the growing concerns over global climate change and dependence on fossil-based products [1, 2]. Domestic biomass resources can substitute a significant portion of petroleum utilized to produce fuels and chemicals, thus creating jobs across the country in various economy sectors [3]. Microalgae is a prospective biomass feedstock in terms of high area productivity, minimized competition with conventional agriculture, capability to grow in various water conditions, close to zero net carbon dioxide emissions and providing valuable co-products [4]. However, an algal biorefinery based on current processing technologies cannot be commercialized unless the energy efficiency of harvesting and dewatering equipment is significantly improved [5-9]. Glycerol is a byproduct when the algal biomass is converted to biodiesel via transesterification reactions and the utilization of glycerol for chemical production has a positive economic impact on biodiesel production [10]. Therefore, our goal is to develop the optimal process design of an integrated algal biorefinery to simultaneously produce biodiesel and glycerol-derived chemicals under both economic and environmental criteria.

In this work, a superstructure of an algal biorefinery with a number of technology alternatives is developed to simultaneously produce biodiesel and several glycerol-derived chemicals. The superstructure begins with an open-pond cultivation system, and then the mature algae product from the open pond is harvested by auto-flocculation and dissolved air flotation sequentially. Later, the algal biomass is concentrated by either pressure filtration or centrifugation. We employ a cell disruption technology from bead beating, microwaving, high pressure homogenization, and sonication, before lipid materials are separated by the solvent extraction system, where the extractant is selected from hexane, isopropanol/hexane, butanol, and supercritical carbon dioxide. Lipid materials are converted to biodiesel through four transesterification technologies, while the remnant is decomposed by anaerobic digestion and the resulting biogas is utilized to generate electricity or synthesize methanol consumed on site. Glycerol obtained from transesterification can be converted to hydrogen, propylene glycol, 1,3-dihydroxyacetone, butyl glycerol either, or poly-3-hydroxybutyrate. Hydrogen production from glycerol includes three alternatives: steam reforming, autothermal reforming, and aqueous phase reforming.

In order to optimize the superstructure and establish an optimal process, we propose a multi-objective mixed-integer nonlinear programming (MINLP) model including four types of constraints. Mass balance constraints describe the mass flow rates of each unit process. Integer variables are chosen to account for the selection of one technology among the alternative options [11]. Energy balance constraints enforce the energy conservation of all equipment by calculating the enthalpy differences between inlet and outlet. The economic and environmental performance is evaluated following the life cycle optimization frame work that integrates the multi-objective superstructure optimization scheme with life cycle assessment (LCA) and techno-economic analysis of the algal biorefinery [5, 11-16]. The life cycle stages of this cradle-to-gate LCA include raw material acquisition, transportation, and production. The life cycle inventories extracted from the process concern both the direct and indirect emissions, which could be aggregated into global warming potential (GWP) to quantify the environment impact to the atmosphere. The model includes two objective functions, which are minimizing the unit biodiesel production cost and the unit GWP. Both objective functions are associated with one gallon gasoline equivalent of the biodiesel produced. The multi-objective MINLP problem is solved using a global optimization framework that integrates the parametric algorithm and branch-and-refine algorithm. The optimal results are plotted in a Pareto-optimal curve which reveals the tradeoffs between the economic and environmental objectives.

References

[1] P. Daoutidis, W. A. Marvin, S. Rangarajan, and A. I. Torres, "Engineering Biomass Conversion Processes: A Systems Perspective," *Aiche Journal,* vol. 59, pp. 3-18, Jan 2013.

[2] D. Yue, F. You, and S. W. Snyder, "Biomass-to-bioenergy and biofuel supply chain optimization: Overview, key issues and challenges," *Computers & Chemical Engineering,* 2013.

[3] D. B. Plan, "Aug 2005. Multi Year Program Plan 2007-2012. US Department of Energy. Office of the Biomass Program," *Energy Efficiency and Renewable Energy*.

[4] J. Ferrell and V. Sarisky-Reed, "National algal biofuels technology roadmap," US Department of Energy College Park, MD2010.

[5] B. H. Gebreslassie, R. Waymire, and F. You, "Sustainable design and synthesis of algae-based biorefinery for simultaneous hydrocarbon biofuel production and carbon sequestration," *AIChE Journal,* vol. 59, pp. 1599-1621, 2013.

[6] J. Gong and F. Q. You, "Optimal Design and Synthesis of Algal Biorefinery Processes for Biological Carbon Sequestration and Utilization with Zero Direct Greenhouse Gas Emissions: MINLP Model and Global Optimization Algorithm," *Industrial & Engineering Chemistry Research,* vol. 53, pp. 1563-1579, Jan 29 2014.

[7] C. Silva, E. Soliman, G. Cameron, L. A. Fabiano, W. D. Seider, E. H. Dunlop, *et al.*, "Commercial-Scale Biodiesel Production from Algae," *Industrial & Engineering Chemistry Research,* 2013.

[8] R. Davis, A. Aden, and P. T. Pienkos, "Techno-economic analysis of autotrophic microalgae for fuel production," *Applied Energy,* vol. 88, pp. 3524-3531, Oct 2011.

[9] R. Davis and A. Aden, *Renewable Diesel from Algal Lipids: An Integrated Baseline for Cost, Emissions, and Resource Potential from a Harmonized Model*: Argonne National Laboratory, 2012.

[10] C. H. Zhou, H. Zhao, D. S. Tong, L. M. Wu, and W. H. Yu, "Recent Advances in Catalytic Conversion of Glycerol," *Catalysis Reviews-Science and Engineering,* vol. 55, pp. 369-453, Oct 2 2013.

[11] B. Wang, B. H. Gebreslassie, and F. Q. You, "Sustainable design and synthesis of hydrocarbon biorefinery via gasification pathway: Integrated life cycle assessment and technoeconomic analysis with multiobjective superstructure optimization," *Computers & Chemical Engineering,* vol. 52, pp. 55-76, May 10 2013.

[12] B. H. Gebreslassie, M. Slivinsky, B. L. Wang, and F. Q. You, "Life cycle optimization for sustainable design and operations of hydrocarbon biorefinery via fast pyrolysis, hydrotreating and hydrocracking," *Computers & Chemical Engineering,* vol. 50, pp. 71-91, Mar 5 2013.

[13] F. Q. You, L. Tao, D. J. Graziano, and S. W. Snyder, "Optimal design of sustainable cellulosic biofuel supply chains: Multiobjective optimization coupled with life cycle assessment and input-output analysis," *Aiche Journal,* vol. 58, pp. 1157-1180, Apr 2012.

[14] F. Q. You and B. Wang, "Life Cycle Optimization of Biomass-to-Liquid Supply Chains with Distributed-Centralized Processing Networks," *Industrial & Engineering Chemistry Research,* vol. 50, pp. 10102-10127, Sep 7 2011.

[15] D. J. Yue, M. A. Kim, and F. Q. You, "Design of Sustainable Product Systems and Supply Chains with Life Cycle Optimization Based on Functional Unit: General Modeling Framework, Mixed-Integer Nonlinear Programming Algorithms and Case Study on Hydrocarbon Biofuels," *Acs Sustainable Chemistry & Engineering,* vol. 1, pp. 1003-1014, Aug 2013.

[16] Q. Zhang, J. Gong, M. Skwarczek, D. Yue, and F. You, "Sustainable process design and synthesis of hydrocarbon biorefinery through fast pyrolysis and hydroprocessing," *AIChE Journal,* vol. 60, pp. 980-994, 2014.

390323 Design of Raceway Ponds for the Production of Algae Biomass

Abstract:

Global warming and scarcity in fossil fuels have driven an interest towards producing biodiesel from algal biomass. However, substantial challenges are encountered in deriving fuels from algae. Primary amongst them is the relatively higher costs of cultivation. Although there are several cultivation techniques that are commercially viable, this presentation focuses on designing optimum outdoor cultivation technology that considers the effect of both sunlight and temperature on algae growth. Raceway ponds are closed loop, oval shaped recirculation channels with paddle wheel that enables circulation and prevent sedimentation. In this work, raceway ponds are designed based on a given geographical location and algae species. Culture properties such as biomass productivity, growth rate, and concentration, and physical properties such as mass flow rate, average velocity, and pond temperature are estimated for a whole year depending on the dynamic behavior of zenith angle, diurnal pattern of solar irradiance, and temperature fluctuations at the location. The designs are modeled as nonlinear programming formulations with about 4100 variables and 3800 constraints. The objective is to meet a specific biodiesel production demand at a minimum net present sink over 10 years. Net present sink includes both equipment and operating costs. Operating costs involve the cost of electricity used for pumping algae biomass. Constraints connect culture properties and raceway pond geometry through experimentally validated growth model [1], temperature model [2], and irradiance model [3]. Maximum specific growth rate of algae biomass is constrained by temperature of the raceway pond, which in turn is constrained by geometry of the raceway pond and surrounding air temperature. Air temperature is calculated based on parameters such as maximum and minimum temperatures of the day along with sunrise time. Biomass concentration is constrained by the available irradiance at a location and geometry of the raceway pond. Solar irradiance at a location is calculated based on the zenith angle. Biomass productivity is calculated from growth rate and biomass concentration. In order to analyze the influence of the type of algae strain and geographical location on these culture properties and geometry of the raceway ponds, case studies with two algae species (*P. Tricornutum* and *I. Galbana*) and four locations (Tulsa, USA; Hyderabad, India; Cape-Town, South Africa; and Rio-de-Janeiro, Brazil) are investigated. The resulting eight Nonlinear Programming (NLP) models: Case 1 – *P. Tricornutum*, Tulsa, USA, and Raceway pond design, Case 2 – *P. Tricornutum*, Hyderabad, India, and Raceway pond design, Case 3 – *P. Tricornutum*, Cape-Town, South Africa, and Raceway pond design, Case 4 – *P. Tricornutum*, Rio-de-Janeiro, Brazil, and Raceway pond design, Case 5 – *I. Galbana*, Tulsa, USA, and Raceway pond design, Case 6 – *I. Galbana*, Hyderabad, India, and Raceway pond design, Case 7 – *I. Galbana*, Cape-Town, South Africa, and Raceway pond design, Case 8 – *I. Galbana*, Rio-de-Janeiro, Brazil, and Raceway pond design, are solved using GAMS 24.2/CONOPT solver. The problems are initialized thousand times using Latin Hypercube Sampling. Preliminary results show that with the model parameters used and a demand of 1 ton/year of biodiesel, Cases 1 through 8 require areas of about 29.99 ha, 25.36 ha, 35.50 ha, 21.92 ha, 31.14 ha, 26.20 ha, 37.03 ha, and 22.89 ha respectively to meet the demand. These results are comparable to the designs presented in [4]. To analyze the impact of algae species over the raceway pond geometry, notice that Cases 1 through 4, where *P. Tricornutum* species is cultured, have lower area than Cases 5 through 8, where *I. Galbana* species are cultured. *P. Tricornutum* species have higher lipid content than *I. Galbana* species and hence requires lower volume to meet the demand. In order to analyze the impact of location on the geometry of raceway pond, notice that Cases 4 and 8, where algae is cultured at Rio-de-Janeiro, Brazil, occupy lower area than other locations. Rio-de-Janeiro, Brazil has higher temperatures and algae growth is favorable at these high temperatures. In this work, a novel mathematical model is developed to estimate the best combination of algae species, geographical location, and raceway pond geometry by combining experimentally validated temperature, irradiance, and algae growth models with optimization. Benefits of such a method include realistic design of raceway ponds that produce the desired amount of biodiesel cost efficiently.

REFERENCES

[1] Yoder JA. Effect of temperature on light-limited growth and chemical composition of skeletonema costatum (bacillariophyceae). Journal of Phycology. 1979;15(4):362-70.

[2] Woodhead T. Simulation of assimilation, respiration and transpiration of crops. By D. T. de Wit et al. Pudoc (Wageningen), 1978. pp. 140. D.F1.22.50. Quarterly Journal of the Royal Meteorological Society. 1979;105(445):728-9.

[3] Ebrahimpour A, Maerefat M. A method for generation of typical meteorological year. Energy Conversion and Management. 2010;51(3):410-7.

[4] M.A. Borowitzka. Commercial production of microalgae: ponds, tanks, tubes and fermenters. Journal of Biotechnology. 1999;70:313-321.

392415 Patterns in Process Safety

Abstract:

Human error is perhaps the most prevalent cause of process accidents as it is in many other technical domains. "Listing human error as one of the causes of an accident is about as useful as listing gravity as the cause of a fall" (Kletz, 1993). Understanding and preventing accidents therefore would require a deeper understanding of human error. Just as patterns such as trees, spirals, waves commonly occur in nature in different contexts and can be used to explain and understand our world, patterns occur regularly in human behavior and therefore in human organizations. Also, a wide variety of sources that reveal human decisions and behaviors are now becoming available. These can be used to observe various aspects of human and organizational behavior – ranging from sustainability reports at the (macro) corporation level, to news and investigation reports at an event level, all the way to plant operators gaze information while operating the distributed control system (DCS) at the micro level. These new sources of data and information can through new data analysis techniques like text mining offer insights into micro and macro behaviors of people and organization. In this paper, we report on our research to identify recurrent patterns that correlate with process plant accidents. Examples based on actual accidents and real human eye tracking data will be presented.

392485 Functional Modelling and Reasoning about Safety in Complex Process Plant

Abstract:

Decision support systems can play a significant role in in the management of safety critical systems by providing facilities for data acquisition, interpretation and information sharing. Knowledge based systems has been proposed by several researchers as tools for implementation of intelligent decision support. Rule bases and qualitative models encoding expert knowledge about the process are here used by inference engines to give advice to decision makers in the analysis of complex problems. Within process safety, such intelligent systems are considered for computer assisted Hazop and for online operator support in fault diagnosis and planning of remedial actions.

Applications of knowledge based technology show that their success depends critically on the nature of the knowledge available and of the problem to be solved. Thus early applications of rule bases showed that the encoding of expert knowledge also should include concepts and relations which could not be formulated in rules but should be expressed in models. The choice of ontologies to represent concepts and relations which are expressed in the models are therefore important in present developments of knowledge based systems, which often include both rules and models.

Modeling and qualitative reasoning about safety problems are typically based on cause-effect relations. For example cause-consequence trees which can be expressed by rules, are explicit representations of the relations between process faults and their causes and consequences. In model based systems for reasoning about process safety, faults have no explicit representation but they are derived from a model of process elements and a rule-base containing generic knowledge about cause-effect relations between the elements.

The paper will present a modeling and reasoning technique based on means-end concepts called Multilevel Flow modelling (MFM). MFM is a technique for functional modeling. Its main focus on relations between means and ends or goals (rather than functions) provides a unique framework for process representation on multiple levels of function and for modeling levels of defense. MFM concepts support qualitative modelling of processes with interacting material, energy and information flows and has been used to model nuclear power plants, power systems and chemical engineering plants. MFM has been applied for cause-consequence analysis in Hazop and for online diagnosis.

The paper will introduce the basic concepts of MFM and present modelling examples from nuclear power and from a recent Hazop related study of an oil and gas separation plant. A short introduction to the MFMStudio, a model based reasoning tool, will also be presented demonstrating the capability of modeling technique and the implemented rule bases for cause-consequence reasoning.

Qualitative modelling and reasoning as implemented with MFM can often with advantage be combined with other types of models in order to increase coverage of the analysis or for validation purposes. In the Hazop study mentioned above, it has been investigated how the reasoning results from an MFM model could be validated by comparing it with simulations using a quantitative model (differential equations). The two models represent different types of knowledge (goals and functions versus physical interdependencies) and support different inferences. Preliminary results and some implications of this validation study will be discussed.

TeCSMART: A New Framework for Modeling and Analyzing Systemic Risk in Socio-technical Systems

Zhizun Zhang and Venkat Venkatasubramanian
Complex Resilient Intelligent Systems Laboratory
Department of Chemical Engineering
Columbia University
New York, NY 10027

Recent systemic failures such as the subprime financial crisis, BP Deepwater Horizon oil spill disaster, and the Northeast Blackout have reminded us, once again, the large-scale risks in complex socio-technical systems. While these are different disasters that happened in different domains, at different facilities, triggered by different events, involve different materials, etc., there are, however, certain common underlying patterns behind such systemic failures. These patterns teach us important fundamental lessons that we had better learn to avoid future disasters. To understand these patterns and learn from them, we need to go beyond analyzing them as independent one-off accidents but in the broader perspective of the potential fragility of all complex engineered systems. We need to study all these disasters from a common systems engineering perspective so that we can thoroughly understand the commonalities as well as the differences in order to better design and control such systems in the future.

With this goal in mind, we have developed a novel modeling framework, called *Teleo-Centric Systems Modeling and Analysis of Risks and Threats (TeCSMART)*. TeCSMART is a seven-layer hierarchical framework that allows to model a complex socio-technical system at different levels of granularity, starting from the equipment and plant level perspectives through the management layer to regulatory and societal perspectives. This facilitates the use of different kinds of models that are appropriate at the different layers. We will discuss the details of this framework as well as the results from a comparative study of BP Deepwater Horizon disaster, the subprime financial crisis, and the Northeast power blackout.

A new system for predicting and rapidly responding chemical and nuclear disasters

Kwanghee Lee[a] and Il Moon[a*]

[a]Department of Chemical and Biomolecular Engineering, Yonsei University, 50 Yonsei-ro, Seodaemun-gu, Seoul 120-749, Korea

Chemical and nuclear disasters involve industrial accidents and chemical and nuclear power plants, incidents from the use of weapons at terror or wars, and accidents at transportation of hazardous materials. Those disasters incur release of severely hazardous materials into atmosphere, and are featured to be very difficult and expensive to forecast as well as highly lethal once an incident occurs. In recent years, many chemical and radiological accidents have taken place in South Korea, for example, the accident of hydrogen fluoride (HF) release at Gu-mi in 2012 killed 5 people and evacuated all Gu-mi residents impossible to sustain their daily business and life. After this disaster, South Korea has made enormous efforts to develop a set of systems for responsively and effectively managing chemical and nuclear disasters. A new software system has been developed by integrating two existing programs designed to predict and maneuver chemical and nuclear disasters in order to set up a response system for dealing with the combined two disasters, as shown in Figure 1, for the first time in Korea. The improved software system is designed to be mainly used by civil defense officers, together with an identification of 625 scenarios of chemical and nuclear disasters. The new system is expected to contribute to upgrade a more systematic regional public protection plan for chemical and nuclear disasters. In addition, it enables all relevant local divisions to share disaster information in real time, resulting in a minimization of possible fatal damages. This new system can potentially be expected to be of great help for the development of a program for managing chemical and nuclear disasters in South Korea. This study was funded by National Emergency Management Agency for two years, and two companies and five Universities are contributed to this project.

* Corresponding author. Department of Chemical and Biomolecular Engineering, Yonsei University, 50 Yonsei-ro, Seodaemun-gu, Seoul 120-749, Korea

E-Mail address: ilmoon@yonsei.ac.kr

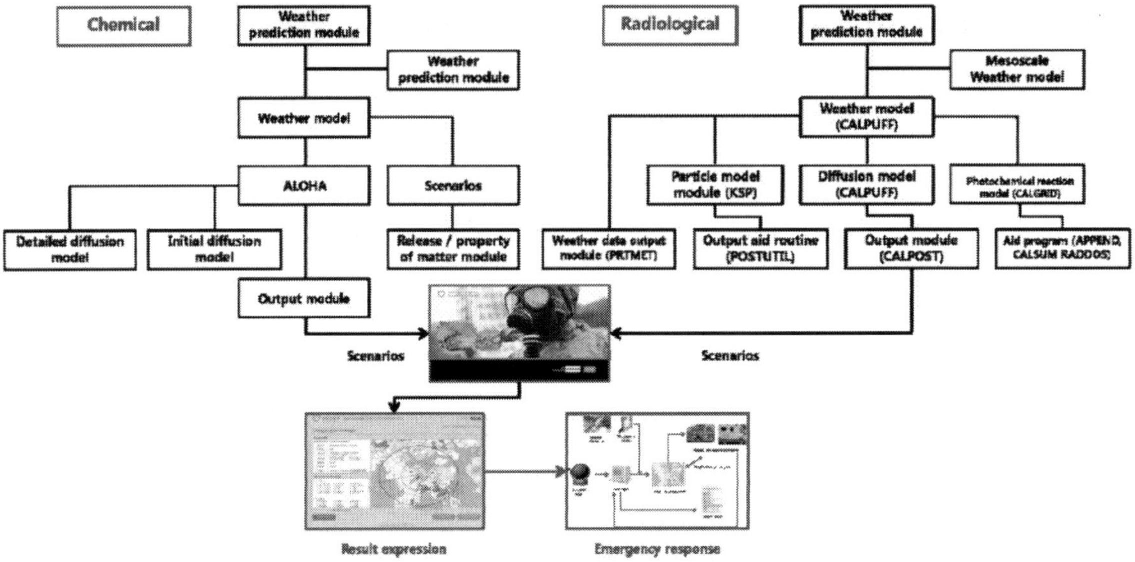

Figure 1 Scheme of integrated program

REFERENCES

[1] Michael M, Bradley. NARAC: an emergency response resource for predicting the atmospheric dispersion and assessing the consequences of airborne radionuclides. Journal of Environmental Radioactivity 2007;96:116-21.

[2] Park JD, Lee YH, Yoon Y, Kim JK, Cho JH, Moon I. Development of COI classification algorithm for chemical terrorism. The Korean Journal of Chemical Engineering 2013:30:559-62.

[3] Han MH, Kim EH, Suh WT, Hwang WT, Jeong HJ. Development of radiological dose assessment systems to support a radiological emergency preparedness in Korea. Annals of Nuclear Energy 2012;43:187-91.

* Corresponding author. Department of Chemical and Biomolecular Engineering, Yonsei University, 50 Yonsei-ro, Seodaemun-gu, Seoul 120-749, Korea

E-Mail address: ilmoon@yonsei.ac.kr

389168 A Safety Knowledge Base and Automatic Synthesis System for the SIS Design of Toxic Gas Facilities

Abstract:

As the use of toxic chemicals is being increased for the manufacturing of high value-added products, the number and loss from related acciedents are being increased globally. The proper design of safety-instrumented system (SIS) is a recommended way as a measure to prevent and safetly handle accidents related to hazadous chemicals. However, the design and validation of the SIS satisfying the required level of risk is not an easy task and requires lots of time and effort of experts. This research proposes an organized safety KB and automatic synthesis system for the SIS layout design for toxic-gas manufacturing, handling, and storage facilities. The safety KB encodes the design rules, regulatory codes, and best practices related to toxic gases. The prototype of developed system is validated against chlorine and ammonia cases, and the KB is being expanded to cover the toxic gases for semiconductor manufacturing.

386667 Global Optimization of Mixed-Integer Nonlinear Optimization Problems

Abstract:

The development of the branch-and-reduce approach to global optimization in the past emphasized constraint propagation and convexification techniques that aimed to provide tight relaxations for continuous optimization problems and relaxations of integer programs [1, 2, 3, 4]. We report recent developments in the integer arsenal of branch-and-reduce and their implementation in the global optimization software BARON. New features include integrality-based techniques, such as automatic constraint classification, preprocessing, probing, active constraint management, branching, and node selection rules. Specific cutting plane generation now produces knapsack cuts, GUB cuts, clique cuts, implication cuts, and flow cuts. Finally, mixed-integer linear programming relaxations are solved to construct new feasible solutions and improve relaxations at nodes of the search tree. Extensive computational results will be presented on problems from a collection of test sets, including process synthesis and operations problems from minlp.org.

1. Ryoo, H. S. and N. V. Sahinidis, Global optimization of nonconvex NLPs and MINLPs with applications in process design, Computers & Chemical Engineering, 19, 551-566, 1995.
2. Sahinidis, N. V., Global optimization and constraint satisfaction: The branch-and-reduce approach, pp. 1-16 in C. Bliek, C. Jermann, and A. Neumaier (eds.), Global Optimization and Constraint Satisfaction, Lecture Notes in Computer Science, Vol. 2861, Springer, Berlin, 2003.
3. Tawarmalani, M. and N. V. Sahinidis, Global optimization of mixed-integer nonlinear programs: A theoretical and computational study, Mathematical Programming, Ser. A, 99, 563-591, 2004.
4. Bao, X., A. Khajavirad, N. V. Sahinidis, and M. Tawarmalani, Global optimization of nonconvex problems with multilinear intermediates, Mathematical Programming Computation, DOI: 10.1007/s12532-014-0073-z, 2015.

390654 Tailored Global Optimization Algorithms for Mixed-Integer Linear Fractional Programming Problems

Abstract:

Mixed-integer linear fractional programming (MILFP) is a class of mixed-integer nonlinear program (MINLP) [1]. It is such a problem that includes both continuous and discrete variables and seeks to optimize an objective function formulated as a ratio of two linear functions subject to linear constraints. MILFP models usually occur in the planning and scheduling optimization for production process or when we apply the concept of functional unit as in the life-cycle assessment (LCA) [2-4]. MILFPs can be computationally intractable for large-scale problems due to its combinatorial nature and pseudo-convexity. In this work, we propose three tailored algorithms to solve this MILFP problem and compare their computation performance.

Normally, MILFPs can be solved by general Mixed-Integer Nonlinear Programming (MINLP) solvers as SBB, DICOPT and BARON [5]. However, for large-scale MILFP problem these general MINLP solvers may be unable to efficiently solve for optimal solution. Based on this situation, we present three tailored algorithms, including the parametric algorithm, reformulation-linearization algorithm and Charnes-Cooper + Branch & Bound algorithm.

Parametric algorithm, which transforms the original MILFP problem to a set of equivalent Mixed-Integer Linear Programming (MILP) subproblems, is proven to be an efficient method for solving MILFP problem [4, 6]. The advantage of parametric algorithm is that the size of each subproblem doesn't change; however the literation times are unpredictable and no gap information will be returned during solution.

Another one is the reformulation-linearization method, which combines the Charnes-Cooper transformation and Glove's linearization method together to transform the original MILFP problem to an equivalent MILP problem by introducing extra auxiliary variables and constraints [7]. The main advantage of reformulation-linearization algorithms is that the equivalent MILP problem only needs to be solved once to get the exact optimal solution for the original problem, and the gap information is available during solution. The drawback is the problem size increases due to the introduction of new variables and constraints.

The last one is Charnes-Cooper + Branch and Bound algorithm, in which we combine the Charnes-Cooper transformation and the branch-and-bound algorithm together [8]. By applying Charnes-Cooper transformation, we get rid of the fractional form of the original objective function. Moreover, we replace the bilinear terms result from Charnes-Cooper transformation with newly introduced variables. By using branch-and-bound algorithm, we guarantee the discrete nature of original binary variables implicitly included in the newly introduced variables. By integrating these two steps, we are able to transform the original MILFP problem into a set of Linear Programming (LP) subproblems, which can be efficiently solved by any LP solver, such as CPLEX. Obviously for each LP subproblem, it's much easier to solve than an MILP, however, due to the nature of branch-and-bound, the actual performance of this algorithm highly rely on the node selecting strategy.

The computation performance for these three solution approaches is demonstrated by some random generated examples and a large-scale MILFP problem derived from an optimal design and operation of water supply chain problem. The results are also compared with those returned by the MINLP solvers, SBB, DICOPT and BARON 12. According to the results, for most cases, parametric algorithm is times faster than other approaches; reformulation-linearization is normally slower but comparable to parametric algorithm; Charnes-Cooper transformation and Branch & Bound, though slower than parametric algorithm and reformulation-linearization, is faster than SBB, and BARON 12. In conclusion, all these three presented algorithms showed their efficiency in in solving large-scale MILFP problems, and parametric algorithm is the most extraordinary one.

References

[1] C. A. Floudas, *Deterministic global optimization*vol. 37: Springer, 2000.

[2] D. Yue, F. You, and S. W. Snyder, "Biomass-to-bioenergy and biofuel supply chain optimization: Overview, key issues and challenges," *Computers & Chemical Engineering,* 2013.

[3] Y. Chu and F. You, "Integration of production scheduling and dynamic optimization for multi-product CSTRs: Generalized Benders decomposition coupled with global mixed-integer fractional programming," *Computers & Chemical Engineering,* vol. 58, pp. 315-333, 11/11/ 2013.

[4] Z. X. Zhong and F. Q. You, "Globally convergent exact and inexact parametric algorithms for solving large-scale mixed-integer fractional programs and applications in process systems engineering," *Computers & Chemical Engineering,* vol. 61, pp. 90-101, Feb 2014.

[5] M. Tawarmalani and N. V. Sahinidis, "A polyhedral branch-and-cut approach to global optimization," *Mathematical Programming,* vol. 103, pp. 225-249, 2005/06/01 2005.

[6] F. You, P. M. Castro, and I. E. Grossmann, "Dinkelbach's algorithm as an efficient method to solve a class of MINLP models for large-scale cyclic scheduling problems," *Computers and Chemical Engineering,* vol. 33, pp. 1879-1889, 2009.

[7] D. Yue, G. Guillén Gosálbez, and F. You, "Global optimization of large scale mixed integer linear fractional programming problems: A reformulation linearization method and process scheduling applications," *AIChE Journal,* vol. 59, pp. 4255-4272, 2013.

[8] A. Charnes and W. W. Cooper, "Programming with linear fractional functionals," *Naval Research Logistics Quarterly,* vol. 9, pp. 181-186, 1962.

385169 Argonaut: Algorithms for Global Optimization of Constrained Grey-Box Computational Problems

Abstract:

Constrained grey-box optimization methods do not employ derivative information of the objective function and/or constraints of the original model for obtaining the global optimum. In typical applications of the above methods, derivative information may be available but deceptive or prohibitively expensive, or even completely unavailable [1, 2]. Development of efficient constrained grey-box methods is of high interest since it enables the optimization of costly simulation models developed to represent industrial processes with a high level of detail, coupling multiple physical, chemical, and mechanical phenomena. In addition, the independency of grey-box methods from derivatives allows the optimization of problems with embedded numerical noise, discontinuities and multiple local optima, which are all reasons for failure of finite-differencing techniques and consequently derivative-based methods [1].

Constrained grey-box methods have a vast pool of application areas ranging from expensive finite-element or partial-differential equation systems and flowsheet optimization to mechanical engineering design, molecular design, material screening, supply chain optimization and pharmaceutical product development, to name a few. In previous work, we have developed a constrained grey-box optimization method to optimize a Pressure Swing Adsorption (PSA) process for CO_2 capture [3, 4] and natural gas purification [5], while current and future applications of interest are Simulated Moving Bed (SMB) processes for separation of p-xylene, o-xylene and m-xylene, H_2 separation and air enrichment using zeolites and natural gas liquefaction using multi-stream heat exchangers.

During the past decades there have been significant developments in the area of derivative-free methods, however, the vast majority of methods have been developed for box-constrained problems, or problems with known linear constraints [1, 6]. Moreover, dimensionality of the original problem and the number of constraints are two major limitations of the performance of existing methods, which are typically tested on a small set of problems with low dimensionality and small number of constraints.

In this work, we present a novel AlgoRithm for Global Optimization of coNstrAined grey-box compUTational problems (ARGONAUT), which is developed to solve constrained grey-box problems with a large number of input variables and constraints. The novel components of the algorithm involve variable selection, model identification for the objective and each of the grey-box constraints using a large pool of possible basis functions, surrogate model fitting and global optimization of the model parameters, global optimization of the grey-box formulation using a global optimization solver [7], clustering of the obtained local and global solutions, and iterative bound refinement until convergence. The capacity of ARGONAUT is shown through a large set of problems from known standard libraries for constrained optimization, a selected set of in-house applications, such as SMB process and protein structure prediction, and several applications from the open literature. In fact, this work aims to formalize a comprehensive test suite for constrained derivative-free algorithms. Finally, ARGONAUT is compared with commercially available constrained derivative-free software.

References:

1. Conn AR, Scheinberg K, and Vicente LN, *Introduction to derivative-free optimization*. MPS-SIAM Series on Optimization. 2009, Philadelphia: SIAM.

2. Forrester AIJ, Sóbester A, and Keane AJ, *Engineering Design via Surrogate Modelling - A Practical Guide*. 2008: John Wiley & Sons.

3. Hasan MMF, et al., *Nationwide, Regional, and Statewide CO2 Capture, Utilization, and Sequestration Supply Chain Network Optimization*. Industrial & Engineering Chemistry Research, 2014; **53**(18): p. 7489-7506

4. Hasan MMF, First EL, and Floudas CA, *Cost-effective CO2 capture based on in silico screening of zeolites and process optimization*. Physical Chemistry Chemical Physics, 2013; **15**(40): p. 17601-17618

5. First EL, Hasan MMF, and Floudas CA, *Discovery of novel zeolites for natural gas purification through combined material screening and process optimization*. AIChE Journal, 2014; **60**(5): p. 1767-1785

6. Rios LM and Sahinidis NV, *Derivative-free optimization: a review of algorithms and comparison of software implementations*. Journal of Global Optimization, 2013; **56**(3): p. 1247-1293.10.1007/s10898-012-9951-y

7. Misener R and Floudas C, *ANTIGONE: Algorithms for coNTinuous / Integer Global Optimization of Nonlinear Equations*. Journal of Global Optimization, In Press, DOI: 10.1007/s10898-014-0166-2, 2014

Logic-based outer-approximation for the global optimization of generalized disjunctive programs

Francisco Trespalacios and Ignacio E. Grossmann

Many optimization problems are represented with algebraic equations, using continuous and discrete variables, giving rise to Mixed Integer (Linear or Nonlinear) Programs (MILP/ MINLP). An alternative way to represent this type of problems is Generalized Disjunctive Programming (GDP) proposed by Raman and Grossmann[1] that involves not only algebraic equations, but also disjunctions and logic propositions. This higher level representation, which is of special importance in engineering, allows us to exploit the logic structure of the problem to obtain better formulations and improved solution methods.

GDP problems are normally reformulated as MILP/MINLP models to exploit the developments of these types of solvers. However, specialized methods have been developed to solve generalized disjunctive programs: disjunctive branch and bound[2], and logic based Outer-approximation[3]. The former can be applied to the global optimization of nonconvex GDPs, as long as each node in the branch and bound is solved to global optimality. The latter, however, can only be applied to convex GDPs (i.e. GDPs in which the objective function and inequality constraints are convex, and the equality constraints are linear).

In this work we present an alternative logic-based outer-approximation algorithm to find the global solution of non-convex GDPs. The general idea of the algorithm is to have a master MILP that overestimates the feasible region of the GDP. This master problem provides a valid lower bound (in a minimization problem), and the selection of only one disjunctive term in each of the disjunctions. With this alternative provided by the master problem, an NLP subproblem is solved to global optimality. This NLP subproblem is smaller and simpler than the continuous relaxation of the MINLP reformulation of the original GDP. After solving this subproblem, infeasibility or optimality integer cuts can be added to the master problem.

This basic algorithm has the advantage of solving only small NLP problems to global optimality, instead of solving a larger MINLP to global optimality from the beginning. Furthermore, by using GDP as framework the NLP subproblem is smaller and simpler than an equivalent method directly applied to the MINLP reformulation. Even with these advantages, the convergence of the algorithm is slow and it may have difficulties finding good or optimal solutions. In order to improve the performance of this logic-based outer approximation, we implement three main features: derivation of additional cuts, partition of the algorithm in two stages, and parallelization of the algorithm.

In the first improvement we develop a novel method to derive a cutting plane in each of the disjunctive terms selected by the master problem. We obtain this cut by solving to global optimality one NLP for each of the selected disjunctive terms. This NLP minimizes the distance between the solution found by the master problem, and a point that lies in the tightest possible convex region that contains the nonconvex region described by the disjunctive term. The idea of solving this NLP to obtain a cut is similar to the well-known concept of the separation problem in MILP[4] and convex MINLP[5].

A second improvement in the algorithm is its partition into two phases. The first phase seeks to find a good feasible solution, while the second phase provides a rigorous lower bound. The first phase tests many alternatives and evaluates the NLP subproblems only for a few seconds. If an NLP subproblem provides a feasible solution, it is a valid upper bound. If the NLP is proven infeasible within that time period, then a feasibility cut is derived, and it is valid for the original MINLP. If the NLP does not find a feasible solution within the given time, an infeasibility cut is derived and used for the first phase. However, this cut is not valid for the original MINLP, so it is discarded in the second phase. Note that the first phase provides valid upper bounds when found, but it does not provide a valid lower bound in general. For the second phase, the invalid cuts found in phase 1 are removed. After this, the rigorous algorithm previously described is solved (without stopping the NLP subproblems until they are solved to global optimality or proven infeasible). In this second phase a stronger MILP relaxation is used to provide good lower bounds.

The third improvement in the algorithm is to parallelize the derivation of cutting planes and solution of subproblems. This parallelization is performed by obtaining a pool of solutions from the master problem. The cut derivation and subproblem solution is then performed in parallel for each of the solutions in the pool.

We illustrate the application of this algorithm with several GDP process networks examples. In some cases, process networks are difficult to solve as MINLPs, but easy NLPs to solve for a given fixed structure. This characteristic makes the logic based outer-approximation an attractive method for solving these type of problems. We compare the solution of these problems using the described algorithm against global optimization solvers. The results show that the algorithm generally finds better solutions faster, and with a smaller optimality gap.

References:

[1] Raman, Ramesh, and Ignacio E. Grossmann. "*Modelling and computational techniques for logic based integer programming*", Computers & Chemical Engineering. 1994; 18(7), 563-578.
[2] Beaumont, Nicholas. "*An algorithm for disjunctive programs*", European Journal of Operational Research.1990; 48(3), 362-371.
[3] Türkay, Metin, and Ignacio E. Grossmann. "*Logic-based MINLP algorithms for the optimal synthesis of process networks*", Computers & Chemical Engineering. 1996; 20(8), 959-978.
[4] Balas, Egon. "*Disjunctive programming*", Annals of Discrete Mathematics, 1979; 5, 3-51.
[5] Stubbs, Robert A., and Sanjay Mehrotra. "A branch-and-cut method for 0-1 mixed convex programming", Mathematical Programming. 1999; 86(3), 515-532.

AICHE 2014 Session: 10C03

Title: Global Optimization of Implicit Functions

Authors: Harry A. J. Watson, Achim Wechsung, Matthew D. Stuber, and Paul I. Barton

Abstract:

The set of variables in an optimization problem can often be partitioned into subsets corresponding to the independent parameters and the dependent variables. Under such a partitioning, the problem's equality constraints are a parametric equation system, the solution of which defines the dependent variables as an implicit function of the independent parameters. As a result, it is only necessary to consider the parameters as decision variables in the problem formulation, if an algorithm is available to calculate values of the implicit function. When the set of parameters is of significantly lower dimension than the set of dependent variables, as is often the case in process optimization problems, this reformulation can be highly advantageous due to the worst-case exponential run-time scaling of all known global optimization algorithms.

Using the well-known branch and bound algorithm for this "reduced-space" reformulation requires a procedure for calculating bounds and/or relaxations on the range of the implicit function in addition to its value at a point in the parameter space. When weak relaxations are used, the branch and bound algorithm will tend to partition boxes in the vicinity of a minimum repeatedly. This tendency is known as the "cluster problem", and is highly dependent on the convergence order of the underestimating scheme used to generate lower bounds [1] [2]. However, it is also known that the cluster effect can disappear when a quadratically convergent lower bounding scheme is used [2]. Unfortunately, the standard machinery for bounding the solution of a system of nonlinear equations, such as the interval Newton method and its variants, exhibits only linear convergence order in the parametric case [3]. Recently, however, methods have been proposed for generating quadratically convergent lower bounds for global optimization with implicit functions.

This paper analyzes the performance of these newly developed methods for obtaining bounds and relaxations on implicitly defined functions in the context of process optimization problems. The methods studied include a sensitivity-based interval bounding method developed in [3], parametric mean-value form McCormick relaxations developed in [4], and a reverse propagation scheme for McCormick objects developed in [5]. These methods (and combinations thereof) are compared with each other and with other widely used algorithms, such as interval Newton and reverse interval propagation methods, on the basis of empirical convergence order and solution speed on a set of process optimization test problems. Comparison is also made with the selective branching approach, implemented using the global optimizer BARON [6]. The results highlight the importance of generating enclosures of the range of implicit functions which are rapidly convergent, inexpensive to calculate, and applicable to large, highly nonlinear problems, as well as the advantages of these methods over selective branching.

[1] K. Du and R. B. Kearfott, "The cluster problem in multivariate global optimization," *Journal*

of Global Optimization, vol. 5, pp. 253-265, 1994.

[2] A. Wechsung, S. D. Schaber and P. I. Barton, "The cluster problem revisited," *Journal of Global Optimization,* vol. 58, no. 3, pp. 429-438, 2014.

[3] A. Wechsung, J. K. Scott and P. I. Barton, "Second-order convergent parametric interval bounds for implicit functions." Submitted.

[4] M. D. Stuber, J. K. Scott and P. I. Barton, "Convex and Concave Relaxations of Implicit Functions." In Press. *Optimization Methods and Software,* 2014.

[5] A. Wechsung, J. K. Scott and P. I. Barton, "Reverse propagation of McCormick relaxations," Submitted to *Journal of Global Optimization.*

[6] N. V. Sahinidis, "BARON 12.1.0: Global Optimization of Mixed-Integer Nonlinear Programs," User's Manual, 2013.

Effective Discretization Strategy for Multi-Component Networks

Elisabet Capón-García, Ignacio E. Grossmann, Konrad Hungerbühler

Process networks comprise a set of process units connected by material flows, which in general consist of a mixture of chemical compounds. The modeling of multi-component networks can be expressed either (i) in terms of individual flows, or (ii) in terms of total flows and compositions. In both cases, non-linear terms appear derived from modeling mixing and splitting conditions in the process units. Hence, the optimization of multi-component process networks involves handling these non-linear terms, which represents a computational challenge especially for large-scale problems. This concern arises in a wide variety of examples, such as the pooling problem (Misener et al., 2011), the multi-period blending problem (Kolodziej et al., 2013b) or the design of water networks (Ahmetovic and Grossmann, 2011).

In the model based on individual flows, non-linear equations are derived from splitting conditions. Specifically, the ratio between individual flows of the streams, that is, the composition, leaving the splitting unit has to be the same as the composition of inlet stream of the splitting unit. This condition leads to the following nonlinear equation:

$$\frac{m_i^{in}}{\sum_{i'} m_{i'}^{in}} = \frac{m_{is}^{out}}{\sum_{i'} m_{i's}^{out}} \qquad \forall i, s$$

where i is the set of components, s is the set of output streams, m_i^{in} is the individual input flow of component i, and m_{is}^{out} is the individual output flow s of component i.

In the literature, a wide number of strategies are presented to tackle the bilinear terms that appear from manipulating the previous constraint, such as the multiparametric disaggregation technique (Kolodziej et al., 2013a) or the linear and logarithmic partitioning schemes (Misener et al., 2011).

This work proposes a discretization strategy of the mass/volume fraction of the individual flows in the process units involving stream splitting for avoiding non-linear terms. The mathematical formulation of the problem corresponds to an MILP that is based on individual flows related to concentration limits in the operating units.

The discretization strategy consists of dividing the stream (or volume) to be split ("initial stream/volume") into a finite number of fractions represented by binary variables. Therefore, for a specific output stream, the output fraction of the total flow and each fraction of the individual flows should be the same for all streams with regard to the "initial stream/volume". Hence, binary variables are related to outlet streams and shared by all individual flows contained in the same stream.

This strategy has been applied to several case studies of the pooling problem, multiperiod blending problem and design of water networks, and compared to other solution approaches. Results have shown that significant improvements can be obtained in terms of solution quality and computational effort.

Bibliography

Ahmetovic, E. and Grossmann I.E., 2011. Global Superstructure Optimization for the Design of Integrated Process Water Networks. AIChE Journal 57, 434–457.

Kolodziej, S.P., Castro, P.M., Grossmann, I.E., 2013a. Global optimization of bilinear programs with a multiparametric disaggregation technique, Journal of Global Optimization 57, 1039–1063.

Kolodziej, S.P., Grossmann, I.E., Furman, K. C., Sawaya, N.W., 2013b. A discretization-based approach for the optimization of the multiperiod blend scheduling problem, Computers and Chemical Engineering 53, 122–142.

Misener, R., Thompson, J.P., Floudas, C.A., 2011. APOGEE: Global optimization of standard, generalized, and extended pooling problems via linear and logarithmic partitioning schemes. Computers and Chemical Engineering 35, 876–892.

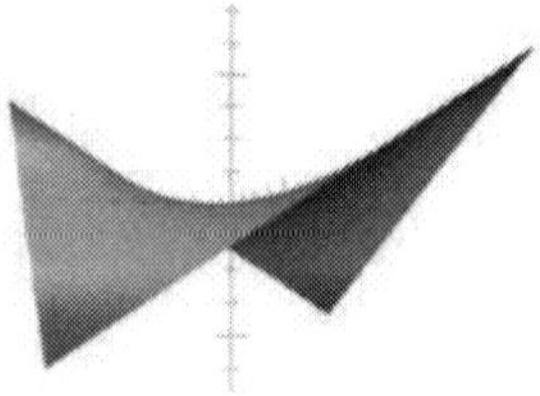

Tightening Piecewise McCormick Relaxations through Partition-Dependent Bounds for Non-Partitioned Variables

Pedro M. Castro (pmcastro@fc.ul.pt)
http://www.researcherid.com/rid/C-3642-2008

Associate Researcher
Centro de Investigação Operacional
Faculdade de Ciências
Universidade de Lisboa

IF INVESTIGADOR FCT

U LISBOA UNIVERSIDADE DE LISBOA

Problem definition (NLP)

- Bilinear program

(P)

$$\min f_0(x)$$

$$f_q(x) \le 0 \quad q \in Q\backslash\{0\}$$

$$f_q(x) = \sum_{(i,j)\in BL_q} a_{ijq} x_i x_j + h_q(x) \quad q \in Q$$

$$0 \le x^L \le x \le x^U$$

$$x \in \mathbb{R}^m$$

Relaxation provides a lower bound

a_{ijq} - parameters
x – vector of continuous variables
BL_q - (i,j) index set
$i \neq j$ – strictly bilinear problems
$i = j$ – can be allowed (quadratic problems)
$h_q(x)$- linear function in x

Introduction

- Bilinear problems occur in a variety of applications
  - Process network problems (Quesada & Grossmann, 1995)
  - Water networks (Bagajewicz, 2000)
  - Pooling and blending (Haverly, 1978)
- Non-convex, leading to multiple local solutions
  - Gradient based solvers unable to certify optimality
- Need for global optimization algorithms
  - What do they have in common?
    - Linear (LP) or mixed-integer linear (MILP) relaxation of (**P**) → LB
      - LP: Standard McCormick envelopes (1976)
      - MILP: Piecewise McCormick envelopes (Bergamini et al. 2005)
      - MILP: Multiparametric disaggregation (Teles et al. 2013; Kolodziej et al. 2013)
    - Solution of (**P**) with fast local solver → UB
      - Using LB as starting point
- Tight relaxation critical to ensure convergence
  - Relative optimality gap (UB-LB)/LB<ε

Piecewise McCormick relaxation $w_{ij} = x_i x_j$

- Domain of divided into N uniform partitions

x_j^L | $n=1$ | $x_{j,1}^U$ | $n=2$ | $x_{j,2}^U$ | . . . | $x_{j,N-1}^U$ | $n=N$ | $x_{j,N}^U$

$x_{j,1}^L$ | $x_{j,2}^L$ | $x_{j,N}^L$ | x_j^U

$$y_{j,1} = true \ \underline{\vee}\ y_{j,2} = true \ \underline{\vee}\ \cdots \ \underline{\vee}\ y_{j,N} = true$$

Optimization will pick single partition

Generalized Disjunctive Program
(Balas, 1979; Raman & Grossmann, 1994)

$$\underline{\bigvee_n} \begin{bmatrix} y_{jn} \\ w_{ij} \geq x_i \cdot x_{jn}^L + x_i^L \cdot x_j - x_i^L \cdot x_{jn}^L \\ w_{ij} \geq x_i \cdot x_{jn}^U + x_i^U \cdot x_j - x_i^U \cdot x_{jn}^U \\ w_{ij} \leq x_i \cdot x_{jn}^L + x_i^U \cdot x_j - x_i^U \cdot x_{jn}^L \\ w_{ij} \leq x_i \cdot x_{jn}^U + x_i^L \cdot x_j - x_i^L \cdot x_{jn}^U \\ x_{jn}^L \leq x_j \leq x_{jn}^U \end{bmatrix}$$

$$x_i^L \leq x_i \leq x_i^U$$
$$x_{jn}^L = x_j^L + (x_j^U - x_j^L) \cdot (n-1)/N$$
$$x_{jn}^U = x_j^L + (x_j^U - x_j^L) \cdot n/N$$

Convex hull →

(**PR-MILP**)

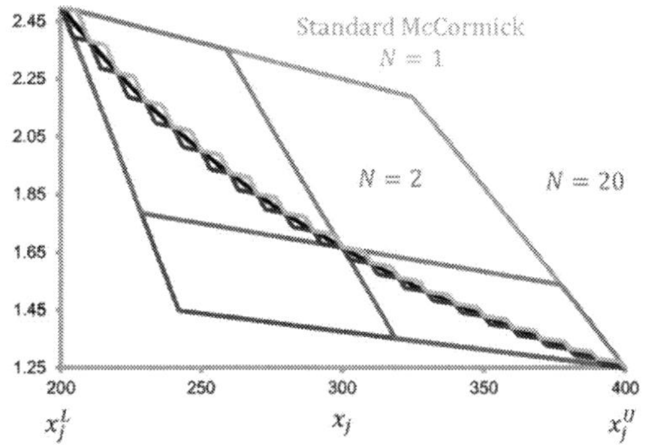

New tighter piecewise relaxation

- Basic idea
  - Use partition-dependent bounds also for variable x_i
    - x_{ijn}^L / x_{ijn}^U lower/upper bound when x_j constrained to n
  - Better bounds $\Rightarrow$ tighter relaxation $\Leftrightarrow$ lower gap

$$\bigvee_n \left[\begin{array}{c} w_{ij} \geq x_i \cdot x_{jn}^L + x_{ijn}^L \cdot x_j - x_{ijn}^L \cdot x_{jn}^L \\ w_{ij} \geq x_i \cdot x_{jn}^U + x_{ijn}^U \cdot x_j - x_{ijn}^U \cdot x_{jn}^U \\ w_{ij} \leq x_i \cdot x_{jn}^L + x_{ijn}^U \cdot x_j - x_{ijn}^U \cdot x_{jn}^L \\ w_{ij} \leq x_i \cdot x_{jn}^U + x_{ijn}^L \cdot x_j - x_{ijn}^L \cdot x_{jn}^U \\ x_{ijn}^L \leq x_i \leq x_{ijn}^U \\ x_{jn}^L \leq x_j \leq x_{jn}^U \end{array} \right] \quad (y_{jn})$$

$$x_{jn}^L = x_j^L + (x_j^U - x_j^L) \cdot (n-1)/N$$
$$x_{jn}^U = x_j^L + (x_j^U - x_j^L) \cdot n/N$$

Convex hull

(PRT-MILP)

$$\min z^R = f_0(x) = \sum_{(i,j) \in BL} a_{ij0} w_{ij} + h_0(x)$$

$$f_q(x) = \sum_{(i,j) \in BL} a_{ijq} w_{ij} + h_q(x) \leq 0 \ \forall q \in Q \backslash \{0\}$$

$$\left. \begin{array}{l} w_{ij} \geq \sum_n (\hat{x}_{ijn} \cdot x_{jn}^L + x_{ijn}^L \cdot \hat{x}_{jn} - x_{ijn}^L \cdot x_{jn}^L \cdot y_{jn}) \\ w_{ij} \geq \sum_n (\hat{x}_{ijn} \cdot x_{jn}^U + x_{ijn}^U \cdot \hat{x}_{jn} - x_{ijn}^U \cdot x_{jn}^U \cdot y_{jn}) \\ w_{ij} \leq \sum_n (\hat{x}_{ijn} \cdot x_{jn}^L + x_{ijn}^U \cdot \hat{x}_{jn} - x_{ijn}^U \cdot x_{jn}^L \cdot y_{jn}) \\ w_{ij} \leq \sum_n (\hat{x}_{ijn} \cdot x_{jn}^U + x_{ijn}^L \cdot \hat{x}_{jn} - x_{ijn}^L \cdot x_{jn}^U \cdot y_{jn}) \\ x_i = \sum_n \hat{x}_{ijn} \end{array} \right\} \forall (i,j)$$

$$\left. \begin{array}{l} x_j = \sum_n \hat{x}_{jn} \\ \sum_n y_{jn} = 1 \end{array} \right\} \forall \{j | (i,j) \in BL\}$$

$$\left. \begin{array}{l} x_{jn}^L = x_j^L + \dfrac{(x_j^U - x_j^L) \cdot (n-1)}{N} \\ x_{jn}^U = x_j^L + \dfrac{(x_j^U - x_j^L) \cdot n}{N} \\ x_{jn}^L \cdot y_{jn} \leq \hat{x}_{jn} \leq x_{jn}^U \cdot y_{jn} \end{array} \right\} \forall \{j | (i,j) \in BL\}, n$$

$$x_{ijn}^L \cdot y_{jn} \leq \hat{x}_{ijn} \leq x_{ijn}^U \cdot y_{jn} \ \forall (i,j) \in BL, n \in \{1, \ldots, N\}$$

How to generate bounds x_{ijn}^L & x_{ijn}^U?

- Optimality bound contraction with McCormick envelopes
  - Solve multiple instances of LP problem (**BC**), two per:
    - Non-partitioned variable x_i
    - Partitioned variable x_j
    - Partition n
  - Focus on regions that can actually improve current UB
    - z' obtained from solving (**P**) with local solver
  - (**BC**) infeasible? Remove partition n^* of j^* from (**PRT-MILP**)
  - Computation can be time consuming

(**BC**)

$$x_{i^* j^* n^*}^L := \min x_{i^*} \quad (x_{i^* j^* n^*}^U := \max x_{i^*})$$

$$f_0(x) = \sum_{(i,j) \in BL_0} a_{ij0} w_{ij} + h_0(x) \leq z'$$

$$f_q(x) = \sum_{(i,j) \in BL_q} a_{ijq} w_{ij} + h_q(x) \leq 0 \quad \forall q \in Q \backslash \{0\}$$

$$\left. \begin{array}{l} w_{ij} \geq x_i \cdot x_j^L + x_i^L \cdot x_j - x_i^L \cdot x_j^L \\ w_{ij} \geq x_i \cdot x_j^U + x_i^U \cdot x_j - x_i^U \cdot x_j^U \\ w_{ij} \leq x_i \cdot x_j^L + x_i^U \cdot x_j - x_i^U \cdot x_j^L \\ w_{ij} \leq x_i \cdot x_j^U + x_i^L \cdot x_j - x_i^L \cdot x_j^U \end{array} \right\} \forall (i,j) \in BL$$

$$x_{j^*}^L = x_{j^* n^*}^L \leq x_{j^*} \leq x_{j^* n^*}^U = x_{j^*}^U$$

$$x^L \leq x \leq x^U$$

New optimization algorithm

- Key features
  - User selects:
    - N partitions
    - Partitioned variables x_j
  - Preliminary bound contraction stage
  - Bounds updated at different levels
  - LB from MILP relaxation
  - UB from NLP (one starting point)
  - Computes an optimality gap

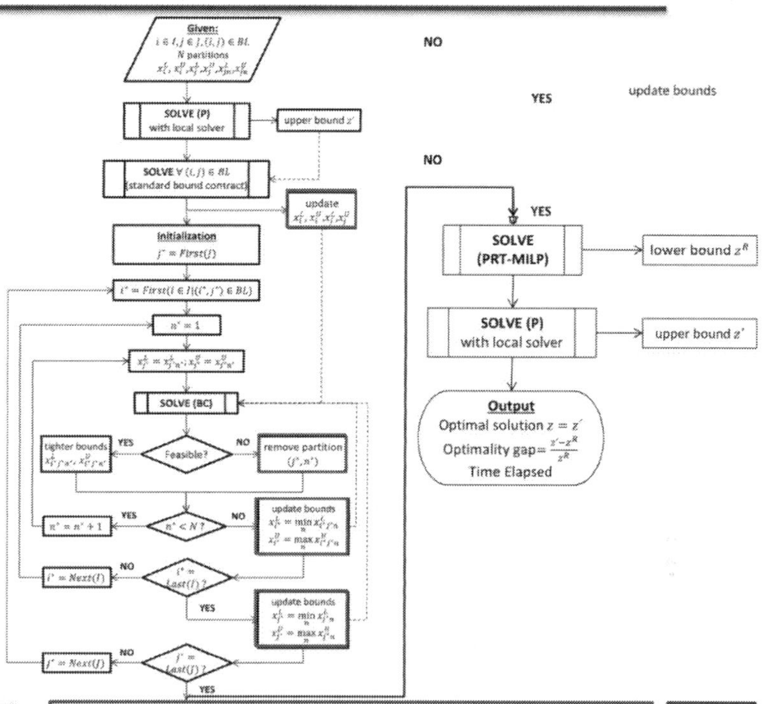

Illustrative example P1

$$\min f_0(x) = -x_1 + x_1x_2 - x_2$$
$$-6x_1 + 8x_2 \leq 3$$
$$3x_1 - x_2 \leq 3$$
$$0 \leq x_1, x_2 \leq 1.5$$

Partitioned variable: x_1
Non-partitioned variable: x_2
$BL = \{(2,1)\}$

Global optimum

$z = f_0(x) = -1.083333$

Standard optimality bound contraction

$$0.357143 \leq x_1 \leq 1.375$$
$$0 \leq x_2 \leq 1.26$$

	Gap
(**PR-MILP**) relaxation	
$z^R = -1.185207$	8.60%
(**PRT-MILP**) relaxation	
$z^R = -1.169619$	7.38%

Partition-dependent bound contraction $N = 3$

Partition n	x_{1n}^L	x_{1n}^U	x_{21n}^L	x_{21n}^U
1	0.357143	0.696429	LP is infeasible	
2	0.696429	1.035714	0	1.137609
3	1.035714	1.375	0.107143	1.195150

How about using more partitions?

- Up to two orders of magnitude reduction in optimality gap
  - Major increase in total computational time
  - Still, new approach performs best!

P1

Relaxation	Partitions (N)	15	150	1500	7500
Standard (PR-MILP)	Total CPUs	0.93	1.60	3.63	302
	Optimality gap	1.67%	0.188%	0.0189%	0.0038%
Tightened (PRT-MILP)	Optimality gap	0.77%	0.010%	0.0002%	
	Total CPUs	3.63	27.8	267	

P2

$$\min 6x_1^2 + 4x_2^2 - 2.5x_1x_2$$

$$x_1x_2 - 8 \geq 0$$

$$1 \leq x_1, x_2 \leq 10$$

Relaxation	Partitions (N)	9	90	900	2700
Standard (PR-MILP)	Total CPUs	0.83	1.2	7.76	478
	Optimality gap	8.41%	0.84%	0.0830%	0.0278%
Tightened (PRT-MILP)	Optimality gap	0.84%	0.02%	0.0002%	
	Total CPUs	3.33	24.0	223	

Results for larger test problems

- 16 easiest water-using design problems
  - New (**PRT-MILP**) approach leads to lower gaps at termination (3600 CPUs)
  - Faster in 56% of the cases (finest partition level)
  - Single failure when finding best-known solution for (**P**)

Subset of problems selected for illustration

	Number of variables			Standard (PR-MILP)			Tightened (PRT-MILP)		
	Non-partitioned $\|I\|$	Partitioned $\|J\|$	Partitions (N)	Solution	Total CPUs	Gap	Gap	Total CPUs	Solution
Ex2	20	16	10	74.4699	18.8	0.1250%	0.0484%	141	74.4699
			50		3609	0.1245%	0.0103%	1536	
Ex6	30	25	10	142.0816	44.7	0.0435%	0.0110%	205	142.0816
			50		3614	0.1379%	0.0041%	1090	
Ex8	30	20	10	164.4898	18.8	0.7134%	0.2132%	207	164.4898
			100		3613	1.9508%	0.0130%	1803	
Ex10	15	9	10	169.1173	6.69	0.0160%	0.0008%	45.3	169.1173
			100		809	0.0023%	0.0000%	323	
Ex14	32	24	10	329.9689	282	1.2259%	1.0499%	447	329.5698
			20	329.9566	3615	2.3368%	0.9206%	4270	
Ex15	40	30	10	361.5177	231	0.7963%	0.7427%	723	361.5177
			20	361.6856	3618	0.9361%	0.7779%	4486	361.6856
Ex16	24	16	10	285.9343	17.4	1.1105%	0.8861%	212	285.9343
			50		3610	1.5755%	0.0526%	2042	

How to determine the number of partitions?

- Quality of relaxation increases with N
  - Careful not to generate intractable MILPs (Gounaris et al. 2009)
- Several works identify most appropriate value for a particular problem
  (Misener et al. 2011; Wittmann-Holhbein and Pistikopoulos, 2013; 2014)
  - No method to estimate N as a function of problem complexity
  - Needed to provide a fair comparison with commercial solvers
- Proposed formula meets important criteria:
  - Returns an integer value
  - Ensures minimum of two partitions
    - So as to benefit from piecewise relaxation scheme
  - Reduces N with increase in problem size
    - Roughly inversely proportional (r^2=0.76)

$$N = 1 + \left\lceil \frac{\alpha}{|I| \cdot |J|} \right\rceil \qquad \alpha = 1.8E4$$

Comparison to commercial solvers

- Performance profiles
  (Dolan & Moré, 2002)
  - KPI: Optimality gap

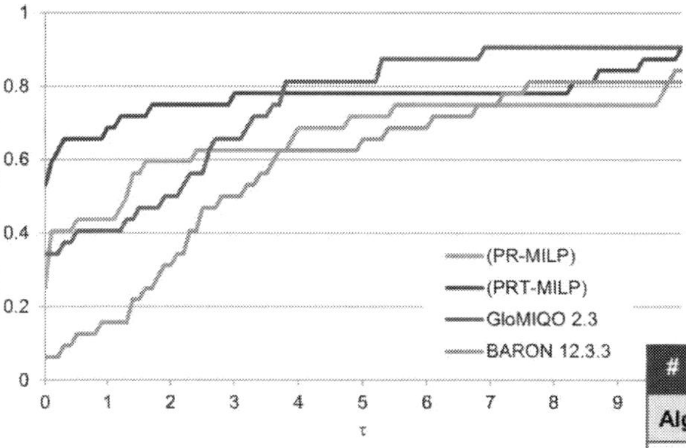

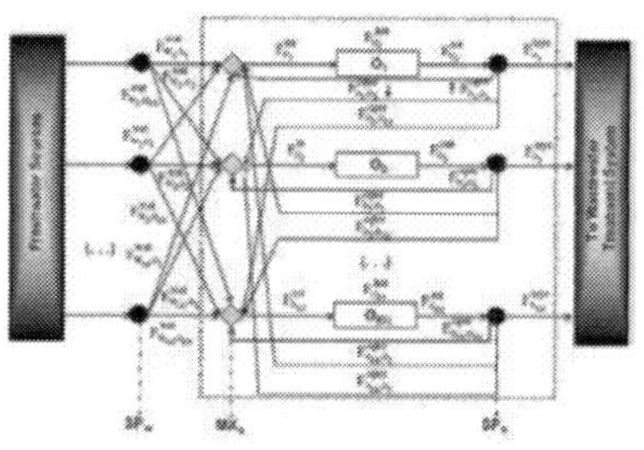

- (**PRT-MILP**) better (**PR-MILP**)
- GloMIQO outperforms BARON

GAMS 24.1.3, CPLEX 12.5.1, GloMIQO 2.3, BARON 12.3.3, Intel Core i7-3770 (3.07 GHz), 8 GB RAM, Windows 7 64-bit
Termination criteria: gap=0.0001%, Time=3600 CPUs

Failures finding best-known solution (34 problems)

Algorithm	GloMIQO	(PRT-MILP)	BARON	(PR-MILP)
Suboptimal	3	5	2	7
No solution	0	0	4	0
Total	**3**	**5**	**6**	**7**

Why not bivariate partitioning?

- Domain of x_i and x_j known a priori for each 2-D partition
  - Slightly better performance than univariate partitioning for P1
  - Gap for 125^2 partitions one order of magnitude larger vs. (PRT-MILP)

(PR-BV-MILP)

$$\bigvee_{n=1}^{N}\bigvee_{n'=1}^{N}\begin{bmatrix} y_{injn'} \\ w_{ij} \ge x_i \cdot x^L_{jn'} + x^L_{in}\cdot x_j - x^L_{in}\cdot x^L_{jn'} \\ w_{ij} \ge x_i \cdot x^U_{jn'} + x^U_{in}\cdot x_j - x^U_{in}\cdot x^U_{jn'} \\ w_{ij} \le x_i \cdot x^L_{jn'} + x^U_{in}\cdot x_j - x^U_{in}\cdot x^L_{jn'} \\ w_{ij} \le x_i \cdot x^U_{jn'} + x^L_{in}\cdot x_j - x^L_{in}\cdot x^U_{jn'} \\ x^L_{in} \le x_i \le x^U_{in} \\ x^L_{jn'} \le x_j \le x^U_{jn'} \end{bmatrix} \forall (i,j)$$

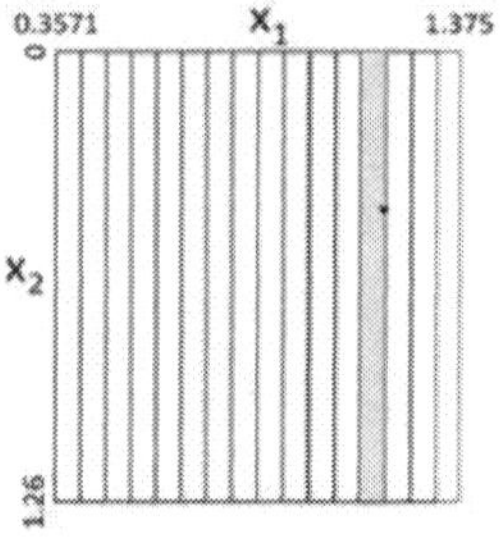

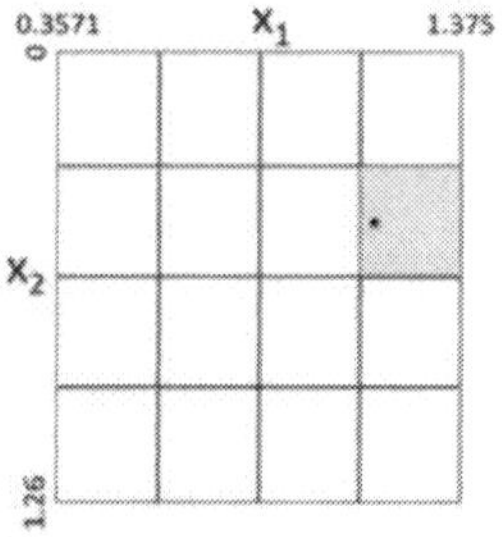

Relaxation	Partitions (N)	16	144	2500	10000	
Univariate (PR-MILP)	Total CPUs	0.88	1.42	9.11	429	
	Optimality gap	1.70%	0.197%	0.0114%	0.0029%	
	N	4	12	50	100	125
Bivariate (PR-BV-MILP)	Optimality gap	1.28%	0.188%	0.0078%	0.0029%	0.0018%
	Total CPUs	0.93	0.91	7.55	217	469

Conclusions

- Three piecewise relaxation (PR) schemes for bilinear problems
- Novel approach with partition-dependent bounds for all bilinear variables
  - Significant improvement in relaxation quality
  - Need for extensive optimality-based bound contraction (BC)
    - Total computational time is sometimes lower
- New algorithm outperforms state-of-the-art global solvers
  - Specific class of water-using network design problems
- PR and BC schemes should be used to greater extent
  - Integrating with spatial B&B subject of future work

- See CACE 2015, 72, 300-311 for further details

- Acknowledgments:
  - Luso-American Foundation (2013 Portugal-U.S. Research Networks)
  - Fundação para a Ciência e Tecnologia (Investigador FCT 2013)

Heterogeneity and reduction for complex network dynamics

Ioannis G. Kevrekidis, Carlo Laing, Alexander Holiday, Tom Bertalan
Chemical and Biological Engineering & PACM, Princeton University, USA

Summary. In the study of dynamic, large-scale, complex networks, coarse-grained (reduced, low-dimensional) descriptions of the network can be useful not only in identifying significant network features, but also in enabling reduced, systems-level analysis of the evolution dynamics. Obtaining such a coarse-grained description for networks is a challenging problem. Additionally, the implementation of coarse-grained computational algorithms requires a systematic approach for translating low-dimensional network features to full-scale network realizations. We discuss recent advances, and experience with applications, in both these directions for complex network evolution.

Complex, large scale networks are often dynamically evolving in time. One can discriminate several different forms for such an evolution: (a) it may be that the connectivity of the network is fixed, but properties of the nodes evolve in time (e.g. concentrations in a complex biochemical reaction network); this we term dynamics *on* networks; (b) it may be that the only thing evolving is the connectivity of the network - edges either form or disappear (alternatively, their weights evolve) in time; this is termed dynamics *of* networks; finally (c) it is possible that *both* properties of the nodes and existence/weights of edges evolve, giving us dynamics *of and on* networks or, as it is also sometimes termed, *adaptive* network dynamics. In all these cases we have complex, large scale evolving dynamical systems, and this work presents several efforts towards their algorithmic model reduction.

We begin by describing and illustrating an approach to coarse-graining the evolution of networks whose connectivity changes dynamically. The approach is formulated within our *equation-free* framework given a full, detailed model of the network evolution dynamics. We assume that we know what the crucial macroscopic network features are, the "dependent variables" in terms of which we can write effective closed evolution equations (e.g. the network degree distribution). We then perform short bursts of detailed network evolution simulations from which we estimate the right-hand-sides (and the actions of the Jacobians) of these unavailable closed evolution equations on the fly. The approach involves repeated use of lifting and restriction operators that translate between actual network realizations and their (appropriately chosen) coarse observables - the more "difficult" of the two is the lifting: creation of networks that have prescribed statistical features. The equation-free framework will be used here to accelerate temporal simulations (through coarse projective integration), and to implement coarse-grained fixed point algorithms (through matrix-free Newton-Krylov GMRES). The approach will be illustrated through a simple network evolution example, for which analytical approximations to the coarse-grained dynamics can be independently obtained, so as to validate the multiscale computational results. The scope and applicability of the approach will be discussed.

So far we have assumed that the appropriate coarse-grained observables (the "right" variables in terms of which coarse network evolution equations can be expressed) are known *a priori*. We now discuss the detection of such variables based on data-mining tools, in particular manifold-learning techniques like Diffusion Maps. We extend these data mining approaches to cases in which every data point in a time series is a "snapshot" of an evolving graph. We are thus able to detect intrinsic low-dimensionality in ensembles of graphs, which is the first step in coarse-graining the detailed graph information; the second step is the detection of the appropriate coarse variables that parameterize the network evolution. One of the main challenges in mining graph evolution data is the definition of a suitable pairwise similarity metric in the space of graphs. We explore two practical solutions to solving this problem: one based on finding subgraph densities, and one using graph spectral information. We demonstrate the data-mining process by detecting and parametrizing low-dimensional families of graphs arising from graph construction algorithms as well as from dynamic graph evolution laws.

The last component of the work addresses heterogeneous networks, and the construction of good coarse-grained observables for such networks. This involves a "technology transfer" from modern uncertainty quantification methods, namely from generalized polynomial chaos techniques and from noninvasive, collocation-type implementations of them. There, the response of a system to a parameter whose value comes from a known probability distribution is expanded in terms of orthogonal polynomials weighted by this distribution. Here, the behavior of a system of heterogeneous units, whose heterogeneity parameter comes from a known probability distribution, is again expanded in terms of orthogonal polynomials weighted by this distribution. This allows the collective behavior of large numbers of coupled heterogeneous oscillators to be described in terms of "just a few" evolving expansion coefficients, in the spirit of spectral methods. What makes the approach particularly appealing, however, is the ability to use it both in cases where the heterogeneity is intrinsic to the nodes (e.g. neurons whose kinetic property values are sampled from a known distribution), as well as in cases where the heterogeneity is characteristic of the network structure (e.g. neurons whose degree in a network is sampled from a known degree distribution). Remarkably, the approach works equally well, at least in principle, when joint intrinsic as well as structural heterogeneities are considered. What now becomes interesting is the case where these heterogeneity distributions are not independent, but rather correlated; then the simple and powerful noninvasive collocation-type methods are not easily applicable any more.

Our illustrative examples range from "textbook" Kuramoto-on-a-network type dynamics to more detailed heterogeneous neuronal network models, where both intrinsic and structural heterogeneity arises. "Lifting" (the construction of networks consistent with one or more statistical properties) in the context of Mixed Integer Linear Programming is discussed and contrasted to stochastic optimization techniques (like simulated annealing); this component of the work has been done in collaboration with C. Floudas (Princeton) and C. Gounaris (CMU).

References

[1] A. C. Tsoumanis, Rajendran, K., Siettos, C. I., and Kevrekidis I. G., Coarse-graining the dynamics of network evolution: the rise and fall of a networked society, N. J. Phys. 14 (083037) (2012)

[2] K. A. Bold, K. Rajendran, B. Rath and I. G. Kevrekidis, An equation-free approach to coarse-graining the dynamics of networks, J. Comp. Dyn. **1**(1) (2014), also http://arxiv.org/abs/1202.5618

[3] K. Rajendran, A. C. Tsoumanis, C. I. Siettos and I. G. Kevrekidis, On the Selection of Collective Variables in the Coarse-graining of Dynamics on Networks, submitted to Phys. Rev. E (2013)

[4] K. Rajendran and I. G. Kevrekidis, Mining data in the form of graphs using pair-wise graph similarity measures, submitted to Phys Rev E (2013)

386919 Multiscale Modeling of Rod-like Structures Using Molecular Dynamics and Continuum Mechanics

Abstract:

Molecular structures can be adequately studied using molecular dynamics (MD), although the utility of MD is limited by the large high computational costs of atomistic simulations. One possible way to reduce these costs is to replace the atomistic description of the system under consideration with a continuum mechanical one.

In this work, we focus in particular on rod-like structures where the extension in one direction is clearly larger than in the remaining two dimensions. Well-known examples of such materials on the nanoscale include carbon nanotubes and macromolecules such as DNA and cellulose bundles. For these materials, the concept of geometrically exact beams introduced by Simo (also known as Reissner-Simo beam theory) is a promising candidate for a continuum formulation.

Using atomistic molecular simulations to parameterize a continuum model, we can obtain numerical results for a variety of geometric configurations and system conditions much more efficiently than using exclusively atomistic MD simulations. To this end, we outline a systematic procedure that estimates the necessary parameters (the initial geometry and strain-stress relationships) at finite, constant temperatures. By using benchmark problems that bridge the molecular and continuum scales, we can demonstrate the validity of the estimated parameters in finite element models of the beam model. We show how this model can be used to derive continuum representations of carbon nanotubes under various conditions.

388982 Continuum/Mesoscale/Molecular Level Hybridization in Complex Fluid Flows

Abstract:

Continuum/Mesoscale/Molecular Level Hybridization in Complex Fluid Flows

Sesha Hari Vemuri, Yu Liu, Myung S. Jhon, and Lorenz T. Biegler

Department of Chemical Engineering, Carnegie Mellon University, Pittsburgh, PA 15213

Considerable effort is being dedicated to understand the range of time and length scale phenomena that take place in a broad range of systems including colloidal suspensions, polymer solutions, micro- and nano hydrodynamic flows, especially in biological systems. These systems are characterized by multiscale interaction between "large and slow" polymer particles, and "small and fast" solvent particles [1]. A detailed modeling of such systems can be provided by a molecular dynamics (MD) simulation where the motion of all the particles is tracked via Newton's equations of motion. However, this approach has severe drawbacks due to the limitations on typical length scale covered in the MD simulations and the computational time scaling as $O(N^2)$, where N is the number of particles. On the other hand, the continuum based theory breaks down for these systems as the confinement size is comparable to or smaller than the molecular/pseudo-particle mean free path.

Mesoscopic models, and notably those arising from kinetic theory, are natural candidates to fill this gap because they operate precisely at intermediate scales between the atomistic and continuum levels. During the past two decades, lattice Boltzmann method (LBM), which originates from Boltzmann transport equation has proven successful to model characteristic features of nano/microscopic flows, such as the occurrence of slip boundary conditions near solid walls [2]. Furthermore, being a pseudo particle simulation methodology, it has the advantages of easily implementing easily the rotational degrees of freedom of the particles [3], and ability of simulating multi-physical phenomena in complex geometries. In this work, we present a hybrid MD/LBM method for the simulation of hydrodynamic phenomena in polymer solutions and colloidalsystems.

This will be achieved by generating a buffer zone at the component interface (e.g., polymer-solvent interface, fluid-boundary interface etc.) where both MD and LBM solutions are valid. Here, the particle velocities and forces obtained are exchanged between particles modeled by LBM and those modeled by MD [3]. The constrained dynamics algorithm forces the instantaneous mean particle velocity to equate the continuum solution at the other boundary. The equations of motion in the MD are also modified to match the external force obtained from the LBM via proper scaling.

To validate our hybrid model, we examine our results of the benchmark complex fluid flow systems including including colloidal solution & flow in a confined geometry and will extend our analysis towards the complex multi-physical phenomena various applications.

[1] A. Chatterji and J. Horbach, *J. Chem. Phys.*, **122**, 184903 (2005)

[2] J. Horbach and S. Succi, *Phys. Rev. Lett.*, **96**, 224503 (2006)

[3] P. Ahlrichs and B. Dünweg, *Int. J. Mod. Phys. C,* **09**, 1429 (1998)

Quantifying the Free Energy Landscape for Curvature Induction and Sensing in Lipid Bilayer Membranes Through a Mesoscale Model

Background: Several proteins recruited to the cell membrane have been found to induce and stabilize curvature in the lipid bilayer. Curvature generation in the cell membrane is an essential part of several key biophysical processes including cell motility, endocytosis, and invadopodia formation. In addition to the colocalization of curvature-inducing proteins to sites of curvature *in vivo*, these membrane proteins have been shown to both migrate along mean curvature gradients and tubulate liposomes *in vitro* [1, 2]. Curvature-inducing membrane proteins are thought to generate curvature through either the insertion of an amphipathic helix, or through the electrostatic scaffolding of the bilayer [3,4,5]. Both these mechanisms create a local deformation in the bilayer. It is not clear how these proteins sense mean curvature gradients with different affinities, and what major determinants impact their segregation, curvature induction, or the morphological phase transitions they drive.

Methods: In this work, the lipid/cell membrane is modeled with a continuum Monte Carlo model evolved with the Helfrich Hamiltonian [6, 7]. In the model, the membrane is described by an infinitesimally thin triangulated sheet, which can access length scales on the order of 100 nm and above. Membrane proteins interact with the membrane through a prescribed intrinsic curvature field and can move along the triangulated mesh. The intrinsic curvature field is approximated as a Gaussian function, with parameters derived from relating the curvature energy to isothermal titration experiments measuring Epsin association energy with the bilayer. Some membrane proteins such as those comprising BAR domains are thought to induce an anisotropic curvature field, and are modeled by an anisotropic two-dimensional Gaussian function. In addition to single proteins, a larger step function is also used to approximate an entire assembly of proteins such as those present in a clathrin coat during clathrin-mediated endocytosis. The excess area present in the membrane can also be controlled in our model. This excess area gives rise to a renormalized tension, which can be quantified through a non-linear fit of the undulation spectrum.

Several free energy methods have been coupled to the membrane model in order to determine a membrane proteins chemical potential. In addition to the Widom particle insertion method [8], two free energy perturbation methods are utilized to track the free energy for the addition of a curvature inducing protein to the bilayer. The Widom method computes the excess chemical potential upon insertion of a membrane protein, but this method is known to fail at high densities. For this reason, both thermodynamic integration and Bennett acceptance methods are used to validate Widom calculations at low densities and access the free energy landscape at higher protein densities in simulation [9,10].

Results: Using our mesoscale membrane model, the chemical potential of curvature inducing proteins is quantified for planar bilayers. The planar geometry represents a 400 by 400 nm patch of the cell membrane. For low protein concentrations, the chemical potential required to bend the membrane to a certain curvature field is shown to depend on both the extent as well as the amplitude of the curvature field. A larger curvature field senses mean curvature gradients to a greater degree. The chemical potential also scales linearly with bending rigidity. Anisotropic curvature inducing proteins are shown to have preferential alignment with background mean curvature fields.

At higher densities, these curvature inducing membrane proteins cluster to form tubules. Our free energy methods show a pitchfork bifurcation in the excess chemical potential at the point of tubulation, which is a signature of a morphological phase transition. This morphological phase transition is further quantified in its dependence on tension, with higher tensions requiring a higher critical concentration of proteins. This model connects the strength of an individual protein curvature field to the mesoscale morphological phase transitions seen *in experiments.*

The free energy of vesicle nucleation in clathrin-mediated endocytosis is also quantified with our model. A free energy barrier is shown to be present during the formation of the vesicle neck. Our results suggest that tension hinders curvature induction, and the formation of the vesicle neck. These results suggest a novel mechanistic interpretation to experimental studies reporting actin localization to some sites of endocytosis, which is necessary to overcome membrane tension.

References

[1] Ford, M., et al. Curvature of clathrin-coated pits driven by epsin. Nature. 2002.

[2] Capraro, B., Yoon, Y., Cho, W., Baumgart, T. Curvature Sensing by the Epsin N-TerminalHomology Domain Measured on Cylindrical Lipid Membrane Tethers. J. Am. Chem. Soc. 2010.

[3] Campelo, F., McMahon, H., and Kozlov, M.The Hydrophobic Insertion Mechanism of Membrane Curvature Generation by Proteins. Biophysical Journal. 2008.

[4] McMahon, H., and Gallop, J. Membrane curvature and mechanisms of dynamic cell membrane remodeling Nature. 2005.

[5] Zimmerberg, J., and McLaughlin, S. Membrane Curvature: How BAR Domains Bend Bilayers Current Biology. 2004.

[6] Helfrich, W. Elastic Properties of Lipid Bilayers - Theory and Possible Experiments. Z.Naturforsch. 1973.

[7] Ramakrishnan, N., Kumar, P. B. and Ipsen, J. H. Monte Carlo simulations of fluid vesicles with in-plane orientational ordering. Phys. Rev. E. 2010.

[8] Widom, B. Some Topics in the Theory of Fluids. J. Chem. Phys. 1963.

[9] Frenkel, D., and Smit, B. *Understanding Molecular Simulation*, 2nd ed. 2001.

[10] Bennett, C. H. Efficient estimation of free energy differences from Monte Carlo data. J. Comp. Phys. 1976.

374538 Rippling Instabilities on the Surface of Stressed Crystalline Solids

Abstract:

Morphological instabilities on surfaces of stressed solids and the resulting pattern formation are of major fundamental and technological significance toward improving materials function and reliability and developing innovative directed assembly processes for nanotechnology. Such a well-known phenomenon is the Asaro–Tiller/Grinfeld (ATG) instability, which leads to the formation of a regular pattern of crack-like grooves emanating from the surface of a stressed elastic solid and deepening by surface diffusion.

We have carried out a systematic analysis of morphological evolution of stressed solid surfaces under the action of an electric field, or a temperature gradient, or the simultaneous action of both and reported that sufficiently strong external fields can inhibit the ATG instability and prevent surface cracking; a linear stability theory predicted fairly accurately the current-induced stabilization of surface morphology in stressed elastic solids. However, complex aspects of surface morphological evolution and pattern formation are not accounted for by the linear theory.

Using self-consistent dynamical simulations according to a well-tested fully nonlinear model of surface morphological evolution, we have studied the surface morphological response of stressed solids both in the absence of and under the simultaneous action of external fields focusing on low-amplitude long-wavelength perturbations from the planar surface morphology. We find that in addition to the ATG instability, a long-wavelength tip-splitting instability may be triggered forming a pattern of secondary ripples on the surface; rippling occurs in the absence of external field application, as well as for weaker-than-critical applied external fields. We have developed a nonlinear wavelength selection theory to analyze the formation of such secondary surface ripples. We have also carried out a comprehensive simulation study to characterize the ripple formation, identify the critical wave-length for rippling, and determine the dependence of the critical wave number for the onset of rippling on the applied external field strength.. Finally, we have examined the effect of the Arrhenius temperature dependence of the surface diffusivity, the strongest temperature dependence of all the relevant material properties, on surface rippling under the action of an externally applied temperature gradient.

382383 Robust Control of Multiscale Process Systems Under Uncertainty: An Application to Epitaxial Thin Film Growth

Abstract:

Multiscale processes are characterized by inherently coupled chemical and physical phenomena occurring over different temporal and spatial scales [1]. In such systems, control of the events evolving at the microscopic scale is essential to product quality while efficient operations require manipulated variables at the macroscopic scale for real-time feedback control [2]. Accordingly, to describe the behavior of these processes, it is common to employ a multiscale model encompassing macro-scale continuum equations (e.g., Partial Differential Equations) embedded with micro-scale simulations (e.g., Kinetic Monte Carlo). Although such a model is capable to provide a fair representation of a system, the microscopic processes are subject to model parameter uncertainty that can significantly affect the control and optimization objectives. As such, to achieve the desired objectives, it is crucial to develop control strategies that are robust to these uncertainties.

This study explores a systematic framework to analyze model parameter uncertainty for the purpose of robust control and optimization in multiscale models. The difficulties in analysing uncertainty in such systems arise due to the computational intensity of the fine-scale simulations (typically KMC simulations). Moreover, the lack of a closed formulation between the process optimization objectives and the model parameters, and absence of sensors to measure microscopic features at fine scales make this analysis even more challenging. Motivated by these observations in this work, Power Series Expansion (PSE), which obtains a simpler representation of the actual system, is employed to analyze model uncertainty propagation [3].

The thin film growth process has been used in this study to effectively exemplify a multiscale process system. Strong dependence of the electrical and mechanical properties of thin films on their microstructure requires accurate control of the product's quality (e.g., film's roughness and thickness) at the fine scale. For uncertainty analysis of the thin film process, the probability distributions of microscopic events are determined using the PSE-based models that rely on the prior distribution of the uncertain parameters. Subsequently, the upper and lower bounds on desired outputs (e.g., film roughness) are determined at a given confidence level of the distributions obtained for the states of the system [4]. Moreover, multiple reduced-order lattices are employed in KMC simulations to reduce the computational costs. The potential application of the proposed method can be illustrated in open-loop optimization or closed-loop purposes. In open-loop optimization, the optimal temperature profile is obtained to optimize the end-point properties of the thin film. For closed-loop purposes, on the other hand, measurements are not available as frequent as required for the development of an effective feedback control strategy. The proposed algorithm can therefore be applied to develop a robust estimator that is able to provide the estimate of the controlled output using online measurements. Furthermore, the feedback control of the film microstructure can be achieved via offline identification of a low-order model from lower and upper bounds on the process outputs in the presence of uncertainty.

[1] Vlachos, D. G., (2005) A review of multiscale analysis: Examples from systems biology, materials engineering, and other fluid-surface interacting systems. Advances in Chemical Engineering, 30, 1-61.

[2] Christofides, P. D., Armaou, A., Lou, Y. & Varshney, A., (2008) Control and optimization of multiscale process systems. Boston: Birkhäuser.

[3] Nagy, Z. K. & Braatz, R. D., (2007) Distributional uncertainty analysis using power series and polynomial chaos expansions. Journal of Process Control, 17, 229-240

[4] Rasoulian, S., Ricardez-Sandoval, L.A. "Uncertainty analysis and robust optimization of multiscale process systems with application to epitaxial thin film growth" Chemical Engineering Science, In Press.

388856 Lattice Boltzmann Method Based Multiscale Modeling of Two-Phase Flows through Microarchitectures with Microfeatures

Abstract:

Lattice Boltzmann Method based multiscale modeling of two-phase flows through microarchitectures with microfeatures

Agnieszka Truszkowska[1] and Goran Jovanovic[1]

[1]Oregon State University, Department of Chemical, Biological, and Environmental Engineering, Corvallis, OR 97330, USA

Emails: truszkoa@onid.orst.edu, goran.jovanovic@oregonstate.edu

Segmented two-phase flows in microscale-based structures display attractive characteristics such as enhanced mixing and intensification of mass and heat transfer, reduced Taylor dispersion, and opportunity to control and manipulate dispersed phase; bubbles or droplets. Two-phase flow in microscale-based structures is commonly used in emulsification and encapsulation operations, microreactors, material synthesis, bioassays and many other applications ([1],[2]). Apart for its purposeful presence, dispersed phase could occur in microarchitectures as a side effect emerging from phase transition, inadvertent inflow with process feed, or through equipment gaps ([3],[4]). Whether part of an intended operation or unwanted operating byproduct, second/discrete phase has to be properly controlled to achieve desired system performance.

Optimization of two-phase processes on microscale level results in numerous opportunities related to both operating conditions and design architecture. However, efficient and comprehensive optimization often requires recurring use of numerical models in addition to experimental data. Additionally, these computational models require high resolution computing due to dominant interfacial transport phenomena. Due to the need for excessive computational resources many researchers reduce modeling effort to a representative small segment of the system, or make overly simplifying assumptions and approximations.

In this paper we present a novel, multiscale-modeling approach for numerical representation of the two-phase flow in microscale-based structures containing micro features such as posts and pillars. Microscale-based architectures containing various micro features are relatively unexplored design options ([5], [6]) with vast potential for optimization due to their extreme design flexibility. We explore their characteristics and provide a novel multiscale modeling approach within the framework of Shan and Chen Lattice Boltzmann modeling method.

Example used in the numerical experiments is an array of cylindrical micropillars with staggered, equilateral arrangement. A multiscale operator in the form of field of discrete vector quantities, forcing terms, is developed. Discrete vector forcing terms are applied on the interface of the dispersed phase, which results in an equivalent average behavior of the dispersed phase in the geometrically simplified scale, just as in the geometrically fully resolved scale. We present several variations of the operator as well as its characteristics and robustness. Finally, we discuss applicability and limitations of the approach based on model system simulations and classical benchmark cases.

[1] Shui, Lingling, Jan CT Eijkel, and Albert van den Berg, "Multiphase flow in microfluidic systems--Control and applications of droplets and interfaces", Advances in Colloid and Interface Science, 133,1 (2007): 35-49

[2] Gunther, Axel, and Klavs F. Jensen, "Multiphase microfluidics: from flow characteristics to chemical and materials synthesis", Lab on a Chip, 6,12 (2006): 1487-1503

[3] Jensen, Mads Jakob, Goran Goranovic, and Henrik Bruus, "The clogging pressure of bubbles in hydrophilic microchannel contractions", Journal of Micromechanics and Microengineering, 14,7 (2004): 876

[4] Litterst, C., et al., "Improved gas bubble mobility in CHIC-type flow channels", Actuator, Messe Bremen, Germany (2004): 541-544

[5] Krishnamurthy, Santosh, and Yoav Peles, "Gas-liquid two-phase flow across a bank of micropillars", Physics of Fluids 19 (2007): 043302

[6] De Loos, S. R. A., et al. "Gas--liquid dynamics at low Reynolds numbers in pillared rectangular micro channels", Microfluidics and nanofluidics, 9,1 (2010): 131-144

371300 Characterizing Tip-Surface Contact Between Monolayer Functionalized Silica Surfaces through Molecular Dynamics Simulation

Abstract:

Chemisorbed alkylsilane monolayer coatings have been shown to reduce friction, adhesion, and wear of surfaces under shearing contact. Exploration of monolayer properties at the nanoscale is commonly performed through atomic force microscopy (AFM). However, inconsistencies in values and trends of friction properties may arise between separate studies as a result of variance in probe tip size. Flater et al. proposed that for tip contact of a monolayer-functionalized surface, plowing of monolayer chains provides an additional contribution to friction force.[1] Molecular dynamics simulations of Harrison and coworkers reveal an increased coefficient of friction during shearing of a buckyball tip compared to shearing of an infinitely flat amorphous surface, supporting the idea of molecular plowing presence and the dependence of shear behavior on tip size.[2] However, it remains unclear under which contact dimensions tip-surface contact behavior begins to deviate from that of infinitely flat (surface-surface) contact. Furthermore, although simulations of surface-surface contact may provide a good approximation to the contact behavior of many AFM tips, system size effects have yet to be fully explored. Using molecular dynamics simulations, we examine system size dependence during surface-surface contact for alkylsilane monolayers and study the influence of substrate size for both flat and curved tip-like contacting agents to help better define the tip-surface contact regime.

For systems of surface-surface contact, we find contact behavior to have a minimal dependence on system size; however, clear trends arise for systems of tip-surface contact. We observe a distinct transition in friction force as a function of substrate size for flat tip contacting agents. We attribute this to size influence on monolayer disorder arising from chain tilting at tip edges. A similar transition is found for contacting agents displaying curvature, where shear behavior is shown to transition from sliding to a plowing-like mechanism. This behavior is investigated through examination of penetration depth and local disorder of the countersurface monolayer, which reveals that less highly curved tips are unable to exhibit sufficient monolayer penetration and may be effectively modeled through surface-surface simulations.

[1] Flater EE, Ashurst WR, Carpick RW, Nanotribology of octadecyltrichlorosilane monolayers and silicon: self-mated versus unmated interfaces and local packing density effects, *Langmuir*, 23, (2008).

[2] Knippenberg MT, Mikulski PT, Harrison JA, Effects of tip geometry on interfacial contact forces, *Modelling and Simulation in Materials Science and Engineering*, 18, (2010).

Visualization and Prediction of Operating Conditions of Blast Furnace in Low-Dimensional Space

Koji Hazama[1], Manabu Kano[1], Shinroku Matsuzaki[2] and Akira Morita[2]

[1]Department of Systems Science, Kyoto University, Kyoto, Japan
[2]Nippon Steel & Sumitomo Metal Corporation, Chiba, Japan

Abstract

A large number of variables are measured in a blast furnace, and the number is increasing rapidly. It is difficult to extract valuable information from a huge amount of data, especially when there is the complex relationship among variables; engineers and operators do not make the most of process data to ensure productive and stable operation. Thus, it is effective to develop a data-based operation support system so that engineers and operators can evaluate operating conditions and make necessary adjustments. The present research is aiming to develop an evaluation method of operating conditions through visualizing and predicting them. To visualize high-dimensional data, supervised dimensionality reduction techniques such as partial least squares (PLS) and supervised distance preserving projection (SDPP) were investigated. In addition, two types of methods for evaluating the predicted future operating conditions were developed. The results of applying the proposed methods to operation data of a blast furnace show that the proposed methods can predict the transition of operating conditions in the future and provide information useful for evaluating the operating conditions.

1 Introduction

In the steel industry, blast furnaces are used to smelt iron ore to produce pig iron. The number of measured variables in a blast furnace and the quantity of stored data are *big* and fast increasing; the relationship among many process variables is too complex. Thus, it is not easy for engineers and operators to fully utilize process data to monitor operating conditions and improve the productivity and the stability of the operation. In fact, only experienced operators can evaluate the operating conditions and take appropriate actions.

Since the stable operation of blast furnaces is especially important to realize efficient operation of the entire plant, the development of a data-based operation support system must be useful. The aim of this research is to develop a method for evaluating operating conditions of a blast furnace through visualization and prediction: how good the process is and will be operated.

In order to visualize high-dimensional data to intuitively grasp the conditions, the dimensionality has to be reduced. A wide variety of methods for dimensionality reduction have been proposed[1)]; they are categorized according to the following two features:

1. linear or nonlinear
2. supervised or unsupervised.

Nonlinear techniques are further divided into two groups based on whether the purpose is to preserve global properties or local properties. Traditional unsupervised linear techniques include principal component analysis (PCA) while linear discriminant analysis (LDA) is a representative supervised linear technique. Examples of nonlinear techniques to preserve global properties are Multidimensional Scaling (MDS)[2], Isomap[3], Stochastic Proximity Embedding (SPE)[4], t-distributed Stochastic Neighbor Embedding (t-SNE)[5], and Autoencoder[6]. Meanwhile, major nonlinear techniques to preserve local properties include Locally Linear Embedding (LLE)[7], Laplacian Eigenmaps (LE)[8], and Linear Tangent Space Alignment (LTSA)[9]. These non-linear techniques are all unsupervised.

The present work is aiming to help engineers and operators evaluate operating conditions by developing a method to visualize and predict them in real time, hence it seems reasonable to use the prior evaluation of the operating conditions by engineers and operators as reference (teacher) signals. Therefore, two supervised dimensionality reduction techniques, i.e., partial least squares (PLS) regression and supervised distance preserving projection (SDPP)[10], were investigated and compared. While PLS is a linear technique, SDPP is a nonlinear technique that preserves local properties of data.

In addition, new methods were developed to predict and evaluate future operating conditions. The proposed methods visualize a target sample on a low dimensional space, evaluate the operating condition, and use past data to derive the expected transition of operating conditions in the future. Hence, the methods can provide operators with useful information to evaluate the operating conditions and judge whether intervention is necessary or not.

This paper is organized in the following manner. In Section 2, two visualization techniques are briefly explained. Section 3 proposes the methods to predict future operating conditions. Application results of these methods to operation data of a blast furnace are described in Section 4. Finally, the summary of this paper is given in conclusions.

2 Dimensionality reduction

PLS is a linear technique for reducing dimensionality with supervised data. On the other hand, SDPP optimizes a projection matrix so that the local properties of projected data can be as close to those of supervised data as possible. PLS is used to construct an initial projection matrix for SDPP in this work.

2.1 PLS

PLS has been widely used in various fields because it can cope with a colinearity problem. Given N samples, measurements of M input variables and an output variable are denoted by $\{\boldsymbol{x}_1, \boldsymbol{x}_2, \ldots, \boldsymbol{x}_N\} \in \Re^M$ and $\{y_1, y_2, \ldots, y_N\}$. Each column of an input matrix $\boldsymbol{X}$ and an output vector $\boldsymbol{y}$ are mean-centered and scaled appropriately. In PLS, $\boldsymbol{X}$ and $\boldsymbol{y}$ are decomposed as follows:

$$\boldsymbol{X} = \boldsymbol{T}\boldsymbol{P}^T + \boldsymbol{E}, \ \boldsymbol{y} = \boldsymbol{T}\boldsymbol{q} + \boldsymbol{f}, \tag{1}$$

where $\boldsymbol{T} \in \Re^{N\times R}$ is the score matrix whose columns are the latent variables $\boldsymbol{t}_r \in \Re^N (r = 1, 2, \ldots, R)$, $\boldsymbol{P} \in \Re^{M\times R}$ is the loading matrix of $\boldsymbol{X}$, and $\boldsymbol{q} \in \Re^R$ is the regression coefficient vector of $\boldsymbol{y}$. R denotes the number of adopted latent variables. $\boldsymbol{E} \in \Re^{N\times M}$ and $\boldsymbol{f} \in \Re^N$ are errors.

2.2 SDPP

SDPP[10] computes a projection matrix from a high dimensional space to a low dimensional space. Suppose that N samples $\{\boldsymbol{x}_1, \boldsymbol{x}_2, \ldots, \boldsymbol{x}_N\} \in \Re^M$ and their corresponding responses $\{\boldsymbol{y}_1, \boldsymbol{y}_2, \ldots, \boldsymbol{y}_N\} \in \Re^L$ are given, and assume that there is a continuous mapping $f : \mathcal{X} \mapsto \mathcal{Y}$. SDPP is designed to compute a low-dimensional subspace $\mathcal{Z}$ of dimensionality r with $r \ll M$ so that the local geometry of the samples in $\mathcal{Y}$ and $\mathcal{Z}$ can be matched as much as possible. SDPP assumes that the subspace $\mathcal{Z}$ can be obtained by a linear transformation of $\boldsymbol{X}$; that is, for a sample $\boldsymbol{x}$, the new representation in the subspace is $\boldsymbol{z} = \boldsymbol{W}^T\boldsymbol{x}$, where the projection matrix is $\boldsymbol{W} \in \Re^{M \times r}$.

$\boldsymbol{W}$ can be computed by minimizing the following criterion:

$$\boldsymbol{J}(\boldsymbol{W}) = \frac{1}{N}\sum_{n=1}^{N}\sum_{\boldsymbol{x}_k \in \boldsymbol{A}(\boldsymbol{x}_n)} (d_{nk}^2(\boldsymbol{W}) - \delta_{nk}^2)^2, \tag{2}$$

where $\boldsymbol{A}(\boldsymbol{x}_n)$ is a neighborhood of $\boldsymbol{x}_n$, and pairwise distances are calculated with the Euclidean distance, i.e., $d_{nk}^2(\boldsymbol{W}) = \|\boldsymbol{z}_n - \boldsymbol{z}_k\|^2$ and $\delta_{nk}^2 = \|\boldsymbol{y}_n - \boldsymbol{y}_k\|^2$. Given a matrix $\boldsymbol{B}$, the element in the nth row and kth column is expressed as $\boldsymbol{B}_{nk}$ in this section. After denoting the neighborhood graph $\boldsymbol{G}_{nk}$ of $\boldsymbol{x}_i$ as

$$\boldsymbol{G}_{nk} = \begin{cases} 1 & \text{if } \boldsymbol{x}_k \text{ is a neighbor of } \boldsymbol{x}_n \\ 0 & \text{otherwise} \end{cases} \tag{3}$$

and the squared pairwise distance matrices as $\boldsymbol{D}_{nk} = d_{nk}^2(\boldsymbol{W})$ and $\boldsymbol{\Delta}_{nk} = \delta_{nk}^2$, $\boldsymbol{J}$ is rewritten as

$$\boldsymbol{J}(\boldsymbol{W}) = \frac{1}{N}\sum_{n=1}^{N}\sum_{k=1}^{K} \boldsymbol{G}_{nk}(\boldsymbol{D}_{nk} - \boldsymbol{\Delta}_{nk})^2, \tag{4}$$

where K is the number of neighbors. In addition, the gradient of $\boldsymbol{J}$ with respect to $\boldsymbol{W}$ is denoted as follows:

$$\begin{aligned} \nabla_{\boldsymbol{W}}\boldsymbol{J} = & \frac{4}{N}\sum_{n=1}^{N}\sum_{k=1}^{K} \boldsymbol{G}_{nk}(\boldsymbol{D}_{nk} - \boldsymbol{\Delta}_{nk}) \times \\ & (\boldsymbol{x}_n - \boldsymbol{x}_k)(\boldsymbol{x}_n - \boldsymbol{x}_k)^T\boldsymbol{W}. \end{aligned} \tag{5}$$

3 Evaluation methods

In this section, two methods for evaluating the operating conditions in the future are proposed. One is an evaluation method based on the frequency of neighbor samples, referred to as a frequency-of-neighbors method. The other is based on the distribution of neighbor samples; this method is referred to as a track-forecast method because it is like a typhoon (tropical cyclone) track forecast. These evaluation methods work in the two dimensional space derived with the above-mentioned visualization techniques. The nearness is measured with the Euclidean distance in the two-dimensional space in this work; however, correlation or other indices can be also used to measure the nearness in the low-dimensional space or in the original high-dimensional space.

3.1 Frequency-of-neighbors method

A new method for evaluating operating conditions based on a visualization result is proposed. The procedure is as follows: 1) prepare a database consisting of process data measured in the past, 2) select K samples $\boldsymbol{x}_k^{(0)} (k = 1, 2, \ldots, K)$ nearest to the target sample, i.e., the current sample $\boldsymbol{x}^{(0)}$, from the database, 3) select their future samples $\{\boldsymbol{x}_k^{(t^*)} | t^* = St(t = 1, 2, \ldots, T)\}$ from the database, where S is the predetermined time interval such as an hour, and T is the number of future conditions to be evaluated, 4) extract the class labels corresponding to the selected samples, and 5) evaluate the operating conditions at present and at elapsed time t by the rate of occurrence of each class label. The results are displayed in a bar chart or a table.

3.2 Track-forecast method

This method provides expected transition of a target sample by drawing circles at each elapsed time on the basis of past data. The result, which is similar to typhoon track forecast, is displayed on the two dimensional space created by a dimensionality reduction technique such as PLS and SDPP. Similar to the frequency-of-neighbors method, K nearest neighbors and their future samples are selected. The selected samples are denoted by $\boldsymbol{x}_k^{(t^*)}$. The center of each circle is determined as follows:

$$\boldsymbol{c}^{(t^*)} = \frac{1}{K} \sum_{k=1}^{K} \boldsymbol{x}_k^{(t^*)} \tag{6}$$

The radius of a circle is set to the distance that is sufficient to include 70% of K nearest neighbors.

4 Case study

The proposed evaluation methods were applied to operation data of a blast furnace at Nippon Steel & Sumitomo Metal Corporation. A total of 177 process variables are measured in the blast furnace, and 146 variables were used. Experienced engineers labeled a part of samples as very good, good, bad, very bad, and low production in advance, and the labels were used as the teacher signals; as a result, the operation data were divided into five classes. Samples with the same label are expected to form a cluster, which is as far from each of other clusters with different labels as possible. In addition, clusters corresponding to ordered labels such as very good, good, bad, and very bad should be in proper order in the two-dimensional space.

In addition to the operating conditions of the blast furnace, the throughput is an important performance indicator. In this work, the blasting flow rate was used as a measure of the throughput.

4.1 Visualization results

To balance the number of samples corresponding to each label, 600 samples were extracted at random from each class, and a total of 3,000 samples were used for modeling. The visualization results of a total of 12,000 samples by using PLS and SDPP are illustrated in Fig. 1, where samples with labels such as very bad, bad, good, very good, and low production are colored red, magenta, cyan, blue, and green, respectively.

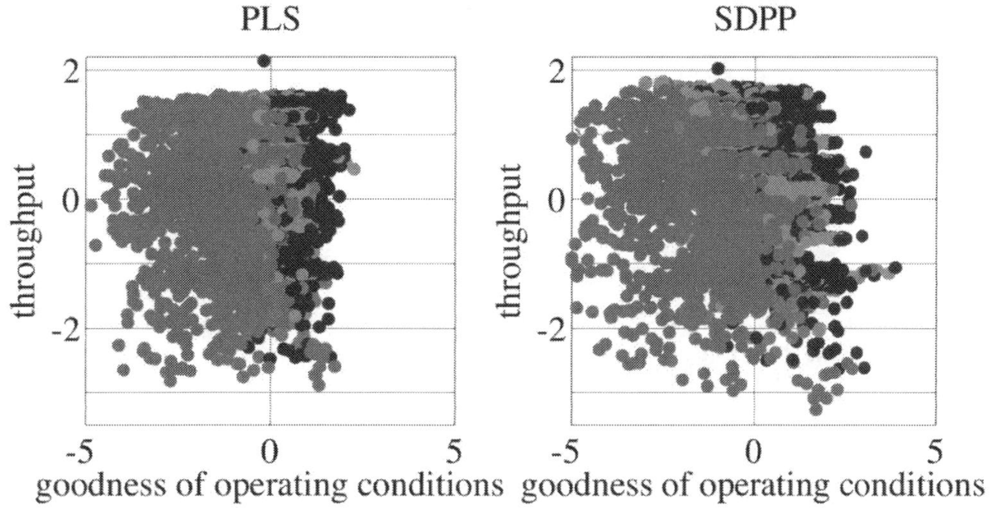

Fig. 1: Visualization results of operation data of the blast furnace. Samples labeled with very bad, bad, good, very good, and low production are colored red, magenta, cyan, blue, and green, respectively.

Each axis of the two-dimensional space has physical meaning. The horizontal axis is corresponding to the goodness of the operating conditions, and the vertical axis is corresponding to the throughput. The visualization results show that the samples are, in general, located according to their labels in proper order in the horizontal axis direction; that is, the operating conditions were successfully visualized. It is also shown that samples labeled as low production are located relatively below the other samples; in other words, the conditions of these samples are not necessarily bad, but the throughput is lower than the other samples. Although samples with the same label are clustered, the clusters overlap each other. In addition, there is not a huge difference between the visualization results of PLS and SDPP.

4.2 Evaluation of operating conditions

To grasp the visualization result intuitively, a mesh plot was employed, where each cell was colored according to the most frequent label within it. The number of cells was predetermined. The evaluation result of the frequency-of-neighbors method is illustrated in Fig. 2. The mesh plot indicates that the operating condition of the target sample is very good, and the throughput is high. The bar chart shows that the samples labeled with very good are the most frequent neighbors of the target sample, and the proportion of the neighbors labeled with good increases as the number of neighbors grows. This confirms the visualization result of the mesh plot.

The prediction of future conditions is shown in Table 1, where K is 100, and samples labeled as good and very good are denoted by "Good", samples labeled as very bad and bad are "Bad", and samples labeled as low production are "Low". Table 1 indicates that the operating conditions tend to deteriorate within 3 hours, although the current state is good.

The evaluation result of the track-forecast method when K is 100 is illustrated in Fig. 3, where the actual data after the elapsed time are displayed as squares. The size of each square becomes large with time. This result shows that the operating conditions would gradually deteriorate in next three hours. Moreover, the result

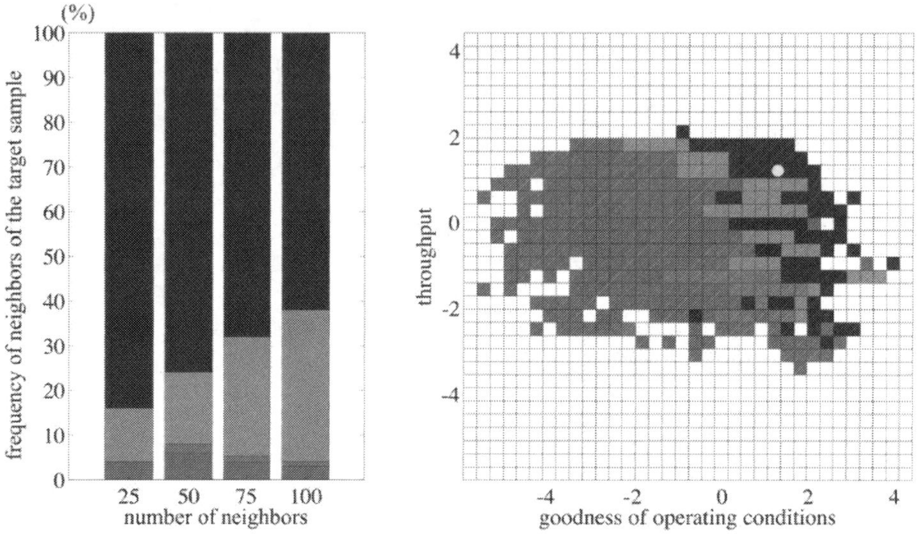

Fig. 2: Result of the frequency-of-neighbors method. In the bar chart, the proportions of samples with labels such as very bad, bad, good, very good, and low production are colored red, magenta, cyan, blue, and green, respectively. The mesh plot is derived with the result of SDPP in Fig. 1. Each cell in the mesh plot is colored according to the label of the most frequent samples within it in a similar way to the bar chart. The target sample is colored yellow in the mesh plot.

Table 1: Evaluation result of the frequency-of-neighbors method.

Conditions	Current	+1h	+2h	+3h
Good	96	96	92	90
Bad	3	4	7	8
Low	1	0	1	2

(actual data) confirms that the operating conditions deteriorated.

5 Conclusions

In this paper, operation data of a blast furnace was visualized by using PLS and SDPP, and two methods evaluating the operating conditions were proposed. The results confirm that the proposed methods can provide engineers and operators with useful information to evaluate the operating conditions and judge whether intervention is necessary or not. Since the results largely depend on predetermined class labels, the clarification of evaluation indices of operating conditions and the successful determination of class labels are the key to the improvement of visualization and evaluation accuracy.

References

1) L. Maaten, E. Postma, and H. Herik, "Dimensionality Reduction: A Comparative Review", Technical Report TiCC-TR 2009-005, Tilburg University, 2009.

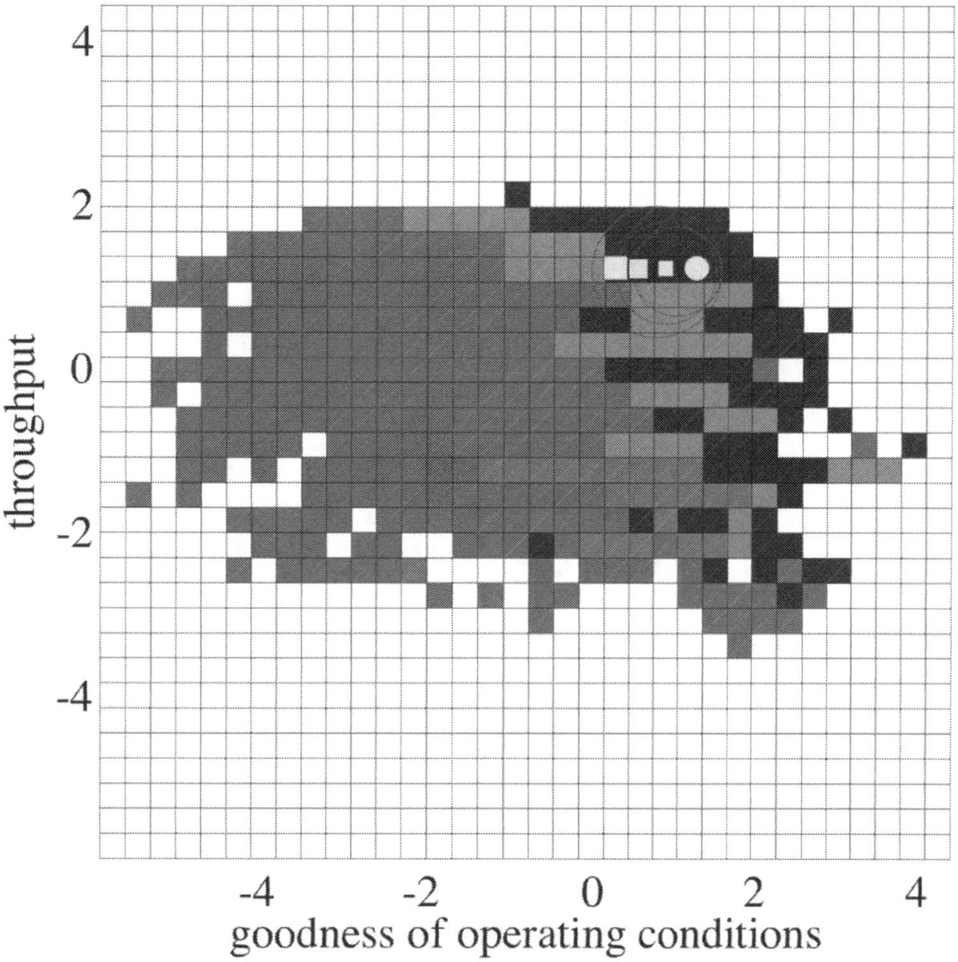

Fig. 3: Result of the track-forecast method. The mesh plot is derived with the result of SDPP in Fig. 1. A cell is colored according to the most frequent label within it, i.e., when the most frequent label is very bad, bad, good, very good, or low production, the color is red, magenta, cyan, blue, or green, respectively. The target sample is a yellow circle, and the actual data after the elapse of a certain period of time are squares. With the lapse of time, the line width of each oval becomes thinner, and the size of a square grows smaller.

2) J. Kruskal, "Multidimensional scaling by optimizing goodness of fit to a non-metric hypothesis", *Psychometrika*, Vol. 29, No. 1, pp. 1-27, 1964.

3) J. Tenenbaum, "Mapping a manifold of perceptual observations", *Advances in Neural Information Processing Systems*, Vol. 10, pp. 682-688, 1998.

4) D. Agrafiotis, "Stochastic Proximity Embedding", *Journal of Computational Chemistry*, Vol. 24, No. 10, pp. 1215-1221, 2002.

5) L. Maaten and G. Hinton, "Visualizing Data using t-SNE", *Journal of Machine Learning Research*, Vol. 9, pp. 2579-2605, 2008.

6) D. Demers and G. Cottrell, "Non-linear dimensionality reduction", *Advances in Neural Information Processing Systems*, Vol. 5, pp. 580-587, 1993.

7) S. Roweis and L. Saul, "Nonlinear dimensionality reduction by Locally Linear Embedding", *Science*, Vol. 290, No. 5500, pp. 2323-2326, 2000.

8) M. Belkin and P. Niyogi, "Laplacian Eigenmaps and spectral techniques for embedding and clustering", *Advances in Neural Information Processing Systems*, Vol. 14, pp. 585-591, 2002.

9) Z. Zhang and H. Zha, "Principal manifolds and nonlinear dimensionality reduction via local tangent space alignment", *SIAM Journal of Scientific Computing*, Vol. 26, No. 1, pp. 313-338, 2004.

10) Z. Zhu, T. Similä, and F. Corona, "Supervised Distance Preserving Projections", *Neural Processing Letters*, Vol. 38, No. 3, pp. 445-463, 2013.

Application of Feasible Generalized Least Squares to Disturbance Identification

Megan A. Zagrobelny
James B. Rawlings

Department of Chemical and Biological Engineering

THE UNIVERSITY of WISCONSIN MADISON

AIChE Annual Meeting, Atlanta, GA
November 19, 2014

Overview

1. Background
   - Motivation
   - Autocovariance least squares method
2. Feasible Generalized ALS Method
3. Industrial Example
4. Conclusions

Motivation

Identification of disturbance models has important applications:

- Model predictive controller performance monitoring
  - Is MPC achieving optimal performance?
  - Compare current performance to ideal benchmark
  - Theoretical benchmark (Zagrobelny et al., 2013) requires
    - ★ Model of process and controller
    - ★ Knowledge of noises affecting the system

  Estimator design

Disturbance Model Identification

- Bayesian estimation (Mehra, 1972; Matisko and Havlena, 2013)
  - Need prior distribution of disturbance covariances
  - Computationally demanding

Disturbance Model Identification

- Bayesian estimation (Mehra, 1972; Matisko and Havlena, 2013)
  - Need prior distribution of disturbance covariances
  - Computationally demanding
- **Maximum likelihood estimation** (Mehra, 1969; Kristensen et al., 2004)

Disturbance Model Identification

- Bayesian estimation (Mehra, 1972; Matisko and Havlena, 2013)
  - Need prior distribution of disturbance covariances
  - Computationally demanding
- **Maximum likelihood estimation** (Mehra, 1969; Kristensen et al., 2004)
- Covariance matching (Myers and Tapley, 1976; Valappil and Georgakis, 2000; Tzou and Lin, 2001)
  - Less computational effort
  - Produce biased results; not guaranteed to converge

Disturbance Model Identification

- Bayesian estimation (Mehra, 1972; Matisko and Havlena, 2013)
  - Need prior distribution of disturbance covariances
  - Computationally demanding
- **Maximum likelihood estimation** (Mehra, 1969; Kristensen et al., 2004)
- Covariance matching (Myers and Tapley, 1976; Valappil and Georgakis, 2000; Tzou and Lin, 2001)
  - Less computational effort
  - Produce biased results; not guaranteed to converge
- **Correlation techniques** (Mehra, 1970, 1972)
  - Includes autocovariance least squares (ALS) method (Odelson et al., 2006; Rajamani and Rawlings, 2009)

Disturbance Model Identification

- Bayesian estimation (Mehra, 1972; Matisko and Havlena, 2013)
  - Need prior distribution of disturbance covariances
  - Computationally demanding
- **Maximum likelihood estimation** (Mehra, 1969; Kristensen et al., 2004)
- Covariance matching (Myers and Tapley, 1976; Valappil and Georgakis, 2000; Tzou and Lin, 2001)
  - Less computational effort
  - Produce biased results; not guaranteed to converge
- **Correlation techniques** (Mehra, 1970, 1972)
  - Includes autocovariance least squares (ALS) method (Odelson et al., 2006; Rajamani and Rawlings, 2009)
- Subspace identification (Ljung, 1999; Van Overschee and De Moor, 1995; Qin, 2006)
  - Primarily designed for model identification
  - Also provides optimal Kalman filter gain and innovation covariance

Noise Model in State Space

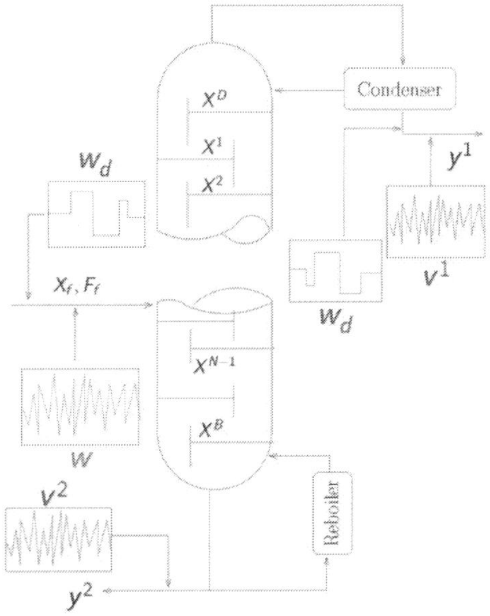

Noise Model in State Space

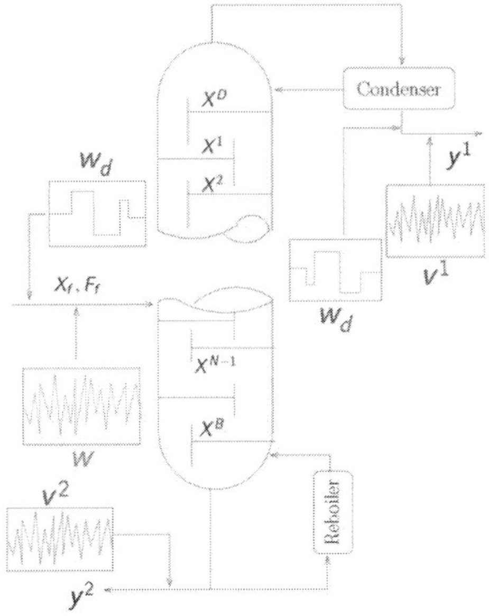

- Process noise: $w(k) \sim N(0, Q_w)$

$$x(k+1) = Ax(k) + Bu(k) + B_d d(k) + Gw(k)$$

Noise Model in State Space

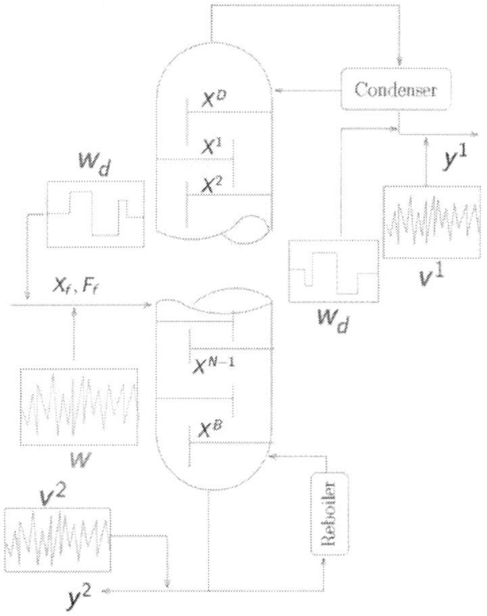

- Process noise: $w(k) \sim N(0, Q_w)$

$$x(k+1) = Ax(k) + Bu(k) + B_d d(k) + Gw(k)$$

Integrated disturbance noise: $w_d(k) \sim N(0, Q_{wd})$

$$d(k+1) = d(k) + G_d w_d(k)$$

Noise Model in State Space

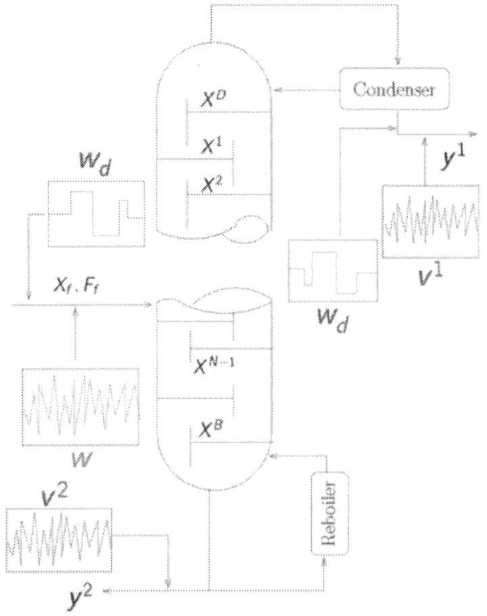

- Process noise: $w(k) \sim N(0, Q_w)$

$$x(k+1) = Ax(k) + Bu(k) + B_d d(k) + Gw(k)$$

Integrated disturbance noise: $w_d(k) \sim N(0, Q_{wd})$

$$d(k+1) = d(k) + G_d w_d(k)$$

Measurement noise: $v(k) \sim N(0, R_v)$

$$y(k) = Cx(k) + C_d d(k) + v(k)$$

Zagrobelny, Rawlings FGLS Disturbance Identification 5 / 20

Noise Model in State Space

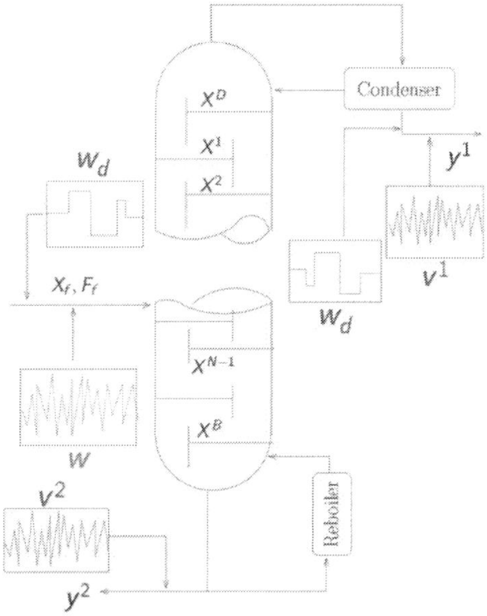

- Process noise: $w(k) \sim N(0, Q_w)$

$$x(k+1) = Ax(k) + Bu(k) + B_d d(k) + Gw(k)$$

Integrated disturbance noise: $w_d(k) \sim N(0, Q_{wd})$

$$d(k+1) = d(k) + G_d w_d(k)$$

Measurement noise: $v(k) \sim N(0, R_v)$

$$y(k) = Cx(k) + C_d d(k) + v(k)$$

How do we find these variances from data?

Distinguishing noise types

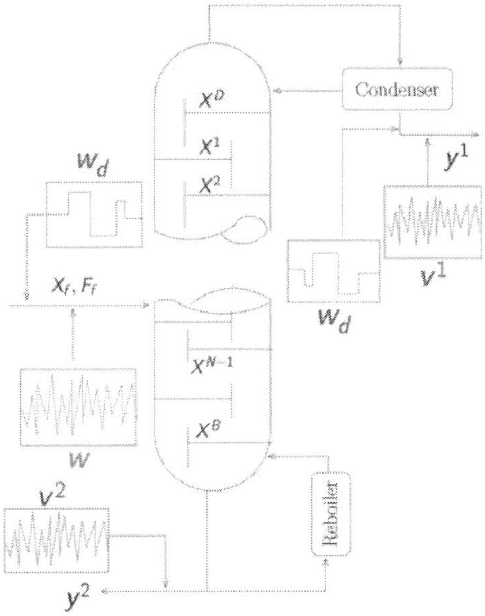

- Process noises **propagate and affect later measurements**

  Measurement noises **affect only current measurement**

$$y(1) = CAx(0) + CGw(0) + v(1)$$

$$y(2) = CA^2x(0) + CAGw(0) + CGw(1) + v(2)$$

Distinguishing noise types

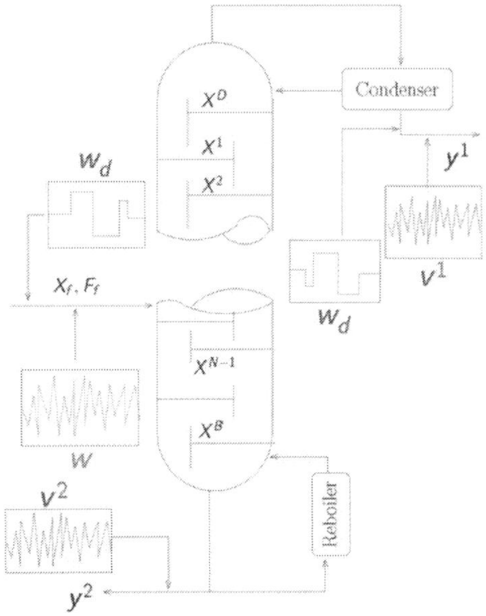

- Process noises propagate and affect later measurements

  Measurement noises affect only current measurement

$$y(1) = CAx(0) + CGw(0) + v(1)$$

$$y(2) = CA^2x(0) + CAGw(0) + CGw(1) + v(2)$$

Process noise and measurement noise leave different footprint on autocovariances of one-step ahead prediction error

Optimization Problem

- Define L-innovation: $\mathscr{Y}(k) = y(k) - C\hat{x}(k|k-1)$

  From data, estimate autocovariances of $\mathscr{Y}$ different lags

  From the model, these autocovariances are linear functions of Q_w and R_v

$$\underbrace{\text{vec}\left(\mathbb{E}\begin{bmatrix} \mathscr{Y}(k)\mathscr{Y}(k)' \\ \vdots \\ \mathscr{Y}(k+N-1)\mathscr{Y}(k)' \end{bmatrix}\right)}_{b} = \mathscr{A}\underbrace{\begin{bmatrix} \text{vec}(Q_w) \\ \text{vec}(R_v) \end{bmatrix}}_{x}$$

Optimization Problem

- Define L-innovation: $\mathscr{Y}(k) = y(k) - C\hat{x}(k|k-1)$
  From data, estimate autocovariances of $\mathscr{Y}$ different lags
  From the model, these autocovariances are linear functions of Q_w and R_v

$$\underbrace{\text{vec}\left(\mathbb{E}\begin{bmatrix}\mathscr{Y}(k)\mathscr{Y}(k)' \\ \vdots \\ \mathscr{Y}(k+N-1)\mathscr{Y}(k)'\end{bmatrix}\right)}_{b} = \mathscr{A}\underbrace{\begin{bmatrix}\text{vec}(Q_w) \\ \text{vec}(R_v)\end{bmatrix}}_{x}$$

$\Longrightarrow$ Least-squares optimization problem: $\min_{Q_w, R_v} \|\mathscr{A}x - \hat{b}\|_W^2$

Optimization Problem

- Define L-innovation: $\mathscr{Y}(k) = y(k) - C\hat{x}(k|k-1)$
  From data, estimate autocovariances of $\mathscr{Y}$ different lags
  From the model, these autocovariances are linear functions of Q_w and R_v

$$\underbrace{\text{vec}\left(\mathbb{E}\begin{bmatrix}\mathscr{Y}(k)\mathscr{Y}(k)' \\ \vdots \\ \mathscr{Y}(k+N-1)\mathscr{Y}(k)'\end{bmatrix}\right)}_{b} = \mathscr{A}\underbrace{\begin{bmatrix}\text{vec}(Q_w) \\ \text{vec}(R_v)\end{bmatrix}}_{x}$$

$\Longrightarrow$ Least-squares optimization problem: $\min_{Q_w, R_v} \|\mathscr{A}x - \hat{b}\|_W^2$

But how should we weight the least squares problem?

Weighting the Least Squares Problem[1]

Model:

$$\hat{b} = \mathscr{A}x + e$$

$\longrightarrow$

Least squares problem:

$$\min_x \|\hat{b} - \mathscr{A}x\|_W^2$$

- Original ALS method uses $W = I$ (Odelson et al., 2006)
  Generalized least squares method uses

$$W = \left(\text{var}(\hat{b})\right)^{-1}$$

$\longrightarrow$ Minimum variance linear unbiased estimates for Q_w and R_v (Rajamani and Rawlings, 2009)

[1]Schmidt 1976, p. 1-18,64-68; Magnus and Neudecker 1999, p. 254-261

Generalized ALS Method

- Need var $\left(\hat{b}\right)$ to weight problem optimally
  - $\hat{b}$ contains sample autocovariances of $y(k) - \hat{y}(k-1)$

Generalized ALS Method

- Need $\text{var}\left(\hat{b}\right)$ to weight problem optimally
  - $\hat{b}$ contains sample autocovariances of $y(k) - \hat{y}(k-1)$

Theoretical formula for this covariance in Rajamani and Rawlings (2009):

- Samples are correlated $\rightarrow$ calculation is intractable for large sample size (even for small models)
- Requires knowledge of unknown Q_w and R_v

Generalized ALS Method

- Need $\text{var}\left(\hat{b}\right)$ to weight problem optimally
  - $\hat{b}$ contains sample autocovariances of $y(k) - \hat{y}(k-1)$

Theoretical formula for this covariance in Rajamani and Rawlings (2009):

- Samples are correlated $\rightarrow$ calculation is intractable for large sample size (even for small models)
- Requires knowledge of unknown Q_w and R_v

Conclusion: generalized least squares is not feasible!

Feasible generalized ALS Method[1]

- Approximate $\hat{\Sigma} = \text{var}(\hat{b})$ from the data (*e.g.* sample variance) and weight problem

$$\min_x \|\hat{b} - \mathscr{A}x\|^2_{\hat{\Sigma}^{-1}} \quad \Rightarrow \hat{x}_{FGLS}$$

[1]Schmidt (1976, p. 69-72)

Feasible generalized ALS Method[1]

- Approximate $\hat{\Sigma} = \text{var}(\hat{b})$ from the data (*e.g.* sample variance) and weight problem

$$\min_x \|\hat{b} - \mathscr{A}x\|^2_{\hat{\Sigma}^{-1}} \quad \Rightarrow \hat{x}_{FGLS}$$

Statistical properties (minimum variance, unbiased) are uncertain

Simulations show reduced variance

[1]Schmidt (1976, p. 69-72)

Zagrobelny, Rawlings FGLS Disturbance Identification 10 / 20

- Weight ALS problem with estimate of $\text{cov}(\hat{b})$ from data
  Simulations consistently show improved results

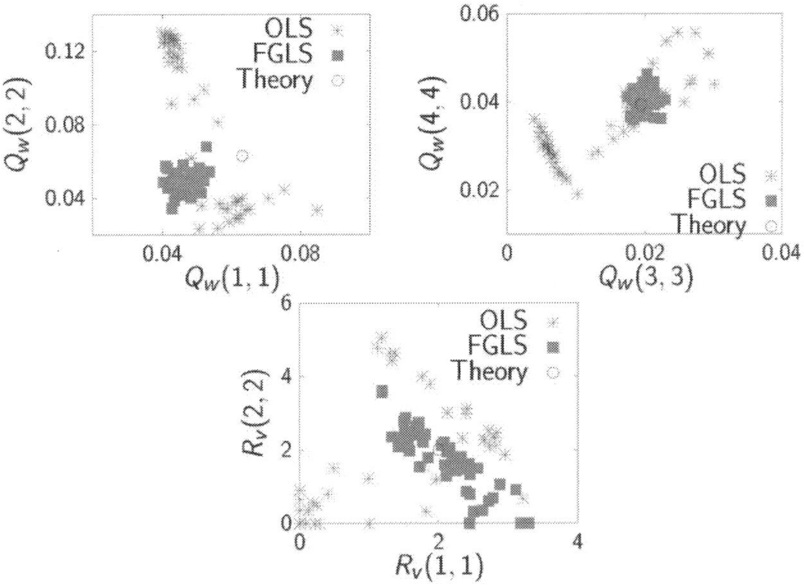

ALS estimates of diagonal elements of Q_w and R_v using OLS and FGLS methods.

- Evaluate ALS results by designing estimator:
  - Calculate innovations: $\mathscr{Y} = y(k) - \hat{y}(k|k-1)$
  - Optimal estimator $\rightarrow$ white innovations
  - Plot autocovariances of innovations: $\mathscr{Y}_a(k)\mathscr{Y}_b(k-j)$ vs. lag j

Optimal estimator behavior

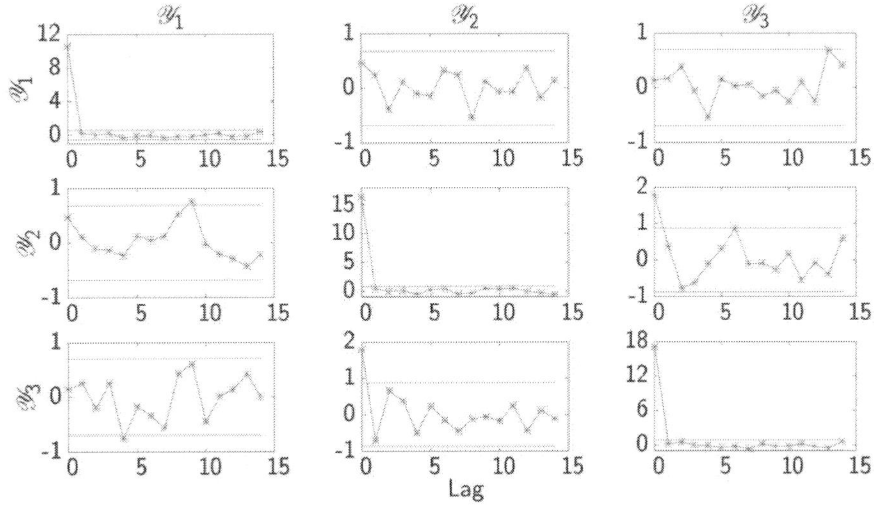

- Evaluate ALS results by designing estimator:
  - Calculate innovations: $\mathscr{Y} = y(k) - \hat{y}(k|k-1)$
  - Optimal estimator $\rightarrow$ white innovations
  - Plot autocovariances of innovations: $\mathscr{Y}_a(k)\mathscr{Y}_b(k-j)$ vs. lag j

Sub-optimal estimator behavior

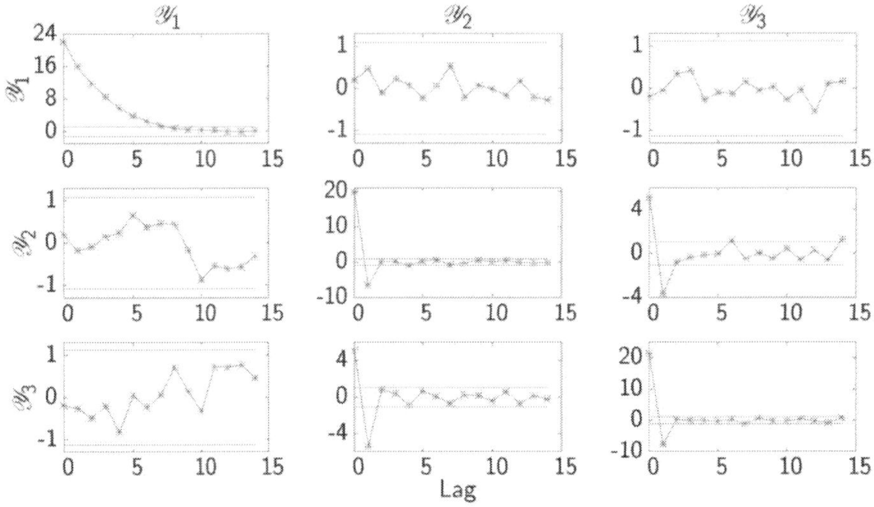

Industrial Data from Air Separation Unit

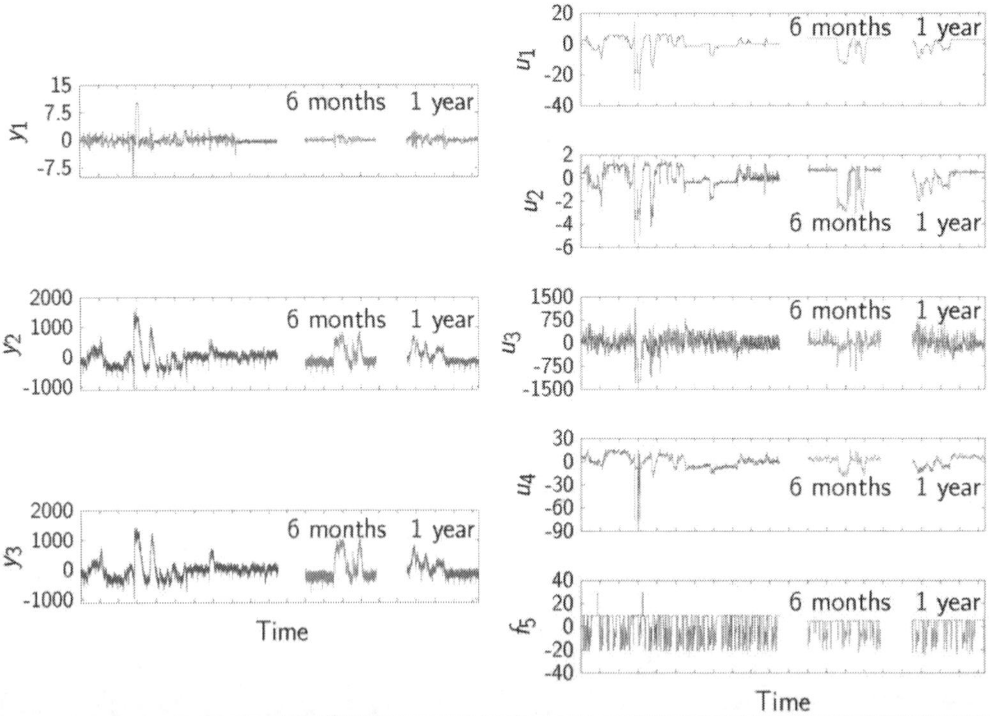

Ordinary vs. Feasible Generalized ALS Method

- ALS estimates $\longrightarrow$ New estimators $\longrightarrow$ Process data

  Optimal behavior using feasible generalized ALS estimates

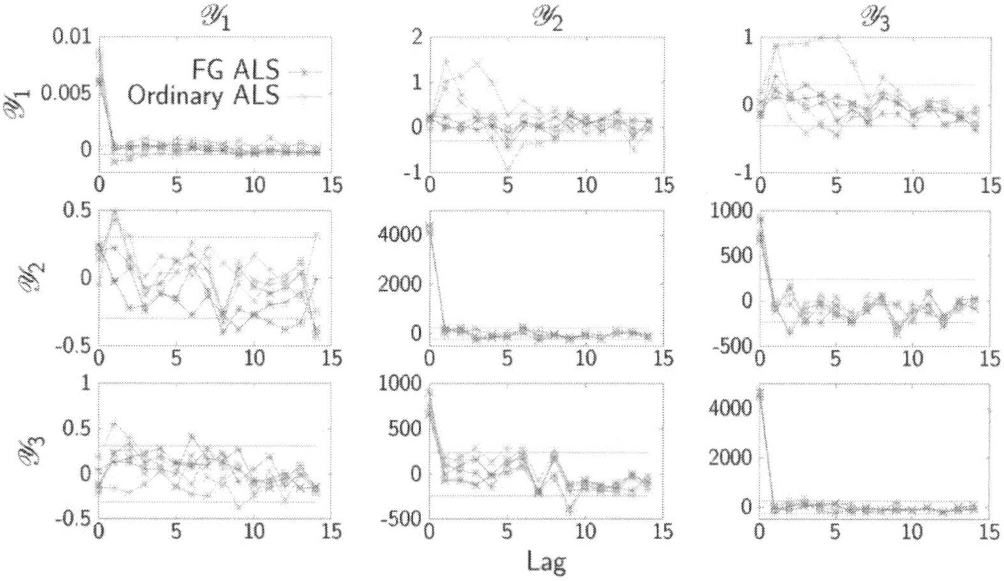

Consistency of ALS Results

- Solve ALS problem using first data set $\longrightarrow$ estimator L_1

  Filter all data sets using L_1 $\longrightarrow$ Nearly perfect results for 65% of the data sets

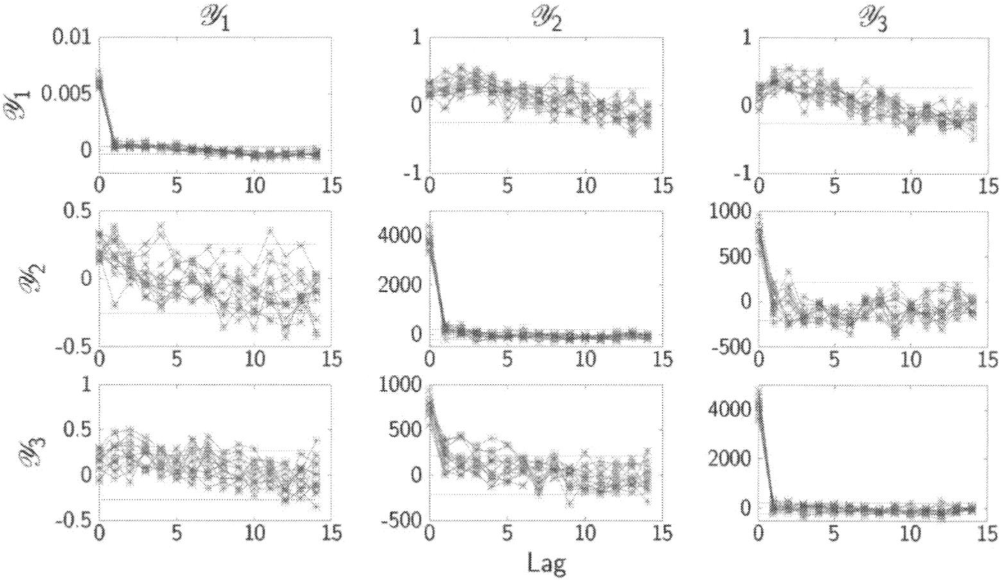

Results: Later Time Periods

- Apply estimator from first data set to data from 6 months and 1 year later
  ⟶ Estimator is nearly optimal

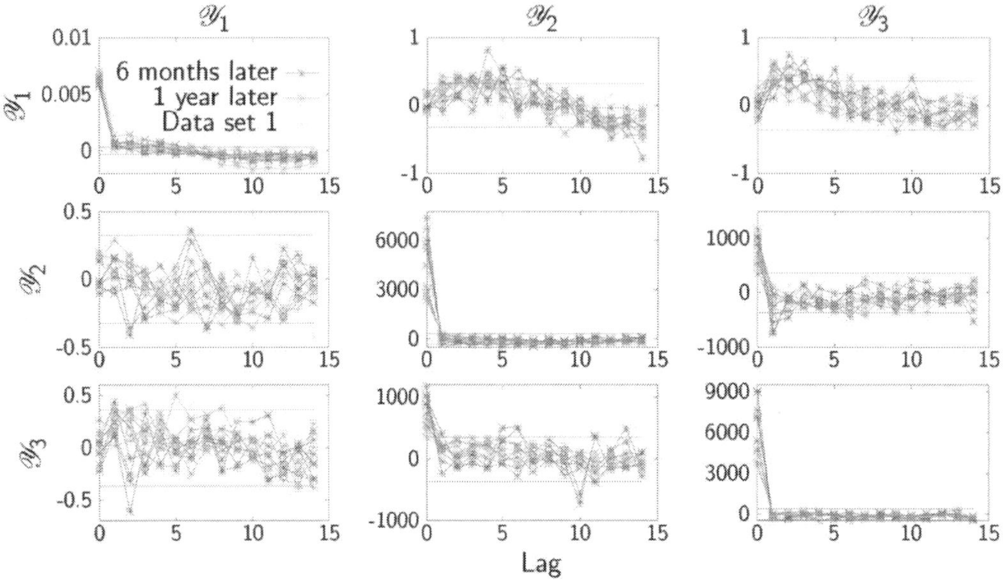

Conclusions

- Feasible generalized least squares should be applied to ALS method
  - Reduces variance of ALS estimates
  - Offers a practical alternative between ordinary ALS (suboptimal) and generalized ALS (impractical)

Conclusions

- Feasible generalized least squares should be applied to ALS method
  - Reduces variance of ALS estimates
  - Offers a practical alternative between ordinary ALS (suboptimal) and generalized ALS (impractical)

  Applied feasible generalized ALS to industrial data
  - Produces estimator with optimal behavior
  - Noise characteristics are consistent over a one year period

Conclusions

- Feasible generalized least squares should be applied to ALS method
  - Reduces variance of ALS estimates
  - Offers a practical alternative between ordinary ALS (suboptimal) and generalized ALS (impractical)

  Applied feasible generalized ALS to industrial data
  - Produces estimator with optimal behavior
  - Noise characteristics are consistent over a one year period

Zagrobelny, Rawlings FGLS Disturbance Identification 17 / 20

Acknowledgments

- Thank you to Drs. Jesus Flores-Cerrillo, Gangshi Hu, and Xiaojing Jiang at Praxair, Inc. for their collaboration and for providing the industrial data.
- Funding: NSF Grant #CTS-1159088, Industrial members of TWCCC

References I

N. R. Kristensen, H. Madsen, and S. B. Jørgensen. Parameter estimation in stochastic grey-box models. *Automatica*, 40:225–237, 2004.

L. Ljung. *System Identification - Theory for the User*. Prentice Hall, New Jersey, second edition, 1999.

J. R. Magnus and H. Neudecker. *Matrix Differential Calculus with Applications in Statistics and Econometrics*. John Wiley, New York, 1999.

P. Matisko and V. Havlena. Noise covariance estimation for Kalman filter tuning using Bayesian approach and Monte Carlo. *Int. J. Adaptive Cont. Signal Proc.*, 27(11): 957–973, 2013.

R. K. Mehra. Identification of stochastic linear dynamic systems. In *1969 IEEE Symposium on Adaptive Processes (8th) Decision and Control*, 1969.

R. K. Mehra. On the identification of variances and adaptive Kalman filtering. *IEEE Trans. Auto. Cont.*, 15(12):175–184, 1970.

R. K. Mehra. Approaches to adaptive filtering. *IEEE Trans. Auto. Cont.*, 17:903–908, 1972.

K. A. Myers and B. D. Tapley. Adaptive sequential estimation with unknown noise statistics. *IEEE Trans. Auto. Cont.*, 21:520–523, 1976.

References II

B. J. Odelson, M. R. Rajamani, and J. B. Rawlings. A new autocovariance least-squares method for estimating noise covariances. *Automatica*, 42(2):303–308, February 2006.

S. J. Qin. An overview of subspace identification. *Comput. Chem. Eng.*, 30:1502–1513, 2006.

M. R. Rajamani and J. B. Rawlings. Estimation of the disturbance structure from data using semidefinite programming and optimal weighting. *Automatica*, 45:142–148, 2009.

P. Schmidt. *Econometrics*. Marcel Dekker, Inc, 1976.

H. K. Tzou and Y. T. Lin. The tracking of a manoeuvring object by using an adaptive Kalman filter. *Proceedings of the Institute of Mechanical Engineers*, 215(2):125–130, 2001.

J. Valappil and C. Georgakis. Systematic estimation of state noise statistics for extended Kalman filters. *AIChE J.*, 46(2):292–308, 2000.

P. Van Overschee and B. De Moor. A unifying theorem for three subspace system identification algorithms. *Automatica*, 31(12):1853–1864, December 1995.

M. A. Zagrobelny, L. Ji, and J. B. Rawlings. Quis custodiet ipsos custodes? *Annual Rev. Control*, 37:260–270, 2013.

Disturbance Covariance Identification Using Maximum Likelihood Estimation

Megan A. Zagrobelny
James B. Rawlings

Department of Chemical and Biological Engineering

THE UNIVERSITY of WISCONSIN MADISON

AIChE Annual Meeting, Atlanta, GA
November 19, 2014

Table of Contents

Why Use MLE to Find Disturbance Covariances?

- Goal: identify process and measurement noise covariances using maximum likelihood estimation
  - Motivated by performance monitoring and estimator design

Why Use MLE to Find Disturbance Covariances?

- Goal: identify process and measurement noise covariances using maximum likelihood estimation
  - Motivated by performance monitoring and estimator design

  Why not use the ALS method?
  - The ALS problem becomes increasingly arbitrary as we start changing the weighting, minimizing the rank, *etc.*
  - Choice of weighting is particularly important
  - ALS problem is more complicated than just least-squares

$$\min_{Q_w, R_v} \|\mathscr{A} \operatorname{vec} \begin{bmatrix} Q_w \\ R_v \end{bmatrix} - \hat{b}\|_W^2 + \rho \operatorname{tr}\left(\begin{bmatrix} I & 0 \\ 0 & 0 \end{bmatrix} Q_w \right) \quad \text{s.t. } Q_w \geq 0, R_v \geq 0$$

Why Use MLE to Find Disturbance Covariances?

- Goal: identify process and measurement noise covariances using maximum likelihood estimation
  - Motivated by performance monitoring and estimator design

  Why not use the ALS method?
  - The ALS problem becomes increasingly arbitrary as we start changing the weighting, minimizing the rank, *etc.*
  - Choice of weighting is particularly important
  - ALS problem is more complicated than just least-squares

$$\min_{Q_w, R_v} \|\mathscr{A} \text{vec} \begin{bmatrix} Q_w \\ R_v \end{bmatrix} - \hat{b}\|_W^2 + \rho \text{tr} \left(\begin{bmatrix} I & 0 \\ 0 & 0 \end{bmatrix} Q_w \right) \quad \text{s.t. } Q_w \geq 0, R_v \geq 0$$

  Maximum likelihood problem has a simpler derivation

Maximum Likelihood Estimation

- Start with:
  - Measurement(s) y
  - Distribution of y in terms of unknown parameter vector, θ

Maximum Likelihood Estimation

- Start with:
  - Measurement(s) y
  - Distribution of y in terms of unknown parameter vector, θ

  Define the likelihood as $p(y;\theta)$ (the probability that we get measurement y when θ is the parameter)

Maximum Likelihood Estimation

- Start with:
  - Measurement(s) y
  - Distribution of y in terms of unknown parameter vector, θ

  Define the likelihood as $p(y;\theta)$ (the probability that we get measurement y when θ is the parameter)

  Maximum likelihood estimation solves the problem

$$\max_{\theta} p(y;\theta)$$

  or equivalently

$$\min_{\theta} \quad -\ln p(y;\theta)$$

MLE Methods in the Literature

- Earliest methods consist of several steps:
  - Deterministic parameter estimates → filter parameters → driving noises (Mehra, 1969, 1972; Kashyap, 1970)

MLE Methods in the Literature

- Earliest methods consist of several steps:
  - Deterministic parameter estimates → filter parameters → driving noises (Mehra, 1969, 1972; Kashyap, 1970)

- More recent methods:
  - Expectation maximization: iterates between estimating the states and optimizing the covariances (Shumway and Stoffer, 1982; Bavdekar et al., 2011; Li and Badgwell, 2014)
  - Grey-box modeling: model structure is known, but some deterministic and/or stochastic parameters are estimated (Bohlin and Graebe, 1995; Kristensen et al., 2004)
  - Direct MLE method assuming independent innovations (Bavdekar et al., 2011)

MLE Methods in the Literature

- Earliest methods consist of several steps:
  - Deterministic parameter estimates $\rightarrow$ filter parameters $\rightarrow$ driving noises (Mehra, 1969, 1972; Kashyap, 1970)

  More recent methods:
  - Expectation maximization: iterates between estimating the states and optimizing the covariances (Shumway and Stoffer, 1982; Bavdekar et al., 2011; Li and Badgwell, 2014)
  - Grey-box modeling: model structure is known, but some deterministic and/or stochastic parameters are estimated (Bohlin and Graebe, 1995; Kristensen et al., 2004)
  - Direct MLE method assuming independent innovations (Bavdekar et al., 2011)

  Contributions of our work:
  - Directly formulate the MLE problem for outputs of a linear system with known model
  - Prove that a solution exists
  - Connect with ALS problem (information about uniqueness)

Setting up the Maximum Likelihood Problem

- Consider our state space system with process and measurement noise:

$$x^+ = Ax + w \qquad\qquad y = Cx + v$$

  We measure N samples of vector y

$$\mathbb{Y} = \begin{bmatrix} y_1 & y_2 & \dots & y_N \end{bmatrix}$$

Setting up the Maximum Likelihood Problem

- Consider our state space system with process and measurement noise:

$$x^{+} = Ax + w \qquad\qquad y = Cx + v$$

We measure N samples of vector y

$$\mathbb{Y} = \begin{bmatrix} y_1 & y_2 & \dots & y_N \end{bmatrix}$$

  - Each y_i is normally distributed
  - y_i and y_j are correlated due to the process noise:

$$y_1 = CAx_0 + Cw_0 + v_1 \qquad y_2 = CA^2x_0 + CAw_0 + Cw_1 + v_2$$

Setting up the Maximum Likelihood Problem

- Consider our state space system with process and measurement noise:

$$x^{+} = Ax + w \qquad\qquad y = Cx + v$$

We measure N samples of vector y

$$\mathbb{Y} = \begin{bmatrix} y_1 & y_2 & \dots & y_N \end{bmatrix}$$

  - Each y_i is normally distributed
  - y_i and y_j are correlated due to the process noise:

$$y_1 = CAx_0 + Cw_0 + v_1 \qquad y_2 = CA^2x_0 + CAw_0 + Cw_1 + v_2$$

Can't separate the likelihood:

$$p(y_1, \dots, y_N; m, P) \neq \prod_{i=1}^{N} p(y_i; m, P)$$

Zagrobelny MLE Disturbance Identification 6 / 18

Distribution of the Measurements

- Write all the measurements in terms of process and measurement noises:

$$y(k) = v(k) + Cw(k-1) + CAw(k-2) + CA^2w(k-3) + \dots$$

Distribution of the Measurements

- Write all the measurements in terms of process and measurement noises:

$$y(k) = v(k) + Cw(k-1) + CAw(k-2) + CA^2w(k-3) + \dots$$

Assuming that $A^k \approx 0, \quad \forall k \geq K$

$$Y = \begin{bmatrix} y(0) \\ y(1) \\ \vdots \\ y(N) \end{bmatrix} = \underbrace{\begin{bmatrix} CA^{K-1} & \dots & C & 0 & \dots & 0 \\ 0 & CA^{K-1} & \dots & C & \dots & 0 \\ & & \ddots & \ddots & \ddots & \\ 0 & \dots & 0 & CA^{K-1} & \dots & C \end{bmatrix}}_{\mathbb{O}} \begin{bmatrix} w(-K) \\ w(-K+1) \\ \vdots \\ w(N-1) \end{bmatrix} + \begin{bmatrix} v(0) \\ \vdots \\ v(N) \end{bmatrix}$$

The vector of outputs, Y, is a linear transformation of the vector of noises

Zagrobelny | MLE Disturbance Identification | 7 / 18

Forming the Likelihood

- The measurements are normally distributed:

$$\begin{bmatrix} y(0) \\ \vdots \\ y(N) \end{bmatrix} | Q_w, R_v \sim \mathcal{N}(0, P) \qquad P = \mathbb{O} \begin{bmatrix} Q_w & & \\ & \ddots & \\ & & Q_w \end{bmatrix} \mathbb{O}' + \begin{bmatrix} R_v & & \\ & \ddots & \\ & & R_v \end{bmatrix}$$

$$\mathbb{O} = \begin{bmatrix} CA^{K-1} & \dots & C & 0 & \dots & 0 \\ 0 & CA^{K-1} & \dots & C & \dots & 0 \\ & & \ddots & \ddots & \ddots & \\ 0 & \dots & 0 & CA^{K-1} & \dots & C \end{bmatrix}$$

Forming the Likelihood

- The measurements are normally distributed:

$$\begin{bmatrix} y(0) \\ \vdots \\ y(N) \end{bmatrix} | Q_w, R_v \sim \mathcal{N}(0, P) \qquad P = \mathbb{O} \begin{bmatrix} Q_w & & \\ & \ddots & \\ & & Q_w \end{bmatrix} \mathbb{O}' + \begin{bmatrix} R_v & & \\ & \ddots & \\ & & R_v \end{bmatrix}$$

$$\mathbb{O} = \begin{bmatrix} CA^{K-1} & \dots & C & 0 & \dots & 0 \\ 0 & CA^{K-1} & \dots & C & \dots & 0 \\ & & \ddots & \ddots & \ddots & \\ 0 & \dots & 0 & CA^{K-1} & \dots & C \end{bmatrix}$$

Simplify negative log likelihood to form optimization problem:

$$\min_{Q_w, R_v} \quad \ln \det P + Y'P^{-1}Y$$

MLE Disturbance Identification Problem

- Optimization problem from negative log likelihood:

$$\min_{Q_w, R_v} \quad \ln \det P + Y' P^{-1} Y$$
$$P = \mathbb{O}\text{diag}(Q_w \dots Q_w)\mathbb{O}' + \text{diag}(R_v \dots R_v)$$

MLE Disturbance Identification Problem

- Optimization problem from negative log likelihood:

$$\min_{Q_w, R_v} \quad \ln \det P + Y' P^{-1} Y$$
$$P = \mathbb{O}\text{diag}(Q_w \dots Q_w)\mathbb{O}' + \text{diag}(R_v \dots R_v)$$

Different from standard problem:

- Finding Q_w, R_v (parameters of P) rather than P itself
- Only have one sample of Y

What can we say about the existence and uniqueness of the solution?

Existence of Solution

- Goal: prove that solution exists

Existence of Solution

- Goal: prove that solution exists

Challenges:

- $Q_w, R_v \geq 0$ is unbounded
- $\phi \to \infty$ on boundary

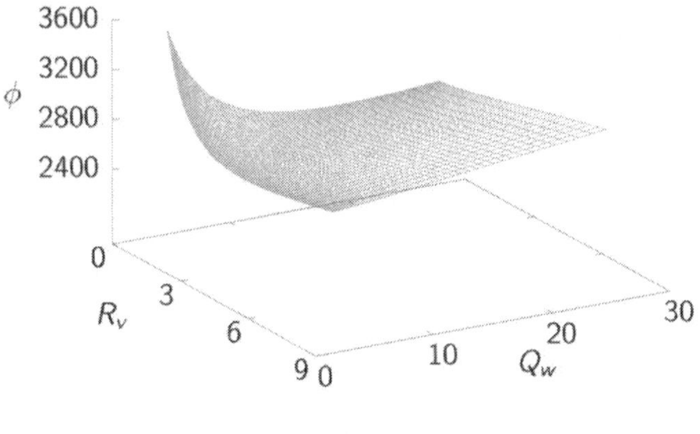

Zagrobelny MLE Disturbance Identification 10 / 18

Proof for Existence of Solution

- Show that $\phi \not\to -\infty$ on boundaries

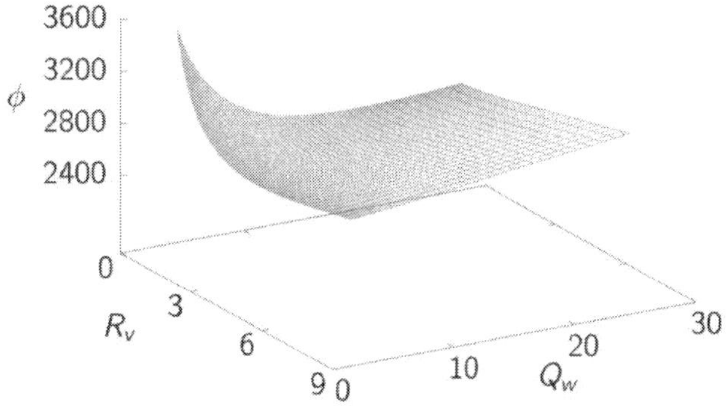

Proof for Existence of Solution

- Show that $\phi \not\to -\infty$ on boundaries

  $\phi(Q_1, R_1) = \phi_1$ for some (Q_1, R_1)

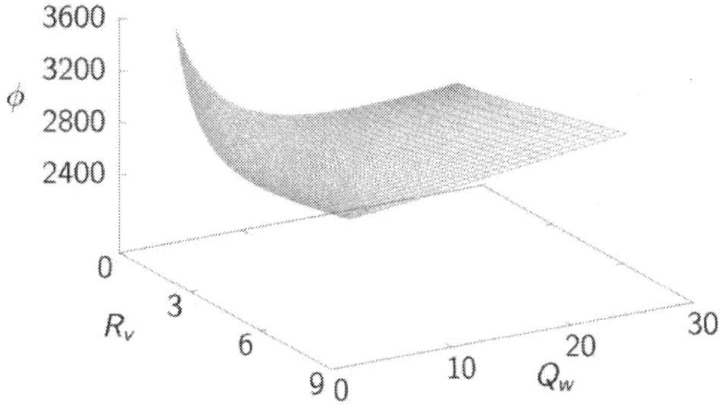

Proof for Existence of Solution

- Show that $\phi \not\to -\infty$ on boundaries

  $\phi(Q_1, R_1) = \phi_1$ for some (Q_1, R_1)

  Define new sublevel set

  $$\Omega' := \{(Q_w, R_v) | Q_w \geq 0 \\ R_v \geq 0, \ \phi(Q_w, R_v) \leq \phi_1\}$$

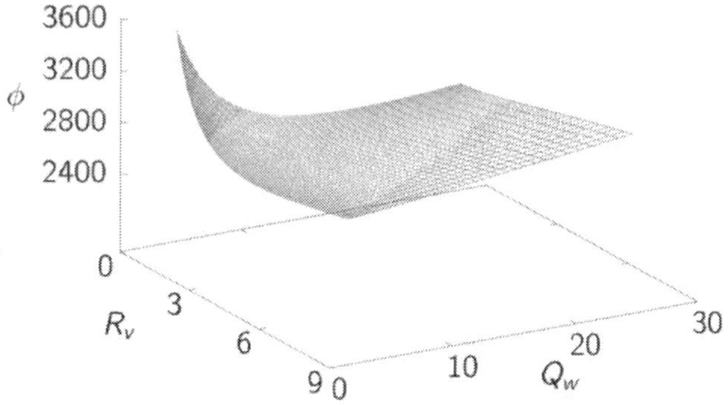

Proof for Existence of Solution

- Show that $\phi \not\to -\infty$ on boundaries

  $\phi(Q_1, R_1) = \phi_1$ for some (Q_1, R_1)

  Define new sublevel set

  $$\Omega' := \{(Q_w, R_v) | Q_w \geq 0 \\ R_v \geq 0, \ \phi(Q_w, R_v) \leq \phi_1\}$$

  Solution exists on Ω' (by Weierstrass theorem) $\rightarrow$ solution exists on $Q_w \geq 0, R_v \geq 0$

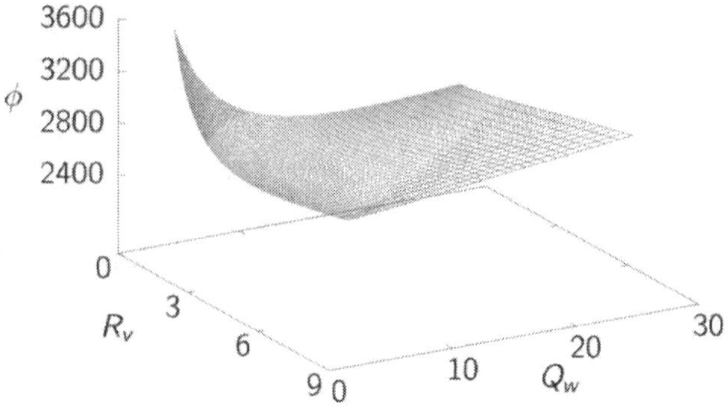

Comparison to ALS Problem

- MLE problem has same first order conditions as optimally weighted ALS problem

  As a consequence:
  - No unique ALS solution $\Rightarrow$ no unique MLE solution
  - Unique ALS solution $\Rightarrow$?

Comparison to ALS Problem

- MLE problem has same first order conditions as optimally weighted ALS problem

  As a consequence:
  - No unique ALS solution $\Rightarrow$ no unique MLE solution
  - Unique ALS solution $\Rightarrow$?

  Necessary conditions for uniqueness when noise-shaping matrix G is unknown:
  - (A, C) observable
  - $\text{rank}(C) = n$
  - $\text{rank}(A) = n$

Zagrobelny MLE Disturbance Identification 12 / 18

Scalar Example

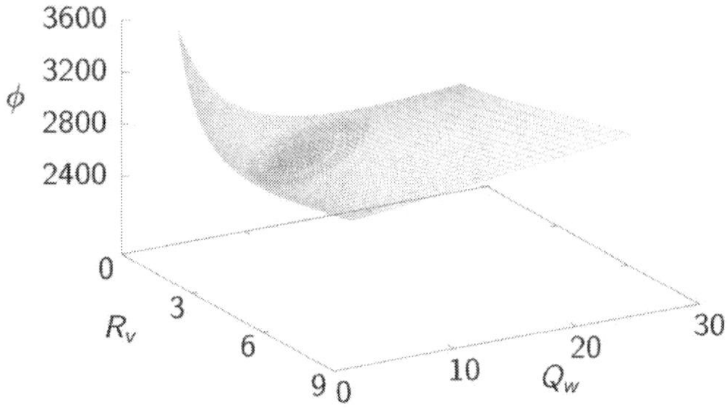

- $Q = 7$ and $R = 3$ used to generate data
- Objective function has unique minimum

Scalar Example

$N = 1000$

	Q_w	R_v	ϕ	Time (s)
MLE	8.69	2.74	2656.85	2.86
ALS	8.66	2.65	2657.09	1.36
True	7	3	2659.4	

- $Q = 7$ and $R = 3$ used to generate data
- Objective function has unique minimum
- MLE solution has maximum likelihood compared to ALS solution or true values

Scalar Example

$N = 1000$

	Q_w	R_v	ϕ	Time (s)
MLE	8.69	2.74	2656.85	2.86
ALS	8.66	2.65	2657.09	1.36
True	7	3	2659.4	

$N = 10000$

	Q_w	R_v	ϕ	Time (s)
MLE	6.76	3.07	26277.25	**178**
ALS	6.64	3.10	26277.39	1.88

- $Q = 7$ and $R = 3$ used to generate data
- Objective function has unique minimum
- MLE solution has maximum likelihood compared to ALS solution or true values
- MLE method slows significantly as more data is used

$n = p = 5$ Example

- MLE solution has lowest objective value for larger example
- Changing initial guess reduces computational time

Method	Estimates	ϕ	**Time (min)**
MLE $Q_0 = I$	$\text{diag}(Q_w) = [8.50 \quad 14.29 \quad 28.58 \quad 11.19 \quad 24.48]$ $\text{diag}(R_v) = [1.53 \quad 1.90 \quad 1.58 \quad 2.99 \quad 0.801]$	**15375**	**34.8**
MLE $Q_0 = \hat{Q}_{ALS}$	$\text{diag}(Q_w) = [8.34 \quad 14.25 \quad 28.60 \quad 11.25 \quad 24.25]$ $\text{diag}(R_v) = [1.52 \quad 1.91 \quad 1.57 \quad 3.00 \quad 0.822]$	**15375**	**20.5**
ALS $Q_0 = I$	$\text{diag}(Q_w) = [6.42 \quad 14.55 \quad 25.01 \quad 8.78 \quad 17.48]$ $\text{diag}(R_v) = [1.38 \quad 1.70 \quad 1.14 \quad 2.33 \quad 0.94]$	**15465**	**0.173**
True Covariances	$\text{diag}(Q_w) = [8.92 \quad 13.07 \quad 25.11 \quad 11.98 \quad 24.20]$ $\text{diag}(R_v) = [1.51 \quad 2.10 \quad 1.39 \quad 3.78 \quad 1.11]$	**15391**	

Conclusions

- We formulated the disturbance model identification as an MLE problem
  - Simpler derivation and more meaningful results compared to the ALS method

Conclusions

- We formulated the disturbance model identification as an MLE problem
  - Simpler derivation and more meaningful results compared to the ALS method

Proved that solution exists

Connected to ALS problem $\rightarrow$ necessary conditions for uniqueness

Conclusions

- We formulated the disturbance model identification as an MLE problem
  - Simpler derivation and more meaningful results compared to the ALS method

Proved that solution exists

Connected to ALS problem $\rightarrow$ necessary conditions for uniqueness

MLE method produces reliable results for small examples

Conclusions

- We formulated the disturbance model identification as an MLE problem
  - Simpler derivation and more meaningful results compared to the ALS method

Proved that solution exists

Connected to ALS problem $\rightarrow$ necessary conditions for uniqueness

MLE method produces reliable results for small examples

Computation time or storage requirements remain a challenge

- Sparsity and other tricks help, but more improvement is needed

Future work: cases with $p < n$ (no unique solution)

Acknowledgments

- Thank you to Drs. T.A. Badgwell and W. Li of ExxonMobil for helpful discussion

  Funding: NSF grant #CTS-1159088, Industrial members of TWCCC

Bibliography

V. A. Bavdekar, A. P. Deshpande, and S. C. Patwardhan. Identification of process and measurement noise covariance for state and parameter estimation using extended Kalman filter. *J. Proc. Cont.*, 21:585–601, 2011.

T. Bohlin and S. F. Graebe. Issues in nonlinear stochastic grey box identification. *Int. J. Adaptive Cont. Signal Proc.*, 9:465–490, 1995.

R. L. Kashyap. Maximum likelihood identification of stochastic linear systems. *IEEE Trans. Auto. Cont.*, 15(1):25–34, 1970.

N. R. Kristensen, H. Madsen, and S. B. Jørgensen. Parameter estimation in stochastic grey-box models. *Automatica*, 40:225–237, 2004.

W. Li and T. A. Badgwell. Structured covariance estimation for state prediction. Accepted for publication in 53rd IEEE Conference on Decision and Control, 2014.

R. K. Mehra. Identification of stochastic linear dynamic systems. In *1969 IEEE Symposium on Adaptive Processes (8th) Decision and Control*, 1969.

R. K. Mehra. Approaches to adaptive filtering. *IEEE Trans. Auto. Cont.*, 17:903–908, 1972.

R. H. Shumway and D. S. Stoffer. An approach to time series smoothing and forecasting using the EM algorithm. *J. Time Series Anal.*, 3:253–264, 1982.

Behavior of ϕ on Boundaries

- If $\phi(Q_w, R_v) \to -\infty$, then a solution doesn't exist

Behavior of ϕ on Boundaries

- If $\phi(Q_w, R_v) \to -\infty$, then a solution doesn't exist

  Prove that $\phi(Q_w, R_v) \not\to -\infty$ on the boundaries
  - Eigendecomposition of P:

$$P = \underbrace{\begin{bmatrix} W_1 & \dots & W_{Np} \end{bmatrix}}_{W} \operatorname{diag}\left(\lambda_1 \dots \lambda_{Np}\right) W'$$

  - W_i, λ_i are functions of Q_w and R_v

Behavior of ϕ on Boundaries

- If $\phi(Q_w, R_v) \to -\infty$, then a solution doesn't exist
  Prove that $\phi(Q_w, R_v) \not\to -\infty$ on the boundaries
  - Eigendecomposition of P:

$$P = \underbrace{\begin{bmatrix} W_1 & \dots & W_{Np} \end{bmatrix}}_{W} \operatorname{diag}(\lambda_1 \dots \lambda_{Np}) \, W'$$

  - W_i, λ_i are functions of Q_w and R_v
  - Rewrite objective in terms of eigenvalues:

$$\phi(Q_w, R_v) = \sum_{i=1}^{Np} \left(\ln(\lambda_i) + \frac{(W_i' Y)^2}{\lambda_i} \right)$$

Behavior of ϕ on Boundaries

- If $\phi(Q_w, R_v) \to -\infty$, then a solution doesn't exist
- Prove that $\phi(Q_w, R_v) \not\to -\infty$ on the boundaries
  - Eigendecomposition of P:

$$P = \underbrace{\begin{bmatrix} W_1 & \dots & W_{Np} \end{bmatrix}}_{W} \operatorname{diag}(\lambda_1 \dots \lambda_{Np}) \, W'$$

  - W_i, λ_i are functions of Q_w and R_v
  - Rewrite objective in terms of eigenvalues:

$$\phi(Q_w, R_v) = \sum_{i=1}^{Np} \left(\ln(\lambda_i) + \frac{(W_i' Y)^2}{\lambda_i} \right)$$

  - As $\lambda_i \to 0$, $\frac{1}{\lambda_i} \to \infty$ faster than $\ln(\lambda_i) \to -\infty$
  - Also require $W_0' Y \neq 0$, where $W_0 = \text{null}(Y)$: true as long as $\begin{bmatrix} y_1 & \dots & y_N \end{bmatrix}$ is full row rank (we assume with probability one)

TOPOLOGICAL PRESERVATION TECHNIQUES FOR NONLINEAR PROCESS MONITORING

Michael C. Thomas, Jose Romagnoli

Department of Chemical Engineering, Louisiana State University, Baton Rouge, LA

American Institute of Chemical Engineers
November 2014

PSE @ LSU
LSU
PROCESS SYSTEMS ENGINEERING AT LOUISIANA STATE UNIVERSITY

Outline

- Introduction: Process monitoring
- Self Organizing Maps
- Case Study: Tennessee Eastman Process
- Results
- Conclusions and Further Studies
  - Industrial Separation Tower

2

Process Monitoring Tasks

- ***GOAL**: detect and assist with the management of abnormal situations*
- *Fault Detection*
  - Recognizes a deviation from the normal operating regime using process measurements
- *Fault Identification*
  - Helps personnel identify the fault by finding the measured variables most related to the fault
- *Fault Diagnosis*
  - Determines the root causes of the fault

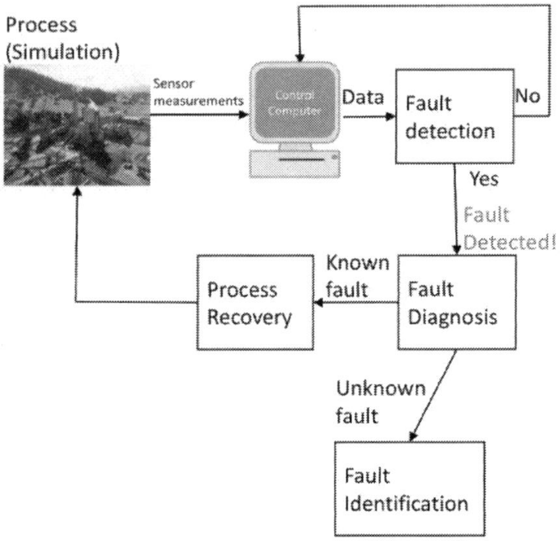

3

Problem Statement

- What is the probability an online data observation belongs to an operating region or "class"?
- Data driven fault detection utilizes the large data sets produced by DCS
- Principal Component Analysis
  - Common, linear data-driven method
  - Possess a suite of tools useful for process monitoring
- Self-Organizing Maps (SOM) summary:
  - It is an ideal tool for data mining and analysis
  - Provides excellent visualization tools to explore the data
- However, **does not** provide all necessary tools for process monitoring tasks.
  - SOM-based discriminant function
  - SOM-based contribution plots
  - SOM-based diagnosis tools → ***MSOM***

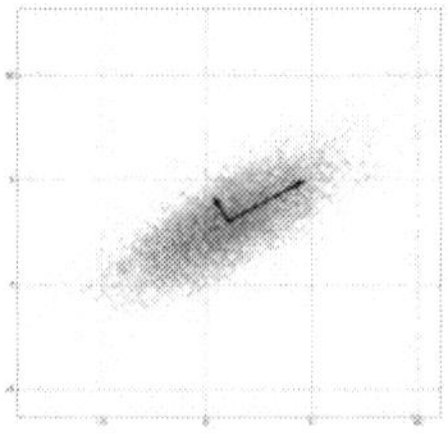

Our Contribution

Validated within a benchmark industrial problem

4

Self Organizing Maps (SOM)

- **Goal**: represent high dimensional data structure
- **Solution**: 2-dimensional representation of data manifold that preserves topology of original data
- Visualizations show:
  - Relationships between clusters of data
  - Relationships between variables
- A data vector's *best matching unit* (BMU) is the map vector most similar to a data vector for classification
  - Used to show location of data on the map

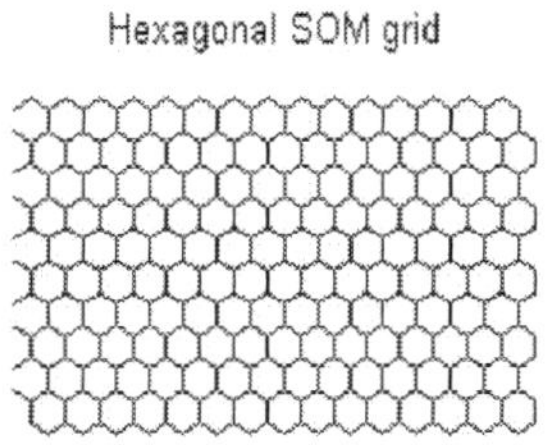

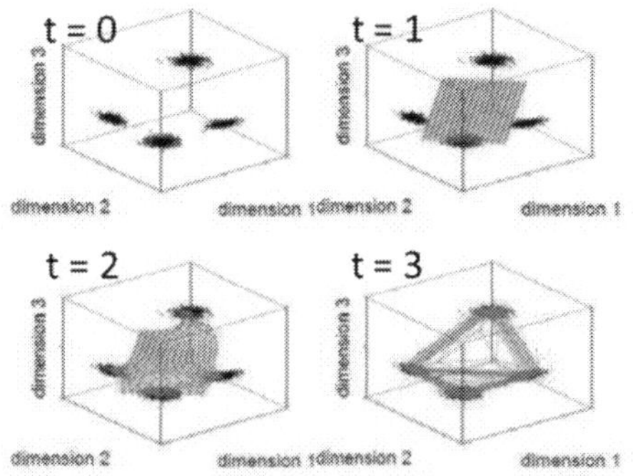

5

SOM Visualizations: Distance Matrix

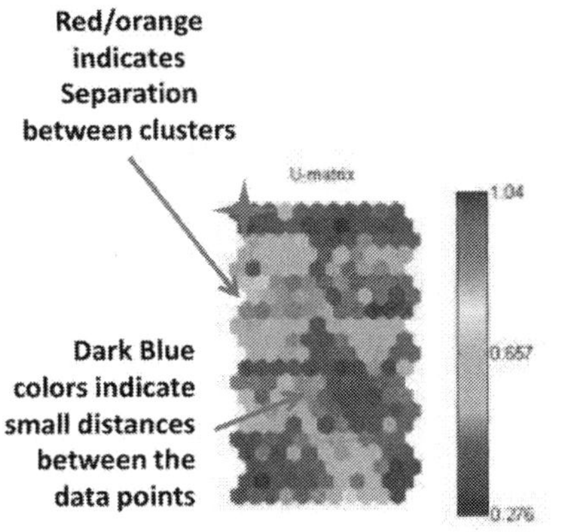

U-Matrix of the same process indicating two clear clusters of operating data

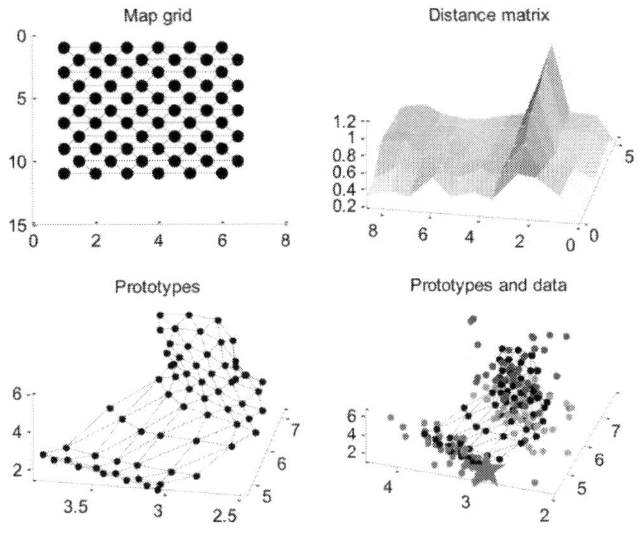

6

SOM Fault Detection

- Kernel Density Estimation
- Gaussian Mixture Modeling
  - SOM map vectors used as the center of kernels
- Kernel function:
  - $\hat{f}(x) = \frac{1}{nh^d} \sum_{i=1}^{n} K\left\{\frac{1}{h}(x - X_i)\right\}$

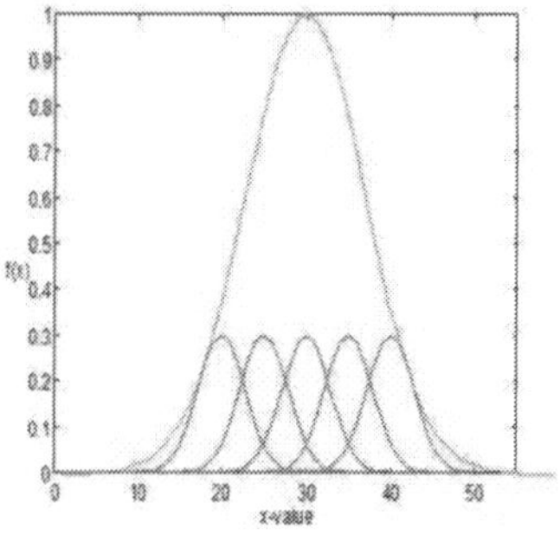

Where

h: is the window width/smoothing parameter

n: is the number of data points

K: is the kernel function- the distribution of the bumps

KDE

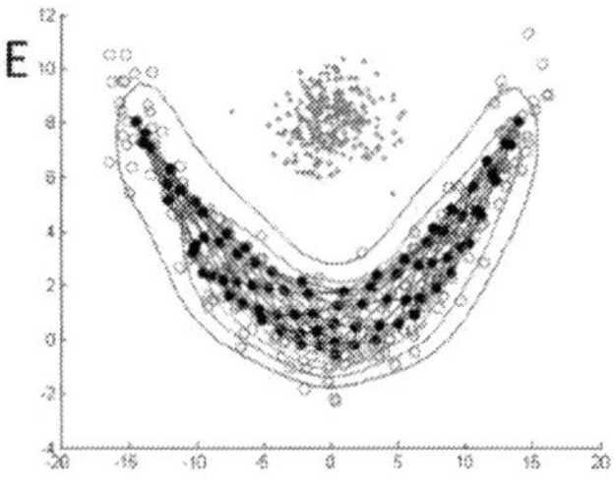

PCA Ellipsoid

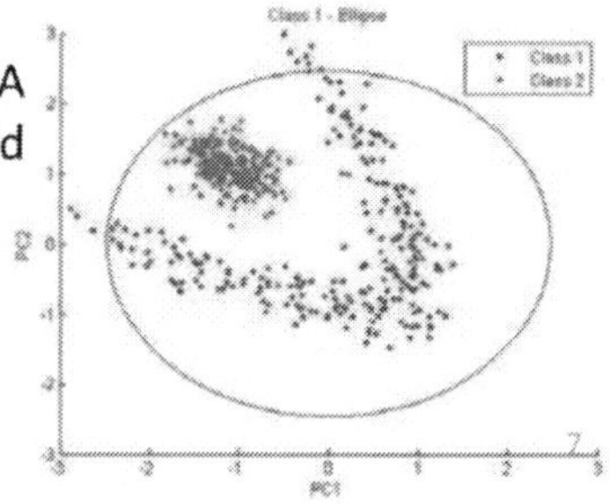

SOM Fault Identification

- Train a SOM to fit each faulty region
- Allows for integration of new faults without affecting other fault representations
- Fault identification
  - Residual is difference between point and *normal* BMU

$$r_f(t) = \left(x_j(t) - m_{c_j}\right)^2$$

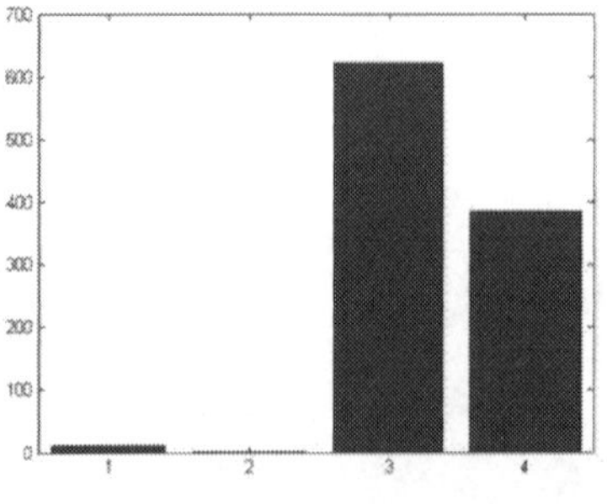

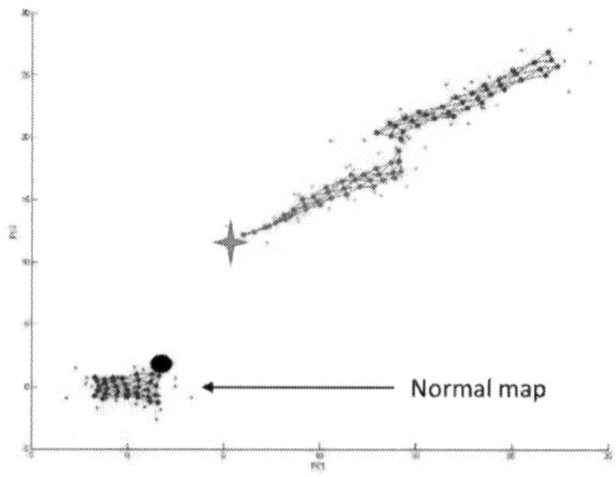

Fault Diagnosis: MSOM

- Train a SOM to fit each faulty region
- Allows for integration of new faults without affecting other fault representations
- Eliminates bordering map units which can cause misclassifications using 1SOM

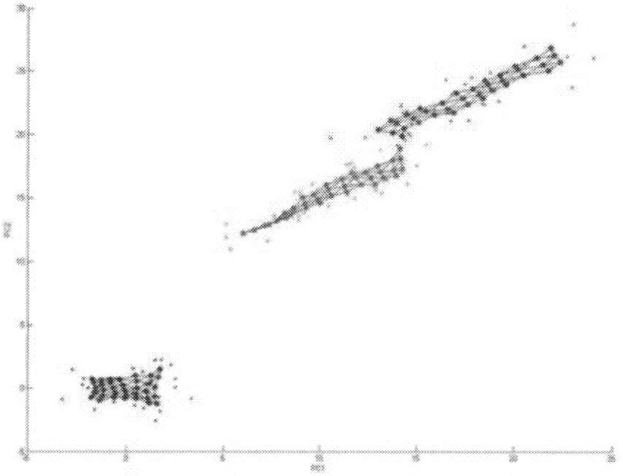

- Feature extraction
  - Projection to each map
- Discriminant analysis
  - Probability a data vector is associated with a map
- Maximum selection
  - Data will be associated with whatever map it's closest to

9

MSOM Process Monitoring

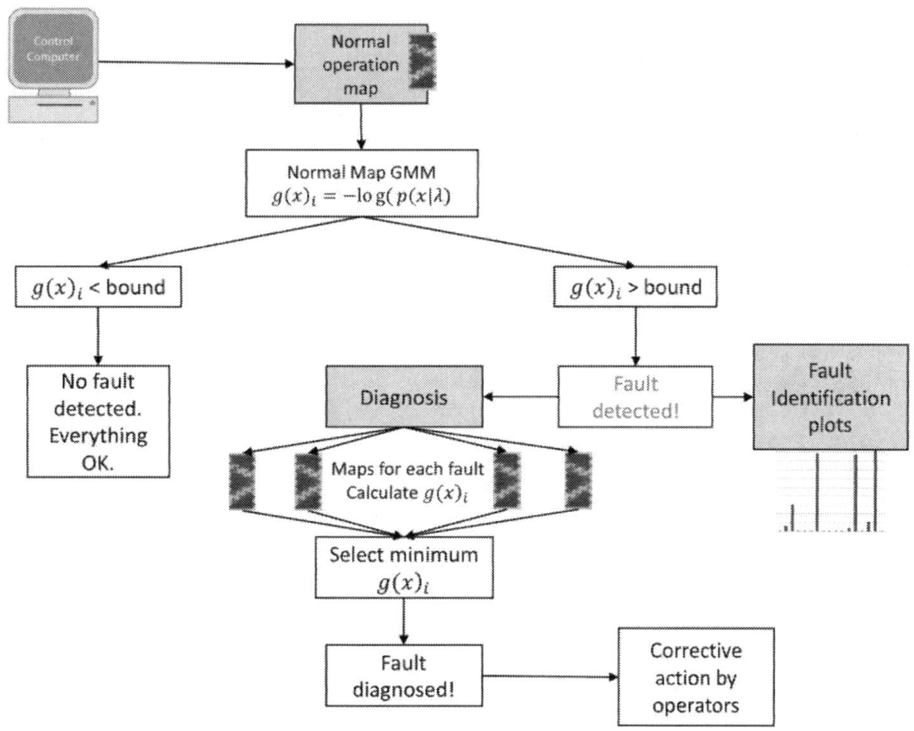

10

Case Study

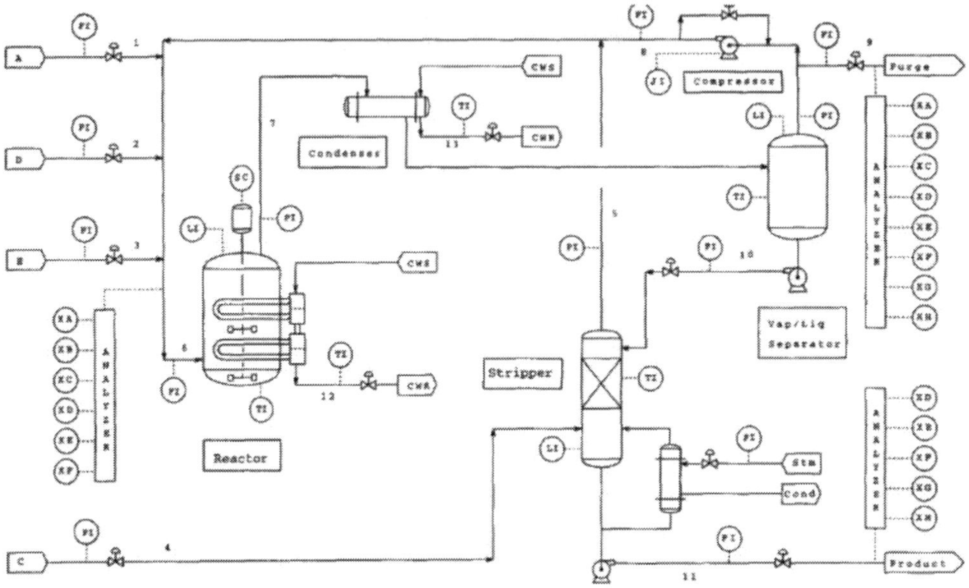

The ***Tennessee Eastman Process (TEP)*** was created by the Eastman Chemical Company to provide a reliable industrial process for evaluating control and monitoring methods

11

Case Study

- The TEP process contain 20 preprogramed faults.
  - Most of these faults are particular feed-step-change faults.
- In this study, in-depth analysis to characterize other more difficult faults such as random variations, sticky valves and slow drift in kinetics is addressed.
- Specific faults considered in this work are:
  - **Fault 1** is a step change in the composition of stream 4.
  - **Fault 7** is a step change in the pressure of stream 4.
  - **Fault 8** is a random variation in the feed composition of stream 4.
  - **Fault 11** imposes a random variation in the cooling water to the reactor, which requires the control system to constantly adjust the cooling water flow to compensate for the reduced or increased cooling capacity of the water used.
  - **Fault 13** is a slow drift in the kinetics of the reaction which requires many adjustments by the control system to accommodate the changing composition of the reactor output.
  - **Fault 14** is a sticky valve in the reactor cooling water. Its effect on the rest of the process is similar to the earlier random variation faults where it creates variations that would not be observed under normal operation.
  - **Faults 14 and 1 (14+1)** is the combination of faults 1 and 14, where the state is disturbed by a step change in the composition of steam 4 while coping with a sticky valve.

Examples of Faults

Fault 1: Step change in feed comp.

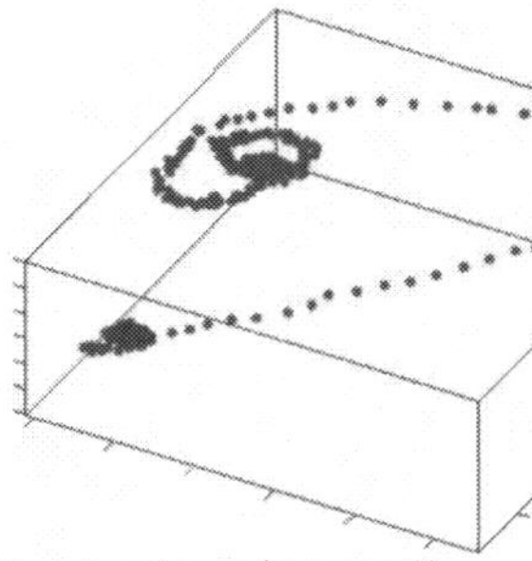

Fault 1 is a step change from one SS area to another.

Fault 14: sticky cooling valve

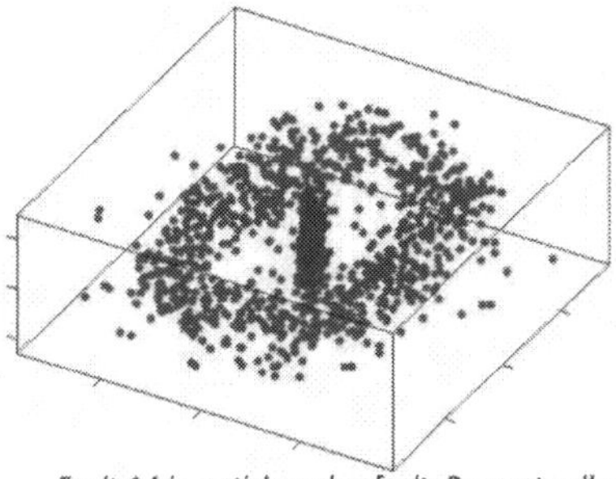

Fault 14 is a sticky valve fault: Prevents all variables from entering the normal operating region simultaneously forming an "orbit" around normal operations

Fault 8: random variation

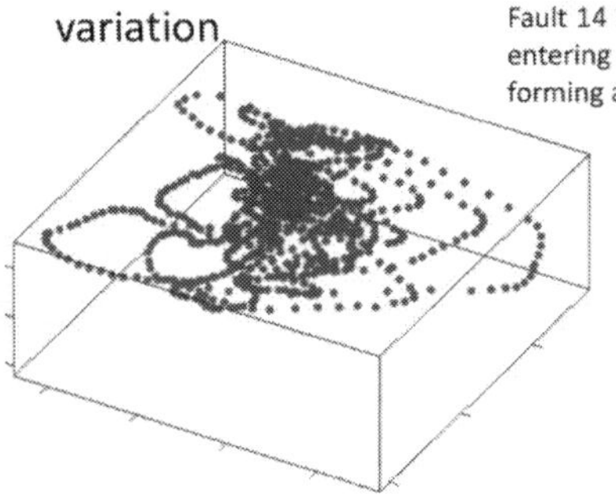

Fault 8 is a random variation with constant adjustment by the control system.

13

Fault Diagnosis

- SOM U-matrix can visualize how the data cluster in multidimensional space

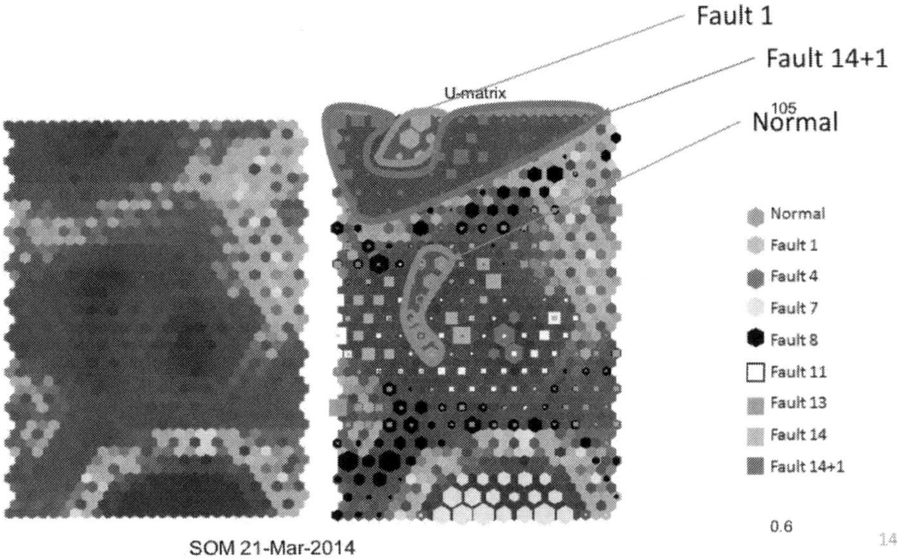

14

Fault Visualization

- Visualize trajectory of faults on SOM U-matrix

Fault 1

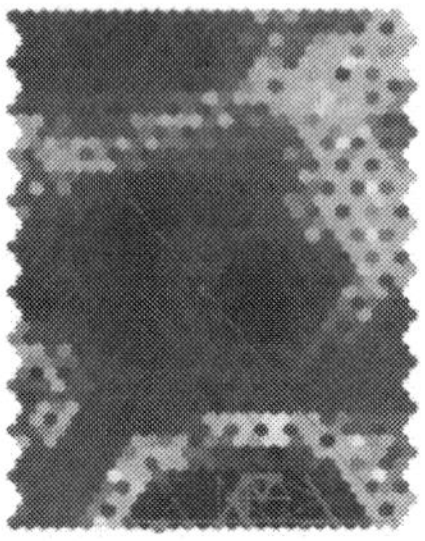

Fault 7

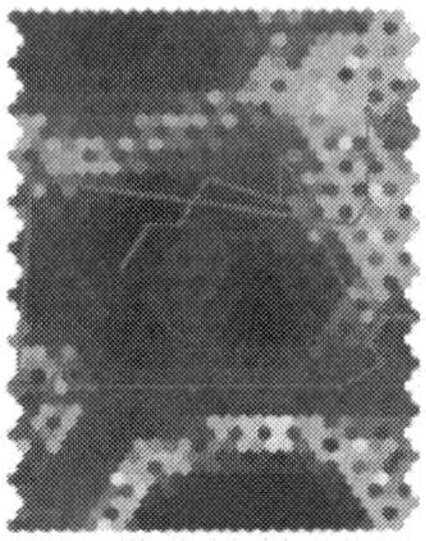

Fault 13

Fault 13: slow drift in the kinetics of the reaction

requires many adjustments by the control system to accommodate the changes

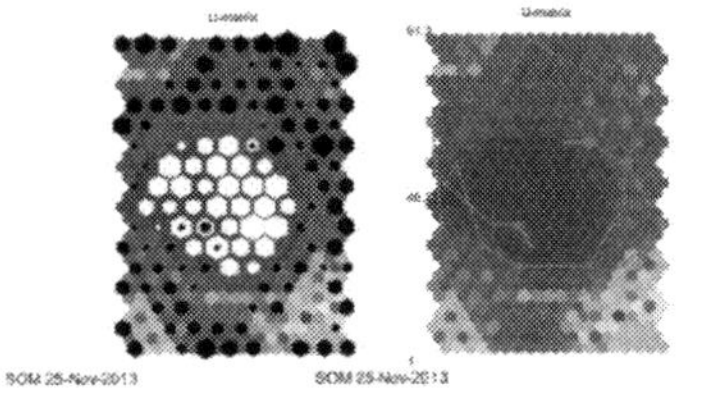

Fault 8 – Random Variation of Stream 4 Composition
Here the effects of the fault are masked by the action of the control system.
SOM manages to separate the faulty data from normal operation, and the trajectory can be seen circling, but not entering the normal region

Fault Identification

Fault 11:
Random variation in reactor cooling water

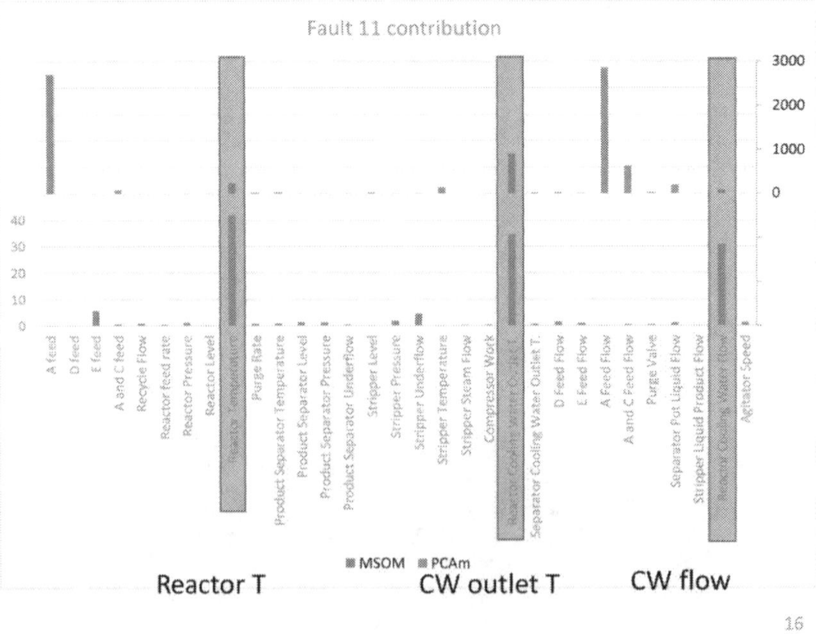

16

Fault Diagnosis

- Fault diagnosis tests the discriminating ability of the different methods
- MSOM offers improvements on SOM
- Both address the shortcomings of linear technique

	Correct Diagnosis Rate (%)		
TEP fault	mPCA	1-SOM	MSOM
1	19.9%	93.0%	99.3%
4	0%	95.0%	97.5%
7	37.2%	99.9%	100.0%
8	79.8%	79.2%	91.4%
11	69.5%	74.7%	94.6%
13	97.4%	84.1%	94.3%
14	21.3%	92.6%	99.3%
14+1	96.8%	90.3%	99.6%

17

Fault Diagnosis – 1-SOM

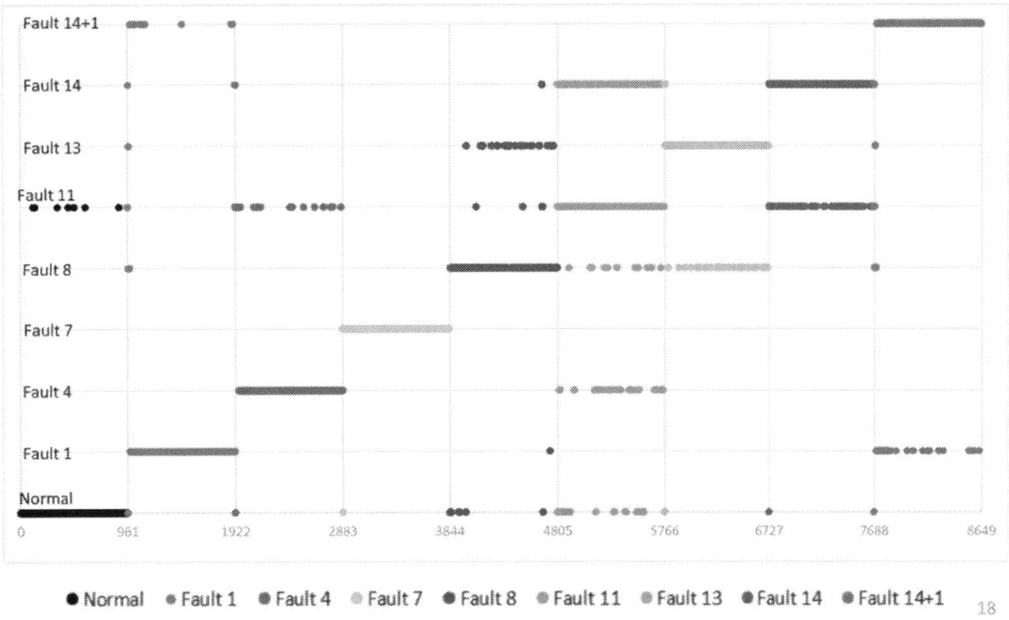

18

Fault Diagnosis - MSOM

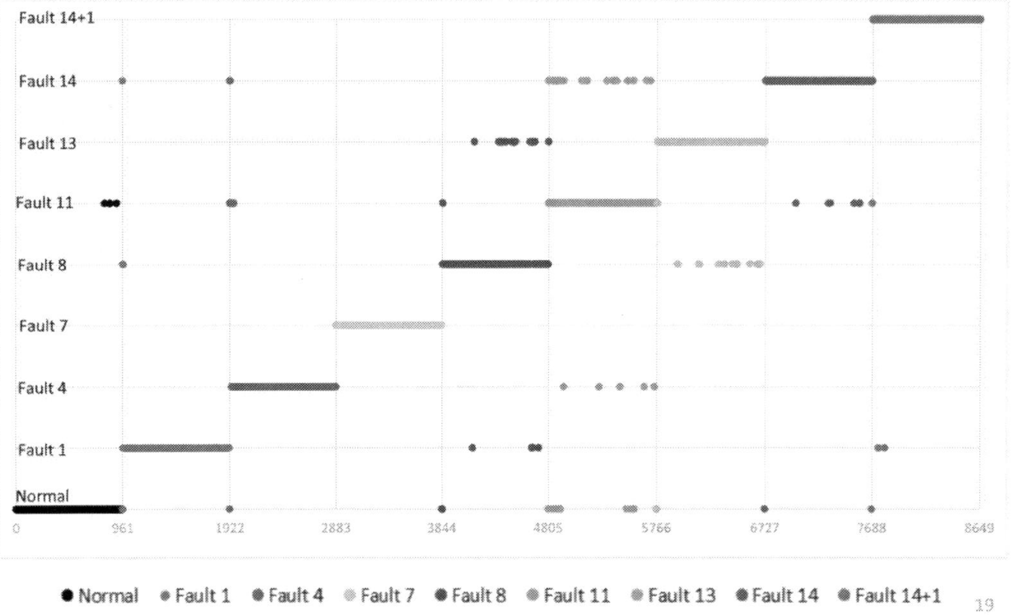

19

Industrial application

Separation system for reactor products

- **Challenging**: Different grades during operation
- 2 faults analyzed: 1 known and an unknown
- Goal: diagnose the known fault, identify/detect the new fault

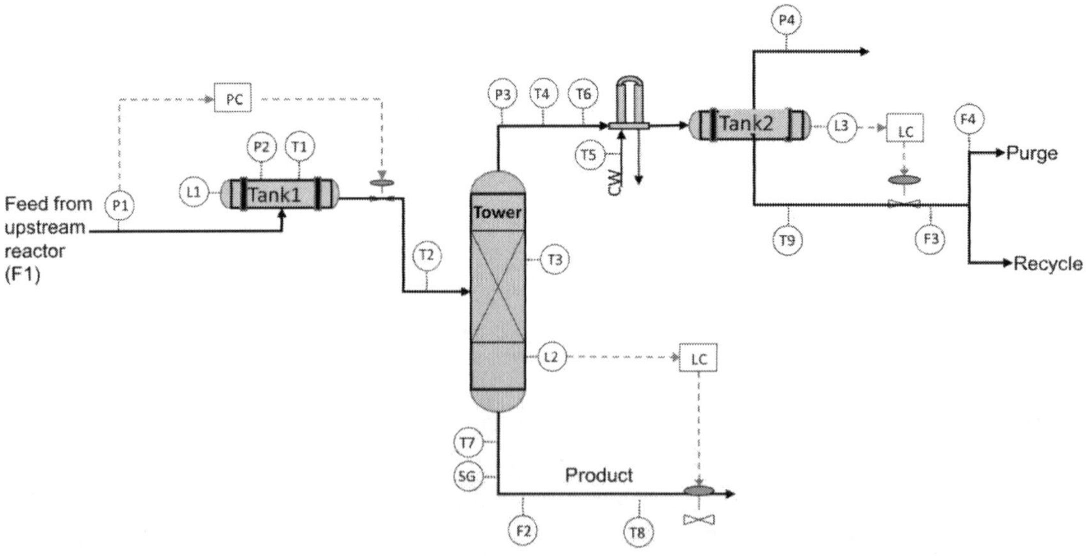

Summary

- In this work, SOMs were applied to all aspects of process monitoring
- **Fault detection** was performed by using a Gaussian mixture model of the normal region.
- For the fault **variable identification**, a novel method was introduced using residuals of the deviation from the normal region.
- **Fault diagnosis** was done by creating a map for each fault (MSOM), which makes incorporating new faults into the scheme much simpler.
- **Applications**:
  - Tennessee Eastman Process (TEP)
    - Previous studies of the TEP have considered particular feed-step-change faults; here, an analysis of dynamic faults such as random variations, sticky valves and a slow drift in kinetics were included.
  - Preliminary results in an industrial system

Results indicate that MSOM is able to improve upon linear distance preservation techniques and standard SOM-based approaches in process monitoring tasks

21

Thanks!

- Questions?

Polymerization processes are an important class of chemical processes. Continuous polymerization reactors are widely used to produce synthetic polymer products such as styrene. The increasing demand for high quality polymers has given impetus to controller designs that provide good control of the polymer product properties and minimize off-spec product during the start-up and the grade transitions, leading to using automatic control techniques [1]. The increased level of automation, however, also makes process control systems susceptible to equipment abnormalities, such as actuator and sensor faults. If not properly handled, they can lead to consequences ranging from failures to meet product quality specifications to plant shutdowns, incurring substantial economic losses and safety hazards. It is desired that faults be detected and the faulty equipment be quickly identified so that corrective control action can be taken before the fault causes a catastrophic failure. This realization has motivated significant research efforts in the area of fault detection and isolation (FDI).

A typical approach to FDI is to utilize the information embodied in the process model to detect and isolate faults (see, e.g., [2, 3] for reviews). In this approach, residuals are generated as fault indicators by exploiting the analytical redundancy provided by the process model. Faults are detected by checking whether or not the residuals breach their thresholds, and isolated using certain isolation logic. Along this line, the problem of FDI has been studied extensively for linear systems, and existing results include the parity space approach, the observer approach, the fault detection filter approach, and the parameter identification approach (see, e.g., [2]). These results, however, may not remain effective for styrene polymerization process owing to the strong nonlinear characteristics including hysteresis, input multiplicities as well as disjoint bifurcations and isola behavior [4] .

Recently, the problem of FDI has also been studied for nonlinear process systems subject to actuator or sensor faults. In [5], a nonlinear FDI filter is designed to generate residuals that are affected by a particular fault and not affected by disturbances and the rest of the faults. The problem of actuator fault isolation is also studied by exploiting the system structure to generate dedicated residuals (see, e.g., [6]). In addition, adaptive estimation techniques are used to explicitly account for unstructured modeling uncertainty for FDI (see, e.g., [7]). The above results rely on the FDI requirements being satisfied at nominal operating conditions. Recently, the idea of active fault isolation has been proposed to enhance fault isolation by driving the process to a desired operating point where residuals can become uniquely sensitive to faults (see [8]).

Compared to actuator faults, fewer results are available for sensor FDI of nonlinear process systems. In the existing results, this problem has been studied by using nonlinear state observers [9], adaptive estimation techniques [11], and a bank of observers [11,12]. The observer gain in [11], however, is obtained through the first order approximation of the nonlinear dynamics. The enhanced applicability of the state observer (see also [13]) aids in the explicit consideration of process nonlinearity in the FDI design. In summary, while there is a plethora of results that have addressed separately the problems of isolating actuator and sensor faults, there exist limited results on distinguishing between and isolating actuator and sensor faults in a unified framework for nonlinear process systems. As both faults are concerned, an actuator (or sensor) fault could result in unexpected behavior of residuals designed for isolation of only sensor (or actuator) faults because the computation of residuals may use incorrect input (or output) information. Recently , a method that uses banks of high-gain observers was proposed to isolate actuator and sensor faults [14].

In this work, we address the problem of isolating actuator and sensor faults in styrene polymerization process. The key idea is to exploit the analytical redundancy in the styrene process through state observer design based on proposed framework in [14]. To this end, out of possible actuator and sensor faults, we consider subsets of faults, and design state observers and associated residuals such that each is only sensitive to faults in the corresponding subset. With the ability of detecting faults in a subset, they can be isolated by checking whether the corresponding residuals breach their thresholds. Simulation results will be presented to show the efficacy of the proposed framework.

References

1. Gandhi, Rahul, Dave Baldwin, and Prashant Mhaskar. Safe-Parking of a Styrene Polymerization Process. Industrial & Engineering Chemistry Research 48.15 (2009): 7205-7213.

2. P. M. Frank. Fault diagnosis in dynamic systems using analytical and knowledge-based redundancy: A survey and some new results. Automatica, 26:459-474, 1990.

3. J. Bokor and Z. Szab. Fault detection and isolation in nonlinear systems. Annu. Rev. Contr., 33:113-123, 2009.

4. Lemoine-Nava, Roberto, Antonio Flores-Tlacuahuac, and Enrique Saldvar-Guerra. Non-linear bifurcation analysis of the living nitroxide-mediated radical polymerization of styrene in a CSTR. Chemical engineering science 61.2 (2006): 370-387.

5. C. De Persis and A. Isidori. A geometric approach to nonlinear fault detection and isolation. IEEE Trans. Automat. Contr., 46:853-865, 2001.

6. P. Mhaskar, C. McFall, A. Gani, P. D. Christofides, and J. F. Davis. Isolation and handling of actuator faults in nonlinear systems. Automatica, 44:53-62, 2008.

7. X. Zhang, M. M. Polycarpou, and T. Parisini. Fault diagnosis of a class of nonlinear uncertain systems with Lipschitz nonlinearities using adaptive estimation. Automatica, 46:290-299, 2010.

8. Du, Miao, and Prashant Mhaskar. Active fault isolation of nonlinear process systems. AIChE Journal 59.7 (2013): 2435-2453.

9. R. Rajamani and A. Ganguli. Sensor fault diagnostics for a class of non-linear systems using linear matrix inequalities. Int. J. Contr., 77:920-930, 2004.

10. X. Zhang. Sensor bias fault detection and isolation in a class of nonlinear uncertain systems using adaptive estimation. IEEE Trans. Automat. Contr., 56:1220-1226, 2011.

11. M. Mattei, G. Paviglianiti, and V. Scordamaglia. Nonlinear observers with H performance for sensor fault detection and isolation: a linear matrix inequality design procedure. Contr. Eng. Prac., 13:1271-1281, 2005.

12. M. Du and P. Mhaskar. Isolation and handling of sensor faults in nonlinear systems. Automatica, 50:1066-1074,2014.

13. R. Findeisen, L. Imsland, F. Allgwer, and B. A. Foss. Output feedback stabilization of constrained systems with nonlinear predictive control. Int. J. Rob. & Non. Contr., 13:211-227, 2003.

14. Du, Miao, James Scott, and Prashant Mhaskar. Actuator and sensor fault isolation of nonlinear process systems. Chemical Engineering Science 104 (2013): 294-303.

Efficient polytopal enclosures for the reachable set of nonlinear control systems

Stuart M. Harwood and Paul I. Barton

Process monitoring and fault detection are critical activities in maintaining optimal, efficient, and above all, safe process behavior. These tasks are especially important when applied to large industrial processes, where failure can cost human lives, and the complexity of the system necessitates automation. In general, model-based techniques for fault detection can involve a nonlinear dynamic model of the state of the system. Measured system outputs can be compared to the model predictions, and excessive deviation determines a fault [4]. A critical part of this procedure is estimating the state of the system. Further, realistic state estimation should take into account uncertain parameters in the model, uncertainties in the measured inputs, and typical process disturbances. The result is then a range of process behaviors; a fault is flagged if the measured outputs deviate outside this predicted range [1].

Estimation of the range of states of the dynamic system that result from the uncertainties, otherwise known as the reachable set, is a challenging task. To make use of the nonlinear dynamic models in real time requires that efficient techniques for the estimation of the reachable set be used. The quickest techniques, based on comparison theorems, yield upper and lower bounds on each state, or an interval enclosure of the reachable set [3, 2]. However, overestimation may occur due to the wrapping of the reachable set within an interval.

This work presents a method for the construction of polytopal enclosures of the reachable set of uncertain nonlinear control systems. It is inspired by the comparison theorem results. Consequently, it has the advantage that its enclosures can be obtained very quickly, through the solution of an initial value problem in ordinary differential equations that is related to the original dynamic system. Therefore established numerical integration methods and efficient codes implementing those methods can be used. Meanwhile, the use of polytopes permits much more flexibility in obtaining a tight enclosure, and indeed significant improvements will be presented.

References

[1] Youdong Lin and Mark A. Stadtherr. Fault Detection in Nonlinear Continuous-Time Systems with Uncertain Parameters. *AIChE Journal*, 54(9):2335–2345, 2008.

[2] Joseph K. Scott and Paul I. Barton. Bounds on the Reachable Sets of Nonlinear Control Systems. *Automatica*, 49(1):93–100, 2013.

[3] Adam B. Singer and Paul I. Barton. Bounding the Solutions of Parameter Dependent Nonlinear Ordinary Differential Equations. *SIAM Journal on Scientific Computing*, 27(6):2167–2182, January 2006.

[4] Venkat Venkatasubramanian, Raghunathan Rengaswamy, Kewen Yin, and Surya N. Kavuri. A review of process fault detection and diagnosis: Part I: Quantitative model-based methods. *Computers & Chemical Engineering*, 27(3):293 – 311, 2003.

377398 Data-Based Fault Identification and Accommodation in Sampled-Data Controlled Particulate Processes

Abstract:

Fault-tolerant control of particulate processes is a fundamental problem encountered in a wide range of industries, including the agricultural, chemical, food, mineral, and pharmaceutical industries. This problem is significant given that malfunctions in the control system or process equipment can negatively impact the particle size distribution of interest and thus harm the desired end product quality. Major bottlenecks in the design of model-based fault-tolerant control systems for particulate processes include the infinite-dimensional nature of the process model as well as the complex and uncertain dynamics of particulate processes. An effort to address these problems was initiated in [1]-[2] where a methodology for the detection, isolation and handling of control actuator faults in particulate processes was developed based on low-order models that capture the dominant process dynamics. A number of subsequent studies were carried out to account for various implementation issues that arise in the design of fault-tolerant control systems, including the discrete and delayed availability of output measurements [3], and the presence of multi-rate sampling and sensor faults [4].

In both studies, fault detection was achieved by designing a fault-free time-varying alarm threshold off-line and later comparing this with values of the residual for the entire duration of the process. However, the scheme for fault detection was stability-based, leaving ``small" malfunctions that do not lead to instability to go undetected. In designing this threshold, there are competing design requirements that need to be considered. For example, there is the need to tighten the threshold for timely fault detection; however, an extremely tight bound may result in false alarms. It was also assumed in those studies that a fault identification scheme was already in place which was able to determine the nature and location of the fault. This information was utilized in determining the appropriate response for fault accommodation. After each fault, a new alarm threshold for fault detection had to be calculated and used since the closed-loop system will have different stability properties after each fault accommodation event.

In this work, our aim is to address some of these limitations by integrating within the fault-tolerant control methodology a fault identification mechanism that allows for immediate detection of faults and/or malfunctions while determining its location and magnitude. One key element of the proposed scheme is that it can still be used for fault identification even after fault accommodation. This allows for timely fault detection in the event of consecutive system faults. This is an advantage over the previous detection schemes where a new alarm threshold had to be calculated after every fault accommodation event. This recalculation may result in delays in the fault detection preceding a fault. Timely or even instantaneous fault identification is important even for faults that do not immediately result in an unstable behavior since these malfunctions may later on result in poor process performance or even instability. In addition, rapid detection will also allow for systematic scheduling of plant maintenance and equipment repair or replacement.

Motivated by the above considerations, we develop in this study a model-based framework for the integrated detection, identification and accommodation of actuator faults in sampled-data particulate processes described by complex population balance equations. Initially, model reduction techniques are applied to derive a finite-dimensional model to be used in designing a stabilizing sample-and-hold state feedback controller. This controller uses past values of the state measurements in between sampling times. The controller then utilizes updated state measurements when sensor readings are received at discrete times. Through a stability analysis, an explicit characterization of the behavior of the closed-loop system is obtained as a function of the controller design parameters, the update time, and the actuator health. This characterization is then used as a metric in determining the appropriate post-fault response once a fault is detected. Fault identification is carried out by solving a data-based moving-horizon optimization problem. Data from the fault identification scheme are used in the fault accommodation which involves modifying the controller design parameters based on the stability plots generated from the stability analysis. Finally, the proposed fault-tolerant control framework is applied to a simulated model of a non-isothermal continuous crystallizer and is shown to effectively handle simultaneous and consecutive faults.

References:

[1] N. H. El-Farra and A. Giridhar, "Detection and management of actuator faults in controlled particulate processes using population balance models," *Chemical Engineering Science*, **63**: 1185-204, 2008.

[2] A. Giridhar and N. H. El-Farra, "A unified framework for detection, isolation and compensation of actuator faults in uncertain particulate processes," *Chemical Engineering Science*, **64**: 2963-2977, 2009.

[3] T. Napasindayao and N. H. El-Farra, "Fault detection and accommodation in particulate processes with sampled and delayed measurements,"
Industrial & Engineering Chemistry Research, **52**:12490-12499, 2013.

[4] T. Napasindayao and N. H. El-Farra, "Sensor fault accommodation strategies in multi-rate sampled-data control of particulate processes," In *Proceedings of the 10th IFAC Symposium on Dynamics and Control of Process Systems*, Mumbai, India, pp. 379-384, 2013.

358604 Design of Sampled Data Nonlinear Observers for Processes Monitoring

Abstract:

State observers provide a powerful tool for model-based process monitoring. On-line estimation of unmeasured states provides critical information for the early detection of operational problems that could affect safety or product quality.

Chemical processes typically evolve continuously in time, but process measurements are discrete in time and the implementation of the observer is realized in discrete time using the sampled data. Therefore, time discretization issues naturally arise in specific applications. If the sampling period of all measurements is relatively small, one possible approach is to design the observer in continuous time and discretize the observer equations at the implementation stage. Another possibility is to discretize the process dynamics model at the beginning and perform the observer design in discrete time. Both approaches are reasonable and there no clear superiority of one over the other in the case of linear systems.

For nonlinear processes, time discretization of a continuous-time observer, or a priori discretization of the process dynamics with time step equal to the sampling period, generally involves numerical errors, whose effect on the accuracy of the state estimates needs to be evaluated.

In the presence of infrequent measurements with large sampling periods, special care is needed in the design of the observer, to account for system dynamics in between sampling instants. In particular, inter-sample behaviour of the system states should be predicted and accounted for in the observer calculations ([1]).

The present paper will explore all three approaches in the design of sampled data observers for nonlinear processes. In order to have a common basis of comparison, the same observer synthesis point of view will be applied: nonlinear Luenberger observer based on exact error linearization with eigenvalue assignment ([2], [3]).

After a discussion on the general theoretical issues involved, the paper will proceed with a detailed evaluation and comparison in a batch reactor case study. In particular, a chemical reaction system following the Lotka-Volterra mechanism will be considered. One of the concentrations will be assumed to be measured and the other will need to be estimated on line. This system is chosen for two reasons: First, it is an oscillatory system, so that discretization errors have a pronounced effect on observer performance. Second, it is a system where the synthesis of a nonlinear Luenberger observer based on exact error linearization and eigenvalue assignment can be performed analytically.

The results for the Lotka-Volterra reaction system have shown that both the discretized continuous-time observer and the discrete observer design based on a priori discretization, lead to accurate state estimates under fast sampling. However, estimation accuracy decreases quickly with increasing size of the sampling period. Moreover, discrete observer design based on a priori discretization suffers from poor robustness to errors in the sampling period. The ultimate solution for the case of large sampling periods and in the presence of possible errors in the sampling schedule is to use the continuous-time observer in conjunction with an inter-sample output predictor, to predict the evolution of the output during the time period in between two consecutive measurements ([1]).

References:

[1] I. Karafyllis and C. Kravaris, "From continuous-time design to sampled-data design of observers", IEEE Transactions on Automatic Control, Vol. AC-54, pp.2169-2174, 2009.

[2] C. Kravaris, J. Hahn and Y. Chu, "Advances and selected recent developments in state and parameter estimation", Computers & Chemical Engineering, Vol. 51, pp.111-123, 2013.

[3] N. Kazantzis and C. Kravaris, "Nonlinear observer design using Lyapunov's auxiliary theorem", Systems & Control Letters, Vol. 34, pp.241-247, 1998.

364276 The Virtual Product-Process Design Lab

Abstract:

The Virtual Product-Process Design Laboratory (VPPD-Lab) has been developed as a computer aided tool for product-process development. Rather than design chemicals based products through an experimental-based trial and error approach, the powerful methodologies and tools in VPPD-Lab help to replace some of the experimental steps by performing virtual experiments using model-based techniques. VPPD-Lab offers a quick search of active ingredients, solvents, and additives to determine a set of promising formulations that match the desired (target) product properties (functions). This is followed by experimental-based validation, and evaluation of the most promising candidates for eventual selection as the final product, reducing thereby, valuable time and resources at the preliminary design stage.

The earlier version of VPPD-Lab consisted of templates for the design of microcapsules, pesticide uptake, and homogeneous liquid formulated products such as paint formulation, sunscreen lotion, and insect repellent lotion. The algorithms in VPPD-Lab can perform ingredients selection and testing, solubility tests, and property measurements using its available library of property models and chemicals databases. VPPD-Lab is also able to directly access ICAS (Integrated Computer Aided System) as a support tool to (i) consult several other chemicals databases, (ii) predict property values which are missing in the databases predicted through the ICAS property prediction tool (ProPred), and (iii) perform process design using process models available in ICAS MOT (Modeling testbed) to design the processes. However, the main limitations of property models available in VPPD-Lab is that the application range of the models is not large enough to include a wider spectrum of chemicals based products, such as, emulsified formations, blended synthetic fuels and lubricants, and many more. Covering a wider spectrum of products requires the use and management of a wide range of property models, calculation procedures and a very large amount of data. A systematic framework is needed to manage the data and models are necessary as well as the modules for a better communication between VPPD-Lab and support tools.

This work proposes a systematic framework to improve and extend the application range of VPPD-Lab. New property prediction models have been developed and integrated to the different steps of the product design algorithms. The chemicals databases have been extended and integrated, and property models needed for each type of products are managed through a knowledge base representation system (ontology) enabling a flexible and efficient communication between the product-design templates and supporting tools (databases, models, and connected software). The new version of VPPD-Lab has included new templates for the design of blended and emulsified products with the new user interface of a generic workflow to design products and processes. Therefore, VPPD-Lab is now generic, user friendly, and widely applicable with options for a wide range of chemicals based liquid products.

Two case studies dealing with tailor-made design of jet-fuels and an emulsified formulation product is highlighted to illustrate the new features and options available in VPPD-Lab.

368170 Oscillatory Amplitudes Regulation Method in 1,3-PD Production By Klebsiella Pneumoniae

Abstract:

Oscillatory amplitudes regulation method in 1,3-PD production by *Klebsiella pneumoniae*

Hangzhou Wang[a], Tong Qiu[b], Jinsong Zhao[b],Bingzhen Chen[b*]

[a]Memorial University, St. John's, N.L., Canada

[b]Tsinghua University, Beijing, China

1,3-propanediol (1,3-PD) is an important chemical raw material, and it can be used to produce polyesters, polyether and polyurethanes with excellent properties. Microbial fermentation production has many advantages [1, 2] over chemical synthesis of 1,3-PD. But oscillatory behavior was observed in both experiments and mathematical simulation during the continuous fermentation of *Klebsiella pneumoniae*, in which biomass, product and substrate oscillations exist under certain operating conditions. It was reported that these oscillations were caused by the existence of Hopf bifurcations in the system[3]. To design a more stable process and achieve high product quality, these operating condition points corresponding to the Hopf singularity ones should be avoided. And in some particular situations, the singularity operation point is located exactly at the maximum concentration point; the oscillation is unavoidable or is prone to generate. In these situations, it is a better way to regulate the oscillatory phenomenon, that is, to reduce the oscillatory amplitude so as to increase process stability to some extent. In this paper, a regulation method of oscillatory amplitudes is proposed so that the oscillatory amplitude can be attenuated when the oscillation itself is unavoidable. The detailed steps for this regulation method of oscillatory amplitudes as follows:

1. Calculate all steady state solutions for the dynamic system[4].
2. Identify all Hopf singularity points in the steady state curve[5, 6].
3. Simulate the oscillatory phenomenon and find the amplitudes under different operation conditions.
4. Establish the relationship model between the amplitudes and operation conditions(such as feed dilution rate, initial concentration of substrate).
5. Use the relationship model to regulate oscillatory amplitudes in process design.

According to the proposed method, with the results of oscillation simulation around the singularity operation condition points, the product concentration oscillatory amplitudes under different operation conditions can be obtained. These calculated amplitude data can be regressed to a smooth amplitude curve. The curve fits the varying tendency of the process, and can represent the relationship between amplitude and operation conditions. As a result, the characteristics of the oscillatory behavior in dynamic process can be studied quantitatively The oscillation behavior can be predicted and regulated so that a process with better stability can be designed.

Keywords: 1,3-PD, Hopf Singularity points, Oscillatory Amplitudes, Process Design and Regulations

References

1. Wong, C.-L.; Huang, C.-C.; Chen, W.-M.; Chang, J.-S., Converting crude glycerol to 1,3-propandiol using resting and immobilized Klebsiella sp. HE-2 cells. *Biochemical Engineering Journal* **2011,**58–59, (0), 177-183.

2. Menzel, K.; Zeng, A. P.; Deckwer, W. D., High concentration and productivity of 1,3-propanediol from continuous fermentation of glycerol by Klebsiella pneumoniae. *Enzyme and Microbial Technology* **1997,**20, (2), 82-86.

3. Sridhar, L. N., Elimination of Oscillations in Fermentation Processes. *AIChE Journal* **2011,**57, (9), 2397-2405.

4. Wang, H.; Chen, B.; He, X.; Zhao, J.; Qiu, T., Numerical Analysis Tool for Obtaining Steady-State Solutions and Analyzing Their Stability Characteristics for Nonlinear Dynamic Systems. *Journal of Chemical Engineering of Japan* **2010,**43, (4), 394-400.

5. Wang, H.; Zhang, N.; Qiu, T.; Zhao, J.; He, X.; Chen, B., Analysis of Hopf Points for a Zymomonas mobilis Continuous Fermentation Process Producing Ethanol. *Industrial & Engineering Chemistry Research* **2012,**52, (4), 1645-1655.

6. Wang, H.; Zhang, N.; Qiu, T.; Zhao, J.; He, X.; Chen, B., A process design framework for considering the stability of steady state operating points and Hopf singularity points in chemical processes. *Chemical Engineering Science* **2013,**99, (0), 252-264.

---

* Corresponding author. Tel.: +86 10 62781499;fax: +86 10 62770304.

E-mail address: dcecbz@tsinghua.edu.cn (Bingzhen Chen)

388102 Molecular Design of Optimum CO2 Capture Solvents: From Conceptual Screening to SAFT-Based Validation

Abstract:

The wide adoption of chemical absorption/desorption systems for post-combustion CO_2 capture in industry is currently challenged by the high energy penalty in solvent regeneration and the environmental impacts associated with solvents and their derivatives. Intense research efforts reported in recent years are predominantly based on lab and pilot-scale experiments to select solvents which may potentially improve the overall performance of absorption/desorption CO_2 capture. However, this is very challenging due to a) the highly non-ideal solvent-CO_2-water chemical interactions, b) the countless combinations of potential capture solvent and blend candidates and c) the need for combined consideration of numerous thermodynamic, kinetic and sustainability properties as performance criteria prior to selecting solvents with optimum capture features. Computer-aided molecular design methods (CAMD) can help address these challenges and have been successful in supporting the synthesis of molecules with desired physical, chemical and environmental properties in non-CO_2 separations [1]. Despite extensive developments in CAMD methods, few recent works have reported their utilization in the design of CO_2 capture solvents or mixtures for chemical and physical absorption using the Statistical Associating Fluid Theory with potentials of Variable Range (SAFT-VR) [2, 3] and for physical absorption using the Perturbed Chain Polar Statistical Associating Fluid Theory (PCP-SAFT) [4]. While these approaches enable an accurate and reliable determination of solvent-process vapour-liquid equilibria, the set of few solvents screened to date will be further expanded as research efforts extend the rigorous predictive capabilities of SAFT-based models towards additional molecular structures. On the other hand, few CAMD approaches have also been reported [5, 6, 7] that use less accurate, but well-established group contribution methods to screen a much wider set of molecular structures by approximating as solvent performance criteria mainly properties relevant to thermodynamics and toxicity.

In our current work, we propose the use of an optimization-based CAMD method to select post-combustion CO_2 capture solvents of optimum performance in molecular and mixture properties associated with thermodynamics, kinetics and sustainability. The problem is first approached in a fast screening stage where solvent structures are evaluated based on the simultaneous consideration of important pure component properties. For the first time numerous properties are considered as performance criteria reflecting solvent characteristics based on thermodynamic (e.g., vapour pressure, CO_2 solubility etc.), kinetic (solvent basicity, steric hindrance etc.) and sustainability (e.g., health and safety hazard, life cycle assessment etc.) behaviour. The simultaneous consideration of properties selected to capture the molecular chemistry effects on the absorption/desorption process compensates for the initial utilization of simpler models and ensures the selection of fewer but more effective solvents. A few highly-performing solvents are further evaluated using the SAFT-VR and SAFT-γ equations of state to predict accurately the very non-ideal solvent-water-CO_2 mixture vapour-liquid equilibrium behaviour. Different functionalities of the employed CAMD method are used both to design optimum, novel molecular structures and to screen a dataset of commercially available amine solvents suitable for CO_2 capture. The obtained results reveal interesting structure-property trade-offs and point to commercial molecules, which have very recently been considered or have yet to be employed as capture solvents.

Acknowledgements

Funding from the European Commission under grant FP7-ENERGY-2011-1-282789-CAPSOL is gratefully acknowledged.

Cited References

[1] Adjiman, C.S., Galindo, A., 2010, Process Systems Engineering: Volume 6: Molecular Systems Engineering, Pistikopoulos, E.N., Georgiadis, M.C., Dua, V. (Eds).

[2] Mac Dowell N., Pereira F.E., Llovell F., Blas F.J., Adjiman C.S., Jackson G., Galindo A., 2011, Transferable SAFT-VR models for the calculation of the fluid phase equilibria in reactive mixtures of carbon dioxide, water, and n-alkylamines in the context of carbon capture, Journal of Physical Chemistry B, 115, 8155-8168.

[3] Pereira, F.E., Keskes, E., Galindo, A., Jackson, G., Adjiman, C.S., 2011, Integrated solvent and process design using a SAFT-VR thermodynamic description: High-pressure separation of carbon dioxide and methane, Computers and Chemical Engineering, 35 (3), 474-491.

[4] Bardow, A., Steur, K., Gross, J., 2010. Continuous-molecular targeting for integrated solvent and process design. Ind. Eng. Chem. Res. 49, 2834–2840.

[5] Juan Salazar, Urmila Diwekar, Kevin Joback, Adam H. Berger, Abhoyjit S. Bhown, Solvent Selection for Post-Combustion CO2 Capture, Energy Procedia, Volume 37, 2013, Pages 257-264

[6] Bommareddy, S., Chemmangattuvalappil, N.G., Solvason, C.C., Eden, M.R. Simultaneous solution of process and molecular design problems using an algebraic approach (2010) Computers and Chemical Engineering, 34 (9), pp. 1481-1486.

[7] Chemmangattuvalappil, N.G., Eden, M.R. A novel methodology for property-based molecular design using multiple topological indices (2013) Industrial and Engineering Chemistry Research, 52 (22), pp. 7090-7103.

389735 Structure Generation of Optimal Reactants and Products Using Signature Descriptors

Abstract:

Selection of optimal chemicals and maximization of process efficiency has successfully been carried out by utilizing process design techniques that integrate product design considerations in a reverse problem formulation. The combined process-product design problem is solved for design targets in terms of properties that are later used to design optimal products. However, most of these techniques have been implemented in non-reactive systems. Different efforts to address the paucity of techniques for reactive systems have been restricted to single unknown reactants. Also, utilization of linear property models and solution schemes that do not have a provision for treating property constraints are among some of the other restrictions. In this work an algorithm has been developed for the design of optimal reactants and products that is not restricted by the number of unknown reactants and products. Nonlinear property models can be incorporated in the algorithm and the solution scheme considers property constraints. The products of the reaction(s) are constrained by property bounds arising from the process to ensure the optimality of the process. An optimization problem is formulated for each of the products in order to select the best chemical structures that satisfy the property bounds and have the best dominant property. The recently developed molecular signature descriptors have been utilized to perform molecular design. The signature descriptor is a systematic codification system over an alphabet of atoms describing the extended valence of the atoms of a molecule. The signature of a molecule can be expressed as a linear combination of its atomic signatures. Topological indices of molecules that form various property models can be linearly represented in terms of molecular signature descriptors [1]. Different quantitative structure activity/property relationships (QSARs/QSPRs) have been applied in the developed algorithm to estimate properties from molecular structures. QSARs/QSPRs can be expressed in terms of different topological indices. Thus the target properties can be expressed in terms of signatures. In the developed method, the problem is formulated in terms of the number of occurrences of each signature, i.e. the number of times each signature appears in the molecular structure. The dominant property to be optimized and the property constraints can thus be ultimately expressed in terms of the number of occurrences of the signatures in the product molecule. After identifying the occurrence values from each of the formulated problems, the molecular structure of products is generated by using an algorithm developed by Chemmangattuvalappil et al. [2]. Next, using the chemical equation to track the migration of groups from reactants to products, the structure of the reactants is determined by considering the appropriate addition and/or removal and/or rearrangement of groups. Thus, candidate reactant structures can be generated irrespective of the type and number of reactions as the introduced algorithm focuses on the design of products first and then uses the chemical equations to identify the reactant structures. Graph theory principles have been utilized to track the signatures and to avoid the generation of infeasible molecular structures. To solve the resulting nonlinear integer optimization problem(s), MI-LXPM, a real coded genetic algorithm (RCGA) developed by Deep et al. [3] has been incorporated in the solution scheme. RCGAs are general purpose population based search techniques that mimic the principles of natural selection and natural genetics laid down by Charles Darwin and encode the solution (represented as chromosomes) in real numbers. Property functions that may be discontinuous, non-differentiable and/or highly non-linear can be handled by genetic algorithms (GAs). Real coding is generally more efficent than binary and gray coding and it helps avoid the "Hamming-Cliff" problem. MI-LXPM is a robust and efficient GA for solving nonlinear integer programming problems. It utilizes the recently developed and self adaptive Laplace crossover (LX) operator and a tunable power mutation (PM) operator [4]. A special truncation procedure is utilized to handle integer restrictions on variables instead of rounding off carried out in previous algorithms. One may not begin with an array of feasible points since a parameter free constraint handling procedure dependent on tournament selection is used. Thus search of feasible points is not a constraining factor. This contribution will illustrate the developed methods and highlight their use through a case study.

[1] J.L. Faulon, D.P. Visco Jr., R.S. Pophale, 2003, Journal of Chemical Information and Computer Sciences, 43, 707-720

[2] N.G. Chemmangattuvalappil, M.R. Eden, 2013, Industrial & Engineering Chemistry Research, 52, 7090-7103

[3] K. Deep, K.P. Singh, M.L. Kansal, C. Mohan, 2009, Applied Mathematics and Computation, 212, 505-518

[4] K. Deep, M. Thakur, 2007, Applied Mathematics and Computing, 193, 211-230

390512 An Application of Computer-Aided Molecular Design (CAMD) Using the Signature Molecular Descriptor: Designing New Water-Reducing Admixtures for Cement through Cement Paste Rheology Evaluation

Abstract:

Concrete is a material that forms the literal foundation of our modern society. Concrete is no longer simply cement, water and aggregates, but rather, it also typically contains a complex array of chemical admixtures that modify its many properties to produce a cost effective, durable, formable and mechanically sound material. Water-reducing admixtures (WRAs) have been developed to reduce the required water content of concrete for a given workability or increase the workability for a given water content.

The development of new chemical admixtures for concrete is normally an experimental approach wherein one uses prior knowledge to suggest potential effective compounds. This approach is time-consuming, incremental and typically expensive. To accelerate the process, we propose a computer aided molecular design (CAMD) approch using the Signature molecular descriptor as a powerful alternative. CAMD is the application of computer-implemented algorithms that are utilized to identify and design new molecules with optimally predicted properties such that they can be tested and evaluated for efficacy. In this work, we are interested in the flow properties of cement paste, described by its rheological properties. Accordingly, a parallel-plate rheometer was used to measure the effect of eighteen low molecular weight organic chemicals belonging to different classes (sugars, sugar alcohols, and sugar acids) on rheology by directly obtaining yield stress and viscosity data. Workability, measured by a simple industry standard flow (mini-slump cone) test, was also obtained. Setting time, the time required for cement paste to stiffen, was measured since a higher workability is normally accompanied by an extended setting time, which is an undesired side-effect in some applications. After creating and refining four different quantitative structure–property relationship (QSPR) models for our four properties of interest (yield stress, plastic viscosity, workability, and setting time), a structure enumeration algorithm wass employed to generate structures outside of the original training set that have optimally predicted properties such as high workability, low yield stress and viscosity, and short setting time. Over 670 structures were newly generated and screened against a non-commercial chemical database, ZINC, to identify commercially available compounds. As a result, 6 compounds were purchased and tested for their effect on the flowability (i.e. workability, yield stress, and viscosity) and the setting time of cement paste for evaluation. Experimental results validated the previously developed QSPR models. Results confirm that the CAMD strategy is a powerful tool for generating novel and non-intuitive structures with the ability to increase flowability of cement paste with controlled setting time.

Graph Based Database Development for Gasoil Reactive Systems

AIChE Abstract
Charles Solvason and Jorge Martinis

Sessions:

- 10A03 Product and Molecular Design

In the petroleum refining industry, the accurate design and optimization of hydrocarbon conversion processes for gasoil feedstocks demand a growing effort in developing rigorous and reliable kinetic models (Van Geem *et al.*, 2002). Existing techniques rely on the approximation of gasoil stream compositions using a set of pseudo-components called "oil cuts" characterized by a set of lumped kinetic models. Although the lumped models are easy to construct and implement, they lack molecular information which makes them feedstock dependent and of limited use in process simulation and design. Developments by Neurock *et al.* (1994), Trauth *et al.* (1994), and Stark *et al.* (1995) address this limitation using molecular reconstruction techniques to approximate the individual components of the gasoil streams via stochastic optimization methods coupled with Monte Carlo kinetic models. The application of stochastic methods is, however, limited by the computational intensity imposed by the large sampling frequency needed to achieve statistical significance (Van Geem 2007). For specific classes of gasoils, Hudebine and Verstraete (2011) utilize a hybrid stochastic method that maximizes an entropy function with Lagrangian parameters associated with analytical constraints in order to reduce the computational burden. However, the method strongly depends on the initial set of molecules which must be chosen and built in relation to the type of gasoil studied.

An alternative, deterministic approach to molecular reconstruction has been developed by Bryan Research & Engineering (BRE) for the process simulator, ProMax®. The approach replaces the stochastic methods with a parallelized SQL call to a molecular database that contains all chemical constituents important to gasoil products and streams. The method is dependent on two competing factors, the completeness of the database and the size of the database. Unfortunately, most gasoil reconstructions require representations of chemical constituents of up to 40 atoms in a hydrogen suppressed, molecular graph. Enumerating all chemical constituents up to this size is cumbersome for even state-of-the-art graph generation algorithms and would result in a database too large to be efficiently searched (Peironcely *et al.*, 2012).

To address these limitations, a hybrid method is proposed that limits the size of the graph database using a set of constraints provided by molecular group and point group theory. In the first step of the method, a modified set of molecular functional groups described by Joback and Reid (1987) are used as building blocks to construct a complete set of vertex subgraphs that contain all double bonds, aromatic ring structures, and non-aromatic ring structures necessary to model gasoil systems. Next, these vertices are perturbed using point group symmetry operations to create a set of isomeric vertex graphs. Of these graphs, constitutional isomeric vertices are selected and separated into terminal and internal vertex sets. The internal vertices are constructed using graph generation algorithms and the terminal vertices are added at the final step to create complete molecular graphs. By utilizing coarse-grained sub-graphs of vertices, specific isomer classes common to gasoil systems can be created, without overloading the database, thus ensuring the capability of the SQL call to search the database in a timely manner. In addition, the database generation routing is constructed such that it is expandable by toggling various stereoisomer point groups and vertex rotations. The molecular library will be utilized to illustrate the database generation method as it is applied to a diesel hydrotreater case study.

Van Geem, K.M., Hudebine D., Reyniers M.F., Wahl F., Verstraete J.J., Marin, G.B. (2007): "Molecular reconstruction of naphtha steam cracking feedstocks based on commercial indices." *Computers & Chemical Engineering,* 31, pp. 1020-1034.

Hudebine D. and Verstraete J.J. (2011): "Reconstruction of Petroleum Feedstocks by Entropy Maximization. Application to FCC Gasolines." *Chemical Reaction Modelling of Refining Processes,* 66(3), pp. 461-477.

Joback K.G. and Reid R.C. (1987): "Estimation of pure-component properties from group contributions." *Chemical Engineering Communications*, 57(16), pp. 233-243.

Peironcely, J.E., Rojas-Chertó M., Fichera D., Reijmers T., Coulier L., Faulon, J. and Hankemeier T. (2012): "OMG: Open Molecule Generator." *Journal of Cheminformatics* 4(21), pp. 1-13.

Neurock, M., Nigam, A., Trauth, D., and Klein, M.T. (1994): "Molecular representation of complex hydrocarbon feedstocks through efficient characterization and stochastic algorithms." *Chemical Engineering Science*, 49, pp. 4153-4177.

Stark S.M., Neurock M., Klein M.T. (1995): "Comparison of MIMD and SIMD Strategies for Monte Carlo Modeling of kinetically coupled reactions." *Computers& Chemical Engineering*, 19(6), pp. 719 -742.

Trauth D.M., Stark S.M., Petti T.F., Neurock M., and Klein, M.T. (1994): "Representation of the molecular structure of petroleum resid through characterization and Monte Carlo modeling. *Energy & Fuels*, 8(3), pp. 576-580.

386157 Determining Optimal Groups for Group Contribution Methods

Abstract:

Given the sheer number of chemical compounds in use today, determining a new or untested compound's properties is a problem beyond current laboratory resources. This necessitates the need for quantitative structure activity relationships (QSARs), which are models used to predict certain properties based solely on the chemical structure of a compound. Often, QSARs are in the form of group contribution methods, which assume that a property of a compound can be modelled based on the number of occurrences of molecular subunits, or "groups". Many group contribution methods are derived from regression over large datasets, but many of these regression problems require first specifying the identity of these groups, a requirement which introduces some bias in the model building. While many group contribution methods have seen a substantial degree of success [1], many new methods are still based to some degree on the original UNIFAC groups [2]. Some recent approaches have applied best subset selection methods to alleviate this bias [3,4], but these methods still must specify a larger set from which to select the best components. We propose a method that aims to investigate the entire space of subgroups and select the set of groups which provides the best predictive accuracy from a given dataset. This is essentially optimal group selection for a given model. This method has promise in improving group contribution methods for properties that have historically been difficult to predict and design for as well as determining important substructures for more complicated properties like toxicity or bioactivity.

[1] Group-contribution based estimation of pure component properties. Jorge Marrero and Rafiqul Gani.Fluid Phase Equilibria. 2001. 183, 183-208.
[2] Group-contribution estimation of activity coefficients in nonideal liquid mixtures. Aage Fredenslund, Russel L. Jones, and John M. Prausnitz. AIChE Journal. 1975. 21(6), 1086-99.
[3] Choosing Feature Selection and Learning Algorithms in QSAR. Martin Eklund, Ulf Norinder, Scott Boyer, and Lars Carlsson. Journal of Chemical Information and Modeling 2014 54 (3), 837-843
[4] Towards Optimal Descriptor Subset Selection with Support Vector Machines in Classification and Regression. Holger Fröhlich, Jörg K. Wegner and Andreas Zell. QSAR & Combinatorial Science. 2004. 23(5), 311-318

Optimal Constrained State Estimation of Parabolic PDE Systems

Xiaodong Xu
Supervisors: Stevan Dubljevic and Biao Huang

19 November 2014

2014 AIChE Conference

Department of Chemical and Materials Engineering
University of Alberta

Outline

- System Description
- System Decomposition
- Optimal Constrained State Estimation
- Estimation Result

System Description

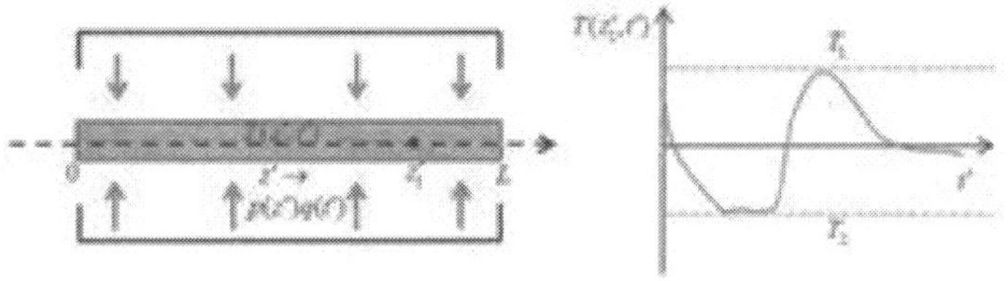

Figure: A long thin rod being heated in a multizone furnace

$$\rho C_p \frac{\partial T'}{\partial t'} = k\frac{\partial^2 T'}{\partial z'^2} + p(z')q(t')$$

$$\frac{\partial T'}{\partial z'} = 0, \; z' = 0, L$$

System Description

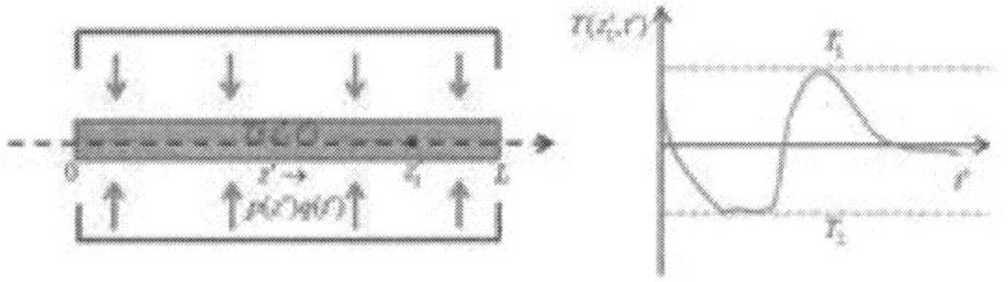

Figure: A long thin rod being heated in a multizone furnace

$$\rho C_p \frac{\partial T'}{\partial t'} = k\frac{\partial^2 T'}{\partial z'^2} + p(z')q(t')$$

$$\frac{\partial T'}{\partial z'} = 0, \; z' = 0, L$$

$$t = \frac{t'k}{\rho C_p L^2}, \; z = \frac{z'}{L}, \; B(z)u(t) = \frac{p(z')q(t')L^2}{kT_0}, \; x(z,t) = \frac{T(z',t')}{T_0}$$

$$\boxed{\begin{array}{l} \frac{\partial \bar{x}(z,t)}{\partial t} = \frac{\partial^2 \bar{x}(z,t)}{\partial z^2} + B(z)u(t), z \in [0,1] \\ \frac{\partial \bar{x}(0,t)}{\partial z} = 0, \frac{\partial \bar{x}(1,t)}{\partial z} = 0 \end{array}}$$

System Description

- Describing a wide of family of problems including diffusion-convection-reaction processes
- Involve spatial differential operator satisfying the spectrum decomposition assumption

$$\frac{\partial \bar{x}(z,t)}{\partial t} = \frac{\partial^2 \bar{x}(z,t)}{\partial z^2} + B(z)u(t) + \xi(t), \quad \bar{x}(z,0) = x_0(z)$$

$$\frac{\partial \bar{x}(0,t)}{\partial z} = 0, \quad \frac{\partial \bar{x}(1,t)}{\partial z} = 0$$

$$y(t) = \bar{x}(z_1,t) + \eta(t)$$

$$\text{State Constraint}: \quad \mathcal{X}_{\min} \leq \bar{x}(z_1,t) \leq \mathcal{X}_{\max}$$

$$\text{Assumption}: \quad \|\xi(\mathrm{t})\| \leq \mathrm{u_n}$$

System Description

- Defining the state function $x(t)$ on the state space $L_2(0,1)$ as $x(t) = \bar{x}(z,t)$.

$$\dot{x}(t) = \mathcal{A}x(t) + \mathcal{B}u(t) + \mathcal{G}\xi(t), \quad x(0) = x_0 \in D(\mathcal{A})$$

$$y(t) = \mathcal{C}x(t) + \eta(t) \ t > 0$$

$\mathcal{A} = \frac{\partial^2}{\partial z^2}$ with $D(\mathcal{A}) = \left\{h(z) \in X, h(z)\frac{dh(z)}{dz} \text{ are a.c, and } \frac{dh(0)}{dz} = \frac{dh(1)}{dz} = 0\right\}$,
$\mathcal{B} = B(z)$,
$\mathcal{G} = 1$,
$\mathcal{C}h(z) = h(z_1)$.

System Description

◇ Defining the state function $x(t)$ on the state space $L_2(0,1)$ as $x(t) = \bar{x}(z,t)$.

$$\dot{x}(t) = \mathcal{A}x(t) + \mathcal{B}u(t) + \mathcal{G}\xi(t),\ x(0) = x_0 \in D(\mathcal{A})$$

$$y(t) = \mathcal{C}x(t) + \eta(t)\ t > 0$$

$\mathcal{A} = \frac{\partial^2}{\partial z^2}$ with $D(\mathcal{A}) = \left\{ h(z) \in X, h(z)\frac{dh(z)}{dz} \text{ are a.c, and } \frac{dh(0)}{dz} = \frac{dh(1)}{dz} = 0 \right\}$,
$\mathcal{B} = B(z)$,
$\mathcal{G} = 1$,
$\mathcal{C}h(z) = h(z_1)$.
The spectrum of $\mathcal{A}$ can be obtained by solving the following eigenvalue problem:

$$\mathcal{A}\phi_n = \frac{d^2\phi_n}{dz^2} = \lambda_n\phi_n$$

with

$$\frac{d\phi_n(0)}{dz} = \frac{d\phi_n(1)}{dz} = 0$$

$$\lambda_n = -(n\pi)^2, n = 0, 1, 2, \cdots$$

$$\phi_n(z) = \begin{cases} 1, n = 0 \\ \sqrt{2}\cos(n\pi z), n = 1, 2, 3, \cdots \end{cases}$$

System Decomposition

◇ The spectrum of the operator $\mathcal{A}$, i.e. $\sigma(\mathcal{A}) = \{\lambda_0, \lambda_1, \cdots\}$ satisfy

$$\lambda_0 \geq \lambda_1 \geq \lambda_2 \geq \cdots \geq \lambda_n \geq \cdots, \quad \lim_{n\to\infty} \lambda_n = -\infty$$

Given a positive number γ, it is possible to choose m that, for all $n > m$,

$$\lambda_n \leq -\gamma$$

therefore, $\sigma(\mathcal{A})$ is the union of two parts: $\sigma(\mathcal{A}) = \sigma_1(\mathcal{A}) \bigcup \sigma_2(\mathcal{A})$. where $\sigma_1(\mathcal{A}) = \{\ \lambda_0 \quad \cdots \quad \lambda_m\ \}$ and $\sigma_2(\mathcal{A}) = \{\ \lambda_{m+1} \quad \lambda_{m+2} \quad \cdots\ \}$.

System Decomposition

◇ The spectrum of the operator $\mathcal{A}$, i.e. $\sigma(\mathcal{A}) = \{\lambda_0, \lambda_1, \cdots\}$ satisfy

$$\lambda_0 \geq \lambda_1 \geq \lambda_2 \geq \cdots \geq \lambda_n \geq \cdots, \quad \lim_{n\to\infty} \lambda_n = -\infty$$

Given a positive number γ, it is possible to choose m that, for all $n > m$,

$$\lambda_n \leq -\gamma$$

therefore, $\sigma(\mathcal{A})$ is the union of two parts: $\sigma(\mathcal{A}) = \sigma_1(\mathcal{A}) \bigcup \sigma_2(\mathcal{A})$. where $\sigma_1(\mathcal{A}) = \{\ \lambda_0 \quad \cdots \quad \lambda_m\ \}$ and $\sigma_2(\mathcal{A}) = \{\ \lambda_{m+1} \quad \lambda_{m+2} \quad \cdots\ \}$.

$$\boxed{\begin{aligned} &P_s(\cdot) = \tfrac{1}{2\pi j} \int_\Gamma R(\lambda; A)(\cdot)d\lambda \\ &P_f = I - P_s \end{aligned}}$$

$$\begin{aligned} \dot{x}_s(t) &= \mathcal{A}_s x_s(t) + \mathcal{B}_s u(t) + \mathcal{G}_s \xi(t), \ x_s(0) = x_{s0} = P_s x_0 \\ \dot{x}_f(t) &= \mathcal{A}_f x_f(t) + \mathcal{B}_f u(t) + \mathcal{G}_f \xi(t), \ x_f(0) = x_{f0} = P_f x_0 \\ y(t) &= \mathcal{C}_s x_s(t) + \mathcal{C}_f x_f(t) + \eta(t) \end{aligned}$$

System Decomposition

$$x_f(t) = T_{\mathcal{A}_f}(t)x_{f0} + \int_0^t T_{\mathcal{A}_f}(t-\tau)\left(\mathcal{B}_f u(\tau) + \mathcal{G}_f \xi(\tau)\right)d\tau$$

We denote by $T_{\mathcal{A}_f}$ as the C_0-semigroup of generator $\mathcal{A}_f$. Then $T_{\mathcal{A}_f}$ is exponentially stable, i.e. $\left\|T_{\mathcal{A}_f}\right\| \leq Me^{-\gamma t}$, M is a positive constant.

Theorem: For the infinite dimensional states x_f in equation (6), $\exists M > 0$ and $\gamma \leq -\lambda_n$, we have the following inequality

$$\|x_f\| \leq Me^{-\gamma t}\|x_f(0)\| + \frac{M}{\gamma}\|\mathcal{B}_f u(t)\| + \frac{M}{\gamma}\|\mathcal{G}_f\|u_n$$

System Decomposition

$$x_f(t) = T_{\mathcal{A}_f}(t)x_{f0} + \int_0^t T_{\mathcal{A}_f}(t-\tau)\left(\mathcal{B}_f u(\tau) + \mathcal{G}_f \xi(\tau)\right)d\tau$$

We denote by $T_{\mathcal{A}_f}$ as the C_0−semigroup of generator $\mathcal{A}_f$. Then $T_{\mathcal{A}_f}$ is exponentially stable, i.e. $\left\|T_{\mathcal{A}_f}\right\| \le Me^{-\gamma t}$, M is a positive constant.

Theorem: For the infinite dimensional states x_f in equation (6), $\exists M > 0$ and $\gamma \le -\lambda_n$, we have the following inequality

$$\|x_f\| \le Me^{-\gamma t}\|x_f(0)\| + \frac{M}{\gamma}\|\mathcal{B}_f u(t)\| + \frac{M}{\gamma}\|\mathcal{G}_f\| u_n$$

$$\boxed{x_f(t) \approx Me^{-\gamma t}\|x_f(0)\| + \frac{M}{\gamma}\|\mathcal{B}_f u(t)\| + \frac{M}{\gamma}\|\mathcal{G}_f\| u_n}$$

$$\begin{aligned}
\dot{x}_s(t) &= \mathcal{A}_s x_s(t) + \mathcal{B}_s u(t) + \mathcal{G}_s \xi(t),\ x_s(0) = x_{s0} \\
x_f(t) &\approx Me^{-\sigma t}\|x_f(0)\| + \frac{M}{\sigma}\|\mathcal{B}_f u(t)\| + \frac{M}{\sigma}\|\mathcal{G}_f\| u_n \\
y(t) &= \mathcal{C}_s x_s(t) + \mathcal{C}_f x_f(t) + \eta(t) \\
\text{State Constraint} &: \mathcal{X}_{\min} \le \mathcal{C}_s x_s(t) + \mathcal{C}_f x_f(t) \le \mathcal{X}_{\max}
\end{aligned}$$

Optimal Constrained State Estimation Formulation

$$\begin{aligned}
\dot{x}_s(t) &= \mathcal{A}_s x_s(t) + \mathcal{B}_s u(t) + \mathcal{G}_s \xi(t),\ x_s(0) = x_{s0} \\
x_f(t) &\approx Me^{-\sigma t}\|x_f(0)\| + \frac{M}{\sigma}\|\mathcal{B}_f u(t)\| + \frac{M}{\sigma}\|\mathcal{G}_f\| u_n \\
y(t) &= \mathcal{C}_s x_s(t) + \mathcal{C}_f x_f(t) + \eta(t) \\
\text{State Constraint} &: \mathcal{X}_{\min} \le \mathcal{C}_s x_s(t) + \mathcal{C}_f x_f(t) \le \mathcal{X}_{\max}
\end{aligned}$$

At every estimation time instant:

$$\boxed{\min \quad J(\hat{x}_s, t, u(\cdot))}$$

The performance index is given by

$$\begin{aligned}
J &= \tfrac{1}{2}[\hat{x}_s(0) - x_{s0}]^T P_0^{-1}[\hat{x}_s(0) - x_{s0}] \\
&+ \tfrac{1}{2}\int_0^T \left\{\left[\dot{\hat{x}}_s(\tau) - \mathcal{A}_s\hat{x}_s(\tau) - \mathcal{B}_s u(\tau)\right]^T \mathcal{G}_s^T R \mathcal{G}_s \left[\dot{\hat{x}}_s(\tau) - \mathcal{A}_s\hat{x}_s(\tau) - \mathcal{B}_s u(\tau)\right]\right\}d\tau \\
&+ \tfrac{1}{2}\int_0^T \left\{\left[y(\tau) - \hat{y}_f(\tau) - \mathcal{C}_s\hat{x}_s(\tau)\right]^T Q \left[y(\tau) - \hat{y}_f(\tau) - \mathcal{C}_s\hat{x}_s(\tau)\right]\right\}d\tau
\end{aligned}$$

where $\hat{y}_f(t) = \mathcal{C}_f\hat{x}_f(t)$. P_0, R, and Q are based on the statistics of the problem.

Constrained Optimal State Estimation Formulation

Defining $V(t) = \dot{\hat{x}}_s(t) - \mathcal{A}_s x_s(t) - \mathcal{B}_s u(t)$, the performance index is given by

$$
\begin{aligned}
&J = \tfrac{1}{2}[\hat{x}_s(0) - x_{s0}]^T P_0^{-1} [\hat{x}_s(0) - x_{s0}] \\
&+\tfrac{1}{2}\int_0^T \left\{ V(\tau)^T G_s^T R G_s V(\tau) \right\} d\tau \\
&+\tfrac{1}{2}\int_0^T \left\{ \left[y(\tau) - \hat{y}_f(\tau) - \mathcal{C}_s\hat{x}_s(\tau)\right]^T Q \left[y(\tau) - \hat{y}_f(\tau) - \mathcal{C}_s\hat{x}_s(\tau)\right] \right\} d\tau
\end{aligned}
$$

Minimization Problem:

$$
\begin{aligned}
\min_{V(t)} \quad & J(\hat{x}_s, t, V(\cdot)) \\
\text{s.t.} \quad & \dot{\hat{x}}_s(t) = \mathcal{A}_s\hat{x}_s(t) + \mathcal{B}_s u(t) + V(t) \\
& \mathcal{X}_{\min} \leq \mathcal{C}_s\hat{x}_s(t) + \hat{y}_f(t) \leq \mathcal{X}_{\max}
\end{aligned}
$$

System Description
Optimal Constrained State Estimation
State Estimation Results

◇ Unconstrained Minimization Problem:

$$
\begin{aligned}
\min_{V(t)} \quad & J(\hat{x}_s, t, V(\cdot)) \\
\text{s.t.} \quad & \dot{\hat{x}}_s(t) = \mathcal{A}_s\hat{x}_s(t) + \mathcal{B}_s u(t) + V(t)
\end{aligned}
$$

◇ Unconstrained Minimization Problem:

$$\begin{aligned} \min_{V(t)} \quad & J(\hat{x}_s, t, V(\cdot)) \\ \text{s.t.} \quad & \dot{\hat{x}}_s(t) = \mathcal{A}_s \hat{x}_s(t) + \mathcal{B}_s u(t) + V(t) \end{aligned}$$

$$\boxed{\mathcal{X}_{\min} \leq \mathcal{C}_s \hat{x}_s(t) + \hat{y}_f(t) \leq \mathcal{X}_{\max}}$$

System Description
Optimal Constrained State Estimation
State Estimation Results

◇ Unconstrained Minimization Problem:

$$\begin{aligned} \min_{V(t)} \quad & J(\hat{x}_s, t, V(\cdot)) \\ \text{s.t.} \quad & \dot{\hat{x}}_s(t) = \mathcal{A}_s \hat{x}_s(t) + \mathcal{B}_s u(t) + V(t) \end{aligned}$$

$$\boxed{\mathcal{X}_{\min} \leq \mathcal{C}_s \hat{x}_s(t) + \hat{y}_f(t) \leq \mathcal{X}_{\max}}$$

◇ Constrained Minimization Problem $\mathcal{X}_{\min} \leq \mathcal{C}_s \hat{x}_s(t) + \hat{y}_f(t)$

$$\begin{aligned} \min_{V(t)} \quad & J(\hat{x}_s, t, V(\cdot)) \\ \text{s.t.} \quad & \dot{\hat{x}}_s(t) = \mathcal{A}_s \hat{x}_s(t) + \mathcal{B}_s u(t) + V(t) \\ & \mathcal{C}_s \hat{x}_s(t) + \hat{y}_f(t) = \mathcal{X}_{\min} \end{aligned}$$

◇ Constrained Minimization Problem $\mathcal{C}_s \hat{x}_s(t) + \hat{y}_f(t) \leq \mathcal{X}_{\max}$

$$\begin{aligned} \min_{V(t)} \quad & J(\hat{x}_s, t, V(\cdot)) \\ \text{s.t.} \quad & \dot{\hat{x}}_s(t) = \mathcal{A}_s \hat{x}_s(t) + \mathcal{B}_s u(t) + V(t) \\ & \mathcal{C}_s \hat{x}_s(t) + \hat{y}_f(t) = \mathcal{X}_{\max} \end{aligned}$$

Constrained Optimal State Estimation Formulation

$$\begin{aligned}\min_{V(t)} \quad & J(\hat{x}_s, t, V(\cdot)) \\ \text{s.t.} \quad & \dot{\hat{x}}_s(t) = \mathcal{A}_s \hat{x}_s(t) + \mathcal{B}_s u(t) + V(t) \\ & \mathcal{C}_s \hat{x}_s(t) + \hat{y}_f(t) = \mathcal{X}_{\min}\end{aligned}$$

$$\Downarrow$$

$$\begin{aligned}\min_{V(t)} \quad & J(\hat{x}_s, t, V(\cdot)) \\ \text{s.t.} \quad & \dot{\hat{x}}_s(t) = \mathcal{A}_s \hat{x}_s(t) + \mathcal{B}_s u(t) + V(t) \\ & \mathcal{C}_s \mathcal{A}_s \hat{x}_s(t) + \mathcal{C}_s \mathcal{B}_s u(t) + \mathcal{C}_s V(t) + \dot{\hat{y}}_f(t) = 0 \\ & \mathcal{C}_s \hat{x}_s(t) + \hat{y}_f(t) = \mathcal{X}_{\min}\end{aligned}$$

Constrained Optimal State Estimation Formulation
Hamiltonian:

$$\begin{aligned} H(t) &= \tfrac{1}{2}\left[V(t)^T G_s^T R G_s V(t)\right] + \\ &+ \tfrac{1}{2}\left\{\left[y(t) - \hat{y}_f(t) - \mathcal{C}_s \hat{x}_s(t)\right]^T Q \left[y(t) - \hat{y}_f(t) - \mathcal{C}_s \hat{x}_s(t)\right]\right\} \\ &+ \lambda(t)^T \left[\mathcal{A}_s \hat{x}_s(t) + \mathcal{B}_s u(t) + V(t)\right] \\ &+ \mu(t)^T \left[\mathcal{C}_s A_s \hat{x}_s(t) + \mathcal{C}_s \mathcal{B}_s u(t) + \mathcal{C}_s V(t) + \dot{\hat{y}}_f(t)\right]\end{aligned}$$

(C.1). $\lambda(T) = 0,\ \hat{x}_s(0) = x_{s0} - P_0 \lambda(0)$

(C.2). $\frac{d\lambda(t)^T}{dt} = -\left(\frac{\partial H}{\partial \hat{x}_s}\right)$

(C.3). $\frac{\partial H}{\partial V} = 0$

(C.4). $\mathcal{C}_s \hat{x}_s(t) + \hat{y}_f(t) = \mathcal{X}_{\min}$

$$\boxed{\begin{aligned}\dot{\hat{x}}_s(t) &= f(\hat{x}_s(t), \lambda(t)) \\ \dot{\lambda}(t) &= g(\lambda(t), \hat{x}_s(t))\end{aligned}}$$

Numerical Example

Consider the following systems:

$$\frac{\partial x(z,t)}{\partial t} = \frac{\partial^2 x(z,t)}{\partial z^2} + B(z)u(t) + \xi(t), \quad z \in [0,1]$$

$$\frac{\partial x(0,t)}{\partial z} = 0, \ \frac{\partial x(1,t)}{\partial z} = 0, \ x(z,0) = sin(2\pi z)$$

$$y(t) = Cx(z,t) + \eta(t)$$

$$\mathcal{X}_{\min} \le Cx(z,t) \le \mathcal{X}_{\max}$$

where $\mathcal{X}_{\min} = -0.4$, $\mathcal{X}_{\max} = \infty$, B is a shaping function around the control point $z_c = 0.9995$, C is a point observation operator at sensing point $z_o = 1$:

$$B(z) = \frac{1}{2\varepsilon} 1_{[z_c-\varepsilon, z_c+\varepsilon]}(z)$$

$$Cx(t) = x(z_o, t)$$

where $\varepsilon = 0.0005$, $1_{[z_c-\varepsilon,z_c+\varepsilon]} = \begin{cases} 1, z_c - \varepsilon \le z \le z_c + \varepsilon \\ 0, \text{otherwise} \end{cases}$

Plant State Profile

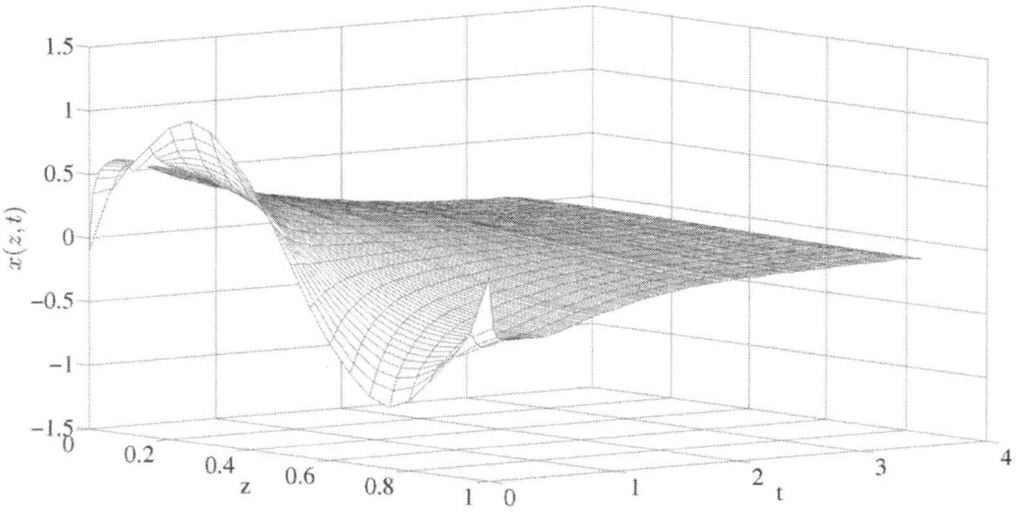

Estimation Results

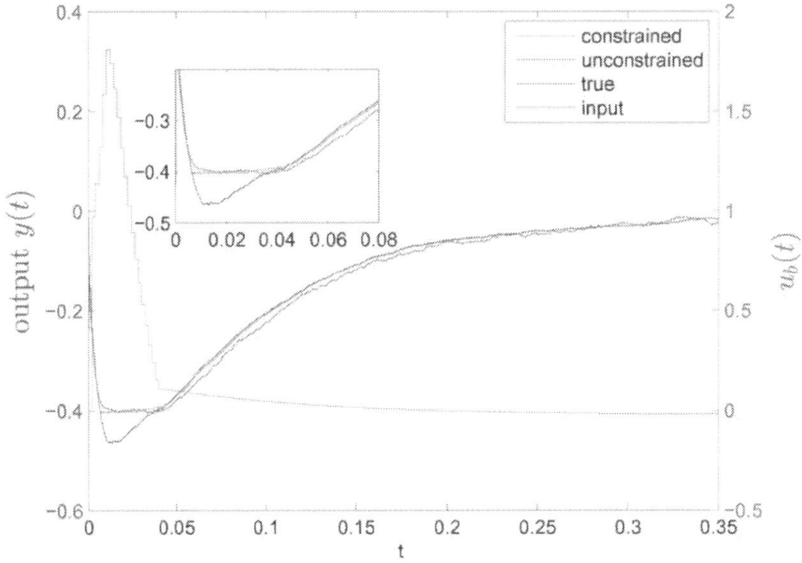

Estimation Results

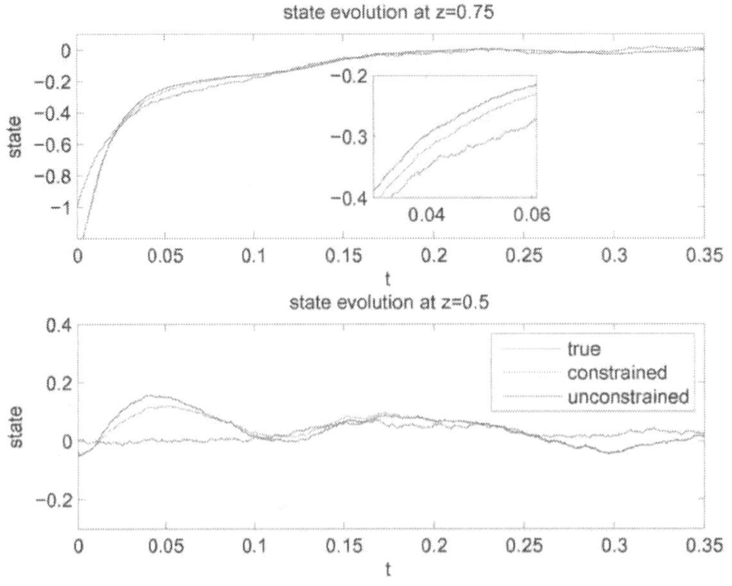

Optimal Constrained State Estimation

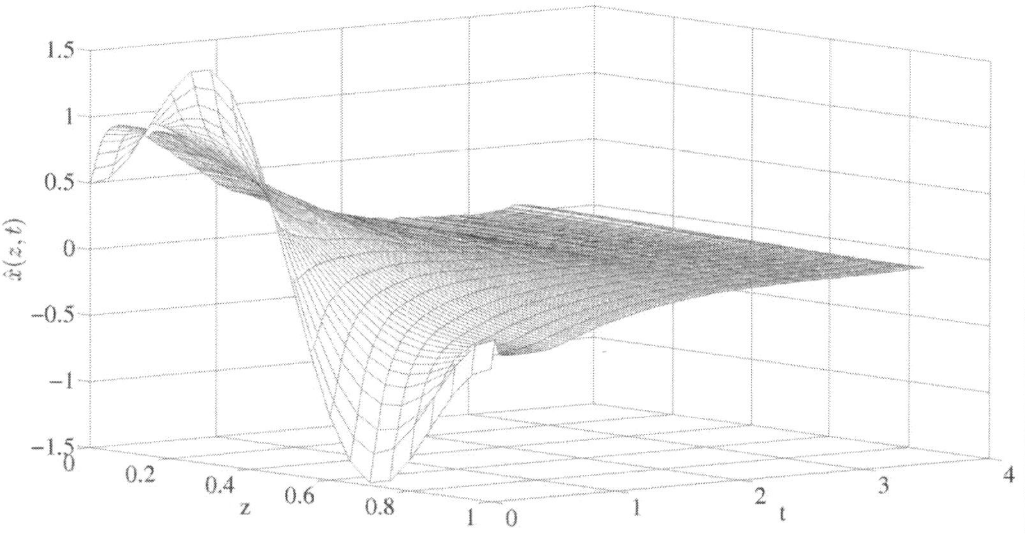

Optimal Unconstrained State Estimation

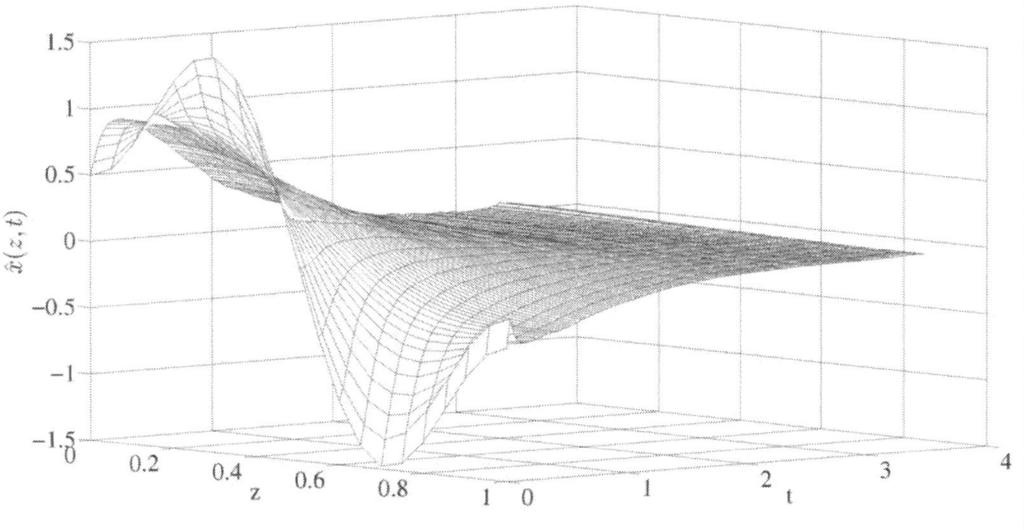

Conclusion

Constraints in control systems:

$$\mathcal{U}_{\min} \leq u \leq \mathcal{U}_{\max}$$
$$\mathcal{X}_{\min} \leq x \leq \mathcal{X}_{\max}$$

- The original system is decomposed into two sub-systems;

  The optimal constrained state estimation problem is converted to the dual optimal control problem.

  The proposed state estimation method improves the estimation accuracy.

*Liu, Liu, Biao Huang, and Stevan Dubljevic. "Model predictive control of axial dispersion chemical reactor." Journal of Process Control (2014).

Acknowledgements

- Distributed Parameter System Lab
- Natural Sciences and Engineering Research Council of Canada (NSERC)
- Department of Chemical and Materials Engineering, University of Alberta

370161 Handling State Constraints and Economics in Feedback Control of Transport-Reaction Processes

Abstract:

Economic model predictive control (EMPC) is a practical optimal control-based technique that has recently gained widespread popularity within the process control community and beyond because of its unique quality of effectively integrating process economics and feedback control (e.g., [1]-[2]; see, also, [3] for an overview of recent results and references). However, most of EMPC systems have been designed for lumped parameter processes described by linear/nonlinear ordinary differential equation (ODE) systems. In our previous work [4]-[5], EMPC systems with a general economic cost function for parabolic PDE systems were proposed which operate the closed-loop system in a dynamically optimal fashion. Specifically, the EMPC schemes were developed with a reduced-order model (ROM) (i.e., low-order nonlinear ODE model) derived through Galerkin's method using analytical eigenfunctions [4] and empirical eigenfunctions derived by proper orthogonal decomposition (POD) [5], respectively. To achieve high accuracy of the ROM derived from the empirical eigenfunctions of the original PDE system, the POD method usually needs a large ensemble of solution data (snapshots) to contain as much local and global process dynamics as possible. Constructing such a large ensemble of snapshots becomes a significant challenge from a practical point of view; since currently, there is no general way to realize a representative ensemble. Based on this consideration, an adaptive proper orthogonal decomposition (APOD) methodology was proposed to recursively update the ensemble of snapshots and compute on-line the new empirical eigenfunctions in the on-line closed-loop operation of PDE systems [6]-[7]. The ROM accuracy is indeed limited by the number of the empirical eigenfunctions adopted for the ROM; in practice, when a process faces state constraints, the accuracy of the ROM based on a limited number of eigenfunctions may not be able to allow the controller to avoid a state constraint violation.

For hyperbolic PDE systems (e.g., convection-reaction processes where the convective phenomena dominate over diffusive ones), it is common that the eigenvalues of the spatial differential operator cluster along vertical or nearly vertical asymptotes in the complex plane, and thus, many modes are required to construct finite-dimensional models of desired accuracy. Considering this, the Galerkin's method would be computationally expensive for this case compared to finite difference method utilizing a sufficient large number of spatial discretization points. However, no work on applying EMPC to hyperbolic PDE systems as been completed.

Motivated by the above considerations, we address two open issues in the context of applying EMPC to transport-reaction processes. First, we focus on a system of nonlinear parabolic PDEs and propose a novel EMPC design integrating adaptive proper orthogonal decomposition (APOD) method with a high-order finite-difference method to handle state constraints. The computational efficiency and constraint handling properties of this scheme are evaluated using a tubular reactor example modeled by two nonlinear parabolic PDEs. Second, we formulate an EMPC system that accounts for both manipulated input and state constraints for a system of first-order hyperbolic PDEs. Various closed-loop simulations scenarios are presented to demonstrate the overall effectiveness of this EMPC scheme using a plug flow reactor example.

[1] Angeli D, Amrit R, Rawlings JB. On average performance and stability of economic model predictive control. IEEE Transactions on Automatic Control. 2012;57:1615-1626.
[2] Amrit R, Rawlings JB, Angeli D. Economic optimization using model predictive control with a terminal cost. Annual Reviews in Control. 2011;35:178-186.
[3] Ellis M, Durand H, Christofides PD. A tutorial review of economic model predictive control methods. Journal of Process Control, in press.
[4] Lao L, Ellis M, Christofides PD. Economic model predictive control of transport-reaction processes. Industrial \& Engineering Chemistry Research, in press.
[5] Lao L, Ellis M, Christofides PD. Economic model predictive control of parabolic PDE systems: Addressing state estimation and computational efficiency. Journal of Process Control. 2014;24:448-462.
[6] Ravindran SS. Adaptive reduced-order controllers for a thermal flow system using proper orthogonal decomposition. SIAM Journal on Scientific Computing. 2002;23:1924-1942.
[7] Varshney A, Pitchaiah S, and Armaou A. Feedback control of dissipative PDE systems using adaptive model reduction. AIChE Journal.

377317 Analysis and Control of Networked Distributed Processes with Sensor and Actuator Errors

Abstract:

With the significant growth in communication and networking capabilities in recent years, process operations have become increasingly reliant on sensor and control systems that are accessed over shared communication networks rather than dedicated links. While this transition has been motivated by the economic and operational benefits of shared networks, it also introduces a number of control challenges that need to be addressed. Examples include the inherent limitations on the transmission and processing capabilities of the communication medium, in the form of network resource constraints, real-time scheduling constraints, data losses and delays, which may degrade the overall closed-loop performance and even lead to instability if not accounted for in the control system design. This realization has motivated significant research work on the analysis an design of networked control systems, and the literature on this topic is quite extensive (see, for example, [1]-[2] for some results and references in this area).

Yet, a careful examination of the available results shows that, compared with the significant attention that lumped parameter systems (modeled by ordinary differential or difference equations) have enjoyed, networked control of spatially distributed systems (modeled by partial differential equations) has received only limited attention. Efforts to address sensor-controller communication constraints in the context of spatially distributed control systems were made in a number of recent studies (see, for example, [3]-[6]) which focused mainly on highly-dissipative infinite-dimensional systems and led to a number of finite-dimensional networked control design methodologies that enforce closed-loop stability with minimal sensor-controller communication requirements.

The design methods in these studies are based on the consideration that the state measurements used for model updates are exactly known at the update time, and the assumption that the control action computed by the controller is also transmitted flawlessly and accurately to the control actuator. In many practical situations, however, these assumptions need to be re-examined in light of the possible degradation in the quality and precision of the transmitted sensor and actuator data over the network. Possible sources for the presence of measurement and actuation errors include interference with other signals in the field in the case of wireless networks, errors due to noisy sensor readings which are subject to drift over time, the use of discrete sensors (e.g., binary sensors) and discrete actuators (on/off valves) in the control system which cannot be varied continuously, the use of data fusion and aggregation techniques, and the need to conserve network bandwidth by reducing data precision. The end result in all these situations is that only inexact values of the state and/or input variables will be available, and the resulting errors may cause a significant deterioration in the closed-loop performance, if not outright instability, if not handled properly in the control system design.

Motivated by these considerations, we present in this work a methodology for the design and analysis of networked controllers for a class of spatially distributed process systems with uncertain low-order dynamics, discrete sensor-controller communication, and sensor and actuator errors. Initially, a model-based networked controller that enforces exponential closed-loop stability in the absence of sensor and actuator errors is synthesized using an approximate finite-dimensional model of the infinite-dimensional system. Lyapunov techniques are then used to analyze the implementation of the model-based controller in the presence of bounded measurement and actuation errors. Precise conditions for stability and ultimate boundedness of the closed-loop trajectories are derived and used to obtain an explicit characterization of the interdependence between the achievable ultimate bound on the one hand, and the sensor/actuator error bounds, the sensor data transmission rate, the size of process-model mismatch and the choice of sensor/actuator spatial configurations, on the other. It is shown that by judicious selection of these parameters, closed-loop stability with minimal performance degradation can be attained. Finally, the implementation of the networked controller on the infinite-dimensional system is analyzed, and the results are illustrated using a diffusion-reaction process example.

References:

[1] J. P. Hespanha, P. Naghshtabrizi, and Y. Xu, "A survey of recent results in networked control systems," *Proceedings of the IEEE*, **95**:138-162, 2007.

[2] P. D. Christofides, J. Liu, and D. Muoz de la Pena, *Networked and Distributed Predictive Control: Methods and Nonlinear Process Network Applications*. London: Springer-Verlag, 2011.

[3] Y. Sun, S. Ghantasala, and N. H. El-Farra, "Networked control of spatially distributed processes with sensor-controller communication constraints," in *Proceedings of American Control Conference*, St. Louis, MO, 2009, pp. 2489-2494.

[4] Z. Yao and N. H. El-Farra, "On model-based networked control of nonlinear spatially distributed process systems," in *Proceedings of 18th Mediterranean Conference on Control and Automation*, Marrakech, Morocco, 2010, pp. 898-903.

[5] Z. Yao, Y. Sun, and N. H. El-Farra, "Resource-aware scheduled control of distributed process systems over wireless sensor networks," in *Proceedings of American Control Conference*, Baltimore, MD, 2010, pp. 4121-4126.

[6] Z. Yao and N. H. El-Farra, "Networked model predictive control of spatially distributed processes," in *Proceedings of American Control Conference*, Washington, DC, 2013, pp. 2062-2067.

380202 Data-Driven Methods for the Analysis of Multiscale Stochastic Systems

Abstract:

Complex dynamic engineering models often contain dynamical processes evolving over broad ranges of time and length scales. Separation of timescales can pose problems in modeling and simulation, as accurately resolving the fast modes on timescales which are relevant for the slow modes requires significant computational power. Reducing the model to only contain the slow modes is often based on *a priori* knowledge or analytic transformations of the variables, which may not be immediately obvious for complex systems. We propose data-driven techniques which automatically uncover the slow variables from simulation data. Using nonlinear independent component analysis [1, 2], a nonlinear dimensionality reduction technique, we show that we can recover the slow variables, even when the system dynamics are complex and nonlinear. We demonstrate our methods through examples from chemical reaction networks and molecular simulations.

[1] Singer, Amit, and Ronald R. Coifman. "Non-linear independent component analysis with diffusion maps." *Applied and Computational Harmonic Analysis*25.2 (2008): 226-239.

[2] Dsilva, Carmeline J., et al. "Nonlinear intrinsic variables and state reconstruction in multiscale simulations." *The Journal of chemical physics* 139.18 (2013): 184109.

Dynamic Reduced Order Models for a Bubbling Fluidized Bed Adsorber

Mingzhao Yu and Lorenz T. Biegler

Chemical Engineering, Carnegie Mellon University, Pittsburgh, PA,

Distributed parameter first-principle models give an accurate physical description of chemical processes, but lead to large-scale models whose numerical solution is challenging and computationally expensive. The large amount of time required to solve the large-scale detailed model limits its use for process control and optimization tasks, where a large number of simulations need to be run. As a result, dynamic reduced order models with reasonable accuracy are required for advanced process control and other model based real-time applications.

In this study, we aim to build a dynamic reduced model for a bubbling fluidized-bed (BFB) adsorber, which is the essential component of an integrated carbon capture system for fossil fuel power plants. In this one-dimensional, three-region model, hydrodynamic behavior is described by partial differential and algebraic equations (PDAEs), constructed from mass and heat conservation and hydrodynamic correlations. Detailed equations for steady-state and dynamic BFB model can be found in [1, 2]. Besides, the reaction kinetic model for adsorption of CO_2 onto an amine-impregnated sorbent is developed in [3]. After spatial discretization, the model is converted to a differential and algebraic equation (DAE) system consisting of nearly 30000 equations. Solving such large-scale DAE system is computationally expensive. Thus, a reduced model of BFB adsorber is needed for simulation-based optimal design and control of the carbon capture system.

In the BFB model, where states are involved in different phases and regions, the time scale difference in differential states may lead to a high stiffness ratio and thus more integration steps. Singular perturbation is a useful tool for the reduction of models showing multiple time scale behavior. The underlying idea is the quasi-steady state approximation for fast states, which converts the differential equations describing the states with fast dynamics to algebraic equations. Eigenvalue analysis technique is used to locate fast and slow eigenvalue groups and unit perturbation spectral resolution (UPSR) matrix [4] is used to obtain a quantitative measure of the levels of association between all eigenvalues and states. Besides, we also apply the nullspace projection method [5] to the reaction kinetics of adsorption process to take advantage of the scale difference of reaction rates. The reformulated dynamic BFB model can be solved much faster while maintaining good accuracy.

Moreover, huge number of differential and algebraic equations, especially after spatial discretization, also increases the simulation time. The proper orthogonal decomposition (POD) is a powerful model reduction tool to create an accurate low dimensional approximation of large spatially distributed models. The key idea of POD is to find a low dimensional set of orthogonal basis functions, in which the spatial distribution information is captured, to represent the original high dimensional system. So POD technique is applied to identify the dominant spatial features inside the BFB adsorber and method of snapshots is used to generate POD empirical basis functions. Preliminary result shows that, instead of discretization over 100 points, only 6 to 7 basis functions are needed for each state, with an average projection error less than 0.1% , which shows great potential in further reducing the number of equations and computational cost.

[1] A. Lee and D. C. Miller. "A One-Dimensional (1-D) Three-Region Model for a Bubbling Fluidized-Bed Adsorber". In: Ind. Eng. Chem. Res. 52 (2013), pp. 469–484.

[2] S. Modekurti, D. Bhattacharyya, and S. E. Zitney. "Dynamic Modeling and Control Studies of a Two-Stage Bubbling Fluidized Bed Adsorber-Reactor for Solid-Sorbent CO2 Capture". In: Ind. Eng. Chem. Res. 52 (2013), pp. 10250–10260.

[3] A. Lee et al. "A model for the adsorption kinetics of CO2 on amine-impregnated mesoporous sorbents in the presence of water". In: 28th International Pittsburgh Coal Conference, Pittsburgh, PA, USA. 2011.

[4] I. T. Cameron and A. M Walsh. "Unravelling complex system dynamics using spectral association methods". In: Elsevier B.V., 2004. Chap. The Integration of Process Design and Control, pp. 126–145.

[5] Y. Nie, L. Biegler, C,Villa, and J, Wassick. "Reactor modeling and recipe optimization of polyether polyol processes: polypropylene glycol". AIChE J. 2013, 59, 2515–2529

386460 Adaptive Control of Dissipative Distributed Parameter Systems Via Recursively Updated Reduced Order Models

Abstract:

In dissipative distributed parameter systems (DPSs) with underlying physical and chemical spatiotemporal phenomena such as chemical reactions, diffusion, convection and dispersion, physical properties and reaction parameters are often unknown. Such systems exemplified by flow reactors, chemical vapor deposition systems and microelectronic fabrication processes, can be mathematically modeled by dissipative partial differential equations (PDEs) with unknown parameters [1]. Thus, we may need to identify the unknown parameters during process evolution to design the control structure for such systems.

Adaptive control and system identification methods have been used extensively in estimation and control of linear DPSs over regular domains [2], [3]. One of the limitations of current techniques is that they can not directly be applied to neither general linear DPSs over irregular domains nor nonlinear DPSs. Such limitations may be circumvented via model reduction. In this work, we focus on dissipative systems whose infinite dimensional representation of the PDE dynamics can be approximated by reduced order finite dimensional models [1]. The recursively updated reduced order model (ROM) of the system is constructed by discretization of the PDEs using Galerkin's method. The set of basis functions needed for discretizing the system is initially computed and then recursively updated using adaptive proper orthogonal decomposition [4], [5]. Then the resulting recursively updated ROM that includes unknown parameters due to unknown reaction and diffusion terms in the PDE system, is considered as the basis for Lyapunov-based adaptive control design. The effectiveness of the proposed control structure is successfully illustrated on regulation of the spatiotemporal dynamic of temperature in a catalytic reactor with unknown reaction and diffusion parameters.

[1] P. D. CHRISTOFIDES, Nonlinear and robust control of PDE systems, Birkhauser, New York, 2000.

[2] M. BOHM, M. A. DEMETRIOU, S. REICH, AND I. G. ROSEN, Model reference adaptive control of distributed parameter systems, SIAM Journal on Control and Optimization, 36 (1998), pp. 33–81.

[3] A. SMYSHLYAEV AND M. KRSTIC, Adaptive control of parabolic PDEs, Princeton Univ. Press, Princeton, NJ, 2010.

[4] D. BABAEI POURKARGAR AND A. ARMAOU, Design of APOD-based switching dynamic observers and output feedback control for a class of nonlinear distributed parameter systems, J. Proc. Cont., to appear (2014).

[5] D. BABAEI POURKARGAR AND A. ARMAOU, Modification to adaptive model reduction for regulation of distributed parameter systems with fast transients, AIChE J., 59(12) (2013), pp. 4595–4611.

Nonlinear control of a differential algebraic equations system using modified POD method

Manda Yang, Antonios Armaou

In many chemical industry processes, nonlinear PDEs need to be solved due to spatial distributed nature of the system dynamics. In general, these PDEs can't be solved analytically. One of the methods is to decompose the function into spatial part and temporary part using Galerkin's method. Compared with standard Galerkin's method, which provides optimal basis functions in linear system, in non-linear system, proper orthogonal decomposition method (POD) can provide a better set of basis to approximate the function we need to solve [1]. Basis functions generated by POD are optimal with respect to that the average square of distance between the function and basis is minimized. Data from open loop process is used to approximate the solution of the minimizing problem. This method, which is called a combination of POD and method of snapshots, has been used extensively in analyzing the dynamics distributed system. But this method also has some limitations. Firstly, it requires the process is fully excited so that all the possible spatial modes are contained in the data we use to generate POD basis functions. However, there is no method to properly excite the system [2]. Besides, sampling too few snapshots may also result in basis functions that can't represent the dynamics of the closed-loop system accurately.

To relax the requirements of POD method, we develop a new method to generate basis function. For a time-varying system, the optimal basis functions to capture the characteristics of the process may change as time goes by [2]. Therefore, the first step is to use new snapshots to recalculate the optimal basis functions. This step will add new important basis function(s) and discard unimportant ones based on the energy ratio the present basis function captures. It will make the result more accurate and reduce unnecessary computational load. This method is called adaptive proper orthogonal decomposition (APOD).

After basis functions (ψ) are identified using APOD, the inner product of ψ and $A\psi$ is calculated to generate new basis functions, where A is an operator representing the dynamics of the process. In this step, we further decrease the number of modes without losing too much information of the system. Finally, we will use the new basis functions to design nonlinear controller in differential algebraic equations system. By decreasing the number of modes we need, this method is computationally cheaper, which is important for continuous estimation of optimal basis functions in APOD method.

Other than the method proposed here, there are other approaches in literature trying to improve POD method. One of them is the work of Sirovich in which he created a competing method to POD using $\dot{v}$ [3]. Compared to this method, the proposed method has the following advantages: the basis functions generated by this method are orthonormal and orthogonal to process dynamics; because of this property, subsequent analysis of system dynamics and controller design can be simplified when calculating inner product. Besides, this method reduced the number of modes needed and therefor is more computationally efficient; which is a significant feature when implemented on line.

REFERENCES

[1] Sirovich, L. (1987). Turbulence and the dynamics of coherent structures: part 1 coherent structures.

Quarterly of Applied Mathematics, 561–571.
[2] Pitchaiah, S., & Armaou, A. (2010). Output feedback control of distributed parameter systems using adaptive proper orthogonal decomposition. Industrial & Engineering Chemistry Research, 49, 10496–10509.
[3] Sirovich, L., Knight, B., & Rodriguez, J. (1990). Optimal low-dimensional dynamical approximations. Quarterly of Applied Mathematics, 535–548.

388998 Backstepping Observer and Controller Design for Parabolic PDE with Time-Varying Domain

Abstract:

In this work a PDE backstepping based observer and control law for one-dimensional unstable heat equatio with time-varying spatial domain are developed. The underlying parabolic partial differential equation (PDE) with time-varying domain is the model emerging from process control applications such as crystal growth. In backstepping observer and control law synthesis [1], a characteristic feature is that the PDE describing the transformation kernel of the associated Volterra integral is time-dependent. In this work, the kernel PDE is solved numerically and both observer and state-feedback controller are simulated for the application of temperature regulation in the Czochralski crystal growth process.

[1] M. Krstic and A. Smyshlyaev, Boundary control of PDEs: A course on backstepping designs. Philadelphia: SIAM, 2008.

387911 Initialization Strategies for Global Optimization in Refinery Planning

Abstract:

Refinery planning is the problem of determining the optimal crude slate for a refinery in order to meet all product demands and specifications while satisfying all operational constraints. Refinery planning is typically represented as a generalized pooling problem resulting in a mixed-integer, nonlinear optimization problem. The very large number of bilinear and trilinear terms can lead to local optima, which can translate into millions or even billions of lost revenues for the refinery operators over the course of a year. Despite recent developments in global optimization, many of the new techniques are impractical due to problem size and complexity as well as the solution time constraints present in a refinery environment. Commonly used techniques (such as successive linear programming, SLP, and successive quadratic programming, SQP) require that all variables be initialized prior to optimization.

In this paper we explore a convex relaxation-based generation of an initial solution for the refinery planning problem. The algorithm uses recent enhancements in global optimization techniques applied to a large-scale industrial problem while employing specialized bounding algorithms to further improve accuracy and performance. Combined with other techniques and heuristics, the end result is the creation of a consistent initialization strategy to help reduce the occurrence of sub-optimal solutions in refinery planning problems.

Electric Energy Procurement for Large Industrial Consumers under Uncertainty in Electricity Price and Product Demand

In general, large industrial electricity consumers procure their required electricity through bilateral contracts or in the pool (day-ahead or real-time market). The main purpose of signing bilateral contracts is to hedge against uncertainty in electricity prices, whereas buying electricity in the pool provides the opportunity of demand side management, i.e. shifting load to low-price hours. Thus, there is a trade-off between these two purchasing options, and the challenge is to find the optimal energy procurement strategy.

This problem has been studied by Conejo et al. (2005), Carrión et al. (2007), and Zare et al. (2010). While Conejo et al. (2005) solve a deterministic mid-term energy procurement problem considering a set of bilateral contracts and hourly changing spot prices, Carrión et al. (2007) apply Stochastic Programming to account for uncertainty in the electricity spot prices. Uncertainty is also the focus of the work by Zare et al. (2010), who solve a similar problem by applying Information Gap Decision Theory. However, in these works, the production profile is fixed, which leaves no room for load shifting.

In this work, we solve the energy procurement problem considering detailed production scheduling decisions, which increases operational flexibility and allows us to shift load in order to take advantage of varying electricity prices. Based on the work by Mitra et al. (2012), we develop a mixed-integer linear programming (MILP) model that incorporates electricity contract options into the production scheduling model, and we apply a two-stage stochastic programming model to take uncertainty into account. Here, we not only consider uncertainty in the electricity prictes but also in the product demand, which is typically the main source of uncertainty in production planning.

In order to reduce the computational effort of solving the MILP stochastic program, we first apply scenario reduction techniques (Dupacova et al., 2003) to decrease the number of scenarios while retaining a good approximation of the probability distribution. Since the problem is still difficult to solve as already the deterministic problem involves up to 5,000 binary variables, 15,000 continuous variables and 12,000 constraints since it is defined on an hourly basis over a one-week horizon, we further develop a tailored solution method based on Benders Decomposition. We apply the proposed framework to a real-world industrial case study provided by Praxair and show the value of the stochastic solution.

References

Carrión, M., Philpott, A. B., Conejo, A. J. & Arroyo, J. M., 2007. A stochastic programming approach to electric energy procurement for large consumers. *IEEE Transactions on Power Systems,* 22(2), pp. 744-754.

Conejo, A. J., Fernadez-Gonzalez, J. J. & Alguacil, N., 2005. Energy procurement for large consumers in electricity markets. *IEE Proceedings - Generation, Transmission and Distribution,* 152(3), pp. 357-364.

Dupacova, J., Gröwe-Kuska, N. & Römisch, W., 2003. Scenario reduction in stochastic programming: An approach using probability metrics. *Mathematical Programming,* Volume Ser. A 95, pp. 493-511.

Mitra, S., Grossmann, I. E., Pinto, J. M. & Arora, N., 2012. Optimal production planning under time-sensitive electricity prices for continuous power-intensive processes. *Computers & Chemical Engineering,* Volume 38, pp. 171-184.

Zare, K., Moghaddam, M. P. & Sheikh El Eslami, M. K., 2010. Electricity procurement for large consumers based on Information Gap Decision Theory. *Energy Policy,* 38(1), pp. 234-242.

A bi-level heuristic for steel plant scheduling under complex time-sensitive price structures

Hubert Hadera[1,2], Iiro Harjunkoski[1*], Ignacio E. Grossmann[3], Guido Sand[1], Sebastian Engell[2]

[1]ABB Corporate Research, Wallstadter Str. 59, 68526 Ladenburg, Germany

[2] Technical University of Dortmund, Emil-Figge-Straße 70, 44221 Dortmund, Germany

[3]Carnegie Mellon University, 5000 Forbes Ave., Pittsburgh, PA 15213, USA

*Corresponding author, Email: iiro.harjunkoski@de.abb.com

Demand response of industrial production processes can support the transformation process of today's energy supply systems by enabling an active load management. For energy-intensive industries this can be seen as an opportunity to reduce costs taking advantage of process flexibility. This is especially important in the times of the economic crisis when the energy-intensive industry is facing under-utilization which creates valuable process flexibility. As a new way of operating cost reduction, production plants may enter liberalized time-sensitive energy markets. Under special arrangements with a supplier, a plant may be also obliged to commit to certain electricity consumption levels. These aspects create potential for energy-driven production schedule optimization.

With our work we would like address a continuous-time scheduling problem of a Melt Shop section in a stainless steel plant. The plant is assumed to pursue one day schedule with the goal of satisfying complex production rules, minimizing the earliest task completion time and energy bill. The bill is a subject of time-sensitive complex pricing and load deviation penalties. Multiple hourly-varying contracts, onsite generation plant with special constraints and possibility of selling surplus electricity to the grid is considered in this work.

Although the topic of energy-aware scheduling is gaining attention in the literature, considerations of such complex cost structures within continuous-time batch scheduling framework has not been thoroughly investigated. Recent work introduces volatile electricity markets in scheduling of continuous processes (Castro et al. 2009, Castro et al. 2011, Mitra et al. 2012). In Castro at el. (2013) an RTN approach has been successfully used to deal with single price curve and load deviation problem in a stainless steel plant. However, the solution uses a discrete time approach. For the continuous-time driven models, Hait, A., & Artigues (2011) use CP with MILP formulation to arrive at a heuristic solving detailed scheduling of Electric Arc Furnace stage and human operator availability. In Hadera & Harjunkoski (2013) Melt Shop schedule with regard to a single price and load deviation problem is not tractable in reasonable computational times. Tracking of pre-agreed electricity load profiles has been tackled by Nolde & Morari (2010), however without any price signals. Some more complex pricing tariffs have been reported in Ashok (2006) and Hadera et al. (2014).

To tackle the problem of complex price structures and load deviation we develop a monolithic MILP model utilizing the benefit of exact continuous-time approach. The scheduling model is expanded with energy-awareness using event binaries. The binaries account for the fact whether a task started/finished at a given time interval that is related to energy pricing problem. Next, having the knowledge concerning electricity consumption, we employ optimization of both energy purchase structure and deviation penalties. Since the resulting formulation gives rise to an intractable problem, here we propose a bi-level heuristic. The upper level is an approximation of the original problem scheduling only a part of the available equipment. In the lower level, the rough solution from the upper level problem is fed and evaluated under certain level of flexibility. The resulting schedule of the entire process is then used to update assignment on the most important Electric Arc Furnace stage. This in turn is used to solve the upper level problem in the next iteration.

The model is tested on various problem instances. The results show that the heuristic helps to find good quality solutions in reasonable time. In addition, we show the benefit of energy-aware scheduling compared to more traditional makespan driven approaches. The main limitation of the model that should be addressed in further work is computational performance for even larger scale instances. Also the deterministic nature of the input data concerning the process disturbances and pricing might be tackled with robust optimization or stochastic programming approaches.

References:

S. Ashok, 2006, Peak-load management in steel plants, App. Energy, vol. 83, no. 5, pp. 413-424

P.M. Castro, I. Harjunkoski and I.E. Grossmann, 2009, New continuous-time scheduling formulation for continuous plants under variable electricity cost, Industrial and Engineering Chemistry Research, vol. 48, no. 14, pp. 6701-6714

P.M. Castro, I. Harjunkoski and I.E Grossmann, 2011, Optimal scheduling of continuous plants with energy constraints, Computers and Chemical Engineering, vol. 35, no. 2, pp. 372-387

P. Castro, L. Sun, I. Harjunkoski, 2013, Resource–Task Network Formulations for Industrial Demand Side Management of a Steel Plant, Ind. Eng. Chem. Res., 52, 36, 13046–13058

H. Hadera, I. Harjunkoski, 2013, Continuous-time Batch Scheduling Approach for Optimizing Electricity Consumption Cost, In: Andrzej Kraslawski and Ilkka Turunen, Editor(s), Computer Aided Chemical Engineering, Elsevier, 32, pp. 403-408

H. Hadera, I. Harjunkoski, I. E Grossmann, G. Sand, S. Engell, 2014, Steel production scheduling optimization under time-sensitive electricity cost, Computer Aided Chemical Engineering, Elsevier, accepted

A. Hait and C. Artigues, 2011, A hybrid CP/MILP method for scheduling with energy costs, European Journal of Industrial Engineering, vol. 5, no. 4, pp. 471-489

S. Mitra, I. E. Grossmann, J. M. Pinto and N. Arora, 2012, Optimal production planning under time-sensitive electricity prices for continuous power-intensive processes, Computers and Chemical Engineering, vol. 38, pp. 171-184

K. Nolde and M. Morari, 2010, Electrical load tracking scheduling of a steel plant, Computers & Chemical Engineering, 34, pp. 1899-1903

379539 A Conceptual Design of Shale Gas Condensate Recovery Process for Maximum Energy Savings

Abstract:

Shale gas condensate refers to the light liquid hydrocarbons recovered from lease separators at natural gas or shale gas wells. It primarily consists of pentanes and heavier hydrocarbons. By definition, condensate is lighter than crude oil, but heavier than natural gas liquids. The API gravity of condensate is typically higher than 50 degrees. It is a valuable mixture that can be converted to different petroleum products like naphtha, jet fuel, and diesel etc. Plenty of new condensates show up at the Gulf Coast from production in the Eagle Ford and other shale basins. Forecasters predict that total Eagle Ford condensate output will reach 250,000 and 400,000 barrels per day (BPD) by 2020. However, increasing the production from shale gas condensate is not easy for existing refineries, especially those in the Gulf Coast of Mexico. This is because condensates contain too many components belonging to the light naphtha range that can overwhelm most Gulf Coast refineries that are mostly designed to process heavier crudes. Thus, a novel conceptual design for shale gas condensate recovery is developed in this paper to enable the production activity and meanwhile to maximize energy savings during the production.

A four-columns process system including pre-flash column, debutanizer, condensate splitter and vacuum distillation column is investigated for condensate recovery process. To optimize the process, a superstructure is constructed first considering many alterative operation conditions. Next, a simulation-assisted mixed-integer linear programming (MILP) model is developed and solved to derive the optimal process synthesis. In order to maximize the energy savings, heat exchanger network (HEN) design is taken into account in conjunction with the superstructure optimization. Finally, rigorous plant-wide simulations are conducted to validate the feasibility and capability of the entire conceptual design coupling separation and heat integration.

Generalized Disjunctive Programming Model for Scheduling a Pulp Plant Under Energy Constraints

Pedro M. Castro
Djêide Rodrigues
Henrique A. Matos

INVESTIGADOR FCT
LNEG
LISBOA
UNIVERSIDADE DE LISBOA

Industrial case study

- Cooking process of a sulphite pulp plant
  - 4 parallel batch digesters
  - Ordered sequence processing tasks
    - Charge with wood chips and fill with chemicals
    - Heat contents with steam
    - Cook, relief pressure and discharge
  - Shared equipment
    - One per stage (except cooking)

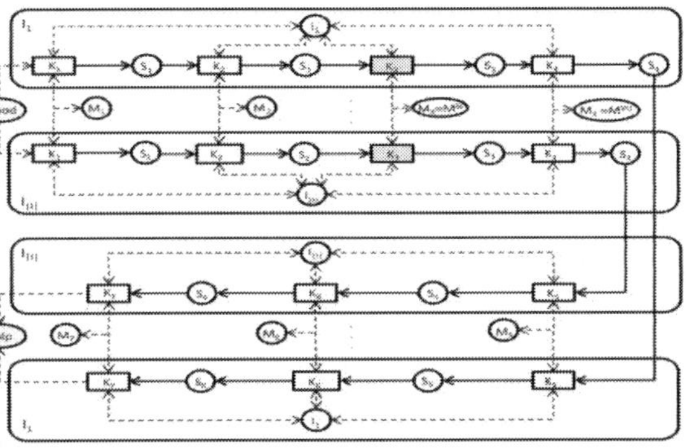

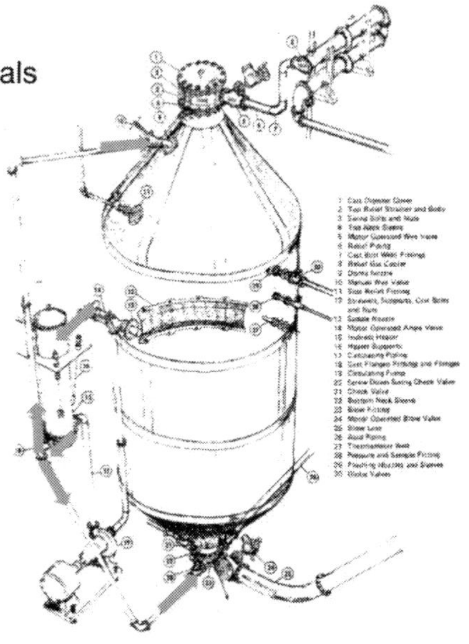

Heating is the bottleneck stage

- Two digesters being heated simultaneously
  - Complex interaction due to different heating rates between key temperatures

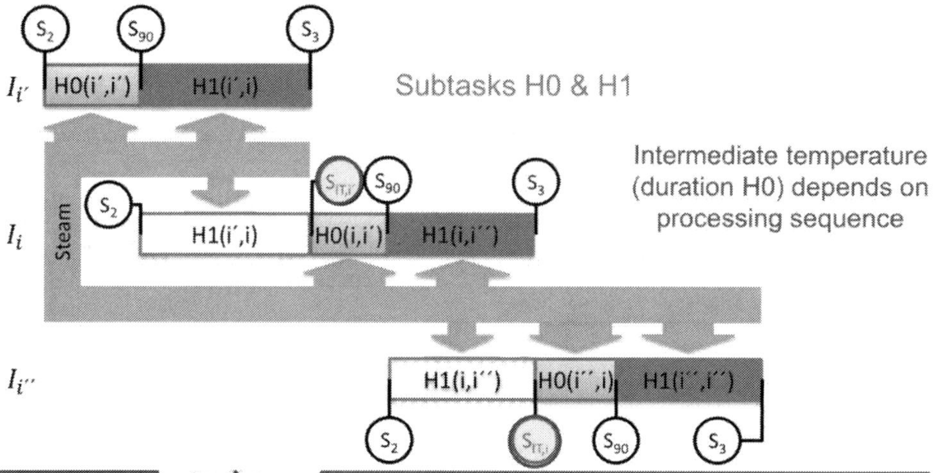

November 19, 2014 | 14AIChE | Industrial Applications in Design and Operations I | 3

Constraints involving continuous variables

- Wrap-around operator (Shah et al. 1993)
  - $\Omega(t) = \begin{cases} t + |T| & ,t < 1 \\ t & ,1 \le t \le |T| \\ t - |T| & ,t > |T| \end{cases}$
    - Used also for index m
- Timing constraints (Wu & Ierapetritou 2004)
  - Consecutive time slots
    - $Ts_{\Omega(t+1),m} \geq Te_{t,m} - H|_{t=|T|} \forall m \notin M^{VU}, t$
      - Cooking tasks allowed to overlap

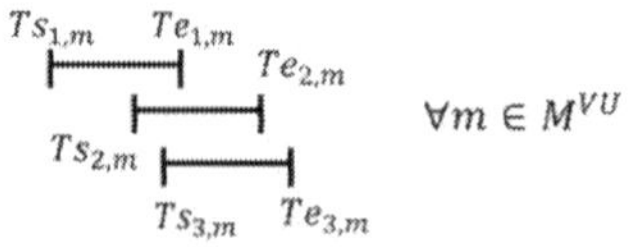

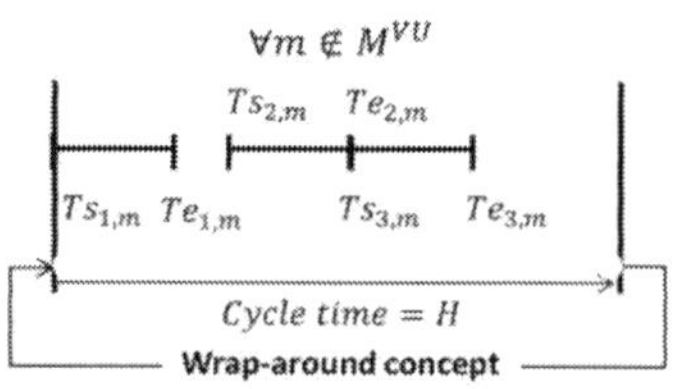

  - Consecutive units (stages)
    - $Ts_{t,\Omega(m+1)} = Te_{t,m} - H|_{m=|M|} \; \forall m \in M^{ZW}, t$
    - $Ts_{t,\Omega(m+1)} \geq Te_{t,m} - H|_{m=|M|} \; \forall m \notin M^{ZW}, t$

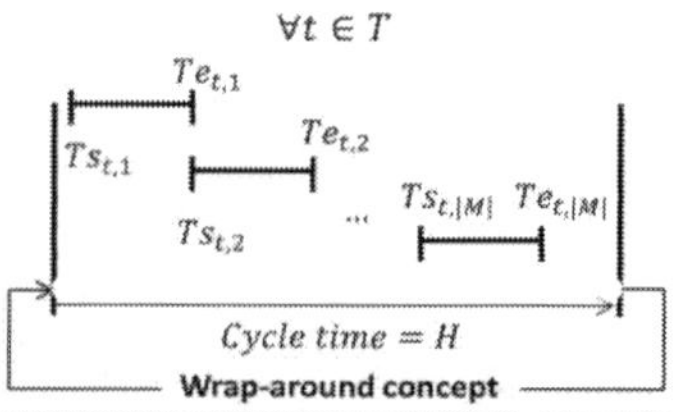

November 19, 2014 | 14AIChE | Industrial Applications in Design and Operations I | 4

Constraints with binary variables

- GDP formulation
  - Single slot t per digester i

$$\bigvee_i \begin{bmatrix} Y_{i,t} \\ Te_{t,m} = Ts_{t,m} + p_{i,m} \end{bmatrix} \; \forall m \notin M^{SL}, t$$

$$\bigvee_t Y_{i,t} \; \forall i \in I$$

  - Heating-stage ($m \in M^{SL}$) constrains are more complex

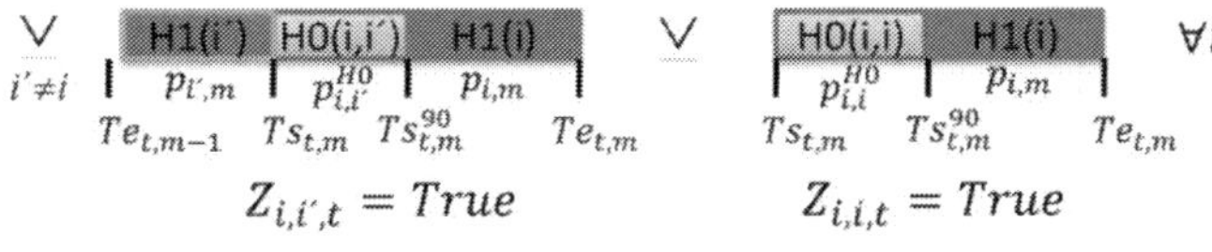

$$\bigvee_i \begin{bmatrix} Y_{i,t} \\ Te_{t,m} = Ts^{90}_{t,m} + p_{i,m} \\ \underline{\bigvee}_{i' \neq i} \begin{bmatrix} Z_{i,i',t} \\ Ts^{90}_{t,m} = Ts_{t,m} + p^{H0}_{i,i'} \\ Te_{t,m-1} \le Ts_{t,m} - p_{i',m} \end{bmatrix} \underline{\bigvee} \begin{bmatrix} Z_{i,i,t} \\ Ts^{90}_{t,m} = Ts_{t,m} + p^{H0}_{i,i} \end{bmatrix} \end{bmatrix} \; \forall m \in M^{SL}, t \in T$$

$$\underline{\vee}_{i' \neq i} Z_{i',i,\Omega(t+1)} \Rightarrow Y_{i,t} \; \forall i \in I, t \in T$$

From GDP to MILP model

- Convex hull reformulation of disjunctions (Castro & Grossmann 2012)
  - $Te_{t,m} = Ts_{t,m} + \sum_{i \in I} p_{i,m} \cdot Y_{i,t} \; \forall m \notin M^{SL}, t$
  - $Te_{t,m} = Ts^{90}_{t,m} + \sum_{i \in I} p_{i,m} \cdot Y_{i,t} \; \forall m \in M^{SL}, t$
  - $Ts^{90}_{t,m} = Ts_{t,m} + \sum_{i \in I} \sum_{i' \in I} p^{H0}_{i,i'} \cdot Z_{i,i',t} \; \forall m \in M^{SL}, t$
  - $Te_{t,m-1} \le Ts_{t,m} + \sum_{i \in I} \sum_{i' \in I, i' \neq i} p_{i',m} \cdot Z_{i,i',t} \; \forall m \in M^{SL}, t$
- Logic propositions converted into (in)equalities
  - $\sum_{i \in I} Y_{i,t} = 1 \; \forall t$
  - $\sum_{t \in T} Y_{i,t} = 1 \; \forall i$
  - $Y_{i,t} \ge \sum_{i' \in I, i' \neq i} Z_{i',i,\Omega(t+1)} \; \forall i, t$ $\underline{\vee}_{i' \neq i} Z_{i',i,\Omega(t+1)} \Rightarrow Y_{i,t} \; \forall i \in I, t \in T$
  - $Y_{i,t} = \sum_{i' \in I} Z_{i,i',t} \; \forall i, t$
- Objective function: minimize cycle time
  - $\min H$

Alternative solution approaches

- Current work
  - Continuous-time model with multiple time grids
  - Sequential, multistage multiproduct batch plant
    - Each digester can be viewed as a production order
    - Tasks of production recipe define the stages
    - Single unit per stage (shared equipment)
  - Challenging constraints derived from Generalized Disjunctive Programming (Balas 1979, Raman & Grossmann 1994)
    - Integrated heating tasks
- Previous work (Castro et al. 2003)
  - RTN-based periodic scheduling formulations
    - Multipurpose plant
    - Tractable and accurate with discrete-time
    - Continuous-time with single time grid requires too many event points
      - Returns suboptimal solutions
    - 4+4 subtasks H0 & H1 explicitly defined per digester

Problem data

- Duration $p_{i,m}$ non-heating tasks (min)

Stage/Digester	I3	I4	I5	I6
Wood-chips filling (M1)	45	54	60	62
Acid filling (M2)	50	50	44	43
Cooking (M4)	152	142	120	120
High pressure degassing (M5)	30	34	36	36
Low pressure degassing (M6)	34	36	48	46
Pulp blowing (M7)	20	30	37	34

- Duration heating tasks (min), stage M_3
  - Two scenarios with constant steam availability (13 and 14 t/h)
    - Data from dynamic simulation model in gPROMS (Castro et al. 2002)

H0(i,i') ≡ $p_{i,i'}^{H0}$	I3		I4		I5		I6	
F_{steam}^{total} (t/h)	13	14	13	14	13	14	13	14
I3	46	46	34	28	40	34	39	31
I4	42	36	66	66	60	55	59	52
I5	32	25	44	35	55	50	48	38
I6	36	28	48	39	53	44	59	54
H1(i) ≡ $p_{i,M3}$	90	88	99	96	107	104	107	104

  - Capacity (10^3 kg) I3<I4<I5≈I6
    - The larger the digester i' the lower the temperature reached by i after steam sharing and thus the longer the duration of H0(i, i')

Optimal periodic schedules

- Shared equipment view
- Digesters view

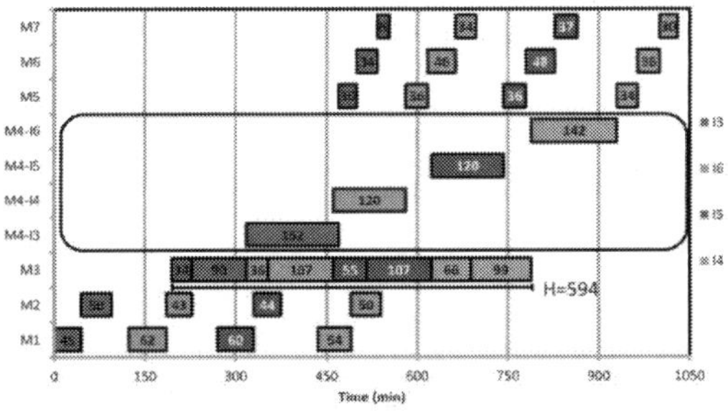

13 t/h

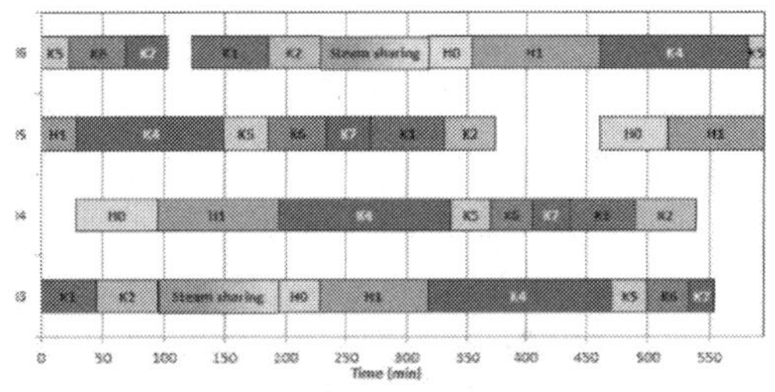

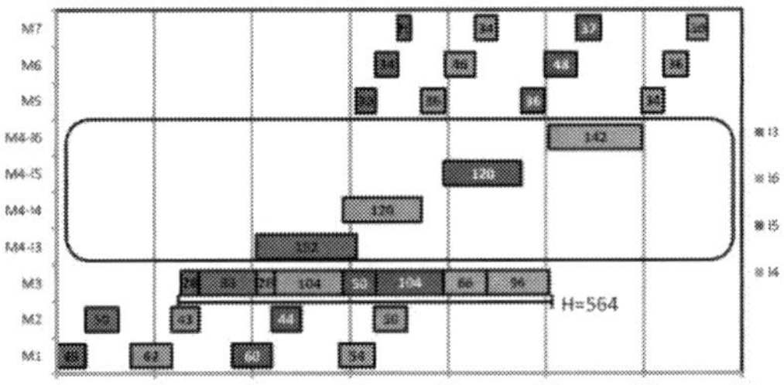

14 t/h

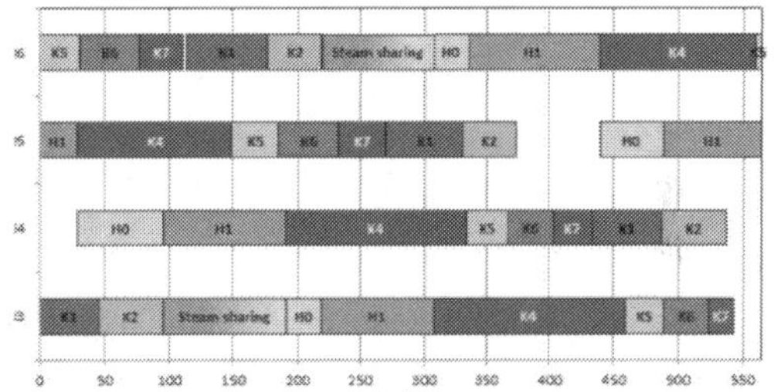

Results for different approaches

Formulation	Discrete-time	Continuous-time		Discrete-time	Continuous-time	
		Single grid	Multiple grids		Single grid	Multiple grids
F_{steam}^{total} (t/h)	13			14		
Integer variables	33263	2688	79	31583	3584	79
Total variables	64154	3327	141	60914	4435	141
Constraints	30932	3331	132	29372	4435	132
Cycle time H (min)	**594**	606	**594**	**564**	571	**564**
Total CPU (s)	563	576	0.11	217	25252	0.21

- New multiple time grid model performs best
  - Two orders of magnitude reduction in problem size and total CPUs vs. discrete-time
    - Even better performance vs. single grid continuous-time
  - Does not require iterative search procedure over # slots
  - Much simpler to include more digesters
    - No need for further tasks and intermediate states
      - RTN-based models

Capacity expansion studies

- Estimate productivity increase from extra digester
  - 1/10 value from ratio of batch sizes b_i
    - Doubles from 13 to 14 t/h
  - Investment should be accompanied by a new boiler

		F_{steam}^{total} (t/h)	13			14		
Digester sequence	$\sum_i b_i$ (10^3 kg)	Estimated impact (%)	H (min)	P (10^3 kg/h)	A.I. (%)	H (min)	P (10^3 kg/h)	A.I. (%)
I3-I6-I5-I4	1087	-	594	109.80	-	564	115.64	-
I3-I5-I6- I3*-I4	1295	19.1	693	112.12	2.1	644	120.65	4.3
I3-I4- I6-I5-I4*	1359	25.0	734	111.09	1.2	665	122.62	6.0
I3-I5-I6-I5*-I4	1388	27.7	737	113.00	2.9	683	121.93	5.4
I3-I6-I5-I6*-I4	1393	28.2	740	112.95	2.9	684	122.19	5.7

How far can we go in terms of size?

- New formulation very robust
  - Problems with 60 digesters solved <12 min
    - Over 200,000 integer variables
- $H_{|I|=8} < 2 \cdot H_{|I|=4}$
  - Better not to repeat pattern -(I3-I6-I5-I4)-
    - I3-I5-I6-I5*-I6*-I4-I3*-I4* 13 t/h
    - I3-I4-I3*-I4*-I5-I6-I5*-I6* 14 t/h
  - Goes against periodic scheduling concept!

				F_{steam}^{total} (t/h)	13		14	
Digesters	|I|	Integer variables	Total variables	Constraints	H (min)	CPU (s)	H (min)	CPU (s)
I3+I4+I5+I6	4	79	141	132	594	0.11	564	0.21
2·I3+2·I4+I5+I6	6	251	343	222	846	0.11	786	0.12
2·I3+2·I4+2·I5+2·I6	8	575	697	328	1161	0.35	1076	0.24
4·I3+4·I4+4·I5+4·I6	16	4351	4593	912	2321	0.61	2152	0.46
5·I3+5·I4+5·I5+5·I6	20	8399	8701	1300	2901	1.42	2690	0.66
10·I3+10·I4+10·I5+10·I6	40	65599	66201	4200	5801	42.4	3920	2.34
15·I3+15·I4+15·I5+15·I6	60	219599	220501	8700	8701	708	8070	199

Conclusions

- Cyclic scheduling problem of a batch pulp plant
- Main challenge: Model complex interaction between two digesters being heated simultaneously
- Easier with GDP formulation (multiple time grids)
  - Retains underlying logic structure
  - Links different constraints to alternative operational decisions
- Orders of magnitude faster than RTN-based formulations
  - Discrete and continuous-time
- Future work
  - Handle variable profiles for heating utilities
- See IECR 2014, 53, 17098-17111 for further details
- Acknowledgments:
  - Financial support from FEDER (Programa Operacional Factores de Competitividade –COMPETE) and Fundação para a Ciência e Tecnologia
    - FCOMP-01-0124-FEDER-020764 and Investigador FCT 2013

382822 Decomposition Method for Mutiperiod Blending Scheduling

Abstract:

The efficient blending of liquid fuels and crude oils to meet both technical and environmental specifications has been a growing research area in recent years. The multiperiod blending problem seeks to maximize the profits associated with this process. Specifically, multiple liquid streams with various properties enter supply tanks. Then, they are fed into blending tanks where they are mixed in some proportion to meet a set of specifications. Finally, they are fed into demand tanks. The goal is then to select the flows that maximize the overall profit of the blending process. In the process, supply and demand may vary with time. Furthermore, operational constraints do not allow having flow coming in and going out of a blending tank simultaneously.

The modelling of this process makes use of binary variables to represent if there is flow between two tanks at any given time period, making the problem highly combinatorial. The description of the mixing constraints requires the use of bilinear terms, so global optimization tools are needed to find the optimal solution. These two issues make the mixed-integer nonlinear program (MINLP) difficult to solve.

Kolodziej et al.[1] use a discretization-based approach to solve the multiperiod blending problem to global optimality. The method used in their work discretizes some continuous variables with a certain precision, using the multiparametric disaggregation technique[2]. The algorithm they present shows good performance in comparison to commercial global optimization tools. The main downside is that it takes a long time to find feasible solutions in some cases. In practice, finding good feasible solutions fast is of utmost importance.

In this work we first present an alternative formulation to the multiperiod blending problem. This new formulation provides a structured problem that is easy to decompose. We then present a decomposition approach to the multiperiod blending problem. The proposed algorithm decomposes the MINLP model in two levels: a master problem and a subproblem. The master problem is a mixed-integer linear program (MILP) relaxation of the original MINLP, obtained through the multiparametric disaggregation technique. This problem provides an upper bound to the MINLP. The subproblem is an MINLP in which a subset of discrete variables is fixed, yielding a reduction in the number of binary variables and bilinear terms. These problems are solved successively until the gap between the upper and lower bounds is closed.

For the solution of the MINLP we use a similar approach to the one presented by Kolodziej et al.[1]. This approach uses two MILP formulations derived from the multiparametric disaggregation technique. One of the MILPs is a relaxation; therefore, it provides an upper bound. The other MILP guarantees that, if a solution is found, then that solution is feasible for the original MINLP. The second MILP provides a lower bound. The algorithm starts by using a low precision discretization and solves both MILPs. If the gap between the upper bound and the lower bound is smaller than a specified parameter the algorithm stops. If it is larger, then the algorithm uses more precision in the discretization and solves the two MILPs again. This is repeated iteratively until the gap is smaller than the specified parameter.

We illustrate the performance of the algorithm with over 40 instances of the multiperiod blending problem. In general, the decomposition algorithm finds better solutions faster than the algorithm by Kolodziej et al.[1] and than commercial global optimization solvers. We also present a parallelization of the algorithm, showing the improvement in solution times.

[1] Kolodziej, S. P., Grossmann, I. E., Furman, K. C., and Sawaya, N. W. "A discretization-based approach for the optimization of the multiperiod blend scheduling problem." Computers & Chemical Engineering. 2013; 53, 122-142.

[2] Kolodziej, Scott, Pedro M. Castro, and Ignacio E. Grossmann. "Global optimization of bilinear programs with a multiparametric disaggregation technique." Journal of Global Optimization 2013; 57(4), 1039-1063.

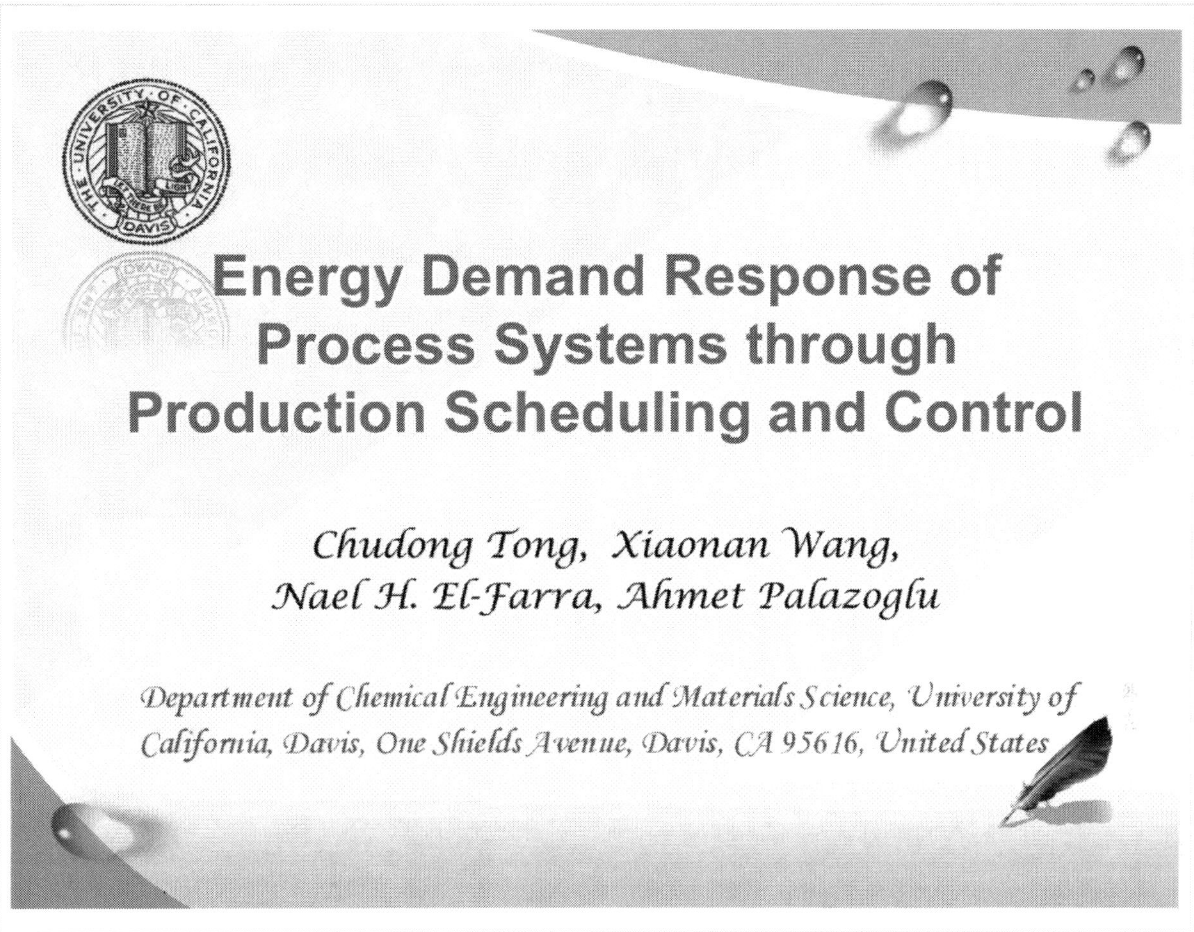
Energy Demand Response of
Process Systems through
Production Scheduling and Control
Chudong Tong, Xiaonan Wang,
Nael H. El-Farra, Ahmet Palazoglu
Department of Chemical Engineering and Materials Science, University of
California, Davis, One Shields Avenue, Davis, CA 95616, United States

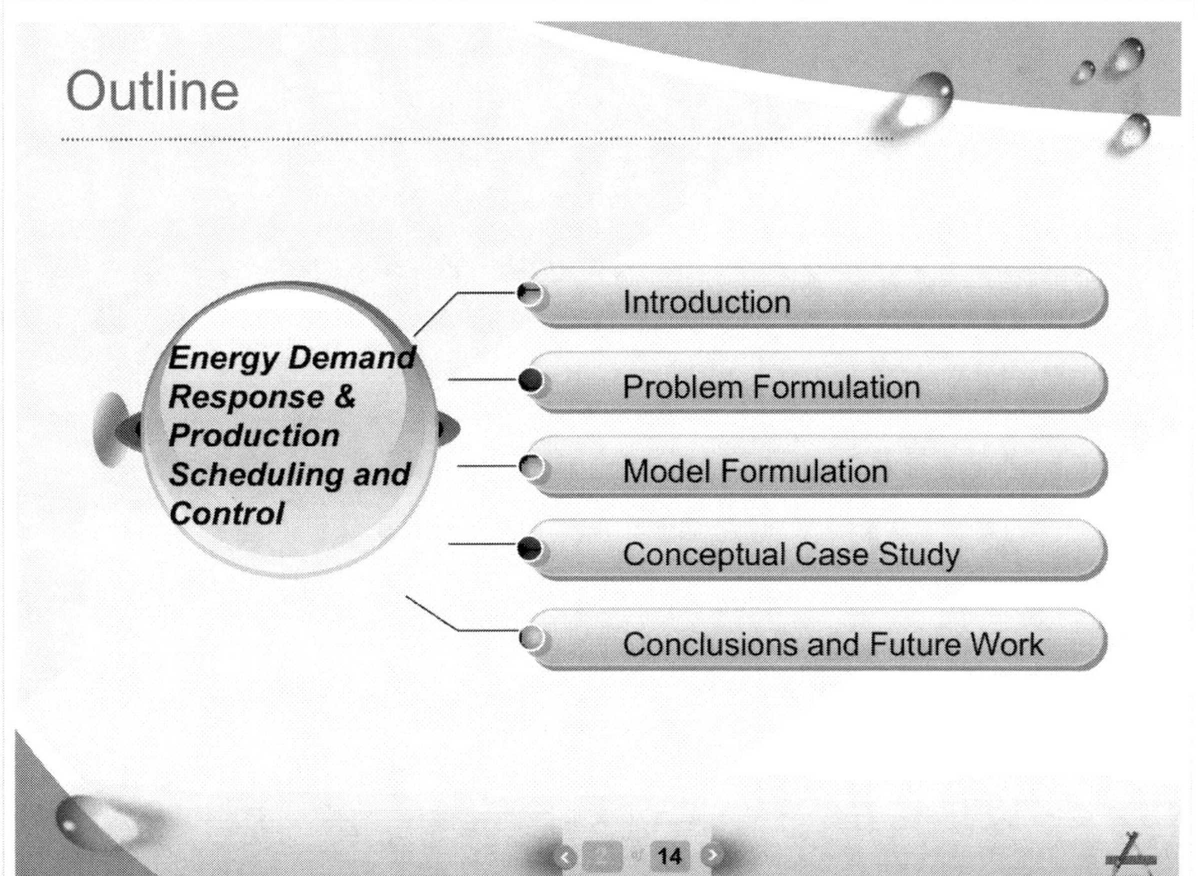
Outline
Energy Demand Response & Production Scheduling and Control
Introduction
Problem Formulation
Model Formulation
Conceptual Case Study
Conclusions and Future Work
14

Introduction

Demand Response

- Time-varying electricity pricing
- Improving economic performance
- Residential & Industrial

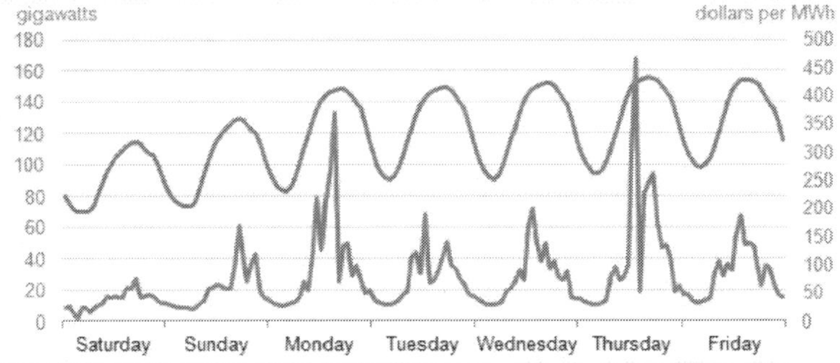

Hourly real-time energy prices and electricity demand in July, 2013

U.S. Energy Information Administration based on PJM data.

Introduction

Demand Response

Scheduling & Control

- Sequential - Simultaneous

Changing of Operating Conditions

↓

Transition Process ?

Related Work

On Demand Response

- Transition behavior problem has not been adequately addressed in the literature
- Minimum stay constraints for describing ramp-up transition and rate of change constraints for restricting transitions between operating points

  S. Mitra, I. E. Grossmann, J. M. Pinto, and N. Arora, "Optimal production planning under time-sensitive electricity prices for continuous power-intensive processes," Comp. Chem. Eng., 38, 171–184, 2012.

- The manufacturing facility is assumed to be operated at continuous production levels with the transitional behavior ignored

  D.I. Mendoza-Serrano and D. J. Chmielewski, "Demand response for chemical manufacturing using economic MPC. In Proc. 2013 American Control Conf, Washington DC, pp. 6655–6660.

On Scheduling & Control

- Some early attempts to tackle simultaneous scheduling and control problems without explicitly considering energy consumption
- The incorporation of energy constraints related to electricity pricing and availability increases the complexity of plant scheduling and control problems

Problem Formulation

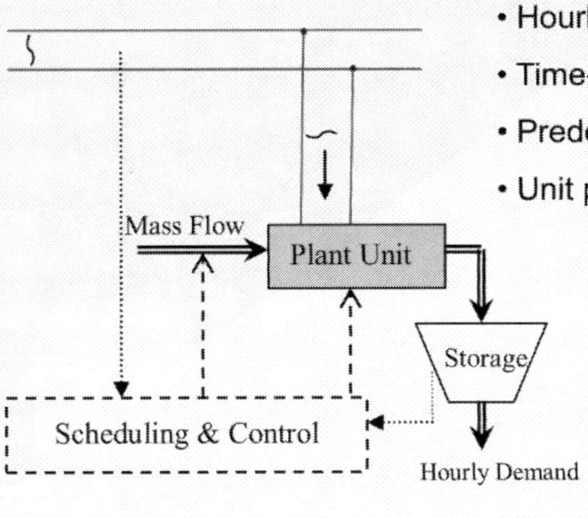

Constraints

- Hourly demand satisfied
- Time-varying electricity prices known a priori
- Predefined discrete steady-state operating modes
- Unit prices of inventory and raw material costs

Objective

- Simultaneous determination of operating modes at each time instant, and the control profile for the production changes
- Minimization of total cost

Solution Overview

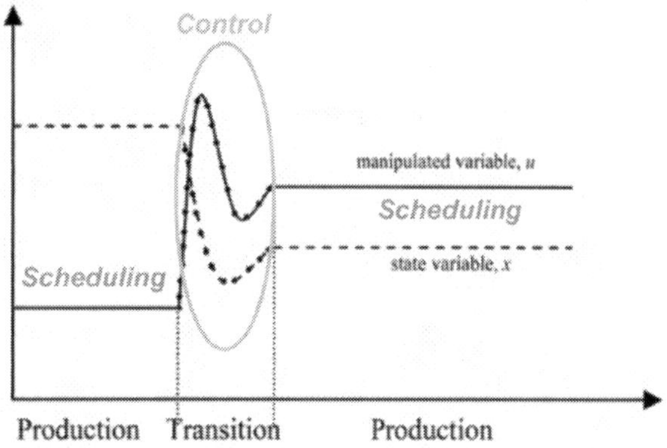

Assumptions

- Energy demand is assumed to be roughly proportional to the material flow
- Process operated under steady-state conditions generates on-spec product, while off-spec product is vented during transition

Incorporation of dynamic process model → MIDO problem

Pre-defined discrete operating modes ↔ Off-line calculation of control parameters and profiles for all possible transitions → MINLP problem

Model Formulation

Mode Assignment

$m \in M$: a set of discrete operating modes corresponding to different production levels

$\sum_{m \in M} y_m^h = 1 \quad \forall h$ → Only one mode can be active within each *h*

$\tilde{y}_m^h = y_m^{h-1} \quad \forall m, h \neq 1$ → Defines a backward binary variable

$\tilde{y}_m^1 = y_m^0 \quad \forall m$ → Assigns the backward binary variable with an initial value

Transition Logic Constraints

$$\sum_{m' \in M} z_{m',m}^h = y_m^h \quad \forall h$$

$$\sum_{m' \in M} z_{m,m'}^h = \tilde{y}_m^h \quad \forall h$$

→ If $z_{m,m'}^h = 1$, a transition from mode *m* to *m'* occurs from time step *h*-1 to *h*

Model Formulation

Storage Balance

The storage level in the storage unit at end of each time step is

$$s^h = \tilde{s}^h + \sum_{m \in M} (y_m^h \cdot p_m) \cdot \underbrace{\left[1 - \sum_{m',m \in M} (z_{m',m}^h \cdot t_{m',m}) \right]}_{\text{transition loss}} - d^h$$

$$\tilde{s}^h = s^{h-1} \qquad \forall h \neq 1$$

$$\tilde{s}^1 = s^0$$

Capacity limits

$$0 \le s^h \le s_{\max} \qquad \forall h$$

To avoid depleting the storage at the last time step, we set

$$s^H \ge s^0$$

d^h Hourly demand specified as a constant rate

p_m Production level associated with operating mode *m*

$t_{m',m}$ The transition time from *m'* to *m*, calculated offline

$t_{m,m} = 0, \ \forall m \in M$ self-transition

Model Formulation

Electricity Cost

Production Period + Transition Period

$$\Delta E^h = \alpha \cdot \sum_{m \in M} \left[y_m^h \cdot (\bar{u}_m^1 + \bar{u}_m^2 + \cdots + \bar{u}_m^n) \right] \cdot \left[1 - \sum_{m',m \in M} (z_{m',m}^h \cdot t_{m',m}) \right]$$

$$\Delta e_i^h = \alpha \cdot \sum_{m',m \in M} \left[z_{m',m}^h \cdot (u_{m',m}^{i,1} + u_{m',m}^{i,2} + \cdots + u_{m',m}^{i,n}) \right] \cdot \Delta t$$

$\bar{u}_m^1, \bar{u}_m^2, \cdots, \bar{u}_m^n$ represent the steady-state values of manipulated variables, and *n* is number of manipulated variables

$u_{m',m}^{i,1}, u_{m',m}^{i,2}, \cdots, u_{m',m}^{i,n}$ are sampled values of manipulated variables at each sampling step *i*, where $i = 1, 2, \cdots, N_{m',m}$ and $N_{m',m} \cdot \Delta t = t_{m',m}$

α is ratio of energy required to material flow rates

Model Formulation

Additional Transition Cost

$$C_{raw}^h = \vartheta_r \cdot \sum_{m',m\in M} \left(z_{m',m}^h \cdot t_{m',m} \cdot y_m^h \cdot p_m \right)$$

ϑ_{el}^h time-varying electricity prices on hourly basis

ϑ_s unit inventory cost

ϑ_r unit price of raw material

Objective Function

$$J = \min\{\Phi_1 + \Phi_2 + \Phi_3\}$$ → total cost minimum

$$\Phi_1 = \sum_{h=1}^{H} \left(\Delta E^h \cdot \vartheta_{el}^h \right)$$ → the cost of purchased electricity during production period

$$\Phi_2 = \sum_{h=1}^{H} \sum_{i=1}^{N_{m'm}} \left(\Delta e_i^h \cdot \vartheta_{el}^h \right) + \sum_{h=1}^{H} C_{raw}^h$$ → the transition cost that consists of wasted energy and raw material

$$\Phi_3 = \sum_{h=1}^{H} s^h \cdot \vartheta_s$$ → inventory cost

Conceptual Case Study

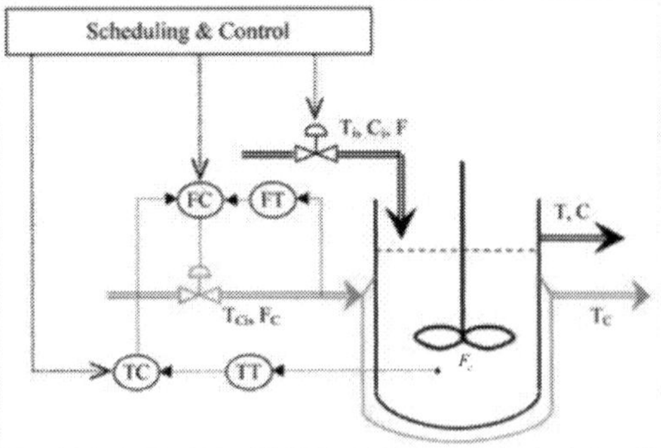

$$\frac{dC}{dt} = \frac{F}{V}(C_i - C) - k_0 \cdot e^{-E/RT} \cdot C$$

$$\frac{dT}{dt} = \frac{-\Delta H \cdot k_0 \cdot e^{-E/RT} \cdot C}{\rho C_p} + \frac{F}{V}(T_i - T) + \frac{UA_c \cdot (T_c - T)}{\rho C_p \cdot V}$$

$$\frac{dT_c}{dt} = \frac{F_c}{V_c}(T_{ci} - T_c) + \frac{UA_c}{\rho_c C_{pc} \cdot V_c}(T - T_c)$$

Operating modes	Production rates F (ft^3/h)	Jacket flow F_c (ft^3/h)
6	50 (1.416 m^3/h)	58.227 (1.648 m^3/h)
5	45 (1.274 m^3/h)	54.040 (1.530 m^3/h)
4	40 (1.133 m^3/h)	49.747 (1.409 m^3/h)
3	**35 (0.991** m^3/h)	**45.327 (1.284** m^3**/h)**
2	30 (0.850 m^3/h)	40.752 (1.154 m^3/h)
1	25 (0.708 m^3/h)	35.985 (1.019 m^3/h)

Comparison

(i) the conventional case ignoring transition costs

(ii) the proposed model with transition costs included

(iii) no demand response (only the nominal operating mode is active)

Simulation Results

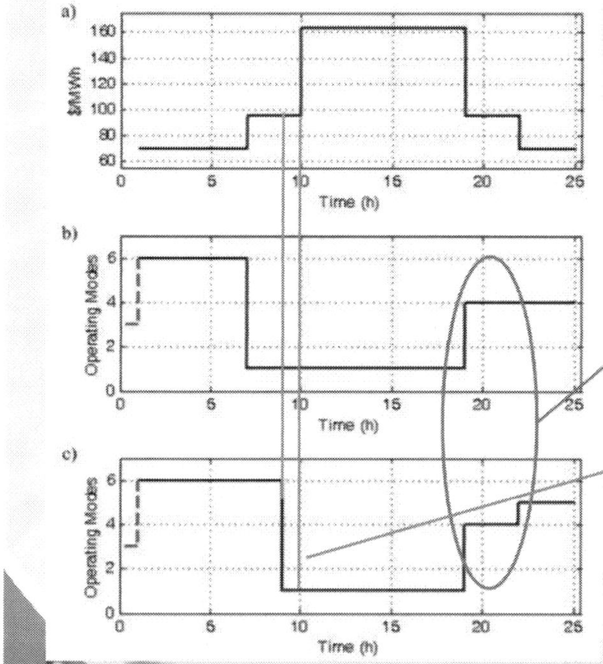

- The process switches to a lower production level when the electricity price rises up and vice versa
- Less frequent mode switching occurs when no transition costs considered
- The process tends to trigger mode switching when the electricity price is relatively low in order to reduce the electricity cost during the transition period

Simulation Results

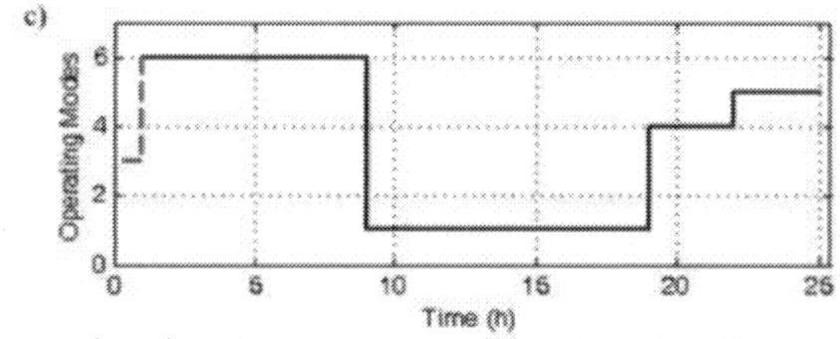

How the product demand is met from the storage unit when the process operates at the low production level

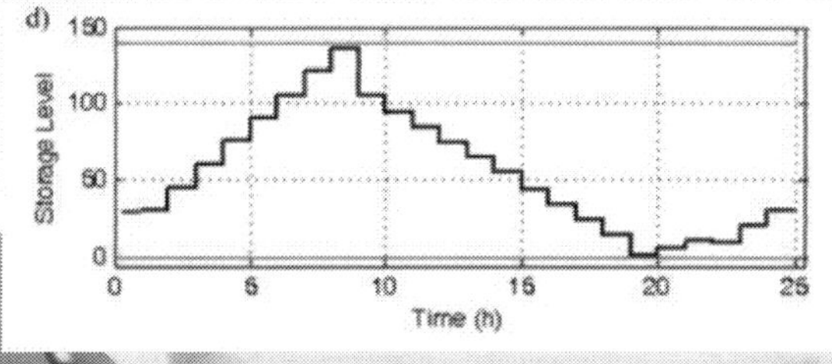

- (i) achieves the minimum cost due to the absence of transition cost
- With the proposed energy demand management, the total cost (ii) is reduced compared with the solution (iii)

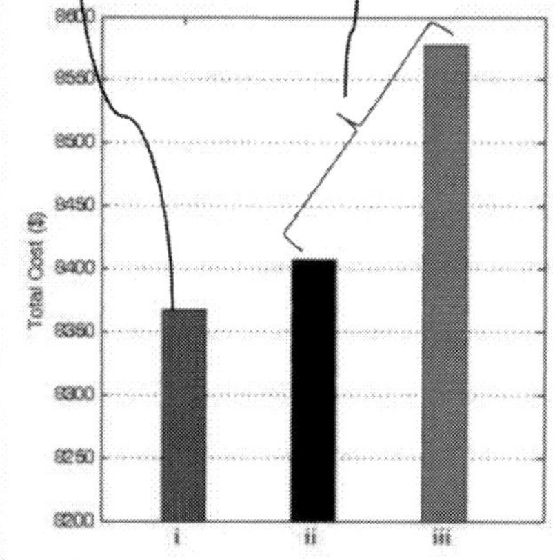

Conclusions and Future Work

- Energy demand management considering transition behavior and time-varying electricity prices is investigated.
- The formulated optimization model is capable of realizing demand response in terms of operating modes and transitional behavior.
- The scheduling and control problem is addressed in a closed-loop manner.
- Future work will focus on the integration of scheduling and control.

Thank you!

Support from China Scholarship Council (CSC) and UCDavis CHMS Department are gratefully acknowledged!

Reference

J. Aghaei, and M. I. Alizadeh, “Demand response in smart electricity grids equipped with renewable energy sources: A review,” *Renew. Sust. Energ. Rev.*, vol. 18, pp. 64–72, Feb. 2013.

S. Mitra, I. E. Grossmann, J. M. Pinto, and N. Arora, “Optimal production planning under time-sensitive electricity prices for continuous power-intensive processes,” *Comput. Chem. Eng.*, vol. 38, pp. 171–184, Mar. 2012.

Y. Wang and L. Li, “Time-of use based electricity demand response for sustainable manufacturing systems,” *Energy*, vol. 63, pp. 233–244, Dec. 2013.

H.G. Kwag and J.O. Kim, “Optimal combined scheduling of generation and demand response with demand resource constraints,” *Appl. Energy*, vol. 96, pp. 161–170, Aug. 2012.

M. Paulus and F. Borggrefe, “The potential of demand-side management in energy-intensive industries for electricity markets in Germany. *Appl. Energy*, vol. 88, pp. 432–441, Feb. 2011.

D.I. Mendoza-Serrano and D. J. Chmielewski, “Demand response for chemical manufacturing using economic MPC. *In Proc. 2013 American Control Conf*, Washington DC, pp. 6655–6660.

R. Mahadevan, F. J. Doyle, and A. C. Allcock, “Control-Relevant Scheduling of Polymer Grade Transitions,” *AIChE J.*, vol. 48, pp. 1754–1764, Aug. 2002.

X. Wang, A. N. Palazoglu, N. H. El-Farra, "Operation of residential hybrid renewable energy systems: Integrating forecasting, optimization and demand response", in: Proceedings of American Control Conference, 2014.

377680 Bittorrent-Inspired Distributed Energy Resource Management: A P2P Framework for Utility 2.0

Abstract:

This work examines a peer-to-peer (P2P) approach to management and control of distributed energy resources (DER) on an electric power distribution grid based on a modified version of the BitTorrent filesharing protocol[1]. It is proposed that customers with DER - defined as small (<50 kW) generation, storage, and/or flexible loads - be treated as active peers in a power distribution swarm coordinated via the Internet. The sheer number of possible future IP-enabled devices of this category (the so-called 'Internet of things') motivates the need for a decentralized management strategy, as centralized approaches are not well suited to this scale and complexity of nodes.

The proposed management strategy is inspired by the ability of the BitTorrent filesharing protocol to efficiently distribute large files over bandwidth-constrained networks, while minimizing network capacity requirements[2]. Our modified protocol applies the principles of BitTorrent file distribution to the problem of electric power grid ancillary services provision by making an analogy between data and energy flows, in which DER can upload or download energy to or from grid peers at a certain power, respectively. In this framework, simple distributed client controllers autonomously dispatch local DER as needed, while attempting to satisfy both local and swarm objectives using common advanced control and optimization techniques such as model predictive control. This approach can both reduce transmission congestion by increasing local resource utilization and decreasing peak demand as well enable resilient and efficient operation of microgrids post-emergency (e.g. Hurricane Sandy).

To test the modified protocol, an open source python BitTorrent client is modified to manage a small swarm of window air conditioning units via infrared using a Raspberry Pi Model B single-board linux machine[3] and a custom IR transceiver circuit. The A/C units function as physical devices participating in an otherwise virtual simulation. Real time power use of each A/C unit is collected every 2 seconds by Tweet-a-watt[4] modified Kill-a-watt power meters using ZigBee communication. This small-scale swarm demonstrates the potential of the BitTorrent-inspired framework to reduce peak demand and enhance efficiency through integrating local execution of real-time models and optimization with P2P dispatch of network DER.

References:

1) B. Cohen, Proc. First Workshop on Economics of Peer-to-Peer Systems, (2003)

2) A. R. Bharambe, C. Herley, V. N. Padmanabhan, SIGMETRICS'05, June 6–10, (2005)

3) Pi, Raspberry. "An ARM GNU/Linux box for $25." Take a byte (2012)

4) Flahiff, Daniel. TWEET-A-WATT! A power monitor that twitters your energy use. Inhabitat. Retrieved on March 05, 2009 from: http://www.inhabitat.com/2009/02/04/tweet-a-watt-by-ladyada/ (2009)

UNIVERSITY OF MINNESOTA

Nonlinear Deadtime Compensated Control of Steam Power Plants

Nahla Alamoodi and Prodromos Daoutidis
University of Minnesota
Minneapolis, MN

ID# 373610

2014 AIChE Annual Meeting

Atlanta, GA

1

Conventional power plants and projected energy landscape

New challenges for conventional power plant control :

- ✧ Large variations of load demand due to introduction of renewable power
- ✧ Coal preprocessing a source of deadtime
- ✧ Integrating different types of fuel

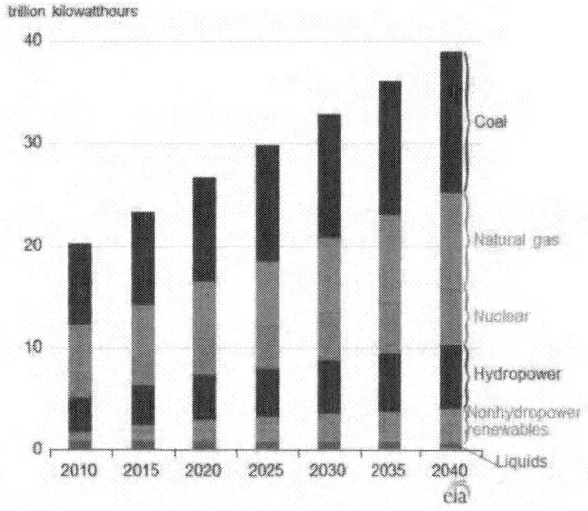

World net electricity energy by source, 2010-2040*

*US Energy Information Administration, *International Energy Outlook 2013.*

2

Steam power plant

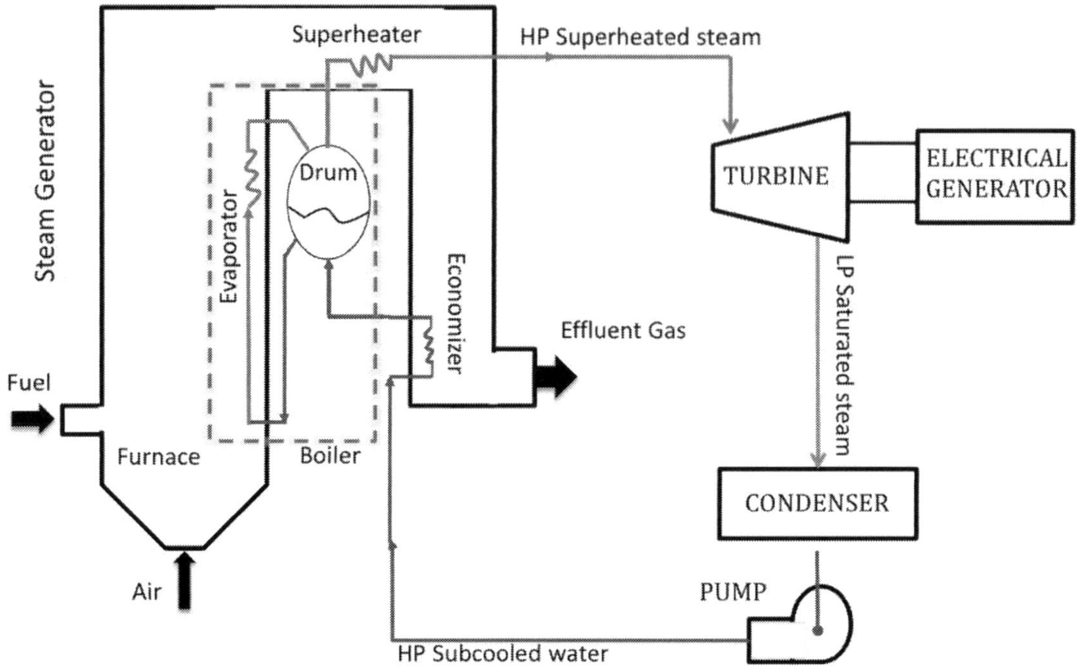

3

Objective: Control of power generation in turbine-generator system and pressure in boiler

- Strong interactions between power generation and pressure in the boiler
- Introduction of dead time by coal
- Nonlinear process

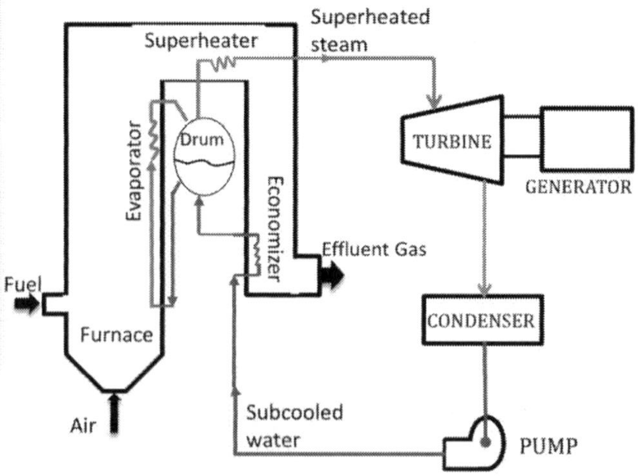

4

Current state-of-the-art technology, previous and present work

- Current technology:
  - PID controllers with compensation for time delays, nonlinearities, interactions and I/O constraints[1]
- Previous studies:
  - SISO Coordinated Control Strategies based on PID controllers[2]
  - Adaptive[3] and linear predictive control[4]
- Present study:
  - MIMO nonlinear model-based controller with deadtime compensation

1-US department of Energy, Plant process control workshop, 2006.
2- Garduno-Ramirez, R., Lee K. Y., *Control Engineering Practice*, 2005, 13.
3-Liu, H., Li.,S., Chai., T., *Electrical Power and Energy Systems*, 2003, 25.
4--Liu, X., Guan, P., Chan, C.W., *IEEE Transactions on control systems technology*, 2010, 18 (5).

5

Process modeling

Boiler model

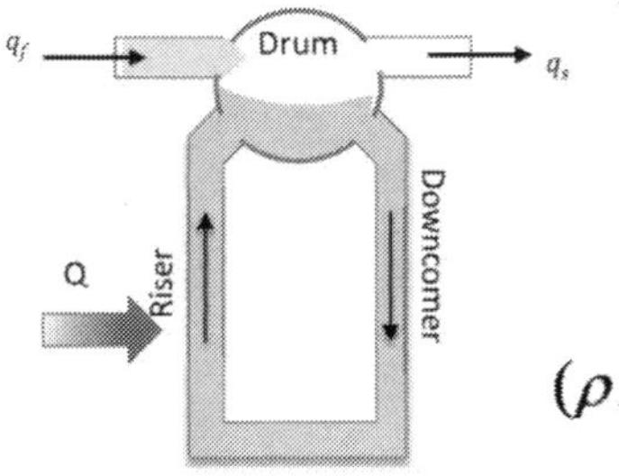

Mass and energy balance:

$$\frac{d\rho}{dt}=\frac{\partial\rho}{\partial P}\frac{dP}{dt}, \quad \text{and } V_t=V_w+V_s$$

$$(\rho_w-\rho_s)\frac{dV_w}{dt}+\left[V_w\left(\frac{\partial\rho_w}{\partial P}\right)+V_s\left(\frac{\partial\rho_s}{\partial P}\right)\right]\frac{dP}{dt}=q_f-q_s$$

$$\left[V_s\left(\rho_s\frac{\partial h_s}{\partial P}\right)+V_w\left(\rho_w\frac{\partial h_w}{\partial P}\right)-V_t+m_tC_p\frac{\partial T^{sat}}{\partial P}\right]\frac{dP}{dt}+(h_s-h_w)\frac{d(\rho_sV_s)}{dt}=q_f(h_f-h_w)-q_sh_c+\eta_b\Delta Hq_e$$

$$q_s=\frac{kP}{\sqrt{T^{\text{sup}}}}u_s$$

Astrom, K., Bell, R., *Automatica*, 2000, 36.

6

Process modeling
Turbine-generator model

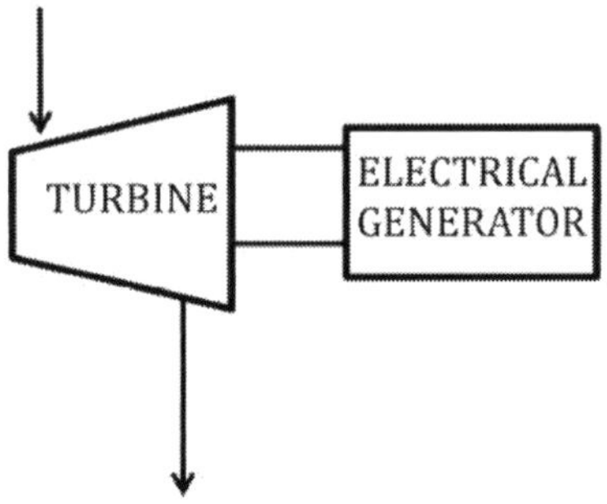

- *Approximate lumped model*

$$\tau \frac{dE}{dt} = \eta_T q_s \left(h^{\sup}\left(T^{\sup}, P_b^{sat}\right) - h^{sat}\left(P_{cond}^{sat}\right)\right) - E$$

Steam and water properties are evaluated using the standard formulas given by the International Association for the Properties of Water IAPWS-IF97

–

7

Controller design

Control problem consists of:

✧Two manipulated variables q_e and u_s

✧Two state variables *E* and *P*

✧Deadtime associated with the fuel flow rate θ

✧ Parameters adopted from P16-G16 – A 160 MW coal-fired power plant located in Sweden*

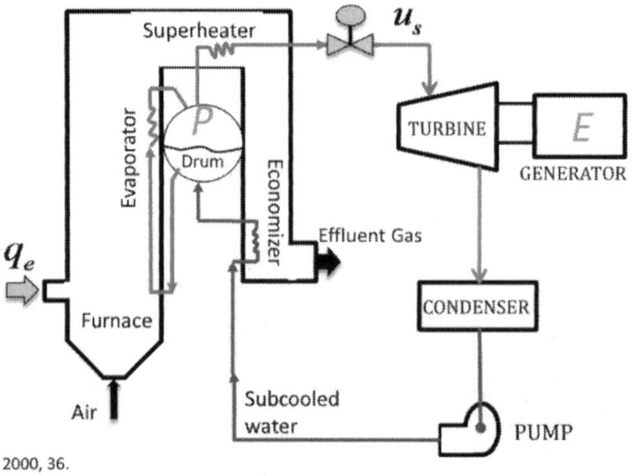

*Astrom, K., Bell, R., *Automatica*, 2000, 36.

8

Model state feedback structure* with deadtime compensation**

$$\frac{dx}{dt} = f(x) + g_1(x)u_1 + g_2(x)u_2$$
$$y_1 = h_1(x)$$
$$y_2 = h_2(x)$$

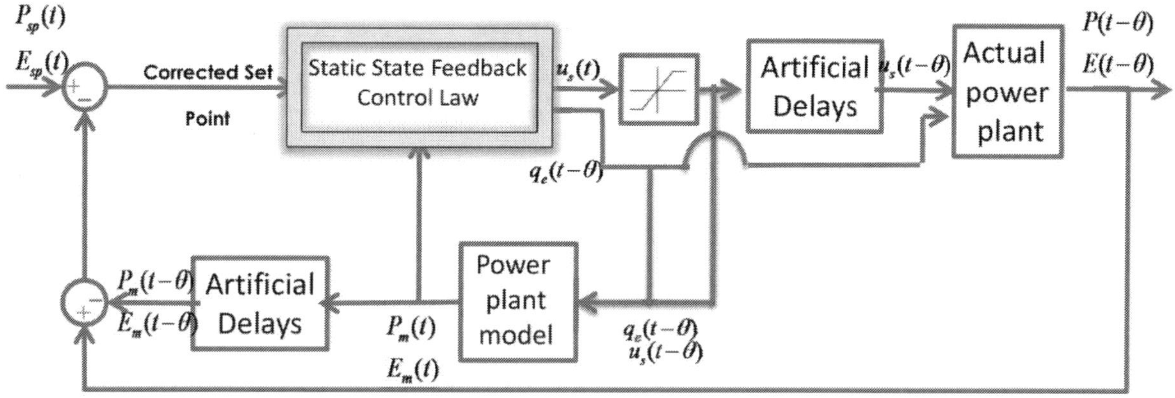

*Kravaris, C., Daoutidis, P., Wright, R., *Chemical Engineering Science*, 1994, 49, 13
**Kravaris C., Wright, R., *Chemical Engineering Science*, 2003, 58.

9

Linearizing state feedback in the absence of deadtime

If all r_i are finite, and the decoupling matrix

$$C(x) = \begin{bmatrix} L_{g_1}L_f^{r_1-1}h_1(x) & L_{g_2}L_f^{r_1-1}h_1(x) \\ L_{g_1}L_f^{r_2-1}h_2(x) & L_{g_2}L_f^{r_2-1}h_2(x) \end{bmatrix}$$ is nonsingular

The state feedback control law:

$$u = C(x)^{-1} \begin{bmatrix} \dfrac{v_1 - \sum_{k=0}^{r_1} \beta_{1k} L_f^k h_1(x)}{\beta_{1r_1}} \\ \dfrac{v_2 - \sum_{k=0}^{r_2} \beta_{2k} L_f^k h_2(x)}{\beta_{2r_2}} \end{bmatrix} \Rightarrow$$

Decoupled linear input/output closed-loop system:

$$\begin{bmatrix} \sum_{k=0}^{r_1} \beta_{1k} y_1^{(k)}(t) \\ \sum_{k=0}^{r_2} \beta_{2k} y_2^{(k)}(t) \end{bmatrix} = \begin{bmatrix} v_1 \\ v_2 \end{bmatrix}$$

$\beta_{1k} \cdots \beta_{2k}$ are tunable parameters

v_1, v_2 are reference inputs

10

Model state feedback structure* with deadtime compensation**

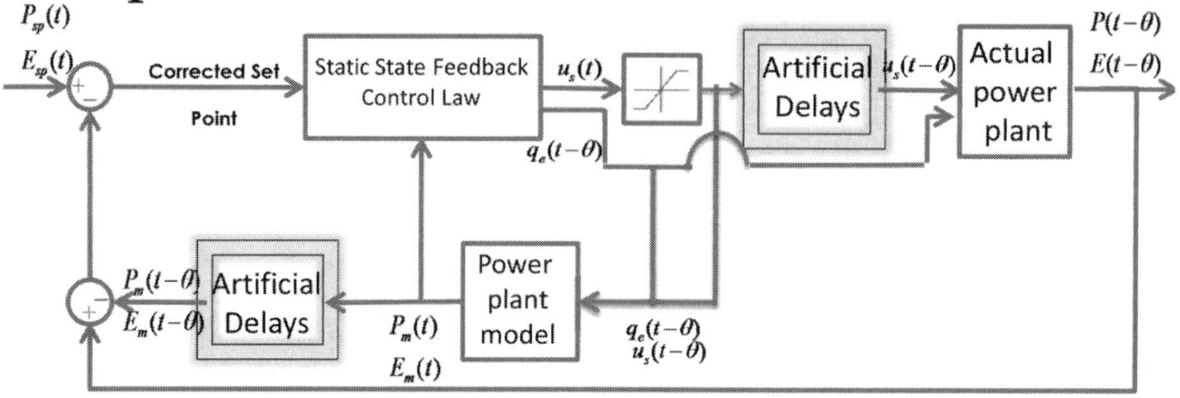

Closed-loop linearized decoupled system:

$$P(t)+\beta_{11}\frac{dP(t)}{dt}=P_{sp}(t-\theta)$$

$$E(t)+\beta_{21}\frac{dE(t)}{dt}=E_{sp}(t-\theta)$$

For performance comparison a PI-Generalized Multi Delay Compensator (PI-GMDC)★ is designed and tested.

*Kravaris, C., Daoutidis, P., Wright, R., *Chemical Engineering Science*, 1994, 49, 13
**Kravaris C., Wright, R., *Chemical Engineering Science*, 2003, 58.
★Jerome, N., Ray, W., *AIChE Journal*, 1986, 32, 6.

11

Tests & Results

A-Narrow range control: Setpoint tracking of power demand

Multiple step changes of 80→86.8→82.5→78.5 MW in the setpoint of power demand

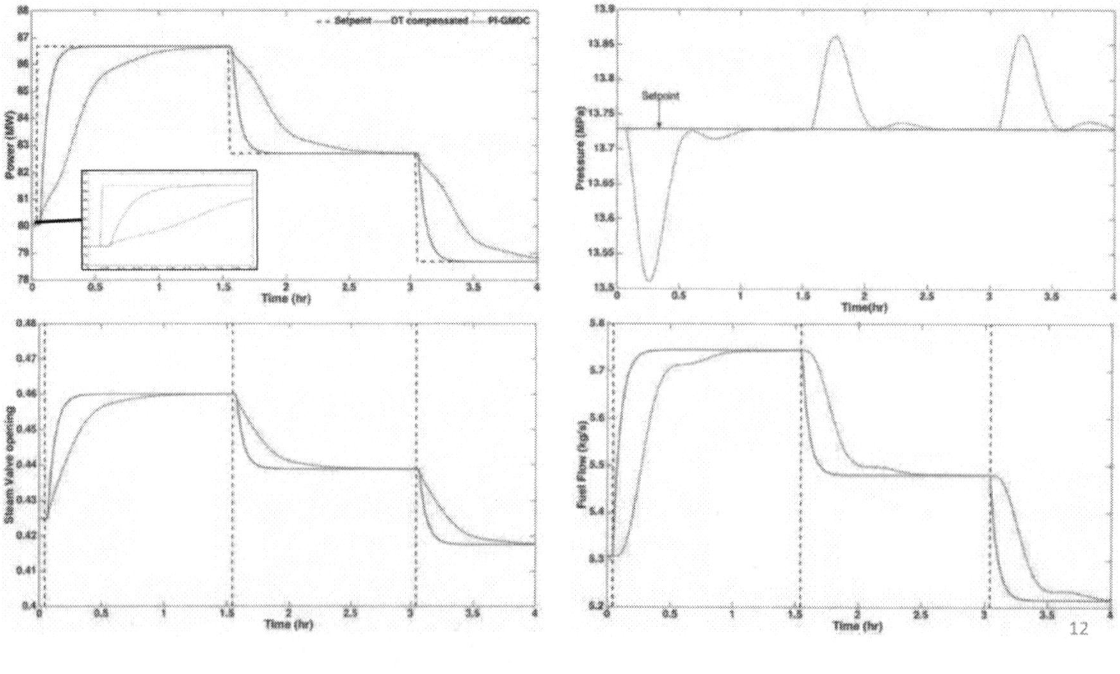

12

Narrow range control: Setpoint tracking of power demand

Multiple step changes of 80→86.8→82.5→78.5 MW in the setpoint of power demand

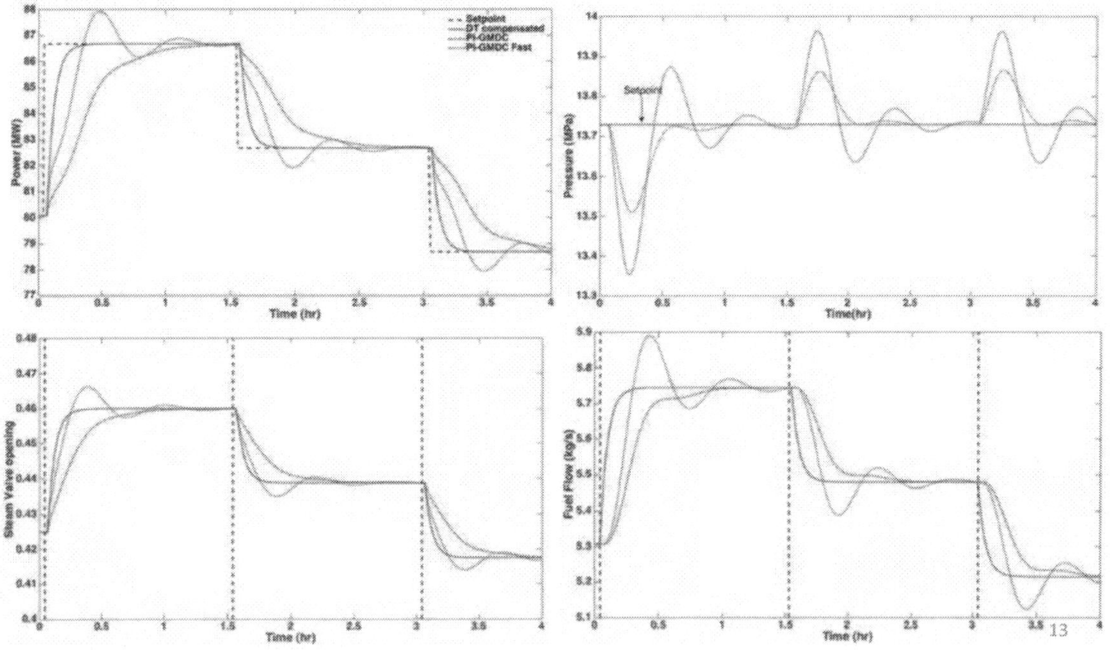

B- Disturbance rejection for a 5% decrease in the boiler efficiency

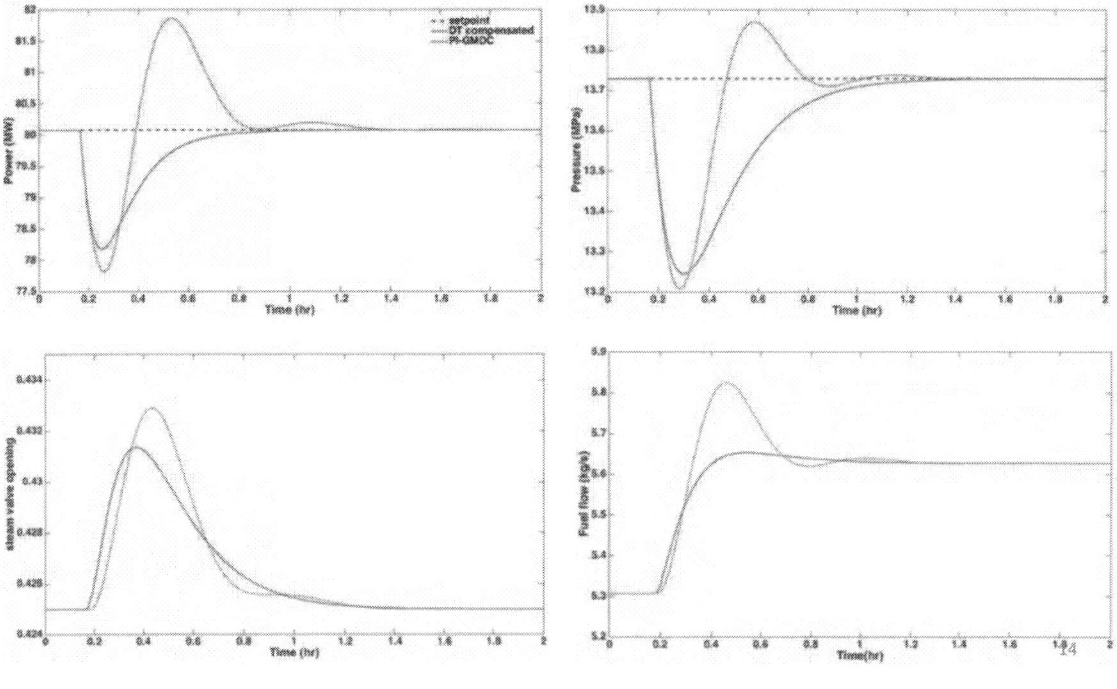

C- Wide range control: Tracking of ramping power setpoint at 2% /min*

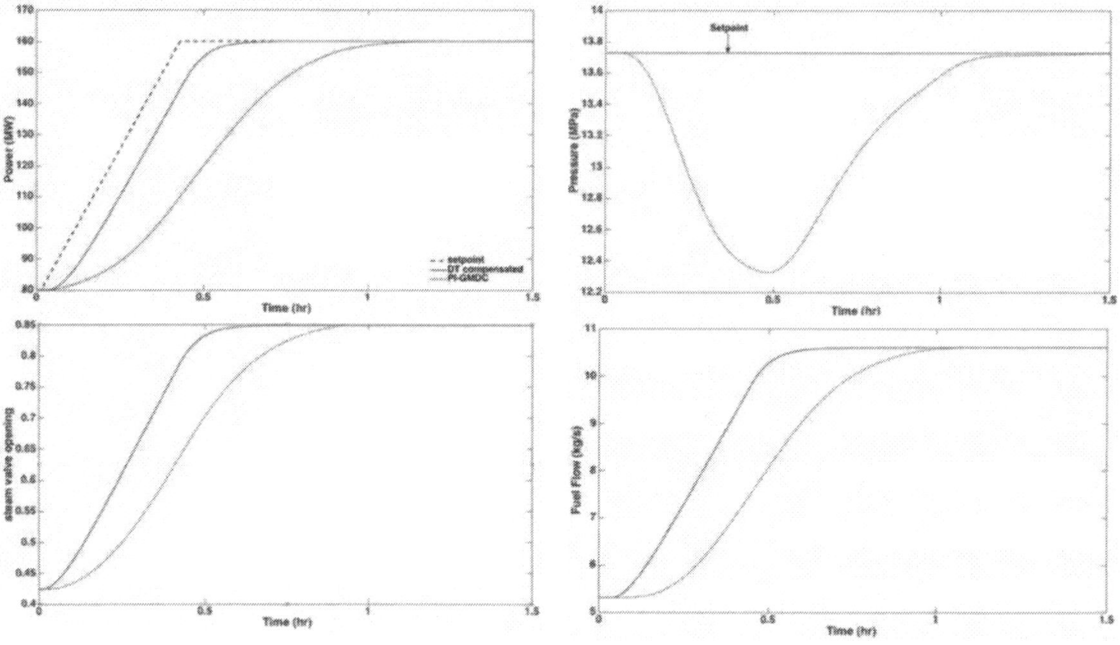

*H. Spliethoff, *Power Generation from Solid Fuels*, Springer, 2010.

15

D- Wide range control: Tracking of large stepping of power setpoint ---"worst case" scenario

- ±20% MCR Step changes to power setpoint at 20 minutes intervals
- Random step changes 90-120s* in the deadtime associated with fuel flow

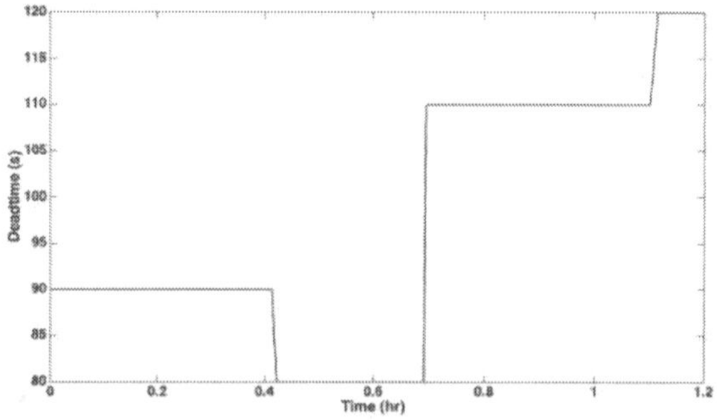

* J. M. Blaazer, "Advanced process control for power plants: Improving overall performance through control of internal process variables", M.S. thesis, 3mE, TUDelft, Netherlands, 2010.

16

Wide range control: Tracking of large stepping of power setpoint

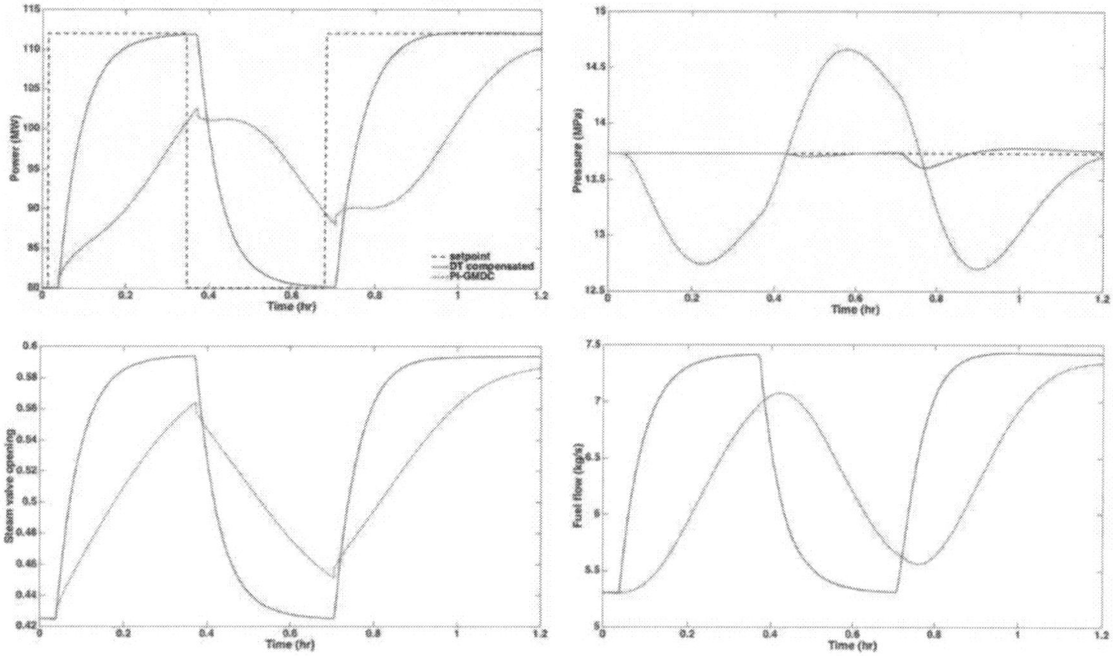

17

Concluding remarks

- High performance of nonlinear controller in tracking changes in power and rejecting disturbances
- Strong candidate to face challenges of renewable power introduction to power industry
- Future study will include water level dynamics inside the drum

18

Acknowledgement

Daoutidis Research Group

Sponsors:

أدنــوك

ADNOC

شركة بترول ابوظبي الوطنية

19

Thank you

20

Abstract Title:
Theoretical & Experimental Investigation of the Effects of Polymer Chemistry on Polymer-Electrolyte Dye Sensitized Solar Cell Performance

Author:
Yuriy Y. Smolin, Siamak Nejati, Mona Bavarian, Kenneth K.S. Lau, and Masoud Soroush*

Keywords:
First-Principles Macroscopic Modeling, Polymer Electrolyte, Dye Sensitized Solar Cell, Initiated Chemical Vapor Deposition

Section:
Modeling and Control of Energy Systems (10B11)

Abstract:
Dye sensitized solar cells are commonly regarded as one of the most promising next generation solar cell technology[1], because they have good performance and can be made relatively inexpensively without intensive purification and fabrication steps. However, the liquid electrolyte of conventional DSSCs are prone to leakage and evaporation—hindering DSSC applications, durability, and thermal stability. Novel polymer electrolyte DSSCs can overcome shortcomings and can enhance the cell I-V behavior. Recent experimental studies have demonstrated that polymer electrolytes are promising alternatives to liquid electrolytes in DSSCs[2-4]. Currently, there is a need for better understanding how polymer electrolyte chemistry affects cell performance. Accurate first-principles models can provide the better understanding, and the needed insight and guidance. By properly describing the physical and electrochemical processes occurring inside the cells, it is possible to elucidate the competing effects of DSSC components on the overall cell performance and guide experimental work.

Continuing our previous work[5] on mathematical modeling of a liquid-electrolyte DSSC, in this work theoretically and experimentally we investigate the effects of polymer-electrolyte chemistry on the performance of polymer-electrolyte DSSCs. Furthermore, to gain a complete understanding of the photochemical processes inside the DSSCs, a first-principles macroscopic model is developed and integrated with experiments using initiated chemical vapor deposition (iCVD) to synthesize and incorporate the polymer electrolytes within the mesoporous TiO_2 photoanode. The model is used to elucidate the effects of polymer chemistry on the cell performance, and to determine how and why the

photochemical processes inside such solar cells are altered by poly(2-hydroxyethyl methacrylate)

(PHEMA), poly(glycidyl methacrylate) (PGMA), and poly(4-vinylpyridine) (P4VP). By varying the polymer electrolyte chemistry, this work show that DSSC cell characteristics, including open circuit voltage, short circuit current density, and fill factor can be tuned. Parameter estimation using the macroscopic model indicates that a shift in the conduction band of TiO_2 and a suppression of the back electron transfer at the dye-TiO_2-electrolyte interface are induced by the methyl ester of PGMA, the pyridine group of P4VP, and the hydrophilic pendant group of PHEMA. The estimated differences between the TiO_2 conduction band edge and the redox potential of the electrolyte are 0.873 eV, 0.994 eV and 1.059 eV for P4VP, PGMA, and PHEMA, respectively. Estimates of the recombination rate constant for PVP and PGMA are, respectively, 65% and 39% lower than that for PHEMA.

* Corresponding Author: ms1@drexel.edu

(1) Mari, O.; Kei, O.; Ai, S.; Hideyuki, T.; Nozomu, H.; Akira, E.; Hiromitsu, T.; Momoji, K.; Akira, M. *Japanese Journal of Applied Physics* **2010**, *49*, 04DP10.
(2) Nogueira, A. F.; Durrant, J. R.; De Paoli, M. A. *Advanced Materials* **2001**, *13*, 826.
(3) Wang, P.; Zakeeruddin, S. M.; Moser, J. E.; Nazeeruddin, M. K.; Sekiguchi, T.; Gratzel, M. *Nat Mater* **2003**, *2*, 402.
(4) Nejati, S.; Lau, K. K. S. *Nano Letters* **2010**, *11*, 419.
(5) Bavarian, M.; Nejati, S.; Lau, K. K. S.; Lee, D.; Soroush, M. *Industrial & Engineering Chemistry Research* **2013**.

Cyber-Physical-Energy Systems Testbed: A Distributed Solar Power Case Study

Authors: Elaine Hale, Bryan Palmintier, Jin Wei, Roisin Mossop, Julieta Giraldez, David Biagioni, Wesley Jones, Bri-Mathias Hodge

The future electricity system is a distributed control problem on an enormous scale: millions of generation and load devices requiring coordinated but decentralized decision making. The interactions between a residential rooftop PV system and the household appliances will feed up into the local distribution system. The interactions among the different residential and commercial systems feed into a distribution level supply or demand that will be further fed up into the transmission system. At the transmission level coordination is necessary to be able to provide ancillary services or to link sub-systems where supply and demand are mismatched with those that currently have the opposite power balance due to fluctuating conditions. The cyber (information flow) layer architecture formalizing the interactions between these different systems needs to be developed while respecting that the physical electricity interactions are dictated by the physics of the electricity and the communication systems.

Historically in the transmission and distribution networks have been modeled independently of one another due to computational limitations. However, this does not allow for an understanding of the impact of distributed devices on bulk system operations, nor of the impact of transmission level policies on distributed generation. In this work we have developed a simulation platform that ranges from the appliance-level to the Independent System Operator (ISO) level which allows for the investigation of these impacts. The open source software GridLab-D is utilized for distribution level system modeling, and has been linked with the FESTIV transmission level model to represent both ends of the power system operations spectrum. As most transmission level substations have multiple distribution feeders, and the typical timescale for feeders is much lower than for bulk system operations, thousands of GridLab-D instances must be run to supply the information for each FESTIV time step. These simulations are only possible to be completed in realistic timescales due to the use of NREL's Peregrine high performance computing system.

The testbed has been utilized to study the impacts of high penetration distributed solar power on transmission level system operations. Distributed solar power installations are currently growing very quickly in a number of areas of the United States, and are beginning to impact transmission system operations. For example, while system operators generally do not have any visibility into distributed solar installations or their power output, and only have visibility into their output based on deviations from their load forecasts. The case study examined includes the impact of high penetration distributed solar power on the unit commitment and economic dispatch process, bus voltage levels, and distributed solar power forecasting, among other options.

Dynamic Optimization of Cryogenic Carbon Capture with Large-scale Adoption of Renewable Power

Seyed Mostafa Safdarnejad
Trent Hall
John Hedengren
Larry Baxter

Chemical Engineering Department
Brigham Young University (BYU)
November 2014

1

Outline

- Background of the Research
- Challenges for Power Sector
- Scope of the Study
- Results
- Future Work

2

PRISM

Outline

- ➢Background of the Research
- ➢Challenges for Power Sector
- ➢Scope of the Study
- ➢Results
- ➢Future Work

3

PRISM

Increased Electricity Production from Renewable Energy (RE) Sources

- ➢ 733% growth in RE supply from 1990 to 2012
- ➢ 50% of globally added electricity generating capacity from RE (2022)

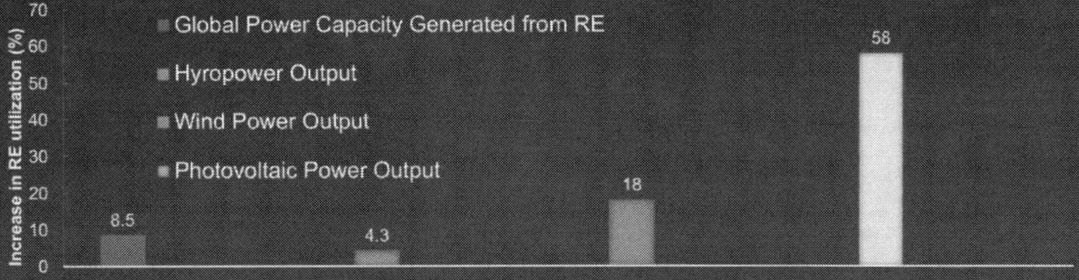

[1] Trends in global CO_2 emissions, PBL Netherlands Environmental Assessment Agency, 2013

4

Increase in CO_2 Emission

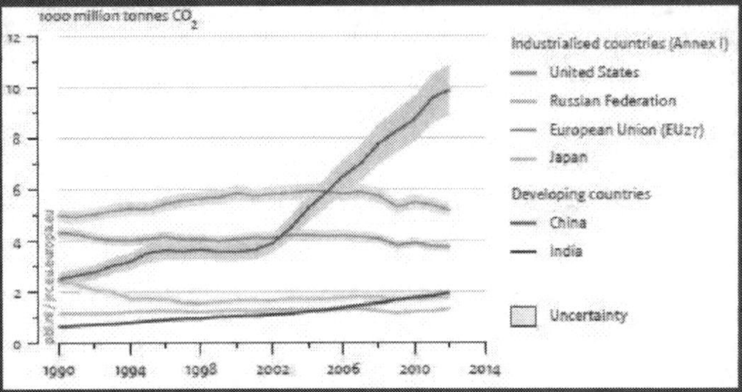

- 1.1 % increase in global emissions from 2011 to 2012
- 38% of the US CO_2 emission from electricity generation

[1] Trends in global CO_2 emissions, PBL Netherlands Environmental Assessment Agency, 2013
[2] http://www.epa.gov/

5

Increasing Restriction for CO_2 Emission

- Clean Air Act CO_2 emission regulations for new power plants (Sep. 2013)
- EPA's Clean Power Plan CO_2 emission regulations for existing fossil – fueled power plants (June 2014)

Solutions:

1- Shift from coal to natural gas (NG)

2- Increased utilization of the RE sources

3- Integrate CO_2 removal system with power plants

→ The main reasons for 4% reduction in CO_2 emission in the US in 2012

6

Outline

- Background of the Research
- Challenges for Power Sector
- Scope of the Study
- Results
- Future Work

7

Challenges for Power Sector

Intermittent RE sources

- Integration of RE sources with conventional power sources
- Storage of RE

Impact of CO_2 removal system on the power plant

- High energy consumption of physical and chemical absorption CO_2 removal systems
- Development of new technologies for CO_2 removal

8

PRISM

Cryogenic Carbon Capture ™ (CCC)

- Separation of CO_2 by desublimation
- Consumes less energy than chemical and physical absorption systems
- Can store energy in the form of the refrigerant (LNG)
- Manages the energy loss of CO_2 removal during peak demand
- Rapid-load-change capability
- Scalability of the proposed storage system to large-scale systems

9

PRISM

Outline

- Background of the Research
- Challenges for Power Sector
- Scope of the Study
- Results
- Future Work

10

Scope:

- Dynamic integration and optimization of CCC with power generation units
- 90% CO_2 capture rate
- Maximize the operational profit of the hybrid system
- Meet the residential and CCC electricity demands

11

Hybrid System of Power Generation and CCC

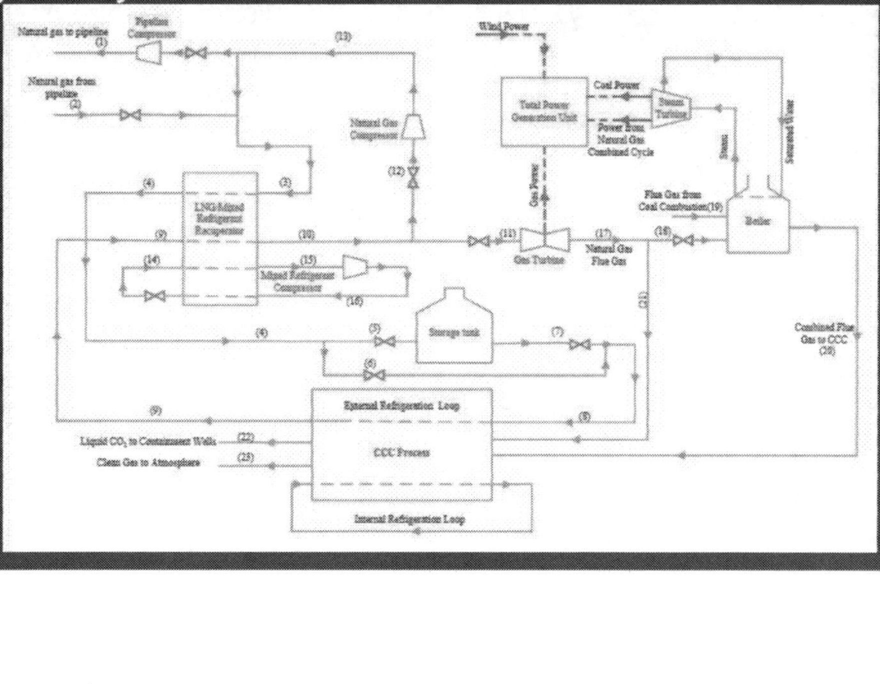

12

BYU PRISM

Energy Storage of Cryogenic Carbon Capture

- Two refrigeration cycles
- Overproduction and storage of LNG during low electricity demand
- Using stored LNG during peak hours
- More power available by ramping down the refrigeration compressor load
- Power production in a gas turbine during peak demand

13

BYU PRISM

Optimization Approach

- Objective function: $\ell 1$-norm

$$\min_{x,y_m,u} \Phi = w_{hi}^T e_{hi} + w_{ho}^T e_{lo} + y_m^T c_y + u^T c_u + \Delta u^T$$

$$s.t.\;\; 0 = f(\dot{x}, x, u, d)$$

$$0 = g(y_x, x, u, d)$$

$$a \geq h(x, u, d) \geq b$$

$$\tau_c \frac{\delta y_{t,hi}}{\delta t} + y_{t,hi} = sp_{hi}$$

$$\tau_c \frac{\delta y_{t,lo}}{\delta t} + y_{t,lo} = sp_{lo}$$

$$e_{hi} \geq (y_m - y_{t,hi})$$

$$e_{lo} \geq (y_{t,lo} - y_m)$$

- Orthogonal collocation on finite elements for DAE to NLP conversion
- Active set Method or Interior Point Optimization Method by using APMonitor Modeling Language

14

Outline

- ➢Background of the Research
- ➢Challenges for Power Sector
- ➢Scope of the Study
- ➢Results
- ➢Future Work

15

Power Production vs. Electricity Demand

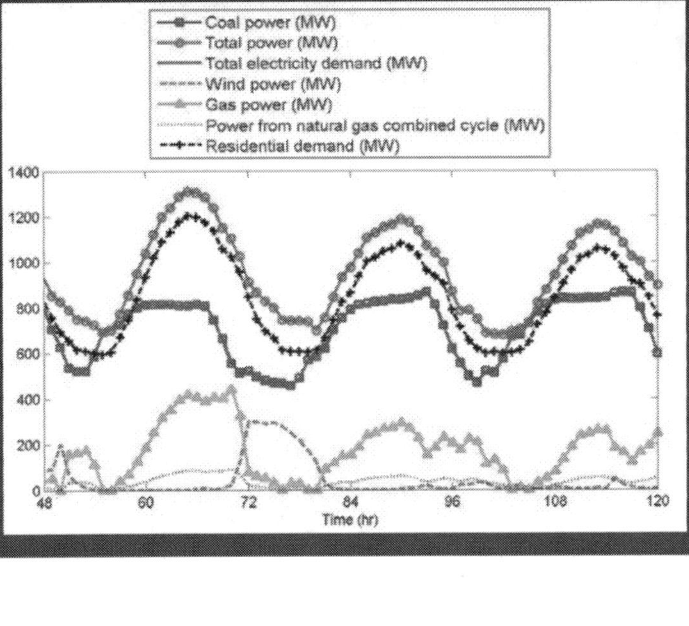

- ➢ Met the total electricity demand
- ➢ No wasted energy
- ➢ 100% utilization of the wind power
- ➢ Energy storage of CCC corresponds with power production from NG
- ➢ Shifting wind power to peak demand by using energy storing CCC

16

Natural Gas (NG) Imported to and Exported from the Plant and LNG Inventory in the Tank

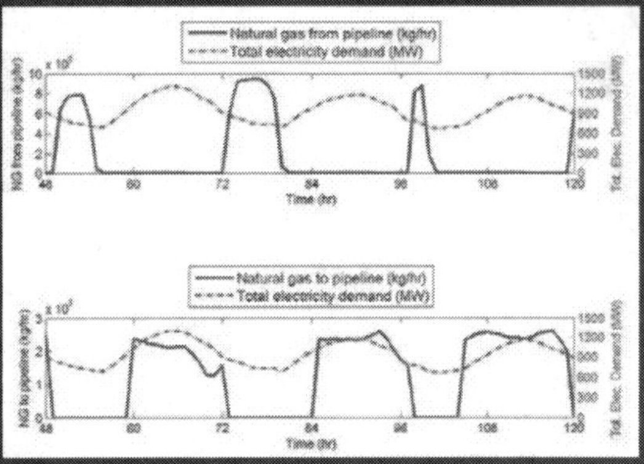

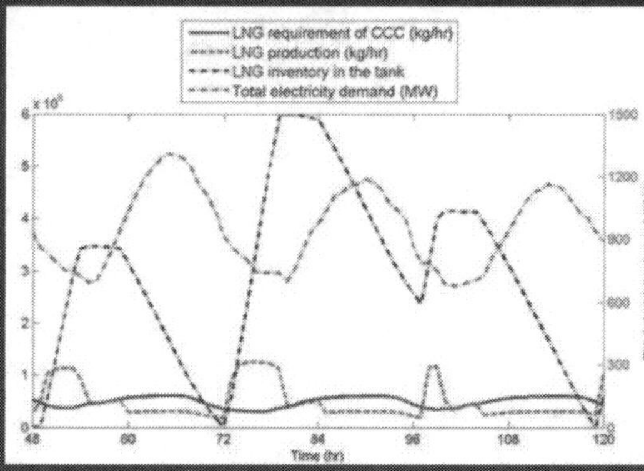

- Import of NG during low electricity demand
- NG export during high electricity load
- LNG supply from tank during peak hours

17

Demand Curves for the Main Electricity Consuming Components of the Hybrid System

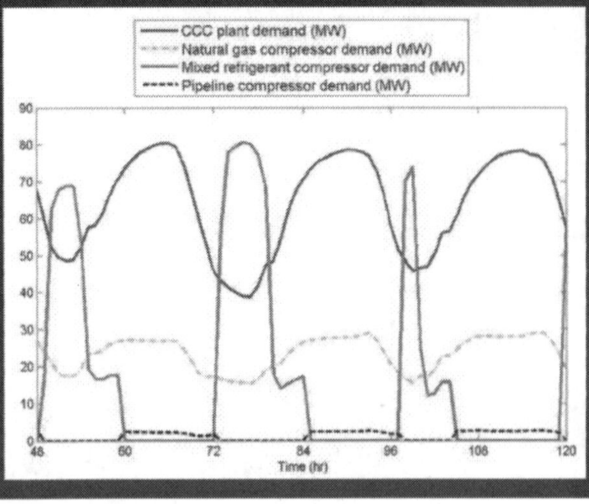

- Dependency of the NG compressor and CCC plant demands on the residential electricity demand
- Decrease of the mixed refrigerant compressor demand during peak hours
- Transfer of the saved electricity to meet the peak demand
- The main energy storing aspect of CCC (meeting the parasitic load of the CCC-based carbon capture with stored LNG)

18

Profitability Comparisons

- $15k/hr average profit without storage tank
- 68% decrease in the profit without NG export
- $33k/hr with storage tank (122% increase in the profitability compared to the non-storing CCC)
- Extra revenue from storage system ≈ 2.1 cent/kWh
- Cost of construction of the CCC plant ≈ 2 cent/kWh
- **Extra revenue from the storage system covers the construction cost of the CCC plant**

19

Outline

- Background of the Research
- Challenges for Power Sector
- Scope of the Study
- Results
- Future Work

20

PRISM

Future Work

- Consideration of the baseline steam boilers for the hybrid system
- Application of the proposed system for smart grid integration
- Exploring the uncertainty in prices and wind power data

21

PRISM

Acknowledgement

- Sustainable Energy Solutions (SES)
- Graduate and undergraduate students at Brigham Young University

22

PRISM

THANKS FOR YOUR ATTENTION

Q & A?

23

382809 Modeling and Nonlinear Operability Analysis of a Membrane Reactor for Direct Methane Aromatization

Abstract:

This presentation addresses the computation of feasible regions for the design and control of the direct methane aromatization (DMA) conversion process of natural gas to fuels and chemicals as an alternative to petroleum. In this case, we focus on the equilibrium-limited production of H_2 and C_6H_6 from CH_4 using a catalytic membrane reactor. Membrane reactors (MRs) are emerging systems that enable process intensification by combining two unit operations (reaction and separation) in one process, resulting in higher efficiencies and lower costs. MRs allow higher conversions than conventional packed-bed reactors due to the reaction equilibrium shift towards the products caused by the selective removal of one of these products through the membrane[1]. Specifically, in the DMA process, the reaction equilibrium is shifted towards products as H_2 is removed through a H_2-selective membrane (ion transport-based). Restrictions on process target specifications, such as reaction conversions and stream purities, as well as multiple challenges associated with controlling temperatures and compositions, make the problem of designing and controlling MRs for the DMA process a challenging area of research. In this project, the design of MRs employing process operability tools[2] enables the mapping of the feasible regions between process input and output variables using a developed MR model. The calculated output constraints can also be used for the design of model predictive controllers[3].

The MR simulation set up considers a shell and tube reactor, in which the CH_4 feeds the reaction side (tube packed with catalysts) and the sweep gas (He) flows in the permeation side (shell) assuming cocurrent and countercurrent flow configurations. The H_2 produced in the tube permeates to the shell through the membrane layer that is placed on the surface of the tube wall. The produced outlet streams from the tube (retentate) and shell (permeate) are rich in C_6H_6

and H_2, respectively. For the modeling studies, a nonlinear model that characterizes the 2-step reaction mechanism for the non-oxidative conversion of methane is developed, including the calculation of thermodynamics and kinetic properties[4]. Also, the membrane flux expression assumes a relationship proportional to the transmembrane partial pressure gradient of the species with a ¼ dependence[5]. Using the developed model, a process operability mapping is then obtained considering the input and output process operating regions. In particular, for the operability calculations, the variables that define the available input set (AIS) for the reactor are the diameter, length, temperature, and pressure; and for the membrane are selectivity and permeance. Based on the defined input specifications, the operability calculations provide an achievable output set (AOS) for the following performance variables: methane conversion, benzene production rate, hydrogen permeation through the membrane, selectivity of benzene/hydrogen, and concentration of the undesired intermediate, ethylene, in the retentate. The obtained preliminary results demonstrate the capabilities of the nonlinear operability approach to provide guidelines for experimental research. These guidelines include the set of operating conditions for process control and specifications that the design variables associated with the reactor and membrane should have in the laboratory so that desired process output targets can be achieved.

References

1. Lima F.V., Daoutidis P., Tsapatsis M., Marano J.J. Modeling and optimization of membrane reactors for carbon capture in Integrated Gasification Combined Cycle units. *Ind. Eng. Chem. Res.* **2012**, 51, (15), 5480–5489.

2. Lima F.V., Georgakis C. Input-output operability of control systems: the steady-state case. *J. Proc. Cont.* **2010**, 20, (6), 769-776.

3. Lima F.V., Georgakis C., Smith J.F., Schnelle P.D., Vinson D.R. Operability-Based Determination of Feasible Control Constraints for Several High-Dimensional Nonsquare Industrial Processes. *AIChE Journal* **2010**, 56, (5), 1249-1261.

4. Li L., Borry R.W., Iglesia E. Design and optimization of catalysts and membrane reactors for the non-oxidative conversion of methane. *Chem. Eng. Sci.*, **2002**, 57, (21), 4595-4604.

5. Li J.L., Yoon H.S., Wachsman E.D. Hydrogen permeation through thin supported $SrCe_{0.7}Zr_{0.2}Eu_{0.1}O_{3-\delta}$ membranes; dependence of flux on defect equilibria and operating conditions. *J. Membrane Sci.* **2011**, 381, (1-2), 126-131.

Supervisory Energy Management of an Integrated Fuel Cell / Battery System for Vehicular Applications

C. Ziogou[1], D. Giaouris[1], C. Yfoulis[2], F. Stergiopoulos[2], P. Seferlis[3], S.Voutetakis[1], S. Papadopoulou[1,2]

[1]Chemical Process and Energy Resources Institute (CPERI), Centre for Research and Technology Hellas (CERTH), PO Box 60361, 57001, Thessaloniki, Greece,
{cziogou, giaouris,paris}@cperi.certh.gr

[2] Department of Automation Engineering, Alexander Technological Educational Institute of Thessaloniki, P.O. Box 14561, 54101, Greece, {cyfoulis,fstergio,shmira}@autom.teithe.gr

[c] Department of Mechanical Engineering, Aristotle University of Thessaloniki, PO Box 484, 54124 Thessaloniki, Greece, seferlis@cperi.certh.gr

This work presents an energy management strategy (EMS) for an integrated fuel cell and battery system for vehicular applications. A challenging issue when integrating fuel cells to such applications is the handling of the dynamics that arise by their use. Additionally when a fuel cell is combined with a battery then various decisions should be made in order for the integrated system to operate in an optimum way, while fulfilling the frequently and rapidly changing load demand (Rodatz et al., 2005). In that context the resulting hybrid system should be complemented by a control system that will be able to handle the complex dynamics and the discrete decisions of the overall system (Arce et al., 2009). At the core of the considered control system an EMS is developed that dictates the appropriate decisions for each subsystem at various operating conditions. The goal of this work is to explore the behavior of the hybrid system under a realistic operation scenario which will be controlled by an EMS that relies on a Hybrid Timed Automaton (HTA) (Hahn et al., 2013). Besides the fulfillment of the load demand, the objectives of the EMS include the optimal use of the available hydrogen, the protection of the battery's lifetime and the preservation of the fuel cell's durability.

The design and development of an efficient and modular EMS is motivated by a system that considers a lithium-ion battery stack, a Polymer Electrolyte Membrane (PEM) fuel cell system and a programmable electronic DC load that simulates the behavior of an actual load. Based on the chosen topology, the main energy providers are the battery and the fuel cell. For each subsystem a dynamic mathematical model is developed and a set of operation scenarios will be performed in order to produce the appropriate amount of data. Subsequently a formal parameter estimation technique that involves mainly the battery and the fuel cell subsystem will be applied. A sensitivity analysis will be performed for the determination of the major variables and parameters that influence the behavior and response of the integrated system (e.g. State of Charge (SOC) of the battery, excess ratio levels of hydrogen and air of the fuel cell etc). By this analysis a set of variables/parameters will be determined that will guide the levels of operation of each subsystem.

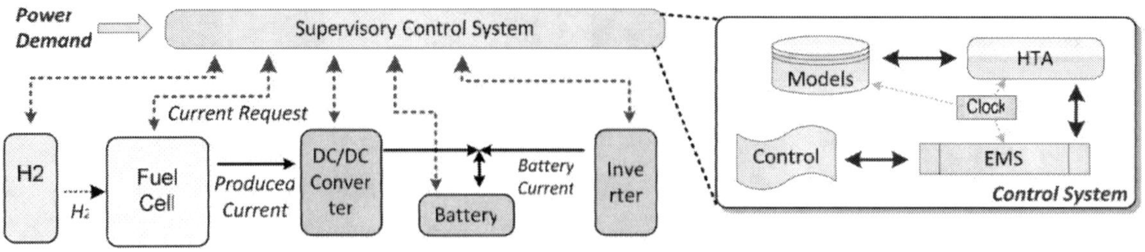

Figure 1: Integrated subsystems and components of the control system

In order to supervise the operation of the subsystems an EMS is necessary. The purpose of the EMS is twofold, to supervise the status of each subsystem and to be able to adjust to system operational modes (Ziogou et al., 2013). In this work the proposed EMS will take into account the main requirements and constraints which are posed by the vehicle's operation. One of the most challenging tasks for an effective EMS is to consider the interactions among the electronic and the electrochemical subsystems. Therefore, an approach that can handle this challenge is a hybrid timed automaton (HTA) that switches between the system's operational modes and sets the set-points for the lower-level control loops. More specifically an HTA is a dynamic approach that describes the evolution in time of a set of discrete and continuous state variables. The interaction between discrete and continuous operating states motivates the use of the hybrid approach and thus the hybrid automaton appears as a formulation that includes the semantics for the realization of a generic EMS. In this context a modular strategy is proposed which uses the automaton based structure while the system requirements are translated into a propositional based logic representing a set of operating rules. It consists of a series of switching actions in order to drive the system from any initial condition to an operating mode appropriate for every subsystem, taking into consideration the specifications and the constraints.

After the development of the EMS a deterministic optimization problem is formulated in order to examine a set of operation scenarios and to determine the values for the variables that will result from the sensitivity analysis The scope of the optimization framework it to enhance the performance of the system and most importantly to enable a smooth operation of the subsystems in terms of reliability. The utilization pattern of the battery and of the fuel cell determines the frequency of equipment replacement and maintenance and therefore the operating costs, over the lifespan of the overall system. Thus, the optimization problem focuses on these operation parameters, which are considered as decision variables. The optimization framework is evaluated on a case study regarding the optimal operation of the system. The optimal values of the operating parameters are identified and their effect to the overall system operation is discussed. Finally the effectiveness of the EMS we be evaluated using a set of predefined driving profiles described by a time-varying curve of power demand. The results of the case study will focus on the coupling of the fuel cell and the battery subsystems towards the production of the desired power demand. The results will reveal the potential of the proposed EMS, explore the synergy among the various subsystems and demonstrate the response of the integrated system.

Acknowledgement

This work is implemented through the Operational Program "Education and Lifelong Learning", co-financed by the European Union (European Social Fund) and Greek national funds, program "Archimedes III".

References

Arce A., del Real A., Bordons C., 2009, MPC for battery/fuel cell hybrid vehicles including fuel cell dynamics and battery performance improvement, Journal of Process Control, 19, 8, 1289-1304.

Hahn E.M., Hartmanns A., Hermanns H., Katoen J.P., 2013, A compositional modelling and analysis framework for stochastic hybrid systems, Formal Methods in System Design, 43, 2, 191-232.

Rodatz P., Paganelli G., Sciarretta A., Guzzella L., 2005, Optimal power management of an experimental fuel cell/supercapacitor-powered hybrid vehicle, Control Engineering Practice, 13, 1, 41-53.

Ziogou C., Ipsakis D., Seferlis P., Bezergianni S., Papadopoulou S., Voutetakis S., 2013, Optimal production of renewable hydrogen based on an efficient energy management strategy, Energy, 55, 58-67.

Automatic alarm systems are inseparable components of current industrial plants. They are designed such that in case of abnormal events operators are notified and the appropriate precautionary actions are taken in shortest possible time. However, in many cases the number of less informative alarms annunciated is much larger than it could be useful or even be handled. This situation is usually called "alarm flood". On the other hand, alarm thresholds should not be set such that the alarm system fails to announce the occurrence of an abnormal event in an effective way [1].

To address these issues, the design and implementation of efficient alarm systems have been an active research area. The goal has been to decrease the number of false alarms without increasing the number of missed alarms. To this end, a variety of techniques have been proposed. These techniques include multivariate monitoring [2], model-based monitoring [3-6], threshold optimization [7], and alarm signal processing [8].

This paper presents a probabilistic model-based method of improving alarm-system performance. The performance is quantified in terms of number of false and missed alarms. This method uses a novel data-driven probabilistic modeling approach to calculate the differences between measurements and model-predictions of process variables. The model is trained properly to cover all possible states of the process using the process historical data. The calculated differences will be used to identify abnormal situations and trigger appropriate alarms. It is shown that the modeling approach can model highly non-linear and non-monotonic behavior observed in data with minimum number of parameters. To make the calculation of the differences computationally tractable, we use an efficient probabilistic inference technique, which in combination with the model presents a unique framework for probabilistic reasoning particularly when there are a large number of variables to deal with.

References

[1]. Rothenberg, D. H. *Alarm Management for Process Control: A Best-Practice Guide for Design, Implementation, and Use of Industrial Alarm Systems*. New York: Momentum Press, 2009.

[2]. Chen, T. On reducing false alarms in multivariate statistical process control. *Chemical Engineering Research and Design* **2010**, *88* (4), 430-436.

[3]. Ahooyi, T. M.; Soroush, M.; Arbogast, J. E.; Seider, W. D.; Oktem, U. G. Maximum-likelihood maximum-entropy constrained probability density function estimation for prediction of rare events. *AIChE Journal* **2014**, *60* (3), 1013-1026.

[4]. Mohseni Ahooyi, T.; Abrogast, J. E.; Oktem, U.; Seider, W. D.; Soroush, M. Estimation of Complete Discrete Multivariate Probability Distributions from Scarce Data with Application to Risk Assessment and Fault Detection. *Industrial & Engineering Chemistry Research* **2014**, *53* (18), 7538-7547.

[5]. Mohseni Ahooyi, T.; Abrogast, J. E.; Soroush, M. Rolling Pin Method: a Novel Efficient General Method of Joint Probability Modeling. *Industrial & Engineering Chemistry Research, accepted*, DOI: 10.1021/ie503584q.

[6]. Mohseni Ahooyi, T.; Abrogast, J. E.; Soroush, M. Applications of the Rolling Pin Method: I) an Efficient Alternative to Bayesian Network Modeling and Inference. *Industrial & Engineering Chemistry Research*, accepted, DOI: 10.1021/ie503585m.

[7]. Jiang, R. Optimization of alarm threshold and sequential inspection scheme. *Reliability Engineering & System Safety* **2010**, *95* (3), 208-215.
[8]. Rheineck-Leyssius, A. R.; Kalkman, C. J. Advanced pulse oximeter signal processing technology compared to simple averaging. i. effect on frequency of alarms in the operating room. *Journal of Clinical Anesthesia* **1999**, *11* (3), 192-195.

387285 Multivariate Control Charts in Radial Coordinates

Abstract:

Multivariate Control Charts in Radial Coordinates
Ray Wang, Michael Baldea, and Thomas F. Edgar

McKetta Department of Chemical Engineering

The University of Texas at Austin, 1 University Station C0400, Austin, TX 78712

Mark Nixon, Willy Wojsznis, Ricardo Dunia

Emerson Process Management, Austin, TX

Email: mbaldea@che.utexas.edu

10B06 Process Monitoring and Fault Detection (https://aiche.confex.com/aiche/2014/10/sessions/sessioninfopopup.cgi?sessionid=26932)

For the past decades, the chemical industry has established itself as a prime source and collector of "big data" in the form of process information. These data are diligently stored on historian systems, with sample times that are on the order of seconds. However, they often remain dormant and underutilized, placing many branches of the chemical industry in the paradoxical situation of being data rich, but information poor.

In this paper, we focus on process fault detection as one of the key insights to be gained from historical and real-time data. We take a geometric approach based on the radial coordinate representation that we introduced recently [1]. In our previous work, we proposed representing each data sample on a Kiviat ("radar" or "star") plot, with the plots of successive scans stacked along a time axis normal to the plotting plane. In this paper, we use this representation to define a multivariate control chart. We use the centroid of the radial representation of each sample as a condensed metric to define the state of the process at a given time instant. The centroid point can then be used to construct a multivariable control chart in the sense of MacGregor and Kourti [2] (where this concept has been developed for the two-dimensional case). The multivariate control chart is based on a confidence ellipse defined by the centroids computed for the normal operating state of the process. For fault detection, the real-time data samples whose centroids fall within the ellipse can be considered within specification, while data outside the ellipse are indicative of a faulty state.

We apply these developments to an industrial data set concerning the operation of a compressor that is prone to surge as well as to fault data collected from simulating the Tennessee Eastman Challenge Process [3]. We show that the performance of the proposed method is at least comparable in terms of fault detection speed and missed detection rates to other approaches discussed in the literature [4,5]. In addition, we show that the method provides unique capabilities in detecting process-wide abnormal events, e.g., column flooding and compressor surge. Subsequently, we discuss an extension of our method to assessing the occurrence of faults during process transitions between operating states.

References:

[1] Ray C. Wang, Ricardo Dunia, Michael Baldea, Thomas F. Edgar, Visual Data Mining and Fault Detection in Radial Coordinates, American Institute of Chemical Engineers Annual Meeting, San

[2] J.F. MacGregor, T. Kourti, Statstical process control of multivariate processes, Control Eng. Pract. 3 (1995) 403–414.

[3] J.J. Downs, E.F. Vogel, A plant-wide industrial process control problem, Comput. Chem. Eng. 17 (1993) 245–255.

[4] E.L. Russell, L.H. Chiang, R.D. Braatz, Fault detection in industrial processes using canonical variate analysis and dynamic principal component analysis, Chemom. Intell. Lab. Syst. 51 (2000) 81–93.

[5] Y. Zhang, Fault Detection and Diagnosis of Nonlinear Processes Using Improved Kernel Independent Component Analysis (KICA) and Support Vector Machine (SVM), Ind Eng Chem Res. 47 (2008) 6961–6971.

380520 A Methodology for Leak Detection in Natural Gas Pipelines Under Different Consumer Usage Patterns

Abstract:

Model based fault detection is one of the most widely used approaches for industrial process fault diagnosis [1]. For leak detection in a natural gas pipeline, software methods such as model-based methods or statistics based methods have been applied [2-5]. The model-based leak detection strategy is based on the pressure and flow rate measurement at the inlet and outlet of the pipeline and an estimation algorithm from the natural gas flow model. Estimation algorithms proposed for a single natural gas pipeline leak detection include Unscented Kalman Filter, particle filter, and Extended Kalman Filter [6-9]. However, leak detection in natural gas pipeline networks involving different consumer usages has not been studied.

In this paper, we consider the effects of different consumer usages in a natural gas pipeline network. Two different consumer usages are considered which are constant industrial usage and household consumer. The household consumer usage varies based on temperature and the time of the day and can't be predicted precisely. The effect of leak on the flow rate of the pipeline network is simulated using the non-isothermal modelling. An estimation algorithm is developed for natural gas pipeline fault detection (leak size and location) in the presence of different consumer usages. Furthermore, due to the length of the pipeline, there is a time delay of the effect of consumer usage at both ends of the pipeline. Based on the non-linear isothermal model of the natural gas flow in a pipeline and boundary condition of constant pressure at both ends, a simple model was derived showing the effect of consumer usage on the inlet and outlet flow rate at steady state. With a random leak treated as an unknown input, an unknown input observer with time delay is designed based on the above model for leak detection. Two residuals are generated for the leak size and location identification. An alarm of leak is triggered according to the magnitude of the residuals.

[1] Isermann R. Model-based fault-detection and diagnosis–status and applications. Annual Reviews in control.

2005;29:71-85.

[2] Geiger IG. Principles of leak detection. 2005.

[3] Billmann L, Isermann R. Leak detection methods for pipelines. Automatica. 1987;23:381-5.

[4] Wang S, Carroll J. Leak Detection for Gas and Liquid Pipelines by Transient Modeling. International Oil & Gas Conference and Exhibition in China 2006.

[5] Benkherouf A, Allidina A. Leak detection and location in gas pipelines. Control Theory and Applications, IEE Proceedings D: IET; 1988. p. 142-8.

[6] Liu M, Zang S, Zhou D. Fast leak detection and location of gas pipelines based on an adaptive particle filter. International Journal of Applied Mathematics and Computer Science. 2005;15:541.

[7] Reddy HP, Narasimhan S, Bhallamudi SM, Bairagi S. Leak detection in gas pipeline networks using an efficient state estimator. Part-I: Theory and simulations. Computers & Chemical Engineering. 2011;35:651-61.

[8] Wan J, Yu Y, Wu Y, Feng R, Yu N. Hierarchical leak detection and localization method in natural gas pipeline monitoring sensor networks. Sensors. 2011;12:189-214.

[9] Tang W. Modeling, Estimation, and Control of Nonlinear Time-Variant Complex Processes: Texas Tech University; 2013.

383600 Robust Virtual Metrology of Plasma Etch Tools Using Growing Structure Multiple Models Systems with Dynamic Partial Least Squares Models

Abstract:

Robust virtual metrology of plasma etch tools using Growing Structure Multiple Models Systems with Dynamic Partial Least Squares models

Bo Lu[a], John Stuber[b], Thomas F. Edgar[a]

[a] McKetta Department of Chemical Engineering, [b] Texas Instruments Inc.
The University of Texas at Austin, 1 University Station C0400, Austin, TX 78712
Email: edgar@che.utexas.edu

Traditional semiconductor manufacturing factories have relied on run-to-run process control with external metrology. Only the most critical processes have metrology on every wafer; even in these cases, excursions in product quality are not detected until many more wafers have been processed. Virtual metrology aims to predict end-of-batch outputs using fault diagnostic trace data and other end-point readings. Based on predictions made by virtual metrology, control engineers can then make the decision on whether to perform additional metrology sampling or apply feed-forward adjustment in subsequent processing steps. However, semiconductor processes experience shifts and drifts in their operation, making it difficult to develop data-driven models with high fidelity predictions. To increase the robustness of these models, model updates such as recursive model update or moving-window updates can be used (Khan et al. 2007, Lu et al. 2014). However, adaptive models do not predict well across discontinuity or major events (such as preventive maintenance) and are sensitive to disturbances in update data. On the other hand, Mixture modeling by combining the predictions of more accurate local models has also been proposed; but this approach requires large amount of training data to initialize and introduces additional complexity in both the modeling and the interpretation of results. Consequently, there exists a need to develop a robust virtual metrology framework capable of maintaining prediction accuracy across event boundaries and over longer periods of process drifts and disturbances.

In this study, a novel divide-and-conquer approach combining growing structure multiple model systems (GSMMS) with Dynamic Partial Least Squares (DPLS) has been proposed. GSMMS models based on simple least squares models have been shown to be effective in approximating nonlinearity in dynamic systems with two or three input variables (Liu et al. 2009). The GSMMS utilizes a growing self-organizing map (GSOM) to adaptively allocate future incoming data to the appropriate node. In cases of process discontinuities or disturbances that are previously unseen, the GSOM "grows" a new node to accommodate the new behavior. This flexible partitioning scheme reduces the need for large amount of initial training data. Efficient dynamic partial least squares (DPLS) models provide estimates of local model parameters for each GSOM node for much larger space of input variables (Galicia et al. 2011). The proposed method is applied to an industrial dataset of a metal etch process from Texas Instruments. Comparisons against the moving-window adaptive PLS model and the mixture modeling with K-means clustering demonstrates the robustness of the GSMMS-DPLS model over current state-of-art alternatives.

References

Galicia, H.J., He, Q.P. & Wang, J., 2011. A reduced order soft sensor approach and its application to a continuous digester. *Journal of Process Control*, 21(4), pp.489–500.

Khan, A., Moyne, J. & Tilbury, D., 2007. An approach for factory-wide control utilizing virtual metrology. ... , *IEEE Transactions on Semiconductor Manufacturing*, 20(4), pp.364–375.

Liu, J. et al., 2009. Growing Structure Multiple Model Systems for Anomaly Detection and Fault Diagnosis. *Journal of Dynamic Systems, Measurement, and Control*, 131(5), p.051001.

Lu, B., Stuber, J. & Edgar, T.F., 2014. Integrated Online Virtual Metrology and Fault Detection in Plasma Etch Tools. *Industrial & Engineering Chemistry Research*, 53(13), pp.5172–5181.

377696 Input Design for Active Fault Diagnosis Using Zonotopes

Abstract:

In chemical processes, aerospace systems, and many other applications, the trend toward increasing complexity and automation has made component malfunctions and other abnormal events (i.e., faults) increasingly frequent and difficult to diagnose. Without corrective action, such faults lead to performance degradation and potentially critical situations. Thus, maintaining safe and profitable operation requires fast and accurate determination of whether a fault has occurred (fault detection) and which component has failed (fault diagnosis).

Approaches to automatic fault diagnosis can be classified as either active or passive. In the passive approach, input-output data are collected in real-time and faults are diagnosed based on comparisons with a process model or historical data. In contrast, the active approach involves injecting a signal into the system to improve the diagnosability of potential faults. Compared to passive methods, research on active fault diagnosis has been very limited.

This presentation discusses a deterministic, set-based approach for active fault diagnosis. The process of interest, under nominal and various faulty conditions, is described by a set of linear discrete-time models subject to bounded disturbances and measurement noise. Faults are modeled by discrete switches between these models. Given deterministic bounds on the disturbances and measurement noise, the objective is to compute an input that is guaranteed to lead to outputs that are consistent with at most one model. First, a method is presented for computing the set of all such inputs using computations with zonotopes (a class of centrally symmetric convex polytopes). These computations are very efficient and scalable to high dimensions. Second, a method is presented for choosing an input from this set that is minimally harmful with respect to other control objectives. Although this leads to a difficult nonconvex optimization problem, it is shown to be equivalent to a mixed-integer quadratic program that can be solved efficiently. The resulting input design method is significantly more efficient than other existing methods that provide a guarantee of diagnosis. Finally, the design of active inputs within a closed-loop framework is discussed using a moving horizon approach. All methods are demonstrated for examples with multiple fault models.

386302 Effective Importance Sampling in Sequential Monte Carlo Moving Horizon Estimation for Nonlinear Dynamic Systems

Abstract:

State estimation from noisy measurements is a fundamental problem common to many fields of applied science such as engineering, meteorology and remote sensing, as well as in widely different fields such as agricultural sciences and economics. Adequate knowledge of the state of a dynamic system is crucial for the tasks of forecasting, control or optimization of the system dynamics in all these fields. For general nonlinear dynamic systems characterized by non-Gaussian measurement and model uncertainties, analytical solutions to state estimation problems do not exist. In recent decades, there has been a great deal of interest in solving the nonlinear state estimation problem. The focus has turned to sampling based methods that can approximately implement the recursive Bayesian solution with significant computational efficiency, these methods are broadly known as the sequential Monte Carlo (SMC) numerical approach. Instead of a closed form representation of the state conditional probability density function (pdf), the SMC approach obtains a dynamic set of samples drawn according to the conditional pdf, from which point estimates of the state trajectory can be easily estimated.

The SMC approach has been recently used to implement a sequential Monte Carlo moving horizon estimation (SMCMHE)[1]. The estimation of the most probable state trajectory conditioned on a moving window of measurements is achieved by maximizing the joint conditional pdf of the states in the trajectory. The resulting optimization problem is posed over an evolving discrete grid in state space over a window of sampling instances inside a chosen horizon. Samples drawn from the joint conditional pdf naturally form a non-uniform and randomized dynamic grid, which is a suitable candidate for implementing a dynamic programming approach for solving the optimization problem. Such a dynamic grid of samples, evolving according to state dynamics, effectively eliminates "the menace of expanding grids", which is a commonly encountered limitation in applications of dynamic programming. The samples are automatically clustered in high probability regions of the state space. Furthermore, a fixed sample size is independent of the number of state variables, which protects the approach from "the curse of dimensionality". There are many statistical tests available to assess an effective sample size. The Viterbi algorithm has been effectively used for forward and backward sweep searches of the objective function values on the grid. The main computational cost is on maintaining a grid of objective function values by appending new values at each sample time and dropping the earliest values as the horizon moves forward.

In this paper the focus is on effective and efficient sampling of the conditional pdf to provide a randomized non-uniform grid on which a maximum a posterior (MAP) state estimation problem is approximately solved. Since the actual conditional pdf is not available for sampling, one must resort to other approaches to draw samples such as importance sampling. In the literature, many approaches have been proposed to identity suitable importance density functions, many of which are tailored to specific classes of problems. Examples include the use of approximate nonlinear filters such as the extended Kalman filter (EKF) and the unscented Kalman filter (UKF). These are generally known as local linearization particle filters because each sample is drawn from a different importance density function, typically a bank of Gaussians propagated by linearized dynamics. They entail a considerable cost of computation due to the burden of implementing a large bank of nonlinear filters in the background. In this paper a single non-Gaussian importance density is proposed, which is known analytically up to a constant. Two key features of this importance density can be readily established, its mode and its support. This involves an optimization problem and the solution to a nonlinear equation. Such knowledge of the importance density makes it amenable to effective sampling by established methods such as the Metropolis-Hastings algorithm, by centering the sampling process around the mode. This approach is readily extended to state estimation problems involving constraints on state variables. Both the mode and support of the importance density can be constrained, hence the randomized grid is automatically limited to the constraint hyper-surface in state space.

The presentation will include numerical examples of bench mark nonlinear problems from literature to demonstrate the efficacy of the proposed approach. Results include mean squared error of estimation and computational times averaged over multiple realizations. Comparisons will be provided with implementation of traditional moving horizon estimation with arrival cost computed by the EKF.

References:

[1] Ungarala, S., Sequential Monte Carlo moving horizon estimation, AIChE Annual Meeting, Pittsburgh, PA, 2012.

On the Use of Extents for Process Monitoring and Fault Diagnosis

S. Srinivasan, J. Billeter and D. Bonvin

Laboratoire d'Automatique, Ecole Polytechnique Fédérale de Lausanne
EPFL – Station 9, 1015 Lausanne, Switzerland
sriniketh.srinivasan@epfl.ch

Process monitoring and fault diagnosis are broadly used to control quality and enforce safety compliance in industrial processes. Processes are commonly monitored by online spectroscopy, with either PCA or calibration techniques such as PLS being used to predict abstract or physical process variables. By comparing these variables to historical data measured under normal operating conditions, possible faults are detected based on deviations from statistical thresholds [1]. One then tries to identify the causes of these faults, identification being easier when monitoring involves a calibration that predicts physical process variables and when fault detection uses a model to relate controlled and manipulated variables [2].

Exploiting the structure of balance equations, a transformation can separate multivariate data into decoupled variant/invariant states, which can be investigated individually to identify rate laws and reconstruct unmeasured quantities [3]. A convenient linear transformation uses the generalized concept of extents [4, 5], which coincides with a time-invariant transformation used to model rank-deficient spectroscopic data [6]. This transformation requires only limited process information, namely, the reaction stoichiometry, the species transferring between phases, the composition of inlet flows and the initial conditions. Moreover, this transformation was adapted to handle calorimetric and spectroscopic data [7, 8].

This contribution addresses the applicability of the transformation to extents for process monitoring and fault diagnosis. By comparing the extents computed from measurements of the current batch with either their prediction or the extents computed from previous batches, significant deviations can point at possible faults and provide a systematic way of identifying their causes.

References:

[1] V. Venkatasubramanian, R. Rengaswamy, S. N. Kavuri, K. Yin, Comp. Chem. Eng. 27 (2003) 327
[2] V. Venkatasubramanian, R. Rengaswamy, K. Yin, S. N. Kavuri, Comp. Chem. Eng. 27 (2003) 293
[3] S. Srinivasan, J. Billeter, D. Bonvin, IFAC Proceedings Vol. 1 (2013) 102
[4] M. Amrhein, N. Bhatt, B. Srinivasan, S. Bonvin, AIChE J. 56 (2010) 2873
[5] N. Bhatt, M. Amrhein, D. Bonvin, Ind. Eng. Chem. Res. 49 (2010) 7704
[6] J. Billeter, Y.-M. Neuhold, K. Hungerbühler, Chemom. Intell. Lab. Syst. 95 (2009) 170
[7] S. Srinivasan, J. Billeter, D. Bonvin, Chem. Eng. J. 207-208 (2012) 785
[8] J. Billeter, S. Srinivasan, D. Bonvin, Anal. Chim. Acta 767 (2013) 21

Predictive Control + Parallel Computing for Complex Processes with Multiscale Objectives

Antonios Armaou

Department of Chemical Engineering, The Pennsylvania State University, University Park, PA 16802

Stevan Wilson

Andritz Automation, Decatur, GA 30030

The performance of production process for many advanced and biological materials is measured both in terms of yield and quality. The latter term is usually a combination of desired aspects of the product microstructure. Direct, in process, sensor measurement at the microstructure level is either impractical or infeasible, necessitating soft sensors, calculated from multiscale process modeling, for controller synthesis. Multiscale models are traditionally used to quantify process evolution across all relevant length scales and characterize product behavior within current computational limitations. Such models, however, pose significant challenges for controller design due to a) the unavailability of closed form equations of the molecular-level description components and b) the high computational demands of the resulting system, both of which often prevent direct implementation inside the control structure.

An industrially relevant example for this situation is the thin film deposition processes widely used by the microelectronics and solar energy industries (such as the production of photovoltaic systems). Motivated by the complex and nonlinear process dynamics, strict quality requirements, high product value, and high production costs, a significant amount of research has been applied to this problem. While earlier work focused on designing structures that allow the use of standard feedback controllers by directly using simulations in lieu of measurements, more recent results have been focused on combinations of model-based controllers with observers. These structures are based on models derived by using system identification methods. Unfortunately, the use of predictive controllers adds more computational demands due to their mathematical complexity.

To address both the computationally demanding issues of process modeling and predictive control, the authors investigated the combination of parallel computing with system identification methods. This resulted in both increases the computational power available, and also in a controller synthesis method that can properly utilize the results from parallel computations. The method is able to provide a computationally tractable online system identification scheme based on a combination of subspace and adaptive system identification methods. This scheme allows for approximate identification of the drift and diffusion terms of the underlying stochastic differential equations. This model was incorporated into a predictive control scheme that drives the process towards desired objectives. By employing optimization algorithms that account for the nonlinear and uncertain nature of the underlying dynamics, the predictive controller was capable of efficiently identifying near optimal control action profiles that also lead to reduced process variance. Discussion of each algorithm and the total system are presented along with results for the process under control.

363557 Stochastic Model Predictive Control of High-Dimensional Systems: An End-to-End Continuous Pharmaceutical Manufacturing Case Study

Abstract:

Probabilistic uncertainties are ubiquitous in complex dynamical systems and can impair closed-loop performance of model predictive control (MPC) approaches, which are widely used to realize high-performance operation of complex systems. Stochastic MPC provides an alternative approach for conventional robust MPC strategies [1], which design the optimal control law with respect to worst-case system uncertainties that can have vanishingly small probability of occurrence. Stochastic MPC approaches shape the predicted probability density functions of system states and outputs in an optimal manner over a finite prediction horizon [2,3,4,5]. Stochastic MPC also enables considering state constraints in a probabilistic sense (so-called chance constraints) to alleviate the inherent conservatism of robust MPC approaches that merely deal with deterministic worst-case perturbations. The key challenge in stochastic MPC is however the propagation of probabilistic uncertainties. The commonly used methods (e.g., Monte Carlo and Markov Chain Monte Carlo methods [6]) for probabilistic uncertainty analysis are prohibitively expensive for real-time control. Additional complexity in stochastic MPC arises from chance constraints, which entail the computation of multivariable integrals [7].

In this talk, a fast stochastic MPC algorithm with chance constraints is presented for high-dimensional stable linear systems with time-invariant probabilistic uncertainties (e.g., uncertainties in initial conditions and system parameters). Generalized polynomial chaos (PC) theory [8,9] is used to enable efficient uncertainty propagation through the high-dimensional system model. In the PC approach, the implicit mappings between the uncertain variables/parameters and system outputs are replaced with an expansion of orthogonal polynomials, whose statistical moments can be determined efficiently from the expansion coefficients. To determine the expansion coefficients, Galerkin projection [9] is adapted to construct PC expansions for a general class of linear differential algebraic equations (DAEs), so that the SMPC algorithm is applicable to both regular and singular/descriptor systems. The quadratic dynamic matrix control

(QDMC) algorithm [10] is used to formulate an input-output framework for stochastic MPC with output constraints. The resulting probabilistic input-output framework is independent of the state dimension to circumvent the prohibitive computational costs of control of uncertain systems with a large state dimension.

The proposed probabilistic input-output MPC approach with chance constraints is used for control of an end-to-end continuous pharmaceutical manufacturing process (with approximately 8000 states) [11] in the presence of time-invariant, probabilistic parametric uncertainties. The process has nine inputs and three outputs---the active pharmaceutical ingredient (API) dosage of tablets, the impurity content of tablets, and the production rate of the process [12]. The critical quality attributes (CQAs) of the manufactured tablets consist of the API dosage and impurity content of the tablets, which should be regulated consistently in a stochastic setting. The closed-loop simulation results reveal that the stochastic MPC approach leads to a much tighter distribution (i.e., lower variance) of the API dosage around its desired setpoint. In addition, the chance constraint imposed on impurity content ensures that the tablets meet the stringent drug specifications in the presence of system uncertainties. These results indicate the promising potential of the proposed fast stochastic MPC approach for application to control of high-dimensional systems (e.g., pharmaceutical manufacturing), in which stringent requirements on robust closed-loop performance should be met.

[1] A. Bemporad and M. Morari. Robust model predictive control: A survey. In Robustness in Identification and Control (A. Garulli and A. Tesi, eds.), 207-226, Springer Berlin, 1999.

[2] A. Schwarm and M. Nikolaou. Chance-constrained model predictive control. AIChE Journal, 45:1743-1752, 1999.

[3] D. Bernardini and A. Bemporad. Scenario-based model predictive control of stochastic constrained linear systems. In Proceedings of the 48th IEEE Conference on Decision and Control, 6333-6338, Shanghai, 2009.

[4] G. C. Calafiore and L. Fagiano. Robust model predictive control via scenario optimization. IEEE Transactions on Automatic Control, 58:219-224, 2013.

[5] A. Mesbah, S. Streif, R. Findeisen, and R. D. Braatz. Stochastic nonlinear model predictive control with probabilistic constraints. In Proceedings of the American Control Conference, To Appear, Portland, 2014.

[6] M. Vidyasagar. Randomized algorithms for robust controller synthesis using statistical learning theory. Automatica, 37:1515-1528, 2000.

[7] F. Oldewurtel, D. Sturzenegger, P. M. Esfahani, G. Andersson, M. Morari, and J. Lygeros. Adaptively constrained stochastic model predictive control for closed-loop constraint satisfaction. In Proceedings of the American Control Conference, 4681-4688, Washington, 2013.

[8] N. Wiener. The homogeneous chaos. American Journal of Mathematics, 60:897-936, 1938.

[9] D. Xiu and G. E. Karniadakis. The Wiener-Askey polynomial chaos for stochastic differential equations. SIAM Journal of Scientific Computation, 24:619-644, 2002.

[10] C. E. Garcia and A. M. Morshedi. Quadratic programming solution of dynamic matrix control (QDMC). Chemical Engineering Communications, 46:73-87, 1986.

[11] B. Benyahia, R. Lakerveld, and P. I. Barton. A plant-wide dynamic model of a continuous pharmaceutical process. Industrial Engineering & Chemistry Research, 51:15393-15421, 2012.

[12] J. A. Paulson, A. Mesbah, S. Streif, R. Findeisen, and R. D. Braatz. Fast stochastic model predictive control of high-dimensional systems. Submitted to IEEE Conference on Decision and Control, Los Angeles, 2014.

A framework for model reduction State Estimation and Robust Multi-Parametric Model Predictive Control in Anaesthesia

Ioana Nascu[a], Romain S. C. Lambert[a], Efstratios N. Pistikopoulos[a*]

[a]Dept. of Chemical Engineering, Centre for Process Systems Engineering (CPSE), Imperial College London SW7 2AZ, Lodnon, U.K

*e.pistikopoulos@imperial.ac.uk

Keywords: Moving Horizon Estimation, Explicit/Multi-Parametric Model Predictive Control, Anaesthesia

Multi-parametric/explicit model based control is a technology that enables instantaneous computation of the optimal values of the control inputs using simple function evaluations. This allows faster, more robust and reliable solutions compared to the ones obtained via on line optimization techniques (Pistikopoulos, Dua et al. 2002). However, complex process models need to be simplified in order to be used in multi-parametric studies. Therefore, we rely on model order reduction techniques to obtain simpler yet descriptive models. The main idea is that high-dimensional state vectors may be transformed so that they belong to lower-dimensional subspaces. The implementation of explicit/multi-parametric MPC, and in general, MPC is that it relies on the assumption of full space information and noise-free data availability. However, in reality, state information needs to be inferred from the available, likely noisy, output measurements using a state estimator. Dealing with system constraints entails the simultaneous use of moving horizon estimation (MHE) which can be implemented in a multi-parametric fashion (Darby and Nikolaou 2007, Voelker, Kouramas et al. 2010).

A framework for model reduction state estimation and robust multi-parametric model predictive control is developed and applied to both volatile and intravenous anaesthesia processes. The framework comprises of four essential consecutive steps. The first step is model reduction followed by the linearization step, essential in the case of complex non-linear processes. The first two steps are interchangeable and can be applied using different methodologies. For the state estimation step different strategies are implemented and compared. The final step of the framework is the development of robust multi-parametric model predictive control strategies. All the aforementioned steps of the framework are implemented and tested for different processes, using different methodologies and strategies.

General anaesthesia is an important part of surgery and intensive care unit (ICU) and requires critical evaluation of the induced quantities of drugs into the patient (Dumont, Martinez et al. 2009).To ensure the patient's safety and maintain his vital function before, during and after surgery, the role of the anaesthesiologist has become more and more complex and indispensable. The anaesthesiologist must estimate the drug effect in the patient's body and calculate the corresponding drug infusion rates. Consequently, the automation of routine actions constitutes a potential great improvement for the anaesthesiologists, granting them more time to allocate on safety critical issues that can threaten the safety of the patient and to avoid excessive over-dosages and under-dosages.

In order to control the depth of sedation based on predictive control strategies, a model that captures the dynamical response of the patient is required. The pharmacokinetic (PK) – pharmacodynamic (PD) blocks denote compartmental models, used to represent the distribution of drugs in the body, i.e. mass balance. They rely on a conservation principle applied to the exchange of chemicals among coupled macroscopic systems called compartments (central compartment, fast and slow equilibrating peripheral compartments).

In each compartment the drug concentration is assumed to be uniform, assuming perfect and instantaneous mixing. The transport rate of the drug leaving the compartment is assumed to be proportional to the drug concentration. The PK-PD models most commonly used for Propofol are the 4th order compartmental model described by Schnider, Minto et al. (1998), Schnider, Minto et al. (1999) and Minto, Schnider et al. (1997).

Control of anaesthesia is, however, a challenging task and is characterized by several challenges, such as: inter- and intra-patient variability, multivariable characteristics, variable time delays, dynamics dependent on hypnotic agent and stability issues (Absalom, De Keyser et al. 2011). During the recent decades, many PID tuning techniques have been studied. The main problem with this type of controllers is that they fail to anticipate the response of the patient and they do not consider prior knowledge of the pharmacometrics drug metabolism. Therefore their performance is inherently rendered sub-optimal. As a result, model based strategies using fuzzy, predictive and adaptive control algorithms have been developed and used in clinical trials (Ionescu, Keyser et al. 2008, Krieger, Panoskaltsis et al. 2014).

In this study, we use model order reduction for the simultaneous design of robust constrained controllers and observers. The use of such multi-parametric reduced-order observers is not only more computationally attractive but also more efficient by avoiding poor observability of some of the states in the full order model. Furthermore, we incorporate another important aspect by considering the simultaneous implementation of robust mp-MPC together with different state estimators for the regulation of the depth of anaesthesia (DOA). We perform the simulations on a set of 12 virtual generated patients under constraints and the presence of noise in the induction and maintenance phase. During the anaesthesia process, the state values which are different for each patient, cannot be directly measured. As a result they have to be obtained from the system output measurements using advanced state estimation. This will lead to more accurate state values for each individual patient. Some of the important aspects of this application are the high inter- and intra-patient variability and the presence of noise that has a high impact on the estimations and also on the performance of the controllers. Furthermore, we consider the parameter estimation of the nonlinear part of the system, given by the Hill curve. Thus, instead of using parameter-scheduling techniques in order to compensate the nonlinearity, by implementing the inverse of the hill curve using the nominal patients values we use the actual estimated values for each patient. All the designed controllers using the advanced estimation techniques are implemented and tested against one another and their performances are being evaluated. Using estimation for the parameters of the nonlinear part of the system will bring great improvements to the performance of the controllers even under the effect of disturbances and the existence of constraints.

References

Absalom, A. R., R. De Keyser and M. M. R. F. Struys (2011). "Closed loop anaesthesia: are we getting close to finding the Holy Grail?" Anesth Analg **112**(3): 516-518.

Darby, M. L. and M. Nikolaou (2007). "A parametric programming approach to moving-horizon state estimation." Automatica **43**(5): 885-891.

Dumont, G. A., A. Martinez and J. M. Ansermino (2009). "Robust control of depth of anesthesia." International Journal of Adaptive Control and Signal Processing **23**(5): 435-454.

Ionescu, C. M., R. D. Keyser, B. C. Torrico, T. D. Smet, M. M. R. F. Struys and J. E. Normey-Rico (2008). "Robust predictive control strategy applied for propofol dosing using BIS as a controlled variable during anesthesia." IEEE Transactions on Biomedical Engineering **55**(9): 2161-2170.

Krieger, A., N. Panoskaltsis, A. Mantalaris, M. C. Georgiadis and E. N. Pistikopoulos (2014). "Modeling and analysis of individualized pharmacokinetics and pharmacodynamics for volatile anesthesia." IEEE Transactions on Biomedical Engineering **61**(1): 25-34.
Minto, C. F., T. W. Schnider, T. D. Egan, E. J. Youngs, H. J. Lemmens, P. L. Gambus, V. Billard, J. F. Hoke, K. H. Moore, D. J. Hermann, K. T. Muir, J. W. Mandema and S. L. Shafer (1997). "Influence of age and gender on the pharmacokinetics and pharmacodynamics of reminfetanil. I. Model development." Anesthesiology **86**: 10-23.
Pistikopoulos, E. N., V. Dua, N. A. Bozinis, A. Bemporad and M. Morari (2002). "On-line optimization via off-line parametric optimization tools." Computers and Chemical Engineering **26**(2): 175-185.
Schnider, T. W., C. F. Minto, P. L. Gambus, C. Andresen, D. B. Goodale, S. L. Shafer and E. J. Youngs (1998). "The influence of method of administration and covariates on the pharmacokinetics of propofol in adult volunteers." Anesthesiology **88**(5): 1170-1182.
Schnider, T. W., C. F. Minto, S. L. Shafer, P. L. Gambus, C. Andresen, D. B. Goodale and E. J. Youngs (1999). "The influence of age on propofol pharmacodynamics." Anesthesiology **90**(6): 1502-1516.
Voelker, A., K. Kouramas and E. N. Pistikopoulos (2010). Simultaneous constrained moving horizon state estimation and model predictive control by multi-parametric programming.

Acknowledgments

Financial support from EPSRC (EP/G059071/1, EP/I014640), the European Research Council (MOBILE, ERC Advanced Grant, No: 226462) and the European Commission (OPTICO/G.A. No.280813) is gratefully acknowledged.

Integral controllability (IC) is necessary and sufficient to achieve robustness of decoupling multivariable control with integral action, in the sense defined by (Garcia and Morari 1985). IC refers to the fact that the steady-state gain matrix model $\hat{\mathbf{G}}$ and actual plant $\mathbf{G}$ must satisfy the eigenvalue inequality

$$-\text{Re}\left[\lambda\left(\mathbf{G}\hat{\mathbf{G}}^{-1}\right)\right] < 0 \tag{1}$$

for all plants $\mathbf{G} \in U$ where U is an uncertainty set. A typical uncertainty set is based on ellipsoidal uncertainty, resulting from standard least-squares estimation:

$$U = \left\{ \mathbf{G} = \begin{bmatrix} \mathbf{g}_1^T \\ \vdots \\ \mathbf{g}_n^T \end{bmatrix} \in \Re^{n \times n} \;\middle|\; \left(\mathbf{g}_i^T - \hat{\mathbf{g}}_i^T\right)\mathbf{M}^T\mathbf{M}\left(\mathbf{g}_i - \hat{\mathbf{g}}_i\right) \le c^2 \right\} \tag{2}$$

Therefore, a plant model $\hat{\mathbf{G}}$ that does not satisfy eqn. (1) is not useful for such a controller, even if $\hat{\mathbf{G}}$ is close to the true plant $\mathbf{G}$. This may easily happen, particularly for ill-conditioned systems. It should be stressed that $\hat{\mathbf{G}}$ might satisfy eqn. (1) for some plants $\mathbf{G} \in U$ but not for all plants $\mathbf{G} \in U$. Therefore, it becomes important to rigorously check whether a model $\hat{\mathbf{G}}$ satisfies eqn. (1) for all plants $\mathbf{G} \in U$ or not. This, however, is challenging.

A sufficient condition for satisfaction of eqn. (1) for all $\mathbf{G} \in U$ was proposed in (Darby and Nikolaou 2009) as

$$J \triangleq \sum_{k=1}^{n} c \frac{\left\|\hat{\mathbf{u}}_k\right\|_1}{\hat{\sigma}_k} \sqrt{\hat{\mathbf{v}}_k{}^T (\mathbf{M}^T\mathbf{M})^{-1} \hat{\mathbf{v}}_k} < 1 \tag{3}$$

where

$$\hat{\mathbf{G}} = \sum_{k=1}^{n} \hat{\sigma}_k \hat{\mathbf{u}}_k \hat{\mathbf{v}}_k^T \tag{4}$$

is the singular value decomposition of the identified model $\hat{\mathbf{G}}$. Note that eqn. (3) suggests a simple check, and does not need a search over the set U, required by eqns. (1) and (2). However, because eqn. (3) is sufficient but not necessary, it could potentially be conservative, thus possibly rejecting models that would be acceptable. To address this problem we undertook a numerical study using Monte Carlo simulations on a number of cases.

As an example, a test was performed on an ill-conditioned 2x2 distillation column (Skogestad and Morari 1987) with assumed real steady-state gain matrix

$$\mathbf{G} = \begin{bmatrix} 87.8 & -86.4 \\ 108.2 & -109.6 \end{bmatrix} \tag{5}$$

and identified steady-state gain matrix

$$\hat{\mathbf{G}} = \begin{bmatrix} 88.79 & -86.72 \\ 109.88 & -109.63 \end{bmatrix} \tag{6}$$

resulting from a simulated identification experiment with PRBS inputs.

Figure 1 shows whether each plant $\mathbf{G}$ among 10^6 instances in the uncertainty ellipsoid combined with the model $\hat{\mathbf{G}}$ satisfies eqn. (1). Each arrow in this plot starts at the point corresponding to $\mathbf{g}_1^T - \hat{\mathbf{g}}_1^T$ and ends at the point corresponding to $\mathbf{g}_2^T - \hat{\mathbf{g}}_2^T$. Yellow and red arrows indicate satisfaction and violation of IC, respectively. Figure 2, which summarizes the results of Figure 1, shows that even a slight increase in the value of J above 1 accounts for increase in IC violation percentage, which suggests that eqn. (3) is not conservative.

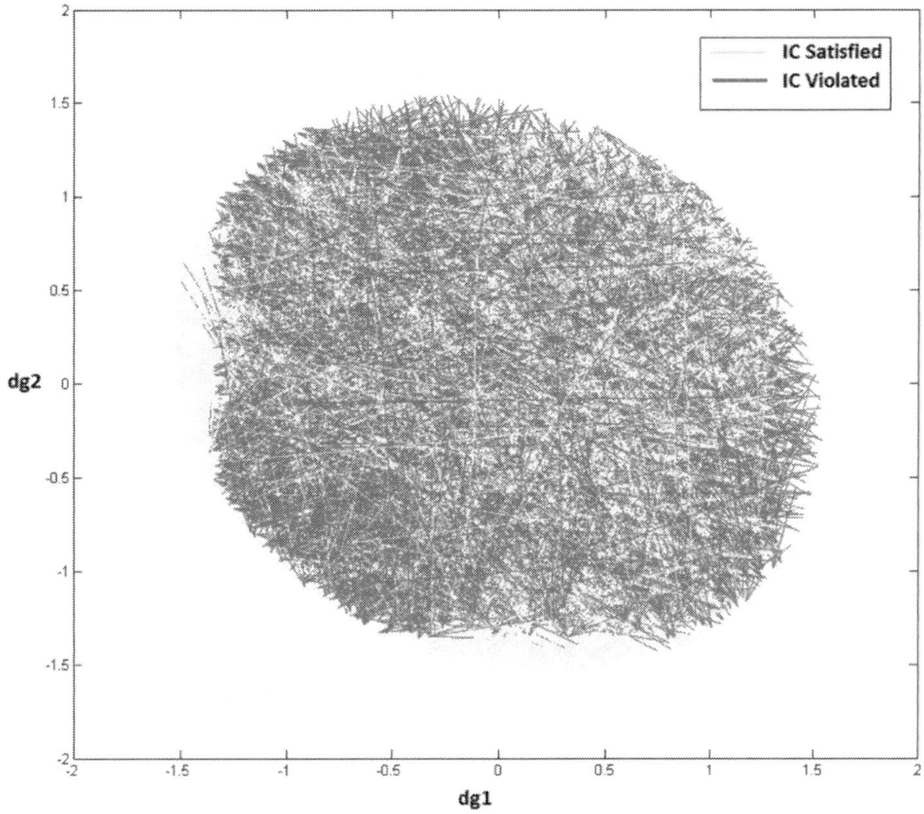

Figure 1: Monte Carlo simulation using model $\hat{\mathbf{G}}$ and plants $\mathbf{G}=(\mathbf{g}_1 \quad \mathbf{g}_2)^T \in \mathbf{U}$

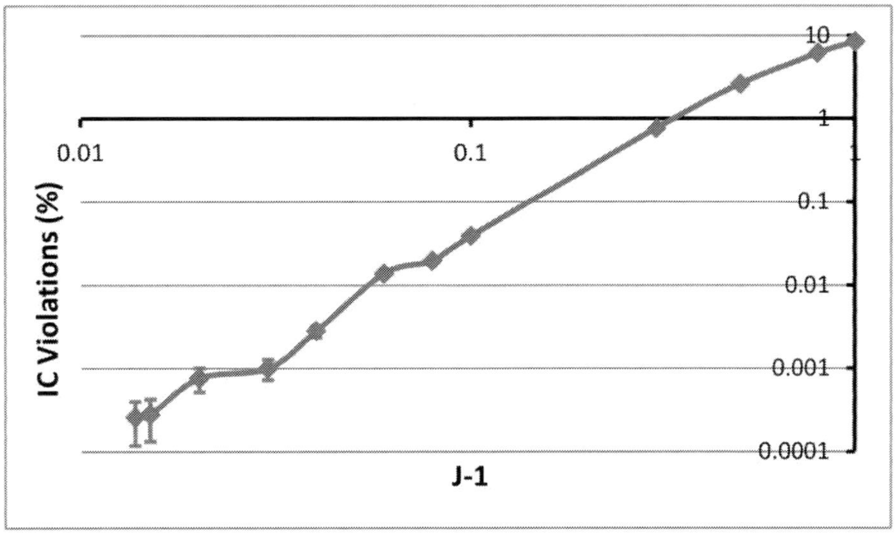

Figure 2: Percentage of IC violation varying with $(J-1)$

Additional cases studies with 3x3, 4x4, and 5x5 systems will be presented.

References

Darby, M. L. and M. Nikolaou (2009). "Multivariable system identification for integral controllability." Automatica **45**(10): 2194-2204.

Garcia, C. E. and M. Morari (1985). "Internal model control: 2. design procedure for multivariate systems." Ind. Eng. Chem. Process Des. Dev. **24**: 472-484.

Skogestad, S. and M. Morari (1987). "Control Configuration Selection For Distillation-Columns." Aiche Journal **33**(10): 1620-1635.

369922 Accounting for the Regulatory Control Layer in Economic Model Predictive Control of Nonlinear Chemical Processes

Abstract:

Industrial chemical process control architectures typically use two control layers, an upper advanced control layer and a lower regulatory control layer, to maintain stability and robustness of a process. In the advanced or constrained control layer, model predictive control (MPC) is often used to account for coupling of process dynamics, process constraints, and optimal process operation considerations. In the regulatory control layer, many proportional-integral-derivative (PID) control loops are used to implement the control actions computed by the advanced control layer. Recently, many researchers have considered using economic MPC (EMPC) in the advanced control layer since it combines feedback control with process economic optimization by employing an objective function that reflects the process economics in its formulation (e.g., [1]-[4]). Within the context of the theoretical developments of various EMPC systems, a common assumption made is that the control actuators operate with a zeroth-order hold (i.e., perfect actuation). However, control actuators have their own dynamics that cause the control actuators to deviate from this ideal behavior. Actual actuation of realistic control actuators is typically modeled as a first-order system with dead time. The parameters of the first-order system (i.e., gain, dead time, and time constant) may be fit to the response data of a given actuator. After this is completed, bounds on the stabilizing gains and time constants of the PID control loops can be established as well as robust stability guarantees can be made while explicitly accounting for the dynamics of each layer of the control architecture. Additionally, appropriate constraints limiting the change of the inputs between sampling periods may be imposed in the EMPC problem to ensure the EMPC solution may be implementable in the regulatory layer.

The main contribution of this work is to investigate the stability properties of control structures with EMPC accounting for the regulatory control layer and to propose a regulatory control layer monitoring methodology by taking advantage of the EMPC solution. To accomplish the goal of deriving conditions where robust stability guarantees can be made accounting for the regulatory control layer, a rigorous theoretical treatment of the overall control framework is completed. Constraints limiting the change of the inputs between sampling periods are imposed in the EMPC layer to ensure that the computed input profile may be tracked by the regulatory control layer (thus, ensuring the existence of a sufficient time-scale separation between the control layers). The closed-loop system (i.e., the system in closed-loop with both layers of the control architecture) may be written as a standard singular perturbation model (e.g., [5]-[6]). Using singular perturbation arguments, conditions for guaranteed closed-loop stability are derived while accounting for the supervisory and regulatory control layers, the actual dynamics of the control actuators, and process economic optimization. Furthermore, utilizing the computed input profile from the EMPC and an appropriate fault-detection and isolation filter (FDI) (e.g., [7]-[8]), a monitoring methodology is proposed for the regulatory control layer to monitor the performance of the PI control loops and to automatically re-tune the gain and time constant of the PI control loops that are performing poorly. The overall approach is demonstrated using several chemical process examples.

[1] Angeli D, Amrit R, Rawlings JB. On average performance and stability of economic model predictive control. IEEE Transactions on Automatic Control. 2012;57:1615-1626.
[2] Diehl M, Amrit R, Rawlings JB. A Lyapunov function for economic optimizing model predictive control. IEEE Transactions on Automatic Control. 2011;56:703-707.
[3] Heidarinejad M, Liu J, Christofides PD. Economic model predictive control of nonlinear process systems using Lyapunov techniques. AIChE Journal. 2012;58:855-870.
[4] Huang R, Harinath E, Biegler LT. Lyapunov stability of economically oriented NMPC for cyclic processes. Journal of Process Control. 2011;21:501-509.
[5] Kokotovic P, Khalil HK, O'Reilly J. Singular Perturbation Methods in Control: Analysis and Design. London, England: Academic Press, 1986.
[6] Christofides PD and Teel AR. Singular perturbations and input-to-state stability. IEEE Transactions on Automatic Control, 1996;41:1645--1650.
[7] Leosirikul A, Chilin D, Liu J, Davis JF, and Christofides PD. Monitoring and retuning of low-level PID control loops. Chemical Engineering Science. 2012;69:287-295.
[8] Mhaskar P, Liu J, Christofides PD. Fault-Tolerant Process Control: Methods and Applications. London, England: Springer-Verlag, 2013.

378328 Proactive Quasi-Decentralized Networked Model Predictive Control

Abstract:

The fundamental and practical challenges associated with control of networked process systems have been the focus of significant research work in process control over the past two decades, and have motivated many research studies in this area. Examples include results on traditional centralized and decentralized control, distributed model predictive control (e.g., [1]-[2]), passivity-based control (e.g., [3]), agent-based systems (e.g., [4]), and multi-time scale control (e.g., [5]). In addition to these results, research efforts within process control have also begun to address the new challenges that arise from the use of networked control systems in process operations (e.g., resource constraints, real-time scheduling constraints, etc.). Efforts to address some of these issues were initiated in [6]-[8] and focused on the development of resource-aware networked plant-wide control, estimation and scheduling strategies with explicitly characterized closed-loop stability properties.

These results were subsequently extended in [9] to incorporate performance optimization considerations through the use of model predictive control techniques. Specifically, an adaptive quasi-decentralized MPC framework, where each subsystem is controlled by a Lyapunov-based MPC controller, was developed, and a forecast-triggered communication strategy was used to minimize the cross communication of measurements between the distributed subsystems. This communication strategy exploits the stability properties of each subsystem and generates forecasts of the future behavior of each subsystem locally. When the forecast indicates a potential significant loss of control performance for a particular subsystem, its corresponding control system sends out requests to all the other subsystems for measurements which are used to update the model states within the local MPC controller. In this manner, a balance can be maintained between the performance of each subsystem and the utilization of network resources. Given that this adaptive (i.e., request-based) strategy requires that the local control system broadcast requests to all the other subsystems simultaneously and then await the arrival of measurements from the entire plant within a possibly short period of time, this could potentially lead to unnecessary network utilization which might lead to communication delays and data losses.

To address this problem, we present in this work a quasi-decentralized MPC framework that combines proactive and adaptive communication strategies. A set of local MPC controllers are initially designed based on sampled-data distributed bounded Lyapunov-based controllers, and their closed-loop stability and performance properties are characterized in terms of local (subsystem-specific) and global (plant-wide) thresholds. A hybrid communication strategy with complementary proactive and adaptive components is then devised to meet the control and communication objectives at the local and plant-wide levels. In both components, each subsystem forecasts the future evolution of the local state within the next sampling interval (relative to the corresponding stability/performance thresholds) and then decides whether to establish or suspend communication with the rest of the plant. In the proactive strategy, the forecasted breach of the specified threshold triggers each subsystem to broadcast its measurements to update the model states within the other subsystems to maintain the overall plant stability; while in the adaptive strategy the subsystem responds to the local threshold breach by requesting measurements from the other subsystems to update the states of the models embedded within the subsystem itself to maintain the local subsystem stability.

The proactive broadcast-based communication strategy is implemented when the process state of the entire plant is far from the nominal equilibrium point and its corresponding control objective is to stabilize the overall system and drive its state to some neighborhood of the equilibrium point. As the process state converges to this neighborhood, the contribution of the local behavior to the overall plant stability becomes difficult to discern, and thus the proactive strategy becomes infeasible. Under such circumstances, all control systems switch to the adaptive request-based communication strategy where the control objective becomes the practical stabilization of every subsystem. A simulated chemical process network example is used to illustrate the implementation of the proposed approach.

References:

[1] B. T. Stewart, S. Wright and J. B. Rawlings, "Cooperative distributed model predictive control for nonlinear systems," *J. Process Control*, **21,** 698–704, 2011.

[2] P. D. Christofides, J. Liu and D. M. de la Pena. *Networked and Distributed Predictive Control: Methods and Nonlinear Process Network Applications*. Springer-Verlag, London, 2011.

[3] K. R. Jillson and B. E. Ydstie, "Process networks with decentralized inventory and flow control," *J. Process Control*, **17,** 399-413, 2007.

[4] M. D. Tetiker, A. Artel, F. Teymour and A. Cinar, A., "Control of grade transitions in distributed chemical reactor networks: An agent-based approach," *Comp. Chem. Eng.*, **32**,1984-1994, 2008.

[5] S. Jogwar, M. Baldea and P. Daoutidis, "Dynamics and control of process networks with large energy recycle, " *Ind. Eng. Chem. Res.*, **48**, 6087-6097, 2009.

[6] Y. Sun and N. H. El-Farra, "Quasi-decentralized model-based networked control of process systems," *Comp. Chem. Eng.*, **32**, 2016-2029, 2008.

[7] Y. Sun and N. H. El-Farra, "A quasi-decentralized approach for networked state estimation and control of process systems," *Ind. Eng. Chem. Res.*, **49**, 7957-7971, 2010.

[8] Y. Sun and N. H. El-Farra, "Resource-aware quasi-decentralized control of networked process systems over wireless sensor networks," *Chem. Eng. Sci.*, **69**, 93-106, 2012.

[9] Y. Hu and N. H. El-Farra, "Adaptive quasi-decentralized model predictive control of networked process systems," *Distributed MPC Made Easy*, R. Negenborn and P. Maestre (Eds.), Chapter 12, 209-223, Springer-Verlag, Berlin, 2014.

382664 Quality Relevant Model Predictive Control for Batch Processes Via Subspace Identification

Abstract:

Batch processes play an important role in chemical, pharmaceuticals, bio-chemicals and materials industries for the production of low-volume but high-value-added products. The nonlinear and time-varying dynamics over a wide range of operating conditions and absence of equilibrium points preclude the direct application of control strategies designed for continuous systems. The primary control objective in batch processes is to reach a specified product quality by batch termination to consistently produce products of desired quality.

The existing batch process control strategies can be divided into trajectory tracking and inferential quality approaches [1, 2]. In trajectory tracking control, it is not guaranteed to meet to the desired quality even with perfect tracking if there is significant variation in the initial conditions from batch to batch in that disturbances encountered during the new batch could alter the relationship between the product quality and the trajectories of the process variables. In parallel, the data-driven inferential quality control is developed through multivariate statistical process control, such as partial least squares (PLS) regression beginning with batch-wise unfolding of multiway batch data. The multiway analysis based inferential model calls for the entire batch trajectory to predict the quality of the batch. However, measurements are only available up to the current sampling instant during the batch operation, and the future data required to predict the final quality is not measured yet, which is treated as a missing data problem [3]. In reality, the problem is not one of missing data but rather one of dynamic modeling. Therefore, a data-driven dynamic model is recently developed for quality control in batch processes, integrated with multiway PLS [4]. Nevertheless, multiway analysis based inferential model includes process measurement trajectories that only indirectly predict product quality, because it is recognized that the final product quality only depends on the final state of the process.

Motivated by above considerations, a subspace identification based quality relevant control strategy is developed for batch processes in this work. To this end, canonical variate analysis (CVA) method is employed to identify a dynamic state space model for the batch process [5]. Subsequently, an inferential model is constructed to predict the final product quality using the identified final state. A model predictive controller is then designed to achieve desired final product quality. The proposed approach is applied to a penicillin fermentation process and the simulation results indicate that the superior performance of the proposed controller over the traditional trajectory tracking controller.

[1] Cruickshank, S. M., Daugulis, A. J., & McLellan, P. J. (2000). Dynamic modeling and optimal fed-batch feeding strategies for a two-phase partitioning bioreactor. Biotechnology and bioengineering, 67(2), 224-233.

[2] Golshan, M., MacGregor, J. F., Bruwer, M. J., & Mhaskar, P. (2010). Latent Variable Model Predictive Control (LV-MPC) for trajectory tracking in batch processes. Journal of Process Control, 20(4), 538-550.

[3] Flores-Cerrillo, J., & MacGregor, J. F. (2004). Control of batch product quality by trajectory manipulation using latent variable models. Journal of Process Control, 14(5), 539-553.

[4] Aumi, S., Corbett, B., Clarke Pringle, T., & Mhaskar, P. (2013). Data driven model predictive quality control of batch processes. AIChE Journal, 59(8), 2852-2861.

[5] Larimore, W. E. (2013, June). Identification of nonlinear parameter-varying systems via canonical variate analysis. In American Control Conference (ACC), 2013 (pp. 2247-2262). IEEE.

386607 Shaping the Spatiotemporal Dynamics of Transport-Reaction Processes Via Adaptive Proper Orthogonal Decomposition

Abstract:

In recent years, the control of spatiotemporal dynamics of transport-reaction processes has been recognized as an important subject in many industrial chemical processes exemplified by flow reactors and microelectronic fabrication processes. The control problem of such systems which can be mathematically modeled by partial differential equations (PDEs) may typically be addressed via model order reduction (MOR) because the infinite dimensional representation of the PDE dynamics of the systems can be approximated by reduced order finite dimensional models [1].

We focus on shaping the spatiotemporal dynamics of such processes via regulation of error dynamics between the process and a target PDEs. The infinite dimensional representation of the error dynamics is reduced by MOR. Specifically, we apply a combination of Galerkin's method and adaptive proper orthogonal decomposition (APOD) to construct a recursively updated reduced order model (ROM) of the spatiotemporal error dynamics [2], [3]. The ROM is then used as the basis to design the dynamic observer and nonlinear feedback controller for error dynamics regulation. The proposed control structure is successfully applied on shaping the spatiotemporal thermal dynamics in a catalytic reactor.

[1] P. D. CHRISTOFIDES, Nonlinear and robust control of PDE systems, Birkhauser, New York, 2000.

[2] D. BABAEI POURKARGAR AND A. ARMAOU, Design of APOD-based switching dynamic observers and output feedback control for a class of nonlinear distributed parameter systems, J. Proc. Cont., to appear (2014).

[3] D. BABAEI POURKARGAR AND A. ARMAOU, Modification to adaptive model reduction for regulation of distributed parameter systems with fast transients, AIChE J., 59(12) (2013), pp. 4595–4611.

Air Separation with Cryogenic Energy Storage: Optimal Scheduling Considering Electric Energy and Reserve Market Participation

Cryogenic energy storage (CES) is a new technology that is being pioneered in the UK (Harrabin, 2012). The idea of CES is to store energy in the form of liquid gas. When energy is needed at a later time, the liquid gas is pumped to high pressure and evaporated; the high-pressure gas can then be used to drive a turbine to generate electricity. Interestingly, although CES on an industrial scale is a relatively new approach, the technology used for CES is well-known and essentially part of every air separation plant that utilizes cryogenic separation. This raises the question of whether an air separation plant can benefit from its CES capability by, for example, storing energy when the electricity price is low and recovering energy for its own use or to sell it back to the grid when the price is high. Moreover, the ability of such a plant to generate power creates the new opportunity of participating in reserve markets.

In this work, we investigate the potential operational benefits of adding CES to an existing air separation plant. Based on the scheduling model proposed by Mitra et al. (2012), we develop a mixed-integer linear programming (MILP) model that integrates the air separation plant and the CES system, while considering participation in the day-ahead energy market as well as the operating reserve market.

When participating in the reserve market, we sell power generation capacity that can be dispatched when unplanned power generation is required in order to balance real-time electricity supply and demand. This is an intrinsically uncertain process since we do not know in advance when reserve will actually be needed. This means that if we sell reserve capacity for a certain period of time, we have to guarantee that we can dispatch the agreed amount of power at any time during that period of time. To consider this uncertainty, we apply a Robust Optimization approach in which the concept of the "budget of uncertainty" (Bertsimas & Sim, 2004) is used to control the level of conservatism.

We apply the proposed framework to a real-world industrial case study provided by Praxair, which leads to an MILP with up to 20,000 binary variables, 180,000 continuous variables and 110,000 constraints. Despite the large size of the problem, the MILP can be solved with modest computational effort (within minutes to 0.1 % integrality gap). The results show that adding a CES system to an existing air separation plant can significantly reduce the operating costs, particularly when we also consider the interaction with the reserve market.

References

Bertsimas, D. & Sim, M., 2004. The Price of Robustness. *Operations Research,* 52(1), pp. 35-53.

Harrabin, R., 2012. *Liquid air 'offers energy storage hope'* [online]. Available at: http://www.bbc.com/news/science-environment-19785689 [accessed May 2014].

Mitra, S., Grossmann, I. E., Pinto, J. M. & Arora, N., 2012. Optimal production planning under time-sensitive electricity prices for continuous power-intensive processes. *Computers & Chemical Engineering,* Volume 38, pp. 171-184.

373014 Design of Cryogenic Systems of Advanced Power Plants Using Simultaneous Heat Integration and Process Optimization

Abstract:

Heat integration is a very useful tool for reducing of chemical process systems and is essential to design cost-effective cryogenic separation systems. Unfortunately, most heat integration technologies are designed for application after process optimization, and assume stream temperatures and flowrates are fixed. These methods focus on synthesis of the optimal heat integration network and do not consider the interaction between adjusting process operating conditions (temperatures, pressures, flowrates, compositions) and heat integration. See Furman & Sahinidis (2002) for an extensive review of these methods.

In contrast to these approaches, Duran and Grossmann (1986) developed an optimization friendly formulation for calculating minimum utility requirements based on pinch methods. Their approach allows for the utility target portion of the heat integration equations to be embedded into a flowsheet optimization problem. Kamath et al (2012) extended the approach to consider phase changes and applied it to optimize multistream heat exchangers for natural gas liquefaction.

In this paper, we extend the work of Duran, Grossmann and Kamath into a framework for simultaneous heat integration and flowsheet optimization. Emphasis is placed on using equation-based models with accurate second derivatives and efficient, large-scale nonlinear programming (optimization) algorithms. The following key details will be highlighted:
- Reformulation of the Duran-Grossmann model to reduce the number of large elements in the Hessian (2nd derivative) matrix
- Decomposition of heat exchangers to accommodate phase changes and satisfy the constant heat capacity assumption
- An optimization friendly formulation of approximate area targeting formulas

These methods are demonstrated in two case studies: design of an air separation unit and design of a CO2 processing unit and compression train, both for advanced coal oxycombustion power plants. These cryogenic systems are excellent demonstrations of the importance of simultaneous heat integration and process optimization. For example, adjusting the operating pressures of distillation column in the air separation unit can significantly impact the design of the accompanying multistream heat exchanger.

References:
Duran, M. A., & Grossmann, I. E. (1986). Simultaneous optimization and heat integration of chemical processes. AIChE Journal, 32(1), 123–138.

Furman, K. C., & Sahinidis, N. V. (2002). A Critical Review and Annotated Bibliography for Heat Exchanger Network Synthesis in the 20th Century. Industrial & Engineering Chemistry Research, 41, 2335–2370.

Kamath, R. S., Biegler, L. T., & Grossmann, I. E. (2012). Modeling Multistream Heat Exchangers with and without Phase Changes for Simultaneous Optimization and Heat Integration. AIChE Journal, 58(1).

379044 Equation Oriented Modeling and Optimization Framework for Cryogenic Process and Energy Systems

Abstract:

In response to the recent surge in natural gas supply in the United States, several large scale liquefaction (LNG) plants have been proposed for transporting and potentially exporting natural gas [1]. Futhermore, there has been a strong reinvestment in bulk chemicals production; with many processes requiring products of air separation units (ASUs). LNG and ASU processes operate at cryogenic temperatures and high pressures, and the power required for compression is significant: air separation alone accounts for 3.5% of the U.S. manufacturing electricity demand [2]. Tight heat integration is imperative to ensure energy efficiency and profitability.

Both processes make use of multistream heat exchangers (MHEX) to maximize the recovery of refrigeration from cryogenic products. These are either of the plate-fin or coil-wound type, and allow for heat exchange between multiple product streams. Approach temperatures are typically very small, and one or more of the process streams may undergo phase change. Such units are essential to the efficient performance of the plant, and they must be considered at the flowsheet design and optimization stage. However, at present, very limited information is available in the published literature on equation-oriented, optimization-friendly MHEX models.

A MHEX model must be able to account for stream phase transformations, which, in general, are dealt with by incorporating discrete variables in the model equations. In this paper, we propose a different approach for constructing a robust equation-oriented MHEX model, based on the pseudo-transient continuation approach developed by the authors [3,5,6]. Our method bypasses the need for making discrete decisions on the phase of the streams by obtaining explicit transient expressions of stream temperature. We demonstrate that the proposed transient model has the same steady-state solution as a conventional MHEX model.

Subsequently, we integrate the developed MHEX model in the pseudo-transient process modeling and optimization framework introduced in previous work [3,4]. Using the unit operation library in [3], we explore two design case studies: the PRICO natural gas liquefaction process [7], and an air separation unit for nitrogen production [8]. We begin by performing flowsheet optimization in the nominal case, showing significant cost and energy savings compared to results from literature. Next, we explore the design of these processes under uncertainty. In the ASU case, we assume a variable electricity pricing structure, and identify the optimal flowsheet design and operating conditions for peak and off-peak pricing hours. For the LNG process, we consider a non-uniform natural gas composition and find the optimal design for all conditions. Our studies reveal that the MHEX is a potential bottleneck in the design and operation of tightly heat integrated processes that face multiple service conditions, a result that is consistent with the dynamic study reported in [8].

References

[1] Proposed North American LNG Export Terminals, 2014. Available at https://www.ferc.gov/ industries/gas/indus-act/lng/lng-export-proposed.pdf

[2] Zhu, Y.; Legg, S.; Laird, C.D. Optimal design of cryogenic air separation columns under uncertainty. Comput. Chem. Eng. 2010, 34, 1377-1384.

[3] Pattison, R.P.; Baldea, M. A pseudo-transient framework for equation-oriented modeling, simulation, and optimization of process and energy systems. AIChE J. Submitted.

[4] Zanfir, M.; Baldea, M.; Daoutidis, P. Optimizing the catalyst distribution for countercurrent methane steam reforming in plate reactors. AIChE J. 2011, 57, 2518–2528.

[5] R.C. Pattison, M. Baldea, Optimal Design of Air Separation Plants with Variable Electricity Pricing, Foundations of Computer-Aided Process Design (FOCAPD), Cle Elum, WA, July 2014.

[6] R.C. Pattison, M. Baldea, Equation-Oriented Models of Multistream Heat Exchangers for Flowsheet Optimization, 24th European Symposium on Computer Aided Process Engineering – ESCAPE 24,Budapest, Hungary, June 15-18, 2014.

[7] Price, B.C.; Mortko, R.A. PRICO - A simple, flexible proven approach to natural gas liquefaction. In Proceedings of the 17th International LNG/LPG conference, Gastech '96. Austria Centre, Vienna. 1996.

[8] Cao, Y.; Swartz, C.L.E.; Baldea, M. Design for Dynamic Performance: Application to an Air Separation Unit. In American Control Conference, San Francisco, CA, 2683-2688, 2011.

Microgrid power systems are a promising avenue for the integration of distributed power generation units. These microgrids contain both local power generation and load, operate autonomously from the area's macrogrid, and are connected to the macrogrid through only a single point of common coupling. This method for integrating distributed generation minimizes the undesirable interactions with the macrogrid that may disrupt power quality, minimizes the need for expensive overhauls to the macrogrid architecture, and allows for continued operation of distributed units during macrogrid outages. Interest in these microgrids has also been sparked by improving economics of distributed technologies for renewables, energy storage, and thermal units.

The economics of microgrids as an alternative to traditional power supply is a topic of some debate [1-3]. However the overall cost compares, proper design of these microgrid systems is certainly important since the capital cost of distributed units is relatively high. In addition, a poorly designed microgrid may lead to unexpected loss of load, significantly inflated fuel usage, or a higher than expected level of reliance on the macrogrid.

There have been previous literature efforts to determine the optimal design of microgrid systems for various case studies [2-5]. Overall, there has been little analysis of combined heat and power microgrids. This is partially due to the fact that heat and power supply has historically been decoupled. However, the potential benefits of these cogeneration systems, in terms of reduced fossil fuel usage and reduced greenhouse gas emissions, have been qualitatively identified. Moreover, distributed energy design studies have primarily analyzed cases in which the system is isolated (no connection to the macrogrid).

This paper analyzes the economic and environmental performance of combined heat and power microgrids in a macrogrid-connected scenario. Distributed technologies considered include wind turbines, solar photovoltaics, gas-fired microturbines, electric boilers, gas-fired boilers, and lead-acid batteries. A mixed integer linear program optimization problem is formulated to determine the microgrid design, i.e. the included units and their sizes. The optimal design is analysed with respect to the level of independence from the macrogrid. In addition, the impact of several environmental public policy initiatives, i.e. emission taxation and emission caps, is explored in the context of these microgrids.

The results indicate that microgrids are currently more expensive than traditional heat and power supply, but that the inclusion of small amounts of local thermal generation may not significantly increase cost. Such microgrids are promising for the support of critical power systems in locations like hospitals and military installations, and may even be economically viable when the economic losses that result from macrogrid outages are considered. However, the overall greenhouse gas emissions may rise as a result of inefficient utilization of available heat from local thermal units.

A reasonable level of emission taxation is not seen to promote reduced emissions in almost all microgrid cases considered. In contrast, an emissions cap is seen to result in a significant change in microgrid design and emissions levels, but does result in substantial cost increase. For low levels of emission reduction, the optimal microgrid design relies heavily on wind power since it is currently cheaper than solar PV. For high levels of emissions reduction, the optimal microgrid design utilizes a hybrid wind-solar design to take advantage of the complimentary nature of their power production.

[1] P. M. Costa and M. A. Matos, "Economic analysis of microgrids including reliability aspects," in *9th International Conference on Probabilistic Methods Applied to Power Systems*, 2006.

[2] E. D. Mehleri, H. Sarimveis, N. C. Markatos, and L. G. Papageorgiou, "Optimal design and operation of distributed energy systems," in *21st European Symposium on Computer Aided Process Engineering*, pp. 1713-1717, 2011.

[3] O. Hafez and K. Bhattacharya, "Optimal planning and design of a renewable energy based supply system for microgrids," *Renewable Energy*, vol. 45, pp. 7-15, 2012.

[4] S. Mizani and A. Yazdani, "Optimal design and operation of a grid-connected microgrid," in *2009 IEEE Electrical Power & Energy Conference (EPEC)*, pp. 1-6, IEEE, 2009.

[5] E. Koutroulis, D. Kolokotsa, A. Potirakis, and K. Kalaitzakis, "Methodology for optimal sizing of stand-alone photovoltaic/wind-generator systems using genetic algorithms," *Solar Energy*, vol. 80, no. 9, pp. 1072–1088, 2006.

390723 On the Use of Multistage Stochastic Programming for the Design of Smart Grid Coordinated Building Thermal Energy Storage Systems

Abstract:

Energy consumption by Heating Ventilation and Air Conditioning (HVAC) systems is usually heaviest when electricity prices are at their highest. One approach to reducing energy costs is to employ Building Thermal Energy Storage (BTES) to time-shift power consumption away from periods of high energy cost to periods of low cost.

Due to the uncertainty of electricity prices, the problem of optimal equipment sizing must be formulated as a stochastic program. However, rather than being a fairly simple two-stage stochastic program, the dynamics imposed by the storage devices requires the formulation to be of the far more challenging multistage class.

In this work, we propose a novel approximate solution procedure for this class of multistage stochastic programming problems. The approach utilizes the computational efficiency of the recently developed method of Economic Linear Optimal Control (ELOC) and its extension Constrained ELOC. The proposed method occurs in two stages. The first is a global search over the here-and-now variables as well as the parameters of the ELOC policy. However, this first search assumes a statistical constraint enforcement mechanism (similar to chance constrained optimization). In the second step is a gradient based search over the here-and-now variables, but used the Constrained ELOC policy to enforce point-wise-in-time constraints. To address the possible occurrence of an infeasible solution from the Constrained ELOC policy a fairly simple barrier type approach is used to improve convergence properties.

391477 Optimum Management of Energy Decisions Empowered By Internet of Things (IoT) Architecture - Application on a Smart Grid with Renewable Energy Sources and Hydrogen Storage

Abstract:

In a networked environment where various energy sources, including renewable (RES) such as photovoltaic and wind generators, are present, there is an increased need for flexible and adaptive energy management. Besides the varying supply, due to the nature of RES, the demand also varies through time by the end user. The combination of the requirements that arise by the dynamic supply and demand in the grid formulates a highly complex environment that can be modeled as a complex system of interconnecting processes. Such a complex system has a large number of design variables and its decision-makings require real time data collected from devices, systems and processes. Thus design and operation of a multi-source energy network involves numerous decision-making at various levels and domains. As systems become more interconnected and diverse, it is important to be able to design the interactions among the various components in an efficient and robust way, prior to their deployment to a specific environment. To address this challenge it is important to rely on an architecture that considers the heterogeneity and the dynamic behavior of the involved components. The Internet of things (IoT) is considered as a promising networking paradigm that bridges the gap between the cyber and physical world and can be used as an alternative approach to the management of energy decisions in a networked environment. The implementation of an IoT enabled architecture can provide a better insight about the physical systems and their interconnection to the cyber world. In this work a generic scheme is proposed that facilitates the design and deployment of scalable, flexible and easily maintainable modular systems for the energy domain. The proposed scheme is not limited to a particular technical implementation and can be used for a wide range of applications.

The objective of this work is to develop a reliable, sustainable and adaptive architecture that will implement the optimum energy decisions in a complex networked ecosystem. More specifically, this work investigates the impact of emerging IoT on energy networked systems in a smart grid environment. To achieve this objective, the modeling of the IoT objects is discussed in order to identify the requirements of decision support systems in dynamic and distributed environments. Furthermore, the relation of the developed IoT enabled architecture and the energy management decisions are overviewed. Finally, some preliminary results are presented that show the implementation of the proposed scheme to a smart grid system with RES.

In order to implement a generic and context-aware framework that will be able to accommodate the needs of the various devices within the smart grid domain, an architectural reference model is adopted that describes the structure of the system and an application specific model is developed that describes the requirement of the energy domain. The subsystems are modeled as objects or entities of a Unified Modeling Language (UML) diagram. The operations are modeled by a Finite State Machine (FSM) and the rules of operation are represented by propositional logic. The implementation is realized by a middleware, which is situated between the low level device drivers and the upper level industrial automation software.

The developed energy management scheme (EMS), which is empowered by the IoT architecture, is implemented for the operation of a smart grid that involves three autonomous stations (nodes) with local battery storage capabilities, hydrogen generation and long term hydrogen storage options. Each station has a PV array, wind generators and a diesel generator for backup purposes. Also at one of the stations there is a Polymer Electrolyte Membrane (PEM) fuel cell and a PEM electrolyzer for the production of hydrogen. This smart grid is located at Xanthi, Greece and its operation is monitored by a supervisory control station while the Machine to Machine (M2M) communication for the decision making is implemented by the IoT enabled architecture. The performance of the involved nodes is explored and the overall operation of the grid under various operating scenarios is presented. Besides the typical operation, a series of unplanned events are also explored in order to verify the response of the developed EMS in terms of flexibility and robustness of the energy exchange at a network level.

387132 Combined Heat and Power for Residential Neighborhoods: A Data-Driven Modeling and Optimization Study

Abstract:

Combined heat and power (CHP) facilities are a very promising path to reducing CO_2 emissions and increasing efficiency in the power generation sector, especially when combined with residential solar power generation. CHP facilities rely on natural gas, a cleaner fuel than coal, to generate electricity and hot and chilled water. The ability to supply these essential residential utilities in an efficient way on a medium to large scale opens the path for combining district cooling, heating and power generation, and suggests that CHP plants are a very appealing choice for providing integrated utilities for the neighborhood of the future. Yet, there are currently no CHP plants that serve just residential neighborhoods, and published works exploring this possibility are scarce.

Approximately 20% of the electricity produced in the United States is used in the residential sector, with about half of this energy going to heating, ventilation, and air-conditioning (HVAC) systems [1]. Coal-fired power plants, which provide a majority of the electricity, have very high CO_2 emissions and among the lowest efficiencies of all power generation facilities. CHP plants powered by natural gas can reach efficiencies over 80%, more than twice that of coal-powered plants, which are often less than 38% [2]. In combination with district cooling, CHP plants also lower CO_2 emissions by a factor of 10 [3]. In a CHP case study conducted by Austin Energy, the local power utility in Austin, TX, in 2010, Austin Energy's Fayette Power coal-powered plant was only able to convert 35% of the primary fuel into electricity, and an additional 6% was lost in transmission to a large consumer (Dell Children's Medical Center) situated less than 10 miles (16.8km) away. A subsequent investment in a localized CHP facility at Dell Childern's demonstrated that at least 75% of the primary energy source (natural gas) can be converted to power with no transmission losses [4]. Yet- and in spite of these appealing energy efficiency benefits - there are currently no CHP plants in the United States that provide integrated utilities (heating, cooling and power) to residential neighborhoods, and studies exploring such opportunities are not available.

In the paper, we describe the optimal integration of a CHP plant as a utility producer for a residential neighborhood, and the potential for using photovoltaics and energy storage, implemented in a centralized fashion, to alleviate fluctuations in residential demand. Utilizing data collected by Pecan Street Research Inc., a smart-grid demonstration project headquartered at The University of Texas at Austin, residential heating, cooling, and electricity demand are analyzed and evaluated. These data are then used to create a time-resolved energy profile pertaining to residential energy demand. Subsequenlty, we compute an optimal operating strategy for an integrated CHP/solar utility and determine the optimal ratio of generating capacity between CHP and photovoltaic generation, such that the effect of fluctuations in energy demand is minimized. We demonstrate that CHP a viable means for providing district-level cooling and power in a predominantly cooling climate and, furthermore, we study the economic and operational impact of integrating, in a similarly centralized manner, CHP production with various forms of energy storage (e.g., thermal, electrical).

References

[1] U.S. Energy Information Administration, "Annual Energy Review 2011." 27-Sep-2012.

[2] Combined Heat & Power Association, "What is CHP?"

[3] DHC + Technology Platform, "District Heating & Cooling." Mar-2012.

[4] Jim Collins. Combined Heat & Power Conference 2011, "Case Studies: Mueller Energy Center at Dell Children' Medical Center, Austin, TX." 3-Nov-2010.

385301 Global Optimization of Real Time Operation of an Industrial Gas Network

Abstract:

Air Liquide operates a number of industrial gas pipeline networks around the world. Each of them connects several air separation plants and a number of customers. In each of the plants, there can be one or more air separation units (ASU), which separates air into pure oxygen, nitrogen and argon in either gas or liquid phase. Usually, the gaseous oxygen and nitrogen supply different pipeline sections at different pressure levels after being pressurized either in the ASU or by compressors downstream of the ASU.

The operation of such a network of plants, pipelines and customers is complicated due to the rapidly fluctuating electricity prices and customer demands. To efficiently operate this network under these conditions, real-time optimization techniques are necessary. The real-time optimization model was developed by Air Liquide to describe both plant and pipeline operations through mass balance, energy balance and regression on historical data. Nonconvex functional forms are necessary to accurately describe plant characteristics. Also, certain discrete decisions must be made to choose whether certain compressors and air separation units should be turned on or off. Due to these complications, the resulting model is a mixed integer nonlinear programming problem (MINLP), with multiple local optima.

Global optimization techniques are thus essential to exploit this model and to make the best operating decisions. For use in a real-time optimization setting, this model must be solved in a reasonable time. Furthermore, the size and numerical features of the problem pose a challenge in obtaining a reliable and robust solution to the model.

We will present the collaborative effort between Carnegie Mellon University and Air Liquide in addressing the above challenges. We will present the strategies employed to tackle each of the challenges, along with computational results.

Natural Gas Reduction in Industrial Waste Incineration by Optimized Short-term Scheduling

Matteo Abaecherli, Elisabet Capón-García, Stavros Papadokonstantakis and Konrad Hungerbühler

Waste production is a major issue in all branches of the chemical industry. Bulk chemicals have a quite low waste to product ratio, but they are produced in huge quantities, pharmaceuticals and fine chemicals are in contrast produced in low quantities, but they have a large waste to product ratio (Poliakoff et al., 2002). In some countries, such as Switzerland, up to 70 % of the chemical waste is incinerated (BAFU, 2012). Major chemical industrial sites have own in-house incineration units which guarantee an independent waste management. On the one hand, the heat released in the incineration process is generally used for steam production. In order to keep the incineration at steady state, it is necessary a minimum energy input into the system. Hence, in case that the waste does not provide enough energy, additional natural gas is required. On the other hand, the optimal functionality of the incineration treatment and its well thought-out planning and scheduling are of utmost importance to avoid constraining unnecessarily the daily production (Wassick, 2009). However, most decisions in practice are still made on a purely empirical basis.

In order to improve the operation of waste incineration plants, (Anderson et al., 2005) applied a multi-objective evolutionary algorithm. This strategy provided the operators with the ability to adapt some selected parameters of the incineration units to improve the overall treatment performance based on a given input and set of constraints. Recently, Wassick (2009) has presented the benefits of using systematic modeling and optimization in the waste management and plant integration in the Dow Company. Over one year the optimized scheduling of the presented model was compared with the actual schedule of the normal work process. Results indicate that optimal scheduling tools provide schedules with lower costs.

This work aims to provide to waste management operators with a detailed short-term optimized schedule for liquid waste incineration. The optimization objective is to reduce the overall consumption of natural gas through adequate mixing and scheduling of waste streams with different energy content, thus avoiding situations with excess and shortage of energy.

Additionally holdups in the production caused by bottlenecks in the waste treatment units have to be avoided and all legal and technical constraints have to be respected.

The waste management and incineration system considered in this work consists of several storage tanks, a complex piping network, tank wagons for the waste transport, unloading pumps and firing lances of the incineration units. A mixed-integer linear programming (MILP) formulation is introduced to approach the short-term planning and scheduling of liquid waste transfer from different production sites to the incineration units and its disposal. The model is composed of two main parts: (i) waste transfer and mixing issues, (ii) the incineration process itself with its associated energy calculations and constraints. The scheduling horizon was set in the range of one week discretized in uniform time intervals.

The developed optimization tools have been tested and validated in a chemical plant in Switzerland. It could be shown, that, compared to the current performance of waste incineration, the optimized schedule would have reduced the overall natural gas consumption, while respecting all system and regulatory constraints and guaranteeing the provision of enough energy for a continuous and proper incineration.

References

Anderson, S. R., Kadirkamanathan, V., Chipperfield, A., Sharifi, V., & Swithenbank, J. (2005). Multi-objective optimization of operational variables in a waste incineration plant. Computers & Chemical Engineering, 29(5), 1121 - 1130.

BAFU. (2012). Sonderabfallstatistik: Federal Office of the Environement, Switzerland.

Poliakoff, M., Fitzpatrick, J. M., Farren, T. R., & Anastas, P. T. (2002). Green Chemistry: Science and Politics of Change. Science, 297(5582), 807-810.

Scienceindustries. (2012). The Swiss Chemical and Pharmaceutical Industry.

Wassick, J. M. (2009). Enterprise-wide optimization in an integrated chemical complex. Computers & Chemical Engineering, 33(12), 1950 - 1963.

380430 Polymerization Process Reconfiguration from Molecular Weight Distribution

Abstract:

To meet the market requirements, a polymerization process is expected to be capable of producing a number of polymer grades with different end-use properties. Due to the limitations of the conventional process structures, many polymer grades, especially some high-quality grades, may not be available even under a wide range of operating conditions. Hence, a more flexible process configuration needs to be considered and designed. On the other hand, molecular weight distribution (MWD) is at the core in establishing key quality indices for polymers. Toughness, hardness, stiffness, strength, and viscoelasticity are among the properties dependent on MWD. Few research efforts, however, have been made to determine the optimal structure and operating conditions of polymerization processes from different specifications on MWD due to the complexity of MWD modeling and calculation.

Reactor network plays a central role in determining the characteristics of polymerization processes. In our study, we develop and analyze a comprehensive superstructure of series and parallel reactor configurations. First, on the basis of the superstructure model, several steady-state optimization problems with specified MWDs, like polymer productivity maximization, will be studied to obtain the optimal steady-state structure and operating conditions. Next, dynamic reconfiguration problems will be further proposed and solved to show the optimal time-dependent process structure and control trajectories. For polymer grade transitions, a time-varying process configuration is expected to help minimize the transition time or the quantity of waste products. A high-density polyethylene (HDPE) slurry process with several continuous stirred-tank reactors (CSTRs), together with other units including flash drums, coolers, compressors, and pumps, is considered as an example to illustrate the feasibility and advantages of the proposed reactor network synthesis method. The polymerization process reconfiguration from MWD could lead to the development of new polymer grades, the improvement of the end-use properties and productivity of products, and the improvement of process operating policies.

Reference

1. Lakshmanan A, Biegler LT. Synthesis of Optimal Chemical Reactor Networks. *Ind. Eng. Chem. Res*. 1996;35:1344-1353.

2. Schweiger CA, Floudas CA. Optimization framework for the synthesis of chemical reactor networks. *Ind Eng Chem Res*. 1999;38:744-766.

3. Khare NP, Seavey KC, Liu YA, Ramanathan S, Lingard S, Chen CC. Steady-State and Dynamic Modeling of Commercial Slurry High-Density Polyethylene (HDPE) Processes. *Ind Eng Chem Res*. 2002;41:5601-5618.

4. Khare NP, Lucas B, Seavey KC, Liu YA, Sirohi A, Ramanathan S, Lingard S, Song Y, Chen CC. Steady-State and Dynamic Modeling of Gas-Phase Polypropylene Processes Using Stirred-Bed Reactors. *Ind Eng Chem Res*. 2004;43:884-900.

Agent Based Simulations for Synchronous and Asynchronous Decision Making Strategies in Supply Chains

Nihar Sahay[a], Marianthi Ierapetritou[a], John Wassick[b]

[a] Department of Chemical and Biochemical Engineering, Rutgers University, Piscataway, NJ

[b] The Dow Chemical Company, United States

A chemical supply chain is a network of suppliers, production facilities, warehouses and markets designed to acquire raw materials, manufacture and store and distribute products among the markets. In today's complicated economic environment, these individual entities may or may not belong to a single company. Each entity has its own objective function which drives its operations. The entities make decisions based on time-scales appropriate to their own operations. Therefore a supply chain is essentially an asynchronous distributed decision-making system.

Although modern Enterprise Resource Planning (ERP) systems attempt to connect the supply chain members and facilitate synchronized operations, achieving complete synchronization of all the supply chain operations is still a challenge. Each individual operation has its own local dynamics and rate of generating and exchanging data. The individual members take many decisions locally at different points in time and decision-making rates also vary. For example, warehouses may have different review frequencies under periodic inventory review policies which would result in one warehouse modifying its inventory more regularly compared to another.[1] As the decisions and actions taken by the individual members depend on the information available to them, time and rate of exchange of information plays an important role in determining the dynamics of the supply chain network.

In a practical scenario, asynchronous decision making within a supply chain is quite common. There are quite a number of works which cover the distributed and asynchronous nature of decision making. Carle et al.[2] propose an agent-based metaheuristic to solve large-scale multi-period supply chain network design problems. Using the concept of asynchronous agent teams, they propose an efficient hybrid metaheuristic called Collaborative Agent Team. This model integrates design and modeling concepts and can be used to reengineer real-world supply chain networks. Tolone[3] focuses on agility and time-based manufacturing as critical factors for manufacturing enterprises in today's world. He considers a multi-company supply chain planning and execution environment and emphasizes the importance of real-time and asynchronous collaboration technology in allowing manufacturers to increase their supply chain agility. Luh et al.[4] focus on the issue of coordination of activities across a network of suppliers to quickly respond to market conditions. They combine mathematical optimization and the contract net protocol for make-to-order supply network coordination. The overall problem is decomposed into organizational sub-problems. Scheduling of activities by the individual organizations is based on their internal situations and inter-organization prices. Coordination is carried out in a distributed and asynchronous manner. Sohlenkamp and Chwelos[5] present a virtual office environment that supports communication, cooperation and awareness in both synchronous and asynchronous modes. It is modeled after the standard office and abstracts elements of physical offices required for collaborative work like people, rooms, desks and documents.

Simulation models have been widely used to depict supply chain networks. The main idea behind using detailed simulation models is to be able to capture the true dynamics of the network. In a practical scenario, no supply chain operates completely synchronously or asynchronously. It is therefore essential for the simulation models to include the synchronicities and asynchronicities present in the network in order to capture the true dynamics. The way these models are coded is important to get the right set of results. Lormier et al.[6] discuss that the success of individual based models depends on its design and assumptions made during its construction. In particular, they consider synchronous and asynchronous scheduling as two methods for updating object characteristics during interaction. They show that the two methods produce different results for a simple individual based model, particularly for high population densities and increased interaction complexity. Boer et al.[7] investigate how the market dynamics for continuous, asynchronous simulation are different from discrete time simulation. They show that the continuous, asynchronous simulation reveals more information and the market prices are different between the two cases. Androulakis and Reklaitis[1] developed a methodology for decentralized decision making for problems consisting of interacting sub-systems where each is described by local properties and dynamics. Their methodology is based on the solution of a series of sub-problems where each minimizes each local objective while maximizing a common Lagrangean function.

In this work, we investigate how the asynchronous operations of the different members of the supply chain can affect its overall behavior. We consider that the different entities of the network operate based on their timescales. Based on their own operating cycles, they update the relevant information which impacts the information set available to others at any particular point in time. In order to illustrate this, we use rather simple case studies. We use agent based simulation models to capture the dynamics of these networks. To include production scheduling for the manufacturing plants, we use embedded optimization with the agent based simulation models. We also develop models for the synchronous operation and compare the results for the scenarios. The results show that the frequency of interaction between the decision makers and the level of shared information available to decision – making entities can have a significant impact on cost and Supply Chain dynamics. It is observed that synchronous interactions result in lower cost compared to asynchronous decision making. However, since synchronous interactions are not always feasible, the model for the asynchronous scenarios provides useful results regarding the resource utilization in the supply chain.

References

1. Androulakis IP, Reklaitis GV. Approaches to asynchronous decentralized decision making. *Computers & Chemical Engineering.* 2/28/ 1999;23(3):341-355.
2. Carle MA, Martel A, Zufferey N. The CAT metaheuristic for the solution of multi-period activity-based supply chain network design problems. *Int J Prod Econ.* Oct 2012;139(2):664-677.
3. Tolone WJ. Virtual situation rooms: connecting people across enterprises for supply-chain agility. *Comput Aided Design.* Feb 2000;32(2):109-117.
4. Luh PB, Ni M, Chen HX, Thakur LS. Price-based approach for activity coordination in a supply network. *Ieee T Robotic Autom.* Apr 2003;19(2):335-346.
5. Sohlenkamp M, Chwelos G. Integrating communication, cooperation, and awareness: the DIVA virtual office environment. Proceedings of the 1994 ACM conference on Computer supported cooperative work; 1994; Chapel Hill, North Carolina, USA.
6. Caron-Lormier G, Humphry RW, Bohan DA, Hawes C, Thorbek P. Asynchronous and synchronous updating in individual-based models. *Ecological Modelling.* 4/10/ 2008;212(3–4):522-527.
7. Boer K, Kaymak U, Spiering J. From discrete-time models to continuous-time, asynchronous modeling of financial markets. *Comput Intell.* May 2007;23(2):142-161.

Modeling Steam-Methane Reforming for Dynamic Risk Analysis

Ian Moskowitz[1], Warren D. Seider[1], Masoud Soroush[2], Ulku Oktem[3], and Jeffrey Arbogast[4]

1. *Department of Chemical and Biomolecular Engineering, University of Pennsylvania, Philadelphia, PA 19104-6393*
2. *Department of Chemical and Biomolecular Engineering, Drexel University, Philadelphia, PA 19104*
3. *Department of Risk Management and Decision Process Center, University of Pennsylvania, Philadelphia, PA 19104-6393*
4. *Air Liquide, Newark, DE 19702*

Process simulations are used by chemical manufacturers to optimize the design of plants and their associated control systems, but little emphasis has been placed on using simulations to design and operate safety systems. Process simulations designed for safety purposes have the potential to inform operators and plant managers about accident scenarios likely to occur. The simulations can also be used to design and recommend safety systems that are most effective at mitigating the risks of postulated accident scenarios. A major difficulty lies in the development of process safety simulations; what attributes should a simulation designed for safety have? How should a simulation for process safety differ from a simulation for process design or control? How can a safety simulation be integrated into management of daily operations?

This presentation focuses on the development of a safety simulation for a steam-methane reforming (SMR) process. These processes often use pressure-swing adsorption (PSA) to recover H_2 product from off-gases. This inherently-transient unit is responsible for dynamic effects throughout the plant, which are crucial to understand when properly designing safety systems. Another attribute of the SMR process is the furnace box in the SMR-unit. Temperatures in this box are as high as 1,500K, but these are too high for thermocouple measurements. Consequently, in safety models, good temperature predictions are needed to avoid catalyst overheating and degeneration. Safety systems must respond by arresting rapid temperature rises resulting in accident scenarios.

This paper presents a robust method of using a plant dynamic model together with alarm and measurement data to conduct safety-system analysis. As all models contain process-model mismatch, the robust method will allow operators to comfortably make safety decisions for the plant. Issues concern the rigor of furnace models (e.g., in radiation heat transfer – view factor determination; the need for computational fluid dynamics; the need to model each of 400 tubes) and the PSA cycles involving each of 12 adsorption units. For safety purposes, given extensive alarm and process variable data, the need for high precision, involving high computational demands is shown to be relaxed.

Stated differently, this paper shows how simulations together with alarm and measurement data are used robustly to conduct safety-system analysis. Bayesian analysis fits this scheme well, with simulations providing informative <u>prior distributions</u> for the failure probabilities of the safety systems, and the alarm and process data providing the likelihood distributions, with which the <u>posterior</u> failure probabilities are updated. In this fashion, simulations describe the failure probabilities of safety systems using the traditional event-tree formulation. Also, they are used to set alarm threshold bands for the process in which the failure probabilities are the minimized objective function.

371613 Application of a Mixed-Integer Linear Program Based Planning and Scheduling Tool to Reduce Capital Investments in Batch Facility Debottlenecking

Abstract:

Understanding capacity for new products in large scale production of therapeutic Monoclonal Antibodies is a challenging proposition. Production generally proceeds through many sequential unit operations with parallel equipment for the slower cell culture unit operations and a single purification train. Each unit operation may be supported by a variety of solutions which are prepared in a limited number of mixing vessels. Storage and delivery of solutions to the applicable unit operation must respect both the connectivity limitations in the plant design as well as the temporal coordination with the rest of the production process. Additionally, equipment cleaning using a limited set of Clean-In-Place (CIP) skids adds complexity to the planning and scheduling problem. Often, there is not a single piece of equipment that defines the plant bottleneck, but rather a series of steps that may cross multiple pieces of equipment which limits performance.

Traditionally, understanding plant bottlenecks is a challenge even for products the plant has already run. Debottlenecking a new product using back of the envelope calculations and broad assumptions is likely to result in inaccurate investment targets. However, current tools for planning and scheduling support high resolution operational modeling of batch facilities and allow for accurate identification of plant bottlenecks. Here, we describe how a sequential process of (1) identifying bottlenecks using a model and (2) testing alternative engineering solutions to bottlenecks significantly reduced debottlenecking costs by eliminating unnecessary capital project scope and simultaneously identified previously unidentified bottlenecks that would have been overlooked without the insight of a detailed model.

383187 Optimization and Analysis Methods for Maritime Inventory Routing

Abstract:

Inventory routing problem (IRP) appears in various chemical supply chains (SCs), when distribution routing and scheduling decisions are made with considerations of inventories at suppliers/customers. Based on travel distances, real-time inventories and forecasted production/consumption profiles of suppliers and customers, we optimize the routes, schedules, and loading/discharging amounts, while ensuring that the inventory levels are kept within a certain range. Compared with the traditional order call-in scheme, integration of distribution routing and scheduling together with inventory control helps to reduce the distribution cost, which accounts for a large fraction of the total operating cost of the distribution SC. One important type of IRP, which finds applications in the petro-chemical sector, is the maritime inventory routing problem (MIRP), where multiple products are transported between supplier ports and customer ports, so that the upper and lower inventory level constraints of these ports are satisfied. The travel time between ports normally takes several days. Each port in the SC network has its own access windows during which a loading or discharging is allowed, which also adds more complexity to the problem.

In this work, we propose a discrete-time Mixed-Integer-Programming (MIP) model, and develop a suite of methods and tools to aid decision-making in MIRP. First, in addition to travel distance, heterogeneous vessels and access windows, we also consider soft windows, and penalized overflow and underflow. The objective is to minimize the overall cost, which includes travelling cost, material holding cost, overflow/underflow cost and vessel rental cost. Second, we present new methods to accurately calculate rental cost based on rental modes (short-term or long-term). Third, we develop tools that help the user to easily manipulate data as well as the resulting schedules and re-optimize seamlessly. Finally, to give some insights on the SC performance under uncertainty, we present results from an analysis based on different scenarios generated based on (1) different rental modes, (2) uncertainty in consumption/production forecast, (3) uncertainty in vehicle availability, and (4) expanded access windows. We show that the proposed framework can lead to significant savings.

Assessing Flexibility and Risk in Supply Chains using Hybrid Simulation based Optimization

Nihar Sahay and Marianthi Ierapetritou

Department of Chemical and Biochemical Engineering, Rutgers University, Piscataway, NJ

The main goal of an efficient supply chain is to produce acceptable quality products and deliver them to customers at a competitive cost within reasonable delivery times. With increasing uncertainty in product demand, transportation times and production times, it becomes more challenging to achieve these targets thus leading to the introduction of risk in supply chain decision-making. Along with maintaining efficiency, it is also important that there is flexibility within the network so that risk can be controlled. There can be many aspects of flexibility related to the different processes. However, in case of supply chains, the total system flexibility needs to be evaluated and not just flexibility of a particular component.[1] Achieving flexibility incurs costs and therefore, there is a need for assessment of flexibility levels and risk management during the planning of supply chains under uncertainty.

Researchers have studied various dimensions of flexibility in the context of supply chains and have also proposed different strategies for achieving them. Vickery et al.[2] study five key supply chain flexibilities: product flexibility, volume flexibility, launch flexibility, distribution flexibility and responsiveness to target markets. Pujawan[3] presents a framework for assessing flexibility of a supply chain. He identifies the different dimensions of flexibility and assesses their drivers. He presents a framework to assess the flexibility of a supply chain. Winkler[4] discusses the design and use of strategic supply chain networks to improve flexibility. He shows how structural, technological and human potentials of the strategic supply chain network can be managed to gain flexibility. Garavelli[5] studies process flexibility and logistical flexibility and proposes that configurations with limited flexibility perform relatively better compared to no or total flexibility when the trade-offs between costs and flexibility are considered. He also found that achieving flexibility in upstream activities leads to better results than downstream activities.

Under uncertain conditions, usually the expected performance measure is optimized. The obtained solution performs well for all the scenarios on an average. However variability in performance due to uncertainty in conditions is an important aspect that needs to be assessed.

Eppen and Martin[6] use two stage stochastic programming model for production capacity planning for the auto industry. Demand is considered to be uncertain. They use downside risk as the risk measure. Barbaro and Bagajewicz[7] present a methodology to include financial risk management in the framework of two-stage stochastic programming for planning under uncertainty. A probabilistic definition of financial risk is adapted to be used in a two-stage stochastic programming model. You et al.[8] study risk management for mid-term planning of global multi-product chemical supply chain. They consider demand and freight rate to be uncertain. They propose a multi-period planning model and use two-stage stochastic programming approach within it. An algorithm based on multi-cut L-shaped method is proposed to solve the large scale problems. They compare four different risk measures using the proposed framework.

In this work, we study the trade-offs between cost and flexibility of a supply chain. The supply chain is modeled using an agent-based simulation, which helps to capture the supply chain dynamics in a realistic scenario. Therefore, it helps evaluate the flexibility of the network under realistic conditions. Demand is considered to be uncertain. Variability in demand that the supply chain fulfills is used to define how flexible the network is. A multi-period planning problem is solved using a rolling horizon framework where a two stage stochastic programming problem is solved at each period. In order to solve the two stage stochastic programming problem, hybrid simulation based optimization is used. Hybrid simulation based optimization has been shown to provide a few advantages in solving supply chain planning problems.[9,10] The simulation model provides a better representation of the dynamics of the supply chain while the optimization model is used to guide the simulation towards the optimal solution. The framework is also used to study risk management. The different objectives give rise to a multi-objective optimization problem which is solved using the ε-constraint method. The approach has been used to solve small scale case studies in order to demonstrate its effectiveness.

References

1. Upton DM. The Management of Manufacturing Flexibility. *California Management Review.* Winter94 1994;36(2):72-89.

2. Vickery Sn, Calantone R, Dröge C. Supply Chain Flexibility: An Empirical Study. *Journal of Supply Chain Management.* 1999;35(2):16-24.
3. Pujawan IN. Assessing supply chain flexibility: a conceptual framework and case study. *International Journal of Integrated Supply Management.* 2004;1(1):79-97.
4. Winkler H. How to improve supply chain flexibility using strategic supply chain networks. *Logist. Res.* 2009/03/01 2009;1(1):15-25.
5. Garavelli AC. Flexibility configurations for the supply chain management. *Int J Prod Econ.* 8/11/ 2003;85(2):141-153.
6. Eppen GD, Martin RK, Schrage L. A Scenario Approach to Capacity Planning. *Operations Research.* 1989;37(4):517-527.
7. Barbaro A, Bagajewicz MJ. Managing financial risk in planning under uncertainty. *AIChE Journal.* 2004;50(5):963-989.
8. You F, Wassick JM, Grossmann IE. Risk management for a global supply chain planning under uncertainty: Models and algorithms. *AIChE Journal.* 2009;55(4):931-946.
9. Sahay N, Ierapetritou M. Supply chain management using an optimization driven simulation approach. *AIChE Journal.* 2013;59(12):4612-4626.
10. Nikolopoulou A, Ierapetritou MG. Hybrid simulation based optimization approach for supply chain management. *Computers & Chemical Engineering.* 12/20/ 2012;47(0):183-193.

377394 Optimal Dynamic Allocation of Mobile Plants to Monetize Associated or Stranded Natural Gas Under Uncertainty

Abstract:

Associated or stranded natural gas presents a particular challenge to monetize due to its low volume and lack of supporting infrastructure. Recent proposals for deploying mobile, modular plants, such as those which perform gas-to-liquids (GTL) conversion or liquefied natural gas (LNG) on a small scale have been identified as possible attractive routes to gas monetization. However, such modular technologies are yet unproven in the marketplace. In addition, uncertainty about supply, demand and price conditions greatly influence the outcome of investment decisions in the oil and gas industry. To address these issues, we propose a multi-stage stochastic programming framework which determines the most optimal allocation decisions for a decision-maker intending to make use of mobile plants to monetize associated or stranded gas under uncertainty. We then demonstrate how this framework can be applied to a real-world case study of the Bakken shale play in North Dakota, where large amounts of associated gas are currently being flared. The development of this framework paves the way for decision-makers to screen and evaluate unproven modular technologies systematically and also provides a robust strategy for profit.

An Optimization Framework for the Assessment of Shale Gas Resources

Omar J. Guerra[a], Andrés J. Calderón[b], Lazaros G. Papageorgiou[b], Jeffrey J. Siirola[a], Gintaras V. Reklaitis[a,*]

[a] School of Chemical Engineering, Purdue University, West Lafayette IN 47907 USA.

[b] Department of Chemical Engineering, UCL (University College London), London WC1E 7JE, UK

[*] Corresponding author. E-mail address: reklaiti@purdue.edu (G.V. Reklaitis).

Abstract

Unconventional primary energy resources like Shale gas can play an important role in the world's energy production in the decades to come. It is estimated that 32% of the current total natural gas resources corresponds to Shale gas (EIA, 2013) and its importance in the global market is increasing as new technologies for exploration and exploitation are available. Shale gas could be a "game changer" with the potential of lowering prices and expanding supply. The availability of natural gas at low prices can bring new opportunities for chemical and electricity generation industries (Jenner and Lamadrid, 2013; Siirola, 2014). Nevertheless, the production of shale gas resources depends extensively on production costs and productivity where minor changes can make the difference between a resource that is economically viable and one that is unattractive at current market conditions. For instance, the extraction of shale gas is more aggressive in terms of environmental impact when compared to conventional gas. In particular, high water consumption and the potential for ground water contamination represent major concerns, which could hinder the development of Shale gas plays (Cafaro and Grossmann, 2014; Clark et al., 2013; Eaton, 2013; Vidic et al., 2013; Warner et al., 2013).Therefore, the assessment of Shale gas production is a challenging problem where economic and environmental aspects need to be considered at both the single field and supply chain decision levels.

In this work, an optimization framework for the assessment of shale gas resources is presented. The framework is based on a disaggregated approach that includes Shale gas production, water management and supply chain models. First, rigorous simulations for water and shale gas treatment facilities are carried out in order to provide detailed information about process performance under different configurations and operating conditions. Then, an optimization model is proposed to address the optimal development of a shale gas field. The decisions addressed in this level are: well-pad locations; capacities, technologies and location of gas and water treatment facilities. The design of the gas and water transport network from well-pads to treatment facilities is also considered. Based on the results, surrogate models are developed to describe the performance of a shale gas field in terms of Capex, Opex, gas production and water use.

Finally, those models are integrated into a Mixed Integer Linear Programming (MILP) supply chain optimization model in order to deal with strategic decisions such as selection, capacity and location of process technologies on a regional and multi-time period basis. The capabilities of the developed optimization framework are demonstrated using a nationwide case study involving data from Colombia. This optimization framework is used to analyze the trade-off between economic and environmental aspects of Shale gas production and commercialization in the Colombian context.

References

Cafaro, D., Grossmann, I., 2014. Strategic Planning, Design and Development of the Shale Gas Supply Chain Network. AIChE J. 60, 2122–2142.

Clark, C.E., Horner, R.M., Harto, C.B., 2013. Life cycle water consumption for shale gas and conventional natural gas. Environ. Sci. Technol. 47, 11829–11836.

Eaton, T.T., 2013. Science-based decision-making on complex issues: Marcellus shale gas hydrofracking and New York City water supply. Sci. Total Environ. 461-462, 158–169.

EIA, 2013. Technically Recoverable Shale Oil and Shale Gas Resources : An Assessment of 137 Shale Formations in 41 Countries Outside the United States (http://www.eia.gov/analysis/studies/worldshalegas/).

Jenner, S., Lamadrid, A.J., 2013. Shale gas vs. coal: Policy implications from environmental impact comparisons of shale gas, conventional gas, and coal on air, water, and land in the United States. Energy Policy 53, 442–453.

Siirola, J.J., 2014. The Impact of Shale Gas in the Chemical Industry. AIChE J. 60, 810–819.

Vidic, R.D., Brantley, S.L., Vandenbossche, J.M., Yoxtheimer, D., Abad, J.D., 2013. Impact of shale gas development on regional water quality. Science 340, 1235009_1- 1235009_9.

Warner, N.R., Christie, C. a, Jackson, R.B., Vengosh, A., 2013. Impacts of shale gas wastewater disposal on water quality in western Pennsylvania. Environ. Sci. Technol. 47, 11849–11857.

379788 Capacity Expansion with Rational Markets in the Industrial Gas Industry

Abstract:

The standard approach to develop capacity expansion plans leverages optimization tools in order to maximize the profit obtained during a finite time horizon. The problem with this formulation is that it assumes that companies decide what share of the market they satisfy. In reality, buyers select their providers from a pool of potential suppliers based on their own interest. Therefore, it is important to consider market preferences in order to develop reliable expansion plans.

We address the problem of establishing a capacity expansion plan that maximizes the net present value (NPV) of an industrial gas company with multiple facilities, subject to markets that determine the company's income according to their own cost. The capacity expansion problem with a rational market resembles a strategic game in which companies establish their capacity expansion plans, and then the markets decide from which provider their demands are satisfied. We consider the strategic game with perfect information; the suppliers are aware of the decision criterion of the rational markets and the markets observe the actions of the suppliers before their decisions are made.

We formulate the capacity expansion problem with rational markets as a bilevel optimization problem. The upper level problem maximizes the leader's NPV and the lower level problem minimizes the cost paid by the markets. The decisions in the upper level problem are the investment related to capacity expansion, while in the lower level the markets decide from which provider they buy. The lower level decisions are based on the economic interest of the markets and the prices offer by the different providers.

The resulting formulation is a mixed-integer bilevel linear programming (MIBLP) problem in which the upper level variables are mixed-integer, and the lower level variables are all continuous. In order to solve this problem, we develop two solution strategies that transform the bilevel problem into a single-level optimization problem. Both reformulations rely on the conditions met by linear programs (LPs) at the optimal solution.

The first strategy replaces the lower level problem with its Karush-Kuhn-Tucker (KKT) optimality conditions. KKT conditions ensure optimality of the lower level problem by satisfying stationarity, primal feasibility, dual feasibility, and complementary slackness of the solution. The main difficulty with this reformulation is the nonlinearities introduced with the complementary slackness conditions. We reformulate the nonlinear complementary slackness conditions as big-M constraints using the active-set approach. The final formulation is a single-level MILP problem that can be solved directly. However, the use of big-M constraints introduces a large number of binary variables that complicate the solution of the problem.

The second strategy leverages the strong duality property of linear programs. In this case, the lower level LP is replaced by both its primal and dual constraints; strong duality is enforced with a constraint that equates the primal and dual objective functions. The complication of this approach comes from the nonlinearities that appear in the dual objective function of the lower level problem. These nonlinearities arise because the upper level variables are treated as parameters in the lower level. Therefore, the single-level reformulation includes the multiplication of upper level variables and lower level dual variables. However, the nonlinearities can be reformulated using exact linearization when all the upper level variables are discrete.

The KKT and the Primal-Dual strategies are implemented in an instance of the capacity expansion problem with rational markets. The results show that both reformulations yield exactly the same solution as expected. However, the primal-dual reformulation is computationally much more efficient. The significant difference in solution times is a consequence of the large number of binary variables introduced in the KKT strategy. The KKT reformulation introduces one binary variable for each lower level inequality constraint, whereas the primal-dual reformulation does not introduce any new binary variables.

The expansion plan obtained from the bilevel formulation is also compared with the plan obtained from the standard single-level capacity expansion formulation. The results show that the single-level formulation yields higher investment cost because it overestimates the attainable market share. In comparison to the bilevel approach, such an expansion plan yields significantly lower NPV when evaluated in a rational market environment. These results clearly demonstrate the importance of considering markets' behavior to develop capacity expansion plans that are more robust in the dynamic conditions of the markets.

Rolling Horizon Approach for Production - Distribution Coordination of Industrial Gases Supply-Chains

Miguel Zamarripa[1], Pablo A. Marchetti[2], Ignacio E. Grossmann[1],
Lauren Cook[3], Pierre-Marie Valton[4], Tejinder Singh[3], Tong Li[3], and Jean André[4]

[1]Dept. of Chemical Engineering, Carnegie Mellon University, Pittsburgh, PA 15213
[2]INTEC (UNL-CONICET), Güemes 3450, Santa Fe, Argentina.
[3]American Air Liquide Inc., Delaware Research and Technology Center, Newark, DE 19702
[4]Air Liquide, Claude & Delorme R&D Center, 78350 Jouy-en-Josas, France

An industrial gases supply-chain coordinates production/distribution decisions at multiple plants/depots while satisfying several customer demands. Plants operate air separation units, which feature high electricity consumption, to produce gaseous and liquid products. Distribution is carried out by pipeline for gaseous products and by truck for liquid products. For truck distribution, multiple depots are considered. Moreover, trucks at a given depot can deliver product from multiple sources. In order to ensure customer storage replenishments, product can be purchased from alternative sources. As the number of sources and customers increase, the selection of the possible routes to be included in the formulation becomes a critical issue.

Production decisions include production modes and rates at each plant, with varying efficiencies and power consumptions. To handle electricity price fluctuations during the day a peak/off-peak pricing scheme is considered. The main distribution decisions include the association of vehicles with routes at each time period. Besides, distribution capacity depends on the inventory levels of liquid products at each plant. A detailed distribution management including availability of trucks, trailers, and drivers, and truck load, travel, and delivery times can be indirectly considered. At the supply-chain level, production and distribution decisions must be coordinated to fulfill several customer demands, local demands, and guarantee a safety stock in the inventory levels while minimizing the overall cost. Fixed customer demands based on forecast data or consumption profiles are considered. Consequently, inventory management plays a key role to guaranty the replenishment of storage tanks at customer locations. Main model assumptions: i) two time periods per day (peak/peak off) are considered; ii) trucks do not visit more than 4 customers in a single delivery; iii) maximum and minimum inventory levels; iv) truck availabilities and capacities.

This work further elaborates on a mixed-integer linear programming (MILP) formulation to minimize the overall cost of production and distribution of industrial gases supply-chains. While previous work provides optimal production-distribution decision making[1], the computational effort required to solve the problem is still an issue to consider when industrial size examples are studied. This issue is mainly due to the combinatorial nature of the problem at the distribution side, where multiple trucks are used to pick up products and visit multiple locations at each time period. This work proposes a rolling horizon (RH) framework to enhance the solution of production-distribution coordination problems for large-scale industrial gases supply chains.

The case studies considered result in large-scale optimization problems. Being an effective decomposition technique, the rolling horizon approach has been widely studied in literature.[2,3] It provides a series of small-scale optimization models that reduce the computational burden of the full-size problem.[4] The RH method divides the time horizon into a "detailed time block" and an "aggregated time block" and solves a set of sub-problems reducing the complexity of the model. The "detailed time block" uses the representation of the original model, while a simplified model must be provided for the "aggregated time block". RH has been applied in a number of supply-chain management problems. Ierapetritou et al. (2010)[3] proposed a new heuristic method to improve the solution quality of the rolling horizon method for integrated planning and scheduling optimization. The heuristic method is based on the development of a production capacity model through a parametric programming approximation. Another interesting approach to reduce the computational effort in strategic supply-chain planning models is presented by Kostin et al. (2011).[5] Based on the RH approach they developed a new heuristic method where some of the integer variables are considered as continuous, finding that the relaxation of the integer variables is tight. This means that the solution of the relaxed model is very close to the optimal solution of the original problem.

This work describes a rolling horizon framework in which two approaches for the "aggregate time block" are studied. The first one relies on a relaxation of the integrality condition of binary variables. The second one consists of aggregating the detailed distribution by considering direct shipping with unlimited trucks availability to satisfy customer demands. Both approaches, which provide rigorous lower bounds, are studied under a forward RH scheme. The results of the proposed framework show a significant reduction in the computational burden for up to a factor of 7 times faster, while improved solutions are obtained compared to a full space solution which cannot be solved to optimality. These solutions provide higher coordination among plants/depots to satisfy a common set of shared customer demands. Additionally, a comparison between the rolling horizon heuristic and a Lagrangean Decomposition of the full-size problem is presented.

[1] Marchetti, P. A., Gupta, V., Grossmann, I. E., Cook, L., Valton, P. M., Singh, T., Li, T., and André J. (2014). Simultaneous Production and Distribution of Industrial Gas Supply-Chains. In press

[2] Dimitradis, A. D., Shah, N., and Pantelides, C. C. (1997). RTN-based Rolling Horizon Algorithms for Medium Term Scheduling of Multipurpose Plants. *Computers Chemical Engineering*, 21, S1061-S1066.

[3] Li, Z., and Ierapetritou M. G. (2010). Rolling horizon based planning and scheduling integration with production capacity consideration. *Chemical Engineering Science*, 65, 5887-5900

[4] Castro, P. M., Harjunkoski, I., and Grossman I.E. (2010) Rolling-Horizon Algorithm for Scheduling under Time-Dependent Utility Pricing and Availability. *Computer Aided Chemical Engineering*, 28, 1171-1176.

[5] Kostin, A.M., Guillén-Gosálbez, G., Mele, F.D., Bagajewicz, M.J., and Jiménez, L. (2011). A novel rolling horizon strategy for the strategic planning of supply chains Application to the sugar cane industry of Argentina. *Computers and Chemical Engineering*, 35, 2540-2563.

382719 Mixed-Integer Programming Methods for Inventory Routing

Abstract:

To reduce the distribution cost in manufacturing supply chains (SCs), customers and vendors adopt vendor managed inventory (VMI) policies, where the vendor decides when and how much to serve each customer (as long as the inventories of customers are kept within an acceptable range). To implement a VMI policy, the vendor has to repeatedly solve the inventory routing problem (IRP), which, in essence, is the integration of inventory control and distribution scheduling – the vendor decides the vehicle routes and schedules simultaneously with delivery amounts and visiting times. A wide range of constraints, including customer access windows, maximum working and driving time limits, and heterogeneous fleet, makes IRP a very hard problem. Recently, we proposed the first mixed-integer programming (MIP) model that accounts for all restrictions and characteristics of practical IRP problems[1]. While general and easily extendable to handle other features, the proposed model is computationally expensive. To address this challenge, in this work, we develop advanced solution methods, including (1) preprocessing methods and (2) a decomposition algorithm.

First, in the preprocessing phase, we eliminate some nodes and arcs in the SC network that are unlikely to be visited or used in (an optimal solution for) the current planning horizon, since IRP is a detailed scheduling problem with relatively short horizon. Based on inventory level information, consumption profiles, and geographical information, our preprocessing methods generate a much smaller SC network which is then used to formulate smaller MIP models. We also present preprocessing methods that allow us to identify infeasible or suboptimal routes which are excluded from the current formulation.

Second, we design a decomposition algorithm which considers a vehicle routing problem at the higher (first) level, and a detailed scheduling problem at the lower (second) level. For the higher level vehicle routing problem, we find the optimal routes and route-vehicle pairings to minimize transportation cost while ensuring that the minimum demand of each customer is satisfied (i.e., the inventory at the end of horizon is above a minimum terminal level). The first level solution gives a lower bound on the overall objective (of the minimization problem) as it is essentially a relaxation of the full IRP problem. Using the solution of the first level subproblem as input, we solve the detailed lower level distribution scheduling problem with customers, vehicles, drivers and all related constraints. The lower level solution gives an upper bound, since it is a feasible solution for the IRP. The two subproblems are solved iteratively until the gap between the two bounds is closed. We also present methods for the reduction of iterations.

Finally, we present computational results based on real instances. We show that the proposed methods yield very good solutions fast, which means that they can be used in practice. While the case studies we discuss are from an industrial gas company (Praxair), the models and solution methods we propose are applicable in SCs in various sectors.

[1] Dong, Y.; Sundaramoorthy, A.; Pinto, J.M.; Maravelias, C. T. A MIP Model for Inventory Routing in Industrial Gases Supply Chain. *Industrial & Engineering Chemistry Research*, submitted for publication.

385479 Game-Theoretic Modeling and Optimization of Decentralized Supply Chains

Abstract:

When entering a business, the manufacturer is often encountered with questions such as where to locate the plants, how big the plants should be, which conversion technology to choose, how much to produce, how to set the prices, etc. [1]. Although there is a large body of literature on the modeling and optimization of supply chain design and operations in the process system engineering community and the operations research field, most of these works viewed the supply chain from a centralized perspective and integrated the various components of the supply chain into a monolithic model [2, 3]. Under this approach, it was implicitly assumed that the decision maker has full control over the entire supply chain so that all the strategic and operational decisions can be implemented successfully. However, the management over a supply chain is very often decentralized in practice [4, 5]. In other words, the different entities in a supply chain may be under the charge of different stakeholders, who may even have conflicting interests. These entities would strive to maximize their own benefits and compete with their peers, thus leading to a non-cooperative supply chain [6, 7].

The goal of this work is to develop a novel game-theoretic modeling and optimization framework to assist the manufacturers in optimal design and strategic planning in a three-echelon supply chain system. The first echelon consists of a set of individual suppliers. The second echelon includes a set of candidate sites for building the manufacturing facilities. The third echelon consists of a set of individual distributors. The manufacturer is assumed as the leader in this problem, who will take actions first. The suppliers and distributors are assumed as the followers, who will then react to the leader's decisions. We formulate a bilevel program to model the Stackelberg leader-follower relationship [8]. We model the competitions among the suppliers and distributors under the generalized Nash equilibrium assumption [9]. We also address the selection of the location, sizing, and technology of manufacturing facilities, strategic plans on material flows, as well as the determination of transfer prices. Since bilevel programs cannot be directly handled by the existing solvers, we reformulate the problem into an equivalent single-level mixed-integer nonlinear program (MINLP) by replacing the followers' problems with the corresponding Karush-Kuhn-Tucker (KKT) conditions. However, the resulting MINLP involves bilinear and concave terms in the objective function, which makes the problem nonconvex. To facilitate the solution process, we propose an improved branch-and-refine algorithm which solves the nonconvex MINLP problem efficiently by solving a sequence of mixed-integer linear programming (MILP) problems by iteration [10, 11].

To illustrate the application of the proposed game-theoretic modeling and optimization framework, a case study on the design and strategic planning of a biofuel supply chain is presented [12-18]. The biorefinery investor tends to maximize his total profit by optimizing the location, capacity, and technology selection of the biorefineries, the strategic plan on harvesting and production, as well as the setting of transfer prices. Given the biorefinery investor's and the competitors' decisions, the individual supplier tends to maximize his own profit by optimizing the amount of biomass shipping to every installed biorefinery, and the individual distributor tends to maximize his own profit by optimizing the amount of biofuel to purchase from each of the installed biorefineries. Optimization results of the non-cooperative supply chain are obtained and comparison with the centralized design is presented. The proposed solution strategy is shown to be much more efficient than the existing general-purpose MINLP solvers.

References

[1] I. Grossmann, "Enterprise-wide optimization: A new frontier in process systems engineering," *AIChE Journal,* vol. 51, pp. 1846-1857, 2005.

[2] N. Shah, "Process industry supply chains: Advances and challenges," *Computers & Chemical Engineering,* vol. 29, pp. 1225-1236, 5/15/ 2005.

[3] L. G. Papageorgiou, "Supply chain optimisation for the process industries: Advances and opportunities," *Computers & Chemical Engineering,* vol. 33, pp. 1931-1938, 12/10/ 2009.

[4] J. Gjerdrum, N. Shah, and L. G. Papageorgiou, "Fair transfer price and inventory holding policies in two-enterprise supply chains," *European Journal of Operational Research,* vol. 143, pp. 582-599, 12/16/ 2002.

[5] J. Gjerdrum, N. Shah, and L. G. Papageorgiou, "Transfer Prices for Multienterprise Supply Chain Optimization," *Industrial & Engineering Chemistry Research,* vol. 40, pp. 1650-1660, 2001/04/01 2001.

[6] K. Yeh, J. H. Lee, C. Whittaker, and M. J. Realff, "Two Stage Bilevel Programming Approach for Representation of Biorefinery Investment Decision Making in a Pre-Established Timberlands Supply Chain," in *8th International Conference on Foundations of Computer-Aided Process Design*, Cle Elum, Washington, USA, 2014.

[7] Y. Bai, Y. Ouyang, and J.-S. Pang, "Biofuel supply chain design under competitive agricultural land use and feedstock market equilibrium," *Energy Economics,* vol. 34, pp. 1623-1633, 9// 2012.

[8] J. F. Bard, *Practical bilevel optimization: algorithms and applications* vol. 30: Springer, 1998.

[9] F. Facchinei and C. Kanzow, "Generalized Nash equilibrium problems," *4OR,* vol. 5, pp. 173-210, 2007/09/01 2007.

[10] D. Yue and F. You, "Planning and scheduling of flexible process networks under uncertainty with stochastic inventory: MINLP models and algorithm," *AIChE Journal,* vol. 59, pp. 1511-1532, 2013.

[11] J. Gong and F. You, "Optimal Design and Synthesis of Algal Biorefinery Processes for Biological Carbon Sequestration and Utilization with Zero Direct Greenhouse Gas Emissions: MINLP Model and Global Optimization Algorithm," *Industrial & Engineering Chemistry Research,* vol. 53, pp. 1563-1579, 2014/01/29 2014.

[12] D. Yue, F. You, and S. W. Snyder, "Biomass-to-bioenergy and biofuel supply chain optimization: Overview, key issues and challenges," *Computers & Chemical Engineering,* 2014.

[13] F. You and B. Wang, "Life Cycle Optimization of Biomass-to-Liquid Supply Chains with Distributed–Centralized Processing Networks," *Industrial & Engineering Chemistry Research,* vol. 50, pp. 10102-10127, 2011/09/07 2011.

[14] F. You, L. Tao, D. J. Graziano, and S. W. Snyder, "Optimal design of sustainable cellulosic biofuel supply chains: Multiobjective optimization coupled with life cycle assessment and input–output analysis," *AIChE Journal,* vol. 58, pp. 1157-1180, 2012.

[15] K. Tong, J. Gong, D. Yue, and F. You, "Stochastic Programming Approach to Optimal Design and Operations of Integrated Hydrocarbon Biofuel and Petroleum Supply Chains," *ACS Sustainable Chemistry & Engineering,* vol. 2, pp. 49-61, 2014/01/06 2013.

[16] K. Tong, M. J. Gleeson, G. Rong, and F. You, "Optimal design of advanced drop-in hydrocarbon biofuel supply chain integrating with existing petroleum refineries under uncertainty," *Biomass and Bioenergy,* vol. 60, pp. 108-120, 1// 2014.

[17] D. Yue, M. A. Kim, and F. You, "Design of Sustainable Product Systems and Supply Chains with Life Cycle Optimization Based on Functional Unit: General Modeling Framework, Mixed-Integer Nonlinear Programming Algorithms and Case Study on Hydrocarbon Biofuels," *ACS Sustainable Chemistry & Engineering,* vol. 1, pp. 1003-1014, 2013/08/05 2013.

[18] B. H. Gebreslassie, Y. Yao, and F. You, "Design under uncertainty of hydrocarbon biorefinery supply chains: Multiobjective stochastic programming models, decomposition algorithm, and a Comparison between CVaR and downside risk," *AIChE Journal,* vol. 58, pp. 2155-2179, 2012.

381994 Quantitative Operability Analysis of Process Supply Chains

Abstract:

The North American forest products industry (FPI) is primarily commodity-based, and is facing challenges from low-cost competitors, leading to a sustained decline in demand and over-capacity of equipment (Thorp, 2005). As a consequence, capital and operational costs, as well as R&D activities have been reduced, resulting in obsolete pulp and paper mills. A possible strategy to improve the struggling business model is through revenue diversification, in which value-added specialty products are produced along with conventional pulp and paper products (Orzechowska, 2005). In order to achieve this, new process technologies need to be incorporated into existing pulp and paper mills, and essentially transform them into integrated forest biorefineries (IFBR). The obsolete mills can be retrofitted and transformed into more sustainable IFBRs to promote product variety. The additional capital cost incurred is justified by having improved profitability and greater robustness against market volatility (Wising and Stuart, 2006).

A key consideration in the transformed FPI paradigm is that product, process and supply chain designs have adequate flexibility and robustness to changes in market conditions. This is applicable both at the strategic design level, where market changes sustained over long periods are considered, as well as at the tactical/operation level, where responsiveness to shorter-term variability is required.

Process plants operate in a continually changing environment and are required to operate satisfactorily under sustained changes as well as short-term fluctuations. Operability of a chemical process reflects the ability of a system to perform satisfactorily under the conditions away from the nominal operating and/or design conditions (Grossmann and Morari, 1984). Two considerations of operability are flexibility and dynamic responsiveness. Flexibility reflects the ability of a system to remain feasible in the face of sustained changes, while dynamic responsiveness refers to the ability to respond rapidly to short-term fluctuations. Systematic and quantitative analyses for process operability have been proposed (Grossmann and Morari, 1984; Swaney and Grossmann, 1985), as well as the incorporation of operability considerations in optimization-based plant design (Mohideen *et al*., 1996; Baker and Swartz, 2004).

Analogous to chemical processes, supply chain processes are required to satisfactorily meet customer demand or transition between operating policies rapidly to exploit external opportunities. The above considerations have motivated an investigation into an optimization based framework for the design of operable supply chains. A quantitative framework to analyze supply chain flexibility was developed and demonstrated through case studies (Wang *et al*., 2013). Quantitative measures of responsiveness have also been established and included in supply chain design and analysis (You *et al*., 2008; Mastragostino and Swartz, 2014). The focus of the present work is to combine both flexibility and responsiveness considerations within a single supply chain optimization framework. This is particularly useful within the IFBR context, in which demand variation in commodity and value-added products occurs at different frequencies. Sustained uncertain demand levels in the former are handled through flexibility considerations, while supply chain dynamics are taken into account with consideration of short-term fluctuations in value-added product demand.

The proposed framework offers insights when examining the relations and trade-offs between flexibility, responsiveness, and profitability. It can be applied not only to the design of new process supply chain networks, but also to existing supply chains for performance assessment based on economics and operability criteria. While motivated by FPI transformation, the methodology outlined is widely applicable to general process supply chain networks.

References

Baker, R. and Swartz, C.L.E. (2004). Simultaneous solution strategies for inclusion of input saturation in the optimal design of dynamically operable plants. *Optimization and Engineering*, 5, 5-24.

Grossmann, I.E. and Morari, M. (1984). Operability, resiliency and flexibility: Process design objectives for a changing world. In *Proc. Int. Conf. on Foundations of Computer-Aided Process Design*, Westerberg, A.W. and Chien, H.H., eds, 931-1030.

Mastragostino, R. and Swartz, C.L.E. (2014). Dynamic operability analysis of process supply chains for forest industry transformation. Accepted for publication, *Ind. Eng. Chem. Res.*

Mohideen, M.J., Perkins, J.D. and Pistikopoulos, E.N. (1996). Optimal design of dynamic systems under uncertainty. *AIChE Journal*, 42, 2251-2272.

Orzechowska, A. (2005). Saving the Canadian Industry: Discussing Possible Solutions. *Pulp and Paper Canada*, 106, 25-27.

Swaney, R. and Grossmann, I. (1985). An index for operational flexibility in chemical process design. Part I: Formulation and theory. *AIChE Journal*, 31, 621-630.

Thorp, B. (2005). Biorefinery Offer Industry Leaders Business Model for Major Change. *Pulp and Paper*, 79, 35-39.

You, F. and Grossmann, I.E. (2008). Design of responsive supply chains under demand uncertainty. *Comp. Chem. Eng.*, 32, 3090-3111.

Wang, H., Mastragostino, R. and Swartz, C.L.E. (2013). Flexibility analysis of process supply chains. Paper 518g, AIChE Annual Meeting, San Francisco, CA.

Wising, U. and Stuart, P.R. (2006). Identifying the Canadian Forest Biorefinery. *Pulp and Paper Canada*, 107, 25-30.

Soft Regulation: Coordinating Distributed Self-interested Agents in Sociotechnical Systems

Yu Luo[*†], *Garud Iyengar*[‡] *and Venkat Venkatasubramanian*[*†‡]

Complex Resilient Intelligent Systems Laboratory
Department of Chemical Engineering
Department of Industrial Engineering and Operations Research
Columbia University
New York, NY 10027

Regulating new industries such as fracking is challenging, sometimes even controversial. On the one hand, under-regulation may result in safety threats to plant personnel, surrounding communities, and the environment. On the other, over-regulation may hinder innovation, progress, and economic growth. Since one typically has limited understanding of, and experience with, the novel technology in practice, it is difficult to accomplish a properly balanced regulation. How do we promote economic growth yet ensure safety under such circumstances? This is further complicated by the presence and actions of humans who tend to act in a self-interested manner which may not benefit the system overall. Controlling such a sociotechnical system with conflicting objectives and behaviors is a challenging problem.

In this work, we will discuss a new type of control policy, called soft regulation, that attempts to strike the right balance. In contrast to hard regulation - laws that forbid and punish, soft regulation encourages and guides the agents by creating appropriate incentives. Soft regulation resolves one of the major hurdles we face in systems involving self-interested agents, i.e., agents that only pursue maximal individual benefit, with individual interests that do not align with those of the system. In other words, agents take actions that maximize individual payoff (or minimize individual risk); but by doing so, they cause systemic instability.

We propose that such a framework can be effective for newborn industries which satisfy the following criteria: (i) We have a limited understanding of the technology - the data are incomplete and noisy, and the models are inaccurate; (ii) The resources are relatively abundant since the technology is new, and the system is weakly coupled; (iii) Distributed self-interested agents are the dominant decision makers who pursue their own interests.

An overview of our soft regulation algorithm is the following. There is a central regulator who lets the agents decide what is best for themselves and act accordingly. The regulator collects aggregate and privacy-preserved information of the system behavior statistics, shares the aggregate picture with the agents, and provides recommendations. This creates a collective, distributed, learning environment for both the regulator and the agents he oversees. The agents have the flexibility to accept, reject, or partially accept the recommendations from regulator based on their own self interests.

We will discuss simulation results that demonstrate that this soft regulation benefits both the overall system and the individual agents by stabilizing the system and expediting the process of reaching optimum. The mechanism is incentive compatible as the agents voluntarily choose to cooperate while maximizing their own payoffs.

---

[*] Complex Resilient Intelligent Systems Laboratory
[†] Department of Chemical Engineering
[‡] Department of Industrial Engineering and Operations Research

374818 On-Line Determination of Appropriate Control Loop Configuration Using Directed Spectral Decomposition of Process Data

Abstract:

An important part of designing effective decentralized multivariable control schemes is determining control loop configurations that minimize control loop interactions. Traditionally, the most appropriate control loop configuration is often determined using the relative gain array of a process obtained at a single operating condition, with the configuration unchanged for the duration of subsequent process operation. For systems with significant non-linear characteristics the appropriate control loop configuration may change with operating conditions, leading to poor control system performance and potential closed loop instability. For such systems, good control system performance can be maintained and closed loop instability avoided over a variety of operating conditions if the control loops can be reconfigured appropriately on-line using informative data collected during process operation.

Our goal is to develop and evaluate a procedure for on-line data-driven determination of the most suitable control loop configuration using directed spectral decomposition (DSD)—a spectral analysis technique that uses informative process data to determine connectivity strength between process variables. The procedure involves two steps: 1) quantifying connectivity strength between process variables using DSD; and 2) using connectivity strengths to ascertain the control loop configuration that minimizes control loop interactions at the current operating conditions. The procedure has been tested on a simulated example 3-input, 3-output non-linear stirred mixing tank under varying operating conditions. This presentation will focus on the development of the proposed procedure and the specific results obtained by applying the procedure to the stirred mixing tank.

376030 Lyapunov Stability Analysis of Economic NMPC

Abstract:

The ultimate goal of any operation strategy for a process plant is to make profit. Traditionally this goal could be achieved by a two-layer control system, where the upper layer solves an optimization problem aiming at maximizing profit to get the optimal setpoints for the controlled variables in the layer below. The lower layer then keeps the controlled variables at their given setpoints. However, there are some problems with this two-layer structure. For example, the setpoints calculated by the upper layer might not be reachable by the lower layer. One of the solutions is to combine these two layers and include the economic criterion directly into the cost function of the lower layer controller when an optimization-based controller such as MPC is used. This approach is often referred to as Economic MPC, and has been gaining increased interest in recent years.

In this study we would like to analyze Lyapunov stability properties of Economic NMPC. Lyapunov stability properties of setpoint tracking NMPC problems have been well established with the assumption that the stage cost is minimal at the desired equilibriums points so that the objective function could be used as a Lyapunov function. However, since Economic NMPC controller generally does not attain a minimum at the steady state optimal operation point, the objective function cannot be used as a Lyapunov function directly.

As described in previous studies [1, 2], we consider a modified analysis through a transformed rotated stage cost. This is generated by first subtracting the optimal steady state from the variables and stage cost, and then augmenting the transformed cost to form the rotated stage cost. As shown in [1, 2] the modified problem has the same solution as the original problem, if the modified stage and terminal costs are strongly convex. Moreover, strong convexity can always be ensured through the addition of quadratic regularization terms. In this talk we establish stability results for three different terminal criteria: (1) fixed terminal point with finite cost function, (2) fixed terminal point with infinite horizon cost function, and (3) terminal set and a terminal cost for a finite horizon cost function. Moreover, we provide a simple constructive method based on Gershgorin bounds for regularization that ensure strongly convex stage costs where stability is guaranteed. We demonstrate these concepts on a detailed case study with two distillation columns in sequence.

1. M. Diehl, R. Amrit, J. B. Rawlings, A Lyapunov function for economic optimizing model predictive control, *IEEE Transactions on Automatic Control* 56 (2011) 703–707

2. J. Jäschke, X. Yang and L. T. Biegler, Fast Economic Model Predictive Control Based on NLP-Sensitivities, *Journal of Process Control*, in press (2014)

CARLEMAN LINEARIZATION-BASED ANALYTICAL MODEL PREDICTIVE CONTROL ACTIONS OF NONLINEAR SYSTEMS

Yizhou Fang and Antonios Armaou,
Department of Chemical Engineering,
The Pennsylvania State University, University Park, PA

Model Predictive Control (MPC) is developed as a powerful control method applied in chemical, pharmaceutical, and petroleum industries over the past two decades. Since control signals are computed by repeatedly solving receding-horizon optimization problems, the control actions evolve as the system dynamics carry on, which enables MPC to reject external disturbances and tolerate model inaccuracy while following process constraints. The computation of MPC is more complex than that of classical controllers. Complex systems with high nonlinearity require significant amount of computation. When constraints are considered, the amount of computation is even higher and it is impossible to avoid this issue by deriving analytical solutions. Developing an advanced and systematic MPC synthesis methodology for nonlinear complex systems is motivated. [1]

The objective is to derive search algorithms that reduce the amount of computation and remove computational delay in resolving dynamic optimization problems for nonlinear complex systems. The search algorithm computed from an algebraic nonlinear optimization problem (NLP) with analytically computed sensitivity reduces computation and removes computational delay in online operations. Based on Carleman linearization, the method proposed enables analytical solutions to the optimal control problem, while keeping the nonlinearity of system dynamics and constraints. Control actions are designed in bilinear expressions which combine feedback control laws and constant control signals.

The proposed method was computationally efficient by allowing standard, gradient-based, search algorithms. Compared with nonlinear MPC, the proposed method takes less time to compute the optimal policy. The work is illustrated by an unstable Continuous Stirred Tank Reactor with high nonlinearity and unmeasured disturbances. With the proposed method, the closed loop system remained high practical stability under disturbances and the deviation from set-points maintained less than 1%.

References

1. A. Armaou and A. Ataei, "Piece-wise constant predictive feedback control of nonlinear systems," J. Process Control, vol. 24, no. 4, pp. 326–335, Apr. 2014.

Model predictive control (MPC) has played a key role in several advanced process control system in chemical process industry for the past three decades. Every MPC algorithm is based on a predictive mechanism which can reasonably predict the plant outputs in the future given a change in the inputs, based on a linear or nonlinear model of the plant. Its ability to honor plant constraints both soft and hard, helps to maintain product properties with more precision, which means product giveaways can be minimized without violating specifications, thus leading to more profitable operation.

Uncertainty is ubiquitous in every model based control, both linear and nonlinear. Recent advances [1] have utilized Lyapunov-based techniques to design robust MPC formulations which (for certain classes of systems) allow an analytical solution resulting in a formulation computationally faster than the min-max optimization techniques traditionally used to handle uncertainty. This formulation has also been shown to possess an explicitly characterized region of robust stability. This essentially rely (or use) some information about the bounds on the uncertainty (as well as a structure of the uncertainty), and in general do not estimate the uncertain parameter (and therefore do not need to assume a specific form of the uncertainty).

In another set of results, the model in the controller is augmented with some form of 'disturbance' states or parameters which are in turn estimated on-line, effectively turning the controller into an adaptive control design. In one approach, the current measured process output is compared with the corresponding prediction from the model and the difference is used as a bias term which is added to the model predictions used within the MPC [2]. This type of additive feedback is also referred as a constant output disturbance. Other approaches handle the measurement noise associated with the bias correction term by 'filtering' the measurements/bias term in some form. These set of results are typically referred to as offset-free MPC designs, where parameter/state estimation techniques such as Kalman filter [3] and Luenberger Observer [4], have been utilized.

Notwithstanding the available designs, there exist a lack of comprehensive tuning guidelines for MPC formulations in general. Existing results dictate MPC tuning by approximating the MPC problem as an unconstrained LQR, with process variables normalized by range, Q and R are diagonal matrices close to Identity. Motivated by these considerations, in this work we present a comprehensive tuning framework for the MPC. Specifically, the focus will be first on linear MPC, where given the dynamic matrices and constraints, values of the MPC tuning constants will be prescribed to achieve a certain prescribed response of the controlled variables. The prescribed response will be characterized as a first order response to a set point change, and quantified through the time constant. In conjunction with the tuning rules, a two-level MPC formulation will be proposed. At the first level (that will be implemented infrequently), achievable, but optimal, first order time constants will be computed. At the second level (on line MPC implementation), the aforementioned MPC will be implemented to achieve the desired closed loop behavior. Subsequently, the formulation will be generalized to consider process models with explicit time delays, and MPC formulations that include the offset-free mechanism. Chemical process examples will be presented to illustrate the tuning mechanism.

References

[1] P. Mhaskar, "Robust Model Predictive Control Design for Fault-Tolerant Control of Process Systems," *IECR*, vol. 45, pp. 8565–8574, 2006.

[2] S. J. Qin and T. A. Badgwell, "A survey of industrial model predictive control technology," *Control engineering practice*, vol. 11, no. 7, pp. 733–764, 2003.

[3] K. Muske and T. Bagwell, "Disturbance models for offset-free linear model predictive control." *Journal of Process Control*, vol. 12, pp. 617–632, 2002.

[4] M. Wallace, B. Das, P. Mhaskar, J. House, and T. Salsbury, "Offset-free model predictive control of a vapor compression cycle," *Journal of Process Control*, vol. 22, pp. 1374?–1386, 2012.

A Novel Approach to Updating Soft Sensors in Industrial Processes

Kuilin Chen[a], Jie Yu[a], Ivan Castillo[b] and Leo H. Chiang[b]

[a]Department of Chemical Engineering

McMaster University, 1280 Main Street West, Hamilton, Ontario, L8S 4L8, Canada

[b]Analytical Tech Center

The Dow Chemical Company, 2301 Brazosport Blvd., Freeport, TX 77541, USA

The partial least squares (PLS) method is widely used in the process industry to generate soft sensors to predict difficult-to-measure variables utilizing easy-to-measure variables. However, the prediction accuracy of soft sensors tends to degrade after a period of online operation due to a variety of factors, including process fouling, abrasion of mechanical components, drifted operating conditions, process faults, and changes in the external environment.. Though some dynamic and recursive modeling approaches are developed to overcome this degradation issue, these approaches are not desirable in industry because of their complex implementation and a lack of information that could lead to overfitting the soft sensor model.

In this study, a novel model mismatch index is developed to monitor the performance of PLS soft sensor models in industrial applications. Based on estimated regression parameters, the Kalman filter (KF) based model mismatch index is developed and employed to detect the mismatch between the current soft sensor model and process dynamics. The predictions of the soft sensor model are recognized as uncertain if the model mismatch index overpasses the predefined control limit. In order to avoid a biased update for soft sensor models under faulty operations, the decision of the model update can be made based on the model mismatch index with the assistance of traditional T^2 and SPE process monitoring indexes. Furthermore, a contribution plot for the model mismatch index is developed in order to isolate the leading abnormal regression coefficients that are causing the degradation of the soft sensor model. After updating the model, the regression coefficients are estimated through PLS regression using samples of the training set and the current process conditions. Consequently, the proposed method can handle changes in process dynamics and lead to accurate quality predictions under various process conditions.

The proposed approach is validated using simulated and industrial case studies, indicating that the calculated index can capture the model mismatch of the soft sensor model earlier, which in turn results in more accurate predictions. In addition, the comparison with the recursive partial least squares (RPLS) method demonstrates that the proposed approach leads to stable model structures, which is more desirable in industrial applications.

Development of a Systematic Approach for Plant-Wide Control System Design

Dustin Jones[a], Debangsu Bhattacharyya[a], Richard Turton[a], and Stephen E. Zitney[b]

[a] Department of Chemical Engineering, West Virginia University, Morgantown, WV, USA.
[b] National Energy Technology Laboratory, Morgantown, WV, USA

Traditionally, control structures are designed based upon heuristic methods that do not use the process knowledge in a systematic manner. In this work, a systematic, model-based approach for control structure design has been developed by significantly enhancing the methodology proposed by Skogestad[1]. The approach is a two-stage method starting with a top-down analysis and ending with a bottom-up design. In the first stage, a top-down approach is taken to generate a list of manipulated, control, and disturbance variables considering a scalar operation objective and other process constraints. In the second stage, a bottom-up approach is used for designing the control system taking into account the results from the previous stage.

In the first stage of the plant-wide control system design approach, controlled variables are selected for their self-optimizing control performance and controllability while ensuring satisfactory performance in terms of dead-time and closed loop interactions. This work has considered both self-optimizing and control performance as well as addressed issues related to loop-interactions and superstructure constraints. Our proposed method involves a three-stage procedure: *a priori* analysis, controlled variable selection, and *a posteriori* analysis. The developed methodology can be deployed to large-scale process systems. This approach results in a large-scale, constrained, mixed-integer multi-objective optimization problem. For solving this problem, a parallelized, bi-directional branch and bound algorithm with dynamic search strategies has been developed to solve the problem on large computer clusters.

In the next stage, the focus is on the development of a rigorous, model-based approach for the selection of secondary controlled variables. A partially controlled plant analysis is undertaken to determine the optimal secondary controlled variables. Here a systematic three-stage approach is proposed that addresses issues of controllability and economic performance of the control system. The first stage involves the generation of an initial set of candidate secondary controlled variables and the generation of selection constraints. During the second stage, secondary controlled variables are selected to minimize integral absolute errors (IAEs) of the primary controlled variables subject to minimal loop interactions as determined by a relative gain array analysis. Finally, during the third stage, control performance of the secondary controlled variables is evaluated at off-design operations using a nonlinear process model.

Finally, the design of the supervisory control layer is completed. Gramian-based interaction measures are used to determine the optimal form of the supervisory control that balances controller complexity with control performance. Finally, model-based advanced controllers are designed for supervisory control.

The methodology is applied to an integrated gasification combined cycle (IGCC) power plant with CO_2 capture. This plant should be operated optimally in the face of disturbances without violating operational and environmental constraints. The process models are developed in Aspen Engineering Suite while the algorithms for control structure selection and the controllers are implemented in Matlab.

1. Skogestad, S., "Control structure design for complete chemical plants", *Computers and Chemical Engineering*, 28, 219-234, 2004

Valuation and Operation of Energy Commodity Storage –paramter effects and extended model

Fanghui Fan, Fernando Garcia and Jeffrey Kantor*

May 12, 2014

Abstract

Supply, demand and other exogenous events contribute energy commodity price fluctuations that incur financial risks for energy facility owners and operators. Energy commodity storage is an effective way to mitigate price fluctuation and lower the financial risks.

In previous work we proposed a storage model for the case of a single commodity with stochastic price serving a known and fixed demand. Convenience yield is included in the model. A 2-dimensional (2D) Hamilton-Jacobi-Bellman (HJB) equation was solved to obtain the optimal storage strategy [1]. In this paper the control policy is further investigated. The valuation result is compared with a no control policy and with threshhold control policy [3]. The effects of the convenience yield and the demand are investegeted. The problem is further extended by including stochastic demand. A 3-dimentional (3D) HJB equation with embedded optimization problem is formulated and solved. Similar derivation process of the 3D HJB equation can be found in [4], [5], [6], and [7]. First order convergence of of the 3D compuatation is verified numerically.

In 2D problem, the optmial control ends up to be sliding mode control. The chattering can be reduced or elimiated by reducing the amplitude of the signum function or replacing the signum function by a high-slope saturation function [2]. The valuation result with optmical control is much higher than those without any control policy executed and with threshold method. The convenience yield is positively related with the valuation result while the demand is the opposite. That's because the bigger convenience yield is, the more benefit it can bring by storing the enenrgy commodity; the higher the demand, the more cost is spent to satisfy it, and therefore we have lower valuation result. Finite difference method with the first-order accuracy is employed to solve the 3D HJB equation. The calculated numerical convergence order is approximately 1, which is consistent with the numerical scheme order we used to solve the HJB equation. Therefore we say our 3D computation converges.

This extended 3D model can be easily adapted to stochastic convenience yield, stochastic volatility or another stochastic commodity price.

*Author to whom correspondence should be addressed: Kantor.1@nd.edu

Proper boundary conditions need to be chosen to solve the corresponding 3D problem.

References

[1] Fanghui Fan and Jeffrey Kantor. Valuation and optimal operation of energy commodity storage. In *AIChE*, 2013.

[2] Hassan K. Khalil. *Nonlinear Systems*. Upper Saddle River, NJ: Prentice Hall, third edition, 2002.

[3] J. Mattingley, Yang Wang, and S. Boyd. Receding horizon control. *Control Systems, IEEE*, 31(3):52–65, June 2011.

[4] Robert C. Merton. Option pricing when underlying stock returns are discontinuous. *Journal of Financial Economics*, 3:125–144, 1976.

[5] Robert C. Merton. *Continuous time finance*. Blackwell publishers, 1990.

[6] Matt Thompson, Matt Davison, and Henning Rasmussen. Valuation and optimal operation of electric power plants in competitive markets. *Operations Research*, 52(4):546–562, July–August 2004.

[7] Matt Thompson, Matt Davison, and Henning Rasmussen. Natural gas storage valuation and optimization: A real options application. *Naval Research Logistics*, 56(3):226–238, 2009.

372643 Generalized Modeling Framework for the Optimal Synthesis of Solvent-Based Post-Combustion CO2 Capture Processes

Abstract:

Solvent-based chemical absorption/desorption systems are the most mature among industrial CO_2 capture applications because the technology is well established, the capture conditions are relatively mild, and the process can be easily retrofitted onto existing plants. Despite this maturity, a major challenge hindering their widespread adoption in industry is the very high cost that is introduced to the operation of CO_2 emitting plants, mainly due to the thermal demands for solvent regeneration. Reported research findings indicate that modifications in the structural and operating characteristics of conventional absorption/desorption flowsheets may result in significant energetic and efficiency improvements. Most reported CO_2 capture process design approaches focus on the use of rigorous absorption/desorption simulation models supporting the identification of realistic design solutions, but mainly rely on empirically identified flowsheet configurations [1]. Empirical knowledge is clearly useful, yet the proposed improvements are bound to be limited unless an extensive number of structural and operating parameters are taken into account in optimum flowsheet design. This requirement may be well supported by process synthesis methods which combine generic representations of process layouts with optimization algorithms to systematically investigate an enormous range of potentially optimum process options. However, the utilization of such methods in CO_2 capture process design is very limited. Their systematic consideration with rigorous process models to support the identification of both realistic and highly optimum solutions has yet to be considered.

This work proposes the implementation of a generic process modeling framework [2] in the context of a flexible and inclusive synthesis model to support the optimum design of absorption/desorption CO_2 capture flowsheets. The framework serves as a mathematical tool able to reproduce any potentially favorable representation of solvent based CO_2 capture processes supported by an underlying superstructure. The proposed superstructure consists of modules representing generic process tasks (e.g., reactive and nonreactive separation, heat transfer, mixing) and interconnecting streams emulating material flows. Each module may be independently assigned with a process model representing a particular task, the type of equipment utilized, the desired operating conditions and so forth. Many different modules may be connected in the same flowsheet using an inclusive set of streams (e.g., recycle, bypasses and so forth). The proposed generic tasks account for (a) reaction, mass and heat exchange options between different phases within each module, and (b) stream mixing and splitting options to enable distribution of materials among different modules. Each module may be independently diversified in the same flowsheet into one of the three structural blocks considered in this work; namely the column section, the heater associated with phase change and the heat exchanger.

In particular, the behavior of column sections for reactive or non-reactive separation is described using an equilibrium based (EQ) model, where orthogonal collocation on finite elements approximation techniques are employed (OCFE). EQ/OCFE model formulation enables a robust and flexible representation of the column section that accommodates high resolution in describing the occurring phenomena in a compact in terms of number of required modeling equations form. Furthermore, OCFE formulation offers a competitive advantage by treating the column section height, a decision variable in the design problem, as a continuous variable thus greatly facilitating the numerical solution in non-linear optimization. In this way complex column configurations with multiple side feed and product draw streams are effectively represented and simulated. Column section size is deduced by the aggregate effect of multiple finite elements with adjustable length towards the achievement of separation targets set by the design specifications.

Using the proposed synthesis approach a number of novel absorption/desorption flowsheet configurations is generated and evaluated systematically. Such flowsheets are designed using various combinations of recycle split streams acting as multiple feed and product draw streams towards different column sections, aiming at the maximization of the driving forces inside the separation columns. Optimal design using the modular flowsheet supported by the EQ/OCFE model enables the calculation of the main design variables such as the heat duties for the required condenser and reboiler units in the stripping columns, the operating temperatures and pressures as well as structural features such as column section packing heights, and heat exchanger area for heat integration. Process design is based on the minimization of a suitable objective function comprised of the sum of the total capital and process costs in each flowsheet. Several solvents are considered including Monethanolamine (MEA), 2-amino-2methyl-1-propanol (AMP), and Diethanolamine (DEA). Vapor-liquid equilibrium calculations are performed using statistical associating fluid theory for potentials of variable range (SAFT-VR) [3-4].

Acknowledgements

The authors would like to thank Prof. Claire Adjiman, Prof. Amparo Galindo, Prof. George Jackson, and Dr. Alexandros Chremos, of Imperial College London for providing the SAFT-VR thermodynamic models. Funding from the European Commission under grant FP7-ENERGY-2011-1-282789-CAPSOL is gratefully acknowledged.

Cited References

[1] Le Moullec Y., Kanniche M., 2010, Description and evaluation of flowsheet modifications and their interaction for an efficient monoethanolamine based post-combustion CO_2 capture, Chemical Engineering Transactions, 21, 175-181.

[2] Damartzis T., Papadopoulos A.I., Seferlis P., 2014, Optimum synthesis of solvent-based post-combustion CO_2 capture flowsheets through a generalized modeling framework, Clean Technologies and Environmental Policy, DOI: 10.1007/s10098-014-0747-2.

[3] Rodriguez J., Mac Dowell N., Llovell F., Adjiman C.S., Jackson G., Galindo A., 2012, Modelling the fluid phase behaviour of aqueous mixtures of multifunctional alkanolamines and carbon dioxide using transferable parameters with the SAFT-VR approach, Molecular Physics, 110, 2012, 1325-1348.

[4] Mac Dowell N., Pereira F.E., Llovell F., Blas F.J., Adjiman C.S., Jackson G., Galindo A., 2011, Transferable SAFT-VR models for the calculation of the fluid phase equilibria in reactive mixtures of carbon dioxide, water, and n-alkylamines in the context of carbon capture, Journal of Physical Chemistry B, 115, 8155-8168.

Benchmarking Process Alternatives for Post-Combustion CO_2 Capture from a Sustainability Perspective

Sara Badr[a], Stavros Papadokonstantakis[a], Konrad Hungerbühler[a], Theodoros Damartzis[b,c], Panos Seferlis[b,c],

[a] ETH Zurich, Institute of Chemical and Bioengineering, Wolfgang-Pauli Strasse 10, 8093 Zurich
[b] Chemical Process and Energy Resources Institute, Centre for Research and Technology-Hellas, 57001 Thermi, Greece
[c] Department of Mechanical Engineering, Aristotle University of Thessaloniki, 54124, Thessaloniki, Greece

A multi-criteria sustainability assessment is carried out for a number of process scheme alternatives for post-combustion CO_2 capture using amine scrubbing. Environmental, health and safety hazard assessment and a life cycle analysis are used to compare the performance of five process configurations and to create a benchmark for further steps of process optimization. The tested configurations compare the conventional flowsheet to others with features such as absorber intercooling, multiple feed absorbers and strippers for three standard amine solvents: Monoethanolamine (MEA), Diethanolamine (DEA) and 2-Amino-2-methyl-1-propanol (AMP).

Process hazard assessment is expressed through the EHS method, the Inherent Safety Index (ISI) method and the Hazard Identification and Ranking (HIRA) method. It is strongly influenced by the inventories of chemicals within the process units, which are functions of equipment sizes, required solvent flow rates and solvent degradation. Process conditions also influence the accumulation of hazardous degradation products such as nitrosamine compounds within the process.

Process emissions and energy consumption are the basis for the life cycle assessment, where the Cumulative Energy Demand (CED), Global Warming Potential (GWP), Eco-Indicator 99 (EI99) and ReCiPe are used as indicators to characterize the environmental impact of process operation. Solvent degradation and losses determine the solvent makeup rates, the impact of which is represented through the adopted cradle-to-gate approach for indicating impacts of solvent production.

Interest in CO_2 capture is mainly environmental and it is therefore crucial to identify process and solvent characteristics that influence the sustainability of the capture process. The feedback of sustainability assessment to process design algorithms is therefore essential in that respect.

Acknowledgements

Funding from the European Commission under grant FP7-ENERGY-2011-1-282789-CAPSOL is gratefully acknowledged.

391945 Seamless integration of solvent-based carbon capture processes into natural gas combined cycle (NGCC) plants: rigorous modelling, optimization and operability analysis

Abstract:

Solvent-based carbon capture can potentially mitigate carbon emissions associated with fossil fuel consumptions. However, minimizing the penalties associated with the costs of carbon capture requires detailed understanding of the undelaying physical and chemical phenomena. The present study exploits a rigorous modelling methodology developed in our research group at Centre for Process Systems Engineering (CPSE), Imperial College and in collaboration with Process Systems Enterprise (PSE). The methodology benefits from rate-based distributed modelling of the underlying transport phenomena according to the so-called film theory in addition to accurate modelling of the thermodynamic properties based on statistical associating fluid theory (SAFT), which are implemented in the gPROMS platform.

Such advanced modelling technologies provided a compact and rigorous representation of the physical system, which was validated based on the experimental data from the UKCCSRC pilot plant at Sheffield for the case of monoethanolamine (MEA) solvent.

The validated model was then optimized based on key performance indicators (KPIs) such as the energy consumptions per ton of captured CO2, the solvent slippage per ton of captured CO2, and the required packing per ton of captured CO2 and so on.

Finally, the validated and optimized model was applied in order to study the dynamic behaviour of the carbon capture process and its interaction with the upstream natural gas combined cycle (NGCC) plant, subject to dynamic changes in the load and the feed composition.

389184 Optimization of the Various Modes of Flexible Operation for Post-Combustion CO2 Capture Plant

Abstract:

Abstract

Adding a post-combustion CO_2 capture plant to a power plant comes with a substantial energy penalty as significant amounts of steam must be used to strip off the absorbed CO_2 and regenerate the solvent. This leads to a decrease in the power output of the plant, typically by ~30 %. One option to mitigate the adverse effect of power plant output loss is to operate the CO_2 capture plant in flexible modes in response to the varying electricity market price. This way the overall profit can be improved while meeting the peak electricity demand. Flexible operation can be done, for example, by reducing the capture level (through CO_2 venting, by-passing a part of flue gas, CO_2 capture level reduction, etc.), or by reducing the amount of regenerated spent solvent on a periodic or temporary basis, i.e., storing the CO_2 rich solvent in a tank for a time while ensuring a constant CO_2 capture level by using the already regenerated solvent stored in another tank. The objective of this work is to simulate, optimize, and analyze the economics of the various modes of flexible operation including the CO_2 capture level reduction, solvent storage, and the both options simultaneously. An equilibrium based model for complete absorption/regeneration system has been used and its optimizations are performed in the *gPROMS* interface.

There are three novel features in this work when compared to the previous works [1-5]: 1) A more detailed, accurate process flow sheet model is used for the analysis and optimization rather than first-order approximations more suitable for quick screening purposes. 2) Many previous works treated captured CO_2 as products of market value. They then posed the problem as optimizing the flexible operation in light of varying energy price and CO_2 market price. In contrast, we consider penalties on the CO_2 emission. We assume that captured CO_2 holds no market value as the market for CO_2 is likely to be highly limited in view of the huge CO_2 quantity once CO_2 capture becomes ubiquitous. 3) Both options of dynamically varying the capture level and storing the solvent (with varying storage capacities and different input energy cost profiles) are considered simultaneously. Therefore, the simultaneous operation can have advantages in optimizing both capture reduction level and regeneration level for the solvent storage. The Simultaneous mode of the flexible operation has been shown to offer the greatest advantage over the inflexible operation.

References

1. Cohen, S.M., et al., *Comparing post-combustion CO 2 capture operation at retrofitted coal-fired power plants in the Texas and Great Britain electric grids.* Environmental Research Letters, 2011. **6**(2): p. 024001.

2. Cohen, S.M., G.T. Rochelle, and M.E. Webber, *Optimizing post-combustion CO2 capture in response to volatile electricity prices.* International Journal of Greenhouse Gas Control, 2012. **8**(0): p. 180-195.

3. Wiley, D.E., M.T. Ho, and L. Donde, *Technical and economic opportunities for flexible CO2 capture at Australian black coal fired power plants.* Energy Procedia, 2011. **4**(0): p. 1893-1900.

4. Cohen, S.M., G.T. Rochelle, and M.E. Webber, *Optimal operation of flexible post-combustion CO2 capture in response to volatile electricity prices.* Energy Procedia, 2011. **4**(0): p. 2604-2611.

5. Chalmers, H., et al., *Valuing flexible operation of power plants with CO2 capture.* Energy Procedia, 2009. **1**(1): p. 4289-4296.

372693 A Process Design and Operability Framework for Post-Combustion CO2 Capture Systems and Solvents: Integrated Assessment and Selection

Abstract:

Post-combustion CO_2 capture technologies are predominantly based on the use of solvents that enhance the separation efficiency in absorption/desorption systems by selectively dissolving CO_2. A large part of published research is focused on the identification of new solvents with improved thermodynamic, reactivity and sustainability characteristics [1] to help reduce the high operating costs and environmental impacts associated with absorption/desorption systems. As solvents are inherent components of the separation systems in which they are utilized, the consideration of different solvents in the optimization of absorption/desorption characteristics such as process structure and size, topology of recycle streams and solvent feed flowrates [2], to name a few, is receiving significant attention in CO_2 capture research to further improve the process economics. However, CO_2 capture processes are essentially dynamic environments susceptible to variations in operating parameters (e.g., temperature, pressure, and composition, solvent degradation and losses) and influence from exogenous disturbances (e.g., input flue gas composition). Absorption/desorption systems designed for optimality assuming constant operating conditions will divert from optimal operation if any of these parameters vary in time, as is often the case in industrial practice. This issue has received very limited attention in CO_2 capture research, where the effects of variability are largely overlooked during absorption-desorption design as it is generally assumed that a properly designed control system will eventually compensate for such effects. On the other hand, the effects of different CO_2 capture solvents in process design under operating variability that explicitly considers the controllability properties and the operability performance of the overall process system have yet to be addressed systematically.

The current work addresses the economic performance simultaneously with the static and dynamic controllability properties of amine based CO_2 capture processes. Amine solvent candidates are introduced into a unified framework that enables the optimal absorption/desorption design while considering variations in process parameters and disturbances in process operation. The proposed framework consists of a process design and a nonlinear sensitivity analysis stages that evaluate process design and control structure configurations under the influence of multiple simultaneous parameter variations and/or exogenous disturbances. The absorption/desorption design stage involves a process synthesis approach employing reactive and non-reactive separation column modules of sufficient modelling detail and flexibility acquired through the implementation of an orthogonal collocation on finite elements (OCFE) approximation technique [2]. The nonlinear sensitivity analysis stage investigates the steady-state and dynamic effects process parameters and disturbances impose on the system for a given control structure. The control scheme categorizes control objectives and preferences in the utilization of manipulated variables in a multivariable centralized fashion [3]. As a result, an extended sensitivity surface around the optimal operating points is developed that reflects the variability of process manipulated and controlled variables in response to changes in process parameters and disturbances. It is within this framework that alternative solvents of different chemical structure and physical properties are introduced in order to study their effects in optimal design and operability performance under operating conditions variations. Each solvent constitutes an additional design feature in the formulated optimization problem, demonstrating alternating process design directions under the imposed variability. The performed sensitivity analysis generates useful insights regarding the control structure selection and the range of parameter variations within which the solvents demonstrate optimum performance. Furthermore, it allows the adaptive modification of the process design characteristics, inclusive of the available solvent options, in order to enhance process operability.

The proposed developments are illustrated in case studies considering amine solvents such as Monethanolamine (MEA), 2-amino-2methyl-1-propanol (AMP) and Diethanolamine (DEA) which are commonly used for CO_2 capture. The thermodynamic behavior of the CO_2-water-amine mixtures is modelled through pressure-loading relationships extracted from the Statistical Associating Fluid Theory for potentials of variable range (SAFT-VR) [4, 5]. Different absorption/desorption structures are evaluated including recycle split streams, muti-pressure cascades etc. in view of the considered solvents.

Acknowledgements

The authors would like to thank Prof. Claire Adjiman, Prof. Amparo Galindo, Prof. George Jackson, and Dr. Alexandros Chremos, of Imperial College London for providing the SAFT-VR thermodynamic models. Funding from the European Commission under grant FP7-ENERGY-2011-1-282789-CAPSOL is gratefully acknowledged.

Cited References

[1] Papadopoulos A.I., S. Badr, A. Chremos, E. Forte, T. Zarogiannis, P. Seferlis, S. Papadokonstantakis, C. S. Adjiman, A. Galindo, G. Jackson, 2014, Efficient screening and selection of post-Combustion CO_2 capture solvents, Chemical Engineering Transactions, 39, ISBN 978-88-95608-30-3; ISSN 2283-9216.

[2] Damartzis T., A.I. Papadopoulos, P. Seferlis, 2014, Optimum synthesis of solvent-based post-combustion CO_2 capture flowsheets through a generalized modeling framework, Clean Technologies and Environmental Policy, DOI: 10.1007/s10098-014-0747-2.

[3] Seferlis, P., Grievink, J., 2004, "Process Design and Control Structure Evaluation and Screening using Nonlinear Sensitivity Analysis", Computer Aided Chemical Engineering, 17, 326-351.

[4] Rodriguez J., Mac Dowell N., Llovell F., Adjiman C.S., Jackson G., Galindo A., 2012, Modelling the fluid phase behaviour of aqueous mixtures of multifunctional alkanolamines and carbon dioxide using transferable parameters with

the SAFT-VR approach, Molecular Physics, 110, 2012, 1325-1348.

[5] MacDowell N., Pereira F.E., Llovell F., Blas F.J., Adjiman C.S., Jackson G., Galindo A., 2011, Transferable SAFT-VR models for the calculation of the fluid phase equilibria in reactive mixtures of carbon dioxide, water, and n-alkylamines in the context of carbon capture, Journal of Physical Chemistry B, 115, 8155-8168.

379421 Increasing the Operating Range of an IGCC Plant with Carbon Capture through the Integration of a CaO-Based Thermal Storage System

Abstract:

Using pre-combustion CO_2-capture, IGCC plants show significant potential for efficient power generation with carbon capture. The gasification and gas processing steps however have multiple temperature and flow constraints which severely limit the flexibility of IGCC plants to meet the dynamic demands of the current grid [1]. To address this issue, the authors recently proposed a CaO-based energy storage system increasing the load range of a base IGCC plant by about ±20-25% with respect to its nominal output without the need to cycle the gasifier island and while maintaining its 90% CO_2-capture rate [2-3].

The storage system relies on the reversible decomposition of $CaCO_3$ into CaO and CO_2 (calcination) to store excess energy during low electricity demand periods. This excess energy originates from the combustion of excess syngas produced in the gasifier and is stored chemically and thermally as hot CaO. During high demand periods, the stored energy can be recovered by recombining the CaO with CO_2 (carbonation) extracted from the IGCC plant. This recombination results in high grade reaction heat (at 650 °C) suited for steam and electricity production and hence results in a temporary increase of the IGCC power output. In this process, CO_2 is temporarily stored as $CaCO_3$, which allows to turndown the primary carbon capture system leading to a reduction of the parasitic load of the primary carbon capture system and hence a further increase in the net output of the IGCC plant.

This presentation will give an overview of the proposed energy storage system and explore its integration into an IGCC plant equipped with CCS. Two different storage configurations are considered herein, using either an indirectly fired air-blown calciner or a directly fired oxygen-blown calciner. The system's round-trip efficiency was found to vary between 60 and 70%, depending on the storage operating conditions and configuration [4]. The loss associated with the purge and replacement of the solid inventory required to maintain a sufficiently high $CaO/CaCO_3$-particle reactivity was found to be the dominant factor limiting the round-trip efficiency. The combustion of the syngas in the calciner instead of the combined cycle of the IGCC plant further results in increased heat losses as the efficiency of the Rankine steam cycle employed to convert the thermal energy in the calciner and carbonator into electric power is lower than the efficiency of the combined cycle. A high degree of internal heat recuperation minimizing the calciner syngas demand was therefore found to be essential to achieve a high round-trip efficiency. Based on efficiency and storage size, the energy storage system was found to be competitive with other storage solutions proposed in the literature.

[1] IEAGHG, Operating Flexibility of Power Plants with CCS; June 2012

[2] Vandersickel, A., et al., *Integration of a CaO-based energy and CO2 storage system in an IGCC plant with carbon capture*, in *ECOS 2013*. 2013: Guilin, China

[3] Vandersickel, A., et al., *A CaO-based energy and CO2 storage system for the flexibilization of an IGCC plant with carbon capture*. sumitted to Industrial & Engineering Chemistry Research, 2014

[4] Vandersickel, A., et al., *Integration of a CaO-based thermal storage szstem in an IGCC plant with carbon capture,* submitted to IMECE 2014, Paper Nr. IMECE2014-38113

Design and Optimization of Low-CO_2-Emitting Organic Rankine Cycle for Liquefied Natural Gas Cold Exergy Recovery

Ung Lee[1], Kyeongsu Kim[1], Chonghun Han[1*]

[1]*School of Chemical and Biological Engineering, Seoul National University,*

The depletion of fossil fuel due to increasing energy consumption has been accelerating, thus the energy crisis is becoming more severe. Moreover, the need to reduce fossil fuel consumption has been attracting more attention recently due to concerns related to global warming. Pulverized coal (PC) power plants are one of the largest consumers of fossil fuel, and they produce a significant amount of CO_2. Therefore, much research has been focused on improving the efficiency of PC power plants and reducing their CO_2 emission in order to address both fossil fuel depletion and greenhouse gas reduction. The installation of the bottoming cycle in conventional power plants has recently gained attention. The combined cycle is able to produce additional power without consuming fossil fuel through utilization of waste heat generated from the steam cycle. Although many studies have focused merely on improving the efficiency of the topping cycle, it is also important to improve the bottoming cycle's efficiency in order to produce more power from power plants. Because the waste heat from the steam cycle (the temperature of which is too low to be utilized for other purposes within the steam cycle) is used, the combined cycle can produce more power without generating additional CO_2. There have been other noteworthy studies performed to recover low-grade heat using liquefied natural gas (LNG) as a heat sink. In this way, the temperature difference between the heat source and the heat sink is increased because the condensation of the working fluid takes place at a point lower than ambient temperature. Consequently, the bottoming cycle can produce more power with the same heat source utilization.

In this work, a multicomponent combined Rankine cycle utilizing LNG cryogenic exergy was proposed. Since LNG is a multicomponent mixture comprised of more than eight components, the evaporation process of LNG is not isothermal. Therefore, the exergy recovery of the LNG through the ORC working fluid can be significantly improved by utilizing multicomponent working fluids, which have a nonisothermal condensing nature. Also, the proposed ORC is designed to recover the entire evaporation enthalpy of LNG by using a multicomponent working fluid to maximize the power generation within the cycle. To illustrate the performance of the ternary mixture ORC quantitatively, ORCs with single and binary components were also examined in this study.

Another important point of this study is the integration of the ORC with a PC power plant with a CO_2-capture unit. The proposed power cycle utilized low-pressure steam from a PC plant to enhance the total power generation. In addition, captured CO_2 from the PC power plant was utilized to preheat the working fluid. In this manner, the proposed power cycle can preheat the working fluid and reduce the energy needed for liquefaction of CO_2.

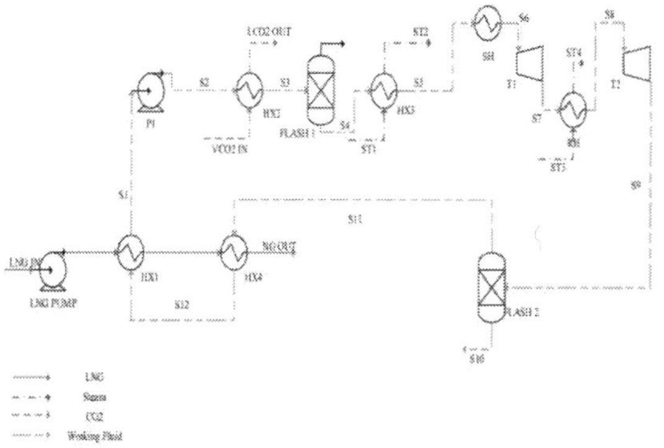

Figure. 1 Process flow diagram of ORC using ternary mixture working fluid.

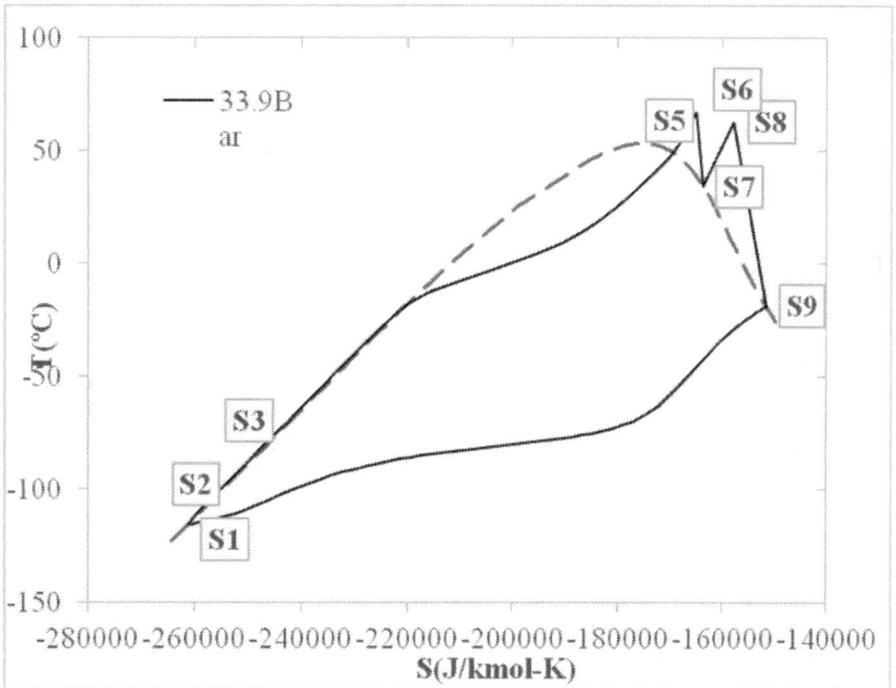

Figure. 2 T-S diagram of ORC using ternary mixture working fluid.

Figure 1 and 2 give a principal process scheme of ternary mixture working fluid and corresponding T-S diagram. The ORC uses R601-R23-R14 ternary mixture as its working fluid and is integrated with a steam cycle as a bottoming cycle. By utilizing the hot and cold exergy of low-pressure steam and LNG that are initially wasted, the ORC is able to generate additional power without consuming fossil fuel. The CO_2 generated from the combined cycle is liquefied by utilizing its latent heat as a heat source for the working fluid. Consequently, the energy consumption for the CO_2-liquefaction process is reduced by more than 50% in this study. The nonisothermal condensing nature of the ternary mixture working fluid can reduce the exergy loss of the system, and the consequent power generation and efficiency of the ORC are significantly improved. Power generation from the ternary mixture ORC is increased by about 56% and 59% as compared with the pure and binary mixture ORCs, respectively. Important design parameters such as pump discharge pressure, working fluid composition, and turbine inlet and outlet pressure are also optimized to recover the maximum power from the ORC.

Acknowledgement

This research was supported by the second phase of the Brain Korea 21 Program in 2014, by Institute of Chemical Processes in Seoul National University, by MKE and grant from the LNG Plant R&D Center funded by the Ministry of Land, Transportation and Maritime Affairs (MLTM) of the Korean government, by the Energy Efficiency & Resources Core Technology Program of the Korea Institute of Energy Technology Evaluation and Planning(KETEP) granted financial resource from the Ministry of Trade, Industry & Energy, Republic of Korea (No. 2010201020006D), (No. 20132010201760) and (No. 20132010500050).

387239 A Time Scale-Bridging Approach for Integrating Production Scheduling and Process Control

Abstract:

The control and operation hierarchy of the chemical supply chain comprises several layers of functionality, including regulatory and supervisory control, and production scheduling and planning. These functions act over different time horizons, ranging from seconds, in the case of regulatory control, to months, in the case of production planning). Since they all serve the common goal of improving process efficiency, profitability and safety, it is conceivable that carrying out several of these functions in an integrated fashion will yield further economic and/or operational benefits. Indeed, extensive literature studies have been dedicated to the interplay between regulatory and supervisory control, and to the integration of planning and scheduling.

Conversely, the integration of scheduling and control has received far less attention. This is partly due to organizational challenges [1]: the two functions are typically the responsibility of different parts of an organization, with planning and scheduling being carried out by a business arm, and control belonging to an engineering/operations division. However, integrating these two functions also poses significant technical challenges: their time horizons and execution frequencies differ considerably. Moreover, scheduling must account for discrete decisions, in addition to using continuous variables as is typical in control calculations [2].

In this work, we present a novel framework for integrating short-term scheduling and nonlinear control of continuous chemical processes. We draw our inspiration from scale-bridging techniques used in multi-scale simulation, whereby a reduced-order model of the system characteristics in a faster time scale (or smaller lenghtscale) is used to improve the accuracy of the system model at a higher level in the time/space scale hierarchy. In a similar vein, we develop a low-dimensional dynamic model of the closed-loop input-output behavior of the process under nonlinear control [3]. This low order time scale-bridging model is then incorporated into the scheduling layer as a dynamic constraint. The schedule is computed in the form of a time-varying setpoint signal for the process controller [4].

We demonstrate the implementation of these concepts using multivariate input-output linearizing feedback control, which imposes a linear and decoupled closed loop behavior on the systems of interest. We then present two illustrative case studies, a multiple product reactor [5] and a polymerization reactor [6]. We show that the economic performance of the proposed scheduling mechanism is comparable to that of scheduling formulations that rely on the full process model when a perfect model is available, and emphasize the benefits of our approach in the presence of plant-model mismatch.

References:

[1] S. Engell and I. Harjunkoski. Optimal operation: Scheduling, advanced control and their integration. Comput. Chem. Eng., 47:121–133, 2012.

[2] M. Baldea and I. Harjunkoski. A systematic review of the integration of production

scheduling and process control. Comput. Chem. Eng., submitted.

[3] M. Baldea, P. Daoutidis. Dynamics and Nonlinear Control of Integrated Process Systems, Cambridge University Press, 2012.

[4] J. Du, J. Park, I. Harjunkoski, and M. Baldea. Integrating Production Scheduling and Process Control using Internal Coupling Models. ESCAEP 24, Budapest, Hungary.

[5] A. Flores-Tlacuahuac and I.E. Grossmann. Simultanous cyclic scheduling and control of a multiproduct CSTR. Ind. Eng. Chem. Res., 45:6698–6712, 2006

[6] P. Daoutidis, M. Soroush, and C. Kravaris. Feedforward/feedback control of multivariable

nonlinear processes. AIChE J., 36(10):1471–1484, 1990.

Theoretical Framework for Formulating MIP Scheduling Models with Multiple and Non-Uniform Discrete-Time Grids

Sara Velez and Christos T. Maravelias
Department of Chemical and Biological Engineering, University of Wisconsin – Madison
1415 Engineering Dr., Madison, WI 53706, USA

The modeling of time plays a key role in the formulation of mixed-integer programming (MIP) models for scheduling, production planning, and operational supply chain planning problems. It affects the size of the model, the computational requirements, and the quality of the solution. While many types of time grids have been proposed in the literature for continuous-time models, discrete-time models had always been thought of as having a single grid with equally spaced time points. However, in this work we show that models based on multiple and nonuniform discrete-time grids can in fact be formulated. Specifically, we develop algorithms for the generation of separate grids for each task, unit, and material, based on problem data. The proposed algorithms are executed in just a few seconds and therefore do not introduce significant overhead compared to the time it takes to solve the MIP scheduling model. The resulting MIP model has significantly smaller number of variables and constraints. We build upon our previous work to generate separate grids for utilities; and modify the task and unit grids to account for maintenance, changeovers, limited intermediate storage, and utilities. Importantly, we prove five properties which ensure that our multi-grid formulation will not cutoff the best possible solution.

Utility consumption can vary throughout the execution of a task and the amount of utility that is available can change as well. Capturing the variations in utility consumption and availability requires a grid with a small time step. If a single grid is being used, it is necessary to add more binary variables as time points are added to capture the utility consumption. Another option when using a single-grid is to approximate the utility profiles, but this may lead to suboptimal solutions. We propose to use a separate grid for the utility that can be made as fine as desired without affecting the task/unit grids or the number of binary variables.

The proposed framework leads to substantial reductions in the size of the formulations. Computational results showed that our multi-grid models: (1) obtain better solutions than a model based on a single coarse grid (which is the practice today), and (2) obtain better solutions and are solved faster than single-grid models based on a sufficiently fine grid.

Beyond the computational enhancements, the proposed framework has four additional advantages. First, the algorithms for the generation of the grids require only problem data as input so they can be readily applied for a range of problems. Second, the proposed models have exactly the types of constraints present in single-grid models: unit utilization, batch capacity, material balance, and resource constraints. No new types of constraints are needed to synchronize the different grids. Third, additional processing characteristics, such as consumption and production of materials during the execution of tasks, can be considered using the same ideas and methods. Fourth, it allows us to formulate models that enforce resource constraints at a frequency higher than the frequency at which tasks are allowed to start, a highly desirable feature that has been overlooked in the academic literature (e.g., model utilization of cleaning-in-place and cleaning-out-of place resources in batch manufacturing facilities).

385398 Optimal Short-Term Scheduling of Turnarounds in an Integrated Site

Abstract:

Integrated sites involve material flow in process networks. The disruption of flow due to planned maintenance turnarounds on some process units affects production rates in both upstream as well as downstream units. Due to this, long-term maintenance planning often recommends the alignment of a subset of the turnarounds around the same time in order to make use of synergistic effects.

In this paper, we consider a short-term maintenance scheduling setting, where we focus on a time scale of a few months within which a number of turnarounds are scheduled to take place. A major concern during the time leading up to a set of turnarounds is the planning for inventory build-up in anticipation of a turnaround, so as to try and maintain normal operation and satisfy demands during these long outages. Another issue is the allocation of maintenance manpower among various subtasks, which is a scarce resource that has been negotiated for months in advance.

The goal of this work is to provide an optimization formulation that will take into account operational level detail and suggest optimal maintenance schedules of these units while negotiating resource constraints, within a time interval that spans the recommended maintenance durations of a subset of the production units. In particular, we study the short-term scheduling of detailed maintenance tasks and subtasks, ramping up and down of unit production, allocation of available maintenance manpower, use of intermediate storage, and the effect on material flow in the entire site network. Not only maintenance task and manpower work schedules, but also site-wide short-term production plans and inventory policy decisions fall out of the optimization framework. We use an industrial-size site network example to illustrate the results.

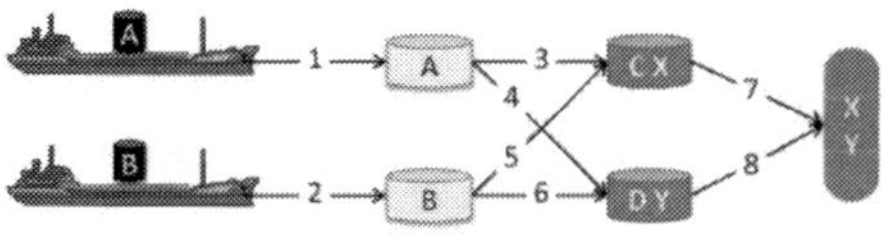

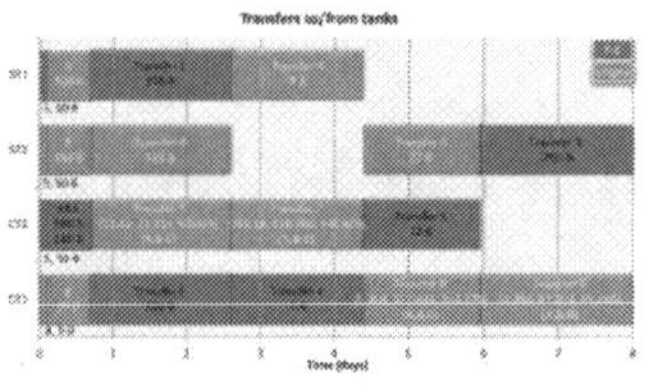

Resource-Task Network Continuous-Time Formulation for Crude Oil Blending Operations

Pedro M. Castro

INVESTIGADOR FCT

LISBOA

Ignacio E. Grossmann

Carnegie Mellon

Introduction

Carnegie Mellon

- Improved crude oil blend scheduling can reach M$/year (Kelly and Mann, 2003)
  - Capitalize on purchase opportunities from spot market
  - Reduce variation in mixture quality & flowrate charged to distillation columns
- Alternative models have been proposed
  - Discrete-time (Lee et al. 96)
  - Continuous-time, unit-specific, STN-based (Ierapetritou & co-workers 03, 04, 07; Li et al. 07, 12; Yadav & Shaik 12)
  - Continuous-time with priority slots (Mouret et al. 09)
- MINLP scheduling problem
  - Bilinear blending constraints guarantee no composition discrepancies
    - Between tank & its outlet streams
  - Favored approach: two-stage MILP-NLP solution procedure
    - MILP generation:
      - Remove bilinear blending constraints (Jia & Ierapetritou, 03)
      - Relax them with global (Lee et al. 96) or piecewise McCormick envelopes (Li et al. 12)
    - NLP generation:
      - Fix binary variables using solution from MILP

Problem main features

- 1. Marine vessels characterized by arrival time at_{mv} and volume of crude to discharge $v_{mv,cr}^{in}$
- 2. Single docking station available

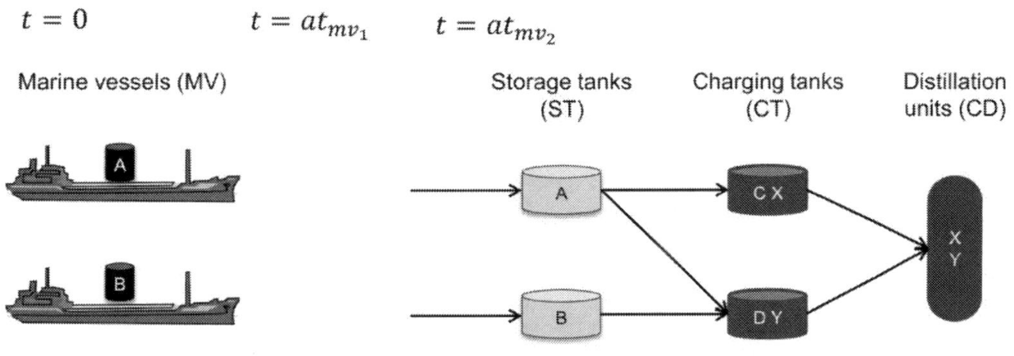

- 3. Tanks hold a single or multiple crudes

Main features (cont.)

- 4. Tanks have storage capacity v_{tk}^{min}, v_{tk}^{max} and must respect composition limits $c_{tk,pr}^{min}$, $c_{tk,pr}^{max}$
- 5. It is assumed that crude properties blend linearly
- 6. Final product demands $\in [d_{cr}^{min}, d_{cr}^{max}]$
- 7. No simultaneous flow in & out in blending tanks

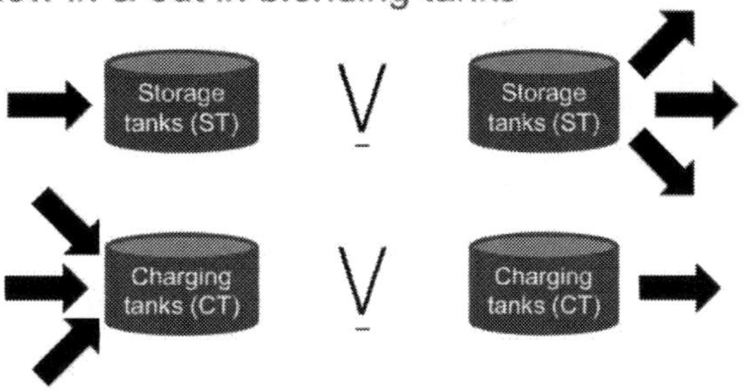

- 8. A charging tank only feeds one CDU at a time
- 9. A CDU can be charged by a single CT at a time
- 10. CDUs are always operating

RTN process model

Carnegie Mellon

- Allows for multiple streams in/out tanks

Problem 1

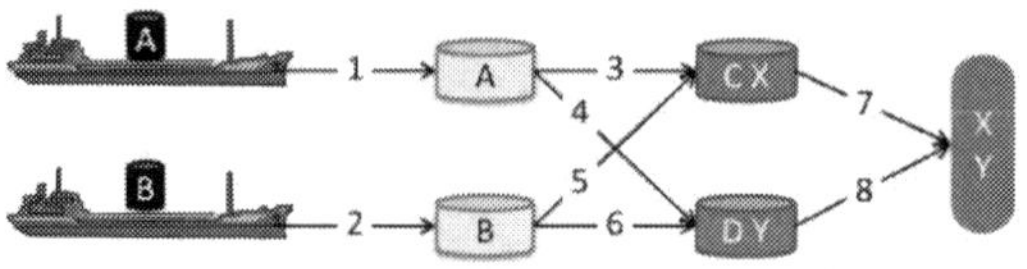

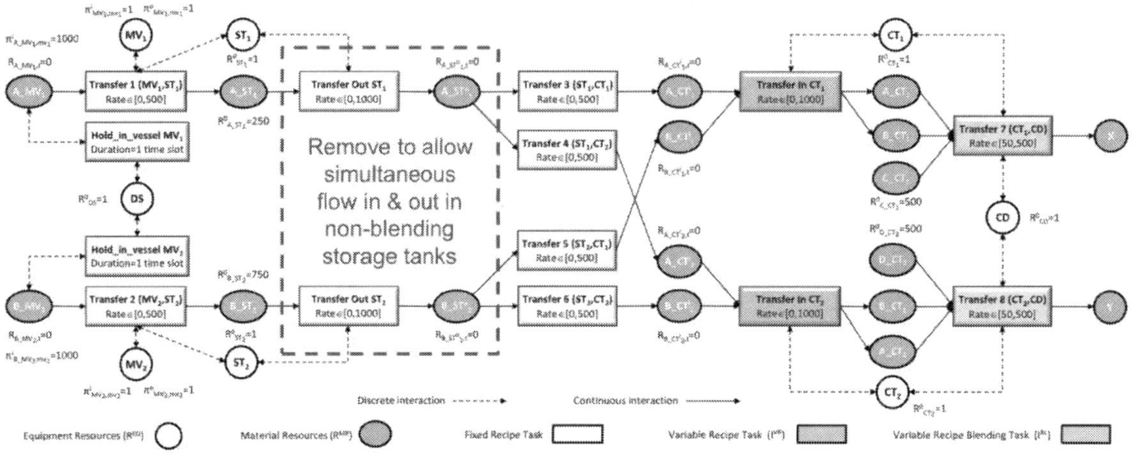

Mathematical formulation (MINLP)

Carnegie Mellon

- Consists of two parts:
  - Scheduling constraints (MILP)
  - Blending constrains & objective function (NLP)
- Time representation
  - Continuous-time with single time grid

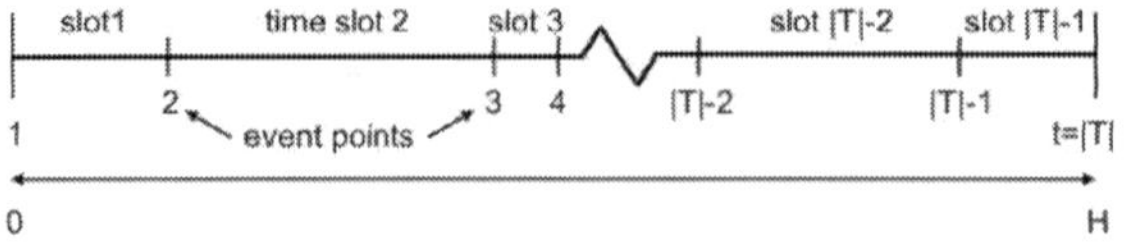

Timing variables

$$T_1 = 0$$

$$T_{|T|} = H$$

- Why not discrete-time?
  - Worse performance for continuous plants (Castro and co-workers 09,10)
  - Time grids with more slots => more bilinear terms & constraints
- Why not multiple time grids (a.k.a. unit-specific)?
  - Fewer time slots but larger number of inefficient big-M constraints
    - » Low-quality relaxation detrimental for global optimization

Arrival & departure of marine vessels

Carnegie Mellon

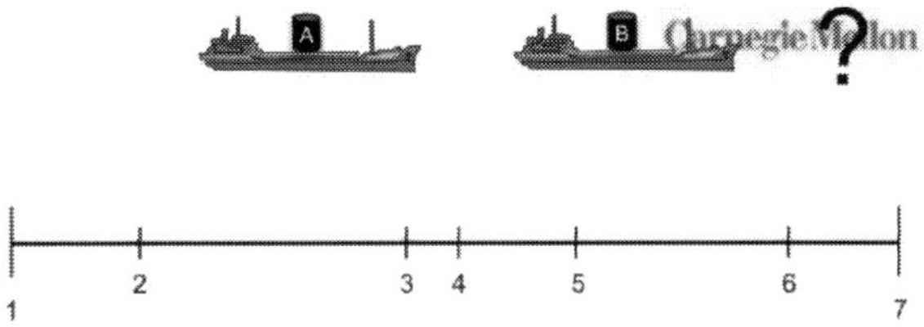

- Each event point with single incoming vessel

$$\bigvee_{mv\in MV}\begin{bmatrix} Z^{i}_{t,mv} \\ T_t \geq at_{mv}\end{bmatrix} \bigvee \begin{bmatrix} Z^{no\ i}_{t} \\ T_t \geq 0\end{bmatrix} \ \forall t \neq |T|$$

- And single departure vessel

$$\bigvee_{mv\in MV}\begin{bmatrix} Z^{o}_{t,mv} \\ T_t \geq at_{mv} + \sum_{cr} v^{in}_{mv,cr}/\rho^{max}_{mv,st_*}\end{bmatrix}$$

$$\bigvee \begin{bmatrix} Z^{no\ o}_{t} \\ T_t \geq 0\end{bmatrix} \ \forall t \neq 1$$

- Arrival/departure occurs once

$$\bigvee_{t\neq|T|} Z^{i}_{t,mv} \ \forall mv \in MV \qquad \bigvee_{t\neq 1} Z^{o}_{t,mv} \ \forall mv \in MV$$

- Convex hull reformulation of GDP
(Grossmann & co-workers 94, 12)

$$T_t \geq \sum_{mv\in MV} Z^{i}_{t,mv} \cdot at_{mv} \ \forall t \neq |T|$$

$$T_t \geq \sum_{mv\in MV} Z^{o}_{t,mv} \cdot \left(at_{mv} + \frac{\sum_{cr} v^{in}_{mv,cr}}{\rho^{max}_{mv,st_*}}\right) \ \forall t \neq 1$$

$$\sum_{t\neq|T|} Z^{i}_{t,mv} = 1 \ \forall mv \in MV$$

$$\sum_{t\neq 1} Z^{o}_{t,mv} = 1 \ \forall mv \in MV$$

Alternative objective functions

Carnegie Mellon

- Maximize gross margin
  - $\max \sum_{cr\in CR}\sum_{r\in R^{CT}_{cr}}\sum_{t\neq|T|}\sum_{i\in I^{VR},\bar{\lambda}_{r,r,i}=-1} \bar{\xi}_{r,i,t} \cdot p_{cr}$
    - Upper bound on number of distillation runs
      - $\sum_{cd\in CD} ND_{cd} \leq nd$
- Minimize operating cost
  - $\min \sum_{mv\in MV}(WS_{mv} c^{wsea}_{mv} + HB_{mv} c^{harb}_{mv}) + \sum_{tk\in TK}\sum_{t\neq|T|} AI_{tk,t} \cdot DT_t \cdot c^{inv}_{tk} + \sum_{cd\in CD}(ND_{cd} - 1)\cdot c^{chg}$ (Bilinear term: $AI_{tk,t} \cdot DT_t$)
    - Sea waiting
      - $WS_{mv} \geq T_t - at_{mv} - (H - at_{mv})\cdot(1 - Z^{i}_{t,mv}) \ \forall mv \in MV, t \neq |T|$
    - Vessel harboring
      - $HB_{mv} \geq T_{t'} - T_t - \left(H - \frac{\sum_{cr} v^{in}_{mv,cr}}{\rho^{max}_{mv,st_*}}\right)\cdot(2 - Z^{i}_{t,mv} - Z^{o}_{t',mv}) \forall mv \in MV, t' > t$
    - Inventory at charging and storage tanks
      - $AI_{tk,t} = \sum_{r\in R^{MR}_{tk}} \frac{R_{r,t} + R^{end}_{r,t}}{2} \ \forall tk \in TK, t \neq |T|$
      - $DT_t = T_{t+1} - T_t \ \forall t \neq |T|$
    - Crude changeovers in distillation columns

Nonlinear blending constraints

Carnegie Mellon

- Outlet stream with same composition as blend inside tank
  - $\sum_{r' \in R_r^{CR}} \bar{\xi}_{r',i,t} = \zeta_{i,t} \cdot R_{r,t} \; \forall i \in I^{BL}, r \in R_i^{BL}, t \neq |T|$ (Bilinear term)
    - Volume fraction leaving tank during slot t: $\zeta_{i,t}$
    - Amount of resource inside the tank: $R_{r,t}$
    - Amount of resource consumed by task: $\bar{\xi}_{r',i,t}$

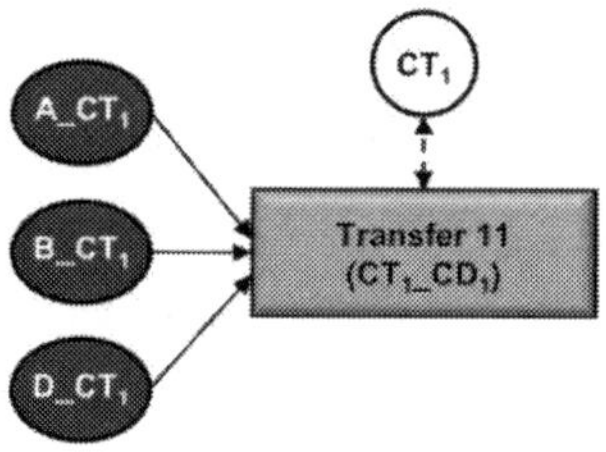

- McCormick (76) LP relaxation
  - $\sum_{r' \in R_r^{CR}} \bar{\xi}_{r',i,t} = W_{r,i,t} \; \forall i \in I^{BL}, r \in R_i^{BL}, t \neq |T|$
    - New set of continuous variables $W_{r,i,t}$ and constraints
      - $W_{r,i,t} \geq 0 \; \forall i \in I^{BL}, r \in R_i^{BL}, t \neq |T|$
      - $W_{r,i,t} \geq R_{r,t} + R_r^{max} \cdot (\zeta_{i,t} - 1) \forall i \in I^{BL}, r \in R_i^{BL}, t \neq |T|$
      - $W_{r,i,t} \leq R_r^{max} \cdot \zeta_{i,t} \; \forall i \in I^{BL}, r \in R_i^{BL}, t \neq |T|$
      - $W_{r,i,t} \leq R_{r,t} \; \forall i \in I^{BL}, r \in R_i^{BL}, t \neq |T|$

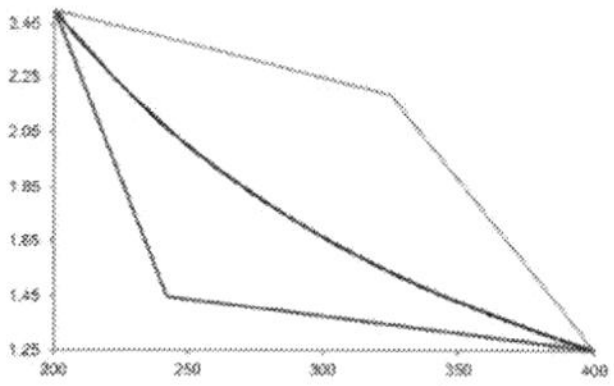

Alternative relaxation approach (MILP)

Carnegie Mellon

- MDT: Multiparametric disaggregation (Teles, et al. 13)
  - Discretize variables $\zeta_{i,t}$ over set of powers $k \in \{\psi, \ldots, \eta\}$
    - $\zeta_{i,t} \leq 1$, thus $\eta = \lfloor \log_{10} 1 \rfloor = 0$
    - $\psi \leq -1 \Rightarrow$ tighter than McCormick
      - Relaxation quality improves as ψ decreases
    - Add residual variable $0 \leq \ddot{\zeta}_{i,t} \leq 10^{\psi}$ to obtain continuous domain
      - $\zeta_{i,t} = \sum_{k=\psi}^{\eta} \sum_{j=1}^{9} j \cdot 10^k \cdot Y_{i,t,j,k} + \ddot{\zeta}_{i,t} \; \forall i \in I^{BL}, t \neq |T|$

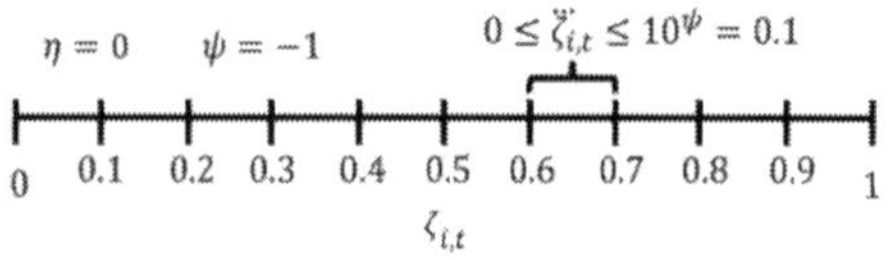

  - Binary variables $Y_{i,t,j,k}$ identify digit $j \in \{0, \ldots, 9\}$ for position k
    - $W_{r,i,t} = \sum_{k=\psi}^{\eta} \sum_{j=1}^{9} j \cdot 10^k \cdot \hat{R}_{r,i,t,j,k} + \ddot{W}_{r,i,t} \; \forall i \in I^{BL}, r \in R_i^{BL}, t \in T$
    - $R_{r,t} = \sum_{j=0}^{9} \hat{R}_{r,i,t,j,k} \; \forall i \in I^{BL}, r \in R_i^{BL}, t \neq |T|, k \in \{\psi, \ldots, \eta\}$
    - $\sum_{j=0}^{9} Y_{i,t,j,k} = 1 \; \forall i \in I^{BL}, t \neq |T|, k \in \{\psi, \ldots, \eta\}$
    - $0 \leq \hat{R}_{r,i,t,j,k} \leq R_r^{max} \cdot Y_{i,t,j,k} \; \forall i \in I^{BL}, r \in R_i^{BL}, t \neq |T|, k \in \{\psi, \ldots, \eta\}, j \in \{0, \ldots, 9\}$

Test problems

Carnegie Mellon

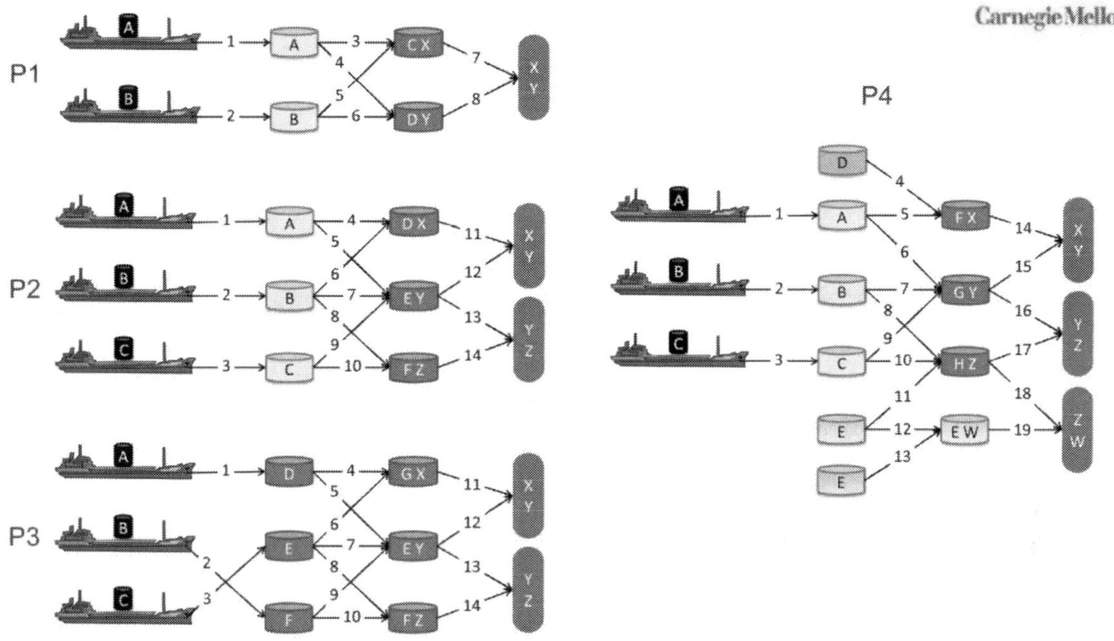

Results for gross margin maximization

Carnegie Mellon

- Vs. Mouret et al. (09)
  - New best solutions for P1 and P2
    - We now allow for interruption of marine vessels discharge
  - Global optimality proven for all but P3
    - Lower optimality gap (0.37% vs. 2.2%)
  - Higher CPUs due to MDT relaxation
    - McCormick solves in a few seconds

P2

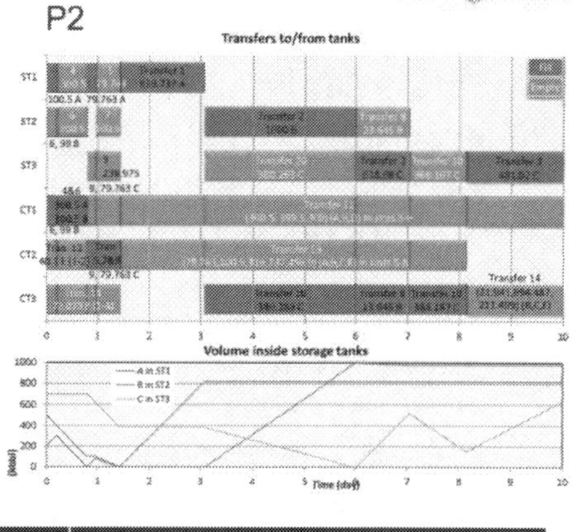

Model type	Priority slots (Mouret et al. 09)				RTN, single time grid (this work)			
Upper bounding strategy (MILP)	MINLP without blending constraints				Blending constraints relaxed with MDT			
Problem	# slots	Gross margin [k$]	Gap	CPUs	# slots	Gross margin [k$]	Gap	CPUs
P1	5	7975	0%	0.35	6	7982.5	0%	1.02
P2	6	10117.5	0%	0.93	9	10246.1	0%	55.6
P3	9	8544.9	2.2%	5.96	7	8544.9	0.37%	3600
P4	4	13254.8	0%	0.47	6	13254.8	0%	8.15

Results for cost minimization

Carnegie Mellon

- Vs. Discrete-time model of Lee et al. (96)
  - Better solutions in all problems
    - Consistent with previous results for continuous plants
  - Continuous-time model rigorous relaxation of discrete-time model (for the same # slots)
    - Event location is chosen by optimization rather than being fixed

Reference	Lee et al.[3]		This work		
Time representation	Discrete		Continuous		
Problem	# slots	Cost [k$]	# slots	Cost [k$]	Reduction
P1	8	217.667	7	210.538	3.3%
P2	10	352.55	7	320.496	9.1%
P3	12	296.56	7	287.000	3.2%
P4	15	420.99	7	365.088	13.3%

Comparison of optimization algorithms

Carnegie Mellon

- Operating cost minimization harder objective function
  - Quality of MILP relaxation much worse
  - No problem could be solved to global optimality
- GloMIQO 2.3 and MILP-NLP MDT strategy tied for best solution found
  - MDT optimality gap up to 4 orders of magnitude smaller
  - GloMIQO and MDT with two suboptimal outcomes.
- BARON 12.3.3 could not find a feasible solution in 12/22 cases

		$\|T\| = (5,5)$					$\|T\| = (8,8)$		
	Approach	Discrete variables	Total equations	Cost [$]	Gap	CPUs	Cost [$]	Gap	CPUs
P2	McCormick	76	1015	382600	78.8%	0.37	360208	81.9%	1.38
	MDT ($\psi = -1$)	356	2783		0.46%	25.4	320496	16.2%	3600
	MDT ($\psi = -2$)	556	3683	361800	0.045%	50.8	-	-	-
	MDT ($\psi = -4$)	956	5483		0.0006%	1185	-	-	-
	GloMIQO				9.6%	3600	320504	37.0%	3600
	BARON	76	695	374488	126%	3600	no sol.	-	3600
	DICOPT			361800	-	0.67	329400	-	6.03
P3	McCormick	88	1676	338400	50.4%	0.59	292168	62.3%	5.38
	MDT ($\psi = -1$)	704	5040		0.26%	42.6	288867	48.1%	3600
	MDT ($\psi = -2$)	1184	6892	331700	0.026%	555	-	-	-
	GloMIQO				0.031%	3600	287000	45.8%	3600
	BARON	88	1020	354155	215%	3600	no sol.	-	3600
	DICOPT			351267	-	0.74	300000	-	2.35

Conclusions

Carnegie Mellon

- Resource-Task Network continuous-time MINLP formulation can capture most important constraints
- Very efficient, can solve problems to global optimality
  - GloMIQO
  - Two-stage MILP-NLP strategy using MDT relaxation

- Future work
  - Spatial branch-and-bound algorithm with proposed strategy
- Further details in IECR 2014, 53, 15127-15145

- Acknowledgments
  - 2013 Portugal-U.S. Research Networks
  - PTDC/EQU-ESI/118253/2010

FUNDAÇÃO LUSO-AMERICANA

FCT Fundação para a Ciência e a Tecnologia

COMPETE

368873 Simultaneous Scheduling and Control Using Fast MPC

Abstract:

In chemical process operations simultaneous consideration of scheduling and control helps to enhance the overall process profitability but meanwhile increases the complexity of modeling and thus the computational efficiency of the solution approaches (Terrazas-Moreno, et al., 2008) (Engell and Harjunkoski, 2012). Online integration of scheduling and control requires updating operation solutions for both scheduling and control levels at real time in the presence of disturbance, thus the online integration requires a repetitive solution of the integrated problem at each time interval (Zhuge and Ierapetritou, 2012).

Model Predictive Control (MPC) is an online optimization technique that involves repetitively solving optimization problems over a future time horizon. To reduce the computation burden of conventional MPC, multi-parametric Model Predictive Control (mp-MPC) was proposed to solve for the explicit control law offline and thus the online optimization is reduced to simple function evaluations (Pistikopoulos, 2009). However, as the problem size increases in terms of state dimension and prediction horizon, the number of polyhedral regions in the state partition increases exponentially and this limits the applicability of mp-MPC to small and medium-sized control problems (Richter, et al., 2012).

Fast MPC on the other hand is capable in handling large scale problems, and therefore can be used to facilitate the efficient integration of scheduling and control of large-scale processes. Fast MPC transforms the MPC problem into a convex quadratic programming for which efficient nonlinear programming methods and computational tools can be used to speed up the computation by exploring the problem structure. Among the existing work in the literatures three solution approaches can be identified: active set method (Ferreau, et al., 2008), interior point method (Rao, et al., 1998) (Wang and Boyd, 2010) and Fast gradient method (Richter, et al., 2009) (Richter, Jones and Morari, 2012).

In this study we propose a cascade control strategy that involves two control loops for the online integration of scheduling and control. In the outer loop we approximate the original process dynamics using a piece-wise affine (PWA) model and incorporate it with the scheduling level. This leads to an integrated problem that is subject to linear constraints. The primary MPC at the outer loop solves the integrated problem and generates both the production scheduling and the control solution. However, only the scheduling solution is transferred to the inner loop where the secondary MPC (a fast MPC) treats the scheduling solution as parameters and computes the exact control solution online. Note that these two loops correspond to different models. The outer loop uses the integrated model while the inner loop uses the process dynamic model. Following this approach the primary MPC is able to achieve an overall optimalility for both scheduling and control levels efficiently and the secondary MPC is able to respond quickly to the local disturbance. Results of case studies show the feasibility and efficiency of the proposed approach.

References:

(1) Terrazas-Moreno, S.; Flores-Tlacuahuac, A.; Grossmann, I. E. Simultaneous design, scheduling, and optimal control of a methyl-methacrylate continuous polymerization reactor. *AIChE Journal* **2008,**54, (12), 3160-3170.

(2) Engell, S.; Harjunkoski, I. Optimal operation: Scheduling, advanced control and their integration. *Computers & Chemical Engineering* **2012,**47, 121-133.

(3) Zhuge, J.; Ierapetritou, M. G. Integration of Scheduling and Control with Closed Loop Implementation. *Industrial & Engineering Chemistry Research* **2012,**51, (25), 8550-8565.

(4) Pistikopoulos, E. N. Perspectives in multiparametric programming and explicit model predictive control. *AIChE Journal* **2009,**55, (8), 1918-1925.

(5) Richter, S.; Jones, C. N.; Morari, M. Computational Complexity Certification for Real-Time MPC With Input Constraints Based on the Fast Gradient Method. *Automatic Control, IEEE Transactions on* **2012,**57, (6), 1391-1403.

(6) Ferreau, H. J.; Bock, H. G.; Diehl, M. An online active set strategy to overcome the limitations of explicit MPC. *International Journal of Robust and Nonlinear Control* **2008,**18, (8), 816-830.

(7) Rao, C. V.; Wright, S. J.; Rawlings, J. B. Application of Interior-Point Methods to Model Predictive Control. *Journal of Optimization Theory and Applications* **1998,**99, (3), 723-757.

(8) Wang, Y.; Boyd, S. Fast Model Predictive Control Using Online Optimization. *Control Systems Technology, IEEE Transactions on* **2010,**18, (2), 267-278.

(9) Richter, S.; Jones, C. N.; Morari, M. Real-time input-constrained MPC using fast gradient methods. *Decision and Control, 2009 held jointly with the 2009 28th Chinese Control Conference.*, **2009**, pp 7387-7393.

Industrial aspects on production scheduling models and solution methods

Iiro Harjunkoski, Christos T. Maravelias, Pedro M. Castro

ABB Corporate Research, Wallstadter Str. 59, 68526 Ladenburg, Germany
University of Wisconsin, Madison, WI 53706, United States
Centro de Investigação Operacional, Faculdade de Ciências, Universidade de Lisboa, 1749-016 Lisboa, Portugal

Abstract

This presentation provides a brief review on existing scheduling methodologies developed for process industries, focusing on the industrial aspects of scheduling, and discusses the main characteristics of existing approaches, including their strengths and weaknesses. Even if optimization tools of today can in many cases effectively support plant level production planning, there is still clear potential for improvements, especially in transferring academic results into industry. Usability, interfacing and integration are some central aspects that will be discussed in the presentation. After the introduction and problem classification, emphasis is given to lessons learned from industry including some general guidelines and examples on the modeling and solution of industrial problems.

The paper presents the main findings of an extensive review paper (Harjunkoski et al., 2014) and aims at reducing the gap between theory and practice. As scheduling problems arise in almost any type of industrial production facilities (Pulp and Paper, Metals, Oil and Gas, Chemicals, Food and Beverages, Pharmaceuticals, Transportation, Service, Military, etc.), it is important to ensure that the re-usability of models can be maximized. This is a challenge since the production of various processes must be appropriately planned to ensure that the equipment, material, utilities, personnel and other resources are available at the plant when they are needed to realize the production tasks. Apart from economic benefits, good schedules can also contribute to reducing the environmental load, energy demand, violations of various regulations and help coping more efficiently with uncertainties, both in production as well as in customer order levels.

The presentation puts special focus on deployability issues – something so far neither discussed much within the scheduling literature – nor in reviews. The practical value and benefits of using optimization approaches for planning and scheduling will be presented through a number of examples and main emerging industrial challenges will be discussed.

References

Iiro Harjunkoski, Christos Maravelias, Peter Bongers, Pedro Castro, Sebastian Engell, Ignacio Grossmann, John Hooker, Carlos Méndez, Guido Sand and John Wassick (2014). Scope for Industrial Applications of Production Scheduling Models and Solution Methods. *Computers and Chemical Engineering*, 62, pp. 161-193

383661 Optimal Scheduling of Converter Aisle Operation in a Nickel Smelting Plant

Abstract:

The Converter Aisle of a Nickel Smelting plant represents a significant challenge in terms of achieving efficient planning and scheduling. The scheduling is complicated by several factors including hybrid dynamics due to the interaction of discrete logical decisions and operational restrictions. The complexity of this multivariable system greatly complicates the derivation of a schedule by means of inspection, thus necessitating a systematic methodology by which optimal scheduling can be achieved.

The process configuration includes one or more furnaces which transfer ladles of furnace matte via cranes to several converters which follow a standard batch recipe. Furnaces are characterized by a continuous inlet flow of Dry Solid Charge (DSC) with a discrete drawing of the Flash Furnace Matte (FFM), referred to as tapping. The standard batch recipe on the converters includes iteratively charging the converters with FFM, followed by high purity oxygen blows that reject iron, sulfur, and impurities, thereby purifying the FFM and producing Bessemer Matte (BM). Byproducts resulting from the oxygen blows include SO_2 and a layer of liquid impurities referred to as Converter Slag (CS). Oxygen blows must be followed by a removal of the CS to avoid re-contamination of the BM by means of a Converter Skim [1]. The blow, skim, and charge operations must be performed several times to ensure that the BM will meet the purity tolerances, whereas the number of FFM ladles added to each converter determine the cast size and thus the production level.

The objective of this work is to design a centralized optimization-based scheduling algorithm that will maximize production by coordinating multiple furnaces and several converters operating in parallel by making FFM allocation decisions to satisfy matte demand based on converter cycle progression. Matte allocation decisions are subject to restrictions such as shared crane usage, furnace capacity constraints, and a range of operating furnace flow rates. The scheduling optimization must account for the semi-continuous furnace operation due to the continuous inlet flow and discrete removal. Converters are characterized by both batch and cyclical operation where casting times must be offset so as to account for the limited casting availability. Furthermore, converter blowing must be sequenced and timed to achieve compliance with SO_2 emission tolerances which limit the number of converters operating in parallel. This work builds on an initial simplified model [2], and contributes toward the development of a real-time decision support tool to optimize production and systematically handle constraints.

Continuous-time formulations have been successful in solving industrial problems [3],[4]. As such, the scheduling optimization is formulated as a continuous-time mixed integer linear program (MILP). The formulation is modeled in **AMPL** and is solved using **CPLEX 12.5**. A series of symmetry-breaking and tightening constraints are developed and employed to improve the computation efficiency of the optimization to facilitate the use of this technology in real-time applications. A reactive scheduling mechanism is included to allow rescheduling for when process disturbances impact the operating schedule. A methodology for reducing radical scheduling changes due to the optimization during reactive scheduling is presented using a tiered optimization approach that progressively increases the degrees of freedom available as required, in order to achieve a feasible production schedule.

The optimization formulation will be presented, and its performance demonstrated through case studies simulating real plant operating conditions. Challenges encountered will be identified, and strategies for addressing them discussed.

[1] Warner. A. E. M., Liu. J., Javor. F., Lawson. R., Shellshear. W., Hoang. T., and Falcioni. R. (2005). *Developments in Pierce-Smith Converting at Inco's Copper Cliff Smelter During the Last 35 Years*. pp. 27-43, San Francisco, California.

[2] Ewaschuk. C. M., Swartz. C. L. E., and Zhang. Y. (2013). *Optimal Converter Aisle Scheduling in a Nickel Smelting Plant*. In Proceedings of the IFAC MMM, pp. 202-207, San Diego, California.

[3] Floudas. C.A. and Lin. X. (2004). *Continuous-time versus discrete-time approaches for scheduling of chemical processes: A review*. Computers & Chemical Engineering, 28, pp. 4341-4359.

[4] Harjunkoski. I., Maravelias. C. T., Bongers. P., Castro. P. M., Engell. S., Grossmann. I.E., Hooker. J., Méndez. C., Sand. G., Wassick. J. (2014). *Scope for industrial applications of production scheduling models and solution methods*. Computers & Chemical Engineering, Volume 62, pp. 161-193.

370072 Economic Model Predictive Control Using Nonlinear Empirical Models

Abstract:

Significant strides have been made in the last five years to unify economic process optimization and feedback control. Economic model predictive control (EMPC) is a feedback control technique that attempts to tightly integrate economic optimization and feedback control since it is a predictive control scheme that is formulated with an objective function representing the process economics. Key theoretical developments on stability, performance, and robustness of EMPC have been presented as well as numerous demonstrations of improved economic performance of closed-loop systems under EMPC [1]-[3] (for an overview of recent results on EMPC see, also, [4]). As its name implies, EMPC requires the availability of a dynamic model to compute its control actions. However, in industrial processes, it may be difficult to obtain an accurate first-principles model of the process. Furthermore, EMPC may dictate a possibly time-varying operating policy, that is, it may not force convergence of the closed-loop state to a pre-defined operating steady-state. A consequence of such operating strategy is that the operating region in state-space under EMPC may be sufficiently large such that the effect of nonlinearities is significant (i.e., a linear model cannot adequately capture the process dynamics in the region of operation). For the cases where the development of a sufficiently accurate first-principles dynamic model is not possible, system identification techniques may need to be used to obtain an accurate empirical input/output model of the process dynamics. In particular, nonlinear autoregressive moving average with exogenous inputs (NARMAX) models may be one type of nonlinear system identification techniques employed and used to construct an empirical model [5].

In this work, EMPC schemes using NARMAX models are investigated and discussed. In this direction, EMPC schemes are formulated with NARMAX models and the applicability of using these models with EMPC is discussed. In the first part, a nonlinear system identification methodology is presented that can be used to identify model structure and parameters to construct a NARMAX model capable of adequately describing the process dynamics. Since EMPC may dictate a time-varying operating strategy, being able to construct a NARMAX model that can describe the process transients is especially important. In the second part, the advantages and disadvantages of the using various NARMAX models in the context of EMPC are discussed. Finally, the EMPC schemes are demonstrated through several chemical process examples.

[1] Angeli D, Amrit R, Rawlings JB. On average performance and stability of economic model predictive control. IEEE Transactions on Automatic Control. 2012;57:1615-1626.
[2] Huang R, Harinath E, Biegler LT. Lyapunov stability of economically oriented NMPC for cyclic processes. Journal of Process Control. 2011;21:501-509.
[3] Heidarinejad M, Liu J, Christofides PD. Economic model predictive control of nonlinear process systems using Lyapunov techniques. AIChE Journal. 2012;58:855-870.
[4] Ellis, and M, Durand H, Christofides PD. A tutorial review of economic model predictive control methods. Journal of Process Control, in press.
[5] Billings SA. Nonlinear System Identification: NARMAX Methods in the Time, Frequency, and Spatio-Temporal Domains. Chichester, West Sussex, United Kingdom: John Wiley & Sons, 2013.

A Comparison of PLS and Subspace Identification for Batch Model Predictive Quality Control

Brandon Corbett Kuilin Chen Prashant Mhaskar

Batch processes are a class of chemical processes characterized by the reaction of ingredients over a finite duration. These processes are industrially relevant for the production of high value and quality critical products such as pharmaceutical and high quality polymers. The traditional control objective of stabilizing operation around a desired operating condition is not applicable to batch processes because they are inherently transient, encountering a wide range of operating conditions. Instead, the control objective for these processes is to reach the desired product quality by batch termination. This is achieved by manipulating inputs, such as heating, cooling, and ingredient addition, to influence the terminal quality. However, arriving at a control policy is complicated by a number of factors. Unlike continuous processes, which can be modeled adequately by a linear approximation around the desired operating condition, the nonlinearities inherent to chemical processes must be explicitly handled for successful batch control. Further complicating the control problem is that online measurements are typically sparse in batch operations. Approaches to address these problems can generally be classified in two categories: trajectory tracking control and inferential quality control.

Trajectory tracking control for batch processes involves using closed-loop control to track pre-determined trajectories of measurable process variables. These pre-determined trajectories are often taken from historically successful trajectories. Industrially, tracking is usually achieved using PI control (possibly with some gain scheduling to address the nonlinear process dynamics). Additionally, literature contributions have highlighted the improved trajecotry tracking using model predictive control [2, 3, 6, 7]. However, even if perfect tracking is achieved, the desired terminal quality may not be met. This is because the predefined trajectories fail to account for disturbances in initial conditions and throughout the batch.

The key idea behind inferential quality control is to use process measurements, as they become available, to estimate the terminal quality. Then, using the estimated quality, control of the terminal quality is achieved by altering the input trajectories. In practice this is achieved by implementing the inferential model in a shrinking-horizon, model predictive control framework where the objective is to reduce deviation of the estimated quality from the desired quality.

The ability of inferential quality model predictive control to drive the batch the desired quality is dependent on the accuracy of the inferential model. For this reason, the control problem becomes a model identification problem. This model identification problem has a number of interesting features. First, because of the wide operating range of the process, the nonlinearities of the process must be considered. In addition, the sparse measurements associated with batch processes mean that full state information is rarely available. Furthermore, the necessity for low process development costs often rules out development of a detailed first principals model. For similar reasons, it is usually necessary to minimize the need for identification batches. The problem is thus defined as follows: identify a model that can accurately predict end point quality from historically available process measurements with few or no identification runs. In this work we compare different approaches to this problem including an analysis of partial least squares (PLS) methods and a new approach based on subspace identification.

Partial least squares (PLS) models have been used extensively in literature as inferential models for inferential quality control [4, 5, 1]. In this approach, historical process data for a number of batches is arranged into a regressor and response matrix. The data is arranged such that each row of the regressor matrix contains the process measurements and inputs from a completed batch. Likewise, each row of the response matrix contains the corresponding terminal quality. Then a linear model relating the process measurements and terminal qualities can be obtained by regression. Note that the columns of the regressor matrix will naturally be highly correlated because of the time-series nature of the measurements it contains. To resolve this problem, the regressor and response matrix are projected onto corresponding orthogonal subspaces of a reduced dimension. The values of the regressor matrix after the projection are referred to as the scores while the projection matrix itself is referred to as the loadings. PLS methods have been proven effective in many literature studies and industrial cases. An understanding of PLS methods from a state-space perspective serves to provide further insights into the strengths and limitations of the approach.

In a state-space dynamic model, the state of the system captures its instantaneous operating condition. Therefore, the terminal quality of a batch must depend on its terminal state. With this understanding, we return to our analysis of PLS models. In the PLS approach, the scores are used to calculate the terminal quality. Therefore, we conclude that the scores are, in some sense, an estimate of the terminal state. Furthermore, the loadings are a linear model that relates the process measurements throughout the batch to the terminal state. Because of the batch-wise unfolding used in these methods, each column of the regressor matrix contains a time-indexed process measurement. The loading associated with a given column is the linear contribution that the time-indexed measurement makes to the score. Therefore, the loadings can be thought of as a time-varying linear model relating the process outputs to the terminal state. As a result, when the correlation between process measurements and the terminal state changes, the validity of a PLS model is reduced. One circumstance in which this occurs is when a new control scheme is introduced. Since the purpose of developing the PLS model, in this case, is implementation in a new control scheme, this underlines an inherent discrepancy in the approach.

As an alternative to PLS models, we propose the use of a state space model in conjunction with state estimation. In this approach, past measurements are used in conjunction with the model to arrive at a state estimate. The model can then be used to predict the terminal state for a given input trajectory. The approach capitalizes on the fact that the (nonlinear) relationship between the terminal state and the desired quality variable is often known. Therefore, from a prediction of the terminal state, the terminal quality can be easily obtained. Furthermore, since state-space models are invariant under different control schemes, the approach is reliable for implementation in model predictive control. The greatest challenge to this approach is to identify a state-space model from historical data. To do this, we adapt subspace identification methods. A detailed analysis of PLS-based and new results on subspace based methods is presented and illustrated using simulation results.

References

[1] Siam Aumi, Brandon Corbett, Tracy Clarke-Pringle, and Prashant Mhaskar. Data-driven model predictive quality control of batch processes. *AICHE JOURNAL*, 59(8):2852–2861, AUG 2013.

[2] Siam Aumi, Brandon Corbett, Prashant Mhaskar, and Tracy Clarke-Pringle. Data-based modeling and control of nylon-6, 6 batch polymerization. *IEEE TRANSACTIONS ON CONTROL SYSTEMS TECHNOLOGY*, 21(1):94–106, JAN 2013.

[3] Siam Aumi and Prashant Mhaskar. An adaptive data-based modeling approach for predictive control of batch systems. *CHEMICAL ENGINEERING SCIENCE*, 91:11–21, MAR 22 2013.

[4] J Flores-Cerrillo and JF MacGregor. Control of batch product quality by trajectory manipulation using latent variable models. *JOURNAL OF PROCESS CONTROL*, 14(5):539–553, AUG 2004.

[5] J Flores-Cerrillo and JF MacGregor. Iterative learning control for final batch product quality using partial least squares models. *INDUSTRIAL & ENGINEERING CHEMISTRY RESEARCH*, 44(24):9146–9155, NOV 23 2005.

[6] J Flores-Cerrillo and JF MacGregor. Latent variable mpc for trajectory tracking in batch processes. *JOURNAL OF PROCESS CONTROL*, 15(6):651–663, SEP 2005.

[7] Masoud Golshan, John F. MacGregor, Mark-John Bruwer, and Prashant Mhaskar. Latent variable model predictive control (lv-mpc) for trajectory tracking in batch processes. *JOURNAL OF PROCESS CONTROL*, 20(4):538–550, APR 2010.

385868 ARX-Based Approach for Steady-State Identification Analysis of Industrial Processes

Abstract:

For the past decades, process systems engineering technologies, such as model predictive control (MPC) and real time optimization (RTO), have been employed to bring economic benefits to chemical processes. Along those lines, steady-state identification (SSI) approaches can help enhancing chemical processes by providing data for updating lumped parameters in stationary models, and by complementing other frameworks. For example, a classic steady-state RTO scheme crucially depends on SSI methodologies for its calculations, in order to update the parameters of the stationary model that represents various process operating conditions. The development and implementation of SSI methods have attracted much attention in numerous investigations[1], but present several challenges, including processing noisy irregular process data, errors Type I and II. In this work, a new SSI method based on the ARX (Auto-Regressive model with eXogenous inputs) technique is proposed and compared against other techniques from the literature via process systems examples.

Steady-state analyses have been performed with different strategies in the literature. For example, Cao and Rhinehart[2] proposed a methodology, denoted as F-like test, based on the comparison of two variances for the determination of steady states. Bhat and Saraf[3] extended the F-like test to consider the early determination of steady state and gross error detection in a crude distillation unit by means of tuning the critical values, implementing a linear Kalman filter, and performing data reconciliation by least-square techniques. Kelly and Hedengren[4] developed a method for SSI to detect non-stationary drifts in chemical processes. Flehmig and Marquardt[5] reported a technique which can infer trends in unmeasured states by using a linear process model. Jiang and coworkers[6] derived a method that can find steady-state points employing wavelet theory. Yao and coworkers[7] implemented a SSI methodology based on principal component analysis specific for batch processes. Finally, Le Roux et al.[8] and Tao et al.[9] proposed a technique based on a polynomial equation assuming a specified degree, and a moving window of data. Such techniques were employed for SSI by using the polynomial slope as index or critical value. Our proposed method in this presentation provides an alternative that permits the SSI with reduced tuning. This method considers a typical ARX model, in which the singularity of the model matrices is used as an index for steady-state determination.

In this contribution, the F-like test[2], the polynomial-based approach[9] and our proposed technique are implemented for SSI of two different cases. The first case considers a simulated data set in steady state and it is performed to evaluate the techniques under simple conditions. The second case corresponds to a non-linear CSTR described in Pannocchia and Rawlings[10] when operated at different steady states. During steady-state transitions, a non-linear model predictive control scheme is implemented to track set-point changes due to the non-linear characteristics of the system. For each case, we assume two white noise levels (with standard deviations equal to 3 and 5 % of the output value) that are typical of real industrial operating conditions. Results of these implementations show that the tuning of the F-like test and the polynomial-based methods is the most critical step for good performance of these approaches. Moreover, these techniques rely on an index that is sensitive to a user-defined threshold. On the other hand, our proposed approach depends on a mathematical index that characterizes the identifiability properties of the system and therefore can be implemented with reduced tuning. The conclusions of the analyses performed in these examples will facilitate the selection and implementation of the appropriate method for a depropanizer column owned by Petrobras S.A. with real industrial data.

(1) Korbel, M.; Bellec, S.; Jiang, T.; Stuart, P. Steady State Identification for On-line Data Reconciliation Based on Wavelet Transform and Filtering. *Computers & Chemical Engineering* **2014**, *63*, 206.

(2) Cao, S.; Rhinehart, R. R. An Efficient Method for On-line Identification of Steady State. *Journal of Process Control* **1995**, *5*, 363.

(3) Bhat, S. A.; Saraf, D. N. Steady-State Identification , Gross Error Detection , and Data Reconciliation for Industrial Process Units. *Industrial & Engineering Chemistry Research* **2004**, *43*, 4323.

(4) Kelly, J. D.; Hedengren, J. D. A Steady-state Detection (SSD) Algorithm to Detect Non-stationary Drifts in Processes. *Journal of Process Control* **2013**, *23*, 326.

(5) Flehmig, F.; Marquardt, W. Inference of Multi-variable Trends in Unmeasured Process Quantities. *Journal of Process Control* **2008**, *18*, 491.

(6) Jiang, T.; Chen, B.; He, X.; Stuart, P. Application of Steady-state Detection Method Based on Wavelet Transform. *Computers & Chemical Engineering* **2003**, *27*, 569.

(7) Yao, Y.; Zhao, C.; Gao, F. Batch-to-Batch Steady State Identification Based on Variable Correlation and Mahalanobis Distance. *Industrial & Engineering Chemistry Research* **2009**, *48*, 11060.

(8) Le Roux, G. A. C.; Santoro, B. F.; Sotelo, F. F.; Teissier, M.; Joulia, X. Improving Steady-state Identification. *Proceedings of the 18th European Symposium on Computer Aided Process Engineering – ESCAPE 18* **2008**, *25*, 459.

(9) Tao, L.; Li, C.; Kong, X.; Qian, F. Steady-state Identification with Gross Errors for Industrial Process Units. *Proceedings of the 10th World Congress on Intelligent Control and Automation* **2012**, 4151.

(10) Pannocchia, G.; Rawlings, J. B. Disturbance Models for Offset-free Model-predictive Control. *AIChE Journal* **2003**, *49*, 426.

Modeling of an Industrial Diesel Hydro-processing Plant by the Continuous Lumping Approach

Ayşe Dilan Çelebi[a], Erdal Aydın[a], Hasan Şıldır[a], Yaman Arkun[a,*]

Ümmühan Canan[b], Onur Kartal[b], Fatih Demir[b], Murat Erdoğan[c]

[a]*Department of Chemical and Biological Engineering, Koc University, Rumeli Feneri Yolu, 34450 Sariyer, Istanbul, TURKEY*

[b]*TUPRAS R&D Department, Kocaeli, TURKEY*

[c]*TUPRAS Kirikkale Refinery, Kirikkale, TURKEY*

*Corresponding author: yarkun@ku.edu.tr; Tel: +90-212 338 1313, Fax: +90-212 338 1548

In refineries, sulfur compounds are undesirable as they poison catalysts and cause corrosion problems in refining equipments. Sulfur emissions also pose environmental and health problems. Increasing demand for ultra-low sulfur diesel calls for improved design and operation of hydro-processing plants.

In this study the industrial hydro-processing plant consists of catalytic hydro-desulfurization (HDS) and hydro-cracking (HC) reactors as shown in Fig.1. The feed to the reactors consists of a blend of diesel and light vacuum gas oil. HDS reactions occur at high temperatures in the presence of hydrogen to produce higher-value hydrocarbon products with low sulfur levels. Reactions are exothermic due to hydrogenation. In the first two HDS beds, the organic sulfur compounds are converted to hydrogen sulfide. Hydrocracking (HC) occurs in the last two beds where heavier hydrocarbons are cracked to lower molecular weight petroleum fractions. Although cracking reactions are endothermic, the overall hydro-cracking process is exothermic since hydrogenation is highly exothermic. Inter-stage cooling by quench hydrogen is used in both reactors to control the bed temperatures. Reactor effluent is next separated into useful end products such as naphta and diesel. The main operational objective is to maximize the diesel production and keep its sulfur content below its specification (in this case 5 ppm) by running the reactor under optimal operating conditions.

We developed non-isothermal, pseudo-homogeneous plug flow steady-state and dynamic models for catalytic hydro-desulfurization and hydro-cracking reactors by using the continuous lumping approach. Continuous lumping considers the complex reactive mixture as a continuum and uses an intrinsic property to characterize its properties [1]. In this work, true boiling point (TBP) was used to characterize the reaction mixture and the model consists of a set of integro-differential and algebraic equations. The distribution of sulfur in the feed and the critical desulfurization and cracking rate constants are modeled as continuous function of the true boiling point as well. The resulting kinetic and other model parameters were estimated by using real plant data. Model predictions of the reactor bed temperatures, product distribution (naphta and diesel) and the sulfur amounts are in excellent agreement with real plant data. The developed models are planned to be used for optimization and control purposes.

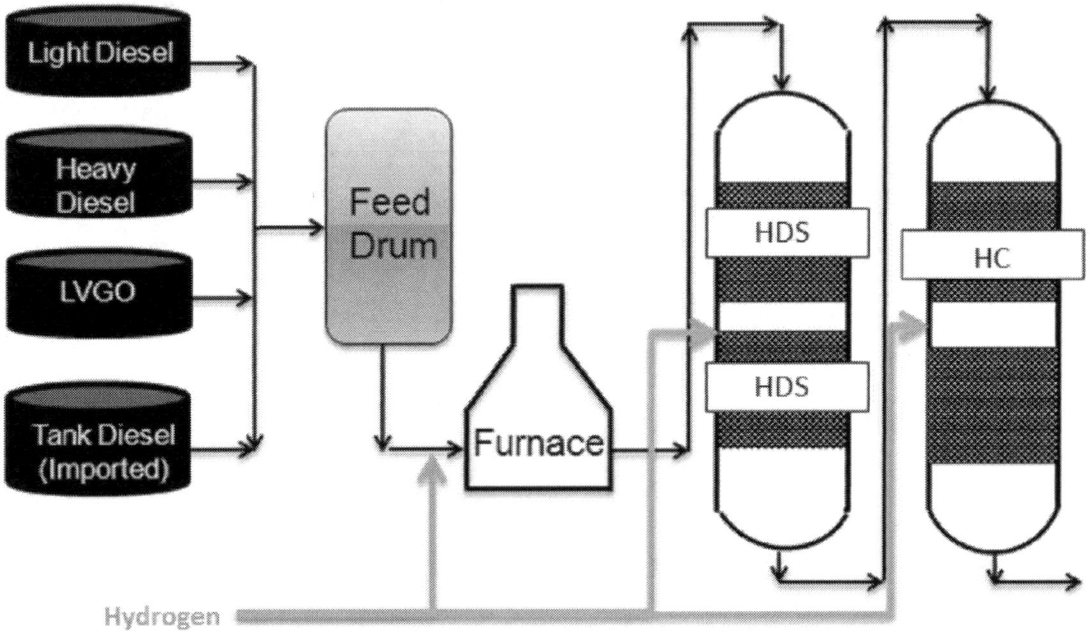

Fig.1. Hydro-processing Plant.

[1] Sildir, H., Y. Arkun, B. Cakal, D. Gokce & E. Kuzu (2012) A dynamic non-isothermal model for a hydrocracking reactor: Model development by the method of continuous lumping and application to an industrial unit. Journal of Process Control, 22, 1956-1965.

Keywords: Hydrodesulfurization, hydrocracking, continuous lumping, reactor modeling, parameter estimation.

Reactor Modeling of Gasoil Systems with Graph-Based Databases

AIChE Abstract
Jorge Martinis and Charles Solvason

Sessions:

- 10B03 Process Modeling and Identification

In the petroleum refining industry, the accurate design and optimization of hydrocarbon conversion processes for gasoil feedstocks demand a growing effort in developing rigorous and reliable kinetic models (Van Geem et al. 2007). Existing techniques rely on the approximation of gasoil stream compositions using a set of pseudo-components called "oil cuts" characterized by a set of lumped kinetic models. Although the lumped models are easy to construct and implement, they lack the necessary fundamental molecular information needed to model the kinetics of the process at the elementary step level.

To address this issue, Bryan Research & Engineering (BR&E) has developed a Process Reactors Suite (PRS) for the process simulator, ProMax® that when provided a set of user selected species, is capable of automatically generating all reactions sets and their corresponding rate expressions via the application of strict chemical rules, Lagrangian maximization functions, and optimization algorithms. At its basic level, PRS relies on molecular reconstruction techniques that transform characterization data into molecular structures suitable for reaction kinetics modeling. Unlike most molecular reconstruction approaches that utilize computationally intensive stochastic optimizations at runtime (Van Geem et al. 2007), PRS uses a deterministic molecular reconstruction approach to create a molecular database network that contains all chemical constituents important to gasoil products and streams prior to runtime. When needed, the database is called and returns the species requested. The effectiveness of the method is dependent on two competing factors, the completeness of the database and size of the database. Since most gasoil reconstructions require representations of molecular structures of up to 40 atoms in a hydrogen suppressed, molecular graph, enumerating all chemical constituents would strain the capability of state-of-the-art graph generation algorithms (Peironcely et al. 2012). Furthermore, it would result in a database too large to be efficiently searched. The PRS addresses these concerns in two ways. First, a hybrid method is proposed that limits the size of the database using a set of constraints provided by molecular group, point group, and graph theories without compromising its completeness. Second, levels of resolution are introduced such that compound species (e.g. C8 aromatics) are manipulated by the end user, while the chemical structures in the database (e.g. o-xylene, m-xylene, p-xylene, and ethylbenzene) are utilized in the reactor modeling. If desired, the user may toggle between the levels of resolution, but all kinetic modeling proceeds at the highest possible resolution.

This paper utilizes a case study on a diesel hydrotreater to illustrate the motivations and challenges in developing the graph-based molecular library and the Process Reactors Suite.

Van Geem, K.M., Hudebine D., Reyniers M.F., Wahl F., Verstraete J.J., Marin, G.B. (2007): "Molecular reconstruction of naphtha steam cracking feedstocks based on commercial indices." *Computers & Chemical Engineering,* 31, pp. 1020-1034.

Peironcely, J.E., Rojas-Chertó M., Fichera D., Reijmers T., Coulier L., Faulon, J. and Hankemeier T. (2012): "OMG: Open Molecule Generator." *Journal of Cheminformatics* 4(21), pp. 1-13.

Natural Gas

Liquefaction Models

Introduction

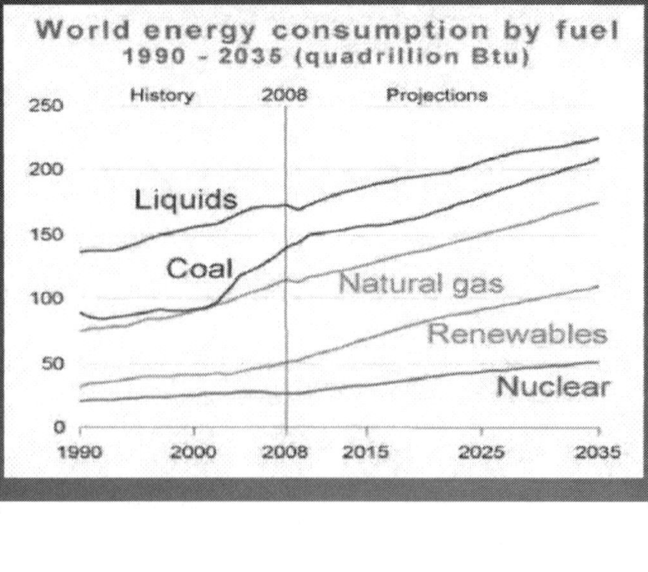

Cryogenic Carbon Capture (CCC)

- This process could used on coal power plants
- Cleans the CO_2 emissions from the flue gas
- Needs a steady flow of Liquefied Natural Gas (LNG)
- LNG is used as a refrigerant in the CCC Process

Cryogenic Carbon Capture

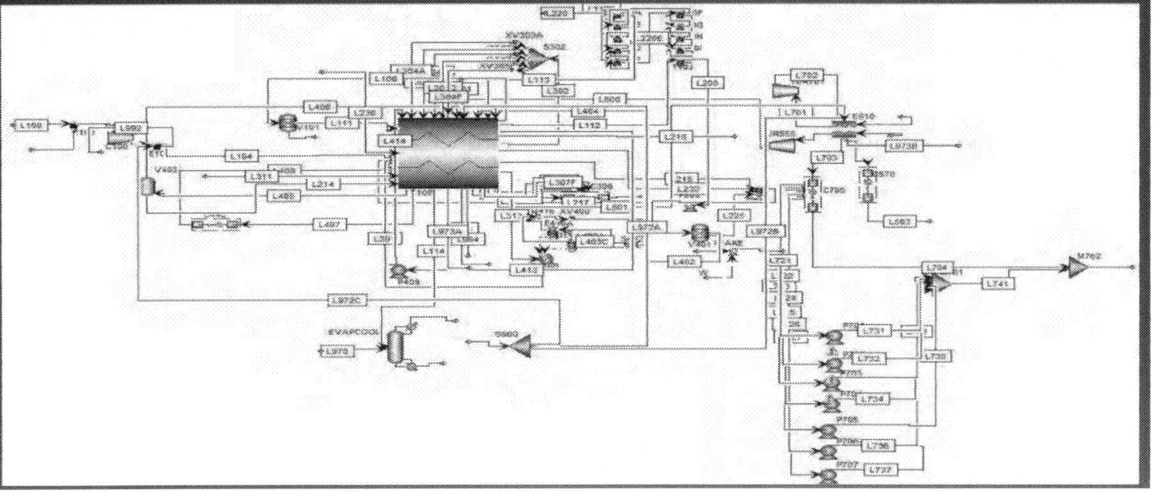

Cryogenic Carbon Capture

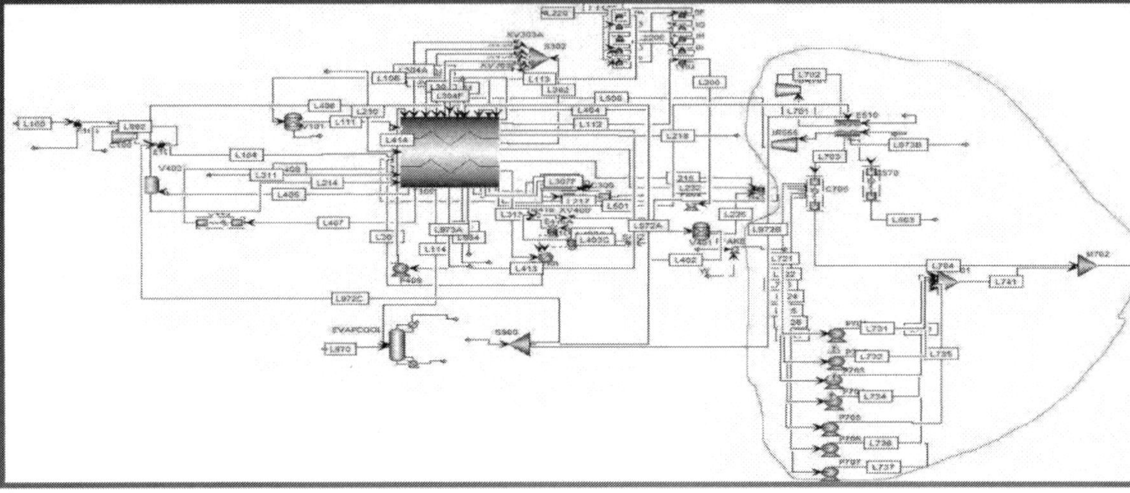

Power Plant Demand

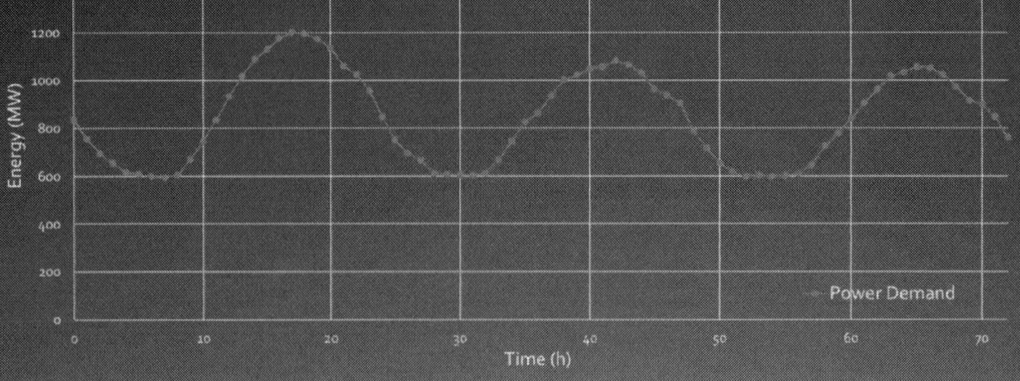

Power Plant Demand
Energy (MW)
1400
1200
1000
800
600
400
200
0
0
10
20
30
40
50
60
70
Time (h)
Power Demand
Coal Power Plant

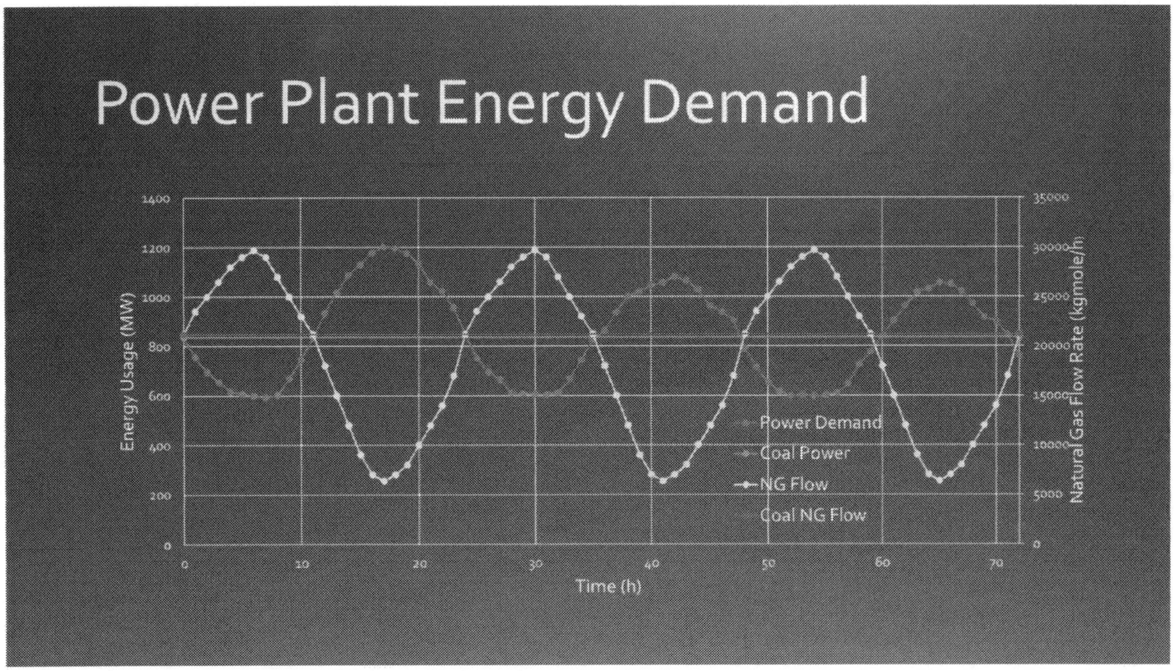
Power Plant Energy Demand
Energy Usage (MW)
1400
1200
1000
800
600
400
200
0
0
10
20
30
40
50
60
70
Time (h)
Natural Gas Flow Rate (kgmole/h)
35000
30000
25000
20000
15000
10000
5000
0
Power Demand
Coal Power
NG Flow
Coal NG Flow

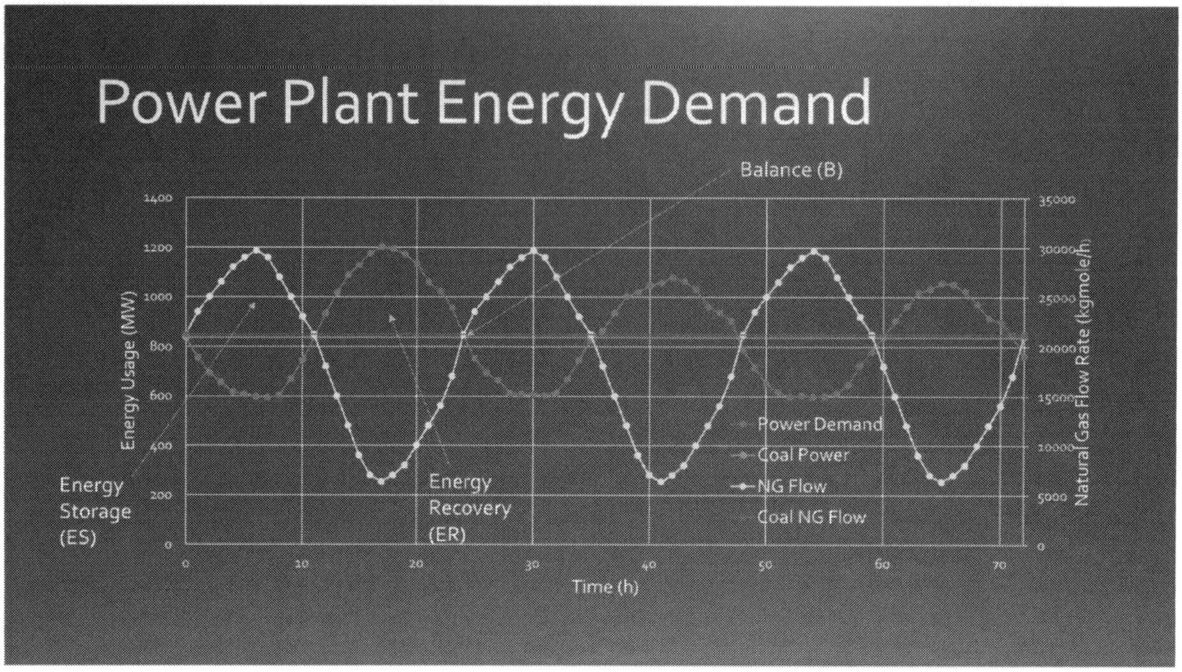
Power Plant Energy Demand
Balance (B)
Energy Usage (MW)
1400
1200
1000
800
600
400
200
0
Natural Gas Flow Rate (kgmole/h)
35000
30000
25000
20000
15000
10000
5000
0
0
10
20
30
40
50
60
70
Time (h)
Energy Storage (ES)
Energy Recovery (ER)
Power Demand
Coal Power
NG Flow
Coal NG Flow

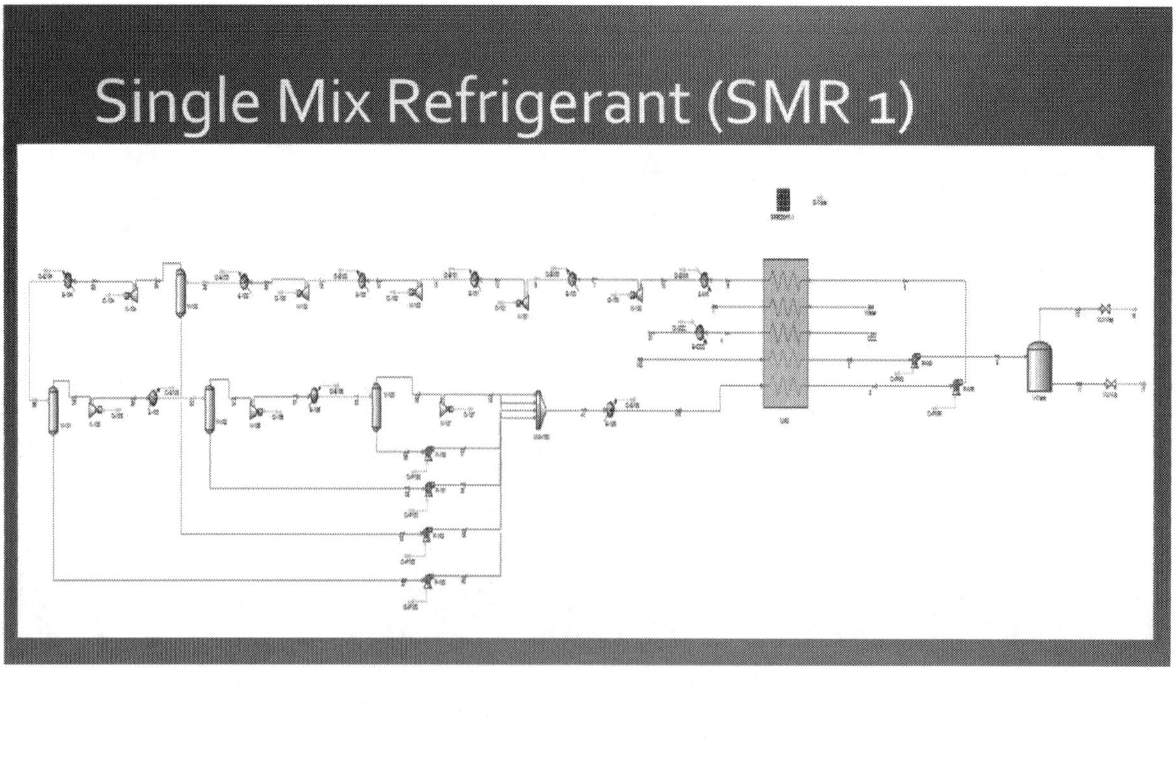
Single Mix Refrigerant (SMR 1)

SMR1 Temperature Profile

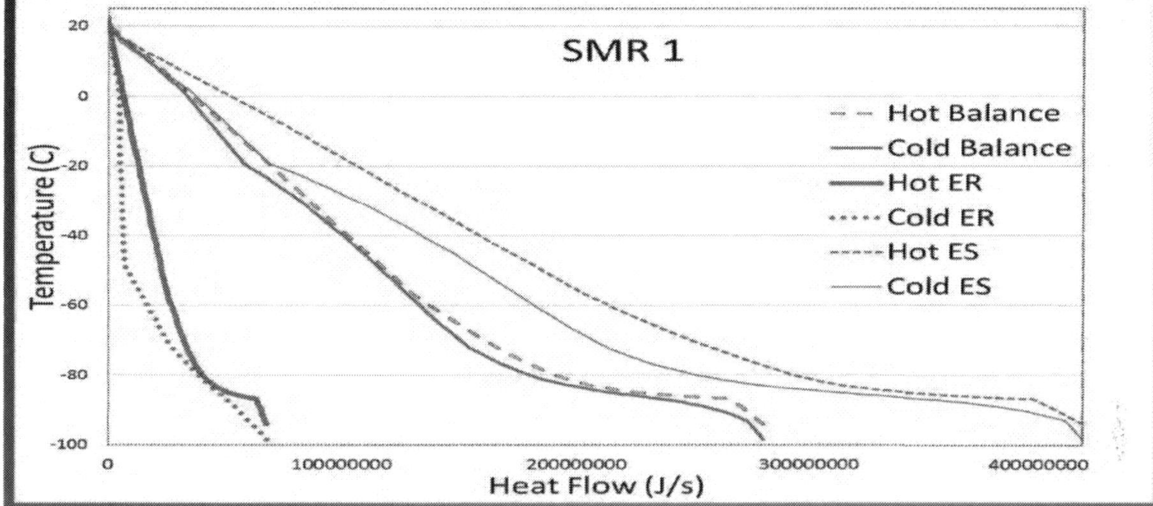

Single Mix Refrigerant (SMR 2)

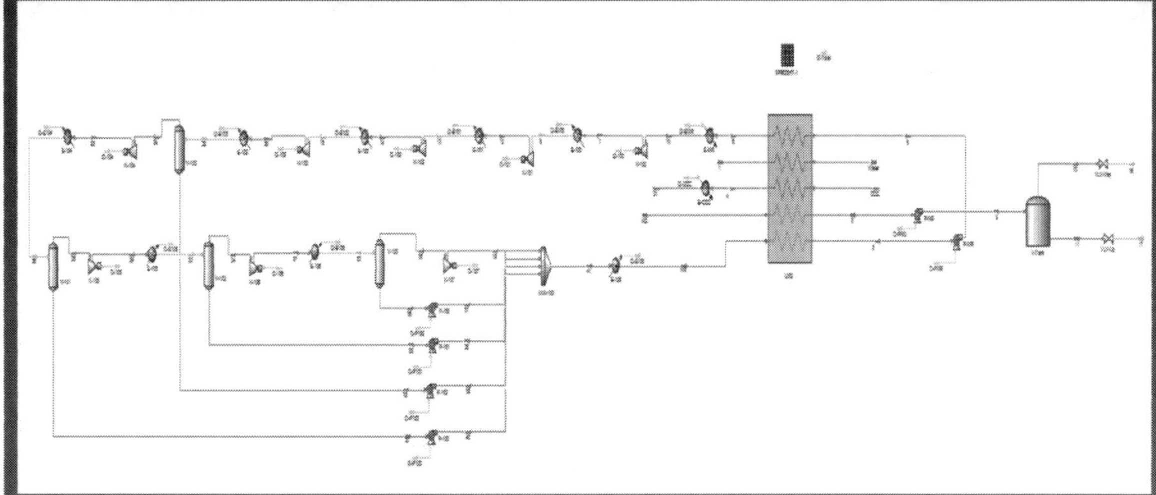

SMR 2 Temperature Profile

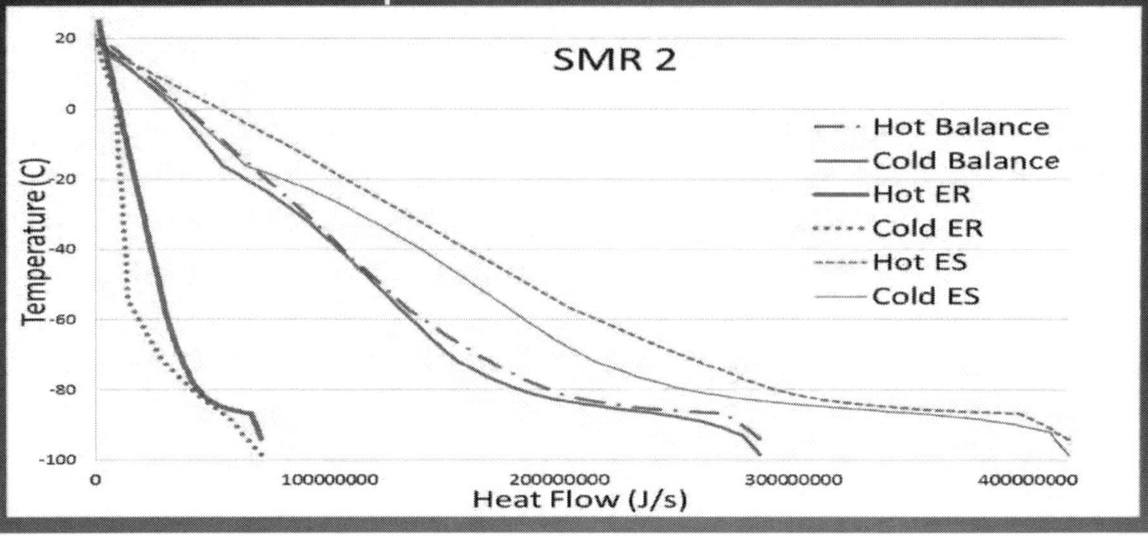

C3_MR

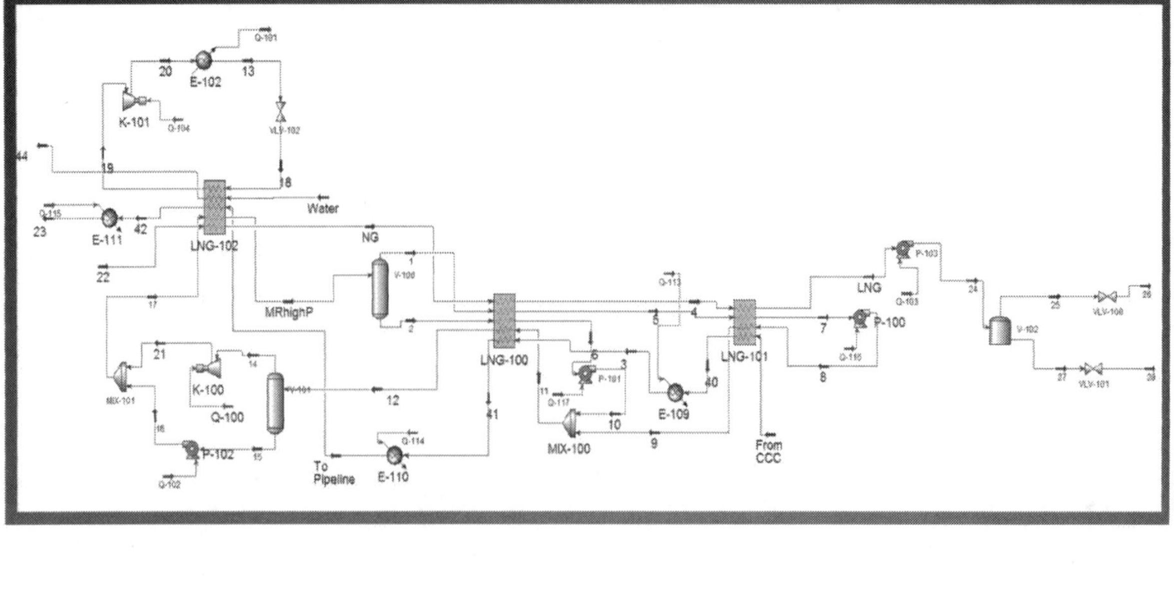

C3_MR Temperature Profile

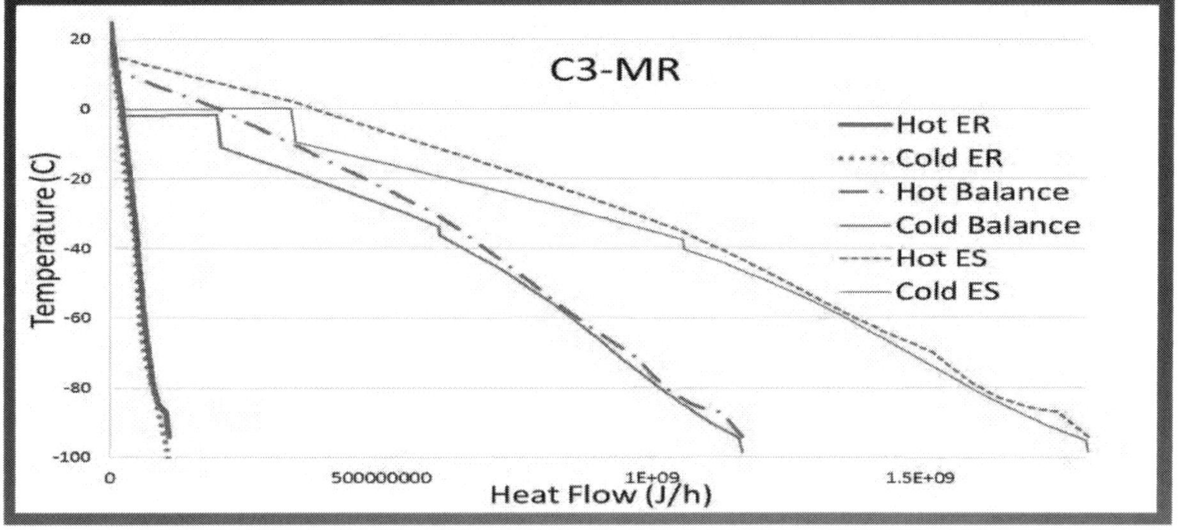

Modified Dual Mix Refrigerant (MDMR)

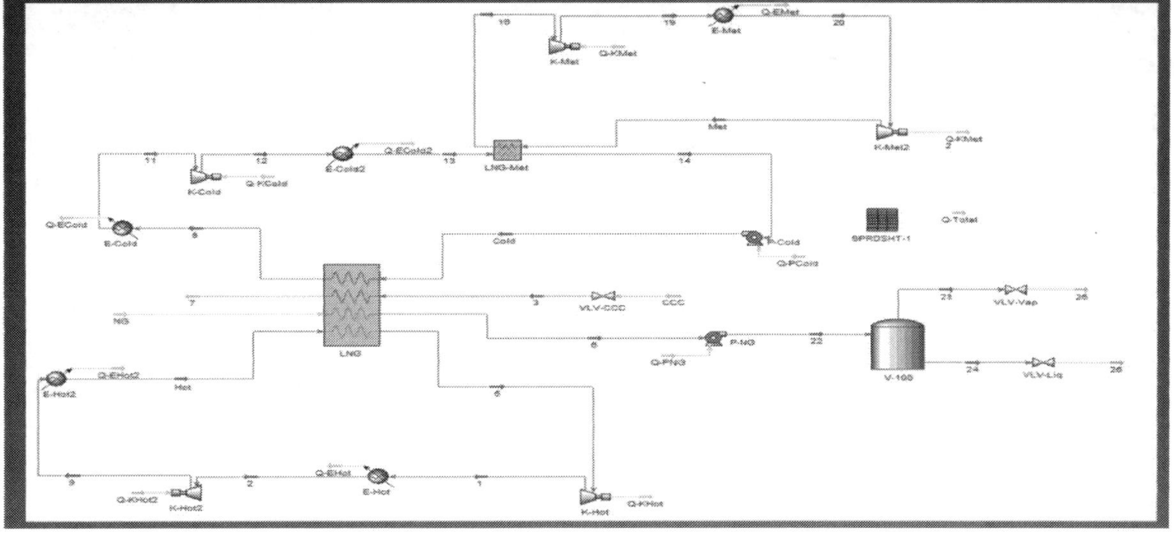

MDMR Temperature Profile

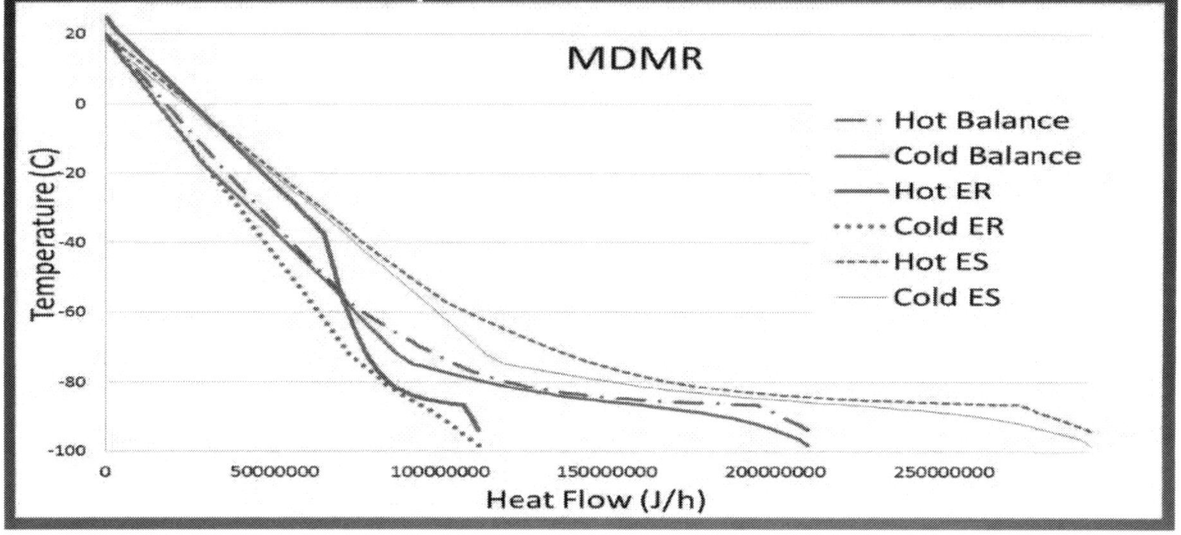

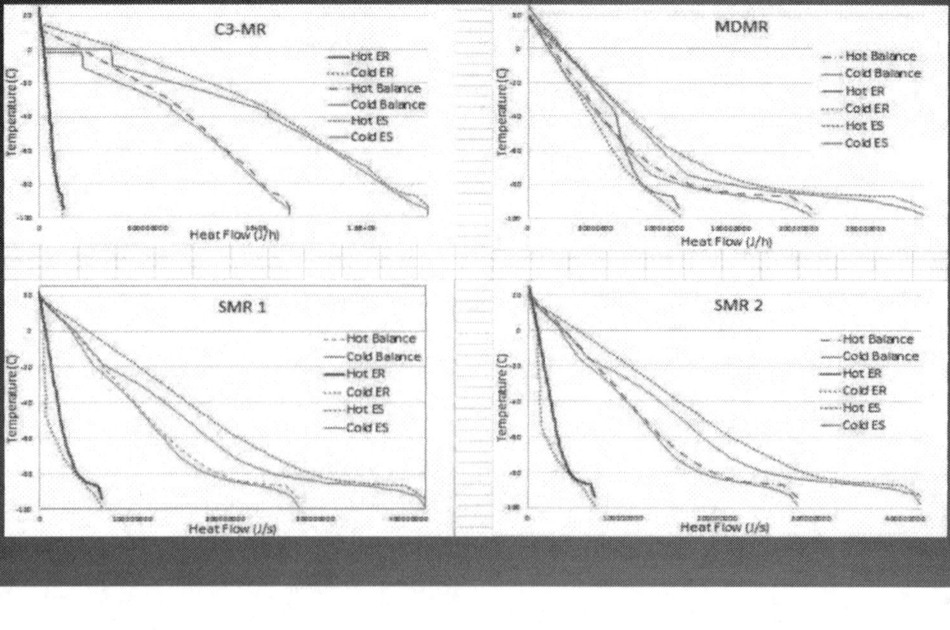

Energy Consumption

- Total energy consumption
- $\text{Wnet} = \sum Wcompressor - \sum Wexpander$

Energy Consumption (MW)

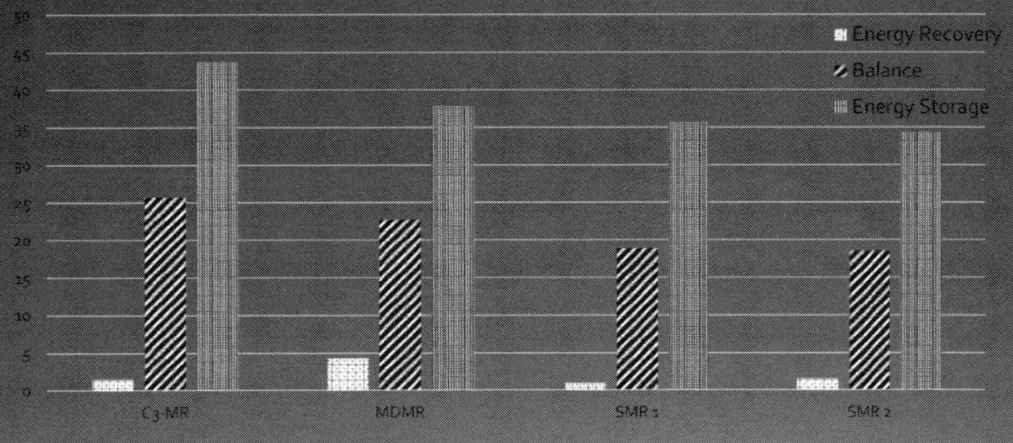

Capital Cost

- Used Aspen HYSYS Economic analysis.
- Capcost (Excel)

Capital Cost (USD $)

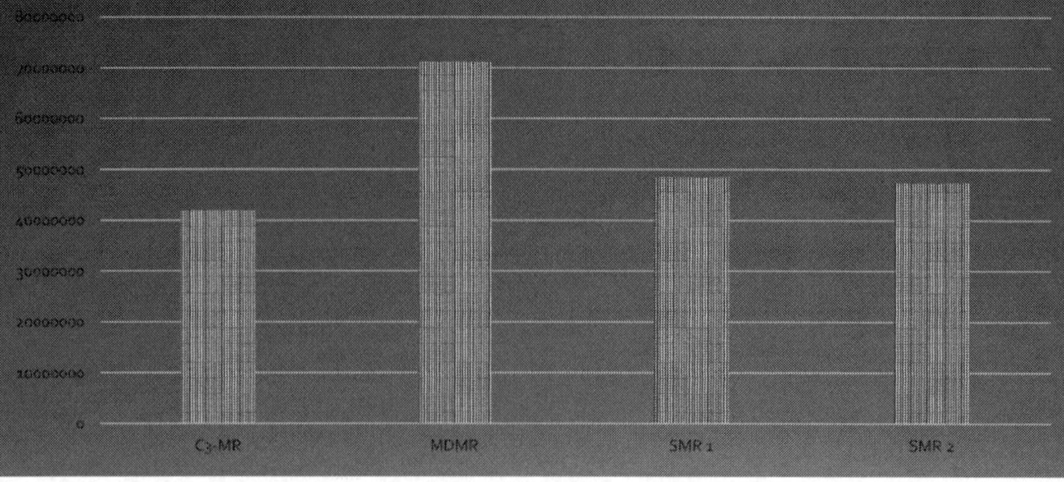

Exergy Loss

Equipment	Exergy loss equation
Compressor	$\Delta E_x = E_{in} - E_{out} + W_c$
Expander	$\Delta E_x = E_{in} - E_{out} - W_c$
LNG heat exchanger	$\Delta E_x = \sum E_{in} - \sum E_{out}$
Water cooler	$\Delta E_x = E_{in} - E_{out}$
Valve	$\Delta E_x = T_0 (S_{out} - S_{in})$

Exergy Loss (MW)

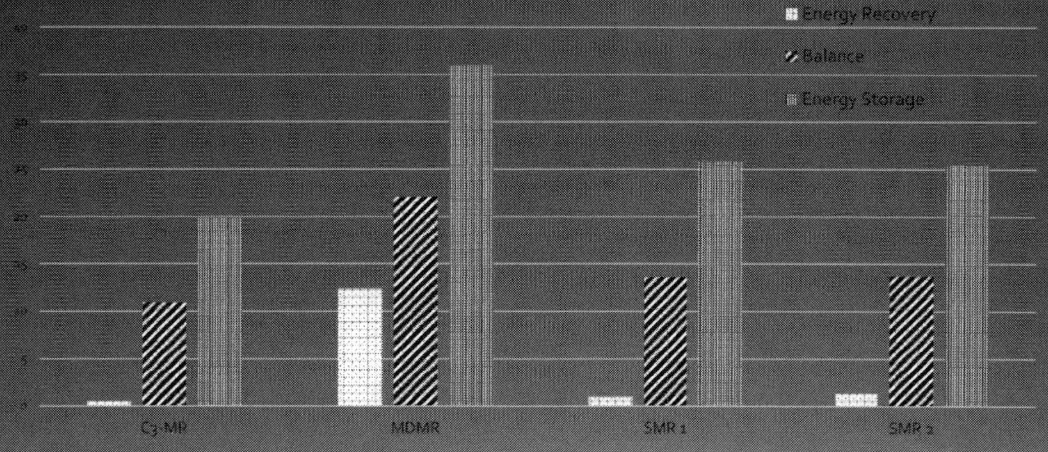

Heat Exchanger Efficiency

- $\Delta S_{mix} = \sum S_{out} - \sum S_{in}$
- $\Delta S = \sum S_{out} - \sum S_{in}$
- $\eta = \frac{(\Delta S_{mix} - \Delta S)}{\Delta S_{mix}}$
- S = Entropy (kJ/h)

Heat Exchanger Efficiency

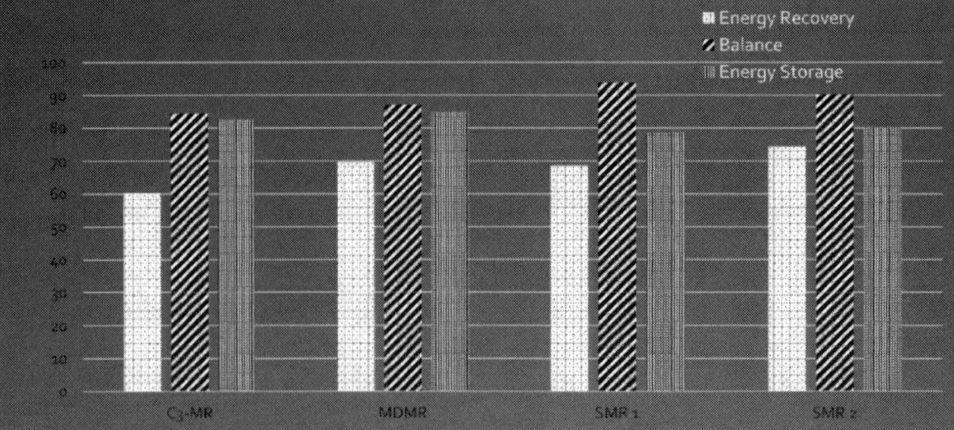

Dynamic Model

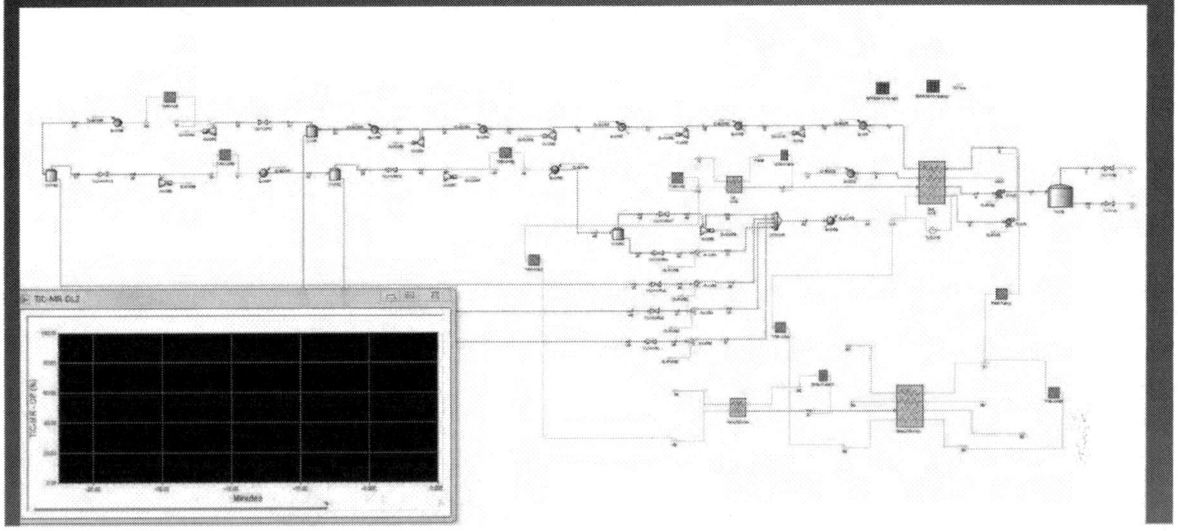

LNG Heat Exchanger Efficiency

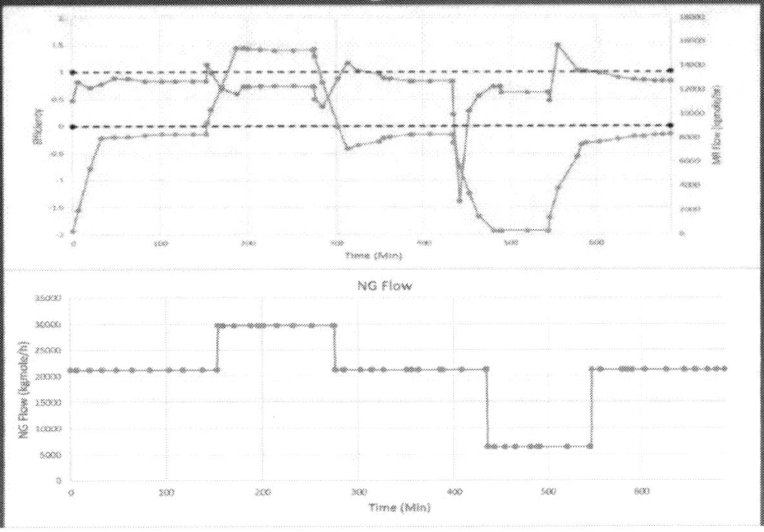

Conclusion

- CCC process use to clean CO_2 out of flue gas
- LNG process follows the power demand curve
- All four models have their strengths

389105 Global Optimality Properties of Total Annualized and Operating Cost Problems for Compressor Sequences

Abstract:

In this work, the minimum total annualized cost problem is studied for a series of non-isentropic compressors and coolers that brings a gas with constant compressibility factor from a specified initial pressure and temperature to a specified final pressure and the same temperature. It is established analytically that at the global optimum, the cooler outlet temperatures are equal to the minimum allowable temperature. For constant heat capacity, constant compressibility factor gases, additional properties of the globally optimal compressor sequence are analytically established for the minimum operating cost case. The aforementioned properties permit development of a solution strategy that identifies the globally-minimum operating cost. Several case studies are presented to illustrate the developed theorems and solution strategies.

1964 2014 50th ANNIVERSARY UCDAVIS CHEMICAL ENGINEERING AND MATERIALS SCIENCE

Proactive Optimization-Based Reconfiguration of Heat-Exchanger Super Networks

Xiaonan Wang, Ahmet Palazoglu and Nael H. El-Farra
xnwang@ucdavis.edu

2014 AIChE Annual Meeting
Atlanta, GA
November 20, 2014

UCDAVIS

Outline

Introduction

Heat-Exchanger Networks/ Proactive Reconfiguration

Problem Statement

- Framework
- Formulation

Demonstrative Example

- Supervisory Optimization
- Dynamic Control

Summary and Future Work

2

Introduction

Heat-Exchanger Network (HEN)

- ➢ Utilizing internal flow streams to satisfy outlet temperature targets
- ➢ Minimize operating cost

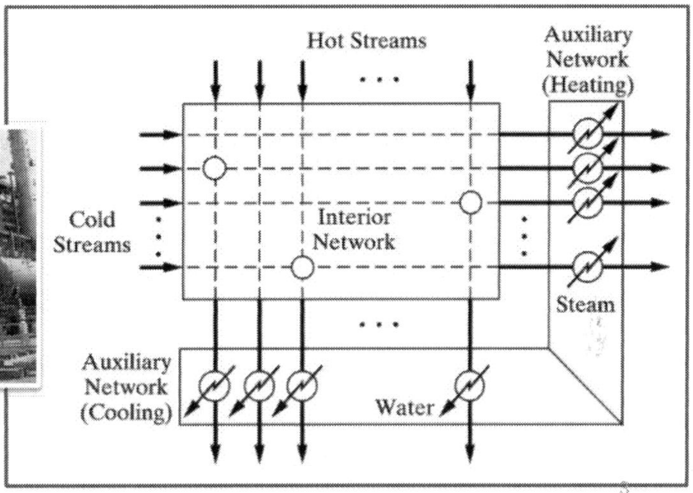

3

Seider, Warren D., Junior D. Seader, and Daniel R. Lewin. PRODUCT & PROCESS DESIGN PRINCIPLES: SYNTHESIS, ANALYSIS AND EVALUATION. John Wiley & Sons, 2009.

Introduction

Emerging energy challenges

High renewable penetration

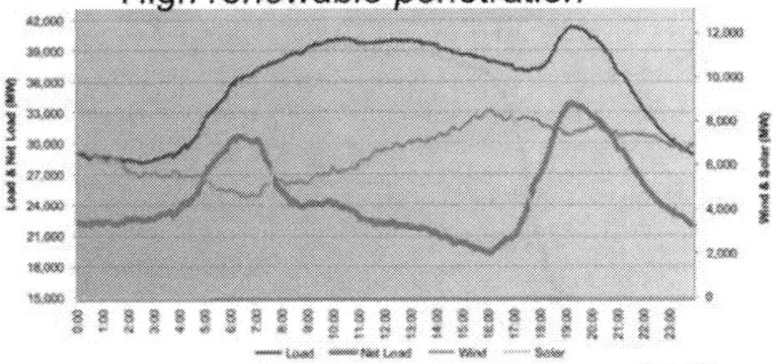

Projected load and renewable profiles in April 2020

33% | 2020 | CA

Industrial plant powered by renewable cogeneration
- ✓ Deal with Variations and Uncertainties

- ✓ Improve economic performance

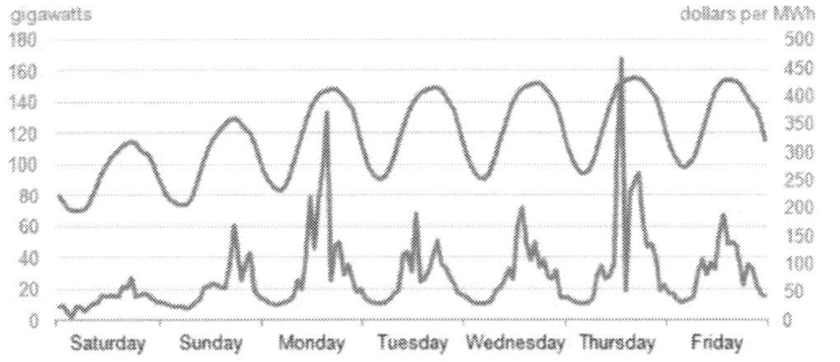

Hourly real-time energy prices and electricity demand in July, 2013

4

Operating flexibility analysis for R.12-03-014, Mark Rothleder, Shucheng Liu, and Clyde Loutan, CPUC workshop, June 4, 2012.
U.S. Energy Information Administration based on PJM data.

Problem Statement

HEN Synthesis & Design

- Process energy demand
- Recovery & Utilities
- Trade-off: Capital vs Operating cost

Proactive HEN Reconfiguration

- RT Optimization to minimize cost
- Dynamic transition is realized
- Benefits-Penalties

HEN Operation & Control

- Existing HEN
- Flexible control
- Energy efficiency

Research groups such as Floudas; Grossmann; CANMET Energy Technology Centre and others

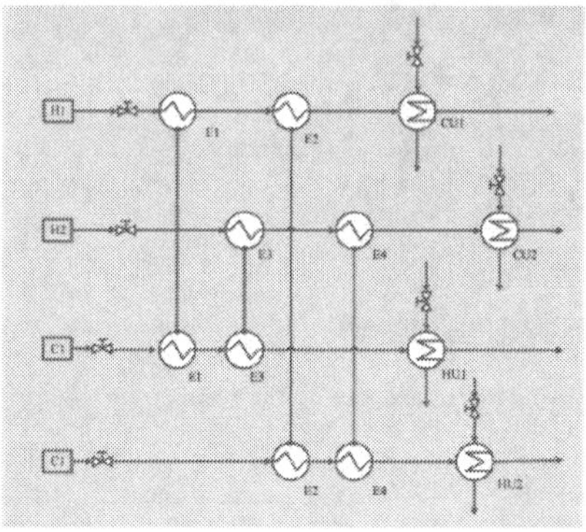

Research groups such as Marchetti; Gundersen; Skogestad and others

5

Problem Statement

HEN Synthesis & Design

- Process energy demand
- Recovery & Utilities
- Trade-off: Capital vs Operating cost

Proactive HEN Reconfiguration

- RT Optimization to minimize cost
- Dynamic transition is realized
- Benefits-Penalties

HEN Operation & Control

- Existing HEN
- Flexible control
- Energy efficiency

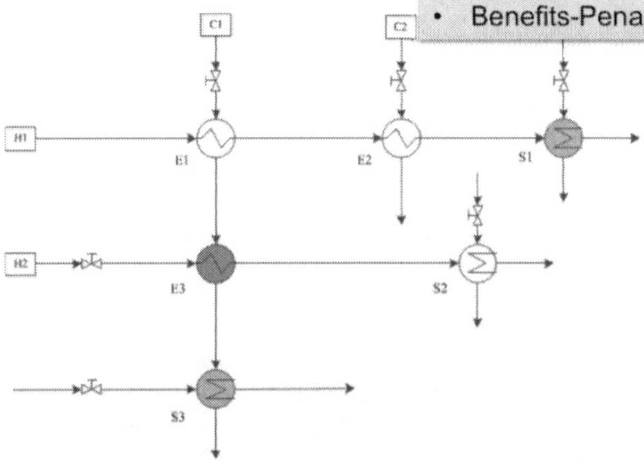

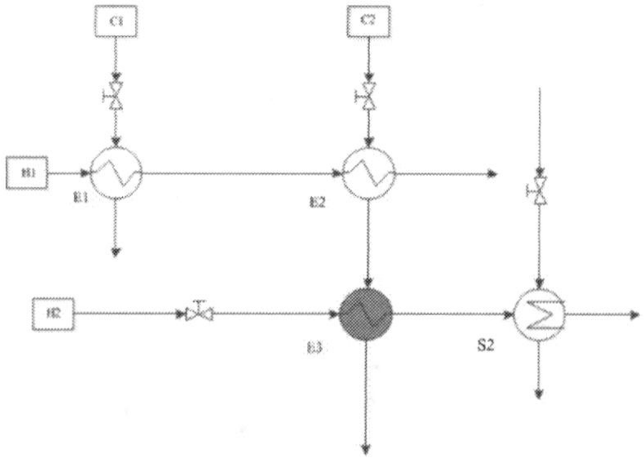

Overview of Methodology

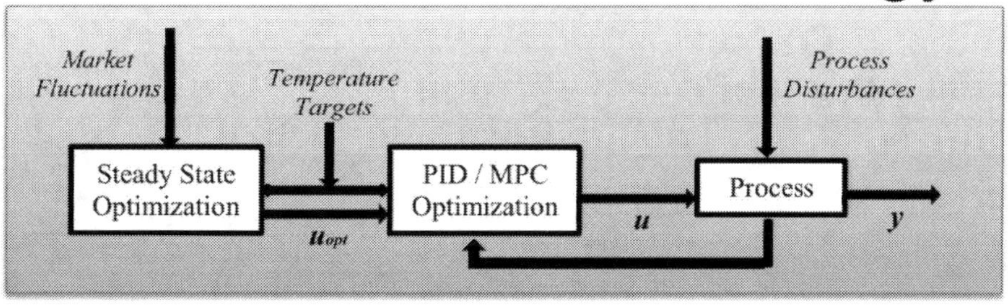

Steady-state optimization decides real-time operation rules under the market and energy resource fluctuations

Local controllers realize the optimal operation rules and transitions to satisfy temperature targets

Explicit process model demonstrates dynamic behavior

7

Problem Formulation

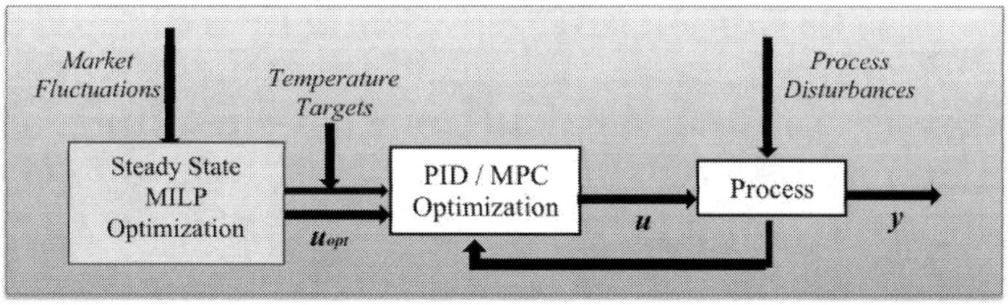

$$\min J = \alpha\left(\sum_{i=1}^{N_H} C_{CU} C_i + \sum_{j=1}^{N_C} C_{HU} H_j\right) - \beta\left(C_{ele\_sell} P_{sell}\right) + \lambda\left(\sum_{k=1}^{N_E} C_{HE} Q_k + \sum_{i=1}^{N_H} C_{HEh} C_i + \sum_{j=1}^{N_C} C_{HEc} H_j\right)$$

Utility Cost | Profit from Sold Electricity | Discounted Capital Investment of HEs

$$C_{HU} = \begin{cases} C_{re}, & when \sum_{j=1}^{N_C} H_j \le P_{renewable}\eta_{re} \\ C_{ele\_buy}/\eta_{ele}, & when \sum_{i=1}^{N_H} H_j \ge P_{renewable}\eta_{re} \end{cases}$$

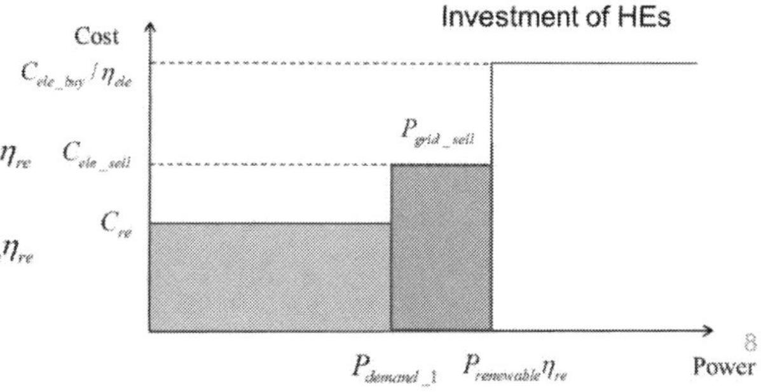

8

Problem Formulation

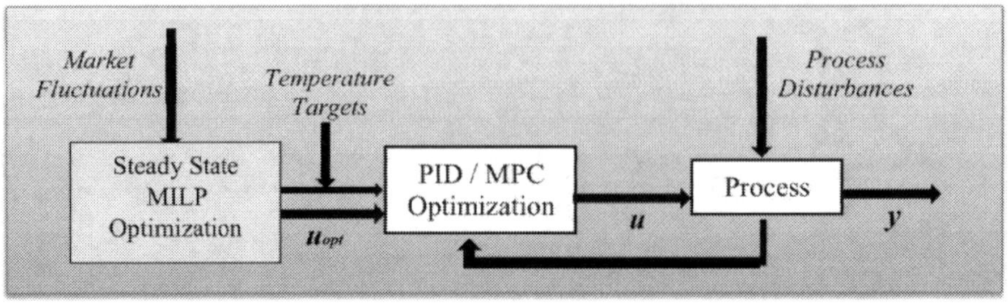

$$\min J = \alpha\left(\sum_{i=1}^{N_H} C_{CU} C_i + \sum_{j=1}^{N_C} C_{HU} H_j\right) - \beta\left(C_{ele\_sell} P_{sell}\right) + \lambda\left(\sum_{k=1}^{N_E} C_{HE} Q_k + \sum_{i=1}^{N_H} C_{HEh} C_i + \sum_{j=1}^{N_C} C_{HEc} H_j\right)$$

Utility Cost — Profit from Sold Electricity — Discounted Capital Investment of HEs

$$C_{HU} = \begin{cases} C_{re}, & when \sum_{j=1}^{N_C} H_j \le P_{renewable}\eta_{re} \\ C_{ele\_buy}/\eta_{ele}, & when \sum_{i=1}^{N_H} H_j \ge P_{renewable}\eta_{re} \end{cases}$$

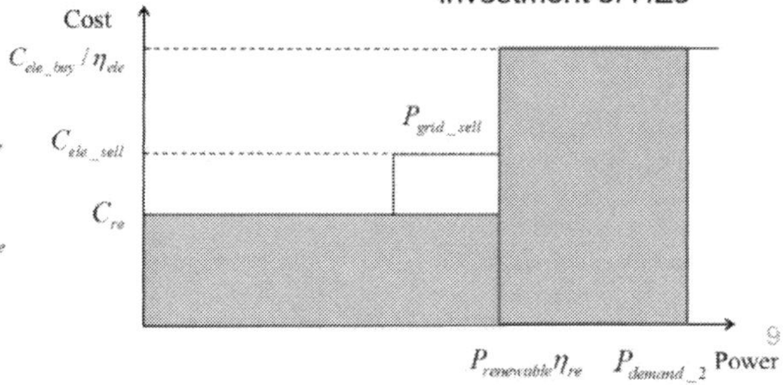

9

MILP Formulation

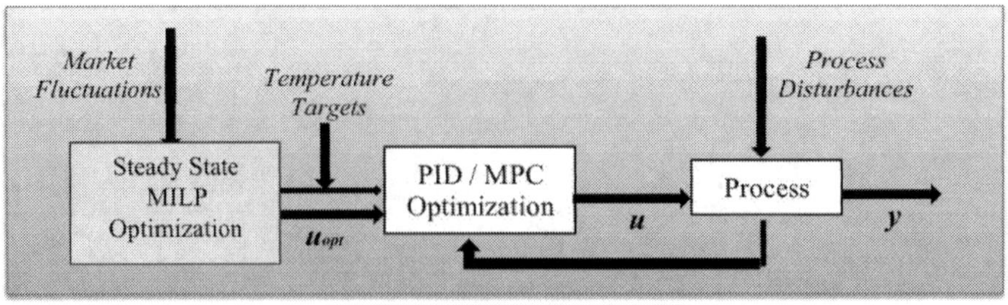

Heat/ Energy balance

Overall for each stream

$$(mC_p)_i(T_i^{in} - T_i^{out}) = \sum_{j=1}^{N_C} Q_{i,j,1} + \sum_{j=1}^{N_C} Q_{i,j,2} \dots + \sum_{j=1}^{N_C} Q_{i,j,Ns} + C_i, i \in N_H$$

$$(mC_p)_j(T_j^{out} - T_j^{in}) = \sum_{i=1}^{N_H} Q_{i,j,1} + \sum_{i=1}^{N_H} Q_{i,j,2} \dots + \sum_{i=1}^{N_H} Q_{i,j,Ns} + H_j, j \in N_C$$

For each stream at each stage

$$for\ l = 1, 2, \dots, N_S$$

$$(mC_p)_i(T_{i,l}^{in} - T_{i,l}^{out}) = \sum_{j=1}^{N_C} Q_{i,j,l}, i \in N_H$$

$$(mC_p)_j(T_{j,l}^{out} - T_{j,l}^{in}) = \sum_{i=1}^{N_H} Q_{i,j,l}, j \in N_C$$

Utility matches

$$(mC_p)_i(T_i^{Ns} - T_i^{out}) = C_i, i \in N_H$$

$$(mC_p)_j(T_j^{out} - T_j^{Ns}) = H_j, j \in N_C$$

10

MILP Formulation

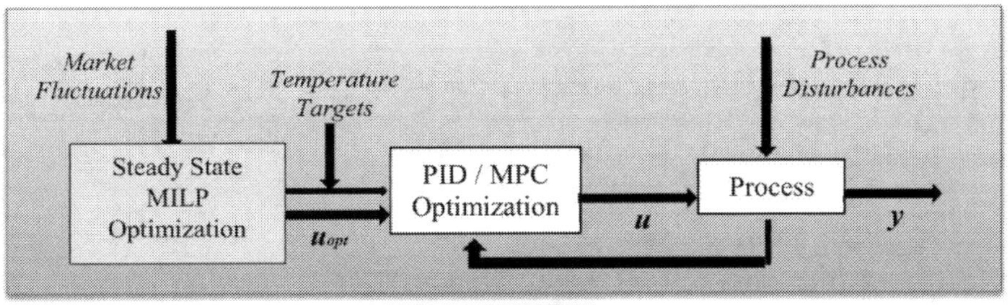

Temperature

Assignment of inlet temperatures

$$T_i^{out} = T_i^{target}, i \in N_H$$
$$T_j^{out} = T_j^{target}, j \in N_C$$

Targets of outlet temperatures

$$T_{i,0} = T_i^{in}, i \in N_H$$
$$T_{j,0} = T_j^{in}, j \in N_C$$

Feasibility of temperatures

$$T_i^{out} \le T_{i,Ns} \le \ldots \le T_{i,1} \le T_{i,0}, i \in N_H$$
$$T_j^{out} \ge T_{j,Ns} \ge \ldots \ge T_{j,1} \ge T_{j,0}, j \in N_C$$

Capacity

- Availability from all utility resources
- Maximum and Minimum heat exchanger capacity limits

$$0 \le Q_{i,j,l} \le Q_{i,j,l,\max}, i \in N_H, j \in N_C, l \in N_S$$

Logical constraints

- Matches that take place denoted by 0-1 binary variables

Determine the topology of the HEN according to active matches and heat duties with respect to each stage.

11

Control-based Reconfiguration

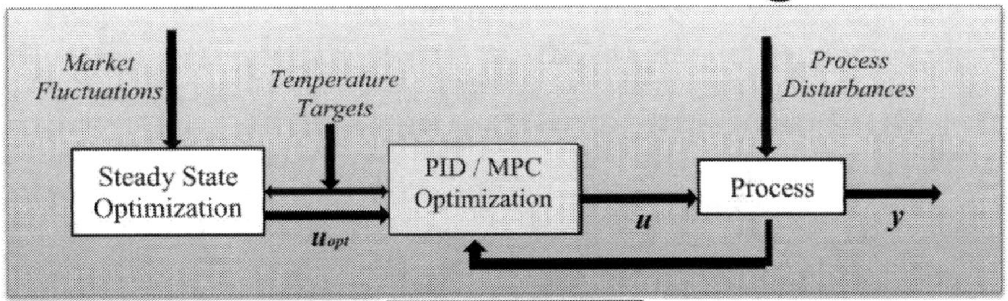

Dynamic Model

Proportional-Integral (PI) controllers

Internal Model Control (IMC) tuning (First-order lag plus deadtime model)

Control variables in an HEN:
(a) process stream bypasses around heat heat-exchangers
(b) utility stream flow rates in service units
(c) splits of process streams

12

Illustrative Example

2-HOT-2-COLD Process streams

Stream	mC_p (kW/K)	T_{in} (K)	T_{out} (K)	Enthalpy (kW)
H1	0.7	443	333	77
H2	1.0	423	313	110
C1	1.7	293	353	-102
C2	1.3	278	348	-91

Pinch point and first-law analysis

Pinch temperatures:
313K (H)
293K (C)
Minimum utility:
4.1 kW (H)
0 kW (C)

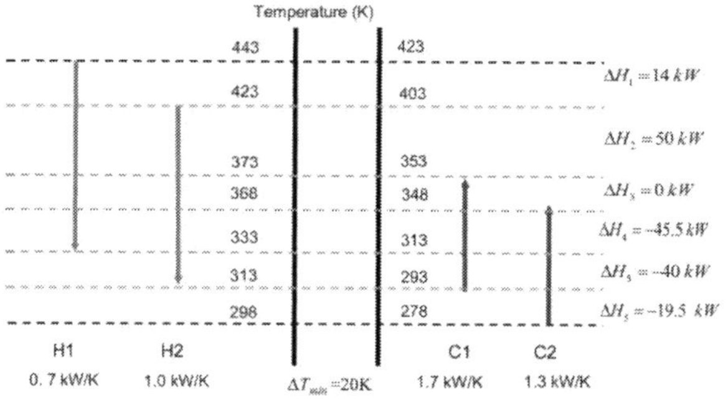

Energy Supply and Cost

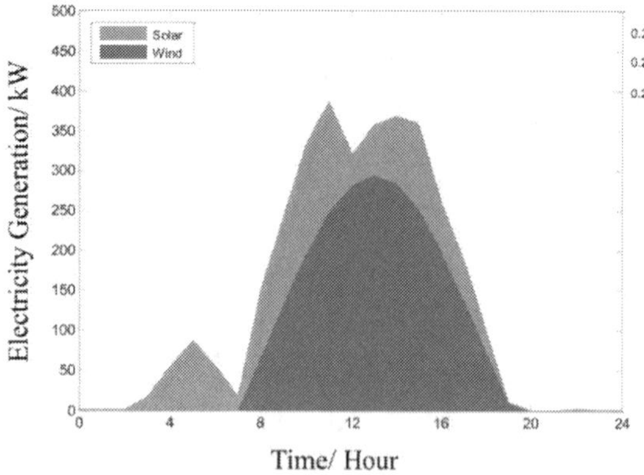

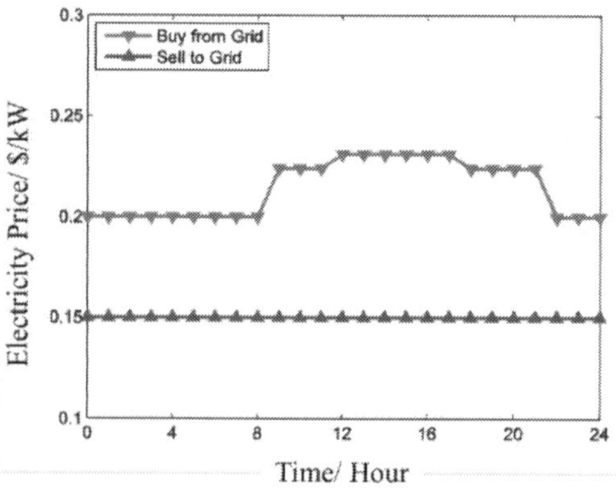

- Renewable generation over a typical summer day at Sacramento Valley.
- Real-time forecasting and update of available resources and electricity price.

- Time-varying electricity price to buy from and sell to the grid over a day.
- Compensate renewable generation.

Results

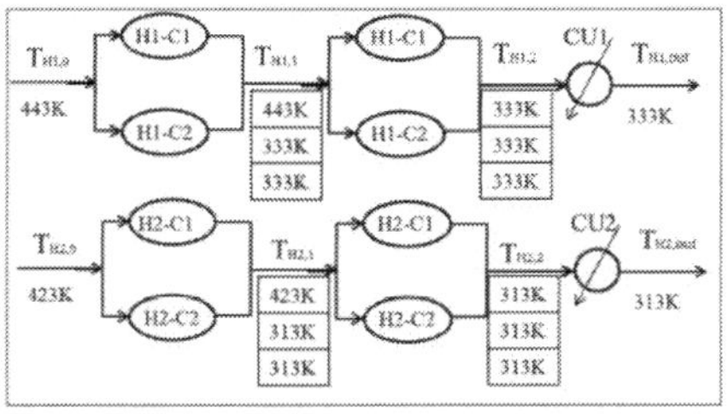

Temperature of HOT streams

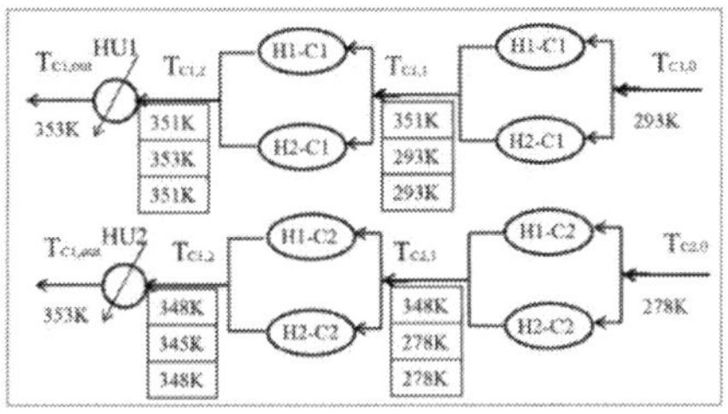

Temperature of COLD streams

Multiple-Stage ($Ns = 2$)

Heat duty (kW)	0:00am – 11:00am	11:00 am – 14:00pm	14:00pm – 17:00pm	17:00pm – 24:00pm
Q_H1_C1_1	0	78.8	78.8	0
Q_H1_C1_2	78.8	0	0	78.8
Q_H1_C2_1	0	0	0	0
Q_H1_C2_2	0	0	0	0
Q_H2_C1_1	0	23.4	20.5	0
Q_H2_C1_2	20.5	0	0	20.5
Q_H2_C2_1	0	88.5	91.5	0
Q_H2_C2_2	91.5	0	0	91.5
Q_H1_W	0	0	0	0
Q_H2_W	0	0	0	0
Q_C1_S	2.94	0	2.94	2.94
Q_C2_S	0	2.94	0	0

15

Results

Configuration 1 *(Off-peak hours)*

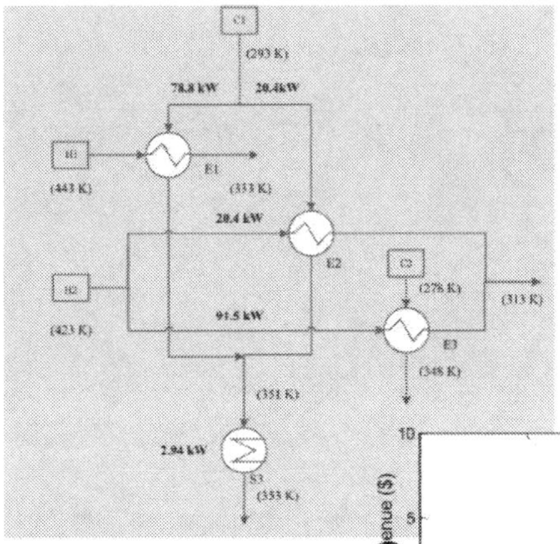

0:00 am – 11:00 am
14:00 pm – 17:00 pm
17:00 pm – 24:00 pm

Optimal real-time economic performance over a day.

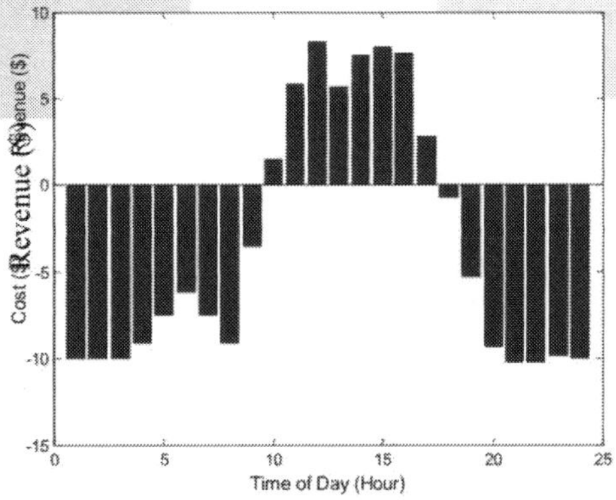

Configuration 2 *(Peak hours)*

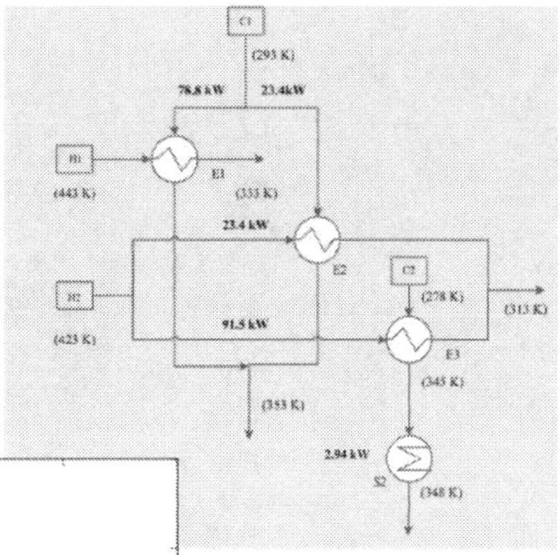

11:00 am – 14:00 pm

Selling electricity back to the grid brings a profit.

16

17

Dynamic Reconfiguration

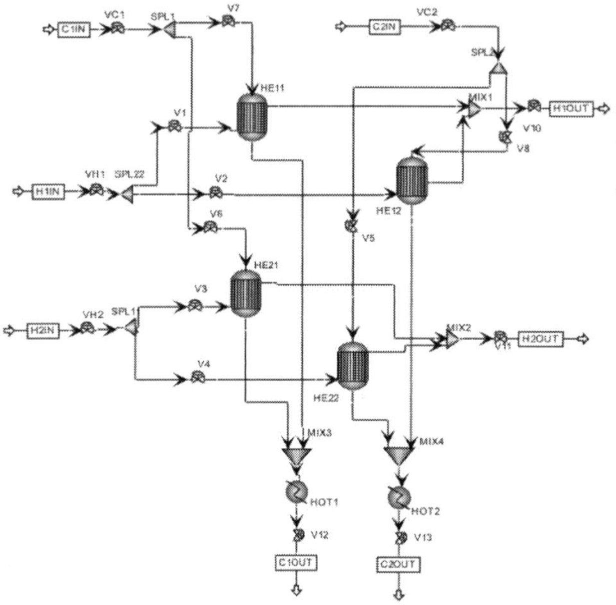

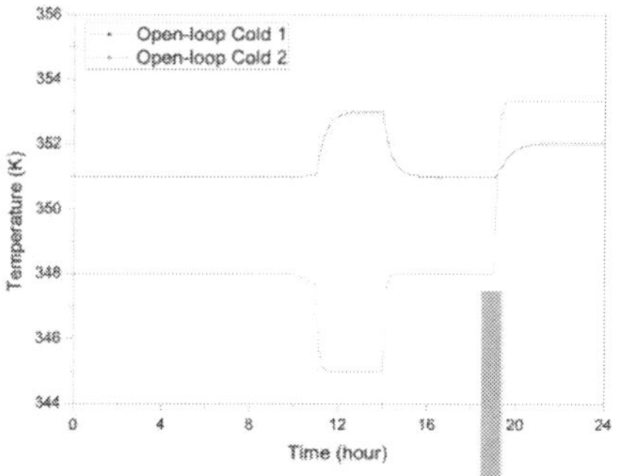

Process disturbance:
Sudden change of hot stream 2 inlet temperature

17

17

Dynamic Reconfiguration

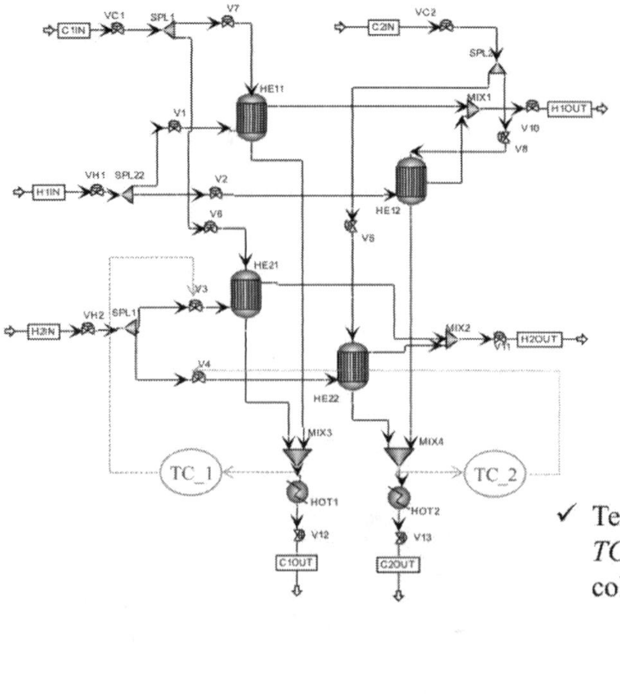

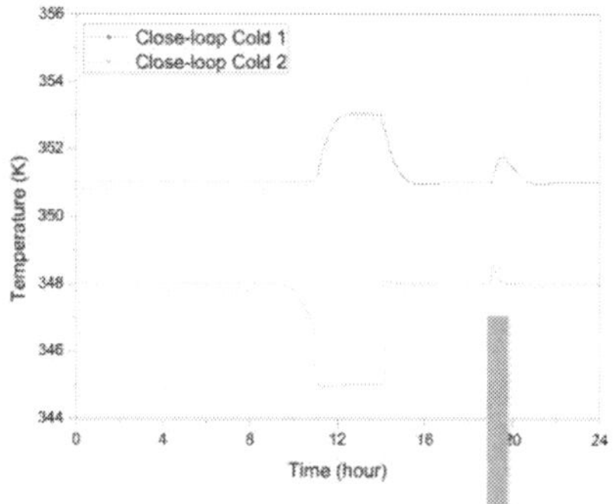

Process disturbance:
Sudden change of hot stream 2 inlet temperature

- ✓ Temperature controllers *TC_1* and *TC_2* to drive the outlet temperature of cold streams 1 and 2 to the targets.

18

Dynamic Reconfiguration

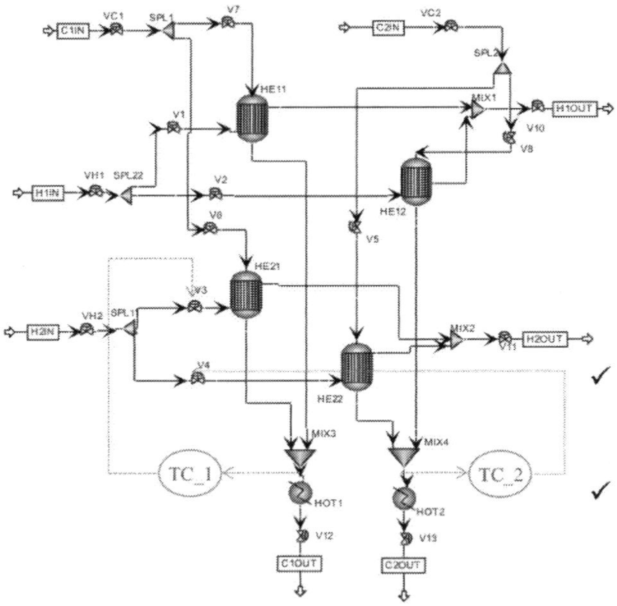

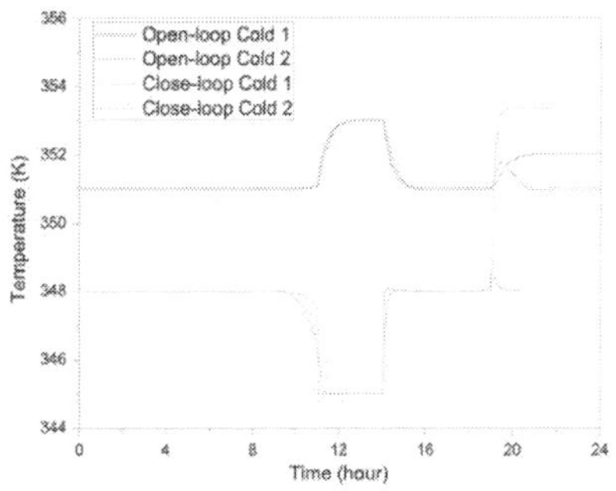

- ✓ Temperature controllers *TC_1* and *TC_2* to drive the outlet temperature of cold streams 1 and 2 to the targets.
- ✓ Settling time shortened compared with the open-loop system at the two switching times, and disturbances not preventing set-points tracking.

19

Conclusions

- **Summary**

- A proactive methodology to re-configure in real-time the optimal operation and control strategies of a HEN within a super structure.
- Steady-state optimization decides real-time operation rules under the varying renewable generation and electricity pricing.
- Local controllers orchestrate the required structural changes to realize the optimal strategy and achieve economic benefits.

- **Future work**

- Explicit equipment cost at design stage and transition cost in the dynamic operation can be incorporated into the cost function.
- Model predictive control can be used to link the supervisory optimization and structural reconfiguration

20

Thank you!

Any Questions?

xnwang@ucdavis.edu

Financial support from California Energy Commission, UC Davis CHMS Department, Phillips 66 Fellowship and are gratefully acknowledged.

21

Reference

Aguilera, N., & Marchetti, J. L. (1998). Optimizing and controlling the operation of heat-exchanger networks. *AIChE journal*, 44(5), 1090-1104.
Ciric, A. R., & Floudas, C. A. (1991). Heat exchanger network synthesis without decomposition. *Computers & Chemical Engineering*, 15(6), 385-396.
Dipama, J., Teyssedou, A., & Sorin, M. (2008). Synthesis of heat exchanger networks using genetic algorithms. *Applied Thermal Engineering*, 28(14), 1763-1773.
Floudas, C. A. (1995). *Nonlinear and mixed-integer optimization: fundamentals and applications*. Oxford University Press, New York.
Glemmestad, B., Skogestad, S., & Gundersen, T. (1999). Optimal operation of heat exchanger networks. *Computers & chemical engineering*, 23(4), 509-522.
González A H, Odloak D, Marchetti J L. (2006). Predictive control applied to heat-exchanger networks. *Chemical Engineering and Processing: Process Intensification*, 45(8): 661-671.
Gorji-Bandpy, M., Yahyazadeh-Jelodar, H., & Khalili, M. (2011). Optimization of heat exchanger network. *Applied Thermal Engineering*, 31(5), 779-784.
Gundepsen, T., and Naess, L. (1988). The synthesis of cost optimal heat exchanger networks: an industrial review of the state of the art. *Computers & Chemical Engineering*, 12(6), 503-530.
Lersbamrungsuk, V., Srinophakun, T., Narasimhan, S., & Skogestad, S. (2008). Control structure design for optimal operation of heat exchanger networks. *AIChE Journal*, 54(1), 150-162.
Linnhoff, B., & Flower, J. R. (1978). Synthesis of heat exchanger networks: I. Systematic generation of energy optimal networks. *AIChE Journal*, 24(4), 633-642.
López-Maldonado, L. A., Ponce-Ortega, J. M., & Segovia-Hernández, J. G. (2011). Multiobjective synthesis of heat exchanger networks minimizing the total annual cost and the environmental impact. *Applied Thermal Engineering*, 31(6), 1099-1113.
Wang, X., Teichgraeber, H., Palazoglu, A., & El-Farra, N. H. (2014). An economic receding horizon optimization approach for energy management in the chlor-alkali process with hybrid renewable energy generation. *Journal of Process Control*, 24(8), 1318-1327.
Yee, T. F., & Grossmann, I. E. (1990). Simultaneous optimization models for heat integration—II. Heat exchanger network synthesis. *Computers & Chemical Engineering*, 14(10), 1165-1184.
Zamora, J. M., & Grossmann, I. E. (1998). A global MINLP optimization algorithm for the synthesis of heat exchanger networks with no stream splits. *Computers & Chemical Engineering*, 22(3), 367-384.

22

370166 Handling Preventive Sensor Maintenance in Economic Model Predictive Control of Nonlinear Chemical Processes

Abstract:

Traditionally, fault detection schemes are used to identify faulty sensors [1]-[2]. Essential to the objectives of next-generation advanced manufacturing is the development of tightly integrated components of process systems (i.e., process control, process economics, scheduling) which are proactive instead of reactive. One potential area of interest is to integrate process economics, control, and preventive maintenance. Preventive maintenance plays an important role in maintaining a high degree of reliability of a process system and thus, reducing the changes of a failure or fault. For example, as process sensors age and becomes sluggish (e.g., due to gradual damage), it becomes necessary to conduct preventive maintenance (e.g., recalibration, verification, or replacement) on sensors to avoid process faults or failures. The driving force behind integrating control and maintenance is to avoid the need for a process shutdown required to complete a preventive maintenance task. To accomplish this goal, the process control system should explicitly account for a preventive maintenance task which may take process control components (e.g., actuator or sensor) off-line for a period of time. In a previous work, we considered integrating preventive maintenance of control actuators, process control, and process economics through the design of an economic model predictive control (EMPC) scheme [3]. However, we did not consider handling sensor maintenance and thus, the consideration of sensor maintenance is an open problem.

This work focuses on developing a state-estimation-based Lyapunov-based economic model predictive control (LEMPC) method (e.g., [3]-[4]) that can handle preventive sensor maintenance. To complete the sensor maintenance task, the sensor is taken off-line until the maintenance task is completed. Since a measurement from the sensor is not available during this time, a certain observability assumption is imposed on the process with the remaining sensors. A state estimator is used to estimate the current state when the sensor is taken off-line and provide the state estimation to the LEMPC. For guaranteed closed-loop stability as well as to provide an accurate state estimate to the EMPC, a robust moving horizon estimator (RMHE) is used [4]. The approach is applied to a chemical process example under the proposed LEMPC to demonstrate that the LEMPC is able to maintain stability of the process when a sensor is taken off-line for preventive maintenance.

[1] Qin SJ, Li W. Detection and identification of faulty sensors in dynamic processes. AIChE Journal. 2001;47:1581-1593.
[2] Mehranbod N and Soroush M and Panjapornpon C. A method of sensor fault detection and identification. Journal of Process Control. 2005;15:321-339.
[3] Lao L, Ellis M, Christofides PD. Smart Manufacturing: Handling preventive actuator maintenance and economics using model predictive control. AIChE Journal. 2014;60:2179-2196.
[4] Ellis M, Zhang J, Liu J, Christofides PD. Robust moving horizon estimation based output feedback economic model predictive control. Systems \& Control Letters. 2014;68:101-109.

375086 Economic Model Predictive Control of the Electric Arc Furnace Using Data-Driven Models

Abstract:

Electric arc furnaces (EAF) are widely used in the steel industry to produce molten steel from scrap metal. The EAF is used for approximately one-third of the total steel produced in the world and its use is expected to increase. Although the technology for recycling steel is more economical and flexible than large blast furnaces that consume vast amounts of iron ore and coal, steel manufacturing is generally an energy intensive process with harsh operating conditions that is prone to dramatic shifts in demand and price due to the cyclical nature of the industry. In addition, steel production is a complicated process with complex physical and chemical phenomena that are not well understood. Therefore, the EAF involves relatively low levels of automation and relies heavily on operator involvement [1]. Nevertheless, the continual rapid growth in the power of silicon chips, digital sensors and high-bandwidth communications can potentially improve steelmaking and make the control and optimization of the EAF a crucial component of modern steel mills.

The development of optimal operating policies for the electric arc furnace aimed at improving the corporate profitability, energy efficiency and environmental sustainability of steelmaking is essential for successful control and optimization. One of the main challenges in the EAF operation is that the measurements related to the steel quality are often unavailable during the batch and only made through off-line laboratory analysis. Moreover, the EAF exhibits strong nonlinear, transient operating behavior and is characterized by stages with varying dynamics that makes conventional system identification approaches, where a single linear model is identified, ill-suited for developing an accurate temporal model that is capable of characterizing the time-varying nonlinear dynamics [2].

Recent efforts have recognized the importance of including the economic objective function in standard model predictive control (MPC) formulations, and analyzed the stability properties of economic MPC formulations [3, 4]. In the context of batch processes, standard MPC regulation is equivalent to tracking predefined process set-point trajectories. It has been recognized that for batch processes, where the end-point quality is the variable of importance, it remains meaningful to penalize only the final quality variable deviation in the objective function, while incorporating input restriction as constraints in the optimization problem [5]. One way to optimally compute the inputs is to solve another problem where the final quality (obtained as a solution of the first problem) is imposed as a constraint with the objective function only containing other terms of economic relevance [6].

Motivated by the above considerations, an economic model predictive control (EMPC) framework for the EAF is developed in this work. The design utilizes the multi-model approach to identify local linear models and an appropriate weighting scheme to capture the nonlinear nature of the EAF using only limited process measurements [5], and notions of economic MPC to compute the best input profile that meets the desired quality specifications. To this end, a database of historical process measurements is initially clustered into a number of operating regions and a weighting scheme is devised for the training data to appropriately weight the local models. Subsequently, local linear models are estimated simultaneously using an appropriate latent variable regression method, namely the partial least squares (PLS) regression technique. The resulting model is used to design a two-level predictive controller that enables minimization of an economic objective function subject to end-point and path constraints. The proposed economic MPC is benchmarked against existing trajectory tracking approaches and the simulation results demonstrate the superior performance of the economic MPC over the traditional trajectory tracking approaches.

References

[1] R.D.M. MacRosty and C.L.E. Swartz. Dynamic modeling of an industrial electric arc furnace. Ind. Eng. Chem. Res., 44:8067-8083, 2005.

[2] M. Golshan, J.F. MacGregor, M.-J. Bruwer, and P. Mhaskar. Latent variable model predictive control (LV-MPC) for trajectory tracking in batch processes. J. Proc. Cont., 20:538-550, 2010.

[3] M. Heidarinejad, J. Liu, and P.D. Christofides. State-estimation-based economic model predictive control of nonlinear systems. Sys. Cont. Let., 61:926-935, 2012.

[4] R. Amrit, J.B. Rawlings, and L.T. Biegler. Optimizing process economics online using model predictive control. Ind. Eng. Chem. Res., 58:334-343, 2013.

[5] S. Aumi, B. Corbett, T. Clarke-Pringle, and P. Mhaskar. Data-driven model predictive quality control of batch processes. AIChE J., 59:2852-2861, 2013.

[6] Z. Chong and C.L.E. Swartz. Optimal operation of process plants under partial shut-down conditions. AIChE J., 59:4151-4168, 2013.

390187 Application of Economic MPC to a Hybrid Fuel Cell Vehicle

Abstract:

The notion of a fuel cell powered vehicle has captured the imagination of many as a clean/efficient alternative to the internal combustion engine. However, the capital cost of a fuel cell power unit is significant, and threatens to be a show-stopping hurtle. As such, many have looked to hybridize the fuel cell with lower cost energy storage devices; a rechargeable battery and/or super-capacitor. If given a hybrid vehicle configuration, a fundamental question concerns the coordination of power output from each device. How much power should each device provide in response to a demand? How fast should a storage device be recharged and to what level? If the system contains more than one storage technology - each with unique power/energy density characteristics - then the issue becomes even more complicated. Central to the power coordination question is the physical limits of each device. Clearly, a battery or super-capacitor can hold only a finite amount of energy and cannot output power if the reserve is fully depleted. In addition, the rate of charging and discharging of these devices should be limited to observe heat dissipation related safety concerns. Similarly, a fuel cell will have a maximum power output limit and cannot accept any power. In addition, one may wish to limit the ramp rate of fuel cell power output, in an effort to reduce degradation rates.

Clearly the prominent role of equipment limitations, with respect to energy storage capacity and maximum power, suggests the use of predictive control for constraint enforcement. In this work we investigate the use of Economic MPC for the hybrid vehicle application. In this novel formulation the economic objective is defined as the instantaneous energy loss from the storage devices. Unfortunately, the resulting on-line optimization problem turns out to be a computationally slow quadratically constrained quadratic programming problem. Thus, the main result of the paper is to show that an approximation of the EMPC problem will lead to a computationally efficient quadratic programming implementation with virtually zero loss in performance.

391598 Economic Model Predictive Control with Triggered Evaluations: State and Output Feedback

Abstract:

In recent years, there are strong calls for the development of economic model predictive control (EMPC) to maximize the economic profit from a process. Unlike the traditional two-layer real-time optimization structure, EMPC addresses the economic consideration of a process within the framework of model predictive control (MPC) with a general economic cost function which in general is not quadratic. It has been shown that EMPC typically leads to time-varying process operation instead of steady-state operation. In the literature, several results of EMPC have been developed. In [1], an EMPC design with a terminal constraint for nonlinear systems was proposed. In this type of design, the first step is to determine the economically optimal steady states; and then an EMPC scheme is designed to optimize a general economic cost function which requires that the closed-loop system state settles to a steady state at the end of the prediction horizon.An application of EMPC to cyclic processes as well as a closed-loop stability analysis was also discussed. The above results have demonstrated that EMPC can lead to significant economic profits. However, the computational complexity of an EMPC optimization problem is typically much higher than a traditional MPC optimization problem due to the generality of the cost function.

In a recent work [2], a two-layer design for optimal time-varying operation of nonlinear systems was proposed. In this two-layer design, an EMPC is used in the upper layer to compute dynamic economic optimization policies for process operation and a traditional Lyapunov-based MPC is used to ensure the closed-loop system state follows the optimal time-varying trajectories computed by the upper layer. In order to improve the computational efficiency, the upper-layer EMPC is evaluated at the beginning of each finite-time operating window. This two-layer design provides an approach to achieve improved economic performance while reducing the computational burden. However, this two-layer design requires the re-design of the control system structure and may not be applicable to other existing EMPC designs.

In this work, the objective is to develop a more general approach to reduce the computational burden of EMPC designs. Specifically, event-triggered approach will be adopted in this work to reduce the number of evaluations of an EMPC optimization problem. Event-triggered approaches have been widely used in the design of control systems that have shared communication and computation resources and have been used in state estimation in wireless sensor networks to reduce information transmission in the network while maintaining stability and performance. However, event-triggered evaluation has not been studied within the framework of EMPC. The adoption of event-triggered evaluation in EMPC is not a trivial task. The time-varying operation of EMPC needs to be considered in the design of the triggering condition; the potential open-loop operation needs to be considered in the design of the EMPC and the interplay between the threshold of the triggering condition and the stability of the closed-loop system needs to be carefully studied.

In this work, we will develop event-triggered approaches in the framework of EMPC. The developed approaches can be applied to other existing EMPC designs with appropriate re-designs. First, we consider the case that state feedback is available and revise the LEMPC design in [3] to have triggered evaluations. The triggering condition is designed based on the difference between the actual system state and its predicted value. At a sampling time, if the triggering condition is satisfied, the EMPC is re-evaluated. Subsequently, we consider the case that only output feedback is available and integrate triggered evaluation to an output feedback LEMPC. In this case, a robust moving horizon estimator is used to reconstruct the state information from output measurements and the corresponding triggering condition is based on the difference between the measured and predicted output as well as its time derivatives. In both cases, revised implementation strategies for the event-triggered EMPC are provided and sufficient conditions that ensure the closed-loop stability are derived. A chemical process is used to illustrate the effectiveness of the proposed designs.

[1] M. Diehl, R. Amrit, and J. B. Rawlings. A {L}yapunov function for economic optimizing model predictive control. IEEE Transactions on Automatic Control, 56:703--707, 2011.

[2] M. Ellis and P. D. Christofides. Optimal time-varying operation of nonlinear process systems with economic model predictive control. Industrial & Engineering Chemistry Research, in press.

[3] M. Heidarinejad, J. Liu, and P. D. Christofides. Economic model predictive control of nonlinear process systems using lyapunov techniques. AIChE Journal, 58:855--870, 2012.

370158 Selection of Control Configurations for Economic Model Predictive Control Systems

Abstract:

Control structure design (i.e., the selection of manipulated, controlled, and measured variables or outputs) has been the subject of extensive research within the process control community for many years resulting in many methods for input-output and control configuration selection (e.g. [1]-[2]). Most of the control structure selection methodologies have been developed using linear steady-state or dynamic process models with the assumption that the system is to be operated at steady-state (i.e., the main control objective is to force the system to the desired operating steady-state and maintain operation at this state in the presence of disturbances). Within the context of dynamic operation of nonlinear systems, fewer results and methodologies on control structure selection exist that explicitly consider the process dynamics and nonlinearities. A simple and potentially effective method for evaluating control configurations of multivariable nonlinear systems is to employ a relative degree analysis which may be useful since the relative degree is essentially a measure of the direct effect or physical closeness between an input and an output [3]. Recently, economic model predictive control (EMPC), which is a nonlinear predictive control scheme that optimizes an objective function describing the process economics a control scheme, has been proposed to integrate economic optimization and process control (e.g., [4]-[6]). EMPC may operate the system in a possibly dynamic fashion (i.e., forced dynamic operation) to optimize the process economics. For EMPC, not all of the possible manipulated inputs must have a direct effect on the economic cost of the EMPC since it is not derived from traditional control objectives. As a consequence, control configuration selection for EMPC (i.e., which inputs to manipulate for a given EMPC cost) is an open and relevant problem. Since EMPC may dictate a dynamic operating policy, the system may be operated in a larger region of operation (i.e., the effect of nonlinearities in the process may become significant) compared to traditional/conventional control schemes which force the system to operate in a small neighborhood of the steady-state. Thus, traditional methods that evaluate control structures on the basis of steady-state operation using linear or linearized models may not provide sufficient results within the context of EMPC.

Owing to the aforementioned considerations, a methodology for control configuration selection for EMPC is proposed. Treating the economic cost function as the output, a relative degree analysis is completed to determine which inputs have the most direct dynamic effect on the economic cost. The choice of inputs that are controlled by EMPC are the inputs that have a low relative degree with respect to the cost function (typically, one or two). The remaining possible inputs are partitioned to the set of inputs controlled by EMPC and the set of remaining inputs that are not controlled by EMPC on the basis of a sensitivity analysis and a relative degree analysis of any known disturbances. Furthermore, the set of inputs selected for EMPC is ensured to be a stabilizing one. The remaining inputs not controlled by EMPC may be held constant if the control configuration selected has a sufficient degree of robustness or they may be manipulated through other control systems (i.e., outside of EMPC). An evaluation and analysis of the control configuration selection methodology is provided using a chemical process example.

[1] Heath JA, Kookos IK, Perkins JD. Process control structure selection based on economics. AIChE Journal. 2000;46:1998-2016.
[2] Skogestad S. Self-optimizing control: the missing link between steady-state optimization and control. Computers \& Chemical Engineering. 2000;24:569-575.
[3] Daoutidis P, Kravaris C. Structural evaluation of control configurations for multivariable nonlinear processes. Chemical Engineering Science. 1992;47:1091-1107.
[4] Angeli D, Amrit R, Rawlings JB. On average performance and stability of economic model predictive control. IEEE Transactions on Automatic Control. 2012;57:1615-1626.
[5] Heidarinejad M, Liu J, Christofides PD. Economic model predictive control of nonlinear process systems using Lyapunov techniques. AIChE Journal. 2012;58:855-870.
[6] Huang R, Harinath E, Biegler LT. Lyapunov stability of economically oriented NMPC for cyclic processes. Journal of Process Control. 2011;21:501-509.

381814 Effects of Closed-Loop Dynamics in Dynamic Real-Time Optimization

Abstract:

In the multilayer plant decision-making hierarchy, real-time optimization (RTO) optimizes plant economics in response to low frequency disturbances affecting profit. RTO interacts with the lower level process automation activities in a cascade fashion by providing economically optimal set-points for the underlying regulatory control system - typically MPC as an advanced, multivariable-constrained control strategy. Most industrial applications use a high fidelity, fundamental steady-state process model in the RTO formulation (Marlin and Hrymak, 1997; Darby et al., 2011). However, for systems that undergo relatively frequent transitions, the use of a dynamic model at the RTO level would be more appropriate, hence transforming the steady-state RTO to dynamic RTO (D-RTO) (Tosukhowong et al., 2004; Würth et al., 2011). An alternative to the two-layer architecture is a single-layer economic MPC based on a dynamic fundamental model (Amrit et al., 2013), which in some cases considers a hybrid objective of plant economics and control tracking (Zanin et al., 2002).

D-RTO in the two-layer framework typically uses an open-loop model prediction, which does not recognize the presence of the plant control system. In this study, the effects of the closed-loop behaviour of the MPC control system are considered at the upper level D-RTO calculations. We consider an industrial type, input-constrained MPC controller based on a state-space formulation at the lower level. Both the D-RTO and controller levels are implemented based on a moving horizon approach, with a time-scale separation. We compare the performance of open-loop and closed-loop D-RTO formulations. The open-loop D-RTO formulation optimizes process input moves to minimize operating cost or maximize production profit based on the dynamic process model and operational constraints. In this case, set-points to the lower level MPC controller are essentially the open-loop input/output trajectories. On the other hand, the closed-loop D-RTO problem considers a rigorous inclusion of the MPC closed-loop dynamics in its formulation, thus optimizing the controller set-point trajectories directly. The overall closed-loop

D-RTO problem structure is in the form of a multilevel programming problem as it considers a sequence of MPC controller optimization problems over the D-RTO optimization horizon. A simultaneous solution approach is employed in which the convex MPC-QP subproblems are reformulated using their equivalent Karush-Kuhn-Tucker (KKT) optimality conditions, initially derived in Baker and Swartz (2008). This permits the original problem to be transformed to a single-level mathematical program with complementarity constraints (MPCC) that can be handled using an exact penalty formulation or an interior point approach.

The formulations are applied to multi-input, single-output (MISO) and multi-input, multi-output (MIMO) systems for steady-state transition problems with an objective to minimize the transition cost. Re-optimization at the D-RTO level is executed periodically based on a predefined sample time, which is larger than that of MPC. To gauge the benefit of using the closed-loop formulation over the open-loop formulation, a cost performance measure is introduced to compute a cumulative plant profitability during the transient period. Compared to the open-loop formulation, the closed-loop formulation is shown to give better transition economics. It also helps aid feasibility through set-point trajectories that are appropriately backed-off from output constraints, hence minimizing the degree of output constraint violation during control execution. The problem formulations are presented, challenges discussed, and avenues for future work identified.

References

Amrit, R., Rawlings, J. B., and Biegler, L. T. (2013). Optimizing process economics online using model predictive control. Computers & Chemical Engineering, 58(0), 334 – 343.

Baker, R. and Swartz, C. L. E. (2008). Interior point solution of multilevel quadratic programming problems in constrained model predictive control applications. Industrial & Engineering Chemistry Research, 47(1), 81–91.

Darby, M. L., Nikolaou, M., Jones, J., and Nicholson, D. (2011). RTO: An overview and assessment of current practice. Journal of Process Control, 21(6), 874 – 884.

Marlin, T. E. and Hrymak, A. N. (1997). Real-time operations optimization of continuous processes. In Kantor, J., Garcia, C., and Carnahan, B. (Eds.), AIChE Symposium Series: Proceedings of the 5th International Conference on Chemical Process Control, Vol. 93, pp. 156 164. AIChE and CACHE.

Tosukhowong, T., Lee, J. M., Lee, J. H., and Lu, J. (2004). An introduction to a dynamic plant-wide optimization strategy for an integrated plant. Computers & Chemical Engineering, 29(1), 199 – 208.

Würth, L., Hannemann, R., and Marquardt, W. (2011). A two-layer architecture for economically optimal process control and operation. Journal of Process Control, 21(3), 311 – 321.

Zanin, A., de Gouva, M. T., and Odloak, D. (2002). Integrating real-time optimization into the model predictive controller of the FCC system. Control Engineering Practice, 10(8), 819 – 831.

370155 Safe Economic Model Predictive Control

Abstract:

Maintaining safe operations of process systems is of the highest priority. In fact, it has been argued that dealing with the challenges associated with achieving safe operations should be viewed as a process control problem as process control and automation are ultimately responsible for (safely) regulating a process in the presence of uncertainty [1]. One potential control methodology that may help to integrate safety and control is economic model predictive control (EMPC) [2]-[5] because it is a unique predictive control technique that uses a general cost function in its formulation (i.e., typically, a non-quadratic cost function). Owing to the fact that the objective function does not need to be positive definite with respect to a given set-point, it may incorporate other considerations (outside of feedback control considerations) into the design of the EMPC scheme through the addition of appropriate constraints imposed in the EMPC optimization problem and the modification of the cost function to explicitly account for these other considerations. For instance, an EMPC scheme was formulated for a handling preventive maintenance of control actuators in [5]. Integrating safety and process control is an open and challenging area as new safety metrics, process monitoring methodologies, and control schemes that incorporate safety need to be introduced.

To integrate EMPC with safety considerations (i.e., handle safety through feedback control), a novel control and monitoring methodology is proposed. First, the safeness of a region of operation with respect to a possible failure (i.e., determine if stability of the process can be maintained for any point in the region of operation if a failure occurs) is characterized. Second, a statistical analysis is used to evaluate the probability of a possible failure. To this end, it is important to point out that it might be best with respect to the process economics to operate a process in specific region of operation. However, this region of operation must also be acceptable with respect to the process safety. Even if it is safe to operate in this region, the reliability of process components decreases and the probability of failure increases with time in general [6], and the safeness of this region of operation may decrease with time. Thus, the probability of a potential failure is computed with time. If the probability of a failure exceeds a threshold and the process cannot be safely operated in the region of operation if a failure occurs, the region of operation is shifted to a safer region of operation regardless of the fact that the process economics performance may decrease in this region of operation. To transition between regions of operation, appropriate constraints are added to the EMPC optimization problem. Through a rigorous stability analysis, recursive feasibility and closed-loop stability are analyzed. The proposed control and monitoring technique, which effectively integrates process control, process economics, and safety considerations, is demonstrated with a chemical process example.

[1] Leveson NG, Stephanopoulos G. A system-theoretic, control-inspired view and approach to process safety. AIChE Journal. 2013;60:2-14.
[2] Angeli D, Amrit R, Rawlings JB. On average performance and stability of economic model predictive control. IEEE Transactions on Automatic Control. 2012;57:1615-1626.
[3] Huang R, Harinath E, Biegler LT. Lyapunov stability of economically oriented NMPC for cyclic processes. Journal of Process Control. 2011;21:501-509.
[4] Heidarinejad M, Liu J, Christofides PD. Economic model predictive control of nonlinear process systems using Lyapunov techniques. AIChE Journal. 2012;58:855-870.
[5] Lao L, Ellis M, Christofides PD. Smart Manufacturing: Handling preventive actuator maintenance and economics using model predictive control. AIChE Journal. 2014;60:2179-2196.
[6] Barlow RE, Proschan F. Mathematical Theory of Reliability. Philadelphia, PI: Society for Industrial and Applied Mathematics, 1996.

A New Multi-Period MINLP Model for Optimal Long-Term Shale Gas Development Planning

Markus G. Drouven[1] and Ignacio E. Grossmann[2]
Department of Chemical Engineering
Carnegie Mellon University
Pittsburgh, PA 15213

[1]mdrouven@cmu.edu, [2]grossmann@cmu.edu

ABSTRACT

The production of shale gas from unconventional resource plays is transforming the U.S. natural gas supply. Advances in production technologies, notably the dual application of horizontal drilling and hydraulic fracturing, allow the extraction of vast deposits of trapped natural gas that, until recently, were uneconomic to produce. The Energy Information Administration predicts that shale gas will account for 50% of total U.S. natural gas production by 2040 [1]. Natural gas demand is also expected to increase in the electric power and nearly all other industrial sectors. Furthermore, shale gas supplies light hydrocarbons such as ethane, propane and butane that are used as basic feedstock in the chemical industry. In light of the current projections the "shale gas revolution" is expected to impact world economics for up to a century or longer [2].

Future development of shale gas resources requires an extensive expansion of the existing gas production, transmission, and processing infrastructure. Virtually all stages of the shale gas supply chain need to be expanded and upgraded to match the ever growing natural gas supply and demand [3]. Since the necessary capital investments for drilling rigs, pipelines, boosting stations and midstream processing facilities are substantial, the long-term planning of upstream production and natural gas transmission is a key challenge.

Yet, so far the shale gas development planning problem has received little attention from researchers in academia. Past publications have focused primarily on conventional and offshore gas field planning [4-5]. In comparison, unconventional gas production is characterized by the development of a large number of candidate well pads, which calls for tailor-made planning models and solution algorithms capable of identifying optimal development strategies within acceptable solution times. Currently however, there are virtually no computer tools available to support decision-makers in the development of the shale gas value chain.

Recently, Cafaro and Grossmann [6] proposed a large-scale nonconvex mixed-integer nonlinear programming (MINLP) model to identify the optimal shale gas supply chain extending from well pads over the gathering system beyond midstream processing plants to interstate pipelines. The nonlinearities in the proposed model arise from nonconvex power law expressions capturing economies of scale applying to investment costs. The authors develop a decomposition approach based on successively refining a piecewise linear approximation of the objective function. The model is based on the assumption that the shale gas composition is independent of well locations and offers only limited capability to handle composition variations.

In this work we address the deterministic, multi-period shale gas development planning problem. The scope of the problem extends from upstream production at well pads over the construction of flow and gathering lines to midstream processing facilities. As part of this problem, investment and operating decisions have to be made throughout the planning horizon. These decisions include: (a) the selection of a set of well pads for development and production, (b) pipeline routing and sizing, (c) the selection and dimensioning of gathering and boosting stations, and (d) the identification of suitable freshwater reservoirs for drilling and fracturing operations. The objective is to maximize the net present value over a planning horizon of 10 years.

The optimal transmission system is determined by means of a superstructure-based optimization approach. The proposed superstructure is derived from real gathering systems and accounts for characteristics common in shale gas transmission networks, such as pad-to-pad connections. We show that the recently proposed MINLP shale gas development planning model by Cafaro & Grossmann can be reformulated as a mixed-integer linear programming (MILP) model by taking advantage of discrete equipment sizes. The discrete nature of pipeline diameters and standardized compressor sizes is exploited by introducing a large number of binary variables and mixed-integer linear constraints. While the reformulated linear model is significantly larger than its nonlinear counterpart, the key advantages of the reformulation are: (a) the simplified treatment of discrete equipment sizes as continuous variables is no longer necessary, (b) the economic scaling effects that are dominant in the oil and gas industry are still accounted for, (c) linearity in the constraints and objective function is preserved, and (d) global optimality can be guaranteed.

The reformulated model is also extended to explicitly account for variable gas compositions at distinct well pads. The motivation for this extension is that in practice the shale gas composition can vary significantly within a shale play. In the Marcellus play for example, the gas composition ranges from "wet gas", which is rich in heavier hydrocarbons such as ethane, butane or propane, to "dry gas", which is nearly pipeline quality gas. The major challenge in accounting for variable gas compositions is that the corresponding mass balances contain a large number of bilinear terms, rendering the model an MINLP. In this work we investigate several solution strategies, ranging from efficient initialization and bounding procedures over Benders and Lagrangean decomposition to multi-parametric disaggregation algorithms. We compare their computational performance and highlight individual limitations and advantages in detail.

The proposed optimization planning model is applied to a set of real-world case studies. Initial results reveal non-intuitive well development strategies, the effects of capital expenditure constraints, and demonstrate the value of rigorous mathematical optimization for this particular application.

References

[1] U.S. Energy Information Administration (EIA). Annual Energy Outlook with Projections to 2040. April 2013.

[2] Siirola, J. J. The Impact of Shale Gas in the Chemical Industry. *AIChE J.* 2014, *60*, 810-819.

[3] Goellner, J. F. Expanding the Shale Gas Infrastructure. *AIChE CEP* August 2012, 49-52.

[4] Goel, V.; Grossmann, I.E. A Stochastic Programming Approach to Planning of Offshore Gas Field Developments under Uncertainty in Reserves. *Comput. Chem. Eng.* 2004, *28* (8), 1409-1429.

[5] Tarhan, B.; Grossmann, I.E.; Goel, V. Stochastic Programming Approach for the Planning of Offshore Oil or Gas Field Infrastructure under Decision-Dependent Uncertainty. *Ind. Eng. Chem. Res.* 2009, *48*(6), 3078-3097.

[6] Cafaro, D. C.; Grossmann, I. E. Strategic Planning, Design, and Development of the Shale Gas Supply Chain Network. *AIChE J.* 2014. doi: 10.1002/aic.14405.

386110 Integrated Oil Field Management through Optimal Placement and Scheduling of Drilling and Infrastructural Installations and Activities

Abstract:

High capital expenditures of oil field development, increasing global oil demand, limited oil resources, and multifaceted interactions between production infrastructures have alerted the oil production industry to move towards integrated field management and avoid peacemeal and myopic decisions. In this framework, any decision demands a wholistic analysis of the complex interactions between multiple reservoirs, wells, surface infrastructure network, and energy market. The outcome of this analysis sets several goals and instructions for the operators and contractors in different parts of the production chain, and these optimal decisions have substantial economic impacts. Infrastructure placement and allocation are amongst the most important design decisions over the long planning horizons that demand such analyses. The combinatorial, nonlinear and spatio-temporal dynamic nature of this problem poses several challenges and usually leads to large models with considerable computational costs. Once production and installation scheduling are also integrally tackled with the aforementioned topics, the new problem is even more challenging as the problem size increases much further. However, drilling and installation scheduling is a must for field development projects; e.g. there are always limited drilling rigs that should be scheduled to drill many wells in a field, and all drillings cannot be done simultaneously. Therefore, in this study, we target an integrated study to consider installation scheduling as well. In our recent studies [1, 2], we addressed joint well and infrastructure placement, allocation and production scheduling in oil fields with the shared surface network. We formulated dynamic nonconvex MINLP models and modified an outer approximation algorithm (based on the works of Grossmann and coworkers [3]) to solve them. In both of these studies, we assumed that all drillings and installations are done at the beginning of the production horizon. In this study, we extend our previous works to address the integrated problem of surface network design and placement with production and installation scheduling, which has not been addressed so far in the literature. In this problem, the search space is substantially large due to two components: design, and scheduling. Hence, we solve this joint problem in two stages. In the first stage, we obtain the potential and promising locations and allocations for the design problem using our rigorous dynamic MINLP model. This uses detailed governing equations and employs advanced mathematical techniques to account for the numerous interactions in the field. In the second stage, we solve the drilling and installation scheduling problem to determine the time and order of each activity. In both stages, we employ rigorous dynamic NLPs to model the multiphase flow inside the reservoir. Although the final solution can be sub-optimal, it still can provide much improvement in the financial objective value. In fact, the numerical results show significant improvement in the net present value of a literature problem with two reservoirs. The best solution progresses to the highest NPV of *MM$* 240.4 and suggests drilling 6 new wells, establishing 8 new connections and installing two new manifolds over almost 1200 days. The second stage of the solution had a key role in improving the final NPV.

REFERENCES

1. Tavallali, M.S., I.A. Karimi, A. Halim, D. Baxendale, and K.M. Teo, *Well Placement, Infrastructure Design, Facility Allocation, and Production Planning in Multi-Reservoir Oil Fields with Surface Facility Networks.* Industrial & Engineering Chemistry Research, 2013. **Under review**.

2. Tavallali, M.S., I.A. Karimi, K.M. Teo, D. Baxendale, and S. Ayatollahi, *Optimal producer well placement and production planning in an oil reservoir.* Computers & Chemical Engineering, 2013. **55**: p. 109-125.

3. Viswanathan, J. and I.E. Grossmann, *A combined penalty function and outer-approximation method for MINLP optimization.* Computers and Chemical Engineering, 1990. **14**(7): p. 769-782.

383379 Long- and Short-Term Optimization Model for Shale Gas Water Management

Abstract:

With the advancement in directional drilling and hydraulic fracturing, shale gas is predicted to provide almost half of the United States natural gas supply in twenty years. Since water use is associated with each step of the drilling and production process, it is an important aspect in shale gas development. The challenge with water management is that a large volume of it is required in a short-period of time. Water sources can be described as interruptible and non-interruptible sources. Whereas withdrawal from a non-interruptible source remains constant, withdrawal from an interruptible source can be affected by seasonal variation in water availability. As a result, even in water-rich regions, water use can be under strain due to the high demand. To this end, impoundments can be installed close to the water sources in order to serve as buffer tanks to store water when there is high freshwater supply in the sources. In order to avoid hauling freshwater to shale gas wellpads through trucks, overland or buried pipeline could be installed to transport freshwater from water sources to the wellpads. While permanent pipeline can be buried underground when the operators set up gas pipelines, overland pipelines are temporary and can be rented. Flowback water can be reused at the next frac through adequate treatment and blending with freshwater. Thus, based on the flowback water composition, different treatment techniques can be adopted.

We present a new mixed-integer linear programming (MILP) model for optimizing daily and long-term decisions in water use through a discrete-time representation of the State-Task Network. The proposed model extends the previous work by the authors (Yang et al., 2004), which dealt only with operations, to include capital investment decisions. Specifically, the objective is to minimize the overall cost including capital cost of impoundment, piping, and treatment facility, and operating cost including freshwater, pumping, and treatment. Given are the potential freshwater source location and withdrawal data, potential impoundment location, wellpad storage, location, and total number of stages, treatment unit capability and location, and the number of frac crews available. The goal is to determine the location and capacity of impoundment, the type of piping, treatment facility locations and removal capability, as well as the frac schedule, and the water sources to obtain freshwater. In addition, we examine the impact truck hauling restriction has on the overall cost and frac schedule. The problem is optimized over a long planning horizon, which increases the computational difficulty for solving the MILP model. A real-world case study in the Utica Shale is optimized to illustrate the application of the proposed formulation.

Reference

Yang, L., J. Manno and I.E. Grossmann, "Optimization Models for Shale Gas Water Management," submitted for publication (2014).

382301 Natural Gas to Liquid Transportation Fuels, Olefins, and Aromatics (GTL+C2_C4+C6_C8)

Abstract:

The United States faces major challenges towards the production of hydrocarbon fuels and high value chemicals. These major challenges comprise of high crude oil prices, volatility of the global oil market, and excessive greenhouse gas (GHG) emissions. A recent review has shown that these challenges can be addressed using single or hybrid feedstocks such as coal, biomass, and natural gas towards the production of liquid transportation fuels [1]. However, high value petrochemicals, such as C2 to C4 olefins and C6 to C8 aromatics, are an important part of refineries since they can substantially increase the profitability of these plants. Therefore, hybrid energy systems that produce liquid fuels from alternative feedstocks should also consider the production of C2 to C4 olefins and C6 to C8 aromatics. The primary C2 to C4 olefins are ethylene, propylene, and butadiene, whereas the key C6 to C8 aromatics include benzene, toluene, and the xylene isomers. Moreover, recent progress in the shale gas industry provides cheap and abundant natural gas feedstock that can be utilized towards fuels and chemicals production. Recent work has already shown that the production of liquid fuels from natural gas (GTL) can be competitive with typical petroleum processes [2]. Therefore, this work considers the production of liquid fuels concurrently with C2 to C4 olefins and C6 to C8 aromatics, starting from a natural gas feedstock.

A techno-economic and environmental assessment of a novel refinery that converts natural gas to liquid fuels, C2 to C4 olefins, and C6 to C8 aromatics (GTL+C2_C4+C6_C8) will be performed using an optimization-based framework. The process synthesis superstructure will contain multiple natural gas conversion pathways. Synthesis gas (syngas) produced from the reformers will be directed to either the Fischer-Tropsch refining section or to the methanol production section. The Fischer-Tropsch hydrocarbons will be upgraded to fuel-grade quality products in either a ZSM-5 reactor or through a series of treatment units. The methanol will either be converted into gasoline or olefins. The olefins may be directed to the olefin purification section or further processed to produce gasoline and distillate.

Aromatics will either be separated from naphtha produced from the refinery or through conversion of liquefied petroleum gases (LPG). Any CO2 produced in the plant can be vented, sequestered, or recycled back in the refinery.

The optimization-based framework consists of process design, process synthesis, and global optimization strategies to determine the optimum plant topology under different scenarios [2-13]. Simultaneous heat, power, and water integration is present in the model to keep the utilities costs at minimum levels. This framework ensures the selection of the process that produces the liquid fuels, C2 to C4 chemicals, and C6 to C8 aromatics in the most preferable way (lowest cost or highest profit). The superstructure proposes different methods of production of syngas, hydrocarbons, and chemicals; therefore the tradeoffs between each method for the proposed refinery will be investigated. Different case studies are presented to explore the effect of the plant capacity and olefins/aromatics production levels. The key foundations, topological decisions, economical and environmental aspects are discussed.

[1] Floudas, C. A.; Elia, J. A.; Baliban, R. C. Hybrid and single feedstock energy processes for liquid transportation fuels: A critical review. Computers & Chemical Engineering 2012, 41 (6), 24-51.
[2] Baliban, R. C.; Elia, J. A.; Floudas, C. A. Novel Natural Gas to Liquids Processes: Process Synthesis and Global Optimization Strategies. AIChE Journal 2013, 59 (2), 505-531.
[3] Baliban, R. C., Elia, J. A., Floudas, C. A. Toward novel biomass, coal, and natural gas processes for satisfying current transportation fuel demands, 1: Process alternatives, gasification modeling, process simulation, and economic analysis. Industrial & Engineering Chemistry Research 2010, 49, 7343-7370.
[4] Elia, J. A., Baliban, R. C., Floudas, C. A. Toward novel biomass, coal, and natural gas processes for satisfying current transportation fuel demands, 2: Simultaneous heat and power integration. Industrial & Engineering Chemistry Research 2010, 49, 7371-7388.
[5] Baliban, R. C., Elia, J. A., Floudas, C. A. Optimization framework for the simultaneous process synthesis, heat and power integration of a thermochemical hybrid biomass, coal, and natural gas facility. Comp. Chem. Eng. 2011, 35, 1647-1690.
[6] Baliban, R. C., Elia, J. A., Floudas, C. A. Simultaneous process synthesis, heat, power, and water integration of thermochemical hybrid biomass, coal, and natural gas facilities. Comp. Chem. Eng. 2012, 37, 297-327.
[7] Baliban, R. C., Elia, J. A., Misener, R., Floudas, C. A. Global Optimization of a MINLP Process Synthesis Model for Thermochemical Based Conversion of Hybrid Coal, Biomass, and Natural Gas to Liquid Fuels. Comp. Chem. Eng. 2012, 42, 64-86.
[8] Baliban, R. C., Elia, J. A., Weekman, V., Floudas, C. A. Process Synthesis of Hybrid Coal, Biomass, and Natural Gas to Liquids via Fischer-Tropsch Synthesis, ZSM-5 Catalytic Conversion, Methanol Synthesis, Methanol-to-Gasoline, and Methanol-to-Olefins/Distillate Technologies. Comp. Chem. Eng. 2012, 47 (12), 29-56.
[9] Baliban, R. C.; Elia, J. A.; Floudas, C. A. Biomass to liquid transportation fuels (BTL) systems: process synthesis and global optimization framework. Energy Environ. Sci. 2013, 6 (1), 267-287.
[10] Baliban, R. C.; Elia, J. A.; Floudas, C. A. Biomass and Natural Gas to Liquid Transportation Fuels: Process Synthesis, Global Optimization, and Topology Analysis. Industrial & Engineering Chemistry Research 2013, 52 (9), 3381-3406.
[11] Baliban, R. C.; Elia, J. A.; Floudas, C. A.; Gurau, B.; Weingarten, M. B.; Klotz, S. D. Hardwood Biomass to Gasoline, Diesel, and Jet Fuel: 1. Process Synthesis and Global Optimization of a Thermochemical Refinery. Energy & Fuels 2013, 27 (8), 4302-4324.
[12] Baliban, R. C.; Elia, J. A.; Floudas, C. A.; Xiao, X.; Zhang, Z.; Li, J.; Cao, H.; Ma, J.; Qiao, Y.; Hu, X. Thermochemical Conversion of Duckweed Biomass to Gasoline, Diesel, and Jet Fuel: Process Synthesis and Global Optimization. Industrial & Engineering Chemistry Research 2013, 52 (33), 11436-11450.
[13] Niziolek, A.; Onel, O.; Elia, J. A.; Baliban, R. C.; Xiao, X.; Floudas, C. A.; Coal and Biomass to Liquid Transportation Fuels: Process Synthesis and Global Optimization Strategies. Industrial & Engineering Chemistry Research 2014, DOI: 10.1021/ie500505h.

Introduction and Motivation

Sustainable production of environmentally friendly electricity has become a major topic of discussion and research worldwide in recent years. One promising technology for this purpose is the solid-oxide fuel cell (SOFC) power generation system [1]-[5]. An SOFC utilizes a carbonaceous fuel gas (synthesis gas, for example) and an oxidant (typically air) to efficiently produce electrical power through an electrochemical reaction across a solid oxide barrier. There are key advantages to using SOFCs for power production. For example, the anode exhaust is mainly H_2O and CO_2, which are easily separated if the anode and cathode streams are not mixed downstream, and the cathode exhaust is mostly N_2, which can be used for additional power generation or heat recovery and then safely vented. Moreover, SOFCs are fuel flexible, quiet, and highly efficient [6]. Recent studies have shown that carbon capture and sequestration (CCS) can be added to bulk-scale SOFC systems at very low costs and parasitic energy penalties, and that these methods are capable of recovering essentially 100% of direct CO_2 emissions [1]-[2]. However, the total cradle-to-grave life-cycle impact of bulk SOFC systems has not yet been performed; upstream fuel processing, plant commissioning and decommissioning and other sub-processes in the electricity supply chain must be considered to conclude if SOFC plants are truly a sustainable method of generating electricity from fossil fuels.

Methodology

This work performs a comprehensive life-cycle analysis (LCA) of a bulk-scale SOFC power plant fueled by natural gas (NG) and compares the results to a current state-of-the-art natural gas combined cycle (NGCC) plant. Sub-processes such as plant commissioning (including resources and energy consumed during the fabrication of the SOFC stacks themselves and any relevant balance-of-plant), upstream fuel acquisition and processing, downstream electricity distribution networks, and the plant itself are all considered within the cradle-to-grave life-cycle boundary. This study employs the *ReCiPe 2008* LCA method, which defines a set of standardized metrics to categorize the impacts of products (electricity in this case) at the mid-point (such as climate change, ecotoxicity or fossil fuel depletion) and end-point (such as loss of human life and ecosystem degradation) levels, as shown in Figure 1 [7].

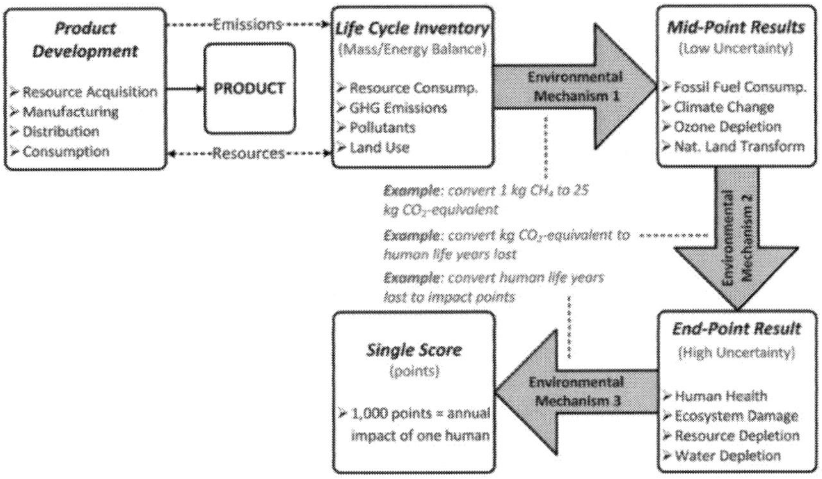

Figure 1: Example of a mid-point to end-point life-cycle assessment strategy. Adapted from [7]

End-point results are normalized to a unified point system to obtain a description of the product's entire life-cycle impact so that it may be compared to other products. It is important to note that it is a subject of debate as to which of the many LCA methods currently in use is the most accurate, but provided that a comparative LCA is performed with the same method for each product, meaningful comparisons may be obtained. All calculations for both the NGCC and SOFC processes were scaled to a basis unit of 1 MW-h of delivered electricity. In order to accommodate the impacts of capital and infrastructure development, typical plant lifetimes and operating capacities from literature were used.

Preliminary Results and Conclusions

The LCA simulation results that will be presented show that the manufacturing stage of 1 kW of SOFC cells and their associated balance-of-plant contributes a noticeable portion to the entire life-cycle impact of the SOFC process. Most emissions from SOFC manufacturing are the result of consuming electricity during the manufacturing process, which was assumed to be supplied by the average electricity mix in the United States. With regards to end-point impacts, it will be shown that the manufacturing of 1 kW of SOFCs has the impact of approximately 9% of the impact of a human being over one year.

The addition of CCS to the NGCC process reduces direct CO_2 emissions by 90%, but lowers global CO_2 emissions by only 81% due to increased upstream and downstream emissions resulting from lower NGCC plant efficiencies. Moreover, the decreased efficiency of the NGCC plant with CCS results in increases to NG and water consumption. Uncaptured direct emissions increase due to increased fuel consumption as well. Although adding CCS to the NGCC plant is partially offset by increases to resource consumption and uncaptured emissions, it will be shown to decrease the total life-cycle impact of the NGCC process by approximately 21%.

The addition of CCS to the SOFC process reduces direct CO_2 emissions by 100% and decreases global CO_2 emissions by approximately 93%. However, adding CCS to a SOFC process has a very low parasitic energy penalty and hence does not facilitate the need for significantly more NG. As such, the impacts of CCS on upstream NG acquisition and processing are small. The SOFC process without CCS is found to have an end-point impact significantly lower than the NGCC plant without CCS on a per-MW-h basis. The end-point reduction of adding CCS to the SOFC plant is significant, yielding an overall decrease of approximately 35%.

From an economically incentivized perspective, it is possible to combine the results of this study with that of previous techno-economic analyses to determine the added cost of CO_2 and eco-point reductions for the SOFC and NGCC cases. The LCA results that will be presented indicate that adding CCS to a SOFC plant is nearly 20 times more cost effective than for a NGCC plant on a per-unit of CO_2 avoided or on a per-eco-point avoided basis. The addition of CCS to a SOFC plant is therefore found to be both economically and environmentally preferable to that of a NGCC plant even when the entire life-cycle of both plants is considered.

References

[1] Adams T.A. II, Barton, P.I. High-efficiency power production from natural gas with carbon capture. *Journal of Power Sources.* **2010**, *195*, 1971.

[2] Adams, T.A. II, Barton P.I. High-efficiency power production from coal with carbon capture. AIChE J. **2010**, *56*, 3120.

[3] Duan, L.; Yang, Y.; He, B.; Xu, G. Study on a novel solid oxide fuel cell/gas turbine hybrid cycle system with CO2 capture. Int. Journ. of Energy Res. **2012**, *36*, 139-152.

[4] Becker, W.L.; Braun, R.J.; Penev, M.; Melaina, M. Design and technoeconomic performance analysis of a 1 MW solid oxide fuel cell polygeneration system for combined production of heat, hydrogen, and power. Journal of Power Sources. **2012**, *200*, 34-44.

[5] Nease, J, Adams TA II. Systems for peaking power with 100% CO_2 capture by integration of solid oxide fuel cells with compressed air energy storage. J. Power Sources. **2013**, *228(15)*, 281-293.

[6] T.A. Adams II, J. Nease, D. Tucker, P. Barton, Energy conversion with solid oxide fuel cell systems: short and long term outlooks. Ind. Eng. Chem. Res. **2013**, *52*, 3089-3111.

[7] M Goedkoop, R. Heijungs, M. Huijbregts, A. De Schryver, J. Struijs, R. van Zelm. *Recipe 2008 First Edition (revised, version 1.08).* Ministerie van Volkshuivesting, Ruimtelijke Ordening en Milieubeheer: 2013.

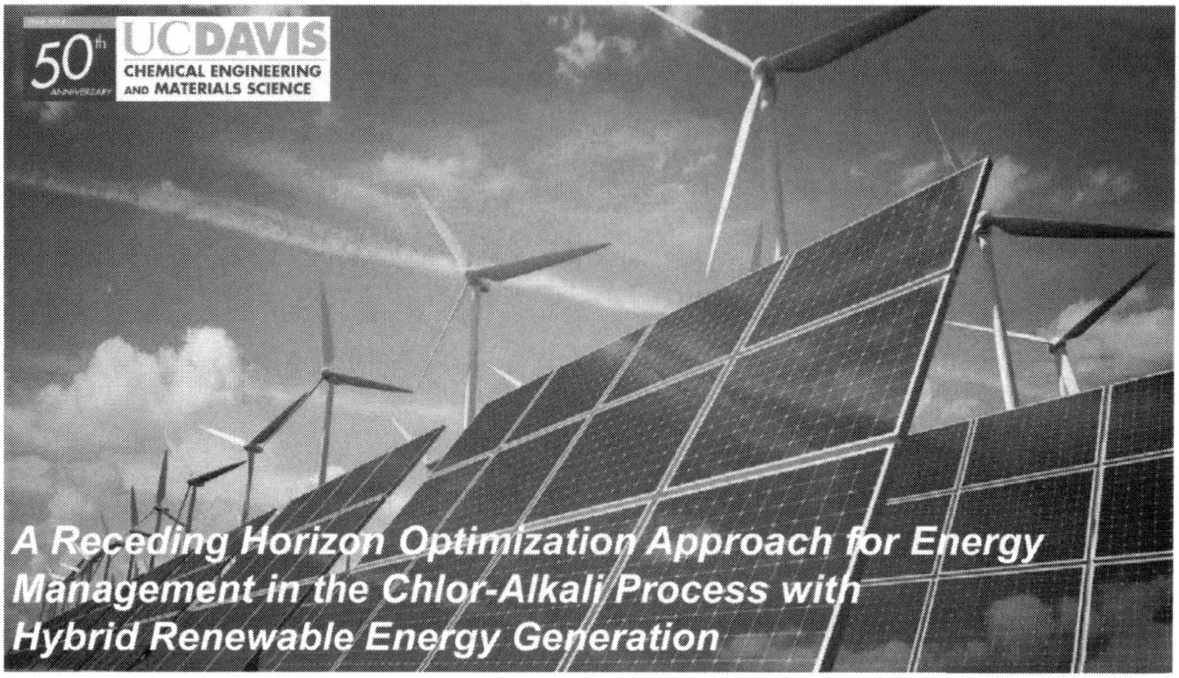

A Receding Horizon Optimization Approach for Energy Management in the Chlor-Alkali Process with Hybrid Renewable Energy Generation

Xiaonan Wang, Ahmet Palazoglu and Nael H. El-Farra
xnwang@ucdavis.edu

2014 AIChE Annual Meeting
Atlanta, GA
November 20, 2014

UCDAVIS

Outline

Energy Management Chlor-Alkali Process

- Introduction
- Component Modeling
  - Renewable Energy Generation
  - Chlor-Alkali Plant
- Operational Optimization
  - Receding-Horizon Optimization
  - Formulation
- Results and Discussion
- Conclusions

2

Introduction

Motivation for research on ENERGY

- *Generation* Side
  - Rising rate of consumption of fossil fuels
  - Benefits of using alternative energy resources

"The Stone Age did not end for lack of stone. And the Oil Age will end long before the world runs out of oil."

Sheikh Zaki Yamani
Former Oil Minister, Saudi Arabia.

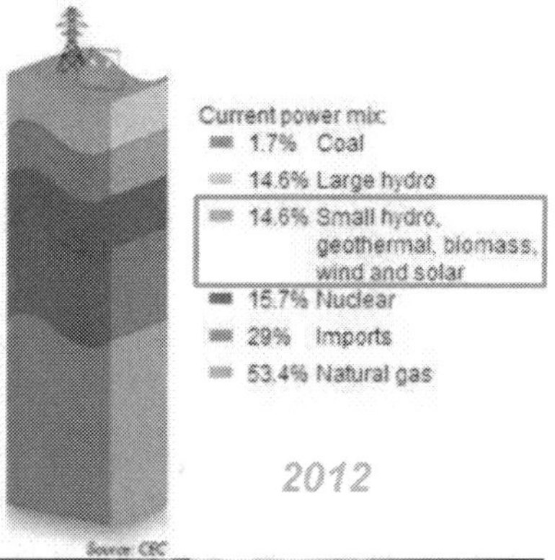

Goal of California Energy Commission
33% from Renewables by 2020

3

http://www.energy.ca.gov/renewables/renewable_links.html

Introduction

Motivation for research on ENERGY

- *Generation* Side
- *Demand* Side
  - Energy intensive process
  - Demand-Responsive loads

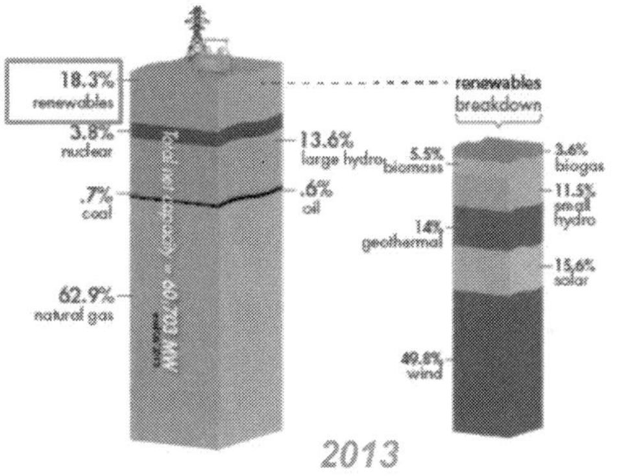

Time-varying electricity price

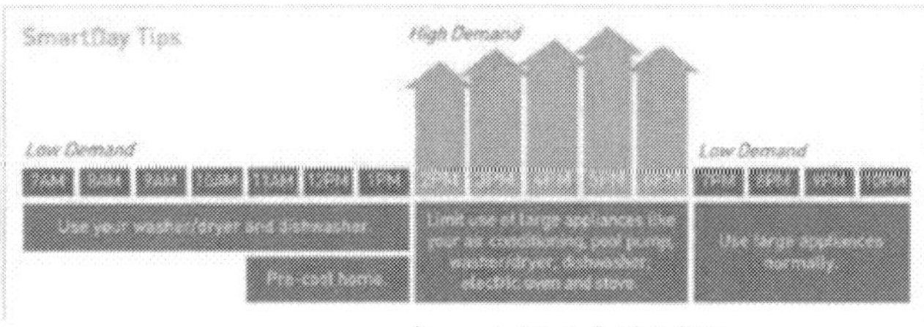

Smart Rate| PG&E

4

http://www.caiso.com/informed/Pages/CleanGrid/default.aspx

Introduction

Produce chlorine, caustic soda and hydrogen

- Optimal dispatch policy for a grid connected industrial plant

Energy intensive process

- Renewable supply – wind/solar/fuel cell
- Grid connected – buying and selling power
- Demand Management

Hydrogen "Economy"

- Byproduct recovery
- Energy recovery

5

Modeling: Solar Energy

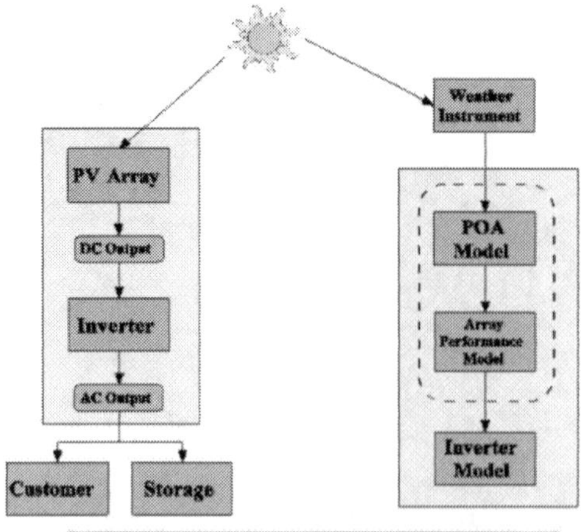

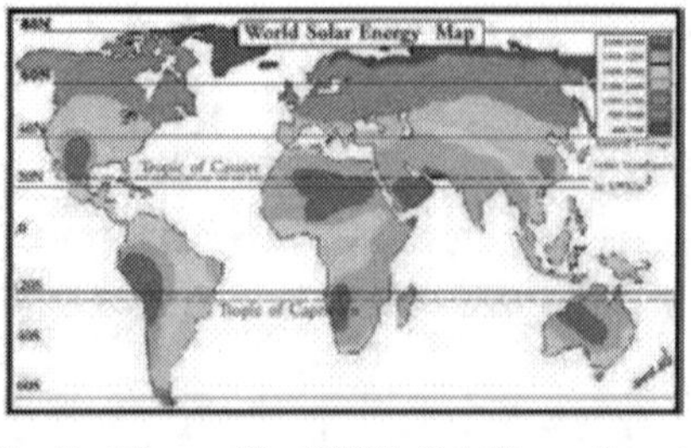

http://is.njit.edu/competition/2009/Cat2_2_Winner_Group142/

Photovoltaic/Thermal hybrid solar system (PVT system)

- **Plane-of-Array (POA) model:**
- **PV performance model**:
  - Input: global irradiance, temperature

$$I_L - I_S\{\exp[\alpha(v_{pv} + R_S i_{pv})] - 1\} - \frac{v_{pv} + R_S i_{pv}}{R_{sh}} - i_{pv} = 0$$
$$p = v_{pv} i_{pv}$$

5-parameter model (W. Soto et al., 2005)

- **Thermal collector**

$$\eta_{el} = \frac{V_{mpp} I_{mpp}}{G A_{pv}} \qquad \eta_{th} = \frac{\dot{m} C_p (T_{out} - T_{in})}{G A_{pv}}$$

$$\eta_{th} = \tau_\alpha - \tau \eta_{el}$$

6

Modeling: Wind Energy

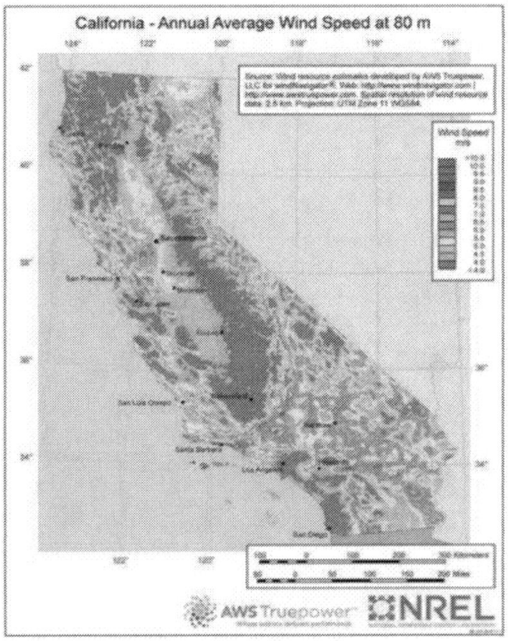

- **Efficiency:**
  - Less than **59.3%**

  Conversion of kinetic energy of the wind to mechanical energy of the turning rotor
  - Only **10%-30%**

  A complete wind turbine system, considering strength and durability
- **Power:** $P_t = \frac{1}{2} C_p(\lambda) \rho A v_w^3$

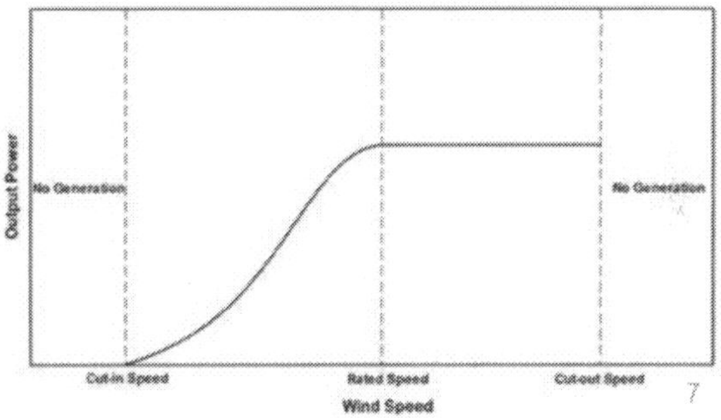

7

Modeling: Storage

STORAGE: Satisfy unmet demand due to intermittent supply

- **Short-term variation**
  - Rechargeable battery technologies
- **Long-term variation**
  - Electrolyzer–Hydrogen storage–Fuel cell

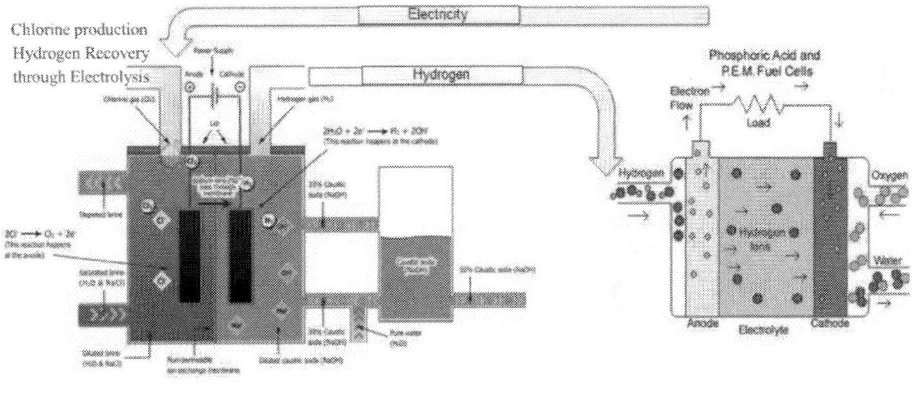

8

http://americanhistory.si.edu/fuelcells/basics.htm

Modeling: Plant Diagram

Long-term variation Chlor-Alkali Electrolysis – Hydrogen storage– Fuel cell

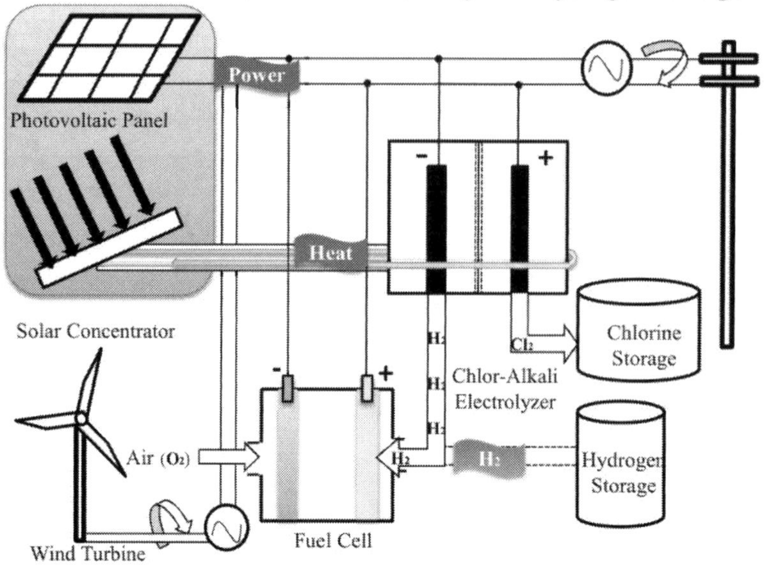

Produce chlorine, caustic soda and hydrogen through electrolysis

9

Wang, X., Teichgraeber, H., Palazoglu, A., & El-Farra, N. H. (2014). An economic receding horizon optimization approach for energy management in the chlor-alkali process with hybrid renewable energy generation. Journal of Process Control, 24(8), 1318-1327.

Modeling: Chlor-Alkali Process

Reaction $2Na^+(aq) + 2Cl^-(aq) + 2H_2O \rightarrow 2Na^+(aq) + 2OH^-(aq) + Cl_2(g) + H_2(g)$

Anode $2Cl^-(aq) \rightarrow Cl_2(g) + 2e^-$

Cathode $2Na^+(aq) + 2H_2O + 2e^- \rightarrow H_2(g) + 2Na^+(aq) + 2OH^-(aq)$

Cell Voltage $U_{rev} = \frac{\Delta G}{zF}$ $U_{tn} = \frac{\Delta H}{zF}$ $U_{el} = U_{rev} + \frac{r_1 + r_2 T_{el}}{A} I_{el} + S \ln\left(\frac{t_1 + t_2/T_{el} + t_3/T_{el}^2}{A} I_{el} + 1\right)$

Reversible/ Thermoneutral/ Electrochemical

Energy Balance $C_t \frac{dT}{dt} = \dot{Q}_{gen} - \dot{Q}_{loss} - \dot{Q}_{cool}$

Efficiency $\eta_F = \frac{(I/A)^2}{f_1 + (I/A)^2} f_2$ $\eta_e = \frac{U_{tn}}{U_{el}}$ $\eta_{el} = \frac{P_{out}}{P_{in}} = \frac{\dot{n}\, HHV(H_2)}{P_{FC}}$

Faraday efficiency/ Energy efficiency/ Overall electrolyzer efficiency

H_2 Flow Rate $\dot{n}_{H_2} = \eta_F \frac{n_c I_{el}}{zF}$

10

Modeling: Fuel Cells

Proton Exchange Membrane Fuel Cell (PEMFC)

$$U_{FC}(t) = U_{FC,0} + E_1 T_{FC,0}(t) + E_2 \log\left(\frac{I_{FC}(t)}{I_{FC,0}}\right) + \frac{R_{FC}}{T_{FC}(t)} I_{FC}(t)$$

$$\dot{n}_{H_2} = N_{cell,FC} \frac{p\dot{V}_{H_2}}{RT_{FC}} \qquad \eta_{FC} = \frac{P_{out}}{P_{in}} = \frac{P_{FC}}{\dot{n}\,HHV(H_2)}$$

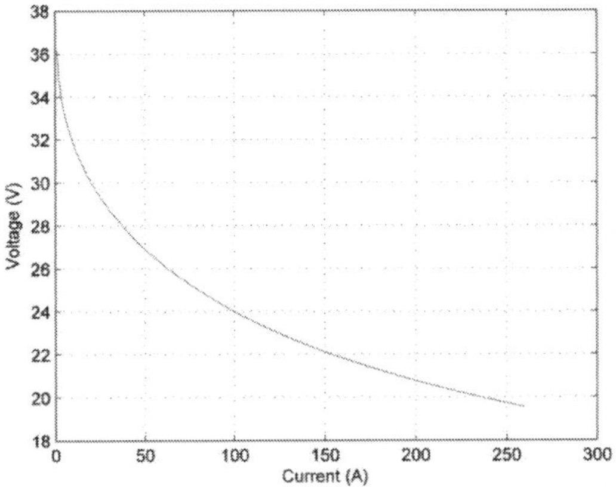

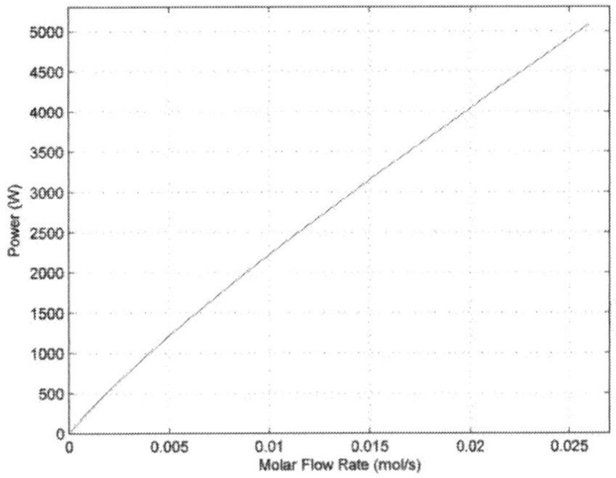

11

Receding-Horizon Optimization

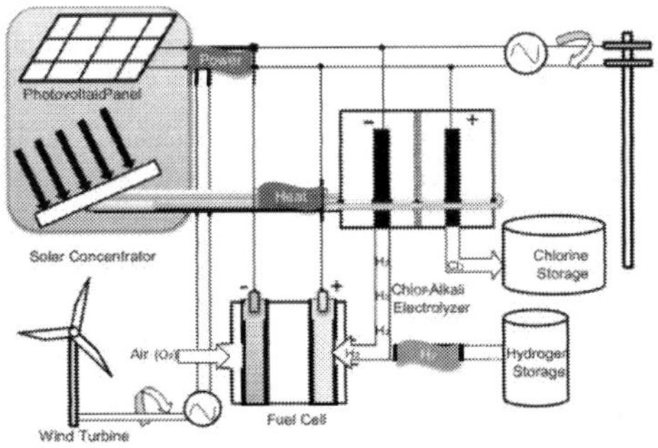

Receding-Horizon Optimal Dispatch

12

Receding-Horizon Optimization

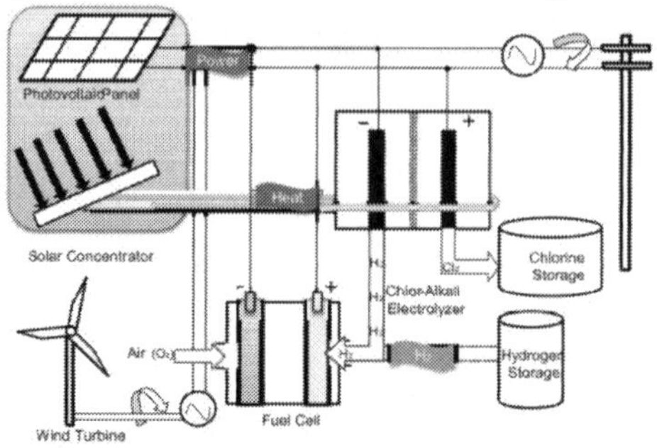

Receding-Horizon Optimal Dispatch

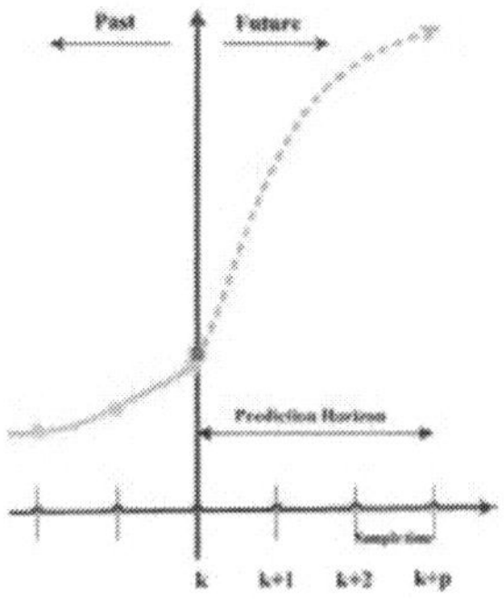

12

Receding-Horizon Optimization

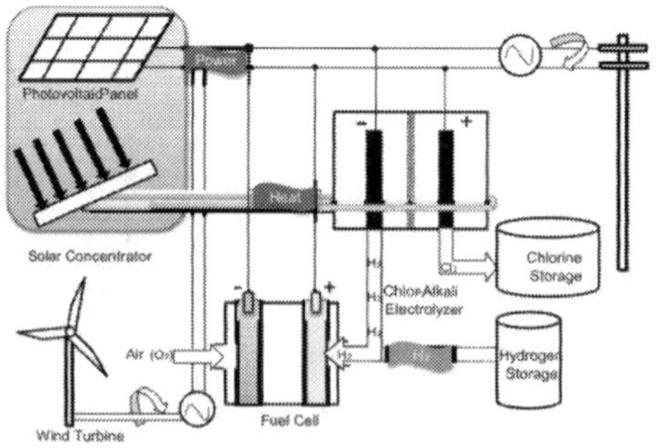

Receding-Horizon Optimal Dispatch

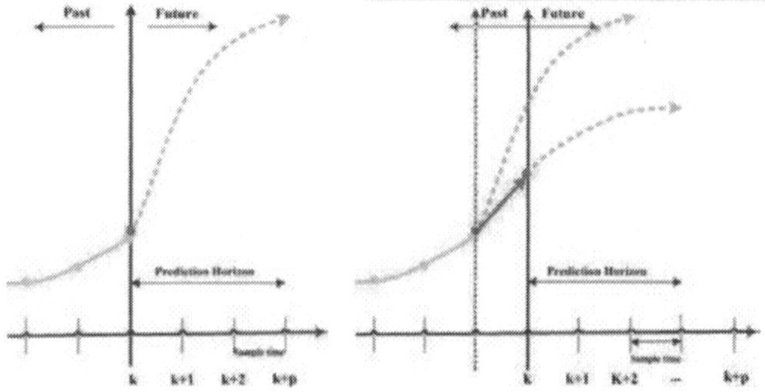

12

Receding-Horizon Optimization

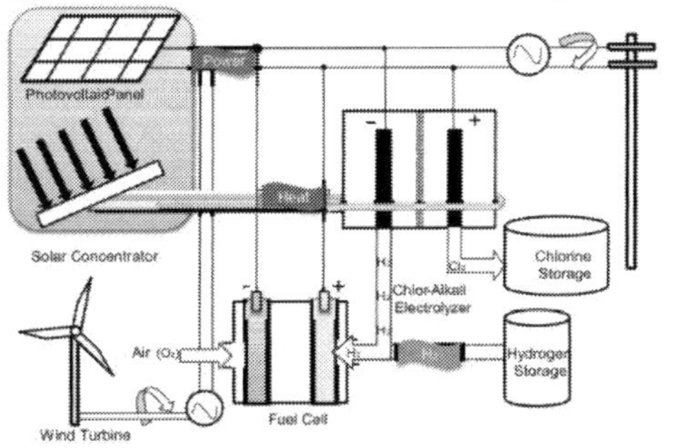

Receding-Horizon Optimal Dispatch

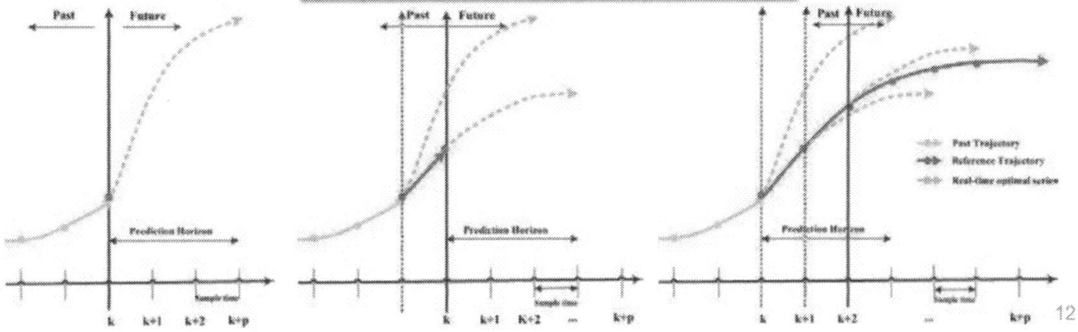

Optimization Formulation

Economic Objective

Cost (Operation & Environmental) Profit (Cogeneration & Product)

$$\min \sum_{t=j}^{j+N\Delta} \alpha \left(\sum_i P^t_{i,ref} C^t_i \right) + \beta \left(\sum_i P^t_{i,ref} Cen^t_i \right) - \lambda_1 \left(P^t_{grid\_sell,ref} C^t_{grid\_sell} \right) - \lambda_2 \left(Q^t_{product} C^t_{product} \right)$$

$$w.r.t.\ P^j_{PV,ref}, P^j_{wind,ref}, P^j_{FC_{in},ref}, P^j_{FC_{out},ref}, P^j_{grid\_buy,ref}, P^j_{grid\_sell,ref}$$

$$s.t. \forall t \in [j\Delta, (j+N)\Delta)$$

$$P^t_{required} = f(Q^t_{product})$$

$$P^t_{required} \le \sum_i P^t_{i,ref} - P^t_{grid\_sell,ref}$$

$$P^t_{i,\min} \le P^t_{i,ref} \le P^t_{i,\max}$$

$$|P^{t+\Delta}_{i,ref} - P^t_{i,ref}| \le dP_{i,\max}$$

$$i = \{PV, wind, FC_{in}, FC_{out}, grid\_buy\}$$

Receding-horizon optimization

- Modeling of individual components
- Required power depends on production level
- Decision variables are the power outputs/exchanges
- Weighting factors emphasize priorities
- Linear convex optimization problem.

Optimization Results

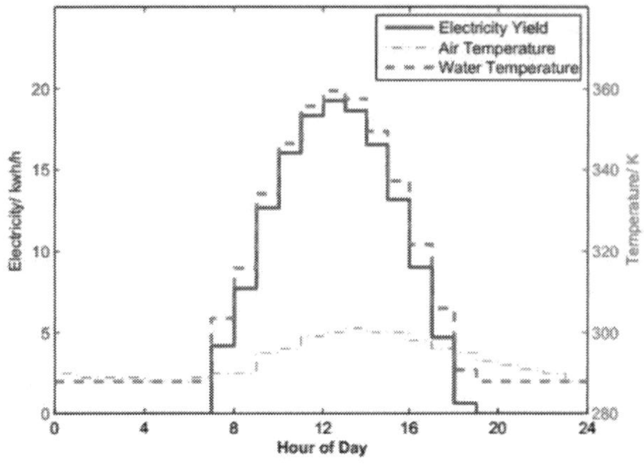

(a) PV power and thermal yield

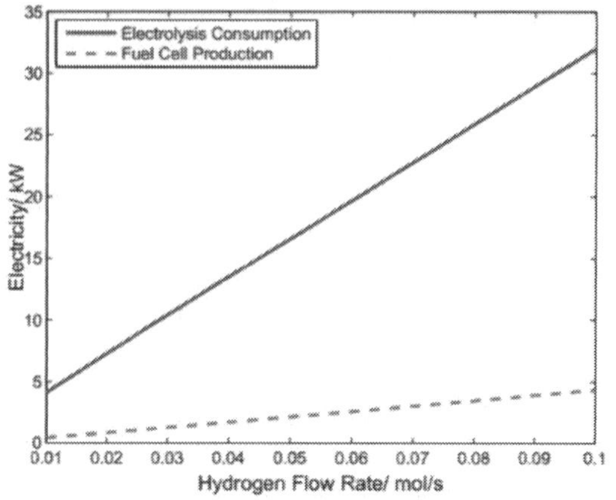

(b) Power of electrolysis and fuel cells

- Electricity generation and water temperature obtained by the heating from solar combi-panels.
- Solar irradiation and temperature have key influence on the energy conversion.

- Relations between electricity power (consumed by electrolysis and produced by fuel cells) and hydrogen flow rates.
- Recovery efficiency is less than 13.5% (temperature matters).

14

Optimization Results

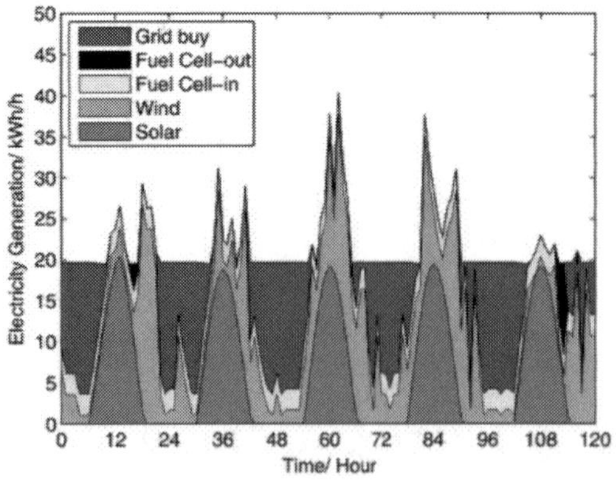

(a) Overall electricity dispatch

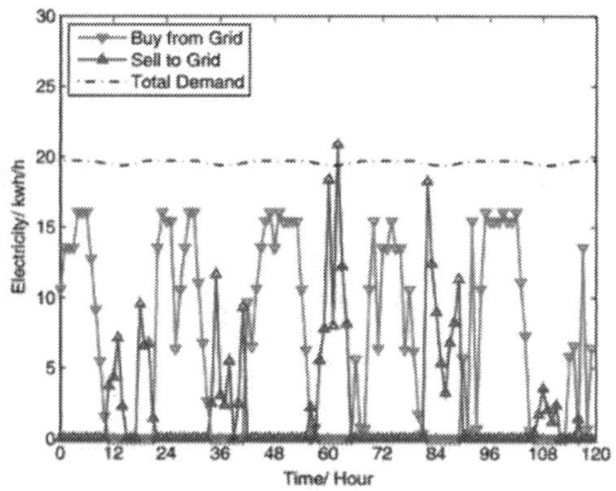

(b) Electricity bought from and sold to the grid compared with total demand

- One can see the complementary relations between all resources.
- Purchased hydrogen used less due to cost.

- Electricity is bought during evening hours.
- Electricity is sold when generation exceeds demand.

15

Demand Response

$$\min \sum_{t=j}^{j+N\Delta} \alpha\left(\sum_i P^t_{i,ref} C^t_i\right) + \beta\left(\sum_i P^t_{i,ref} Cen^t_i\right) - \lambda_1\left(P^t_{grid\_sell,ref} C^t_{grid\_sell}\right) - \lambda_2\left(Q^t_{product} C^t_{product}\right)$$

$$\frac{Q_{product}(t) = \delta_1(t) Q_{high} + \delta_2(t) Q_{normal} + \delta_3(t) Q_{low}}{t = j, j+\Delta, ..., j+N\Delta}$$

$$s.t. \sum_{t=j}^{j+N\Delta} Q_{product}(t) \geq Q_{total} \qquad P^t_{required} = f\left(Q^t_{product}\right)$$

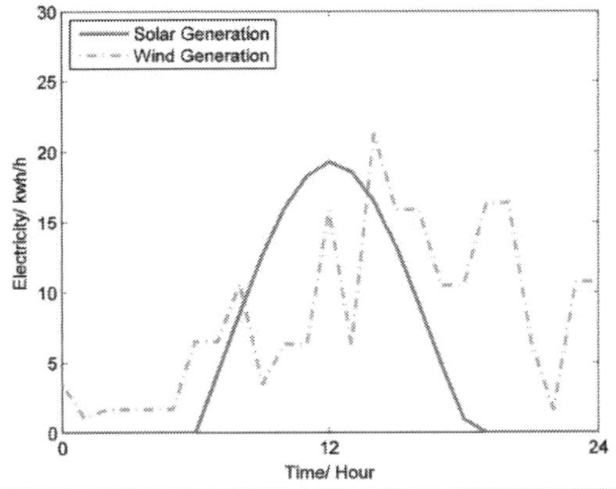

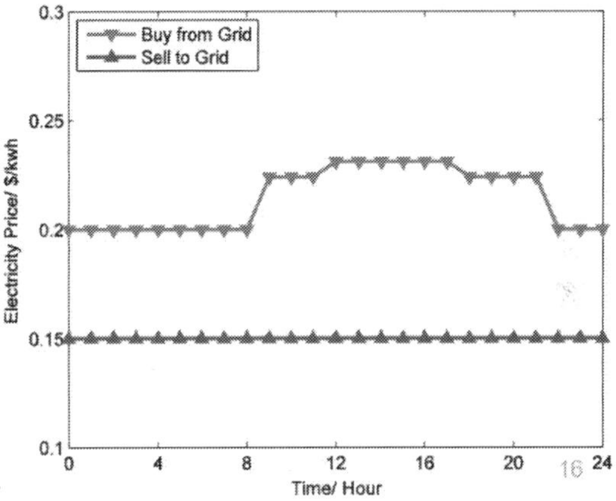

16

Demand Response

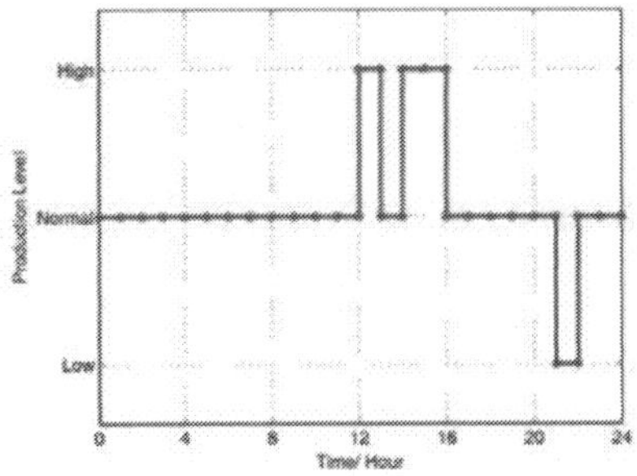

Production level

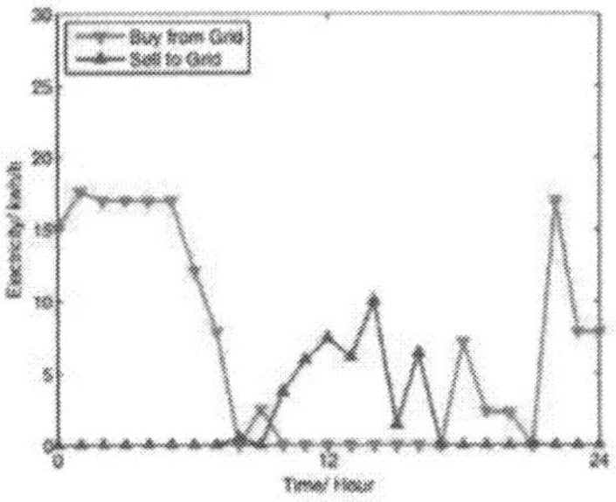

Electricity exchanged with the grid

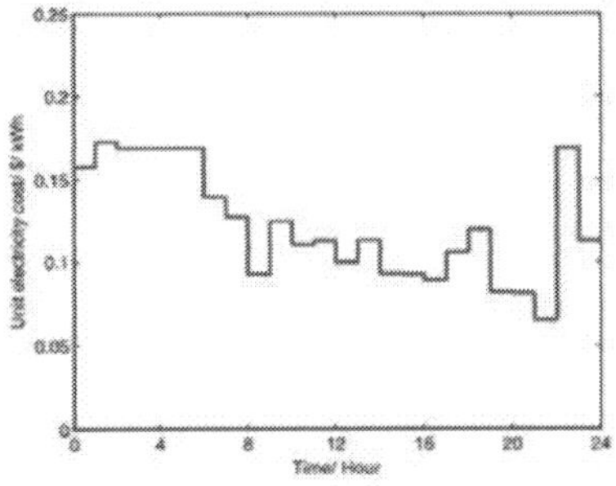

Hourly electricity cost

- Binary variables are used to select operating level. This results in an *MILP*.
- A minimum daily production level is imposed.
- Operation is switched based on cost of electricity and the availability of renewable resources.

17

Day-ahead Scheduling

$$P^j_{i,ref} = arg\min \sum_{t=j}^{j+N\triangle} \| y^t_i - y^{t,r}_i \|^2_Q + \| \ell(P^t_{i,ref}) \|^2_R$$

$$y_i^{t,r}$$

A day-ahead contract generated between the users and the grid operators

✓ how much electricity to be purchased

✓ how much to be sold-back

$$\sum_{t=j}^{j+N\Delta} \alpha\left(\sum_i P^t_{i,ref}C^t_i\right) + \beta\left(\sum_i P^t_{i,ref}Cen^t_i\right) - \lambda_1\left(P^t_{grid\_sell,ref}C^t_{grid\_sell}\right) - \lambda_2\left(Q^t_{product}C^t_{product}\right)$$

Same as previous cost function

18

Day-ahead Scheduling

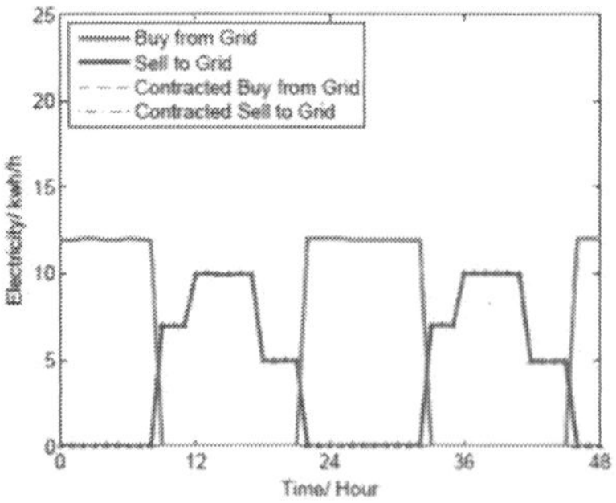

(a)

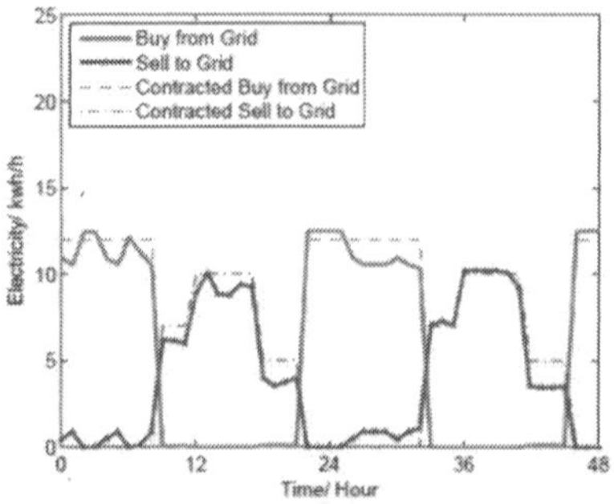

(b)

$$P^j_{i,ref} = arg\min \sum_{t=j}^{j+N\triangle} \| y^t_i - y^{t,r}_i \|^2_Q + \| \ell(P^t_{i,ref}) \|^2_R$$

- Deviations on both sides from the contracted buying and selling curve induce certain penalties.
- Trade-off between maximizing profits and following the scheduled electricity trade.
- Based on how to penalize:

(a) no allowance of deviation;
(b) fines for deviation (flexibility).

- Daily profits increased by 4%

19

Summary

- ✓ Optimal design and operation of a grid-connected HRES that combines PVT systems, wind energy conversion systems and fuel cells, to supply power to a chlor-alkali plant
- ✓ Combination of real-time forecasting, optimization and demand response can improve the system efficiency
- ✓ Production cost can be reduced

20

Thank you!

Any Questions?

xnwang@ucdavis.edu

Financial support from California Energy Commission, Phillips 66 Fellowship and UC Davis CHMS Department are gratefully acknowledged.

21

Reference

[1] E. Worrell, D. Phylipsen, D. Einstein, N. Martin, "Energy use and energy intensity of the US chemical industry", Technical Report, Lawrence Berkeley National Lab., CA (US), 2000.
[2] J. Chlistunoff, "Advanced chlor-alkali technology", Technical Report, Final Technical Report, LAUR 05-2444. Los Alamos, NM, USA, 2005.
[3] M. Paulus, F. Borggrefe, "The potential of demand-side management in energy-intensive industries for electricity markets in Germany", Applied Energy 88 (2011) 432–441.
[4] R. Santorelli, A. Schervan, A. Delfrate, "Energy production from hydrogen co-generated in chlor-alkali plants by the means of PEM fuel cell systems", Nuvera Fuel Cells Europe (2009).
[5] M. Broeren, D. Saygin, M. Patel, "Forecasting global developments in the basic chemical industry for environmental policy analysis", Energy Policy 64 (2014) 273–287.
[6] S. Kelouwani, K. Agbossou, R. Chahine, "Model for energy conversion in renewable energy system with hydrogen storage", Journal of Power Sources 140 (2005) 392–399.
[7] K. Zhou, J. Ferreira, S. De Haan, "Optimal energy management strategy and system sizing method for stand-alone photovoltaic-hydrogen systems", International Journal of Hydrogen Energy 33 (2008) 477–489.
[8] X. Wang, A. N. Palazoglu, N. H. El-Farra, "Operation of residential hybrid renewable energy systems: Integrating forecasting, optimization and demand response", in: Proceedings of American Control Conference, 2014.

In energy research and policy there is a big effort to achieve a large penetration of renewable energy generation, predominantly wind and solar. These intermittent sources in general increase uncertainties in energy supply and thus could, along with variations of energy demand, deteriorate the stability of the whole energy system. Demand side management with time-variable operation of energy-intensive industrial plants has been proved helpful to solve this problem, e.g. [1]. Another complementary approach is implementing sufficient energy storage capacity.

Many chemicals are good energy carriers in the sense of high energy density and ability for long-term storage. Due to their amenable physical properties, it has been widely recognized that chemicals-based energy storage could be a promising solution on the GWh-level [2].

Analog to batteries, one of intuitive configurations of chemicals-based energy storages is encapsulating chemical conversion processes in a closed-loop unit [3]. Although this configuration offers flexibility in selection of construction sites, the difficulty in heat management is immense in practice. There is another possible configuration that has not yet received enough awareness. Herein the processes realizing the energy storage function are integrated into a chemical manufacturing plant. The product can either be sold on the chemicals market or be used as energy storage medium. In comparison to the former configuration, the effect of the integration on process design is twofold. On the one hand, the process design becomes more challenging, because the synergy of both functional parts should be maximized. On the other hand, one enjoys more freedom in selecting proper processes and conducting heat integration.

In this presentation, taking ammonia production as an example, two processes in different configurations are considered. The processes are synthesized taking into account specialties of each configuration. The corresponding flowsheets are simulated in Aspen Plus® with rigorous steady-state models. Based on the simulation results, the benefits of the integrated configuration pertaining to the performance and the heat management are highlighted.

References:

[1] Ghobeity, A.; Noone, C.J.; Papanicolas, C.N.; Mitsos, A. Optimal time-invariant operation of a power and water cogeneration solar-thermal plant. Solar Energy (85) 2295-2320, 2011.

[2] Lewis, N.S.; Nocera, D.G. Powering the Planet: Chemical challenges in solar energy utilization. Proceedings of the National Academy of Sciences (43) 15729-15735, 2006.

[3] Al-musleh, E.I.; Mallapragada, D.S.; Agrawal, R. GWh level renewable energy storage and supply using liquid CO2. In Kraslawski, A. and Turunen, I. edited: Proceedings of the 23rd European Symposium on Computer Aided Process Engineering – ESCAPE 23, 415-420, 2013.

376488 Simulation and Control of Plasma-Enhanced Chemical Vapor Deposition (PECVD) Process for the Manufacturing of Silicon Thin Film Solar Cells

Abstract:

Solar energy is currently one of the most promising clean and sustainable energy sources, and plasma enhanced chemical vapor deposition (PECVD) is an emerging deposition technique used for the manufacturing of silicon thin film solar cells which are the most widely used solar cell systems. In a PECVD process, the silane and hydrogen gases are activated by collisions with plasma ions, and those activated species are transported to the substrate where thin film growth through surface reactions occur. Furthermore, the key benefit of using PECVD is in the low operational temperature allowing the deposition of temperature sensitive films and relatively high deposition rate reducing the processing time [1]. However, the widespread application of PECVD is limited by the significant nonuniformity in the deposition rate along the spatial coordinate, particularly for a large-scale chamber. This effect leads to the production of thin film solar cells with low conversion efficiency due to undesired surface morphology characterized by aggregate surface roughness and slope. Motivated by this, the modeling of a PECVD process has been necessitated in an effort to control the surface morphology of thin films to desired constraints for enhanced solar cell conversion efficiency.

This work focuses on the simulation of a silicon thin film deposition process through PECVD and the design of a model predictive control scheme to improve solar cell performance. A sinusoidal grated wafer is used to achieve surface roughness within the range of the visible light spectrum, which is desired for optimal light trapping [2]. Initially, a kinetic Monte Carlo (kMC) model is presented based on the solid-on-solid assumption to simulate the thin film growth process which consists of 4 microscopic processes associated with silyl radicals including physisorption, chemisorption, surface diffusion, and H-abstraction [3]. The kMC simulation is regarded as a representation of a real PECVD process, and the system parameters used in the simulation are identified through the calibration of the simulation results with experimental data. Then, a reduced-order model is introduced to predict the dominant system dynamics of the silicon thin film growth computed by the kMC simulation, and is used for the design of a model predictive controller (MPC) [4]. Finally, the closed-loop simulation results show that the production of the silicon thin film with a desired surface morphology that leads to improved light trapping efficiency is achieved by regulating the spatially distributed deposition rate through the proposed MPC scheme.

[1] Armaou A, Christofides PD. Plasma enhanced chemical vapor deposition: Modeling and control. Chem. Eng. Sci. 1999;54:3305-3314.

[2] Campa A, Isabella O, Erven R, Peeters P, Borg H, Krc J, Topic M, Zeman M. Optimal design of periodic surface texture for thin-film a-Si:H solar cells. Prog. Photovolt. 2010;18:160-167.

[3] Tsalikis DG, Baig C, Mavrantzas VG, Amanatides E, Mataras D. A hybrid kinetic Monte Carlo method for simulating silicon films grown by plasma-enhanced chemical vapor deposition. The J. of Chem. Phys. 2013;139:204706.

[4] Huang J, Hu G, Orkoulas G, Christofides PD. Simulation and control of porosity in a three-dimensional thin film solar cell. Ind. & Eng. Chem. Res. 2013;52:11246-11252.

388164 Effect of Drill Pipe Whirling on Cementing Integrity in Horizontal Shale Gas Wells: A Simulation Study

Abstract:

Shale gas production from horizontal wells faces potential problems related to gas leakage from various zones of a well into the air and water reserves. Avoiding such leaks, and thus achieving better environmental friendliness as well as well productivity, is one of the biggest challenges for shale gas production. To avoid gas communication between well zones in a rock formation penetrated by a well (i.e. to achieve zonal isolation) the integrity of the cement placed in the annulus between the metal casing and the borehole wall must be ensured at all times. While the cement may be placed and set properly in the annulus, subsequent drilling operations that extend the well depth, may affect the strength of the cemented annulus.

During drilling of subsequent well sections, the drill pipe often experiences a whirling motion and impacts on the casing of already cemented sections. As a result of this impact, excessive stresses are generated in the casing and cemented annulus, which compromises the strength of the cement sheath. While this hypothesis is reasonable, it is difficult to confirm it through direct measurement of stresses developed as a result of whirling-induced impact forces. Therefore, we undertook a simulation study, to estimate such stresses. This exercise will help in assessing the need for better control of drill pipe whirling and in designing cement slurries that can sustain impact forces of such magnitude. Comparisons with experimental observations will be discussed.

388292 Combination of Electrolysis and Catalytic Oxidation to Simultaneously Produce Hydrogen and Chlorine Using Waste Hydrogen Chloride Gas

Abstract:

Hydrogen chloride is widely produced as a byproduct of processes where chlorine is used as a primary raw material. The recovery of hydrogen chloride presents a significant opportunity in terms of economic yields and environmental impact.

A number of processes are available for producing Cl_2 from HCl. Two traditional approaches are [1]:

(1) Electrolysis of HCl with production of Cl_2 and H_2.

(2) Oxidation of HCl by air or O_2 in the presence of a catalyst, e.g., Deacon-type processes.

Direct electrolysis requires extensive electricity and the energy consumed to produce hydrogen exceeds the energy released by the same hydrogen when used as a fuel. The strength of this path is that both hydrogen and chlorine can be produced at the same time. On the other hand, Deacon-type processes have a simple flowsheet and low power requirements, while only chlorine is achieved as the product. A combination of these two approaches is feasible to achieve multiple products as well as achieving some energy savings. Furthermore, using renewable energy sources to power electrolysis reactions and to realize thermochemical splitting of HCl is a promising dierction, especially when solar energy, as a renewable resource, applied to supply heat and electricity at the same time from one integrated system, called photovoltaic/thermal hybrid solar systems (or PVT systems) [2].

In the present work, we propose a combination of electrolysis and catalytic oxidation to simultaneously recover hydrogen and chlorine through waste HCl gas. Thermochemical decomposition of hydrogen chloride generally involves three distinct reaction steps: hydrogen production, oxygen production and materials regeneration. We examine various design schemes and present a thermodynamic analysis. The proposed process is tested using advanced process simulations tools (Aspen Plus®). A preliminary set of reaction steps are given below:

Block	Description	Process
S1H2	Step1 H_2 Production	$2\ Cu\ (s) + 2\ HCl\ (g) \rightarrow 2\ CuCl\ (l) + H_2\ (g)$ 450°C
S2ELE	Step 2 Electrolysis	$2\ CuCl\ (s) + H_2O\ (l) \rightarrow CuCl_2\ (aq) + Cu\ (s) + H_2O\ (l)$ 25°C
S3CL2	Step 3 Cl_2 Production	$CuCl_2 + 1/2O_2 \leftrightarrow CuO + Cl_2$ 25°C
S4FB	Step 4 Fluidized Bed Reactor	$2CuCl_2(aq) + H_2O(g) \rightarrow 2CuCl(l) + 2HCl(g) + 1/2O_2(g)$ 450°C
S5	Step 5 $CuCl_2$ recovery	$CuO + 2HCl \leftrightarrow CuCl_2 + H_2O$ 100°C

Further developments of the reaction cycles are being conducted to improve efficiency and facilitate eventual commercialization.

[1] Pan H Y, Minet R G, Benson S W, et al. Process for converting hydrogen chloride to chlorine. Industrial & engineering chemistry research, 1994, 33(12): 2996-3003.

[2] Chow T T. A review on photovoltaic/thermal hybrid solar technology. Applied Energy, 2010, 87(2): 365-379.

378927 Thermal Conversion of Biomass: Pyrolysis Reaction Networks for Optimal Reactor Design

Abstract:

Biochar is charcoal generated for intentional soil amendment by pyrolyzing sustainable biomass feedstocks. In addition to providing a method for carbon sequestration, properly engineered charcoals can increase the water holding and cation exchange capacities of soils, improving the ability of plants to survive under drought conditions and reducing fertilizer runoff into watersheds. Fertilizer runoff has become a serious problem, because as much as 70% of fertilizer applied to crop fields is leached into the groundwater or lost to streams and rivers, eventually leading to large hypoxic "dead" zones in the world's oceans (including the Gulf of Mexico).

One desirable quality of reactors used for the thermochemical conversion of biomass is that they be "feedstock-blind": capable of producing consistent products even when the feedstocks or the feedstock properties vary. To achieve a satisfactory level of reproducibility, we must be able to accurately control the operation of a reactor so that each biomass load will be processed with the temperature program required to produce biochar with the desired properties.

Biomass pyrolysis involves a complicated series of parallel and sequential reactions that include dehydration, depolymerization and decomposition of the primary biomass components (hemicellulose, cellulose, lignin), devolatilization, and condensation. Therefore, the distribution of the pyrolysis products as well as the chemical and physical properties of the produced biochar will depend on the complete temperature history of the biomass particles.

Clearly, the design of optimal reactors for biomass pyrolysis requires a detailed kinetic model for every feedstock. To address this challenge, we will first present a systematic procedure that can be used to develop lumped models for biomass pyrolysis. While many earlier studies based their analysis on thermogravimetric experiments with constant heating ramps, our procedure uses a sequence of step-isothermal experiments to identify the potential pseudo-components of each biomass feedstock. Dynamic simulations and nonlinear least-squares methods are then used to test several reaction networks and select the candidate that best fits the experimental data.

We applied this method to develop pyrolysis models for slash pine and eucalyptus wood obtained from a managed forest that belongs to Rice University. Our results showed that a network of first-order parallel and sequential reactions with as many as five solid pseudo-components must be used to accurately describe the pyrolysis dynamics of these woods. The kinetic parameters of the individual reactions and their heats of reaction were also estimated using thermogravimetric and differential scanning calorimetry data.

The pyrolysis reaction networks were then used to simulate the transient operation of fixed-bed and rotary kiln or auger reactors used for slow pyrolysis of biomass. Reactors of the second type are usually characterized by (a) biomass residence times that may be three or more orders of magnitude longer than the residence times of gas, and (b) very high solid to gas mass ratios, conditions that allow for simplification of their modeling equations. Transient mass balances were developed for all the gaseous and solid species. The solid phases were assumed to be continuous media that exchange heat with the flowing gas, decompose and release volatile products into the gas phase. Fixed-bed reactor models assumed a stationary solid phase (where spatial variations are induced solely by temperature gradients) and two-dimensional PDEs with axial convection and two dispersion terms for the gases. Two-dimensional PDEs were used to describe the transient energy balances for the solid and gas phases, with heat exchange through the reactor wall since this is the most common method for heating these reactors. Finally, the moving bed reactor models considered plug flow of the solid phases.

Due to the significant heat effects of pyrolysis reactions, numerical solutions of the modeling equations revealed that large and prolonged axial and radial temperature gradients develop in these reactors. As a result, biochar particles located in different locations of a fixed-bed reactor may experience widely different temperature histories and, thus, achieve different conversions. Even prolonged exposure to high pyrolysis temperatures may not be enough to eliminate these differences and guarantee that the produced biochar will have uniform properties.

A lab-scale fixed-bed reactor was used to validate the fixed-bed reactor models. The reactor had a total volume of 1 liter, was heated by placing it inside a furnace and was equipped with four thermocouples to monitor the temporal and spatial temperature distributions. Axial and radial temperature gradients often exceeding 100°C per inch were recorded in this small reactor when the furnace temperature was set to 500°C. Different spatial and temporal temperature patterns were observed for runs with pine and eucalyptus wood particles, a strong indication that pyrolysis models should be feedstock-specific.

We will close this presentation by discussing the implications of these results for the design and optimal operation of "feedstock-blind" reactors.

A COUPLED LAGRANGIAN-EULERIAN MODEL FOR NUMERICAL SIMULATION OF WIND TURBINES PERFORMANCE UNDER RAINY CONDITIONS

Ahmad Cheshmberah And Hamid Arastoopour

Wanger Institute for Sustainable Energy Research (WISER), Department of Mechanical, Materials and Aerospace Engineering (MMAE), Illinois Institute of Technology (IIT), Chicago, Illinois 60616

Abstract

Wind Turbines power output is constantly influenced by their environmental conditions, including raining and icing. Therefore, understanding the effect of rain is necessary to enhance the efficiency of the wind turbines used in regions with considerable number of rainy days.

To obtain this understanding, we developed a new multiphase computational fluid dynamics (CFD) modele by coupling the conventional Lagrangian Discrete Phase Model (DPM) and the Eulerian Volume of Fluid (VOF) models to be able to simulate the rain and capture the resulting water layer formation over the turbine blades. In this study, we used National Renewable Energy Laboratory's (NREL) S809 airfoil profile of a horizontal-axis wind turbine. In addition, we included the effect of surface tension as well as the surface property of the airfoil.

We used our model to run several simulations in order to identify the effect of each of the following factors: the momentum that rain droplets add to the water layer in addition to their mass, surface tension, and contact angle between water and airfoil.

Our simulation showed that, in general, the rain causes significant increase in drag coefficient compared to the no-rain condition. The more non-wettable the surface is, the larger this unfavorable increase in drag coefficient will be. In case of the lift coefficient, it was observed that for the most part, the rain favorably increases the lift coefficient and this increase was significant for wettable surfaces. We also found out that, the effect of surface tension between water and air on the lift and the drag coefficients is considerably more that the effect of the momentum of the rain droplets.

Based on our results, we can claim that a wind turbine with a wettable surface will suffer the least drag from a rainy atmosphere and, in certain conditions, might even benefit from it through the increase in lift force.

386368 Multiphase Flow Modeling and CFD Analysis during Managed Pressured Drilling (MPD) in Oil Reservoirs

Abstract:

Managed Pressure Drilling (MPD) addresses a multitude of technical problems encountered in carbonate reservoirs of numerous oil fields. The solid cuttings transport flowrate, stability and cost of cuttings transport from the carbonate rock formation towards the surface must be monitored accurately, because drilling safety and efficiency is of paramount importance. A profound understanding of the effect of the geometry and rheology on the pressure profile along the wellbore is essential, so appropriate rheological models (Rooki et al., 2012) of the drilling fluids (derived from viscometric studies) must be used toward technically relevant and numerically reliable CFD modeling studies.

Drilling fluids perform several simultaneous tasks during MPD, offering hydrostatic pressure, cooling the bit, transporting cuttings to surface, maintaining wellbore stability and preventing contamination of the formation. Thus, they must possess acceptable stability in terms of density and rheology over a range of external conditions. Industrial demand is trending towards advanced drilling fluids with tailor-made properties, which can alleviate problems as they arise. Current high-performance requirements foster the development of smart drilling fluids, which are stable at downhole conditions, can perform the required tasks and have a much lower environmental footprint.

Computational Fluid Dynamics (CFD) modeling of cuttings transport during MPD drilling operations can potentially have a strong impact on exploring numerically the attainable envelope of MPD drilling operations, and it has been attempted in very few cases (Nakagawa et al., 1999; Li & Kuru, 2003; Rooki et al., 2013). Research studies for gas-solid-liquid flows (the fundamental cuttings transport mechanism) have been published in the past, but most are not concerned with the elucidation of flow patterns and cuttings distributions, particularly for smart drilling fluids. Modern commercial and/or open-source CFD codes can tackle these problems successfully, yielding detailed state variable (pressure, slip velocity, solids loading) profiles as a function of drilling configuration and pipe position. This paper aims to present relevant case studies for non-Newtonian (e.g. Herschel-Bulkley) drilling fluids, which can be very efficient in MPD implementations.

LITERATURE REFERENCES

1. Li, Y., Kuru E., Numerical modeling of cuttings transport with foam in horizontal wells, *Journal of Canadian Petroleum Technology* 42: 54-61 (2003).

2. Nakagawa E.Y., Silva V., Boas M.B.V., Silva P.R.C., Shayegi S., Comparison of aerated fluids/foam drilling hydraulics simulators against field data, *SPE Paper 54319 presented at the SPE Asia Pacific Oil and Gas Conference and Exhibition*, Jakarta, Indonesia (1999).

3. Rooki, R., Doulati Ardejani, F., Moradzadeh, A., Mirzaei, H., Kelessidis, V.C., Maglione, R., Nourozi, M., Optimal determination of rheological parameters for Herschel–Bulkley drilling fluids using genetic algorithms, *Korea Australia Rheology Journal* 24: 163-170 (2012).

4. Rooki, R., Doulati Ardejani, F., Moradzadeh, A., Norouzi, M., Simulation of cuttings transport with foam in deviated wellbores using computational fluid dynamics, *Journal of Petroleum Exploration & Production Technology* 3(3): 1-11 (2013).

381824 Analytical Model for the Electrochemical Reduction of CO2 in a Microfluidic Cell

Abstract:

Globally, CO_2 emission through the combustion of fossil fuels has increased by about 1.6 times between 1990 (the Kyoto Protocol reference year) and 2013, with approximately 9.9 GtC added to the atmosphere in 2013.[1] To limit the potential effect of climate change caused by carbon emission, many strategies are being explored, including carbon capture and sequestration (CCS) and increasing the use of carbon-neutral energy sources such as nuclear, wind, solar and hydroelectricity. When it is powered by the carbon-neutral electricity source, electrochemical conversion of CO_2 not only provides an economic way of utilizing the CO_2 captured, but also can serve as a storage mechanism for storing excess energy from intermittent renewable sources into fuels.[2]

Many early studies on electrochemical conversion of CO_2 are experimental in nature, focusing on the possible mechanisms for the various products of CO_2 electro-reduction, or exploring different types of electrodes and catalysts to improve performance[3]. As an alternative approach to the laborious experiments, first-principles modeling of electrochemical reactors can complement the current experimental work by elucidating the complex transport and electrochemistry particularly in porous electrodes, and help in designing and optimizing such reactors. Currently, there is a lack of detailed full cell model and no analytical model for the aqueous electrochemical reduction of CO_2 in a microfluidic cell, which has been demonstrated to be an effective reactor and a versatile analytical tool[4].

Here, we will present a full steady-state isothermal model and the approximate analytical solutions for an electrochemical microfluidic cell that reduces CO_2 to CO. Conversion of CO_2 into CO is attractive due to the versatility of CO (with H_2) as a feedstock in Fischer-Tropsch synthesis to a variety of products including liquid hydrocarbon fuels. We will begin our analysis with a full model that takes into account all the significant physics and electrochemistry in the cell, such as the transport of species and charges, momentum and mass conservations, and electrochemical reactions. The full model consists of partial differential equations (PDEs) is solved using finite element method. It is calibrated and validated using experimental data obtained for various inlet flow rates and compositions. A narrow-gap approximation and scale arguments allow for a significant reduction in the mathematical complexity of the full model; that is, the PDEs reduce to a set of ordinary differential equations and Poisson's equations. Integration, Taylor-series expansions, homogenization and separation of variables then allow for approximate analytical solutions. The approximate analytical solutions are verified with the numerical results obtained from the full model; good agreement is found. As the approximate analytical solutions are more computational efficient, it can be suitably extended for stack design and optimization.

1. T. A. Boden, G. Marland and R. J. Andres, ed. O. R. N. L. Carbon Dioxide Information Analysis Center, Oak Ridge, Tenn., U.S.A., 2010.

2. D. T. Whipple and P. J. A. Kenis, *The Journal of Physical Chemistry Letters*, 2010, **1**, 3451-3458.

3. H.-R. M. Jhong, S. Ma and P. J. A. Kenis, *Current Opinion in Chemical Engineering*, 2013, **2**, 191-199.

4. D. T. Whipple, E. C. Finke and P. J. A. Kenis, *Electrochemical and Solid-State Letters*, 2010, **13**, B109.

384880 Development of a Mathematical Model for Microbial Desalination Cells

Abstract:

Conventional waste water treatment processes, especially the aeration process, typical require a lot of energy input. In particular, wastewater treatment consumes 2% of global power capacity at an annual cost of $40 billion worldwide. Microbial fuel cells (MFCs) can convert the organic compounds in the waste water to electrons and protons by microorganisms growing on the anode surface. The electrons can be used by external electrical circuit as power supply, while the protons move to the cathode through the ion exchange membrane and interact with oxygen to produce water on the cathode surface. Since MFCs can produce power from the organic compounds in waste water, MFCs may provide a sustainable way for water treatment. Microbial desalination cells (MDCs) derive from MFCs by adding a third compartment between the anode and the cathode, separated by anion or cation exchange membranes for a new function of desalination. Given complex desalination processes and strong interactions between biological, electrochemical, and engineering factors in MDCs, a proper mathematic model will be essential for the optimization and the scaling up of MDCs. While several mathematical models exist for MFCs or microbial electrolysis cells (MECs) [1-8], no model has been developed for MDCs. Based on an existing MFC model [4], a mathematic MDC model was developed in this work to quantify MDC desalination performance under dynamic loadings (for both substrate and saline water) and different external resistance.

The developed MDC model considers the desalination process driven by both the electric potential between an anode and a cathode electrodes and the salt diffusion process due to a concentration gradient across the ion exchange membranes. The Nernst-Monod equations were used to quantify substrate consumption and bacterial growth. The model was calibrated using experimental data obtained from a lab-scale MDC upon the change of substrate flowrates, and validated by the data from the experimental conditions with different substrate concentrations, salt concentrations, and external electrical resistance. The validated model was then used to predict the performance of the MDC affected by either single or multiple operating parameters. Optimal operation parameters such as influent acetate feed concentration and flow-rate, influent salt feed concentration and flow-rate, external electrical resistance, and the ratio of the volumes of the anode and the salt compartments, were determined from the simulations. To the best of our knowledge, this is the first mathematic model for MDCs.

References:

[1] Peng, S.K., Liang, D.W., Diao, P., Liu, Y.Y., Lan, F., Yang, Y.H., Lu, S.F., Xiang, Y. Nernst-ping-pong model for evaluating the effects of the substrate concentration and anode potential on the kinetic characteristics of bioanode. *Bioresour. Technol.* **2013,** *136*, 610-616.

[2] Picioreanu, C., Head, I.M., Katuri, K.P., van Loosdrecht, M.C.M., Scott, K. A computational model for biofilm-based microbial fuel cells. *Water Res.* **2007,** *41*, 2921-2940.

[3] Zeng, Y.Z., Choo, Y.F., Kim, B.H., Wu, P. Modelling and simulation of two-chamber microbial fuel cell. *J. of Power Sources.* **2010,** *195*, 79-89.

[4] Pinto, R. P.; Srinivasan, B.; Manuel, M. F.; Tartakovsky, B. A two-population bio-electrochemical model of a microbial fuel cell. *Bioresour. Technol.* **2010,** *101* (14), 5256-5265.

[5] Pinto, R. P.; Srinivasan, B.; Escapa, A.; Tartakovsky, B., Multi-Population Model of a Microbial Electrolysis Cell. *Environ. Sci. Technol.* **2011,** *45*, 5039-5046.

[6] Marcus, A. K.; Torres, C. I.; Rittmann, B. E. Conduction-based modeling of the biofilm anode of a microbial fuel cell. *Biotechnol. Bioeng.* **2007,** *98* (6), 1171-1182.

[7] Marcus, A.K., Torres, C.I., Rittmann, B.E. Evaluating the impacts of migration in the biofilm anode using the model PCBIOFILM. *Electrochim. Acta.* **2010,** *55*, 6964-6972.

[8] Picioreanu, C.; van Loosdrecht, M. C. M.; Curtis, T. P.; Scott, K. Model based evaluation of the effect of pH and electrode geometry on microbial fuel cell performance. *Bioelectrochemistry.* **2010,** *78* (1), 8-24.

379819 Inventory Optimization for Production Network Flexibility

Abstract:

The industrial production of chemicals is characterized by complex networks of unit operations that transform raw materials into products. In these networks, material flows are constrained by the production capacity of the processing units, and the performance of the whole system often depends on the capacity of some limiting processes. The rigorous design of production networks is intended to coordinate the production capacity across all units under normal operation, but the variability in processing conditions often creates temporal bottlenecks that affect network performance.

It is a common practice in industry to allocate storage units at different stages of the production network in order to decouple the production of successive sections. The role of inventory is to buffer temporal mismatches among supply availability, processing rates, and demand. Different types of inventories are used for each purpose: raw material inventories hedge against variability in supply, intermediate inventories protect against variability in processing capacity, and final product inventories hedge against variability in demand. The importance of intermediate inventories resides in their ability to reduce processing units' interdependence by delaying the formation of bottlenecks and increasing capacity utilization. In this framework, inventory management strategies can have a significant impact in the performance of the production network.

Inventory management includes all decisions related to the replenishment and depletion of inventories. In the context of production networks, inventory decisions are closely related to production decisions because most processing units are simultaneously internal suppliers and consumers. Furthermore, inventory replenishment requires using available capacity in the upstream processing units, which reinforce the role of capacity constraints in production networks.

A realistic representation of variability in production networks implies adopting a probabilistic description of the corresponding process parameters. The resulting mathematical model integrates the stochastic processes describing variability, the production decisions limited by capacity constraints, and the history dependent levels of inventory. The characterization of these stochastic inventory levels in capacitated production networks is a difficult task. They depend on the other random processes involved in the system and on the inventory management decisions.

The optimal inventory management strategy for a single storage unit subject to capacitated replenishment and uncertain demand has been proved to follow a modified base-stock policy [1]. The policy consists of a set of rules intended to maintain the inventory at the base-stock level; however, this is not always possible because inventory replenishments are constantly limited by upstream production capacity. Using a base-stock policy in integrated production networks not only determines the inventory management, but also the production strategy that maximizes utilization when the inventory is below the base-stock level.

Adopting base-stock policies reduces the inventory management problem to finding the optimal base-stock level for each inventory location and to determine replenishment priorities among them. However, the resulting formulation is very challenging even for the simplest probabilistic descriptions of the stochastic parameters. In essence, the formulation includes stochastic differential equations that describe the rate of change of the inventory levels as functions of stochastic parameters and the base-stock policies.

In order to overcome the difficulties associated with such a complex formulation, a discrete-time version of the model is used. The model is based on the principles of discrete-event simulation, which is a widely used method to evaluate inventory management strategies. The idea is to include the discrete time model together with base-stock policies in an optimization problem that determines base-stock levels and replenishment priorities. In this framework, the variability in supply, processing capacity, and demand can be represented with sample-paths of the corresponding stochastic processes during long time horizons.

The sample-path optimization approach has the advantage that it provides an approximation of the optimal solution of the stochastic model using as a basis the deterministic problem given by random samples. Additionally, it has the flexibility to analyze production networks of complex topology and arbitrary characterizations of the stochastic processes.

The proposed methodology is implemented in two examples with variability in supply, production capacity, and demand. The first example is a single production-inventory system; it is used to illustrate the implementation of the base-stock policy and the solution method. The second example is a production network with processing units in series and parallel, and multiple inventory locations. The results show the importance of inventory management in the performance of production networks and the effectiveness of the proposed approach.

References

[1] Federgruen, A., and P. Zipkin. 1986. An Inventory Model with Limited Production Capacity and Uncertain Demands I. The Average-Cost Criterion. *Mathematics of Operations Research,* 11, 193.

367801 Plant-Wide Coordination for the Energy-Efficient Scheduling of an Integrated Steel Production Process

Abstract:

In this work, we present a coordination method to synchronize the schedules of two consecutive production sections in an integrated steel plant: meltshop and hot rolling mill. The meltshop produces steel slabs, which are stored in a slab yard, and then have to be reheated before being processed by the hot rolling mill. A well-coordinated schedule allows some slabs from the melt shop to be charged directly into the reheating furnace of the hot rolling mill. This operational technique, called hot charging, leads to substantial energy savings and productivity increase but is challenging in terms of production planning and scheduling.

A review detailing the different hot charging techniques and corresponding integrated production scheduling approaches can be found in Tang [5]. The problem of generating an overall production schedule satisfying both meltshop and hot strip mill constraints has been tackled using AI methods (expert systems, random heuristic search and constraint satisfaction problems), multi agent methods [4] or operations research methods. Our work is based on the approach proposed by Xu [6][7], consisting of a bi-level coordination framework in which the lower-level optimizers are the meltshop and hot rolling mill schedulers while the upper-level focuses only on the bottleneck of the two processes. The advantage of this approach is that it reuses two existing independent optimization modules for the meltshop and the hot rolling mill, without having to develop a monolithic model which would be intractable. The Meltshop Optimizer (MSO) is based on the decomposition strategy presented in Harjunkoski & Grossmann [3] which includes multiple MILP optimization steps using continuous-time formulation and global precedence variables. The Hot Strip Mill Optimizer (HSO) is based on the approach presented by Biondi & Saliba [1] which uses a combination of optimization steps and algorithmic rules in order to build and sequence rolling programs. The upper-level coordinator considers only the bottleneck of the process, namely the continuous casting and hot strip mill stages, and is therefore called intersection model. The coordination objective is to maximize the hot charging ratio by reducing the slab storage times in the slab yard.

An important challenge of this coordination framework is that the intersection model considers only the production stages of the bottleneck of the process. For this reason it is not guaranteed that the schedule proposed by this simplified model would be feasible when considering the constraints of all upstream and downstream production stages. Xu [6][7] proposed an iterative framework in which logical and integer cuts are progressively added to the intersection model, in order to remove from the search space the job sequences and machine assignments that are not feasible for the lower level schedulers. In this work we show that the potential inability of the sub-optimizers to follow the desired master schedule is not caused by infeasible job sequences but instead by an unrealistic timing of the jobs that doesn't correspond to the normal process operation and throughput capacity. The problem comes from the difficulty to generate a realistic continuous casting schedule without considering explicitly the upstream steel-making stages. Our approach is to use the information contained in an initial overall solution obtained by a first call of the MSO and HSO schedulers. From this feasible solution, we extract knowledge about the meltshop process in terms of load balancing of the casters and casting production capacity. Based on this data we then generate automatically additional constraints to the intersection model, which lead to a more realistic master schedule.

For a better evaluation of hot charging, the inner structure of rolling programs and casting sequences are considered in the intersection model. The hot charging condition for each slab (whether the waiting time of the slab is lower than a certain time limit) can be represented by a binary variable. This model extension doesn't add much complexity to the earlier heatgroup/program based model since the assignment, precedence and timing decision variables are maintained. The shifting algorithm, a linear shift of the hot rolling mill schedule in order to guarantee feasibility, is also modified to take into account this deeper granularity of the mathematical model. The coordination framework has been tested on several real-world large-scale instances, and the results show the benefits of the proposed model developments in terms of hot charging ratio.

References:

(1) Biondi, M., Saliba, S. and Harjunkoski, I., 2011, Production Optimization and Scheduling in a Steel Plant: Hot Rolling Mill, 18th IFAC World Congress Milano (Italy)

(2) Cowling, P. I. and Rezig, W. (2000) Integration of continuous caster and hot strip mill planning for steel production. Journal of Scheduling, 3 (4), 185-208.

(3) Harjunkoski and I. E. Grossmann, 2001, A Decomposition Approach for the Scheduling of a Steel Plant Production, Computers & Chemical Engineering, 25, pp. 1647-1660

(4) Ouelhadj D. 2003, A Multi Agent system for the integrated dynamic scheduling of steel production, PhD Thesis, University of Nottingham.

(5) Tang, L., Liu, J., Rong, A. and Yang, Z., 2001, A review of planning and scheduling systems and methods for integrated steel production, European Journal of Operational Research, Vol.133, Issue 1, pp.1-20

(6) Xu, C., 2013, Coordination and Decomposition of Large-Scale Planning and Scheduling Problems with Application to Steel Production, Thesis, Technische Universität Dortmund.

(7) Xu, C., Sand, G., Harjunkoski, I., Engell, S., 2012, A new heuristic for plant-wide schedule coordination problems: The intersection coordination heuristic, Journal of Computers & Chemical Engineering, Vol. 42, 11 July 2012, pp. 152–167

375253 Mathematical Modeling and Global Optimization for Entire Petrochemical Planning Operations

Abstract:

In the last twenty years, the petrochemical industry has succeeded by creating markets and supplying them with suitable products used to create goods such as plastics, cosmetics, lubricants, and paints. Petrochemical production begins in a refinery that separates crude oils mainly into lighter components such as naphtha, light naphtha, top oil, and liquid fuels including gasoline, diesel and jet fuel. The lighter components are further processed into various petrochemicals such as ethylene, propylene, butadiene, benzene, toluene, xylol and some other high-valued products via cracking, butadiene extraction, hydrotreating, etherification, and polymerization processes. Nowadays, tighter competition, strict environmental regulations, and lower-margin profits, drive the petrochemical industry to apply new technologies to improve their planning operations.

Integrated petrochemical operations include both refinery and chemical production operations, such as crude oil blending and processing, production unit operations, and product blending and distribution [1-2]. Planning of integrated petrochemical operations involves optimization of crude blending, processing amounts for production units, unit production modes, flow connections between production units, and pooling and blending operations to satisfy quality requirements of production units, intermediates and final products. Mathematical modeling of production units, pooling and blending operations introduces bilinear, trilinear terms, and higher order terms, which result to a mixed integer non-convex nonlinear optimization problem.

The refinery planning problem has received considerable attention since the introduction of linear programming in 1950s. Research focused on developing different models and algorithms to solve large-scale industrial problems, leading to commercial software such as RPMS (Refinery and Petrochemical Modeling System) [3], PIMS (Process Industry Modeling System) [4], and GRTMPS (Haverly Systems) [5]. Commercially available software can be extended to model and optimize integrated petrochemical processes, however, inaccuracy caused by non-rigorous linear models and approximate algorithms may reduce the overall profitability or sacrifice product quality. Moreover, no global optimality can be guaranteed. On the other hand, nonlinear models and specialized algorithms have also been proposed for refinery planning problems. For instance, Pinto and Moro [6] developed a nonlinear planning model for production planning which allows the implementation of nonlinear process models as well as blending relations. Li et al.[7] presented a refinery planning model that utilizes simplified empirical nonlinear process models with considerations for crude characteristics, product yields, and qualities. Alhajri et al.[8] developed a nonlinear model to address the refinery planning problem. Alattas et al. [9] developed a fractionation index based nonlinear model for crude distillation units and integrated it into the linear refinery planning model, solving it with NLP solvers without guaranteeing global optimality. A comprehensive review on refinery planning can be found in Shah et al. [1]. It can be concluded that very few theoretical developments and results have been reported for global optimization of integrated refinery and petrochemical planning problems.

In this presentation, we first propose detailed nonlinear models to predict product yields and properties in production units including crude distillation unit, vacuum distillation unit, hydrocracking units, catalytic cracking units, ethylene-cracking units and other processing units present in a petrochemical plant. The yield and property prediction models for crude distillation and vacuum distillation units are developed using swing cut theory based on crude assay data. Empirical non-linear models are developed for other processing units, including bilinear, trilinear, quadratic, polynomial, and exponential terms. Parameter estimation for all of the developed models is performed by globally minimizing the least squares error between the parametric models and real production data. Moreover, property indices in blending units are linearly additive and calculated on weight or volume basis, which introduce bilinear and trilinear terms. We also introduce binary variables to denote different operation modes for several production units.

The entire planning model is a non-convex mixed integer nonlinear optimization model (MINLP). To solve this large-scale nonlinear model, we propose an optimization-based procedure to obtain the tightest lower and upper bounds for variables especially the variables involving nonlinear terms. Then, we incorporate those tightest lower and upper bounds into commercial solver ANTIGONE [10] to obtain e-global optimality. Finally, a user-friendly platform is developed to allow the user to modify the planning model by updating model parameters when new data is available, product demands and specifications, cost parameters and many more. Several large-scale industrial examples are solved to illustrate the efficiency of our proposed model and global optimization approach.

References

[1] Shah, N. K.; Li, Zu; Ierapetritou, M. G. Petroleum Refining Operations: Key Issues, Advances, and Opportunities, *Industrial and Engineering Chemistry Research*, 2011, *50*, 1161-1170.

[2] Shah, N. K.; Ierapetritou, M. G. Short-Term Scheduling of a Large-Scale Oil-Refinery Operaitons: Incorporating Logistics Details. *AIChE Journal*2011, 57, 1570-1584.

[3] RPMS (Refinery and Petrochemical Modeling Systems): A System Description; Bonner and Moore: Houston, TX, 1979.

[4] Aspen PIMS System Reference (v7.2), Aspen Technology Inc.; Burlington, MA, 2010.

[5] GRTMPS (Haverly Systems), http://www.haverly.com/main-products/13-products/9-grtmps.

[6] Pinto, J. M.; Moro, L. F. L. A Planning Model for Petroleum Refineries. *Braz. J. Chem. Eng.* 2000, *17*, 575-585.

[7] Li, W. K.; Hui, C. W.; Li, A. X. Integrating CDU, FCC, and Product Blending Models into Refinery Planning. *Computers and Chemical Engineering* 2005, *29*, 2010-2028.

[8] Alhajri, I.; Elkamel, A.; Albahri, T.; Douglas, P. L. A Nonlinear Programming Model for Refinery Planning and Optimization with Rigorous Process Models and Product Quality Specifications. International Journal of Oil, Gas, and Coal Technology 2008, 1, 283-307.

[9] Alattas, A. M.; Grossmann, I. E.; Palou-Rivera, I. Integration of Nonlinear Crude Distillation Unit Models in Refinery Planning Optimization. *Industrial and Engineering Chemistry Research* 2011, *50*, 6860-6870.

[10] Misener, R.; Floudas, C. A. ANTIGONE: Algorithms for continuous/Integer Global Optimization of Nonlinear Equations. Journal of Global Optimization, In press, 2014.

389669 Integrated Planning, Scheduling, and Dynamic Optimization of Continuous Manufacturing Processes

Abstract:

Planning, scheduling, and dynamic optimization of chemical processes are highly interconnected [1-5]. The aim of planning is to determine the production assignment of the manufacture processes from the enterprise level over a long time horizon. In each planning periods, scheduling is used to determine the production sequence and time, task assignment, resource allocation, etc. Dynamic optimization is to determine the optimal transition processes when different products are manufactured in a processing unit sequentially. Though the three problems are able to be solved sequentially, the integration of these problems will result in a better performance compared to sequential approaches and has gathered increasing attention recently [6-16].

However, most of the previous studies concentrate on only parts of the integrated problem. Some works focus on the integrated problem of planning and scheduling, where the transition times and the transition costs are assumed to be known parameters [12]. Some other works concentrate on the integration of scheduling and dynamic optimization considering variable transition times and transition costs, which are determined by time-dependent trajectories [7, 17-20]. The integrated problem is formulated into a mixed-integer dynamic optimization (MIDO) model, which is then reformulated into a mixed-integer nonlinear programming (MINLP) full space model by discretizing the differential equations [7, 21, 22]. The full space model characterizes all the detailed information of planning, scheduling, and dynamic optimization, however, a large amount of computational time is needed to solve the full space model due to its large-scale [23].

In this work, we propose a novel method for solving the integrated planning, scheduling, and dynamic optimization problem for a multi-product continuous process. We first reformulate the integrated MIDO problem into a large-scale MINLP full space problem. To reduce the computational complexity, we propose an efficient flexible recipe method which can approximate the full space model into an online planning and scheduling model and offline dynamic optimization models. The key is that the planning and scheduling model is linked to the dynamic models via transition times and transition costs. The relationship between a transition time and the corresponding transition cost can be approximated by a set of discrete points, which are obtained by solving the dynamic optimization problems offline. The candidate transition times and transition costs are then appended to the planning and scheduling level problem, replacing the nonlinear equations concerning the dynamic models. The flexible recipe method ends up to be a mixed-integer linear program (MILP), which is easier to solve than the original MINLP full space problem. A bi-level decomposition method is then introduced to further improve the computational efficiency of the flexible recipe method, which formulates the upper level planning problem and lower level detailed problem. The Bi-level method then solves the upper and lower level problems iteratively until a pre-determined stopping condition is met.

To demonstrate the applicability of the proposed full space modeling framework, flexible recipe method and bi-level decomposition algorithm, we investigate a methyl methacrylate (MMA) polymerization process with azobisiso-butyronitrile initiator and a toluene solvent. The dynamic model of the MMA process is described by a set of differential equations. Different products with different molecular weights are produced during the steady state production time period. Through the proper arrangement of the production manufacturing, the process industry will be able to increase their profits. An integrated model is formulated to contain decision making across production assignment, production sequence and time and transition process strategy according to specific order demand. In a small scale problem, there are 3 products and 3 24-hour-long planning periods. The result reveals that: both the flexible recipe method and the bi-level decomposition algorithm can solve the problem within 2 seconds with the MILP solver CPLEX 12, while the full space method takes 3,000 seconds computational time to solve the problem with the MINLP solver SBB. In a large scale problem, there are 5 products and 4 168-hour-long planning periods. The bi-level decomposition algorithm can solve the problem within 2 minutes and the flexible recipe method can solve the problem with about 40 min (both methods use the MILP solver CPLEX 12), while the full space method fails to obtain a feasible solution within 30 hours with the MINLP solver SBB. From the result of the case study, the computational efficiency of the proposed solution methods is demonstrated.

References

[1] J. Kallrath, "Planning and scheduling in the process industry," *Or Spectrum,* vol. 24, pp. 219-250, Aug 2002.

[2] I. Harjunkoski, R. Nyström, and A. Horch, "Integration of scheduling and control—Theory or practice?," *Computers & Chemical Engineering,* vol. 33, pp. 1909-1918, 2009.

[3] E. Capon-Garcia, M. Moreno-Benito, and A. Espuna, "Improved Short-Term Batch Scheduling Flexibility Using Variable Recipes," *Industrial & Engineering Chemistry Research,* vol. 50, pp. 4983-4992, May 4 2011.

[4] P. M. Verderame, J. A. Elia, J. Li, and C. A. Floudas, "Planning and Scheduling under Uncertainty: A Review Across Multiple Sectors," *Industrial & Engineering Chemistry Research,* vol. 49, pp. 3993-4017, May 5 2010.

[5] C. T. Maravelias and C. Sung, "Integration of production planning and scheduling: Overview, challenges and opportunities," *Computers & Chemical Engineering,* vol. 33, pp. 1919-1930, Dec 10 2009.

[6] I. Grossmann, "Enterprise-wide optimization: A new frontier in process systems engineering," *AIChE Journal,* vol. 51, pp. 1846-1857, Jul 2005.

[7] Y. F. Chu and F. Q. You, "Integration of scheduling and control with online closed-loop implementation: Fast computational strategy and large-scale global optimization algorithm," *Computers & Chemical Engineering,* vol. 47, pp. 248-268, Dec 20 2012.

[8] Y. F. Chu and F. Q. You, "Integrated Scheduling and Dynamic Optimization of Complex Batch Processes with General Network Structure Using a Generalized Benders Decomposition Approach," *Industrial & Engineering Chemistry Research,* vol. 52, pp. 7867-7885, Jun 12 2013.

[9] J. J. Zhuge and M. G. Ierapetritou, "Integration of Scheduling and Control with Closed Loop Implementation,"

Industrial & Engineering Chemistry Research, vol. 51, pp. 8550-8565, Jun 27 2012.

[10] S. Terrazas-Moreno, A. Flores-Tlacuahuac, and I. E. Grossmann, "Lagrangean heuristic for the scheduling and control of polymerization reactors," *AIChE Journal,* vol. 54, pp. 163-182, Jan 2008.

[11] Z. K. Li and M. G. Ierapetritou, "Integrated production planning and scheduling using a decomposition framework," *Chemical Engineering Science,* vol. 64, pp. 3585-3597, Aug 15 2009.

[12] M. Erdirik-Dogan and I. E. Grossmann, "A decomposition method for the simultaneous planning and scheduling of single-stage continuous multiproduct plants," *Industrial & Engineering Chemistry Research,* vol. 45, pp. 299-315, Jan 4 2006.

[13] Y. Chu, F. You, J. M. Wassick, and A. Agarwal, "Integrated planning and scheduling under production uncertainties: Bi-level model formulation and hybrid solution method," *Computers & Chemical Engineering,* 2014.

[14] D. B. Birewar and I. E. Grossmann, "Simultaneous Production Planning and Scheduling in Multiproduct Batch Plants," *Industrial & Engineering Chemistry Research,* vol. 29, pp. 570-580, Apr 1990.

[15] D. J. Yue and F. Q. You, "Planning and Scheduling of Flexible Process Networks Under Uncertainty with Stochastic Inventory: MINLP Models and Algorithm," *AIChE Journal,* vol. 59, pp. 1511-1532, May 2013.

[16] A. Flores-Tlacuahuac and I. E. Grossmann, "Simultaneous cyclic scheduling and control of a multiproduct CSTR," *Industrial & Engineering Chemistry Research,* vol. 45, pp. 6698-6712, Sep 27 2006.

[17] Y. F. Chu and F. Q. You, "Integrated Scheduling and Dynamic Optimization of Sequential Batch Processes with Online Implementation," *AIChE Journal,* vol. 59, pp. 2379-2406, Jul 2013.

[18] Y. F. Chu and F. Q. You, "Integration of production scheduling and dynamic optimization for multi-product CSTRs: Generalized Benders decomposition coupled with global mixed-integer fractional programming," *Computers & Chemical Engineering,* vol. 58, pp. 315-333, Nov 11 2013.

[19] Y. F. Chu and F. Q. You, "Integration of Scheduling and Dynamic Optimization of Batch Processes under Uncertainty: Two-Stage Stochastic Programming Approach and Enhanced Generalized Benders Decomposition Algorithm," *Industrial & Engineering Chemistry Research,* vol. 52, pp. 16851-16869, Nov 27 2013.

[20] A. Flores-Tlacuahuac and I. E. Grossmann, "Simultaneous Scheduling and Control of Multiproduct Continuous Parallel Lines," *Industrial & Engineering Chemistry Research,* vol. 49, pp. 7909-7921, Sep 1 2010.

[21] V. Bansal, V. Sakizlis, R. Ross, J. D. Perkins, and E. N. Pistikopoulos, "New algorithms for mixed-integer dynamic optimization," *Computers & Chemical Engineering,* vol. 27, pp. 647-668, May 15 2003.

[22] R. J. Allgor and P. I. Barton, "Mixed-integer dynamic optimization I: problem formulation," *Computers & Chemical Engineering,* vol. 23, pp. 567-584, May 1 1999.

[23] L. T. Biegler, *Nonlinear programming: concepts, algorithms, and applications to chemical processes* vol. 10: SIAM, 2010.

Integrated Refinery Planning and Scheduling

Pedro Castillo Castillo, Vladimir Mahalec
Department of Chemical Engineering, McMaster University, Hamilton, ON, Canada

Petroleum refineries produce multiple products be separating crude oils into fractions which are then processed further by various reaction and separation units to produce components which are blended into final products. In order to accommodate seasonal variations in demand for various products, major refinery units (e.g. crude distillation, FCC, catalytic reforming) change their operation from one operating mode to another (e.g. maximum gasoline or maximum diesel). Refinery long term production plan (typically 12 months) computes how much to produce by each process unit in each mode of operation during each month over the planning horizon. Industry practice is to employ month by month multiperiod model to compute the production via successive LP (e.g. Aspen PIMS or Haverly) or via NLP (e.g. Aspen PIMS AO). Results from the production plan become constraints for refinery scheduling, which is most often carried out via interactive simulation (e.g. Aspen ORION). Since monthly multi-period model is guaranteed to be feasible only at the period boundaries, the constraints imposed by the production plan may sometimes lead to schedules which are intra-period infeasible.

This work introduces inventory-pinch based algorithm for integration of long term production plans with short term planning and scheduling. Instead of using month by month multi-period model, we use a model where periods are delineated by the inventory pinch points for various product pools (e.g. gasoline and diesel). This reduces drastically the number of periods in the long term planning model and enables use of nonlinear, more accurate refinery models as a basis for long term planning. In order to increase likelihood that the constraints for short term planning model lead to feasible schedules, we decompose the planning problem into two levels: first level solves for optimal operating conditions via multiperiod NLP models, while the second level computes via multiperiod MILP both the long term plan and a short term plan; the latter is comprised of daily and weekly periods for the first month of the planning horizon. Constraints from the short term planning model are used to formulate continuous time scheduling models for refinery subsystems. Overall refinery schedule is computed by a decomposition algorithm which aims to minimize the number of switches while operating the process units at their optimal operating states.

Integrated Planning of Oil Products Blending and Multimodal Distribution

Diego C. Cafaro, INTEC(UNL-CONICET), Santa Fe, ARGENTINA
Agustín F. Montagna, Chem. Eng. School, UNL, Santa Fe, ARGENTINA
Jaime Cerdá, INTEC(UNL-CONICET), Santa Fe, ARGENTINA

Summary

Planning blending operations has become a critical task for petroleum refiners. The physical mixture of refined hydrocarbons permits to produce finished products under increasingly stringent specifications. Blending of gasoline and distillates is accomplished by mixing proportionate amounts of different oil fractions. Moreover, additives including octane enhancers, metal deactivators, anti-oxidants, anti-knock agents, gum and rust inhibitors, are added to provide specific properties not inherent in hydrocarbons [1]. For this purpose, the refinery infrastructure usually comprises component and product tanks as well as blending equipment with multiple operational features, like limited production rates and finite capacities.

In turn, refined products distribution is a very challenging problem involving multiple transportation modes. Oil pipelines are extensively used by petroleum refiners for carrying large quantities of refined petroleum products over long distances on land [2]. Ships are the other major mode for the primary transportation of oil products from refineries to distribution centers or harbors. Then, minor pipelines, trains or barges deliver oil products from distribution centers to petrochemical plants, airports or hubs, while trucks are the most common choice for the last-mile distribution, from transportation hubs to gas stations or small customers.

Scheduling pumping and delivery operations in multiproduct pipelines is a complex logistic problem. Limited tank capacities, delivery dates and refinery production plans are problem conditions to be taken into account. But one of the most challenging issues is the interface generation. As different products are shipped through the same line, usually without separation devices, a product mixture is formed in the interface of two consecutive batches. The product contamination strongly depends on the ordered pair of species put in contact. Optimizing the size and sequence of product batches transported through pipelines requires accurate models and efficient computational tools [3].

Besides, the optimal planning of other transportation modes (ships, trains and trucks) seeks to determine the set of delivery routes for a fleet of vehicles, commonly housed at depots [4]. Every vehicle route must start and finish at the assigned depot, each customer is to be visited by at least one of the vehicles, and vehicle capacities must not be exceeded. The objective is usually the minimization of the travel cost, preventing from delays in demand fulfillment.

To reduce refinery operation costs, blending operations should be effectively integrated to final product distribution, in order to fulfill customer demands with the right product at the right time. The short-term planning of both blending and delivery operations becomes a complex task due to the need of managing a large number of orders, delivery time-windows, multiple transportation modes, alternate blending equipment, limited number of product tanks, quality specifications, and variable product recipes. Manual and non-exact planning generates inefficient schedules that result in higher production costs, quality giveaways, larger inventories, and delivery demurrages. Optimization techniques such as mixed-integer linear programming (MILP) models can be used to generate better schedules that aim to minimize operation costs and improve customer satisfaction. Recently, MILP approaches based on unit-specific, slot-based representations that mimic the production sequence have been proposed. However, even simplifying the delivery plan representation, the CPU time required to obtain feasible solutions is extremely large, and the usefulness of such tools in a real-world environment is actually low [5,6,7].

To overcome this limitation, we develop a comprehensive strategy based on a more efficient MILP formulation [8]. The new computational tool is able to find an integrated short-term plan of blending and distribution tasks, simultaneously optimizing the operation of all the transportation modes. The proposed model makes use of efficient sequencing and inventory tracking constraints, which permits to optimally determine: (a) the number of batches of each product to be processed, the operation times and production rates; (b) the recipe for each batch; (c) the batch-blender allocation, sequencing and timing; (d) the assignment of product batches to storage tanks; and (e) the multimodal delivery plan to timely fulfill customer orders over the planning horizon.

The major elements of the model are: (i) blending components and tanks, (ii) blenders, (iii) storage tanks for the blended products, and (iv) single-product orders to be fulfilled over the planning horizon and their delivery time windows. The model is based on the following assumptions:

(a) Each order should be served through a certain transportation mode, connected to a limited number of finished product tanks.
(b) Blending components and final products have known quality specifications given by the property indices, such as octane number and sulfur index. Each blend component is characterized by constant properties, while blended products properties should be within the desired interval.
(c) Blenders operate in a semi-continuous mode, mixing several component flows in different proportions to get a batch of blended product.
(d) Similar to previous works [6], linear mixing correlations on either volume or weight basis are assumed to determine the product properties.
(e) Since non-identical blenders are available, the number of batches of each product to be processed is initially unknown [9].
(f) Component tanks can receive and discharge flows at the same time.
(g) A single product can be hold in each tank during the time horizon.
(h) The delivery of a product batch from the assigned tank can start only if its production run has finished and is completely unloaded into the tank.

Linear model constraints are destined to: (1) allocate product batches to blenders, (2) determine the sequence of product batches processed in each blender, (3) fulfill customer orders, (4) coordinate product deliveries, (5) monitor inventories of components and finished products at refinery tanks, (6) control refined product properties according to specifications, and (7) identify product changeovers at each blender. Finally, the objective function aims to minimize the sum of production, demurrage and transitions costs.

Compared to previous works, results show drastic reductions in the CPU time, of up to three orders of magnitude in real-world problems of the oil industry. This alternative formulation is able to find the optimal planning of blending and distribution operations at modest CPU times, thus satisfying a challenging requirement of the modern refining industry. The model efficiency relies on very tight relaxations, with integrality gaps below 1%.

References

1. "Blending in petroleum refining" (2006). http://www.eoearth.org/view/article/150699. Accessed: May 5, 14.
2. Cafaro, D.C. and J. Cerdá (2012). Comput. Chem. Eng., 38, 185-203.
3. Cafaro, V.G., Cafaro, D.C., Méndez, C.A. and J. Cerdá (2012). Comput. Chem. Eng., 30, 252-256.
4. Cocccola, M., Cafaro, V.G., Méndez, C.A. and D.C. Cafaro (2014), Ind. Eng. Chem. Eng., In press.
5. Li, J., Karimi, I.A. and R. Srinivasan (2009), AIChE J., 56, 2, 441-465.
6. Li, J. and I.A. Karimi (2011), Ind. Eng. Chem. Res., 50, 15, 9156-9174.
7. Li, J., Li, W.K., Karimi, I.A. and R. Srinivasan (2007), AIChE J., 53, 2659-2680.
8. Montagna, A.F., Cafaro, D.C. and J. Cerdá (2013). Comput. Aided Chem. Eng., 32, 649-654.
9. Marchetti, P.A., Méndez, C.A. and J. Cerdá (2010), Ind. Eng. Chem. Res., 49, 6482-6498.

Integration of Procurement Planning and Short-term Scheduling in Petroleum Refineries under Consideration of Crude Delivery Uncertainty

Shujing Zhang and Qiang Xu

Dan F. Smith Department of Chemical Engineering
Lamar University, Beaumont, Texas 77710, USA

Abstract

Petroleum refineries is always confronted with various risks that affect their normal operations and profitability. Risk factors include volatile crude market, demand changes of petroleum products, and increasingly strict environmental regulations. Especially, the availability of new crude sources will necessitate the evaluation of refinery assets to leverage refinery procurements in order to maximize their profits. Facing tremendous challenges on multiple fronts, refiners need to examine different aspects of supply chain components and continuingly improve their bottom line. Long-term planning and short-term scheduling constitute two important decision elements for petroleum refineries. Planning is responsible for creating profitable plans exploiting the constantly changing market opportunities, while scheduling is required to execute feasibly and optimally as much as possible according to obtained plans. Unfortunately, due to their different time interests and process constraints considered, discrepancies always exist, which lead to inherent gap between generated plans and schedules. It is estimated that bridging actual scheduled operations closer to the optimal plan while respecting the process constraints for refineries could yield refining companies up to 10 cents per barrel. Therefore, two-way communications between planning and scheduling need to be enhanced in order to validate the generated plans and make any necessary adjustments for the plans to be feasible under real operation settings.

Literature review suggests that limited attention has been devoted to the integration of long-term procurement planning and detailed crude scheduling. Existing publications are either focused on an aggregated planning problem without considering realistic operating practices and connection decisions between units [1], or neglecting the ever-changing crude supplies/product demands, variable shipment sizes and demurrage costs [2]. Most importantly, the aforementioned works disregard the continuous operation of crude distillation units. The reason for these sacrifices lies in the fact that that long planning horizon imposes extreme difficulties on solving a full-space problem with integrated formulation of planning and scheduling. Therefore, decomposition strategies are broadly deployed [3]. In this paper, an iterative framework integrating refineries crude procurement and detailed scheduling under consideration of crude delivery uncertainty is proposed. The long-term planning is decomposed into sub-periods representing different scheduling horizons. Hierarchically, the planning problem serves as a master sub-problem while the detailed scheduling is a salve sub-problem. The crude purchasing and refining plans from master problem are provided as inputs to the slave problem. Under the circumstances that any planned operations are impossible with detailed processing constraints or delivery uncertainty, a feedback will be sent from the slave problem to the master problem in the form of extra constraints. A forward rolling-time horizon framework will be created in such a way that once a portion of the time horizon has been scheduled, no additional modifications are allowed for the given period of time horizon. Thus, the number of iterations required to schedule the entire time horizon will be minimized. Accordingly, the finalized refining profits will be compromised due to consistent refinement from detailed scheduling problem. A case study will be demonstrated to validate the capability of the proposed integration method.

References

1. Zhang, J., Y. Wen, and Q. Xu, Simultaneous optimization of crude oil blending and purchase planning with delivery uncertainty consideration. Industrial & Engineering Chemistry Research, 2012. 51(25): p. 8453-8464.
2. Oddsdottir, T.A., M. Grunow, and R. Akkerman, Procurement planning in oil refining industries considering blending operations. Computers & Chemical Engineering, 2013. 58: p. 1-13.
3. Maravelias, C.T. and C. Sung, Integration of production planning and scheduling: Overview, challenges and opportunities. Computers & Chemical Engineering, 2009. 33(12): p. 1919-1930.

Model-Free Control of Fractional Crystallization

The Department of Energy must extract, vitrify, and encase the hazardous constituents contained in 56 million gallons of nuclear waste currently stored at the Hanford site in the state of Washington. Separating non-radioactive components from the waste can reduce costs and expedite cleanup efforts (Herting, 1996; Nassif et al., 2008). In particular, sodium salts—which are non-radioactive and innocuous by themselves—make up a significant portion of the unprocessed waste and potentially can be removed with a crystallization-separation operation and treated as low-activity waste. However, as nuclear waste solutions contain many solutes that can crystallize, methods for reliably monitoring and controlling *multi-solute* crystallizations are needed. The operation in such a circumstance is referred to as fractional crystallization.

In this work we explore the use of process analytical measurement techniques to implement model-free methods for controlling fractional crystallizations of a model two-solute system (sodium nitrate and sodium sulfate in water) with the objective of improving the crystal size distribution to facilitate separation. Model-free control strategies—such as supersaturation or concentration control (Fujiwara et al., 2005)—have been successfully applied to control single-solute crystallization (Nagy and Braatz, 2012), but have not been extended to multicomponent crystallizations. The presence of multiple solutes affects solution behavior and can convolute measurement signals. This introduces complications in the supersaturation control strategy, which relies on solubility information and concentration measurement feedback (via infrared or other spectroscopic techniques). In considering these complications for the model two-solute solution we developed methodologies for quickly gathering the solubility of a multicomponent solution and also extracting the concentration of multiple components from the complex solution infrared absorbance spectra. Applying these techniques we are able to apply supersaturation control during fractional crystallization.

Fujiwara, M., Nagy, Z. K., Chew, J. W. and Braatz, R. D. 2005. First-Principles and Direct Design Approaches for the Control of Pharmaceutical Crystallization. *Journal of Process Control* 15, 493-504

Herting, D. L. 1996. Clean Salt Process - Final Report. Environmental Management

Nagy, Z. K. and Braatz, R. D. 2012. Advances and New Directions in Crystallization Control. *Annual Review of Chemical and Biomolecular Engineering, Vol 3* 3

Nassif, L., Dumont, G., Alysouri, H. and Rousseau, R. W. 2008. Pretreatment of Hanford Medium-Curie Wastes by Fractional Crystallization. *Environmental Science & Technology* 42, 4940-4945

Cooling Crystallization within Dual-Impinging-Jet Mixers: Mathematical Modeling, Theory Analysis, and Experimental Validation

Mo Jiang and Richard D. Braatz
Department of Chemical Engineering, Massachusetts Institute of Technology

An objective in pharmaceutical crystallization is to control crystal size distribution (CSD), to improve process efficiency and product quality. An effective way to improve the control of CSD is through combining continuous seeding and growth with concentration control [1]. One well-studied approach for continuous seeding is to combine solution and anti-solvent streams in a dual-impinging-jet (DIJ) mixer (e.g., [1]–[3]). Rather than anti-solvent crystallization as commonly used, this DIJ mixer combines hot and cold saturated solutions to generate small uniform seed crystals [4], which appears to be the first time a cooling DIJ nucleation has been reported. The product crystals were less than 10 microns in length, placing them in the correct size range for direct application in inhalers [5].

Theoretically, if the mixing was perfect (that is, concentration and temperature were completely mixed at the molecular scale in the mixer), the average supersaturation level would be too low to nucleate crystals in a cooling DIJ mixer. The surprising result of nucleating crystals by combining hot and cold saturated solutions in a DIJ mixer motivated the theoretical analysis of the system, which showed that nucleation was enabled by the much faster energy transfer than mass transfer rates near the interface. This difference in rates enabled the temperature of the hot solution to drop to approximately the average temperature of the two solutions before its solution concentration had significantly changed, resulting in a supersaturation sufficiently high to nucleate crystals.

The theoretical analysis is supported with process simulations. The two-dimensional energy and molar balances are solved using COMSOL input with an analytical solution for the velocity field, to generate spatial distributions of temperature, concentration, and supersaturation near the hot-cold interface within a cooling DIJ mixer. In the most important spatial region of importance for characterization of nucleation, the two-dimensional fields are shown to be very close to analytical solutions derived from a one-dimensional approximation of the energy and molar balances. This simplification applied the derivation of design criteria to facilitate quick assessment of whether any particular pharmaceutical-solute combination will nucleate crystals in a cooling DIJ mixer, based on the physicochemical properties. These criteria could save time and material by avoiding or reducing trial-and-error experiments, which is helpful at the early stage of pharmaceutical process development. The process simulations are then expanded to include the full solution of the spatially-varying population balance model [1] to calculate the crystal size distributions throughout the spatial domain. The mathematical models and methodologies are validated for the crystallization of L-asparagine monohydrate (LAM) in aqueous solution.

References

[1] X. Y. Woo, R. B. H. Tan, and R. D. Braatz, "Precise tailoring of the crystal size distribution by controlled growth and continuous seeding from impinging jet crystallizers," *CrystEngComm*, vol. 13, no. 6, pp. 2006–2014, 2011.

[2] B. K. Johnson and R. K. Prud'homme, "Chemical Processing and Micromixing in Confined Impinging Jets," *AIChE J.*, vol. 49, no. 9, pp. 2264–2282, 2003.

[3] M. Midler Jr., E. L. Paul, E. F. Whittington, M. Futran, P. D. Liu, J. Hsu, and S. H. Pan, "Crystallization Method to Improve Crystal Structure and Size," U.S. Patent 5,314,506A, May 24, 1994.

[4] M. Jiang, M. H. Wong, Z. Zhu, J. Zhang, L. Zhou, K. Wang, A. N. Ford Versypt, T. Si, L. M. Hasenberg, Y. E. Li, and R. D. Braatz, "Towards achieving a flattop crystal size distribution by continuous seeding and controlled growth," *Chem. Eng. Sci.*, vol. 77, pp. 2–9, 2012.

[5] T. G. D. Capstick and I. J. Clifton, "Inhaler technique and training in people with chronic obstructive pulmonary disease and asthma," *Expert Rev. Respir. Med.*, vol. 6, no. 1, pp. 91–103, Feb. 2012.

383475 Crystallization Kinetics in an Airlift Crystallizer: A Parameter Estimation Study

Abstract:

Batch solution crystallization is one of the oldest downstream processes, which involves the contribution of several phenomena such as growth, nucleation, agglomeration, dissolution etc. Among those kinetic processes the growth of seeds is often the only desirable one, whereas primary and secondary nucleation and agglomeration give rise to problems in both process and product control leading to variability in the product quality and unwanted crystal size distributions (CSD), shapes and impurity uptakes.

Crystallization in traditional stirred-vessel crystallizers, in which especially secondary nucleation is difficult to avoid, often leads to a wide multimodal CSD. Recently it has been shown [1,2] however that when crystallization is performed in a recently developed air lift crystallizer a strong suppression in the amount of fine crystals can be obtained compared to a stirred crystallizer; probably due to the suppression of the rate of secondary nucleation by attrition.

To prove this suppression in the rate of secondary nucleation and to get more insight in the kinetics and thus in the performance in the airlift crystallizer a more quantitative study is necessary in which the nucleation and growth phenomena are modelled and analyzed in both type of crystallizers.

The mixing strategy is the major distinction between an air lift and a stirred vessel crystallizer. In an airlift crystallizer, injected air bubbles are responsible for the mixing and the suspension of the crystals in an airlift, whereas the impeller fulfils these tasks in a stirred vessel crystallizer. It is most likely that the differences in hydrodynamics between the two types of crystallizer could have a strong impact on the crystallization kinetics. Especially secondary nucleation might be affected due to the absence of the high energetic collisions between the impeller and the crystallizer which are dominant in a stirred vessel crystallizer. [3-6].

In this contribution, the growth and nucleation behavior during batch operation in an air lift crystallizer for L-ascorbic acid/water system has been analyzed. Model verification and parameter estimation was performed based on experiments under a range of process conditions. A comparison is made between the results of the eighteen liter air lift crystallizer and that of a five liter stirred tank crystallizer to get more insight in dominating crystallization mechanisms in the airlift; allowing a further optimization of this novel type of crystallizer.

Crystallization experiments, described earlier [1], have been used with a number of different supersaturation profiles, batch times and seed loadings to estimate the parameters of the crystallization models in a model verification and parameter estimation procedure. The measured supersaturation profile and CSD of seeds and final product from these experiments were used to select the most appropriate models for nucleation and growth and to estimate the relevant kinetic parameters using a variety of models implemented in gPROMS Model Builder.

Both size dependent and size independent growth models are used along with different models for primary and secondary nucleation. Different secondary nucleation mechanisms were taken into account ranging from attrition based to activated nucleation mechanism [2,3,5,6,7].

The results show indeed a strong suppression of the secondary nucleation by attrition in the airlift crystallizer compared to stirred vessel crystallizer under all studied process conditions, leading to larger crystals and a reduced number of fines. However, the supersaturation profile and CSD in the airlift crystallizer cannot be simply described by a reduction in attrition rate. The parameter estimation study revealed that a different, activated nucleation model is needed to describe the experimental results properly. Apparently in the absence of the impeller in the airlift crystallizer, which avoids the attrition caused by the crystal-impeller and crystal-baffle collisions, an alternative secondary nucleation mechanism becomes dominant. An activated surface nucleation model, in which it is assumed that preordered clusters or pre-nuclei are formed in the immediate vicinity of the crystal surface or on the crystal surface [7] gives the best fit for the experimental data of the air lift crystallizer.

This means that although secondary nucleation by attrition is suppressed in airlift crystallizer, secondary nucleation is not completely absent in this type of crystallizer for the studied process conditions. The indication of the presence of an alternative nucleation mechanism is important as it opens possibilities to optimize the production of high quality crystals in this type of crystallizer.

Acknowledgments: *This work was supported by the European Commissions Framework 7 program through the OPTICO consortium*

References:

1. *R. Lakerveld, J.J.H van Krochten, H.J.M.. Kramer, An airlift crystallizer can suppress secondary nucleation at a higher supersaturation compared to a stirred crystallizer, submitted (2013)*
2. *Soare, A., Lakerveld, R., van Royen, J., Zocchi, G., Stankiewicz, A.I., Kramer, H.J.M., 2012. Minimization of Attrition and Breakage in an Airlift Crystallizer. Ind Eng Chem Res 51, 10895-10909.*
3. *J. W. Mullin, M. Chakraborty, K. Mehta, Nucleation and growth of ammonium sulphate crystals from aqueous solution, Journal of Applied Chemistry, 20, 12 (1970)*
4. *A. Soare, S. A. Pérez Escobar, A. I. Stankiewicz , M. Rodriguez Pascual , H. J. M. Kramer, 2-D Flow and Temperature Measurements in a Multiphase Airlift Crystallizer,* Ind. Eng. Chem. Res., *52, 34 (2013)*
5. *S.K. Bermingham, H.J.M. Kramer, and G.M. van Rosmalen, Towards on-scale crystalliser design using compartmental models. Computers & Chemical Engineering, 22 (1998)*
6. *Kramer, H.J.M., S.K. Bermingham, and G.M. van Rosmalen, Design of industrial crystallisers for a given product quality. Journal of Crystal Growth,. 198 (1999)*
7. *A. Mersmann, B. Braun, M. Loffelmann, Prediction of crystallization coefficients of the population balance, Chemical Engineering Science, 57, 20 (2002)*

376312 Protein Crystal Shape and Size Control Using a Plug Flow Crystallization Configuration

Abstract:

Crystallization plays a key role in the context of separation and purification methods for the production of therapeutic drugs. Considering the fact that crystal size and shape distributions have significant influence on the bioavailability of drugs such as the dissolution rate, filterability in the subsequent downstream process, and stability as a carrier to the target site, it becomes of particular interest to the pharmaceutical industry to produce crystals with desired size and shape distributions [1]-[2]. Traditionally, batch crystallization processes have been widely used in the pharmaceutical industry. However, owing to a few well-known potential drawbacks OF batch processes such as batch-to-batch variability, the difficulty in the scale-up, mixed suspension mixed product removal (MSMPR) crystallization process has received growing attention, and many efforts have been made in order to produce crystals from the MSMPR process with a higher production rate and desired product quality [3]-[4]. However, due to the presence of back-mixing, those crystals nucleated at a later stage during the crystallization process will reside a relatively short amount of time in the crystallizer and thus they will end up leaving the crystallizer with undesired size and shape distributions. To this end, plug flow crystallizer (PFC) has been proposed to produce crystals with desired size and shape distributions [5]-[7].

Motivated by the above considerations, the present work focuses on modeling of a PFC with 5 segments for the production of tetragonal HEW lysozyme crystals through kinetic Monte Carlo (kMC) simulation methods in the way described in [8] using the rate equations originally developed by [9]. A seeding strategy is used to decouple the nucleation from the crystal growth processes [10]. Furthermore, a constraint on the supersaturation level is imposed so that the system is enforced to stay in the metastable regime where the degree of primary nucleation is negligible [11]. Then, a population balance equation (PBE) is presented to describe the spatio-temporal evolution of the crystal volume distribution, and by applying the method of moments (MOM) to the PBE, a reduced-order moment model is derived because kMC models are not readily available in a closed form. Along with the mass and energy balance equations, the leading moments are used for the estimation of the spatio-temporal evolution of the crystal size and shape distributions in an optimization problem. More specifically, the crystallizer jacket temperatures at each segment and the superficial flow velocity are chosen as the decision variables in the optimization problem and the objective function is to minimize the squared deviation of the average size and shape of crystals produced throughout the PFC. Besides, dynamic models are developed and used for the design of a feed-forward control (FFC) strategy for the production of crystals with desired size and shape distribution properly suppressing the undesired effect caused by disturbances in the inflow solute concentration and crystal fines distribution.

[1] Patience DB, Rawlings JB. Particle shape monitoring and control in crystallization process. AIChE J. 2001;47:2125-2130.

[2] Wang L, Lee MH, Barton J, Hughes L, Odom TW. Shape-control of protein crystals in patterned microwells. J. Am. Chem. Soc. 2008;130:2142-2143.

[3] Griffin DW, Mellichamp DA, Doherty MF. Reducing the mean size of API crystals by continuous manufacturing with product classification and recycle. Chem. Eng. Sci. 2010;65;5770-5780.

[4] Alvarez A, Singh A, Myerson AS. Crystallization of cyclosporine in a multistage continuous MSMPR crystallizer. Crysal Growth & Design. 2011;11:4392-4400.

[5] Eder R, Schmitt E, Grill J, Radl S, Gruber-Woelfler H, Khinast J. Seed loading effects on the mean crystal size of acetylsalicylic acid in a continuous-flow crystallization device. Cryst. Res. Technol. 2011;3:227-237.

[6] Ridder B, Majumder A, Nagy Z. Population balance model based multi-objective optimization of a Multi-Segment Multi-Addition (MSMA) continuous plug flow antisolvent crystallizer. Ind. & Eng. Chem. Res. 2014;53:4387-4397.

[7] Alvarez A, Myerson A. Continuous plug flow crystallization of pharmaceutical compounds. Crysal Growth & Design. 2010;10:2219-2228.

[8] Kwon JS, Nayhouse M, Christofides PD, Orkoulas G. Modeling and control of crystal shape in continuous protein crystallization. Chem. Eng. Sci. 2014;107:47-57.

[9] Durbin SD, Feher G. Simulation of lysozyme crystal growth by the Monte Carlo method. J. Cryst. Growth. 1991;110:41-51.

[10] Liu JJ, Ma CY, Hu YD, Wang XZ. Effect of seed loading and cooling rate on crystal size and shape distributions in protein crystallization - a study using morphological population balance simulation. Comp. & Chem. Eng. 2010;34:1945-1952.

[11] Shi D, Mhaskar P, El-Farra NH, Christofides PD. Predictive control of crystal size distribution in protein crystallization. Nanotechnology. 2005;16:S562-S574.

384909 Crystallization Process Design: Accelerating Process Design through Modelling and Control

Abstract:

Pressure to deliver robust, high-performing processes at scale with compressed timelines and greater process understanding is increasing. Advanced modelling and process control represent one possible approach to accelerating process development while delivering on process understanding and performance. These approaches compliment and support the experimental programs undertaken during process development.

In this talk, the application of novel crystallization process models, continuous crystallization operation and advanced process control to the design of batch crystallization processes will be discussed. The deployment of these tools allows for the systematic mapping and interrogation of the design space, streamlining the development of a robust process at scale. A number of case studies applying this approach to process design will be presented.

Title

Steady-state and Startup Optimization for Combined Cooling and Antisolvent Crystallization (CCAC) in Continuous Mixed Suspension Mixed Product Removal (MSMPR) Cascade Crystallizers

Authors

Yang Yang and Zoltan K. Nagy, Chemical Engineering, Purdue University, West Lafayette, IN

Abstract

Crystallization is widely used as an important separation and purification technique in pharmaceutical industry [1]. Traditionally, most crystallization processes are performed in batch systems due to the high operating flexibility and short development time. However, in recent decades continuous crystallization is of great interest due to its enhanced reproducibility, reduced operating cost, simplified scale up and improved efficiency of process and equipment as compared with the traditionally used batch crystallization [2]. Similarly to the batch crystallization, either cooling or antisolvent addition could be used to generate supersaturation in a continuous crystallization process. Alternatively, combined cooling and antisolvent crystallization (CCAC) is also possible in continuous crystallization. Because in batch systems the CCAC approach could result in higher yield and larger control freedom on the multiple crystal properties like crystal size, shape and crystalline form as compared with cooling or antisolvent addition only crystallization [3].

Mixed suspension mixed product removal (MSMPR) crystallizer is one of the most important continuous crystallizers with its benefits demonstrated in the literature [2][4-7].

In continuous manufacturing, the control of startup is always a very crucial problem, due to the considerable amount of time product that can be wasted during this time. Methods to shorten the startup duration time and reduce the amount of waste in continuous crystallization are desirable.

This work presents a comprehensive study on the various control and optimization approaches for both the steady-state and startup behaviors of a continuous MSMPR cascade crystallizer through a model-based framework. Both cooling and antisolvent addition are applied to generate supersaturation at all cascade stages in order to provide the largest degree of freedom that can lead to the maximum steady-state crystal size achievable through an offline open-loop control approach. The optimal operating trajectory is obtained and compared with the optimal batch trajectory under similar scenario. Such comparison proves that the batch system can achieve larger crystals as compared with the cascade system. However, the optimal trajectories of the two systems show great similarities on both temperature and antisolvent dimensions.

In addition, different methods to shorten the startup duration time and reduce the waste amount have been discussed and proved to be feasible using a same open-loop optimal control framework. These methods include changing the initial solution compositions, applying dynamic antisolvent profiles, applying dynamic temperature profiles, and applying dynamic combined antisolvent/temperature profiles. The most efficient and practical way to achieve the goal turns out to be using dynamic antisolvent profiles only during startup. The proposed approach could reduce with approximately 50% of the startup duration and the waste amount while having no impact on the final steady-state product qualities.

References:

[1] Myerson, A. S. Handbook of Industrial Crystallization, 2nd ed.; Butterworth-Heinemann: Boston, 2002.

[2] Alvarez, A. J.; Singh, A.; Myerson, A. S. Crystallization of cyclosporine in a multistage continuous MSMPR crystallizer. *Cryst. Growth Des.* **2011**, 11, 4392-4400.

[3] Yang, Y.; Nagy, Z. K. Model-based systematic design and analysis approach for unseeded combined cooling and antisolvent crystallization (CCAC) systems. *Cryst. Growth Des.* **2014**, 14, 687-698.

[4] Wong, S. Y.; Tatusko, A. P.; Trout, B. L.; Myerson, A. S. Development of continuous crystallization processes using a single-stage mixed-suspension, mixed-product removal crystallizer with recycle. *Cryst. Growth Des.* **2012**, 12, 5701-5707.

[5] Quon, J. L.; Zhang, H.; Alvarez, A.; Evans, J.; Myerson, A. S.; Trout, B. L. Continuous crystallization of aliskiren hemifumarate. *Cryst. Growth Des.* **2012**, 12, 3036-3044.

[6] Zhang, H.; Quon, J.; Alvarez, A. J.; Evans, J.; Myerson, A. S.; Trout, B. Development of continuous anti-solvent/cooling crystallization process using cascaded mixed suspension, mixed product removal crystallizers. *Org. Process Res. Dev.* **2012**, 16, 915-924.

[7] Vetter, T.; Burcham, C. L.; Doherty, M. F. Regions of attainable particle size in continuous and batch crystallization processes. *Chem. Eng. Sci.* **2014**, 106, 167-180.

AUTHOR INDEX

AUTHOR INDEX

AUTHOR INDEX

AUTHOR INDEX

AUTHOR INDEX

AUTHOR INDEX

AUTHOR INDEX

AUTHOR INDEX

AUTHOR INDEX